CAPTURE GAMMA-RAY SPECTROSCOPY

AIP CONFERENCE PROCEEDINGS 238

CAPTURE GAMMA-RAY SPECTROSCOPY

PACIFIC GROVE, CA 1990

EDITOR:
RICHARD W. HOFF
LAWRENCE LIVERMORE
NATIONAL LABORATORY

American Institute of Physics New York

Authorization to photocopy items for internal or personal use, beyond the free copying permitted under the 1978 U.S. Copyright Law (see statement below), is granted by the American Institute of Physics for users registered with the Copyright Clearance Center (CCC) Transactional Reporting Service, provided that the base fee of $2.00 per copy is paid directly to CCC, 27 Congress St., Salem, MA 01970. For those organizations that have been granted a photocopy license by CCC, a separate system of payment has been arranged. The fee code for users of the Transactional Reporting Service is: 0094-243X/87 $2.00.

© 1991 American Institute of Physics.

Individual readers of this volume and nonprofit libraries, acting for them, are permitted to make fair use of the material in it, such as copying an article for use in teaching or research. Permission is granted to quote from this volume in scientific work with the customary acknowledgment of the source. To reprint a figure, table, or other excerpt requires the consent of one of the original authors and notification to AIP. Republication or systematic or multiple reproduction of any material in this volume is permitted only under license from AIP. Address inquiries to Series Editor, AIP Conference Proceedings, AIP, 335 East 45th Street, New York, NY 10017-3483.

L.C. Catalog Card No. 91-57923
ISBN 0-88318-830-9
DOE CONF-901057

Printed in the United States of America.

CONTENTS

Preface xvii

NUCLEAR STRUCTURE MODELS AND THEORY

Talks

Microscopic Self-Consistent and Collective Model Description
of Nuclear Structure 3
   K. Heyde, C. De Coster, D. Van Neck, and M. Waroquier
Structure of Nonrotational States and Distribution of $E\lambda$
Strength in Deformed Nuclei 16
   V. G. Soloviev
Pseudo-Spin Symmetry and Nuclear Structure.. 30
   J. P. Draayer
The Pseudo-L Quantum Number in Odd-A Nuclei... 46
   D. D. Warner
Collective Excitation Spectra of Transitional Even Nuclei................................ 55
   P. Quentin, I. Deloncle, J. Libert, and J. Sauvage
The Residual Proton–Neutron Interaction and Nuclear Collectivity 67
   R. F. Casten
Is Collective Motion Symmetric in the Neutron–Proton
Degrees of Freedom?.. 82
   J. N. Ginocchio, A. Leviatan, and M. W. Kirson
Octupole and Dipole Collectivity in Nuclei 90
   D. Kusnezov
Systematic Behavior of Low Energy Octupole States.. 98
   P. D. Cottle
$M1$ Strength and Scissors Strength in Rare-Earth Nuclei................................ 105
   C. De Coster and K. Heyde
$M1$ Excitation Scheme in Deformed Nuclei... 114
   T. Otsuka and I. Morrison
Nuclear Shell Model Calculations with Non-Local Interactions 122
   S. A. Moszkowski, S. D. Bloom, and D. A. Resler

Posters

Shell Model Study on Mixed-Symmetry 2^+ State in ^{56}Fe 131
   H. Nakada, T. Otsuka, and T. Sebe
Multiparticle–Multihole Configuration Mixing Within
the Neutron–Proton Interacting Boson Model... 134
   A. F. Barfield and B. R. Barrett
Excitation Modes in Non-Axial Nuclei 137
   A. Leviatan and J. N. Ginocchio

Systematics of Effective Charges of Proton and
Neutron Bosons 140
    A. Wolf and R. F. Casten

A Consistent Description of the Mixing of Intruder and
Normal States in ^{114}Cd 143
    J. Jolie and K. Heyde

Recent Progress in the Application of Extended Supersymmetries
to Odd–Odd Nuclei in the Au Region 146
    J. Jolie

Evidence for Core Polarization Interaction in the Single
Closed Shell Region 150
    Zs. Dombrádi, S. Brant, and V. Paar

Signature Dependence and Inversion in Two Quasi-Particle
Rotational Bands of Even-Even Nuclei 153
    A. K. Jain and A. Goel

Phase Reversal in Odd–Even Staggering in K^+ and K^- Bands
of Even–Even 2QP Rotational Spectra 156
    A. K. Jain

Completeness of Two-Particle Spectra of Deformed Nuclei 159
    P. C. Sood, R. K. Sheline, R. W. Hoff, and A. K. Jain

Application of the VMI Model to Rotational Bands of Odd–Odd
Rare-Earth Nuclei 162
    A. K. Jain and A. Goel

A Comparison of the Princeton and Beijing Energy Expressions
for Rotating Nuclei 165
    R. A. Naumann, Y. Y. Sharon, and A. Adarkar

Construction of Multi-Cluster States for Nuclear Cluster Models 168
    A. Novoselsky and J. Katriel

EXPERIMENTAL NUCLEAR STRUCTURE

Talks

Gamma Ray Induced Doppler Broadening and the Determination
of Lifetimes of Excited Nuclear States 173
    H. G. Börner, J. Jolie, S. J. Robinson, E. G. Kessler, S. Ulbig,
    R. F. Casten, M. S. Dewey, G. L. Greene, R. D. Deslattes, K. P. Lieb,
    B. Krusche, and J. A. Cizewski

Analysis of Recoil Doppler Profiles Obtained with Ultra High
Resolution (n,γ) Spectroscopy 189
    J. Jolie, H. G. Börner, and S. J. Robinson

Precise Absolute Gamma-Ray Wavelength Measurements 199
    E. G. Kessler, M. S. Dewey, G. L. Greene, R. D. Deslattes, and H. Börner

Mixed Symmetry States in Nuclei Near Closed Shells 210
    S. J. Robinson, J. Jolie, and J. Copnell

Probing Collective Excitations with the $(n,n'\gamma)$ Reaction 218
    S. W. Yates

The First Submagic Nucleus: ^{96}Zr 227
    G. Molnár

Photon Scattering Experiments in the Rare Earth Region—
A New Approach to Low Lying Dipole Modes 234
    P. von Brentano, A. Zilges, R. Jolos, A. Richter, R. D. Heil,
    U. Kneissl, H. H. Pitz, and C. Wesselborg

Energy Repulsion and Width Attraction in Two Resonance Mixing 244
    P. von Brentano

Investigation of Odd Ba Nuclei from Neutron Induced Reactions 249
    V. A. Bondarenko, I. L. Kuvaga, P. T. Prokofjev, V. A. Khitrov,
    Yu. V. Kholnov, L. H. Khiem, Yu. P. Popov, and A. M. Sukhovoj

Octupole Deformation in Odd–Odd Nuclei ^{154,156}Eu
    A. V. Afanasjev, T. V. Guseva, J. J. Tambergs, and M. K. Balodis

Features of the Level Schemes of $^{164-166}$Dy 257
    F. Hoyler, K. Föhl, H. G. Börner, B. Krusche, S. J. Robinson,
    and P. Schillebeeckx

Intrinsic Excitations in ^{192}Ir 263
    J. Kern, M. K. Balodis, W. Beer, S. Brant, R. F. Casten,
    A. Chalupka, C. Coveca, J.-Cl. Dousse, R. Eder, T. von Egidy,
    D. G. Gardner, M. A. Gardner, P. Giacobbe, R. L. Gill,
    E. Hagn, R. W. Hoff, M. A. Hungerford, I. A. Kondurov,
    I. V. Kononenko, N. D. Kramer, V. A. Libman, Yu. E. Loginov,
    A. V. Murzin, V. Paar, P. T. Prokofjev, A. Raemy,
    H. J. Scheerer, H. H. Schmidt, W. Schwitz, L. I. Simonova,
    P. A. Sushkov, and E. Zech

Prompt and Delayed γγ Coincidences in the ^{191}Ir$(n,\gamma)^{192}$Ir Reaction 271
    I. A. Kondurov, Yu. E. Loginov, and P. A. Sushkov

Nuclear Structure of ^{190,192,194}Ir by Charged-Particle Spectroscopy 279
    D. G. Burke and P. E. Garrett

Distribution of Photon Strength in Nuclei by a Method
of Two-Step Cascades 287
    F. Becvar, P. Cejnar, R. E. Chrien, and J. Kopecky

Excitation Modes and Decay Channels in Compound States 298
    Yu. P. Popov

Neutron Capture in Perspective 308
    W. Gelletly

Non-Yrast Spectroscopy of Tellurium Nuclei 323
    J. A. Cizewski, R. G. Henry, and C. S. Lee

Mesonic Effects in Nuclei Near ^{208}Pb Deduced from β Decay 331
    E. K. Warburton

Single Particle States in the Heaviest Known Nuclei 339
    I. Ahmad, R. R. Chasman, A. M. Friedman, and S. W. Yates

Spectroscopic Studies Near the Proton Drip Line 347
    K. S. Toth, D. M. Moltz, J. M. Nitschke, P. A. Wilmarth,
    and J. D. Robertson

Nuclear Structure of Neutron-Rich Fission Fragments 355
    W. R. Phillips

Level Lifetimes in the PS Region: Spherical and Deformed
Structures at $A \sim 100$ 367
    H. Ohm, M. Liang, M. Büscher, U. Paffrath, B. De Sutter,
    K. Sistemich, and G. Molnár

A New Method for Picosecond Lifetime Measurements Using
Electronic Timing: Nuclear Structure Applications 375
R. L. Gill

Reliability of Short Lifetimes Measured by the Doppler Shift
Attenuation Method 383
J. Keinonen

Posters

Precision Gamma-Ray Measurements in ^{25}Mg Following Thermal
Neutron Capture in ^{24}Mg 393
S. Michaelsen, K. P. Lieb, L. Ziegeler, and T. von Egidy

Grid Lifetime Study of the 1762 keV State in ^{49}Ti Populated
via Cascade Feeding 396
S. Ulbig, J. Jolie, H. G. Börner, M. S. Dewey, K. P. Lieb,
and S. J. Robinson

Level Density in ^{51}V at Spin 9/2–15/2 Measured
with the ^{50}V(n,γ) Reaction 399
S. Michaelsen, K. P. Lieb, and S. J. Robinson

The Lowest $J^\pi = 1^+$ Analogue and Antianalogue in ^{56}Fe 404
Z. Guo, C. Alderliesten, C. van der Leun, and P. M. Endt

Grid Lifetime Measurements and Shell Model Calculations in ^{59}Ni 407
S. Ulbig, K. P. Lieb, H. G. Börner, S. J. Robinson,
and J. G. L. Booten

Measurement of Conversion Electrons from the ^{75}As(n,e^-) Reaction—
A Test of the Description of ^{76}As in the $U_\nu(6/12) \otimes U_\pi(6/12)$ Scheme 412
J. Jolie, F. Hoyler, and B. Krusche

Measurement of the Natural Line Shape of Krypton Conversion
Electrons from Gaseous ^{83m}Kr 415
D. J. Decman and W. Stoeffl

$E2$ and $M1$ Strengths and Strong Sub-Shell Closure Effects
in Neutron-Rich $A \sim 100$ Nuclei 418
F. K. Wohn, H. Mach, G. Molnár, K. Sistemich, J. C. Hill,
M. Moszyǹski, R. L. Gill, W. Krips, D. S. Brenner,
and R. F. Casten

The Level Scheme of ^{110}Ag from the (n,γ) Reaction 421
I. A. Kondurov, P. A. Sushkov, M. Bogdanović, J. Simić
H. Seyfarth, H. A. Baader, D. Breitig, R. Koch, G. Barreau,
H. G. Börner, R. Brissot, S. Kerr, T. D. Mac Mahon,
T. Mitsunari, H. Faust, and K. Schreckenbach

Interacting Boson–Fermion–Fermion Description of Odd–Odd
Indium Nuclei 425
Zs. Dombrádi, T. Fényes, S. Brant, and V. Paar

New Features in the Systematics of Low-Spin States
in Even $^{106-120}$Cd 428
J. Kumpulainen, R. Julin, J. Kantele, A. Passoja, W. H. Trzaska,
E. Verho, and J. Väärämäki

Levels in ^{129}Te Populated in the ^{128}Te(n,γ) Reaction 431
    C. A. Stone, B. E. Zimmerman, C. E. Ford, P. F. Mantica, Jr.,
    and W. B. Walters

Doppler Shift Lifetime Measurements at Low Recoil Velocities
and Mixed-Symmetry States in ^{134}Ba 434
    T. Belgya, B. Fazekas, G. Molnár, Á. Veres, R. A. Gatenby,
    E. M. Baum, E. L. Johnson, and S. W. Yates

Tests of Octupole Band Structures 437
    P. D. Cottle, S. M. Aziz, J. W. Holcomb, T. D. Johnson,
    K. W. Kemper, M. L. Owens, E. L. Reber, K. A. Stuckey,
    S. L. Tabor, P. C. Womble, J. D. Brown, E. R. Jacobsen,
    Y. Y. Sharon, S. G. Buccino, and F. E. Durham

Detailed Spectroscopy of ^{144}Nd 440
    S. J. Robinson, H. G. Börner, S. Judge, J. Jolie, and P. Schillebeeckx

Extraction of Spins and Mixing Ratios from Directional
Correlation Data 443
    S. J. Robinson

Experimental Evidence for Octupole Deformation from Observation
of Parity Doublets in ^{152}Eu, ^{154}Eu, ^{156}Eu
Neutron Capture Studies 446
    M. K. Balodis, P. T. Prokofjev, and A. V. Afanasjev

Nuclear Structure Investigations and Lifetime Measurement
in ^{156}Gd 455
    J. Klora, H. G. Börner, T. von Egidy, H. Hiller, S. Judge,
    B. Krusche, V. A. Libman, H. Lindner, L. L. Litvinsky,
    U. Mayerhofer, A. V. Murzin, and S. J. Robinson

Investigation of Gamma Transitions Populating the 1094 keV
Isomeric State in the ^{167}Er(n,γ) Reaction 458
    M. Bogdanović and J. Simić

Study of Odd–Odd ^{196}Au via the (p,d) Reaction 461
    G. Rotbard, G. Berrier, M. Vergnes, J. M. Maison, S. Fortier,
    J. Vernotte, J. Kalifa, L. Rosier, and P. Van Isacker

The Level Scheme of ^{198}Au Studied with (d,p), (n,γ), and (n,e^-)
Reactions and the Level Scheme of ^{196}Au Studied with (d,t)
and $(^3\text{He},\alpha)$ Reactions 464
    U. Mayerhofer, T. von Egidy, H. Lindner, H. Hiller, J. Klora,
    H. Trieb, and A. Walter

High Resolution Spectroscopy Using Transfer Reactions 467
    H. Lindner, T. von Egidy, H. Hiller, J. Klora, U. Mayerhofer,
    H. Trieb, and A. Walter

High Resolution Study of the ^{222}Ra Exotic Decay 470
    M. Hussonnois, J. F. Le Du, L. Brillard, J. Dalmasso,
    and G. Ardisson

High-Spin States of ^{238}Pu Using a Heavy-Ion One-Neutron
Transfer Reaction 473
    M. A. Stoyer, J. O. Rasmussen, A. A. Shihab-Eldin, D. Cline,
    K. Helmer, A. E. Kavka, W. J. Kernan, B. Kotlinski, E. Vogt,
    C. Y. Wu, C. Bingham, M. W. Guidry, X. L. Han, R. W. Kincaid,
    X. T. Liu, H. Schechter, M. L. Halbert, and D. Hensley

Nuclear Structure of ^{241}Pu from (n,γ) and (n,e) Reaction
Measurements 476
    D. H. White, R. W. Hoff, H. G. Börner, G. Colvin, F. Hoyler,
    and K. Schreckenbach

**Polarization Measurements in Nuclear Resonance Fluorescence
Experiments** 479
    R. D. Heil, A. Degener, H. Friedrichs, A. Jung, U. Kneissl,
    S. Lindenstruth, J. Margraf, H. H. Pitz, H. Schacht, B. Schlitt,
    U. Seemann, R. Stock, C. Wesselborg, P. von Brentano,
    A. Zilges, G. Müller, and M. Schumacher

**Potentialities of a Multidetector System to Study Cascade
γ Decay of a Compound State of Complex Nuclei** 482
    V. A. Khitrov, Yu. P. Popov, and A. M. Sukhovoj

**On the Construction of a Complex Gamma-Decay Scheme
on the Basis of the Spectroscopic Data from Reactions
$(n,2\gamma)$ and (n,γ)** 485
    S. T. Boneva, V. A. Khitrov, Yu. V. Kholnov, V. D. Kulik,
    L. H. Khiem, P. D. Khang, Yu. P. Popov, A. M. Sukhovoj,
    and E. V. Vasilieva

**Dependence of Two-Step Cascade Intensities on the Excitation
Energy of Intermediate Levels for Three Nuclei** 488
    S. T. Boneva, V. A. Khitrov, A. M. Sukhovoj, and A. V. Vojnov

**On Estimates of Radiative Strength Functions of Soft Primary
Transitions in ^{181}Hf** 491
    V. A. Bondarenko, I. L. Kuvaga, P. T. Prokofjev, G. L. Rezvaya,
    L. I. Simonova, V. A. Khitrov, Yu. V. Kholnov, L. H. Khiem,
    P. D. Khang, Yu. P. Popov, and A. M. Sukhovoj

Cascade γ Decay of the ^{187}W Compound State 494
    S. T. Boneva, V. A. Khitrov, L. A. Malov, Yu. P. Popov,
    A. M. Sukhovoj, E. V. Vasilieva, M. R. Beitins, P. T. Prokofjev,
    G. L. Rezvaya, and L. I. Simonova

SUPERDEFORMATION IN NUCLEI

Talks

Superdeformation in the Murcury Region 499
    R. R. Chasman

Shape Isomers: Mean-Field Description and Beyond 511
    P. Bonche, S. J. Krieger, M. S. Weiss, J. Dobaczewski, H. Flocard,
    P. H. Heenen, and J. Meyer

Superdeformation in the Hg–Tl–Pb Region 523
    E. A. Henry, J. A. Becker, S. W. Yates, T. F. Wang, W. Kuhnert,
    M. J. Brinkman, J. A. Cizewski, M. A. Deleplanque, R. M. Diamond,
    F. S. Stephens, F. Azaiez, W. Korten, and J. E. Draper

Global Systematics of Superdeformation 533
    R. K. Sheline and P. C. Sood

Posters

Interacting Boson Model for Superdeformation 543
    M. Honma and T. Otsuka
Analysis of High-Multiplicity Gamma-Ray Events 546
    M. J. Brinkman, J. A. Cizewski, D. R. Manatt, J. A. Becker,
    E. A. Henry, N. Roy, R. M. Diamond, F. S. Stephens,
    M. A. Deleplanque, C. W. Beausang, and J. E. Draper
Pattern Recognition in Gamma–Gamma Coincidence Data Sets 549
    D. R. Manatt, F. L. Barnes, J. A. Becker,
    J. V. Candy, E. A. Henry, and M. J. Brinkman

CAPTURE REACTION MECHANISMS, PROTON CAPTURE, AND RESONANCES

Talks

Direct Neutron Capture and Related Mechanisms 555
    J. E. Lynn and S. Raman
Gaussian Distribution for Spacings of Simple Neutron Resonances 572
    G. Rohr
The Real Optical– and Shell– Model Potentials 579
    R. D. Lawson, S. Chiba, P. T. Guenther, and A. B. Smith
Tensor Polarized Deuteron Capture Reactions and D-State Effects in Light Nuclei 587
    H. R. Weller, G. Feldman, M. J. Balbes, L. H. Kramer,
    J. Z. Williams, and D. R. Tilley
Isovector Giant Quadrupole Resonance in $(\vec{p},\gamma)$ 595
    G. Feldman, L. H. Kramer, H. R. Weller, E. Hayward
    and W. R. Dodge

Posters

Resonances in the Direct Radiative Capture of a Light Nucleus by a Light Nucleus 604
    A. Mondragón and E. Hernández
Calculations of Capture Cross Sections and Gamma-Ray Spectra with Different Strength Function Models 607
    J. Kopecky and M. Uhl
Statistical Description of the γ-Ray Cascades after Thermal Neutron Capture 610
    B. Krusche
Calculated Neutron Radiative Capture Cross Section Using Modifying Exciton Model 614
    Z. H. Lu, C. Z. Mo, and Y.-K. Ho

Polarized Neutron–Proton Capture for Neutron Energies
from 19 to 50 MeV 617
 G. Fink, P. Doll, S. Hauber, M. Haupenthal, H. O. Klages,
 H. Schieler, F. Smend, and G. D. Wicke

Systematics of 3.5 to 100 MeV ^{2}H(γ,n) Data.. 620
 A. Wolf, S. Kahane, and Y. Birenbaum

Capture Gamma Rays from Broad Neutron Resonances
in the Proton-Odd Nuclei ^{14}N, ^{19}F, and ^{27}Al 624
 M. Igashira, H. Kitazawa, S. Kitamura, H. Anze,
 and M. Horiguchi

Electromagnetic Transitions from Broad s-Wave Neutron
Resonance in the sd-Shell Nuclei ^{24}Mg, ^{28}Si, and ^{32}S.................................... 627
 H. Kitazawa, M. Igashira, Y. Achiha, Y. Lee, N. Mukai,
 K. Muto, and T. Oda

The Measurement of the ^{12}C(n,γ) ^{13}C Reaction in the Pygmy
Resonance Region 630
 Z. Huang, L. Zhu, L. Ho, X. Shi, and D. Ding

Gamma-Ray Intensities Following Thermal Neutron Capture
in ^{117}Sn 633
 V. R. Skoy and E. I. Sharapov

Study of the ^{91}Zr(n,α) ^{88}Sr and ^{187}Os(n,α) ^{184}W Reactions
on Resonance Neutrons... 636
 J. Andrzejewski, Yu. M. Gledenov, M. P. Mitrikov, Yu. P. Popov,
 P. V. Sedyshev, I. Chadraabal, and L. Ho Bom

Asymmetries in Nucleon Radiative Capture Caused by
Nuclear Structure... 639
 R. Guidotti, F. Saporetti, G. Maino, and A. Ventura

STATISTICAL PROPERTIES OF NUCLEAR LEVELS

Talks

Precision and Completeness... 645
 C. van der Leun

Chaos in Nuclear Level Schemes 655
 J. F. Shriner, Jr., G. E. Mitchell, and T. von Egidy

Scars and the Order to Chaos Transition 665
 H. Frisk

New Formula for Spin-Dependent Level Density... 672
 V. Paar, S. Brant, D. K. Sunko, M. G. Mustafa,
 and R. G. Lanier

Poster

Level Densities, Expectation Values and Strengths with
Interactions: Convolution Forms and Applications 685
 V. K. B. Kota

HIGH-ENERGY NEUTRONS, GAMMA-RAY PRODUCTION

Talks

Neutron-Induced Gamma-Ray Production 697
R. O. Nelson, D. M. Drake, R. C. Haight, C. M. Laymon,
S. A. Wender, P. G. Young, M. Drosg, A. Pavlik, H. Vonach,
and D. C. Larson

**Hard Photon Production in Nuclear Reactions at Low
and Intermediate Energies** 705
Y.-K. Ho

Radiative Capture and Preequilibrium γ Emission 714
P. Obložinský

**Neutron–Proton Bremsstrahlung Studies Using the White Neutron
Source at the LAMPF/WNR** 723
S. A. Wender, R. O. Nelson, M. E. Schillaci, and M. Blann

**High Resolution Measurement of Radiative Proton Capture
and Pair Production at Intermediate Energies** 733
B. Höistad, E. Nilsson, J. Thun, S. Dahlgren, S. Isaksson,
G. S. Adams, and C. Landberg

Poster

**A Study of Reaction Mechanisms for Gamma Production
in Fast-Nucleon Induced Reactions** 742
A. Höring, H. A. Weidenmüller, F. S. Dietrich, M. Herman,
and G. Reffo

FUNDAMENTAL PHYSICS WITH NEUTRONS

Talks

**Parity and Time Reversal Symmetry Violation
in Neutron–Nucleus Scattering** 747
C. R. Gould, J. E. Bush, C. M. Frankle, D. G. Haase, G. E. Mitchell,
C. D. Bowman, J. D. Bowman, J. Knudson, S. Penttilä, S. J. Seestrom,
J. J. Szymanski, S. H. Yoo, V. W. Yuan, N. R. Roberson, X. Zhu,
P. P. J. Delheij, and H. Postma

**The Measurements of Parity Violation in Resonant
Neutron-Capture Reactions** 756
E. I. Sharapov, S. A. Wender, H. Postma, S. J. Seestrom, C. R. Gould,
A. Wasson, Yu. P. Popov, and C. D. Bowman

**Radiative Capture in Few-Nucleon Systems and
Exchange Currents** 764
K. Abrahams, M. W. Konijnenberg, and R. Wervelman

**Current Results and Future Prospects for a Neutron Lifetime
Determination Using Trapped Protons** 774
M. S. Dewey

Results of Neutron Lifetime Measurements with Gravitational
UCN Trap 787
    V. P. Gudkov, A. G. Kharitonov, V. V. Nesvizhevsky, A. P. Serebrov,
    S. O. Sumbaev, R. R. Taldaev, V. E. Varlamov, A. V. Vasilyev,
    V. P. Alfimenkov, V. I. Lushikov, V. N. Shvetsov, and A. V. Strelkov

Parity Violation in Resonant Neutron Reactions 797
    A. Müller, H. L. Harney, and E. D. Davis

Posters

Neutron–Antineutron Oscillation Experiment: Results
and Perspectives 805
    T. Bitter, F. Eisert, P. El-Muzeini, M. Kessler, U. Kinkel, E. Klemt,
    W. Lippert, R. Werner, D. Dubbers, K. Gobrecht, M. Baldo-Ceolin,
    F. Bobisut, D. Gibin, A. Guglielmi, M. Laveder, F. Mattioli,
    M. Mezzetto, G. Puglierin, A. Sconza, M. Vascon, L. Visentin,
    P. Benetti, E. Calligarich, R. Dolfini, M. Genoni, A. Gigli Berzolari,
    F. Mauri, A. Piazzoli, A. Rappoldi, G. L. Raselli, and D. Scannicchio

Study of Parity Violation in the ^{6}Li(n,α) ^{3}H Reaction
with Polarized Neutrons 808
    J. Andrzejewski, A. D. Antonov, Yu. M. Gledenov, M. P. Mitrikov,
    Yu. P. Popov, I. S. Okunev, B. G. Peskov, E. V. Shul'gina,
    and V. A. Vesna

P-Even Effects in a Direct γ Transition Following Neutron
Capture in ^{113}Cd Around $Ep = 7$ eV Resonance 811
    V. P. Alfimenkov, S. B. Borzakov, Yu. D. Mareev, L. B. Pikelner,
    V. R. Skoy, A. S. Khrykin, and E. I. Sharapov

New Measurements of the Electron–Neutron Spin Asymmetry
in the Neutron Beta Decay 814
    B. G. Erozolimskii, I. A. Kuznetsov, I. V. Stepanenko, I. A. Kuida,
    and Ju. A. Mostovoi

NUCLEAR ASTROPHYSICS

Talks

Nuclear and Astronomical Constraints on the Site for R-Process
Nucleosynthesis 827
    G. J. Mathews, G. Bazan, and J. J. Cowan

Capture Reactions on ^{14}C in Nonstandard Big Bang
Nucleosynthesis 840
    M. Wiescher, J. Görres, and F. K. Thielemann

^{176}Lu—An s-Process Thermometer 850
    N. Klay, F. Käppeler, H. Beer, and G. Schatz

Changes in Nuclear Decay Rates as a Result of High Stellar
Temperature: ^{148}Pm and ^{176}Lu 860
    K. T. Lesko

Second Generation New Theoretical Data of Beta Decay Far
from Stability and of Double Beta Decay and Implications
for Astrophysics 870
    H. V. Klapdor-Kleingrothaus
Some Limitations of Detailed Balance for Inverse Reaction
Calculations in the Astrophysical p–Process 882
    D. G. Gardner and M. A. Gardner
Cross-Section Measurements on Radioactive Samples 892
    P. E. Koehler and H. A. O'Brien

Posters

Cross Section of the ^{7}Li$(n,\gamma)^8$Li Reaction at Stellar Energy 900
    Y. Nagai, M. Igashira, N. Mukai, K. Takeda, F. Uesawa, T. Ohsaki,
    T. Ando, H. Kitazawa, S. Kubono, and T. Fukuda
Capture Rate of the ^{12}C$(n,\gamma)^{13}$C Reaction in Inhomogeneous
Big-Bang Models and Stellar Evolution 903
    Y. Nagai, M. Igashira, K. Takeda, N. Mukai, F. Uesawa,
    S. Motoyama, H. Kitazawa, annd T. Fukuda
The 30 keV Averaged Neutron Capture Cross Sections
of ^{56}Fe and ^{60}Ni 906
    F. Corvi, G. Fioni, A. Mauri, and T. Babeliowsky
Beta Decay of Some FP Shell Nuclei for Presupernova Stars............................ 909
    K. Kar, S. Sarkar, and A. Ray
Precision Measurement of the Half-Life of ^{56}Co 911
    D. E. Alburger and C. Wesselborg
Importance of $E2$ Transitions in ^{175}Lu(n,γ) Isomer Production
Calculations... 914
    M. A. Gardner and D. G. Gardner
The ^{180m}Ta$(\gamma,\gamma')^{180}$Ta Cross Section at 1.33 and 4.0 MeV and
Its Astrophysical Consequences.. 917
    Zs. Németh, F. Käppeler, and G. Reffo

NEW FACILITIES, APPLICATIONS, AND RELATED TOPICS

Talks

The Advanced Neutron Source 923
    S. Raman and J. B. Hayter
Capture Gamma-Ray Spectroscopy Using Cold Neutron
Beams 929
    C. A. Stone, D. F. R. Mildner, R. Zeisler, and D. C. Cranmer
Considerations in Upgrading Intermediate Flux Reactors
by the Addition of Cold Neutron Beams 936
    D. D. Clark
Experiments with (n,γ)–(γ,e^+e^-) Converters 943
    B. Krusche

Experimental Assessment of the Performance of a Proposed Lead
Slowing-Down Spectrometer at WNR/PSR 953
    M. S. Moore, P. E. Koehler, A. Michaudon, A. Schelberg, Y. Danon,
    R. C. Block, R. E. Slovacek, R. W. Hoff, and R. W. Lougheed

Use of N-Capture γ-Rays for Studies in Condensed Matter Physics 961
    R. Moreh

Internal Oxidation of Sb and In in Silver Studied with Nuclear
Reaction Analysis/Channeling Combined with HFI
Measurements 972
    D. O. Boerma

Thin Film Analysis by Means of Resonant Proton Capture
Reactions 983
    K. P. Lieb, W. Bolse, T. Corts, T. Kacsich, A. Kehrel,
    and M. Uhrmacher

Elemental Mapping of Planetary Surfaces Using Gamma-Ray
Spectroscopy 994
    R. C. Reedy

Posters

Neutron Focusing Using Converging Guides 1003
    D. F. R. Mildner

Boron Neutron Capture Therapy 1006
    F. Stecher-Rasmussen, R. L. Moss, and M. W. Konijnenberg

Design and Application of *In Situ* Prompt-Gamma Probe
as a Salinometer 1008
    J.-H. Chao and C. Chung

Photoexcitation of ^{50}Cr, ^{53}Cr and the Debye Temperature
of Chromium 1011
    D. Levant, R. Moreh, and O. Shahal

Elemental Analysis of Coal Based on Spectrometry of Prompt
Gamma Rays from Neutron Induced Reactions 1014
    S. Pospíšil, Z. Janout, J. Koníček, and M. Vobecký

Participants List 1015

Author Index 1026

PREFACE

The Seventh International Symposium on Capture Gamma-Ray Spectroscopy and Related Topics was held at the Asilomar Conference Center, Pacific Grove (Monterey peninsula) California, on October 14–19, 1990. This symposium is part of a series of meetings whose thematic material has included capture gamma-ray spectroscopy and which were held in Studsvik, Sweden (1969), Petten, The Netherlands (1974), Brookhaven, USA (1978), Grenoble, France (1981), Knoxville, USA (1984), and Leuven, Belgium (1987). The 1990 Symposium was sponsored by the American Physical Society, the U.S. Department of Energy, and the Lawrence Livermore National Laboratory.

The five-day Symposium was attended by 167 participants from 23 countries. The Symposium program was designed to continue the general themes of the earlier conferences, with emphasis on recent experimental and theoretical developments in low-energy nuclear physics as determined with a variety of nuclear probes including capture reactions. Topics discussed during the Symposium were the following: neutron and proton capture (high-resolution gamma spectroscopy, short lifetimes, gamma-ray production, reaction mechanisms), nuclear structure (collective and single-particle phenomena, superdeformation, proton-neutron interactions), statistical properties of nuclear levels (chaotic behavior), nuclear astrophysics (nucleosynthesis, chronometry), fundamental physics with neutrons (neutron lifetime), new facilities, new instrumentation, new neutron sources, and applications (elemental mapping of planetary surfaces, analysis of near-surface solid-state reactions).

The general scope of the Symposium program was developed in consultation with the Advisory Committee. The Program Committee members were instrumental in developing the program themes and selecting the invited speakers, again with the assistance and advice of the Advisory Committee. The members of these committees are listed below.

Advisory Committee:

K. Abrahams (The Netherlands), C. D. Bowman (USA), R. F. Casten (USA), R. E. Chrien (USA), T. von Egidy (W. Germany), W. Gelletly (UK), F. Iachello (USA), J. Kern (Switzerland), H. V. Klapdor-Kleingrothaus (W. Germany), J. E. Lynn (UK), R. A. Meyer (USA), C. H. Poppe (USA), P. T. Prokofjev (USSR), S. Raman (USA), O. W. B. Schult (W. Germany), R. K. Sheline (USA), V. G. Soloviev (USSR), P. H. M. Van Assche (Belgium).

Program Committee:

I. Ahmad (USA), H. G. Börner (W. Germany), P. von Brentano (W. Germany), R. R. Chasman (USA), J. Cizewski (USA), F. S. Dietrich (USA), J. P. Draayer (USA), D. G. Gardner (USA), W. D. Hamilton (UK), K. Heyde (Belgium), Y.-K. Ho (China), F. Käppeler (W. Germany), J. Keinonen (Finland), I. Kondurov (USSR), K. P. Lieb (W. Germany), G. J. Mathews (USA), W. C. McHarris (USA), E. Norman (USA), T. Otsuka (Japan), R. Piepenbring (France), V. Paar (Yugoslavia), Yu. Popov (USSR), P. Quentin (France), C. W. Reich (USA), K. Schreckenbach (W. Germany), G. L. Struble (USA), C. van der Leun (The Netherlands), D. D. Warner (UK), M. S. Weiss (USA), S. Wender (USA), S. W. Yates (USA).

The technical program was comprised of 76 invited papers and 77 contributed papers. The invited talks were presented in eight half-day sessions during mornings and afternoons and were organized according to topic. A single morning session was held each day. In the afternoons

(Monday, Tuesday, and Thursday), two parallel sessions of oral presentations were held. The contributed papers were presented as posters. Two evenings (Monday and Tuesday) were devoted to viewing the posters. The session chairmen who presided during the oral presentations were the following: K. Abrahams, C. D. Bowman, R. E. Chrien, F. S. Dietrich, T. von Egidy, E. Norman, S. Raman, J. O. Rasmussen, C. W. Reich, G. L. Struble, M. Vergnes, F. K. Wohn, and J. L. Wood. Concurrent with the poster sessions, industrial exhibits from the Bicron Corporation and the Canberra Nuclear Products Group were displayed. The National Nuclear Data Center of Brookhaven National Laboratory also participated with a display.

The social program during the Symposium included a welcoming reception for symposium participants on Sunday evening, a banquet in the dining hall at Asilomar on Wednesday evening, and a visit to the Monterey Bay Aquarium for all participants on Thursday evening.

These proceedings contain the papers of the Symposium which were received in the form of camera-ready copy; the contents of these original contributions have not been edited. The papers in this volume have been arranged into nine sections, according to topic. Although this arrangement is not identical to the order in which they were presented at the Symposium, the logical sequences of papers on a given topic during the oral presentations have been retained.

The organization of this technical conference was an effort where many people contributed. Thus, I want to acknowledge their contributions and to thank many of them individually for their work. The members of the Advisory Committe and the Program Committe provided invaluable assistance in planning the general scope of the program and in the selection and acceptance of invited and contributed papers. I thank them for these efforts and for the many useful suggestions they provided with respect to conference organization. I was extremely fortunate to have the participation of the organizers of the five most recent symposia in this series. The dedication of these scientists (K. Abrahams, R. E. Chrien, T. von Egidy, S. Raman, and P. H. M. Van Assche) to the continuing success of this symposium series is very much appreciated.

The planning for the many aspects of the Symposium beyond the question of technical program was the responsibility of the Organizing Committee, whose members are listed below:

Organizing Committee:

J. A. Becker, H. M. Blann, D. J. Decman, M. A. Gardner, E. A. Henry, R. W. Hoff (Chairman), R. G. Lanier, R. W. Lougheed, L. G. Mann, C. K. McGregor, M. Mustafa, M. N. Namboodiri, W. Stoeffl, H. I. West, D. H. White, R. M. White.

I want to thank them for their hard work and attention to detail that was so essential to the success of the Symposium. This Committee was assisted by many LLNL staff members whose contributions and dedication to the administrative aspects of the Symposium are very much appreciated. The administrative staff, under the direction of Sharon Blackwell performed beautifully and put in long hours of work before, during, and after the actual event. We thank Sue Frumenti, Roberta Moeller, Joyce Plis, Nancy Henry and Sue Garber for all of their help. In addition, we were ably assisted by members of the LLNL Technical Information Department. The symposium logo and posters were designed by Pamela Allen and Ellen Baldwin. The editing and paste-up of the papers required for the production of the Proceedings was done in superb style by Catherine Williams and Coralyn McGregor. We thank all of these people for their fine contributions.

The companion's program, those extra activities that were available to the 32 spouses and friends who accompanied scientists to Asilomar, was ably organized by Patricia Hoff. The companions toured historical houses in Monterey, visited the village of Carmel and the Carmel mission, and hiked on Point Lobos. We thank Patricia for developing this fine program and for her advice and help on many operational aspects of the Symposium.

Financing was an important question whose resolution was essential to the success of the Symposium. We were very fortunate to obtain financial support from three different organizational elements within the Lawrence Livermore National Laboratory, namely, the Nuclear Test and Experimental Science directorate through our own Nuclear Chemistry Division, the Physics Department, and the Chemistry Department. I thank C. H. Poppe, R. W. Kuckuck, M. S. Weiss, J. D. Anderson, C. B. Tarter, and C. Gatrousis at LLNL for providing this support. Our thanks, in turn, go to the U.S. Department of Energy for making this Laboratory support possible. Also, we thank the Bicron Corporation and the Canberra Corporation for their financial support.

Lastly, I thank all of the speakers, the authors of the invited and contributed papers, the session chairmen, and everyone who attended the Symposium for their contributions to the Symposium. After all, a truly successful Symposium can happen only with the support and enthusiastic participation of its clientele. Those of you who attended the Symposium at Asilomar in 1990 provided all of this in good measure, and for this I am grateful.

Plans are already underway for the next Symposium in this series which is expected to be held somewhere in Europe in 1993. At this time, a firm decision as to location has not been reached.

Richard W. Hoff
May 13, 1991

Nuclear Structure Models and Fields

MICROSCOPIC SELF-CONSISTENT AND COLLECTIVE MODEL DESCRIPTION OF NUCLEAR STRUCTURE

K.Heyde, C.De Coster, D.Van Neck and M.Waroquier

Institute for Theoretical Physics and Institute for Nuclear Physics
Proeftuinstraat 86 - B 9000 Gent (Belgium)

ABSTRACT

We show that, starting from density-dependent nucleon-nucleon interactions (extended Skyrme forces), one is able to construct an interaction that leads to correct saturation properties in the nucleus and also behaves as a good effective interaction in describing low-lying excitations. This is illustrated in various mass regions.

In those regions where many valence protons and neutrons determine the nuclear structure, one has to resort on collective model descriptions. We shortly point out the importance of symmetry considerations (the Interacting-Boson model). More in particular, we concentrate on those modes of motion where the interplay of protons and neutrons is dominant : the M1 scissor mode, intruder 0^+ excitations and shape coexistence. Results for the Cd region, the Pb region and the $N = 20$ nuclei are presented.

INTRODUCTION

The atomic nucleus, as a quantal system, can be well described as a collection of A interacting nucleons, interacting with a short-range interaction that contains mainly two-body effects but where higher-order terms can be present too. It is the aim to describe both low-lying nuclear excitations (bound states $E < 0$) as well as scattering properties ($E > 0$) starting from a single nucleon-nucleon interaction. This is a very ambitious task, in the light of the rich spectrum of modes of motion appearing in the nucleus.

In the present article we shall discuss (i) the interaction of external fields (p, n, e, γ, ...) with the nucleus **as a test** of mean field properties ; (ii) study excitations in nuclei with only a few valence nucleons **as a test** of the nuclear

4 Description of Nuclear Structure

shell-model methods ; (iii) study excitations in nuclei with open proton and neutron shells **as a test** of collective model descriptions. Here we shall concentrate on some typical proton-neutron excitations (M1 scissor, shape coexistence in various mass regions).

MEAN-FIELD SELF-CONSISTENT CALCULATIONS

Starting from density-dependent nucleon-nucleon interactions[1,2], we have derived average nuclear properties such as the Hartree-Fock field. In the example of ^{208}Pb, the proton- and neutron average fields have been derived and are illustrated in Fig.1. Once the fields $U^{(p)}(r)$ and $U^{(n)}(r)$ are determined, starting from the Hartree-Fock orbitals $\varphi_{\alpha_i}(\vec{r})$ ($\alpha_i \equiv$ all quantum numbers), one can calculate various average properties such as the total binding energy (E), the mass and charge density distributions and the various radii. The extended Skyrme force calculations provide a good description of most of these nuclear quantities for spherical nuclei as is illustrated in Fig.2 for a number of

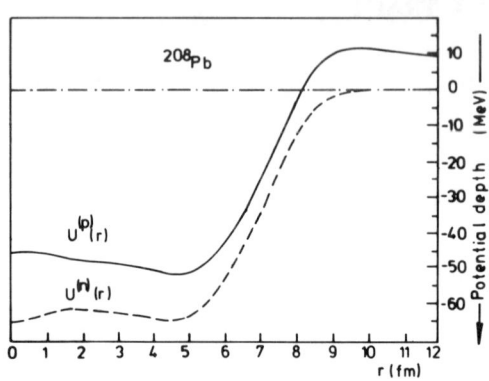

Fig. 1. The calculated self-consistent proton and neutron potentials $U^{(p)}(r)$ and $U^{(n)}(r)$ for ^{208}Pb as determined from an effective Skyrme type interaction.

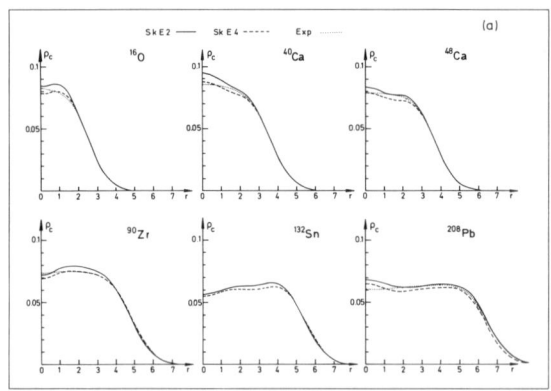

Fig. 2. Charge densities for ^{16}O, ^{40}Ca, ^{48}Ca, ^{90}Zr, ^{132}Sn and ^{208}Pb using self-consistent Hartree-Fock calculations and extended Skyrme forces[1,2] (SkE2, SkE4), compared with the experimental data.

charge density distributions[1].

Even though the concept of nucleon motion in an average mean field is widely accepted, it is still good to carry out, both experimentally and theoretically, stringent tests for this concept.

TEST OF AVERAGE FIELD PROPERTIES

The ideal probe to test the nucleonic motion is the electromagnetic interaction. This can be carried out using electron scattering off the nucleus. The electron now acts as a microscope to study the nucleons moving inside the nucleus. Modifiying the energy ($\hbar\omega$) and momentum ($\hbar\vec{q}$), various details going from the nuclear bulk properties down to details of the nucleonic motion can be scrutinized. The process can be depicted as one where transfer of a certain amount of energy $\hbar\omega$ and momentum $\hbar\vec{q}$ to the nucleus is performed. By varying, at a certain energy, the magnitude of $\hbar\vec{q}$ and measuring the scattered electron (scattering angle, energy, momentum), one can determine the charge density in the nucleus. This process is depicted in Fig.3 in a schematic way. By varying $\hbar\omega$ in the energy region of $\hbar\omega \simeq 100$ MeV, the nucleus will mainly react by emitting a single nucleon out of the nucleus (quasi-elastic regime). Such experiments have been done over the recent years, in particular at Bates (MIT), NIKHEF (Amsterdam) and Saclay.

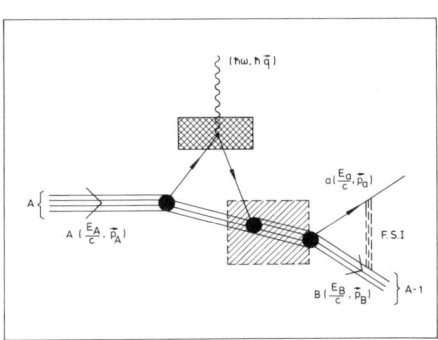

Fig.3. *Electromagnetic interaction, exchanging energy $\hbar\omega$ and momentum $\hbar\vec{q}$ with a nucleus consisting of A nucleons (with energy E_A/c and momentum $\vec{p}_A$). In the above process, a nucleon is emitted from the nucleus (with energy E_a/c and momentum $\vec{p}_a$) leaving an A-1 system (energy E_B/c, and momentum $\vec{p}_B$). The notation FSI denotes eventual final-state interactions between the emitted nucleon and the remaining A-1 nucleons.*

We have studied this process, using the self-consistent Hartree-Fock method, in order to determine the average field, the internal nucleon wave functions as well as the wave functions describing the ejected nucleon[3,4] (Fig.4). By calculating the response of the nucleus to the electromagnetic field,

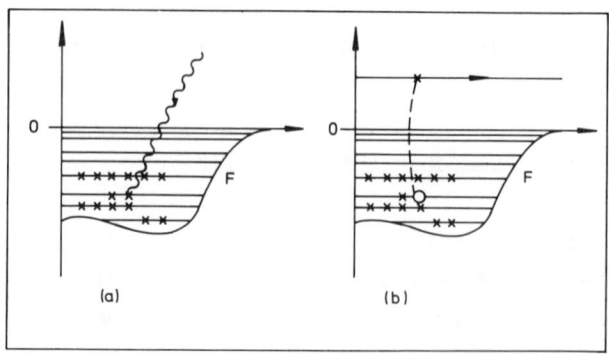

Fig.4. *The self-consistent Hartree-Fock + RPA method used to calculate one-nucleon emission processes. Here, a photon is absorbed by a bound nucleon (part a), leaving the nucleus into a single-hole configuration with one nucleon in a continuum state that can escape, leaving an A-1 nucleus as residual nucleus (part b).*

cross-sections for (e,e'p) and (γ,n), reactions have been calculated. In the case of (e,e'p) reactions e.g. from studying the proton spectra in coïncidence with the scattered electron (exclusive scattering), one can determine the nuclear spectral functions i.e. the Fourier transform of the nucleon single-particle wave function in the nucleus. If the independent particle shell-model picture has to have some sense, the spectral function (or the velocity distribution function (Fig.5)) should be well described by the Hartree-Fock single-particle wave functions. Here, indeed, rather good agreement results.

One can even go one step beyond by calculating the coupling of the single-particle motion in the nucleus with polarization and correlation effects of the nuclear medium. Such calculations were carried out recently, again in a fully self-consistent way[5], which we illustrate for the proton-hole states in ^{89}Y.

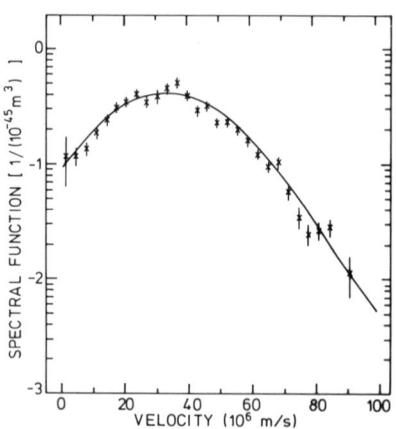

Fig.5. *The experimental spectral function for a $1p_{3/2}$ proton in ^{16}O. The theoretical distribution is also given.*

TEST OF THE NUCLEAR SHELL MODEL NEAR CLOSED SHELLS

Besides an explanation of the global aspects of the nuclear characteristics, one should be able to describe at the same time and using the same nucleon-nucleon interaction, the finer (local) details of the nuclear many-body system. Such a challenging set of calculations was carried out some time ago starting from Skyrme forces[6] (SkE). In studying the finer details of the nuclear excitations one has the advantage that the larger part of all nucleons form a rather inert core determined by the nearest closed shells and thus by the average field characteristics. So, only a very limited number of nucleons determines most of the low-lying nuclear characteristics. One has to solve the Schrödinger equation in a limited model space. Such calculations are quite feasible at an near to closed-shell configurations. Provided the effective interaction so used shows the correct short-range pairing properties (Fig.6), one can test the nuclear shell-model methods nicely. Extensive tests for doubly-closed shell nuclei (^{16}O, ^{40}Ca, ...), nuclei with two-particle or two-hole configurations outside a closed shell (^{18}O, ^{42}Ca, ..., ^{134}Te, ...) and even for single-closed shell nuclei (Sn nuclei) have been carried out [1,6]. We illustrate the state of the art self-consistent shell-model calculations for the nucleus ^{116}Sn (Fig.7) where good agreement is obtained ; clearly as good as quasi-particle calculations specially adapted for this particular mass region (s.p. energies, two-body forces, model space, ...) do.

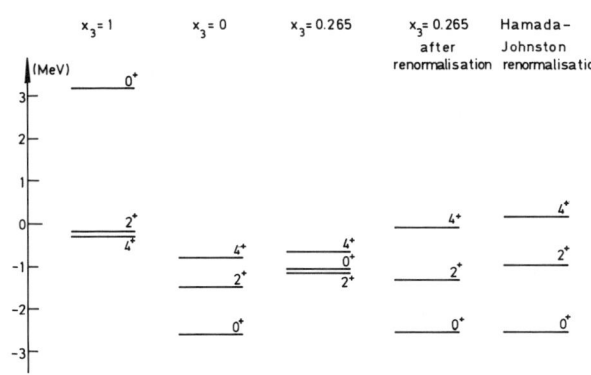

Fig.6. *Antisymmetrized and normalized neutron two-body matrix elements for the $(1d_{5/2})^2$ configuration in ^{18}O, evaluated with the SkE4 force (ref. 1,2). Various repartitions between a density-dependent two-body force ($x_3=0$) and a zero-range three-body force ($x_3=1$) are illustrated as well as the optimal fit with $x_3=0.265$ (before and after renormalization). Comparison with the Hamada-Johnston interaction is also presented.*

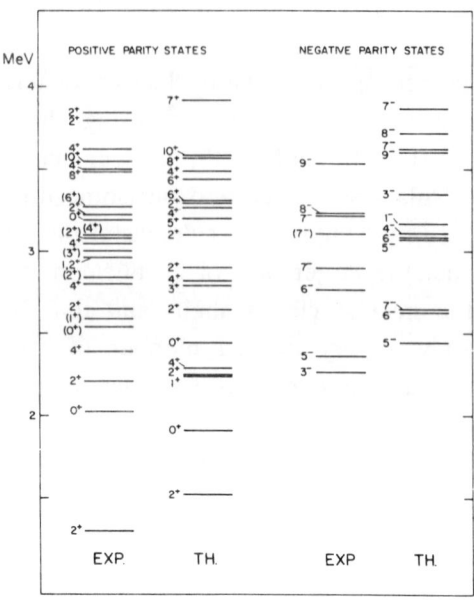

Fig.7. *Experimental and calculated low-lying positive and negative parity levels in* ^{116}Sn. *Levels belonging to the intruder band based on a 2p-2h proton excitation are not included in the figure (ref.6).*

Large-scale shell-model calculations when many more valence nucleons are present (Wildenthal, Brown[7,8]) e.g. in the full sd-model space show some important shortcomings in the particular mass region (N = 20; Z≃11,12). This poses serious questions that point towards possible effects caused by nuclear deformation setting in at various regions of the nuclear mass table. To this end, we examine what a study of collective modes of motion can highlight about the nuclear dynamics in those regions.

TEST OF NUCLEAR COLLECTIVE MOTION : OPEN-SHELL NUCLEI

The shell-model is not well suited to describe modes of motion when many valence protons and neutrons are present at the same time. Nuclear excitations quite often contribute in a coherent way : collective vibrations, rotations, ... of the nuclear density distributions then result.

It was shown, in particular by Arima and Iachello[9], in recent years, that specific truncations to the huge shell-model space, making use of symmetry properties of the nuclear many-body problem, gave rise to an elegant treatment of nuclear collective motion. Taking into account only the 0^+ and 2^+ coupled nucleon pairs that are largely favoured by the short-range character of the residual nucleon-nucleon interaction (called S and D pairs, respectively) and mapping these fermion pairs to s and d boson states, the Interacting Boson model (IBM) is obtained. This IBM model, describing the various group chain reductions of the SU(6) group, contains nuclear collective excitations such as anharmonic vibrations (U(5) limit), rotations (SU(3) limit) and γ-unstable motions (O(6) limit). Various extensions, such as the study of odd-A nuclei making use of Boson-Fermion symmetries[10,11] and extensions treating the proton and neutron degrees of freedom (IBM-2)[12,13] explicitely have been highly successful.

Here we shall concentrate on some typical proton-neutron excitations that give rise to interesting new modes of motion : the proton-neutron non-symmetric modes of motion and intruder 0^+ excitation at low-energy serving as the band-head of nuclear collective bands.

a. Proton-neutron non-symmetric motion

When one considers proton and neutron excitations, the resulting combined modes of motion are not always symmetric in the charge variables. This can be the case for both vibrations and rotational excitations (Fig.8) where, in the latter case, a relative scissor M1 mode of motion results in simple collective models. Suggested to occur theoretically near $E_x \approx 3$ MeV in deformed nuclei as $J^\pi = 1^+$ states with B(M1;↑) values of 2 to 3 μ_N^2, such

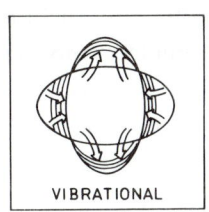

VIBRATIONAL

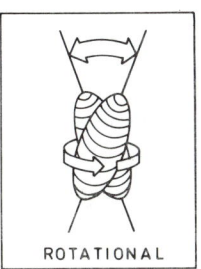

ROTATIONAL

states were indeed shown to exist, via (e,e') scattering[14]. Detailed experimental and theoretical studies in recent years have pointed towards a more complex description where the underlying shell-model properties cannot all be averaged out into purely collective motion. A number of contributions in the present conference concentrate on this topic in more detail[15-19]. Recently some interesting new results, relating to both the nature of the orbital M1 strength around $E_x \approx 3$ MeV and the spin strength at higher energies, have been obtained. First, a close correlation between the summed M1 strength (which is mainly orbital in character up to $E_x \approx 3.5$ MeV) and the

Fig.8. Schematic illustration of non-symmetric collective proton-neutron modes. In the upper part, an isovector quadrupole vibrational excitation is shown whereas in the lower part, an isovector rotational mode of motion (called scissor mode) is illustrated.

nuclear ground-state equilibrium quadrupole deformation was observed[16,20]. Secondly, the unperturbed spin-flip 1^+ strength relating to $\ell,j=\ell+1/2 \rightarrow \ell,j=\ell-1/2$ excitations and located between 4 and 8 MeV is dramatically changed, through the residual interactions. Spin strength is pushed up and separates out two groups of 1^+ spin-flip excitations. Results for ^{154}Sm will be discussed by C.De Coster in more detail during this conference.

b. Proton-neutron interaction : shape coexistence and intruder excitations

A quite unexpected consequence of the strongly attractive proton-neutron interaction is the observation of very low-lying $J^\pi = 0^+$ excitations near single-closed shell nuclei. Normally, closed shells form an inert core which stabilizes the motion of the remaining valence nucleons. If now, single-closed shell nuclei are studied[21-23] (Sn nuclei, Pb nuclei, ...) in which the number of valence nucleons is maximal, the closed shell becomes unstable against 2p-2h excitations across the closed shell. Even though the unperturbed energy for such excitations is very high and of the order of 7-10 MeV in medium-heavy and heavy nuclei, the proton-neutron interaction modifies the nucleon motion in an important way so as to give rise to very low-lying $J^\pi = 0^+$ excitations. A most dramatic illustration of these "intruder" excitations was shown to exist in the Pb region by the LISOL group in Leuven[23]. Such excitations correspond also to states with a large quadrupole deformed shape which quite often can give rise to shape coexistence.

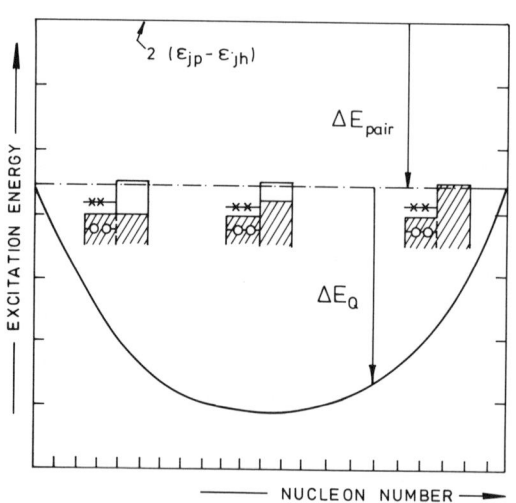

Fig.9. Schematic picture containing the various binding energy terms that contribute to the low-lying $J^\pi = 0^+$ 2p-2h intruder excitations. The unperturbed energy $2(\varepsilon_{jp} - \varepsilon_{jh})$, the pairing gain ΔE_{pair} and the proton-neutron interaction energy ΔE_Q are given as a function of the number of valence nucleons.

The detailed mechanism for describing these intruder excitations has been discussed in ref.24. The basic effect, though, is illustrated in Fig.9, in a schematic way. Starting from an unperturbed energy $2(\varepsilon_{j_p} - \varepsilon_{j_h})$ needed to form the 2p-2h excitation, which is taken constant over a given mass region at a single-closed shell, various energy corrections have to be considered. The first correction, ΔE_{pair}, takes into account the extra pairing correlation energy amongst the 0^+ coupled particle and hole pair. This energy gain is also taken as constant over the given mass region. Apart from typical shell-model effects, affecting the single-particle energies with changing nucleon number (relative self-energy corrections or monopole energy shift ΔE_M), that are not changing the 2p-2h 0^+ excitation energy in a major way, the attractive proton-neutron force gives the dominant energy correction. Due to polarization effects of the proton-neutron force, changing 0^+ coupled pairs into 2^+ coupled pairs, both the ground-state and intruder pair distributions become modified as follows

$$|\widetilde{0^+_\pi \otimes 0^+_\nu}> \Rightarrow |0^+_\pi \otimes 0^+_\nu> + \alpha |2^+_\pi \otimes 2^+_\nu> + ... \quad . \tag{1}$$

Using now a residual quadrupole-quadrupole proton-neutron interaction and 0^+ ground-state and intruder wave functions that are approximated by SU(3) wave functions, one derives the quadrupole binding energy gain ΔE_Q as

$$\Delta E_Q \cong 2\kappa \, \Delta N_\pi \cdot N_\nu \quad . \tag{2}$$

It is now in particular the latter term, through the $\Delta N_\pi \cdot N_\nu$ dependence ($\Delta N_\pi = 2$ for a proton 2p-2h 0^+ excitation and N_ν the number of valence neutron pairs), which is causing the intruder 0^+ state to occur at such low energies as is observed in the various single-closed shell regions e.g. the Z=50 (Sn), the Z=82 (Pb), the N=20 nuclei.

Since the above procedure only describes te variation in band-head excitation energy for the 0^+ 2p-2h configuration, one has no description of the nuclear dynamics of the band-structure that develops on top of this. Using configuration mixing within the IBM-2, one can (i) calculate the collective contribution to the excitation energy $\Delta E_{coll.}$ (which simplifies into the term ΔE_Q if only a quadrupole-quadrupole proton-neutron force would be used) and; (ii) describe the band-structure in both the (N_π, N_ν) and $(N_\pi + 2, N_\nu)$ system (for

12 Description of Nuclear Structure

proton 2p-2h excitations) and their mixing. Recent calculations for ^{114}Cd have been performed that give a consistent description of all the low-lying levels up to $E_x \approx 2$ MeV as well as most of the electromagnetic E2 static and decay properties. These results are discussed in detail by J.Jolie (ref.25) and are presented as a poster at the conference.

In those cases where the proton-neutron attractive force is able to overcome the shell-gap energy needed to create the 2p-2h configuration, the ground state can completely change its character from a spherical into a deformed configuration. Illustrations of this extreme case are :

(i) the increase in the nuclear radii for the odd-mass Au nuclei for $A \leq 186$ where the intruding proton configuration from across the $Z = 82$ closed shell becomes the ground-state configuration (Fig.10) and,

(ii) the $N = 20$ neutron rich nuclei around $Z = 11, 12$ defining a small region of increased binding energy relative to the other nuclei within the sd-shell-model space. In the latter region, 2p-2h neutron excitations across the $N = 20$ closed shell modify the ground-state nuclear structure. Both calculations using the simple intruder picture[26] as well as large-scale shell-model calculations[27,28] explain these properties caused mainly by the attractive proton-neutron interaction (Fig.11) between the valence protons and the extra neutron 2p-2h configuration.

A final comment centers around the question where proton (or neutron) 4p-4h, ... np-nh excitations in medium-heavy and heavy nuclei could result[29]. Analyzing the basic contributions to the

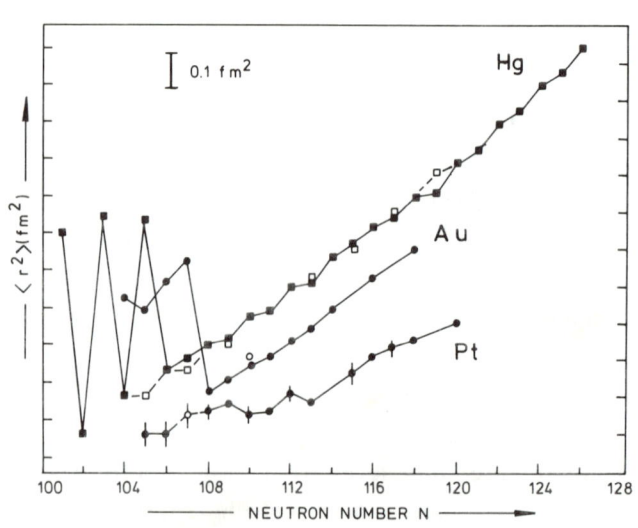

Fig.10. Variation (for both the ground-state (full symbols) and isomeric state (open symbols)) of the nuclear radius for the Hg, Au and Pt nuclei towards the N = 104 mid-shell region.

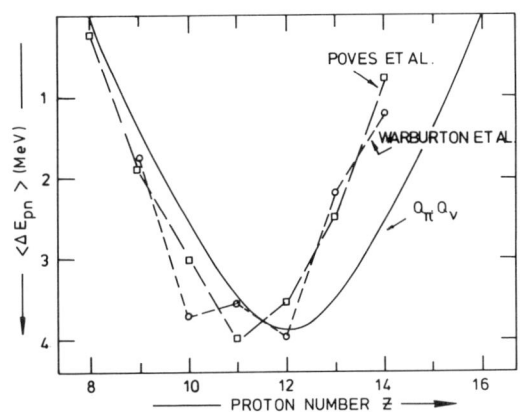

Fig.11. Interaction energy expressing the extra gain in binding energy originating from the proton-neutron force using various shell-model calculations[26-28], as a function of the proton number at $N=20$.

energy of the intruder configurations and comparing with the lowest-lying 2p-2h excitations one observes that both the unperturbed energy, the monopole correction ΔE_M and the proton-neutron interaction energy ΔE_Q scale exactly with the number of particles and holes. Only the pairing energy correction ΔE_{pair} for $\frac{n}{2}$ 0^+ coupled pairs reduces over the energy gain $-G\Omega$ for a single-pair times the number of pairs $\frac{n}{2}$, according to the expression

$$\Delta E_{pair} = -\frac{n}{2} G\Omega \left(1 - \frac{n}{2}\cdot\frac{1}{\Omega} + \frac{1}{\Omega}\right) \quad , \tag{3}$$

with Ω the shell degeneracy and $\frac{n}{2}$ the number of pairs. Combining the above results, one gets as a quite general result that

$$E_{np-nh}(0^+) \simeq \left(\frac{n}{2}\right) E_{2p-2h}(0^+) \quad , \tag{4}$$

where the approximate equal sign is best fulfilled for large Ω and quite small n values. The results for ^{116}Sn and ^{192}Pb are presented in Fig.12.

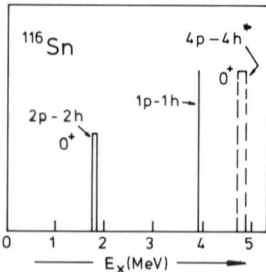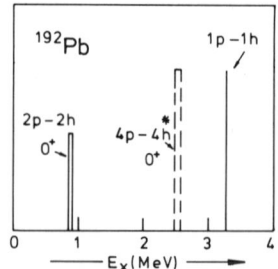

Fig. 12. Calculated energy for the lowest intruder 0^+ 4p-4h proton configurations in ^{116}Sn and ^{192}Pb. The observed 0^+ 2p-2h intruder energy and a typical value for the lowest-lying 1p-1h excitations are also given for comparison.

CONCLUSION

In the present article, we have discussed the results obtained from performing calculations for both the average nuclear properties as well as for the detailed characteristics in nuclear excited states, all starting from a single density-dependent nucleon-nucleon interaction. We point towards various testing grounds for nuclear structure aspects originating from the many-body problem in the nucleus. We also discuss those specific collective modes of motion where the residual proton-neutron interaction plays a dominant role : study of non-symmetric proton-neutron excitations such as the nuclear scissor mode and study of low-lying intruder $J^\pi = 0^+$ excitations and its relation to nuclear shape coexistence. In particular, recent applications to the Cd nuclei, the Pb region and the N = 20 neutron rich nuclei have been discussed.

The work presented here is the result of the theoretical nuclear physics research carried out over the last years in the Institute for Nuclear Physics and the Institute for Theoretical Physics in Gent. We are grateful to J.Ryckebusch, J.Moreau and J.Van Maldeghem for the use of material highlighting these research topics. Much of the results have also been obtained in collaboration with many groups. Here, we mention O.Scholten, A.E.L.Dieperink at the KVI, Groningen; M.Huyse and P.Van Duppen at the IKS, Leuven; J.L.Wood from Georgia Tech. in Atlanta and his various co-workers. We also like to thank A.Richter and U.Kneissl and co-workers for much discussions on $J^\pi = 1^+$ states in nuclei.

REFERENCES

1. M.Waroquier, K.Heyde, G.Wenes, Nucl.Phys. A404, 269; 298 (1983)
2. M.Waroquier, Hoger Aggregaatsthesis, University of Gent, unpubl., 1987
3. J.Ryckebusch, Ph.D.Thesis, University of Gent, unpubl., 1988
4. J.Ryckebusch et al., Nucl. Phys. A476, 237 (1988)
5. D.Van Neck, J.Ryckebusch, M.Waroquier, Phys. Lett., to be publ.
6. M.Waroquier et al., Phys. Repts. 148, 249 (1987)
7. B.H.Wildenthal in : Progr. Part. nucl. Phys. 11, 5 (1984)
8. B.A.Brown, R.Radhi and B.H.Wildenthal, Phys. Repts. 101, 313 (1983)
9. F.Iachello and A.Arima, The Interacting Boson Model (Cambridge Univ. Press, 1988)
10. P.Van Isacker, Hoger Aggregaatsthesis, University of Gent, unpubl., 1983
11. J.Jolie, Ph.D.Thesis, University of Gent, unpubl., 1985
12. P.Van Isacker, K.Heyde, J.Jolie and A.Sevrin, Ann. Phys. (N.Y.) 171, 253 (1986)
13. O.Scholten et al., Nucl. Phys. A438, 41 (1985)
14. D.Bohle et al., Phys. Lett. 137B, 27 (1984)
15. J.Ginocchio, proc. of this conf.
16. C.De Coster and K.Heyde, proc. of this conf.
17. T.Otsuka, proc. of this conf.
18. S.Robinson, proc. of this conf.
19. M.Harder, proc. of this conf.
20. A.Richter, private communication
21. K.Heyde et al., Phys. Repts. 102, 291 (1983)
22. J.Bron et al., Nucl. Phys. A318, 335 (1979)
23. P.Van Duppen et al., Phys. Rev. Lett. 52, 1974 (1984) and Ph.D.Thesis, university of Leuven, unpubl., 1985
24. K.Heyde et al., Nucl. Phys. A466, 189 (1987); A484, 275 (1988)
25. J.Jolie and K.Heyde, Phys. Rev. C, to be publ.
26. K.Heyde and J.L.Wood, J.Phys. G, to be publ.
27. E.Warburton et al., Phys. Rev. C41, 447 (1990)
28. A.Poves and J.Retamosa, Phys. Lett. 184B, 311 (1987)
29. K.Heyde et al., to be publ.

STRUCTURE OF NONROTATIONAL STATES AND DISTRIBUTION OF Eλ STRENGTH IN DEFORMED NUCLEI

V.G. Soloviev

Joint Institute for Nuclear Research, Dubna, USSR

ABSTRACT

The specific properties of the quasiparticle-phonon nuclear model with a phonon operator consisting of the electric and magnetic parts are discussed. It is asserted that magnetic states in deformed nuclei cannot be consistently treated without interactions of the electric type. The structure of nonrotational states of doubly even deformed nuclei is discussed. Distribution of E2 and E3 strength among low-lying states is exemplified by ^{238}U.

1. INTRODUCTION

The energies and wave functions of two-quasiparticle and one-phonon states in doubly even deformed nuclei were calculated in 1960-1975. Good enough description (see refs. [1-5]) of experimental data available at that time was obtained and predictions were made which were later confirmed experimentally. In all subsequent years much progress was achieved in studying high-spin states which moved aside researches on low-spin states in deformed nuclei. Nevertheless, a great amount of new experimental data is needed for a further development of the theory. Advances are obvious in describing in the IBM the rotation bands based on the ground, β and γ vibrational states. The important role is played by the sdg IBM [6] and sdf IBM [7]. New microscopic calculations of nonrotational states of deformed nuclei are made within the quasiparticle-phonon nuclear model (QPNM) [8-12].

2. DESCRIPTION OF NONROTATIONAL STATES IN DEFORMED NUCLEI IN THE QPNM

The QPNM Hamiltonian for nonrotational states of deformed nuclei consists of the average field of a neutron and a proton system in the form of the axial-symmetric Woods-Saxon potential, monopole pairing, isoscalar and isovector particle-hole (ph) and particle-particle (pp) multipole, spin-multipole and tensor interactions. Effective interactions between quasiparticles are taken in a separable form. As is known, separable interactions of the rank $n_{max} > 1$ are widely used in describing nucleon-nucleon interactions and studying three-body nuclear systems and light nuclei. In refs.[13,14], the QPNM has been generalised for separable interactions of any rank n_{max} whereas in refs.[11,12,15] simple interactions with $n_{max} = 1$ were used.

Then, we make the Bogolubov transformation

$$a_{q\sigma} = u_q \alpha_{q\sigma} + \sigma v_q \alpha^+_{q-\sigma}$$

and introduce the RPA phonons $Q_{Ki\sigma}$ and $Q_{Ki\sigma}$. Here $q\sigma$ are quantum numbers of single-particle states, $\sigma = \pm 1$, q equals K^π and asymptotic quantum numbers $Nn_z \Lambda \uparrow$ at $K = \Lambda + 1/2$ and $Nn_z \Lambda \downarrow$ at $K = \Lambda - 1/2$.

In spherical nuclei one-phonon states of the electric type are described by the multipole $\lambda \mu$ and spin-multipole $\lambda' L K$ with $\lambda' = L$ interactions and of the magnetic type are described by the spin-multipole $\lambda' L K$ with $\lambda' = L \pm 1$ and tensor interactions. If in deformed nuclei one introduces independent phonons of the electric and magnetic types, then the number of states will be doubled. Consider, for example, the $K^\pi = 2$ state shown in fig. 1. It can be treated as a one-phonon octupole states with $\lambda \mu = 32$ for which E3 transition from the $I^\pi K = 3^- 2$ level will be enhanced. This state can

18 *Eλ* Strength in Deformed Nuclei

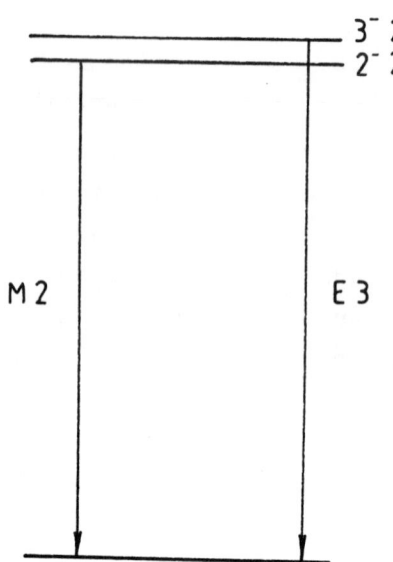

Fig. 1

The first $K^\pi=2^-$ state described either as quadrupole magnetic state with enhancement of M2 transition or as octupole electric state with enhancement of E3 transition.

also be treated as a one-phonon quadrupole magnetic state with $\lambda'LK$ =122 and M2 transition from the $I^\pi K=2^-2$ level can be enhanced.

To avoid the above doubling, the phonon operator with fixed value of $K^{\tilde{\pi}}$ has been introduced in ref. [14] which consists of the electric and magnetic parts in the form

$$Q^+_{Ki_2\sigma} = \frac{1}{2}\sum_{qq'}\{\psi^{Ki_2}_{qq'}[f^{\lambda K}(qq')u^{(+)}_{qq'}A^+(qq';K\sigma) + if^{\lambda'LK}(qq')u^{(-)}_{qq'}\mathcal{O}^+(qq';K\sigma)] - \varphi^{Ki_2}_{qq'}[f^{\lambda K}(qq')u^{(+)}_{qq'}A(qq';K-\sigma) + if^{\lambda'LK}(qq')u^{(-)}_{qq'}\mathcal{O}(qq';K-\sigma)]\}. \quad (1)$$

Here $f^{\lambda K}(qq')$ and $f^{\lambda'LK}(qq')$ are the matrix elements of the multipole and spin-multipole with $\lambda'=L\pm 1$ interactions;

$$A^+(q_1q_2;K\sigma) = \sum_{\sigma'}\delta_{\sigma'(K_1-K_2),\sigma K}\,\sigma'\alpha^+_{q_1\sigma'}\alpha^+_{q_2-\sigma'}$$

or

$$\delta_{K_1+K_2,K}\,\alpha^+_{q_2\sigma}\alpha^+_{q_1\sigma},$$

$$\mathcal{A}^+(q_1,q_2;K\sigma) = \sum_{\sigma'} \delta_{\sigma'(K_1-K_2),\sigma K} \alpha^+_{q_1,\sigma'} \alpha^+_{q_2,-\sigma'}$$

or

$$\delta_{K_1+K_2,K} \, \sigma \alpha^+_{q_2,\sigma} \alpha^+_{q_1,\sigma} \quad ,$$

$U^{(\pm)}_{qq'} = u_q v_{q_1} \pm u_{q_1} v_q$, and $i = 1,2,3,\ldots$ is the root number of the RPA secular equation.

The RPA one-phonon state is described by the wave function $Q^+_{K i_2 \sigma} \Psi_0$, where Ψ_0 is the ground state wave function of a doubly even nucleus, determined as the phonon vacuum. The RPA secular equations including all the above-mentioned interactions have the form of a determinant of the rank $24 \cdot n_{max}^{14}$. With inclusion of separable forces of the rank $n_{max} > 1$ the rank of the determinant increases n_{max} times. The RPA equations for 0^+ states are given in ref. [16].

In the QPNM calculations the exited state wave function is written in the form

$$\Psi_\nu(K_0^{\pi_0}\sigma_0) = \Big\{ \sum_{i_0} R^\nu_{i_0} Q^+_{K_0 i_0 \sigma_0} + \sum_{\substack{K_1 i_1 \sigma_1 \\ K_2 i_2 \sigma_2}} \frac{1}{2} P^\nu_{K_1 i_1} Q^+_{K_1 i_1 \sigma_1} Q^+_{K_2 i_2 \sigma_2} \Big\} \Psi_0 \quad (2)$$

where $\nu = 1,2,3,\ldots$ is the number of a state with given $K_0^{\pi_0}$. The secular equation for the energies E_ν is

$$\det \Big\| (\omega_{K_0 i_0} - E_\nu) \delta_{i_0 i'_0} - \sum_{K_1 i_1 \leq K_2 i_2} (1 + \delta_{K_1 i_1, K_2 i_2})^{-1}$$

$$\frac{U^{K_0 i_0}_{K_1 i_1, K_2 i_2} \, U^{K_0 i'_0}_{K_1 i_1, K_2 i_2} \, [1 + \mathcal{H}^{K_0}(K_1 i_1, K_2 i_2)]}{\omega_{K_1 i_1} + \omega_{K_2 i_2} + \Delta\omega(K_1 i_1, K_2 i_2) - E_\nu} \Big\| = 0 \quad . \quad (3)$$

Here ω_{Ki} are the energies of the RPA equations, the functions $U^{K_0 i_0}_{K_1 i_1, K_2 i_2}$ denote the quasiparticle-phonon interactions, the functions $\mathcal{H}^{K_0}(K_1 i_1, K_2 i_2)$ and $\Delta\omega(K_1 i_1, K_2 i_2)$ are responsible the effect of the Pauli principle in two-phonon terms of the wave function (2). The form of these functions depends on the interactions taken into account. The rank of the determinant (3) equals the number of one-phonon terms in the wave function (2). It is important that eq.(3) coincides in form with the equations from refs. 10-12 taking into account only multipole ph interactions at $n_{max}=1$. The form of eq.(3) is independent of what multipole, spin-multipole and tensor interactions are included and of the rank n_{max} of separable interactions.

It can be concluded that a mathematical apparatus of the QPNM constructed in ref. [14] can serve as a basis for calculations of many characteristics of low- and high-lying states of deformed nuclei.

The necessity of introducing the phonon operator (1) consisting of the electric and magnetic parts can be illustrated by the matrix element of $M\lambda$ transition between one-phonon states. According to [14] it has the form

$$(\Psi_o^* Q_{K_4 i_4 \sigma_4} \mathcal{M}(M\lambda\mu) Q^+_{K_o i_o \sigma_o} \Psi_o) =$$

$$= i \sum_\sigma \delta_{\sigma_o, K_o + \sigma\mu, \sigma_4 K_4} \sum_{\tau=p,n} \sum_{q_1, q_2, q_3}^{\tau} \Gamma(M\lambda\mu; q_1, q_2) \vartheta^{(+)}_{q_1, q_2} \cdot \quad (4)$$

$$\cdot (\Psi^{K_o i_o}_{q_2 q_3} \Psi^{K_v i_4}_{q_3 q_1} + \varphi^{K_o i_o}_{q_2 q_3} \varphi^{K_v i_4}_{q_3 q_i})(f^{\lambda K_o}(q_2 q_3) U^{(+)}_{q_2 q_3} \times$$

$$\times f^{\lambda' L K}(q_3 q_1) U^{(-)}_{q_3 q_1} + \lambda \longrightarrow \lambda' L) ,$$

where $\Gamma_\tau(M\lambda\mu; q_1, q_2)$ is the matrix element of $M\lambda$ transition. It contains the product of electric and magnetic matrix elements, i.e. it equals zero if only magnetic-type interactions are used. This means that in contrast with spherical nuclei in deformed nuclei one cannot describe magnetic states separately, i.e. without electric-type interactions. Electric states and $E\lambda$ transitions can consistently be described without magnetic-type interactions.

The mathematical apparatus developed in [14] can be used to describe M2 and M3 strength distributions and to describe the probabilities of M1, M2 and E2 transitions between excited states.

3. THE STRUCTURE OF NONROTATIONAL STATES IN DOUBLY EVEN NUCLEI

Vibrational states were calculated with single-particle energies and wave functions of the Woods-Saxon potential with monopole and quadrupole pairing, isoscalar and isovector ph and pp multipole interactions at $n_{max}=1$ with the radial dependence $\partial V(r)/\partial r$ where $V(r)$ is the central part of the Saxon-Woods potential. To decrease the number of free constants, the constants of isovector ph interactions $\varkappa_1^{\lambda\mu}=-1.5\varkappa_0^{\lambda\mu}$ and the constants of pp interactions $G^{\lambda\mu}=0.9\varkappa_0^{\lambda\mu}$. The constants of isoscalar ph interactions were chosen in the interval $\varkappa_0^{\lambda\mu} = 0.012-0.022\,fm^2 MeV^{-1}$; they increased with λ and μ and decreased with A.

The states with $K^\pi \neq 0^+$ have been calculated for ^{168}Er, ^{170,172,174}Yb and ^{178}Hf in ref. [15]. The states with any K^π including 0^+ have been calculated in [17] for the isotopes of Th, U, Pu, ^{250}Cf, ^{256}Fm and other nuclei. Good enough description of the relevant experimental data was obtained and predictions were made.

The study of vibrational states with $K^\pi \neq 0^+$ in well deformed doubly even nuclei has shown that the energy and structure of each state are mainly determined by the single-particle energies and wave functions of the Woods-Saxon potential, monopole pairing and ph isoscalar multipole interaction. The multipole ph isovector interaction, quadrupole pairing and multipole pp interaction play an auxiliary role. Inclusion of pp interaction improves the description of vibrational states. Moreover, it proves the RPA to be applicable for the description of states with an energy less than 1 MeV.

Nonrotational states with $K^\pi = 0^-$, 1^-, 2^+, 3^+ and 4^+ with energies up to 2.5 MeV have the dominating one-phonon components. For states with an energy up to 2 MeV the dominating contribution to the wave function normalisation from the one-phonon component is more than 80% and from the two-phonon component not more than 10%. If the single-particle energy of the intruder state $p\,541\!\uparrow$ is lowered by 1 MeV according to the calculations [18], then one gets that in ^{168}Er the calculated energies and $B(E\lambda)$ values are equal to $K^\pi = 3^+_1$, $E_1 = 1.63$ MeV, $B(E4)=0.8$ s.p.u. and $K^\pi = 4^+_1$, $E_1 = 2.13$ MeV, $B(E4) = 0.8$ s.p.u. A rough estimation of the effect of three-phonon terms added to the wave function (2) provides that $B(E2; 4^+_1 \to 2^+_1) = 0.6$ s.p.u.

In describing 0^+ excited states the role of pp interactions is essential as the change of G^{20} entails the change of several low-lying poles of the RPA secular equation. Based on the calculations [17] we can state that introduction of pp interaction improves the description of 0^+ states but this improvement is not sufficient.

Collective vibrational states are not limited by quadrupole and octupole states. It has been shown in [17,19] that in some cases multipole interactions with $\lambda = 5-9$ lead to the mixing of two-quasiproton and two-quasineutron configurations in the states with large K. This mixing (calculated in refs. [17,19]) is exemplified in table 1. In ref. [20], this mixing has been calculated with inclusion of an additional neutron-proton interaction. Perhaps the necessaty of including high-multipolarity interactions is related to the inclusion [21] of higher multipolarity deformations with $\lambda = 5, 6$ and 7 which were found out to be important in the regions of barium and radium.

Table 1

Calculated mixing of two-quasiproton and two-quasineutron configurations

Nuclei	K_ν^π	λ_o^μ $^2\mathcal{K}_o^\mu{}_{-1}$ fm MeV^{-1}	E_ν MeV	Structure (%) nn two-quasi-neutron		Structure (%) pp two-quasi-proton	
^{168}Er	4_1^-	54 0.020	1.0	633↑+521↓	81	411↓+523↑	18
	4_2^-		1.6	− " −	18	− " −	80
^{174}Yb	5_1^-	55 0.024	1.9	624↑+521↓	27	411↓+514↑	72
	5_2^-		2.2	− " −	70	− " −	28
^{176}Hf	6_1^+	66 0.024	1.35	512↑+514↓	73	404↓+402↑	26
	6_2^+		1.75	− " −	27	− " −	71
^{178}Hf	8_1^-	98 0.024	1.1	624↓+514↓	75	404↓+514↑	25
	8_2^-		1.4	− " −	24	− " −	75
^{238}U	5_1^-	55 0.020	1.61	631↓+734↑	93	642↑+523↓	2.2
	5_2^-		1.99	622↑+752↑	93	− " −	5.8
	5_3^-		2.01	− " −	3.4	− " −	87
^{250}Cf	4_1^-	54 0.018	1.2	734↑−620↑	72	660↑+514↓	9
	4_2^-		1.8	− " −	26	− " −	32
^{256}Fm	5_1^+	65 0.020	1.1	613↑+622↓	69	521↑+514↑	21
	5_2^+		1.4	− " −	27	− " −	70
	7_1^-	77 0.020	1.4	622↓+725↑	88	633↑+514↑	10
	7_2^-		1.5	− " −	10	− " −	88
^{260}104	5_1^-	55 0.020	1.0	620↑−725↑	74	624↑+521↓	23
	5_2^-		1.2	− " −	24	− " −	76

4. DISTRIBUTION OF E2 AND E3 STRENGTH AMONG LOW-LYING STATES

The collectivity of the first quadrupole and octupole states and its absence in higher-lying states up to giant resonances undelie phenomenological models including IBM. In ref.[11], for the first time it was shown experimentally that the most collective is not the first but higher lying states $K_\nu^\pi = 3_4^-$ in ^{168}Er and the largest part of E2 strength is concentrated in ^{172}Yb not in the first and second quadrupole states but in the energy interval 2-3 MeV. Table 2 shows the examples of such a nonstandard distribution of E3 strength calculated in refs. 15,17. It would be interesting to get experimental confirmation of distribution like that.

Distribution of E2 and E3 strength among low-lying states is exemplified by ^{238}U in table 3 in the form of sums $\sum_{\mu i} B(E\lambda,\mu;i)$ and $\sum_{\mu i} \omega_{\lambda\mu i} B(E\lambda,\mu;i)$ in the energy interval of 1 MeV. For the states with energy more than 5.5 MeV $B(E2)$ and $B(E3)$ values are very small. The states lying above the first quadrupole ones contain 58% and the states lying above the first octupole states contain 37% of the total strength up to 6 MeV. For the states with energy up to 6 MeV contributions to the energy weighted isoscalar sum rules amount to 15% for the quadrupole and 28% for the octupole states.

Distribution of E3 strength in ^{238}U for each state with $K^\pi = 0^-, 1^-, 2^-, 3^-$, calculated in ref. [17] in the RPA is shown in fig. 2 as $B(E3,i)$-value. It is seen from it that apart from the first there is a great number of collective octupole states, for instance, at energies 3.5, 4.8 and even at 5.4 MeV. The calculations with the wave function (2) have shown that the wave functions of a large number of states in the interval 2-5 MeV contain a large number of two-quasiparticle and two-phonon compo-

Table 2
Nonstandard distribution of E3 strength calculated in QPNM

Nuclei	$I=3$ K_ν^π	E or ΔE, MeV	B(E3) s.p.u.	Nuclei	$I=3$ K_ν^π	E or ΔE, MeV	B(E3) s.p.u.
^{168}Er	3_1^-	1.6	0.14	^{178}Hf	0_1^-	2.0	2.0
	3_2^-	2.1	0.6		0_2^-	2.4	4.0
	3_3^-	2.2	0.3		1_1^-	1.4	0.5
	3_4^-	2.4	2.0		1_2^-	1.5	0.3
					1^-	2.0–3.0	9.5
^{170}Yb	1_1^-	1.5	0.7				
	1_2^-	2.2	1.2	^{238}U	3_1^-	1.6	0.2
	2_1^-	1.6	2.4		3_2^-	1.9	2.0
	2_2^-	1.9	2.0				
	3_1^-	1.6	0.1	^{250}Cf	0_1^-	1.4	3
	3^-	2.0–2.6	3.3		0_2^-	1.8	4
					1_1^-	1.15	8
^{172}Yb	1_1^-	1.2	1.4		1_2^-	1.5	2.0
	1_2^-	2.2	2.2		1_3^-	1.8	2.5
^{174}Yb	0_1^-	1.7	1.0				
	0^-	2.0–3.0	5.0	^{256}Fm	2_1^-	0.9	5
	1_1^-	1.8	0.1		2^-	1.5–2.5	5
	1^-	1.9–3.2	4.2				
	2_1^-	1.2	1.6				
	2_2^-	2.7	1.7				

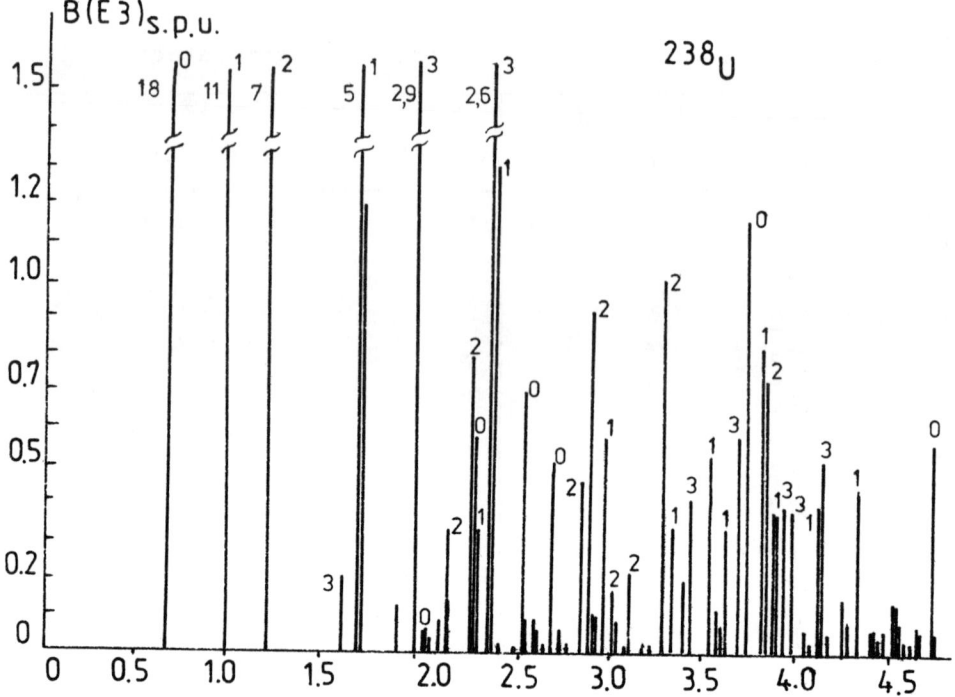

Fig. 2

Distribution of E3 strength among the low-lying states in ^{238}U. Notation: B(E3) are given in single-particle units, large B(E3) values are given on the left from the straight lines and K=0,1,2,3 are given on the right for B(E3) > 0.2 s.p.u.

Table 3.
Distribution of E2 and E3 strength in ^{238}U

	ΔE, MeV					
	1-2	2-3	3-4	4-5	5-6	1-6
$\sum_{\mu=0,2} \sum_{i} B(E2,\mu;i)_{s.p.u.}$	5.4	2.7	0.9	2.4	1.5	12.9
$\sum_{\mu=0,2} \sum_{i} \omega_{2\mu i} B(E2,\mu;i)_{s.p.u.}$	6.1	6.4	3.0	10.5	7.8	33.8
$\sum_{\mu=1,2,3} \sum_{i} B(E3,\mu;i)_{s.p.u.}$	45.2	10.2	9.3	4.6	1.7	71.0
$\sum_{\mu=0,1,2,3} \sum_{i} \omega_{3\mu i} B(E3,\mu;i)_{s.p.u.}$	46.0	25.6	33.7	21.0	7.3	134.0

nents. It can be concluded that experimental study of the $E\lambda$ strength distribution in doubly even deformed nuclei is of great interest.

5. OTHER PROBLEMS

The theme of the symposium allows one to dwell upon two problems concerning neutron resonances. The first is the problem of many-quasiparticle components of the wave functions of neutron resonances which has been posed for the first time in ref.[23]. With increasing excitation energy the structure of states becomes more complex. The contribution to the normalisation of the wave functions of many-quasiparticle components increases whereas the contribution of one-quasiparticle components decreases up to $10^{-4} - 10^{-7}$. Our knowledge of the structure of states, except for high-spin ones, decreases exponentially with increasing excitation energy. The following question is to be answered: Are there low-spin high-lying states whose wave functions contain large, definite many-quasiparticle components? There are many-quasiparticle isomers with high spins whose large life-time is due to the absence of few-quasiparticle components. This indicates a small fragmentation of these states. Low-spin states are fragmented more strongly than high-spin ones. How strong is this fragmentation? Even in the strength distribution of deep hole states pronounced maxima were observed experimentally.

A large contribution of the many-quasiparticle configuration to the normalisation of the neutron resonance wave function will enhance E1, M1 and E2 transitions from neutron resonances to the states lying 1-2 MeV below. Experiments of this type are to be performed with new detectors. The cardinal problem is to be answered: What is the structure of highly excited states: chaos or large many-quasiparticle configurations or both?

The second problem is experimental determination of the dependence of density in highly excited states on spins. For this purpose, the isomer ^{178}Hf with $K^{\pi}=16^+$, energy 2.5 MeV and half-life of 31 years should be used as a target for $(n\gamma)$ reaction. This makes it possible to measure the density of states with spins $31/2^+$ and $33/2^+$ and to obtain the s-wave neutron strength function.

In conclusion I would like to thank L.A.Malov, A.V.Sushkov and N.Yu.Shirikova for help discussions.

REFERENCES

1. C.J.Gallagher, V.G.Soloviev, Mat.Fys.Skr.Den.Vid.Selsk. $\underline{1}$, №11 (1962).
2. V.G.Soloviev, Nucl.Phys. $\underline{69}$, 1 (1965).
3. V.G.Soloviev, Theory of complex nuclei. (Pergamon Press, Oxford, 1976).

4. E.P.Grigoriev, V.G.Soloviev. Structure of even deformed nuclei (Nauka, Moscow, 1974).
5. S.P.Ivanova, A.L.Komov e.a. Part.Nucl. 7, 450 (1976).
6. N.Yoshinaga, Y.Akiyata, A.Arima. Phys.Rev. C38, 419 (1988).
7. A.F.Barfield, B.R.Barrett e.a. Ann.pPhys. 182, 344 (1988).
8. V.G.Soloviev, Part.Nucl. 9, 580 (1978).
9. V.G.Soloviev, Prog.Part.Nucl.Phys. 19, 107 (1987).
10. V.G.Soloviev, Theory of atomic nuclei. Quasiparticle and Phonons. (Energoatomizdat, Moscow, 1989).
11. V.G.Soloviev, N.Yu.Shirikova, Z.Phys. A301, 263 (1981); Yad.Fyz. 44, 1443 (1982).
12. V.O.Nesterenko, V.G.Soloviev e.a. Yad.Fyz. 44, 1443 (1986).
13. V.G.Soloviev, Fizika 22,   (1990).
14. V.G.Soloviev, Preprint JINR E4-90-119, Dubna, 1990.
15. V.G.Soloviev, N.Yu.Shirikova, Z.Phys. A334, 149 (1989).
16. V.G.Soloviev, Z.Phys. A334, 143 (1989).
17. V.G.Soloviev, A.V.Sushkov., N.Yu.Shirikova, Preprint JINR E4-90-449.
18. D.A.Arseniev, S.I.Fedotov e.a. Phys.Lett. 40B, 305 (1972).
19. V.G.Soloviev, A.V.Sushkov, J.Phys. G16, L57 (1990).
20. H.Massman, J.O.Rasmussen e.a., Phys.Rev. C9, 2312 (1974).
21. A.Sobiczewski, Z.Patyk e.a., Nucl.Phys., A485, 16 (1988).
22. I.M.Govil, H.W. Fulbright e.a. Phys.Rev. C33, 793 (1986); C36, 1442 (1987).
23. V.G.Soloviev, Phys.Lett. 42B, 409 (1972).

PSEUDO-SPIN SYMMETRY AND NUCLEAR STRUCTURE

J. P. Draayer
Department of Physics and Astronomy
Louisiana State University, Baton Rouge, LA 70803-4001

ABSTRACT

Consequences of the validity of the pseudo-spin concept, which follows from the fact that the strength of the spin-orbit interaction in the single-particle shell-model hamiltonian is approximately four times the strength of the orbit-orbit term, are reviewed. In particular, it follows that the pseudo-LS-coupled scheme is the natural choice for a many-particle shell-model theory of heavy nuclei. Going beyond the pseudo-LS scheme to situations where deformation inducing terms dominate the residual two-body interaction, as is the case for deformed nuclei of the rare earth and actinide regions, the pseudo-SU(3) scheme and its multi-$\hbar\omega$ extension, the pseudo-symplectic model, are good many-particle shell-model theories. Some preliminary results for ^{238}U are presented. The contribution of intruder states to enhanced quadrupole collectivity in deformed nuclei is also considered.

INTRODUCTION

The three-dimensional isotropic harmonic oscillator, H_0, augmented with the usual one-body spin-orbit ($l \cdot s$) and orbit-orbit (l^2) interactions,

$$H = H_0 + Cl \cdot s + Dl^2, \qquad (1)$$

is known to be a good approximation for the nuclear single-particle hamiltonian.[1,2] The l^2 term, with $D < 0$, pushes high angular momentum states down relative to those with lower l values, a feature that occurs automatically when a more realistic interaction like a Woods-Saxon form is used for the central potential, while the phenomenological $l \cdot s$ term with $C < 0$, which couples space and spin degrees of freedom, is required to achieve shell closures at the magic nucleon numbers 2, 8, 20, 50, 82, 126 and 184. Unfortunately, the required value for C is so large that the spin-orbit interaction destroys the underlying SU(3) symmetry of the oscillator for all but light ($A \lesssim 28$) nuclei, thereby rendering it of little apparent value in attempts at unraveling the structure of heavy ($A \gtrsim 100$) systems. Specifically, for heavy nuclei the $j = n+1/2$ orbital of the n-th oscillator shell, which includes levels with $j = l \pm 1/2$ with $l = n, n-2, \cdots, 1$ or 0, is pushed down among the orbitals of the next lower shell. This yields new shells with normal parity $j = 1/2, 3/2, \cdots, n-1/2$ orbitals plus a $j = n+3/2$ unique parity intruder from the shell above.[3]

In this paper this seemingly unfavorable situation is shown to give way to a much more favorable one because for heavy nuclei $C \approx 4D$ or the Nilsson parameter $\mu = 2D/C \approx 0.5$. First of all, this condition insures that at the $0\hbar\omega$ level pseudo-spin is a good symmetry because the level splitting generated by the $l \cdot s$ and l^2 interactions can be duplicated by a pseudo-oscillator hamiltonian plus a pseudo l^2 interaction, with at most a very small breaking of the pseudo-spin symmetry by a pseudo $l \cdot s$ term.[4,5] Secondly, when the residual deformation driving quadrupole-quadrupole interaction dominates over the pseudo l^2 term, which must be the case when well-developed rotational bands are present, each pseudo-spin symmetry will have associated with it an yrast band that is dominated by its leading irreducible representation (irrep) of pseudo SU(3), the symmetry group of the pseudo oscillator.[6] And finally, going beyond pseudo SU(3) to its pseudo-symplectic extension, which incorporates major shell ($2n\hbar\omega$, $n = 1, 2, ...$) mixing into the picture, reenforces the goodness of the pseudo SU(3) picture.[7] As a consequence, strongly deformed configurations, especially superdeformed bands, are expected to be simple when expressed in the framework of the pseudo SU(3) model and its pseudo symplectic extension.

PSEUDO-SPIN SYMMETRY

A straightforward way to gain an appreciation for the significance of the pseudo-spin concept is shown in Figure 1 where eigenvalues of H are plotted as a function of the Nilsson parameter $\mu = 2D/C$ that measures the relative strength of the l^2 and $l \cdot s$ terms. For the special value $\mu = 0.5$, the orbital pairs with $j = l + 1/2$ and $j = (l+2) - 1/2$ are degenerate for all l values. Furthermore, the splitting of these degenerate pairs follows a $\tilde{l}(\tilde{l}+1)$ rule where $\tilde{l}$ is the average l of the pair, $\tilde{l} = [l+(l+2)]/2 = l+1$. This association can be characterized as a special "normal $\leftrightarrow$ pseudo" unitary transformation that makes this degeneracy that of pseudo spin-orbit partners, $\tilde{j} = \tilde{l} + \tilde{s}$ where $j = \tilde{j}$, $\tilde{l} = l+1$ and $\tilde{s} = 1/2$. For the single-particle basis states this transformation has the form,

$$|\tilde{n}(\tilde{l},\tilde{s})\tilde{j}\tilde{m}\rangle = U_{njm,\tilde{n}\tilde{j}\tilde{m}}(l,\tilde{l}) |n(l,s)jm\rangle,$$

$$U_{njm,\tilde{n}\tilde{j}\tilde{m}}(l,\tilde{l}) = \delta_{n-1,\tilde{n}} \delta_{j,\tilde{j}} \delta_{m,\tilde{m}} \delta_{l \pm 1/2, \tilde{l} \mp 1/2}. \quad (2)$$

From the structure of $U_{njm,\tilde{n}\tilde{j}\tilde{m}}(l,\tilde{l})$ it is clear that the transformation is simply a relabeling of the basis states with all levels of the n-th shell, except the one with $j = n+1/2$, associated with levels of the $\tilde{n}$-th shell of another oscillator, dubbed a "pseudo" oscillator by its originators even though its algebraic properties are identical to those of the "normal" oscillator, where $\tilde{n} = n - 1$. Under this mapping the single-particle hamiltonian transforms as follows:

$$H_0 + Cl \cdot s + Dl^2 \quad \underset{\text{— pseudo} \rightarrow}{\overset{\leftarrow \text{normal} —}{\text{————————}}} \quad \tilde{H}_0 + (4D-C)\tilde{l} \cdot \tilde{s} + D\tilde{l}^2 + (\hbar\omega + 2D - C). \quad (3)$$

Since the $(\hbar\omega+2D-C)$ term is a constant, the pseudo form for the interaction, $\tilde{H} = \tilde{H}_0 + \tilde{C}\tilde{l}\cdot\tilde{s} + \tilde{D}\tilde{l}^2$, has the *same* excitation spectrum as the normal one, $H = H_0 + Cl\cdot s + Dl^2$, when $\hbar\tilde{\omega} = \hbar\omega$, $\tilde{C} = (4D-C)$ and $\tilde{D} = D$. The transformation is physically meaningful because for real (A ≳ 100) systems $C \approx 4D$ so $\tilde{C} \approx 0$. Specifically, as indicated in the figure, $\mu_\nu \approx 0.4$ and $\mu_\pi \approx 0.6$. Here ν denotes neutrons and π protons. This places heavy nuclei very close to the exact pseudo-spin limit ($\mu = 0.5$) of the theory.[8] In particular, the average value for μ is almost exactly 0.5. The familiar single-particle shell-model hamiltonian for heavy nuclei can therefore be replaced by a less familiar but equivalent pseudo form which is inherently simpler because it has a much smaller spin-orbit term.

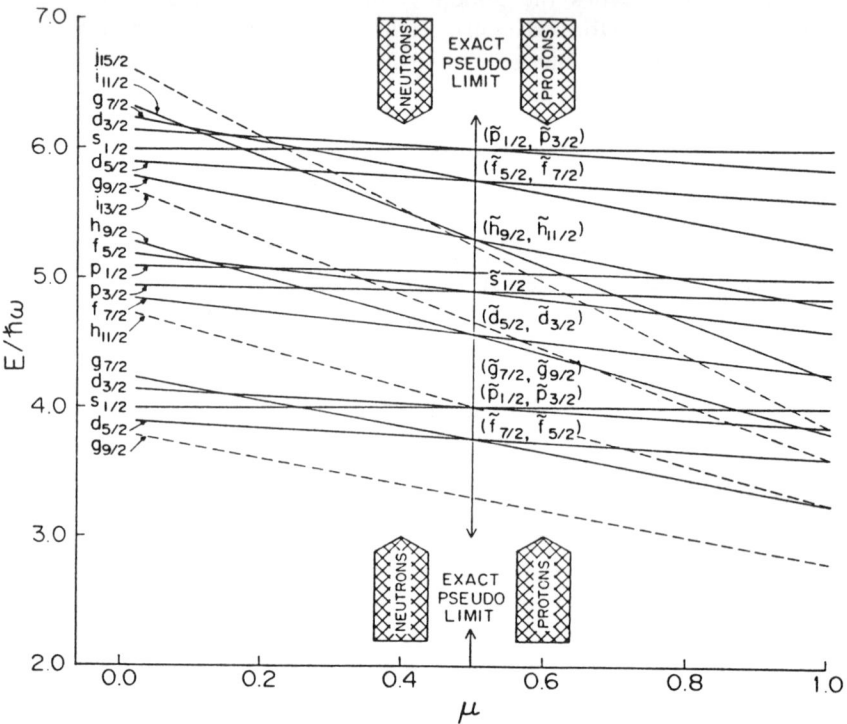

FIGURE 1. Plot of the eigenvalues of the reduced single-particle hamiltonian given by $H/\hbar\omega = n - \kappa(2l\cdot s + \mu l^2)$, where $\mu = 2D/C$ and $\kappa = -C/2\hbar\omega$, for the specific value $\kappa = 0.05$ and $0.0 \leq \mu \leq 1.0$. Notice that the $j = (l+2)-1/2$ and $j = l+1/2$ levels are degenerate for $\mu = 0.5$. Indeed, as shown in the text, the $\mu = 0.5$ spectrum can be duplicated by a simpler hamiltonian $H/\hbar\tilde{\omega} = \tilde{n} - \kappa\mu\tilde{l}^2$ where $\hbar\tilde{\omega} = \hbar\omega$ and $\tilde{n} = n-1$ with $j = l + s = \tilde{l} - \tilde{s}$ where $\tilde{l} = l+1$ and $\tilde{s} = 1/2$. In addition, in each pseudo shell there is the abnormal parity intruder level, shown as dashed, with $j = (n+1)+1/2 = n+3/2$ from the shell above. Since heavy nuclei fall near the $\mu = 0.5$ limit (~0.6 for protons and ~0.4 for neutrons) a many-particle description in terms of basis states of the pseudo oscillator is expected to be superior to one that uses normal oscillator basis states.

An operator that transforms $Cl \cdot s + Dl^2$ into $(4D-C)\tilde{l}\cdot\tilde{s} + D\tilde{l}^2 + (2D-C)$ has been identified, namely, $U = \exp(i\pi h) = 2ih$ where $h = \tilde{s}\cdot\hat{r}$ is the helicity.[9] It is important to note, however, that unlike $U_{njm,\tilde{n}\tilde{j}\tilde{m}}(l,\tilde{l})$ this U does not carry H_0 into $\tilde{H}_0 + \hbar\omega$. So while it is true that the helicity form U is simple and attractive because it transforms the one-body spin-orbit and orbit-orbit terms in H into their pseudo-spin counterparts and, furthermore, leaves operators that depend only on spatial coordinates invariant, like interactions that generate deformation and the electric multiples, it does not commute with the kinetic energy operator and as a consequence it does not effect the "normal $\leftrightarrow$ pseudo" transformation that is an integral part of the pseudo-spin concept. In particular, the helicity transformation does not lead to a realization of the pseudo-LS, pseudo SU(3), and pseudo symplectic symmetries that, as is shown below, are important parts of a many-particle shell-model theory of heavy nuclei.

PSEUDO LS-COUPLED SHELL MODEL

As indicated above, the pseudo scheme organizes the normal parity $j = 1/2$, $3/2, \cdots, n-1/2$ levels of the n-th oscillator shell into a pseudo oscillator shell with $\tilde{n} = n-1$. For example, the $(4s_{1/2}, 2d_{3/2}, 2d_{5/2}, 0g_{7/2})$ normal parity levels of the $n = 4$ shell are mapped onto the $(3\tilde{p}_{1/2}, 3\tilde{p}_{3/2}, 1\tilde{f}_{5/2}, 1\tilde{f}_{7/2})$ orbitals of a $\tilde{n} = 3$ shell. In fact, this mapping of single-particle orbitals defines the pseudo coupling scheme. To grasp its full significance, recall that the dynamical symmetry group for the usual many-particle generalization of the single-particle theory, with particles distributed among the lowest available single-particle levels, is U(NM), the unitary group in (N by M) dimensions, where $N=(n+1)(n+2)/2$ is the spatial degeneracy of the n-th oscillator shell and M=2 or 4 for a spin or spin-isospin formulation of the theory. The U(N) $\otimes$ U(M) direct product subgroup of this U(NM) group separates the full (N by M) dimensional space into its spatial and spin (M=2) or spin-isospin (M=4) parts. Irreps of U(N), which are labeled by a Young pattern $[f] = [f_1, f_2, ..., f_N]$, specify the space symmetry while irreps $[f^c] = [f_1^c, f_2^c, ..., f_M^c]$ of U(M), which must be related to the [f] of U(N) by row-column interchange to insure overall antisymmetry in U(NM) as required by the Pauli Exclusion Principle, label the complementary spin or spin-isospin symmetry, as appropriate.[10] The breakup of the full model space into irreps of U(N) $\otimes$ U(M) for M = 2 and subgroups of these symmetries is shown schematically in Figure 2.

An important difference between the two one-body interactions is that the $l \cdot s$ term couples different spatial ([f]) symmetries whereas l^2 does not. When the strength of all terms like $l \cdot s$ that couple different spatial symmetries is small relative to others like l^2 that do not, the [f] and therefore [f^c] will be good quantum numbers. Whereas this is so for the many-particle extension of $\tilde{H}$, see below, it certainly is not for H. This feature is important because the full model space can then be partitioned into disjoint (pseudo) subspaces $\{[\tilde{f}] \& [\tilde{f}^c]$ of $\tilde{U}(\tilde{N}) \otimes \tilde{U}(\tilde{M})\}$ that have much smaller dimensions than can be realized with the normal scheme. In addition, as is known to be the case for the surface delta interaction[4] and, as is demonstrated below for the quadrupole-quadrupole interaction, if the residual two-body interaction is a pseudo-space scalar operator it actually reenforces the goodness of the pseudo $[\tilde{f}]$ and $[\tilde{f}^c]$ symmetries. The end result is that instead of dealing with the full model space, one can deal with a much smaller subspace comprised of a collection of a few important spatial symmetries.

For heavy nuclei the valence protons and neutrons occupy different shells so an identical particle ($\tilde{M}=2$) formulation must be applied to each. Specifying the $[\tilde{f}]$ and $[\tilde{f}^c] = [\tilde{f}_1^c, \tilde{f}_2^c]$ labels is then equivalent to specifying the total number of normal parity particles and their pseudo spin, $\tilde{m} = \tilde{f}_1^c + \tilde{f}_2^c$ and $\tilde{S} = (\tilde{f}_1^c - \tilde{f}_2^c)/2$. Each normal parity $\tilde{m}$-particle space, with $\tilde{m} = \tilde{m}_\pi$ for protons and $\tilde{m} = \tilde{m}_\nu$ for neutrons, is partitioned into subspaces with $\tilde{S} = 0, 1, 2, 3, ..., \tilde{S}_{max}$ for $\tilde{m}$ even and $\tilde{S} = 1/2, 3/2, 5/2, ..., \tilde{S}_{max}$ for $\tilde{m}$ odd, where $\tilde{S}_{max}$ is the minimum of $\tilde{m}/2$ and $\tilde{N}-\tilde{m}/2$. This means that $\Delta\tilde{S}$ is always an integer. Also, to each $\tilde{S}$ there is a complementary set of spatial configurations. To the extent the pseudo-spin symmetry is good, one therefore expects to observe sets of states, like rotational sequences, that differ in *total* angular momenta ($J = \tilde{L} + \tilde{S}$) by integer (even-A compared with even-A) or half-integer (odd-A with even-A) amounts. Collective states that differ by unit alignment are therefore a natural consequence of good pseudo-spin symmetry.[11]

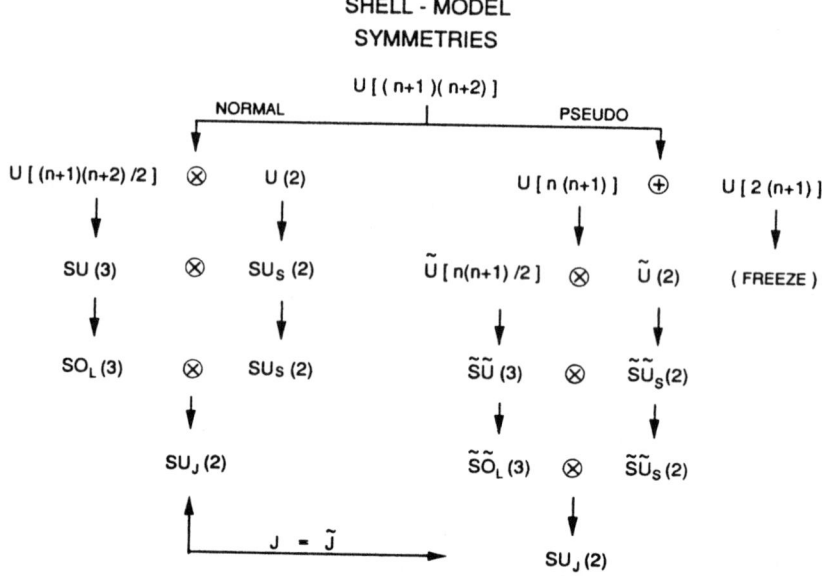

FIGURE 2. Schematic diagram showing both normal and pseudo symmetries for identical nucleons in the n-th major shell of a three-dimensional isotropic harmonic oscillator potential. The total degeneracy is $\Sigma_j(2j+1) = (n+1)(n+2)$. Under the pseudo-spin decomposition this breaks up into two parts, a $(2j_{max}+1) = 2(n+1)$ subspace that is frozen out because the j_{max} level dips below the Fermi level and the remaining $\Sigma_j'(2j+1) = (n+1)(n+2) - 2(n+1) = n(n+1)$ valence space that can be mapped onto an oscillator shell of one less quanta, $\tilde{n} = n-1$.

In considering the alignment, it is very important to understand that it can be either proton or neutron in origin or a combination of the two because the basis states are coupled configurations: $|\Psi^J\rangle = |[(\tilde{\alpha}_\pi \tilde{L}_\pi \tilde{\alpha}_\nu \tilde{L}_\nu)^{\tilde{L}} \times (\tilde{S}_\pi \tilde{S}_\nu)^{\tilde{S}}]^J\rangle$, where $\tilde{\alpha}_\kappa$ labels multiple occurrences of the $\tilde{L}_\kappa$ values and $\kappa = (\pi,\nu)$. The symmetry group of this combined π-ν system is $[\tilde{U}_\pi(\tilde{N}_\pi) \otimes \tilde{U}_\pi(2)] \otimes [\tilde{U}_\nu(\tilde{N}_\nu) \otimes \tilde{U}_\nu(2)]$. This direct product structure can be reordered as for $|\Psi^J\rangle$ so the pseudo-space and pseudo-spin associations are made first, $[\tilde{U}_\pi(\tilde{N}_\pi) \otimes \tilde{U}_\nu(\tilde{N}_\nu)] \otimes [\tilde{U}_\pi(2) \otimes \tilde{U}_\nu(2)]$. In this expression, $\tilde{N}_\kappa = (\tilde{n}_\kappa+1)(\tilde{n}_\kappa+2)/2$ is the pseudo-space degeneracy of the $\kappa = (\pi,\nu)$ subshell. This argument demonstrates that the observation of simple alignment is consistent with good total pseudo-spin symmetry, $\tilde{S} = \tilde{S}_\pi \times \tilde{S}_\nu$, provided the π-ν interaction like the π-π and ν-ν terms conserves $\tilde{S}$. As is demonstrated below, the real quadrupole-quadrupole, $Q^\pi \cdot Q^\nu$, which does connect configurations with different $\tilde{S}$ symmetry, but only weakly as compared to the symmetry preserving couplings, is such an interaction.

PSEUDO SU(3) COUPLING SCHEME

The importance of the SU(3) model for light nuclei follows from the dominance of the quadrupole-quadrupole interaction, Q·Q, over the one-body l·s and l^2 terms as well as over all other two-body forms.[12,13] Even though the spin-orbit interaction is relatively strong, yrast states of nuclei like ^{20}Ne and ^{24}Mg are typically 60-80% pure leading SU(3) representations.[14] This feature can be understood as follows: First of all, SU(3) is a subgroup of U(N) with Casimir invariant $C_2 = (Q \cdot Q + 3L^2)/4$. Hence Q·Q conserves spatial symmetry. In addition, since the expectation value of Q·Q is proportional to the square of the deformation, Q·Q further subdivides each U(N) irrep [f] into (λ,μ) irreps of SU(3) with the most deformed of these lying lowest and the least deformed highest.

The amount and sharpness of the separation, first into irreps of U(N) and then by SU(3), depends on the relative strength of the symmetry preserving and symmetry breaking interactions. And this in turn depends upon whether or not the available space supports strongly deformed configurations. In the ds-shell case, systems like ^{20}Ne and ^{24}Mg have leading (λ,μ)'s with relatively large deformation so Q·Q overpowers the other interactions and yrast states have good [f] and (λ,μ) quantum labels. In a restricted space like the $d_{5/2}$ subshell of the ds shell, however, the same hamiltonian will not display the same level of quadrupole collectivity and other interactions like pairing might very well appear to dominate. The success of the SU(3) model in the ds-shell shows that the many-particle dynamics can promote (quadrupole) collectivity over single-particle and other noncollective effects.

We now argue that the pseudo SU(3) scheme,[6] with $\widetilde{SU}(3)$ standing in the same relationship to $\tilde{U}(\tilde{N})$ as SU(3) does to U(N), see Figure 2, provides a similar explanation for observed quadrupole collectivity in heavy deformed nuclei. There are, of course, some differences: 1) the valence neutrons and protons occupy different major shells, 2) the normal Q·Q interaction is not the quadratic invariant of $\widetilde{SU}(3)$, and 3) whereas for the ds shell the coefficient D of l^2 is positive, for heavy nuclei it is always negative. Regarding the first of these issues, a very simple calculation has been done which demonstrates that the protons and neutrons being in different shells makes very little difference so long as they interact, albeit even weakly, through their quadrupole fields. Specifically, even if the separate proton and neutron interactions are dominated by pairing, a small $Q_\pi \cdot Q_\nu$ interaction between the two suffices to drive the whole system towards the strong coupled pseudo SU(3) limit of the theory.[15]

To dispense with the second matter, recall that the pseudo-spin scheme is an excellent starting point for a many-particle description of heavy nuclei, whether they are deformed or not. In particular, the $\tilde{l}\cdot\tilde{s}$ interaction is weak relative to $\tilde{l}^2$ so the pseudo-spin symmetry is good. Also, since the surface delta interaction, which is known to be a reasonably good effective interaction for many applications, is a pseudo-spin scalar operator, the residual two-body interaction is not expected to change this picture very much by inducing additional mixing among pseudo-spin symmetries.[4] And even more importantly, although Q·Q is not an invariant of $\widetilde{SU}(3)$, under the "normal ↔ pseudo" mapping it transforms into its pseudo counterpart plus small corrections,[16]

$$Q\cdot Q = \kappa \tilde{Q}\cdot\tilde{Q} + \cdots . \qquad (4)$$

Indeed, within the leading pseudo-space symmetry, the sum of all correction terms has been shown to induce less than a 1% change in calculated yrast states.[17]

To explore the third and final question, which concerns the relevance of the pseudo SU(3) scheme for strongly deformed nuclei for which D < 0, which is a condition that helps to destroy SU(3) in the fp-shell, consider the many-body problem (tildes now suppressed for simplicity) with hamiltonian,

$$H = H_0 + C\sum_i l\cdot s + D\sum_i l_i^2 - \frac{1}{2}\chi Q\cdot Q = H_0 + C\sum_i l\cdot s + D\sum_i l_i^2 - \frac{1}{2}\chi(4C_2 - 3L^2). \qquad (5)$$

The last form for H follows because within a major oscillator shell Q·Q = $4C_2 - 3L^2$. Here C_2 is the quadratic Casimir invariant operator of SU(3) which has eigenvalue $\lambda^2+\lambda\mu+\mu^2+3(\lambda+\mu)$ in the (λ,μ) representation and L^2 is the square of the total angular momentum with eigenvalue L(L+1). Since the C_2 and L^2 interactions are diagonal in an SU(3) basis they split but do not break the oscillator symmetry. And because χ is always positive, the SU(3) representations with the largest eigenvalue for C_2 and hence the greatest intrinsic deformation, since $\beta^2 \sim \langle Q\cdot Q\rangle = \langle 4C_2\rangle$ in L=0 states, lie lowest.

Energies of calculated 0^+ states of this hamiltonian with C = 0 and D < 0 are plotted in Figure 3 as a function of χ for the simple but representative case $(ds)^4[f]=[4]$ with $(\lambda\mu)$'s = (8,0), (4,2), (0,4), and (2,0). This pseudo ^{20}Ne case, which differs from the real ^{20}Ne nucleus in that D < 0 instead of being positive and C = 0 instead of being nonzero, shows that under conditions very similar to those for pseudo SU(3) applications in rare earth and actinide nuclei the symmetry breaking *decreases* sharply as the strength of the deformation inducing quadrupole-quadrupole interaction *increases*. A χ of 0.06-0.07 is probably realistic for both a real and pseudo ^{20}Ne application of the theory. In the figure the dashed curve is the $0^+_1(8,0)$ intensity when a spin-orbit interaction (C < 0) with one-tenth its normal strength is included in H. This too is an appropriate for pseudo SU(3) applications. Yrast states of heavy deformed nuclei are therefore expected to be dominated to at least the 80% level by the leading pseudo SU(3) symmetry.

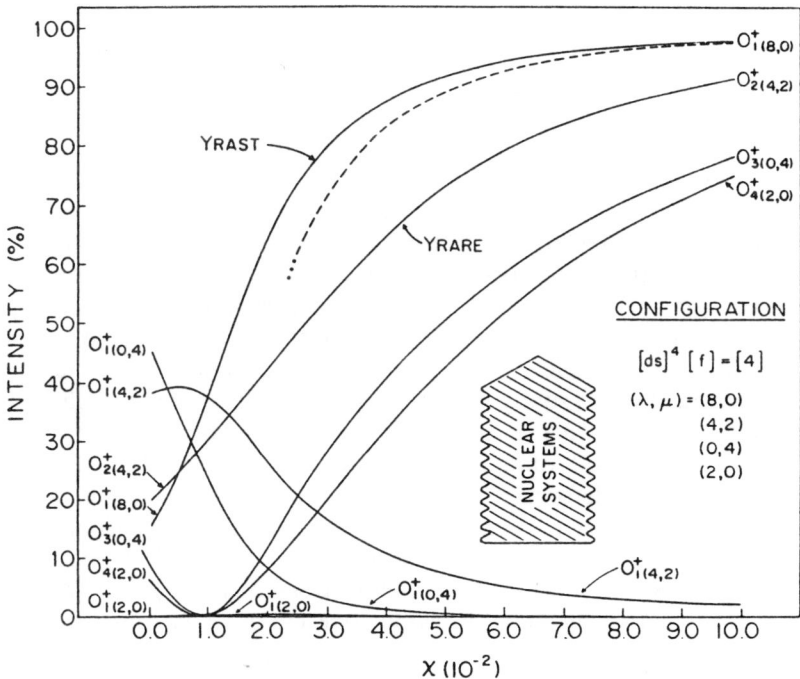

FIGURE 3. Intensity of SU(3) representations in calculated $L^\pi = 0^+$ states of the model hamiltonian (4) in the space $(ds)^4[f]=[4]$ with $(\lambda\mu)$'s = (8,0), (4,2), (0,4), and (2,0). The parameter D was set at -0.2 (Mev) while χ was assigned values between 0.0 and 0.1 (Mev). These parameter values for a pseudo ^{20}Ne system with *no* spin-orbit splitting, which differ significantly from the values required for a real ^{20}Ne application (C ≈ -2.0 and D ≈ +0.2 for a μ ≈ -0.2), simulate an application of the theory to deformed nuclei of the rare earth and actinide regions. A full breakdown of the intensity is given for the 0^+_1 yrast state while results for the main SU(3) component only are shown for each of the remaining states, 0^+_α, α = 2, 3, & 4. Note that SU(3) symmetry breaking *decreases* sharply as the strength of the deformation inducing quadrupole-quadrupole interaction *increases*, particularly for the yrast state. A realistic value for χ is about 0.06 - 0.07. Even at half this value, the yrast state is more than 80% pure $(\lambda,\mu) = (8,0)$. The dashed curve just below the $0^+_1(8,0)$ solid line is the corresponding result when a spin-orbit interaction is included with a strength C = -0.2, which is down by a factor of ten from the actual value but again a representative strength for pseudo SU(3) applications.

From the results of this simple representative calculation several important conclusions can be drawn. First of all, because $\tilde{C} \approx 0$ for nuclei of the rare earth and actinide regions, mixing among different pseudo-space symmetries is expected to be small. This means that the normal parity valence space can be truncated down to a reasonable size, for example the [f] = [4] symmetry in the pseudo ^{20}Ne case just considered, so simple yet realistic calculations can be carried out. Secondly, it is important to again stress that the breaking of the SU(3) symmetry *decreases* very sharply as the strength of the deformation inducing quadrupole-quadrupole interaction *increases*. More specifically, the spreading out of the various SU(3) irreps within [f] symmetries by Q·Q reduces the SU(3) symmetry breaking induced by the l^2 term. This means, for example, that the normal parity contribution to yrast states of strongly deformed configurations should be dominated by the leading pseudo SU(3) symmetry.

It is important to understand that the leading pseudo SU(3) symmetry is the most deformed configuration available in the space under consideration. Going beyond normally deformed configurations to superdeformed and perhaps even hyperdeformed structures, therefore implies shifting particles into higher-lying configurations. For example, in the pseudo ^{20}Ne example one candidate for a superdeformed band is the configuration obtained by lifting two particles out of the p shell into the fp shell. This leads to pseudo SU(3) irreps contained in the product (8,0) x (0,2) x (6,0). The leading irrep in this case is $(\lambda,\mu)=(14,2)$. Since in L=0 states the square of the deformation is proportional to the expectation value of C_2, under the action of the *same* hamiltonian this arrangement of nucleons will have a deformation that is nearly twice that of the leading (8,0) configuration: $<C_2(14,2)>/<C_2(8,0)> = 3.14$ which yields $\beta(14,2)/\beta(8,0) = 1.77$. Since the hamiltonian does not change, configurations like the one containing the (14,2) are expected to display even less representation mixing than the one containing the (8,0) since for this arrangement of particles the dominance of the Q·Q term in the energy matrix will be even more pronounced. To summarize, based on the results of this representative study, superdeformed bands are expected to be better pseudo SU(3) nuclei and hence better rotors than normally deformed ones.[18] This is consistent with recent experimental results.[19]

PSEUDO-SYMPLECTIC MODEL

The symplectic model, which can be used to describe both low-lying *and* giant resonance states of light (A $\leq$ 28) nuclei that are strongly deformed and which is sometimes also called the microscopic collective model, extends the SU(3) model to include multiple $2\hbar\omega$ shell-model excitations of the monopole ($l=0$) and quadrupole ($l=2$) type.[7] This multi-$\hbar\omega$ feature means the action of the real quadrupole operator, not just its intra-shell $0\hbar\omega$ preserving part, can be fully accomodated. The symmetry algebra of the scheme generates a realization of the noncompact symplectic group Sp(3,R), which has the Elliott SU(3) as its maximal compact subgroup. Associated with each SU(3) shell-model irrep is an infinite dimensional Sp(3,R) representation. To reiterate, the important point is that these symplectic spaces are complete with respect to the action of the real quadrupole operator. This means intraband and interband E2 transition strengths between low-lying as well as giant resonance configurations can be reproduced within the framework of the model without the use of proton and neutron effective charges.[20]

The pseudo symplectic model extends the pseudo SU(3) picture by allowing for inter-shell $2\hbar\omega$ excitations of the monopole and quadrupole type among the normal parity orbitals just like the symplectic scheme does for the normal (ds-shell) SU(3) model.[21] In this case, however, the underlying symmetry group is pseudo Sp(3,R) with pseudo SU(3) as a subgroup. Since the symplectic extension means that multiple $2\hbar\omega$ excitations in the proton and neutron normal parity spaces are included, it would seem that unique parity excitations of this type should also be considered, as well as mixing between the two. In this first application of the theory, however, these couplings are not included. The argument that is usually given to support this narrowing in the scope of the problem is that for states in even-even nuclei that lie below the backbending region the unique parity parts of the wavefunctions are dominated by pairing correlations which differ from quadrupole correlations in that they do not require couplings to higher shells for strength enhancement. The validity of this bold assumption can certainly be challenged, and, indeed, is currently under further investigation, see below.

The model hamiltonian is taken to be the sum of the proton and neutron pseudo harmonic oscillator hamiltonians plus a real quadrupole-quadrupole interaction. And as indicated above, since the current application of the theory is restricted to low-lying states of even-even nuclei, contributions to the dynamics from protons and neutrons in the unique parity levels can and will be suppressed in what follows. Also, to insure that the mean field of the harmonic oscillator is conserved under perturbation by the quadrupole-quadrupole interaction, $Q^c \cdot Q^c$ is replaced by $Q^c \cdot Q^c - (Q^c \cdot Q^c)_{shell}$ where $(Q^c \cdot Q^c)_{shell}$ is an operator that reproduces single-shell traces of $Q^c \cdot Q^c$,[22]

$$H = \hbar\omega \tilde{N} - \frac{1}{2}\chi[Q^c \cdot Q^c - (Q^c \cdot Q^c)_{shell}] + H_r, \qquad (6)$$

$$\tilde{N} = \tilde{N}_\pi + \tilde{N}_\nu \text{ and } Q^c = Q^c_\pi + Q^c_\nu.$$

The superscript "c" in these equations denotes the real "collective" quadrupole operator, $Q^c = \sqrt{16\pi/5}\, r^2 Y_2(\hat{r})$, as opposed to the symmetrized "algebraic" one of Elliott, $Q^a = \sqrt{16\pi/5}\,[b^4 p^2 Y_2(\hat{p}) + r^2 Y_2(\hat{r})]/2$, where b is the oscillator length parameter. (This Q^a is the same as the Q used in the previous section.) While the matrix elements of Q^c and Q^a are identical within an oscillator shell, Q^c has nonvanishing matrix elements between shells that differ by two quanta whereas Q^a does not. The major shell separation energy $\hbar\omega$ is given by the usual empirical rule $41A^{-1/3}$[Mev]. The hamiltonian includes a residual interaction, H_r, that allows the inertia and band splitting features of the spectrum to be easily reproduced.[23]

The real mass quadrupole operators Q^c_π and Q^c_ν can be expressed in terms of the generators of symplectic algebras: $Q^c_\alpha = Q^a_\alpha + \sqrt{6}/2[B^+_{2\alpha} + B_{2\alpha}]$ where $\alpha = \pi$ & ν. The operators $B^+_{lm\alpha}$ ($B_{lm\alpha}$) with $l = 0$ & 2 are the $2\hbar\omega$ raising (lowering) generators of the $Sp_\alpha(3,R)$ algebras and the Q_α^a are quadrupole generators of the respective $SU_\alpha(3)$ subalgebras. Through a second quantization formalism these operators ($O = Q^a, B^+, B^-$) can be expanded in terms of pseudo SU(3) tensors, $O = x\tilde{O} + \cdots$, where $\tilde{O}$ has the same tensorial character as O, see (4). The constant x in this expansion is always greater than unity, ranging from a high of about 1.4 for $O = Q^a(\tilde{n} = 0)$ to a low of about 1.1 for $O = B_{2m}^+(\tilde{n} = 6)$. An average x value for actinide nuclei is 1.14. In pseudo SU(3) applications the correction terms were found to yield less than a one percent change in calculated energies and electromagnetic transition rates. Therefore O was replaced by $\tilde{O}$ in this study.

Neglecting all higher order effects, the hamiltonian (6) can be rewritten as,

$$H = \hbar\omega\tilde{N} - \frac{1}{2}\chi[\tilde{Q}^c \cdot \tilde{Q}^c - (\tilde{Q}^c \cdot \tilde{Q}^c)_{shell}] + a\tilde{L}^2 + b\tilde{X}_3^a + c\tilde{X}_4^a. \qquad (7)$$

In this expression, an average x value has been absorbed into χ and the last three terms are an explicit form for H_r. The operator $\tilde{L}$ is the total pseudo angular momentum while the $\tilde{X}_3^a$ and $\tilde{X}_4^a$ terms are third and fourth order rotational scalars: $\tilde{X}_3^a = [\tilde{L}x\tilde{Q}^a x\tilde{L}]$ and $\tilde{X}_4^a = [(\tilde{L}x\tilde{Q}^a)^1 x (\tilde{Q}^a x\tilde{L})^1]$. This special form for the residual interaction means the moment of inertia and band splitting of the low-lying states can be adjusted without otherwise affecting the dynamics.[24] Note that matrix elements of this H_r vanish in 0^+ states. An estimate for the strength χ of the quadrupole-quadrupole interaction can be made by equating the difference in the expectation value of H in the 0^+ state of the $0\hbar\omega$ (λ,μ) irrep and the so-called stretched $2\hbar\omega$ $(\lambda+2,\mu)$ irrep to the excitation energy, $80A^{-1/3}$, of the giant monopole resonance.[25] This estimate for χ can be improved upon by using it to calculate eigenfunctions of the hamiltonian, (7), with a=b=c=0 in the 0^+ and 2^+ spaces and calculating the $2_1^+ \rightarrow 0_1^+$ reduced E2 transition probability. If the E2 strength is smaller (bigger) than the experimental result a bigger (smaller) value of χ should be chosen. The assumption of a linear relation between E2 values and χ suffices in interpolating between results. The residual interaction can be ignored in this process because even though it changes the excitation energy of the 2_1^+ state, it has virtually no effect on the calculated $2_1^+ \rightarrow 0_1^+$ transition strength. A similar procedure can be used to determine the parameters a, b and c of H_r.[21]

Now consider ^{238}U. It has 10 valence protons in the Z = 82 - 126 shell and 20 neutrons in the N = 126 - 184 valence space. The distribution of these particles between normal and unique parity orbitals is made by selecting a reasonable deformation, say $\beta \approx 0.25$, and filling each level of the appropriate Nilsson diagram with a pair of particles. This procedure yields 6 normal parity protons and 12 normal parity neutrons for the most probable occupancies of the pseudo $n_\pi = 4$ and $n_\nu = 5$ shells. The corresponding occupancies for the unique parity parts are 4 protons and 8 neutrons in the $i_{13/2}$ and $j_{15/2}$ orbitals, respectively. The normal parity configurations give rise to the leading pseudo SU(3) irreps (18,0) for protons and (36,0) for neutrons. The strong coupled SU(3) irreps are given by the product of these two irreps, $(18,0) \otimes (36,0) \equiv \{(\lambda,\mu)\}$. Of this set of irreps the one with the maximum eigenvalue of C_2 and hence the largest deformation, which follows because $\beta^2 \sim \langle Q^a \cdot Q^a \rangle$, is expected to dominate the low-lying structure. This irrep, $(\lambda,\mu)=(54,0)$, is the leading $2\hbar\omega$ SU(3) symmetry.

Applying these considerations to ^{238}U yields a value of 0.00138(Mev) for χ. With this χ, the calculated $2_1^+ \rightarrow 0_1^+$ transition rate is $2.48 e^2 b^2$, which is nearly within the error bars of the adopted experimental number, $2.42\pm0.04\ e^2 b^2$.[26] Since the leading pseudo SU(3) irrep for ^{238}U is (54,0), it is like ^{20}Ne of the ds-shell in that the theory predicts the absence of a $K^\pi=2^+$ (gamma) band. This is consistent with the fact that in ^{238}U a second 2^+ level has not been found below the first excited 0^+ level. Because $\mu = 0$, the b and c parameters were set equal to zero. The predicted value for a is then 0.00541. A best fit to the experimental energy spectrum and E2 transition strengths for the $L^\pi = 0^+ - 12^+$ states was obtained with $\chi = 0.00135$(MeV) and $a = 0.00465$. In the calculation a full $20\hbar\omega$ symplectic basis was used, with dimensionalities ranging from 489 for the $L^\pi = 0^+$ space up to 4069 for $L^\pi = 12^+$. While the best fit value for χ is within 5% of the initial estimate, the final value for the parameter "a" is about 15% below the simplest prediction. This larger difference for "a" can be attributed to the fact that $Q^c \cdot Q^c$ is more effective in generating rotations than $Q^a \cdot Q^a$ suggests.

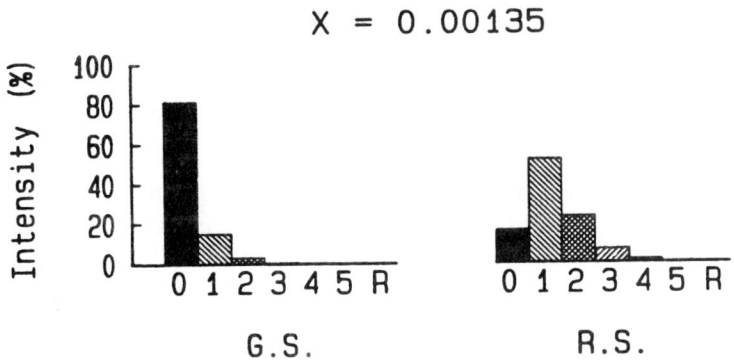

FIGURE 4. Calculated intensities for the ground state ($0_1^+ \rightarrow$ G.S.) and resonant mode ($0_2^+ \rightarrow$ R.S.) of the pseudo symplectic hamiltonian in a full $20\hbar\omega$ symplectic basis with the best-fit value for the strength of the $Q^c \cdot Q^c$ interaction, $\chi = 0.00135$. The abscissa labels the number of $2\hbar\omega$ excitations. For $\chi=0.0$ the ground state and resonant state are pure SU(3) configurations with (λ,μ) values (54,0) and (56,0), respectively. In agreement with this simple picture, the $(\lambda,\mu)=(54,0)$ $0\hbar\omega$ irrep makes up 81.0% of the intensity profile of the ground state while 15.2% comes from the $2\hbar\omega$ space, 3.2% from the $4\hbar\omega$ space, 0.5% from the $6\hbar\omega$ space, and 0.1% from the $8\hbar\omega$ space. Similarly, the dominant contribution to the resonant mode is the $(\lambda,\mu)=(56,0)$ irrep with 45.5% intensity. The $(\lambda,\mu)=(52,2)$ adds an additional 5.1% to this for a total of 50.6% in the $2\hbar\omega$ space. The remaining strength includes strong mixing from the shells above ($4\hbar\omega$, 23.6%; $6\hbar\omega$, 7.0%; $8\hbar\omega$, 1.6%; $10\hbar\omega$, 0.3%) and below ($0\hbar\omega$, 16.8%) with some intensity (0.1%) way out at $12\hbar\omega$.

Rather than show a near perfect fit to the experimental spectrum, only calculated intensities of the ground (0_1^+) and resonant (0_2^+) states are given in Figure 4 for the final best-fit χ value. Recall that for $\chi = 0.0$ the ground and giant monopole states are pure $0\hbar\omega$ [$(\lambda,\mu)=(54,0)$] and $2\hbar\omega$ [$(\lambda,\mu)=(56,0)$] excitations, respectively. An important feature is the amount of vertical mixing that is required to reproduce the observed E2 transition strengths. The $(\lambda,\mu)=(54,0)$ $0\hbar\omega$ configuration makes up 81.0% of the intensity profile of the ground state with slightly over a 15% contribution from the $2\hbar\omega$ space [11.3% $(\lambda,\mu)=(56,0)$ and 4.0% $(\lambda,\mu)=(52,2)$] and under 5% from the $4\hbar\omega$ space [1.4% $(\lambda,\mu)=(58,0)$, 1.3% $(\lambda,\mu)=(54,2)$ and 0.4% $(\lambda,\mu)=(50,4)$], etc. The resonant mode, on the other hand, has significant multi-$\hbar\omega$ admixtures extending out to $12\hbar\omega$, but in agreement with the most naive picture, the dominant contribution is the $(\lambda,\mu)=(56,0)$ configuration with 45.5% intensity. The $(\lambda,\mu)=(52,2)$ irrep adds an additional 5.1% for a total of 50.6% in the $2\hbar\omega$ space. The $0\hbar\omega$ and $4\hbar\omega$ spaces contribute 16.8% and 23.6%, respectively, to the structure of the resonant mode. At each level, $n = 1, 2, ...$, the stretched SU(3) irrep, $(\lambda+2n, \mu)$, was found to be the most important. Detailed results for excitation energies and electric quadrupole transition strengths are given in Table I. Results for the pseudo SU(3) scheme and the collective model are included in the table for comparison.[27,28] Note that the pseudo SU(3) B(E2) strengths saturate at $J_i = 6$ while the collective model results show no such trend. The pseudo symplectic B(E2) results, on the other hand, show saturation near to where the effect is observed experimentally, $J_i = 12\text{-}16$.

Table I. Experimental and calculated excitation energies [E_i (Mev)] and electric quadrupole transition strengths [(B(E2) (e^2b^2)] in ^{238}U. B(E2) values are quoted for the pseudo SU(3) and collective model (CM) theories in addition to those for the pseudo symplectic scheme [Sp(3,R)] introduced in this paper. The pseudo SU(3) results were renormalized to the adopted experimental B(E2: $2_1 \rightarrow 0_1$) number, $2.42 e^2 b^2$.

E_i (Mev)				B(E2) (e^2b^2)			
Exp	Sp(3,R)	J_i	J_f	Exp	Sp(3,R)	SU(3)	CM
0.0449	0.0435	2	0	2.42	2.45	2.42	2.34
0.1487	0.1451	4	2	3.51	3.48	3.37	3.40
0.3072	0.3048	6	4	3.87	3.81	3.53	3.85
0.5178	0.5225	8	6	3.57	3.96	3.42	4.16
0.7757	0.7982	10	8	4.21	4.02	3.14	4.43
1.0765	1.1320	12	10	4.33	4.03	2.73	4.68

INTRUDER LEVELS

The pseudo scheme(s) all remove the highest j state associated with the n-th major shell of the oscillator, $j = n+1/2$, from active consideration because it is pushed down among levels of the shell immediately below by the spin-orbit interaction. Furthermore, the $j = n+3/2$ level that penetrates down from the $(n+1)$-st shell into the active model space is normally assumed to play a passive role in the dynamics of low-lying excited states. Specifically, the usual assumption is that for states below the backbending region this unique parity intruder level, under the action of the hamiltonian, only supports paired (J=0) states and therefore can contribute additional binding energy to the system but nothing to the dynamics.

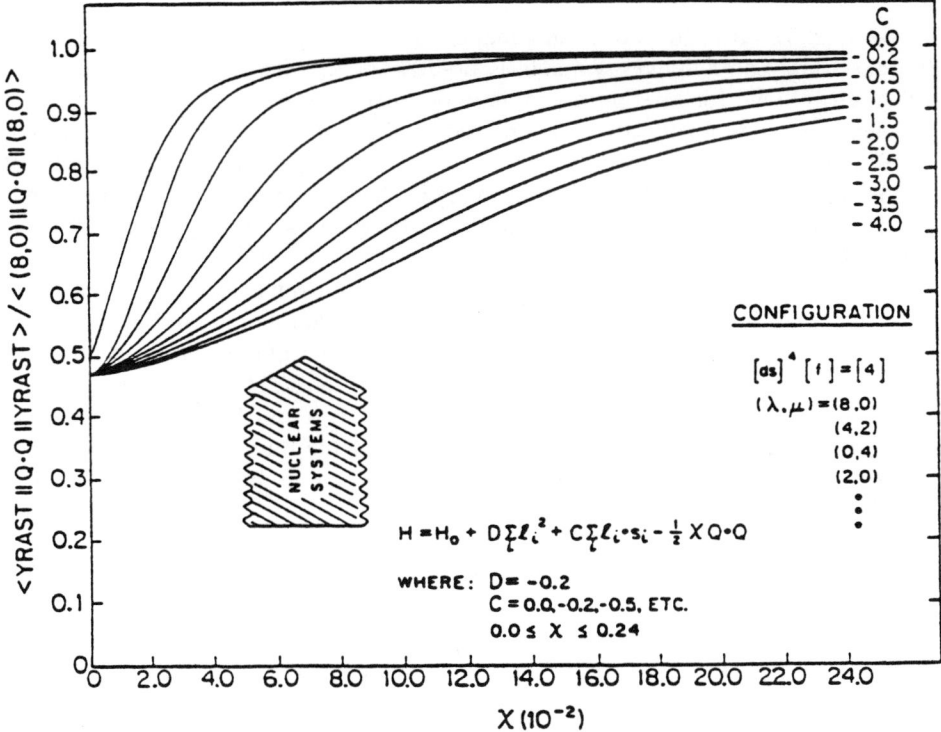

FIGURE 5. Normalized expectation values of Q·Q in calculated ground states of the (ds)^4 system for hamiltonian (4), with D = -0.2 and a spin-orbit term with C values as indicated, are plotted as a function of the strength χ of the quadrupole-quadrupole interaction. The normalization factor is the maximum eigenvalue of Q·Q in the (ds)^4 space which is just the expectation of Q·Q in the leading $(\lambda,\mu) = (8,0)$ irrep of SU(3). For the results to be representative of a pseudo SU(3) application, C ≈ -2.5 and, as indicated, χ ≈ 0.06 - 0.07. It follows from these results that even for a very strong spin-orbit splitting, the yrast state can achieve as much as 60-70% of its maximum total quadrupole collectivity.

Through and beyond the backbending region, however, alignment sets in and the intruder can no longer be considered to be passive. We now want to report on preliminary results that challenge this simple assumption. In particular, the results suggest that the coupling of the intruder to its natural partners, even though these may be as much as a full major shell ($1\hbar\omega$) away, is strong and leads to a sizable contribution to the quadrupole moment and therefore E2 strengths of the many-particle system.

In Figure 5 normalized expectation values of Q·Q in calculated ground states of the $(ds)^4$ system for hamiltonian (5) with D = -0.2 and C values as indicated are plotted as a function of the strength of the quadrupole-quadrupole interaction. The normalization factor is the maximum eigenvalue of Q·Q in the $(ds)^4$ space which is just the expectation of Q·Q in the leading $(\lambda,\mu) = (8,0)$ irrep of SU(3). For the results to be representative of a pseudo SU(3) application, $C \approx -2.5$ and, as above, $\chi \approx 0.06 - 0.07$. It follows from this that even for a very strong spin-orbit splitting, the yrast state can achieve as much as 60-70% of its maximum total quadrupole collectivity. Since for rare earth and actinide nuclei the number of particles in the intruder level is typically about 1/3 of the total number of valence particles, and since the intruder level comes from the $(n+1)$-st oscillator shell, a rough estimate for the ratio of the contribution of particles in the unique parity intruder levels to that of those in the normal parity orbitals can be given: $<Q\cdot Q>_{unique}/<Q\cdot Q>_{normal} \approx 0.6 \times C_2(m(n+1)/3,0)/C_2(2mn/3,0) \approx 0.6 \times [(n+1)/(2n)]^2 \approx 0.2$. This analysis assumes the $(\lambda,\mu) = (m(n+1)/3,0)$ and $(2mn/3,0)$ irreps are representative of those that are allowed by the exclusion principle for the unique and normal parity spaces, respectively. This suggests that particles in the unique parity orbitals can be expected to contribute to the quadrupole moment of a deformed system roughly in proportion to their number with a strength that is about half the strength with which the normal parity particles contribute. Since the number of particles in the unique parity space is typically not small, they are important in determining quadrupole moments and E2 strengths.

CONCLUSION

A key feature in the development of the pseudo-spin scheme and the SU(3) and Sp(3,R) derivative models is the empirical result that the Nilsson parameter $\mu = 2D/C \approx 0.5$ for heavy nuclei. Actual estimates for μ are (0.60 & 0.65) for protons with (50 < Z < 82 & Z > 82) and (0.42 & 0.33) for neutrons with (82 < N < 126 & N > 126), respectively. These numbers are close enough to the exact pseudo-spin limit, $\mu = 0.5$, that even with a relatively weak quadrupole-quadrupole interaction the pseudo SU(3) and Sp(3,R) limits of the theory apply. In contrast with this, for the real ds-shell the value for the parameter $\mu \approx -0.2$ (note the sign) and, as is well-known, the $(s_{1/2}, d_{3/2})$ levels do not form a near degenerate pair nor is the $d_{5/2}$ orbital far removed from them. The analysis underscores the need for a deeper understanding of the C ≈ 4D result. In particular, is the neutron-proton difference ($\mu_\nu \approx 0.4$ versus $\mu_\pi \approx 0.6$) more that a Coulomb effect and is there fundamental physics behind the average $\mu \approx 0.5$ result?

Discussions with colleagues O. Castaños and P. O. Hess of the University of México, P. Rochford, a research associate, and students C. Bahri, S. Park and J. Escher, from Louisiana State University are gratefully acknowledged. The nuclear theory effort at Louisiana State University is supported in part by grants from the U. S. National Science Foundation, INT-88-01337 and PHY-89-22550.

REFERENCES

[1] M. G. Mayer, Phys. Rev. 75, 1969 (1949); 78, 16 (1950).
[2] O. Haxel, J. H. D. Jensen, and H. E. Suess, Phys. Rev. 75, 1766 (1949); Z. Physik 128, 295 (1950).
[3] S. G. Nilsson, Kgl. Danske Videnskab. Selskab Mat. Fys. Medd. 29 (1955).
[4] K. T. Hecht and A. Adler, Nucl. Phys. A137, 129 (1969).
[5] A. Arima, M. Harvey, and K. Shimizu, Phys. Lett. B30, 517 (1969).
[6] R. D. Ratna Raju, J. P. Draayer, and K. T. Hecht, Nucl. Phys. A202, 433 (1973).
[7] G. Rosensteel and D. J. Rowe, Phys. Rev. Lett. 38, 10 (1977); Ann. Phys. 126, 343 (1980).
[8] A. Bohr and B. R. Mottelson, *Nuclear Structure*, Vol. I & II (Benjamin, Reading, Mass., 1975).
[9] A. Bohr, I. Hamamoto, and B. R. Mottelson, Phys. Scr. 26, 267 (1982).
[10] K. T. Hecht, in *Annual Review of Nuclear Science* (Academic Press, New York, 1973) Vol. 23, pp. 123-161.
[11] F. S. Stephens et al., Phys. Rev. Lett. 64, 2623 (1990); 65, 301 (1990).
[12] J. P. Elliott, Proc. Roy. Soc. London A245, 128, 562 (1958).
[13] M. Harvey, in *Advances in Nuclear Physics*, ed. M. Baranger and E. Vogt (Plenum Press, New York, 1968), Vol 1, pp. 67-182.
[14] A. Arima and D. Strottman, *A Compilation of SU(3) Wave Functions Obtained from Realistic Matrix Elements*, Oxford University Report No. 46, 1973.
[15] J. P. Draayer and K. J. Weeks, Ann. Phys. 156, 41 (1984).
[16] J. P. Draayer, K. J. Weeks, and K. T. Hecht, Nucl. Phys. A381, 1 (1982).
[17] O. Castaños, J. P. Draayer, and Y. Leschber, Ann. Phys. 180, 290 (1987).
[18] Y. Leschber and J. P. Draayer, Phys. Lett., B190, 1 (1987).
[19] W. Nazarewicz, P. J. Twin, P. Fallan, and J. D. Garrett, Phys. Rev. Lett. 64, 1654 (1990).
[20] C. Bahri, J. P. Draayer, O. Castaños, and G. Rosensteel, Phys. Lett. B234, 430 (1990).
[21] O. Castaños, P. O. Hess, J. P. Draayer, and P. Rochford, Nucl. Phys. A (in press).
[22] O. Castaños and J. P. Draayer, Nucl. Phys. A491, 349 (1989).
[23] O. Castaños, J. P. Draayer, and Y. Leschber, Z. Phys. A329, 33 (1988).
[24] H. A. Naqvi and J. P. Draayer, Nucl. Phys. A516, 351 (1990).
[25] F. E. Bertrand, Nucl. Phys. A354, 129C (1981).
[26] S. Raman et al., Atom. Data and Nucl. Data Tables, 42, 1 (1989).
[27] E. Grosse et al., Phys. Scr. 24, 338 (1981).
[28] P. O. Hess et al., Zeit. Phys. A296, 147 (1980).

THE PSEUDO-L QUANTUM NUMBER IN ODD-A NUCLEI

D.D. Warner
SERC Daresbury Laboratory, Warrington WA4 4AD, U.K.

ABSTRACT

Use of the pseudo-spin transformation in the shell model scheme leads to the approximate validity of the pseudo-L quantum number in odd-A nuclei. The empirical signatures of this symmetry are discussed and it is shown that the transformation gives rise to a new set of intensity relationships for E2 transitions in strongly deformed systems.

INTRODUCTION

It was pointed out many years ago [1-3] that it is possible to transform the normal parity shell model states to a basis in which they are described in terms of pseudo-orbital and pseudo-spin angular momenta, with the result that the effective spin-orbit splitting is greatly reduced. The set of normal parity states in each shell N (which excludes the state of largest j for each N) then forms the complete set of states belonging to the pseudo-oscillator shell $\tilde{N} = N - 1$. The effect of the pseudo-spin transformation on the angular momentum variables for the N = 4 and 5 shells is illustrated in fig. 1. The validity of the concept, and the relevance of the pseudo SU(3) scheme which follows from it, have been discussed in the previous presentation [4]. Moreover, this approach has recently attracted a considerable degree of renewed interest in terms of its application to phenomena observed at higher rotational frequencies [5-7]. It is the purpose of this paper to show that, in the context of a core-particle coupling framework, the pseudo-spin symmetry leads to the appearance of a "new" quantum number in the description of odd-A nuclei whose near-validity leads to certain characteristic features in a number of observables.

THE PSEUDO-L QUANTUM NUMBER

One aspect which is common to essentially all approaches to the description of collective structure in odd-A nuclei is that they can be written in terms of a core-particle Hamiltonian,

$$H = H_c + H_p + H_{cp} \qquad (1)$$

The total angular momentum of the system is, of course, simply $J = R_c + j$, R_c and j being the core and single-particle angular momenta, respectively. Although j itself is not affected by the pseudo-spin transformation, the single-particle orbital and spin angular momenta do change (see fig. 1) so that now $j = \tilde{\ell} + \tilde{s}$, where the tildas denote the pseudo-scheme variables. Then, in the limit in which the pseudo-spin-orbit coupling strength tends to zero, the Hamiltonian of eq.(1) becomes independent of the pseudo spin $\tilde{s}$ (assuming no spin dependent terms in H_{cp}) with

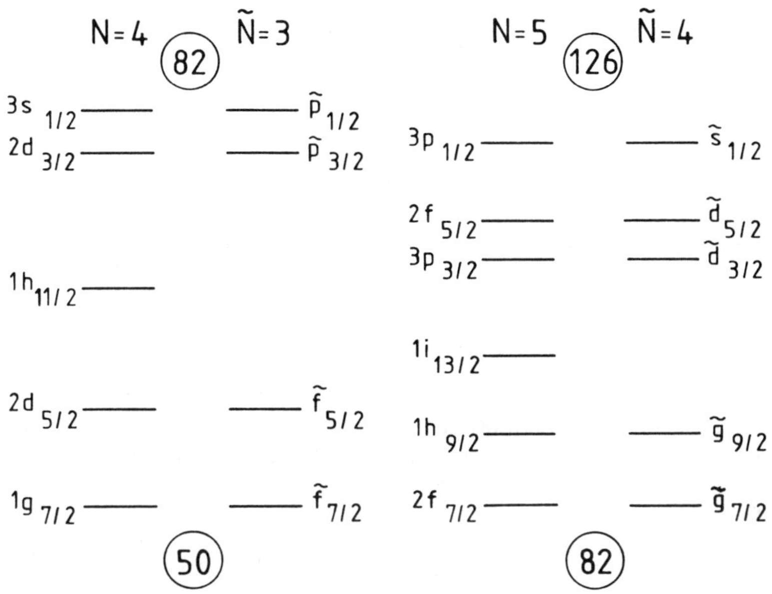

Fig. 1. Pseudo-spin transformation for N = 4 and 5. Pseudo quantum numbers are distinguished by a tilda.

the result that a new quantum number, L, emerges, viz,

$$L = R_c + \tilde{\ell} \quad (2)$$

L is referred to as the total pseudo-orbital angular momentum. Clearly, in the limit of a pseudo-spin independent Hamiltonian, states of the form $|\alpha, L; J = L \pm 1/2\rangle$ will be degenerate (α here denotes other quantum numbers of the system) although such a degeneracy will be lifted by a (diagonal) rotational term of the form J^2.

Before proceeding to consider the observable consequences of the pseudo-L symmetry, it is worth mentioning that such a symmetry has always been an inherent ingredient of the dynamical symmetry limits of the Interacting Boson Fermion Model (IBFM)[8]. The bose-fermi symmetries incorporate the pseudo-spin transformation ab initio and the L quantum number emerges in each of the possible group chain decompositions as the label of the irreducible representations of the group O^{B+F} (3). In addition, this quantum number remains good if the Hamiltonian is taken to be an arbitrary mixture of Casimir operators of higher subgroups from different symmetry limits. That is, the pseudo-L symmetry is maintained in the most general IBFM Hamiltonian, being dependent only on the validity on the initial pseudo-spin decomposition.

SIGNATURES OF PSEUDO-L SYMMETRY

1. Energies

As mentioned above, the most obvious empirical signature of pseudo-L symmetry is the appearance of couplets of levels with $J = L \pm 1/2$. In fact, the levels in the odd-A nucleus can be thought of as arising from coupling the pseudo spin to the states of an even-even nucleus with angular momentum L. An example based on rotational band structure is useful to illustrate this point, and is shown in fig. 2. Three band structures are shown labelled by $K_L = 0$, 1 and 2 where K_L is the projection of L on the symmetry axis of the deformed nucleus. Referring to eq. (2), K_L is identical to the quantum number $\tilde{\Lambda}$ of the pseudo-Nilsson scheme[3,4]. Each band contains L values appropriate to its K_L-value, and each L value gives rise to two degenerate states with $J = L \pm 1/2$. The result for $K_L = 0$ is, in more familiar terms, a $K = 1/2$ band with a decoupling parameter of unity. For $K_L = 1$ or 2 it is, in each case, two degenerate bands with $K = 1/2, 3/2$ and $3/2, 5/2$. This is exactly what would be expected for a vanishing $\tilde{\ell}.\tilde{s}$ strength in the Nilsson scheme. The K_L (or $\tilde{\Lambda}$) = 1 and 2 structures represent pseudo-spin-orbit partners. At low spin and excitation energy, many of these pairs of bands are known. Some examples of the Nilsson orbits involved are given in Tables 1a and 1b along with some of the nuclei in which they have been identified. The nuclei ¹⁸⁵W and ¹⁸⁷Os in particular both exhibit the near-degenerate $K_L = 1$ structure of fig. 2 at their ground states, the Nilsson orbitals involved being the 1/2⁻ [510] and 3/2⁻ [512] which become the 1/2⁻ and 3/2⁻ [4̃1̃1̃] in the pseudo-Nilsson scheme. Indeed, ¹⁸⁵W has already been the focus of a study based on the SU(3) limit of the IBFM [9].

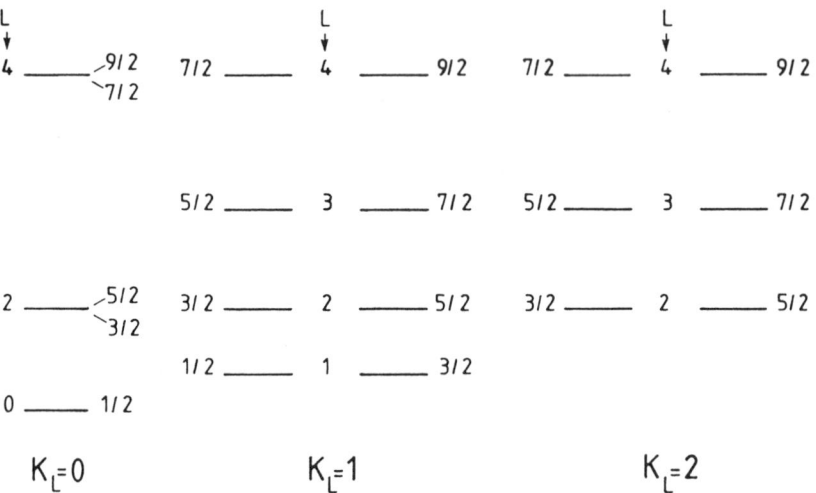

Fig. 2. Rotational band structures in the pseudo-L scheme. States are labelled by L and J.

Table 1a. Examples of pseudo-spin doublets for odd-N nuclei.

Nilsson scheme	Pseudo-spin scheme	Nuclei
3/2[512], 1/2[510]	3/2[$\overline{411}$], 1/2[$\overline{411}$]	^{185}W, ^{187,189}Os
9/2[505], 7/2[503]	9/2[$\overline{404}$], 7/2[$\overline{404}$]	^{189}Os
7/2[514], 5/2[512]	9/2[$\overline{404}$], 7/2[$\overline{404}$]	^{179,181}W, ^{175,177}Hb, ^{173,175}Yb
5/2[523], 3/2[521]	5/2[$\overline{422}$], 3/2[$\overline{422}$]	^{161,163}Er, ^{159,161}Dy, ^{159}Gd

Table 1b. Examples of pseudo-spin doublets for odd-Z nuclei.

Nilsson scheme	Pseudo-spin scheme	Nuclei
1/2[400], 3/2[402]	1/2[$\overline{301}$], 3/2[$\overline{301}$]	^{185,187}Re, ^{187,189}Ir
5/2[402], 7/2[404]	5/2[$\overline{303}$], 7/2[$\overline{303}$]	^{175,177}Ta, ^{177}Lu
3/2[411], 5/2[413]	3/2[$\overline{312}$], 5/2[$\overline{312}$]	^{153,155}Eu

For K = 1/2 bands, the decoupling parameter a is given by the asymptotic expression

$$a = (-1)^{\tilde{N}} \quad ; \quad \tilde{\Lambda} = 0 \qquad (3)$$
$$= 0 \quad ; \quad \tilde{\Lambda} = 1$$

Thus the $K_L = 0$ example of fig. 2 is appropriate for $\tilde{N}$ even; for $\tilde{N}$ odd, the L values would be 1, 3, 5 (mimicking a $K = 0^-$ band in even-even nuclei) producing degeneracies for J = 1/2 - 3/2, 5/2 - 7/2, etc., corresponding to a = - 1. The greatly reduced spin-orbit splitting in the pseudo-spin scheme results in the new label $\tilde{\Lambda}$ for the deformed single particle states being far closer to a good quantum number at realistic deformations, so that the asymptotic values of quantities such as the decoupling parameter provide a far better description of the data than in the "old" scheme. There are many empirical examples which show the near-validity of these values, both for low lying bands in the rare earth region [3] and, more recently, for superdeformed bands [7].

The relevance of the pseudo-L scheme is not limited to well deformed nuclei. Figure 3 shows a comparison of the levels in ^{195}Pt compared to the predictions of the O(6) symmetry in the IBFM. The only quantum numbers of relevance here are those on the immediate left of the levels which denote the L values. Again the characteristic level sequence of the even-even nucleus can be picked out in the sequence of L values, L = 0, 2, 2, 4. Similar examples exist for vibrational nuclei[8].

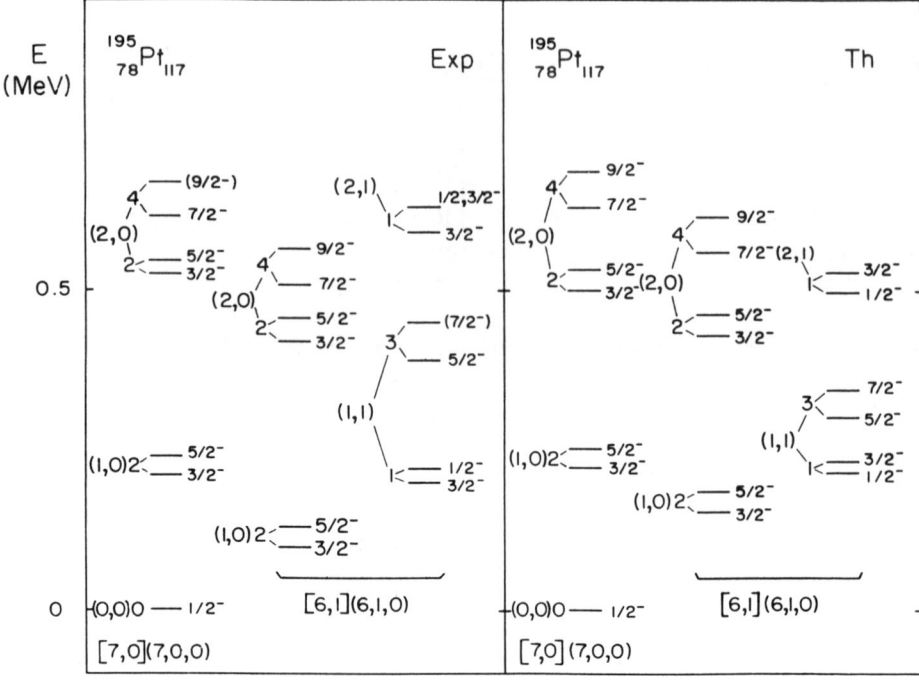

Fig. 3. States in ^{195}Pt versus an IBFM calculation [10] in the O(6) limit.

It is, of course, evident from figs. 2 and 3 that some sequences of L values occur which are not characteristic of even-even nuclei of the same structure. The $K_L = 1$ band of fig. 2 and the L = 1 states of fig. 3 are obvious examples. In the low-lying states of even-even nuclei, such features are absent because of symmetry constraints. In the odd-even nucleus, however, the orbital motions of the core and particle may couple symmetrically or non-symmetrically, the latter case giving rise to the "abnormal" L values. Such states are entirely analogous to the so-called "mixed symmetry" states of the neutron-proton IBM. The crucial difference is that, in the odd-A case, there is no symmetry energy to raise the excitation energy of the states. On the contrary, they can, and frequently do, constitute the ground state of the system, their position relative to the symmetric states being dependent on the Fermi level.

2. Single-particle structure coefficients

The pseudo-L symmetry also has an effect on the matrix elements of the single-particle transfer operator. It has long been recognized that, in deformed nuclei, the predictions for these quantities based on the single-particle structure of the Nilsson orbits are strongly modified by Coriolis coupling. The measured structure factors for the normal-parity states can be understood in terms of the selection rule imposed by the L-symmetry, namely, $\Delta L = \ell$. An example is given in Table 2. The quantities listed are the "effective" C_{jl} coefficients which are those used in the spherical expansion of Nilsson states, modified by the inclusion of Coriolis mixing. The Nilsson model values [11] are compared with the pseudo-L values from an SU(3) description [12] of the same bands. The selection rule cited above leads to vanishing values for states with L odd, since the single particle states in this example have $\ell = 0, 2, 4$. Moreover, values for pseudo-L doublets differ only by a factor $\sqrt{(2J+1)}$. Note that these features are maintained in other regions of nuclei also, such as the Os-Pt transition to γ-soft structure [13].

Table 2. Effective spherical single particle amplitudes for [185]W in Nilsson [11] and pseudo-L [12] schemes.

j	Nilsson $[Nn_3\Lambda],\Omega$	Pseudo-L $[\widetilde{N}\widetilde{n}_3\widetilde{\Lambda}]$, K	L
	[512], 3/2	[411], 3/2	
3/2	0.14	0.00	1
5/2	1.00	0.83	2
7/2	0.12	0.00	3
9/2	0.43	0.69	4
	[510], 1/2		
1/2	0.03	0.00	1
3/2	0.78	0.68	2
5/2	0.15	0.00	3
7/2	0.47	0.61	4
9/2	0.03	0.00	5
	[521], 1/2		
1/2	0.51	0.52	0
3/2	0.31	0.14	2
5/2	0.47	0.17	2
7/2	0.38	0.55	4
9/2	0.45	0.62	4

3. Rotational intensity rules for E2 transitions

The E2 transition rates for bands of good K are governed by the Clebsch-Gordan coefficients $\langle J_i K_i\, 2\, K_f-K_i | J_f K_f \rangle^2$. This dependence gives rise to the well known intensity rules [14] for relative E2 strengths when the initial and final intrinsic states are the same. It has recently been shown [15] that the pseudo-L scheme gives rise to a new expression, viz

$$B(E2; J_i L_i K_{L_i} \to J_f L_f K_{L_f}) = (\frac{5}{16\pi}) e^2 Q_0^2 (2J_f + 1) (2L_i + 1) \qquad (4)$$

$$<L_i K_{L_i} 2 K_{L_f} - K_{L_i} \mid L_f K_{L_f}>^2 \begin{Bmatrix} L_i & L_f & 2 \\ J_f & J_i & 1/2 \end{Bmatrix}^2$$

where Q_0 is the intrinsic quadrupole moment. This expression yields identical results to the good-K basis for $K_L = 0$, $K = 1/2$. For $K_L \neq 0$ the results differ. Referring to fig. 2, the predictions for crossover transitions between bands with $K_1 = K_L - 1/2$ and $K_2 = K_L + 1/2$ are given by

$K = K_1$:
$$B(E2; J \to J-2)_{K_L} = \frac{(2J-3)(J+K+1)}{(2J+1)(J+K-1)} B(E2)_K$$

$K = K_2$:
$$B(E2; J \to J-2)_{K_L} = \frac{(2J+1)(J+K-2)}{(2J-3)(J+K)} B(E2)_K \qquad (5)$$

and similarly for stopover transitions

$K = K_1$:
$$B(E2; J \to J-1)_{K_L} = \frac{(2K+1)^2 (J-1)^2 (J+K+1)}{K^2 (4J^2-1)(J+K)} B(E2)_K$$

$K = K_2$:
$$B(E2; J \to J-1)_{K_L} = \frac{(2K-1)^2 (J+1)^2 (J+K-1)}{K^2 (4J^2-1)(J+K)} B(E2)_K \qquad (6)$$

where $B(E2)_K \equiv B(E2; KJ_i \to KJ_f)$ is the Alaga rule rotational B(E2) value.

The two sets of predictions are compared with the available data [16] for the low lying $K = 1/2$ and $3/2$ ($K_L = 1$) bands in ^{187}Os in Table 3. Of particular interest is the appearance of interband transitions in the pseudo-L scheme, of collective strength. This feature stems from the implicit K-mixing inherent in this scheme. Unfortunately, the data are of insufficient accuracy to provide a convincing test. This is a deficiency common to odd-A nuclei in general.

As mentioned earlier, the relevance of the pseudo-spin scheme for super-deformed states is currently a topic of some interest [7]. Although for $\Delta J = 2$ transitions, eq. (5) indicates that the B(E2) values approach the normal rotational values for large J, significant differences remain for $\Delta J = 1$ transitions.

Table 3. Relative B(E2) values in ^{187}Os.

J_iK_i	J_fK_f	Relative B(E2; $J_i \to J_f$)		
		Expt.	Pseudo L	Good K
5/2 1/2	1/2 1/2	100 (9)	100	100
	3/2 1/2	135 (50)	86	29
	3/2 3/2	41 (21)	29	0
	5/2 3/2	24 (12)	21	0
5/2 3/2	1/2 1/2	65 (7)	29	0
	3/2 3/2	100 (37)	100	100
7/2 3/2	3/2 3/2	100 (12)	100	100
	5/2 3/2	122 (66)	75	150

CONCLUSIONS

As pointed out initially, the pseudo-spin scheme, and the associated quantum numbers L and K_L (Λ) is only exactly valid in the limit of zero pseudo-spin-orbit splitting. Nevertheless the actual residual splitting which remains after the transformation is sufficiently small to render these quantum numbers approximately valid over a broad range of deformation, spanning $\varepsilon_2 \sim 0.2 - 0.6$, as indicated by the empirical information on both normal and superdeformed states. In this sense, the constraints on pseudo-spin symmetry are considerably less than those on achieving the full pseudo-SU(3) symmetry, since the latter also requires a vanishing ℓ^2 interaction in the shell model Hamiltonian. Of course, at sufficiently large deformations, both schemes must break down because of the influence of the neglected states of maximal j in each major shell. Thus, at some point, the normal asymptotic Nilsson quantum numbers become more valid, although it seems likely that this occurs at extremes of deformation which will not be realised in actual nuclei.

At low spin, the L symmetry is most easily broken. This was demonstrated in a recent study [17] in which a $\ell \cdot \tilde{s}$ perturbation was added to an IBFM SU(3) Hamiltonian and shown to correctly reproduce the change in structure in the low lying states of $^{183-187}$W. It can also be inferred qualitatively by noting that states of differing L within a band structure are separated only by rotational energy spacings, whereas bands with differing K_L are separated by single particle spacings. However, the quality of the L quantum number should improve with increasing rotational frequency, since the Coriolis coupling acting on the deformed pseudo-spin partner orbits then becomes large compared to their residual splitting.

At normal deformations the useful increase in spin is limited by the onset of alignment. It is therefore of particular interest to further explore the relevance of the pseudo-L scheme in the second (superdeformed) minimum.

In fact, recent results[18] in the region of Hg nuclei appear to provide a fascinating example in this regard. The authors compare the lowest-lying superdeformed band in ^{192}Hg with an excited (two quasiparticle) superdeformed band in ^{194}Hg and extract the contribution to the aligned angular momentum from the last two neutrons in ^{194}Hg. They find the somewhat astonishing result that, for states corresponding to $\omega \geq 0.2$ MeV/$\hbar$, this alignment is constant and takes the value $1.00 \pm 0.04\,\hbar$. The existence of such exactly quantized aligned angular momentum is, *a priori*, totally unexpected. However, the Coriolis coupling implicit in the pseudo-L scheme, alluded to above, has been shown[19] to arise from a term of the form $J_+ \tilde{s}_- + J_- \tilde{s}_+$ involving stepping operators acting only on the pseudo-spin, rather than the total single particle angular momentum. Moreover, this is consistent with an earlier study[5] which showed that, even at relatively low rotational frequencies, the projection of the pseudo spin on the rotational axis becomes a good quantum number. Thus, it was suggested[18] that the observation of unit alignment in the ^{194}Hg band might be understood in terms of an appropriate pair of pseudo orbitals in which the pseudo-spin alignments add, while the pseudo-orbital angular momenta remain coupled to the deformation axis. If correct, this interpretation would appear to confirm the applicability of the pseudo-L scheme in normal-parity, superdeformed bands. Further examples of quantized alignment in the Hg region have recently been published[20].

REFERENCES

1. A. Arima, M. Harvey and K. Shimizu, Phys. Lett. 30B, 517 (1969).
2. K.T. Hecht and A. Adler, Nucl. Phys. A137, 129 (1969).
3. R.D. Ratna Raju, J.P. Draayer and K.T. Hecht, Nucl. Phys. A202, 433 (1973).
4. J.P. Draayer, contribution to these Proceedings.
5. A. Bohr, I. Hamamoto and B.R. Mottelson, Phys. Scr. 26, 267 (1982).
6. J. Dudek et al, Phys. Rev. Lett. 59, 1405 (1987).
7. W. Nazarewicz et al, Phys. Rev. Lett. 64, 1654 (1990).
8. F. Iachello and P. van Isacker, "The Interacting Boson Fermion Model", Cambridge University Press, to be published.
9. D.D. Warner and A.M. Bruce, Phys. Rev. C30, 1066 (1984).
10. A. Mauthofer et al, Phys. Rev. C34, 1958 (1987).
11. R.F. Casten et al, Matt. Fys. Med. Dan. Vid. Selsk. 38, No. 13 (1972).
12. R. Bijker and V.K.B. Kota, Ann. Phys. (N.Y.) 187, 148 (1988).
13. D.D. Warner et al, Phys. Rev. Lett. 54, 1305 (1985).
14. G. Alaga et al, Matt. Fys. Med. Dan. Vid. Selsk. 29, No. 9 (1955).
15. D.D. Warner and P. van Isacker, Phys. Letts. 247B, 1 (1990).
16. Y.A. Ellis-Akovali, Nucl. Data Sheets, 36, 559 (1982).
17. P. van Isacker, J.P. Elliott and D.D. Warner, Phys. Rev. C36, 1229 (1987).
18. F.S. Stephens et al, Phys. Rev. Lett. 64, 2623 (1990).
19. A. Frank et al, Phys. Lett. 182B, 233 (1986).
20. F.S. Stephens et al, Phys. Rev. Lett. 65, 301 (1990).

COLLECTIVE EXCITATION SPECTRA OF TRANSITIONAL EVEN NUCLEI [+]

P. Quentin

*Department of Physics,
Lawrence Livermore National Laboratory,
Livermore, CA 94550, USA*

and

*C.S.N.S.M.[†] , Bat. 104,
91405 Orsay-Campus, France*

I. Deloncle, J. Libert

*C.S.N.S.M., Bat. 104
91405 Orsay-Campus, France*

J. Sauvage

*Division de Recherche Expérimentale[‡], Institut de Physique Nucléaire,
91405 Orsay-Campus, France*

I. INTRODUCTION

This talk is dealing with the nuclear low energy collective motion as described in the context of microscopic versions of the Bohr Hamiltonian[1]. Two different ways of building microscopically Bohr collective Hamiltonians will be sketched in Section II: one within the framework of the Generator Coordinate Method, the other using the Adiabatic Time-Dependent Hartree-Fock-Bogolyubov approximation. A sample of recent results will be presented in Section III which pertains to the description of transitional even nuclei and to the newly revisited phenomenon of superdeformation at low spin.

We will only consider here the five quadrupole degrees of freedom which are (after a well-known transformation from the lab frame to a body fixed inertial frame) three Euler angles and the usual β and γ parameters[1].

[†] Laboratoire propre du CNRS
[‡] Laboratoire associé au CNRS

The classical Bohr Hamiltonian H is the sum of a potential energy V and of a kinetic energy T splitted into two parts, a vibrational energy T_v and a rotational energy T_r defined with obvious notation as:

$$T_v = \frac{1}{2} \sum_{i,j=1}^{2} B_{ij} \dot{q}_i \dot{q}_j \qquad (1)$$

(with q_1, q_2 standing for β, γ) and

$$T_r = \frac{1}{2} \sum_{i=1}^{3} \Im_i \omega_i^2 \qquad (2).$$

In order to get a quantal description of the nuclear collective modes, one should quantize H, a task for which there are no well-defined prescriptions in the most general case. Using for instance the Pauli prescription. one would get for the vibrational kinetic energy operator acting on a collective wavefunction:

$$\hat{T}_v = -\frac{\hbar^2}{2} \sum_{i,j=1}^{2} (\det B)^{-1/2} \frac{\partial}{\partial q_i} (\det B)^{1/2} (B^{-1})_{ij} \frac{\partial}{\partial q_j} \qquad (3)$$

whereas the rotational kinetic energy would be simply given as

$$\hat{T}_r = \frac{1}{2} \sum_{i=1}^{3} \hat{I}_i^2 / \Im_i \qquad (4).$$

When actually dealing with the Bohr Hamiltonian, one has to take proper care of the symmetries inherent to the problem[1,2] and then one merely considers one sextant in the (β, γ) plane.

The diagonalization of the most general Bohr Hamiltonian is currently performed either by a finite difference treatment of the derivatives which are associated to T_v [2,3] or by projection onto a suitably chosen (and symmetrized) basis[4].

Apart from the quantization problem, the main theoretical task will consist in evaluating microscopically the ingredients of the classical/quantal Bohr Hamiltonian. There are indeed, seven scalar functions of β and γ which need to be determined: the potential V, three moments of inertia ($\Im_1, \Im_2, \Im_3$) and three mass parameters ($B_{\beta,\beta}, B_{\beta,\gamma}, B_{\gamma,\gamma}$).

II. MICROSCOPIC DERIVATIONS OF BOHR COLLECTIVE HAMILTONIANS

II.1 DERIVATION FROM THE GENERATOR COORDINATE METHOD

The Generator Coordinate Method (GCM) starts from a set of Slater determinants or BCS wavefunctions $| \Phi_q >$ depending on a coordinate q (in what follows we will restrict for the sake of clarity, to the simple case of a single such coordinate). The trial wavefunction $|\Psi_\alpha\rangle$ to be used in a variational approach is built as a linear combination of such wavefunctions

$$|\Psi_\alpha\rangle = \int dq\, f_\alpha(q) |\Phi_q\rangle \tag{5}$$

The mixing amplitudes $f_\alpha(q)$ are determined through the variational principle by the following Hill-Wheeler[5] equation

$$\int dq'\, \langle \Phi_q | H - E_\alpha | \Phi_{q'} \rangle\, f_\alpha(q') = 0 \tag{6}$$

or in a condensed vector form as

$$\mathcal{H} f_\alpha = E_\alpha \mathcal{N} f_\alpha \tag{7}$$

where

$$\mathcal{H}(q,q') = \langle \Phi_q | H | \Phi_{q'} \rangle$$
$$\mathcal{N}(q,q') = \langle \Phi_q | \Phi_{q'} \rangle \tag{8}$$

Whereas f_α represents in some way the amount of correlations associated with the degree of freedom which the set $(|\Phi_q\rangle)$ is supposed to describe, its interpretation as a collective wavefunction suffers from the fact that eqs. (6-7) are not standard eigenvalue equations due to the use of non-orthogonal $|\Phi_q\rangle$ states. One therefore currently switches to an alternate representation[6]. For that purpose, one may notice that the matrix $\mathcal{N}$ being real (if the $|\Phi_q\rangle$ states are even with respect to time-reversal) and symmetrical, it can be diagonalized. As a norm matrix moreover, it has non-negative eigenvalues. One first rejects eigenvectors with zero eigenvalues whose existence reflects the non-linearly independent character of the initial set of $|\Phi_q\rangle$ states. This allows one therefore, to consider $\mathcal{N}^{-1}$. One also can evaluate $\mathcal{N}^{1/2}$ and $\mathcal{N}^{-1/2}$ in a straightforward fashion from the so-restricted eigensolutions of $\mathcal{N}$.

The collective wavefunction will then be defined as

$$g_\alpha = \mathcal{N}^{1/2} f_\alpha \tag{9}$$

or more explicitly

$$g_\alpha(r) = \int dq\, (\mathcal{N}^{1/2})_{qr}\, f_\alpha(q) \tag{10}$$

It is then possible to orthonormalize the g_α functions and the Hill-Wheeler equation (7) will become

$$h g_\alpha = E_\alpha g_\alpha \tag{11}$$

where the Hamiltonian matrix h is given in terms of the original $\mathcal{H}$ by

$$h = \mathcal{N}^{-1/2} \mathcal{H} \mathcal{N}^{-1/2} \tag{12}$$

Now starting from the modified Hill-Wheeler equation (11) we will sketch the derivation of the underlying Bohr hamiltonian eigenvalue equation. There exists quite a broad litterature on this subject[7-12]. Here we will rather follow the presentation of reference (11).

Let us first Wigner-transform the Hamiltonian matrix $h(r,r')$ defined in equation (12), into h_w namely

$$h_w(R, \hbar k) = \int ds\, e^{-iks}\, h(R+\tfrac{s}{2}, R-\tfrac{s}{2}) \quad (13)$$

where

$$R = (r+r')/2 \quad (14).$$

Then, one can expand h_w up to second order in $\hbar k$, which corresponds to an expansion on the range of the non locality (rather than to a semiclassical expansion):

$$h_w(R, \hbar k) = h_w(R, 0) + \frac{\hbar^2}{2} k^2 F(R) \quad (15)$$

where

$$p = \hbar k, \quad F(R) = \left.\frac{\partial^2 h_w}{\partial p^2}\right|_{(R,0)} \quad (16).$$

Upon inverse Wigner-transforming h_w

$$h(r,r') = \frac{1}{2\pi}\int dk\, e^{-ik(r-r')}\, h_w\left(\tfrac{r+r'}{2}, \hbar k\right) \quad (17)$$

one gets with the truncated h_w of equation (15):

$$h(r,r') \simeq V\!\left(\tfrac{r+r'}{2}\right)\delta(r-r') - \frac{\hbar^2}{2} F\!\left(\tfrac{r+r'}{2}\right)\delta''(r-r') \quad (18)$$

where $V(r)$ is the diagonal matrix element $h(r,r)$.

Inserting the matrix $h(r,r')$ of equation (18) into equation (11) (the eigenvalue equation for g_α), one gets the Schrödinger equation:

$$\left\{-\frac{\hbar^2}{2}\frac{\partial}{\partial r} F(r)\frac{\partial}{\partial r} + \left[V(r) - \frac{\hbar^2}{8}\frac{\partial^2 F(r)}{\partial r^2}\right] - E_\alpha\right\} g_\alpha = 0 \quad (19).$$

It clearly corresponds to a quantized version of the Bohr collective Hamiltonian upon considering $F(r)$ as an inverse mass and $V(r)$ as a potential. One notices the existence of a corrective term to the potential, proportional to the second derivative of $B(r)$. The latter is usually referred to as a Zero Point Energy (ZPE) corrective term. Now this Bohr hamiltonian eigenvalue equation is defined in terms of h (or h_w). To connect this equation with what is primarily calculated (i. e. the matrix $\mathcal{H}$) one generally relies on the Gaussian Overlap Approximation (GOA) of the GCM. We will not treat that point here and merely refer to the previously quoted litterature.

To summarize this Sub-Section, one should mention that with the preceding approach, one gets a <u>quantized</u> Bohr Hamiltonian with well defined ZPE quantal corrections. On the other hand one should also add that a part of the approach relies on the validity of the GOA to the GCM, for which some elements of assessment are now available[12]. Moreover, the GCM is known[13] to yield mass parameters that are not correct unless double GCM calculations are performed.

II.2 DERIVATION FROM THE ATDHFB METHOD

Among the many derivations of a Bohr collective Hamiltonian from the Adiabatic Time-Dependent Hartree-Fock-Bogolyubov (ATDHFB) method we will concentrate on the approach initiated by Baranger and Vénéroni[14,15]. Here also for the sake of clarity we will consider only one single collective variable even though it is easily extended to more. Similarly we will skip the pairing correlations and briefly present merely the ATDHF version (no pairing included) of this derivation. A generalisation in order to treat these correlations is possible and has indeed been achieved[16].

One starts from the TDHF equation of motion

$$[h, \rho] = i\hbar \dot{\rho} \qquad (20)$$

where the one-body Hamiltonian h is understood as $h(\rho)$, i.e. defined in terms of the one-body reduced density matrix ρ. Now one makes an expansion in some velocities (see references (14,15) for more details):

$$\rho = \rho_0 + \rho_1 + \rho_2 + \cdots \qquad (21).$$

Conversely one can make the same expansion for $h(\rho)$ and get:

$$h = h_0 + h_1 + h_2 + \cdots \qquad (22).$$

All quantities with an even (odd resp.) subscript are even (odd resp.) with respect to the time-reversal operator. Separating the time-even and time-odd parts of the equation of motion (20) one gets two equations. The first (the even one) is usually referred to as the path equation. Its solution will be assumed to be well represented by making the following approximations:

i) all the time-dependence of ρ is contained in a single collective variable q, i. e.

$$\rho(t) = \rho[q(t)] \quad , \quad \dot{\rho} = \dot{q} \frac{\partial \rho}{\partial q} \qquad (23)$$

and therefore one has

$$\rho_0 \propto \dot{q}^0 \; , \; \rho_1 \propto \dot{q}^1 \; , \; \rho_2 \sim \dot{q}^2 \qquad (24),$$

ii) the family $\rho_0(q)$ is defined from Constrained Hartree-Fock (CHF) calculations, namely

$$[h_0 - \lambda Q, \rho_0] = 0 \; , \quad q = tr(\rho_0 Q) \qquad (25).$$

Then it can be shown[15] that the time-odd equation takes the form of a doubly Constrained Hartree-Fock equation

$$[h_0 + h_1 - \lambda Q - \dot{q} P, \rho_0 + \rho_1] = 0 \qquad (26)$$

where one adds to the constraint on $<Q>$, a constraint on the expectation value of an operator P which is univocally defined in terms of $\rho_0(q)$ (and its derivative with respect to q) and which can be shown to be conjugate (in a classical sense) of the operator Q:

$$tr(\rho [Q,P]) = i\hbar \qquad (27).$$

Having solved the double CHF problem, one may compute the energy associated with the solution $\rho = \rho_0 + \rho_1 + \rho_2$ as an expansion again in $\dot{q}$

$$E[\rho] = E[\rho_0] + \frac{1}{2} M(q) \dot{q}^2 \qquad (28).$$

In the above equation $E[\rho_0]$ is the potential energy, as resulting from the simple CHF equation (i.e. with a mere constraint on $<Q>$) and $M(q)$ is a mass parameter which in general depends on q and is proportional to $tr(\rho P/\dot{q})$

It is therefore clear that we have been able to yield a classical Bohr Hamiltonian. The derivation which we have sketched here, provides a generalization of the Thouless-Valatin formalism[17] (or routhian formalism) to collective modes other than pure rotations. As opposed to the GCM, one obtains here good mass parameters [14] and furthermore full self-consistency corrections (beyond the Inglis cranking formula[18]) are appropriately incorporated.

On the other hand this classical Hamiltonian must obviously be quantized and a choice has to be made on the ZPE corrective terms. For those operations, as already pointed out , one lacks a priori theoretical guidance within the mere ATDHFB framework.

III. A SAMPLE OF RESULTS

III.1 TRANSITIONAL NUCLEI

The calculations reported in this Sub-Section have been performed within the ATDHFB framework described in Sub-Section II.2. The Bohr Hamiltonian has been quantized according to the Pauli prescription with no ZPE corrections included. The CHF calculations have used the Skyrme SIII effective nucleon-nucleon interaction[19]. Pairing correlations have been taken care of by means of a BCS approach with constant pairing matrix elements. Whereas self-consistent P operators have been computed and used, one has merely here evaluated Inglis cranking mass parameters. Moreover the so-called Expectation Value Method has been substituted to the full HF plus BCS approach for the static calculations, making use of self-consistent semiclassical solutions (including terms up to the fourth order in $\hbar$ in an expansion à la Wigner-Kirkwood) associated with the considered two-body force. Some of the results discussed here, have already been published elsewhere[21].

Two different regions of transitional nuclei will be considered here ($A \sim 70$ and $A \sim 180$).

The low-lying experimental spectra of ^{74}Ge and ^{76}Se are qualitatively well-reproduced (see reference (21), figure 2). An interesting output of such calculations is obtained when plotting the probability density in the collective plane (β,γ), namely plotting $|\phi_\alpha(\beta,\gamma)|^2$. In that respect, one should bear in mind that in order to have directly interpretable density plots, one generally considers cartesian integration measures for the density functions. To that effect the eigenfunctions must be multiplied by a factor proportional to $(\Im_1\Im_2\Im_3)^{1/2}$ which identically vanishes for $\gamma = n\pi/3$, $(n = 1,...,6)$. Therefore, a nucleus which should be considered as rather prolate (oblate resp.) will have a maximum of the density near the $\gamma = 0$ axis ($\gamma = \pi/3$ resp.) but will of course never have such a maximum on this axis.

In the ^{74}Ge nucleus, one has obtained (see reference (21), figure 3) density distributions almost unaltered at low spin within the ground-state band (no centrifugal stretching or anti-stretching) and a one-phonon state in the so-called β-band which exhibits the expected feature of a single node near the $\gamma = 0$ axis. (In the absence of results on transition probabilities, band assignments have been made by mere energy considerations confirmed, if needed, by a comparison of the K-component content.)

Systematical calculations in the platinum region are currently performed. The excellent result obtained for the ^{186}Pt has been already reported (see reference (21), figure 2). We have extended such calculations to the ^{190}Pt and ^{176}Pt nuclei. Whereas for the former - as seen on figure 1 - the agrement with experimental energies is very satisfactory, it is found to be hardly the case for the ^{176}Pt where the first 2^+ level is significantly too low, missing thus the occurence of a new transitional region below ^{178}Pt as suggested by the data.

The density probability contours of the ^{186}Pt exhibit a very interesting phenomenon. Whereas the yrast band, starts with a rather oblate shape for

$I = 0, 2$, this shape becomes prolate for $I \geq 4$. Conversely, the so-called β-band starts as a one-phonon state for $I = 0$ and gradually corresponds to an oblate solution similar to what had been obtained for the ground state. This is thus a quantitative illustration of the shape transition often advocated in this region.

III.2 SUPERDEFORMATION AT LOW SPIN

The calculations which will be briefly reported here, have been obtained by two of us (I. D. and J. L.) in collaboration with J.P. Delaroche and M. Girod. The results at spin zero have been already published[22]. They correspond to the approach discussed in Sub-Section II.1. Static calculations have been performed within the HFB framework using the D1S Gogny effective force[23] which is not a corrected zero-range force as the Skyrme force and reproduces well at the same time HF-like and BCS-like matrix elements. To approximately take care of the problem of the GCM masses, the mass parameters which have been included, are of the Inglis type (making use however of simplified non self-consistent P operators).

The existence of superdeformed local minima in the potential energy surfaces in the region $A \sim 190$ has been predicted long ago both in Strutinsky -like calculations[24] or in HF plus BCS calculations[25]. It has been recently revisited theoretically[26-28] and the theoretical predictions have been confirmed afterwards experimentally[29] (see the contribution of E. Henry to this Conference for an update of this fast growing experimental domain[30]).

The characteristic feature of potential energy curves (for axially symmetrical shapes) in the region of nuclei with $A \sim 190$ is the existence of a pocket in the ascending part of the fission barrier which corresponds roughly to a mass quadrupole moment of about 45 barns. The relevant energy parameters are the excitation (in a purely static sense here, with the inclusion of ZPE corrective terms though) energy E^* and the inner barrier energy E_B (measured from the ground state). Upon including ZPE corrective terms these energies have been calculated to be (see reference (22), figure 1) in ^{190}Hg, ^{192}Hg and ^{194}Hg:

$E^* \sim 1.5, 3.0, 5.0$ (MeV)

$E_B \sim 5.5, 7.0, 9.5$ (MeV).

Inserting full potential energy surfaces as well as the calculated mass parameters and moments of inertia, one gets eigensolutions of the corresponding Bohr Hamiltonians which present the following features for e. g. the 0^+ states: i) ground states typical of transitional (slightly) oblate nuclei, ii) at values of excitation energies of 4.4, 5.4, 6.9 MeV for the ^{190}Hg, ^{192}Hg and ^{194}Hg nuclei, one gets states that are mostly located in the isomeric well (see reference (22), figure 4). To be specific, if one considers the part of the probability in the collective space corresponding to $\beta \geq 0.4$, one gets 91%, 98% and 96% for each of the above listed nuclei in this particular state. These conclusions have recently been extended to finite spins. Indeed rotational bands built on the previous 0^+ states are found. In ^{190}Hg, the corresponding superdeformed band crosses the yrast (normal) band around I = 10,12. This spin value seems to be too low in view of recent experimental data[30] and can reflect the fact that the static excitation energy (E^*) values in these calculations are found significantly lower than in other dynamical calculations[12,31].

IV. CONCLUSION

Two different derivations in a microscopic fashion of collective Bohr Hamiltonians have been recalled. None of them is without problems. It is convenient at this point to answer the following question: after all why not go on and perform double GCM calculations which would be purely quantal and free from some mass problems? It is unfortunate that at this time, one is at best able to make two dimensional GCM calculations[12]. On the other hand, to treat correctly the quadrupole motion which carries most of the low-energy collective motion, one needs, as it is well known and recalled here, to consider five degrees of freedom. This is the reason why we must go through the above mentionned difficulties.

It is however clear that in the present status such calculations are able to yield a very good qualitative reproduction of data. How quantitative the qualitative agreement is indeed, time will tell. It is particularly appropriate in this context to emphazise the crucial role played by pairing correlations for which such calculations constitute a very demanding test. The obtained reproduction of experimental spectra for rough that it may be deemed, is nevertheless very significant in view of the absence of any ad-hoc parameter fitting except for the force parameters chosen many years ago in a different context, i. e. to reproduce essentially saturation properties (let us also recall that apart from two pairing parameters, the Skyrme SIII force has only six parameters for the whole chart of nuclides). Finally, the obvious point where some effort should be devoted is undoubtedly an improvement of the mass parameters in use which are presently somewhat too crude.

REFERENCES

+ This work was supported in part by the U.S. Department of Energy under Engineering Contract NW 7405-ENG-48.
1. Å. Bohr, Mat.Fys. Medd. Dan. Vid. Selsk.$\underline{26}$ (1952) # 14.
2. K. Kumar and M. Baranger, Nucl. Phys. $\underline{A92}$ (1967) 608.
3. K. Kumar , Nucl. Phys.$\underline{C5}$ (1974) 189.
4. J. Libert and P. Quentin, Z. Phys. $\underline{A306}$ (1982) 315.
5. D.L. Hill and J.A. Wheeler, Phys. Rev. $\underline{89}$ (1953) 1102.
6. see e. g. P. Ring and P. Schuck, The Nuclear Many-Body Problem, Springer Verlag, Berlin (1980).
7. D. M. Brink and A. Weiguny, Nucl. Phys. $\underline{A120}$ (1968) 59.
8. B. Giraud and B. Grammaticos, Nucl. Phys. $\underline{A233}$ (1974) 373.
9. P.G. Reinhard, Nucl. Phys. $\underline{A252}$ (1975) 120.
10. M. Girod and B. Grammaticos, Nucl. Phys. $\underline{A233}$ (1979) 40.
11. J.F. Berger,Thesis, Orsay (1985), unpublished.
12. P. Bonche, J. Dobaczewski, H. Flocard, P.-H. Heenen and J. Meyer, Nucl. Phys. $\underline{A510}$ (1990) 466 .
13. R. E. Peierls and D. J. Thouless, Nucl. Phys. $\underline{38}$ (1962) 154.
14. M. Baranger and M. Vénéroni, Ann. Phys. (New York) $\underline{114}$ (1978) 123.
15. M.J. Giannoni and P. Quentin, Phys. Rev. $\underline{C21}$ (1980) 2060.
16. L. Bennour, J. Libert and P. Quentin, to be published.
17. D.J. Thouless and J.G.Valatin, Nucl. Phys. $\underline{31}$ (1962) 211.
18. D.R. Inglis, Phys. Rev. $\underline{96}$ (1954) 1059; $\underline{103}$ (1956) 1796.
19. M. Beiner, H. Flocard, Nguyen Van Giai and P. Quentin, Nucl. Phys. $\underline{A238}$ (1975) 29.
20. I. Deloncle, K. Bencheikh, L. Bennour, M. Brack, J. Libert, J Meyer and P. Quentin, to be published.
21. I. Deloncle, J. Libert, L. Bennour and P. Quentin, Phys. Lett. $\underline{B233}$ (1989) 16.
22. J.P. Delaroche, M. Girod, J. Libert and I. Deloncle, Phys. Lett. $\underline{B232}$ (1989) 145.
23. J. Dechargé and D. Gogny, Phys. Rev. $\underline{C21}$ (1980) 1568; J.F.Berger, M. Girod and D. Gogny, Nucl. Phys. $\underline{A428}$ (1984) 23c.
24. R.R. Pashkevitch, preprint JINR (Dubna) P4-4383 (1969), unpublished.
25. M. Cailliau, J. Letessier, H. Flocard and P. Quentin, Phys. Lett. $\underline{B46}$(1973) 11.
26. P. Bonche, S.J. Krieger, P. Quentin, M.S. Weiss, J. Meyer, M. Meyer, N. Redon, H. Flocard, and P.-H. Heenen, Nucl. Phys. $\underline{A500}$ (1989) 308.
27. M. Girod, J.P. Delaroche, D. Gogny and J.F. Berger, Phys. Rev. Lett. $\underline{62}$ (1989) 2452.
28. R.R. Chasman, Phys. Lett. $\underline{B219}$ (1989) 227.

29. E.F. Moore, R.V.F. Janssens, R.R. Chasman, I. Ahmad, T.L. Khoo, F.L.H. Wolfs, D. Ye, K.B. Beard, U. Garg, M.W. Drigert, Ph. Benet, Z.W. Grabowski and J.A. Cizewski, Phys. Rev. Lett. 63 (1989) 360.
30. E.A. Henry, invited talk to this Symposium.
31. P. Bonche, S.J. Krieger, M.S. Weiss, J. Dobaczewski, H. Flocard, P.-H. Heenen and J. Meyer, invited talk to this Symposium.

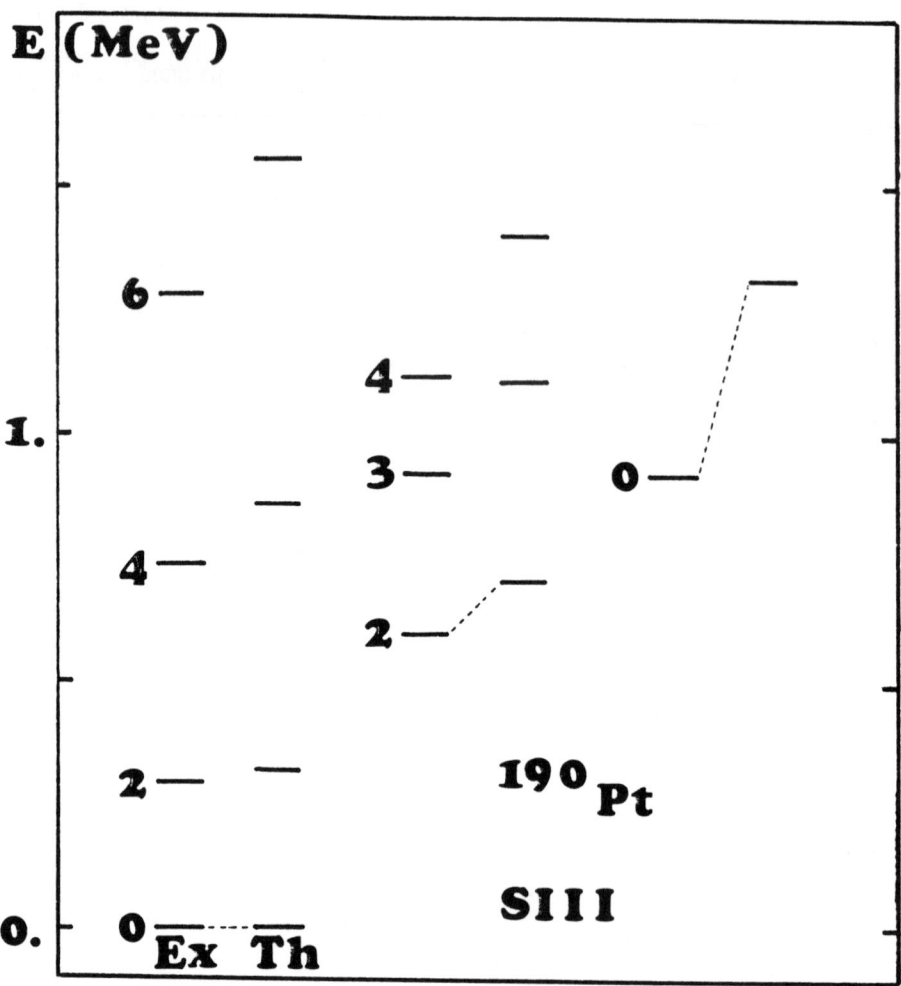

Fig. 1. Comparison of Experimental (Ex) and Theoretical (Th) low-energy spectra (all energies in MeV) of the ^{190}Pt nucleus, calculated with the Skyrme SIII effective interaction.

THE RESIDUAL PROTON-NEUTRON INTERACTION AND NUCLEAR COLLECTIVITY

R. F. Casten
Brookhaven National Laboratory, Upton, New York, 11973, USA

ABSTRACT

The essential role of the valence, residual p-n interaction in the development of collectivity, though long known in general terms, has recently become increasingly apparent. A brief review of the p-n interaction is given, including some very basic nuclear data that illustrate its effects and the phenomenological $N_p N_n$ scheme and the P-factor. This is followed by a discussion of recent experimental extractions of p-n matrix elements throughout the periodic table and theoretical efforts to understand them, in terms of both Shell and Nilsson models.

INTRODUCTION

Nearly four decades ago de Shalit and Goldhaber[1] suggested that the p-n interaction could play an important role in the onset of deformation in nuclei. Since the 1960's Talmi[2] has repeatedly emphasized the critical role of the p-n interaction and discussed its importance in facilitating single nucleon configuration mixing in the Shell Model. In the early 1970's a rapid new region of deformation was discovered near A=100. Later, Federman and Pittel[3] interpreted the onset of deformation as due, primarily, to enhanced p-n interactions in spin-orbit partner single particle states, in particular $p1g_{9/2}$-$n1g_{7/2}$. This work was generalized to heavy nuclei, in particular to the rare earth region by Casten et al.[4] More recently, Hartree Fock calculations[5] for the A=100 region have confirmed the importance of the p-n interaction but stress that other (particularly high-j unique parity) orbits play at least as important a role.

This brief overview will highlight various aspects of the p-n interaction, starting with some basic nuclear data that reveal its effects in a very simple way. The $N_p N_n$ scheme will be briefly reviewed and, then, empirical p-n matrix elements throughout the periodic table, and their interpretation, will be discussed in some detail. Much of this paper should be viewed as embodying the results from recent collaborative papers[6,7] to which the reader is referred for further details and discussion. This review and summary of some of our recent work is similar to one given recently at the Predeal International Summer School[8].

MANIFESTATIONS OF THE p-n INTERACTION IN NUCLEAR STRUCTURE

Despite the long litany of work recognizing that the valence p-n interaction is critical to the development of collectivity, it has, on the whole, been rather neglected as an explicit ingredient in the history of nuclear structure calculations. It is the contention of this review, and of

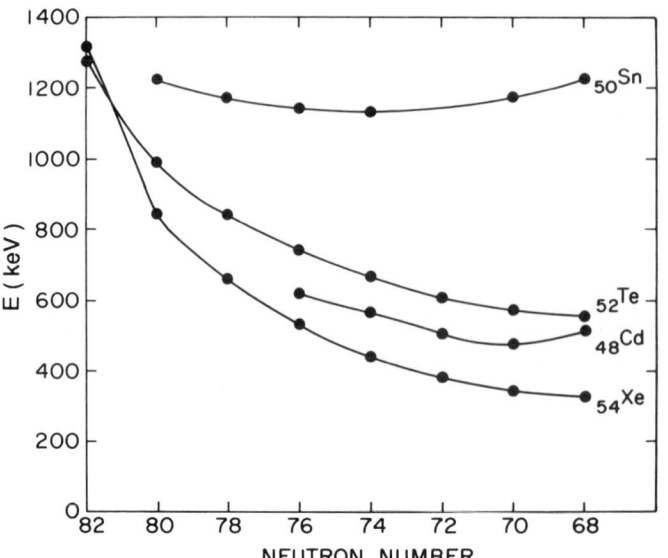

Fig. 1. $E(2_1^+)$ as a function of neutron number for the singly magic nucleus Sn and for nearby elements with a few valence protons.

the work it summarizes, that this is an oversight whose correction may help enrich our understanding of nuclear structure and its evolution. Indeed, evidence for the pivotal role of the p-n interaction is hardly difficult to find: evidence for it is manifest by some of the simplest and best known nuclear data existing. Before discussing more recent work on the p-n interaction, it is perhaps worthwhile to briefly focus on some of this data since its relation to the p-n interaction is often forgotten. This is the purpose of the present section.

Perhaps the most straightforward evidence of the importance of the p-n interaction is provided by the energies of low lying levels in even-even nuclei. The energies of the 2_1^+ levels of nuclei in the Sn region are shown in Fig. 1. Note the constancy of $E(2_1^+)$ in Sn and the rapid decrease for non-singly magic nuclei. Since low 2_1^+ energies are characteristic of the onset of collectivity and deformation, the key role of the p-n interaction is obvious: the *valence* p-n interaction by definition vanishes for Sn as long as we assume the shell closure is intact, but not for the other nuclei. It is easy to understand the Sn results. For simplicity, assume only a single shell model orbit, nlj, is involved. Then, if seniority is a good quantum number, as it must be for any interaction (e.g., a δ force) that can be written in terms of odd tensors[2,9], excitation energies of the j^n configuration are independent of n. The striking feature of Fig. 1 then is the very different behavior for nuclei near Sn with valence protons as well as neutrons. As Talmi[2] has elegantly argued, T=0 residual interactions are much more effective than T=1 interactions in inducing single nucleon configuration mixing, and hence in breaking the seniority scheme. This mixing must lead to non-uniform m-substate distributions and hence is tantamount to deformation, which is characteristically associated with the observed kind of drop in 2_1^+ energies. Thus, the empirical behavior in Fig. 1 is rather

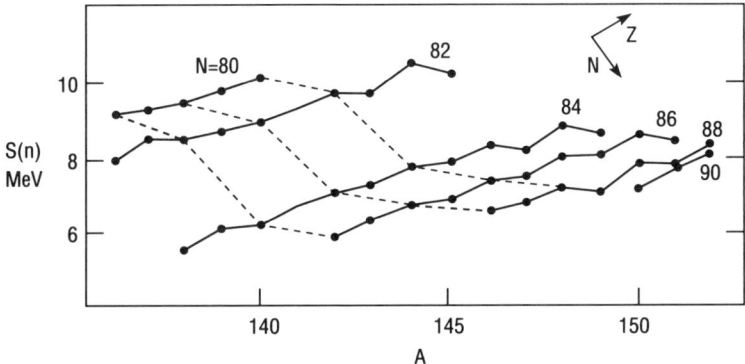

Fig. 2. Single neutron separation energies.

clear evidence of the role of the p-n interaction and this role is consistent with simple properties of residual 2-body interactions.

Nuclear separation energies for the last proton or neutron are some of the simplest data available. They are extremely well known and yet their deep implications for the p-n interaction, though mentioned rather early by Talmi[2], are surprising little recalled. Figure 2 shows some of their systematics. Such plots, though simple and familiar, contain an interesting clue to the p-n interaction and an interesting contrast between its effects and those of the like-nucleon interaction. Inspection of Fig. 2 will show that the separation energy of a given type of nucleon *increases* with the addition of nucleons of the other type and *decreases* with the addition of nucleons of the same type. Thus, for example, S(n) increases with increasing proton number but decreases with increasing neutron number: that is, adding neutrons to a nucleus makes the last neutron less bound while adding protons makes it more bound. Partly, this is related to shell structure effects, which account, for example, for the sharp drop is S(n) after N=82. However, the pervasive pattern shows that, on average, the like nucleon interaction must be repulsive and that the p-n interaction is attractive. Moreover, since the latter consists of both T=1 and T=0 components, and since the T=1 component must be identical (i.e., repulsive) to the p-p and n-n forces, the net attractive character of the p-n interaction implies not only that the T=0 interaction is attractive but also that it must be stronger than the T=1 interaction.

In a multipole expansion of the p-n interaction, the monopole and quadrupole components are often dominant. These two components have complementary effects. It has been shown[10] that the effect of the monopole p-n interaction is to shift single particle energy levels of one kind of particle as a function of the number of nucleons of the other. A dramatic example is seen in Fig. 3. The most striking change is the rapid descent of the $1g_{7/2}$ orbit between ^{91}Zr and ^{131}Sn. Between these two elements, with Z=40 and 50, respectively, the proton $1g_{9/2}$ orbit is filling. The monopole p-n interaction (constant over all space) depends only on the radial overlaps of the proton and neutron wave functions, being largest for similar orbits. [In fact, though rather crude, the

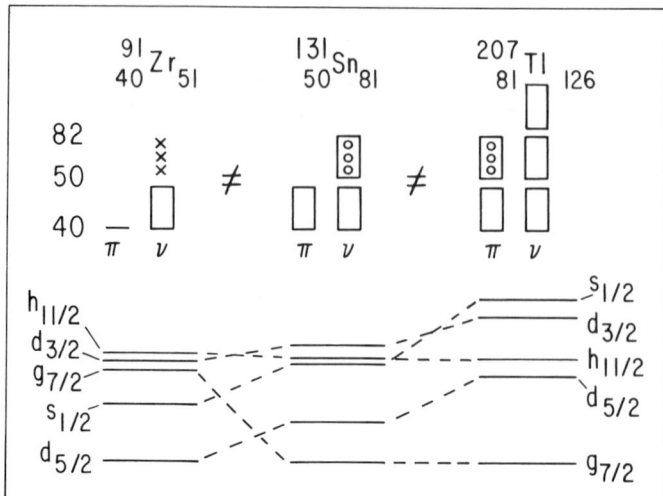

Fig. 3. Single particle energy levels in nuclei with one valence nucleon. The cartoons above each scheme indicate the filled levels in each nucleus. From Ref. 9.

approximation[11] that the monopole p-n matrix elements scale as $1/(\Delta n+\Delta l+1)$ is not terrible[9] for small Δn and Δl.] Being attractive, the favored $p1g_{9/2}$–$n1g_{7/2}$ interaction between Zr and Sn lowers these single particle energies, as clearly seen in Fig. 3. The trend from Sn to Tl can be similarly understood, at least qualitatively. It is interesting to comment in passing that, despite this clear qualitative explanation, a quantitative understanding of the monopole p-n interaction is not yet available. Work currently in progress[12] is attempting to remedy this by developing a reliable and consistent semi-empirical set of two-body matrix elements that characterize the monopole p-n interatction across an entire shell.

The quadrupole component depends primarily on the *angular separation* of the proton and neutron orbits. Its strength varies as $P_2(\cos\theta)$ or $\cos^2\theta$ where θ is the angle between the proton and neutron orbital planes. Viewed in a deformed, or Nilsson, picture, the quadrupole interaction will be strongest for Nilsson orbits with similar Ω or K values such as two downsloping orbits. The quadrupole p-n interaction also depends on the *absolute* angle of each orbit relative to the nuclear symmetry axis or nuclear equator for reasons founded in the basic properties of even tensor interactions in the single-j seniority scheme where matrix elements of an even tensor operator are negatives of each other about mid-shell and, consequently, vanish at mid-shell. The basic effect persists[13] in the multi-j Nilsson scheme where flat Nilsson orbits have very small quadrupole moments. This variation of the quadrupole p-n interaction with orbit angle will have important consequences later.

With this background, we can now look at the phase transitional region near A=150 and see another striking manifestation of the monopole p-n interaction. Figure 4 shows the systematics of $E(2_1^+)$ for the N=88 and 90 isotones. The development of collectivity toward mid-shell would suggest that $E(2_1^+)$ should decrease from Ba to Gd and this is exactly what happens for N=90. In contrast, however, the N=88 isotones display

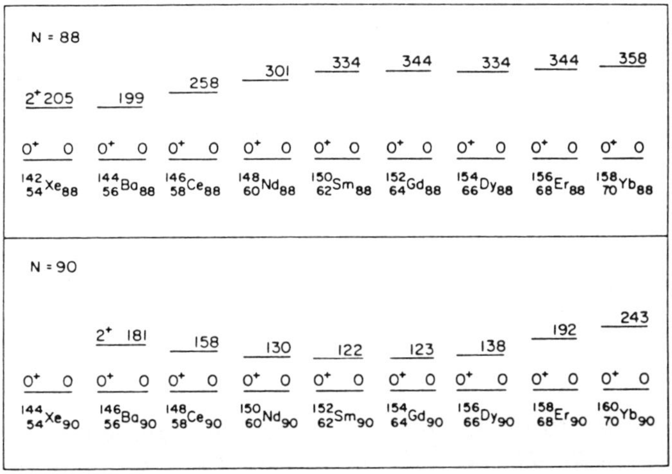

Fig. 4. $E(2_1^+)$ values for N = 88 and 90.

exactly the opposite behavior with a maximum near Gd. It is hard to reconcile this with the expected behavior across a normal major shell. A simple explanation[4], however, is that, at the beginning of the Z=50-82 shell there is a substantial subshell gap[14] at Z=64, but that this gap disappears as neutrons fill the $1h_{9/2}$ orbit near N=90 and the strong attractive monopole interaction with the $p1h_{11/2}$ orbit lowers the energy of the latter. On account of this, for N near 82, the Z=64 gap is present and a nucleus such as Sm with Z=62 actually has, in effect, *fewer* (2) valence protons than Ba (Z=56), and, therefore, quite naturally, a higher $E(2_1^+)$. At N=90 the disappearance of the Z=64 gap leads to a recovery of the normal proton shell 50-82 and the "normal" behavior of $E(2_1^+)$.

THE N_pN_n SCHEME

If the p-n interaction is truly central to the evolution of structure, it should be possible to view the nuclear systematics in terms of some parameter that, at least crudely, reflects its importance. Such an approach is embodied in the N_pN_n scheme[11] in which nuclear observables are plotted against the product of the number of *valence* protons, N_p, times the number of *valence* neutrons, N_n. The rationale behind the N_pN_n scheme is that, if one assumes that the p-n interaction is orbit independent, then its integrated strength should scale as N_pN_n. Examples of N_pN_n plots are shown in Figs. 5 and 6. The simplication wrought by this scheme is evident. The systematics of each observable fall on a smooth curve throughout a half shell. There is no need to repeat here the extensive discussions of the N_pN_n scheme in the literature [see, e.g., refs. 11, 15-17]. It may, however, be useful to cite, without discussion, some of the areas in which the N_pN_n scheme has been applied:

- evolution of collectivity
- subshell effects
- intruder states and shape coexistence

- high spin states
- simplification of collective model calculations
- interpretation of HFB calculations
- r-process nuclei
- monopole radiation and nuclear radii
- isovector M1 excitations, F-spin
- $N_p N_n$ multiplets
- odd-mass nuclei (odd-particle blocking effects)
- heavy ion fusion reactions
- energy weighted sum rules
- saturation of collectivity in deformed nuclei
- monopole and quadrupole p-n interactions

One example may be useful to illustrate since it concerns predictions for nuclei far from stability which are such a focus of current interest.

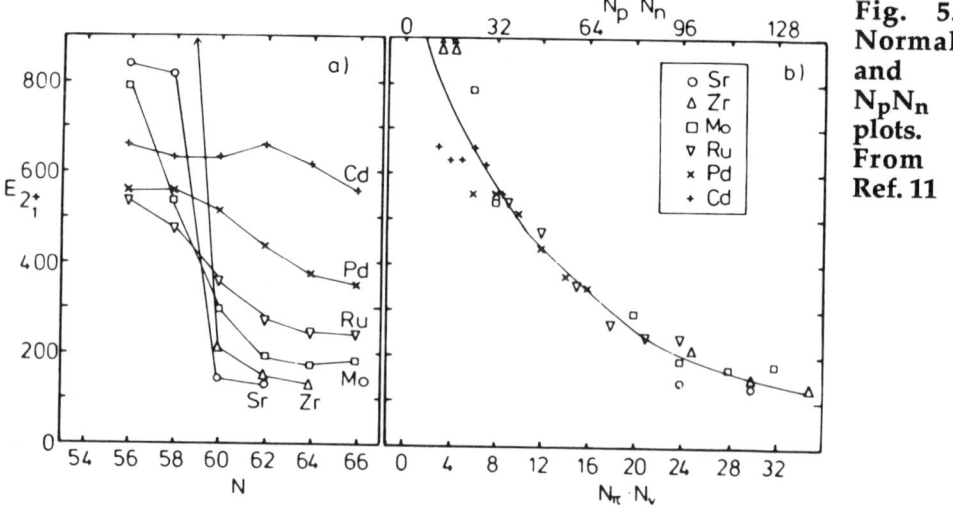

Fig. 5. Normal and $N_p N_n$ plots. From Ref. 11

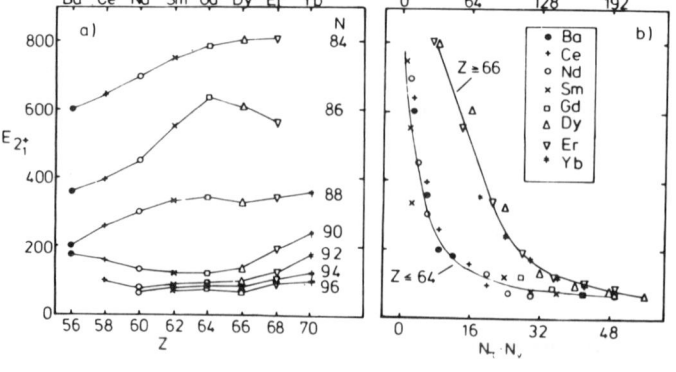

Fig. 6. Normal and $N_p N_n$ plots. From Ref. 11.

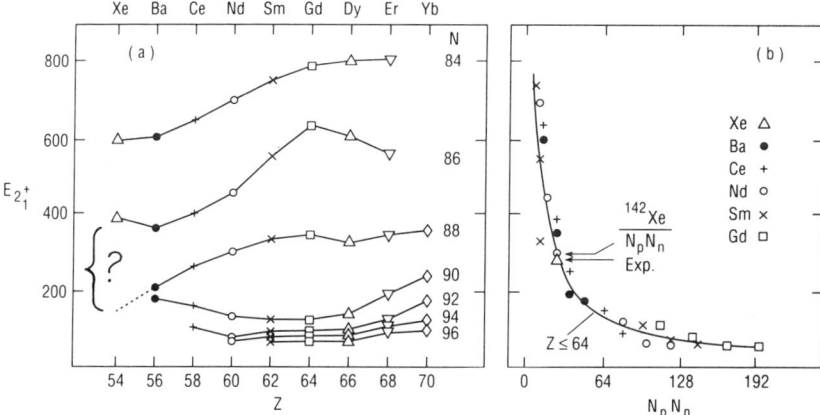

Fig. 7. Normal and N_pN_n predictions for $E(2_1^+)$ in the A = 150 region. A crude extrapolated guess for $E(2_1^+)$ in ^{142}Xe is indicated by the bracket on the left. On the right, the N_pN_n prediction is obtained simply by reading off the curve at $N_pN_n = 24$. From Ref. 18.

Normally, the prediction of properties of nuclei far from stability is a process of *extrapolation* in N, Z or A with all its attendant risks. The N_pN_n scheme is useful here because it frequently converts the normal process of extrapolation to one of *interpolation* since N_pN_n values for many nuclei far from stability are in the same range as those for known nuclei in the same region[16]. The idea is illustrated by the recently studied example of ^{142}Xe, whose level scheme has been deduced from fission product studies[18]. Figure 7 shows a normal and a N_pN_n plot for $E(2_1^+)$ in the rare earth region. If one tries to predict the 2_1^+ energy for ^{142}Xe using normal systematics, an immediate ambiguity arises. The first, most tempting, approach is to follow the N=88 curve. However, this simple extrapolation ignores the fact that, with decreasing Z in this region, one is approaching the Z=50 shell closure and that, at some point, the rigidity associated with the magic numbers should lead to a rise in $E(2_1^+)$, rather than a continued drop. Since, a priori, there is no simple way to anticipate when this will happen in a normal plot, there is considerable uncertainty in the actual value for $E(2_1^+)$: one can only guess that it should fall somewhere between 150-350 keV. The N_pN_n scheme, in contrast, *automatically* balances the competition between increased collectivity and the approach to magicity since this is, in fact, controlled by the p-n interaction. The N_pN_n prediction (see Fig. 7, right) is almost exactly correct.

Clearly, the N_pN_n scheme is useful in considering the systematic behavior of nuclear structure in a given region: that is, *changes in N_pN_n* are correlated with *changes* in structure. However, N_pN_n is a relative quantity, and this complicates the comparison of different mass regions. To facilitate such comparisons, it is useful to introduce[15] a normalized form of N_pN_n, the P factor, defined as

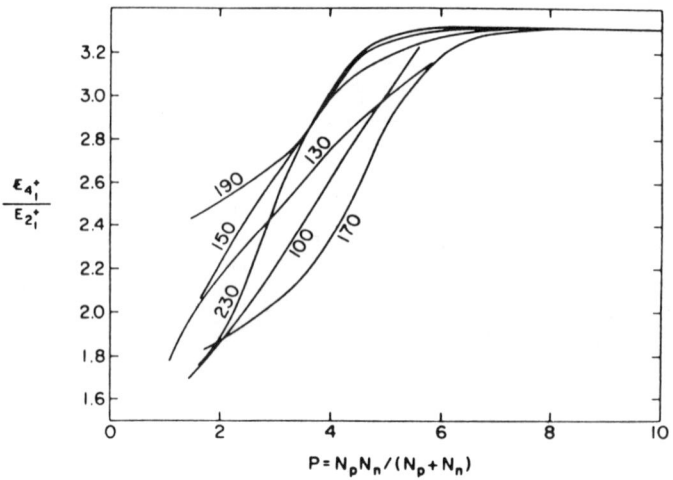

Fig. 8. Plot of $E(4_1^+)/E(2_1^+)$ against the P-factor for six mass regions of heavy nuclei. From Ref. 6.

$$P = \frac{N_p N_n}{N_p + N_n} \qquad (1)$$

The P factor has an obvious physical interpretation, namely as the number of valence p-n interactions per like-nucleon interaction or, equivalently, the average number of p-n interactions per valence nucleon. Normally, the behavior of different mass regions appears rather different when observables are plotted against N, Z or A. Given the importance of the p-n interaction, though, one might expect an underlying similarity in different transition regions. This similarity emerges when the same data are plotted against the P factor. This is illustrated in Fig. 8 which shows that, not only does each region behave very similarly to the others (with the slight exception of the actinides), but all regions also fall within a relatively narrow envelope and each passes through a phase transition between P=4-5. This critical value of P is quite revealing. This is evident if P is expressed, as above, as the ratio of the number of p-n interactions to the number of like-nucleon interactions and if it is recalled that typical p-n interaction strengths are 200-300 keV while the pairing interaction is around 1 MeV. Thus, P_{crit}=4-5 corresponds to the region in a shell where there are 4-5 p-n interactions for each pairing interaction and therefore to just the point where the p-n interaction begins to dominate the pairing interaction. This result[15] highlights once again the intimate connection between the p-n interaction and collectivity on the one hand, and the utility of $N_p N_n$ and the P factor as measures of the integrated p-n strength on the other.

EMPIRICAL VALIDATION OF THE $N_p N_n$ SCHEME

[Note: Much of the following sections is taken from Refs. 6, 7 and 19. I am grateful to my collaborators D. S. Brenner, J.-Y. Zhang, W.-T. Chou, D. D. Warner, and C. Wesselborg for this material.]

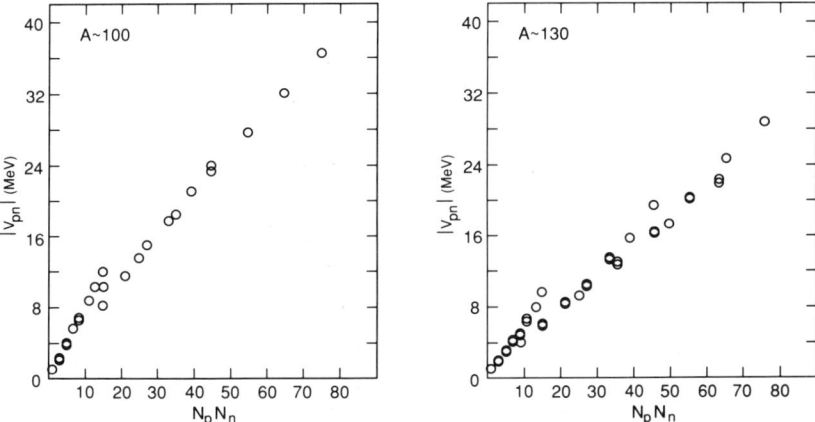

Fig. 9. Plot of integrated empirical p-n interactions V_{pn}, obtained by summing Eq. 2, against $N_p N_n$. From Ref. 7.

The $N_p N_n$ scheme was originally motivated by the recognition of the importance of the valence p-n interaction and is justified by its practical utility. It is based on the assumption that the integrated p-n interaction strength is a simple function of $N_p N_n$. To test this empirically, one needs to extract empirical values for specific p-n matrix elements and demonstrate their approximate orbit independence or, at least, a scaling of their integral with $N_p N_n$. This has recently been done. Zhang et al.[7] have discussed the following double difference of binding energies that isolates the interaction, denoted δV_{pn}, of the last valence proton with the last valence neutron.

$$\delta V_{pn} = \frac{1}{4} [(B(Z+2,N+2) - B(Z+2,N)) - (B(Z,N+2) - B(Z,N))] \quad (2)$$

where $B(Z,N)$ is the (negative) binding energy of the nucleus Z,N. This double difference cancels mean field and like-nucleon interactions. It is discussed more extensively in Ref. 7 where its antecedents in existing mass equations are also discussed. Below, we shall present plots of individual δV_{pn} values over the entire Periodic Table. First, we consider *sums* of δV_{pn} over all valence protons and neutrons giving the integral of δV_{pn}, denoted V_{pn}, which is plotted against $N_p N_n$ in Fig. 9. The result is striking: there is an almost exact linearity in the relationship. [The one exception to this trend are the single points at $N_p N_n = 15$ but these are truly exceptions which prove the rule since these nuclei corresponds to extreme N_p and N_n values of 1 and 15 in comparison to nuclei with N_p and N_n = 3 and 5.] Of particular note in Fig. 9 are the nearly overlapping circles occurring for the same $N_p N_n$ values. These correspond to nuclei

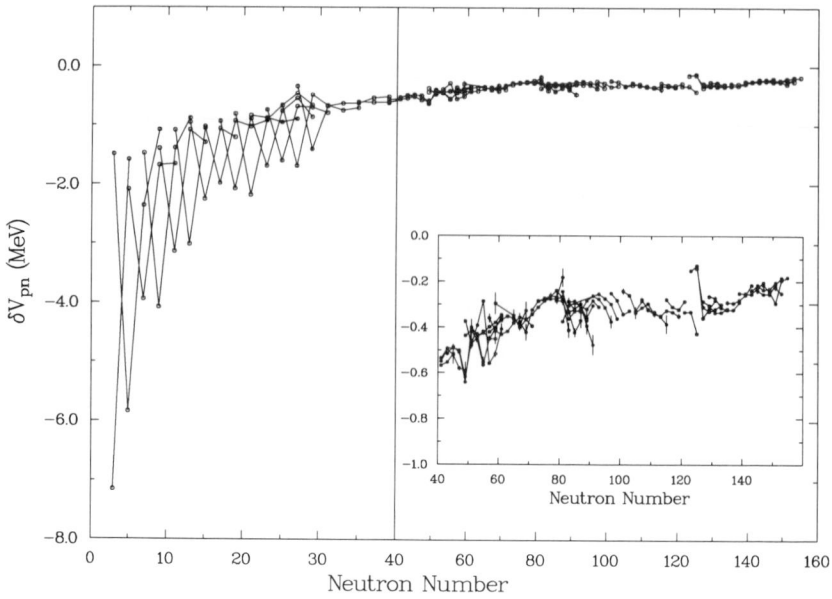

Fig. 10. δV_{pn} data for all even-even nuclei. From Ref. 6.

with different separate values of N_p and N_n but the same valence product. The virtually identical values of V_{pn} strongly support the idea of the approximate linearity of integrated p-n strength with $N_p N_n$ at least up to about the one-third filled part in the shell. These results provide an empirical, microscopic, underpinning to the rationale behind the $N_p N_n$ scheme and give it a conceptual credence beyond its mere practical utility.

EMPIRICAL P-N INTERACTION MATRIX ELEMENTS

Empirical p-n matrix elements of the last proton with the last neutron are plotted for all even-even nuclei in Fig. 10 and show a number of fascinating features[6]. To discuss these, note first that larger (attractive) p-n interaction strengths correspond to *lower* values on the plot. Ignoring, for a moment, the obvious singularities that occur for certain light nuclei, we see that, globally, there is a rather smooth trend toward smaller and smaller values with increasing mass. This has an obvious, and well known, explanation: the p-n matrix elements are sensitive to the overlaps of the respective orbits. For a residual two-body interaction of constant radius, the matrix elements decrease as the radii of the respective Shell Model orbits increase and the wave functions become more spread out in space. This is reinforced by the neutron excess in heavy nuclei which further increases the difference in proton and neutron radii.

Besides this secular trend there are two obvious structural features of the plot. The most dramatic is the set of enormous spikes in δV_{pn} for certain light nuclei. Inspection of the data shows that, in each case, these are specifically those nuclei with N=Z. An explanation for this apparently

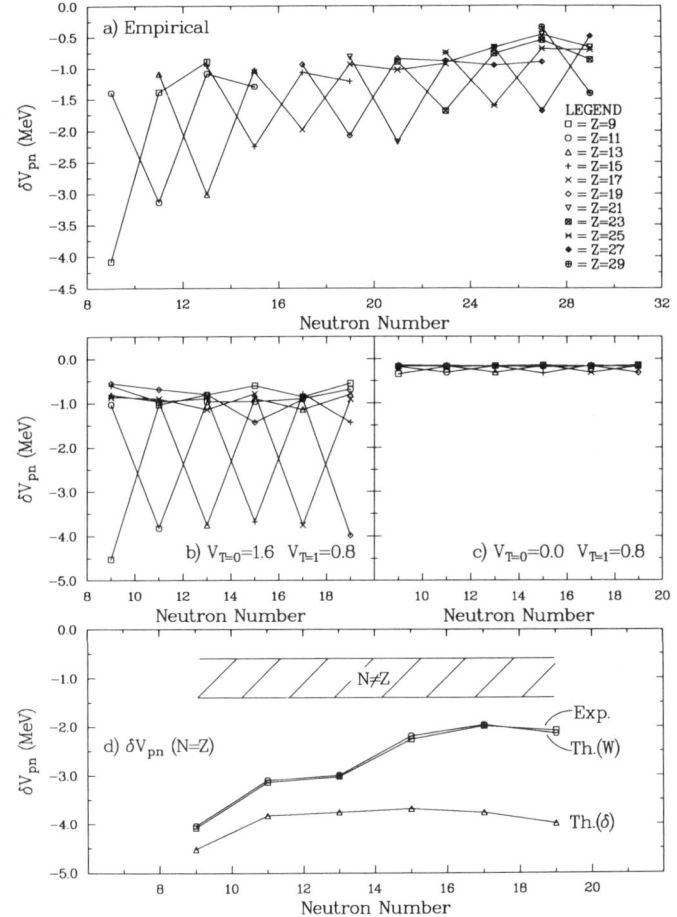

Fig. 11. δV_{pn} values in the region of light N = Z nuclei. Top: Expanded view of the empirical δV_{pn} values of Fig. 10. Middle: Calculations with a surface δ interaction. Bottom: Comparison of the locus of N = Z δV_{pn} values with calculations using a surface δ interaction [Th(δ)] and the Wildenthal-USD interaction [Th(W)]. From Ref. 6.

anomalous behavior is readily at hand[6]. It is implicitly contained[2] in typical parameterizations of nuclear masses which contain a T(T+1) term, which gives an extremum for T=0 characterizing N=Z nuclei. A more microscopic interpretation can be obtained[6] with simple Shell Model calculations. In the middle panels of Fig. 11 are the results of two Shell Model calculations, carried out in the 2s-1d shell (N=8-20) with a surface δ interaction. The single particle energies were chosen to approximately fit the spectra of ^{17}O and ^{17}N while the strength of the T=1 part of the two-body interaction approximately reproduces the spectra of ^{18}O and ^{18}Ne. Calculations are shown for two values of the T=0 strength. On the right the results are given for $V_{T=0} = 0$: clearly these calculations cannot reproduce the empirical results. This is not surprising: indeed, we argued earlier that the T=0 strength is substantially greater than the T=1 strength. The calculation on the left uses a more realistic T=0 strength, namely, $V_{T=0} = 2V_{T=1}$. Even though the space is highly restricted and the force is a

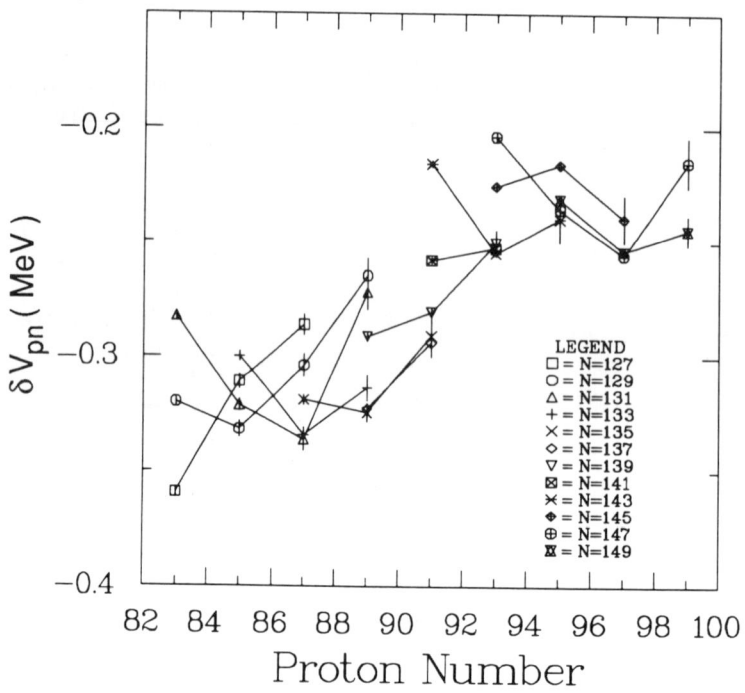

Fig. 12. Experimental δV_{pn} values for the actinides. From Ref. 6.

schematic surface δ interaction, these calculations reproduce the data remarkably well yielding spikes at N=Z of roughly the right magnitude. The only discrepancy is that the spikes are roughly constant in magnitude whereas the empirical spikes decrease in magnitude with increasing A.

The bottom panel shows a calculation using the more realistic Wildenthal universal 2s-1d interaction[20] (USD). In this panel, the δV_{pn} values for nuclei with N≠Z are schematically indicated by the diagonally marked band while the locus of N=Z spikes is indicated by the points lying below this band. The calculations with surface δ and Wildenthal interactions are indicated. The latter exactly reproduces the empirical results. Though the USD interaction is already well established[20], these results provide a particularly sensitive test since δV_{pn} represents the rather small resultant of a double difference of binding energies from four adjacent even-even nuclei. The appearance of N=Z singularities in both the highly schematic calculations and those with a realistic interaction reflects its origin in very basic features of T=0 interactions in two-nucleon wave functions, in particular the strong p-n interactions in N=Z nuclei with protons and neutrons in equivalent orbits characterized by enhanced spatial symmetry of their wave functions.

Despite the approximate orbit independence of the p-n interaction seen in Fig. 10 and implied by Fig. 9, the microstructure in heavier nuclei, that is, the detailed fluctuations in δV_{pn}, is no less interesting. As one example, Fig. 12 shows a decrease of δV_{pn} toward midshell and a levelling off thereafter. It seems likely that this behavior reflects the presence of

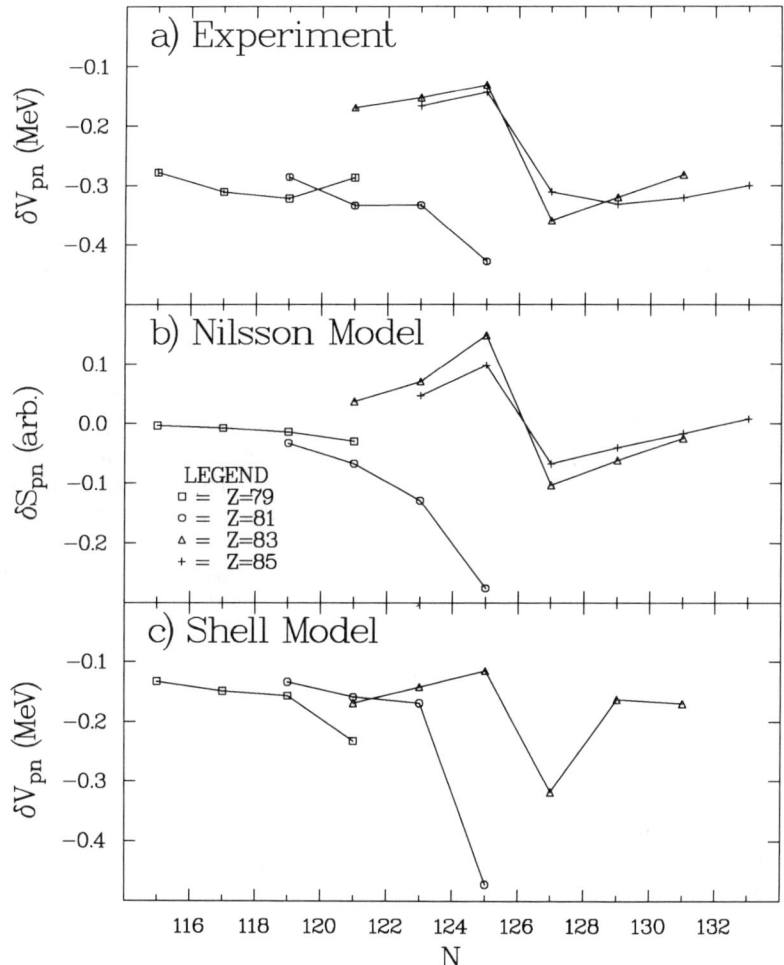

Fig. 13. p-n interaction strengths near ^{208}Pb. Top: Expanded view of the experimental δV_{pn} values from Fig. 10. Middle: Nilsson Model calculations of the quadrupole interaction, δS_{pn}, of the last proton with the last neutron. Bottom: Shell Model calculations of δV_{pn}. From Ref. 19.

both monopole and quadrupole components near the beginning of a shell and the decrease of the quadrupole component toward mid-shell, leaving the dominant monopole component in the latter region. If this is true, the figure suggests an average value of ~ -250 keV for this component.

Other microstructure is both more dramatic and more easily interpretable. The most obvious case is the Pb region shown in expanded form in Fig. 13 along with two sets of calculations[19]. The empirical results show large changes in δV_{pn} on different sides of ^{208}Pb. Specifically, when protons and neutrons are either both above or both below the N=82 and Z=126 magic numbers, the p-n interactions are large. When the proton and neutron numbers are on opposite sides of these two magic numbers, the p-n interactions drop markedly. As noted, this has a simple "overlap" explanation. Single particle Shell Model orbits, labelled by the quantum numbers (n,l,j), at the end of a major shell are typified by relatively high n and low l and j whereas, at the beginning of a shell, the single particle states are of lower n and high l and j. It is clear that the overlap of a pair of orbits at the same end of a major shell will be higher than for orbits on opposite ends of a shell.

This qualitative explanation of the fluctuation patterns in p-n matrix elements near ^{208}Pb, and in other doubly magic regions, can be quantified and substantiated by both Nilsson and Shell Model calculations. The idea behind the Nilsson Model calculations is that the proton and neutron quadrupole moments, and hence the quadrupole p-n interactions, are largest just below and above closed shells and therefore the fluctuations in p-n matrix elements near closed shells should be largely controlled by this component. The calculations for these nearly spherical nuclei were done with a (nearly arbitrary) deformation parameter of $\varepsilon = 0.01$. This arbitrariness only means that the p-n matrix elements are calculated on an arbitrary scale: the calculated *fluctuations* in δV_{pn}, however, are nearly independent of ε. The calculations were done[19] with standard pairing and Nilsson parameters. It is clear that they reproduce the empirical values extraordinarily well. Not only are the sudden jumps around ^{208}Pb reproduced, but also many of the fine details of the microstructure are represented in the calculations.

It is interesting to compare these calculations with those in the Shell Model since the two approaches are rather complementary. The latter were done[19] with a highly truncated space and a surface δ interaction. They are presented in the bottom panel of Fig. 13. Below the Pb magic numbers, the basis states for each kind of nucleon include three low j orbits while, above, a single orbit is used.

There are two interesting conclusions. First, as with the Nilsson Model, the Shell Model nicely reproduces most of the effects seen experimentally. The exception is that the gap in δV_{pn} across Z=82 for N<126 is much smaller in the calculations than in the data. Secondly, as opposed to Nilsson calculations, the Shell Model automatically includes all relevant multipoles in the space considered and, remarkably, even the absolute calculated values between -200 and -400 keV are in excellent agreement with the data.

The fact that the Nilsson Model, with extended space and schematic force, and the Shell Model, with restricted space and more realistic interaction, both reproduce the data and give very similar results, strongly suggests that the origin of the fluctuations in δV_{pn} in the Pb region resides not in small details of the wave functions but in the substantial changes in proton and neutron wave function overlaps as the doubly magic region is traversed. This argues for the essentially simple nature of the underlying physics behind the fluctuations and microstructure in p-n matrix elements in heavy nuclei.

CONCLUSIONS

A number of issues relating to the p-n interaction in nuclear structure have been discussed, including its manifestations in very simple nuclear data, the different roles of its separate multipoles, the N_pN_n scheme and the P factor, their validation through empirical integrated valence p-n interaction strengths, the global and regional behavior of individual p-n matrix elements, including singularities for N=Z nuclei, the saturation of quadrupole p-n interactions near mid-shell, and strong fluctuations in closed shell regions, along with their interpretation in terms of Nilsson and Shell Model calculations.

Research has been performed under contract No. DE-AC02-76CH00016 with the USDOE. I am grateful to my collaborators in much of this work, D. S. Brenner, W.-T. Chou, J.-Y. Zhang, D. D. Warner, and C. Wesselborg. I would also like to thank K. Heyde, I. Talmi, and E. K. Warburton for many useful discussions.

REFERENCES

1. A. de Shalit and M. Goldhaber, Phys. Rev. **92**, 1211 (1953).
2. I. Talmi, Rev. Mod. Phys. **34**, 704 (1962); *Progress in Particle and Nuclear Physics, Collective Bands in Nuclei*, Vol. 9, ed. D. Wilkinson (Pergamon, Oxford, 1983), p. 27.
3. P. Federman and S. Pittel, Phys. Lett. **69B**, 385 (1977) and **77B**, 29 (1978).
4. R.F. Casten et al., Phys. Rev. Lett. **47**, 1433 (1981).
5. J. Dobaczewski et al., Phys. Rev. Lett. **60**, 2254 (1988).
6. D.S. Brenner et al., Phys. Lett. **B243**, 1 (1990).
7. J.-Y. Zhang, R.F. Casten, and D.S. Brenner, Phys. Lett. **B227**, 1 (1989).
8. R.F. Casten, Predeal, Romania, Sept. 1990 (to be published).
9. R.F. Casten, *Nuclear Structure from a Simple Perspective* (Oxford, New York, 1990), Chapters 4 and 5.
10. K. Heyde et al., Nucl. Phys. **A466**, 189 (1987).
11. R.F. Casten, Phys. Lett. **152B**, 145, and Phys. Rev. Lett. **54**, 1991 (1985).
12. P. Federman, G. Arenas-Peris, and S. Pittel, private communication.
13. R.F. Casten, K. Heyde, and A. Wolf, Phys. Lett. **B208**, 33 (1988).
14. M. Ogawa et al., Phys. Rev. Lett. **41**, 1480 (1978); Y. Nagai et al., Phys. Rev. Lett. **47**, 1259 (1981).
15. R.F. Casten, D.S. Brenner, P.E. Haustein, Phys. Rev. Lett **58**, 658 (1987).
16. R.F. Casten, Phys. Rev. **C33**, 1819 (1986).
17. K. Heyde and J. Sau, Phys. Rev. **C33**, 1050 (1986).
18. A.S. Mowbray et al., Phys. Rev. **C42**, 1126 (1990).
19. W.-T. Chou et al., to be published.
20. B.A. Brown and B.H. Wildenthal, Ann. Rev. Nucl. Sci. **38**, 29 (1988); B.H. Wildenthal, private communication to E.K. Warburton.

IS COLLECTIVE MOTION SYMMETRIC IN THE NEUTRON-PROTON DEGREES OF FREEDOM?

Joseph N. Ginocchio[1,2] and Amiram Leviatan[1]
[1]Theoretical Division, Los Alamos National Laboratory,
Los Alamos, New Mexico 87545
[2]Physics Department, University of Washington,
Seattle, Washington 98195

Michael W. Kirson
Nuclear Physics Department, Weizmann Institute of Science,
76100 Rehovot, Israel

ABSTRACT

The purity of intrinsic states of nuclei with respect to a proton-neutron boson symmetry (F-spin) is shown to be largely determined by the difference between proton and neutron deformations and not by whether the Hamiltonian is an F-spin scalar. Upper and lower bounds on F-spin mixing in the ground state band of ^{165}Ho are estimated using recent pion single-charge-exchange data.

INTRODUCTION

Shortly after the introduction of the interacting-boson model[1] as a phenomenologically successful description of low-lying collective spectra of medium to heavy nuclei, it was proposed that a microscopic, shell-model- based interpretation of the model required distinguishing between neutron and proton bosons.[2-3] The resulting proton-neutron interacting-boson model (called IBM-2 to distinguish it from the original IBM-1 with only a single type of boson) contains a natural additional algebraic structure, based on the two-valued proton-neutron degree of freedom. This SU(2) structure, denoted F-spin,[2] holds out the possibility of an additional, possibly-conserved quantum number in nuclear spectroscopy. Although the same microscopic arguments which led to the introduction of IBM-2 support a model Hamiltonian which is not an F-scalar, apparently implying that nuclear collective states do not have well-defined F-spin, data on M1 transitions between low-lying collective states have been argued to require a high degree of F-spin purity in these nuclear levels.[4,5] It will be shown here that the intrinsic-state picture recently developed for interacting-boson models,[6] and the associated geometrical description of nuclear spectra in terms of intrinsic deformations with superimposed collective motion,[7-8] may be capable of bridging these two superficially incompatible claims, and can open the way to the use of additional experimental information to pin down the F-spin purity of nuclear energy levels.

IBM-2 AND F-SPIN

The basic building blocks of IBM-2 are monopole and quadrupole proton ($\rho = \pi$) and neutron ($\rho = \nu$) creation and destruction operators: $s_\rho^\dagger, s_\rho$ ($J^\pi = 0^+$), $d_{\rho\mu}^\dagger$, $\tilde{d}_{\rho\mu} \equiv (-)^\mu d_{\rho-\mu}$ ($J^\pi = 2^+$). The boson number operator $\hat{N}_\rho$ has eigenvalue N_ρ

where $N_\pi(N_\nu)$ is one-half the number of valence protons (neutrons). The basic F-spin doublets are $(s_\pi^\dagger, s_\nu^\dagger)$, and $(d_{\pi\mu}^\dagger, d_{\nu\mu}^\dagger)$, with the convention that proton bosons have F-spin projection $+1/2$, neutron bosons $-1/2$. The F-spin raising and lowering operators are

$$\hat{F}_+ = s_\pi^\dagger s_\nu + \sum_\mu d_{\pi\mu}^\dagger d_{\nu\mu} \,, \quad \hat{F}_- = (\hat{F}_+)^\dagger \tag{1}$$

and the operator defining the z-component of F-spin is $\hat{F}_0 = \frac{1}{2}(\hat{N}_\pi - \hat{N}_\nu)$. These three operators form an SU(2) Lie algebra analogous to isospin. In a given nucleus, with fixed N_π, N_ν, all states have the same value of $F_0 = \frac{1}{2}(N_\pi - N_\nu)$, while the allowed values of the F-spin quantum number F range from $|F_0|$ to $F_{MAX} \equiv \frac{1}{2}(N_\pi + N_\nu)$ in unit steps. F-spin measures the extent to which the states are symmetric in the neutron and proton degrees of freedom. The states with maximum F-spin, F_{MAX}, have the highest symmetry and are the lowest in the energy spectrum in general.

It has been shown[9] that for states with good F-spin, there are no M1 transitions between states with $F = F_{MAX}$. This, together with the observed weakness of B(M1)'s between low-lying collective states in nuclei, is the basis of the claim that F-spin breaking must be small in these nuclei,[4,5] though it has subsequently been realized that selection rules due to symmetries other than F-spin (such as SO(5) and SU(3)) can also lead to vanishing M1 matrix elements.[10]

The general Hamiltonian of IBM-2 has many free parameters, and different forms have been used to fit spectra and E2 transition rates in nuclei. Forms based on microscopic arguments, and hence not F-scalars, can produce F-spin mixing as high as 18% in states of the ground band while fitting spectra and E2 transitions well.[11] (The goodness of the fit to B(M1) depends critically on the poorly-determined boson g-factors.) On the other hand, equally good fits are obtained with near F-scalar Hamiltonians or with interactions which push up in energy states with lower F-spin, resulting in F-spin mixing of 2-4% which suffices to reproduce the B(M1)'s.[12] It is clearly desirable to find alternate experimental criteria to try to pin down more precisely the F-spin purity of nuclear collective states.

AN INTRINSIC STATE

A geometrical description of IBM-2 states can be based on the concept of an intrinsic state which is a condensate of deformed bosons, which do not have well-defined angular momenta.[6-8] The most general such condensate, in which proton and neutron boson numbers are separately conserved, can be written in the form

$$|c> \equiv (N_\pi! N_\nu!)^{-1/2} [b_{c\pi}^\dagger(\beta_\pi, \gamma_\pi)]^{N_\pi} [b_{c\nu}^\dagger(\beta_\nu, \gamma_\nu, \Omega)]^{N_\nu} |0> \tag{2a}$$

where $|0>$ is the boson vacuum,

$$b_{c\rho}^\dagger(\beta_\rho, \gamma_\rho) = (1+\beta_\rho^2)^{-1/2}[s_\rho^\dagger + \beta_\rho \cos\gamma_\rho \cdot d_{\rho 0}^\dagger + \beta_\rho \sin\gamma_\rho \cdot \frac{1}{\sqrt{2}}(d_{\rho 2}^\dagger + d_{\rho,-2}^\dagger)] \tag{2b}$$

$$b_{c\nu}^\dagger(\beta_\nu, \gamma_\nu, \Omega) = R(\Omega) b_{c\nu}^\dagger(\beta_\nu, \gamma_\nu) R^{-1}(\Omega), \tag{2c}$$

and where $R(\Omega)$ rotates the condensate boson through the Euler angles ϕ, θ, ψ which are collectively denoted as Ω. This condensate depends on seven intrinsic shape parameters: the total deformations β_ρ, the axiality parameters γ_ρ, and the Euler angles ϕ, θ, ψ of relative orientation. The values of the shape parameters $\beta_\rho, \gamma_\rho, \Omega$ are determined by minimizing the expectation value of the Hamiltonian in the condensate $|c>$. Whenever $\beta_\rho > 0$, the resulting equilibrium condensate is deformed, and a rotational ground-state band is projected from this intrinsic state. The IBM boson deformation parameter β_ρ differs in numerical value from the familiar Bohr and Mottelson definition[13] since the boson model involves a truncation of nuclear collective motion[1–3,6] to the shell model valence space of monopole and quadrupole pairs of fermions.

The intrinsic state will have well-defined F-spin, F, when $|c>$ is an eigenstate of $\hat{F}^2 = \hat{F}_+\hat{F}_- + \hat{F}_0^2 - \hat{F}_0$, with eigenvalue $F(F+1)$. Since $|c>$ is, by construction, an eigenstate of $\hat{F}_0$, the condition for it to have good F-spin is $\hat{F}_+\hat{F}_-|c> \propto |c>$. Because the effect of $\hat{F}_-$ on the deformed proton boson $b^\dagger_{c\pi}(\beta_\pi, \gamma_\pi)$ is to turn it into a deformed neutron boson $b^\dagger_{c\nu}(\beta_\pi, \gamma_\pi)$, while $\hat{F}_+$ turns $b^\dagger_{c\nu}(\beta_\nu, \gamma_\nu, \Omega)$ into $b^\dagger_{c\pi}(\beta_\nu, \gamma_\nu, \Omega)$, $\hat{F}_+\hat{F}_-|c>$ will have one component proportional to $|c>$, and a second component in which one proton condensate boson and one neutron condensate boson have exchanged expansion coefficients. Thus $\hat{F}_+\hat{F}_-|c>$ can be proportional to $|c>$ only if the expansion coefficients are equal. Since the representation of the expansion coefficients in terms of shape parameters β, γ and Euler angles ϕ, θ, ψ is unique, this means that $\beta_\pi = \beta_\nu, \gamma_\pi = \gamma_\nu$, and the relative Euler angles must vanish in order for $|c>$ to be an eigenstate of $\hat{F}^2$. It is then easy to show that the corresponding eigenvalue is $F_{MAX}(F_{MAX} + 1)$.

These conditions ($\beta_\pi = \beta_\nu, \gamma_\pi = \gamma_\nu, \Omega = 0$) translate into relatively mild constraints on the parameters of the IBM-2 Hamiltonian <u>which need not be an F-scalar</u>. The members of the ground-state band are expected to be well approximated by the projections of states of good angular momentum from the deformed equilibrium condensate. Since the angular-momentum projection operator is an F-scalar, these projected states will also have $F = F_{MAX}$. Thus it is possible for a non-F-scalar IBM-2 Hamiltonian to have a ground-state band of states with nearly pure F-spin, $F = F_{MAX}$, provided only that its equilibrium intrinsic state consists of aligned proton and neutron condensates, of equal deformation.

F-SPIN ADMIXTURES

In order to explore the consequences of different proton and neutron deformations for F-spin breaking, it is convenient to restrict attention to the most likely situation, namely aligned, axially-symmetric but differently-deformed condensates (2) with $\beta_\pi \neq \beta_\nu$, $\gamma_\pi = \gamma_\nu = \Omega = 0$, which we denote by $|c_a>$. It is straightforward to evaluate

$$< c_a|\hat{F}^2|c_a> = F_{MAX}(F_{MAX} + 1) - N_\pi N_\nu (\beta_\pi - \beta_\nu)^2/(1 + \beta_\pi^2)(1 + \beta_\nu^2), \quad (3)$$

from which it is again evident that $F = F_{MAX}$ for $\beta_\pi = \beta_\nu$. This condensate can, in general, be expanded in terms of states of well-defined F-spin, as $|c_a> = \sum_F \alpha_F |F>$, where $|F>$ is the component with the F-spin quoted and α_F is the amplitude of that component in the condensate $|c_a>$. Using the normalization condition $\sum_F |\alpha_F|^2 = 1$ and (3), it follows that

$$NP_{MAX} \equiv N_\pi N_\nu \frac{(\beta_\pi - \beta_\nu)^2}{(1 + \beta_\pi^2)(1 + \beta_\nu^2)} = \sum_{F=|F_0|}^{F_{MAX}-1} |\alpha_F|^2 [F_{MAX}(F_{MAX} + 1) - F(F+1)] \quad (4)$$

where $N = N_\pi + N_\nu$. Both upper and lower bounds on the admixtures of lower F-spin in the intrinsic condensate can be derived by putting $F = F_{MAX} - 1$ and $|F_0|$, respectively, in the sum in (4). That is

$$P_{MIN} = \frac{N}{N_<(N_> + 1)} P_{MAX} \leq 1 - |\alpha_{F_{MAX}}|^2 \leq P_{MAX} \tag{5}$$

where $N_<(N_>)$ is the lesser (greater) of (N_π, N_ν). Thus by knowing the IBM deformation parameters β_ρ an upper and lower bound on the F-spin admixtures can be obtained. For large N_ρ these parameters can be determined from the measured neutron and proton quadrupole moments or $B(E2)$ transition rates.

An explicit form for the F-spin decomposition of the condensate $|c_a>$ can be obtained in the limit of large N_ρ. Introducing a fixed reference value $\overline{\beta}$ of the deformation parameter, the condensate boson $b_{c\rho}^\dagger$ (with $\gamma_\rho = \Omega = 0$) can be expressed as a linear combination of the mutually-orthogonal condensate boson and β-vibration boson for that value, as follows:

$$B_{c\rho}^\dagger \equiv (1+\overline{\beta}^2)^{-1/2}(s_\rho^\dagger + \overline{\beta}d_{\rho 0}^\dagger), \quad B_{\beta\rho}^\dagger \equiv (1+\overline{\beta}^2)^{-1/2}(d_{\rho 0}^\dagger - \overline{\beta}s_\rho^\dagger) \tag{6a}$$

$$b_{c\rho}^\dagger = [(1+\beta_\rho^2)(1+\overline{\beta}^2)]^{-1/2}[(1+\overline{\beta}\beta_\rho)B_{c\rho}^\dagger + (\beta_\rho - \overline{\beta})B_{\beta\rho}^\dagger]. \tag{6b}$$

The condensate $|c_a>$ can then be expanded in terms of the form $(B_{c\rho}^\dagger)^{N_\rho-K_\rho}(B_{\beta\rho}^\dagger)^{K_\rho}$. Using the facts that for any n_π, n_ν, $(B_{c\pi}^\dagger)^{n_\pi}(B_{c\nu}^\dagger)^{n_\nu}$ has well defined F-spin, equal to $\frac{1}{2}(n_\pi + n_\nu)$, with projection $\frac{1}{2}(n_\pi - n_\nu)$, while $B_{\beta\pi}^\dagger$ and $B_{\beta\nu}^\dagger$ have F-spin 1/2, with projections $+1/2$ and $-1/2$ respectively, and applying SU(2) vector-coupling coefficients, each term in the expansion is expressed in terms of components of well-defined F-spin. The leading terms resulting from this procedure are $(B_{c\pi}^\dagger)^{N_\pi}(B_{c\nu}^\dagger)^{N_\nu}$ with $F = F_{MAX}$, and $B_{\beta\pi}^\dagger(B_{c\pi}^\dagger)^{N_\pi-1}(B_{c\nu}^\dagger)^{N_\nu}$, $B_{\beta\nu}^\dagger(B_{c\pi}^\dagger)^{N_\pi}(B_{c\nu}^\dagger)^{N_\nu-1}$, which contain components $F = F_{MAX}$ and $F = F_{MAX} - 1$. If, at this point, the value of $\overline{\beta}$ is chosen so that the coefficient of the F_{MAX} component of the $B_{\beta\rho}^\dagger$ terms vanishes, then only the $F_{MAX} - 1$ component survives, with coefficient

$$x = \sqrt{N_\pi N_\nu/N}\,(\beta_\pi - \beta_\nu)(1+\overline{\beta}^2)/(1+\beta_\pi\overline{\beta})(1+\beta_\nu\overline{\beta}) \tag{7}$$

relative to the leading F_{MAX} term. The equation defining $\overline{\beta}$ is

$$(N_\pi\beta_\nu + N_\nu\beta_\pi)\overline{\beta}^2 + N(1 - \beta_\pi\beta_\nu)\overline{\beta} - (N_\pi\beta_\pi + N_\nu\beta_\nu) = 0 \tag{8}$$

and has one positive root. For almost-equal deformations, $|\beta_\pi - \beta_\nu| \ll 1$, $\overline{\beta} = (N_\pi\beta_\pi + N_\nu\beta_\nu)/N + O((\beta_\pi - \beta_\nu)^2)$; i. e., a boson-number weighted deformation. By inspection of the remaining terms in the expansion, it can be seen that, with this choice of $\overline{\beta}$, the condensate can be written approximately in the form

$$|c_a> \approx e^{-x^2/2}\sum_{k=0}^\infty (x^k/\sqrt{k!})\,|F = F_{MAX} - k>, \tag{9}$$

for large N_ρ.

At this point it is clear that in well-deformed, rotational nuclei, the degree of F-spin mixing in the states of the nucleus is dependent mainly on the degree of similarity of proton and neutron deformations, regardless of whether the IBM-2 Hamiltonian is F-scalar or not. It should be pointed out that non-F-scalar Hamiltonians with equal proton and neutron deformations are not hard to come by. As one example, consider the popular choice

$$H = \epsilon_\pi \hat{n}_{d\pi} + \epsilon_\nu \hat{n}_{d\nu} + \kappa Q_\pi \cdot Q_\nu + \kappa' L_\pi \cdot L_\nu + \alpha_2(s_\pi^\dagger d_\nu^\dagger - s_\nu^\dagger d_\pi^\dagger) \cdot (s_\pi \tilde{d}_\nu - s_\nu \tilde{d}_\pi)$$
$$+ \sum_{L=1,3} \alpha_L (d_\pi^\dagger d_\nu^\dagger)^{(L)} \cdot (\tilde{d}_\nu \tilde{d}_\pi)^{(L)}. \qquad (10a)$$

Here

$$Q_\rho = s_\rho^\dagger \tilde{d}_\rho + d_\rho^\dagger s_\rho + \chi_\rho (d_\rho^\dagger \tilde{d}_\rho)^{(2)}, \qquad (10b)$$

$\hat{n}_{d\rho}$ are the d_ρ-boson number operators, superscripts indicate coupling to well-defined spherical-tensor rank and dots denote scalar products. This is not an F-scalar. The necessary conditions for $\beta_\pi = \beta_\nu = \beta$, $\gamma_\pi = \gamma_\nu = 0$, $\Omega = 0$ to determine a ground-state condensate with well-defined $F = F_{MAX}$ are

$$\epsilon_\pi/N_\nu - \epsilon_\nu/N_\pi - \kappa\sqrt{\frac{2}{7}}(\chi_\pi - \chi_\nu)\beta = 0 \qquad (11a)$$

$$\epsilon_\pi(1+\beta^2)/N_\nu + 2\kappa(1-\beta^2) - \kappa\sqrt{\frac{2}{7}}\beta[\,\chi_\nu(1-\beta^2) + 2\chi_\pi\,] + \frac{2}{7}\kappa\chi_\pi\chi_\nu\beta^2 = 0, \quad (11b)$$

which may be regarded as one equation for determining β, and one constraint equation relating some of the nine parameters of the Hamiltonian. As long as band mixing is weak, the ground state band generated from this condensate will have good F-spin.

The class of Hamiltonians

$$H = \sum_\rho \{A_\rho [d_\rho^\dagger \cdot d_\rho^\dagger - \beta^2(s_\rho^\dagger)^2][H.\ c.\ ] + A'_\rho[\beta s_\rho^\dagger d_\rho^\dagger + \sqrt{\frac{7}{2}}(d_\rho^\dagger d_\rho^\dagger)^{(2)}] \cdot [H.\ c.\ ]\}$$

$$+A_0[d_\pi^\dagger \cdot d_\nu^\dagger - \beta^2 s_\pi^\dagger s_\nu^\dagger][H.\ c.\ ] + \sum_{L=1,3} A_L(d_\pi^\dagger d_\nu^\dagger)^{(L)} \cdot (\tilde{d}_\nu \tilde{d}_\pi)^{(L)} + A_2[s_\pi^\dagger d_\nu^\dagger - s_\nu^\dagger d_\pi^\dagger] \cdot [H.\ c.\ ]$$

$$+A'_2[\beta s_\pi^\dagger d_\nu^\dagger + \sqrt{\frac{7}{2}}(d_\pi^\dagger d_\nu^\dagger)^{(2)}] \cdot [H.\ c.\ ] + A''_2[\beta s_\nu^\dagger d_\pi^\dagger + \sqrt{\frac{7}{2}}(d_\pi^\dagger d_\nu^\dagger)^{(2)}] \cdot [H.\ c.\ ], \qquad (12)$$

where H. c. means Hermitian conjugate, have the remarkable property that the aligned prolate condensate with $\beta_\pi = \beta_\nu = \beta$ is an eigenstate, i.e. there is no band mixing and the ground-state band consists of states of pure $F = F_{MAX}$. (Specific examples of such Hamilitonians were given in the SO(6) and SU(3) dynamical symmetry limits[14] of IBM-2.) However, the condensate is an eigenstate for the Hamiltonian (12) for any choice of the parameters $A_\rho, A'_\rho, A_0, A_1, A_2, A'_2, A''_2, A_3$, but the Hamiltonian is an F-scalar only when $A_\pi = A_\nu = \frac{1}{2}A_0$ and $A'_\pi = A'_\nu = A'_2 = A''_2$.

Similar techniques can be employed to investigate the F-spin decomposition of excited intrinsic bands, including symmetric and antisymmetric β- and γ-vibrations and scissors modes ($\gamma_\rho = \Omega = 0, \beta_\pi = \beta_\nu$ guarantee $F = F_{MAX} - 1$ for the scissors band). The F-spin mixing depends again on the proton and neutron deformations, and generally also on the mixing parameters which determine the relative weights of proton and neutron bosons in the intrinsic normal modes.[8] When the good F-spin of the ground-state band is not due to a symmetry of the Hamiltonian, the mixing in the excited bands can differ from that of the ground band, and from band to band. Similar methods can also be used to obtain the dependence of the F-spin mixing on $\gamma_\pi - \gamma_\nu$ when the axiality constraint is released. The fact that detailed IBM-2 calculations with non-F-scalar Hamiltonians, which have considerable F-spin mixing in the eigenstates, show rather similar F-spin decompositions for different members of the same band[11] strengthens the contention that the F-spin impurity is largely a function of the intrinsic structure underlying the rotational bands.

ESTIMATION OF F-SPIN ADMIXTURES

Recent pion single-charge-exchange experiments on ^{165}Ho at the Clinton P. Anderson Meson Physics Facility at Los Alamos (LAMPF) have measured the difference between the differential cross-section to the analog for the target oriented perpendicular to the pion beam and the differential cross section to the analog for an unoriented target.[15] From this orientation asymmetry in the cross sections the authors extract the ratio of Bohr-Mottelson neutron to proton deformations[15] for ^{165}Ho to be 0.84(8). Multiplying this ratio of deformations by N/Z for ^{165}Ho, the ratio R of the neutron quadrupole moment to the proton quadrupole moment is $R=1.23(12)$. Using the measured proton quadrupole moment of $q_\pi = 3.49(3)$ eb,[17] the neutron quadrupole moment is then R times this moment.

Since the IBM is a model based on the truncation of the shell model space, the effective proton and neutron boson quadrupole transition operator $\hat{T}_\rho$ has a contribution from both the proton and neutron boson quadrupole operators Q_π and Q_ν given in (10b),

$$\hat{T}_\rho = e_B [(1 + \delta) Q_\rho + \delta\, Q_{\tilde{\rho}}], \qquad (13)$$

where $\tilde{\rho} = \pi(\nu)$ for $\rho = \nu(\pi)$, and δ is the effective charge correction which is usually taken to have the value $\delta \simeq 0.5$ in the shell model. The boson strength e_B is taken to be equal to the phenomenologically determined value[18] $e_B = 0.15$ eb, which happens to be twice the quadrupole single-particle unit.

The estimation of the F-spin admixtures is very sensitive to the renormalization coefficient δ. The larger the value of δ taken, the larger the estimation of F-spin admixtures because the larger value of δ means that the transition operator (13) becomes more like an F-spin scalar, and hence a given difference in measured quadrupole moments must be explained by a larger difference in β_π and β_ν thereby implying larger F-spin admixtures in the ground state band. Sometimes very large δ's are used in the IBM[1,19], for example as large as $\delta = 2.7$, but this may be due to the fact that only electromagnetic transition rates are used to determine δ. A better way to determine δ is to measure a collective proton and neutron transition rate. This can be done by measuring a collective transition for both π^+ and π^- inelastic scattering. The best determination of δ comes from fitting both $\pi^\pm$ scattering to the first 2^+ state in the Pd isotopes.[20] In this analysis the value $\delta = 0.69$ was extracted in IBM-1, independent of the IBM-1 Hamiltonian. This value is slightly larger than the shell model value of $\delta = 0.5$, but the use of IBM-1 may have the effect of making δ effectively larger. Since we are using IBM-2 in this paper, we expect a value of δ more consistent with the shell model to be the correct one to use.[21]

In the large N limit, the IBM quadrupole moments are given by a numerical factor relating the moment operator to the transition operator times a Wigner 3j symbol squared times an intrinsic matrix element of the IBM transition operator,

$$q_\rho = 4\sqrt{\frac{\pi}{5}}(2J+1)\begin{pmatrix} J & 2 & J \\ J & 0 & -J \end{pmatrix}^2 \langle c_a|\hat{T}_\rho|c_a\rangle, \qquad (14a)$$

where the intrinsic matrix element of the boson quadrupole operators is

$$\langle c_a|\hat{T}_\rho|c_a\rangle = e_B[(1+\delta)N_\rho\beta_\rho + \delta N_{\tilde{\rho}}\beta_{\tilde{\rho}}] \qquad (14b)$$

where we have used[18] $\sqrt{\frac{2}{7}}\chi_\rho\beta_\rho = 1 - \beta_\rho^2$ and have ignored the contribution of the odd proton to the quadrupole moment.

Comparing these IBM moments to the measured values, we determine the IBM β_π, β_ν to be 1.0 and 1.3 respectively if we take $\delta = 0.5$, the shell model value. With these β_ρ's, we estimate that the maximum of F-spin admixture in the ground state band of ^{165}Ho ($K = J = 7/2$) is $P_{MAX} = 8\%$ and the minimum is $P_{MIN} = 2\%$ with P_{MAX} and P_{MIN} defined in (4) and (5). On the other hand if we use a large value of $\delta = 2.7$, the β_π, β_ν are 0.13 and 0.57, respectively. These β's imply very large F-spin admixtures, between 54% and 13%. However, for these latter relatively small IBM deformations, the intrinsic state may not produce an accurate approximation to the states of the ground band.

SUMMARY

We have shown that in deformed nuclei an F-spin violating Hamiltonian can produce states with pure F-spin provided the IBM neutron and proton deformations are equal, and, if they are not equal, upper and lower bounds on the F-spin admixtures can be estimated from the quadrupole properties of nuclei, as opposed to the M1 properties which are more sensitive to the underlying single-particle structure.

More experiments on a wide range of nuclei are needed to study the difference in proton and neutron quadrupole moments and transition rates in order to establish the quality of F-spin as a quantum number and to set constraints on IBM-2 and on other models of nuclear structure. Although ^{165}Ho is special as a target since it can be aligned, the same information can be obtained by pion single-charge exchange on collective unaligned even-even targets to the 2^+ analog state in the residual nuclei. This cross section is sensitive to the difference between the neutron and proton quadrupole transition densities. These transitions will be possible to detect with the new neutral meson spectrometer being constructed at LAMPF.

In short, we need AFFIRMATIVE ACTION for the other half of the nuclear matrix elements, the neutron matrix elements, so that they are determined to the same quality as the proton matrix elements.

ACKNOWLEDGMENTS

This work was supported in part by the U. S. Department of Energy. One of us (JNG) thanks the Physics Department of the University of Washington for their hospitality during the completion of this manuscript.

REFERENCES

1. F. Iachello and A. Arima, The Interacting Boson Model (Cambridge Univ. Press, Cambridge, 1987).
2. A. Arima, T. Otsuka, F. Iachello, and I. Talmi, Phys. Lett. B66, 205 (1977)
3. T. Otsuka, A. Arima, F. Iachello, and I. Talmi, Phys. Lett. B76, 139 (1978).
4. D.D. Warner, Phys. Rev. C34, 1131 (1986).
5. P.O. Lipas, P. von Brentano, and A. Gelberg, to be published in Reports on Progress in Physics (1990).
6. J. N. Ginocchio and M. W. Kirson, Nucl. Phys. A350, 31 (1980).
7. A. Leviatan, Ann. Phys. 179, 201 (1987).
8. A. Leviatan and M. W. Kirson, Ann. Phys. 201, 13 (1990).
9. P. Van Isacker, K. Heyde, J. Jolie, and A. Servin, Ann. Phys. 171, 253 (1986).
10. P. Van Isacker, H. Harter, A. Gelberg and P. Von Brentano, Phys. Rev. C36, 441 (1987).
11. A. Novoselsky and I. Talmi, Phys. Lett. B160, 13 (1985).
12. P. Von Brentano, et al., J. Phys. G14, S129 (1988) and references therein.
13. A. Bohr and B. R. Mottelson, Nuclear Structure (Benjamin, Reading, MA 1975), Vol. II.
14. A. E. L. Dieperink and I. Talmi, Phys. Lett. B131, 1 (1983).
15. J. N. Knudson et al., LANL Report LA-UR-90-881, submitted to Phys. Rev. Lett.
16. H. C. Chiang and M. B. Johnson, Phys. Rev. Lett. 53, 1996 (1984).
17. P. Raghavan, Atomic Data and Nuclear Data Tables 42, 189 (1988).
18. R. F. Casten and D. D. Warner, Rev. Mod. Phys. 60, 398 (1989); In the course of the present work we noticed that the phenomenological values of e_B given in this reference are close to twice the quadrupole single-particle unit.
19. G. Wenes, T. Otsuka, and J.N. Ginocchio, Phys. Rev. C37, 1878 (1988).
20. J. N. Ginocchio and P. Van Isacker, Phys. Rev. C33, 365 (1986).
21. A. Wolf and R.F. Casten, this Proceedings.

Octupole and Dipole Collectivity in Nuclei

Dimitri Kusnezov
NSCL and Department of Physics and Astronomy,
Michigan State University, East Lansing, MI 48824-1321

Abstract

The role of octupole and dipole degrees of freedom in the description of the large B(E1) rates measured in the $A \sim 100$ and 150 regions is reviewed. We find that algebraic methods offer a simple picture and a good description of most observed phenomena.

Introduction

The question I would like to address is to what degree can a simple model of interactions of symmetric and reflection asymmetric shapes describe the large E1 rates observed in many even-even nuclei in the $A \sim 100$ and 150 regions. There are several objectives of such an exercise. For example, it is interesting to understand to what degree the large E1 rates can be attributed to collective behavior; can an operator with smoothly varying parameters describe the often seemingly random E1 systematics? Further, a simple tractable model can not only offer an intuitive picture of the physics but lead to some predictions as well, as we find in the study of double octupole states below. It is convenient, although not at all necessary, to deal with quantized shapes algebraically. A model to describe the rotations and vibrations of spherical, quadrupole and octupole shapes was introduced over a decade ago,[1] in which an f boson was added to the sd boson space. Unfortunately, it is quite simple to reproduce energy level systematics with very different Hamiltonians, and while a fit to energy spectra is a necessary step, it is not sufficient to guarantee that one has included all the important physics. One way to distinguish models is in the study of transition rates which are more sensitive to the wavefunctions. It is also important that the model be kept as simple as possible. This is a failure of the sdf model which lacks a simple explanation of E1 rates, even if one allows many f bosons into the model space.[2] A p boson degree of freedom was introduced shortly after by Han et al [3] and later by Engel and Iachello [4] to describe the quasi-molecular spectra seen in the light actinides. The addition of the p boson not only allows a rich subalgebra structure including an $SU(3)$ limit not present in the sdf model, but also simple phenomenology of negative parity states. Arguments concerning the physical origin of the p boson can be found in Ref. 2. I would like to review here some of the recent results of the $spdf$ model.

Algebraic Aspects of Octupole Deformation

The $spdf$ model, which describes the interactions of spherical, dipole, quadrupole and octupole deformations, can be realized in terms of a $U(16)$ interacting boson model. Although the algebra $U(16)$ might seem forbiddingly large, one can nevertheless retain a level of simplicity close to that of the $U(6)$ IBM-1. Recently, all the dynamical symmetries of $U(16)$ that reduce to the $O(3)$ algebra were identified, and all the generators, quadratic invariants, as well as many of the quantum number branching rules were constructed. [5-6] A great reduction of these 165 dynamical symmetry limits of $U(16)$ can be made. From calculations using both the sdf and $spdf$ models, the p boson appears to be an essential ingredient. Thus we can consider only those limits of $U(16)$ which do not decouple p and f bosons. By further requiring that boson angular momentum is a generator in these subalgebras, the 165 limits reduces to only 7, 6 of which are given in Table 1. (The seventh limit, included for completeness in Refs. 5-6 is not as relevant.) One further simplification can be made, since the generators $\mathcal{G}_i$ of these algebras can be classified according to whether they separately conserve the total number of positive and negative parity bosons: $[\mathcal{G}_i, \hat{N}_\pm] = 0$ (i.e., subalgebras of $U_{sd}(6) \oplus U_{pf}(10)$, for which sd and pf boson spaces are decoupled). If they do, we refer to them as class A. If they do not, then there are two additional classes: class B which has generators of good parity, and class C which has generators of mixed parity. Since in class A all the negative parity states can be moved in energy relative to the positive parity states without effecting the wavefunctions, Hamiltonians written in terms of these generators cannot accommodate octupole deformation. It is only Hamiltonians with terms from classes B and C which can be used to generate octupole deformation. Of the algebras in Table I, only two are class C, $SU_{pdf}(6)$ and $SU_{pdf}(3)$, and eight are class B, $SU(4) \oplus SU(4)$, $Sp(4) \oplus Sp(4)$, $O_{spdf}(16)$, $O_{pdf}(15)$, $O_{spf}(11)$, $O(10)$, $O_{spdf}(4)$ and $SU_{pdf}(4)$. It remains now to study the individual properties of each of the limits in Table I, including their generic level spectra and electromagnetic transition selection rules, in order to determine whether or not these symmetries are actually realized in nuclei.

$B(E1)$ Strengths in the $A \sim 150$ Region

Octupole deformation has been argued to exist in the neutron-rich rare-earth nuclei, manifesting itself in enhanced $B(E1)$ rates and interleaved positive and negative parity rotational (quasi-molecular) bands. We can pose the question of how much of the observed phenomena in the $Ba - Sm$ region ($N > 82$, $Z = 56 - 62$) can be explained in the absence of octupole deformation. These nuclei and the transition between spherical ($N \lesssim 88$) and quadrupole deformation ($N \gtrsim 88$) has been exhaustively studied in the past, using the conventional $IBM - 1$ Hamiltonian

$$\hat{H} = \varepsilon_d \hat{n}_d + \kappa \hat{Q}_{sd} \cdot \hat{Q}_{sd} + \kappa' \hat{L}_d \cdot \hat{L}_d + \kappa'' \hat{P}^\dagger_{sd} \cdot \tilde{\hat{P}}_{sd}, \qquad (1)$$

The simplest extension of this familiar Hamiltonian to study octupole/dipole excitations can be obtained by replacing the $U_{sd}(6)$ generators $\hat{L}$, $\hat{Q}$, $\hat{P}^\dagger$ in (1)

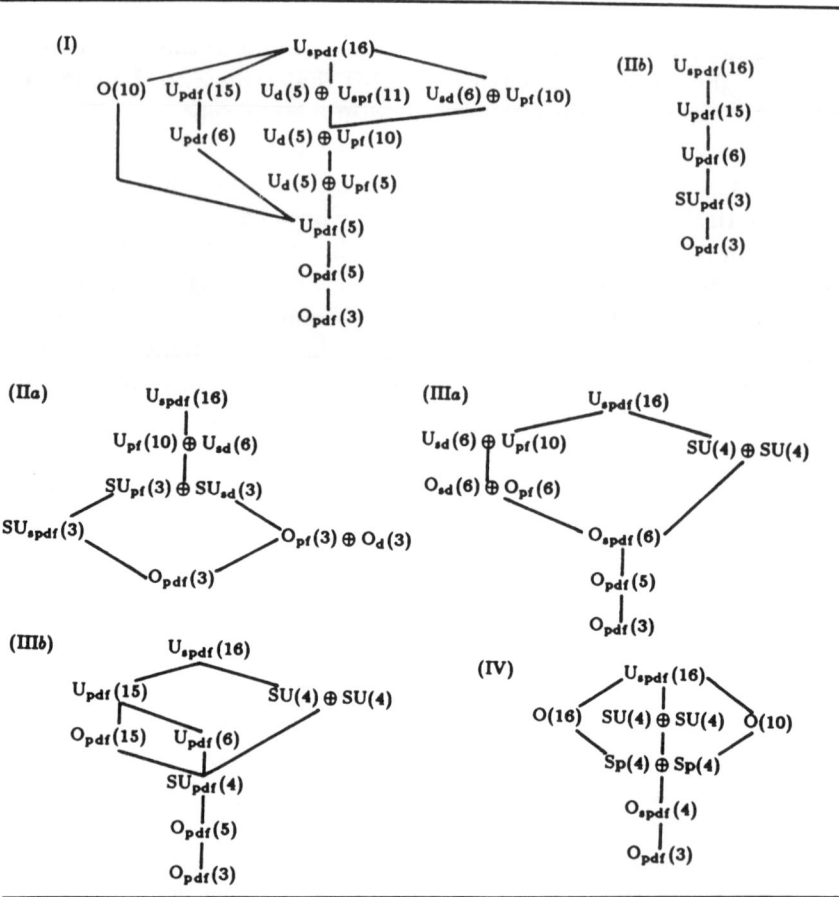

Table 1. pdf dynamical symmetry limits of the $U(16)$ spdf model.[5-6]

with their $U_{spdf}(16)$ counterparts:

$$\hat{H} = \varepsilon_p \hat{n}_p + \varepsilon_d \hat{n}_d + \varepsilon_f \hat{n}_f + \kappa \hat{Q}_{spdf} \cdot \hat{Q}_{spdf} + \kappa' \hat{L}_{pdf} \cdot \hat{L}_{pdf} + \kappa'' \hat{P}^\dagger_{spdf} \cdot \tilde{\hat{P}}_{spdf}. \quad (2)$$

This Hamiltonian treats all bosons on an equal footing. In such a treatment, the parameters $\kappa, \kappa', \kappa''$ and ε_d are fit to the positive parity spectra. As a result, the negative parity structure is completely specified, aside from ε_p and ε_f which are determined from the 1_1^- and 3_1^- states. The only term that can lead to octupole deformation is the ultimate term in (2). For the values of κ'' needed to describe the data, there is effectively no octupole deformation in this parameterization. (The diagonalization of (2) in a full basis and in a one p or f boson basis yields identical results.) For E2 transitions, we choose $\hat{Q}$ as a transition operator, taking $\hat{T}(E2) = e_2 \hat{Q}$, where e_2 is fixed at 0.1eb. The E1 operator, which has the general form

$$\hat{T}(E1) = e_1 \left(\chi [s^\dagger \tilde{p} + p^\dagger \tilde{s}]^{(1)} + [p^\dagger \tilde{d} + d^\dagger \tilde{p}]^{(1)} + \chi'[d^\dagger \tilde{f} + f^\dagger \tilde{d}]^{(1)} \right) \qquad (3)$$

has three free parameters: e_1, χ and χ'. By assuming that χ and χ' have a neutron number (and only weak proton number) dependence, it is possible to describe the E1 systematics measured in the Ba, Ce, Nd and Sm isotopes. This model provides good quantitative agreement to known $B(E1)$, $B(E2)$, $B(E1)/B(E1)$ and $B(E2)/B(E2)$ rates. As a result, we can examine the predictions for the $B(E1)/B(E2)$ ratios, shown in Fig. 1. A more detailed account can be found in Ref. 2. One of the features that such a parameterization can reproduce is the staggering effect as a function of J. This staggering is due to the E1 transition matrix elements, as the yrast B(E2) deexcitations from the positive parity states are close in value to those from the negative parity states.

One of the more interesting nuclei is ^{146}Ba which has peculiar $B(E1)$ properties. [8,10] This parameterization can reproduce the $B(E1)/B(E2)$ ratios in ^{146}Ba, but it over predicts the E1 moment by an order of magnitude. [10] If one believes the collective nature of the octupole band, in view of the rotational positive parity structure, then one approach to modify the E1 moment is to include octupole deformation in the model in a stronger manner, as discussed in Ref. 10. It seems plausible that ^{146}Ba is a candidate for octupole deformation.

OCTUPOLE-INTRUDER CONFIGURATION MIXING

We can consider more complex physical situations, such as octupole – ground state (gs) – intruder (I) configuration mixing. In ^{96}Zr, the I states have an important effect on the $B(E1)$ deexcitations from the octupole two-phonon states (OTP), while in ^{110}Cd, a similar fragmentation of the $B(E1)$ strength from the octupole states is observed. We have a simple description of the intruder-OTP mixing in $spdf$-IBM, [11] in the spirit of the configuration mixing neutron-proton IBM. [9] In order to simulate the structure of the OTP states and their mixing with other collective states in ^{96}Zr, we will treat the ground state and OTP states as vibrational excitations. The $spdf$ Hamiltonian for this first (ground state) configuration is

$$\hat{H}_{gs} = \epsilon(\hat{n}_d + \hat{n}_f) + \epsilon_p \hat{n}_p + \kappa \hat{Q}^2_{spdf} + \kappa' \hat{L}^2_{pdf} + \kappa'' \hat{D}^2_{spdf}. \qquad (4)$$

The origin of the dipole operator $\hat{D}$ is twofold. It serves to fragment the OTP E1 strength among neighboring states since it is the only operator of odd parity, and it can be used for the E1 transition operator. The intruder configuration is

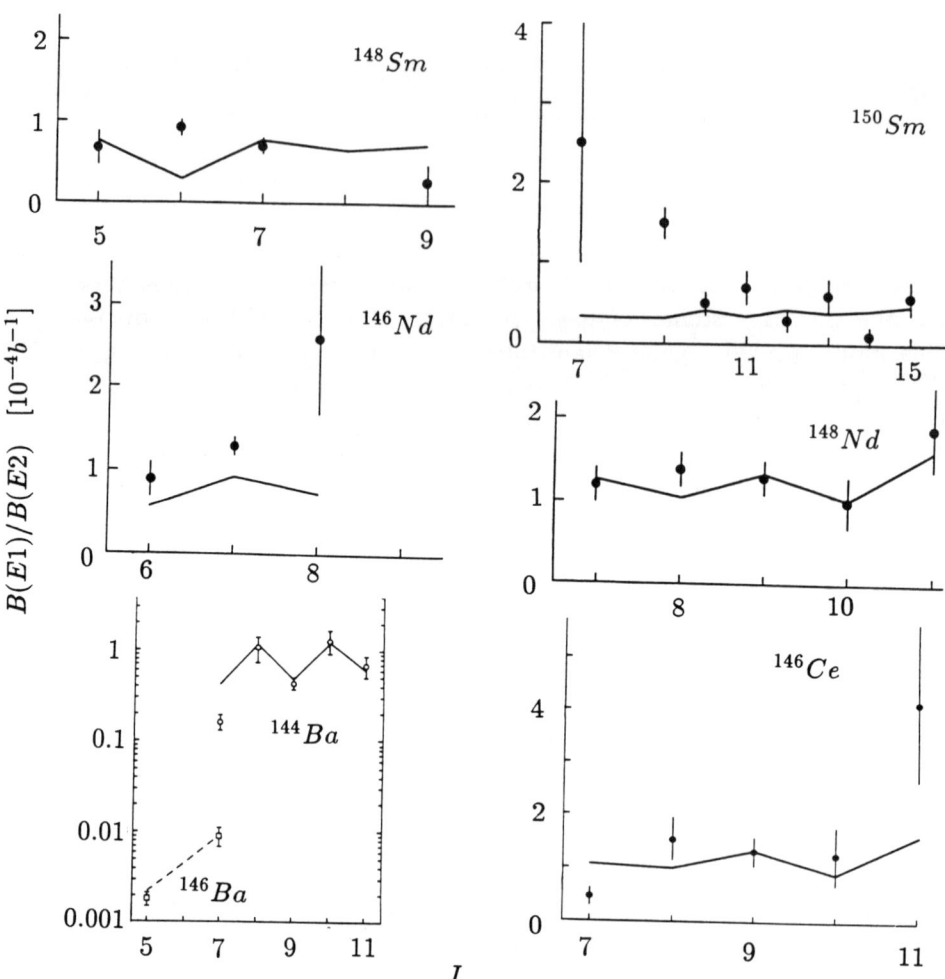

Figure 1. B(E1)/B(E2) ratios in the neutron rich rare-earth nuclei. [2] Experimental data is from Ref. 14.

described by the Hamiltonian

$$\hat{H}_I = \epsilon_2 \hat{n}_d + \kappa_2 \hat{Q}^2. \tag{5}$$

The intruder Hamiltonian is diagonalized in the sd boson space since the inclusion of negative parity states in this configuration has no effect on the results.

The lowest eigenstates of each configuration are then mixed together with a mixing Hamiltonian $\hat{H}_{mix}$, and an additional energy Δ is introduced to reflect the additional energy needed to create the intruder configuration:

$$\hat{H}_{mix} = \alpha[s^\dagger s^\dagger + ss + d^\dagger d^\dagger + \tilde{d}\tilde{d}]_0^{(0)}. \tag{6}$$

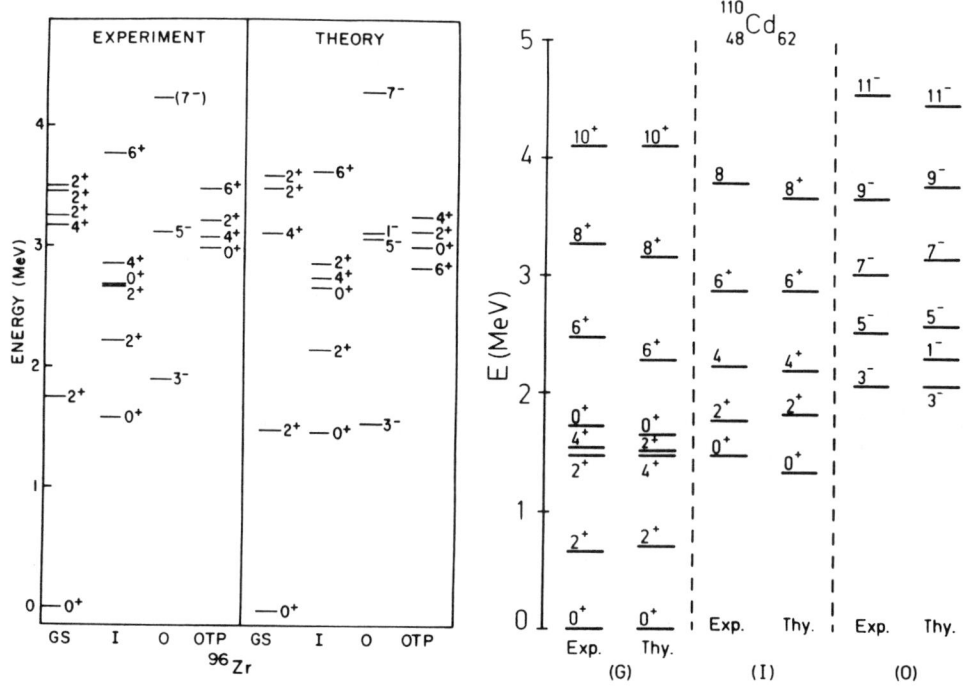

Figure 2. Comparison of experimental ground state (gs), octupole two phonon (OTP), intruder (I) and octupole (O) bands with theory [11,12] for ^{96}Zr and ^{110}Cd.

The results of the calculation are shown in Fig. 2. The resulting wavefunctions for the 1_1^- and OTP states contain less than 2% p-boson composition, indicating rather pure octupole excitations. Further, the double octupole character of the OTP states is found to be obfuscated by the strong fragmentation of the E1 strength,[11] which can be seen in Table 2. The effects of small p-boson admixtures are only manifested in the E1 transition rates.

The E1 and E2 electromagnetic transition operators have been taken in the form $\hat{T}(E1) = e_1(\hat{D}_{gs}^{(1)} + \hat{D}_I^{(1)})$ and $\hat{T}(E2) = e_2(\hat{Q}_{gs}^{(2)} + \hat{Q}_I^{(2)})$. The subscripts gs and I indicate the contribution to the operator from the ground state and intruder configurations, respectively. The only free parameters are the two effective charges e_1 and e_2, which we fix at values used in the previous rare-earth calculations: $e_1 = 0.01 eb^{\frac{1}{2}}$ and $e_2 = 0.1 eb$. Since the states are largely octupole, it is convenient to choose the E3 operator that Barfield *et al* used to describe

Table 2. Experimental and theoretical transition rates in ^{96}Zr and ^{110}Cd. [11-12]

$B(E1)$ $[10^{-5}e^2b]$	Experiment	Theory	$B(E1)/B(E2)$ $[10^{-3}b^{-1}]$	Experiment	Theory
$B(E1: 2^+_{OTP} \to 3^-_1)$	0.95	0.51	$6^+_{OTP} \to 5^-_1/6^+_{OTP} \to 4^+_{OTP}$	1.38	37^a
$B(E1: 4^+_{OTP} \to 3^-_1)$	≤ 0.14	0.01	$6^+_I \to 5^-_1/6^+_I \to 4^+_I$	0.0045	0.0034
$B(E1: 4^+_{3176} \to 3^-_1)$	0.54	2.8	$4^+_I \to 3^-_1/4^+_I \to 2^+_I$	0.0046	0.062
$B(E2)$ $[10^{-2}e^2b^2]$			$B(E2)/B(E2)$		
$B(E2: 5^-_1 \to 3^-_1)$	3.7	3.4	$0^+_{OTP} \to 2^+_{gs}/0^+_{OTP} \to 2^+_I$	0.18	0.38
$B(E2: 2^+_{OTP} \to 0^+_{gs})$	0.06	0.003	$6^+_I \to 4^+_{OTP}/6^+_I \to 4^+_I$	0.05	0.04
$B(E2: 2^+_{OTP} \to 2^+_{gs})$	2.1	0.2	$4^+_I \to 2^+_{gs}/4^+_I \to 2^+_I$	0.40	0.73
$B(E2: 2^+_{3249} \to 2^+_I)$	4.2	2.0	$2^+_I \to 0^+_{gs}/2^+_I \to 0^+_I$	0.008	0.007
$B(E2: 2^+_{3249} \to 0^+_{gs})$	0.068	0.01	$2^+_I \to 2^+_{gs}/2^+_I \to 0^+_I$	0.09	0.01
$B(E2: 4^+_{OTP} \to 2^+_I)$	≤ 0.52	1.6	$6^+_{OTP} \to 4^+_I/6^+_{OTP} \to 4^+_{OTP}$	0.29	0.02
$B(E2: 4^+_{OTP} \to 2^+_{gs})$	≤ 0.08	1.2	Branching Ratio		
$B(E2: 4^+_{3176} \to 2^+_{gs})$	0.1	1.5	$6^+_{OTP} \to 3^-_1/6^+_{OTP} \to 5^-_1$	< 0.05	10^{-4}
			$0^+_{OTP} \to 3^-_1/0^+_{OTP} \to 2^+_I$	–	10^{-3}
$B(E3)$ $[10^{-2}e^2b^3]$					
$B(E3: 3^-_1 \to 0^+_{gs})$	2.9	2.0			

^{110}Cd: $B(E1)$	Experiment	Theory ($\chi = 0.38, \chi' = -0.12$)
$3^+_1 \to 2^+_I$	210	196
$3^+_1 \to 2^+_2$	$\equiv 100$	100
$3^+_1 \to 2^+_1$	50	36
$B(E1)/B(E2)$ $[10^{-4}b^{-1}]$		
$5^- \to 4^+/5^- \to 3^-$	0.85	.02
$7^- \to 6^+/7^- \to 5^-$	0.25	.02

rare-earth nuclei.[13] The comparisons are given in Table 2. Overall, this simple *spdf* parameterization of ^{96}Zr and fixed transition operators are able to account for the qualitative structure of the OTP states, their mixing with intruder configurations and their deexcitations.

In ^{110}Cd, we have used a similar approach to describe the influence of the intruder configurations on octupole band deexcitations. The fit to the *gs*, *I* and *O* bands are shown in Fig. 2, and some branching ratios in Table 2. Since the energy of the 1^-_1 is not known, it has been placed between the 5^-_1 and 3^-_1. The location of this level has an influence on the $B(E1)$ rates. As it stands, we find that the minimal model under predicts the $B(E1)/B(E2)$ ratios, while providing a good description of other $B(E2)$ and $B(E1)$ rates.[12]

Better agreement can be obtained by adjusting the parameters in the E1 operator: a mass dependence of these parameters has been suggested in the above rare-earth studies as well as in the light-actinides.[7] However, we have been interested in the ability of this model to replicate the general features of ^{96}Zr and ^{110}Zr using a simple parameterization. The calculations in ^{96}Zr show that the OTP multiplet members, for which intuitively we might expect to observe deexcitation by E3 transitions, may not deexcite by observable E3 transitions. Equally important is the fact that experimental evidence indicates strong mixing of the OTP excitations with the positive parity levels that occur near in energy. In almost all other nuclei except ^{96}Zr there is a high density of positive parity levels with J^π values the same as the OTP members. With strong mixing, the distinguishing characteristics of the two-phonon octupole multiplet becomes fragmented among all the positive parity levels, and the octupole two-phonon multiplet cannot be identified solely with individual levels.

References

1. A.Arima and F.Iachello, Ann. Phys. **111** (1978) 201.
2. D.Kusnezov, *Doctoral Dissertation*, Princeton University (1988), unpublished.
3. C.S.Han et al, Phys. Lett. **B163** (1985) 295.
4. J.Engel and F.Iachello, Phys. Rev. Lett. **54** (1985) 1126.
5. D.Kusnezov, J. Phys. A: Math and Gen. **22** (1989) 4271.
6. D.Kusnezov, J. Phys. A: Math and Gen., (in press) (1990).
7. T.Otsuka and M.Sugita, Phys.Lett. **B209** (1988) 140.
8. D.Kusnezov and F.Iachello, Phys. Lett. **B209** (1988) 421.
9. P.Duval and B.Barrett, Phys. Lett. **B100** (1981) 223.
10. H.Mach et al, Phys. Rev. **C41** (1990) R2469.
11. D. Kusnezov, E.A.Henry and R.A. Meyer, Phys. Lett. **B228** (1990) 11.
12. J.Kern et al, Nuc. Phys. **A512** (1990) 1.
13. A.F.Barfield et al, Phys. Rev. **C34** (1986) 2001.
14. E.Hammerén et al, Nuc. Phys. **A321** (1979) 71; Z.Sujkowski et al, Nuc. Phys. **A291** (1977) 365; W. Urban et al, Phys. Lett. **B185** (1987) 331; W. R. Phillips et al, Phys. Rev. Lett. 57 (1986) 3257; W. Urban et al, Phys. Lett. **B200** (1988) 424; L.K.Peker, Nucl. Data Sheets **60** (1990) 953.

SYSTEMATIC BEHAVIOR OF LOW ENERGY OCTUPOLE STATES

P.D. Cottle
Physics Department, Florida State University
Tallahassee, Florida 32306 USA

ABSTRACT

Fragmentation of the low energy octupole state and its implications for studies of the systematic behavior of 3_1^- states are discussed. The use of octupole centroids for study of the systematic behavior of octupole states is introduced and applied to studies of the A=150 and 190 regions. In addition, a survey of the dependence of octupole fragmentation on quadrupole deformation is presented. This survey demonstrates that 3_1^- state behavior is generally not strongly affected by fragmentation when $\beta_2 < 0.3$; however, fragmentation effects in the Pt isotopes seem to strongly violate this general rule.

Several characteristics of the behavior of the low energy octupole state (LEOS) have made it an inviting subject for systematic studies[1-4]. First, in much of the Periodic Table the low energy octupole strength is concentrated in a single state, the 3_1^- state. Second, the systematic behavior of the energies of these 3_1^- states is generally quite orderly and straightforward to interpret. The 3_1^- states in the Z=36-50, N=50-82 region (Fig. 1) demonstrate this behavior quite well[5]. In general, a low energy octupole state can be simply understood as the coherent sum of a number of one-particle-one-hole (1p-1h) excitations to the unique parity orbit in a valence shell from the common parity orbit having both orbital and total angular momentum $3\hbar$ less than the unique parity orbit. The proton and neutron valence shells each contain one such pair (a $\Delta j=3$ pair) of orbits, and their contributions add coherently. The octupole state energy reaches a minimum

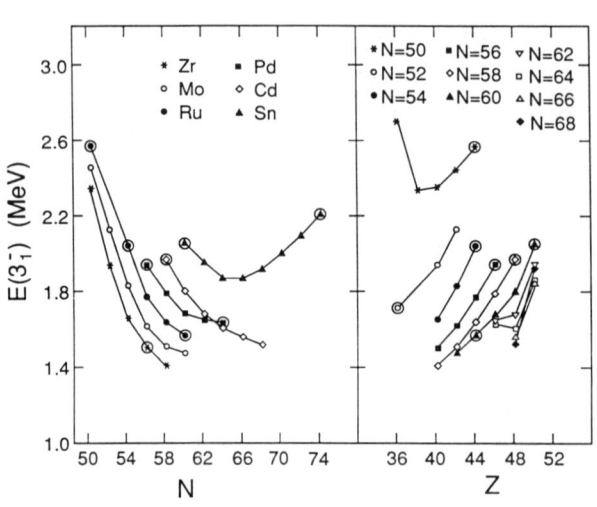

Fig. 1. $E(3_1^-)$ vs. N and Z for the Z=36-50, N=50-82 region. Tentative assignments are denoted by circled data points.

when the lower energy orbit of the $\Delta j=3$ pair is full and the higher energy orbit is empty.

The behavior shown in Fig. 1 can be explained in this context. The $\Delta j=3$ pairs for the proton and neutron valence shells are $p_{3/2}$-$g_{9/2}$ and $d_{5/2}$-$h_{11/2}$, respectively. In the plot vs. Z, the 3_1^- state energies reach a minimum when Z=36-38, where the $p_{3/2}$ orbit is full and the $g_{9/2}$ is empty. The N dependence is similar: the minimum near N=64 occurs as the $d_{5/2}$ orbit is full and the $h_{11/2}$ orbit is empty. In such an orderly plot, the anomalous behavior is easily identified. In Fig. 1, the wayward trend of the Cd isotopes[2] is immediately obvious. Furthermore, Fig. 1 would direct a search for statically octupole deformed nuclei in this region toward Z=36-38 and N=64. This point is discussed more fully in Ref. 5.

It should be cautioned, however, that analyses such as that in the previous paragraph depend on the assumption that the low energy octupole state is concentrated in the 3_1^- state and not fragmented among several 3^- states. If significant fragmentation does occur, then 3_1^- state energies do not give accurate readings of the energies of the low energy octupole states, and the centroids of B(E3; $0_{gs}^+ \to 3^-$) strength must be used instead. Such fragmentation generally occurs in deformed nuclei, where a number of 3^- states corresponding to different alignments of the octupole phonon (K=0, 1, 2 and 3) with respect to the symmetry axis of the nucleus occur. Octupole fragmentation is characteristic of the deformed rare earth and actinide nuclei.

A survey of the low energy octupole states in the A=150 region using centroid energies was recently performed[6] in order to examine the effect of the collapse of the Z=64 subshell gap at N=90 on the behavior of octupole states. The methods we adopted for treating the data serve to illustrate the technical problems associated with such a study. Fig. 2a shows the dependence of 3_1^- state energies in the Z=58-78, N=82-110 region on N. The orderly trends which occur in Fig. 1 are clearly obscured in this region by fragmentation effects. This is obviously a case in which a study of LEOS centroids would be more useful. In order to track the behavior of the LEOS accurately, <u>all</u> of the low energy 3^- strength must be found. Not only is it difficult to find all of this strength, but it is also problematic to determine whether one has found all of it or even a large fraction of it. For the A=150 study, we used a condition based on the energy-weighted sum rule (EWSR) to judge whether a sufficient fraction of the octupole strength were known in a nucleus in order to give a reasonable result for the LEOS energy. Kirson[7] found that for A>60 nuclei the LEOS contains an amount of energy-weighted strength approximately equal to 5% of the EWSR. We chose to limit our A=150 region survey to those nuclei in which more then 2.5% of the EWSR was observed. These "filtered" data meeting the EWSR requirement are shown in Fig. 2b. In this figure a minimum at N=88 becomes obvious, as does the trend of increasing LEOS energy above N=88.

The other obvious characteristic of Fig. 2b is the lack of data. Since the publication of Ref. 6, we[8] have added another point to this plot (^{144}Nd) with the use of proton inelastic scattering,

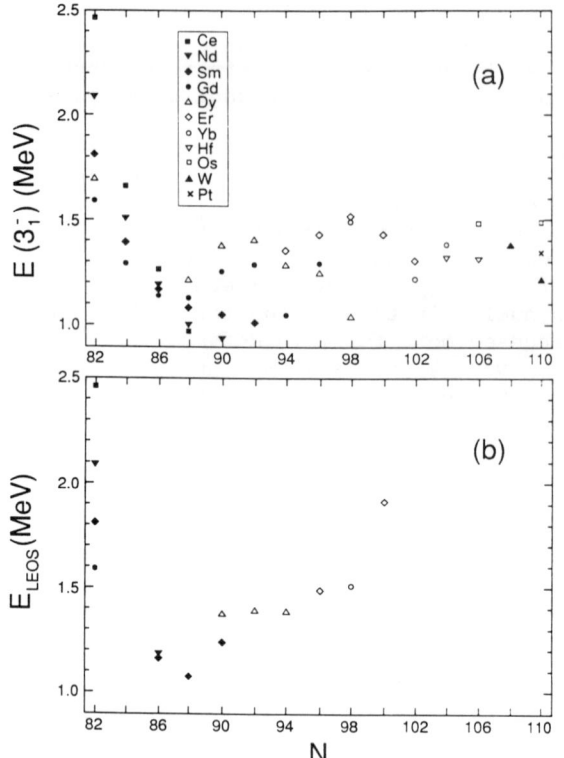

Fig. 2. (a) Energies of 3_1^- states, $E(3_1^-)$, vs. N for nuclei with $58 \leq Z \leq 78$ and $82 \leq N \leq 110$. (b) Centroids of the LEOS, E_{LEOS}, vs. N.

which is an ideal experimental tool for searching for octupole strength in low energy (<4 MeV) states. We are continuing these experiments.

Fragmentation has also been found to be responsible for a very striking anomaly that occurs in the octupole behavior of the A=190 region. The systematic behavior of 3_1^- states in the Pt and Hg nuclei of the A=190-200 region displays[2] a discontinuity that is unique in the Periodic Table. The 3_1^- states of the N=110-120 Pt isotopes lie in the range 1.3-1.7 MeV, while the lowest known 3⁻ states in the N=118-124 Hg isotopes, only two protons away from Pt, occur above 2.4 MeV, nearly 1 MeV above the Pt isotones. This anomaly is illustrated in Fig. 3, where the 3_1^- state energies for Pt, Hg and Pb isotopes are shown. Results from an in-beam γ-ray experiment by Yates et al.[9] suggested the existence of a 3⁻ state in ^{198}Pt at 2.603 MeV, closer to the energies of the Hg states. This result suggested that the low energy octupole states in Pt were strongly fragmented; however, more evidence was needed to support this hypothesis.

We found supporting evidence[10] by examining the results of Deason et al.[11] from studies of the ^{194,196,198}Pt(p,p′) reactions at 35 MeV. We concluded that a total of eleven states that had been found in these three nuclei at energies between 2.1 and 2.9 MeV could be assigned $J^\pi=3^-$ on the basis of their angular distributions. Furthermore, these states contain a substantial amount of E3 strength. In fact, when these results were used to calculate octupole centroids for the Pt isotopes (the open shapes in Fig. 3), and these centroids were used as the LEOS energies instead of the 3_1^- energies, the apparent anomaly in the octupole behavior was resolved.

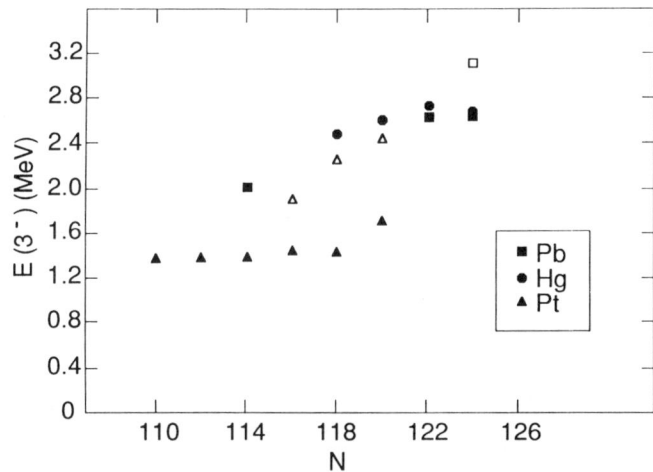

Fig. 3. Energies of 3_1^- states (full shapes) and octupole centroids (open shapes) for N=110-124 isotopes of Pt, Hg, and Pb.

However, the resolution of this problem raised new questions about the Pt isotopes. The Pt isotopes are not as well deformed as the rare earth and actinide nuclei with which octupole fragmentation is usually associated. I discuss this problem further below.

It must be mentioned here that there is some disagreement regarding the above "solution" of the Pt problem. I refer the reader to the Ph.D. thesis of C.S. Lim[12] for a dissenting opinion.

The work which has already been described here suggests several questions. First, in which nuclei is the centroid of the LEOS significantly removed from the 3_1^- state? In such nuclei one would not expect that 3_1^- state surveys would be useful. Second, is the fragmentation observed in the Pt isotopes unusual or even unique? In order to address these questions, as well as to satisfy our curiosity regarding the dependence of octupole fragmentation on quadrupole deformation, we performed[13] a survey of octupole fragmentation in A>60 even-even nuclei. Our "measure" of fragmentation was simply the difference of the energies of the observed LEOS centroid and the 3_1^- state. We denoted this difference by ΔE_3.

The data compiled for this survey included results of experiments using inelastic scattering of protons, deuterons and alpha particles, as well as Coulomb excitation. For most nuclei, the calculation of centroid energies used data on all 3^- states for which $B(E3;0_{gs}^+ \to 3^-)$ strengths were measured. Fragments of the low energy octupole resonance are sometimes observed in extensive studies of spherical nuclei. In order to eliminate these states, we have systematically excluded all 3^- states observed above an energy of $(20 \text{ MeV})/A^{1/3}$, which is approximately the lowest energy found for a low energy octupole resonance in Kirson's compilation[7].

Fig. 4 illustrates the range of ΔE_3 values. All of the values fall below 0.50 MeV, except for ^{196,198}Pt. Already we see that the fragmentation observed in these two isotopes is quite large. This result is even more striking when it is considered that fragmentation is generally associated with large quadrupole deformation, and that these Pt isotopes have relatively small

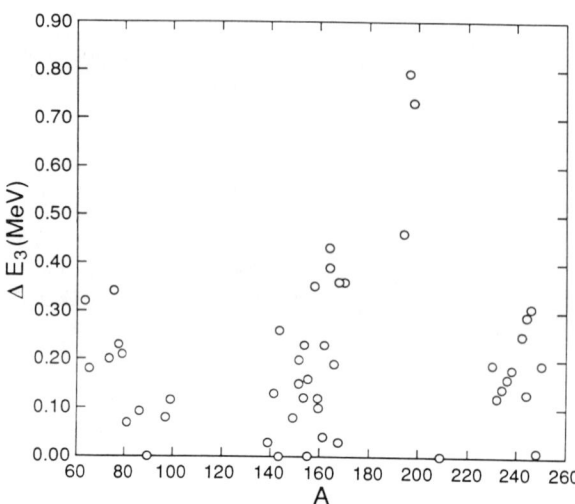

Fig. 4. Values of ΔE_3 vs. A.

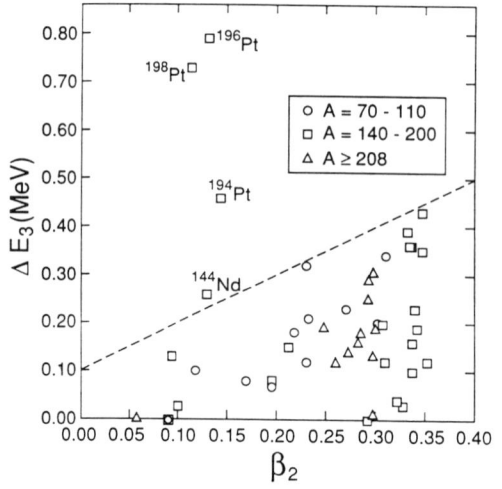

Fig. 5. Values of ΔE_3 vs. β_2.

deformation ($\beta_2 < 0.14$).

The dependence of fragmentation on quadrupole deformation is generally regular, however. Fig. 5 shows the dependence of ΔE_3 on β_2. The line drawn on the plot, $\Delta E_3 = (\beta_2 + 0.10)$ MeV, represents an upper limit for fragmentation for the vast majority of nuclei shown, and demonstrates quite clearly that fragmentation is generally much stronger in well deformed nuclei than it is in spherical nuclei. The primary reason for this is the effect of the Coriolis interaction. The fragmentation of strength among 3^- states with different K quantum numbers is strongly affected by the Coriolis interaction between these 3^- states because this interaction tends to concentrate the octupole strength in the lowest lying 3^- state. The strength of this force is inversely proportional to the moment of inertia of the nucleus, which increases as the quadrupole deformation increases. Consequently, the octupole strength tends to be more strongly fragmented in well deformed nuclei than in spherical nuclei.

The ΔE_3 line also allows us to estimate the β_2 value for which the octupole fragmentation is large enough to render surveys of 3_1^- states unreliable. In general, anomalies in 3_1^- state energy trends become identifiable when the states deviate by 400 keV or so from the energies at which they are expected to occur. Using the $\Delta E_3 = (\beta_2 + 0.10)$ MeV rule, we conclude that surveys of 3_1^- state behavior are not badly distorted by fragmentation effects when

β_2=0.3 or less. Therefore, the only broad regions where such surveys would be unreliable are the deformed rare earth and actinide regions. This is, of course, the result which we expect.

Four exceptional cases appear in Fig. 5. One, ^{144}Nd, which is the result of our recent experiment[8], lies quite close to the line. The others are ^{194,196,198}Pt. In fact, this graph emphasizes the unique nature of the fragmentation in the Pt isotopes.

Despite the puzzling appearance of the Pt isotopes, a calculation predicting the large fragmentation found in ^{194,196}Pt was reported by Engel[14]. Engel's work involved adding an f-boson (carrying three units of angular momentum, and generally associated with an octupole excitation) to an Interacting Boson Approximation-1 calculation which corresponds to the O(6) dynamical symmetry in this model. Nuclei displaying the O(6) symmetry are best described as gamma-soft or triaxial in the geometrical picture. The calculation predicted a single isolated 3_1^- state carrying approximately 7 W.u. near 1.5 MeV in each nucleus, and two more states carrying a total of 15 W.u. near 2.3 MeV. The experimental and theoretical results are compared in Fig. 6. Although the experiment shows the high

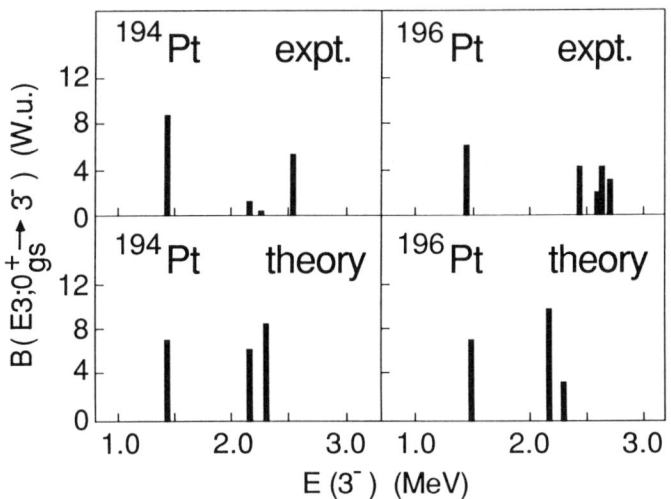

Fig. 6. Comparison of distribution of octupole strength in ^{194,196}Pt to those predicted by Engel (Ref. 14).

lying octupole strength fragmented among more than two states in each nucleus (three in ^{194}Pt, four in ^{196}Pt), the theoretical and experimental distributions are qualitatively similar. The additional fragmentation in the data probably results from the presence of quasiparticle states near 2.5 MeV having J^π=3$^-$ which mix with the octupole states. The quasiparticle states are not included in the IBA, so their influence on fragmentation would not appear in the calculations.

The importance of this result argues for further experimental confirmation of the fragmentation in Pt. One promising line of inquiry involves expanding the investigations that Yates et al. have begun[9] using (p,p'γ) and similar reactions to assign spins and parities to high lying states in Pt nuclei and their neighbors.

In addition, the results on the Pt isotopes suggest that other nuclei which are well described by the O(6) symmetry of the IBA should be tested for severe fragmentation, both experimentally and theoretically. One such region is the Xe-Ba region near A=130. Unfortunately, these nuclei are not stable and the number of experimental probes available to find octupole states there is quite limited.

In summary, I have discussed how the fragmentation of the low energy octupole state can affect the systematic behavior of 3_1^- states and in which regions caution should be exercised when surveying 3_1^- state energies. A survey of octupole fragmentation also reveals that the apparent severe fragmentation in the Pt isotopes is unique. It suggested that experiments be performed to confirm this fragmentation.

This work has been supported by the National Science Foundation and the State of Florida.

REFERENCES

[1] P.D. Cottle and D.A. Bromley, Phys. Rev. C35, 1891 (1987).
[2] P.D. Cottle, K.A. Stuckey and K.W. Kemper, Phys. Rev. C38, 365 (1988).
[3] N.V. Zamfir, R.F. Casten and P. von Brentano, Phys. Lett. B 226, 11 (1989).
[4] R.H. Spear, At. Data and Nucl. Data Tables 42, 55 (1989).
[5] P.D. Cottle (in press, Phys. Rev. C).
[6] P.D. Cottle and M.L. Owens, Phys. Rev. C40, 2904 (1989).
[7] M.W. Kirson, Phys. Lett. 108B, 237 (1982).
[8] P.D. Cottle, S.M. Aziz, K.W. Kemper, M.L. Owens, E.L. Reber, J.D. Brown, E.R. Jacobsen and Y.Y. Sharon (to be published).
[9] S.W. Yates, R. Julin, J. Kumpulainen, and E. Verho, Phys. Rev. C37, 2877 (1988).
[10] P.D. Cottle, K.A. Stuckey, and K.W. Kemper, Phys. Rev. C38, 2843 (1988).
[11] P.T. Deason, C.H. King, R.M. Ronningen, T.L. Khoo, F.M. Bernthal, and J.A. Nolen, Jr., Phys. Rev. C23, 1414 (1981).
[12] C.S. Lim, Ph.D. Dissertation, Australian National University (1989).
[13] P.D. Cottle, M.A. Kennedy, and K.A. Stuckey (in press, Phys. Rev. C).
[14] J. Engel, Phys. Lett. 171B, 148 (1986).

M1 STRENGTH AND SCISSORS STRENGTH IN RARE-EARTH NUCLEI

C.De Coster* and K.Heyde

Institute for Theoretical Physics and Institute for Nuclear Physics
Proeftuinstraat 86, B-9000 Gent, Belgium

ABSTRACT

The M1 strength distribution is studied in rare-earth nuclei, starting from the Nilsson model and using the Quasi-Particle Random Phase Approximation (QRPA), including monopole as well as quadrupole pairing, Coriolis mixing and residual quadrupole and spin-isospin interactions. For the nucleus ^{164}Dy the importance and effect of the different terms in the Hamiltonian are studied.
Results are compared with experimental observations. Special attention is given to the higher-lying spin strength in view of a recent (p,p') experiment performed at TRIUMF on ^{154}Sm.

INTRODUCTION

In analogy to the E1 giant resonance, described by Goldhaber and Teller[1] as due to out-of-phase translational oscillations of protons versus neutrons, Lo Iudice and Palumbo[2] predicted the existence of a low-lying magnetic dipole state, originating from out-of-phase rotational oscillations of protons versus neutrons and denoted as a "scissors" excitation. Independently, the proton-neutron interacting boson model[3], treating valence nucleons in pairs as bosons and thereby taking the charge degree of freedom explicitly into account, described "mixed-symmetry states" which, in the geometrical limit correspond to out-of-phase motion of protons versus neutrons. Other collective models were used to reach similar conclusions, such as the Giant Angle Dipole model (GAD)[4], and the Generalized Bohr-Mottelson model (GBMM)[5].

The experimentally discovered low-lying 1^+ state around 3 MeV in ^{156}Gd in Darmstadt in 1984[6] was therefore at first interpreted in terms of these collective models, but theoretical predictions of excitation energy and M1 strength seemed much at variance with the experimental observations. This then was the onset of a large variety of experiments, in search of the real nature of the M1 excitations, which have been observed in nuclei ranging from ^{46}Ti to ^{238}U[7]. The comparison of inelastic electron scattering cross sections with those obtained in (p,p') experiments revealed the mainly orbital nature of the low-lying excitations, in support of the interpretation as a convection current of protons versus neutrons. Complementary high-resolution experiments, using Nuclear Resonance Fluorescence (NRF) techniques, indicated that the M1 strength around 3-4 MeV, was much more fragmented, thereby ruling out

---

* aspirant of the N.F.W.O.

an interpretation in terms of a single strong collective state. Therefore, most recent theoretical calculations start from a microscopic basis: Cranked Hartree-Fock Bogoliubov (CHFB), QRPA and large scale shell-model calculations[8].

Two more experiments highlighted other interesting aspects of magnetic dipole strength in rare-earth nuclei in general: in a proton pick-up experiment on ^{165}Ho evidence was found for a "pure" two-quasi-particle (2qp) 1^+ rotational band around 2.5 MeV in ^{164}Dy[9]. In recent (p,p') experiments performed at TRIUMF a higher-lying 1^+ spin resonance was found in ^{154}Sm, ^{158}Gd and ^{168}Er[10].

Starting from a microscopic model, we will address the following questions, brought up by experiment:
- What is the real nature of the orbital M1 excitations around 3 MeV, and to what extent can they be interpreted as "collective scissors-like" excitations, in view of the observed fragmentation ?
- Can the observed spin resonance be reproduced theoretically and can the observed structure in the cross-section be explained ?
- Whereto can the possible existence of rather pure 2qp states at lower energy be ascribed ?

Before coming to these points, we briefly sketch the model used.

THE MODEL

In order to describe the deformed rare-earth nuclei, we start from a Nilsson Hamiltonian with stretched coordinates. The ground state quadrupole deformation is taken from Möller and Nix[11] and the κ, μ parameters are taken from Nilsson et al[12], as well as the monopole pairing gap. The strength of the short-range quadrupole pairing correlations is obtained from the ratio of the monopole versus quadrupole contribution to the short-range δ-force matrix elements[13]. Influence of different choices of this strength parameter is discussed in the next section.

We include both monopole and quadrupole pairing in the BCS formalism and then construct correctly antisymmetrized 2qp $K^\pi = 1^+$ intrinsic states

$$|K^\pi = 1^+>_\rho \equiv \alpha^+_{N_1\Omega_1,\rho} \alpha^+_{N_2\Omega_2,\rho} | \tilde{0}> \quad , \quad \rho = \pi,\nu \; , \tag{1}$$

where the Nilsson orbits are characterized by the major shell quantum number N and the projection quantum number on the intrinsic symmetry axis Ω. It is then possible to calculate the magnetic dipole transition probabilities in the unified rotational model with as a result

$$B(M1;0^+ \to 1^+) = \frac{3}{4\pi}(u_1v_2 - v_1u_2)^2 |<N_1\Omega_1|g_\ell \ell_+ + g_s s_+|\overline{N_2\Omega_2}>|^2 \; , \tag{2}$$

where $|\overline{N\Omega}>$ denotes the time-reversed state. The pairing factor acts as a quenching factor, selecting those contributions corresponding with 2qp excitations nearby the Fermi level, as was pointed out previously[14].

Starting from this "unperturbed" picture, we next include Coriolis mixing between 1^+ intrinsic excitations. The coupling matrix elements are calculated using a transformation of the strong coupling wave functions to the weak coupling scheme and given by

$$<I'M',K' = \Omega'_1 + \Omega'_2|H_c|IM,K = \Omega_1 + \Omega_2>$$

$$= \delta_{I,I'} \delta_{M,M'} \frac{1}{2I+1} \sum_{J,J_c \text{even}} 2E_c(2J_c+1) <JKJ_c0|IK> <JK'J_c0|IK'>$$

$$\sum_{\ell_1 j_1} \sum_{\ell_2 j_2} <j_1\Omega_1 j_2\Omega_2 |JK> c\ell_{1j_1}(N_1\Omega_1) c\ell_{2j_2}(N_2\Omega_2)$$

$$\left[\delta_{N_1,N'_1} \delta_{N_2,N'_2} c\ell_{1j_1}(N_1\Omega'_1) c\ell_{2j_2}(N_2\Omega'_2) <j_1\Omega'_1 j_2\Omega'_2 |JK'> \right.$$

$$(u_1 u'_1 + v_1 v'_1)(u_2 u'_2 + v_2 v'_2)$$

$$- \delta_{N_1,N'_2} \delta_{N_2,N'_1} c\ell_{1j_1}(N_1\Omega'_2) c\ell_{2j_2}(N_2\Omega'_1) <j_1\Omega'_2 j_2\Omega'_1 |JK'>$$

$$\left. (u_1 u'_2 + v_1 v'_2)(u_2 u'_1 + v_2 v'_1) \right] \quad , \tag{3}$$

where E_C denotes the core energy, for which the experimental ground-band rotational energies of the $^{A-2}_{Z-2}X_N$ ($^{A-2}_{Z}X_{N-2}$) nucleus are taken to calculate the coupling between proton (neutron) 2qp states.

We also include residual quadrupole-quadrupole and spin-spin separable interactions of the following form

$$H_{\text{res}} = -\kappa_{\pi\pi} Q_\pi \cdot Q_\pi - \kappa_{\nu\nu} Q_\nu \cdot Q_\nu - \kappa_{\pi\nu}(Q_\pi \cdot Q_\nu + Q_\nu \cdot Q_\pi)$$

$$- \tau_{\pi\pi} \sigma_\pi \cdot \sigma_\pi - \tau_{\nu\nu} \sigma_\nu \cdot \sigma_\nu - \tau_{\pi\nu}(\sigma_\pi \cdot \sigma_\nu + \sigma_\nu \cdot \sigma_\pi) \quad , \tag{4}$$

which hence contains an isoscalar and an isovector part. The strength parameters are estimated following a method proposed by Bohr and Mottelson[15] starting from the harmonic approximation

$$\kappa_{\pi\pi} = \kappa_{\nu\nu} = 86.12 A^{-5/3} MeV \quad , \quad \kappa_{\pi\nu} = 388.34 A^{-5/3} MeV \quad , \tag{5}$$

or based on quenching of magnetic moments for heavy nuclei due to first order spin core polarization

$$\tau_{\pi\pi} = \tau_{\nu\nu} = -57.41 A^{-1} MeV \quad , \quad \tau_{\pi\nu} = 1.25 A^{-1} MeV \quad . \tag{6}$$

Due to rotational symmetry breaking, a non-physical excitation occurs, corresponding with a rotation of the nucleus as a whole. It is properly eliminated following a method outlined by Waroquier[16], putting the spurious state at zero energy and having all other eigenstates orthogonal to the spurious one.

In order to evaluate the macroscopic "scissors" picture put forward by collective models, we calculated the overlap of the resulting QRPA 1^+ excitations with the microscopic counterpart of a contra-rotation of protons versus neutrons

$$|1^+, SCISSORS> \equiv (\alpha J_{\pi,+} - \beta J_{\nu,+})|\widetilde{0}> , \qquad (7)$$

where α and β are determined such as to normalize and to orthogonalize this 1^+ excitation with respect to the spurious isoscalar rotational state.

We choose the nucleus ^{164}Dy for a detailed study since it is thus far the only nucleus for which besides fragmented M1 strength around 3 MeV, a pure 2qp excitation around 2.5 MeV is observed. The effect of the different forces included in the model is studied.

LOW-LYING M1 EXCITATIONS AND THE SCISSORS PICTURE

In Fig.1 a and b the low-lying total M1, orbital and spin strength distribution is given for the unperturbed 2qp $K^\pi = 1^+$ states as well as including different forces acting between them. In the unperturbed picture without quadrupole pairing, M1 strength is carried mainly by $(\ell,j)^2$ type 1^+ excitations, where ℓ,j denote the spherical orbit to which the Nilsson state reduces at zero quadrupole deformation. For protons, the most important configurations are $(1h_{11/2})^2$, $(1g_{7/2})^2$, $(2d_{5/2})^2$, while for neutrons the $(1h_{9/2})^2$, $(2f_{7/2})^2$ and $(1i_{13/2})^2$ configurations carry most strength, as is expected since these are the levels closest to the Fermi level[14].

As quadrupole pairing is included, there is a global shift towards higher energies as the strength increases, and the relative position of the 2qp states is changed. Therefore, it is expected that quadrupole pairing influences the couplings induced by the residual interaction and therefore the concentration/fragmentation of M1 strength. Still, in final QRPA results, the formation of a strong excitation with a largely admixed wave function is a constant feature.

Coriolis mixing favors the formation of one strong excitation which can be interpreted as collective in the sense that many 2qp states contribute coherently to the M1 strength. In other words, the amplitudes of proton and neutron 2qp components have opposite sign as is expected for a scissors-like excitation, but all components contribute coherently to the M1 matrix element. This can be seen from Fig.2 where the M1 strength distribution and the scissors overlap are given for a calculation with and without Coriolis mixing. Moreover, in earlier nuclear structure studies it has proven already to be important for close-lying states, as is the case for the 2qp 1^+ excitations.

The repulsive spin interaction pushes the spin strength up to higher energies, giving the low-energy excitations a dominantly orbital character. It also brings the states down in energy, hence separating the low-energy orbital mode, originating from the $(\ell,j)^2$ 2qp 1^+ states, from the higher-lying spin excitations, which will be discussed in the following section.

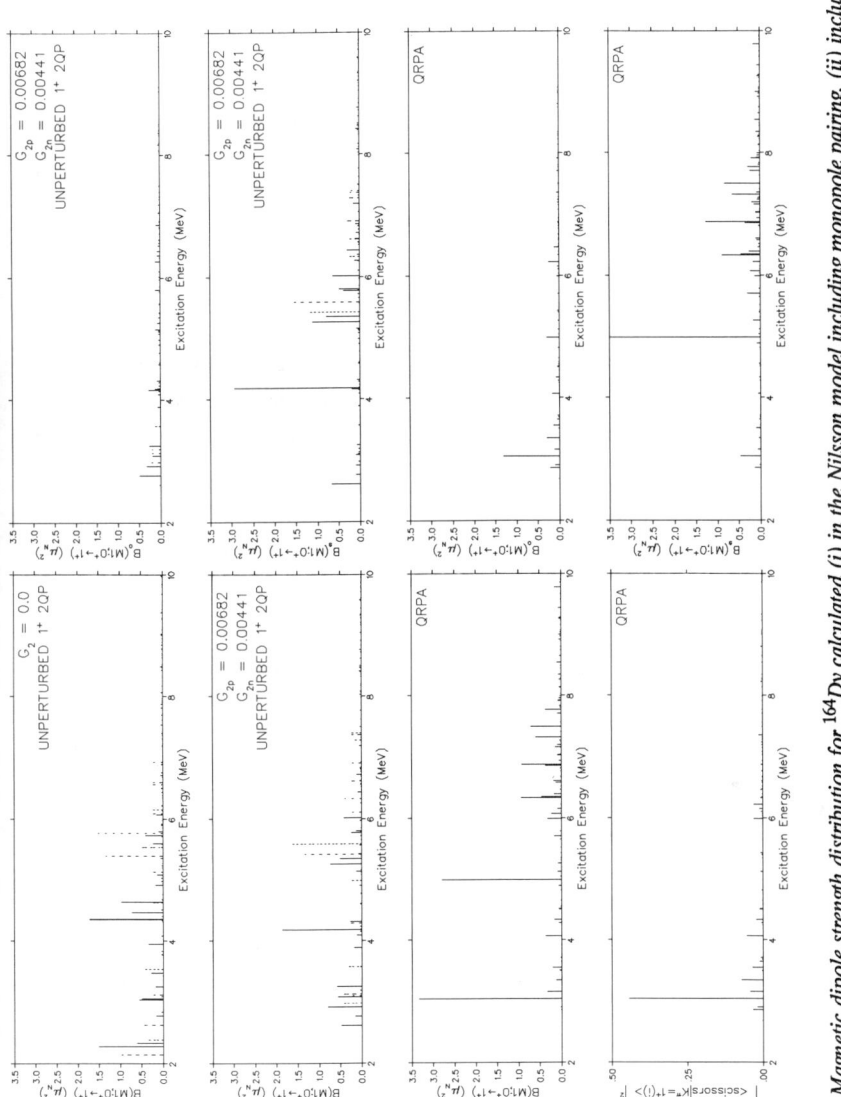

Fig.1. (a) Magnetic dipole strength distribution for ^{164}Dy calculated (i) in the Nilsson model including monopole pairing, (ii) including quadrupole pairing as well. The strengths are indicated on the figure. On both plots the proton (neutron) 2qp excitations are indicated with full (dashed) lines. (iii) M1 strength distribution and (iv) scissors overlap for a full QRPA calculation, performed on the unperturbed basis (ii), including Coriolis mixing as well as separable quadrupole and spin forces.
(b) Orbital and spin contributions to the M1 strength for ^{164}Dy. Both the unperturbed and the QRPA results are given.

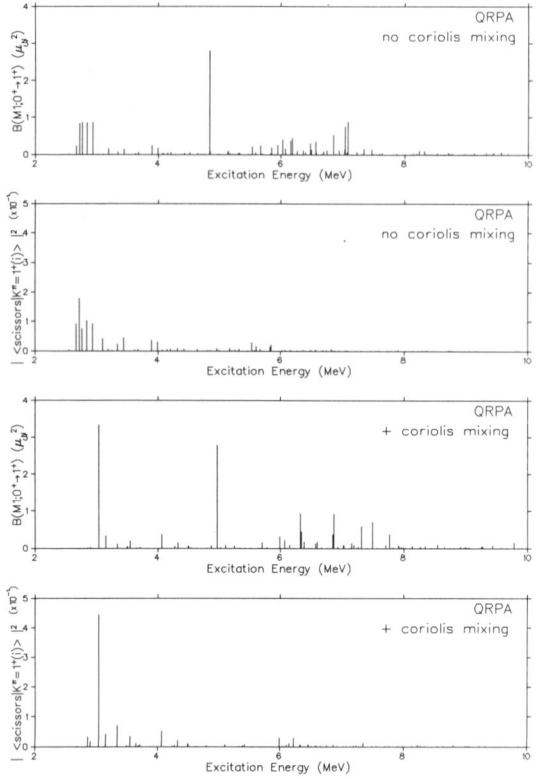

Fig.2. M1 strength distribution and scissors overlap for ^{164}Dy in a QRPA calculation, (i) including everything but Coriolis mixing (ii) including Coriolis mixing as well.

In conclusion, **including all interactions**, a strong 1^+ excitation is predicted at 3.04 MeV with B(M1)=3.33 μ_N^2, which has a dominant orbital character. The spin strength at low energy amounts to 0.46 μ_N^2. Comparing these results with experimental observations from (e,e') and (p,p') experiments, we find that the agreement is satisfactory : the summed strength of three peaks around 3.14 MeV is 3.06 ± 0.19 μ_N^2 and the spin strength is found to be 0.50 ± 0.07 μ_N^2. However, from (γ,γ') experiments fragmentation into three smaller excitations is observed, which is not reproduced by the calculations. We noted earlier that the fragmentation is very sensitive to the unperturbed 2qp energies, thus in view of the discrepancy between empirical and Nilsson single-particle energies, as well as the uncertainty on the quadrupole pairing strength, one cannot expect agreement on the level of fine details. Yet these results can decide about the nature of the excitation, which is indeed to be seen as a contra-rotational excitation of protons versus neutrons. Still, the degree of collectivity is not that high, which explains a scissors overlap of no more than 45%.

One should note here that (γ,γ') shows evidence for more M1 excitations around 2.5 MeV. In our calculations 1^+ states below 3 MeV are predicted, however carrying very little strength.

ONE NUCLEON TRANSFER AND LOW-LYING 1^+ STATES

It is believed that in the pick-up reaction ^{165}Ho(t,α)^{164}Dy a 1^+ rotational band is seen for which the fingerprint pattern fits nicely with a pure 2qproton configuration $[(1h_{11/2})5/2^-$ - $(1h_{11/2})7/2^-]1^+$ in ^{164}Dy. In our "unperturbed" calculations, starting from the Nilsson model including monopole as well as quadrupole pairing, this state is predicted around 2.5 - 2.9 MeV dependent on the strength G_2 and carries appreciable M1 strength. A realistic calculation

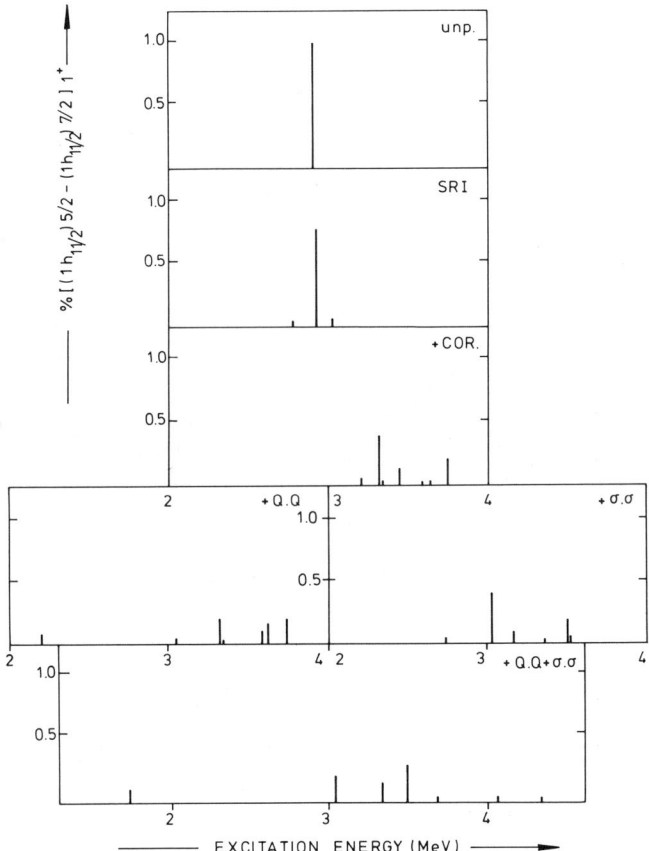

Fig.3. Fragmentation of the $[(1h_{11/2})5/2^- - (1h_{11/2})7/2^-]1^+$ 2qp component over different QRPA 1^+ excitations as interactions are turned on gradually: restoration of rotational symmetry, Coriolis mixing, quadrupole OR spin separable force, both quadrupole AND spin forces.

includes residual interactions as well, and we studied to what extent this configuration is expected to remain pure, turning on the residual forces. This is illustrated in Fig.3. Gradually including symmetry restoration, Coriolis mixing, quadrupole and spin separable forces, it is clear that fragmentation of this 2qp configuration sets in. (it should be mentioned that this qualitative behavior is independent of the parameter choice) The largest fragment corresponds with the strongest M1 excitation, which has a collective nature, so from our calculations there is no clear evidence for a rather pure $(1h_{11/2})^2$ configuration. Thus, only the occurrence of a low-lying well-isolated $(1h_{11/2})^2$ configuration in the unperturbed 2qp 1^+ basis would explain the lack of coupling with the other higher-lying 2qp configurations.

We pointed out before that the initial 2qp energies in the unperturbed picture might deviate appreciably from empirical values. In that sense, one-nucleon transfer reactions provide an important test on the adequacy with which the unperturbed 2qp $K^\pi = 1^+$ states can be described in the Nilsson + BCS model.

HIGHER-LYING M1 STRENGTH : A SPIN EXCITATION

In the unperturbed picture, we find besides the $(\ell,j)^2$ 1^+ configurations below 4 MeV essentially, spin-flip 1^+ excitations $(\ell,j = \ell + 1/2) \rightarrow (\ell,j = \ell - 1/2)$ between 4 and 8 MeV. Including residual interactions, couplings between these states are induced, and thereby the spin separable force is most important. Due to its repulsive nature, it pushes the spin strength up and separates out two groups of excitations, as can be seen from Fig.1. There, the strong state at 4.98 MeV is a proton excitation, while the states above 6 MeV are dominantly neutron excitations, although into some states proton configurations are admixed. This separation between proton and neutron excitations can be explained in view of the low ratio $\frac{\tau_{\pi\nu}}{\tau_{\pi\pi}}$ so that couplings between like nucleon excitations are more likely to occur.

To illustrate this more clearly, we studied the nucleus ^{154}Sm, which has been investigated recently in a (p,p') experiment[10]. In Fig.4, the experimental spectrum is compared with the theoretical prediction for the spin strength and the corresponding unperturbed picture. The separation of proton and neutron centroïds, present already in the latter, is clearly enhanced by the interaction and a double-humped structure is formed which agrees nicely with the experimental result. Also quantitatively the predicted summed spin strength of 13.25 μ_N^2 agrees very well with the experimental value of 11 ± 2 μ_N^2.

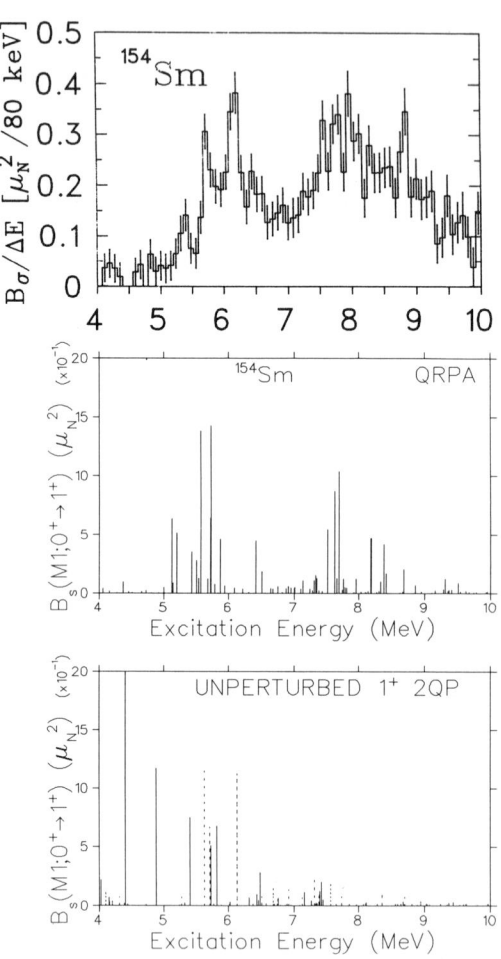

Fig.4. Higher-lying spin strength distribution for ^{154}Sm. The experimental spectrum is compared with the result of a full QRPA calculation and the unperturbed 2qp 1^+ picture (including quadrupole pairing). In the latter, proton (neutron) excitations are indicated by full (dashed) lines.

One remark should be made here. For the extraction of the $B_\sigma(M1)$ values from the experimental data a $\nu(1h_{9/2}\text{-}1h_{11/2})$ configuration was assumed. Looking at the theoretical wave functions, other configurations seem important as well, such as $\nu(1i_{11/2} - 1i_{13/2})$, $\pi(1g_{7/2}\text{-}1g_{9/2})$ and $\pi(1h_{9/2}\text{-}1h_{11/2})$. Hence, the empirical value could differ when including these configurations.

CONCLUSION

Starting from a QRPA calculation for the nucleus ^{164}Dy, which is a representative example for the rare-earth region, we studied the nature of the M1 strength observed in different energy regions. We found that the orbital strength around 3 MeV can indeed be ascribed to a convective current corresponding with contra-rotation of protons versus neutrons. The resulting strong M1 excitation has a complex wave function and can in that sense be denoted as "collective". The much lower B(M1) value compared to the prediction of the collective two-rotor model as well as the scissors overlap of no more than 45%, are due to the fact that not all 2qp configurations contribute to this state: the strongest components correspond to the 2qp states nearest by the Fermi level. This valence property was discussed previously when studying systematics of summed M1 strength over the whole rare-earth region[14].

The higher-lying M1 strength corresponds with a spin excitation. In ^{154}Sm, the double-humped structure could be explained as due to separation of proton and neutron excitations. It is of interest to investigate whether this is a general feature throughout the rare-earth region. Such work is in progress.

REFERENCES

1. M.Goldhaber and E.Teller, Phys.Rev. 74, 1046 (1948).
2. N.Lo Iudice and F.Palumbo, Phys.Rev.Lett. 41, 1532 (1978).
3. F.Iachello and A.Arima, The Interacting Boson Model, Cambridge Monographs on Mathematical Physics, ed. P.V.Landshoff, W.H.Mc Crea, D.W.Sciama, and S.Weinberg (Cambridge University Press, New York, 1987) and references therein.
4. R.R.Hilton, Journal de Physique 'Semiclassical Methods in Nuclear Physics', eds. R.W.Hasse, R.Arvieu, P.Schuck, Colloque C6, Tome 45, 255 (1984).
5. A.Faessler and R.Nojarov, Phys.Lett. 166B, 367 (1986).
6. D.Bohle et al., Phys.Lett. 137B, 27 (1984).
7. for a review, see K.Heyde, Int.Journ.Mod.Phys. A4, 2063 (1989).
8. for a review, see A.Richter in Contemporary Topics in Nuclear Structure Physics, eds. R.F.Casten, A.Frank, M.Moshinsky and S.Pittel (World Scientific, Singapore 1988) 127 and references therein.
9. S.J.Freeman et al., Phys.Lett. 222B, 347 (1989).
10. D.Frekers et al., Phys.Lett. 244B, 178 (1990).
11. P.Möller and J.R.Nix, preprint LA-UR-86-3983, Los Alomos Nat.Lab. (1986).
12. S.G.Nilsson et al., Nucl.Phys. A131, 1 (1969).
13. K.F.Pál, N.Rowley and M.A.Nagarajan, Nucl.Phys. A470, 285 (1987).
14. C.De Coster and K.Heyde, Phys.Rev.Lett. 63, 2797 (1989) and to be publ.
15. A.Bohr and B.Mottelson, Nuclear Structure, Vol.II, (W.A.Benjamin, INC. 1975).
16. M.Waroquier, Proefschrift voorgelegd tot het verkrijgen van de graad van Geaggregeerde van het Hoger Onderwijs, Appendix K, (Gent 1982, unpublished).

M1 EXCITATION SCHEME IN DEFORMED NUCLEI

TAKAHARU OTSUKA and IAIN MORRISON

Department of Physics, University of Tokyo, Hongo, Bunkyo-ku,
Tokyo, 113, Japan
and
School of Physics, University of Melbourne, Parkville, Victoria, Australia

ABSTRACT

We present the M1 excitation scheme in even-even deformed nuclei from the sum-rule viewpoint based on the Nilsson+BCS approach. The sum-rule states are introduced for the Scissors, spin and spin-flip modes. The functional form of the B(M1) sum rule of the Scissors mode is obtained, and its actual value is shown to be 4~6 (μ_N^2). The spin excitation B(M1) turns out to be 10~15 (μ_N^2) including the spin-flip transitions. The total B(M1) is 15~20 (μ_N^2). The effect of the SD and SDG pair truncation is studied to test IBM-2 for M1 excitations. The SDG truncation reproduces very well the calculation without truncation. The SD truncation reproduces the orbital excitation, whereas yields some deviations for the spin excitation.

§1. Introduction

The M1 excitation scheme in deformed even-even nuclei[1] will be presented from a simple perspective. In most theoretical approaches[2] on the 1+ M1 excitations in the energy region of Ex=2 ~ 8 MeV, J^π=1+ or K^π=1+ excited states are obtained first, and then M1 matrix elements from the ground state are calculated. We take an alternative approach in this work. We first introduce several basic excitation modes of K^π=1+ which include the Scissors, spin and spin-flip modes. The sum-rule states of these modes are introduced so that they exhaust the corresponding M1 transitions from the ground state. The sum-rule states are not eigenstates in general. In fact, these K^π=1+ sum-rule states can be fragmented. Proceeding this sum-rule approach, we present a simple and intuitive physical picture on the M1 excitation scheme. The relation to IBM-2 is discussed in terms of pair truncation[3].

§2. Outline

We shall begin with a sketch of our formalism[4]. We first carry out the Nilsson+BCS calculation with standard values of the parameters. The particle number is conserved, so that we can see precisely various dependences of M1 excitations on the proton and neutron numbers. The Nilsson+BCS ground state thus obtained is denoted as $|\psi\rangle$. We choose the Nilsson+BCS method, because it is a simple and yet reliable method for determining the intrinsic state.

Once the Nilsson + BCS ground state $|\psi\rangle$ is fixed, several $K^\pi=1^+$ TDA (*i.e.* one-broken pair) states are constructed by acting J_ρ and S_ρ operators, where ρ refers to as π (proton) or ν (neutron), and J and S denote total and spin angular momentum operators, respectively. We then calculate excitation energies and B(M1)'s for these $K^\pi=1^+$ states.

§3. Scissors mode

In costructing the TDA states, we have to remove the rotational spurious state in the $K^\pi=1^+$ TDA space;

$$|K^\pi=1^+, \text{Spur}\rangle \propto (J_\pi + J_\nu)_{K=1}|\psi\rangle \tag{1}$$

This is nothing but the infinitesimal rotation. The Scissors mode TDA state is *defined* as the orthogonal $K^\pi=1^+$ TDA state generated by the J operators;

$$|K^\pi=1^+, \text{Scis}\rangle \propto [\langle J_\nu^2\rangle J_\pi - \langle J_\pi^2\rangle J_\nu]_{K=1}|\psi\rangle \tag{2}$$

where

$$\langle J_\rho^2\rangle = \langle\psi|(J_\rho \cdot J_\rho)|\psi\rangle, \qquad \rho = \pi, \nu \tag{3}$$

with ($\cdot$) being a scalar product. Since $\langle J_\pi^2\rangle \sim \langle J_\nu^2\rangle$ in practice, the contra-rotation picture between protons and neutrons is clear in eq. (2). Note that the orthogonality

$$\langle K^\pi=1^+, \text{Scis}|(J_\pi + J_\nu)_{K=1}|\psi\rangle = 0 \tag{4}$$

The B(M1) value is evaluated by the formula[5]

$$B(M1: 0^+ \to 1^+_{\text{Scis}}) = \frac{3}{4\pi} \times 2 \times \langle K^\pi=1^+, \text{Scis}|M_1|\psi\rangle^2 \tag{5}$$

where M_1 stands for the K=1 component of the M1 operator written as

$$M = \sum_{\rho=\pi,\nu} (g_l^{(\rho)} L_\rho + g_s^{(\rho)} S_\rho) \tag{6}$$

with the g's being g-factors. Using eqs. (2) and (3), one can derive

$$B(M1: 0^+ \to 1^+_{\text{Scis}}) = \frac{3}{4\pi} \times \frac{\langle J_\pi^2\rangle\langle J_\nu^2\rangle}{\langle J_\pi^2\rangle + \langle J_\nu^2\rangle} \times (g_J^{(\pi)} - g_J^{(\nu)})^2, \tag{7}$$

where

$$g_J^{(\rho)} \equiv g_l^{(\rho)} \frac{\langle L_\rho J_\rho\rangle}{\langle J_\rho^2\rangle} + g_s^{(\rho)} \frac{\langle S_\rho J_\rho\rangle}{\langle J_\rho^2\rangle}, \qquad \rho = \pi, \nu \tag{8}$$

$$\langle L_\rho J_\rho\rangle = \langle\psi|(L_\rho \cdot J_\rho)|\psi\rangle \quad \text{and} \quad \langle S_\rho J_\rho\rangle = \langle\psi|(S_\rho \cdot J_\rho)|\psi\rangle \tag{9,10}$$

We point out that eq. (8) corresponds to the generalized Lande's formula[6].

Since $\langle L_\rho J_\rho \rangle \sim 1$ and $\langle S_\rho J_\rho \rangle \sim 0$ practically, one finds $g_J^{(\pi)} \sim 1$ and $g_J^{(\nu)} \sim 0$. This indicates the orbital dominance in the Scissors mode. On the other hand, deviation upto ~ 0.2 can easily be expected due to spin contribution. In all practical cases, the factor $(g_J^{(\pi)} - g_J^{(\nu)})^2$ in eq. (7) can be estimated roughly as unity.

The quantities $\langle J_\pi^2 \rangle$ and $\langle J_\nu^2 \rangle$ are the proton and neutron angular momenta carried by the Nilsson intrinsic state $|\psi\rangle$. If the proton and neutron systems are completely spherical, the Nilsson wave function is the product of a proton state of angular momentum zero and a neutron state of angular momentum zero. Such a state shows $\langle J_\pi^2 \rangle = \langle J_\nu^2 \rangle = 0$. If one of $\langle J_\pi^2 \rangle$ and $\langle J_\nu^2 \rangle$ is vanished and the other is not vanished, B(M1) in eq. (7) is still zero. In order that B(M1) in eq. (7) is not zero, both protons and neutrons should be deformed. As the general trend, $\langle J_\rho^2 \rangle$ increases as δ increases, where δ stands for the deformation parameter.

The valence part of the Nilsson+BCS wave function is the condensate of Cooper pair

$$|\psi\rangle = |\Lambda_\pi^{N_\pi}\rangle |\Lambda_\nu^{N_\nu}\rangle \qquad (11)$$

where the Λ_π (Λ_ν) denote the proton (neutron) Cooper pairs, and the N_ρ's mean the numbers of Cooper pairs of type ρ[7]. In zeroth-order estimate, $\langle J_\rho^2 \rangle$ is given by N_ρ multipled by the one-pair expectation value $\langle \Lambda_\rho | (J_\rho \cdot J_\rho) | \Lambda_\rho \rangle$. The Cooper pair in a strong Nilsson potential is still dominated by the monopole (S) and quadrupole (D) pairs of nucleons[7], and the amplitudes of the S and D pairs are similar to those of the s and d bosons in the intrinsic state of the SU(3) limit of the IBM. Thus, $\langle \Lambda_\rho | (J_\rho \cdot J_\rho) | \Lambda_\rho \rangle \sim \frac{2}{3} \times 6$, which leads to $\langle J_\rho^2 \rangle \sim 4 N_\rho$. We then obtain a simple relation

$$B(M1: 0^+ \to 1^+_{Scis}) \approx \frac{3}{4\pi} \times 4 \times \frac{N_\pi N_\nu}{(N_\pi + N_\nu)} (g_J^{(\pi)} - g_J^{(\nu)})^2 \qquad (12)$$

By identifying $g_J^{(\pi)}$ and $g_J^{(\nu)}$ as the proton and neutron boson g-factors in IBM-2, eq. (12) becomes identical to the IBM-2 formula in the SU(3) limit with good F-spin for $N_\pi, N_\nu \gg 1$[8]. We can thus relate the Nilsson model sum-rule to that in IBM-2.

Table 1. B(M1) from the ground state to the scissors-mode 1^+ state

	^{164}Dy		
spin quenching factor	1.00	0.75	0.00
B(M1) (μ_N^2) (set A)	6.7	5.9	3.9
B(M1) (μ_N^2) (set B)	6.7	6.1	4.3

Coming back to fermions, we shall consider, as an example, the case of ^{164}Dy with $\delta \sim 0.3$ and $\Delta \sim 0.9$ (MeV) where Δ stands for the pairing gap. The free nucleon values are taken for the orbital g-factors, while the spin g-factors are quenched by a factor 0.75. Table 1 shows the B(M1) values calculated by eq. (7). Table 1 includes calculations of the spin quenching factor 1 and 0 for comparison.

Two sets of single particle orbits are used in Table 1. In set A, we take the single particle orbits in the ordinary one major shell; $50 \leq Z \leq 82$ and $82 \leq N \leq 126$. The set B includes, in addition to the orbits in the set A, $0g_{9/2}$ and $0h_{9/2}$ for protons, and $0h_{11/2}$ and $0i_{11/2}$ for neutrons. These added orbits are spin-flip partners of some orbits in the ordinary major shell, $i.e.$, set A.

Table 1 indicates that the B(M1) values are hardly sensitive to the choice of the single particle orbits as far as the value of δ and Δ remain unchanged. Another interesting point is that the IBM-2 SU(3) formula in eq. (12) produces B(M1)=5.5 (μ_N^2) with the g_J's obtained for spin quenching 0.75. This value of B(M1) is in good agreement with the one in Table 1.

§4. Spin mode

We shall next discuss modes other than the Scissors. Since the Scissors mode is basically orbital, these modes should be primarily spin excitation modes, and the corresponding $K^\pi = 1^+$ TDA ($i.e.$ one-broken pair) states are denoted by $|\Sigma_\rho\rangle$ where $\rho = \pi$ (ν) means that the proton (neutron) broken pair is created in $|\Sigma_\rho\rangle$. Because of the orthogonality to the spurious and Scissors states, one has

$$\langle \Sigma_\rho | (J_\rho)_{K=1} | \psi \rangle = 0, \quad \rho = \pi, \nu \tag{13}$$

which is rewritten as

$$\langle \Sigma_\rho | (L_\rho)_{K=1} | \psi \rangle = - \langle \Sigma_\rho | (S_\rho)_{K=1} | \psi \rangle, \quad \rho = \pi, \nu \tag{14}$$

The M1 matrix element can be written as

$$\langle \Sigma_\rho | (M)_{K=1} | \psi \rangle = (g_s^{(\rho)} - g_l^{(\rho)}) \langle \Sigma_\rho | (S_\rho)_{K=1} | \psi \rangle, \quad \rho = \pi, \nu \tag{15}$$

Since $|g_s^{(\rho)}| \gg |g_l^{(\rho)}|$, this matrix element is dominated by g_s, and $|\Sigma_\rho\rangle$ will be referred to as the spin (excitation) mode. Note that g_s and g_l are destructive in eq. (15).

The spin mode TDA states are decomposed into three components; (i) natural-parity spin mode, (ii) unique-parity spin mode, (iii) unique-parity spin-flip mode. The mode (i) is excited by the natural-parity orbit terms in the M1 operator, while the mode (ii) is due to the unique-parity orbit term, for example, $0h_{11/2} \to 0h_{11/2}$ for protons in ^{164}Dy. The mode (iii) is the excitation from the unique-parity orbit to its spin-flip partner, for example, $0h_{11/2} \to 0h_{9/2}$ for protons in ^{164}Dy. Fig. 1 shows all these excitation modes. Note that the proton excitation from $0g_{9/2}$ to $0g_{7/2}$ and the neutron excitation from $0h_{11/2}$ to $0h_{9/2}$ are included in the natural parity spin modes. In Fig. 1, the excitation energies are calculated in the number-projected TDA in the Nilsson+BCS formalism, except the Scissors mode. The excitation

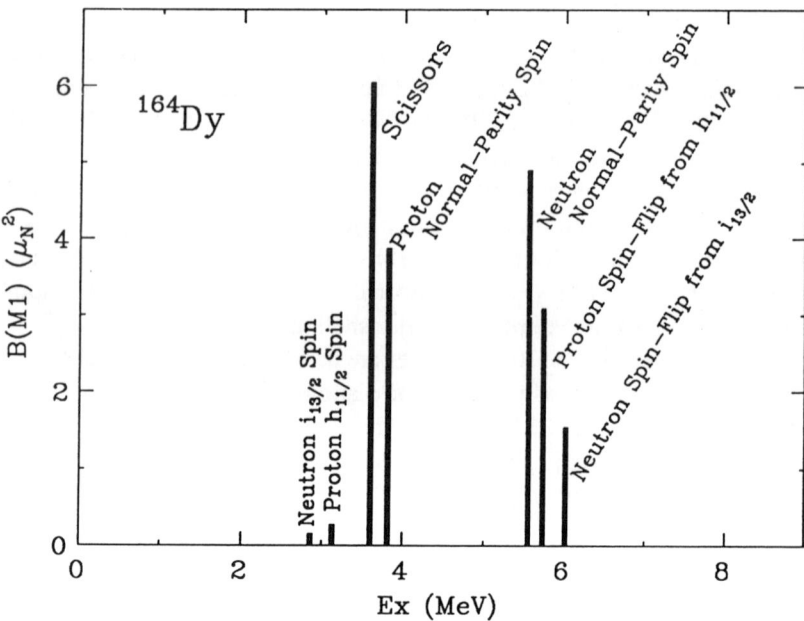

Fig. 1. M1 excitation scheme for ^{164}Dy. The single particle orbit set B is taken.

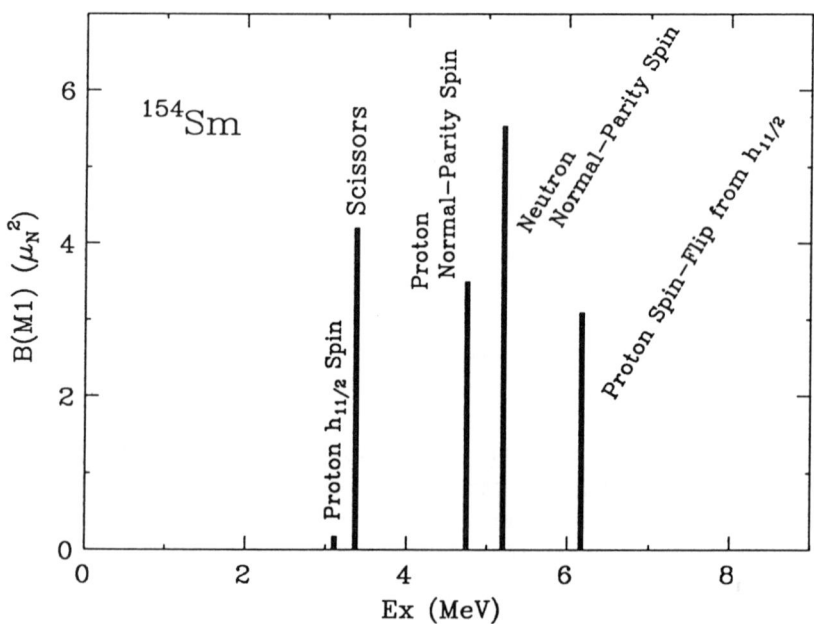

Fig. 2. M1 excitation scheme for ^{154}Sm. The single particle orbit set B is taken.

energy of the Scissors mode is sensitive to the proton-neutron interaction, which is absent in the Nilsson model. We therefore use Ex ~ 66 δ $A^{-1/3}$ (MeV).

The total B(M1) in Fig. 1 is about 20 (μ_N^2) which seems to be in agreement with experiment[9,10]. We point out that the total B(M1) of spin modes is more than twice of the Scissors B(M1). We also emphasize that the proton $h_{11/2}$ unique-parity spin mode at Ex ~ 3 MeV with B(M1) ~ 0.3 (μ_N^2) is in agreement with experimental data[11,12].

The M1 excitation scheme with $g_1^{(\pi)}=g_1^{(\nu)}=0$ is considered next to see the spin contribution. The Scissors mode disappears essentially, as another indication of its orbital character. The proton spin B(M1)'s are increased because the cancellation in eq. (15) disappears due to $g_1=0$. The excitation modes at Ex=4~7 MeV remain with even higher peaks, consistently to spin excitation experiment[9,10].

Fig. 2 shows M1 excitation scheme of ^{154}Sm, where the neutron $i_{13/2}$ spin mode and the neutron spin-flip mode from $i_{13/2}$ are too weak to be visible. The spin excitations at Ex = 5 ~ 6 MeV probably correspond to recent experiment[13].

§5. Pair Truncation

The IBM-2 description of M1 excitation scheme[14] can be examined by introducing the following truncations. The Cooper pair in eq. (11) can be decomposed as

$$A^\dagger = x_0 S^\dagger + x_2 D_0^\dagger + x_4 G_0^\dagger + \ldots \qquad (16)$$

where the x's are amplitudes, and S, D, G, *etc.* denote the pairs of $J^\pi=0^+, 2^+, 4^+$, *etc*[7]. The sdg boson model is the boson image of the SDG-pair truncated space, and is transformed into the sd boson model by renormalization[15].

Since the pair truncation is made in one major shell, we take the orbit set A. Fig. 3 shows the M1 excitation scheme without any truncation, while Fig. 4 shows the result with the truncation to the S, D and G pairs. At first glance, the SDG truncation reproduces essentially the same result as the full calculation, suggesting the usefulness of the sdg boson model for M1 excitation[16]. In the SD truncation, the result of which is not shown because of the length limitation, the excitation energies are shifted more or less, and B(M1) of the Scissors mode is decreased by ~ 20%. We note that, if the spin contribution is neglected ($g_s^{(\pi)}=g_s^{(\nu)}=0$), the SD truncation also reproduces the full result very well, indicating the validity of the naive mapping, SD pairs → sd bosons[2]. This is mainly because the orbital excitation is almost purely collective.

§6. Summary

We summarize this talk as,
(1) Σ B(M1) is 15 ~ 20 (μ_N^2) for rare-earth nuclei with δ ~ 0.3 and Δ ~ 1 MeV. The Scissors B(M1), which is predominantly orbital, is 4 ~ 6 (μ_N^2).
(2) For the Scissors mode, we can derive the B(M1) sum rule (eq. (7)).

(3) The spin modes carry more B(M1) values in total than the Scissors, and appear mostly around Ex = 4 ~ 7 (MeV).
(4) There are similarities and relations between IBM-2 and Nilsson+BCS descriptions of M1 excitation scheme.

Acknowledgement

This work is partly supported by the Japan Society for the Promotion of Science and the Australian Academy of Science under the Bilateral Exchange Program. The numerical calculation is supported financially in part by the Institute for Nuclear Study, University of Tokyo. One of the authors (T.O.) acknowledges the Yoshida Science Foundation for finacial support to his participation to this conference. This work is supported in part by the Grant-in-Aid for General Scientific Research (No. 01540231) by the Ministry of Education, Science and Culture.

References

1. A. Richter, in *Proc. Symp. in Honor of Akito Arima: Nuclear Physics of the 1990's*, ed. by D.H. Feng, J.N. Ginocchio, T. Otsuka and D.D. Strottman, Nucl. Phys. A (1991).
2. C. De Coster, in this proceedings; K. Heyde, in *Nuclear Structure, Reactions, and Symmetries*, ed. by R.A. Meyer and V. Paar (World Scientific, Singapore,1986), p. 288.
3. T. Otsuka *et. al.*, Phys. Lett. **76B** (1978) 139; Nucl. Phys. **A309** (1978) 1.
4. T. Otsuka , Nucl. Phys. **A507**, 129c (1978); T. Otsuka and I. Morrison, to be published.
5. A. Bohr and B.R. Mottelson, in *Nuclear Structure*, (Benjamin, New York, 1975), Vol. 2.
6. A. de-Shalit and I. Talmi, in *Nuclear Shell Theory*, (Academic Press, New York, 1963).
7. T. Otsuka *et. al.*, Phys. Rev. Lett. **48**, 387 (1982).
8. P. van Isacker *et. al.*, Ann. Phys. **171**, 253 (1986).
9. K.-G. Dietrich *et. al.*, Phys. Lett. **220B**, 351 (1989).
10. D. Frekers *et. al.*, Phys. Lett. **218B**, 439 (1989).
11. R. Chapman *et. al.*, in *Contemporary Topics in Nuclear Structure Physics*, ed. by R.F. Casten et. al. (World Scientific, Singapore,1988), p. 165.
12. C. Wesselborg *et. al.*, Phys. Lett. **207B**, 22 (1988).
13. D. Frekers *et. al.*, preprint.
14. F. Iachello, Nucl. Phys. **A358** (1981) 89.
15. T. Otsuka and J.N. Ginocchio, Phys. Rev. Lett. **55**, 276 (1985).
16. I. Morrison *et. al.*, J. Phys. G **15**, 801 (1989).

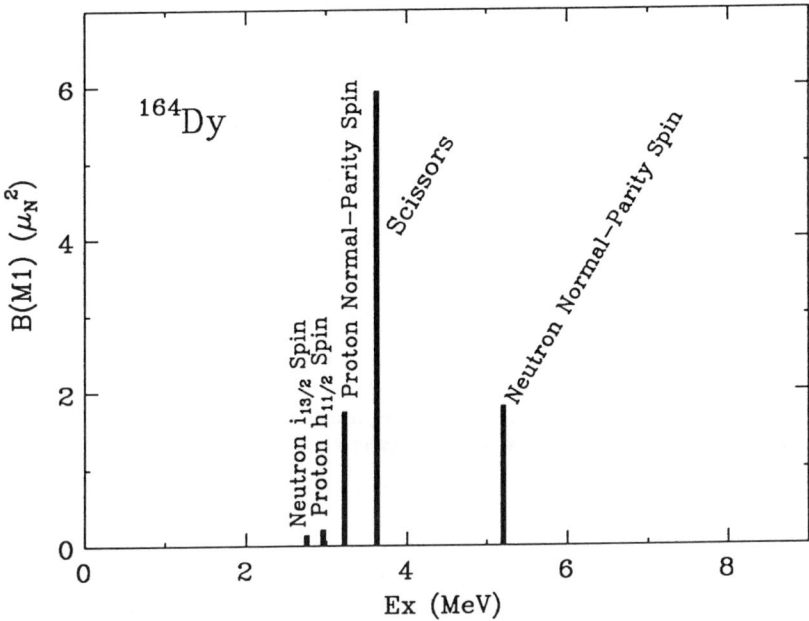

Fig. 3. M1 excitation scheme for ^{164}Dy. The single particle orbit set A is taken.

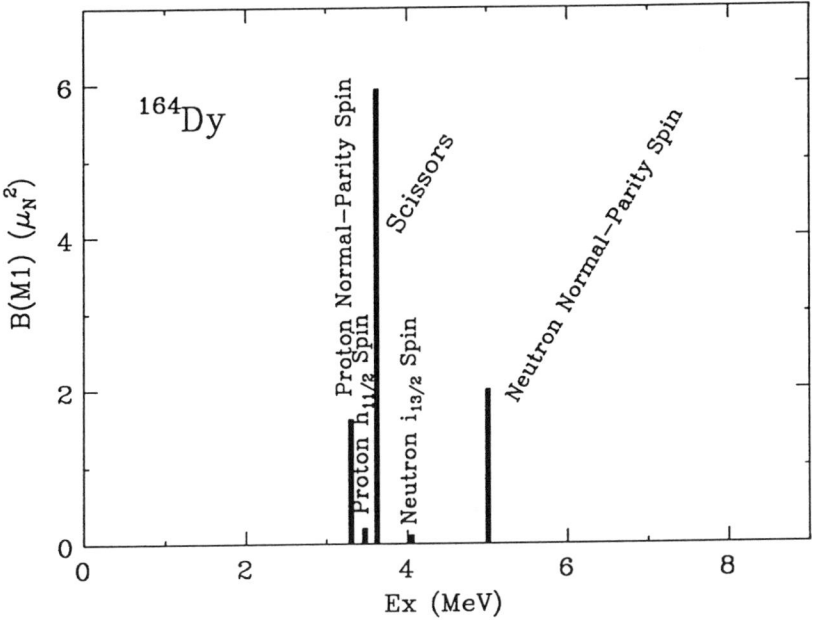

Fig. 4. See the caption of Fig.3, except that the SDG truncated ground state is used.

NUCLEAR SHELL MODEL CALCULATIONS WITH NON-LOCAL INTERACTIONS

S.A. Moszkowski
Department of Physics, UCLA, Los Angeles, CA 90024

S.D. Bloom and D.A. Resler
Lawrence Livermore National Laboratory, Livermore, CA 94550

ABSTRACT

It is becoming clearer with time that non-locality of the nucleon-nucleon interaction can play a significant role in nuclear properties. In this talk we review evidence for such non-locality. Then, using a Gaussian interaction, we discuss the effect of non-locality on two body matrix elements in the nuclear shell model. Finally, we mention some applications. For example, non-locality leads to faster convergence of off-diagonal matrix elements.

I. WHAT IS NON-LOCALITY?
1. Definition

Let us begin with discussing what non-locality is. Basically, the Schroedinger equation for a particle in a conventional (local) potential has the well known form:

$$-(\hbar^2/2m)\psi(\vec{r}) + V(\vec{r})\psi(\vec{r}) = E\psi(\vec{r}). \tag{1}$$

which can also be written as:

$$-(\hbar^2/2m)\psi(\vec{r}) + \int V(\vec{r})\delta(\vec{r}-\vec{r}')\psi(\vec{r}')d^3r' = E\psi(\vec{r}). \tag{2}$$

For a non-local interaction, we have, instead:

$$-(\hbar^2/2m)\psi(\vec{r}) + \int V(\vec{r},\vec{r}')\psi(\vec{r}')d^3r' = E\psi(\vec{r}). \tag{3}$$

i.e. $\vec{r}$ and $\vec{r}'$ don't have to be at the same point.

2. Relation to momentum dependence

Instead of writing the interaction as a non-local operator, we can express it in the form of a local, but momentum dependent interaction. Thus, for example, a non-local Gaussian interaction of the form:

$$V(\vec{r},\vec{r}') = e^{-(\vec{r}+\vec{r}')^2/4a^2} \times e^{-(\vec{r}-\vec{r}')^2/4c^2} \tag{4}$$

is equivalent (apart from a multiplicative constant) to a momentum dependent Gaussian interaction of the form:

$$V(\vec{r},\vec{p}) = e^{-r^2/a^2} \times e^{-p^2c^2} \tag{5}$$

(Actually, this interaction has to be properly symmetrized w.r.t. r and p, so that it is Hermitean.) The connection between non-locality and momentum dependence is more complicated for a non-Gaussian interaction.

3. Relation to energy dependence

Instead of writing the interaction as momentum dependent, we can also try to represent it as local, but energy dependent. However, when we do this, we lose the orthogonality property of wavefunctions of different energy.

II. WHY DO WE NEED NON-LOCALITY?
1. Theory
a. A simplified quark model for non-locality.

Some non-locality of the NN interaction at short distances is expected on account of the quark structure of nucleons, that is to say, due to the nucleon compositeness. To illustrate this point, consider a very crude model, in which each nucleon is replaced by a cluster consisting of two (not three) particles. The two particles have masses M and m, respectively. If there is a harmonic interaction between them, $V = \frac{1}{2}\mu\omega^2 r^2$, where μ is the reduced mass = Mm/(M+m) and r is the spacing between the particles, then the wavefunction of relative motion is a Gaussian proportional to $e^{-\mu\omega r^2/2\hbar} = e^{-r^2/2b^2}$. Now consider the interaction between two clusters whose centers of mass are separated by a distance R. We suppose that the interaction between the particles has a range very long compared to the cluster size b, and that it involves exchanging the light particles on the two clusters. Using standard cluster model methods, it can then be shown that the interaction between the clusters is just of the non-local form discussed above:

$$V(\vec{R},\vec{R}') = e^{-(\vec{R}+\vec{R}')^2/4a^2} \times e^{-(\vec{R}-\vec{R}')^2/4c^2} \tag{6a}$$

with
$$a^2 = 2b^2/[1+\frac{m}{M}]^2 \, , \, c^2 = 2b^2/[1+\frac{M}{m}]^2 \tag{6b}$$

Two particular limits are of interest.
i. For m << M, we obtain $a^2 = 2b^2$, c = 0, i.e. the interaction is local, with a

$$V(\vec{R},\vec{R}') = e^{-\vec{R}^2/2b^2} \times \delta(\vec{R}-\vec{R}') \tag{7}$$

range governed by the overlap of wavefunctions between the two clusters. This case corresponds to the interaction between atoms. Since the atomic nucleus has a much larger mass than an electron, the interactions between atoms are nearly local.

ii. For m = M, which more closely corresponds to the quark model of nucleons, (except for containing 2 quarks instead of 3), we find $a^2 = c^2 = b^2/2$. For this case, the interaction is separable.

$$V(\vec{R},\vec{R}') = e^{-(R^2+R'^2)/b^2} \qquad (8)$$

b. Relativistic Model

Alternatively, some non-locality also arises in relativistic nuclear models, due to relativistic retardation effects.

2. Experimental evidence: Deuteron radius

There are only a few well-known examples of non-local interactions known in nature. The best evidence (to date) for non-locality of the N-N interaction comes from the analysis of the deuteron charge radius[1]. With traditional (basically local) interactions, it is possible to fit N-N scattering phase shifts very well, but one obtains a deuteron radius about 1% larger than the empirical value. This is illustrated in Figure 1. The problem of the deuteron radius can be easily resolved, without destroying the fit to phase shifts, by making the potential non-local[2]. For example, there are simple unitary transformations on the wavefunction which change the short distance behavior, but keep intact both the orthogonality properties of non-degenerate pairs of wavefunction, and the wavefunction at large distances, which is what determines the phase shifts.

III. NON-LOCALITY AND SHELL MODEL CALCULATIONS
1. Two-body interaction matrix elements

It is well known that if we use harmonic oscillator single particle wavefunctions, then the two body matrix elements for Gaussian interactions can be calculated analytically. This also holds for a non-local Gaussian, in fact, the expressions are only slightly more involved. It is readily shown that in the short range limit; i.e., when both the range a and the non-locality range c are small compared to the oscillator length b, then the two body matrix elements reduce to those for a Skyrme interaction (without density dependence). It turns out that non-locality increases the even state matrix elements, but reduces the odd state matrix elements. If a = c, then only matrix elements with relative orbital angular momentum zero survive. In fact, in this case, the interaction is separable. Another case of interest occurs for a long range. Here the interaction effectively reduces to a quadrupole-quadrupole interaction. Finally, for $ac = 2b^2$, all two body matrix elements connecting different oscillator shells vanish identically. An interaction with these (somewhat unrealistic) values of the parameters gives rise to a mean field but no splitting between states belonging to the same irreducible representation of SU(3) but different L values.

We discuss here the Talmi integrals for non-local Gaussian interactions. Suppose we have the following non-local Gaussian interaction:

$$V = -V_0 \exp - \left[\frac{(r+r')^2}{4a^2} + \frac{(r-r')^2}{4c^2} \right] \quad (9)$$

in even states, and ηV in odd states.

Then the integrals involving relative motion can be calculated analytically. For non-local interactions, they are only slightly more complicated that for a local interaction.

We obtain:
$$I_{00} = V_0 \lambda^{3/2} (1-\mu)^{3/2} \quad (10a)$$
where
$$\lambda = \frac{a^2}{a^2+2b^2}, \quad \mu = \frac{c^2}{c^2+2b^2} \quad (10b)$$

Here is an expression for another Talmi integral:
$$I_{11}/I_{00} = \eta(\lambda - \mu) \quad (11)$$

Note that if $\lambda = \mu$, i.e. if $a = c$, (local and non-local ranges are equal), then $I_{11} = 0$. Similarly, all other Talmi integrals with relative angular momentum larger than 0 vanish. This is not surprising, since this case corresponds to a separable interaction, which acts only in relative S-states. Thus, for example:
$$I_{22}/I_{00} = (\lambda - \mu)^2 \quad (12)$$

On the other hand, the Talmi integral I_{20} is finite even for a separable interaction.
$$I_{20}/I_{00} = (\lambda - \mu)^2 + \tfrac{3}{2}(1-\lambda-\mu)^2 \quad (13)$$

However, there is another interesting limit, that where $\lambda + \mu = 1$. This corresponds to the condition $ac = 2b^2$. For this case, we see that $I_{20} = I_{22}$, and, more generally, $I_{N\ell}$ is independent of ℓ. Furthermore, all off-diagonal matrix elements connecting different oscillator shells vanish.
For example,
$$I_{00-20}/I_{00} = \sqrt{\tfrac{3}{2}}(1-\lambda-\mu) \quad (14)$$
which equals 0 in this case.

It is well known that for an infinite range interaction, i.e. $\lambda = 1$, $\mu = 0$, all off-diagonal matrix elements vanish. However, as is shown here, this result holds more generally, even for a finite range interaction, provided the range of the non-locality is chosen equal to $c = 2b^2/a$. In this limit, the interaction provides a mean field, but there is no pairing.

Finally, it should be noted that a finite range space exchange interaction, $a = r_0$, $c = 0$, $\eta = -1$, $(\lambda = \lambda_0, \mu = 0)$ has the same Talmi integrals, and thus two-body matrix elements, as a zero range, but finite non-locality range $a = 0$, $c = r_0$, $\eta = 1$, $(\lambda = 0, \mu = \lambda_0)$ interaction. This indicates that, to some extent the effect of non-locality can be simulated by a local but space exchange range interaction.

2. Connection with Skyrme Interaction

In this talk we will restrict our consideration to a density independent interaction. For a short range interaction, i.e. with both a and c $\ll$ b, we obtain: for the two parameters λ and μ:

$$\lambda = \frac{a^2}{a^2 + 2b^2} \rightarrow \frac{a^2}{2b^2}\left(1 - \frac{a^2}{2b^2}\right) \tag{15a}$$

$$\mu = \frac{c^2}{c^2 + 2b^2} \rightarrow \frac{c^2}{2b^2} \tag{15b}$$

The Skyrme parameters are defined as follows for a local short range interaction:

$$t_0 = \int V(r) \, d^3r = V_0(\pi a^2)^{3/2} \tag{16a}$$

$$t_1 = -\frac{1}{3} \int V(r) r^2 d^3r = -t_0 a^2/2 \tag{16b}$$

$$t_2 = \frac{\eta}{3} \int V(r) r^2 d^3r = \eta t_0 a^2/2 \tag{16c}$$

η is the ratio of odd state to even state interaction.

Then
$$I_{00} = V_0 \lambda^{3/2}(1-\mu)^{3/2} \rightarrow V_0 \left(\frac{\pi a^2}{2\pi b^2}\right)^{3/2} \times \left(1 - \frac{3}{4}\frac{a^2}{b^2}\right) \tag{17}$$

$$I_{no} = \int \psi_n(r) V(r,r') \psi_0(r') d^3r d^3r' \tag{18}$$

For a short range but non-local interaction we must make the following changes:
in t_1, replace a^2 by $a^2 + c^2$, and,
in t_2, replace a^2 by $a^2 - c^2$.
Also, I_{00} is now given by:

$$I_{00} = V_0 \lambda^{3/2}(1-\mu)^{3/2} \rightarrow V_0 \left(\frac{\pi a^2}{2\pi b^2}\right)^{3/2} * \left(1 - \frac{3}{4}\frac{a^2 + c^2}{b^2}\right) \tag{19}$$

Thus non-locality increases the apparent range of the even-state interaction, but decreases that of the odd-state interaction.

In the Skyrme approximation[3], the density independent part of the interaction is written as:
$$V(p,r) = t_0 \delta(r) + \frac{1}{2}[p^2\delta(r) + \delta(r)p^2]t_1 + p\cdot\delta(r)\cdot p t_2 \quad (20)$$
If $t_2 = -t_1$, i.e., $\eta = 1$, then
$$V(p,r) = t_0\delta(r) + \frac{1}{2}[p^2\delta(r) - 2p\cdot\delta(r)\cdot p + \delta(r)p^2]t_1$$
$$= t_0\delta(r) - \frac{1}{2}[\nabla^2\delta(r)]t_1 \quad (21)$$
which is just a local interaction.

The Talmi integrals for relative angular momentum zero are given as follows:
$$I_{no} \to \psi_{no}^2(0)t_0 - \psi_{no}(0)\nabla^2\psi_{no}(0)t_1 \quad (22)$$
Now
$$\nabla^2\psi_{no}(0) = -\frac{n+3/2}{b^2}\psi_{no}(0) \quad (23)$$
Thus
$$I_{no} = \psi_{no}^2(0)\left(t_0 + \frac{n+3/2}{b^2}t_1\right) = \psi_{no}^2(0)t_0\left(1 - \frac{n+3/2}{b^2}(a^2+c^2)\right) \quad (24)$$
Also,
$$I_{11} = [\nabla\psi_{11}(0)]^2 t_2 = \frac{1}{2}\eta t_0 \frac{a^2-c^2}{b^2} \quad (25)$$

It is interesting to list here ratios of the Talmi integrals in the Skyrme approximation:
$$I_{20}/I_{00} = (\lambda-\mu)^2 + \frac{3}{2}(1-\lambda-\mu)^2 \to \frac{3}{2}\times\left(1 - \frac{a^2+c^2}{b^2}\right) \quad (26a)$$
$$I_{22}/I_{00} = (\lambda-\mu)^2 \to \mathcal{O}(a^4) \quad (26b)$$
$$I_{00-20}/I_{00} = \sqrt{\frac{3}{2}}(1-\lambda-\mu) \to \sqrt{\frac{3}{2}}\left(1 - \frac{1}{2}\frac{a^2+c^2}{b^2}\right) \quad (26c)$$
$$I_{11}/I_{00} \quad \eta(\lambda-\mu) \to \frac{1}{2}\eta\frac{a^2-c^2}{b^2} \quad (26d)$$

Finally, we discuss two possible applications of non-local interactions in the nuclear shell model.

3. Overlap integrals

Consider the filling of shells in the A = 100 region. Here we have 40 to 45 protons, filling the p1/2 and g9/2 shells, while there are 55 to 60 neutrons which fill the g7/2 and d5/2 shells. There are, of course, interactions between protons and neutrons, and there is some empirical evidence from study of Zr and Mo isotopes that the interaction between the proton g9/2 and neutron g7/2 shells is especially large.[4] As the neutron g7/2 shell fills, the proton g9/2 shell drops in energy. The apparent special overlap between these shells is larger than calculated for a local zero or finite range interaction. It requires that the interaction is short range in phase space; i.e., the combined r and p space. We discuss here the overlap integrals for a simpler case, but one which has the essential physics, namely particles in the oscillator (s,d) shell.

As can be seen from Table 1, for a local interaction, the overlap integrals are only slightly smaller between s and d than between two particles in the same orbits. However, non-locality enhances the difference, (though not by much unless the non-locality range is large). In the extreme case of a zero range interaction with infinite range non-locality, the overlap integral vanishes unless the two particles are in the same spatial state. It is like a zero range interaction in phase space; i.e., the combined p and r space.

4. Convergence of model space expansion

For a finite non-locality range, the two body matrix elements connecting different oscillator shells disappear faster than for a purely local interaction. This should lead to more rapid convergence of shell model energies as we increase the size of the model space. Table 2 indicates how non-locality increases the rate of convergence for the simple case of ^{4}He, where, in lowest approximation, the nucleons are in the oscillator ground state. Such a faster convergence of the matrix elements can be readily understood. Intermediate states correspond to larger momenta. Now, a non-local interaction can be expressed as a momentum dependent one; i.e., it falls off faster with momentum, (not just momentum transfer, as with a local interaction) then a momentum dependent one. Since the intermediate states have larger momenta, the matrix elements involving such states will be reduced.

IV. SUMMARY

We have discussed the evidence for non-locality in the NN interaction, and the effect of a simple form of non-locality with Gaussians on nuclear shell model calculations. It remains to apply these ideas quantitatively to realistic NN interactions which fit scattering data, as well as the deuteron radius. Of course, this requires consideration of the density dependence in the NN interaction[5].

REFERENCES

1. S.Klarsfeld, J. Martorell, J.A. Oteo, M. Nishimura, and D.W. Sprung, Nucl. Phys. A456, 373 (1986)

2. M.W. Kermode, W. van Dijk, D.W.L. Sprung, M.M. Mustafa, and S.A. Moszkowski, J. Phys. G. in press (1990)

3. D. Vautherin, and D.M. Brink, Phys. Rev. C5, 626 (1972)

4. R. Meyer, "Shell Model and Nuclear Structure", pg 227 - 250, edited by A. Covello, World Scientific, Singapore, 1989

5. S.D. Bloom, D.A. Resler, and S.A. Moszkowski, "Contibuted Papers to the Symposium in Honor of Akito Arima: Nuclear Physics in the 1990's", Santa Fe, NM, May 1-5, 1990, p.5; also LLNL preprint, UCRL-102825, February 1990.

TABLE 1
Illustrating that non-locality leads to larger overlap between particles in the same orbit.

$$\lambda = \frac{a^2}{a^2+2b^2} \; ; \; \mu = \frac{c^2}{c^2+2b^2}$$

V_{ds}/V_{dd}		a	0	$\sqrt{\frac{2}{3}}b$	$\sqrt{2}b$	$\sqrt{6}b$	∞
c	μ	λ	0	0.25	0.5	0.75	1
0	0		0.595	0.891	1.022	1.018	1.000
$\sqrt{\frac{2}{3}}b$	0.25		0.561	0.816	0.971	0.998	1.018
$\sqrt{2}b$	0.5		0.496	0.728	0.921	0.971	1.022
$\sqrt{6}b$	0.75		0.192	0.445	0.728	0.816	0.891
∞	1.00		0.000	0.192	0.496	0.561	0.595

V_{ss}/V_{dd}		a	0	$\sqrt{\frac{2}{3}}b$	$\sqrt{2}b$	$\sqrt{6}b$	∞
c	μ	λ	0	0.25	0.5	0.75	1
0	0		2.440	1.386	1.133	1.040	1.000
$\sqrt{\frac{2}{3}}b$	0.25		2.663	1.336	1.065	1.009	1.040
$\sqrt{2}b$	0.5		3.745	1.993	1.316	1.065	1.133
$\sqrt{6}b$	0.75		4.750	3.219	1.993	1.336	1.386
∞	1.00		5.000	4.750	3.745	2.663	2.440

TABLE 2
Illustrating that non-locality speeds up the convergence of the model space expansion.

$a = \sqrt{\frac{2}{3}}b$; $\lambda = \frac{a^2}{a^2+2b^2} = \frac{1}{4}$; $\mu = \frac{c^2}{c^2+2b^2}$

		0	$\frac{1}{4}$	$\frac{1}{2}$	$\frac{3}{4}$
I_{20}/I_{00}	$\frac{3}{2}(1-\lambda-\mu)^2 + (\lambda-\mu)^2$	0.91	0.38	0.16	0.25
I_{00-20}/I_{00}	$\sqrt{\frac{3}{2}}(1-\lambda-\mu)$	0.92	0.61	0.31	0.00
I_{00-40}/I_{00}	$\sqrt{\frac{15}{8}}(1-\lambda-\mu)^2$	0.77	0.34	0.08	0.00
I_{00-60}/I_{00}	$\sqrt{\frac{35}{16}}(1-\lambda-\mu)^3$	0.62	0.18	0.02	0.00

Nuclear Shell Model Calculations

FIGURE 1
Calculated triplet scattering length a and deuteron radius r_D for realistic NN potentials, and comparison with empirical value. From Ref. 1.

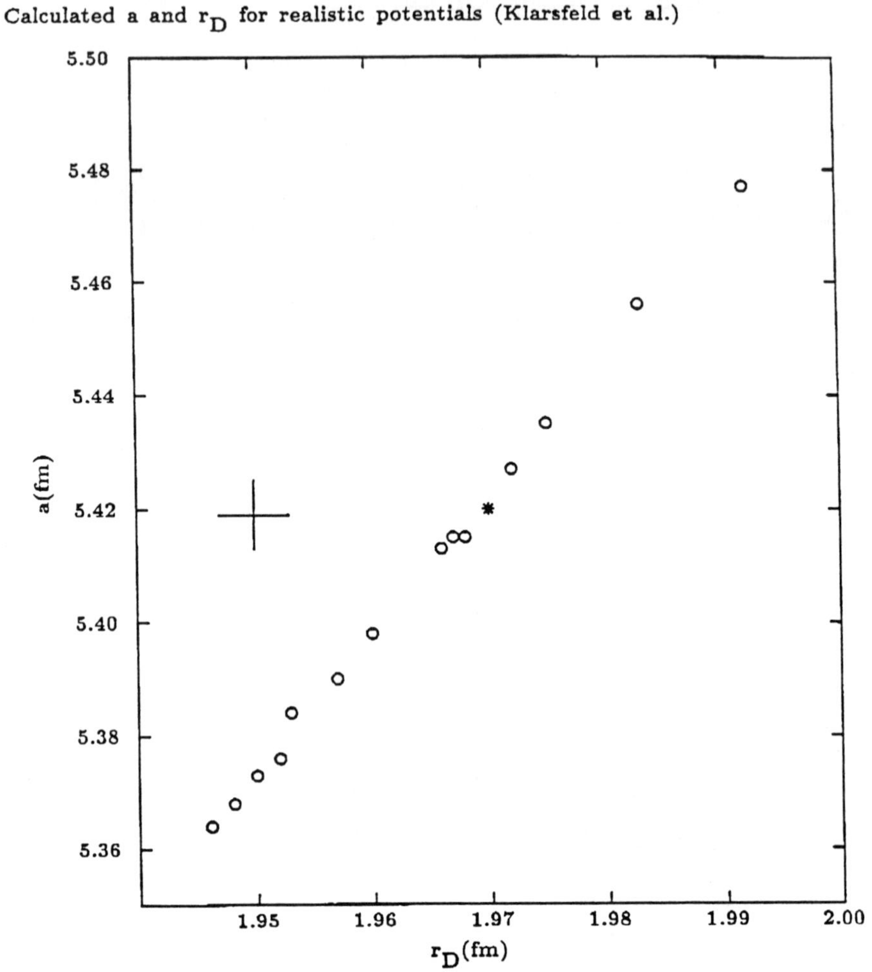

Calculated a and r_D for realistic potentials (Klarsfeld et al.)

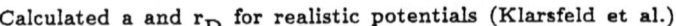

SHELL MODEL STUDY ON MIXED-SYMMETRY 2^+ STATE IN ^{56}Fe

H. NAKADA

Physics Laboratory, School of Medicine, Juntendo University
Hiraga-gakuendai 1-1, Inba-mura, Inba-gun, Chiba 270-16, Japan

T. OTSUKA

Department of Physics, Faculty of Science, University of Tokyo
Hongo 7-3-1, Bunkyo-ku, Tokyo 113, Japan

and

T. SEBE

Department of Applied Physics, College of Engineering, Hosei University
Kajino-cho 3-7-2, Koganei, Tokyo 184, Japan

ABSTRACT

Mixed-symmetric 2^+ state of ^{56}Fe is investigated by shell model in a large configuration space in the pf-shell. We can reproduce experimantal energy levels by using Kuo-Brown interaction. The (e, e') form factors are also reproduced quite well by including the core polarization effect. By analyzing these reliable shell model wavefunctions, we conclude that the 2_2^+ and the 2_4^+ states share the main part of mixed-symmetry state.

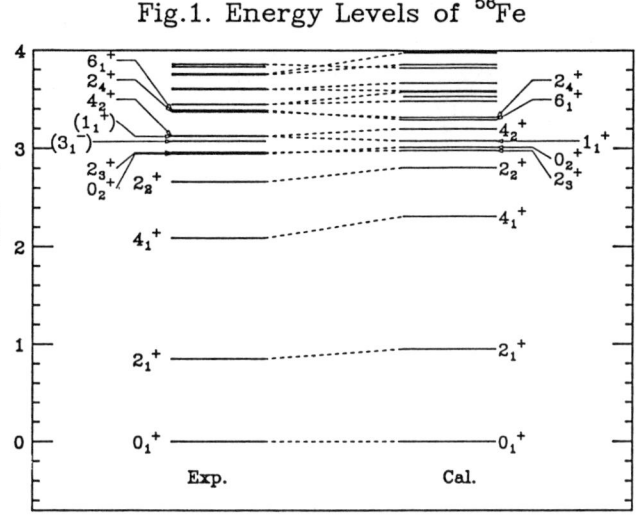

Fig.1. Energy Levels of ^{56}Fe

Mixed-symmetry 2^+ states have been predicted in vibrational nuclei since the birth of the IBM-2[1]. The existence of the 2^+ states of this kind, however, has not been confirmed so far, probably because of too high level density in the relevant energy region. This difficulty could be remedied in lighter nuclei as ^{56}Fe. In fact, experimental research is recently advanced in ^{56}Fe[2-4]. Here we research the mixed-symmetry problem of ^{56}Fe by a realistic shell model calculation.

Assuming ^{40}Ca to be a doubly magic core, ^{56}Fe has 6 valence protons and 10 valence neutrons. We consider configurations as $(0f_{7/2})^{14-k}(0f_{5/2}1p_{3/2}1p_{1/2})^{2+k}$ ($k=0,1,2$). We adopt the Kuo-Brown effective hamiltonian[5], in which single particle energies are determined from experiments at one particle states around ^{40}Ca, and two-body interaction matrix elements are obtained from the G-matrix based on the Hamada-Johnston potential, including $3p$-$1h$ correction by the renormalization. Thus there is no adjustable parameters in the calculation of energy levels. Note that the isospin is conserved, and we restrict ourselves to $T=2$ states. The experimental[6] and calculated energy levels are shown in Fig.1. It is demonstrated that the spectrum seen in experiments is excellently reproduced. Typical value of discrepancy is only 0.1MeV up to $Ex<4$MeV. The B(E2) values are also reproduced with the effective charges of $e_\pi^{eff} = 1.4e$, $e_\nu^{eff} = 0.9e$.

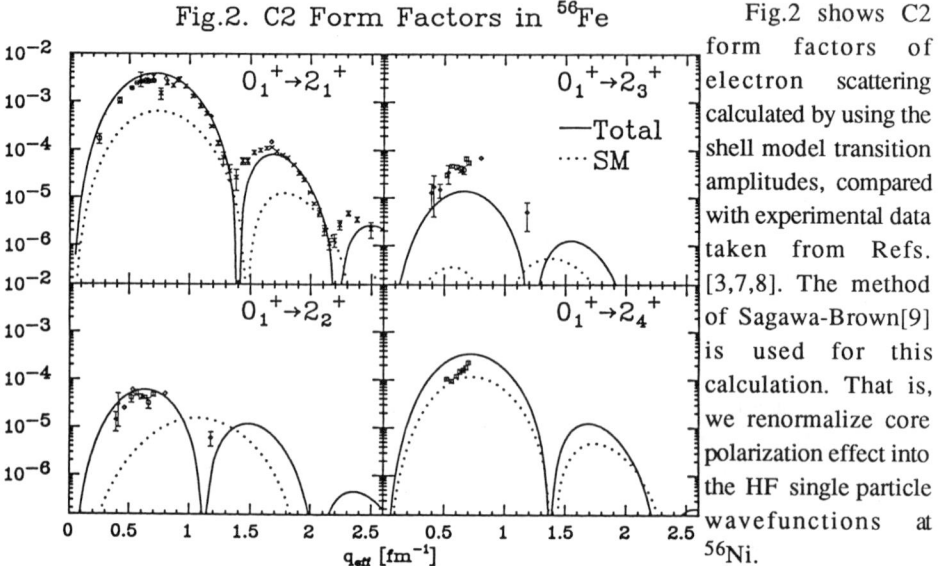

Fig.2. C2 Form Factors in ^{56}Fe

Fig.2 shows C2 form factors of electron scattering calculated by using the shell model transition amplitudes, compared with experimental data taken from Refs. [3,7,8]. The method of Sagawa-Brown[9] is used for this calculation. That is, we renormalize core polarization effect into the HF single particle wavefunctions at ^{56}Ni.

This calculation includes no adjustable parameter again. The form factors of the excitation to the 2_1^+ and 2_2^+ states are reproduced very well. Though there are not so many experimental data of the form factor to the 2_4^+, it should be noticed that the height of the first peak, which is higher than that of the 2_2^+ and the 2_3^+ state, is reproduced. This nature cannot be explained by the wavefunction of Ref.[3] nor [10]. The form factor to the 2_3^+ state is not reproduced so well. This is probably due to that the 2_3^+ state does not have the collectivity. Therefore the form factor of this state may be sensitive to details of its wavefunction.

In order to inspect the present shell model eigenstates in more detail, the wavefunctions of ^{56}Fe are expanded in terms of products of proton and neutron wavefunctions. Table 1 shows overlaps between direct products of the following $\tilde{S}$- or $\tilde{D}$- states and some eigenstates of ^{56}Fe,

$|\tilde{S}_\rho\rangle = \Sigma_i x_{i,\rho}^{(0)} |\varphi_\rho(0_i^+)\rangle$, $|\tilde{D}_\rho\rangle = \Sigma_i x_{i,\rho}^{(2)} |\varphi_\rho(2_i^+)\rangle$,

where $|\varphi_\rho\rangle$ denotes an appropriate proton or neutron basis. The coefficient $x_{i,\rho}^{(J)}$ is determined from the 0_1^+ and the 2_1^+ states of ^{56}Fe. This leads to a generalization of the OAI mapping[1] when we proceed to the correspondence between $\tilde{S}$-, $\tilde{D}$-states and s-, d-boson states. The 0_1^+ is an almost pure $\tilde{S}$-, $\tilde{D}$-state. The 2_1^+ is primarily in the $\tilde{S}\tilde{D}$-subspace, and amplitudes of $|\tilde{S}_\pi\tilde{D}_\nu\rangle$ and $|\tilde{D}_\pi\tilde{S}_\nu\rangle$ components in this state are almost the same. This fact supports the totally symmetric nature of the 2_1^+ state. On the other hand, the 2_2^+ and the 2_4^+ states have large amplitudes for $|\tilde{S}_\pi\tilde{D}_\nu\rangle$ and $|\tilde{D}_\pi\tilde{S}_\nu\rangle$ components with opposite sign.

Table 1. Overlaps between eigenstates and $\tilde{S}$-, $\tilde{D}$-states

state	$|\tilde{S}_\pi\tilde{S}_\nu\rangle$	$|\tilde{S}_\pi\tilde{D}_\nu\rangle$	$|\tilde{D}_\pi\tilde{S}_\nu\rangle$	$|\tilde{D}_\pi\tilde{D}_\nu\rangle$	$\tilde{S}\tilde{D}$-prob.
0_1^+	0.74			0.53	83%
2_1^+		0.54	0.56	0.37	74%
2_2^+		0.38	−0.35	−0.09	28%
2_3^+		0.10	0.03	−0.09	2%
2_4^+		−0.27	0.27	−0.11	16%

It is of interest that the present realistic interaction hardly mix a symmetric state $[|\tilde{D}_\pi\tilde{D}_\nu\rangle]^{(2)}$ into these states. In other words, the realistic interaction seems to conserve the F-spin-like symmetry. Therefore we conclude that the mixed-symmetry 2^+ state appears as one of the basic modes, and that the 2_2^+ and the 2_4^+ states share the mode in ^{56}Fe, contrary to the report in Refs.[2,3]. The remaining half of the mixed-symmetry component should be highly fragmented over wide energy region or go to much higher energy. The 2_3^+ is a non-collective state. It is confirmed that about half of this state is a product of the $\tilde{S}_\pi$ and a neutron non-collective 2^+ state.

In summary, we can understand low-lying 2^+ states of ^{56}Fe in a unified way by large-scale shell model calculation. It is found that the 2_2^+ and the 2_4^+ states share the mixed-symmetry, while the 2_3^+ corresponds to a non-collective mode.

References

[1] T.Otsuka, A.Arima and F.Iachello, Nucl. Phys. A309(1978)1; T.Otsuka, A.Arima, F.Iachello and I.Talmi, Phys. Lett. B76(1978)139; T.Otsuka, Ph.D. Thesis (Univ. of Tokyo, 1979)
[2] S.A.A.Eid et al., Phys. Lett. B166(1986)267; S.P.Collins et al., J. Phys. G15(1989)321
[3] G.Hartung et al., Phys. Lett. B221(1989)109
[4] J.Takamatsu and M. Fujiwara, private communication
[5] T.T.S.Kuo and G.E.Brown, Nucl. Phys. A114(1968)241
[6] H.Junde et al., Nucl. Data Sheets 51(1987)1
[7] R.J.Peterson et al., Nucl. Phys. A153(1970)610
[8] J.Heisenberg et al., Nucl. Phys. A164(1971)353
[9] H.Sagawa and B.A.Brown, Nucl. Phys. A430(1984)84
[10] P.Halse, Phys. Rev. C41(1984)2340

MULTIPARTICLE-MULTIHOLE CONFIGURATION MIXING WITHIN THE NEUTRON-PROTON INTERACTING BOSON MODEL

A. F. Barfield and B. R. Barrett[*]
University of Arizona, Tucson, AZ 85721

ABSTRACT

The IBM-2 + Configuration Mixing method has been expanded, in order to consider several multiparticle-multihole (np-mh) configurations simultaneously. The various np-mh configurations obtained by exciting pairs of protons and/or neutrons across major shell gaps are treated in a consistent manner, along with the normal configuration for a nucleus. As a test, the method is applied to the nucleus ^{192}Hg. Previously determined parameters are utilized for the normal (0p-2h) proton configuration and the expected shape-coexisting (2p-4h) proton configuration. Three other excited np-mh configurations are included, with a consistent set of Hamiltonian parameters.

INTRODUCTION

The two-configuration-mixing method of Duval and Barrett[1], within the neutron-proton Interacting Boson Model (IBM-2), has been successful in describing shape coexistence in nuclei, as seen, for example, in the light isotopes of mercury[1,2]. Recent calculations by Kaup and Barrett[3], using a schematic model, have indicated that shape coexistence may favor multiparticle-multihole (np-mh) excitations over (2p-2h) excitations. In addition, extensive data on superdeformed (SD) bands are now available in several mass regions, and it is of interest to test the plausibility of describing these bands in terms of collective (np-mh) excitations. We have extended the IBM-2 + Configuration-Mixing model in order to treat the ground-state-band (gsb) configuration and several excited (np-mh) configurations within the same formalism.

THEORY

The original two-configuration-mixing method[1] considers the normal configuration and a second configuration in which one pair of protons or neutrons is excited across a major shell gap. In the case of $_{80}$Hg, the shape-coexisting deformed band seen in the light isotopes is built on a (2p-4h) proton configuration, resulting from a proton-pair excitation across the Z = 82 gap. The normal (0p-2h) configuration is described by one proton boson (denoted by $N_\pi = 1$ or 1π), along with the appropriate neutron boson number, N_ν. For simplicity, the (2p-4h) configuration is described by three equivalent proton bosons, denoted by $N_\pi = 3$ (or 3π), with the same neutron boson number. The extended model allows the possible simultaneous

[*]Work supported in part by NSF grant PHY87-23182

Table I. IBM-2 parameters and pair-excitation energies. All values are given in MeV, except χ_ν and χ_π, which are dimensionless. Also, $\alpha = 0.15$, $\beta = 0.10$, $\xi_2 = 0.15$, and $\xi_1 = \xi_3 = -0.30$.

	ϵ	κ	χ_ν	χ_π	$C_{0\nu}$	$C_{2\nu}$	$C_{4\nu}$	Δ
1π	0.68	-0.165	0.4	1.0	0.63	0.05	0.14	
3π	0.35	-0.145	0.4	-1.3	0.00	0.00	0.04	4.0
5π	0.35	-0.12	0.4	-1.8	0.00	0.00	0.00	6.0
7π	0.35	-0.10	0.4	-2.0	0.00	0.00	0.00	8.0
9π	0.35	-0.08	0.4	-2.0	0.00	0.00	0.00	10.0

consideration of other excited proton (and/or neutron) configurations as well, such as 4p-6h ($N_\pi = 5$ or 5π), 6p-8h ($N_\pi = 7$ or 7π), and 8p-10h ($N_\pi = 9$ or 9π).

Standard IBM-2 calculations[4] are done separately for each configuration, and the results are combined with a mixing Hamiltonian. For simplicity, it is assumed that coupling occurs only between configurations that differ by two proton bosons or two neutron bosons. The mixing interaction thus acts only between adjacent configurations, and has the form:

$$V_{mix} = \alpha [s_\rho^\dagger s_\rho^\dagger]^{(0)} + \beta [d_\rho^\dagger d_\rho^\dagger]^{(0)} + h.c. , \qquad (1)$$

where $\rho = \pi$ or ν. The diagonal elements of the Hamiltonian matrix are given by:

$$H_{diag} = \lambda_i + \Delta_{i-1} , \qquad (2)$$

where λ_i is an IBM-2 energy eigenvalue for the ith configuration, and Δ_i is the energy needed to excite i proton (or neutron) pairs.

RESULTS AND CONCLUSIONS

The method is applied to the nucleus ^{192}Hg. Five proton configurations are considered, $N_\pi = 1, 3, 5, 7$, and 9, all with $N_\nu = 7$. The parameters are given in Table I. (The IBM-2 Hamiltonian is given in Refs. 1,2,4.) For simplicity, only the gs bands of each configuration are included in the calculations, up to J = 20.

The 1π (0p-2h) and 3π (2p-4h) parameter values are taken from a previous calculation[5] and result from an overall fit to the even mercury isotopes, $182 \le A \le 198$. The parameters for the remaining configurations are changed as little as possible from the 3π values, consistent with semi-microscopic considerations.

The mixing strengths α and β and the pair-excitation energy Δ_1 are taken from Ref. 5. The multi-pair excitation energies $\Delta_2 - \Delta_4$ are estimated, relative to Δ_1, from the Nilsson diagram for protons, $50 \le Z \le 82$, assuming the normal configuration to have zero deformation and the excited configurations to have deformation $\epsilon \approx 0.27$.

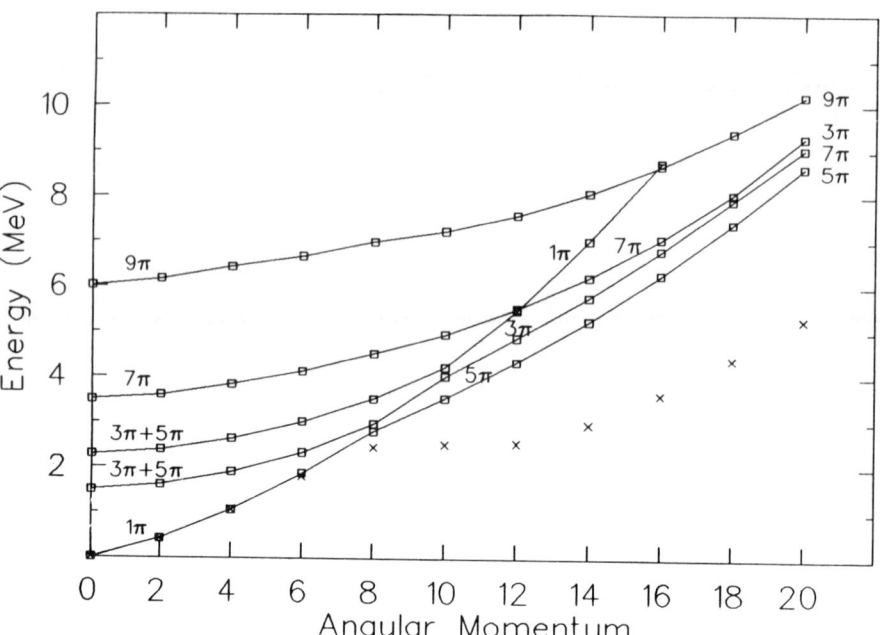

Fig. 1. Energies after mixing for the gsb of five configurations. The calculated bands, denoted by squares, are labelled by the major component(s), and the crosses denote the experimental yrast band[6].

The results are shown in Fig. 1. Although none of the bands have the characteristics of a SD band, it is possible that the inclusion of excited neutron configurations, as well, may result in such a band. In any case, we emphasize that the present calculations are basically a test of the model.

REFERENCES

1. P. D. Duval and B. R. Barrett, Nucl. Phys. **A376**, 213 (1982).
2. A. F. Barfield, et al., Z. Phys. **A311**, 205 (1983).
3. U. Kaup and B. R. Barrett, Phys. Rev. **C42**, 981 (1990).
4. A. Arima and F. Iachello, in "Advances in Nuclear Physics," eds. J. W. Negele and E. Vogt, Vol. 13 (Plenum, N. Y., 1984).
5. A. F. Barfield, Contributed papers, International Conference on Contemporary Topics in Nuclear Structure Physics, Cocoyoc, Mexico, 1988, p. 7.
6. H. Hübel, et al., Nucl. Phys. **1453**, 316 (1986).

EXCITATION MODES IN NON-AXIAL NUCLEI

Amiram Leviatan and Joseph N. Ginocchio
Theoretical Division, Los Alamos National Laboratory Los Alamos, NM 87545

ABSTRACT

Excitation modes of non-axial quadrupole shapes are investigated in the framework of interacting boson models. Both γ-unstable and γ-rigid nuclear shapes are considered for systems with one type of boson as well as with proton-neutron bosons.

Quadrupole distortions in nuclei allow for deviations from axial symmetry. There are some experimental indications for the onset of triaxiality in transitional nuclei and for its relevance for high-spin states. In this work we examine two kinds of non-axial shapes: γ-unstable and γ-rigid. The analysis is carried out in the framework of algebraic models with one type (IBM-1) and two types (IBM-2) of monopole ($s^\dagger$) and quadrupole ($d^\dagger$) bosons.[1] We employ a recently suggested method for identifying the intrinsic and collective components of algebraic Hamiltonians and their normal modes.[2-4]

In IBM-1, the expectation value of the Hamiltonian in a coherent state,[5] which is a condensate of N bosons,

$$|N;\beta,\gamma> \; = \; (N!)^{-1/2}[b_c^\dagger(\beta,\gamma)]^N|0> \tag{1}$$

$$b_c^\dagger \; = \; (1+\beta^2)^{-1/2}[\beta\cos\gamma \cdot d_0^\dagger + \beta\sin\gamma \cdot \frac{1}{\sqrt{2}}(d_2^\dagger + d_{-2}^\dagger) + s^\dagger] \tag{2}$$

defines an energy surface $E(\beta,\gamma)$. The global minimum of E, denoted (β_0,γ_0), determines the equilibrium condensate $|N;\beta_0,\gamma_0>$ which acts as an intrinsic state for the ground band. This intrinsic state serves to separate the Hamiltonian (H) into intrinsic (H_{int}) and collective (H_c) parts. H_{int} is defined to have $|N;\beta_0,\gamma_0>$ as an exact zero-energy eigenstate and to have the same shape for its energy surface as that of the full H. H_c is defined through the relation $H = H_{int} + H_c$. To discuss excitation modes one introduces the following deformed bosons which complement $b_c^\dagger$ (2) in forming a complete orthonormal basis $b_i^\dagger(\beta,\gamma)$ $i = c,\beta,\gamma,x,y,z$

$$b_\beta^\dagger \; = \; (1+\beta^2)^{-1/2}[\cos\gamma \cdot d_0^\dagger + \sin\gamma \cdot \frac{1}{\sqrt{2}}(d_2^\dagger + d_{-2}^\dagger) - \beta s^\dagger] \tag{3a}$$

$$b_\gamma^\dagger \; = \; \cos\gamma \cdot \frac{1}{\sqrt{2}}(d_2^\dagger + d_{-2}^\dagger) - \sin\gamma \cdot d_0^\dagger \tag{3b}$$

$$b_x^\dagger \; = \; \frac{1}{\sqrt{2}}(d_1^\dagger + d_{-1}^\dagger); \quad b_y^\dagger \; = \; \frac{1}{\sqrt{2}}(d_1^\dagger - d_{-1}^\dagger); \quad b_z^\dagger \; = \; \frac{1}{\sqrt{2}}(d_2^\dagger - d_{-2}^\dagger). \tag{3c}$$

Selecting in (2) and (3) the equilibrium deformations (β_0,γ_0) produces the "appropriate" basis for a given H. The deformed bosons in (3) represent excitations of the condensate $|N;\beta_0,\gamma_0>$ (2) which involve β-γ vibrations and x,y,z rotations. Some of these excitations are "spurious" modes (i.e. degenerate with $|N;\beta_0,\gamma_0>$ - the exact ground-state of H_{int}) while others are intrinsic modes. The former are Goldstone bosons associated with spontaneously broken symmetries in $|N;\beta_0,\gamma_0>$ and signal collective motions. The intrinsic modes are obtained by rewritting H_{int} in terms of the "appropriate" basis and preforming a Bogoliubov treatment suitable for large N. The leading order (in N) Bogoliubov image of H_{int} (denoted H^B_{int}) displays harmonic oscillators only in the intrinsic modes which are therefore decoupled from the "spurious" Goldstone modes. The distinction between these two types of modes will become clearer in the discussion below and is elaborated in more detail in refs. [2-4,6].

To describe rigid triaxial shapes ($\beta_0 > 0$, $0 < \gamma_0 < \pi/3$) in the IBM-1, the Hamiltonian is obliged to contain three-body interactions.[6] The energy surface of the corresponding H_{int} exhibits a γ dependence of the form: $\cos^2 3\gamma$ and $\cos 3\gamma$. The leading order Bogoliubov image H^B_{int} takes the form

$$H^B_{int} = \lambda_s^{(\beta\gamma)}\omega_s^\dagger\omega_s + \lambda_a^{(\beta\gamma)}\omega_a^\dagger\omega_a. \qquad (4)$$

There are two intrinsic modes ($\omega_s^\dagger$ and $\omega_a^\dagger$) for a rigid triaxial shape. They are obtained by diagonalizing a 2×2 matrix [6] and involve symmetric (s) and antisymmetric (a) combinations of β- (3a) and γ- (3b) bosons. The three Goldstone bosons missing from (4) are $b_x^\dagger, b_y^\dagger, b_z^\dagger$ (3c) associated with rotations about the x,y,z axes, which are spontaneously broken in the triaxial condensate.

For a γ- unstable shape, the corresponding H_{int} is an O(5) scalar (O(5) is the group generated by $(d^\dagger\tilde{d})^{(L)}$, $L = 1,3$) and the associated energy surface is independent of γ. H^B_{int} for this case takes the form

$$H^B_{int} = \lambda^{(\beta)}b_\beta^\dagger b_\beta. \qquad (5)$$

Only the β- boson survives here as an intrinsic mode. The x-y-z- (3c) and γ- (3a) bosons are all missing from H^B_{int} (5) and correspond to the Goldstone bosons of the spontaneously broken O(5) symmetry.

In IBM-2 one encounters monopole and quadrupole bosons for protons (π) and neutrons (ν). Two sets of deformed bosons $b^\dagger_{i,\rho}$ ($i = c,\beta,\gamma,x,y,z$; $\rho = \pi,\nu$) are introduced.[4] The proton (neutron) basis is obtained from expressions (2)-(3) by inserting β_π,γ_π (β_ν,γ_ν) and using proton (neutron) bosons. The neutron basis is also rotated with respect to the proton basis via three Euler angles $\Omega \equiv (\phi,\theta,\psi)$ to account for the relative orientation between the two quadrupole shapes. The intrinsic state is now a product of π-ν condensates and the intrinsic-collective resolution of the Hamiltonian exists. The energy surface depends in general on seven variables: $\beta_\rho,\gamma_\rho,\Omega$. In the present work we focus on deformed ($\beta_\rho > 0$) non-axial but aligned ($\Omega = 0$) shapes.

In IBM-2, the presence of π and ν degrees of freedom allow for rigid triaxial shapes already with one- and two- body interactions. The possible aligned triaxial shapes obey $0 < \gamma_\pi \neq \gamma_\nu < \pi/3$. The energy surface (for $\Omega = 0$) of the relevant H_{int} displays a γ_ρ dependence of the form: $\cos(\gamma_\pi - \gamma_\nu)$, $\cos^2(\gamma_\pi - \gamma_\nu)$,

$\cos(\gamma_\pi + 2\gamma_\nu)$ and $\cos(2\gamma_\pi + \gamma_\nu)$. The diagonalized form of H^B_{int} is

$$H^B_{int} = \sum_{\alpha=1-4} \lambda_\alpha^{(\beta\gamma)} \Omega_\alpha^\dagger \Omega_\alpha + \lambda_a^{(x)} b^\dagger_{xa} b_{xa} + \lambda_a^{(y)} b^\dagger_{ya} b_{ya} + \lambda_a^{(z)} b^\dagger_{za} b_{za}. \quad (6)$$

There are 7 intrinsic modes for such triaxial shapes. The modes $\Omega_\alpha^\dagger$ ($\alpha = 1-4$) are obtained by diagonalizing a 4×4 matrix involving the bosons $b^\dagger_{\beta\pi}$, $b^\dagger_{\gamma\pi}$, $b^\dagger_{\beta\nu}$, $b^\dagger_{\gamma\nu}$. The 3 remaining intrinsic modes are one-dimensionl scissors modes corresponding to counter-oscillations of protons against neutrons about each of the axes. These modes (denoted by (xa), (ya), (za)) are compsed of antisymmetric combinations of x_ρ, of y_ρ and of z_ρ bosons ($\rho = \pi, \nu$). The 3 symmetric combinations of these bosons are missing from (6) and are the Goldstone bosons of $O(3)$ rotations broken spontaneously in the triaxial π-ν condensate.

As an example of a γ-unstable aligned shape we consider a π-ν configuration with equal but otherwise arbitrary equilibrium values of γ_ρ. The corresponding H_{int} is a scalar under the $O_{\pi+\nu}(5)$ group (generated by a direct sum of the generators of $O_\pi(5)$ and $O_\nu(5)$). Its energy surface (for $\Omega = 0$) exhibits a γ_ρ dependence of the form: $\cos(\gamma_\pi - \gamma_\nu)$ and $\cos^2(\gamma_\pi - \gamma_\nu)$. It should be noted that unlike in IBM-1, in IBM-2 the concept γ-unstable does not mean γ-independence. In π-ν systems, γ-unstable means that the energy surface depends only on a particular combination of the γ_ρ. In the case considered here, the energy surface (for $\Omega = 0$) depends only on $\gamma_\pi - \gamma_\nu$ but not on $\gamma_\pi + \gamma_\nu$. The corresponding H^B_{int} has the form

$$H^B_{int} = \lambda_s^{(\beta)} b^\dagger_{\beta s} b_{\beta s} + \lambda_a^{(\beta)} b^\dagger_{\beta a} b_{\beta a} + \lambda_a \left[b^\dagger_{xa} b_{xa} + b^\dagger_{ya} b_{ya} + b^\dagger_{za} b_{za} + b^\dagger_{\gamma a} b_{\gamma a} \right]. \quad (7)$$

There are 6 intrinsic modes in this case. Two one-dimensional β modes involving symmetric (βs) and antisymmetric (βa) combinations of β_π and β_ν bosons. In addition there are 4 degenerate intrinsic modes (denoted (xa), (ya), (za), (γa)) involving antisymmetric combinations of x_ρ, of y_ρ, of z_ρ and of γ_ρ bosons. The 4 symmetric comninations of the latter bosons are missing from (7) and correspond to the Goldstone bosons of the broken $O_{\pi+\nu}(5)$ symmetry.

The selectivity of E0 transitions may be used to distinguish different shapes. The one-body E0 transition oprator is a scalar under $O(5)$ in IBM-1 and under $O_{\pi+\nu}(5)$ in IBM-2. As such, for a γ-unstable shape it can connect the $J = 0^+$ ground state only with the $J = 0^+$ state belonging to the β band in IBM-1, and in IBM-2, only with the lowest $J = 0^+$ states in the symmetric- and antisymmetric- β bands. In contrast, for triaxial shapes, the rotational states do not have good $O(5)$ symmetry and the previously mentioned selection rules do not apply. This work is supported by the U.S. Department of Energy.

1. F. Iachello and A. Arima, The Interacting Boson Model (Cambridge Univ. Press, Cambridge, 1987).
2. A. Leviatan, Ann. Phys. <u>179</u>, 201 (1987).
3. A. Leviatan Prog. Part. Nucl. Phys. <u>24</u>, 85 (1990).
4. A. Leviatan and M.W. Kirson, Ann. Phys. <u>201</u>, 13 (1990).
5. J.N. Ginocchio and M.W. Kirson, Nucl. Phys. <u>A350</u>, 351 (1980).
6. A. Leviatan and B. Shao, Phys. Lett. <u>B243</u>, 313 (1990).

SYSTEMATICS OF EFFECTIVE CHARGES OF PROTON AND NEUTRON BOSONS

A. Wolf[1] and R. F. Casten[2]
[1]Negev Research Center, Beer-Sheva, Israel
[2]Brookhaven National Laboratory, Upton, New-York, USA

ABSTRACT

Experimental $B(E2; 0_1^+ \rightarrow 2_1^+)$ values for 54 even-even nuclei in the range A = 90 - 200 were used to deduce effective charges of proton, neutron bosons. The results indicate that for deformed nuclei, the ratio e_n / e_p is in general agreement with the schematic Bohr and Mottelson prediction. For vibrational nuclei in the A = 100 and A = 140 regions a very significant enhancement of this ratio is found.

INTRODUCTION

It is well known that the addition of valence nucleons to the closed shells of a nucleus is accompanied by the appearance of a non-spherical field and thus the presence of an additional quadrupole moment of the protons inside the closed shell. This quadrupole core polarization effect can be renormalized as an effective charge of the nucleons added to the core. An estimate of this effect can be obtained following Bohr and Mottelson[1] and noting that the induced quadrupole moment due to the distorted field is of the order $\frac{Z}{A}Q_{sp}$, where Q_{sp} is the single particle quadrupole moment. Thus, the effective charges of the proton, neutron are expected to be :

$$e_p^{eff} = (1 + \frac{Z}{A}) e \qquad (1)$$

$$e_n^{eff} = \frac{Z}{A} e , \qquad (2)$$

respectively. According to this schematic model [1] :

$$e_n^{eff} / e_p^{eff} = Z / (A + Z) \qquad (3)$$

Experimental information on the effective charges of protons, neutrons outside closed shells is of importance for our understanding of the core polarization effect. Such information can be extracted from data on E2 transition probabilities. In this work we present a systematic analysis of $B(E2 ; 0_1^+ \rightarrow 2_1^+)$ experimental data in even-even nuclei in the framework of the IBA-2 model.

RESULTS AND DISCUSSION

We use exact and approximate analytic formulas[2,3] to extract the effective charges e_p, e_n of the proton, neutron bosons respectively. We used 54 B(E2) experimental values for nuclei in the range A = 90 - 200 from the compilation of Raman et al.[4] These values were fitted to the analytical formulas using a least squares fit procedure. In order not to obscure local variations of e_p, e_n as a function of A, we divided the data in 11 groups of neighboring nuclei and the fits were performed separately for each group, using the appropriate analytical formula[2,3] for each group. This procedure is different from that of Raman et al.[5], who tried to find a single e_n / e_p value for a large number of nuclei. In Figure 1 we present as an example the fits for two groups of isotopes. The resulting e_p, e_n as a function of the average mass A in each group are presented in Figure 2. We see that around A = 100, the behavior of the effective charges is rather irregular, while for A > 140 systematic trends can be observed : e_p increases and e_n decreases with mass number. The behavior of e_n / e_p vs. A is presented in Figure 3. For most deformed nuclei, e_n / e_p = 0.3 - 0.5, in agreement with the expected ratio[1] e_n / e_p = Z /(A + Z) ~ 0.3 (eq. (3) above). Vibrational nuclei in the regions A = 100 and A = 140 have e_n / e_p = 1.5 - 2 . The reason for the enhanced e_n / e_p ratio in these nuclei is not completely understood at present, although it is a clear feature of the data. It may be related[6] to the fact that e_n, e_p contain a (length)^2 factor and that in these nuclei the neutrons occupy orbitals significantly farther away from the center of the nucleus than the protons. Thus their effective charge is enhanced by the dimensions of their orbits.

The systematic trends presented in this work warrant detailed theoretical investigation in order to obtain information on the nuclear core polarization effect, which is closely related to the effective charges.

Work supported by the US DOE under Contract DE-AC02-76CH00016.

REFERENCES

1. A. Bohr and B. R. Mottelson, Nuclear Structure (Benjamin, Reading, 1975)vol I , p. 334.
2. J. N. Ginocchio and P. Van Isacker, Phys. Rev. C33, 365 (1986).
3. R. F. Casten and A. Wolf, Phys. Rev. C35, 1156 (1987).
4. S. Raman et al., At. Data Nucl. Data Tables 36, 1 (1987).
5. S. Raman et al., Phys Rev. C37, 805 (1988).
6. W. D. Hamilton et al., Phys. Rev. Lett. 53, 2469 (1984).

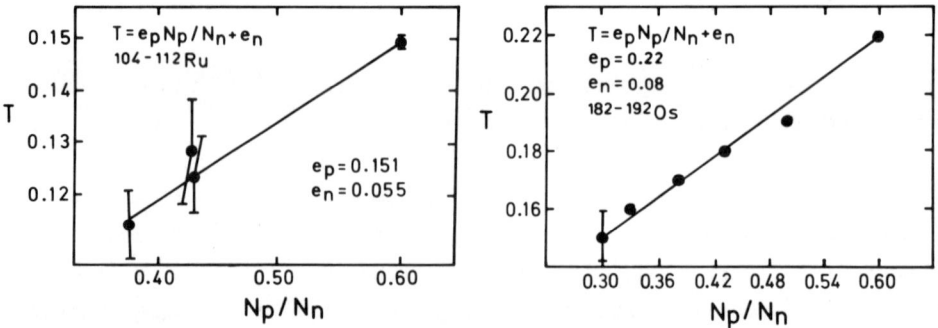

Fig. 1 Fits of experimental B(E2) data to analytical formulas to deduce e_p, e_n for $^{104-112}$Ru and $^{182-192}$Os isotopes.

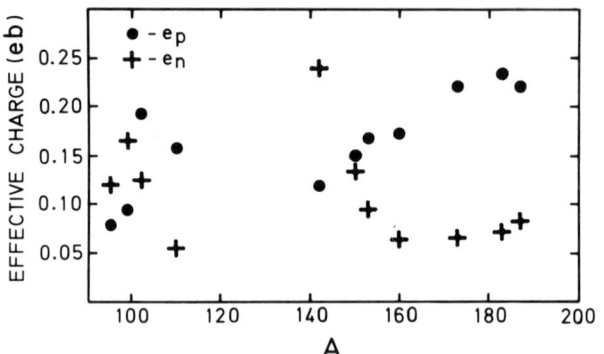

Fig. 2 Systematics of e_p, e_n vs. A. Each symbol represents the e_p or e_n extracted for one of the groups of nuclei and is plotted at the average mass of that group.

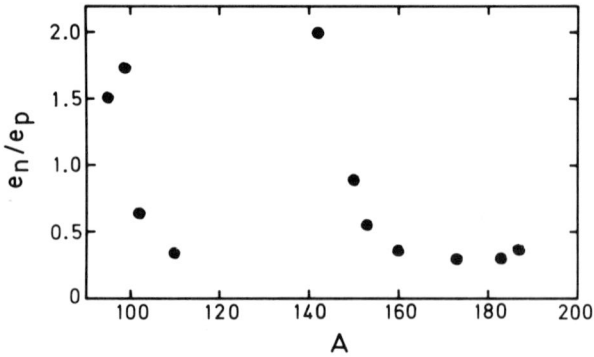

Fig. 3 Systematics of e_n / e_p vs. A. See caption to Fig. 2.

A CONSISTENT DESCRIPTION OF THE MIXING OF INTRUDER AND NORMAL STATES IN ^{114}Cd.

J. JOLIE
Institut Laue-Langevin, 156X, 38042 Grenoble Cedex, France.

K. HEYDE
Institute for Theoretical Physics and Institute for Nuclear Physics, Proeftuinstraat 86, B-9000 Gent, Belgium.

ABSTRACT

Describing the coexistence of normal and intruder states within the framework of the neutron-proton Interacting Boson Model, some inconsistency in the choice of the neutron-proton quadrupole force strength is pointed out. It is shown that this can be removed by rigorously calculating the unperturbed energy of the intruding configuration. This is illustrated by a calculation of the excited levels in ^{114}Cd.

A description of the coexistence of normal and intruder states such as found in nuclei near closed shells should take into account several aspects. It should be able to describe the "normal" spherical states as observed at low energy. Then, it has to describe the structure of the more deformed intruder states and finally it should be able to describe the strong mixing between both. Moreover to have a reliable model it should also be able to explain, on the basis of simple physical arguments, the behavior of the intruder states in a chain of isotopes. These severe constraints limit the possible models but also imply a large experimental effort in order to provide the verification of these models.

For intruder states, observed in medium-heavy nuclei, where collective behavior builts up rapidly, such a model has been developed over the last decade combining shell-model arguments, the particle-core coupling model and the interacting boson model. There the intruder excitations in even-even nuclei are interpreted as a 2p-2h excitation through the closed shell. This is corroborated via the existing (^{3}He,n) reaction studies leading to 0^+ and 2^+ states at low energy in even-even Cd and Sn nuclei. Here e.g. (d,^{3}He) reactions, starting from even-even Cd nuclei into odd-mass In nuclei, show the 1p-1h character of intruder excitations at very low excitation energy ($E_x \leq 1$ MeV). On the basis of shell model arguments one is then able to calculate Δ, the unperturbed excitation energy of the lowest intruder state. The excitation energy of this state is then given by the expression[1]:

$$E_{intr}(0^+) = \Delta + \Delta E_{coll} = 2(\varepsilon_p - \varepsilon_h) - \Delta E_{pairing} + \Delta E_M + \Delta E_{coll}. \quad (1)$$

The first term denotes the unperturbed energy of a 2p-2h excitation. The second term describes the gain in pairing energy and the third the monopole correction. The last term takes account of the extra binding energy gained due to the increase of collectivity and deformation when the number of interacting nucleons increases. This term is dominated by the attractive neutron-proton quadrupole-quadrupole interaction $\kappa Q_\nu.Q_\pi$. To describe the collective excitations of both the normal and intruder states the Interacting Boson Model (IBM-2) is used[2,3]. Then, the normal configurations in ^{114}Cd are described as an interacting system of $N_\nu = 8$ neutron bosons and one $N_\pi = 1$ proton boson describing the 2h states with respect to the Z=50 shell. The intruder configurations are described by $N_\nu = 8$ neutron bosons and $N_\pi = 3$ proton bosons describing the 2p-4h states. The Hamiltonian for both configurations is then diagonalised separately and to the intruder configurations the energy Δ is added. Finally, the lowest states in both configurations are admixed.[3]

The description has been successfully applied to many nuclei. It has, however, a major drawback. This is related to uncertainities occuring as well in the theoretical description as in the experimental knowledge. In general the experimental knowledge of the interaction of intruder and normal configurations is not sufficient to fully test the calculations. Since we are dealing with a complex model for which many interaction strenghts have to be determined, they are in most cases far too many to be able to obtain a quantitative verification of the model. For instance the strength κ can be used to bring the intruder state to its observed energy, since decreasing the kappa of the intruder configuration by 20 keV lowers the intruder configuration by 550 keV. In the past different κ values for both configurations were often used in calculations of intruder states. This should not be the case since, as indicated above, the matrix element of the quadrupole force has a strong and specific dependence on the number of interacting nucleons. By now taking also κ to be strongly dependent on the number of nucleons, part of the argument is lost. Secondly, as shown by Casten, many experimental quantities seem to depend only on the product $N_\pi N_\nu$[4] indicating a rather constant value for κ within a shell.

To investigate these problems we performed a detailed calculation[5] for ^{114}Cd. This nucleus has been investigated by many authors and an almost complete set of E2 reduced matrix elements for the low-lying normal and intruder states is available[6]. In order to avoid a complicated multiparameter fit for this calculation, we intend to obtain as many parameters as possible from the following rules based on physical arguments:

1) The formula (1) is used rigourously in order to calculate Δ the unperturbed energy of the intruder.
2) The collectivity is determined by the number of interacting bosons and thus the intruding configuration will resemble the nucleus with three proton bosons outside the closed shell. Then the parameters used to describe the intruding states can be taken from an earlier study of the Ru isotopes. For the normal states the parameters are obtained from a study of the Cd isotopes.
3) The final adjustment of the parameters should be based on the fact that the trends of the IBM-2 parameters when going from one nucleus to another within a shell are fairly well known on the basis of microscopic studies and phenomenological studies of chains of isotopes, in particular in the region of vibrational nuclei. This means that parameters which are changing rapidly as a function of boson number, such as ε, are allowed to be adjusted. On the other hand parameters which are nearly constant within a shell, such as κ, should be nearly equal for both the $N_\pi=1$ and $N_\pi=3$ configurations.
4) For the description of the electromagnetic decay of the states we determine the parameters for both configurations from experimental quadrupole moments and transitions before they are mixed.

In our calculation of ^{114}Cd we used rule 1 to fix the unperturbed energy of the intruder to be Δ = 4.0 MeV. Then we applied rule 2 to fix all the parameters of the normal and intruding configurations using the parameters derived for ^{114}Cd in ref 2 and for ^{110}Ru in ref 7. Since the mixing of the states is reflected very clearly in their electromagnetic decay, we have to determine the quadrupole operator:

$$T(E2) = e_1[Q_\nu + Q_\pi](N_\pi=1) + e_3[Q_\nu + Q_\pi](N_\pi=3) \qquad (2)$$

with $Q_\rho = (s^\dagger_\rho \tilde{d}_\rho + d^\dagger_\rho s_\rho)^{(2)} + \chi_\rho (d^\dagger_\rho \tilde{d}_\rho)^{(2)}$, which on its own already contains six parameters. Using rule 4 we determine the charges of the quadrupole operator. First the sum $\chi_\nu + \chi_\pi$ for both configurations is fixed by the recently remeasured quadrupole moments $Q(2^+_1)$ for $N_\pi = 1$ and $Q(2^+_2)$ for $N_\pi = 3$. Using furthermore the values obtained for χ_ν in ^{114}Cd for the normal[8] and in ^{110}Ru for the intruder configurations[7], we obtain the values for χ_π. These values are also used to determine the quadrupole operator in the Hamiltonian. The charge e_3 was taken to be 0.103 eb which is the value obtained in the description of the Ru isotopes[7]. Finally the charge e_1 was fitted in order to reproduce the $B(E2;0^+_1 \to 2^+_1)$ and yielded e_1 = 0.086 e.b. The ratio e_3/e_1 = 1.2 obtained in this way is smaller than the one of 1.6 used in ref 2.

Thus using rule 1, 2 and 4 we determined 23 of the 25 parameters in the model before we start the fit. Two parameters left over are the strenghts α and β of the mixing Hamiltonian[3]:

$$H_{mix} = \alpha (s_\pi^\dagger s_\pi^\dagger + s_\pi s_\pi)^{(0)} + \beta (d_\pi^\dagger d_\pi^\dagger + \tilde{d}_\pi \tilde{d}_\pi)^{(0)}. \qquad (3)$$

We took $\alpha = \beta = 0.08$ MeV as obtained in ref 2. Finally we have to chose one parameter $\epsilon(N_\pi=3)$ which is varied in order to reproduce as well as possible the excitation energies and about 25 E2 matrix elements following rule 3. It turns out that one has to lower ϵ from the value of 0.6 MeV, given for ^{110}Ru in ref. 7, to 0.3 MeV. The reduction of ϵ with a factor of 2 is the same as obtained in our description of backbending in the Dysprosium isotopes[9]. When varying around $\epsilon = 0.3$ MeV all features of the quintuplet of states around 1.2 MeV can be reproduced although not exactly for the same value of ϵ. This is due to the fact that all five states are mixing at the same time. In Figure 1 we give, in a schematic way, the different effects of the rules on the excitation energy of the intruder and illustrate the final result by the five highly mixed states near $E_x=1.3$ MeV in ^{114}Cd. More details are given in ref 5.

The calculations we did are most probably the best one can do at present and that can be done for such highly complicated interacting systems which are the Cd nuclei, without doing a physically meaningless multi parameter fit containing 25 parameters to match the 40 data points.

References

1) K. Heyde, J. Jolie, J. Moreau, J. Ryckebush, M. Waroquier, P. Van Duppen, M. Huyse, and J.L. Wood, Nucl. Phys. A466 (1987) 189.
2) K.Heyde, P.Van Isacker, M.Waroquier, G.Wenes, and M.Sambataro, Phys.Rev. C25 (1982) 3160.
3) P. Duval, and B. R. Barrett, Phys. Lett. 100B (1981) 223.
4) R.F.Casten Phys.Lett. 152B (1985) 145, Phys.Rev.Lett. 54 (1985) 1991, and Nucl.Phys. A443 (1985) 1.
5) J. Jolie, and K. Heyde, acc for Phys. Rev. C (1990).
6) C. Fahlander, A. Bäcklin, L Hasselgren, A. Kavka, V. Mittal, L. E. Svensson, B. Varnestig, D. Cline, B. Kotlinski, H. Grein, E. Grosse, R. Kulessa, C. Michel, W. Spreng, H.J. Wollersheim, and J. Stachel, Nucl. Phys. A485 (1988) 327.
7) P. Van Isacker, and G. Puddu, Nucl. Phys. A348 (1980) 124.
8) M. Sambataro, Nucl. Phys. A380 (1982) 365.
9) K. Heyde, J. Jolie, P. Van Isacker, J. Moreau, and M. Waroquier, Phys. Rev. C29 (1984) 1428.

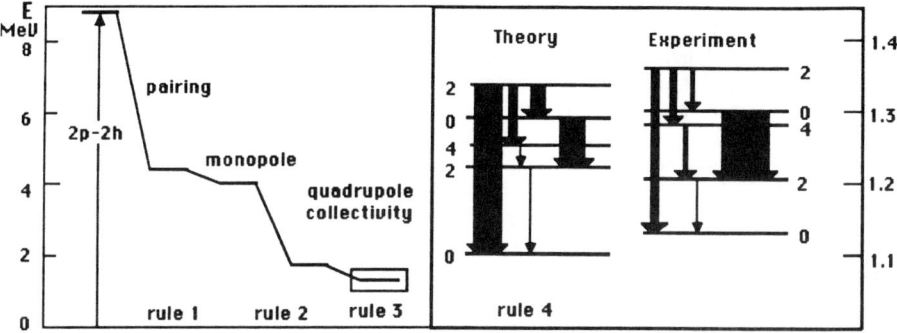

Figure 1: The effect of the different contributions to the energy of the intruder state. On the right the resulting energies and B(E2) values of the highly mixed quintuplet are compared with experiment.

RECENT PROGRESS IN THE APPLICATION OF EXTENDED SUPERSYMMETRIES TO ODD-ODD NUCLEI IN THE AU REGION.

J. JOLIE,
Institut Laue-Langevin, 156X, 38042 Grenoble Cedex, France

The study and understanding of odd-odd nuclei represents a challenging problem to nuclear physicists. The experimental difficulties are mainly connected with the high density of levels in a odd-odd nucleus, typically 20 to 30 levels below 500keV. The theoretical interpretation is complicated because of the interplay between particle and collective degrees of freedom. A model which gives a successful account of the properties of even-even and odd-mass nuclei is the Interacting Boson Model.[1] This model has been shown to be particularly useful and thus seems to provide the ideal framework to tackle the above-mentioned challenge. In extending the IBM to odd-odd nuclei, two different approaches can be followed, in analogy with the treatment of odd-mass nuclei. In the first, a boson-fermion hamiltonian is derived from semi-microscopic arguments; in the second approach, the form of the hamiltonian is restricted by symmetries. Both methods have been extended to odd-odd nuclei. The semi-microscopic IBFFM approach[2] can in principle be applied to many different nuclei but often has to rely on truncations and has also a rather large freedom in the choice of its parameters. The second approach uses symmetry arguments[3]. It has the large disadvantage that it can only be used for very restricted regions of the nuclear mass chart and can only describe those parts of the nuclear excited levels originating from a limited number of shell-model configurations. It has, in contrast, due to the supersymmetric hypothesis, the advantage that it is in a large amount parameter free and analytical soluble. In principle it also allows to test the use of supersymmetry in nuclear physics, since for the odd-odd nucleus it is sometimes possible to fix the complete hamiltonian from the three other members of the supermultiplet. Since the supersymmetric description was proposed five years ago the experimental knowledge of the odd-odd nuclei in the Au region was improved considerably allowing detailed testing of the theoretical description of the odd-odd nuclei in the Au region. To perform real tests of the $U_\nu(6/12) \times U_\pi(6/4)$ extended supersymmetry more wavefunction sensitive observables such as electromagnetic transition rates or transfer reaction amplitudes are derived. These are compare with the experiments.

Most of the details concerning the $U_\nu(6/12) \times U_\pi(6/12)$ supersymmetry can be found in refs 3,4,5. The hamiltonian, function of the Casimir operators of the group chains, has the eigenvalues:

$$E = A [N_1(N_1+5) + N_2(N_2+3) + N_3(N_3+1)] + B [\Sigma_1(\Sigma_1+4) + \Sigma_2(\Sigma_2+2) + \Sigma_3^2] \quad (1)$$
$$+ B' [\sigma_1(\sigma_1+4) + \sigma_2(\sigma_2+2) + \sigma_3^2] + C [\tau_1(\tau_1+3) + \tau_2(\tau_2+1)] + D L(L+1) + E J(J+1).$$

With the reduction rules the excited energy spectrum can then be constructed. Also wavefunctions can be obtained analytically[4,5]. In order to calculated more wavefunction dependent observables, namely transfer reaction amplitudes[6] and electromagnetic transition rates for the odd-odd nucleus, we proceeded partly analytical and partly numerical.

In principle one can obtain for all electromagnetic transition rates an analytic expression. However, in a odd-odd nucleus, the number of possible transitions becomes enormous making the derivation of all the analytic results extremely time consuming. For instance in the theoretical spectrum for ^{198}Au, below 500 keV, one has 157 different B(E2) values and 20 quadrupole to be calculated. Therefore we calculated the electromagnetic decay numerically. The E2 operator for odd-odd nuclei in the $U_\nu(6/12) \times U_\pi(6/4)$ scheme is given by:

$$T^{E2} = e_b(s^\dagger \tilde{d} + d^\dagger s)^{(2)} + e_{f\nu}(-(4/5)^{1/2}(a^\dagger_{1/2\nu}\tilde{a}_{3/2\nu} - a^\dagger_{3/2\nu}\tilde{a}_{1/2\nu})^{(2)}$$
$$-(6/5)^{1/2}(a^\dagger_{1/2\nu}\tilde{a}_{5/2\nu} - a^\dagger_{5/2\nu}\tilde{a}_{1/2\nu})^{(2)}) + e_{f\pi}(a^\dagger_{3/2\pi}\tilde{a}_{3/2\pi})^{(2)} \quad (2)$$

In an odd-odd nucleus the M1 transitions between the excited levels dominate the electromagnetic decay. Unfortunately the magnetic dipole transition rates are not very well described in the interacting boson model. The disagreement with experiment is largest in the IBM-1, since all M1 transitions are forbidden. Despite the use of a F-spin symmetric, in practice an IBM-1 core, one may expect that the description of M1 transitions will be slightly better in the odd-A nuclei and even much better for the odd-odd nucleus as the fermions will largely contribute to the magnetic dipole properties of the nuclear excited levels. We will assume that the creation operator of the ideal fermion $a^\dagger_j$ is in first order the shell-model fermion creation operator $c^\dagger_{nlsj}$. Then we can take the microscopic M1 operator for the fermion part of the M1 operator:

$$T(M1) = (3/4\pi) \{g_b\sqrt{10}(d^\dagger \tilde{d})^{(1)} - \Sigma (3)^{-1/2} <l1/2,j\rho|| g_{l\rho} l_\rho + g_{s\rho} s_\rho$$
$$+ g_{t\rho}(Y_2 \times s_\rho)^{(1)}||l',1/2,j_\rho'>(a^\dagger_{j\rho}\tilde{a}_{j\rho'})^{(1)}(u_{j\rho} u_{j\rho'} + v_{j\rho} v_{j\rho'}) \qquad (3)$$

In (3) the summation runs over the single particle orbits $<l1/2,j\rho|$ considered and over $\rho = \nu, \pi$. The v_j are the BCS occupation probabilities of the considered orbits. This operator includes the tensor part $g_{t\rho}(Y_2 \times s_\rho)^{(1)}$, with $g_{t\rho} \approx 0.01<r^2>g_{s\rho}^{free}$, in order to describe the effect of core polarization.

Recently the excited level scheme of ^{198}Au was studied in detail[7]. This study yielded the branching ratios for the excited levels in ^{198}Au. In an first account of the experiments the gamma-ray intensities were compared[7] with a numerical calculation using the IBFFM approach[2]. Here we calculate the branching ratios in the $U_\nu(6/12) \times U_\pi(6/4)$ using the operators (2) and (3) and the adjusted fit of ^{198}Au as obtained in ref 8. For the E2 effective charges: $e_b = -e_\nu = e_\pi = 0.15$ eb are obtained from previous calculations of ^{194}Pt, ^{195}Pt and ^{197}Au. In order to calculate the magnetic dipole transitions with operator (3) the following g-factors were used: $g_R = 0.4$; $g_{s\nu} = 0.5 g_{s\nu}^{free}$; $g_{s\pi} = g_{s\pi}^{free}$; $g_{l\rho} = g_{l\rho}^{free}$; $g_{t\pi} = 2.3428$ and $g_{t\nu} = -0.8045$. The neutron occupation probabilities were taken from ref 8. The reduction of $g_{s\nu}$ with respect to $g_{s\nu}^{free}$ is determined by the calculation of the known magnetic dipole moment of the 2^- groundstate of ^{198}Au. Using this charges we have calculate the B(M1) and B(E2) values and derive the theoretical branching ratios for the lowest states using the measured energies. In Table 1 we compare the results with experiment and the IBFFM calculation given in 8. The agreement is reasonable with the exception of the excited 0^- states of which the branchings are poorly described. For the first excited state this can be related to the fact that this state in the only one in the IBFFM calculation that has a large contribution from the $\pi s_{1/2}$ orbit[8] which is not considered in the supersymmetric scheme. In contrast to what was found in the IBFFM calculation[8] we find that the experimental $1,2^-$ level at 247keV is 1^-, instead of 2^-. With the branching ratios it is possible to find a one-to-one correspondence between the levels up to 530 keV. Using this assignment refitted the parameters of (1). This yielded A + B= 104, B'=-98.95, C= 38, D=42, and E=-22.75 (all in keV), values very close to the previous ones[8].

From the beginning it was claimed that the best example of $U_\nu(6/12) \times U_\pi(6/4)$ in the Au region should be the quartet of nuclei ^{194,195}Pt and ^{195,196}Au. This due to the fact that the other nuclei in the quartet are good examples of the U(6/12) and the U(6/4) supersymmetries. The main problem with ^{196}Au is the experimental knowledge of its level scheme. Of negative parity only two levels are known the 2^- groundstate and a 12^- state at 595.66 keV. The combination of the lack of experimental knowledge with the expectation that, if it exists, supersymmetry should especially do well for ^{196}Au, makes the investigation of this nucleus a good test of the predictive power of supersymmetry in nuclear structure. Therefore we have predicted the excited levels and transfer reaction amplitudes[9]. In order to obtain a prediction for the negative parity levels of ^{196}Au, we proceeded as follows. From the known level schemes of ^{194}Pt, ^{195}Pt and ^{195}Au we determined all parameters of the Hamiltonian using it's energy eigenvalue expression (1). This is done with a least square fit of (1) to the experimental energies using the quantumnumbers assigned. This fit yielded (all in keV): A = 62.1, B = -2.6, B' = -52.6, C = 49.7, D = 5.0 and E = 6.3.[9] Next we calculated the transfer matrix elements $<L||T||3/2>^2$, for single particle transfer with l=1,3 between the odd-proton groundstate $3/2^+$ and an odd-odd nucleus with the same number of bosons. Then the

J_i	J_f	B(E2)(eb)	B(M1)(n_μ)	BR(exp)	BR(SUSY)	BR(IBFFM)
1_1	2_1	0.0547	0.0485	100	100	100
0_1	2_1	0.031	0	100	0.9	55
	1_1	0	0.3054	77	100	100
1_2	2_1	0.0115	0.0036	45	72	100
	1_1	0.0154	0.0110	19	74	63
	0_1	0	0.0375	100	100	44
4_1	2_1	0.0309	0	100	100	100
3_1	2_1	0.0762	0.0041	100	100	100
	1_1	0.0127	0	15	1.8	0.3
1_3	2_1	0.0445	0.0567	100	100	100
	1_1	0.0444	0.0557	70	46	25
	0_1	0	0.0093	1.6	4	0.6
	1_2	0	0.0186	-	0.35	1
1_4	2_1	0	0.0013	0.4	30	8
	1_1	0	0.0070	-	80	87
	0_1	0	0.0157	100	100	100
	1_2	0.0501	0.0066	8	2.7	0.5
2_2	2_1	0	0.0137	100	100	100
	1_1	0	0.0015	4.4	5.4	93
	0_1	0.0333	0	7.5	1.35	0.
	1_2	0.0499	0.0010	-	0.16	0.5
3_2	2_1	0.0264	0.0252	100	100	100
	1_1	0.0044	0	2.7	0.5	47
	1_2	0.0545	0	-	0.2	0.2
	4_1	0.0402	0.0031	6	0.6	1
	3_1	0.0001	0.0123	-	1.0	0.3
	2_2	0.0081	0.001	-	0.03	0.4
0_2	2_1	0.089	0	-	26	0.9
	1_1	0	0.0465	50	100	100
	1_2	0	0.0279	100	8	12
	1_3	0	0.0021	-	0.15	0.001
	1_4	0	0.0394	-	1.8	2.4
2_3	2_1	0.0259	0.0002	42	100	12
	1_1	0.0111	0.0029	100	89	100
	0_1	0	0	2.3	0	6
	1_2	0	0.0178	5.3	67	4.2
	4_1	0.0005	0	16	0.015	0.02
	3_1	0.0056	0.0022	-	3	0.04
	1_3	0.0014	0.0114	12	11.5	1.0
	1_4	0	0.0006	-	0.45	0.07
	2_2	0	0.0032	-	2	0.5
	3_2	0.0055	0.0406	-	0.3	0

Table 1: Comparison between the experimental branching ratios of the negative parity states ^{198}Au with the ones obtained in the supersymmetric scheme and the ones from the IBFF Also given are the theoretical B(M1) and B(E2) values.

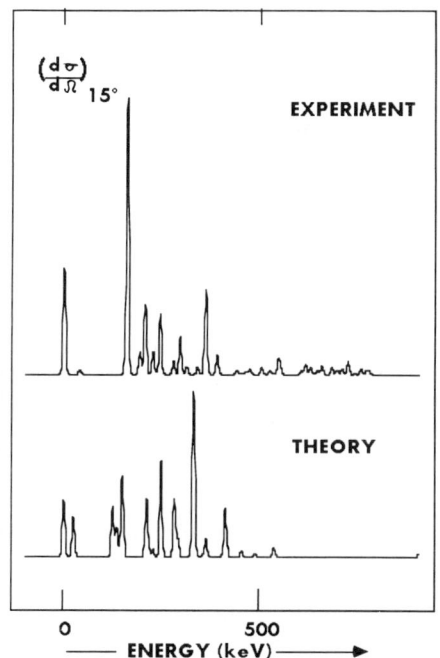

Figure 1: Comparison between the measured and calculated differential cross section $(d\sigma/d\Omega)_\theta$ at 15° for l=1 and 3 transfer.

the pick-up reactions ^{197}Au (p,d) ^{196}Au and ^{197}Au(^{3}He,α)^{196}Au were performed and the differential cross section $(d\sigma/d\Omega)_\theta$ was measured for $\theta = 15°, 30°, 40°$ [6]. Using the theoretical spins and reduced matrix elements we calculated a theoretical value for $d\sigma_l(\theta)$ (l=1,3) which we compare with the experimental one using the theoretical DWBA calculation for $d\sigma_l(\theta)^{D.W.}$ [6]. Since the ground-state has spin 2$^-$ and in theory spin 1$^-$ we here adjust the theory such that the first excited 2$^-$ state at 25.2 keV becomes the groundstate in the theoretical spectrum. This we do in exactly the same way as done in ref 8. We invert the sign of E in (1) and correct the parameter D so that D+E stays the same quantity. This last restriction makes that the Hamiltonian (1) with the adjusted parameters still gives the same results for ^{194}Pt and ^{195}Au.

Having done this we calculate the intensities and give the result in Figure 1 in comparison with the measured intensity at 15 degrees for l=1 and l=3 as obtained in ref 6. One sees that the number of levels excited below 550 agrees very well. While 22 of the 30 levels are excited in theory experimentally 20 levels are observed. Moreover the level density in function of energy seems to be rather well predicted. In detail we do not find a real one-to-one correspondence which is not expected as there is no fitting involved. Maybe it will be possible to obtain a more profound test in the future as more experimental information such as in ref 11, become available.

We concluded that as far as the experiments on ^{198}Au and ^{196}Au are concerned supersymmetry has despite its inherent restrictions predictive power. This is motivated by the fact that the fit after a one-to-one assignment of the levels in ^{198}Au gave very similar parameters for the hamiltonian (1) and that the gross features of the pick-up intensities to ^{196}Au can be predicted.

The author wants to acknowledge, U. Mayerhofer, P. Van Isacker and T. von Egidy for their contributions to this work.

<u>References</u>

1) F.Iachello, and A.Arima,'The Interacting Boson Model', (Cambridge University Press,1987).
2) V.Lopac, S.Brant, V.Paar, O.Schult, H.Seyfarth, A.B.Balantekin,Zeit.Fur.Phys A323(1986) 491.
3) P.Van Isacker, J.Jolie, K.Heyde, and A.Frank, Phys. Rev. Lett. **54** (1985) 653.
4) J.Jolie, PhD thesis (1986) Rijksuniversiteit Gent, unpublished
5) P.Van Isacker, Journ. of Math Phys. 28(1987) 957.
6) J.Jolie, U.Mayerhofer, T.von Egidy, H.Hiller, J.Klora, H.Lindner, H.Trieb, subm to Phys.Rev.C
7) U. Mayerhofer, et al. Nucl. Phys. A492 (1989) 1 and PhD thesis, T.U. Münich, unpublished.
8) D.D. Warner, R.F. Casten, and A. Frank, Phys.Lett.180B (1986) 207.
9) J. Jolie, ILL-internal report 89JO12T (1989).
10) U. Mayerhofer et al., contribution to this conference.
11) M. Vergnes et al., contribution to this conference.

EVIDENCE FOR CORE POLARIZATION INTERACTION IN THE SINGLE CLOSED SHELL REGION

Zs. Dombrádi
Institute of Nuclear Research, 4001 Debrecen, Hungary
S. Brant and V. Paar
Prirodoslovno-matematički fakultet, University of Zagreb,
41000 Zagreb, Croatia, Yugoslavia

ABSTRACT

The multipole structure of the effective proton-neutron interaction was analysed in the odd-odd In nuclei by fitting a dipole plus quadrupole interaction to the splitting of 11 multiplets of $^{110-116}$In. It was found that the quadrupole-quadrupole interaction is about twice as strong as in the doubly magic region. The additional quadrupole-quadrupole interaction strength found is in agreement with the predictions of the particle-vibration theory.[1]

INTRODUCTION

As a second order consequence of the particle-vibration coupling an effective quadrupole-quadrupole proton-neutron interaction was predicted by Bohr and Mottelson.[1] This kind of polarization interaction has been found neither in the double magic [2] nor in the deformed [3] regions. The identification of new proton-neutron multiplet states in the odd-odd In nuclei made possible to extend the search for the core polarization interaction also to the single closed shell region. Although there are not enough complete multiplets known in these nuclei to perform a complete multipole analysis, an approximation to it could be done by truncating the multipole series at its quadrupole member. For sake of comparison also the parameters of the effective interaction characteristic for the doubly magic region were determined.

THE SHORT RANGE INTERACTION

The effective interaction in the doubly magic region was assumed to be a volume delta interaction, which approximates well both other phenomenological interactions and the realistic ones.[4] It was taken in the following form:

$$V_{pn} = V_0 \delta(\vec{r}_p - \vec{r}_n)(1 + \alpha \vec{\sigma}_p \vec{\sigma}_n).$$

The parameters of the interaction were fitted to the somewhat different experimental matrix elements determined by Daehnick[2] and Schiffer and True[5] using harmonic oscillator wave functions for the Slater integral. The oscillator parameter was taken as $b = \sqrt{41.5/(45/A^{1/3} - 25/A^{2/3})}$. This value gives a good approximation both for the light and heavy nuclei.[6]

The fitted parameters of the delta interaction $V_0 = -500$ MeVfm^3 and $\alpha = 0.15$ are close to the values estimated by Zamick[4] from the realistic interactions: $V_0 = -600$ MeVfm^3 and $\alpha=0.125$.

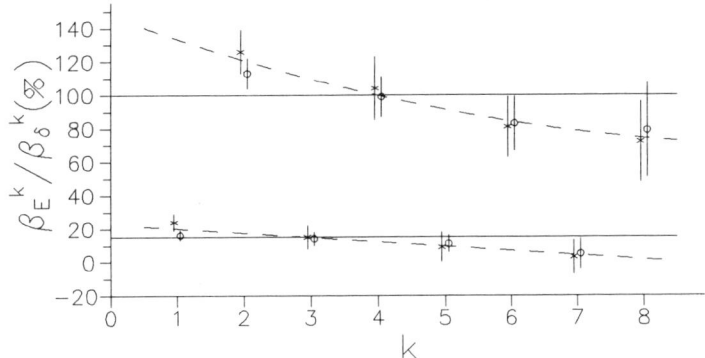

Fig.1. Multipole components of the empirical p-n matix elements compared with the delta interaction components. o : from the data of Schiffer and True [5], x : from Daehnick's [2] data. The solid line: delta interaction, the dotted line shows finite range effect.

The relative multipole coefficients (β^k) [6] determined form the empirical matrix elements are compared with the multipole components of the delta force (with $\alpha = 1$) in Fig.1. As seen in the figure the multipole structure of the empirical interaction is also well approximated by the delta interaction. The systematic deviation of the mean values was described by Schiffer and True [5] using a finite range interaction. The relatively large quadrupole component can also be interpreted as a high frequency core polarization interaction through virtual excitation of giant quadrupole resonances.

CORE POLARIZATION IN ODD-ODD In NUCLEI

To search for the existence of the low-energy core polarization interaction in the single closed region, as an approximation of the multipole decomposition, a dipole plus quadrupole interaction was fitted to the members of 11 multiplets of $^{110-116}$In. The interaction was taken in the following form:

$$V_{pn} = V_0 + V_1[\sigma_p\sigma_n]_0 + V_2 Y_2(p)Y_2(n).$$

The occupation probabilities of the neutron quasiparticles were taken from the systematics of Maldeghem et al.[8] It was found that the strength of the quadrupole interaction is 2.19±0.29 times larger than expected from the above delta interaction. This deviation corresponds to a

$$V_{QQ}^{low} \approx -5\text{MeV} \times Y_2(p)Y_2(n)$$

low-frequency core polarization interaction in addition to the short range one.

Using the IBFM notations, one can estimate the polarization interaction strength as $-10\Gamma_p\Gamma_n/\hbar\omega$, where Γ is the strength of the dynamical interaction and $\hbar\omega$ is the d-boson energy.

The typical dynamical interaction strength in the odd In nuclei is ≈ 1.2 MeV, in the odd tin isotopes $\Gamma_n \approx 0.8$ MeV. In the even tin isotopes $\hbar\omega \approx 1.2$ MeV, leading to an ≈ -8 MeV polarization interaction strength.

The role of the different interactions is shown in Fig.2 through the example of the $\pi g_{9/2}^{-1}\nu h_{11/2}$ multiplet of ^{112}In. It is seen that only half of the splitting is due to the short range interaction.

If the quadrupole-quadrupole interaction found, corresponds to the core polarization interaction, one expects also the presence of the core polarization charge. The quadrupole moments of the long lived states of odd-odd In nuclei can be reproduced well (assuming pure configurations) using $e_n^{eff} = 3.7e$ and $e_p^{eff} = 4.7e$, corresponding to an $e_{low}^{eff} \approx 3.2e$ low-frequency polarization charge. From the particle-vibration theory[1] $e_{low}^{eff} \approx 4e$ is estimated.

The acceptable agreement with the theoretical estimate and the presence of the polarization charge of the correct size indicates that the additional quadrupole interaction is the low-frequency core polarization interaction. The difference between the estimated and found quadrupole strength can be explained by taking into account the smoothing effect of the particle-vibration coupling.[9]

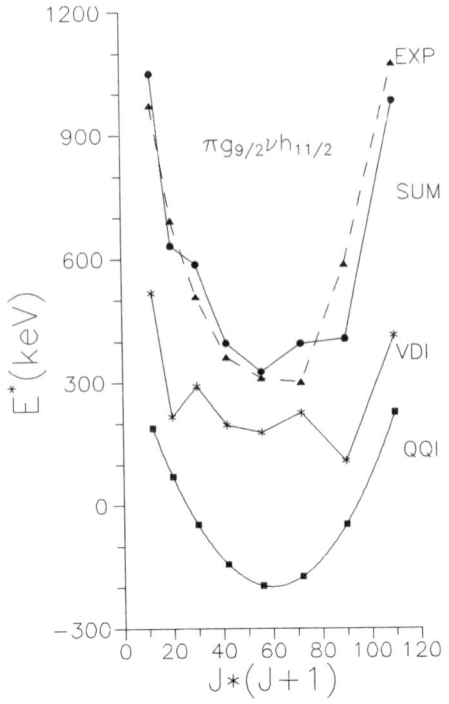

Fig.2. Calculated splitting of the $\pi g_{9/2}^{-1}\nu h_{11/2}$ multiplet of ^{112}In as a function of $J*(J+1)$. EXP: experimental data, VDI: short-range interaction, QQI: core polarization interaction, SUM=VDI+QQI. The multiplets are shifted relative to each other.

REFERENCES

1. A. Bohr and B.R. Mottelson, Nuclear Structure, Benjamin, New York, 1974, vol. 2, chap. 6
2. W.W. Daehnick, Phys. Reports 96, 319 (1983)
3. J.P. Boisson, R. Piepenbring and W. Ogle, Phys. Reports 26C, 101 (1976)
4. L. Zamick, in Proc. Int. Conf. on Nuclear Structure and Nuclear Spectroscopy, Amsterdam, 1974, eds.: H.P. Block and A.E.L. Dieperink, p.24.
5. J.P. Schiffer, W.W. True, Rev. Mod. Phys. 48, 191 (1976)
6. G.F. Bertsch, Practicioner's shell model, North-Holland, Amsterdam, 1972, chap.2.
7. M. Moinester, J.P. Schiffer and W.P. Alford, Phys. Rev. 179, 984 (1969)
8. J. Van Maldeghem and K. Heyde, Phys Rev. C 32, 1067 (1985)
9. Zs. Dombrádi, T. Fényes, S. Brant and V. Paar, in this volume

SIGNATURE DEPENDENCE AND INVERSION IN TWO QUASI-PARTICLE ROTATIONAL BANDS OF EVEN-EVEN NUCLEI

A.K.Jain and Alpana Goel
Physics Department,Univ.of Roorkee,Roorkee-247667,India.

ABSTRACT

The experimental data on two quasi-particle (2qp) rotational bands of even-even rare-earth nuclei reveal significant odd-even staggering in both $K^+=(\Omega_1+\Omega_2)$ and $K^-=|\Omega_1-\Omega_2|$ bands. An empirical rule is proposed for the phase of staggering and is checked with the experimental data. More complexities are exhibited by these rotational bands as compared to odd-odd nuclei which are evidenced in the form of anomalous behaviour seen in the cases where data are known upto higher spins.

INTRODUCTION

We examine the experimental data on 2qp rotational bands of even-even rare-earth nuclei[1] to reveal many new features vis-a-vis the behaviour of rotational bands in odd-odd rare-earth nuclei.The 2qp intrinsic excitations in even-even nuclei differ from their odd-odd counterparts in several ways: (1) the singlet coupling ($\Sigma_1+\Sigma_2=0$) lies Lower in energy than the triplet ($\Sigma_1+\Sigma_2=1$) coupling in contrast to the odd-odd nuclei (2) the parameters of the interaction which split the $K^\pm=|\Omega_1\pm\Omega_2|$ members are entirely different (3) the diagonal odd-even shift (known as the Newby shift in odd-odd nuclei) is different in magnitude and sign from that in odd-odd nuclei.These factors should be taken into account in any discussion of the 2qp rotational bands of even-even nuclei.

DISCUSSION

It was pointed out[2] recently that the rotational bands of odd-odd nuclei exhibit significant odd-even staggering.For high-j configurations of odd-odd nuclei the favoured signature is given by[3]

$$\alpha_f = \tfrac{1}{2}(-1)^{j_p-\tfrac{1}{2}} + \tfrac{1}{2}(-1)^{j_n-\tfrac{1}{2}} \qquad \ldots(1)$$

For $\alpha_f=0$,even spins are favoured and $\alpha_f=1$,odd spins are favored.This rule holds reasonably well for nearly pure-j configurations also.On the other hand an empirical rule proposed by Frisk[4] gives the favored spin I_F in a K=0 band to be $I_F=(j_p+j_n)$ mod2 where the dominating j_p and j_n quantum numbers are to be used.Since the most important mechanism of odd-even shift is the Coriolis coupling to some K=0 band, we realise that the relation (1) immediately follows from this empirical rule and vice-versa.

Recalling that the odd-even shift of K=0 bands in even-even 2qp rotational bands differs in sign, we propose that the favored spin I_F in a K=0 band may be given as

© 1991 American Institute of Physics 153

$I_F=(j_1+j_2-1)$ mod2, and the favored signature for nearly pure-j configurations may be given by

$$\alpha_f = \tfrac{1}{2}(-1)^{j_1-\tfrac{1}{2}} + \tfrac{1}{2}(-1)^{j_2+\tfrac{1}{2}} \qquad \ldots (2)$$

This rule is checked with the data in Table 1. We take the phase of staggering as established provided at least 5 levels are known. The rule seems to work in 8 cases out of the 12 listed cases. It is interesting to note that both K^- and K^+ bands exhibit staggering effect. Sometimes the staggering of the K^+ bands is more pronounced. It is clear that the situation here is more complex than in odd-odd nuclei.

An interesting case where the rule does not work is the $K=4^-$ band in 170Yb; note that its Gallahager partner $K=3^-$ band satisfies the rule. This and two other cases, namely 180Os and 164Er, where levels are known to high spins are shown in Fig.1. The $K=4^-$ band in 170Yb, besides the opposite phase of staggering, shows an anomalous behaviour from I=10 to 12, where the odd-even shift is highly attenuated. Similar behaviour is seen in the $K=7^-$ band in 180Os from I=23 to 25. The $K=5^-$ band in 164Er exhibits a signature inversion at I~10 and again at I~21. An octupole vibrational mixing is also suspected in these cases. A signature inversion seen at low spins in odd-odd nuclei (156Tb and 160Ho) has recently been explained by Hamamoto purely in 2qp+rotor bandmixing calculations in $(h_{11/2})_p (i_{13/2})_n$ space. We have developed a formalism for 2qp+rotor bandmixing calculations in even-even nuclei where a nearly complete Coriolis mixing can be carried out and also the effect of residual interaction taken into account. An investigation of these phenomena is underway.

TABLE 1

S.N.	Nuc.	K^π	Configuration	α_f	Rule valid
1.	152Sm	5$^-$	$\{5/2(413)+5/2(532)\}_p$	0	yes
2.	152Sm	7$^-$	$\{11/2(505)+3/2(651)\}_n$	1	yes
3.	160Dy	1$^+$	$\{5/2(523)-3/2(521)\}_n$	0	yes
4.	162Dy	1$^+$	$\{5/2(523)-3/2(521)\}_n$	0	yes
5.	162Dy	5$^-$	$\{5/2(523)+5/2(642)\}_n$	0	no
6.	164Dy	2$^+$	$\{5/2(523)-1/2(521)\}_n$	0	no
7.	166Er	0$^-$	$\{7/2(523)-7/2(404)\}_p$	0	yes
8.	166Er	1$^+$	$\{7/2(633)-5/2(642)\}_n$	0	no
9.	170Yb	3$^-$	$\{7/2(633)-1/2(521)\}_n$	0	yes
10.	170Yb	4$^-$	$\{7/2(633)+1/2(521)\}_n$	0	no
11.	176Hf	3$^+$	$\{7/2(514)-1/2(521)\}_n$	0	yes
12.	180Os	7$^-$	$\{7/2(633)+7/2(514)\}_n$	0	yes

Financial support from D.A.E.(Govt. of India) is acknowledged.

REFERENCES

1. P.C.Sood, D.M.Headly and R.K.Sheline, At.Data and Nucl. Data Tables (1990).
2. A.K.Jain, J.Kvasil, R.K.Sheline and R.W.Hoff, Phy.Rev. C40, 432 (1989).
3. I.Hamamoto, Phys.Lett. B235, 221 (1990).
4. H.Frisk, Z.Physik A330, 241 (1988).

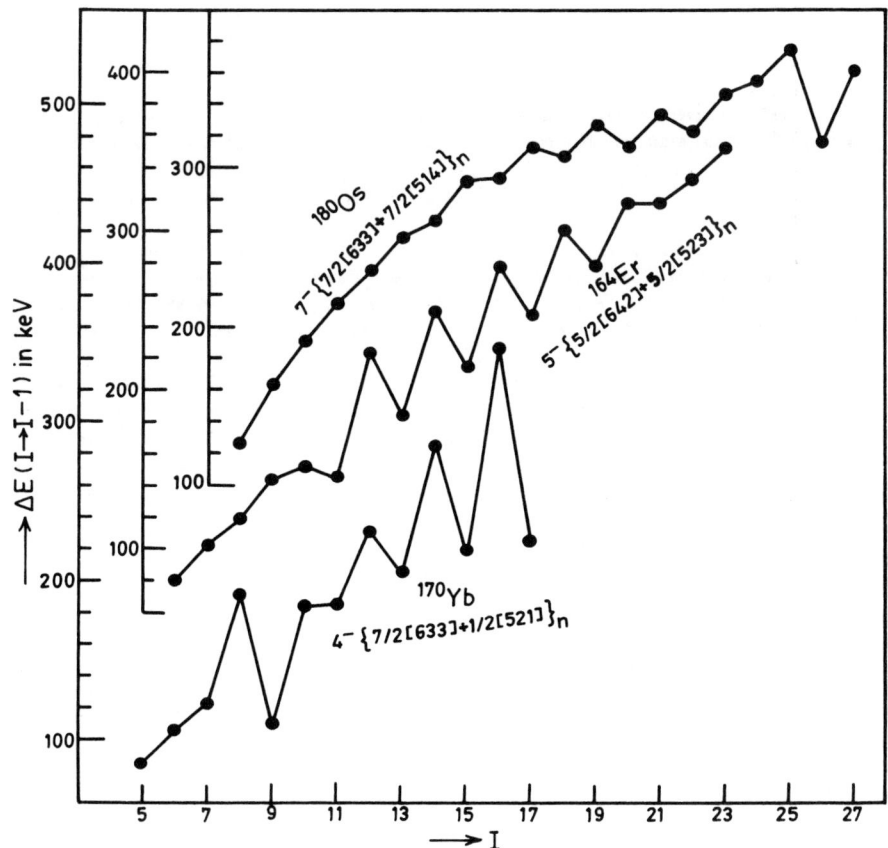

Fig.1. Odd-even staggering and anomalous behaviour of 2qp bands

PHASE REVERSAL IN ODD-EVEN STAGGERING IN K^+ AND K^- BANDS OF EVEN-EVEN 2QP ROTATIONAL SPECTRA

Kiran Jain
Department of Physics, University of Roorkee
Roorkee-247 667 (INDIA)

ABSTRACT

Our analysis of two-quasi-particle rotational bands in even-even nuclei reveals an odd-even staggering in both $K^+ = |\Omega_1 + \Omega_2|$ and $K^- = |\Omega_1 - \Omega_2|$ bands. The phase of odd-even staggering in K^+ and K^- bands is found to be opposite which is in contrast to the behavior of these bands in odd-odd nuclei. In the present paper, we report the preliminary results of our analysis to understand the behavior of these bands.

INTRODUCTION

The intrinsic excitations in an even-even nucleus are observed to occur in the energy region above the pairing gap energy ($\approx$ 1.0 MeV in rare-earth region). The excited particles may be either two protons ($\pi \otimes \pi$) or two neutrons ($\nu \otimes \nu$). These excitations in the nucleus form two-quasi-particle(2qp) doublets with band members $K^{\mp} = |\Omega_1 \pm \Omega_2|$ which split due to residual interaction having singlet state (i.e. $\Sigma = 0$) lower in energy than the triplet state (i.e. $\Sigma = 1$)[1]. These bands are mixed with each other by Coriolis coupling which also leads to the odd-even staggering.

DISCUSSION

The empirical evidence for odd-even staggering in 2qp rotational bands in even-even nuclei is presented in Fig.1 where only representative examples are shown. A strong odd-even shift is present in these bands. It is clear from the figure that both K^+ and K^- bands exhibit this type of behavior. In some cases, the K^+ bands show much greater staggering than the K^- bands. This is just opposite to that observed in doubly-odd nuclei. Recent analysis of Jain et al.[2] shows that most of the K^- bands in odd-odd nuclei are staggered and K^+ bands do not show any staggering. However, few cases are observed in doubly-odd nuclei showing a signature splitting in both bands and some of these are shown in Fig.2.

It is evident from Figs. 1 and 2 that the phase of staggering in K^+ and K^- bands in doubly-even nuclei is

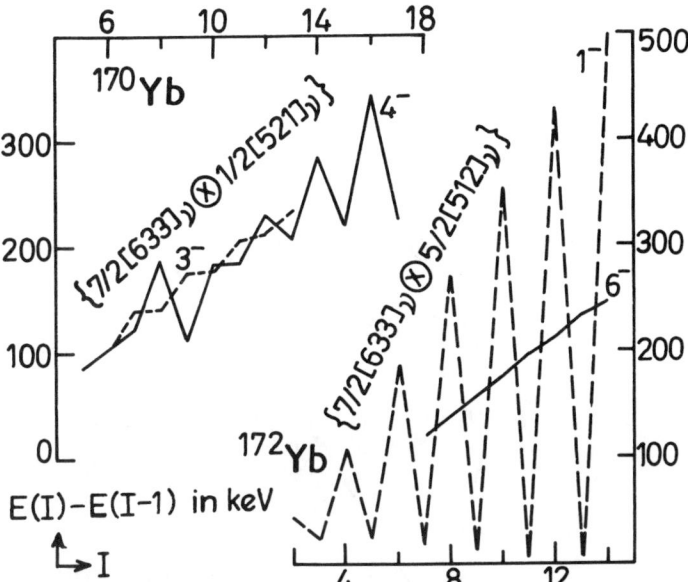

Fig.1: Representative examples showing odd-even staggering in even-even nuclei. The phase of staggering in K^+ (solid lines) and K^- (dashed lines) bands is opposite to each other. Data have been taken from Ref. 3.

opposite while it is same in doubly-odd nuclei. The favored signature in a 2qp rotational band is usually defined as

$$\alpha_F = 1/2(-1)^{j_1-1/2} + 1/2(-1)^{j_2-1/2} \qquad (1)$$

which implies that if $\alpha_F=0$, even-integer spin members are favored and for $\alpha_F=1$, odd-integer spins are favored. We find that this relation is, in genaral, followed by all 2qp rotational bands in odd-odd nuclei showing same phase of staggering in each doublet. The K^+ bands in even-even nuclei do follow this relation but, the favored signature in K^- bands is given by

$$\alpha_F = 1/2(-1)^{j_1-1/2} - 1/2(-1)^{j_2-1/2} \qquad (2)$$

This may be due to the isospin dependence of the residual interaction. Also, most of the 2qp doublets in even-even nuclei occupy one of the high-j orbits and

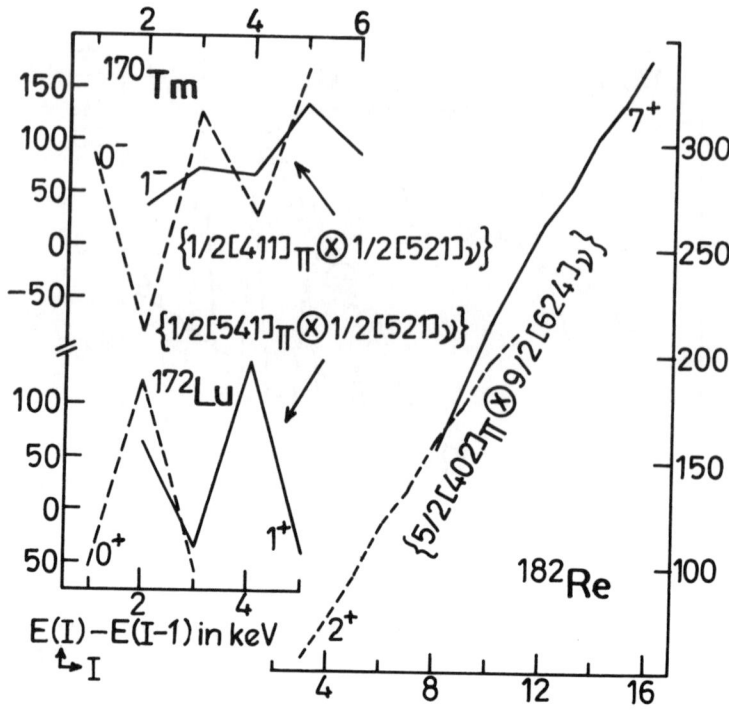

Fig.2: Similar to Fig.1 but for odd-odd nuclei. The phase of staggering in both K^+ and K^- bands is same. Data have been taken from individual references[4].

the staggering from these single-particle orbits is trasmitted to the 2qp bands. A more detailed analysis is underway and will be reported elsewhere.

This work was financially supported by Council of Scientific and Industrial Research, Govt. of India.

REFERENCES

1. C.J. Gallagher, Nucl. Phys. **16**, 215 (1960).
2. A.K. Jain et al., Phys. Lett. **B209**, 19 (1988).
3. P.C. Sood et al., Atomic Data and Nucl. Data Tables, to appear (1990).
4. R.A. Dewberry et al., Phys. Rev. **C24**, 1643 (1981); D. Elmore and W.P. Alford, Phys. Rev. **C14**, 583 (1976); M.F. Slaughter et al., Phys. Rev. **C29**, 114 (1984).

COMPLETENESS OF TWO-PARTICLE SPECTRA OF DEFORMED NUCLEI

P.C. Sood[*+], R.K. Sheline[*], R.W. Hoff[a] and A.K. Jain[*#];
[*]Florida State University, Tallahassee, FL 32306;
[+]Physics Dept., Banaras Hindu University, Varanasi 221005, India; [a]Lawrence Livermore National Laboratory, Livermore, CA 94550; [#]Physics Dept., Roorkee University, Roorkee 247667, India

ABSTRACT

Consideration of the "complete" set of expected two-particle structures in even mass deformed nuclei together with the "complete" set of transitions, connecting states with defined spin and energy ranges and the precisely determined transition energies from crystal spectrometer studies is sought to reveal as yet unidentified levels and band structures involving unplaced transitions. Further a side-by-side study of the odd-odd and the even-even structures points to a possible isospin dependence of the residual interaction. Energy levels of ^{166}Ho and ^{168}Er are used to illustrate these features.

In principle, the average resonance capture (ARC) technique[1,2] can provide an <u>a priori</u> guarantee of the disclosure of all states within certain spin and energy ranges on the premise that the connecting transitions that occur are not likely to be missed. However, in practice, a large number of such transitions remain "not placed" in the deduced level scheme[3-5]. Further, even after establishing the presence of a level, its actual placement within a band with firmly assigned configuration or character can be achieved only in a limited number of cases through particle transfer reaction studies. In many cases, such characterizations are attempted based on an examination of the available configuration space and through recourse to some modelling technique[2,8]. In this paper we illustrate how a careful consideration of the "complete" two-quasiparticle (2qp) structures expected in a deformed even-mass nucleus, taken together with the distinctive features of a specific band, observed in some neighboring nucleus or predicted from theory, may help in revealing as yet unidentified levels[7] by using unplaced transitions through exploiting their precise energies determined from crystal spectrometer investigations. Further we show that a simultaneous description of the 2qp structures in the odd-odd and even-even nuclei points to an isospin dependence of the residual interactions.

As an illustration we discuss the spectroscopy of the odd-odd nucleus ^{166}Ho and its even-even neighbor ^{168}Er. The single particle orbitals of interest are shown in Fig. 1 and are used to construct the 2qp spectra with band members $K=|\Omega_1 \pm \Omega_2|$ in even mass nuclei. These bands constitute (n,p) doublet structures in the odd-odd and (p,p)/(n,n) doublets in the even-even nuclei. We have carried out model calculations[6,8] for the complete 2qp spectra in each case for levels with spins up to 12 and $(E_1 + E_2)$ up to 1200 keV; band-mixings have been included and odd-even staggering is used as a finger-print in specific cases. Results in each case are compared with the available experimental data to deduce configuration assignments and to identify/predict as yet unidentified structures.

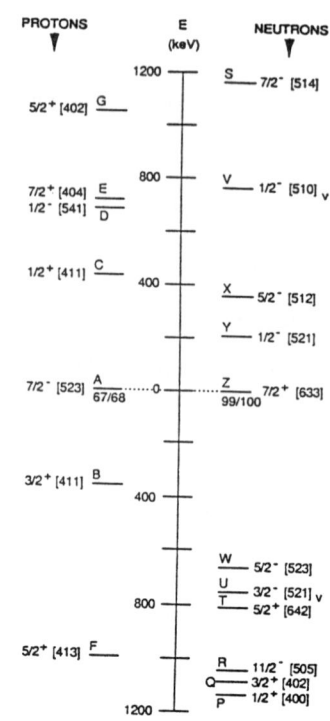

Fig. 1 The proton and neutron single particle orbitals for Z=67/68 and N=99/100 from ^{165}Ho and ^{167}Er respectively.

In ^{166}Ho, ARC studies identify 75 levels below 1MeV, none of which could be assigned unambiguous spin-parity[4]. The deduced level scheme left unplaced 260 low-energy transitions[5], out of a total of 350, observed in n-capture. Our earlier analysis[6] resulted in configuration assignments to 21 bands which mainly involved "A" proton or "Z" neutron orbitals. Careful consideration of beta-feeding rate and constraint led to re-interpretation of the two 1$^+$ bands. In a later study[7], three new bands and 11 additional levels were observed using gamma transitions from n-capture studies. However, our predicted "complete" 2qp set for ^{166}Ho below 1MeV includes ~300 levels with I≤12 constituting over 50 bands with 0≤K≤7; in this set 110 levels have 2≤I≤5, which should exist in ARC data. Further experiments on the lines of ref. 2 and analysis as in ref. 7 aimed at establishing the identity between the two "complete" sets are desired.

As of 1984, capture gamma ray spectroscopy had revealed[9] 36 bands involving 127 levels and 580 transitions in ^{168}Er. Our recent compilation[10] puts the number of identified bands at 41. Over the past decade this nucleus has been the subject of many other experiments and every conceivable theoretical approach. In spite of this tremendous effort, no unambiguous interpretation of these data has emerged except for barely a dozen bands elucidated through particle transfer reactions. Our own effort, which earlier[8] included only K=0 mixing between n-n/p-p bands of same K has since been expanded to cover wide-basis Coriolis mixing. Particularly the extreme staggering observed for the 1359 keV K=1 band is attributed to perturbing effects of $i_{13/2}$ neutron orbitals mainly arising from coupling with the K=0 band whose odd-spin sequence is observed beginning at 1786 keV; description of the observed staggering places the even-spin K=0 band levels starting around 2500 keV, indicative of anomalously large Newby shift in this band. If confirmed, explanation of this anomaly has to be sought. Further, as in the odd-odd case, the "completeness" question needs further effort with the models predicting 56 bands in the explored energy range and over 160 gammas with E up to 1 MeV revealed in n-capture studies still unplaced.

Although the even-even spectra are complicated due to strong admixtures of 2qp structures as well as n-n/p-p admixtures, even at this state of analysis it can be concluded[8] that the doublet splittings in even-even nuclei are on the average about three times larger than those observed in odd-odd nuclei. Interpreted in terms of spin-dependent residual interaction, this feature requires the introduction of isospin dependence of the residual interaction. More detailed studies, particularly of 2qp spectra of even-even nuclei, are needed to establish such a dependence.

These studies have been supported in part by UGC at Varanasi, NSF at Tallahassee, DOE at Livermore and DAE at Roorkee.

1. L.M. Bollinger and G.E. Thomas, Phys. Rev. C2, 1951 (1970).
2. R.W. Hoff et al. Nucl. Phys. A437, 285 (1985).
3. T.J. Kennett, M.A. Islam and W.V. Prestwich, Phys. Rev. C30, 1840 (1984).
4. A.E. Ignatochkin, E.N. Surshikov and Yu.F. Jaborov, Nucl. Data Sheets 52, 365 (1987).
5. V.S. Shirley, Nucl. Data Sheets 53, 223 (1988).
6. P.C. Sood, R.K. Sheline and R.S. Ray, Phys. Rev. C35, 1922 (1987).
7. R.K. Sheline et al. Phys. Rev. C40, 1065 (1989).
8. P.C. Sood, R.K. Sheline and R.S. Ray, Nucl. Instr. Methods B40/41, 462 (1989).
9. D.F. Davidson, W.R. Dixon and R.S. Storey, Can. J. Phys. 62, 1538 (1984).
10. P.C. Sood, D.M. Headly and R.K. Sheline, Atomic Data and Nucl. Data Tables 47 (1990) in press.

APPLICATION OF THE VMI MODEL TO ROTATIONAL BANDS OF ODD-ODD RARE-EARTH NUCLEI

A.K.Jain and Alpana Goel
Physics Department,Univ.of Roorkee,Roorkee-247667,India.

ABSTRACT

The variable moment of inertia model has been applied to the $K^+=(\Omega_p+\Omega_n)$ bands which are relatively free from Coriolis mixing.The inertia parameter shows significant variation with angular momentum which is nearly identical to one of the odd-A rotational bands based on either the neutron or, the proton configuration also involved in the odd-odd band.An explanation in terms of the blocking effect of pairing correlations is presented.

INTRODUCTION

We report the application of the variable moment of inertia model[1] to those rotational bands of odd-odd rare-earth nuclei which are free from Coriolis mixing effects. The $K_>=(\Omega_p+\Omega_n)$ bands are expected to be reasonably unperturbed except for configurations where Ω_p or, $\Omega_n=1/2$ so that a direct mixing with $K_<$ band occurs.

APPLICATION OF THE VMI MODEL

The VMI model is easily extended to odd-odd rotational bands by writing[2]

$$E_K(I)=E_K+\frac{1}{2}C(\mathcal{J}_I-\mathcal{J}_K)^2+\frac{\hbar^2}{2\mathcal{J}_I}[I(I+1)-K(K+1)]$$

where E_K is bandhead energy.The moment of inertia parameter $\mathcal{J}_K$ and the parameter C are determined by a least square fitting to the experimental data.It can be shown that the softness parameter $\sigma=(2K+1)/2C\mathcal{J}_K^3$ for odd-odd nuclei.Reasonably good fits are obtained in the 13 rotational bands analysed by us.

Result of an specific calculation for 178Ta are shown in Fig.1(a) in the form of the moment of inertia parameter and its variation with angular momentum for the three bands.Note that the same neutron configuration namely 7/2 (514)is coupled to different proton configurations in all the three cases.A comparison with the moment of inertia of rotational bands based on the three different proton configurations namely 9/2 (514), 5/2 (402)and 7/2 (404) in the neighbouring 177Ta is revealing.

Fristly, we note that the odd-odd moment of inertia parameter $\mathcal{J}_K$ is significantly greater than that of odd-A rotational bands as expected.Secondly, there is a large increase in the moment of inertia parameter in the odd-odd bands which is quite contrary to our expectations.However, the most remarkable point is that the proton configurations appearing in the odd-odd rotational bands and also seen in the neighbouring odd-A rotational bands have moment of inertia which are placed in the same order with respect to each other; also the variation in the moment of inertia of the odd-odd bands show nearly same behaviour as the odd-A rotational bands.The 7/2 (514) neutron seems to play no role in the variation of the moment of inertia of these bands.It is indeed remarkable that the odd-odd rotational bands exhibit such a large variation in the moment of inertia.

In Fig.1(b) we show a comparison of the same odd-odd rotational band in 176Lu and 178Ta with the odd-proton rotational band in neighbouring odd-A nucleus; again the 7/2(514) neutron is involved in the odd-odd rotational band.The relative ordering of the same odd-odd rotational band in different nuclei is again similar to that in the neighbouring odd-A nuclei.

Similar results are obtained in other cases also where such a comparison is possible.A possible interpretation of this behaviour may be obtained from the blocking effect of the pairing correlations.It is known that the unpaired particles present in the odd-odd system produce the "blocking" effect, resulting in a significant decrease of the pairing correlations.This effect is strongest when the Fermi level (ground state) is blocked.In the case of 178Ta the neutron Fermi level is 9/2 (624) state and the 7/2(514) state turns out to be a hole state.The blocking effect of 7/2(514) neutron therefore appears to be considerably reduced.On the other hand, the three proton configurations involved in 178Ta, namely 7/2(404), 9/2(514) and 5/2(402) lie very close together near the ground state which is 7/2(404) and therefore do affect the pairing correlations.Moreover, the neutrons are known to affect the pairing more than the protons.The individual variations from one configuration to the other e.g. in 178Ta (See Fig.1(a)) can therefore be ascribed as the proton configuration dependence.

We acknowledge the financial support from Department of Atomic Energy (Govt. of India) in the form of a research project.

REFERENCES

1. M.A.J.Mariscotti, G.Schraff - Goldhaber and B.Buck, Phys.Rev. $\underline{178}$, 1864 (1969).
2. Alpana Goel and A.K.Jain, Mod.Phys.Lett. $\underline{A}$ (1990) to appear.

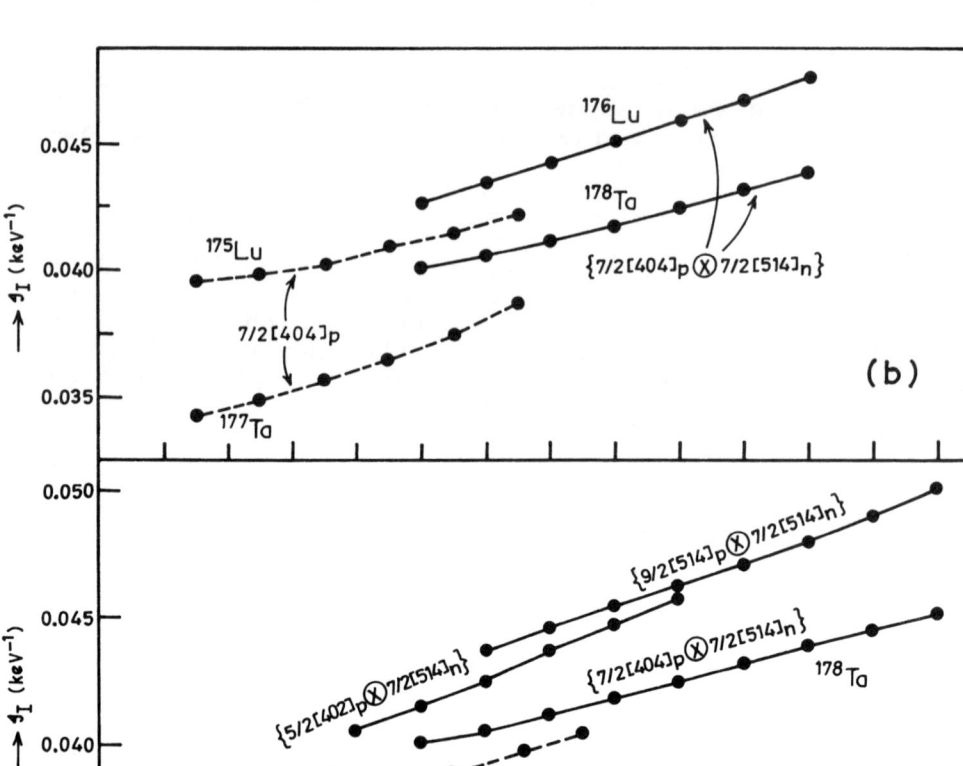

Figure 1.(a) Variation of the moment of inertia parameter $\mathcal{J}_I$ as obtained from the VMI model for 178Ta. A comparison is made with the variation of the moment of inertia of the odd-A rotational bands in neighbouring 177Ta which involve the same proton configurations as in 178Ta.

(b) Same as in (a) but here same band for two different nuclei is shown.

A COMPARISON OF THE PRINCETON AND BEIJING ENERGY EXPRESSIONS FOR ROTATING NUCLEI

R.A. Naumann, Y.Y. Sharon, A. Adarkar
Princeton University

Over the past several years, our group at Princeton has been modelling[1] the extended rotational level sequences now available for nuclei -- especially for the actinides[2], the rare-earths[3] and the recently discovered superdeformed nuclei[4].

Our approach[5] has involved a generalized collective potential:

$$V = (1/n) k q^n + (1/2) A/q^2, \qquad (1)$$

where q is a generalized collective coordinate, A and k are parameters, and n represents the steepness of the long-range part of the potential. This potential leads to an energy eigenvalue expression:

$$E_J = C[(1+A J(J+1))^{n/n+2} - 1], \qquad (2)$$

which gives the energy E_J of an yrast level in terms of its angular momentum J. We have demonstrated that very satisfactory fits are obtained with expression (2) for the experimental data -- at 5 MeV excitation fits within 2 keV are typical.[6]

Recently, Wu and Zeng[7] have reconsidered the Bohr Hamiltonian, $V(\beta,\gamma)$. By making some approximations, they have deduced the simple eigenvalue expression:

$$E_J = C(1+AJ(J+1))^{1/2} - 1). \qquad (3)$$

Obviously, this energy expression suggested by the Beijing group represents a special case (n=2) of our more general approach.

In a still more recent contribution[8], the Beijing group suggested an empirical improvement to their earlier formula of the form

$$E_J = C[1+D(J)(J+1)][(1+A(J)(J+1))^{1/2} - 1] \qquad (4)$$

where D is a small correction factor.

We have carried out extensive least-square fits to the experimental level energies of extended rotational level sequences in the actinide nuclei using expressions (2), (3), and (4). These calculations indicated that for the actinides, the optimized value of our parameter, n, is not 2, but generally lies in the range from 1.6 to 8.

Thus, fits with our expression (2) are generally

superior to fits with the Wu-Zeng expression (3), except for the cases where n≈.5, where these two expressions become identical.

It was very interesting to note that the fits with both our general expression (2) and the extended Chinese expression (4) were not only often both very good, but also frequently led to fits of very similar quality. From our experience, we interpret the Huang-Wu correction term, [1+D(J)(J+1)], of expression (4), as an indication that the long-range part of the potential in expression (1) indeed has an exponent n that is bigger than 2. This accords with the effective collective potential being steeper than the harmonic oscillator (n=2); indeed, the potential's steepness often exceeds even the n=4 dependence implied by the Variable Moment of Inertia Model.

To compare our expression (2) with the Huang-Wu expression (4), we have expanded both analytically[9] in powers of J(J+1) to obtain (B=n/n+2) for expression (2):

$$E_J = CAB[J(J+1)][1+(A/2)(B-1)[J(J+1)] \qquad (2')$$
$$+A^2/6(B-1)(B-2)[J(J+1)]^2 +..]$$

and for expression (4):

$$E_J = C\ (A/2)\ [J(J+1)][1+(1/2)(2D-(A/2))[J(J+1)] \qquad (4')$$
$$+(1/4)A[(A/2)-D][J(J+1)]^2 +..]$$

For very small D (D<<1) and B=1/2, the two expressions (2') and (4') just about coincide! For D=0 and B=1/2, both expressions become C (A/2) [J(J+1)][1-(A/4)J(J+1)+(1/8)A²[J(J+1)]²+...].

Accordingly, it is not surprising that using the optimized values of the parameters for ^{238}U [B=0.5175, D=.000048162] we obtain from (2´) and (4´) closely coincident expressions

$$E_J = 7.4755J(J+1)-.004308[J(J+1)]^2 +.0000051[J(J+1)]^3 \qquad (2*)$$

$$E_J = 7.4750J(J+1)-.004289[J(J+1)]^2 +.0000050[J(J+1)]^3 \qquad (4*)$$

while for ^{234}U, [B=0.7222, D=.0001318], we obtain

$$E_J = 7.3147J(J+1)-.006776[J(J+1)]^2 +.0000193[J(J+1)]^3 \qquad (2+)$$

$$E_J = 7.2770J(J+1)-.005936[J(J+1)]^2 +.0000119[J(J+1)]^3 \qquad (4+)$$

In comparison, with the ^{238}U case, the slightly greater disagreement for the ^{234}U case arises primarily because B is rather different from .5 and secondarily from the larger value of D.

For the even-even actinide nuclei (A≥220), the energy fits provided by these two approaches are usually superior to the other parametrizations previously suggested.

The very close agreement between the two expressions extends also to the ground-state bands of even-even rare-earth nuclei below the back-bend, when $B \cong 0.5$ and D is very small. Thus, for example, for ^{172}Hf [B=.5083, D=.00000292] we obtain from (2´) and (4´):

$E_J = 15.881 J(J+1) - .02460[J(J+1)]^2 + .0000771[J(J+1)]^3$ (2**)

$E_J = 15.875 J(J+1) - .02444[J(J+1)]^2 + .0000755[J(J+1)]^3$ (4**)

For nuclei which are not good rotors, the fits provided by both expressions, and the agreement between the two expressions, are all much worse. For example, for ^{220}Th we obtain from (2´) and (4´), the non-converging and markedly different expansions:

$E_J = 184.05 J(J+1) - 206.82[J(J+1)]^2 + 491.94[J(J+1)]^3$ (2++)

$E_J = 142.55 J(J+1) - 72.52[J(J+1)]^2 + 73.84[J(J+1)]^3$ (4++)

For the good rotors, the coefficient of the $[J(J+1)]^2$ terms are always negative and the coefficients of the $[J(J+1)]^3$ term are positive. This is in agreement with several physical pictures, including a rotation-vibration model. Typically, the relative order of magnitude of the respective coefficients of increasing powers of J(J+1) decrease in both cases by successive powers of about 10^{-3}.

References

1. Princeton University Progress Reports for 1985-1988; contributions by R.A. Naumann, Y.Y. Sharon.
2. Y.Y. Sharon, R.A. Naumann, et al., BAPS 33, 50 (1988).
3. Y.Y. Sharon, R.A. Naumann, A. Adarkar, BAPS 35, 1015 1990).
4. Y.Y. Sharon, R.A. Naumann, G. Loring, BAPS 34, 1169 (1989).
5. R.A. Naumann, Y.Y. Sharon, BAPS 32, 1561 (1987).
6. Y.Y. Sharon, R.A. Naumann, Proc. Conf. on High Spin Nucl. Structure, p. 288 (Argonne, 1988).
7. Wu, Zeng, Comm. Theor. Phys. (Beijing) 8, 51 (1987).
8. Huang, Wu, Zeng, Phys. Rev. C39, 1617 (1989).
9. Xu, Wu, Zeng, Phys. Rev. C40, 2337 (1989).

CONSTRUCTION OF MULTI-CLUSTER STATES FOR NUCLEAR CLUSTER MODELS

A. Novoselsky

Department of Nuclear Physics
The Weizmann Institute of Science, Rehovot 76100, Israel

and

J. Katriel

Department of Chemistry
Technion - Israel Institute of Technology, Haifa 32000, Israel

ABSTRACT

We consider a system of identical particles distributed in several clusters, each one of which is characterized by a Yamanouchi symbol specifying its permutational symmetry. A procedure for coupling the various clusters into an overall well-defined angular momentum and permutational symmetry is presented. The results are immediately applicable to the study of nuclear cluster models as well as to nuclei as quark clusters.

Experimental studies of heavy ion reactions reveal selective excitations which result from cluster correlations. Well known examples are the two nucleon, three nucleon and particularly α-particle transfer reactions. The cross sections of these reactions are very considerably enhanced over those expected from the nuclear shell model without any cluster correlations. One important case in which cluster correlations are significant is when two or more nuclei are fused to form one nucleus and the constituent nuclei keep their identities in the resultant compound nucleus. This situation happens when the attractive nuclear interactions within the clusters are strong while the interactions among them are relatively weak.

The nuclear cluster model which is based on the resonating group method[1] yields a good description of two-cluster phenomena in nuclei. However, serious computational difficulties, resulting from the antisymmetrization procedure, have so far prevented calculations involving more than two clusters.

We present an efficient procedure for coupling a system of identical particles distributed in several clusters into an overall well-defined permutational symmetry. Each constituent cluster is characterized by the sequence of irreps $\Gamma_2, \Gamma_3 \ldots \Gamma_{n-1}, \Gamma_n$ of the permutation group-subgroup chain $S_2 \subset S_3 \subset \ldots \subset S_{n-1} \subset S_n$ which completely determines its permutational symmetry, and which is equivalent to a Yamanouchi symbol Y_n.

The single-cluster wavefunction is denoted by

$$|\Phi_n> \equiv |j^n Y_n \alpha_n J_n> \tag{1}$$

where j is the single-particle angular momentum, J_n is the total angular momentum and α_n labels any remaining degeneracy. The z component of the total angular momentum is suppressed. In a recent paper[2] we presented an efficient algorithm for generating the one-particle coefficients of fractional parentage (cfps). This algorithm can be used to construct the single-cluster wavefunction specified above.

In a general n-particle multi-cluster system the particles are distributed in m clusters such that there are n_α particles in cluster α and $\sum_{\alpha=1}^{m} n_\alpha = n$. The total number of particles in the first $\alpha - 1$ clusters, i.e., $n_1 + n_2 + \ldots + n_{\alpha-1}$, is denoted by q_α. Clearly, $q_1 = 0$. In terms of these symbols the set of indices of the particles in cluster α is $\{n_\alpha\} \equiv \{q_\alpha + 1, q_\alpha + 2, \ldots, q_\alpha + n_\alpha\}$.

In Ref. 3 it was shown that the multi-cluster wavefunctions can be labeled by the Yamanouchi symbols of each cluster, i.e., $Y_{\{n_1\}}, Y_{\{n_2\}}, \ldots, Y_{\{n_m\}}$ as well as by the irrep Γ_n of S_n. In order to construct a multi-cluster state which possesses a total angular momentum J_n, we introduce the multi-cluster cfps, in terms of which we write

$$|\Phi_n^{(m)}> = |j^n \left(\Phi_{\{n_1\}} \Phi_{\{n_2\}} \ldots \Phi_{\{n_m\}}\right) \Gamma_n \alpha_n J_n>$$
$$= \sum_{\substack{\alpha_{n_1} \alpha_{n_2} \ldots \alpha_{n_m} \\ J_{n_1} J_{n_2} \ldots J_{n_m} \\ J^{(2)} J^{(3)} \ldots J^{(m-1)}}} \left[(\ldots((\Phi_{n_1}\Phi_{n_2})J^{(2)}\Phi_{n_3})J^{(3)} \ldots J^{(m-1)}\Phi_{n_m})J_n|\}\Phi_n^{(m)} \right]$$
$$|(\ldots((\Phi_{n_1}\Phi_{n_2})J^{(2)}\Phi_{n_3})J^{(3)} \ldots J^{(m-1)}\Phi_{n_m})J_n> \tag{2}$$

The procedure to construct the multi-cluster wavefunction is based on forming a complete multi-cluster set of states with a particular sequence of m Yamanouchi symbols and some common value of the total angular momentum. An appropriate set of single-cycle class operators of the symmetric group is diagonalized in this basis. The eigenvalues determine the irrep of the symmetric group and the eigenvectors are the multi-cluster coefficients of fractional parentage.

A computation of the eigenvalues of the various single-cycle class operators shows[3] that the two-cycle class operator is sufficient to determine the total symmetry of any multi-cluster state for which the total number of particles satisfies $n \leq 5$ and also for $n = 7$. The two and three-cycle class operators are sufficient for $n \leq 14$, and adding the forth-cycle class operator is sufficient for $n \leq 23$.

The calculation of the matrix elements of the various single-cycle class operators is based on expressing each such operator as a sum of intra-cluster and inter-cluster terms. It will be convenient to introduce the symbol $((\alpha\beta\ldots\epsilon))$ that stands for the sum over all the cycles of a given length whose indices belong to the clusters indicated, in the given order. Thus

$$((\alpha\alpha)) = \frac{1}{2} \sum_{i \in \{n_\alpha\}, j \in \{n_\alpha\}} (i,j) \tag{3}$$

and

$$((\alpha\beta)) = \sum_{i \in \{n_\alpha\}, j \in \{n_\beta\}} (i,j) \tag{4}$$

Using these symbols the class of transpositions becomes

$$[(2)]_n = \sum_{\alpha=1}^{m} ((\alpha\alpha)) + \frac{1}{2} \sum_{\alpha,\beta}^{m}{}' ((\alpha\beta)) \tag{5}$$

The prime indicates that the summation is performed over distinct clusters. Similar expressions are obtained for the three- and four-cycle class operators.

The matrix elements of the intra-cluster operators in the basis specified on the right-hand-side of eq. (2) are trivial because each cluster belongs to a given Yamanouchi symbol. Only the calculation of the inter-cluster matrix elements requires further elaboration[3].

Using the procedure described above we can construct multi-cluster states with *arbitrary* permutational symmetry. This makes it very useful for quark cluster models where each degree of freedom i.e., orbital, spin, color and flavor can belong to any irrep of the symmetric group, provided that the overall state is antisymmetric.

REFERENCES

1. A. Arima, H. Horiuchi, K. Kubodera and N. Takigawa, "Clustering in Light Nuclei" in "Advances in Nuclear Physics," vol. 5, M. Baranger and E. Vogt, Ed., (Plenum, N. Y., 1972).

2. A. Novoselsky, J. Katriel and R. Gilmore, J. Math. Phys. **29**, 1368 (1988).

3. J. Katriel and A. Novoselsky, "Multi-Cluster Wavefunctions with Arbitrary Permutational Symmetry", Ann. Phys. (N.Y.) (submitted).

Experimental Nuclear Structure

GAMMA RAY INDUCED DOPPLER BROADENING AND THE DETERMINATION OF LIFETIMES OF EXCITED NUCLEAR STATES

H.G. Börner[1], J. Jolie[1], S.J. Robinson[1], E.G. Kessler[2], S. Ulbig[3], R.F. Casten[4], S.M. Dewey[2], G. Greene[2], R. Deslattes[2], K.P. Lieb[3], B. Krusche[5], J.A Cizewski[6]

1) Institut Laue Langevin, F38042 Grenoble, France
2) Nat. Inst. of Standards and Technology, Gaithersburg, MD 20899, USA
3) Universität Göttingen, D3400 Göttingen, Germany
4) Brookhaven National Laboratory, Upton, N.Y.,, 11973, USA
5) Universität Giessen, D6300 Giessen, Germany
6) Rutgers University, Physics Dept, New Brunswick, N.J., 08903, USA

Abstract

Measurements of lifetimes of excited states in nuclei yield crucial information for sensitive tests of nuclear models. Here a novel method will be discussed which involves the GRID (Gamma Ray Induced Doppler broadening) technique, in which Doppler broadening is observed in a transition from a nucleus recoiling from the emission of a previous gamma ray. As the recoil energy is extremely small, ultra-high energy resolving power has to be used. To date all such experiments have been carried out at ILL using the GAMS4 double flat crystal spectrometer which is operated in a NIST-ILL collaboration.

The method can be used for all lifetimes below a few picoseconds. The wide range of applicability, together with the very exhaustive set of data often obtained, is an advantage with respect to many other methods. The characteristic features of GRID will be discussed using some selected examples.

1. Introduction

Nuclear spectroscopy is an important tool which allows us to improve our knowledge of the nuclear many-body problem and the role forces play within the nucleus. Many complementary techniques have been developed to elucidate the complex interplay of nucleons and generally it is the combination of several of these techniques (for instance radioactive decay, transfer reactions, neutron capture, inelastic scattering, Coulomb excitation, charged particle reactions and others) which allows the construction and interpretation of detailed nuclear excitation schemes.

The neutron capture γ-ray reaction (in contrast to most other reactions) is known to be non-selective in terms of nuclear structure and, hence, provides the opportunity to determine rather complete sets of low lying nuclear states. A famous example is the identification of all intrinsic excitations below ~2 MeV in ^{168}Er [1]. Such studies were possible due to the high resolving power obtained with bent-crystal spectrometers installed at the High Flux Reactor of the ILL [2]. Such a combination simultaneously permits good resolution and high sensitivity. The results of such measurements provide energies, intensities and branching ratios of γ-transitions connecting nuclear states, important ingredients for comparison with nuclear models.

Within the last three years a new tool has been developed, adding to the field of γ-ray spectroscopy the possibility of measuring the *lifetimes* of short lived nuclear excited states, *populated after neutron capture*. This new technique GRID [3] (Gamma Ray Induced Doppler broadening) is based on the measurement of the extremely small Doppler shift characterizing a gamma ray emitted from a nucleus which is recoiling due to the prior emission of another gamma ray. In the following we will focus on the status of this technique by discussing its application to some selected nuclei.

2. The Instrument

The most accurate method for the determination of gamma-ray wavelengths relies on crystal diffraction from nearly perfect flat crystals of silicon or germanium. The highest precision and resolution, obtained to date, is with a two axis flat crystal spectrometer, GAMS4, installed at the Institut Laue Langevin (ILL), Grenoble, and operated in a NIST/ILL collaboration. The drawback of the extremely low luminosity of such a spectrometer is countered to an acceptable level, by using high mass targets ($\approx$ 10 g) in combination with a high neutron flux, such as available at the throughgoing beam tube H6-H7 at the ILL high flux reactor ($\phi_n = 5 \times 10^{14}$ /cm^{2}s). Details of this spectrometer are discussed elsewhere [4,5]. Here we will discuss a special application of ultra-high resolution work: the measurement of detailed characteristics of line shapes (dispersion) rather than line centroids (precision). An extremely important characteristic of a two-axis flat crystal spectrometer, with a view to the study of lineshapes, is that it can be operated in both a dispersive and a non-dispersive mode. This is illustrated in Fig. 1 and described in detail in refs. [3,4]. A typical measurement procedure involves placing the first crystal at θ_{Bragg} and rocking the second crystal (and detector) so that its orientation is $+\theta_{Bragg}$ or $-\theta_{Bragg}$ from the first diffracted beam. Except for very small ($\approx 10^{-7}$) corrections due to finite vertical divergence, the angular separation of these two diffracted beams is $2\theta_{Bragg}$. At $+\theta_{Bragg}$ the two crystals are parallel to each other. This is a non-dispersive geometry (as the crystal planes are parallel to each other, radiation from a broad energy window determined by the solid angle can be reflected simultaneously) and the profile obtained by rocking the second crystal is the *response function* at angle θ_{Bragg}. At $-\theta_{Bragg}$ the two crystals are aligned at $2\theta_{Bragg}$ with respect to each other. This is a dispersive geometry (only exactly one wavelength λ, within the instrumental resolution, can be successively Bragg-diffracted) and as such one measures the <u>wavelength distribution</u> of the observed γ-rays.

The diffraction angles are measured by polarization sensitive Michelson interferometers [4,5] which have a sensitivity of $\approx 10^{-10}$ radians. The best instrumental resolution obtained is currently in the order of $\Delta E/E \approx 2 \cdot 10^{-6}$ (fig.1)

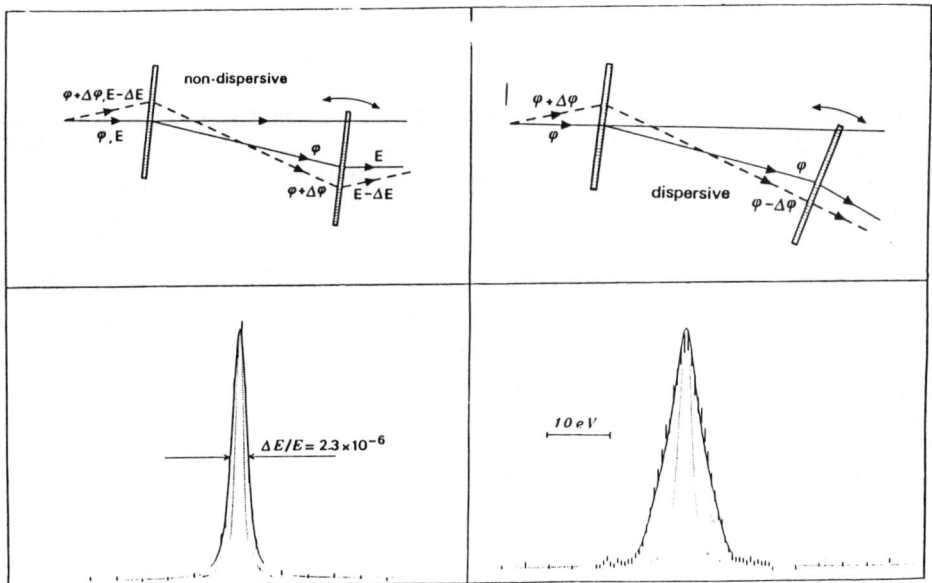

Fig. 1: GAMS4-reflected profiles of a 963 keV transition in ^{152}Sm, emitted after K-capture in ^{152}Eu. In K-capture the nucleus experiences a recoil due to the emission of a neutrino. In this example the recoil is as small as ≈3 eV. Nevertheless, one can clearly discern the Doppler broadening measured in the dispersive mode (right part) by comparison with the lineshape obtained in the non-dispersive mode. A resolution of $\Delta E/E \approx 2 \cdot 10^{-6}$ has been achieved. The dashed line in the non-dispersive data indicates the theoretical lineshape obtained from dynamical diffraction theory calculations for the crystals used (Si, 660 planes, 2.5 mm thickness). The resolution obtained to date is very close to this theoretical limit. The dashed line in the dispersive profile represents again the instrumental response, demonstrating the amount of broadening observed. The spectral distribution of the γ-transition can be obtained by deconvolution of the non-dispersive scan from the dispersive one.

The GRID-Method

For several decades one of the techniques used to study the lifetimes of excited nuclear states has been the Doppler-shift attenuation method (DSAM) [6,7 and refs. therein]. In reactions where accelerated heavy ions interact with target nuclei in a solid medium, the subsequent reaction products can recoil with typical velocities of the order of 0.1 to 10% of the velocity of light. The Doppler shift of gamma rays emitted by the recoil products can be up to several keV and is comparable to the resolution obtained with solid-state detectors. As the shift is attenuated when emission occurs as the nucleus slows down, lifetimes of excited states can then be obtained by comparison with the slowing down time of the recoiling ions in the target. In these reactions one neglects the recoil imparted to the nucleus due to the emission of gamma rays as this leads to recoils which are negligible with respect to the original recoil velocities. However, in the GRID method *it is exactly this small recoil* which gives the *Doppler broadening* we observe, using ultra-high resolution spectroscopy after thermal neutron capture.

When the neutron capture state, at energy E_c, decays via the emission of γ rays this induces a recoil ($v/c = E_\gamma/mc^2$) which leads to velocities $v_R/c \approx 10^{-4}$ to 10^{-6}. The neutron energy of ≈25 meV can be neglected. As these gamma-rays are isotropically emitted this leads to an isotropic distribution of the recoil directions and hence not a Doppler shift, but a Doppler broadening is observed.

The observed Doppler-broadened profile is a function of three separate conditions:

i) the velocity of the recoiling atoms at the moment when the level of interest is populated
ii) the time differential behaviour (slowing down) of the recoiling atoms
iii) the lifetime of the nuclear state which deexcites by the emission of the γ ray being observed.

The principle of the GRID technique is that, provided i) and ii) are known, lifetimes can be extracted.

Fig. 2 summarizes qualitatively different scenarios from which the influence of the velocity distributions (emerging from different population routes) on the expected Doppler profiles can be seen, illlustrating point i.

In all cases we start with the capture of a thermal neutron. The capture state then decays via different routes to an excited state E_x, the lifetime of which we want to measure by determining the Doppler profile of one of its depopulating transitions E_γ. In the simplest case (fig. 2a) the capture state feeds E_x with one primary transition of energy E_p and the nucleus will then recoil with a velocity:

(1) $V_R = E_p/Mc$.

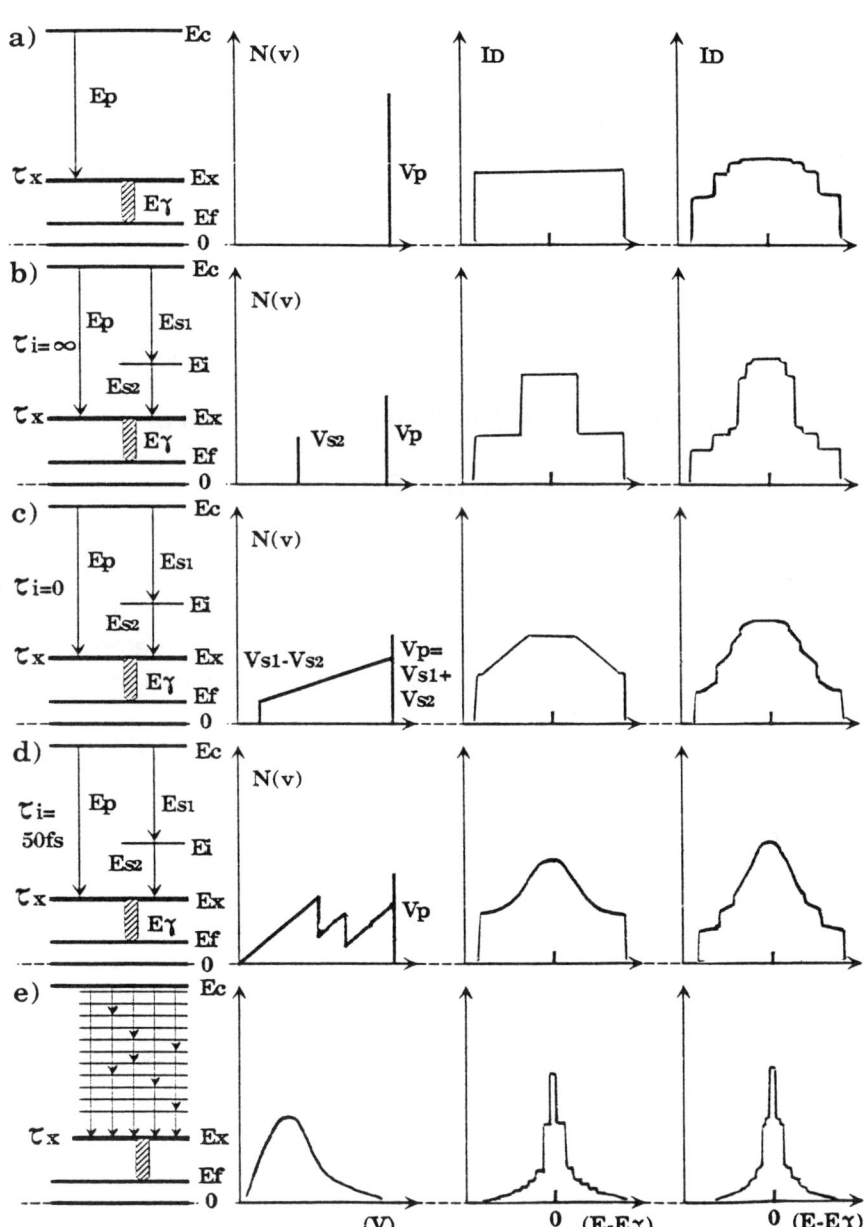

Fig. 2: Five feeding scenarios: Row a): Primary feeding only. b): Two-step cascade with intermediate lifetime $\tau_i = \infty$. c) Two-step cascade with intermediate lifetime $\tau_i = 0$. d) Two-step cascade with finite intermediate lifetime. e) Statistical cascade feeding. τ_i corresponds to the lifetime of the level E_i; v_p, v_{s1} and v_{s2} correspond to the velocity $v/c = E/Mc^2$, induced by the emission of a γ ray with energy E_p, E_{s1} and E_{s2}, respectively.

Then the energy E_γ will be Doppler shifted by:

(2) $\quad \Delta E_D = E_\gamma \cdot V_R/c = (E_\gamma \cdot E_p/Mc^2)\cos\phi$

where ϕ is the angle between directions of observation and recoil. The maximum Doppler shift will be

(3) $\quad \Delta E_D^m = E_\gamma \cdot E_p/Mc^2$

Integrating over all projections we obtain the predicted Doppler profiles. For an infinitely well defined axis of observation, and no slowing down, this leads to a Doppler broadened intensity distribution of:

(4) $\quad I_D(E,v) = $ const, for $(E_\gamma - \Delta E_D) < E < (E_\gamma + \Delta E_D)$
$\quad\quad I_D(E,v) = \quad 0, \quad$ elsewhere

For the case of the two-axis flat crystal spectrometer, with its extremely precise definition of the axis of observation, it only remains to fold these profiles with the instrumental response at E_γ, which has been determined in a non-dispersive scan. The profile discussed above is obtained for a *freely flying* atom.

However, the recoiling nuclei are slowed down in the target material, and this allows us to compare the slowing down time and the lifetime of the excited state E_x:

(5) $\quad \Delta E_D(t) = E_\gamma \cdot v(t)/c$

where $v(t)$ depends on the slowing down process. The decaying intensity of the state E_x at $v(t)$ is a function of the mean time interval between population and decay of this state and, as such, establishes the link between the slowing down time and the lifetime τ_x.

The stopping theory necessary to describe the slowing down depends on detailed information concerning structure and composition of the target material and, hence, the interatomic potentials. Moreover, different recoil velocities might induce completely different mechanisms in the slowing down process. These aspects are reviewed in detail in other contributions [8,9,10].

For the purpose of the discussion here the simplest approach is to approximate the slowing down by discrete binary collisions of the recoiling atoms with other atoms in the bulk of the target (due to the low recoil energy, quantum mechanical scattering effects can be neglected and 'nuclear' scattering is largely dominant). This leads to Doppler profiles summed over the different velocity contributions, which are obtained after collisions with neighbouring atoms:

(6) $$I_D(E) = I_S(E)\exp(-\sum_{k=0}^{N-1} \frac{t_k}{\tau}) + \sum_{n=1}^{N} I_D(E, v_{n-1}) \times$$

$$\times [\exp(-\sum_{k=0}^{n-2} \frac{t_k}{\tau}) - \exp(-\sum_{k=0}^{n-1} \frac{t_k}{\tau})],$$

where $v_k = v_R/2^{k/2}$ is the velocity after the k-th collision and $t_k = r_k/v_k$ is the time between the collisions (k-1) and k. The fraction $I_S(E)$ describes the "stopped peak" due to thermalized atoms [8], while $I_D(E, v_n)$ describes the Doppler profile after the n-th collision. The measured line profiles are then obtained by convolution with the instrumental response function.

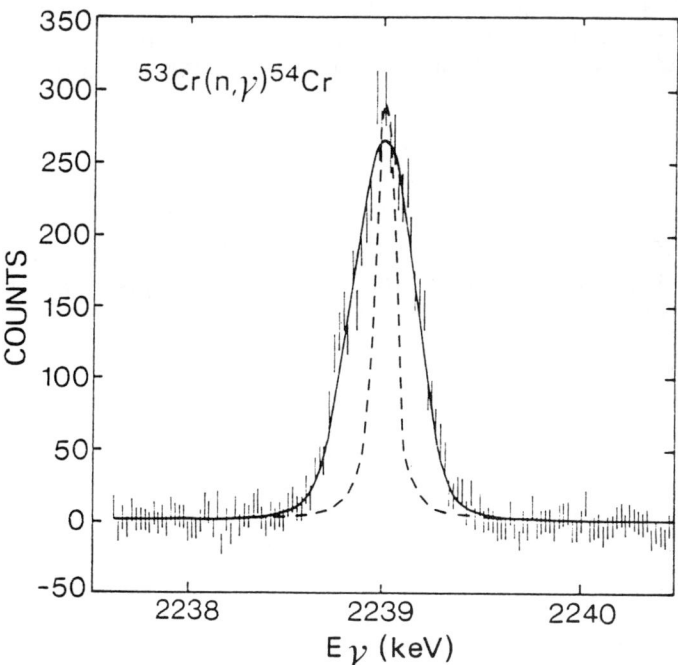

Fig. 3: Doppler-broadened lineshape measured for the transition 3074→835 keV in ^{54}Cr. The dashed line shows the spectrometer response function, the solid line the fitted lineshape.

Examples

I) Primary Feeding

In the example considered above, the level of interest is uniquely fed by one γ-ray (primary). This results in a clearly defined original recoil velocity $v_R/C = E_p/Mc^2$. For the first lifetime measurements [3, 11-13] carried out using GRID, nuclei have been chosen where the levels of interest were fed, to a good approximation, by known, simple patterns. In such a case complications due to unknown feedings and time delays caused by cascade feeding via intermediate states are small, thus allowing a thorough test of the method under transparent experimental conditions. An example of such a measurement is given in ref. [11] where lifetimes of some specific levels in ^{54}Cr have been studied, yielding evidence for a mixed symmetry state at 3074 keV. Fig. 3 shows an example of the line profile obtained for the decay of this 2_3^+ level via the 2239 keV transition to the 2_1^+ state. For the first time a definite lifetime (13(2)fs) was deduced for this level. The figure demonstrates the sensitivity of GRID in this time region. (The characteristics of mixed symmetry states in nuclei near closed shells are reviewed in more detail by S. Robinson et al. in another contribution to this conference).

II) Cascade Feeding

The case of predominant primary feeding, however, is not the standard case in (n,γ), especially in heavier nuclei. More typically, cascade feeding contributes to - or even dominates - the population of a low lying state. The resulting velocity distribution of the recoiling atoms depends on the energies of *all feeding transitions* and the *lifetimes* of the respective *intermediate* levels.

Fig. 2 summarizes some of the most important scenarios which may be involved in different feeding patterns. Starting with the case of predominant primary feeding as discussed above, fig. 2e represents the case of pure statistical population. For each case the corresponding velocity distributions (column 2) and corresponding Doppler profiles (column 3,4) are deduced. Column 3 represents the case for a freely flying particle, which is only observed for very short lifetimes (≈ 1 fs), where the level decays before any slowing down can take place. Column 4 includes, qualitatively, slowing down. This comparison shows that the measured Doppler profiles are quite sensitive to the original velocity distributions.

A model case in ^{49}Ti, representative of the situation cited above, is discussed in another contribution to this conference [14]. In this case the decay of the capture state follows a scenario as depicted in fig. 2d. A primary γ transition feeds (to almost 100%) the 3/2- level at 3260 keV, which in turn decays, via the 1498 keV transition, to the 1762 keV state. We have measured the lifetime of the 3260 keV level to be 17(2) fs. As described in detail in reference [14] one can determine the lifetime of the 1762 keV state as a function of i) the population of the 3260 keV level, ii) the lifetime of the latter and iii) the energy of the directly preceeding transition (1498 keV), as these three parameters allow the determination of the original recoil velocity distribution (including the time differential slowing down) for the decay of the 1762 keV level.

In a more general case one might not know the details of the feeding paths. Can they be reconstructed, and to what extent? To test this, let us assume that we do not know any of the feeding characteristics of the 1762 keV level. Now we fit the measured 1762 keV Doppler profile as a function of a single initial recoil velocity v/c (this corresponds to the scenario described in fig. 2b). We then analyse the respective χ^2 obtained as a function of v/c (fig. 4, curve 1). As the Doppler profiles are sensitive to the recoil velocity, we can correlate the quality of the fit to the velocity distribution N(v). The result is shown in curve 2 of fig. 4 by requiring $N(v) \propto 1/\chi^2$ (and normalizing arbitrarily) Comparison with the velocity distribution calculated from the known decay pattern (curve 3, fig. 4) [14] shows a striking similarity. The mean life of the 1762 keV level, deduced with the known feeding, is 55(5) fs. Using the velocity distribution which results from the shape analysis one obtains ~60 fs, in good agreement. Such a procedure might allow, in a qualitative way, *to test the feeding assumptions* and to verify whether or not important paths are missing.

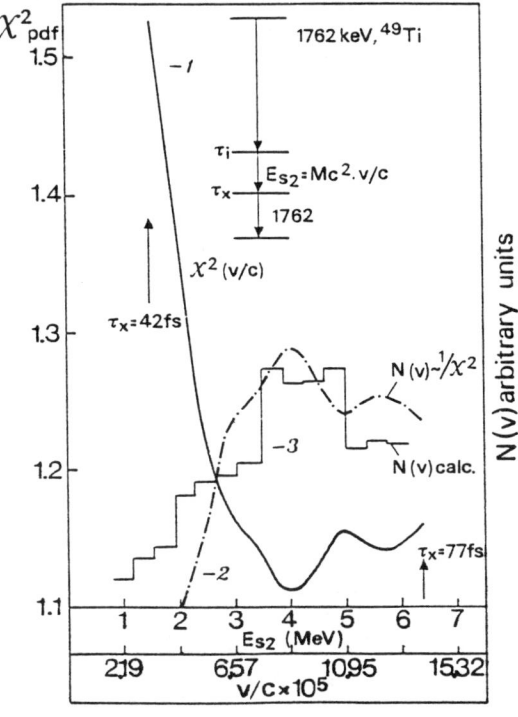

Fig. 4: Investigation of the velocity distribution in a two-step cascade in ^{49}Ti, assuming unknown feeding of the 1762 keV level. N(v) is normalized in an arbitrary manner and therefore reflects a qualitative picture only of the velocity distribution !

III) Statistical Feeding

As mentioned above, cascade feeding procedes via more complex feeding patterns where many intermediate states are involved (fig. 2e). The extension of the treatment from one to many intermediate levels is technically straight forward - provided the respective energies and lifetimes are known. This, however, is generally not the case.

The experimentally missing population strength and the lifetimes of the intermediate states can be approximated by using a statistical model approach. A Monte Carlo code has been developed [15] to simulate these cascades and to deduce the recoil velocity distributions. The level densities are obtained using either a Constant Temperature Fermi gas approach (CTF) [16, 17], or the Bethe parametrisation [16, 17]. The E1, M1 and E2 strengths are obtained using a giant resonance-ansatz. The level distance distribution follows a Wigner distribution (for fixed spins and parities).

Whereas most statistical features depend strongly on the decay branchings, the *time development* is dictated by the decay widths. The recoil distributions are therefore quite sensitive (see fig. 2b-d) to the γ-ray strength functions and may - in turn - provide a method to test them. This, however, will be a rather longstanding effort. An example in ^{144}Nd is described below.

The level scheme of ^{144}Nd had recently been extensively studied [18] by using the ILL-bent crystal spectrometers. We discuss, in the following, one specific state, the 2^+_3 level at 2.073 MeV. The experimental knowledge concerning feeding and decay of this level is summarized in fig. 5a. It appears that ≈85% of the feeding strength has not been observed. This part is simulated by a statistical distribution using the CTF approach. From the measured Doppler profile of the depopulating 1377 keV transition one obtains a mean life $\tau \cong 145^{+40}_{-30}$ fs.

This result is in disagreement with a previous resonance scattering experiment [19] which gave 60(30) fs. If we now wanted to reproduce this latter result we should have to replace the statistical model approach by extreme feeding assumptions with a peaked velocity distribution which corresponds to an energy E_{s2} of ≈1.2 MeV. Fig. 5b shows, in a similar way to the ^{49}Ti case, the χ^2 variation as a function of $v/c=E_{s2}/Mc^2$ and as a function of the deduced lifetime. A lifetime as reported in [19] is incompatible with the above shape analysis. $\tau \approx 60$ fs would require a χ^2 minimum at the corresponding energy. Additionally this is ruled out by the experiment where no strong unplaced γ-transitions are observed in the appropriate energy window. Again, the same interesting feature as observed above emerges from the χ^2 analysis as a function of the variation of v/c: One finds a velocity distribution N(v), which, in this case, is very close to the one calculated from a statistical approach (fig. 5c) and this, inherently, can be seen as a justification of the application of the statistical feeding approach for this state. It should be stressed again that such an analysis is possible due to the very specific shapes of the Doppler profiles. From the slope in fig. 5b, one sees

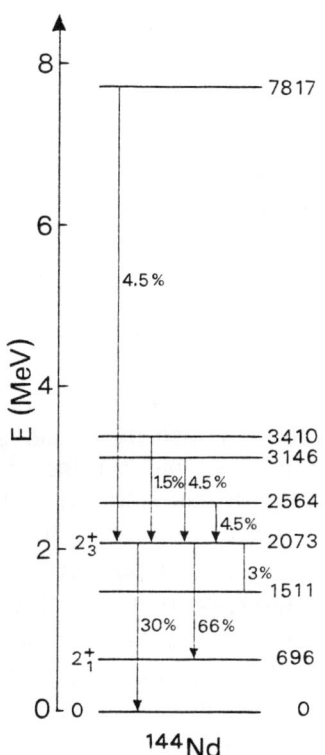

Fig. 5a: Feeding and decay of the 2073 keV, 2^+_3 level in ^{144}Nd

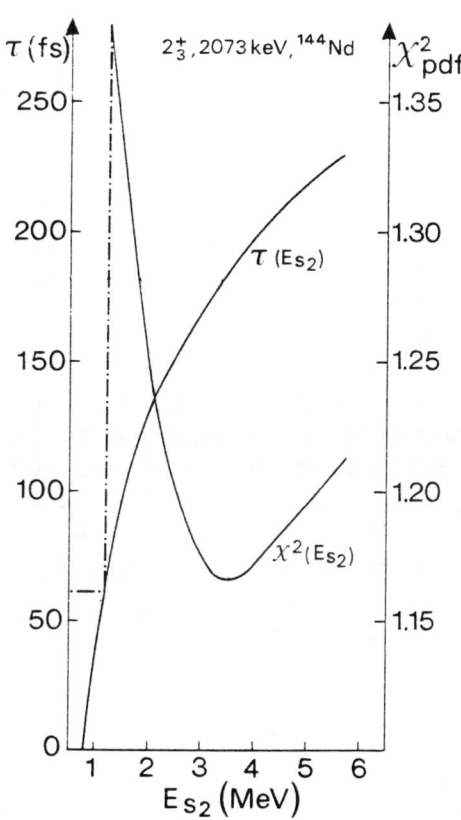

Fig. 5b: Shape analysis for the 1377 keV transition in ^{144}Nd

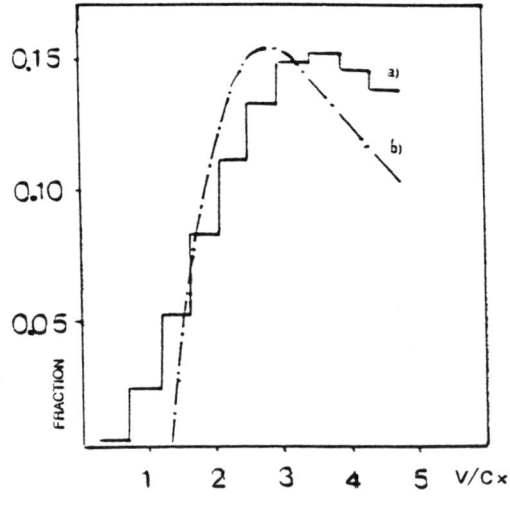

Fig. 5c: Comparison of velocity distributions obtained from
a) a statistical calculation
b) the shape analysis, normalized arbitrarily.

that details of feeding enter more strongly into the Doppler profiles at shorter lifetimes. This leads to two consequences:
a) Uncertainties in the feeding description are less important for longer lifetimes and consequently
b) detailed tests of statistical models can better be done at short lifetimes!

A third case of a reconstruction of the feeding is shown in fig. 6 for a 3⁻, 2022 keV state in ^{168}Er. The shape analysis (χ^2) as a function of v/c yields a recoil distribution N(v) α $1/\chi^2$ which peaks at the velocity which corresponds to essentially primary feeding. Again, this is in agreement with the experiment [1], where one deduces a primary feeding strength of 60%<Ip<95% for this level.

To summarize, we have shown that careful shape analysis of GRID-profiles allows the identification of *trends* in the original recoil distribution which, in turn, are a function of the feeding pattern. These distributions can be compared to those obtained from other inputs, like for instance a statistical model approach. Detailed investigations of this kind are currently in progress. In the examples shown before (fig. 4,5,6) we have also plotted the variation of τ as a function of v/c. One can see that, although the Doppler profiles are very sensitive to the recoil distributions, the τ-values deduced for a given v/c vary in relatively restricted regions, e.g. even for extreme limits of v/c, corresponding to fig2b and 2c, significant lifetime limits can be deduced. A scenario following fig, 2b leads to a *low limit on τ* (depending on the choice of Es2); 2c leads likewise to an *upper limit.*

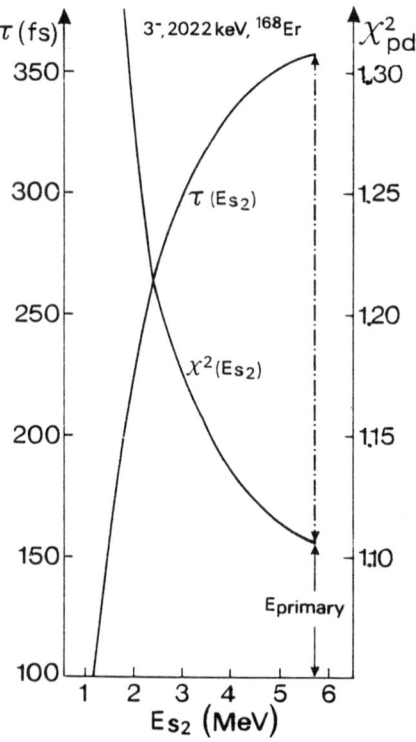

Fig. 6: Shape analysis of the 1942 keV transition in ^{168}Er.

IV) GRID in "heavy nuclei"

In the region of light nuclei many of the lifetimes of levels which can be studied with GRID, can also be investigated using DSAM - at least as long as the lifetimes are not very short. In heavier nuclei, close to or at stability, the situation is different and, generally, only relatively few lifetimes are known, very often studied from Coulomb excitation measurements. Here GRID offers the possibility for more extended studies.

A) ^{196}Pt

^{196}Pt was the first nucleus whose level scheme was interpreted in terms of the O(6) limit of the IBA. It was found that the agreement with predicted O(6) E2 selection rules and branching ratios is excellent. One critical test however, has been lacking to date: The measurement of any absolute transition rates (or limits) for the decay of $\sigma < N$ states, and especially for the 0+ "bandhead" of these families. Fig. 7 shows the experimentally observed decay of the 0_3^+, 1402 keV state (which belongs to the $\sigma = 4$, $\tau = 0$, group) to the 2_1^+ level of the $\sigma = N = 6$, $\tau = 1$, group in ^{196}Pt. In O(6), the selection rules are $\Delta\sigma = 0$, $\Delta\tau = \pm 1$ and therefore the decay of the

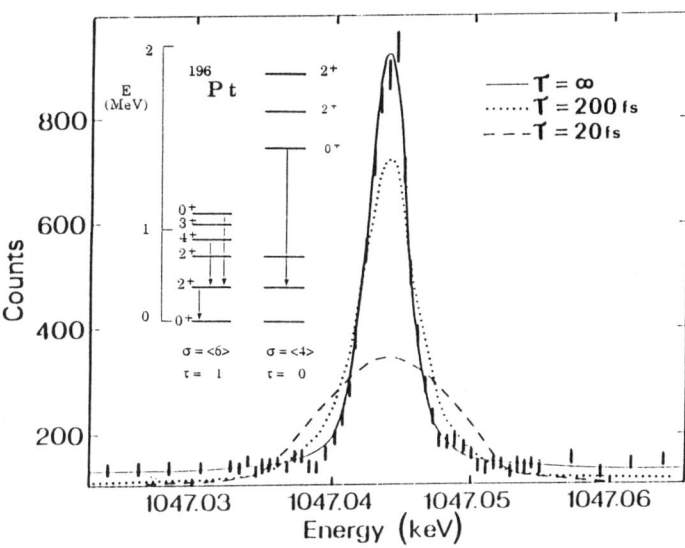

Fig. 7: The Doppler profile observed for the 1047 keV transition in ^{196}Pt, in comparison with the predicted shapes for $\tau = \infty$ (unbroadened shape), 200 fs and 20 fs. It is evident that the peak does not show broadening and thus the *lower limit* of $\tau > 1.86$ ps is obtained.

0_3^+ state should be strongly hindered. The observed decay preserves the stronger τ selection rule ($\tau = 0 \to \tau = 1$) and violates the σ-selection rule. We have obtained results for lifetimes of several levels in ^{196}Pt [20]; however, the principal aim was to obtain a value or limit for the lifetime of the 1402 keV level which decays predominantly by the 1047 keV transition to the 2_1^+ state. Fig 7 shows the data for the 1047 keV transition. A *lower* limit of $\tau > 1.86$ ps is obtained under the assumption of extreme side feeding, following Fig. 2b with E_{S2} corresponding to the highest known secondary feeding transtion. This lifetime limit yields : $B(E2:0_3^+ - 2_1^+) \leq 0.034 e^2b^2$ [20]. This result is in agreement with an order of magnitude hindrance for the σ-forbidden transition in the O(6) limit, since allowed B(E2) values in O(6) are typically $\approx 0.4 e^2b^2$. This provides strong support for the validity of the σ selection rule. Simultaneously it definitely rules out the interpretation of this level in the U(5) limit which predicts a decay 13 times faster than the obtained limiting value.

B) ^{168}Er

For several decades a significant issue in the nuclear structure of deformed nuclei has been whether or not two-phonon collective excitations, such as double gamma vibrations, exist. To date there exist different theoretical approaches treating this problem, which have reached different conclusions. The nucleus ^{168}Er offers an excellent opportunity to test these different models: The 2+ bandhead of the γ vibration lies at the low energy of 821 keV, so that the two-phonon gamma vibration can occur in a relatively clean part of the spectrum. From an extensive earlier study [1] one knew that the lowest possible candidate for a two-phonon $K^\pi = 4+$ gamma vibration was the 4+ bandhead at 2.055 MeV. This is the lowest lying intrinsic $K^\pi = 4+$ excitation and has also the expected energy and decay properties of a dominant E2 branch to the single phonon γ-vibration. However, the absolute strength, $B(E2; 4+ \to 2+\gamma)$, was unknown.

Therefore we have measured the lifetimes of the 4+ and 5+ members of this band [21], at 2055 and 2169 keV, respectively, by measuring the Doppler broadening of the transitions of 1234 keV ($4+ \to 2+\gamma$) and 1277 keV ($5+ \to 3+\gamma$). Fig. 8 shows a summation over 10 individual scans for the 1234 keV transition. The data are also compared to the predicted lineshapes for $\tau = 1$ ps and $\tau = 40$ fs. From the limits for the lifetimes of the 4+ and 5+ levels, we obtain $0.014\ e^2b^2 < B(E2:4+ \to 2+\gamma) < 0.041\ e^2b^2$ and $0.017\ e^2b^2 < B(E2:5+ \to 3+\gamma) < 0.090\ e^2b^2$, respectively. These values are of the same order of magnitude as the known value of $B(E2:2+\gamma \to 0+) = 0.0264(9)\ e^2b^2$ and thus clearly demonstrate that these transitions are collective in nature. These results show, for the first time, definite evidence for the _existence of two-phonon vibrational states_ in a deformed nucleus.

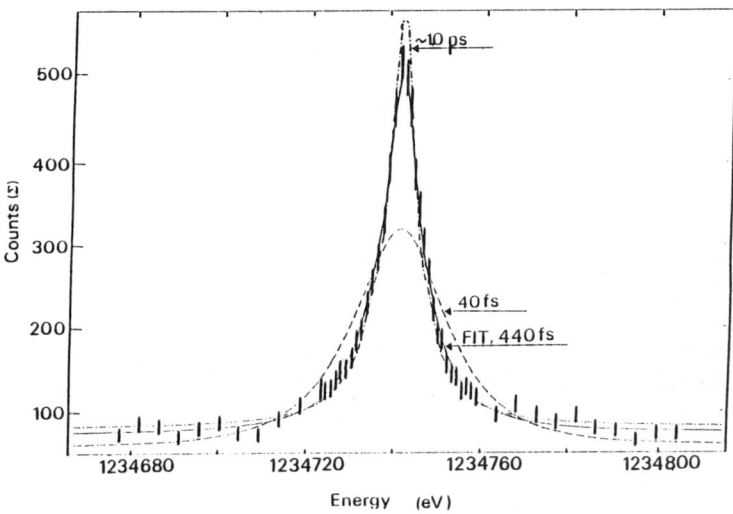

Fig. 8: The Doppler profile measured for the 1234 keV transition in ^{168}Er, shown together with the predicted shapes (dashed lines) for τ = 10 ps and τ = 40 fs.

Conclusion

We have reviewed a new method to measure lifetimes of excited nuclear levels, which uses Gamma Ray Induced Doppler-broadening, obtained after neutron capture and the successive decay of excited states. As the recoil energy is extremely small ultra-high energy resolving power has to be used. To date all such experiments have been carried out at the Institut Laue Langevin using the GAMS4 double flat crystal spectrometer which is operated in a NIST/ILL collaboration. The method can be used for all lifetimes below several picoseconds. It is accessible to all nuclei, throughout the nuclear chart, which can be reached by neutron capture. In cases, where the feeding path has to be simulated by a statistical approach one profits from the fact that the data are not overly sensitive to the details of the statistical model feeding (which in turn complicates the task, if one wants to test these statistical models). Even the limits derived from extreme feeding assumptions define a region of lifetime values which still allows definitive nuclear structure interpretations.

Work has been supported in part under cotract No. DE-AC02-76CH-00016

References

[1] W.F. Davidson, D.D. Warner, K. Schreckenbach, H.G. Börner, J. Simic, M. Stojanovic, M. Bogdanovic, S. Koicki, W. Gelletly, R.F. Casten, G.B. Orr and M.L. Stelts, J. of Phys. G7, 455, 843 (1981)

[2] H.R. Koch, H.G. Börner, J.A. Pinston, W.F. Davidson, R. Roussille, J.C. Faudou, O.W.B. Schult, Nucl. Inst. Meth., 175, 401 (1980)

[3] H.G. Börner, J. Jolie, F. Hoyler, S. Robinson, M.S. Dewey, G. Greene, E. Kessler, R.D. Deslattes, Phys. Lett. B215, 45 (1988)

[4] M.S. Dewey, E.G. Kessler, G.L. Greene, R.D. Deslattes, H.G. Börner, J. Jolie, Nucl. Inst. Meth. A284, 151 (1989)

[5] E.G. Kessler, G.L. Greene, R.D. Deslattes, H.G. Börner, Phys. Rev. C32, 374 (1985)

[6] P.J. Nolan and J.F. Sharpey-Schaffer, Rep. Prog. Phys. 42, 1 (1979)

[7] J. Keinonen, in: Proc. 5th Int. Symp. of Capture gamma-ray spectroscopy AIP Conf. Proc., Vol 125, 557 (1984)

[8] J. Jolie, H.G. Börner, S.J. Robinson, Analysis of Recoil Doppler profiles obtained with Ultra High Resolution (n,γ)-spectroscopy, contribution to this conference

[9] J. Jolie, S. Ulbig, H.G. Börner, K.P. Lieb, S.J. Robinson, P. Schillebeeckx, E.G. Kessler, M.S. Dewey, G.L. Greene, Europhys. Lett. 10, 231 (1989)

[10] A. Kuronen, submitted to Phys. Lett. and J. Keinonen, contribution to this conference

[11] K.P. Lieb, H.G. Börner, M.S. Dewey, J. Jolie, S.J. Robinson, S. Ulbig, Ch. Winter, Phys. Lett. B215, 50 (1988)

[12] S. Ulbig, K.P. Lieb, Ch. Winter, H.G. Börner, J. Jolie, S.J. Robinson, P.A. Mando, P. Sona, N. Taccetti, M.S. Dewey, J.G.L. Booten, F. Brandolini Nucl. Phys. A505, 193 (1989)

[13] S. Ulbig, K.P. Lieb, H.G. Börner, J. Jolie, S.J. Robinson, J.G.L. Booten, M.S. Dewey and Ch Winter, XXV Zakopane School on Physics (1990), World Scientific, in press

[14] S. Ulbig, J. Jolie, H.G. Börner, M.S. Dewey, K.P. Lieb, S.J. Robinson, "Grid lifetime study of the 1762 keV state...", contribution to this conference

[15] B. Krusche et al., to be published, and contribution to this conference

[16] T. v. Egidy, H.H. Schmidt and A.N. Behkami, Nucl. Phys. A481, 109 (1988)

[17] B. Krusche and K.P. Lieb, Phys. Rev. C34, 2103 (1986)

[18] S.J. Robinson et al. , to be published

[19] F.R. Metzger, Phys. Rev., 187, 1700 (1969)

[20] H.G. Börner, J. Jolie, S.J. Robinson, R.F. Casten, J.A. Cizewski, Phys. Rev., in press

[21] H.G. Börner, J. Jolie, S.J. Robinson, B. Krusche, R. Piepenbring, R.F. Casten, A. Aprahamian, J. Draayer, submitted to Phys. Rev. Lett.

ANALYSIS OF RECOIL DOPPLER PROFILES OBTAINED WITH ULTRA HIGH RESOLUTION (n,γ) SPECTROSCOPY.

J. JOLIE, H.G. BÖRNER, S.J. ROBINSON,
Institut Laue Langevin, 156X, 38042 Grenoble CEDEX, FRANCE.

ABSTRACT

The theory used to analyse Gamma-ray Induced Doppler broadening (GRID) is reviewed. The analysis of GRID measurements is done using a collision model which can be used for both monoatomic and polyatomic targets. Very low recoil measurements, as induced by a neutrino (NID), are analysed using a phonon creation model. Comparison with dedicated experiments is discussed.

1. INTRODUCTION.

At the previous capture gamma ray conference, we discussed the first results of Gamma Ray Induced Doppler broadening (GRID) measurements[1,2] performed using the GAMS-4 spectrometer[3,4,5]. In the last three years, the GRID method has been used to measure lifetimes of nuclear excited states in nuclei ranging from ^{37}Cl up to ^{196}Pt. Part of these experiments were dedicated to test and/or extend the method. In this contribution we review these, as well as the method itself. Some experiments dealing more with specific nuclear physics topics are reviewed in another contribution to this conference[6].

The GRID method uses the (n,γ) reaction to produce the nucleus in an excited state. The lifetime of this state is then determined by measuring the Doppler-broadened energy profile of depopulating gamma rays. This broadening may be due to the prior emission of gamma rays, or to a previous beta decay, which give a recoil to the atom, but it can also originate from the thermal motion of the atoms in the target. In contrast to the recoils used in DSAM the recoil velocities used in the GRID method are always very small. Therefore the description of the lineshapes is quite different from that used to describe the large recoils obtained at accelerators. For instance, the recoiling atom slows down within some tens of Ångströms. It is clear that under such conditions no standard backing material can be used. Moreover, the known stopping theories become questionable at such slow velocities. In fact the recoil velocities (0.01% to 0.0001% of c) are much lower than the recoils obtained at accelerators but also higher than the typical binding energies of the solid state. To overcome these problems we have constructed, using simple assumptions, a description of the slowing down process.[2,7,8,9] This description does not attempt to be fully 'microscopic', i.e. it does not include the crystal structure. Instead it is aimed to develop a reliable, yet simple and flexible, method which may be used for many different targets. To this purpose many dedicated experiments have been performed to verify this stopping theory.

2. FORM OF THE LINESHAPE FOR A CONSTANT VELOCITY.

In order to describe the measured profiles we first derive the form of the Doppler broadened lineshape for a transition with energy E_0 coming from a set of recoiling atoms with constant velocity v and observed under a very small solid angle. As GRID can assess very short lifetimes, we take into account the natural linewidth Γ. If the angular distribution between the direction of recoil and the axis of observation is isotropic then the lineshape takes the following form[2]:

$$I_D(E,v) = \frac{Ac}{2E_0\pi v}[\arctan(\frac{2}{\Gamma}(E-E_0(1-\frac{v}{c})))-\arctan(\frac{2}{\Gamma}(E-E_0(1+\frac{v}{c})))] \quad (1)$$

Gamma-gamma correlation between the feeding and decay gamma will lead to an anisotropic distribution between the observed gamma ray and the initial recoil direction. For transitions of multipolarities L = 1 and 2, we have an angular correlation given by $D(\theta) = (1 + a_2P_2(\cos(\theta)) + a_4P_4(\cos(\theta)))$, where the parameters a_2 and a_4 are a function of the multipolarity of the cascading transitions and the spins of the states involved. Then we obtain for the lineshape[9]:

$$I_D(E,v) = (1-a_2/2+3a_4/8)I_0(E,v) + (3a_2/2-30a_4/8)I_1(E,v) + (35a_4/8)I_2(E,v) \qquad (2)$$

with $I_0(E)$ given by (1) and (setting $\beta = v/c$ and $x = \beta E_0$):

$$I_1(E,v) = (A\Gamma/4\pi)\{2x^{-2} + (E-E_0) x^{-3}[\ln((E-E_0(1+\beta))^2 + \Gamma^2/4) \\ -\ln((E-E_0(1-\beta))^2+\Gamma^2/4)]\} + ((E-E_0)^2 - \Gamma^2/4) x^{-2} I_0(E,v) \qquad (3)$$

$$I_2(E,v) = A\Gamma/(6\pi x^2) + 2A\Gamma (E - E_0)^2/(\pi x^4) - ((E-E_0)^2 + \Gamma^2/4)I_1(E,v)x^{-2} \\ + 2(E-E_0)^2((E-E_0)^2 - 3\Gamma^2/4)I_0(E,v)x^{-4} \qquad (4) \\ + A\Gamma(E-E_0)(3(E-E_0)^2 - \Gamma^2/4)[\ln((E-E_0(1+\beta))^2 + \Gamma^2/4) - \ln((E-E_0(1-\beta))^2+\Gamma^2/4)] /(4\pi x^5)$$

This anisotropy is partly lost when the atom scatters off a target atom and completely after some more scatters. In most of the GRID measurements the lifetimes are long enough to neglect the correlation and the much simpler formula (1) can be used. Only for very short lifetimes and for those decays occuring before a collision with any other atom can occur formula (2) should be used instead of (1).

The 3428 keV level in ^{57}Fe is an ideal example of such a very short lived level[10]). In Figure 1 we show the summed profiles of six individual scans of the 2721.3 keV transition, which deexcites this level. Although the nucleus is bound in a solid state target the observed peak corresponds to the 430 eV wide maximal broadening due to free recoil originated by the pure primary feeding from the capture state. The transition is part of a $1/2^- \to 3/2^+ \to 5/2^+$ sequence so only a_2 in formula (2) will determine the correlation. The analysis of the peak allows the extraction of both, the lifetime and the

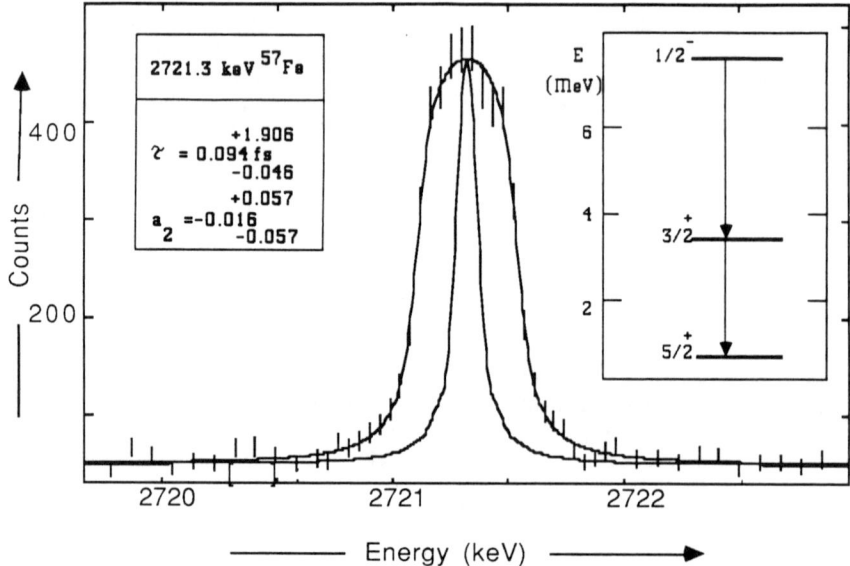

Figure 1: The fully Doppler broadened lineshape of the 2721.3 keV transition in ^{57}Fe. The inserts show the fitted values for τ and a_2, with their 1σ errors, as well as the gamma cascade.

a_2 value, respectively. As a_2 can be used to determine the mixing ratio δ, this allows, in principle, the simultaneous absolute measurement of the B(E2) and B(M1) of the transition. One should, however, keep in mind that the values will have large errors due to the insensitivity of the lineshapes to the lifetime in the region 10^{-15}- 10^{-17}s. The upper limit is given by the absence of any slowing down while the lower is due to the effect of the natural linewidth Γ.

3. DESCRIPTION OF THE LINESHAPES.

After the emission of a primary gamma ray γ_1 the nucleus will start to recoil with an initial velocity $v_R = E\gamma_1/mc$, where m is the mass of the recoiling atom. Since the nucleus is moving in the target it cannot continue to recoil freely with the same velocity v_R. After a time t its velocity will have changed to v'. When it then emits a secondary gamma ray γ_2 the lineshape will be given by formula (1) but now with velocity v'. The total lineshape of γ_2 will thus be a function of the slowing down which is a function of time and the lifetime τ of the level which is deexcited. To obtain the lifetime one needs a description of v(t), the velocity of the recoiling atom in the sample. We will describe this process in a discrete way by splitting it up into a set of averaged distances r(k) in which the atoms will have a mean velocity v(k) (e.g. $v_R=v(0)$) to be calculated. To travel each distance r(k) the atom will need a time t(k). Inserting the radioactive decay law and considering the atom to be thermalised after the N th distance, the total lineshape is then given by[2]:

$$I(E) = N_0 \{ I_S(E) e^{-\Sigma_{k=0}^{N-1} t(k)/\tau} + \Sigma_{n=1}^{N} I_D(E,v(n-1))[e^{-\Sigma_{k=0}^{n-2} t(k)/\tau} - e^{-\Sigma_{k=0}^{n-1} t(k)/\tau}]\}. \quad (5)$$

with $I_D(E,v)$ given by formula (1) [or (2) for $I_D(E,v(0))$]. Once v(k) and r(k) have been determined formula (5) contains two parameters: the lifetime τ and the total number of steps N. N will be determined such that after the last scatter $v < v_{thermal}$ and τ will be fitted. For the stopping peak $I_S(E)$ we use a Doppler broadened lineshape of a Maxwellian velocity distribution around a thermal velocity v_T:

$$I_S(E) = \int_0^{+\infty} 4\pi (3/2\pi \, v_T^2)^{3/2} \exp(-1.5(v/v_T)^2) \, v^2 \, I_D(E,v) dv \quad (6)$$

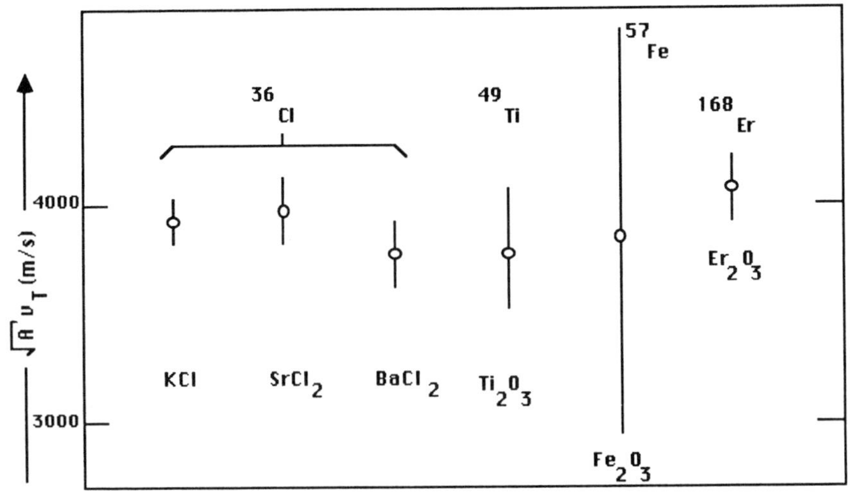

Figure 2: The thermal velocities v_T scaled with $\sqrt{A}$, A being the mass number of the recoiling atom.

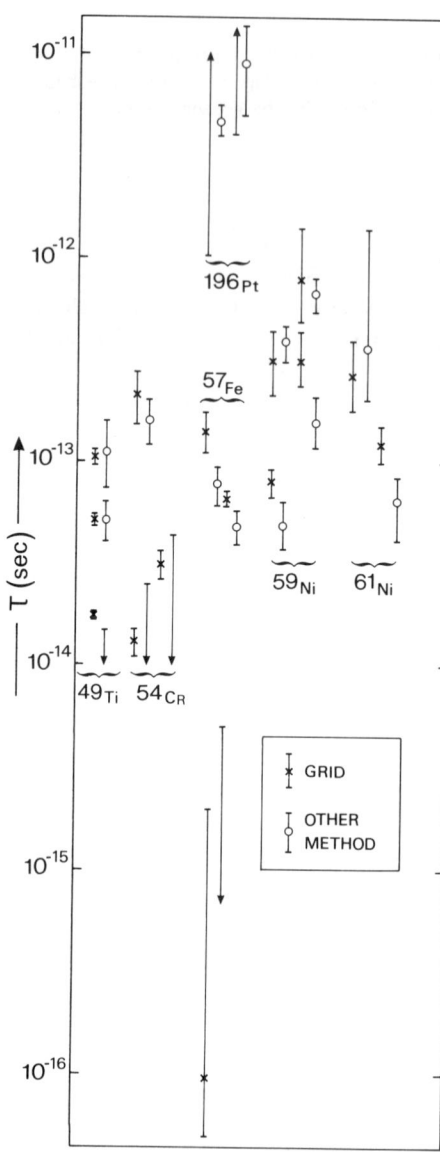

Figure 3: Comparison between the lifetimes obtained with GRID to those obtained by other methods, for monoatomic targets. They are taken from ref 9,11,12,13 and a recent remeasurement of Fe (see also Figure 1).

with $I_D(E,v)$ given by (1). Having no analytic solution this integral has to be solved numerically. That this formula describes the thermal motion of the atoms is nicely illustrated by the values fitted in the different experiments given in Figure 2. We see that they scale as given by $3kT = mv_T^2$ yielding a temperature which is nearly independent of mass and composition of the target.

4. SLOWING DOWN OF ATOMS MOVING IN MONOATOMIC TARGETS

The slowing down arises from collisions of the recoiling atom (which has an energy of a few hundreds of eV) with other atoms in the target (nuclear stopping). The electronic stopping, which is dominant at higher recoils, is small and can be neglected. We will describe these as binary elastic collisions and let the average distance inbetween the successive collisions define the steps k in formula (5). Thus the atom moving in the target with velocity $v(k)$ will encounter another atom after a distance $r(k)$. First we average, for one collision, the energy losses. The target atom can be considered to be at rest since $v(k) \gg v_T$. Neglecting the small difference between the masses as well as the displacement energy, on average the recoiling nucleus will loose half of its kinetic energy in the collision. Then the atom will continue its recoil but now with a mean velocity $v(k+1) = v(k)/\sqrt{2}$. The next step is the determination of $r(k)$. Here we take into account the shape of the repulsive interatomic potential. The best analytical approximation of the real repulsive interatomic potential is the Born-Mayer potential:

$$V(r) = A\, e^{-r/a}. \qquad (7)$$

It is appropriate for the description of the interaction of atoms at separations between 0.5 to 2 Å and is known to describe best the form of the quantum mechanical potential. Moreover its parameters have been fitted to a Thomas-Fermi-Dirac description of pair potentials for all atoms[14]. In order to obtain an estimation for $r(k)$ we define the atomic radius as the distance of closest approach between the two atoms during the collision. This radius $d(E)$ will now depend on the kinetic energy E of the recoiling atom. From $V(2d(E)) = E/2$ we obtain $d(E) = a.\ln(2A/E)/2$. Then $r(k)$ (k referring now to the number of collisions the atom has undergone) is taken to be the mean free path $r(k) = 1/(n\sigma)$, where n is the number of atoms per unit of volume and $\sigma = \pi\, d(E)^2$, and hence:

$$r(k) = V/\pi(a \ln(4A/mv(k)^2))^2 \quad (8)$$

where a and A are the parameters of the potential and V is the volume occupied per atom.

Although this approach is a mixture of different aproximations its application to monoatomic targets gives lifetimes in good agreement with the known lifetimes, as illustrated in Figure 3.

5. DETERMINATION OF THE LIFETIMES WITH POLYATOMIC TARGETS

Since the target temperature at the in beam position is high (~400°C including self heating) the choice of monoatomic targets is limited mainly to metals. In order to study other atomic nuclei the target will have to be made from salts, alloys, oxides etc. which contain the nucleus under study. Therefore we extend our model first to diatomic targets[8] of the form XY. Since we are dealing with targets of the type XY, in which the atom X is recoiling, we will need not only the Born-Mayer potential between the identical atoms but also between the atoms X and Y. This can be obtained from the values for identical atoms using the combination rule of Smith[15]:

$$(A_{xy}/a_{xy})^{2a_{xy}} = (A_x/a_x)^{a_x} (A_y/a_y)^{a_y} \quad (9)$$

which gives the parameters A_{xy} and a_{xy} for the interatomic potential between the atoms X and Y, as a function of the values A_i and a_i (i=x,y) of the interaction between the identical atoms i. Formula (9) contains two unknowns but we can take $a_{xy}= (a_x+a_y)/2$ since the variation of a_i with atomic number is small. Following the same approach as in the preceding paragraph the mean free path r(k) between the collisions of an atom x on an atom y is given as a function of the Born-Mayer parameters A_{xy} and a_{xy} by[8]:

$$r(k) = V/\pi(a_{xy} \ln(2A_{xy}(m_x+m_y)/m_xm_yv(k)^2))^2. \quad (10)$$

with V being the volume occupied by a single atom. After the collision the average velocity is given by $v(k) =[1-2m_xm_y/(m_x+m_y)^2]^{1/2}v(k-1)$. In the diatomic case we now have to define the collision sequence. As all nearest-neighbours of the X atoms are Y atoms, and vice versa, we assume that a recoiling X atom first hits a Y atom and use the formulas derived above. Now being near a Y atom position it will hit preferentially an X atom and we take the mean free path and mean energy loss as derived in section 4. Thus the sequence of collisions is taken to be X on Y, X on X and so on. This double layer approach is clearly an approximation and only holds when the scattering lengths r(k) are less than or equal to the interatomic distance. On the other hand this approach can be seen as a first approximation of the real probability distribution for the collision sequence.

Up to now we have only discussed diatomic targets. In many cases it will be more interesting to use polyatomic targets rather than diatomic. In order to extend the double layer approach for these targets we will use the following rule: The collision sequence is a repetition of a 'principal sequence' that keeps the stochiometric formula of the target material. All partitions of this principal sequence will be taken into account. For instance for Cl recoiling in $SrCl_2$ the principal sequence can be Sr-Cl-Cl, Cl-Sr-Cl or Cl-Cl-Sr. In the last case we then describe the slowing down as a sequence **Cl**-Cl-Sr-**Cl**-Cl-Sr-**Cl**-Cl-Sr-**Cl**-Cl-Sr. The dependence on the sequence is not overly strong, as illustrated in Figure 4 which shows the influence on the lifetime obtained [16] for the 2055 keV level in ^{168}Er. The target used consisted of Er_2O_3.

In order to check the descriptions derived in this paragraph we have performed some dedicated experiments. The first experiment was used to check the extension from monoatomic to diatomic targets by measuring the lineshapes of the short lived states in ^{49}Ti in a pure metallic target and in TiC[9]. Another experiment used ^{36}Cl as recoiling nucleus.[18] Since pure Cl cannot serve as a target we measured the Doppler broadened profiles using four different chlorides: NaCl, KCl, $SrCl_2$, and

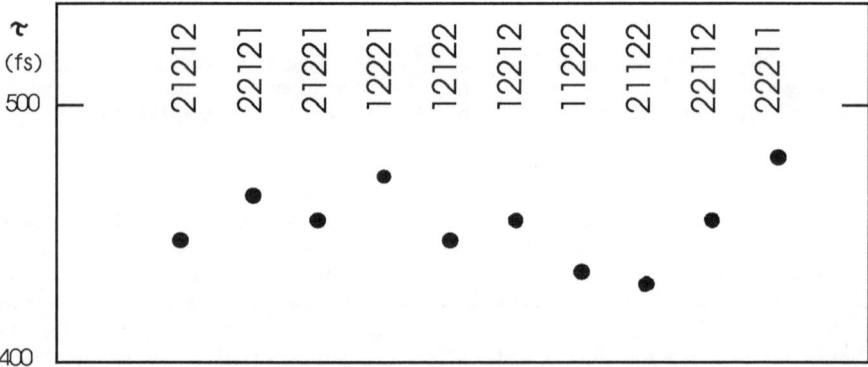

Figure 4: Dependence of the fitted lifetimes on the different collision sequences in Er_2O_3. The lifetimes were obtained using the Bethe formula to describe the statistical feeding.[17] The principal sequence is given above the fitted lifetimes (1 denotes erbium and 2 oxygen).

$BaCl_2$. This experiment was used to test the mass dependence of the slowing down, by changing the mass of the partner atoms in the target and also to test the extension to polyatomic targets. This study confirmed the assumptions made before, as illustrated by Figure 5. In Figure 6 we show all the lifetimes fitted using polyatomic targets in comparison with those measured using other methods. There is good agreement between the lifetimes obtained using GRID amongst themselves and also with the values obtained with other methods. One also notices that the use of targets where the atomic nucleus under study is contained in a chemical compound with lighter atoms, e.g. the oxides A_2O_3, allows the determination of lifetimes in the picosecond region. This is a large advantage when studying medium-heavy and heavy nuclei.

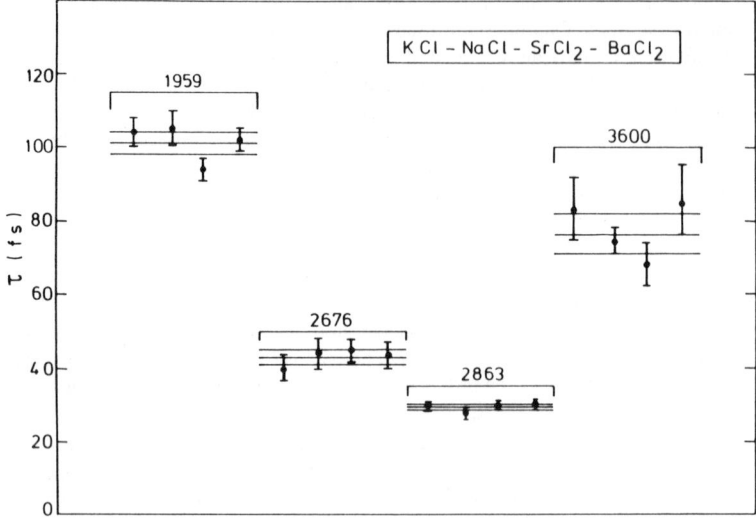

Figure 5: Comparison between the short lifetimes obtained in different Chloride targets.

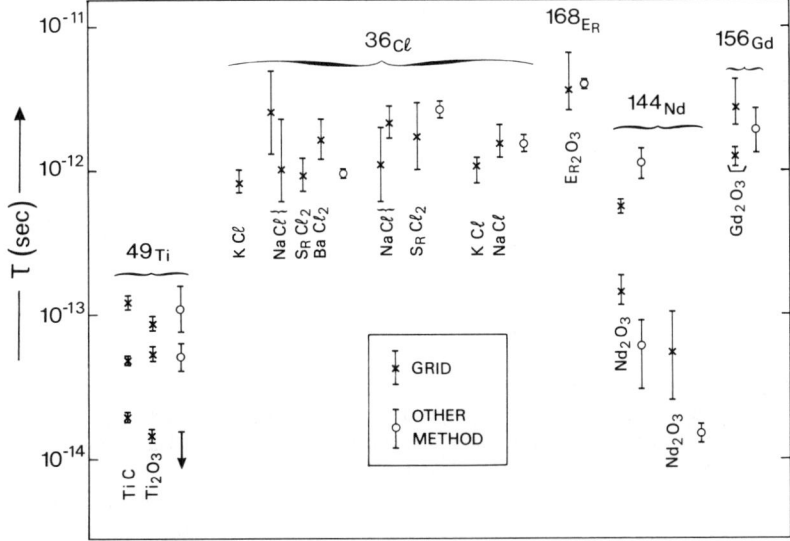

Figure 6: Comparison between the lifetimes obtained with GRID[8,16,18,19,20] for polyatomic targets and those obtained using other methods. The nucleus under study is given and the composition of the target. In some cases two lifetimes are given, corresponding to the values obtained from two different deexciting transitions.

6. GRID AS AN EXPERIMENTAL TOOL TO STUDY INTERATOMIC POTENTIALS.

The GRID method not only allows the study of the lifetimes of the excited states but also the interatomic potential itself[8]). This is done by inverting the method and using the lifetimes as control parameters. When the description of the interatomic potential is good, one should then reproduce the lifetimes of the states. Or better, one measures the same transition but in different chemical compounds. A good slowing down model should then yield consistent lifetimes. In order to really extract the interatomic potential our simple collision model is not appropriate since it is based on statistical averaging of different quantities. Also, its success may be due to positive interference between poor approximations, e.g. defining the atomic radius from the minimum distance in a head-on collision using the calculated Born-Mayer potential parameters.

A more exact calculation can be performed in the following way. Starting from a lattice of atoms and a certain interatomic potential, one can excite one atom to a higher kinetic energy corresponding to the energy of the gamma recoil. In a certain number of time steps one can follow this atom, by minimizing the total energy of all lattice atoms. This clear way, in principle, to describe the slowing-down, is in practice difficult, as one has to recalculate the energy, momentum, and position of all atoms in the lattice under consideration. Moreover GRID observes a random ensemble of recoiling nuclei, and thus one needs to perform this calculation for all possible initial recoil directions. Despite these problems A. Kuronen (University of Helsinki) was able to perform these calculations[21]) for the data we obtained in metallic Ti[7,8]).

He used the combination of molecular dynamics (MD) and Monte Carlo (MC) calculations to simulate the slowing down of low energetic Ti atoms in a lattice containing 216 Ti atoms with a time step of 0.5fs. . About 4000 such events were calculated for each lineshape. The importance in the context of this article lies in the fact that it can be used to check the simple slowing-down model we have derived. Two assumptions of our model were thus tested in detail. First it was found that the electronic energy loss of the recoiling ^{49}Ti atoms had a negligible effect on the lineshapes. Secondly in these calculations the attractive part of the interatomic potential has to be taken into account. It was found that the exact form of this attractive potential did not influence the lineshapes. The correctness of our lifetimes could not yet be tested in detail due to the large choice of interatomic potentials. As yet the consistency of the MD and MC description has not been tested by calculating the lineshapes in a TiC and/or Ti_2O_3 matrix.

In principle this kind of prescription should be the ideal way to analyse the GRID data. However, the large numerical effort argues against a standard use of this method It is desirable that more MD and MC calculations, for a selected number of targets, be used to check the lifetimes obtained in the simple model. In the meantime the simple model can be used, as this leads to a 'phenomenologically' correct and consistent description.

7. VERY SMALL RECOILS.

Up to now we have been using the recoil originating from gamma-ray emission. It is, however, possible to use other nuclear reactions to produce the extra kinetic energy leading to Doppler broadened energy profiles of subsequently emitted gamma rays[4,9]. In this section we will concentrate on a specific kind of beta-decay, namely K-electron capture. There the atom as a whole will, due to the emision of the neutrino, obtain a recoil velocity given by $v_R = Q/m(^AY)c$, with Q the Q-value of the reaction and $m(^AY)$ the mass of the final atom (neglecting secondary effects such as X-ray emission or emission of Auger electrons). This particular beta decay reaction gives, as in the case of gamma emission, a unique value for the initial recoil energy, provided the Q value is less than 1.022 MeV. In contrast to gamma emission it can induce a very small recoil energy of the order of a few eV to the final nucleus.

We now focus our attention on K-electron capture decay from the 0⁻ isomeric state of ^{152}Eu to the 1⁻ level at 963 keV in ^{152}Sm. The neutrino emission gives the final nucleus a recoil energy of 3.0 eV. Initially this reaction was chosen for reasons which are more connected to fundamental physics[22]. However, the electron capture reaction can also be useful for nuclear physics, since it allows a remeasurement, with better precision, of the experiments performed using nuclear resonance fluorescence with the ultra centrifuge technique[23], and thus will yield lifetimes of nuclear excited levels fed by β⁻ decay. The very low initial recoil velocity completely changes the slowing down process due to the fact that the recoil energy does not allow the atom to leave its lattice position. This is most drastically illustrated by the case of Eu_2O_3 and EuF_3. Although for EuF_3 the volume occupied per atom is smaller the observed Doppler broadening is larger, in contrast to what one expects from a collision model.

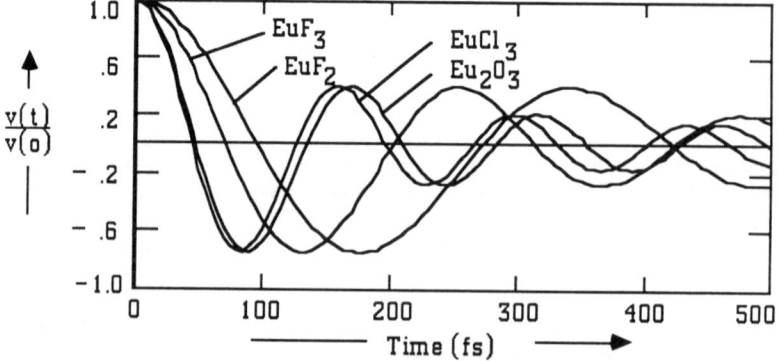

Figure 7: The velocities as function of time extracted from the Doppler broadened lineshapes for four different targets containing Europium with the use of formula (10).

Using the nuclear resonance fluorence technique Langhoff et al. have studied the slowing down of atoms with very small kinetic energies in solids. They concluded that the phonon model (PM) gave the best description for these small recoils[24]. We have used the simplest version of the phonon model, which is the Debye approximation[9]. Then the recoil velocity is given by[24]:

$$v(t) = 3v_0\{(2/\omega_D^2 t^2)\cos(\omega_D t) + ((1/\omega_D t) - (2/\omega_D^3 t^3))\sin(\omega_D t)\} \quad (10)$$

with ω_D the Debye frequency. The Doppler broadened lineshape can then be determined following the method outlined in ref 9). Since the recoil velocities are not much above the thermal velocity we now take the thermal broadening into account as a Gaussian which is folded over I(E)dE. The thermal width σ_T is determined from a measurement of a transition decaying from a level with a known long lifetime . In order to illustrate the slowing down we give, in Figure 7, the v(t) obtained for Sm recoiling in four different target compositions. The fitted Debye frequency has to be seen as an 'effective' frequency or a parameterization of the slowing down process.

8. CONCLUSION

We have reviewed the stopping theory elaborated for the analysis of GRID measurements. Since the initial recoil is of the order of a few hundreds eV/atom or lower, the stopping is dominated by atomic collisions (nuclear stopping) and the exchange effects between the electrons (electronic stopping) are neglected. The description used three years ago[2] has been substantially improved by taking into account the interatomic potential. Since we use tabulated parameters for the Born-Mayer potential no fitting of the slowing-down process for each target is needed. This approach was successfully applied to many monoatomic targets. In order to describe the slowing down in more complex targets we have extended the theory to polyatomic targets[8]. This allows the study of the interatomic potential in different compounds and to test the consistency of the stopping theory. We have also shown how very low energy recoils, as induced by neutrino emission after electron capture, lead to a different slowing down process. These can be described using a phonon creation model based on the Debye approximation. In this case an 'effective' Debye frequency has to be fitted to the observed lineshape. In this contribution we did not mention how the problems related to the feeding of the level whose lifetime one wants to determine are treated. They are discussed in detail in other contributions to this conference.[6,17,25]. Finally a computercode, called GRIDDLE[26], has been developed for the efficient analysis of GRID data.

In conclusion, we have developed a stopping theory to describe the slowing down of recoiling atoms as occuring in GRID and NID measurements. Notwithstanding the crude approximations the theory yields good results and is almost parameter free.

We want to acknowledge E.G. Kessler, M.S. Dewey, J.R. Greene and R.D. Deslattes ; F. Hoyler and P. Schillebeeckx; S. Ulbig and K.P. Lieb; J. Keinonen and A. Kuronen for the many contributions they made to the development of the GRID method.

REFERENCES.

1) H.G. Börner, in IOP Conf. Ser No88 (1988) p.143.
2) J. Jolie, H.G. Börner, F. Hoyler, S. Robinson and M. S. Dewey, in IOP Conf. Ser. No. 88, (1988)p. 586.
3) E.G. Kessler, G.L. Greene, M.S. Dewey, R.D. Deslattes, H.G. Börner and F. Hoyler, in IOP Conf. Ser 88 (1988) p.167.
4) M.S. Dewey, E.G. Kessler, G.L. Greene, R.D. Deslattes, H.G. Börner and J. Jolie, Nucl. Instr. and Meth. A284 (1989) 151.
5) E.G. Kessler et al. contr. to this conference.
6) H.G. Börner et al. contr. to this conference.

7) H.G. Börner, J. Jolie, F. Hoyler, S.J. Robinson and M. S. Dewey, G.L. Greene, E.G. Kessler Jr., R.D. Deslattes, Phys. Lett. B. 215 (1988) 45.
8) J. Jolie, S. Ulbig, H.G. Börner, K.P Lieb, S.J. Robinson, P. Schillebeeckx, E.G. Kessler, M. S. Dewey, and G.L. Greene, Europhys. Lett. 10 (1989) 231.
9) J. Jolie, S.J. Robinson, H.G. Börner, and P. Schillebeeckx, IOP Conf. Ser. 105 (1990) p.179.
10) S. Ulbig, K.P. Lieb, Ch. Winter, H.G. Börner, J. Jolie, S.J. Robinson, P.A. Mando, P. Sona, N. Taccetti, M.S. Dewey, J.G.L. Booten, F. Brandolini, Nucl. Phys. A505 (1989)193.
11) K.P. Lieb, H.G. Börner, M. S. Dewey, J. Jolie, S.J. Robinson, S. Ulbig, and Ch. Winter, Phys. Lett. B. 215 (1988) 50.
12) S. Ulbig et al., contribution to this conference.
13) H.G. Börner, J. Jolie, S.J. Robinson, R.F. Casten, J.A. Cizewski, acc for Phys. Rev. C.
14) A.A. Abrahamson, Phys. Rev. 178 (1969)76.
15) F.T. Smith, Phys.Rev. A 5 (1972) 1708.
16) H.G. Börner, J. Jolie, S.J. Robinson, R. Piepenbring, R.F. Casten, A. Aprahamian, J.P. Draayer, subm to Phys. Rev. Lett.
17) B. Krusche, contribution to this conference.
18) S. Ulbig, J. Jolie, S.J. Robinson, K.P. Lieb, H.G. Börner, and P. Schillebeeckx, to be subm to Phys. Lett A.
19) S.J. Robinson et al., contribution to this conference.
20) J. Klora et al., contribution to this conference.
21) A. Kuronen, preprint
22) H. G. Börner, J. Jolie, S.J. Robinson, P. Schillebeeckx. E.G. Kessler, M. Dewey, and G. Greene, ILL proposal nr. 3.02.421 (1988).
23) F. R. Metzger, Phys. Rev 137 (1965)1415.
24) H. Langhoff, J. Weiss, and M. Schumacher, Z. Physik 226 (1969) 59.
25) S. Ulbig, J. Jolie, H.G. Börner, M.S. Dewey, K.P. Lieb, and S.J. Robinson, contribution to this conference.
26) S.J. Robinson, and J. Jolie, ILL internal report 90RO14T (1990)

PRECISE ABSOLUTE GAMMA-RAY WAVELENGTH MEASUREMENTS

E.G. Kessler, M.S. Dewey, G.L. Greene, R.D. Deslattes
National Institute of Standards and Technology
Gaithersburg, Maryland 20899, USA

H. Börner
Institut Laue Langevin, BP156X, Grenoble, France

ABSTRACT

Gamma-ray wavelengths measured with the joint NIST/ILL GAMS4 facility at the High Flux Reactor, Grenoble, France, are discussed. The primary goal of these measurements is gamma-ray wavelengths which are consistent with the optical wavelength scale and the Rydberg constant with an uncertainty no larger than 0.1 ppm for energies up to 5 MeV. The current status of the Bragg angle and crystal lattice spacing measurements is given. The impact of absolute gamma-ray wavelength measurements on reference energy values, the neutron mass, and the determination of fundamental constants is reviewed. Measurement of structure factors at high energies is also considered.

INTRODUCTION

Absolute gamma-ray wavelength measurements are measurements reported in units of (pico) meters which are consistent with the optical wavelength scale and the Rydberg constant. Absolute gamma-ray wavelengths are most accurately determined by crystal diffraction spectroscopy using flat crystals of high perfection. For the symmetric transmission (Laue) geometry which has been used, the wavelengths are calculated from the relation, $\lambda = 2d\sin\theta$. This equation reveals the experimental quantities which must be measured and the accuracies which must be attained in order to produce gamma-ray wavelengths with an accuracy near 0.1 ppm. It should be noted that the wavelength of a 2 MeV gamma-ray is six orders of magnitude smaller than the wavelength of the optical standard.

The link to the meter and optical wavelength scale is achieved via the measurement of the lattice spacing, d. Current lattice spacing measurement technology is capable of determining the lattice spacing in meters of a crystal used for gamma-ray diffraction with an uncertainty < 0.1 ppm.

The diffraction angles, θ, must be measured with an uncertainty of 0.1 ppm and must be determined in absolute units. Measurements approaching this accuracy (within a factor of 4) have been reported for Bragg angles of approximately 1 degree.[1,2] However, for smaller Bragg angles (0.1 degree) the accuracy decreases to approximately 1 ppm.[3,4] Thus, uncertainty in the determination of Bragg angles limits the accuracy of absolute gamma-ray wavelength measurements. A scheme to more accurately measure the Bragg angles has been developed and will be discussed below.

Precise absolute gamma-ray wavelength measurements are scientifically interesting because:

1). They serve as reference energies for nuclear spectroscopy,
2). The deuteron binding energy which follows from a measurement of the 2.2 MeV n+p capture gamma-ray is needed for a determination of the neutron mass and high energy gamma-ray reference lines derived from atomic mass doublet measurements, and
3). The measurement of a sufficiently large energy interval on the wavelength scale and on the atomic mass scale leads to the combination of fundamental constants, $N_A h/c$.

This last measurement can also be regarded as a measurement of the fine structure constant, α.

In addition to these goals which depend strongly on the high precision absolute nature (line center) of these measurements, there are goals which are more closely associated with the high resolution (line shape rather than line center). These include:

1). Studies of excited state nuclear lifetimes,
2). An experiment to measure the helicity of the neutrino, and
3). Studies concerning the details of diffraction theory at high energies including the measurement of crystal structure factors.

The first item is the subject of other contributions to this conference. A brief discussion concerning the first measurement of structure factors at high energies will be presented below.

In this presentation we will describe some of the features of the diffraction angle measuring spectrometer, GAMS4, including errors associated with the angle interferometers and proposed improvements. We will also discuss the current status of lattice spacing measurements including the accuracy which can routinely be obtained for crystals used for gamma-ray diffraction. Next we discuss the measurements of the deuteron binding energy and large energy intervals. We conclude with a discussion of structure factor measurements made with Ge crystals.

DIFFRACTION ANGLE MEASUREMENTS WITH THE GAMS4 SPECTROMETER

The diffraction angle measurements are made with a double flat crystal spectrometer, GAMS4, operated as a joint NIST/ILL facility at the High Flux Reactor at the Institut Laue-Langevin in Grenoble, France. The spectrometer was designed and built at NIST for the measurement of diffraction angles for radiation < 5 MeV. Because flat crystal spectrometers have a small solid angle of acceptance and thus a low efficiency, only very intense lines can be measured. For gamma-ray energies > 1 MeV, the only available intense lines originate from capture gamma-ray sources which must reside in the reactor during measurement. The ILL with its in-pile source handling system and high neutron flux is the only facility in the world where these measurements can be made.

The description of the GAMS4 facility will be restricted to the current status of the angle measuring spectrometer. For additional details concerning the in-pile prompt gamma source facility see reference 5 and the angle measuring spectrometer see references 2, 6 and 7. The diffraction angles of the two crystals are measured by angle interferometers having a sensitivity of 10^{-10} radians. The angle interferometers are polarization sensitive Michelson interferometers. The plane of polarization of the output light rotates π radians for a change in optical path of one wavelength. Fringe interpolation is achieved by crossing a polarizer with the output light, sensing a null, and reading the polarizer position with an angle encoder. Although fringe reading sensitivity smaller than 1/4000 of a fringe (2×10^{-10} radians) has been achieved, wavelength measurements in the 1 to 2 MeV range are irreproducible at the 1/100 fringe (7.8×10^{-9} radians) level. At 2 MeV and low order reflections, this irreproducibilty translates into $\sim$ 1 ppm uncertainty, a factor of 10 larger than the targeted uncertainty. In Figure 1 a typical narrow gamma-ray rocking curve is shown. Note that the width of the recorded profile is only a fraction of an interferometer optical fringe. Thus the effect of periodic non-linearity with a predominate frequency of 1 cycle/interferometer fringe is not reduced by averaging as would be the case if the gamma-ray profile had a width of several interferometer fringes.

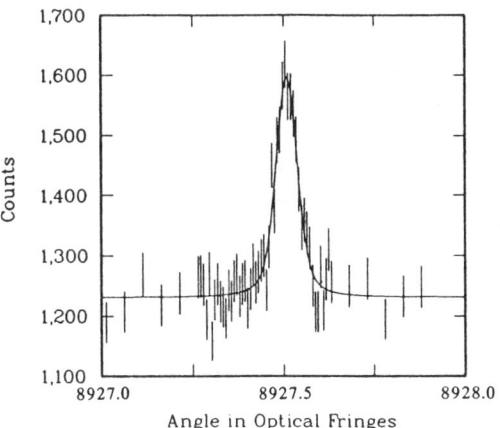

Figure 1. Rocking curve for the 660 reflection from 2.72 mm thick Si crystals. Gamma-ray energy = 1382 keV, θ_B = 0.40° and the displayed angular range = 0.16 s.

At NIST an interferometer identical to the GAMS4 interferometers was tested for non-linearity using pressure scanning. Figure 2 shows a typical scan of the non-linearity of this type of interferometer. The amplitude and frequency of the main non-linearity component are 1/100 of a fringe (7.8×10^{-9} radians) at once per cycle. Changing optical and mechanical components and alignments did not significantly reduce the non-linearity. Additional studies are needed to determine the origins of this non-linearity.

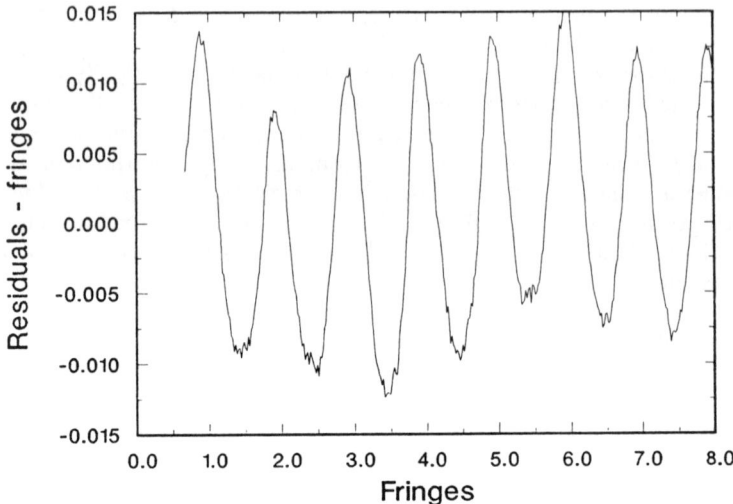

Figure 2. Periodic non-linearity of a rotating polarizer Michelson interferometer (see text). The residuals plotted on the vertical axis are obtained by subtracting a straight line from the recorded fringes.

A heterodyne Michelson interferometer was also tested at NIST for non-linearity. In this interferometer, light from a two-frequency laser whose two frequencies are separated by 1.8 MHz and are orthogonally polarized falls upon a polarization sensitive beam splitter. The beam splitter directs one frequency and polarization into one arm of the interferometer and the other frequency and polarization into the other arm of the interferometer. The two frequencies are recombined at the beam splitter, pass through a polarizer which is oriented at 45 degrees to the two polarizations, and fall upon a detector which responds to the difference frequency. By comparing the phase of the detector signal with the phase of the frequency difference signal of the light emitted directly by the laser, the change in optical path of the interferometer arms can be measured. Investigations of periodic non-linearity of heterodyne Michelson interferometers have been previously reported.[8,9] The non-linearity has been attributed to the mixing of frequencies in the interferometer arms.

Using a commercially available laser, optical elements, and phase measuring electronics, the periodic non-linearities shown in Figure 3 were measured. The crucial components are a laser emitting two frequencies which are orthogonally polarized and a polarization sensitive beam splitter which has a large extinction ratio. The crucial adjustment is the alignment of the laser beam polarization states with the beam splitter polarization axes. Figure 3a shows the non-linearity for a typical "good" alignment of the laser beam and the polarization sensitive beam splitter. By introducing additional polarizers in each of the interferometer arms to further suppress the unwanted frequency, the non-linearity was further reduced as shown in Figure 3b. The non-linearity for the typical heterodyne interferometer is about 1/10 that for the typical rotating polarizer interferometer. The reproducibility of the GAMS4 angle measurements should be significantly improved by using heterodyne interferometry. During the next six months the GAMS4 angle interferometers will be converted to heterodyne angle interferometers. If, after conversion, the GAMS4 angle interferometers have a non-linearity of 0.001 to 0.002 fringes (7.8×10^{-10} to 15×10^{-10} radians), then the non-linearity uncertainty in the angle measurements will be $\sim 5 \times 10^{-7}$ for diffraction angles in the range 0.1 to 0.5° and $\sim 1 \times 10^{-7}$ for larger angles.

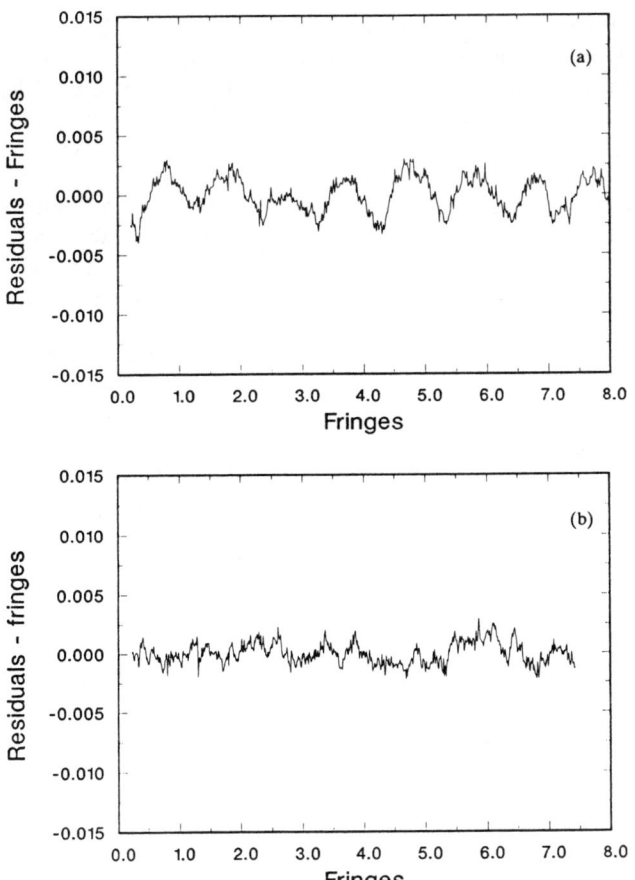

Figure 3. Periodic non-linearity of a two frequency Michelson interferometer (see text). The residuals are obtained by subtracting a straight line from the recorded fringes. Figure 3a shows a typical non-linearity scan, while Figure 3b shows the non-linearity obtained by introducing additional polarizers.

CRYSTAL LATTICE SPACING MEASUREMENTS

The lattice spacing of each set of crystals used on the GAMS4 spectrometer is measured in terms of the wavelength of an iodine stabilized He-Ne laser. The measurements are made in two steps. In the first step the lattice spacing of a particular nearly perfect silicon crystal is compared to the 633 nm wavelength of the stabilized laser. In the second step, the differences in lattice spacing between crystals which will be used for gamma-ray diffraction and the crystal which was compared to the optical wavelength are measured.

A. Absolute Lattice Spacing Measurements

Absolute lattice spacing measurements have been performed at the Physikalisch-Technische Bundesanstalt (PTB) and NIST. Both experiments use a scanning x-ray interferometer of the Laue-Laue-Laue type. The analyzer crystal is mounted on a translation stage and the translation is measured by an optical interferometer. The earlier results from NIST (1973, 1974, 1976)[10,11,12] have an uncertainty of 0.15 ppm while the later result from PTB (1981)[13] has an uncertainty of 0.062 ppm. However the PTB result is smaller than the NIST result by 1.8 ppm, a large discrepancy considering the quoted uncertainty of the individual measurements and an amount much larger than the probable variation in lattice spacings of the two samples. Re-establishment of the NIST x-ray/optical interferometer with several technical improvements led to an understanding of the origin of this large discrepency.[14,15] Measurements have continued and, although a final result is not available yet, preliminary data suggest that the achievable uncertainty will be on the order of 0.01 ppm provided unexplained systematic residue patterns and temporal variations are controlled. Lattice parameter values emerging from available data are, in general, consistent with the PTB lattice parameter value.[16] In summary, absolute lattice parameter determinations with an uncertainty < 0.1 ppm are currently available and it is likely that this uncertainty will be reduced to near 0.01 ppm in the future.

B. Lattice Spacing Comparisons

The difficulty of absolute lattice spacing measurements and the need for a large family of diffraction crystals makes it desirable to have a method to intercompare the lattice spacing of crystal samples. In 1969 Hart suggested a procedure to measure lattice spacing differences having an uncertainty at the 0.01 ppm level.[17] Since then several comparisons have been published based upon similar procedures.[2,16,18,19,20,21,22] The published data clearly demonstrate that lattice comparisons having an uncertainty near 0.01 ppm can be achieved if care is exercised.

The comparison procedure involves x-ray diffraction using a non-dispersive geometry so that there is weaker dependence upon the x-ray wavelength. The unique features of the NIST lattice comparison spectrometer are a precision angle measuring interferometer, rapid interchange of crystal samples, and crystals whose thickness has been chosen so that the x-ray profiles have fine structure oscillations.[23] The fine structure oscillations have a width less than 0.1 of the FWHM of the x-ray profile and serve as a series of sharp convenient references to measure the angular separation of x-ray profiles.

In summary, by using absolute Si lattice parameter measurements followed by lattice comparison measurements, the lattice parameter of crystals used for gamma-ray diffraction can be determined with an uncertainty < 0.1 ppm. At the present time and in the foreseeable future, the uncertainty of gamma-ray wavelengths is and will be dominated by the uncertainty in the diffraction angle measurements and not the uncertainty in the crystal lattice spacing.

THE DEUTERON BINDING ENERGY AND HIGH ENERGY INTERVALS

The extension of the optically based gamma-ray wavelength scale into the several MeV region has the potential of significant scientific dividends. These include reference energies for nuclear spectroscopy, measurement of the neutron mass, and determination of fundamental constants.

A. Reference Energies

The goal here is a more consistent set of reference energies covering a larger energy range for use in nuclear spectroscopy. More than ten years ago a list of gamma-ray energy standards was published entirely based upon a wavelength scale which was established by the above described procedures.[24] Almost all of the standards are for energies < 3.5 MeV and uncertainties of several ppm are typical. In addition to absolute measurements made with calibrated flat crystals, curved crystal spectrometers were used to make relative measurements between the lower energy flat crystal standards and other higher energy lines. The direct measurement of some of the higher energy lines with calibrated flat crystals should result in more accurate standards.

The deuteron binding energy plays a significant role in high energy gamma-ray measurements. It is produced in the reaction $n + {}^1H \rightarrow {}^2H + \gamma$(2.2 MeV) and is crucial in the determination of high energy gamma-ray energies based on atomic mass doublet measurements. Consider the reaction $n + {}^{14}N \rightarrow {}^{15}N + \gamma$(10.8 MeV) and the atomic mass difference, ΔM (in atomic units) between ${}^{14}NDH_2$ and ${}^{15}NH_3$. The sum of the mass difference, ΔM, converted to energy units and the deuteron binding energy is equal to the binding energy of the neutron in ${}^{15}N$. This binding energy then serves as the primary calibration line for the ${}^{15}N$ decay scheme which has a number of intense lines in the 3 to 10 MeV region. Thus the value of the deuteron binding energy measured on the gamma-ray diffraction scale influences energies derived from mass doublet measurements.[25,26]

B. Neutron Mass

Another dividend from a measurement of the deuteron binding energy is a determination of the neutron mass.[4] The deuteron binding energy minus the existing mass spectroscopic data on the interval 2^1H-^2H, leads to the mass difference between the neutron and atomic hydrogen, n-^1H. The neutron mass can then be obtained from the ^1H mass excess.

C. Fundamental Constants

It was noted above that diffraction spectroscopy and atomic mass spectroscopy can both be applied to the MeV region. If the same energy difference can be measured with diffraction spectroscopy (wavelength) and mass spectroscopy (atomic mass units), then a value for $N_A h/c$, the mass-wavelength product can be determined.[27] As an example, again consider the reactions n + $^1H \to {}^2H + \gamma$(2.2 MeV) and n + $^{14}N \to {}^{15}N + \gamma$(10.8 MeV) and the atomic mass difference, ΔM, between $^{14}NDH_2$ and $^{15}NH_3$. The needed wavelength, λ' is determined from $1/\lambda' = 1/\lambda$ (10.8 MeV) - $1/\lambda$ (2.2 MeV) and the relation $\Delta M \lambda' = N_A h/c$ follows from energy equivalence. Note that a precise deuteron binding energy value is again needed for these measurements. Using data from the 1986 adjustment of the fundamental constants,[28] a value of $N_A h/c$ with an uncertainty of 0.089 ppm is derived from the relationship

$$\frac{N_A h}{c} = \frac{M_p \times 10^{-3}}{2R_\infty \alpha^{-2}(m_p/m_e)} \quad (1)$$

where M_p is the proton molar mass (uncertain by 0.01) ppm), m_p/m_e is the proton-electron mass ratio (0.02 ppm), R_∞ is the Rydberg constant (0.001 ppm) and α is the fine structure constant (0.045 ppm).

Conversely, from this equation it can be seen that a measurement of $N_A h/c$ to 0.1 ppm is an independent determination of α at the 0.05 ppm level. Both the mass and the diffraction spectroscopies need to be accurate at the 0.1 ppm level. A recent partial evaluation of the fundamental constants suggests that values of α available from the electron g-factor anomaly and electrical measurements are uncertain at the 0.007 and 0.024 ppm levels, respectively.[29]

ATOMIC STRUCTURE FACTOR MEASUREMENTS AT HIGH ENERGIES

The high resolution and stability of the GAMS4 spectrometer allows accurate recording of line shapes. The independence of the GAMS4 crystals permit profiles to be recorded with the two crystals parallel (nondispersive) and with the two crystals non-parallel (dispersive). In the non-dispersive configuration, the recorded profiles are the instrumental function while in the dispersive configuration, the recorded profiles are a convolution of the instrumental function and the spectral distribution in the incident radiation. For atomic structure factor measurements, the instrumental function (i.e. the crystal diffraction function) is of primary interest. Note that for measurement of lifetimes of excited nuclear states, the spectral distribution of the incident radiation is of primary interest. For nearly perfect crystals and the absence of external influences such as mechanical vibrations and angular drifts, the instrumental function is described by dynamical diffraction theory. Since the shape of the rocking curves for a given reflection and wavelength is extremely sensitive to the structure factor, F, and the crystal thickness, t, a precise recording of the rocking curve and a measurement of the crystal thickness leads to a measurement of F. Similar methods have been used in the x-ray region for precision structure factor determinations.[30,31]

The measurement of the crystal thickness to the desired accuracy is not an entirely trivial matter. For a 0.1% structure factor measurement, the thickness of the diffracting portion of the crystal must also be known to about 0.1%. For crystals with t ~ 2 mm, the thicknesses need to be known to about 2 microns. However, the thickness measurements are less demanding in the gamma-ray region (typical thickness is a few mm) than in the x-ray region (typical thickness is 0.4 to 0.8 mm). In addition, the negligible absorption and minimal polarization of the incident radiation results in a more simple theoretical description of the profiles in the gamma-ray region.

Using Ge crystals with a thickness ~ 1.27 mm and gamma-rays from the reaction ^{48}Ti(n,γ), profiles were recorded for the Bragg reflections (220), (440), (660), and (10,10,0) at 342 keV and for (220) and (440) at 1382 keV. Figure 4 shows a typical profile recording and theoretical fit for the 220 reflection at 1382 keV. Preliminary analysis of the data yields structure factors having an uncertainty of a few tenths percent.

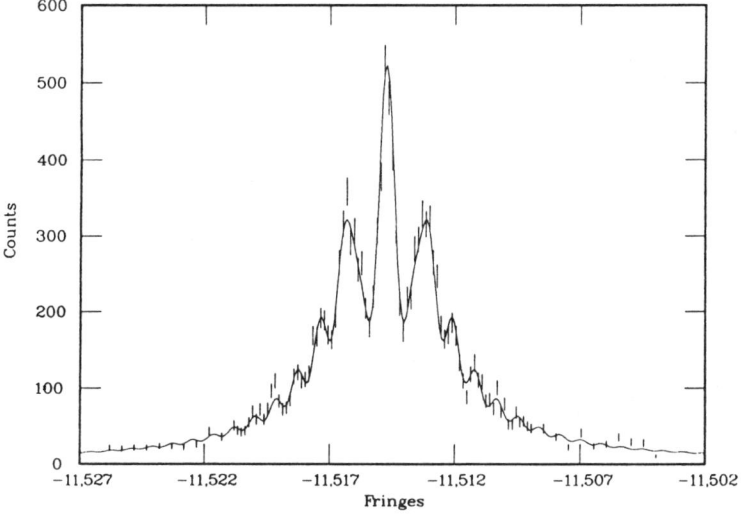

Figure 4. Theoretical (solid line) and experimentally measured rocking curve using the 220 reflection from 1.27 mm thick Ge crystals. Gamma energy = 1381 keV, θ = 0.13° and the displayed angular range = 1s.

In most cases, the periodic non-linearity of the angle interferometers discussed above has minimal influence on the profile measurements whether they are aimed at structure factor or excited state nuclear lifetime determinations. However, the large separation of the two crystals (53 cm) can lead to large angular drifts between the two crystals. Stable environmental conditions are essential for reliable profile measurements.

ACKNOWLEDGEMENTS

The authors acknowledge contributions to the precision measurements from S. Robinson (ILL) and P. Schillebeeckx (formerly ILL). The atomic structure factor measurements are a joint effort with F. Sacchetti and C. Petrillo (Instituto di Struttura della Materia del Consiglio Nazionale delle Ricerche, Frascati) and A. Freund (European Synchrotron Radiation Facility, Grenoble).

REFERENCES

1. E.G. Kessler, R.D. Deslattes, A. Henins, and W.C. Sauder, Phys. Rev. Lett. 40, 171 (1978).
2. R.D. Deslattes, E.G. Kessler, W.C. Sauder, and A. Henins, Ann. Phys. (N.Y.) 129, 378 (1980).
3. E.G. Kessler, Jr., G.L. Greene, R.D. Deslattes, and H.G. Börner, Phys. Rev. C 32, 374 (1985).
4. G.L. Greene, E.G. Kessler, Jr., and R.D. Deslattes, and H.G. Börner, Phys. Rev. Lett. 56, 819 (1986).
5. H.R. Koch, et al., Nucl. Instr. and Meth. 175, 410 (1980).
6. E.G. Kessler, et al., Inst. Phys. Conf. Ser. 88, J. Phys. G: Nucl. Phys. 14 Suppl. (1988).
7. M.S. Dewey, et al., Nucl. Instr. and Meth. A 284, 151 (1989).
8. C.M. Sutton, J. Phys. E: Sci. Instrum. 20, 1290 (1987).
9. N. Bobroff, Appl. Opt. 26, 2676 (1987).
10. R.D. Deslattes and A. Henins, Phys. Rev. Lett. 31, 972 (1973).
11. R.D. Deslattes, et al., Phys. Rev. Lett. 33, 463 (1974).
12. R.D. Deslattes, et al., Phys. Rev. Lett. 36, 898 (1976).
13. P. Becker, et al., Phys. Rev. Lett. 46, 1540 (1981).
14. R.D. Deslattes, et al., IEEE Trans. Instrum. Meas. IM-36, 166 (1987).
15. R.D. Deslattes, in The Art of Measurement, B. Kramer, ed. (VCH Verlagsgesellschaft, Weinheim, FRG, 1988), p. 193.
16. R.D. Deslattes and E.G. Kessler, Jr., CPEM '90, to be published.
17. M. Hart, Proc. R. Soc. London, A309, 281 (1969).
18. M. Ando, D. Bailey, and M. Hart, Acta Cryst. A34, 484 (1978).
19. P. Becker, P. Seyfried, and H. Siegert, Z. Phys. B 48, 17 (1982).
20. D. Windisch and P. Becker, Phil. Mag. A58, 435 (1988).
21. D. Windisch and P. Becker, Phys. Stat. Sol. 118, 379 (1990).
22. D. Hausermann and M. Hart, J. Appl. Cryst. 23, 63 (1990).
23. U. Bonse, et al., Phys. Stat. Sol. (a) 43, 487 (1977).

24. R.G. Helmer, P.H.M. VanAssche, and C. VanDerLeun, At. Data Nucl. Data Tab. 24, 39 (1979).
25. L.G. Smith and A.H. Wapstra, Phys. Rev. C 11, 1392 (1975).
26. A.H. Wapstra, Nucl. Instr. and Meth. A 292, 671 (1990).
27. R.D. Deslattes and E.G. Kessler, Jr., in Atomic Masses and Fundamental Constants-6, J.A. Nolan, Jr. and W. Benenson, eds. (New York, NY: Plenum Press, 1979), p. 203.
28. E.R. Cohen and B.N. Taylor, Rev. Mod. Phys. 59, 1121 (1987).
29. B.N. Taylor and E.R. Cohen, to be published.
30. R. Teworte and U. Bonse, Phys. Rev. B 29, 2102 (1984).
31. M. Deutsch and M. Hart, Phys. Rev. B 31, 3846 (1985).

Mixed Symmetry States in Nuclei Near Closed Shells

S.J. Robinson and J. Jolie
Institut Laue Langevin,156X, 38042 Grenoble Cedex,FRANCE

J. Copnell
Dept. of Physics, Schuster Lab.,University of Manchester,M13 9PL,U.K.

Abstract

We examine proposed mixed symmetry states in both the N=30 and N=84 isotones. In the former case the data are consistent with a mixing of the lowest mixed symmetry state with the two phonon state. This mixing can be described by a perturbation of the strict vibrational limit, due to the introduction of a quadrupole term. This analysis further suggests that no mixing should occur in ^{52}Ti, despite the states being separated by less than 200 keV. The case in the N=84 isotones is less clear due to the influence of two neutron excitations which mix and spread the mixed symmetry strength over several states.

1. Introduction

The experimental study of mixed symmetry states in even-even nuclei is a recent development in nuclear structure physics. Though closely associated with the neutron-proton Interacting Boson Model (IBM-2) [1] they are a natural consequence of any model which includes both neutrons and protons explicitly as separate components [2]. Geometrically one can picture these states as the out of phase motion of collective neutron and proton excitations.

In a spherical nucleus the lowest mixed symmetry state will be a 2^+ level [3] in which the neutron and proton vibrations, which couple in phase to produce the first excited state (2^+_1), now couple antisymmetrically to produce another level (2^+_m). In the absence of any residual interaction between the separate components these two states would be degenerate in energy. However, the introduction of an attractive quadrupole interaction will lower the energy of the symmetric state while having the opposite effect on the antisymmetric level [2]. One can intuitively understand this 'symmetry energy' gap in terms of the extra energy required to spatially separate the two components in the presence of an attractive force.

The antisymmetric form of the wave function means that, for similar E2 transition operators for the two components, their contributions will effectively cancel each other. Therefore, one of the characteristic signatures for such states is weak E2 transitions. On the other hand, the M1 transition between the antisymmetric and symmetric states is expected to be large. Therefore another characteristic signature is a small E2/M1 mixing ratio for the transition between these states [4,5]. This is, however, only a **necessary**, but **not sufficient** condition, for the identification of such states, since single particle transitions will also be predominantly M1 in character. Such judgements should always also take account of absolute transition rates.

2. IBM-2 Formalism

As most investigations of mixed symmetry states have been associated with IBM-2 we will also use this approach here. Since we shall be concerned with spherical nuclei we shall consider only the U(5) limit, and perturbations thereof. In the strict U(5) Hamiltonian there is no 'symmetry energy' gap. Therefore, a Majorana term,

$$M = (s_\nu^\dagger d_\pi^\dagger - d_\nu^\dagger s_\pi^\dagger)^{(2)} \cdot (s_\nu \tilde{d}_\pi - \tilde{d}_\nu s_\pi) - 2 \sum_{k=1,3} (d_\nu^\dagger d_\pi^\dagger)^{(k)} \cdot (\tilde{d}_\nu \tilde{d}_\pi)^{(k)} \quad (1)$$

is introduced, the effect of which is simply to raise the energy of any states with non maximal neutron-proton symmetry (F-spin). The discrepancies between the microscopically derived parameters of the Majorana term [6] and those found in phenomenological fits to real nuclei suggest that this term is far from being well understood.

The simplest general U(5) Hamiltonian is then written as

$$H^{U(5)} = \varepsilon (n_{d_\nu} + n_{d_\pi}) + aM \quad (2)$$

where ε is the d-boson energy (usually assumed to be the same for both neutrons and protons). Within this strict U(5) limit the energies of the two lowest symmetric states and the lowest antisymmetric state are then given by;

$$E(2^+(d)) = \varepsilon, \quad E(2^+(d^2)) = 2\varepsilon, \quad E(2^+_m) = \varepsilon + aN \quad (3)$$

where N is the total boson number (counted from the nearest neutron and proton closed shells). Thus, within any given group of spherical nuclei, for which ε is approximately constant, it would seem that those nearest the closed shells (ie with smaller total boson number N) would provide the easiest opportunity to identify and study the 2^+_m state. We will show later that this is not necessarily the case. However, for the moment we wish to examine in more detail a region which appears to represent a reasonably good example of the simple U(5) case, complicated only by simple two level mixing. (The relevant wave functions, transition operators and reduced matrix elements used in this analysis can be found in refs. [3] and [4].)

3. The N=30 Isotones

Here we consider the simple system represented by the N=30 isotones ^{52}Ti, ^{54}Cr and ^{56}Fe. We shall consider only the lowest three 2^+ levels and assume that only the 2-d boson symmetric and the 1-d boson mixed symmetry states are close enough to show significant mixing. It has already been suggested [7], following the identification of significant M1 decays from both the 2^+_2 (2.66 MeV) and 2^+_3 (2.96 MeV) levels in ^{56}Fe, that these two levels represent a strong mixing of the basis states mentioned above. One of the earliest GRID lifetime measurements [8] showed that, in ^{54}Cr, the M1 strength is concentrated in the 2^+_3 - 2^+_1 transition, from the level at 3.07 MeV. Since, in usual boson counting, ^{54}Cr has one more boson than ^{56}Fe, this was taken [8] as an indication that the mixed symmetry strength is concentrated in the 2^+_3 level, and therefore as a confirmation of the boson number dependence of the 2^+_m excitation energy as given by eqn. (3). Later, the interpretation for ^{56}Fe was confirmed by a study using the (e,e') reaction [9]. The current experimental knowledge of the absolute transition strengths in these nuclei is shown in Fig. 1. Unfortunately only limits can be given for the 2^+_3 level in ^{52}Ti. Further, the values for the other transitions in this nucleus have large errors (see Fig. 2).

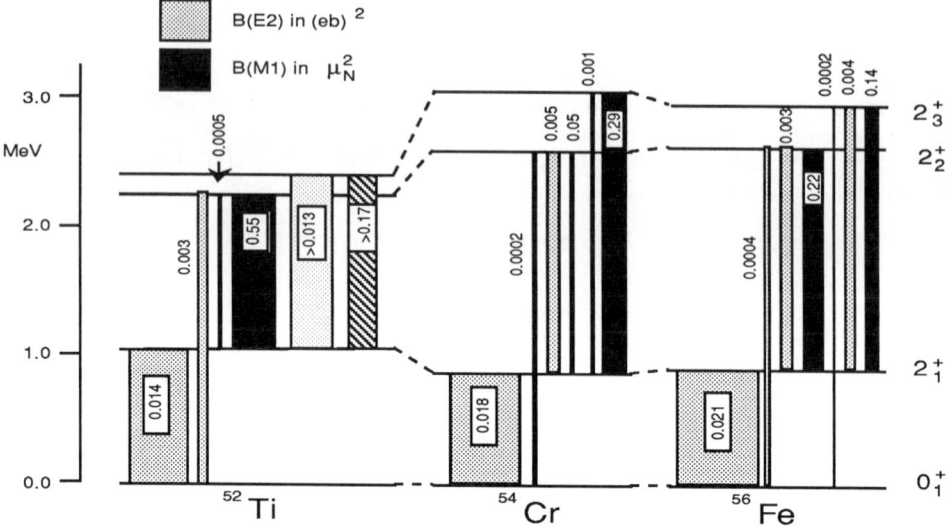

Fig 1. Experimentally determined 2^+ levels and absolute transition strengths for the N=30 isotones. The data are takens from refs [7,8,10,11].

Now, following ref. [7], we write;

$$|2^+_2\rangle = \alpha \, |2^+_m\rangle + \beta \, |2^+(d^2)\rangle$$
$$|2^+_3\rangle = \beta \, |2^+_m\rangle - \alpha \, |2^+(d^2)\rangle \qquad (4)$$

and since the only source of M1 strength to the 2^+_1 state, from either level, is from the $|2^+_m\rangle$ component we can also write

$$\frac{B(M1;2^+_2 \rightarrow 2^+_1)}{B(M1;2^+_3 \rightarrow 2^+_1)} = \frac{\alpha^2}{\beta^2} \qquad (5)$$

Nucleus	N_π	χ_π	e_π(eb)
^{52}Ti	1	-0.5	0.11
^{54}Cr	2	0.0	0.08
^{56}Fe	1	0.5	0.11

Table 1. Parameters used in the level mixing calculations which depend on proton number. For all nuclei the following parameters were kept constant at $N_\nu = 1$, $\chi_\nu = -0.5$, $e_\nu = 0.08$eb, $g_\pi = 1\mu_N$, $g_\nu = 0\mu_N$.

From the data for ^{56}Fe and ^{54}Cr (and taking $\alpha^2 + \beta^2 = 1$) we can therefore say that for ^{56}Fe, $\alpha=0.8$, $\beta=0.6$, and for ^{54}Cr, $\alpha=-0.45$, $\beta=0.9$. Taking the same values for ^{52}Ti as for ^{56}Fe, and taking reasonable values for the other necessary parameters (see Table 1), we can derive values for the E2 and M1 decays from the first three 2^+ levels in these three isotones. These are compared with experiment in Fig 2. It can be seen that the agreement for most decays, and reduced mixing ratios is good.

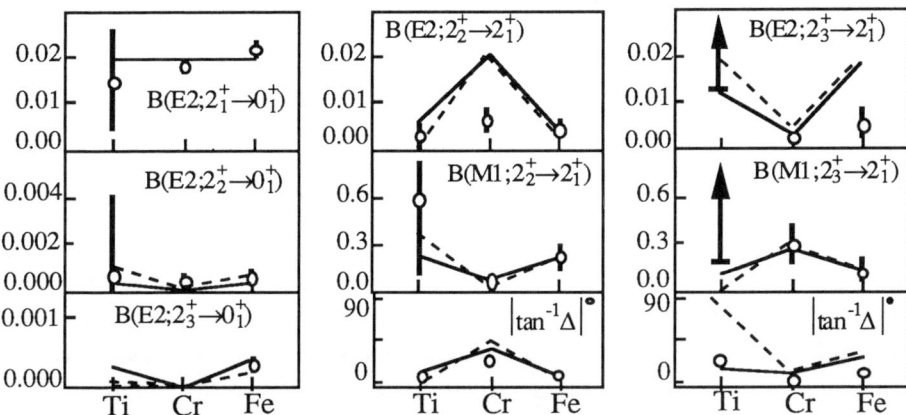

Fig 2. Experimentally determined (open circles with error bars) B(E2) (in e^2b^2) and B(M1) (in μ_N^2) values and reduced mixing ratios in the N=30 isotones. Also shown are the values determined from a semi-empirical two level mixing picture (full lines) and from a calculation using perturbation theory (dashed lines).

4. Perturbation Treatment.

We shall now consider the origin of this mixing since, in the strict U(5) limit, no such mixing can occur. The most natural source for this is a small neutron-proton quadrupole interaction:

$$H_{mix} = \kappa \, Q_\nu . Q_\pi \tag{6}$$

and we can write, to first order in the wave functions

$$|2_i^+\rangle = |(2_i^+)^0\rangle + \sum_{k \neq i} \frac{\langle (2_k^+)^0| H_{mix} |(2_i^+)^0\rangle}{E_i^0 - E_k^0} |(2_k^+)^0\rangle \tag{7}$$

upon renormalisation, and to second order in the energies

$$E(2_i^+) = E^0(2_i^+) + \langle (2_i^+)^0| H_{mix} |(2_i^+)^0\rangle + \sum_{k \neq i} \frac{\langle (2_i^+)^0| H_{mix} |(2_k^+)^0\rangle \langle (2_k^+)^0| H_{mix} |(2_i^+)^0\rangle}{E_i^0 - E_k^0} \tag{8}$$

Upon calculating the appropriate matrix elements, and assuming that only the second and third 2+ levels are close enough to mix appreciably, we then find

$$E(2_1^+) = \varepsilon + \frac{2 \kappa \, N_\nu N_\pi}{N}$$

$$E(2^+(d^2)) = 2\varepsilon + \frac{\kappa \, N_\nu N_\pi}{N(N-1)} \left[4(N-2) - \frac{3}{7}\chi_\nu\chi_\pi \right] - 2\kappa^2 \frac{N_\pi N_\nu (N_\pi \chi_\nu - N_\nu \chi_\pi)^2}{N^2(N-1)(aN-\varepsilon)}$$

$$E(2_m^+) = \varepsilon + aN - \frac{2\kappa N_\nu N_\pi}{N} + 2\kappa^2 \frac{N_\pi N_\nu (N_\pi \chi_\nu - N_\nu \chi_\pi)^2}{N^2(N-1)(aN-\varepsilon)} \quad (9)$$

and for the wavefunctions (for example),

$$|2_i^+\rangle \approx |2^+(d^2)\rangle - \kappa \sqrt{\frac{2N_\pi N_\nu}{(N-1)}} \frac{(N_\pi \chi_\nu - N_\nu \chi_\pi)}{N(aN-\varepsilon)} |2_m^+\rangle \quad (10)$$

The first thing to notice here is that, since we take the χ_ρ equal in ^{52}Ti (which also has $N_\pi = N_\nu = 1$) there is **no mixing** between the two states in question. Now, taking the ε_ρ and χ_ρ parameters from the semi-empirical two level mixing analysis above, we still have to find suitable values for the Hamiltonian parameters ε, a and κ which will, when introduced into these expressions, reproduce the experimental data. In fact this places severe restrictions on the range of suitable values. Finally, by taking $\varepsilon = 1.2$, $a = 0.5$ and $\kappa = -0.2$ (all in MeV) we find that, after renormalisation (and using the convention of eqn. (4) to define α), α(Ti)=1.0 (as explained above), α(Cr)=-0.3 and α(Fe)=0.82, these latter two being in excellent agreement with the semi-empirical determination above. The simplified level scheme deduced from these parameters is shown in Fig. 3 and the transition properties are shown in Fig. 2 as dashed lines.

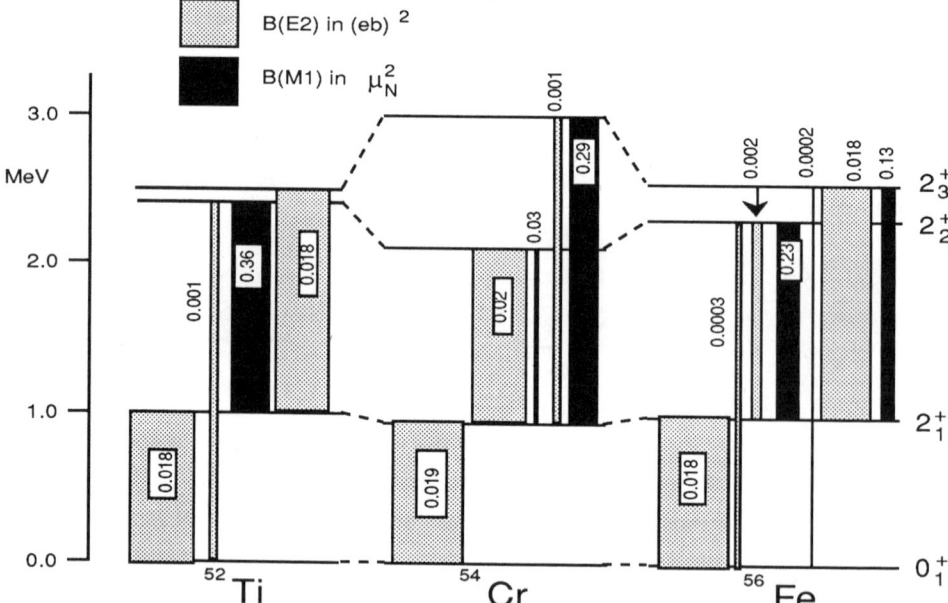

<u>Fig 3</u>. Level energies and transition strengths for the lowest 2+ levels in the N=30 isotones, deduced from a perturbed U(5) analysis.

As can be seen our perturbed U(5) treatment is able to reproduce the experimental data well and thus supports the interpretations of an almost pure mixed symmetry state in ^{54}Cr and a high degree of mixing in ^{56}Fe. The most intriguing result concerns ^{52}Ti in which, because of the parameters involved, no mixing takes place and the two states are almost degenerate. This is in at least qualitative agreement with the known data but the experimental errors are large and only a limit exists for the lifetime of the 2^+_3 level. It would be extremely interesting to have more precise data on this nucleus to examine whether the simple picture our analysis provides, that the two levels represent pure unmixed states only 170 keV apart, can be confirmed.

5. The N = 84 Isotopes

Some of the first cases to be proposed as vibrational mixed symmetry states were 2^+ levels, at around 2 MeV, in the N = 84 isotopes, ^{140}Ba, ^{142}Ce and ^{144}Nd [4]. In a simple analysis mixing ratios and branching ratios were combined to deduce a self-consistent set of transition parameters, rather as in the semi-empirical analysis performed above for the N = 30 isotones. That the neutron effective charge was deduced to be double that of the proton effective charge has been attributed to g-boson renormalisation effects [12].

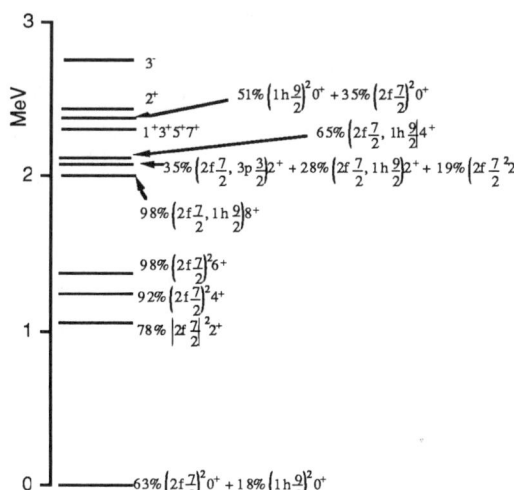

Fig 4. Two neutron spectrum calculated from the PCM parameters used in ^{144}Nd [14]. The major components, in terms of the two coupled fermions, are shown for a number of states

However, the situation in these nuclei is much less clear than the example discussed above. In particular these levels are at an energy where the 2^+ level density is increasing rapidly - there are at least 6 candidate 2^+ levels in the region 2 to 2.5 MeV in ^{144}Nd, and a recent study [13] is likely to reveal even more. It is difficult to imagine that a mixed symmetry excitation would not mix strongly with at least some of these. Indeed this is observed in the wave-functions calculated for the N = 84 isotones [14, 15] using particle core coupling models (PCM) which explicitly couple the two neutrons outside the N =82 closed shell to a system of phonons representing the much larger number of valence protons. In the discussion that follows we shall concentrate on ^{144}Nd, since it is for this nucleus that the most experimental data exist.

The original purpose of our PCM calculations [14] was to test if IBM-2 can be used to describe these nuclei. By mapping the wave functions from the two models (with parameters suitably adjusted to describe the nuclei as well as possible) it has been shown that much of the wave functions of even the low lying states in these nuclei lie outside the IBM-2 model space. In particular the 0^+_2, 2^+_2 and nearly all the 4^+ and 6^+ levels have large components which cannot be described using a system of s and d bosons based on a mapping established using the 0^+_1 and 2^+_1 states. Why this is so can be seen from Fig. 4 which shows the pure two neutron spectrum calculated from the PCM parameters used in ^{144}Nd.

It can be seen that any mapping taken from the first two states will be dominated by $(2f_{7/2})^2$ components, but that at higher energies other components become important. This is simply because it is difficult to build a collective L= 2 boson with this one orbital . For this reason also it is more energetically favourable to create higher spin states purely from the coupling of two neutrons rather than the coupling of an L = 2 neutron pair to the proton phonon system. Thus we must conclude that it will be difficult to describe this nucleus with IBM-2 and any set of parameters which seems to do so is, to some extent, fortuitous.

Now coming to the question of mixed symmetry states. Using the wave function mapping described in ref. 14 we can calculate the overlap of the PCM 2^+ wave functions with the pure U(5) analytic wave functions. This is shown in Fig. 5, from which it can be seen that, although the 2^+_3 state does indeed have some overlap with the mixed symmetry configuration, this accounts for less than 50% of its total strength. Indeed the mixed symmetry strength is spread over several states, even including the 2^+_1 level.

It is interesting to examine the individual proton and neutron transition matrix elements for the 2^+_3 to 2^+_1 transition [14]. Within the PCM they both contribute significantly to the E2 and M1 transitions. However, in the simple IBM-2 picture [4] the E2 part is dominated by the neutron component, whereas the M1 part is dominated by the proton component. This is convincing evidence that the two models are describing very different states.

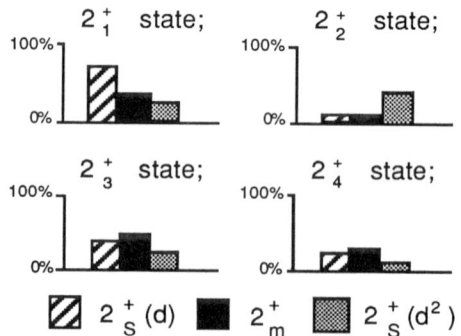

Fig. 5. Overlap of PCM wavefunctions with IBM-2 U(5) analytic wavefunctions for 2^+ levels in ^{144}Nd.

Further evidence comes from recent GRID lifetime measurements [13] which give a lifetime of 145 fs for this state and 550 fs for the 2^+_2. level. This is in disagreement with the previous value, for the 2^+_3 level, of 60 fs [16], and produces a B(E2; $2^+_3 \rightarrow 0^+_1$) of $\approx$ 1 W.u., whereas the usual estimate for such transitions is about 3 W.u. [17]. This will also reduce the values of the absolute transition strengths for the 2^+_3 to

2^+_1 transition and brings everything very close to typical values for single particle transitions. In addition both lifetime values give that $B(M1;2^+_2 \rightarrow 2^+_1) \approx 0.06\ \mu^2$ while $B(M1;2^+_3 \rightarrow 2^+_1) \approx 0.10\ \mu^2$, so that the M1 strength is apparently not concentrated in decays from the 2^+_3 level.

The cumulative effect of all this evidence is to strongly suggest that these states in the N=84 isotones, and in ^{144}Nd in particular, do not represent pure mixed symmetry states. This shows the danger of basing such judgements on mixing ratios and branching ratios alone [4,5] since nothing has changed with regard to the basis of this interpretation.

6. Conclusion

We have shown that a simple two level mixing picture can explain the low lying 2+ level structure in the N=30 isotones. This mixing can be almost perfectly explained by the introduction of a quadrupole perturbation to the simple vibrational Hamiltonian. These calculations suggest that no mixing is present in ^{52}Ti. The situation in the N=84 isotones is much less clear and the experimental and theoretical evidence points to a strong spreading of the mixed symmetry strength in these nuclei.

References

[1] T. Otsuka, A. Arima, F. Iachello and I. Talmi, Phys. Lett. **76B** (1978) 139.
[2] K. Heyde and J. Sau, Phys. Rev. **C33** (1986) 1050.
[3] P. Van Isacker, K. Heyde, J. Jolie and A. Sevrin, Ann. of Phys. **171** (1986) 253.
[4] W.D. Hamilton, A. Irbäck and J.P. Elliott, Phys. Rev. Lett. **53** (1984) 2469.
[5] W.D. Hamilton, J. Phys. G, **16** (1990) 745.
[6] A. Van Egmond and K. Allaart, Nucl. Phys. **A425** (1984) 275.
[7] S.A.A. Eid, W.D. Hamilton and J.P. Elliott, Phys. Lett. **166B** (1986) 267.
[8] K.P. Lieb et al, Phys. Lett. **215B** (1988) 50.
[9] G. Hartung et al, Phys. Lett. **221B** (1988) 109.
[10] S.P. Collins, S.A. Eid, S.A. Hamada, W.D. Hamilton and F. Hoyler, J. Phys. G, **15** (1989) 321.
[11] H. Junde et al, NDS **58** (1989) 677, W. Gongqing et al, NDS **50** (1987) 255 and H. Junde et al, NDS **51** (1987) 1.
[12] O. Scholten et al, Phys. Rev. **C34** (1986) 1962.
[13] S.J. Robinson et al, 'Detailed Spectroscopy of ^{144}Nd', contribution to this conference.
[14] J. Copnell, Ph.D. thesis, University of Sussex (1988).
     J. Copnell, S.J. Robinson, J. Jolie and K. Heyde, Phys. Lett. **222B** (1989) 1, and to be published.
[15] R.A. Meyer, O. Scholten, S. Brandt and V. Paar, Phys. Rev. **C41** (1990) 2386.
[16] F.R. Metzger, Phys. Rev. **187** (1969) 1700.
[17] F. Iachello, Phys. Rev. Lett. **53** (1984) 1427.

PROBING COLLECTIVE EXCITATIONS WITH THE (n,n'γ) REACTION

S. W. Yates
University of Kentucky, Lexington, KY 40506-0055, USA

ABSTRACT

Because the inelastic neutron scattering (INS) reaction can be used to populate both collective and non-collective nuclear states in a statistical manner, it is often possible to probe excited states that cannot be examined with other reactions. Recent advances in Doppler-shift attenuation method (DSAM) measurements following INS have permitted the determination of lifetimes in the femtosecond regime in heavy nuclei. The measured lifetimes can be used to identify low-spin collective excitations. Fast E1 transitions have been observed in two different mass regions and, while these transitions can be offered as evidence for octupole-octupole and quadrupole-octupole excitations, the occurrence of fast E1 transitions is found to be more common than previously thought. The INS reaction has also been employed to examine the well-known M1 scissors mode excitations in the rare earth region and 2^+ mixed-symmetry states in ^{134}Ba.

INTRODUCTION

Neutron-induced reactions have been useful weapons in the arsenal of the nuclear spectroscopist for many years; however, the use of neutron scattering reactions for spectroscopic purposes has not enjoyed the widespread application of some other reactions, e.g., the (n,γ) reaction. The inelastic neutron scattering (INS) or (n,n'γ) reaction has been successfully employed in nuclear structure studies at the University of Kentucky for a number of years. Much of this work has been performed in collaboration with colleagues from the Institute of Isotopes of the Hungarian Academy of Sciences in Budapest. Pulsed-beam time-of-flight methods in neutron and γ-ray spectroscopy have been an area of special emphasis.

Many of the advantages of the INS reaction arise because of the absence of Coulomb effects; thus the nuclear levels can be excited with low incident particle energies--i.e., the nuclei are produced quite "cold". With monoenergetic, accelerator-produced neutrons, low-lying levels can be studied without the attendant complications associated with the presence of radiation from higher-lying levels.

A second advantage of INS is the general non-selectivity of the reaction. Since the population of a level is predominantly determined by the neutron penetrabilities, many differing types of nuclear excitations are accessible. Here the focus will be on some specific examples of collective nuclear states that can be populated with the INS reaction and the wealth of spectroscopic information that can be gleaned from these studies.

EXPERIMENTAL METHODS

The experimental methods employed in both neutron and γ-ray detection measurements at the University of Kentucky have been described previously[1] and will not be repeated here. Instead, the advantages and limitations of these measurements will be examined. It should be noted, of course, that INS measurements are only applicable to stable nuclei. While scattering samples as small as 0.005 mole have been employed for these measurements, the best overall combination of sensitivity and accuracy is achieved with samples of 0.2 to 0.5 moles. The availability of large, isotopically separated samples is thus very important if detailed spectroscopic data are desired.

Following INS, the spectroscopic information can be obtained from either the inelastically scattered neutron or the γ ray deexciting the residual nucleus. Neutron time-of-flight spectroscopy has been employed extensively in a variety of studies in our laboratory, and it has been amply demonstrated[2] that, with pulsed, monoenergetic neutrons, TOF spectroscopy of the inelastically scattered neutrons can also yield valuable spectroscopic information about the lowest-lying nuclear levels. In some cases, the angular distribution of the scattered neutrons can be interpreted to characterize the final states. On the other hand, the resolution for neutron detection is typically poor and such measurements would not be applicable to nuclei with high level densities. These difficulties can be alleviated, however, by observing the γ rays which de-excite the excited levels rather than the inelastically scattered neutrons.

In the γ-ray detection mode, a Ge detector (at present, an n-type spectrometer of 35% relative efficiency surrounded by a 16.5 cm x 20.3 cm BGO anti-Compton annular shield is employed) is used to detect the deexcitation photons, so the resolution of the experiment is greatly improved. It is this mode which is most fruitful for the study of heavy nuclei and, hereafter, the discussion will be limited to γ-ray detection measurements following INS.

Gamma-ray excitation functions. The accurate determination of the γ-ray yield threshold for a particular γ ray uniquely places the level from which the transition arises. When necessary, thresholds can be determined with uncertainties of only a few keV, but threshold determinations to within 30 to 50 keV are generally sufficient for transition placements, even in heavy nuclei. These excitation functions of γ-ray yields (cross sections) are also useful for inferring level spins and parities. Examples of the use of excitation functions to infer level assignments are given in Fig. 1.

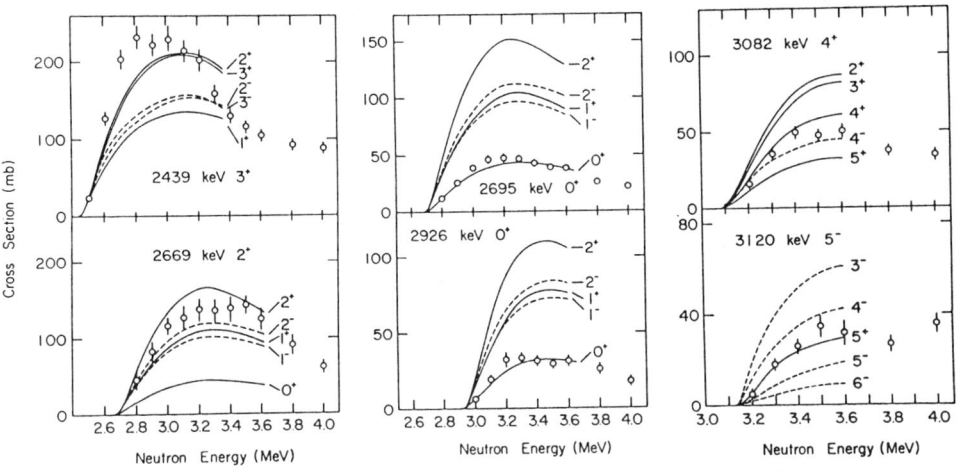

Fig. 1. Inferred (n,n') reaction cross sections (symbols) for selected ^{96}Zr levels compared with Hauser-Feshbach-Moldauer calculations (lines).

Angular distributions. The INS reaction aligns the excited nuclei, so the γ-ray angular distributions from the decays of the excited levels exhibit anisotropies reflecting this alignment, the spins of the levels, and the multipolarities of the transitions. In even-even nuclei it is often possible to find levels which decay by more than one transition--one of known pure multipolarity and others with unknown mixtures of multipolarities. In these cases, the pure multipole transition can be used to deduce the excited-state alignment, and the other distributions can then be analyzed to determine multipole mixing ratios. This thus affords a model-independent method for determining mixing ratios. In practice, the alignments calculated with statistical models are generally sufficient, and tests have shown that mixing ratios determined using these calculated alignments agree well with those determined with the model-independent methods. Figure 2 illustrates the quality of the agreement which can be anticipated for experimental and theoretical γ-ray angular distributions.

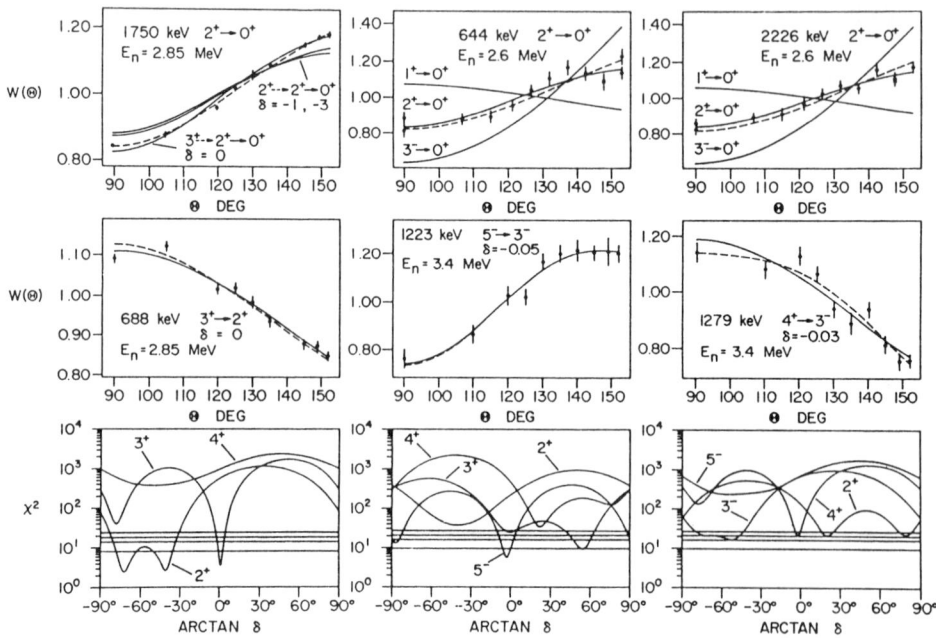

Fig. 2. Angular distributions of ^{96}Zr γ rays from the (n,n'γ) reaction. The dashed lines are fits to the experimental data, and the solid lines were calculated for the indicated spin sequences and mixing ratios.

Lifetimes. In our pulsed-beam experiments, it is generally possible to observe level decay lifetimes greater than 5 ns with conventional timing methods, and searches for lifetimes greater than this are routinely performed. Another time regime can also be investigated in (n,n'γ) measurements by employing the Doppler-shift attenuation method (DSAM). Prior to our measurements[3] on ^{96}Zr, this method with the INS reaction had been confined to nuclei with A < 60. We have shown (e.g., ref. 4) that, given the proper set of conditions, this method can be extended to determine the lifetimes of excited states of much heavier nuclei. While the recoil velocity imparted (v/c ≈ 0.001) in the scattering reactions on even heavy nuclei is sufficient to produce observable Doppler shifts, this method is practically limited to studies of levels which decay by higher energy (>500 keV) transitions--i.e., those levels with lifetimes of less than 1 ps. Examples of Doppler-shift data obtained from the ^{96}Zr(n,n'γ) reaction are presented in Fig. 3.

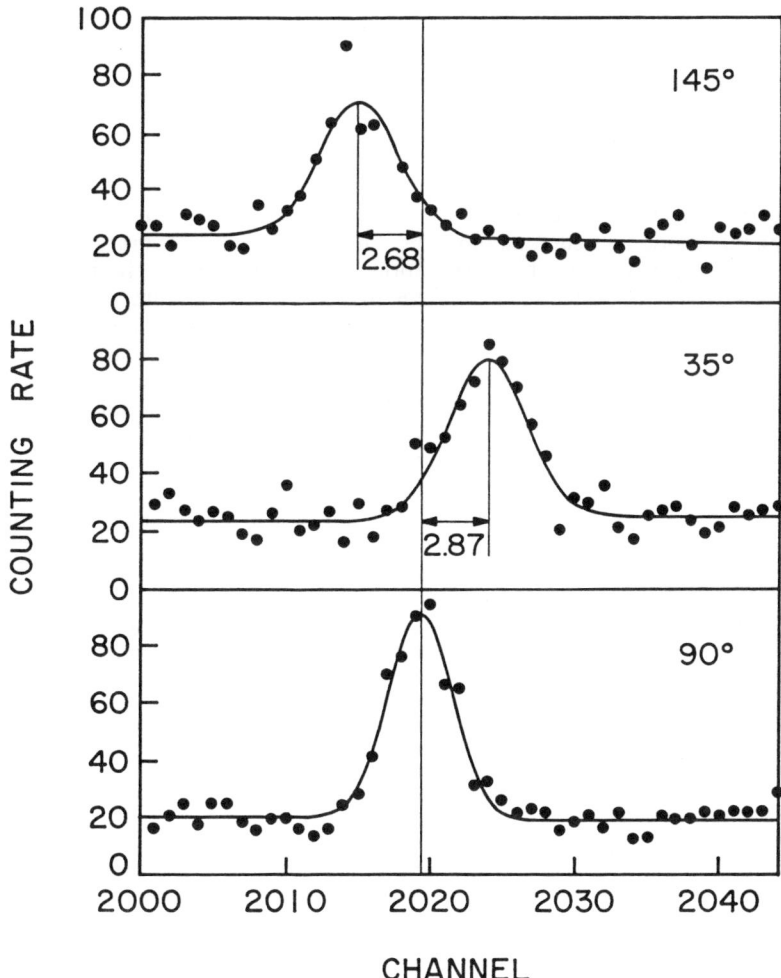

Fig. 3. Energy variation, as a function of angle, of the 3620.67 keV γ-ray peak observed in the INS reaction with 4.3 MeV neutrons. The observed energy shifts are in keV.

The Doppler-shifted γ-ray energy, $E_\gamma(\theta)$, measured at a detector angle of θ with respect to the incident neutrons can be related to E_γ, the energy of the γ ray emitted by a nucleus at rest, by the expression,

$$E_\gamma(\theta) = E_\gamma (1 + F_{exp}(\tau) v_{cm} \cos\theta /c) \quad (1)$$

where v_{cm} is the velocity of the center of mass in the inelastic neutron scattering collision with the atom, and c is the speed of light. $F_{exp}(\tau)$ is the experimental attenuation factor determined from the measured Doppler shift and must be compared with the theoretical attenuation factors to determine the lifetime. In our initial study of the lifetimes of levels in ^{96}Zr, a number of difficulties inherent to our measurements were encountered. The formalism for analyzing these data and the difficulties encountered have been described in ref. 3 and are currently the subject of additional study in our laboratories.

COLLECTIVE EXCITATION MODES

Two-phonon octupole vibrations. While the role of surface vibrational excitations in nuclei has been a subject of study for many years, our knowledge of these fundamental modes remains incomplete. One way in which vibrations in nuclei can be better understood is by the observation of multiphonon excitations. In the quadrupole case, equally spaced, degenerate phonon multiplets are expected, and there are many examples in even-even nuclei near closed shells where the E_{4+}/E_{2+} ratio is near the harmonic value of two, but the expected, closely spaced 0^+, 2^+, 4^+ triplet of two-phonon states is seldom observed. Evidence for three-phonon states was sparse until the reported identification[5] of a complete three-phonon quadrupole vibrational quintet of levels in ^{118}Cd. However, recent measurements[6] of the lifetimes of low-lying excitations in this nucleus indicate that the anharmonicities are larger than originally thought and cast some doubt on this interpretation.

Negative-parity states in many nuclei have been attributed to octupole excitations. In closed-shell nuclei, the octupole vibrations have relatively low excitation energies and compete successfully with the quadrupole mode; in two heavy nuclei, ^{146}Gd and ^{208}Pb, the 3^- state lies lower than the quadrupole phonon and is the first excited state in each. Moreover, these states decay with amazingly similar E3 transition probabilities of about 37 W.u., suggesting that they are indeed collective octupole excitations. These features have stimulated a number of searches for a two-phonon octupole quartet of states with spins and parities of 0^+, 2^+, 4^+, and 6^+ at about twice the energy of the 3^- phonon in these nuclei, but no clear-cut identification has emerged in either nucleus.

Despite the failure of attempts to identify two-phonon octupole excitations in ^{146}Gd, stretch-coupled states of this type have been identified[7,8] in the neighboring nuclei ^{147}Gd and ^{148}Gd. Serendipitously, these states occur as yrast states and decay by characteristic cascades of two E3 transitions because lower multipolarity decays are not possible. Since these states involve the coupling of one or two neutrons to the two-phonon octupole excitation ($vf_{7/2} \times 3^- \times 3^-$ in ^{147}Gd and $v^2 \times 3^- \times 3^-$ in ^{148}Gd), their descriptions are not straightforward, and the search for a complete two-phonon octupole multiplet has continued.

In recently reported measurements[4,9], we have identified two nuclei, ^{96}Zr and ^{144}Sm, with possible octupole-octupole (and also quadrupole-octupole) coupled states based on the observations of fast E1 transitions and characteristic γ-ray branching patterns. The nucleus ^{96}Zr has been shown to display many of the properties of a closed-shell nucleus[9], and recent measurements[10,11] of the lifetime of the first 3^- state yield an E3 transition rate of about 70 W.u., making this $3^- \to 0^+$ transition the fastest known. At somewhat less than twice the one-phonon octupole energy, a quartet of positive parity states which decay to the 3^- phonon with large E1 transition rates was observed. As shown by Kusnezov et al.[12], within the framework of the spdf IBM, large E1 rates can be taken as evidence of transitions from two-phonon octupole states. Unfortunately, mixing between the octupole-octupole states and other quadrupole states can be substantial and the octupole strength can be greatly fragmented, thus these identifications remain tenuous.

The nucleus ^{144}Sm is only two protons removed from ^{146}Gd, which has been shown by Kleinheinz and coworkers[13] to exhibit many of the properties of a doubly closed-shell nucleus, and B(E3;3$^-$ → 0$^+$) has recently been measured[14] to be 38 ± 3 W.u. Based on the DSAM lifetime measurements and γ-ray branchings[15], evidence for a complete two-phonon octupole quartet and a quintet of quadrupole-octupole coupled states was obtained. Only the 4$^+$ level at 3494.1 keV and the 2$^+$ state at 3523.7 keV were observed to decay solely by E1 transitions to the 3$^-$ octupole phonon and can be confidently assigned[4] as members of the two-phonon multiplet. A striking observation (see Figs. 4 and 5) of this work[4] and an accompanying study[15] of the N = 82 isotone ^{142}Nd is that nearly all of the observed E1 transitions are fast. Therefore, E1 transition rates cannot be regarded as unambiguous signatures of octupole character in this mass region.

Collective magnetic dipole states. The discovery of a new class of low-lying collective magnetic dipole excitations observed in inelastic electron and photon scattering experiments (see the excellent review by Richter[16] has stimulated considerable experimental and theoretical interest. This M1 excitation, commonly referred to as the "scissors mode," has been observed in deformed nuclei from ^{46}Ti to ^{238}U and the energies scale as $E_x = 66\delta^{-1/3}$, where δ is the nuclear deformation parameter. Many theoretical approaches, from macroscopic to microscopic, have

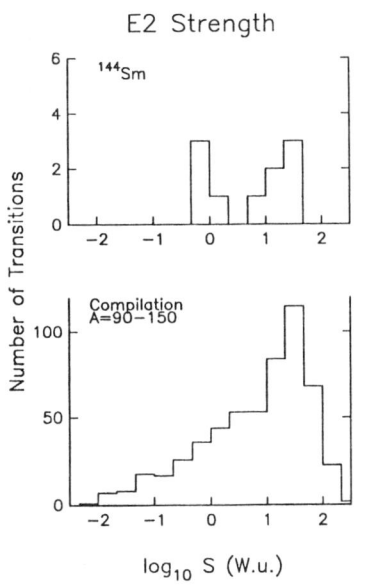
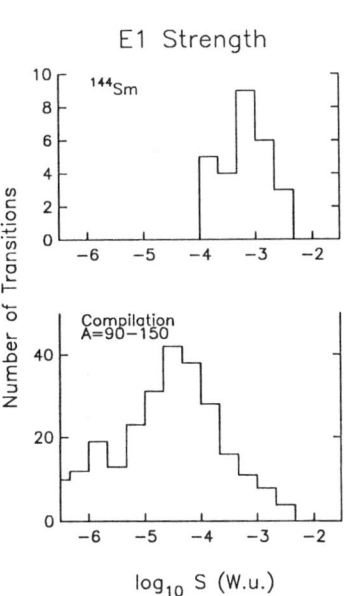

Fig. 4. Comparisons of E1 and E2 γ-ray strengths in the A=90-150 region with measured values in ^{144}Sm.

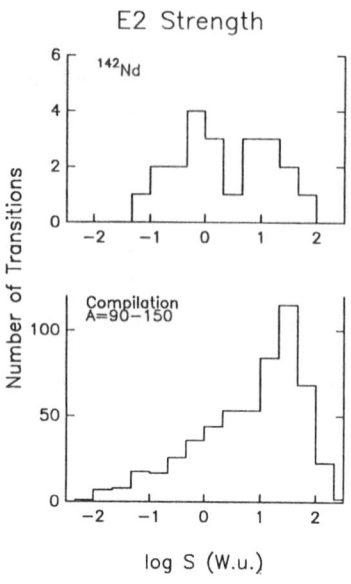 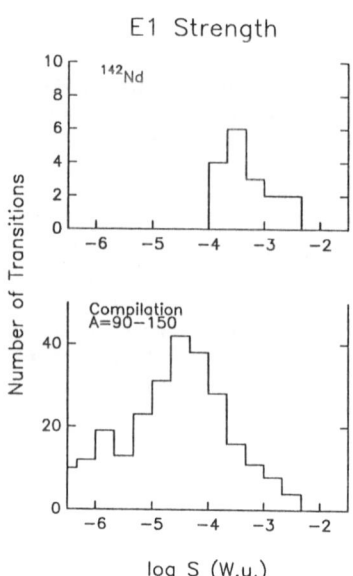

Fig. 5. Comparisons of E1 and E2 γ-ray strengths in the A=90-150 region with measured values in ^{142}Nd.

been developed to explain the M1 scissors mode, and it is now generally agreed that, in any successful description of these states, protons and neutrons must be treated as distinguishable and that this is predominantly an orbital mode.

In a study of two-quasiproton configurations of the form $7/2^-[523] \times 5/2^-[532]$ in ^{164}Dy with the ^{165}Ho(t,α) reaction, Freeman et al.[17] found that the 2539-keV state observed previously in inelastic electron and photon scattering with a large M1 strength (1.67 μ_N^2) is dominated by this two-quasiparticle configuration. This observation appears inconsistent with a collective interpretation of this state. Additional states in ^{164}Dy near 3.1 MeV which were also found in (e,e') and (γ,γ') reactions[18,19] to exhibit large M1 strength (3.15 μ_N^2) were not observed in this proton transfer reaction, a result which could either be taken as evidence that the higher-lying states are the true isovector M1 states or that these states do not involve quasiproton components accessible with the reaction employed. Interestingly, recent intermediate energy proton scattering measurements[20] suggest that ^{164}Dy may be unique in having a significant (15%) spin admixture in the nuclear wavefunctions of the scissors M1 excitations. Clearly, a number of questions remain about the nature of M1 scissors mode states.

In an effort to address the inconsistencies noted above, we have initiated INS studies of ^{162}Dy and ^{164}Dy. In our measurements on these nuclei, we have observed all of the excited 1^+ states suggested as scissors states in these nuclei. The extreme complexity of these spectra, resulting from the non-selective level population noted previously, makes the analysis difficult, particularly when attempting to determine the lifetimes from the small Doppler shifts. At the energies of our measurements, the maximum shifts in the γ-ray energies are about 1.5 keV. On the other hand, the identification of collective M1 transitions should be relatively straightforward, even if we are not able to determine the lifetimes with high precision. In addition to measuring the lifetimes of these states following the INS reaction, we have tentatively (threshold measurements remain to be performed) observed transitions (beyond the known decays to the ground and first 2^+ states) from several scissors mode states. These γ rays have lower energies than the ground-state transitions and are typically obscured in nuclear resonance fluorescence measurements by the intense low-energy bremsstrahlung. We believe our measurements hold the best possibility for observing the complete γ-ray branchings from these isovector M1 excitations.

Most of the studies of collective magnetic dipole (or in the IBM language, mixed-symmetry) states have dealt with deformed nuclei having the SU(3) dynamical symmetry. We have recently examined[21] the mixed-symmetry states in the O(6) nucleus ^{134}Ba where the lowest of these excitations are 2^+ rather than 1^+ states. A description of our low recoil energy (n,n'γ) measurements is described in a separate contribution to this Symposium.[22]

CONCLUSION

Through the examples provided above, some of the capabilities of the INS reaction in studying low-energy collective excitations have been demonstrated. The value of the INS reaction in concert with other probes should be apparent and, in many cases, the information available with INS is unique. The DSAM lifetime measurements provide new opportunities to explore the many fascinating features of nuclear structure.

I wish to acknowledge E. M. Baum, D. P. DiPrete, R. A. Gatenby, E. L. Johnson, J. R. Vanhoy, and D. Wang of the University of Kentucky and T. Belgya, B. Fazekas, G. Molnár, and Á. Veres of the Institute of Isotopes in Budapest for their many contributions. I am also grateful to M. T. McEllistrem, J. L. Weil and many other colleagues for fruitful discussions and their interest in these studies. This work was supported by U. S. National Science Foundation Grants No. PHY-9001465 and INT-8917990 and by the Hungarian Academy of Sciences.

REFERENCES

1. S. W. Yates, Proc. ACS Symposium on Recent Advances in the Study of Nuclei Off the Line of Stability, Chicago, 1985, ACS Symposium Series, **324**, 470 (1986).
2. S. W. Yates, A. Khan, M. C. Mirzaa, and M. T. McEllistrem, Phys. Rev. C**23**, 1993 (1981).
3. T. Belgya, G. Molnár, B. Fazekas, Á. Veres, R. A. Gatenby, and S. W. Yates, Nucl. Phys. **A500**, 77 (1989).
4. R. A. Gatenby, J. R. Vanhoy, E. M. Baum, E. L. Johnson, S. W. Yates, T. Belgya, B. Fazekas, Á. Veres, and G. Molnár, Phys. Rev. **C41**, R414 (1990).
5. A. Aprahamian, D. S. Brenner, R. F. Casten, R. L. Gill and A. Piotrowski, Phys. Rev. Lett. **59**, 535 (1987).
6. H. Mach, M. Moszyński, R. F. Casten, R. L. Gill, D. S. Brenner, J. A. Winger, W. Krips, C. Wesselborg, M. Büscher, F. K. Wohn, A. Aprahamian, D. Alburger, A. Gelberg, and A. Piotrowski, Phys. Rev. Lett **63**, 143 (1989).
7. P. Kleinheinz, J. Styczen, M. Piiparinen, J. Blomqvist, and M. Kortelahti, Phys. Rev. Lett. **48**, 1457 (1982).
8. S. Lunardi, P. Kleinheinz, M. Piiparinen, M. Ogawa, M. Lach, and J. Blomqvist, Phys. Rev. Lett. **53**, 1531 (1984).
9. G. Molnár, T. Belgya, B. Fazekas, Á. Veres, S. W. Yates, E. W. Kleppinger, R. A. Gatenby, R. Julin, J. Kumpulainen, A. Passoja, and E. Verho, Nucl. Phys. **A500**, 43 (1989), and G. Molnár, contribution to this Symposium.
10. H. Ohm, M. Liang, G. Molnár, S. Raman, K. Sistemich, and W. Unkelbach, Phys. Lett **B241**, 472 (1990).
11. H. Mach, S. Ćwiok, W. Nazarewicz, B. Fogelberg, M. Moszyński, J. Winger, and R. L. Gill, Phys. Rev. **C42**, R811 (1990).
12. D. F. Kusnezov, E. A. Henry, and R. A. Meyer, Phys. Lett. **B228**, 11 (1989).
13. P. Kleinheinz, R. Broda, P. J. Daly, S. Lunardi, M. Ogawa, and J. Blomqvist, Z. Phys. A **290**, 279 (1979).
14. A. F. Barfield, P. von Brentano, A. Dewald, K. O. Zell, N. V. Zamfir, D. Bucurescu, M. Ivascu, and O. Scholten, Z. Phys. **A332**, 29 (1989).
15. R. A. Gatenby, Ph.D. Dissertation, University of Kentucky, 1990.
16. A. Richter, Contemporary Topics in Nuclear Structure Physics (World Scientific, Singapore, 1988) pp. 127-164.
17. S. J. Freeman, R. Chapman, J. L. Durell, M. A. C. Hotchkis, F. Khazaie, J. C. Lisle, J. N. Mo, A. M. Bruce, R. A. Cunningham, P. V. Drumm, D. D. Warner, and J. D. Garrett, Phys. Lett. **222B**, 347 (1989).
18. D. Bohle, G. Küchler, A. Richter, and W. Steffen, Phys. Lett. **148B**, 260 (1984).
19. C. Wesselborg, P. von Brentano, K. O. Zell, R. D. Heil, H. H. Pitz, U. E. Berg, U. Kneissl, S. Lindenstruth, U. Siemann, and R. Stock, Phys. Lett. **207B**, 22 (1988).
20. D. Frekers, D. Bohle, A. Richter, R. Abegg, R. E. Azuma, A. Celler, C. Chan, T. E. Drake, K. P. Jackson, J. D. King, C. A. Miller, R. Schubank, J. Watson, and S. Yen, Phys. Lett. **218B**, 439 (1989).
21. G. Molnár, R. A. Gatenby, and S. W. Yates, Phys. Rev. **C37**, 898 (1988).
22. T. Belgya, B. Fazekas, G. Molnár, Á. Veres, R. A. Gatenby, and S. W. Yates, contribution to this Symposium and to be published.

THE FIRST SUBMAGIC NUCLEUS: ^{96}Zr

G. Molnár
Institute of Isotopes, Budapest, H-1525, Hungary

ABSTRACT

Recent in-beam γ-spectroscopic and β-decay experiments as well as inelastic scattering studies reveal the level structure of ^{96}Zr, located at the simultaneous closure of the Z=40 and N=56 subshells, and gave an unambiguous proof of its magic character as well as of the presence of shape coexistence. These properties are reviewed in the present paper. Good agreement with preliminary results from particle-hole RPA calculations is also reported. The recently discovered large octupole vibrational collectivity and the status of the search for two-phonon octupole and quadrupole-octupole states are discussed as well.

INTRODUCTION

Magic nuclei having doubly closed shells are of special interest in nuclear physics because their region is the testing ground of shell model theories. Besides the double major shell closures, there are some known examples of magicity due to a simultaneous occurrence of shell and subshell closures. The nuclei ^{88}Sr (Z=38) and ^{90}Zr (Z=40) have long been observed to have properties resembling those of magic nuclei. More recently, the quasi-magicity of ^{146}Gd has been attributed[1] to the existence of a proton subshell gap at Z=64, influencing the entire chain of N=82 isotones[2]. However, not a single example of a *double* subshell closure has been found.

The nucleus ^{96}Zr has, in fact, long been suspected to be magic. Its first 2^+ state has a high energy of 1750 keV, close to the ^{90}Zr 2^+ energy of 2186 keV. Moreover, single particle transfer data have suggested[3,4] that the $\pi 2p_{1/2}$ proton and the $\nu 2d_{5/2}$ neutron subshells are nearly filled at Z=40 and N=56, respectively. Another unusual feature, characteristic of some magic nuclei like ^{16}O, ^{40}Ca and ^{90}Zr, is that its first excited state is 0^+, at 1582 keV energy. Early shell model calculations[5], however, could not be put to a stringent test because of the poor knowledge of the level scheme at that time. On the other hand, the success of the Pandya transformation when generating ^{96}Nb states has given support to the idea that ^{96}Zr is a good magic core[6].

In the present review, we summarize the results of recent in-beam γ-spectroscopic and β-decay experiments as well as microscopic calculations specifically designed to understand the features of ^{96}Zr and to prove to what extent this nucleus can be considered as magic. In addition, some unexpected features, like the observed large octupole collectivity and possible candidates of two-phonon states involving the octupole phonon, are also discussed.

DOUBLE SUBSHELL CLOSURE AND SHAPE COEXISTENCE

The double subshell closure at Z=40 and N=56 has emerged as one of the most fascinating features of the A~100 mass region. Besides being a unique phenomenon by itself, understanding of this reinforcement of magicity in the transitional region between spherical and deformed structures should also provide a clue to the understanding of the role played by the various single particle orbitals in the development of quadrupole deformation in the region.

According to single-proton transfer reaction studies[3,4], the $\pi 2p_{1/2}$ subshell is nearly filled and the $\pi 1g_{9/2}$ single-particle state lies higher by at least 0.75 MeV at ^{96}Zr. Close examination of the two-proton separation energy[7] differences $\Delta S(2p)$ shows, however, that the subshell gap should be much more comparable to the Z=50 major shell gap (Fig. 1, upper part). It is worth mentioning that the effect is less pronounced for the N=50 isotones than for the N=56 isotones. On the other hand, the reported[4] separation of about 1 MeV between the nearly filled $\nu 2d_{5/2}$ neutron subshell and the next single-particle orbital, $\nu 3s_{1/2}$, is consistent with the fact that $\Delta S(2n)$ has a considerably shallower minimum just above N=56 than above the N=50 major shell closure (Fig. 1, lower part).

Study of the β decay of 0^- ^{96}Y into ^{96}Zr has revealed[8] an extremely fast first-forbidden β transition to the ground state, with a logft value of about 5.6, consistent with a 96±2 % contribution of the $\pi(2p_{1/2})^2$ configuration to the ^{96}Zr ground state. On the other hand,

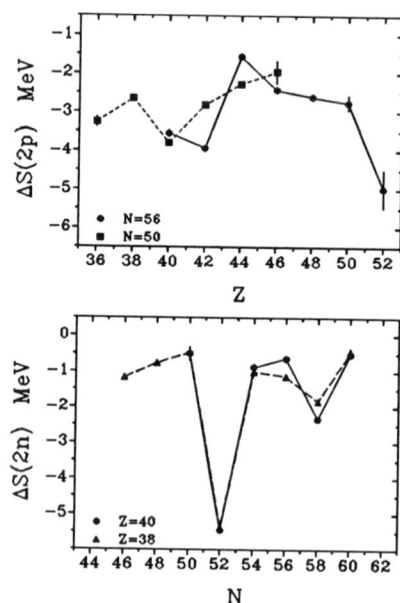

Fig. 1. Two-proton and two-neutron separation energy differences showing the subshell effects at Z=40 and N=56.

extremely weak population of the 1582 keV first excited 0^+ state in β decay confirms that the underlying configuration is the $\pi(1g_{9/2})^2$ configuration. Utilizing the connecting small E0 matrix element value, we have shown[8] that this first excited 0^+ state is associated with a coexisting shape, characterized by a dynamic quadrupole deformation parameter of 0.2, that mixes very weakly with the ground state. The β decay of high-spin $^{96}Y^m$ feeds the 4390 keV 8^+ state, which is identified[9] as the highest-spin member of the $\pi(1g_{9/2})^2$ multiplet. The high energy of this 8^+ state is a further proof of the large proton subshell gap at Z=40.

CLASSIFICATION OF ^{96}Zr EXCITED STATES

Utilization of nonselective in-beam spectroscopic methods like the $(n,n'\gamma)$ or INS reaction[10,11], combined with other in-beam and β decay results as well as with inelastic scattering data gave[11] a nearly complete level scheme of ^{96}Zr. Moreover, the observation of γ decay properties of the individual states, including Doppler-shift lifetime measurements[12], enabled us to classify uniquely most of the lowest-lying states. These classifications are shown in Fig. 2 where the ^{96}Zr excitations are grouped together according to whether their γ-decays favor the first 0^+, 2^+ or 3^- excited states.

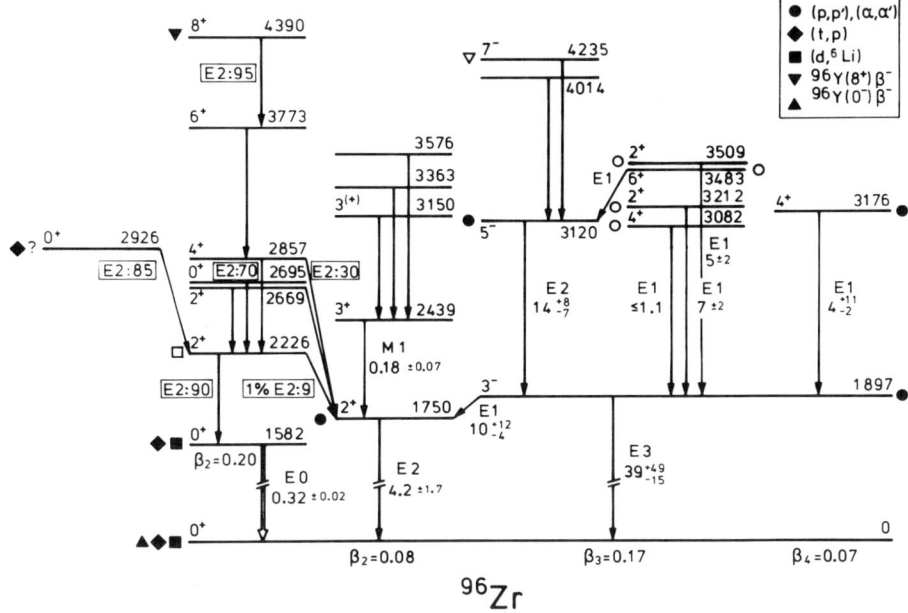

Fig. 2. Classification of ^{96}Zr levels. Relative B(E2) percentages (boxes) and reduced transition probabilities in Weisskopf units (10^{-4} W.u. for E1, $0.5A^{-2/3}$ for E0) are given beside the arrows.

Quadrupole intruder band. The states built on the 1582 keV coexisting 0^+ state were suggested[8,11] to exhibit a moderate quadrupole vibrational collectivity up to the two-phonon level. The 644 keV energy separation between the 0^+ band head and the 2226 keV 2^+ state, consistent with the deformation parameter of 0.2, deduced[8] for the band head, and the $B(E2; 4^+ \rightarrow 2^+)$ rate[11,12] of about 41±24 W.u. have been the main arguments. New lifetime data[13] for the 0^+ and 2^+ states within the band give further support to this conjecture. On the other hand, new lifetime data for the 4390 keV 8^+ level clearly favor the $\pi(1g_{9/2})^2$ two-quasiparticle interpretation as suggested before[9].

Particle-hole states. The recognition[11] that the 1750 keV first 2^+ level is connected to the newly discovered 2439 keV 3^+ state by a fast M1 transition[12] has been a key element in understanding the low energy excitation spectrum and, hence, the size of the neutron subshell gap. The neutron particle-hole configuration $\nu(3s_{1/2}, 2d_{1/2}^{-1})$ has been proposed[11] for this doublet. In order to search for further particle-hole states, (d,d') scattering experiments were carried out with 52 MeV unpolarized[14] and with 22 MeV polarized deuterons[15], yielding a wealth of new information on particle-hole excitations up to about 4.3 MeV.

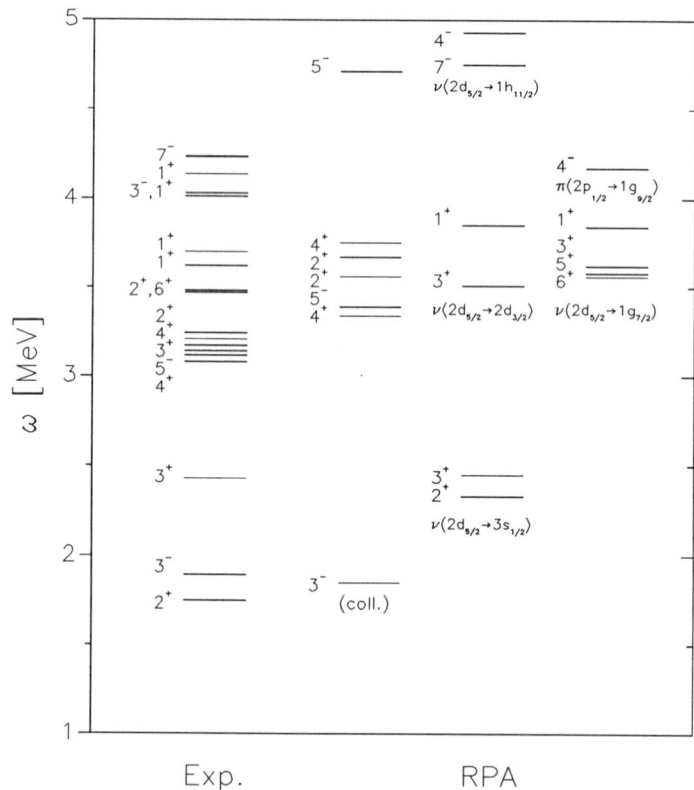

Fig. 3. Comparison of ^{96}Zr levels with the spectrum obtained from particle-hole RPA.

The observed levels are compared with preliminary results[16] of particle-hole RPA calculations in Fig. 3 where the dominant particle-hole configurations are also indicated for each multiplet. The agreement is not unreasonable, especially for the lowest states. However, since we are dealing with relatively weak shell closures, the quasi-particle RPA approach is more reasonable. Such calculations are in progress[17] and they show, in particular, that the inclusion of additional neutron two-quasiparticle components improves greatly the description of the first 2^+ level.

States connected to the 3^- octupole level. There are three groups of levels decaying to the first 3^- state by fast E1 or E2 transitions[11,12], as shown on the right side of Fig. 2. All of them have since been observed[14,15] to be strongly excited in (d,d') scattering. This fact and the close correspondence with the levels predicted by RPA (Fig. 3) may suggest an interpretation in terms of particle-hole multiplets like $\pi(2p_{1/2},1g_{9/2}^{-1})$, $\pi(2p_{3/2},1g_{9/2}^{-1})$ and $\nu(1h_{11/2},2d_{5/2}^{-1})$ for odd and $\nu(1g_{7/2},2d_{5/2}^{-1})$ as well as $\nu(2d_{3/2},2d_{5/2}^{-1})$ for even parity, respectively. On the other hand, the fast E1 rates may signal two-phonon states involving the octupole phonon, as will be discussed in the next section.

STRONG OCTUPOLE MODE AND TWO-PHONON STATES

The blocking of collective quadrupole excitations due to the double subshell closure and the availability of stretched E3 excitations of both proton and neutron type into essentially empty high-j orbitals set favorable conditions for strong octupole collectivity in the neighborhood of ^{96}Zr. Recent lifetime measurements have yielded an even higher octupole transition rate than was expected from the systematics and from earlier measurements[18]. The B(E3) values of 69^{+34}_{-17} W.u. of Ref. 19 and 65±10 W.u. of Ref. 13 are in good agreement with each other and provide the *highest* observed octupole transitional rate for the 1897 keV first 3^- state of ^{96}Zr. Recent RPA calculations[19] were able to reproduce this experimental result and gave the main contributions to the octupole as the $2p_{3/2} \rightarrow 1g_{9/2}$ proton and the $2d_{5/2} \rightarrow 1h_{11/2}$ neutron single-particle transitions, in accordance with the expectations.

A survey of available B(E3) rates[20] has revealed[19] that the first 3^- state exhausts about 10% of the energy-weighted E3 sum rule for all such magic and/or spherical "mirror" nuclei where *both* proton and neutron stretched transitions contribute to the octupole excitation. Surprisingly, the relationship between B(E3) rate and excitation energy has been found to follow the harmonic law as illustrated in Fig. 4 where the fitted solid line has a slope of -1.07±0.13 (Ref. 19). While this is not generally true for octupole excitations the high octupole rate of ^{96}Zr follows directly from this relationship, placing this nucleus in the class of strong harmonic octupole vibrators. It would be of special interest to find cases which confirm the harmonic relationship at even lower octupole energies.

The existence of a strong one-phonon octupole vibration in ^{96}Zr poses the challenge to search for two-phonon states involving the

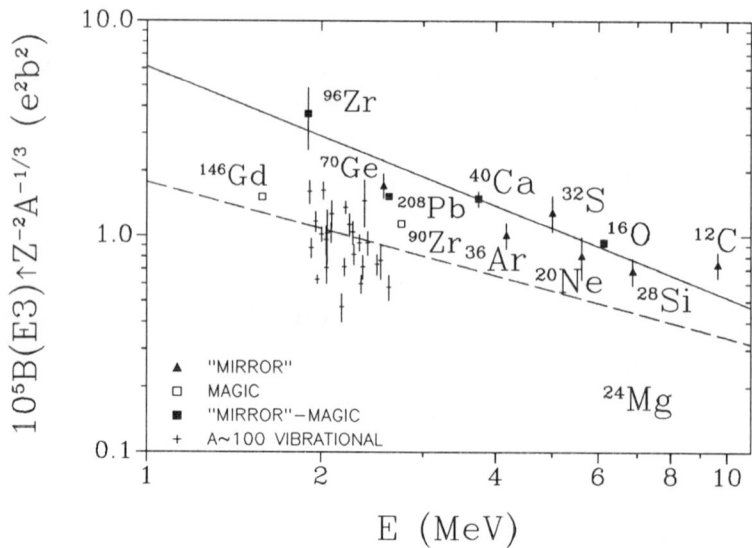

Fig. 4. Scaled B(E3) transition probability against excitation energy for the first 3⁻ states. Fits to strong harmonic octupole vibrators (solid line, filled symbols) and to all quadrupole vibrators (dashed line) are also shown.

octupole phonon. In particular, a two-phonon octupole quartet of 0^+, 2^+, 4^+ and 6^+ states at twice the one-phonon excitation energy and a quintuplet of negative parity states with spins 1-5 at the sum of the quadrupole and octupole energies would be expected for a harmonic case. Based on the observed E2 and E1 transition rates[12] mentioned above it has been proposed[11] that the 3120 keV 5^- state might be a quadrupole-octupole state while the group of positive-parity levels decaying to the 3^- state by strong E1 transitions might be associated with the appropriate members of a two-phonon octupole quartet. In a recent study[21] utilizing the spdf boson version of IBA it has even been shown that the observed decay properties of the candidates for two-phonon octupole states can be reasonably described when mixing with the quadrupole intruder states is included.

To test directly the proposed two-phonon character of the states connected with the 3^- octupole level (d,d') scattering experiments[15] were performed with 22 MeV polarized deuterons. According to a preliminary coupled-channel analysis[15] of the differential cross sections and analyzing powers, the 3120 keV 5^- level is more consistent with a direct excitation. On the other hand, the newly found[14,15] 3749 keV 5^- state and the 3483 keV first 6^+ state might be a quadrupole-octupole or a double octupole excitation, respectively, although the situation is probably more complex and the data require a more refined treatment involving the interference between one-step and two-step processes of various order.

The results presented here have been due to joint efforts of a number of colleagues from several laboratories. I wish to express my gratitude to all co-authors of the cited papers. Valuable discussions with T. Belgya, B. Fazekas, R. A. Gatenby, G. Graw, J. Hebenstreit, E. A. Henry, D. Hofer, P. Kleinheinz, H. Mach, R. A. Meyer, H. Ohm, K. Sistemich, W. Unkelbach and S. W. Yates are especially acknowledged.

REFERENCES

1. P. Kleinheinz et al., Z. Phys. A284, 315 (1978).
2. A. Abbas, Phys. Rev. C29, 1033 (1984).
3. M. R. Cates, J. B. Ball and E. Newman, Phys. Rev. 187, 1682 (1969); B. M. Preedom, E. Newman and J. C. Hiebert, Phys. Rev. 166, 1156 (1968).
4. C. R. Bingham and G. T. Fabian, Phys. Rev. C7, 1509 (1973).
5. D. Gloeckner, Nucl. Phys. A253, 301 (1975).
6. J. R. Comfort, J. V. Maher, G. C. Morrison and J. P. Schiffer, Phys. Rev. Lett. 25, 383 (1970).
7. A. H. Wapstra and G. Audi, Nucl. Phys. A432, 55 (1985).
8. H. Mach, G. Molnár, S. W. Yates, R. L. Gill, A. Aprahamian and R. A. Meyer, Phys. Rev. C37, 254 (1988);
   H. Mach, E. K. Warburton, R. L. Gill, R. F. Casten, J. A. Becker, B. A. Brown and J. A. Winger, Phys. Rev. C41, 26 (1990).
9. M. L. Stolzenwald, G. Lhersonneau, S. Brant, G. Menzen and K. Sistemich, Z. Phys. A327, 359 (1987);
   H. Ohm, M. Liang, M. Büscher, U. Paffrath, B. De Sutter, K. Sistemich and G. Molnár, contribution to this Symposium.
10. S. W. Yates, contribution to this Symposium.
11. G. Molnár, T. Belgya, B. Fazekas, A. Veres, S. W. Yates, E. W. Kleppinger, R. A. Gatenby, R. Julin, J. Kumpulainen, A. Passoja and E. Verho, Nucl. Phys. A500, 43 (1989) and references therein.
12. T. Belgya, G. Molnár, B. Fazekas, A. Veres, R. A. Gatenby and S. W. Yates, Nucl. Phys. A500, 77 (1989).
13. H. Mach, S. Ćwiok, W. Nazarevicz, B. Fogelberg, M. Moszyńsky, J. Winger and R. L. Gill, Phys. Rev. C42, R811 (1990)
14. G. Molnár, J. Hebenstreit, S. Heising, P. Maier-Komor, H. Ohm, D. Paul, P. von Rossen, K. Sistemich and W. Unkelbach, IKP-KFA Jülich Ann. Rep. 1989 (Jülich, 1990), p. 1.
15. D. Hofer, M. Bisenberger, G. Graw, R. Hertenberger, H. Kader, P. Schiemenz and G. Molnár, Understanding the Variety of Nuclear Excitations, Ischia, Italy, May 21-25, 1990, to be published.
16. W. Unkelbach and G. Molnár, IKP-KFA Jülich Ann. Rep. 1989 (Jülich, 1990), p. 110.
17. O. Rosso and W. Unkelbach, private communication.
18. G. Molnár, H. Ohm, G. Lhersonneau and K. Sistemich, Z. Phys. A331, 97 (1988).
19. H. Ohm, M. Liang, G. Molnár, S. Raman, K. Sistemich and W. Unkelbach, Phys. Lett. B241, 472 (1990).
20. R. H. Spear, At. Data Nucl. Data Tables 42, 55 (1989).
21. D. F. Kusnezov, E. A. Henry and R. A. Meyer, Phys. Lett. B228, 11 (1989).

PHOTON SCATTERING EXPERIMENTS IN THE RARE EARTH REGION – A NEW APPROACH TO LOW LYING DIPOLE MODES

P. von Brentano, A. Zilges and R. Jolos *
Institut für Kernphysik, Universität zu Köln, D-5000 Köln 41, Germany

A. Richter
Institut für Kernphysik, Technische Hochschule Darmstadt, D-6100 Darmstadt, Germany

R.D. Heil, U. Kneissl, H.H. Pitz and C. Wesselborg [†]
Institut für Kernphysik, Justus-Liebig-Universität Giessen, D-6300 Giessen, Germany

Abstract

High resolution Nuclear Resonance Fluorescence (NRF) experiments have been carried out to study low lying dipole excitations in rare earth nuclei. In particular, we identified the so called M1 scissors mode near 3 MeV excitation energy. At the same time we found some E1 transitions with strengths up to several mWu. The inspection of the measured decay branching ratios suggests K-mixing for a few states while giving overall evidence for a good K-quantum number at 3 MeV.

INTRODUCTION

Magnetic dipole properties and excitation modes of nuclei have been investigated strongly in terms of various models from the sixties to the eighties [1–6]. One particular feature of these studies has been the prediction of a hitherto experimentally not known orbital magnetic dipole mode [3–6]. In 1983 the Darmstadt group identified this mode by observing a strong M1 transition at 3.075 MeV in ^{156}Gd using electron scattering [7, 8]. This exciting discovery started an intensive work both in the experimental and in the theoretical field. It was found by the Giessen group that the M1-mode can also be seen in high resolution (γ, γ')-experiments [9]. Systematic studies were performed in different regions of the isotopic table ranging from fp-shell nuclei to the actinides with (e,e')-experiments at Darmstadt and by various Nuclear Resonance Fluorescence (NRF) experiments of the Giessen–Darmstadt–Köln–Stuttgart collaboration. We refer the reader to the reviews [10] and [11]. Investigations concerning the strong electric dipole excitations or the measured branching ratios possibly identifying K-mixing have just been started. Their proper examination needs new high precision experiments with different photon endpoint energies and good statistics. In recent proton scattering experiments spin M1

---

*Permanent address: Lab. of Theor. Physics, Joint Inst. for Nuclear Research, Dubna, USSR
[†]Fellow of the A.v. Humboldt Foundation

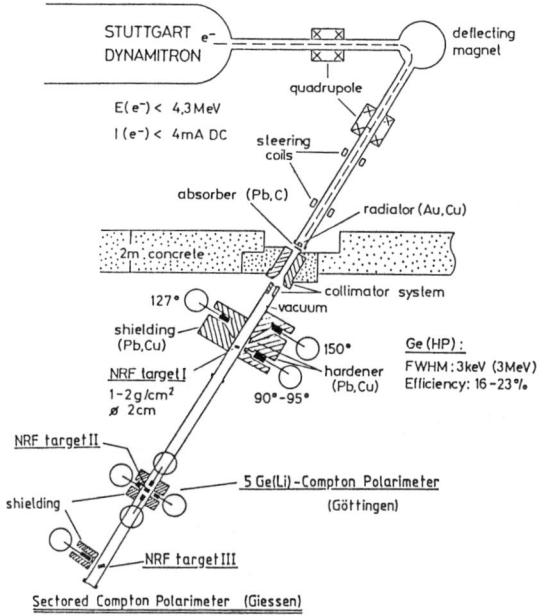

Fig. 1: Experimental set-up at Stuttgart Dynamitron

strength has been observed between 6 and 10 MeV in some rare earth isotopes [12], suggesting that NRF work should be extended to these energies.

Due to the limited space we will restrict ourself in this paper to some high resolution NRF experiments in the rare earth nuclei.

NUCLEAR RESONANCE FLUORESCENCE

The common experimental reactions to explore nuclear structure such as inelastic particle scattering, compound nucleus γ-spectroscopy or neutron capture cover a wide energy and spin range of the excited states. But at higher energies the level density increases and reactions are needed which are very selective in spin and spectroscopic strength in order to study low-spin states with energies above approximately 2 MeV. The method of photon scattering (γ, γ') fulfills these requirements ideally and it offers the possibility to investigate spin one states. In the energy region between 1.5 and 5 MeV one induces dominantly dipole and to less extent quadrupole transitions. Moreover, the mechanism of electromagnetic excitation is well understood and so it is possible to obtain reliable absolute transition strengths from the experimental data.

We are interested in getting an overview of all dipole states up to a certain energy and so it is useful to have a photon source with a continuous energy spectrum like

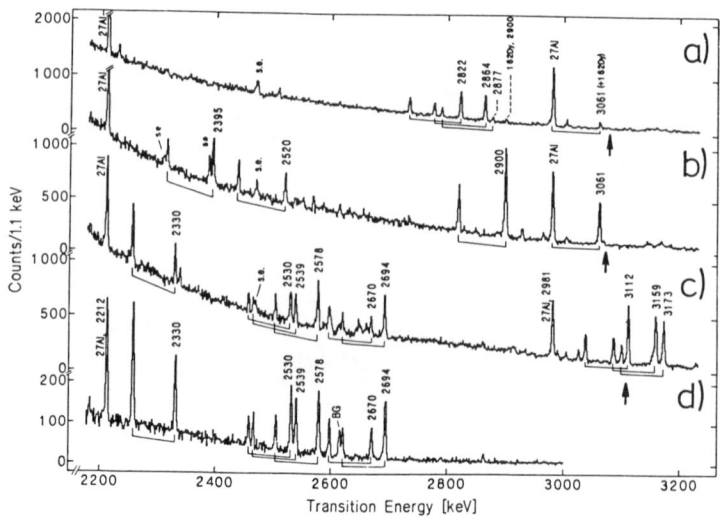

Fig. 2: NRF spectra of ^{160,162,164}Dy (labelled (a)–(c), respectively) measured with 4.1 MeV, and of ^{164}Dy, (d), measured with 2.9 MeV endpoint energy.

e.g. the bremsstrahlung emitted from electrons stopped in a massive radiator target. The deexcitation of the populated states can subsequently be observed with high resolution γ-detectors. This technique of NRF was improved for low energies by Metzger [13] and by the Giessen group [14].

Figure 1 shows a schematic picture of the Stuttgart Dynamitron Photon Scattering Facility. The dynamitron accelerator delivers a dc electron beam with an energy up to 4.3 MeV and a maximum current I_{max}=4 mA. The electron beam is stopped in an Au or Pt target and converted into a continuous bremsstrahlung spectrum. The bremsstrahlung cone is collimated and the low energy part of the spectrum is reduced by lead and graphite absorbers. Three experimental set-ups are operated simultaneously. The first set-up [14] is a three detector arrangement and allows the determination of absolute transition strengths and measurements of angular distributions. The second set-up is a five crystal Compton polarimeter; the third set-up is a single crystal Compton polarimeter both allowing parity determination [15, 16].

Typical NRF-spectra of some Dy-isotopes are shown in fig. 2. The energy resolution is 3–4 keV at 3 MeV excitation energy. A comparison of spectrum c) with spectrum d) shows that a reduction of the photon endpoint energy can improve the peak to background ratio significantly. The spin one states decay both to the 0_1^+ and to the 2_1^+-state. This pair of gamma-rays can be used to identify the level scheme and from the intensity ratio of these transitions one can extract the K-quantum number of the excited states using the Alaga rules.

To obtain the parities of the states one has to compare the photon scattering data with the inelastic electron scattering results from the Darmstadt Linear Accelerator.

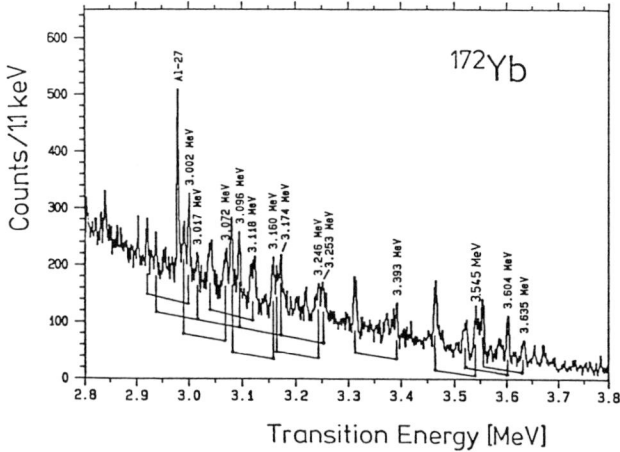

Fig. 3: NRF spectrum of ^{172}Yb between 2.8 and 3.8 MeV.

Additionally, the measurement of the linear polarization of the scattered photons using Compton polarimeters at Stuttgart Dynamitron by the Giessen–Göttingen–Köln collaboration allows model independent parity assignments [15, 16]. Both these experiments are difficult to perform, especially for the weaker transitions, but they are actively pursued.

ΔK=1 TRANSITIONS

In the rare earth region NRF experiments were performed on several Nd- [17], Sm- [18], Gd- [14], Dy- [19], Er- [13, 20] and Yb- [21] isotopes. Figure 3 displays the NRF-spectrum of ^{172}Yb between 2.8 and 3.8 MeV. While the dipole strength in e.g. ^{156}Gd is concentrated in a few transitions, fragmentation as well of the K=1 as of the K=0 strength is generally observed in the heavier nuclei. A fragmentation of the M1 strength can be reproduced by RPA calculations [22, 23].

Due to the fact that we know the parities of only few states we assume in the following discussion that all states with K=1 in an energy window around 3 MeV (where one expects the scissors mode) have positive parity. This assumption has up to now been validated in all cases where the parities have been measured. Clearly it is of utmost importance to measure the parities of all identified spin one states. We refer to the contribution of R.D. Heil in this book.

In order to compare the measurements with the predictions for the scissors mode from various theoretical models (e.g. [24–26]) we plot in fig. 4 the summed K=1 strength around 3 MeV versus the mass number A. The ordinate value is normalized to the IBM-2 estimate for the M1-strength. (Which varies only slightly within the examined nuclei.) One can draw two conclusions from this diagram: The M1-strength up to mass number A=164 is nearly constant and underestimated by the IBM-2 prediction with the g-factors g_π=0.65 and g_ν=0.08. Then one observes a

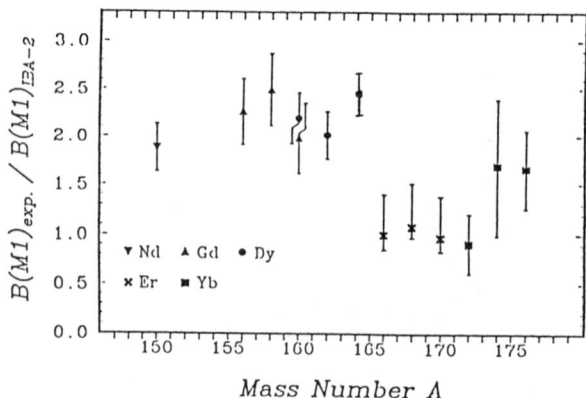

Fig. 4: Systematics of ΔK=1 strength in rare earth nuclei.

certain drop of the strength at ^{166}Er. This anomaly is not yet understood, a new precision measurement on ^{164}Er is clearly warranted.

ΔK=0 TRANSITIONS

States decaying dominantly to the 2_1^+-state correspond to transitions with ΔK=0. The available experimental data on the parities of K=0 spin one states show that all these states have negative parity. Similarly the IBM-2 model does not have $|J^\pi, K\rangle = |1^+, 0\rangle$ states. Thus it is highly favourable to interpret these transitions as E1-transitions.

Figure 5 shows the distribution of K=0 strengths in some Gd-, Dy-, Er- and Yb-isotopes [27]. The ordinate gives the B(E1)↑-strength in $10^{-3} e^2 fm^2$. In the mass region examined here this can be compared with the single particle Weisskopf estimate by 1 mWu $\simeq 2 \cdot 10^{-3} e^2 fm^2$. Thus the strongest transitions have B(E1)-values up to 10 mWu and even some of the weaker transitions above 2 MeV contains several mWu of electric dipole strength.

We will focus on the E1-transitions below 2 MeV excitation energy. The summed K=0 strengths below this energy detected in our NRF-experiments is nearly constant in the examined nuclei. Thus it is natural to assume that these dipole excitations are of collective character as is suggested by various theoretical approaches such as RPA or the IBM with p- or f-bosons [28, 29, 30]. In the vibrational nucleus ^{144}Sm Metzger found comparable collective E1-transitions and interpreted them as a coupling of an octupole and a quadrupole vibration [31]. Barfield et al. [32] gave an equivalent description in the framework of the IBM-1 model with f-bosons.

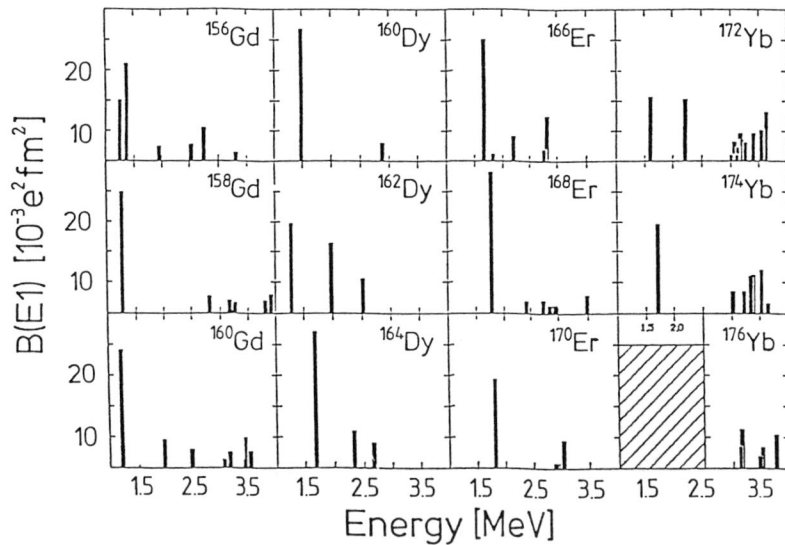

Fig. 5: Distribution of ΔK=0 strength in the rare earth.

BRANCHING RATIOS – A CLUE TO K-MIXING ?

Above we explained that the K-quantum number of the excited states determines the decay branching ratio to different states of the groundstate band. For well deformed even-even nuclei the ratio $R=B(1\to 2_1^+)/B(1\to 0_1^+)$ is 0.5 or 2.0 for a K=1 or K=0 state, respectively. In the (γ,γ')-experiments the ratio is extracted directly from the measured peak areas corrected by the relative energy dependent efficiencies and the angular correlation coefficients. Figure 6 shows an example of the decay of both a J=1, K=0 and a J=1, K=1 state. In the upper part one can see that the peak area arising from the groundstate transition at 1711 keV is smaller than the area arising from the corresponding transition to the 2_1^+-state and so one assigns by inspection a K-quantum number K=0. The opposite is true for the pair of peaks in the lower part of this figure, thus the assigned K-quantum number is K=1.

Now the question arises how well the Alaga rules are fulfilled by the experimental data. From our NRF experiments a data set of about 200 branching ratios in rare earth nuclei was collected. This data set allows us to investigate this question. In order to do this we diplay in fig. 7 the number of states with a certain branching ratio (the frequency of occurence) versus this branching ratio R. The curve shows two prominent maxima: one at $R_{exp} \simeq 0.5$ (as expected for K=1 states) the other at $R_{exp} \simeq 2.0$ (as expected for K=0 states); the number of K=1 states of both parities is roughly three times the number of K=0 states.

The first conclusion from the curve is that more than 80% of the states have decay branching ratios fulfilling the Alaga rules within their error bars. Thus for

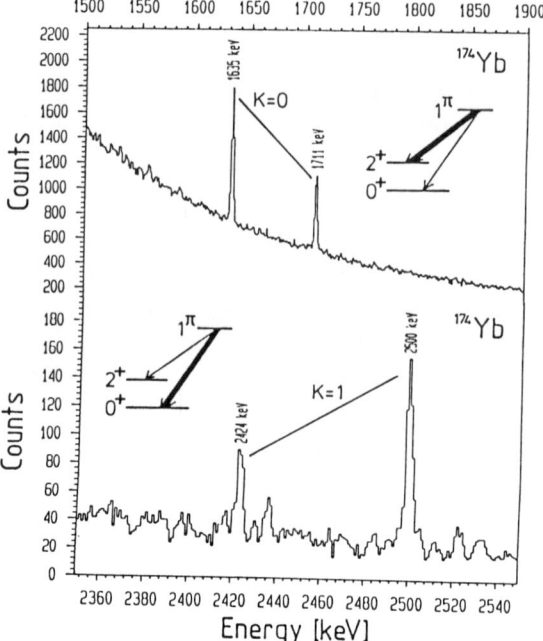

Fig. 6: Two portions of a NRF-spectrum of ^{174}Yb. The upper part shows a $\Delta K=0$, the lower part a $\Delta K=1$ transition.

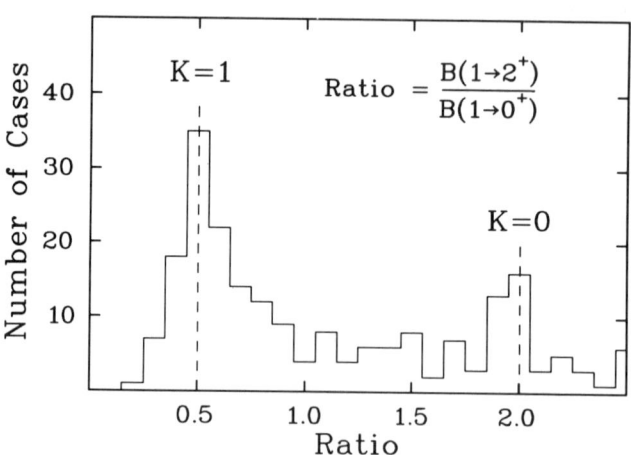

Fig. 7: Frequency distribution P of branching ratios.

most of the states around 3 MeV excitation energy the K-quantum number is still rather good, which is a very interesting finding. Nevertheless there are a few states for which the branching ratios are not compatible with the Alaga rules. One can show that the deviations can not be explained by quadrupole admixtures or other systematic errors in the calculation of the ratios [33]. An inspection shows that in some cases the deviations occur when there are two close-lying spin-one states.

This suggests a possible two state mixing. We started to perform two state mixing calculations, supposing that the measured physical states could be described by the mixture of a pure K=0 with a pure K=1 states. Unfortunately the solution of this two state system is not definite, one gets four different solutions. Due to the fact that we measured in addition to the branching ratios the absolute transition strengths, it is possible to reduce the number of solutions. The degree of reduction depends on the relative experimental errors of the transition strengths. Thus it will be very important to perform new high precision measurements to clarify this point.

An important question arises due to the relative weak absolute E1-transition strengths: Is the K-mixing, determined by such calculations based on the branching ratios, characteristic of the whole wavefunction or only of a small part of it ? If this problem is solved, the observation of branching ratios might become a very useful tool to investigate K-mixing at medium energies. In order to carry out this program one needs to measure parities and one needs also to measure branching ratios with small errors. Fortunately both aims can now be achieved simultaneously in our experiments.

We gratefully acknowledge valuable discussions with R.F. Casten, W. Frank, A. Gelberg, I. Hamamoto, P.O. Lipas, B. Mottelson and P. Vogel. This work was supported by the Deutsche Forschungsgemeinschaft under contract Br 799-32 and Kn 154-18 and partially by the German Federal Minister for Research and Technology (BMFT) under contract Nos. 06OK143 and 06DA184 and by NATO-Grant 5-2-05/RG No. 532/87.

References

[1] W. Greiner, Nucl. Phys. **80**, 417 (1966).

[2] A. Faessler, Nucl. Phys. **85**, 653 (1966).

[3] S.I. Gabrakov, A.A. Kuliev, N.I. Pyatov, D.I. Salamov and H. Schultz, Nucl. Phys. **A182**, 625 (1972).

[4] N. Lo Iudice and F. Palumbo, Phys. Rev. Lett. **41**, 1532 (1978).

[5] F. Iachello, Nucl. Phys. **A358**, 89c (1981).

[6] R.R. Hilton, Zeit f. Phys. **A316**, 121 (1984) and references therein.

[7] A. Richter, Proc. Int. Nuclear Physics Conf., vol.2, ed. P. Blasi and R.A. Ricci, (Florence,1983) p.189.

[8] D. Bohle, A. Richter, W. Steffen, A.E.L. Dieperink, N. Lo Iudice, F. Palumbo, and O. Scholten, Phys. Lett. **B137**, 27 (1984).

[9] U.E.P. Berg, C. Bläsing, J. Drexler, R.D. Heil, U. Kneissl, W. Naatz, R. Ratzek, S. Schennach, R. Stock, T. Weber, H. Wickert, B. Fischer, H. Hollick and D. Kollewe, Phys. Lett. **B149**, 59 (1984).

[10] A. Richter, Nucl. Phys. **A507**, 99c (1990).

[11] U. Kneissl, Part. and Nucl. Phys. **24**, 41 (1989).

[12] D. Frekers, H.J. Wörtche, A. Richter, R. Abegg, R.E. Azuma, A. Celler, C. Chan, T.E. Drake, R. Helmer, K.P. Jackson, J.D. King, C.A. Miller, R. Schubank, M.C. Vetterli and S. Yen, Phys. Lett. **B244**, 178 (1990).

[13] F.R. Metzger, Phys. Rev. **C13**, 626 (1976).

[14] H.H. Pitz, U.E.P. Berg, R.D. Heil, U. Kneissl, R. Stock, C. Wesselborg, and P. von Brentano, Nucl. Phys. **A492**, 411 (1989).

[15] B. Kasten, R.D. Heil, P. von Brentano, P.A. Butler, S.D. Hoblit, U. Kneissl, S. Lindenstruth, G. Müller, H.H. Pitz, K.W. Rose, W. Scharfe, M. Schuhmacher, U. Seemann, Th. Weber, C. Wesselborg and A. Zilges, Phys. Rev. Lett. **63**, 609 (1989).

[16] R.D. Heil, B. Kasten, W. Scharfe, P.A. Butler, H. Friedrichs, S.D. Hoblit, U. Kneissl, S. Lindenstruth, M. Ludwig, G. Müller, H.H. Pitz, K.W. Rose, M. Schuhmacher, U. Seemann, J. Simpson, P. von Brentano, Th. Weber, C. Wesselborg and A. Zilges, Nucl. Phys. **A506**, 223 (1990).

[17] H.H. Pitz, R.D. Heil, U. Kneissl, S. Lindenstruth, U. Seemann, R. Stock, C. Wesselborg, A. Zilges, P. von Brentano, S.D. Hoblit, and A.M. Nathan, Nucl. Phys. **A509**, 587 (1990).

[18] W. Ziegler, C. Rangacharyulu, A. Richter and C. Spieler, Phys. Rev. Lett., in press.

[19] C. Wesselborg, P. von Brentano, K.O. Zell, R.D. Heil, H.H. Pitz, U.E.P. Berg, U. Kneissl, S. Lindenstruth, U. Seemann, and R. Stock, Phys. Lett. **B207**, 2 (1988).

[20] C. Wesselborg, Dissertation, Universität zu Köln, 1988 (unpublished).

[21] A. Zilges, P. von Brentano, C. Wesselborg, R.D. Heil, U. Kneissl, S. Lindenstruth, H.H. Pitz, U. Seemann, R. Stock, Nucl. Phys. **A507**, 399 (1990).

[22] A. Faessler, R. Nojarov and F.G. Scholtz, Nucl. Phys. **A515**, 237 (1990).

[23] J. Speth and D. Zawischa, Phys. Lett. **B211**, 247 (1988); **B219**, 529 (1989).

[24] P. van Isacker, K. Heyde, J. Jolie, M. Waroquier and J. Moreau, Phys. Lett. **B144**, 1 (1984).

[25] C. De Coster and K. Heyde, Phys. Rev. Lett. **63**, 2797 (1989).

[26] T. Otsuka, Nucl. Phys. **A507**, 129c (1990).

[27] P. von Brentano, C. Wesselborg and A. Zilges, in Recent Advances in Nuclear Physics, edited by M.Petrovici and N.V.Zamfir (World Scientific, Singapore,1989), p. 77.

[28] P. Vogel and L. Kochach, Nucl. Phys. **A176**, 33 (1971).

[29] F. Iachello and A. Arima, The Interacting Boson Model, Cambridge University Press, Cambridge 1987.

[30] D. Kusnezov and F. Iachello, Phys. Lett. **B209**, 420 (1988).

[31] F.R.Metzger, Phys. Rev. **C14**, 543 (1976).

[32] A.F. Barfield, P. von Brentano, A. Dewald, K.O. Zell, N.V. Zamfir, D. Bucurescu, M. Ivascu and O. Scholten, Z. Phys. **A332**, 29 (1989).

[33] A. Zilges, P. von Brentano, A. Richter, R.D. Heil, U. Kneissl, H.H. Pitz, C. Wesselborg, Phys. Rev. C, in press.

ENERGY REPULSION AND WIDTH ATTRACTION
IN TWO RESONANCE MIXING

Peter von Brentano

Institut für Kernphysik, Universität zu Köln
5000 Köln 41

Abstract:
It is shown that the effective "perturbed" Hamiltonian H associated with the S matrix of a system of two interacting resonances, which has one strong open reaction channel only and which is timereversal invariant, can be decomposed in a natural way into a diagonal "unperturbed" effective Hamiltonian H° and a real, symmetrical and purely offdiagonal interaction $V = H - H^\circ$. The complex energies of the "perturbed" system exhibit energy repulsion and width attraction as compared to the energies of the "unperturbed" system. This result is the direct generalization of the level repulsion theorem for the bound state case. An example in ^8Be is given.

The quantum mechanical two level system is a very useful concept in physics (1-3). It is of interest to consider not only systems of two bound states but also systems of two unbound states. An interesting property of the bound state system is energy repulsion: the energy difference of two states mixed by a real, symmetrical and offdiagonal interaction is never smaller than the energy difference of the two corresponding unmixed states (1-3). For such an interaction this holds true also for the the two resonance case(4,5).

For a system of two "perturbed" bound states described by the matrix H one finds energy repulsion if the "unperturbed" states, described by the diagonal matrix H°, are mixed by a purely offdiagonal interaction matrix $V = H - H^\circ$ (1-3).

$$H^\circ = \begin{vmatrix} \varepsilon^\circ_1 & , & 0 \\ 0 & , & \varepsilon^\circ_2 \end{vmatrix} , \quad V = \begin{vmatrix} 0 & , & v \\ v & , & 0 \end{vmatrix} \quad (1)$$

Thus the "perturbed" energies show energy repulsion
as compared to the "unperturbed" ones :

$$| \varepsilon^0{}_1 - \varepsilon^0{}_2 | \leq | \varepsilon_1 - \varepsilon_2 | \qquad (2)$$

One notes that a diagonal interaction could change
the energies in both ways and it is therefore not
considered.
If one now considers the mixing of two unbound states
one has to replace the Hamiltonians by effective
Hamiltonians H, H°, which take the decay of the
resonances into account. Thus the "perturbed" and
"unperturbed" energies of the unbound states will
generally be complex.

$$\varepsilon_k = E_k - (i/2)G_k \,,\quad \varepsilon^\circ{}_k = E^\circ{}_k - (i/2)G^\circ{}_k \qquad (3)$$

Particular simple results are obtained if one uses
only internal mixing in the resonance case i.e. a
real, symmetrical and offdiagonal interaction
$V = H - H^\circ$. In this case one finds again
energy repulsion i.e. eq.(4a). An interesting result
in the resonance case with internal mixing is width
attraction (4,5) i.e. eq.(4b):

$$| \text{Re}\,(\varepsilon^0{}_1 - \varepsilon^0{}_2) | \leq | \text{Re}\,(\varepsilon_1 - \varepsilon_2) | \qquad (4a)$$

$$| \text{Im}\,(\varepsilon^0{}_1 - \varepsilon^0{}_2) | \geq | \text{Im}\,(\varepsilon_1 - \varepsilon_2) | \qquad (4b)$$

A simple explanation of width attraction is that for
a sufficiently large interaction V the two states
will become completely mixed and they will thus have
the same lifetimes and total widths.

One must emphasize that the interaction V can in general not be choosen freely but rather it is given.
Namely one knows the true Hamiltonian H, obtained
from the data, and uses a given model Hamiltonian
H°. Then the interaction V is determined as :
$V = H - H^\circ$. It is in general a symmetrical and
complex matrix and corresponds to a combination of
both internal and external mixing.
It will be shown, however, that in the case where
there is only one strong open reaction channel and
under assumption of time reversal invariance there is
a natural choice of the "unperturbed " Hamiltonian
H°, which leads to a real, symmetrical and purely

offdiagonal interaction V.
This choice of the "unperturbed" Hamiltonian is obtained as follows:
The two "perturbed" states with widths G_1, G_2 result from a mixing of one "unperturbed" state with width $G^0{}_1 = G_1 + G_2$ with a second "unperturbed" state with no width i.e. $G^0{}_2 = 0$ by a real, symmetrical and offdiagonal interaction matrix V. These conditions imply the relations (5a,b)

$$\det(H^\circ + V - E) = \det(H - E) \qquad (5a)$$

$$\det(H - E) = (\varepsilon_1 - E)(\varepsilon_2 - E) \qquad (5b)$$

which allow one to obtain from the four real parameters of the "perturbed" Hamiltonian H : $\varepsilon_1 = E_1 - (i/2)G_1$, $\varepsilon_2 = E_2 - (i/2)G_2$, the three parameters of the "unperturbed" Hamiltonian H° : $\varepsilon^\circ{}_1 = E^\circ{}_1 - (i/2)G^\circ{}_1$, $\varepsilon^\circ{}_2 = E^\circ{}_2$ and the one real parameter v of the real, symmetrical and offdiagonal interaction V of eq.(1). The parameters of the "perturbed" Hamiltonian H : ε_1, ε_2 are obtained as the complex pole positions of the S matrix, which is obtained from from a fit of the experimental resonant cross section.

It is remarkable that in the case of two resonances with one open channel only a knowledge of only the two complex resonance positions of the "perturbed" system enables one to obtain an "unperturbed" Hamiltonian and the interaction. The reason is that in the one channel case the total width is equal to the elastic width, which in turn determines a spectroscopic factor of the resonance.
A similar decomposition of the "perturbed" Hamiltonian into an "unperturbed" Hamiltonian and an interaction is also possible in the boundstate case if the corresponding spectroscopic factors are known.

One notes, that even in the case of N resonances and M open channels an effective Hamiltonian can be defined from the S matrix. A special form of the S matrix which defines the effective Hamiltonian has been given by Mahaux and Weidenmüller in their book (6) and it has also been used by a number of other authors (7-10). The special decomposition of the Hamiltonian in the one channel case given above in eq.(5) can be extended from 2 resonances to N resonances as has been shown by Sokolov and Zelevinski(9).

As an example of the level repulsion among resonances the two well known 2^+ resonances in ^{8}Be lying at the excitation energies around 16.6 and 16.9 MeV will be discussed (see refs.(11,12) and refs. contained therein). A beautiful and very detailed investigation of these two resonances in ^{8}Be was done by Hinterberger et. al.(11). They studied the excitation functions of elastic alpha scattering in small energy steps at the Bonn cyclotron and gave also a detailed analysis of the data in terms of the poles of the S matrix, which are given in the column No.1 below. From these "perturbed" energies and using eq.(5) the parameters of the "unperturbed" Hamiltonian were obtained and are given in column No.2 below.

1) "Perturbed complex energies :
   $\varepsilon_1 = (16722-(i/2)\ 108.5(5))$ keV,
   $\varepsilon_2 = (16722 + 288.0(5) - (i/2)\ 73.6(4))$ keV

2) "Unperturbed" complex energies:
   $\varepsilon^0_1 = (16722 + 116.2(6) -i/2)182.1(6))$ keV,
   $\varepsilon^0_2 = (16722 + 171.4(6))$ keV

The obtained mixing matrix element v= 148.3(6) keV agrees with previous work (11).
The fact that one of the "unperturbed" resonances is a bound state in the continuum and has a vanishing width is in the ^{8}Be case due to a selection rule and a quantum number. One notes that the two "unperturbed" levels in ^{8}Be have isospins T = 0 and T = 1, which is in agreement with the vanishing width of the upper "unperturbed" level in the $\alpha\alpha$ channel (11,12).This result gives a physical interpretation for the use of a purely internal mixing representation for the two ^{8}Be resonances.

The parameters of the two 2^+ resonances in ^{8}Be allow an empirical test of the level repulsion theorem for resonances i.e. of eqs.(2),(4).

55.2(9) keV = $|E^0_2 - E^0_1|$ < $|E_2 - E_1|$ = 288.0(5) keV

182.1(6) keV = $|G^0_1 - G^0_2|$ > $|G_1 - G_2|$ = 34.9(9) keV

The first inequality is energy repulsion and the second inequality is width attraction.
There is work in progress in Köln in which these ideas are applied to s - wave neutron resonances.

Acknowledgement: Particular thanks are to Dr. M. Zirnbauer, who has contributed much to this work. I would like to thank G. Böhm, R. F. Casten, W. Frank, A. Gelberg, H. L. Harney, R-D. Herzberg, and V. G. Zelevinski for discussions and to R-D. Herzberg for reading the manuscript. . This work was partly funded by the BMFT under the Contract No. 060 K 143.

References:
1) J. von Neumann and E.P. Wigner, Z. Phys. 30 (1929) 427
2) The Feynman lectures,Vol. 3. Quantum mechanics, R.P. Feynman, R.B. Leighton and M.Sands, Addison-Wesley Publishing Co., London 1970, chapter 9-11
3) C. Cohen-Tannoudji, B. Diu, F. Laloe, Mecanique quantique Herman, Paris (1973) chap. 4.
4) P. von Brentano, Phys. Lett. B.238 (1990),1.
5) P. von Brentano, Phys. Lett. B. Aug.30.(1990).
6) C. Mahaux and H.A. Weidenmüller, Shell-Model approach to nuclear reactions, North-Holland, Amsterdam, (1969), 51
7) J.J.M. Verbaarschot, H.A. Weidenmüller and M.R. Zirnbauer, Phys. Rep. 129 (1985) 369
8) H.Feshbach, Ann.Phys.(N.Y) 5 (1958) 357, 19 (1962) 287
9) V.V. Sokolov and V.G. Zelevinski Nucl. Phys. A 504 (1989) 562
10) V.D. Kirilyuk, N.N. Nikolaev and L.B. Okun, Yad. Fiz. 10 (1969) 1081 [Sov. J. Nucl. Phys. 10 (1970) 617]
11) F. Hinterberger, P.D. Eversheim, P. von Rossen, B. Schüller, R. Schönhagen, M. Thenee, R. Görgen, T. Braml and H.J. Hartmann, Nucl. Phys. A299 (1978) 397
12) F.C. Barker, Phys. Rev. Lett. 35 (1975) 613

INVESTIGATION OF ODD Ba NUCLEI FROM NEUTRON INDUCED REACTIONS

V.A.Bondarenko, I.L.Kuvaga, P.T.Prokofjev
Institute of Physics of the Latvian Academy of Sciences
229021, Salaspils, Latvia
V.A.Khitrov, Yu.V.Kholnov, Le Hong Khiem, Yu. P.Popov, A.M.Sukhovoj
Joint Institute of Nuclear Research, Dubna 141980, USSR

ABSTRACT

Thermal neutron capture reaction on Ba isotopes have been studied. New levels with negative parity are introduced in odd Ba nuclei. They are interpreted in terms of neutron hole state $1h_{11/2}$ coupled with even-even core.

Analysis of the results obtained from the ^{136}Ba $(n,2\gamma)$ reaction by thermal neutron capture was carried out. It was pointed out that the radiative strength function of primary transitions in ^{137}Ba in the wide range of their energies could be described in the framework of the GEDR model taking into account the energy- and temperature-dependence of resonance widths. Deviation from this dependence may be caused by one-particle transitions 4S→3P.

The systematic study of Ba isotope ($135 \leq A \leq 139$) level schemes and radiational properties has been undertaken using thermal neutron capture reaction.

The measurements, performed at the reactor of the Institute of Physics, include both the single spectra measurements and the γ-γ coincidence spectra measurements with two HPGe detectors (10 and 12% relative to 7.6 cm x 7.6 cm NaI). The γ-γ coincidences were measured in the energy region from the threshold at about 150 keV up to the binding energy for each isotope. The information event was stored on magnetic tapes and evaluated afterwards using the traditional method (via the window selection) as well as the method developed in Dubna[1].

The analysis allowed to obtain from spectra the primary transitions with total intensity of about 30% for ^{135}Ba, 76% for ^{137}Ba and 96% for ^{139}Ba. The average excitation region populated by primary transitions is 3 - 4,5 Mev. In ^{137,139}Ba almost all intensity of primary transitions is attributed to E1 transitions of 4s — 3p type. The intensity of the 4s — 3s type M1 transition in ^{137}Ba is comparable to or even exceeds that of the most of E1 transitions. In ^{135}Ba, almost uniform distribution of the strenght of primary transitions is observed. Therefore, the parity of the most 1/2, 3/2 levels populated by primary transitions remains uncertain.

In order to determine the spin values of the lowest levels in ^{135,137}Ba the comparative analysis of transition intensities from (n,γ) and $(n,n'\gamma)$[2] reactions has been performed. In ^{135}Ba the three-step cascade 5302-956-445 populating the $1h_{11/2}$ neutron hole state at 268 keV has been found. This would mean that the primary transition is E1 transition but two others should have E2 multipolarity. In ^{137}Ba we observe only one 7/2 level belonging to the $1h_{11/2}$ configuration. The 1252 keV transition between 7/2

© 1991 American Institute of Physics

and 11/2⁻ states yields 72% of the 11/2⁻ isomer state production in the (n, n'γ) reaction. We have not observed any coincidences with the 1252 keV transition in (n,γ) reaction due to the strong selectivity of 3/2⁻ state population in ^{137}Ba, unlike to ^{135}Ba.

The simple interpretation of negative parity states belonging to $1h_{11/2}$ configuration is given in Fig. 1. The obvious correlation of even-even and odd Ba nuclei energy differences forces to interpret the low-spin branch as an antialignement of $h_{11/2}$ quasiparticle with respect to the even-even core. A somewhat different interpretation can be based on the three-axial rotator model[3]. When the standard parameter procedure is applied, one can find that the position of the lowest $11/2^-$, $7/2^-$ and $15/2^-$ levels in ^{135}Ba is reproduced in good agreement with the experiment by model calculations with $\gamma = 22 \pm 2^0$, $\beta \cdot A^{2/3} = 4.1$, $h^2/2\mathcal{J}_0 = 86 \pm 3$ keV.

Finally, it should be stressed that other models, for example IBFA or "spinor symmetry"[4], can also be used for the interpretation of the negative parity bands in Ba N < 82 region.

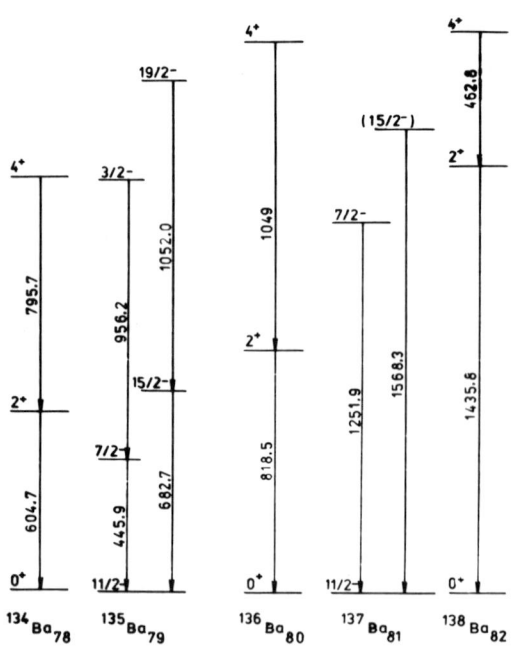

Fig. 1   The energy spacings in the N ⩽ 82 Ba isotopes the high energy negative parity branch in ^{135}Ba is taken from Ref.[5]

Two-step cascades from compound- to ground and three low-lying excited states in ^{137}Ba were studied by capture of thermal neutrons. Forty five intensive cascades were established and 23 of them placed in the decay scheme of ^{137}Ba.

Total intensities of all two-step cascades to 4 final states of ^{137}Ba are given in Table.1.

Table 1.

Energy of cascade (keV)	:Excited :final :level (keV)	Intensity (Percent per decay)	
		Experimental	Calculated
6904	0	15.6 (7)	22
6623	281	53 (3)	17
5004	1900	[2.2]	1.8
4724	2180	[5.5]*	1.1

* The intensity of the cascade 4242+482 keV is added at a level of registration $E_\gamma > 520$ keV for all cascades.

The time resolution of Ge-detectors exceeds the life-time of all the intermediate levels excited in ^{136}Ba (n, 2γ). Then the observed distribution contains both intensities of primary transitions and of secondary ones. Earlier [6] we have developed a method for substraction of cascade intensities corresponding to secondary transitions. The sum of intensities $i_{\gamma\gamma}$ of the decomposed spectrum as a function of the energy of primary transitions is presented in Fig.2 (histogram). For comparison in the same figure there is given a spectrum (points) calculated by the relation:

$$i_{\gamma\gamma}^{cal} = (\langle \Gamma_{\lambda g} \rangle / \Gamma_\lambda) \, (\, \langle \Gamma_{gf} \rangle / \Gamma_g) \langle \rho_g \rangle \Delta E \quad (1)$$

Here $\langle \Gamma_{\lambda g} \rangle$ and $\langle \Gamma_{gf} \rangle$ are the average values of the partial widths of the primary (λ-g) and secondary (g-f) transitions, Γ_λ and Γ_g the total radiative widths of the decaying states λ and g, and $\langle \rho_g \rangle$ the level density in the interval ΔE in the vicinity of the level g. The calculation procedure was analogous to that reported in [7] but the radiative widths of E1-transitions were calculated in the modified model of the giant dipole electric resonance [8].

Intensities of cascades from the compound-state are sufficiently high ($\Sigma i_{\gamma\gamma}^{exp} = 76 \pm 25\%$ per decay, the error includes uncertainty of thermal neutron capture cross section[9] for ^{136}Ba) and one can determine the lower estimate of $\langle \Gamma_{\lambda g} \rangle$ by the relation:

$$\langle \Gamma_{\lambda g} \rangle = i_{\gamma\gamma} \Gamma_\lambda / (\alpha \langle \rho \rangle_g \Delta E) \quad (2)$$

here $\alpha = (\Sigma \Gamma_{gf}/\Gamma_g) \leq 1$ and $\Gamma_\lambda = 125$ meV according to [10]. Then the lower estimate of K(E1+M1) may be obtained by the relation:

$$K(E1+M1) \simeq (i_{\gamma\gamma}\Gamma_\lambda)/(\langle \rho_g \rangle \Delta E\, D_\lambda E_\gamma^3 A^{2/3}) \qquad (3)$$

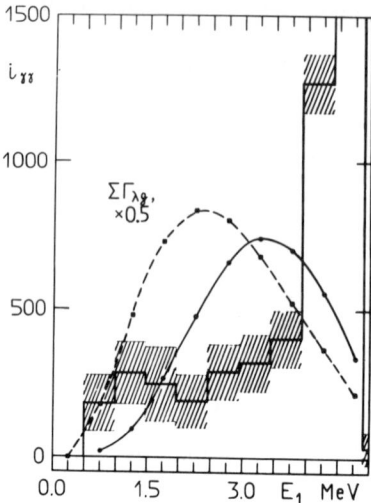

Fig.2. The intensity $i_{\gamma\gamma}$ off cascades (per 10^4 capture events).

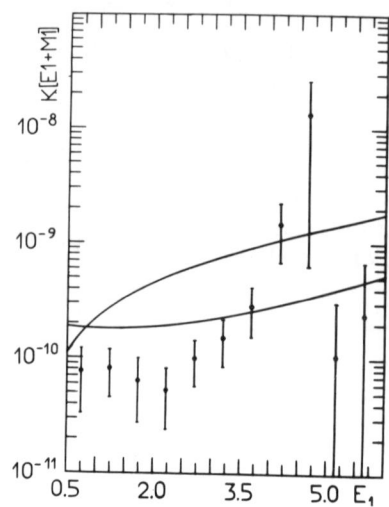

Fig.3. Average experimental values for RSF and two their different predictions.

Average experimental values for RSF over intervals $\Delta E = 0,5$ MeV wide are presented in Fig.3 (points). They are compared with two different predictions: of the model (lower curve) and of the traditional extrapolation of GEDR (upper curve) to the region of low energy transitions. It was assumed that the width of GEDR is constant and the capture cross section for E1-photons changes following the Lorentzian curve. It can be seen from Fig.3 that the experimental values of $K(E1+M1)$ in the wide region of the primary transition energies are in agreement with those predicted by the model [8]. Only for two strong transitions $E_\gamma = 4742$ and 4242 keV which populate $3p_{1/2}$ and $3p_{3/2}$ final states[11] the value of $K(E1+M1)$ is above the estimate from[8].

REFERENCES

1. A.A.Bogdzel et. al. JINR preprint P15-82-706 Dubna 1982
2. V.A.Bondarenko et. al., Izv. Akad.Nauk LatvSSR, ser. Fiz-tch, №6,7 (1983).
3. J.Meyer-Ter-Vehn Nucl. Phys. A249, 111 (1975), A249, 141 (1975)
4. M.M.Michailova J. Phys. 613, L149 (1987)
5. E.Dragulescu et.al. Rev. Roum. Phys. v32, 743 (1987).
6. S.T.Boneva et al., JINR E3-90-45, Dubna, 1990.
7. S.T.Boneva et al., Z.Phys., A330, 153 (1988).
8. S.G.Kadmensky et al., Yad.Fiz., V.37, p.277 (1983).
9. L.Koester et al., Z.Phys., a322, 105 (1985).
10. S.F.Mughabghab, Neutron Cross Section, Vol. 1, Part A, New York, Academic Press, 1984.
11. L.V.Groshev et al., Yad. Fiz., V.10, p.681 (1969).

OCTUPOLE DEFORMATION IN ODD-ODD NUCLEI ^{154,156}Eu

A.V.Afanasjev, T.V.Guseva, J.J.Tambergs, M.K.Balodis
Institute of Physics, Latvian Academy of Sciences,
SU-229021 Riga-Salaspils

The considerable attention is given recently to the possibility of octupole deformation of Eu nuclei. The following arguments favour such an assumption : 1) the calculation of nuclear shape equalibrium deformation stabilization[1] at A~145÷150; 2) the inversion of the odd-even staggering exists in the distribution of Eu isotope charge radii[2] at A~152÷156; 3) the enhanced E1-transitions are observed which are difficult to explain in the framework of unified nuclear model with β_2, β_4-deformation only. The odd nuclei model has been proposed[3] taking into account an axially-symmetric reflection-asymmetric nuclear shape, associated with odd multipolarity deformations β_3, β_5. Owing to the difference between the uncoupled neutron and proton single-particle potential wells, one can write the model Hamiltonian for the odd-odd nuclei in the framework of this model as follows:

$$H = H^p_{sp} + H^n_{sp} + H^p_{pair} + H^n_{pair} + H^p_{core} + H^n_{core} + H_{rot} + H_{rpc} + H_{irot} + H_{ppc} + V_{np} \quad (1)$$

where H^p_{sp}, H^n_{sp} are the single-particle Hamiltonians which do not conserve parity in the reflection-asymmetric mean field of the nucleus; H^p_{pair}, H^n_{pair} are the pairing interaction Hamiltonians; $H_{rot}, H_{rpc}, H_{irot}, H_{ppc}, V_{np}$ are the Hamiltonians analogous to the corresponding terms of the usual nonadiabatical rotational model total Hamiltonian for odd-odd nuclei[4]; the terms

$$H^p_{core} = 1/2 \, E(0^-)(1 - p_n \hat{\pi}_p)$$
$$H^n_{core} = 1/2 \, E(0^-)(1 - p_p \hat{\pi}_n) \quad (2)$$

are responsible for the parity splitting of corresponding parity degenerate nucleon orbitals. As a result we have the doublets of the good parity states with splitting energies ΔE_p, ΔE_n for proton and neutron orbitals correspondingly. The parameter $E(0^-)$ is calculated microscopically or derived from neighbouring even-even nuclei. In (2) $\hat{\pi}_p, \hat{\pi}_n$ are the single-particle parity operators of proton and neutron, respectively; $p_p = \pi_p p$, $p_n = \pi_n p$, where $p = \mp 1$ is the eigenvalue of the good total parity operator for nuclear states and π_p, π_n are the parities of the single-particle states for odd proton and odd neutron, respectively.

Due to the parity splitting of proton as well as neutron orbitals in odd-odd nuclei there should be the parity quadruplets (PQ's) of rotational bands. The different parity states of such PQ are connected via enhanced E1-transitions occuring due to the intrinsic collective E1 moment that results because the proton and neutron centres of mass do not is general coinside in a reflection-asymmetric nucleus. Although, the Nilsson asymptotic quantum numbers except Ω are no longer good quantum numbers in the presence of parity mixing, we shall continue to use them to label the individual bands of PQ:

$K^\pi : \Omega_p[Nn_z\Lambda] \mp \Omega_n[Nn_z\Lambda]$ <———> $K^{-\pi} : \Omega_p[N\mp 1 n_z'\Lambda'] \mp \Omega_n[Nn_z\Lambda]$

$\updownarrow$ $\updownarrow$

$K^{-\pi} : \Omega_p[Nn_z\Lambda] \mp \Omega_n[N\mp 1 n_z'\Lambda']$ <——> $K^\pi : \Omega_p[N\mp 1 n_z'\Lambda'] \mp \Omega_n[N\mp 1 n_z'\Lambda']$

keeping in mind that the indicated configurations are the major components of the given state. The accounting for other interaction modes of the Hamiltonian (1) may result in the considerable distortions of PQ splitting. The parity doublet (PD)[5] in odd-odd nuclei is part of PQ in such a context. The analysis of ^{154}Eu[6,7] level scheme allows us to identify PQ:

p:5/2[413]∓n:3/2[532] <———> p:5/2[532]∓n:3/2[532]

$\updownarrow$ $\updownarrow$

p:5/2[413]∓n:3/2[402] <———> p:5/2[532]∓n:3/2[402]

including the PD's proposed in ref.[5]. Such identification is based on the upper limits of Weisskopf factor F_W-values for E1-transitions (Table 1), having values characteristic for octupole-deformed nuclei[5] $\sim 10^2 \div 10^3$, and on the enhancment of E1-transitions between the PQ's states com-

Table 1. E1-transitions in ^{154,156}Eu and their Weisskopf factors F_W; the last column indicates whether the involved states belong (Y) to the parity doublet (PD) or not (N)

	E_i keV	I_i^π	initial state	E_f keV	I_f^π	final state	F_W	PD
^{154}Eu	129.7	4^-	p5/2[413]+ +n3/2[532]	100.86	4^+	p5/2[413]+ +n3/2[402]	≤5.2+2	Y
	249.4	1^+	p5/2[532]- -n3/2[532]	162.43	1^-	p5/2[413]- -n3/2[532]	≤1.1+4	Y
				180.81	2^-	—— " ——	≤1.1+4	Y
^{156}Eu	87.49	1^-	p5/2[413]- -n3/2[521]	0	0^+	p5/2[413]- -n5/2[642]	9.0+3	N
				22.53	1^+	—— " ——	4.3+4	N
	291.3	1^+	p5/2[532]- -n3/2[521]	87.5	1^-	p5/2[413]- -n3/2[521]	≤5.3+3	Y
				125.5	2^-	—— " ——	≤4.8+3	Y

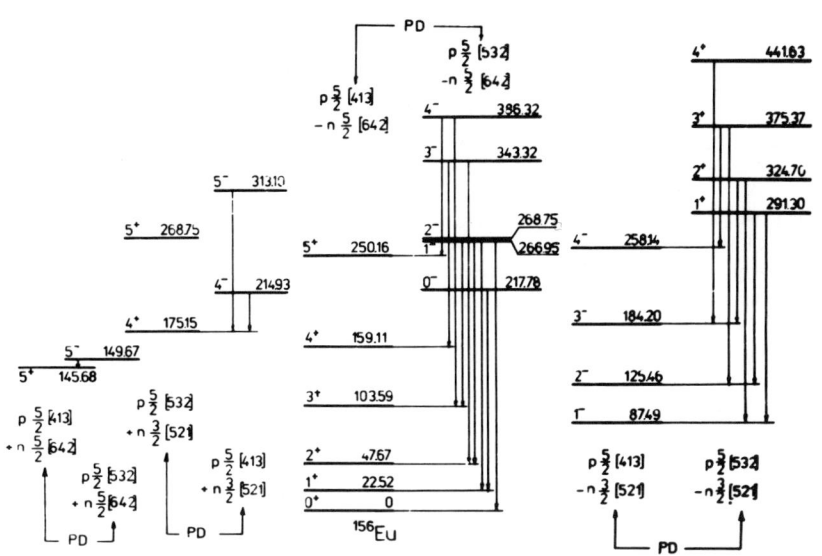

Fig.1. The band structure in the part of ^{156}Eu level scheme; the proposed $K^{\mp}$ parity-doublet bands (labeled PD) are connected; the E1-transitions within PD are presented.

pared to the E1-transitions between the states outside the PQ's. To identify other PQ's in ^{154}Eu is difficult, because the PQ occupies in level scheme the greater energetic interval than PD and also because there is unsufficient information on lifetime of highlying levels.

The same reasons make it difficult to search for analogous PQ's in ^{156}Eu^8. At present, we have been able to identify in the level scheme of ^{156}Eu only the fragments of such PQ's — four PD's with $K^\pi=0^\mp, 1^\mp, 4^\mp, 5^\mp$ (Fig.1), supported by the following considerations:

1. For the $K^\pi=0^\mp$ and $K^\pi=4^\mp$ PD's E1-transitions are favoured in the depopulation of all known levels of the $K^\pi=0^-$ band and of the two known levels of the $K^\pi=4^-$ band
2. E1-transitions inside PD with $K^\pi=1^\mp$ are enhanced if compared with E1-transitions outside this PD (Table 1).
3. The $K^\pi=5^\mp$ bands with 149.67 keV and 145.68 keV bandhead levels form PD, because these levels are, very probably, connected by the 3.99 keV E1-transition although it is not observed because of low energy; other possibility of PD could be the $K^\pi=5^\mp$ bands at 368.57 and 145.68 keV, however, we strongly prefer the former possibility since E1-transition between the 368.57 and 145.68 keV levels is not observed.

REFERENCES

1. S.Cwiok et al, Nucl.Phys.A496,367 (1989)
2. G.A.Alkhasov et al, Sov.J.Nucl.Phys.v.44,734 (1986)
3. G.A.Leander and Y.S.Chen, Phys.Rev.C37,2744 (1989)
4. I.P.Boisson et al, Phys.Reports 26C,100 (1976)
5. R.K.Sheline, Phys.Lett. B219,222 (1989)
6. H.Rotter et al, Nucl.Phys. A417, 1 (1984)
7. M.K.Balodis et al, Nucl.Phys. A472,445 (1987)
8. M.K.Balodis et al, 37th USSR Nucl.spectr.struct. conf. p.117, (1987)

FEATURES OF THE LEVEL SCHEMES OF $^{164-166}$Dy

F. Hoyler, K. Föhl
Phys. Institut, Univ. Tübingen, D-7400 Tübingen, FRG

H.G. Börner, B. Krusche, S.J. Robinson, P. Schillebeeckx
Institut Laue-Langevin, 156X, F-38042 Grenoble Cedex, France

ABSTRACT

The (n_{th},γ) reaction has been measured at the ILL on the target nuclei ^{163}Dy and ^{164}Dy using the Ge pair spectrometer PN4 and the bent crystal spectrometers GAMS. Conversion electrons have been measured using the β-spectrometer BILL, and coincident γ- rays were recorded at the end position of the neutron guide H22. Measurements in different neutron fluxes and the observation of the time dependence of the neutron capture process allowed to assign transitions in ^{166}Dy unambiguously. The multipolarities obtained from the measurement of conversion electrons were used to improve spin and parity assignments for levels in ^{164,165,166}Dy.

INTRODUCTION

The Dy isotopes represent an ideal case for the investigation of a chain of isotopes in the deformed rare-earth region via the (n,γ) reaction due to the large capture cross section of the stable isotopes. Extensive level schemes have been constructed for ^{161,163}Dy by von Egidy et al.[1,2] and a careful investigation of ^{165}Dy has been published recently[3]. The measurements of conversion electrons (CE) at the BILL spectrometer of the ILL allowed the determination of multipolarities and thus a further improvement of the level scheme of ^{165}Dy.

The high cross sections for the ^{164}Dy(n,γ) and the ^{165}Dy(n,γ) reaction allow to investigate the nucleus ^{166}Dy through consecutive neutron capture despite the relative short half-life of only 2.4h for the intermediate nucleus ^{165}Dy provided that an assignment to ^{166}Dy can be made.

The observation of 'collective' M1-transitions in inelastic electron[4] and γ- ray[5] scattering in the rare-earth region led us to search for these states in the spectrum of the ^{163}Dy(n,γ) reaction. Furthermore Burke et al.[6] published recently an investigation of the (t,p) - reaction leading to even-even Dy nuclei and found very large (t,p) -strength to two 0^+ levels at 1655 and 1775 keV excitation energy. Average resonance capture (ARC) studies on ^{162,164}Dy[7] show only two candidates for 2^+ states, which could be members of the K=0^+ bands built onto these 0^+ states. It was therefore another goal of the present investigation to study the decay properties of these K=0^+ bands.

EXPERIMENTAL DETAILS

All experiments reported here have been performed at the HFBR of the ILL Grenoble. γ- rays and conversion electrons have been observed after thermal neutron capture using the curved crystal spectrometers GAMS1,2/3 and the β-spectrometer BILL in the energy range below 2 MeV. Above 2 MeV γ- ray energy data taken with the pair spectrometer PN4 and singles HPGe-spectra have been used. To search for primary transitions populating 1^+ levels in ^{164}Dy coincidences between primary and secondary transitions have been measured at the end position of the neutron guide H22 of the ILL.

A prerequisite for the identification of transitions in ^{164}Dy is the knowledge of lines belonging to ^{165,166}Dy, since, due to the high cross section (2700b) of the ^{164}Dy(n,γ) reaction contaminant lines from this reaction contribute significantly to the total γ- ray intensity.

Special experimental effort has been put into the unambiguous assignment of transitions in ^{166}Dy. Two different approaches have been taken to this end for the observation of γ-rays and CE respectively.

© 1991 American Institute of Physics

The β-decay of ^{165}Dy competes with the double capture process:

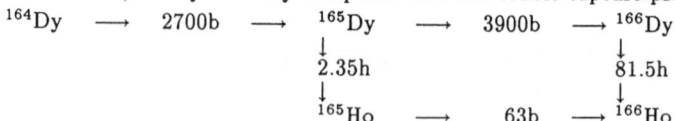

^{164}Dy $\longrightarrow$ 2700b $\longrightarrow$ ^{165}Dy $\longrightarrow$ 3900b $\longrightarrow$ ^{166}Dy
$\downarrow$ 2.35h $\qquad\qquad\qquad$ $\downarrow$ 81.5h
$\downarrow$ $\qquad\qquad\qquad\qquad\qquad$ $\downarrow$
^{165}Ho $\longrightarrow$ 63b $\longrightarrow$ ^{166}Ho

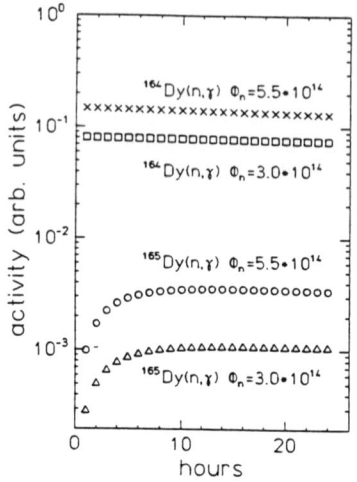

Fig.1 Calculated neutron capture rate for ^{164}Dy

In a constant neutron flux Φ_n, and in saturation of the 2.4h activity, the double to single capture rate is proportional to Φ_n^2. This strong Φ_n dependence implies that it is only feasible to observe ^{166}Dy by irradiation in a high neutron flux, but it offers also the possibility to assign γ-rays to ^{166}Dy in measuring the ^{164}Dy(n,γ) reaction in different neutron fluxes. The in-pile target facility of the ILL allows to place the target in a flux of 5.5×10^{14} n/cm^{2}s at a distance of 50cm from the reactor core or in a flux of appr. 3.4×10^{14} n/cm^{2}s at a distance of 70cm from the core. This is achieved by moving the tube holding the source 50cm towards the GAMS2/3 side [8]. A calculation of the time and flux dependence for the ^{164}Dy(n,γ) and the ^{164}Dy$(2n,\gamma)$ reaction presented in Fig. 1 highlights the relative change in single to double capture rate. We have performed a measurement of the ^{164}Dy(n,γ) reaction in the two target positions. using the existing GAMS3 quartz crystal. Since the reduction in flux is achieved by changing the distance from the target to the crystal the bending radius in this low flux position has to be 23.5m instead of 24m in the standard position. By moving a collimator having a width of appr. 6mm and a height of 15mm in front of the crystal a position was found where practically the same resolution and sensitivity was obtained also for this shorter distance. In Fig. 2 we compare two spectra of the same energy range at the different source positions The relative change in intensity for the two labelled peaks from the double capture process can clearly be seen. On the other hand the figure illustrates also the complexity of the measured spectra. The γ-ray spectrum was measured in the range from 150 to 2000 keV, since the efficiency of GAMS3 decreases strongly for lower energies.

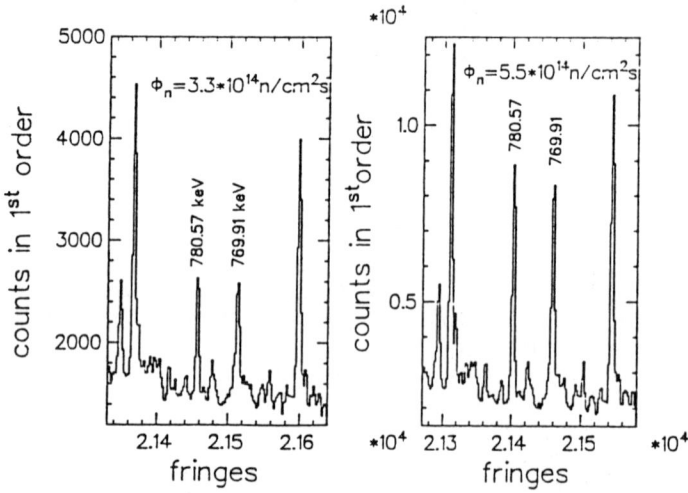

Fig.2. GAMS3 spectrum taken at different neutron fluxes

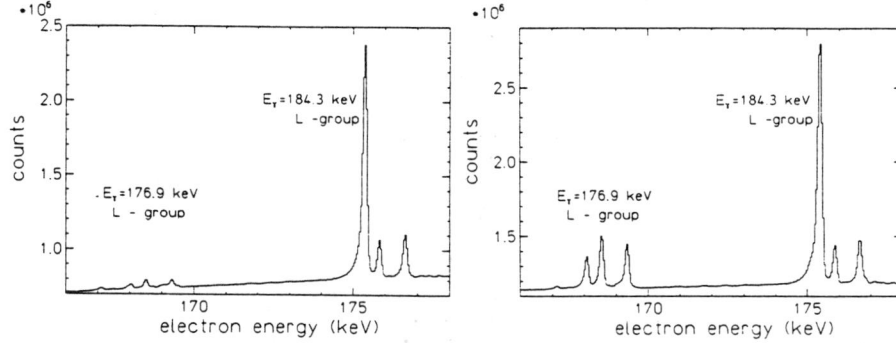

Fig. 3. CE-spectrum taken 2h (left) and 12h (right) after target change (see text)

ii Fig. 1 also shows the second possibility to assign lines from multiple capture - their characteristic grow-in with time. For the ^{164}Dy(n,γ) reaction this time is too fast to be observed with the bent crystal spectrometers due to the time needed for adjusting the targets, but we have observed this grow-in for several transitions measuring the CE-lines with the β-spectrometer BILL in cycling the target in and out from the high flux position. This is shown in Fig. 3 for the L-subshell lines of the E2-transitions at 176 keV in ^{166}Dy.

RESULTS FOR ^{164}Dy

The ARC measurement of Warner et al.[7] provided a complete set of states below 2 MeV in the spin range 2^-,3^- for negative parity and 1^+- 4^+ for positive parity. The band structure proposed in this work could be confirmed for the negative parity states and the level scheme could be extended. For the positive parity states ARC cannot differentiate between 2^+ and 3^+ states. From the (t,p) - measurement [6] two 0^+- bandheads are expected at 1658 and 1774 keV respectively. *No* E0 transition could be assigned in the CE-spectra at the sensitivity limit of the present experiment. Using the Ritz combination principle a level at 1775.8 keV can be constructed, which depopulates to the 2^+ states of the ground and the γ-band. The resulting ratio for the B(E2,0^+-2^+_γ)/B(E2,0^+-2^+_{gs}) of 7 has to be considered as tentative. A γ-ray line of 1581.3 keV can tentatively be assigned to depopulate the 0^+ bandhead at 1654.7 keV. The 2^+ state built on this bandhead can then be identified with the level at 1715.9 keV. The depopulation of this level to the 2^+_{gs} is also observed in coincidences.

Table I: Levels in ^{164}Dy.

E_x(keV)	K,J_i^π	E_x(keV)	K,J_i^π	E_x(keV)	K,J_i^π
0.0	01,0$^+$	1674.905(34)*I	01,1$^-$	1949.658(11)*I	,(2,3)$^-$
73.388(1)*I	01,2$^+$	1686.508(58)	41,5$^-$	1978.611(13)*I	,3$^+$
242.223(1)*I	01,4$^+$	1715.824(44)*I	02,2$^+$	1979.586(19)*I	,2$^+$
501.304(2)	01,6$^+$	1757.969(17)*	01,3$^-$	1985.64(5)*	,(2),3$^+$
761.793(1)*I	21,2$^+$	1775.770(28)?	03,0$^+$	2048.95(9)*	,(2,3)$^+$
828.181(1)*I	21,3$^+$	1796.482(44)*I	22,2$^+$	2053.04(16)*	,(2,3)$^+$
915.960(2)*I	21,4$^+$	1809.439(47)*I	11,1$^-$	2077.84(4)*I	,(2,3)$^+$
976.877(2)*I	21,2$^-$	1840.640(71)*I	11,1$^+$	2112.97(5)*I	,3$^+$
1024.616(3)	21,5$^+$	1846.156(97)*	(11),(2,3)$^-$	2123.94(4)*I	,(2,3)$^+$
1039.267(1)*I	21,3$^-$	1852.640(44)*I	,4$^+$	2152.02(10)*I	,3$^+$
1122.733(1)*I	21,4$^-$	1891.481(23)*I	,4$^+$	2172.91(5)I	,(2-4)$^+$
1225.117(2)*	21,5$^-$	1909.340(41)*I	11,3$^-$	2458.98(8)I	,3$^+$
1588.013(3)*I	41,4$^-$	1921.089(68)*I	(11),2$^+$	2531.11(10)I	12,1$^+$
1654.7(1)?	02,0$^+$	1933.039(44)*I	,3$^+$		

*Level observed in ARC[7], I level observed in primary spectrum
? Level from (t,p)[6].

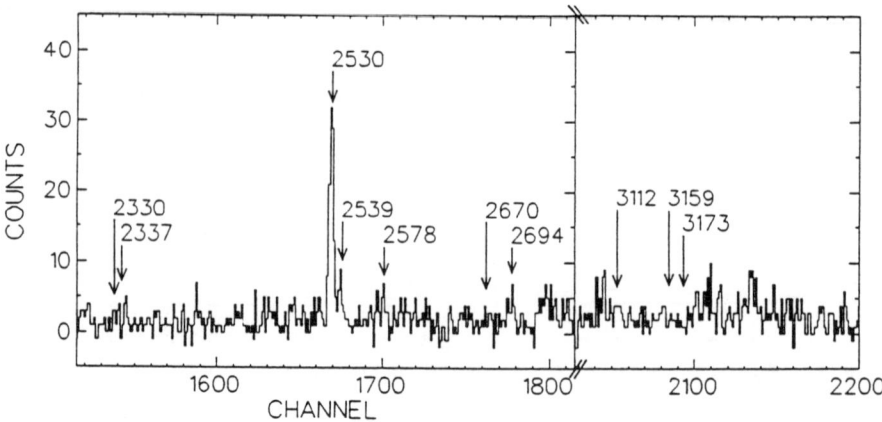

Fig. 4. Spectrum of γ-rays with the coincident energies adding up to the capture state of ^{164}Dy

In Table I excited states of ^{164}Dy as obtained from the present investigation are given with the corresponding spin and parity assignments.

Fig. 4 shows parts of the spectrum obtained for coincident transitions adding up to the binding energy together with the expected energies for 1$^+$levels observed in (γ,γ')[5] The very strongly excited state at 2530 keV corresponds most likely to the K$^\pi$=1$^+$bandhead observed in the ^{165}Ho(t,α) experiment[9], giving further support to the interpretation of this state to be a relatively pure two quasi proton 7/2$^-$[523]×5/2$^-$[532] configuration. Other 1$^+$levels observed in (γ,γ') have been assigned considering the excitation energy and the branching to the first excited 2$^+$state, but no transitions to higher lying bands could be assigned unambiguously.

Table II: E2 transitions in ^{165}Dy. Intensities in arb. units.

Configuration	E_i(keV)	J_i^π	E_f(keV)	J_f^π	E_γ(keV)	I_γ	comment
7/2$^+$[633],2$^+$	538.6	3/2$^+$	0.0	7/2$^+$	538.6	9331	E2
	584.0	5/2$^+$	0.0	7/2$^+$	584.0	3100	E2
			83.4	9/2$^+$	500.6	1453	E2
	649.0	7/2$^+$	0.0	7/2$^+$	649.0	271	E2
			83.4	9/2$^+$	565.6	495	E2
3/2$^+$([651]+7/2$^+$[633],2$^+$)	1108.2	3/2$^+$	0.0	7/2$^+$	1108.2	734	E2
	1158.1	5/2$^+$	0.0	7/2$^+$	1158.1	336	E2
			83.4	9/2$^+$	1074.7	111	E2
1/2$^-$([510]+5/2$^-$[512],2$^+$)	570.3	1/2$^-$	184.3	5/2$^-$	386.0	4396	E2
	605.1	3/2$^-$	184.3	5/2$^-$	420.8	1277	E2
			261.8	7/2$^-$	343.3	464	E2
	658.0	5/2$^-$	184.3	5/2$^-$	473.7	219	E2
			261.8	7/2$^-$	396.2	295	E2
			360.6	9/2$^-$	297.4	52	E2
3/2$^-$([512]+7/2$^-$[514],2$^+$)	1256.5	3/2$^-$	605.1	3/2$^-$	651.4	46	E2
1/2$^-$(3/2$^-$[521],2$^+$)	1080.0	1/2$^-$	158.6	3/2$^-$	921.4	136	E2
	1103.0	3/2$^-$	180.9	5/2$^-$	922.1	202	E2+M1
	1174.9	5/2$^-$	158.6	3/2$^-$	1016.5	175	E2+M1
			180.9	5/2$^-$	994.0	291	M1+E2
	1080.0	1/2$^-$	628.8	5/2$^-$	451.2	76	E2
	1174.9	5/2$^-$	573.6	3/2$^-$	601.4	44	E2

RESULTS FOR ^{165}Dy

As stated in the introduction a careful study of the γ- ray spectrum of the ^{164}Dy(n,γ) reaction has been published recently[3]. The present measurement of the CE-spectrum yielded however new information on the multipolarities of transitions. Especially the observation of E2 admixtures confirmed most of the previously assigned mixings between the vibrational excitations of a lower lying one-quasiparticle state (K[N$n_z\Lambda$],ν) with a one-quasiparticle state K'[N'n'$_z\Lambda$']. The corresponding levels, where such E2 transitions have been observed are given in Table II together with their band assignment.

In contrast to ref.[3] no significant E2 admixture has been observed for the decay of the levels belonging to the K^π=5/2$^-$ band at 533.5 keV (with exception of the observation of the E2 transition to the bandhead of the K^π=1/2$^-$[523] band at 108.2 keV). It has to be concluded that the mixing of the 5/2$^-$(1/2$^-$[521],2$^+$) excitation with the 5/2$^-$[523] configuration has to be rather weak. Also for the K^π=3/2$^-$ band at 1088.0 keV no E2 admixture in the deexciting transitions has been observed.

Out of 890 transitions 387 transitions have been placed in a level scheme comprising 54 levels below 1700 keV excitation energy. For 44 levels spin and parity have been assigned unambiguously. 42 levels have been grouped into 14 rotational bands of which 6 are vibrational excitations of quasi-particle states or do contain significant admixtures of vibrational excitations[3].

RESULTS FOR ^{166}Dy

The level scheme of ^{166}Dy as obtained from our measurements is given in Table III. The intensities are quoted in arbitrary units. The negative parity band with its bandhead at 1095.2 keV has already been proposed by Kaerts in his thesis[10], but the present CE-spectra allowed to determine the multipolarities of the depopulating transitions and thus the parity of the levels.

Table III: Energy levels of ^{166}Dy with their deexciting transitions

E_x(keV)	K, J^π	E_γ(keV)	I_γ	Comment	E_f	K, J_f^π
76.587(1)	0,2$^+$	76.587(1)	113(25)	E2	0.0	0,0$^+$
253.527(2)	0,4$^+$	176.941(1)	801(86)	E2	76.6	0,2$^+$
526.966(3)	0,6$^+$	273.439(2)	208(20)	E2	253.5	0,4$^+$
857.154(6)	2,2$^+$	857.156(11)	231(50)	E2	0.0	0,0$^+$
		780.571(6)	203(40)	E2	76.6	0,2$^+$
928.733(6)	2,3$^+$	852.128(8)	517(103)	E2	76.6	0,2$^+$
		675.218(5)	71(14)	E2	253.5	0,4$^+$
1023.434(6)	2,4$^+$	946.850(15)	122(24)	E2	76.6	0,2$^+$
		769.907(6)	189(39)	E2	253.5	0,4$^+$
1141.261(19)	2,5$^+$	887.734(15)	115(24)	E2	253.5	0,4$^+$
1095.213(6)	2,(2,3)$^-$	238.062(5)	100(10)	E1	857.2	2,2$^+$
		166.479(3)	73(8)	E1	928.7	2,3$^+$
1189.385(6)	2,(3,4)$^-$	260.652(2)	109(10)	E1	928.7	2,3$^+$
		165.950(10)	19(2)		1023.4	2,4$^+$

DISCUSSION

Fig. 5 shows the systematics of the 4$^+$g.s, 2$^+$quasi-γ, 0$^+$quasi-β and the first negative parity state in the even-even Dy isotopes. The ratio of the excitation energies E(4$^+_1$)/E(2$^+_1$) saturates at ^{160}Dy (neutron number 94) for the even-even Dy isotopes at the value of 3.3 for a rigid rotor. It can be observed from Fig. 5 that the compression of the level spacing between g.s. and quasi-γband (and the negative parity band) is accompanied by a strong repulsion of the quasi-β band as the neutron number increases. Maximum deformation (and maximum repulsion for the 0$^+$- band) is reached in ^{164}Dy and a drastic change for the position of the excited bands are found in ^{166}Dy which is before the neutron shell N = 82 to 126 is half-filled.

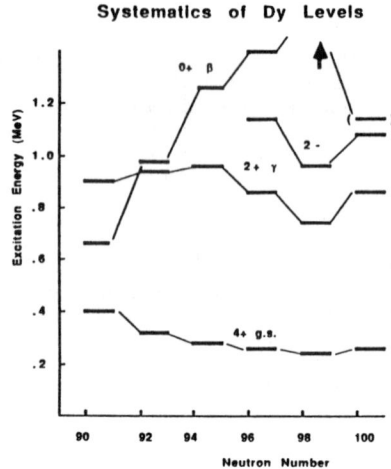

This sudden change may be caused by the energy gap between the 5/2$^-$[523] and the 7/2$^+$[633] neutron orbits as obtained in a deformed potential at neutron number 98. The experimental bandhead energies of the odd-A Dy isotopes have been compared to the most recent theoretical calculations published by Soloviev et al.[11] already in the references [1-3]. It is hoped that the detailed experimental information now available will stimulate new theoretical efforts.

Fig. 5. Systematics of the $4^+_{gs}, 2^+_\gamma, 2^-_1$, and 0^+_β states in the Dy isotopes

REFERENCES

[1] H. H. Schmidt, T. von Egidy, H. J. Scheerer, P. Hungerford, H. G. Börner, S. A. Kerr, K. Schreckenbach, R. F. Casten, W. R. Kane, D. D. Warner, A. Chalupka, M. K. Balodis, T. V. Guseva, P. T. Prokofjev, J. J. Tamberg, Nucl. Phys. **A454**, (1986) 267

[2] H. H. Schmidt, P. Hungerford, T. von Egidy, H. J. Scheerer, H. G. Börner, S. A. Kerr, K. Schreckenbach, F. Hoyler, G. G. Colvin, A. M. Bruce, R. F. Casten, D. D. Warner, I. L. Kugava, V. A. Bondarenko, N. A. Kramer, P. T. Prokofjev, A. Chalupka, Nucl. Phys. **A504**, (1989) 1

[3] E. Kaerts, P. H. M. van Assche, S. A. Kerr, F. Hoyler, H. G. Börner, R. F. Casten, D. D. Warner, Nucl. Phys. **A514**, (1990) 173

[4] D. Bohle, G. Kilgus, A. Richter, C. W. de Jager, H. de Vries, Phys. Lett. **B195**, 326 (1987)

[5] C. Wesselborg, P. von Brentano, K. O. Zell, R. D. Heil, H. H. Pitz, U. E. P. Berg, U. Kneissl, S. Lindenstruth, U. Seemann, R. Stock, Phys. Lett. **B207**, 22 (1988)

[6] D. G. Burke, G. Løvhøiden, T. F. Thorsteinsen, Nucl. Phys. **A483**, (1988) 221

[7] D. D. Warner, R. F. Casten, W. R. Kane, W. Gelletly, Phys. Rev. **C27**, 2292 (1983)

[8] F. Hoyler, H. G. Börner, S. J. Robinson, G. L. Greene, E. Kessler, M. S. Dewey, Inst. Phys. Conf. Ser. No. 88/ J. Phys. G Nucl. Part. Phys. **14 Suppl.**, S161 (1988)

[9] S. J. Freeman, R. Chapman, J. L. Durell, M. A. C, Hotchkis F. Khazaie, J. C. Lisle, J. N. Mo, Phys. Lett. **B222**, 347 (1989)

[10] E. Kaerts, thesis, IKS, Katolieke Universiteit Leuven, 1988

[11] V. G. Soloviev, P. Vogel, G. Jungklausen, Izv. Akad. Nauk SSSR, Ser. Fiz.**31** (1967) 518

INTRINSIC EXCITATIONS IN ^{192}Ir*

Jean Kern
Physics Department, University, CH-1700 Fribourg, Switzerland

M.K. Balodis[8], W. Beer[3], S. Brandt[9], R.F. Casten[2], A. Chalupka[7], C. Coveca[1], J.-Cl. Dousse[3], R. Eder[7], T. von Egidy[7], D.G. Gardner[6], M.A. Gardner[6], P. Giacobbe[1], R.L. Gill[2], E. Hagn[7], R.W. Hoff[6], M.A. Hungerford[7], I.A. Kondurov[4], I.V. Kononenko[5], N.D. Kramer[8], V.A. Libman[5], Yu.E. Loginov[4], A.V. Murzin[5], V. Paar[9], P.T. Prokofjev[8], A. Raemy[3], H.J. Scheerer[7], H.H. Schmidt[7], W. Schwitz[3], L.I. Simonova[8], P.A. Sushkov[4], E. Zech[7].

Bologna-Geel[1]-BNL/Brookhaven[2]-Fribourg[3]-Gatchina/Leningrad[4]-Kiev[5]-LLNL/Livermore[6]-TU Munich[7]-Riga[8]-Zagreb[9]

ABSTRACT

The structure of the odd-odd ^{192}Ir nucleus presents an interesting and challenging problem for both experimentalists and theorists. As a result of the common efforts of nine laboratories, it is possible, for the first time, to propose an extended scheme with 34 levels. The experiments included observation of γ-ray transitions and of conversion electrons emitted after thermal and resonance neutron capture, of direct (d,p) and (d,t) neutron transfer reactions and of the angular distribution of γ-rays from aligned ^{192}Ir nuclei. The results are interpreted in the framework of the rotor-plus-particle and IBFFM models. The nuclear states appear to be strongly mixed. The complex and interesting ground-state configuration is discussed.

---

* Work performed under the auspices of the Swiss National Science Foundation, the Bundesministerium für Forschung und Technologie (BMFT) and the U.S. Departement of Energy.

INTRODUCTION

The ^{192}Ir doubly-odd nucleus lies at the limit of prolate an oblate deformed regions, implying triaxial γ deformation or γ softness and probably coexistence of shapes, in terms of geometrical models, and between the SU(3) and O(6) symmetries, in the algebraic IBM model. It is therefore an intriguing and challenging nucleus from a theoretical point of view. The large level density makes it, however, difficult to investigate experimentally and there was only scarce information hitherto available, athough the nucleus lies close to the stability line. It was only known that a I = 1, 57-keV, 1.4-m isomer decays by an E3 transition[1] to the I = 4 ground state, and that a high-spin 161-keV isomer decays by a transition of probably E5 multipole character[2] to the ground state. The parities of these levels were controversial. We present in this paper results obtained by a collaboration of nine laboratories[3].

THE EXPERIMENTS

The experiments can be best classified according to the type of information they provide. The observation of thermal neutron capture primary and secondary transitions was crucial to establish the level scheme. These experiments were performed at the Paul Scherrer Institute (PSI, formerly EIR) in Würenlingen, using a pair and a curved-crystal spectrometer. By use of the Ritz principle, a scheme including 34 levels was deduced. To avoid the occurence of spurious levels, we set restrictive acceptance conditions. A few spurious transition assignments cannot though be avoided, especially for the higher energy (> 250 keV) transitions which have lower energy precisions.

Several experiments permitted us to assign spin and parities: observation of conversion electrons following thermal neutron capture at Riga (preliminary results from a similar experiment at ILL[4] were also taken into account), of average resonance capture (ARC), using 2-keV and 24-keV neutrons, at BNL and Kiev, of secondary transitions following neutron capture in discrete resonances at Geel and of the angular distribution of γ-rays from oriented nuclei implanted in iron at Munich. Unique spin and parity assignments were obtained for 15 levels and a narrow range of values is proposed for the others.

High resolution ^{191}Ir(d,p) and ^{193}Ir(d,t) direct neutron transfer reactions were observed at Munich using a Q3D spectrograph. These experiments are important to assess the configuration of the observed states. Finally, delayed γ-γ coincidences were measured at the thermal-neutron beam at Gatchina. The isomeric character of a few transitions gives strong arguments for the identification of band heads.

SPINS AND PARITIES OF THE OBSERVED EXCITED STATES

All observed excited states are connected directly or indirectly by decay transitions to the 1.4-m isomeric state. The measured conversion coefficients permit us, in most cases, to determine unambiguously their relative parity. Since the capture states have positive parity, the more intense primary transitions feed the negative parity states. The identification is clear (see Table 1) and the data show the 57 keV isomeric state to have negative parity. This conclusion is supported independently by the recent angular distribution of the ^{193}Ir(d,t) reaction performed by Burke and Garett[5].

Table 1. Comparison of the excitation energies of a few low-lying levels deduced from different experiments.

ARC	Thermal Neutron Capture		Spin
	primaries	secondaries	
56.7(3)	56.9(5)	56.719(5)	1^-
83.0(8)	--	84.274(3)	3^-
105.7(5)	105.05(30)	104.776(5)	1^-
116.2(4) {	115.90(25)	115.563(5)	2^-
	119.5(9)	118.782(2)	3^-
143.3(3)	142.95(40)	143.554(6)	$1^-,2^-$
212.8(3)	212.6(7)	212.805(5)	$1^-,2^-$
228.2(10)	225.7(6)	225.916(5)	3^-

The expected relative ARC intensities were computed according to a procedure presented by Chrien in ref. 6. In Fig. 1 are compared the reduced intensity residues for the 2-keV spectrum, i.e. the differences between experimental and computed values. In the higher part of the γ-ray spectrum, the residues are compatible with

266 Intrinsic Excitations in ^{192}Ir

zero. Between 200 and 300 keV excitation energy we are apparently missing a few levels. Above 400 keV, our level scheme begins to be very incomplete. These findings are not surprising.

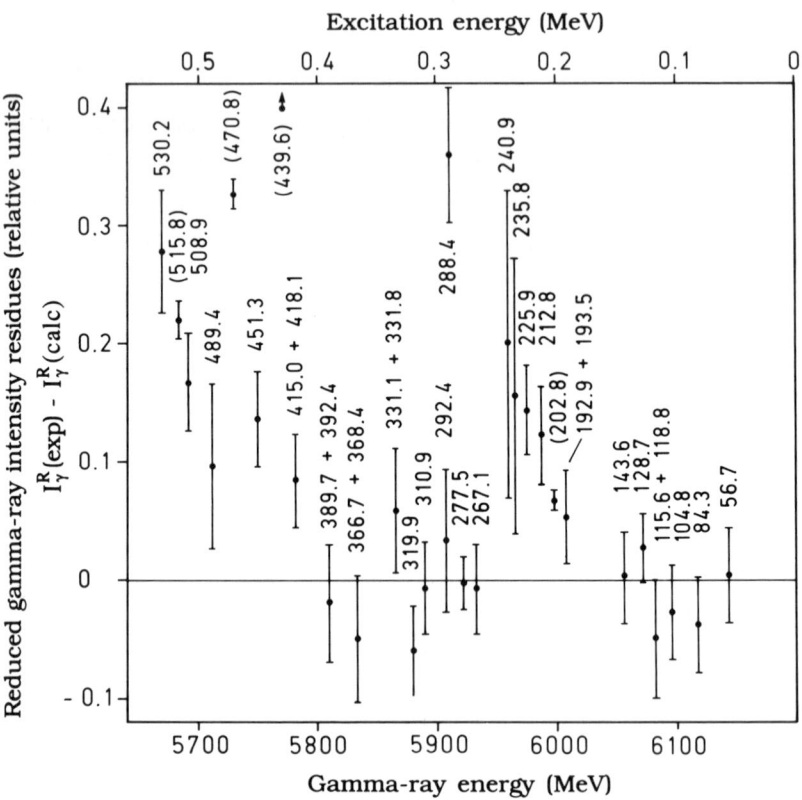

Fig. 1. Difference between experimental and computed reduced intensities $I\gamma/E\gamma^5$ for the 2-keV ARC spectrum (relative units).

THE GROUND STATE SPIN AND PARITY

The ground-state spin I = 4 is firmly established (see ref. 3 and refs. therein). The parity is more difficult to determine. We propose a positive value. No single results leads us unambiguously to this conclusion, but the consideration of a number of converging observations:

1. We assume that the $I^\pi = 1^-$, 1.4-m isomer decays directly to the ground state (Fig. 2.a) by the observed 57 keV E3 transition. It would be possible to consider that the isomeric transition leads to a ~ 10 keV $I^\pi = 4^-$ low-energy excited state (Fig. 2.b). In that case one would expect to observe a ~ 67 keV M3 transition to the ground state. Such transition was not seen in any experiment.

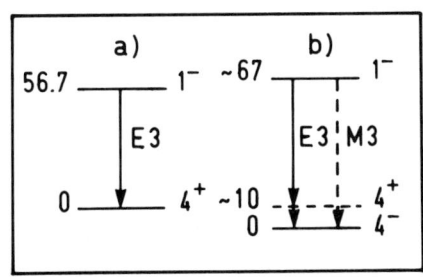

Fig. 2. a) Proposed decay of the 1⁻ isomeric level.
b) Possible alternative decay.

2. From the observation of the angular distribution of γ-rays from oriented ^{192}Ir nuclei, we obtained the value of the disorientation coefficients U_k due to the unobserved β-transitions. These results favour positive parity for ^{192}Ir.

3. The experimental value[7,8] of the ^{192}Ir ground-state magnetic moment is $\mu = +1.924(10)\ \mu_N$. This value can only be explained, in the framework of the rotor-plus-particle model, by a mixed configuration $\{(\pi\ 11/2^-[505\downarrow] - \nu\ 3/2^-[512\downarrow]) + (\pi\ 3/2^+[402\downarrow] - \nu\ 11/2^+[615\uparrow])\}$, mixing induced by the residual V_{pn} interaction.

4. In framework of IBFFM, the only possibility to reproduce the experimental value of μ is to generate a strong mixing between the two-quasiparticle states $(\pi d_{5/2}\ \nu i_{13/2})4^+$ and $(\pi h_{11/2}\ \nu\ \ell_j)4^+$.

As stated above, no one alone of the arguments would be sufficient to prove the positive parity of the ground state, but their combination gives a strong support to our conclusion.

ROTOR-PLUS-PARTICLE INTERPRETATION

We can predict, approximately, by addition of interpolated excitation energies observed in neighbouring odd-A nuclei and using average Gallagher-Moszkowski splittings, the expected level scheme of ^{192}Ir (Fig. 3).

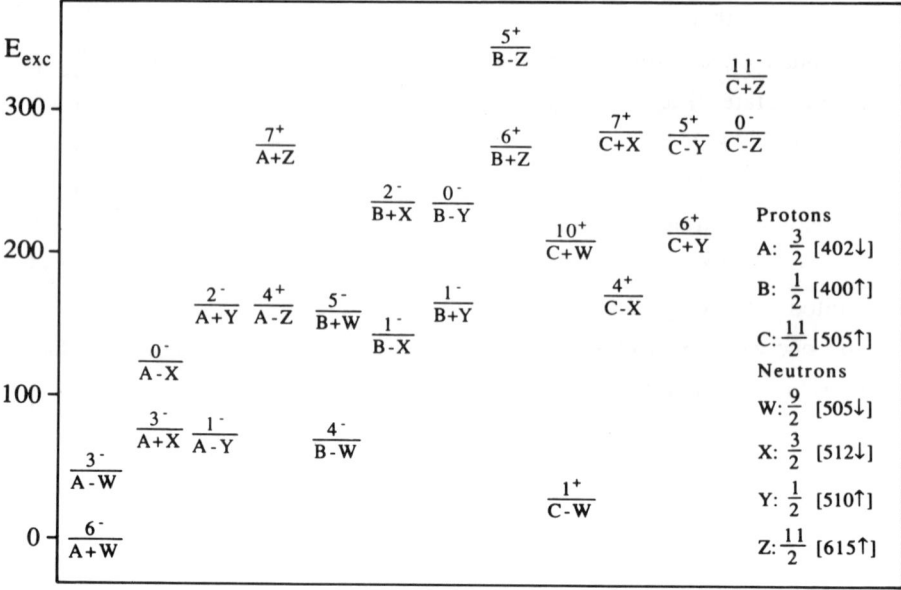

Fig. 3. Predicted ^{192}Ir level scheme.

It appears that most of the low-K bands should have negative parity. It is therefore not surprising that 28 out of 33 excited states we observed have negative parity, states which are, in addition, better populated by primary transitions.

We see that the band heads A-Z and C-X (Fig. 3) are expected to lie at an energy of about 170 keV, whereas their mixed configuration appear to form the ground state. This indicates that our approximation is rather crude, in this case.

We expect the levels that contain the proton in the orbital A, i.e. 3/2 [402↓], to be populated in the neutron direct transfer reactions. In the list of negative-parity levels identified by use of the secondary transitions (Fig. 4), we note that only one level (at 235.5 keV) is not populated in the (d,p) and (d,t) reactions. This points to the fact that all levels contain a certain amount of the 3/2[402↓] configuration and are probably admixed. We believe that such admixtures are due, to a large extent, to Coriolis coupling, which should affect in particular the protons 3/2[402↓] and 1/2[400↑] orbitals and the neutrons 3/2[512↓] and 1/2[510↑] orbitals. Considerable deviations from the pattern of a good rotor are expected also from the γ-deformation or γ-softness, from the presence of states issued from high ℓ orbitals ($h_{9/2}$, $h_{11/2}$ and $i_{13/2}$) and from V_{pn} coupling.

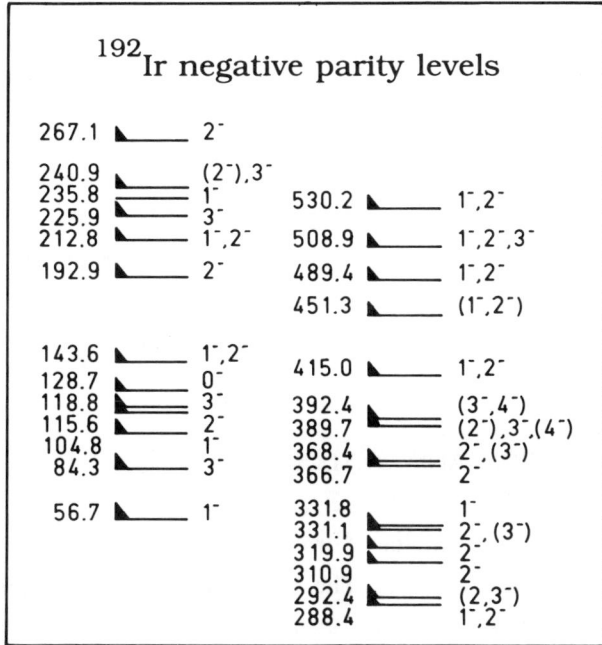

Fig. 4. ^{192}Ir negative parity levels. Black triangles mark the level populated in the neutron direct transfer reactions.

At this stage we can tentatively only propose the main configuration components of a few band heads (Fig. 5).

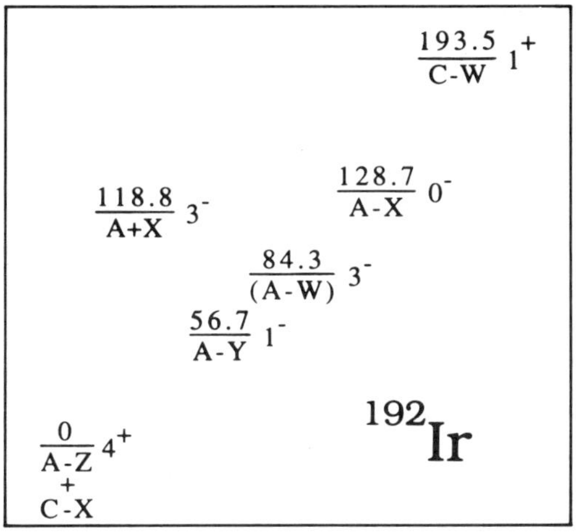

Fig. 5. Tentative assignment of a few band heads, with their dominant configurations. (see Fig. 3 for the orbital labels).

IBFFM CALCULATIONS

Detailed IBFFM calculations were performed[3], yielding interesting insights into the ^{192}Ir level structure. Particularly noteworthy is the mixing between the $(\pi d_{5/2} \nu i_{13/2})4^+$ and $(\pi h_{11/2} \nu p_{3/2})4^+$ configurations needed to explain the ground-state magnetic moment. It is required that the interaction needed to produce this mixing does not affect the lowest position of the 4_1^+ level. This puts stringent conditions on the parameters for the tensor, quadrupole-quadrupole and hexadecapole-hexadecapole interactions.

Another interesting result regards the 161-keV isomer. It is not possible to produce a low-lying 9^- state. In addition, high-spin 11^- and 12^- states have to lie lower in energy. On the other hand, the $(\pi d_{5/2} \nu i_{13/2})9^+$ member of the two-quasiparticle multiplet is expected to lie rather low on the basis of the parabolic rule[9]. This would imply that the decay transition is M5 and not E5, as suggested by Scharff-Goldhaber and McKeown[2]. Such a multipolarity would be compatible with the observed half-life.

CONCLUSION AND OUTLOOK

Good progress was made in unraveling the ^{192}Ir nuclear structure. The long-standing problem of level parities was resolved. The mixed configuration of the states is challenging and provides a good tool to investigate the nuclear interactions in this mass region. Additional results are clearly needed and several new experiments (see e.g. ref. 5) are underway or planned.

REFERENCES

1. J.P. Mize, M.E. Bunker and J.W. Starner, Phys Rev. **96** (1954) 444.
2. G. Scharff-Goldhaber and M. McKeown, Phys. Rev. Lett. **3** (1959) 47.
3. J. Kern, A. Raemy, W. Beer, J.-Cl. Dousse, W. Schwitz, M.K. Balodis, L.I. Simonova, N.D. Kramer, P.T. Prokofjev, R.W. Hoff, R.F. Casten, R.L. Gill, R. Eder, T. von Egidy, E. Hagn, P. Hungerford, H.J. Scheerer, H.H. Schmidt, E. Zech, A. Chalupka, A.V. Murzin, V.A. Libman, I.V. Kononenko, C. Coceva, P. Giacobbe, I.A. Kondurov, Yu. E. Logunov, P.A. Suskov, S. Brant and V. Paar, to be published.
4. J. Kern, H.G. Börner, G.G. Colvin, S. Drissi, T. Von Egidy, K. Kalanga, J.-L. Salicio and K. Schreckenbach, Proc. 6th Conf. on Capt. γ-Ray Spectroscopy, Leuven 1987, Inst. Phys. Conf. Ser. **88** (1988) S571.
5. D.G. Burke and P.E. Garett, Proceedings of this conference.
6. R.E. Chrien, in Capt. γ-Ray Spectroscopy, S. Raman ed., AIP Conf. Ser. No. **62** (1985) 342.
7. E. Hagn, K. Leuthold, E. Zech and H. Ernst, Z. Phys. **A295** (1980) 385.
8. E. Scheidemann, R. Eder, E. Hagn and E. Zech, Proc. Int. Conf. on On-Line Nuclear Orientation, Oxford, U.K. 1988, eds J. Rikovska and N.J. Stone, Hyp. Int. (1988) 493.
9. V. Paar, Nucl. Phys. **A331** (1979) 16.

PROMPT AND DELAYED $\gamma\gamma$ COINCIDENCES IN THE ^{191}Ir(n,γ)^{192}Ir REACTION

I.A.Kondurov, Yu.E.Loginov, P.A.Sushkov

Leningrad Nuclear Physics Institute, Academy of Sciences of the USSR

ABSTRACT

A $\gamma\gamma$-delayed and prompt coincidence experiment on transitions following thermal-neutron capture by ^{191}Ir was performed using a semiconductor and scintillation detectors. The isomeric character of four transitions was established. The half-live was deduced for the following exited states of ^{192}Ir: 66.8 keV 14.6(44) ns; 84.3 keV, 1.9(4) ns; 104.7 keV .7(3) ns; 193.5 keV, 2.7(6) ns. The 118.8 keV level has the half-live $T_{1/2}$ > 15 ns. Using these results and data on prompt coincidences a fragment of ^{192}Ir level scheme is proposed.

INTRODUCTION

Present work is a part of the collaborative investigation of the ^{192}Ir nucleus properties.

The prompt and delayed $\gamma\gamma$-coincidences of some tranitions in the doubly odd $^{192}_{77}$Ir$_{115}$ were investigated using the horizontal channel of WWR-M reactor (LNPI, Gatchina). The arget consisted of 35 mg of metallic iridium powder sotopically enriched to 78% in ^{191}Ir was irradi- ated by he thermal neutrons from the neutron guide with the flux of $\cdot 10^7$ n.$\cdot$cm$^{-2}\cdot$s^{-1}.

PROMPT COINCIDENCES

The prompt $\gamma\gamma$-coincidences were measured by the PGe-Ge(Li) (both are planar) spectrometer. The FWHM of the PGe detector was about 500 eV at the E_γ ~60 keV, that of e(Li) was about 2.5 keV at E_γ ~100 keV. These detectors worked in the common fast-slow measuring system with the FWHM of the TA-converter curve about 35 ns.

The γγ-coincidence spectra were obtained by successively setting of the window of a differentiall discriminator on the lines of the E_γ = 66.8, 77.9, 84.3 and 90.7 keV and the nearby, without lines, parts of the HPGe-spectrum used as background. The spectra of gamma-rays coinciding with these γ-quanta are shown on the fig.1, the discriminator settings are shown in the insert. The qualitative results are presented in the Table 1.

Table 1. Prompt γγ-coincidences result.

γ-line, keV	coincides with E_γ, keV (Ge(Li))
66.8	Ir K x-rays
77.9	Ir K x-rays, 90.7, 134, 156.6, 250.7
84.3	Ir K x-rays, 193.7, 267.4, 284.1, 333.
90.7	Ir K x-rays, 77.9, 88.7, 136.8

The 66.8 kev γ-line is absent in the known list of ^{192}Ir transitions [1] because of strong interference with the 66.831 keV Pt K X-ray line. In a special experiment with shutting down the neutron beam and measuring the long-lived ^{192}Ir ($T_{1/2}$ = 74.8 days) β-decay γ-spectrum the energy of the γ-line was defined as 66.83(2) keV and its intensity as 16.2(12) in units of the preprint [1] or 0.53(7) in γ-rays/100n.

In the coincidence spectrum of the 66.8 keV transition one can see only lines corresponding to Ir K X-rays. The lower limit of intensity of γ-rays whose coincidences with this transition could be yet observed is about 30 units of [1] or ~1 gamma/100 n.

DELAYED COINCIDENSES

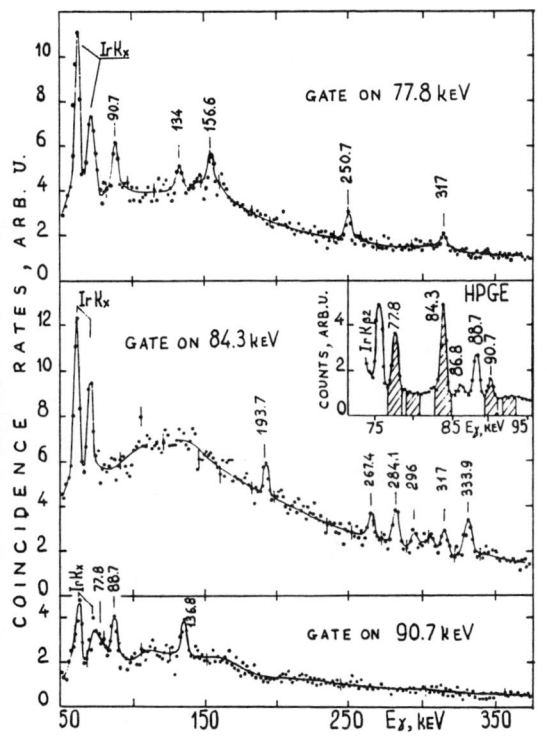

Fig.1. The spectra of prompt γγ-coincidences of the 77.9, 84.3 and 90.7 keV transitions

The lifetime experiments were carried out using the same coincidence spectrometer where the Ge(Li) detector was replaced by the scintillation detector with organic crystal, whose signals corresponding to $E_\gamma > 1$ MeV were used to start the TA-converter. The pulses from HPGe stopped it. The resolving time of this system was about 15 ns at E_γ (HPGe) ~50 keV.

The window of differential discriminator was placed on γ-lynes of interest in HPGe spectrum and on its parts without the lines. The time distributions and gate settings on lines 48.0 and 66.8 keV are shown on the fig.2 and insert 1 of this figure. The values of half lives were determined by the least square fitting of these time distributions.

For the determination by the centroid-shift method the life-times of the levels depopulated by the 77.9, 88.7 and 84.3 keV γ-quanta the target of complex composition - ^{191}Ir + + Gd (Natural) was used.

The centroid position corresponding to the transition

from the lowest excited states in ^{158}Gd (79.5 keV, $T_{1/2}$ = 2.53(2) ns) and ^{156}Gd (88.96 keV, $T_{1/2}$ = 2.19(3) ns) were used for time calibration. The fragment of HPGe γ-spectrum from the mixed target and the gate settings are shown in the insert 2 of the fig.2.

These measurements were resulted in the half-lives values of 2.0(8) ns, 3.6(8) ns and 1.9(4) ns for, correspondingly, γ-transitions 77.8, 88.7 and 84.3 keV. It should be noted that the transitions 77.8 and 88.7 keV are depopulating the same (193.5 keV, 1^+, 2^+) excited state, hence, after meaning procedure the final value of the halflife of this level is $T_{1/2}$ = 2.7(6) ns.

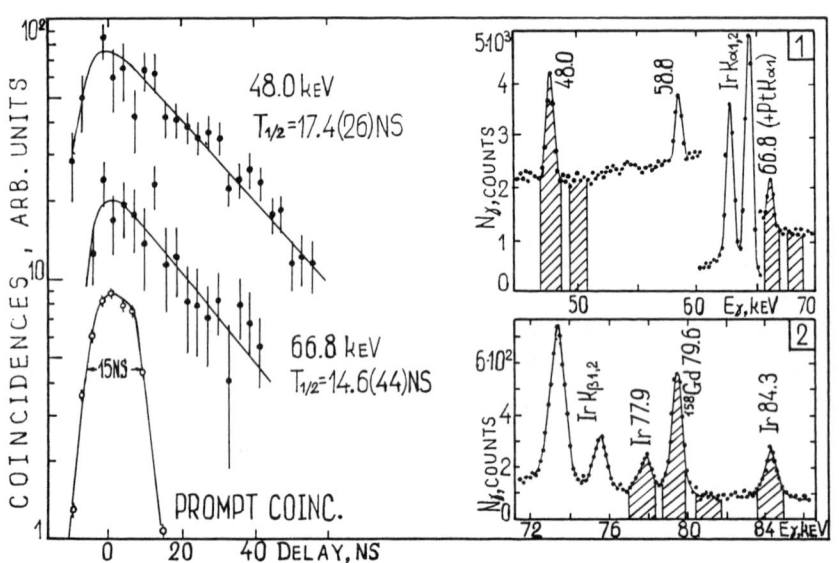

Fig.2 The delayed $\gamma\gamma$-coincidences of the 48.0 and 66.8 keV transitions.

The isomeric character of the γ-line with energy near 118 keV was deduced from the fact of increasing the intensity of this line in the γ-spectrum of delayed (in the time interval of 20 - 40 ns) $\gamma\gamma$-coincidences in comparison with other nonisomeric lines and in accordance with isomeric ones. The appraised value of corresponding half-life is $T_{1/2}$ > 1 ns. In the list [1] of ^{192}Ir γ-lines there is a doublet 118.3(M1)-118.8(E1) kev and one of its components can be delayed, most probably more intensive 118.7818 keV line.

The gammas with the energy ~ 72 keV (the triplet 72.0(M1)- 72.3(M1)- 72.5 keV) is absent both in prompt and delayed γγ-coincidence spectra. Possibly their components (may be the more intensive, 72.0 and 72.3 keV) can depopulate the isomeric states with $T_{1/2}$ > 100 ns.

The results of the lifetime measurements are summarized in the table 2.

Table 2

E level, keV	$T_{1/2}$, ns	E_γ depop., keV	I_i^π	I_f^π	F_w
66.8	14.6(44)	66.83(E1)	$4^-, 5^-$	4^+	$3 \cdot 10^4$
84.3	1.9(4)	84.3 E1	3^-	4^+	$8 \cdot 10^3$
104.7	17(3)	48.0 M1	$1^-, 2^-$	1^-	10^3
118.8	>15	118.8 E1	$3^-, 4^-$	4^+	$>5 \cdot 10^4$
193.5	2.7(6)	77.8 E1	1^+	$1^-, 2^-$	$4 \cdot 10^4$
		88.7 E1	1^+	$1^-, 2^-$	$5 \cdot 10^4$
		136.8 E1	1^+	1^-	$3 \cdot 10^5$

LEVEL SCHEME

Shown on the fig.3 is a fragment of the level scheme of ^{192}Ir deduced from the data presented, the crystal-difraction [1] and internal conversion [2] data.

The spins and parities of the levels with energies of 56.72, 84.27, 104.74, 115.52 and 193.51 keV have been proposed by M.Balodis et al.in [3] and are, correspondingly, 1^-, (3^-), (0^-), $(1)^-$, $(1)^+$. According to the multipolarities of transitions depopulating the 104.74 and 115.52 keV levels we can not exclude the values of $I^\pi = 1^-, 2^-$ for these states.

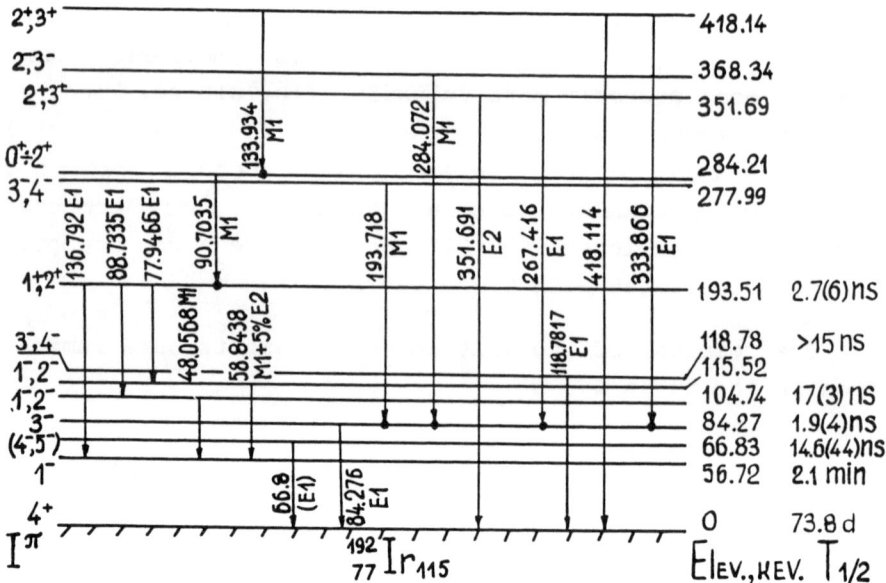

Fig.3 The fragment of the level scheme of ^{192}Ir following from the $\gamma\gamma$-coincidence experiments.

Some new levels can be discussed.

66.83 keV level. The level with the energy 67±1 keV is excited in the ^{193}Ir(dt)-reaction [4] and one can assume that the delayed γ-transition 66.83(2) keV observed in this work depopulates this state. According to its lifetime value the multipolarity of this transition is probably not higher than dipole. The quadrupole character of this transition could permit to observe it in the internal conversion experiment. A weak argument for the parity assignment of this level could be the fact that all another levels excited together with it in the ^{193}Ir(dt)-reaction are of negative parity, and one could assume the same parity of the 66.83 keV level. With this assumption the 66.83 keV transition is of E1 type giving the spin value of this level 3, 4 or 5. The first value must be rejected because of the absence of the primary transition to this state in the thermal and AR-neutrons capture reactions.

118.78 keV level. The state with the energy 118.8(4) keV is excited in the ^{191}Ir(dp)- and ^{193}Ir(dt)-reactions. In the delayed coincidence γ-spectrum the line with the energy 118.65(15) keV is seen and it can be assumed that the γ-transition 118.7818 keV (E1) depopulates this state. In the ARC-experiment the primary γ-transition to this state is observed and it could be assigned $I^{\pi} = 3^-, 4^-$.

277.99 keV level. This level is introduced by the coincidence of 193.7 keV (M1) and 84.3 keV γ-transitions. The most probably assignment of I^{π} is 3^- or 4^-.

284.21 keV level. The 90.7 keV (M1) transition coincides with all three intensive γ-transitions (77.8, 88.7 and 136.8 keV, all are E1) depopulating the 193.51, 1^+ level, and can be placed onto this state. Possible values of I^{π} are $0^+, 1^+$ and 2^+ for the initial, 284.21 keV state.

351.69 keV level. The coincidence of 267.4 keV (E1) - 84.3 keV (E1) tran- sitions and well fulfilled Ritz combination of sum 84.276(3)+ +267.416(10) = 351.692(11) with the E_{γ}= 351.691(5) keV (E2) can be the evidence of existence of excited state of ^{192}Ir with this energy and $I^{\pi}= 2^+$ or 3^+.

368.34 keV level. This level is introduced by the coincidence of 284.1 keV (M1) - 84.3 keV (E1) gammas and such a placing of 284.1 keV transition can be confirmed by the exciting of the levels with energies of 368.0(4) keV and 367.3(3) keV correspondingly in the thermal and AR-neutrons capture; the possible assignments of $I^{\pi} = 2^-$, 3^-.

418.14 keV level. The observed $\gamma\gamma$-cascade 333.9 keV (E1) - 84.3 keV (E1) and Ritz combination of sum 84.276(3) + 333.866(9) = 418.142(10) with E_γ = 418.114(25) keV γ-line allows to accept this excited state and to assign the value $I^{\pi} = 2^+$, 3^+.

CONCLUSIONS

The prompt $\gamma\gamma$-coincidence measurements resulted in the placement in the level scheme of ^{192}Ir three new levels: 277.99 keV, $I^{\pi} = 3^-$, 4^-; 284.21 keV, $I^{\pi} = 0^+$, 1^+, 2^+ and 418.14 keV, $I^{\pi} = 2^+$, 3^+, and confirmed some early accepted excited states, such as 193.51 keV, $I^{\pi} = 1^+$, 2^+ and 351.66 keV, $I^{\pi} = 2^+$, 3^+.

The delayed $\gamma\gamma$-coincidences have shown that almost all low lying states of ^{192}Ir (below 200 keV) are isomeric. Such an isomerism significates the different structure of the states; for example, they can be band heads of various rotational bands with different K.

REFERENCES

[1] A.Raemy, J.Cl.Dousse, J.Kern, W.Schwitz
    IPF-SP-006, Fribourg, 1975.

[2] L.I.Simonova, N.D.Kramer, P.T.Prokofiev
    Preprint of Physics Inst. of Latvian Acad. of. Sc.
    LAFI-010, Salaspils, 1979.

[3] J.Kern, V.Ionescu, A.Bruder, P.Hungerford, H.J.Scheerer,
    A.Chalupka, T.von Egidy, H.H.Schmidt
    Jahresbericht TU Munchen, 1981, s.42.

[4] M.K.Balodis, T.V.Guseva, J.Kern
    Proc. 37th Nucl.Struct.Conf. Jurmala (1987) p.143.

NUCLEAR STRUCTURE OF ^{190,192,194}Ir BY CHARGED-PARTICLE SPECTROSCOPY

D.G. Burke and P.E. Garrett
Physics Department, McMaster University, Hamilton, Ontario
Canada L8S 4K1

ABSTRACT

Single-nucleon transfer reactions have been used to study odd-odd iridium isotopes. Angular distributions of the ^{191,193}Ir(d,t)^{190,192}Ir reactions with 18 MeV deuterons were measured with a typical resolution of 5 keV (FWHM). Angular distributions for the ^{191,193}Ir(d,p)^{192,194}Ir reactions, and spectra at several angles for the ^{189}Os(α,t)^{190}Ir, ^{189}Os(^{3}He,d)^{190}Ir and ^{195}Pt(p,α)^{192}Ir reactions were also measured. Spectroscopic strengths from the ^{193}Ir(d,t)^{192}Ir data support previous rotor-plus-particle interpretations for some ^{192}Ir levels, but the 66 keV level is not readily explained by the models. In ^{190}Ir, the large number of levels populated in both the single-neutron transfer and single-proton transfer reactions implies appreciable mixing of the configurations.

INTRODUCTION

Nuclear structure in the $190 \leq A \leq 198$ region has been difficult to classify in terms of nuclear models. The nuclear shape does not appear to be well-defined, as there is a transition from prolate to oblate ground state deformations[1], with possible triaxiality and γ-softness. In the interacting boson model there is a transition between the SU(3) and O(6) symmetry limits. Single-neutron transfer studies of odd-mass nuclides in this region[2-7] showed that the Nilsson model worked qualitatively for ^{189,191,193}Os, less well for ^{195}Pt, and was unsuccessful for ^{197}Pt. Casten and coworkers showed that there were more $I^{\pi}=\frac{1}{2}^-$ and $\frac{3}{2}^-$ levels in ^{195}Pt and ^{197}Pt than predicted by the Nilsson model, and interpreted these nuclei in terms of the U(6/12) supersymmetry model.[8,9] This description has been refined and extended,[10] but recent tests with single-neutron transfer reactions[7] show that it describes the spectroscopic strengths only qualitatively. Thus a more general treatment, such as an IBFM calculation between the different dynamical symmetry limits, may be more appropriate.

Additional, and perhaps even more stringent, tests of these models may be made by examining odd-odd nuclei, and this is the motivation for the present study of ^{190,192,194}Ir levels. Very little nuclear structure information was previously published for these nuclides, but for some years a large collaboration involving at least 9 laboratories has performed many experiments in an attempt to understand the level structure of ^{192}Ir. This group has now established[11] spins and parities for many levels and made tentative assigments in terms of the IBFFM (Interacting boson-fermion-fermion model) and rotor-plus-particle models. Angular distribuions from the present work yield ℓ-values and spectroscopic strengths which can test these assignments. The population of ^{190}Ir by both neutron-transfer and proton-transfer reactions locates specific two-quasiparticle configurations involving the odd nucleons of both the ^{189}Os and ^{191}Ir targets.

EXPERIMENTS

All experiments were performed at the McMaster University Tandem Accelerator Laboratory using the Enge split-pole magnetic spectrograph, with photographic plates as detectors. Angular distributions of the ^{191,193}Ir(d,t)^{190,192}Ir and ^{191,193}Ir(d,p)^{192,194}Ir reactions were measured with 18 MeV and 16 MeV deuterons, respectively. Because of the high level density, the best possible resolution was desired. For the (d,t) experiments the resolution was typically ~5 keV (FWHM) but the best (d,p) spectra obtained thus far have ~8 keV FWHM. The ^{193}Ir(d,t)^{192}Ir resolution of this study is not quite as good as the value of 3.4 keV reported for the best spectrum measured by Kern et al[11], but the complete angular distributions of the present work provide important new information on ℓ-values and spectroscopic strengths. Spectra at several angles from each of the ^{189}Os(α,t)^{190}Ir and ^{189}Os(^{3}He,d)^{190}Ir reactions were also obtained, using 30 MeV α and 28 MeV ^{3}He beams. Finally, some preliminary measurements of the ^{195}Pt(p,α)^{192}Ir reaction have been made, as previous studies[12] show a good correlation between the (p,α) cross sections and those from the single-proton-pickup (t,α) reaction. A representative (d,t) spectrum is shown in Fig. 1.

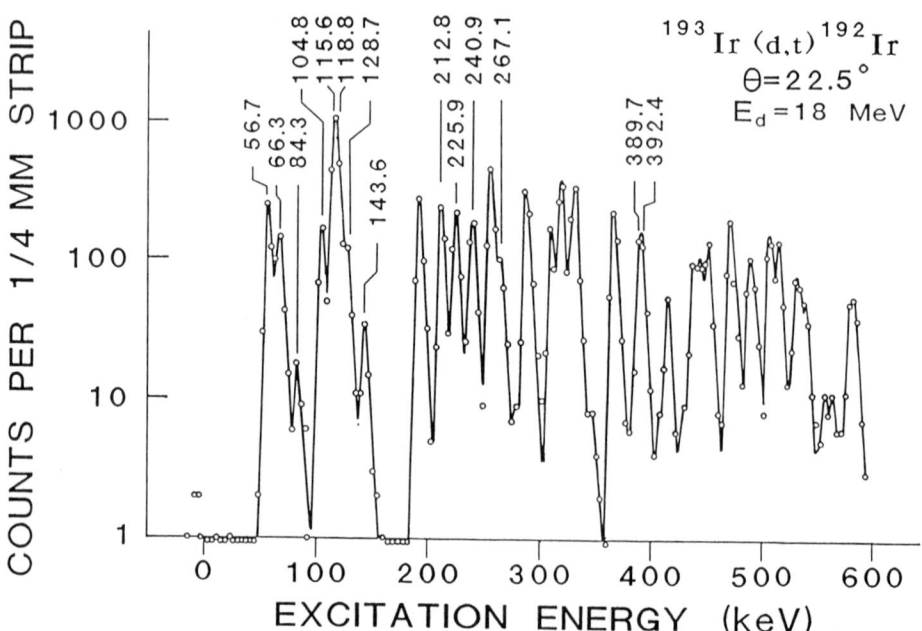

Fig. 1. Triton spectrum from the ^{193}Ir(d,t)^{192}Ir reaction at $\theta=22\frac{1}{2}°$

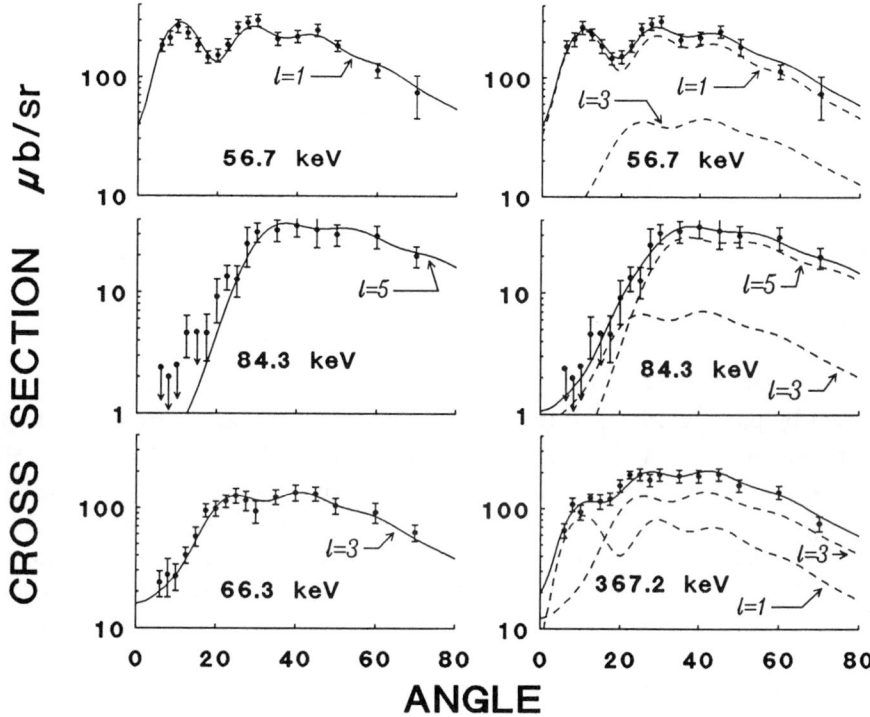

Fig. 2. Angular distributions for several levels populated in the ^{193}Ir(d,t)^{192}Ir reaction. The curves are DWBA predictions for the ℓ-values indicated. The right half shows fits to mixtures of ℓ-values.

Figure 2 shows (d,t) angular distributions for some levels in ^{192}Ir. The curves are DWBA predictions for the ℓ-values indicated, using the same optical model parameters as in ref. 7. Although angular momentum considerations would permit each transition to have several different ℓ-components, many of them are found to be dominated by a single ℓ-value, as seen in the left half of Fig. 2. The right half of this figure shows how a least-squares fitting procedure was used to determine the individual strengths, in mixed-ℓ transitions, which give the best fit to the experimental angular distributions. The strengths, S_ℓ, are defined by the expression for the differential cross section

$$\frac{d\sigma}{d\Omega}(\theta) = \sum_\ell S_\ell N [\frac{d\sigma}{d\Omega}(\theta, \ell)]_{DW}$$

N is a normalization factor for the DWBA cross section, $[\frac{d\sigma}{d\Omega}(\theta,\ell)]_{DW}$, and for the (d,t) reaction it has the value N=3.33. The strengths, S_ℓ, from these data will be used in the following section to test and extend the proposed model interpretations of Kern et al.[11]

DISCUSSION

LEVELS IN ^{192}Ir

For the ^{191,193}Ir(d,t)^{190,192}Ir reactions most of the strong transitions are found to be dominated by $\ell=1$ and $\ell=3$ transitions. This provides independent confirmation of negative parity for a series of previously-established[11] levels starting at 57 keV in ^{192}Ir, and thus supports the proposed[11] positive parity for the ^{192}Ir ground state.

The tentative assignments in the rotor-plus-particle description of ^{192}Ir by Kern et al.[11] are summarized in Fig. 3. The $I^{\pi}=4^{+}$ ground state is not expected to have an observable cross section, and was not seen. The $K^{\pi}=1^{-}$ band starting at 57 keV was assigned as the $\frac{3}{2}^{+}[402]_{\pi} - \frac{1}{2}^{-}[510]_{\nu}$ configuration, which can be populated by picking up a $\frac{1}{2}^{-}[510]$ neutron since the ^{193}Ir target ground state has a $\frac{3}{2}^{+}[402]$ proton. Fig. 4 shows a comparison of observed and predicted strengths for the spin 1 to 4 members of this band as tentatively assigned by Kern et al.[11] The calculated values were obtained from a Nilsson calculation with $\kappa=0.0636$, $\mu=0.393$, $\epsilon_2=0.17$ and $\epsilon_4=0.05$, and assumed a reasonable value $V^2=0.8$ for the fullness parameter due to pairing. The agreement is good, considering that effects of Coriolis mixing in neither the target nor final nucleus have been taken into account. Preliminary calculations of Coriolis mixing effects in ^{192}Ir, including only configurations with the $\frac{3}{2}^{+}[402]$ proton, show that some S_ℓ values can be changed by 10-30%.

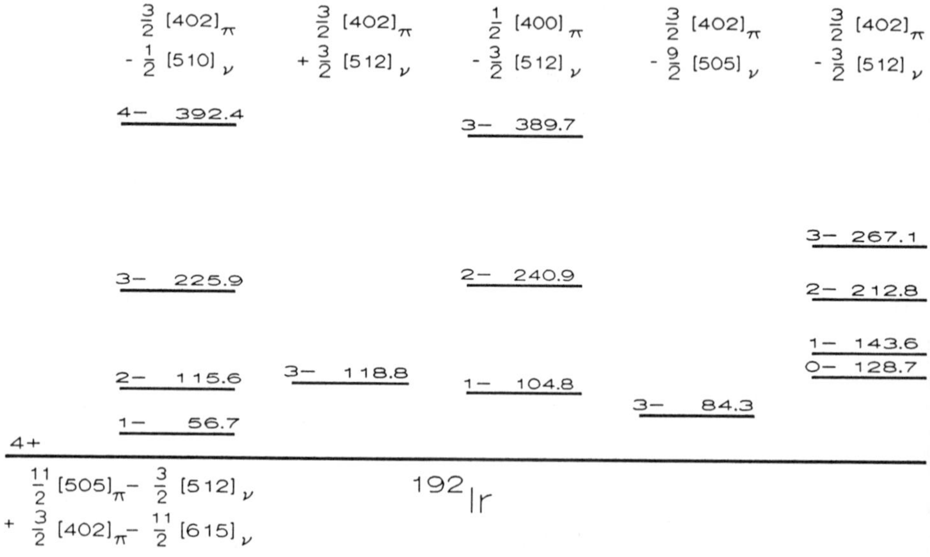

Fig. 3. Tentative assignments of ^{192}Ir levels from Kern et al[11]

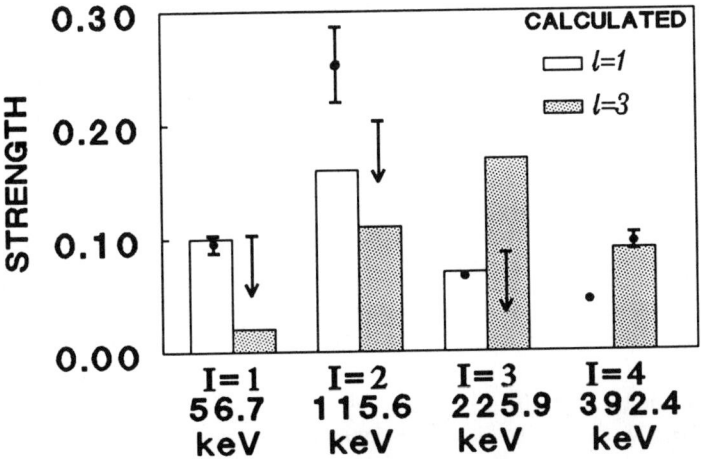

Fig. 4. Comparison of (d,t) strengths for members of the proposed $K^\pi=1^-$, $\frac{3}{2}^+[402]_\pi - \frac{1}{2}^-[510]\nu$ band in ^{192}Ir. Data points shown as arrows are upper limits to possible strengths. The bars represent calculated strengths. The $\ell=1$ strength shown for the 392.4 keV level is believed due to a nearby unresolved level.

IBFFM wavefunctions for the 1^- bandhead at 57 keV have been presented by Kern et al,[11] and include several components which could be populated by $\ell=1$ and $\ell=3$ transitions. However, the ^{193}Ir ground state wave functions used in this description are not yet available so a quantitative comparison of strengths is not yet possible.

The 3^- level at 84 keV is a very interesting case because the (d,t) population appears to be a pure $\ell=5$ transition (see Fig. 2), even though $\ell=1$ and $\ell=3$ transfers would satisfy the angular momentum conservation principle. Very small amounts of the $\ell=1$ and 3 transfers would have been detected, because they have much larger intrinsic cross sections, particularly at forward angles. From the present experiments, upper limits can be set at $S_1 \leq 0.002$ and $S_3 \leq 0.02$. Kern et al.[11] have interpreted this level as the $\frac{3}{2}^+[402]_\pi - \frac{9}{2}^-[505]_\nu$ bandhead, which is expected at low energy since the $\frac{9}{2}^-[505]$ neutron orbital forms the ^{191}Os ground state. The predicted strength for this configuration is $S_5=0.5$, assuming $V^2=0.7$, in excellent agreement with the observed value of $S_5=0.45 \pm 0.1$. The spin 4 member of this band has not been assigned. It is expected to be populated by a much weaker $\ell=5$ transition that could easily be obscured in the experimental data. The IBFFM wavefunctions[11] for the lowest 3^- level include several large components that could be populated by $\ell=1$ and $\ell=3$ transitions, and one that would require $\ell=5$. Thus, this calculation would not reproduce the $\ell=5$ transfer for the lowest observed 3^- level at 84 keV. Furthermore, the $\ell=5$ strength predicted by this model is smaller than the experimental value by a factor of ~7. It therefore appears that the wavefunction for the lowest predicted 3^- level does not correspond with the lowest observed 3^- level. As the calculated energy was ~200 keV too high, and other predicted 3^- levels were only slightly higher, it is quite possible that a different level from the model should be associated with the 84 keV level.

Kern et al.[11] have assigned the 119 keV level as the 3^-, $\frac{3}{2}^+[402]_\pi + \frac{3}{2}^-[512]_\nu$ state, which would have predicted (d,t) strengths of $S_1=0.14$ plus $S_3=0.20$, assuming $V^2=0.8$. These are in reasonable agreement with the experimental values of $S_1 \sim 0.10$ and $S_3 \sim 0.12$. The spin 4 member of this band has not been assigned, but is expected to have a large $\ell=3$ strength, $S_3=0.34$. The only level observed to have such a large $\ell=3$ strength is the one at 257 keV, with $S_3=0.5$.

Four members of the $K^\pi=0^-$, $\frac{3}{2}^+[402]_\pi - \frac{3}{2}^-[512]_\nu$ band have been tentatively assigned[11], starting with the 0^- level at 129 keV. The predicted and observed strengths are compared in Fig. 5. For the 0^- bandhead the results are in very good agreement, but for the other three members the observed values conflict seriously with expectations, indicating that other configurations must have large admixtures in this band.

Another interesting level is the one at 66 keV, for which the (d,t) angular distribution suggests a pure $\ell=3$ transition. It was not populated in the (n,γ) average resonance capture experiments[11], which indicates that the spin is greater than 3. This restricts the spin-parity to 4^- or 5^-. The 4^- value is favored since decay of the spin 9 isomer at 161 keV proceeds directly to the $I^\pi=4^+$ ground state. The IBFFM does not predict any 4^- or 5^- levels below 400 keV, and none

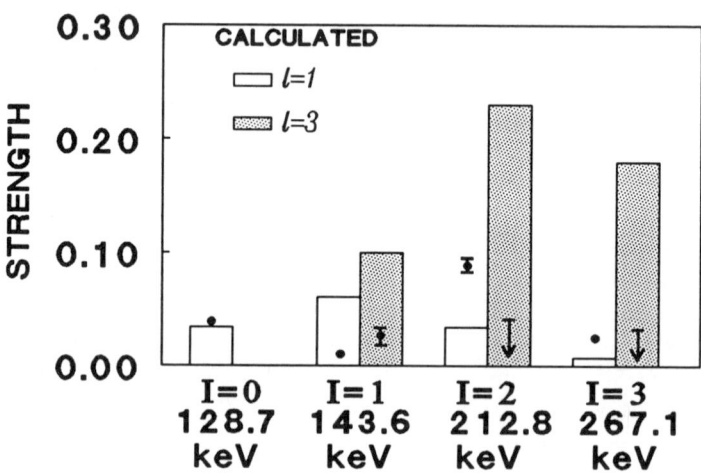

Fig. 5. Comparisons of (d,t) strengths for members of the proposed $K^\pi=0^-$, $\frac{3}{2}^+[402]_\pi - \frac{3}{2}^-[512]_\nu$ band in ^{192}Ir. See caption to Fig. 4.

of the ones predicted by the rotor-plus-particle model below this energy should have strong $\ell=3$ populations. The lowest two- quasiparticle state which would be expected to have a large $\ell=3$ strength to a 4^- or 5^- bandhead is $K^\pi=5^-$, $\frac{3}{2}^+[402]_\pi + \frac{7}{2}^-[503]_\nu$, which should be found at $\geq \frac{1}{2}$ MeV excitation.

LEVELS IN ^{190}Ir

It is not possible at this time to establish an absolute scale of excitation energies for the ^{190}Ir spectra, because the 3 previously proposed ^{190}Ir levels[13] are not expected to have observable cross sections in these reactions. Furthermore, the ^{190}Ir ground state mass has an uncertainty of 200 keV in the most recent mass evaluation[14]. We therefore designate the unknown excitation energy of the first ^{190}Ir level populated in the (d,t) reaction as X. The measured (d,t) Q-value for this level is -1768 ± 2 keV. The ^{189}Os(α,t)^{190}Ir Q-values are known to a similar precision. Thus, although the zero point for the excitation energy scale is not well known, the levels populated in (d,t) can be placed relative to those from (α,t) within ±3 keV. In this way, it is found that 11 out of the first 13 levels in the (α,t) spectrum coincide in energy with ones in the (d,t) spectrum.

Since ^{191}Ir and ^{193}Ir both have $\frac{3}{2}^+[402]_\pi$ ground states, the ^{190}Ir levels populated in the ^{191}Ir(d,t)^{190}Ir reaction should involve various neutron states coupled to a $\frac{3}{2}^+[402]$ proton, as for the ^{193}Ir(d,t)^{192}Ir case. Indeed, there is a similarity in the distributions of ℓ-values and strengths for (d,t) spectra to the two nuclides. For example, the first three levels with observable populations have almost pure $\ell=1$, $\ell=3$, and $\ell=5$ transitions, of similar strengths. In ^{192}Ir these are the 57 keV (S_1=0.10), 66 keV (S_3=0.21), and 84 keV (S_5=0.45 ± 0.1) levels, which appear to correspond to the ^{190}Ir levels at X keV (S_3=0.13), X + 26 keV (S_1=0.09), and X + 38 keV (S_5=0.33) respectively. The similarities continue with higher levels.

For the ^{189}Os(^{3}He,d)^{190}Ir and ^{189}Os(α,t)^{190}Ir studies the target ground state is predominantly the $\frac{3}{2}^-[512]$ neutron, so the ^{190}Ir levels populated should involve various protons coupled to $\frac{3}{2}^-[512]_\nu$. Thus, the ^{190}Ir levels expected to be populated in both the proton-transfer and neutron-transfer processes are the $K^\pi=3^-$ and 0^-, $\frac{3}{2}^+[402]_\pi \pm \frac{3}{2}^-[512]_\nu$ bands. At most, 4 or 5 levels would be expected to have appreciable cross sections in both processes. Experimentally, about twice this many are observed, implying significant mixing in the target ground states or the final states, or both.

Analyses similar to those above for ^{192}Ir can be used to describe some ^{190}Ir levels in terms of the rotor-plus-particle model, with qualitative agreement. A more detailed analysis, including Coriolis mixing in the target ground states as well as the final configurations, is planned. This is probably important because the $\frac{1}{2}^-[510]_\nu$ amplitude mixed into the $\frac{3}{2}^-[512]_\nu$ ground state of ^{189}Os is estimated to be ~0.3. Comparable $\frac{1}{2}^+[400]_\pi$ admixtures in the $\frac{3}{2}^+[402]_\pi$ ground states of the ^{191,193}Ir targets are also likely.

CONCLUDING REMARKS

Charged-particle reaction studies have been used to populate levels in odd-odd iridium isotopes. For ^{192}Ir, in which spin values were previously known for many levels, spectroscopic strengths from these experiments provided tests of the proposed model interpretations and led to a better understanding of the nuclear structure. This demonstrates once again the usefulness of applying different complementary experimental techniques to a complex nuclear structure problem.

Some ^{192}Ir levels appear to be described quite well in the rotor-plus-particle model, e.g., the $I^\pi=3^-$, $\frac{3}{2}^+[402]_\pi - \frac{9}{2}^-[505]_\nu$ state at 84 keV and the $K^\pi=1^-$, $\frac{3}{2}^+[402]_\pi - \frac{1}{2}^-[510]_\nu$ band starting at 57 keV. This is remarkable and somewhat surprising in view of the difficulties encountered in explaining the structures of neighbouring odd-mass nuclei in this transition region. On the other hand, neither the rotor-plus-particle model nor the IBFFM predict low-lying 4^- or 5^- states that could be assigned to the observed level at 66 keV. It will be interesting to see whether the assumption of triaxiality or more complex shapes will explain this interesting level. It is hoped that transition strengths calculated from the IBFFM wave functions will soon be available for comparison with the experimental results.

We are very grateful to J. Kern for permission to use results from his collaboration before their publication, and for useful discussions. Financial support was provided by the Natural Sciences and Engineering Research Council of Canada.

REFERENCES

1. G.J. Gyapong et al., Nucl. Phys. A458, 165(1986).
2. D. Benson, Jr. et al., Phys. Rev. C14, 2095(1976).
3. D. Benson, Jr., et al., Z. Phys. A281, 145(1977).
4. D. Benson, Jr., et al., Z. Phys. A285, 405(1978).
5. Y. Yamazaki and R.K. Sheline, Phys. Rev. C14, 531(1976).
6. Y. Yamazaki, R.K. Sheline and E.B. Shera, Phys. Rev. C17, 2061(1978).
7. D.G. Burke and G. Kajrys, Nucl. Phys. A, in press.
8. D.D. Warner et al., Phys. Rev. C26, 1921(1982).
9. R.F. Casten, et al., Phys. Rev. C27, 1310(1983).
10. A. Mauthofer et al., Phys. Rev. C39, 1111(1989).
11. J. Kern et al., to be published, and also proceedings of this conference.
12. M.A.M. Shahabuddin et al., Nucl. Phys. A307, 239(1978).
13. B. Singh, Nucl. Data Sheets, in press.
14. A.H. Wapstra, G. Audi and R. Hoekstra, At. Data Nucl. Data Tables 39, 281(1988).

DISTRIBUTION OF PHOTON STRENGTH IN NUCLEI BY A METHOD OF TWO-STEP CASCADES

F. Becvar,[1] P. Cejnar[1], R. E. Chrien[2], and J. Kopecky[3]

1. Charles University, Faculty of Mathematics and Physics, 180 00 Prague 8, Czech and Slovak Federal Republic
2. Brookhaven National Laboratory, Upton, New York, 11973, USA
3. Netherlands Energy Research Foundation ECN, P.O. Box 1, 1755 ZG Petten, The Netherlands

ABSTRACT

The applicability of sum-coincidence measurements of two-step cascade γ-ray spectra to the determination of photon strength functions at intermediate γ-ray energies (3 or 4 MeV) is discussed. An experiment based on thermal neutron capture in Nd was undertaken at the Brookhaven National Laboratory High Flux Beam Reactor (BNL HFBR) to test this model. To understand the role of various uncertainties in similar experiments a series of model calculations was performed. We present an analysis of our experimental data which demonstrates the high sensitivity of the method to E1 and M1 photon strength functions. Our experimental data are in sharp contradiction to those expected from an E1 photon strength distributed according to the classical Lorentzian form with an energy invariant damping width. An alternative distribution of Kadmenskij et al., which violates Brink's Hypothesis, is strongly preferred.

INTRODUCTION

In the past three decades a large amount of information has been accumulated from the decay of highly excited nuclear states, especially those states populated in slow neutron capture. This information concerns mainly hard primary γ rays that correspond to the transitions from the capturing states to low-lying levels. On the other hand, very little has been learned on soft primary γ rays with energies $E_\gamma \leq 4$ MeV, although such information is crucial for a better understanding of the deexcitation mechanism.

The radiative strength is usually treated in terms of the conventionally defined photon strength functions $S_\gamma(E_\gamma)$ for various multipolarities of γ radiation[1]. The richest information is available for the E1 photon strength function. In order to calculate the sizes of total radiative widths of neutron resonances, Brink[2] used the principle of detailed balance and the concept of an electric Giant Dipole Resonance (GDR) built on the ground and each excited state. His values were typically a factor of three larger than those observed experimentally. Later, Axel[3] used these assumptions to explain the energy variation of the $S_\gamma^{E1}(E_\gamma)$.

McCullagh, Stelts, and Chrien[4], in their review of E1 strength functions, pointed out that an energy-dependent damping width in the Lorentzian could improve the representation of the strength functions for low-energy γ rays. Subsequently, Kadmenskij, Markushev, and Furman[5] pointed out that for the cases of spherical and transitional nuclei the limit for $S_\gamma^{E1}(E_\gamma)$ as $E_\gamma \to 0$ should be non-zero. Using the Fermi liquid model these authors also predicted a specific dependence of width Γ_G of the electric GDR on the γ-ray energy, as well as a dependence of the shape of the GDR on the nuclear temperature of the final state on which the GDR is built. These considerations lead to a drastic modification of Brink's hypothesis and rule out a purely Lorentzian shape of the electric GDR.

In later papers, Kopecky and Chrien[6] and Kopecky and Uhl[7] used photoneutron data and slow and fast neutron capture data in ^{94}Nb, ^{106}Pd, ^{144}Nd, and ^{198}Au to support the above-outlined predictions of Kadmenskij et al.[5] about the shape and size of $S_\gamma^{E1}(E_\gamma)$. On the other hand, conclusions in Ref. 7, concerning the existence of the non-zero limit of $S_\gamma(E_\gamma)$, need verification that would lean on more direct information on photon strength at low energies. We note in this connection that an attempt to understand behavior of E1 and M1 photon strengths for low-energy primary transitions was made by Aldea and Seyfarth[8] and Furman et al.[9], who studied γ-α cascades in the ^{143}Nd(n,γα)^{140}Ce reaction.

In this paper we demonstrate that the method of Two-Step Cascade (TSC) γ-γ transitions, devised by Hoogenboom[10] more than 30 years ago, makes the needed information available. We present results based on an experimental study of the ^{143}Nd(n,γ)^{144}Nd reaction and compare them with analogous results following from model calculations. We discuss the sensitivity of the proposed method to various models used for photon strength functions, and present conclusions concerning the best choice from these models for the reaction studied.

THE EXPERIMENT

The experiment is based on a two-detector coincident measurement of two-step cascades that follow thermal neutron capture and end at a final state of known spin and parity. By fixing an energy sum a final state can be selected and energy spectrum of all the transitions involved in TSC deexcitation can be obtained. This spectrum, called hereafter a TSC γ-ray spectrum, may be divided into two components: a set of discrete, well-resolved lines and an unresolved quasi-continuum. The discrete lines correspond to cascades proceeding via intermediate states with excitation energies below ~ 3 MeV. The shape and size of the TSC spectrum carry information about the photon strength functions that govern the emission of γ rays in the deexcitation process. As the TSC spectra are expected to display a gross structure with a maximum at approximately 3 or 4 MeV, the behavior of the photon strength functions at these energies can be studied. A set of the TSC spectra can be obtained, one for each final state.

We have undertaken the above-described measurement for the thermal ^{143}Nd(n,γ)^{144}Nd reaction. A sample of natural Nd was irradiated in a beam of thermal neutrons at the BNL HFBR for about 300 hours. Two back-to-back HPGe detectors were used, having measured overall photopeak efficiencies of 1.95 and 2.30% at energy of 696 keV, including the effect of solid angle. Each event, represented by the deposited energies of the coincident γ rays and the associated detection time difference, was recorded on a magnetic tape. This information was later scanned off-line to yield the needed γ-ray spectra.

In the special scanning procedure used, a region that surrounds the peak corresponding to a preselected final state at which two-step cascades terminate is defined in the energy sum-time plane. The region is used to provide a background correction for a given TSC spectrum. This guarantees that the resulting TSC spectra represent a real effect, free from background, and that these spectra behave as if they were yielded by a spectrometer whose energy-response function is confined to a narrow single peak. The energy resolution achieved is represented by FWHM of 2.8 and 4.9 keV at energies of 600 and 7100 keV, respectively, including long-term instabilities.

The spectra obtained were corrected for energy variation of detector efficiencies and converted into spectra of *absolute* TSC intensities. A separate thermal (n,γ) single-spectrum measurement with a mixed Nd+Cl target has been undertaken to determine an absolute intensity for a primary transition to the $J^\pi = 4^+$ state at 1314 keV in ^{144}Nd. The result, $I_\gamma = 7.36\%$, has been used as a reference value in the normalization.

An example of a TSC spectrum is given in Fig. 1. It corresponds to all cascades terminating at the first $J^\pi = 2^+$ level in ^{144}Nd at 696 keV. The spectrum displays a symmetry that follows from the impossibility to distinguish between the primary and the secondary γ rays. A large number of discrete lines can be seen in Fig. 1. For some of them, energies of the corresponding intermediate ^{144}Nd states are given. Intensity fluctuations of these lines are very broad: the TSC intensity of the strongest line in the spectrum, corresponding to the $J^\pi = 4^+$ intermediate state at 1314 keV, reaches a peak value of 0.012 keV^{-1}, while a large number of still well-separated lines are weaker by a factor of 100.

The most important constituent of the spectrum is the quasi-continuum that shows up as a smear in a vicinity of the base line. In order to minimize the uncertainty due to the obvious fluctuations, an integral over a restricted energy interval, situated in the middle of the spectrum, can be chosen as a measure of the size of the TSC spectrum. We selected an energy interval of 2.4 MeV. In this case, the integral could be determined with a statistical accuracy of 2.6%. The integral of the TSC spectrum thus represents a positive, statistically-significant effect.

The integrals of TSC spectra were additionally corrected for effects of pair production and positron annihilation, as well as for backscattering of γ rays. These corrections are small: the overall change is about 6%. We also made corrections for effects of vetoing, caused by detection of those γ rays that were emitted after two-step cascades reached the fixed final state.

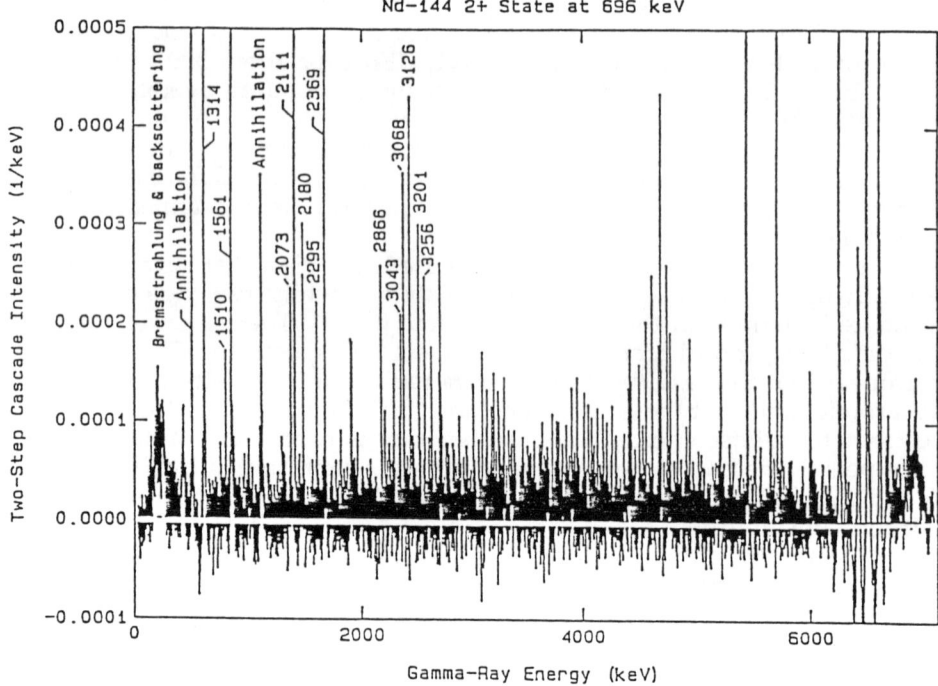

Fig. 1. The TSC γ spectrum for the $J^\pi = 2^+$ final state of ^{144}Nd at 696 keV. Only a small fraction of the range of the TSC intensity is shown.

Vetoing correction factors range from 0.73 to 1.39. No corrections were done for bremsstrahlung. Its contribution to the useful part of a TSC spectrum is expected to be small, not higher than 2%.

All TSC spectra obtained are subject to a common systematic error of 16%, associated with uncertainties of the above-outlined corrections and normalization.

In a similar manner, TSC γ-ray spectra were obtained also for the $J^\pi = 0^+$ ground state and for the states at 1314, 1510, and 1561 keV with J^π equal to 4^+, 3^-, and 2^+, respectively. The integrals of these spectra could be determined with a statistical accuracy of 9.7, 6.8, 7.2, and 12.4%, respectively.

MODELING

In order to test various hypotheses about photon strength functions we modeled the TSC process and compared the resulting spectra with those observed experimentally.

Let us assume a cascade $EJ\pi \to E'J'\pi' \to E''J''\pi''$. Here, $EJ\pi$ stands for an excitation energy, spin and parity of the initial state. Similarly, $E'J'\pi'$

and E"J"π" belong to the intermediate state and the fixed final state, respectively. Following the definition of the photon strength function the expectation value of a partial radiative width for the first step $EJ\pi \rightarrow E'J'\pi'$ can be expressed as[1]

$$< \Gamma_\gamma(EJ\pi \rightarrow E'J'\pi') > = E_{\gamma 1}^{2L+1} S_\gamma^{XL}(E_{\gamma 1})/\varrho(E,J) \qquad (1)$$

where XL stands for multipolarity of the transition $EJ\pi \rightarrow E'J'\pi'$, i.e. XL = E1, M1, E2, etc., $E_{\gamma 1}$ is a γ-ray energy of this transition and ρ(E,J) - density of levels with spin J at an excitation energy E.

The expectation value of the total radiative width for the state $EJ\pi$ is

$$< \Gamma_{\gamma EJ\pi} > = \sum_{J'\pi'} \int_{E_0}^{E} \varrho(E',J') < \Gamma_\gamma(EJ\pi \rightarrow E'J'\pi') > dE'$$
$$+ \sum_k < \Gamma_\gamma(EJ\pi \rightarrow E_k J_k \pi_k) > . \qquad (2)$$

The first term of this expression includes the continuum of states with E' > E_0 that are described by level density ρ. The second term belongs to transitions to the remaining discrete states, labeled by subscript k. It is assumed that below the energy E_0 a full set of states is known from experiments. Analogous expressions can be written for $<\Gamma_\gamma(E'J'\pi' \rightarrow E"J"\pi")>$ and $<\Gamma_{\gamma E'J'\pi'}>$.

The expectation value of a TSC intensity, per unit energy interval, can be expressed via expectation values of one-step intensities:

$$< I_{\gamma\gamma}(E_{\gamma 1}) > = \sum_{J'\pi'} < I_\gamma(EJ\pi \rightarrow E'J'\pi') >$$
$$< I_\gamma(E'J'\pi' \rightarrow E"J"\pi") > \varrho(E',J'). \qquad (3)$$

Here $E' = E - E_{\gamma 1}$ and

$$< I_\gamma(EJ\pi \rightarrow E'J'\pi') > = < \Gamma_\gamma(EJ\pi \rightarrow E'J'\pi') > / < \Gamma_{\gamma EJ\pi} > . \qquad (4)$$

A similar expression can be written for $<I_\gamma(E'J'\pi' \rightarrow E"J"\pi")>$. Equation 3 is valid for primary γ-ray energies $E_{\gamma 1}$ < $E-E_0$. The introduced quantity $<I_{\gamma\gamma}(E_{\gamma 1})>$ is an energy density, i.e. it represents an intensity per unit of energy.

Examples of calculated TSC spectra are given in Fig. 2. Besides the values of $<I_{\gamma\gamma}(E_{\gamma 1})>$ the values of individual terms of the right-hand side of Eq. 3 for various J'π' are plotted. The spectra in Fig. 2 have been calculated using a level density formula of Kataria et al.[11] with parameters, recommended by Kopecky and Uhl[7]. The photon strength function S_γ^{E1} used in calculations was deduced from the model of Kadmenskij et al.[5].

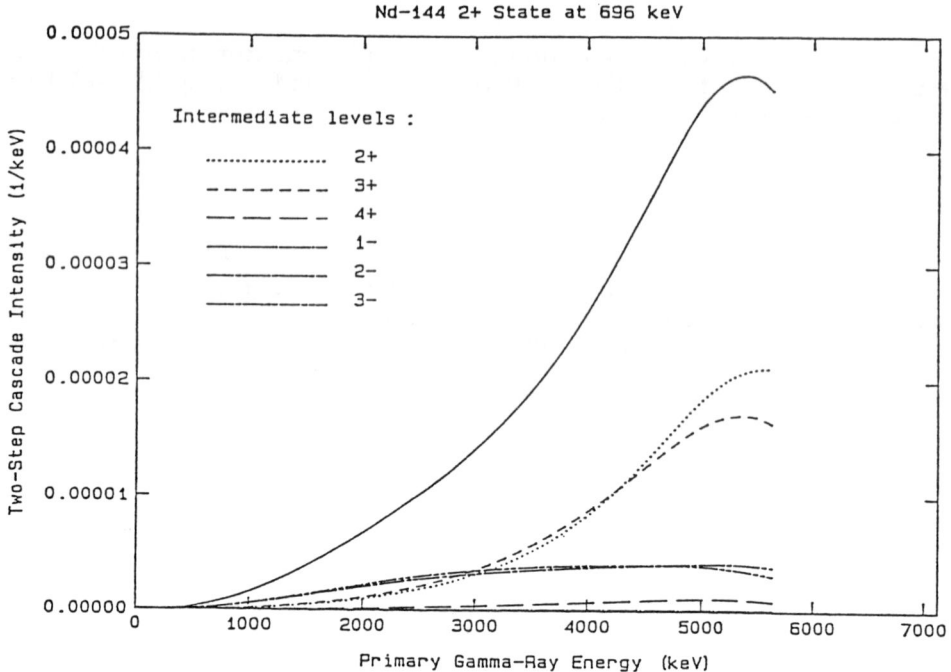

Fig. 2. Expectation values of the TSC intensity vs γ-ray energy for the $J^\pi = 2^+$ state of ^{144}Nd at 696 keV (solid line). Contributions of intermediate states with individual values of J^π are also shown.

For $S_\gamma M1$ and $S_\gamma E2$ single-particle approximations were assumed. The role of M2, E3, and higher multipolarities was neglected.

In order to study the influence of Porter-Thomas fluctuations of partial radiative widths we modified Eq. 3 to yield fluctuating values of TSC intensity. Specifically, in Eq. 3 we introduce additional factors $\theta_{E'J'\pi'}^{(1)}$, and $\theta_{E'J'\pi'}^{(2)}$, that are responsible for Porter-Thomas fluctuations of one-step intensities. The fluctuating TSC intensity can be expressed as

$$I_{\gamma\gamma}^{(fluct)}(E_{\gamma 1}) = \sum_{E'J'\pi'} < I_\gamma(EJ\pi \to E'J'\pi') >< I_\gamma(E'J'\pi' \to E''J''\pi'') >$$
$$\times \Theta_{E'J'\pi'}^{(1)} \Theta_{E'J'\pi'}^{(2)} \delta(E_{\gamma 1} - E + E'). \qquad (5)$$

Summation in Eq. 5 is assumed over a full set of all the intermediate states, whose values of excitation energy E' are distributed in accordance with the adopted level-density formula.

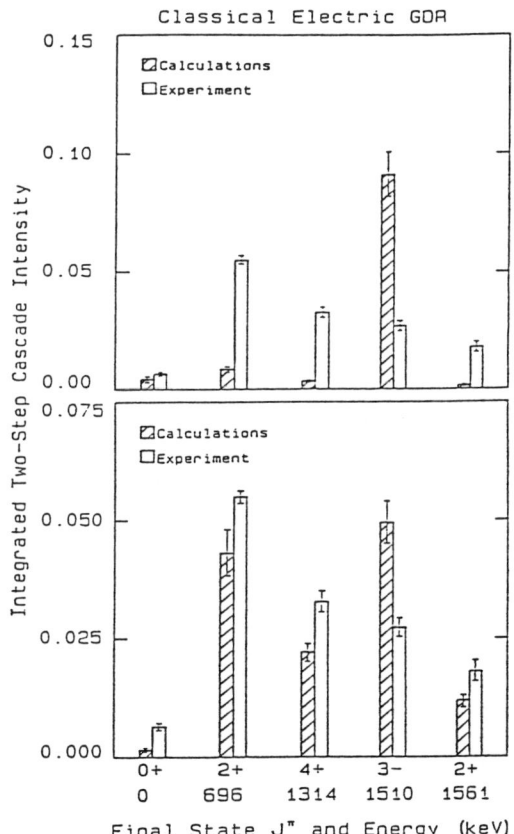

Fig. 3. Calculated and measured TSC intensities for the individual states of ^{144}Nd. The calculated values were obtained using the choice (a) for $S_\gamma^{E1}(E_\gamma)$, see text. The upper part: a magnetic GDR, $\langle\Gamma_\gamma\rangle = 157$ meV. The lower part: a constant strength $S_\gamma^{M1} = 1.2\times10^{-8}$ MeV^{-3}, $\langle\Gamma_\gamma\rangle = 195$ meV.

It can be shown that products $\theta_{E'J'\pi'}^{(1)} \theta_{E'J'\pi'}^{(2)}$ behave as completely independent statistical variables, obeying a distribution which can be easily reproduced by the method of Monte Carlo. This makes it possible to model the values of $I_{\gamma\gamma}^{(fluct)}$.

In order to estimate the role of Porter-Thomas fluctuations we used Eq. 5 and modeled TSC spectra for various final states. All TSC intensities $I_{\gamma\gamma}^{(fluct)}$ we obtained exhibit violent fluctuations, even when they are represented as histograms with a bin width of 3 keV to resemble spectra obtained by a typical Ge-detector. Nevertheless, if one integrates the values of $I_{\gamma\gamma}^{(fluct)}$ over a reasonably wide energy interval, fluctuations of the resulting integral can be kept low. Specifically, for the case of the earlier-mentioned energy interval of 2.4 MeV the residual fluctuations of integrated TSC intensities are represented by rms values of 23, 11, 10, 9, and 11% for cascades populating the ground state and the states at 696, 1314, 1510, and 1561 keV, respectively. These rms values were obtained using the same level-density formula and the same explicit expressions for

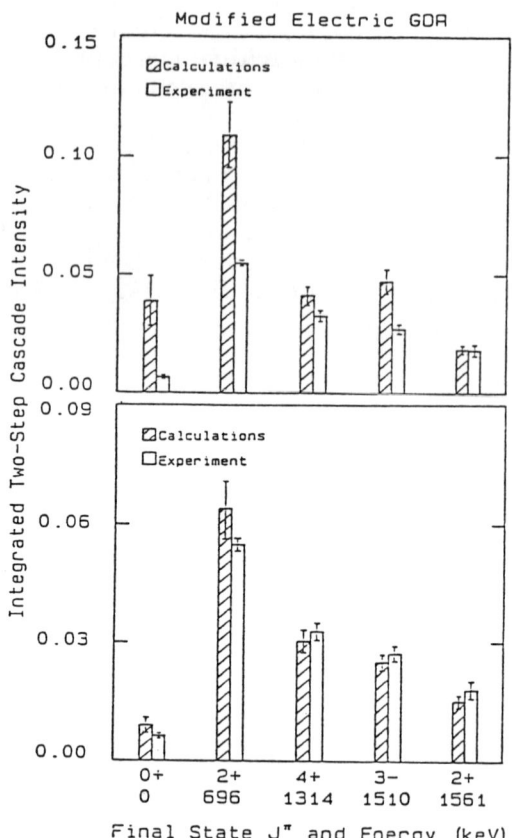

Fig. 4. Calculated and measured TSC intensities for the individual states of ^{144}Nd. The calculated values were obtained using the choice (b) for $S_\gamma^{E1}(E_\gamma)$, see text. The upper part: a magnetic GDR, $\langle\Gamma_\gamma\rangle = 17$ meV. The lower part: a constant strength $S_\gamma^{M1} = 0.3\times10^{-8}$ MeV^{-3}, $\langle\Gamma_\gamma\rangle = 25$ meV.

photon strength functions as in the case mentioned in connection with Fig. 2. We see that the quoted experimental uncertainties in the determination of the quasi-continuum are comparable to, or even lower than those following from the Porter-Thomas fluctuations.

Quantities $\langle I_{\gamma\gamma}\rangle$ and $I_{\gamma\gamma}^{(fluct)}$ were modeled for the ^{144}Nd ground state and the levels at 696, 1314, 1510, and 1561 keV. Using Eq. 2 we also calculated expectation values of total radiative widths for an excitation energy E adjusted at the neutron threshold and for the value $J\pi = 3^-$ that belongs to the dominating capturing state in the reaction studied. The results are presented in a condensed form in Figs. 3-5, where the integrated TSC intensities are plotted for individual levels and compared with analogous quantities deduced from the measured TSC spectra. An energy interval of 2.4 MeV was selected for integration. In Figs. 3-5 experimental errors of the TSC intensities are plotted as well as uncertainties associated with Porter-Thomas fluctuations.

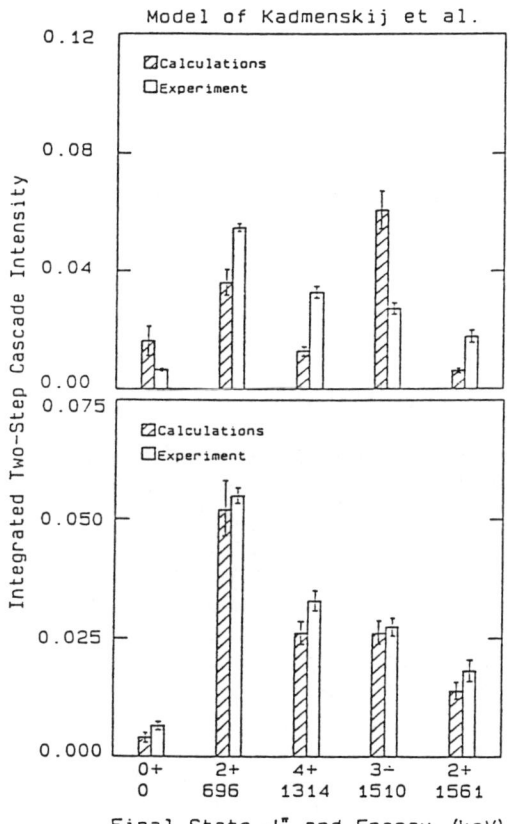

Fig. 5. Calculated and measured TSC intensities for the individual states of ^{144}Nd. The calculated values were obtained using the choice (c) for $S_\gamma^{E1}(E_\gamma)$, see text. The upper part: a magnetic GDR, $\langle \Gamma_\gamma \rangle = 53$ meV. The lower part: a constant strength $S_\gamma^{M1} = 0.6 \times 10^{-8}$ MeV^{-3}, $\langle \Gamma_\gamma \rangle = 72$ meV.

The presented results are based on the use of the level-density formula of Kataria et al.[11] and the following three approximations for the E1 photon strength function:

a) The expression based on a classical electric GDR of Lorentzian shape.

b) The expression (a) modified by inclusion of an energy- and temperature-dependent damping width[6], specifically

$$\Gamma_G = \Gamma_G^0 (E_\gamma^2 + 4\pi^2 T^2)/E_G^2, \quad (6)$$

where T is the nuclear temperature of the level on which an electric GDR is built, Γ_G^0--a constant factor and E_G--the GDR energy.

c) The expression, representing a low-energy approximation within the model of Kadmenskij et al.[5], i.e.

$$S_\gamma^{E1}(E_\gamma) = \frac{(1 + 2f'_1/3)^{1/2} \sigma_0 \Gamma_G^0 \Gamma_G E_G}{3\pi^2 \hbar^2 c^2 (1 + 2f')^{1/2}(E_\gamma^2 - E_G^2)^2}, \quad (7)$$

where σ_0 is the photoabsorption cross section at peak of the GDR, Γ_G is given by Eq. 6 and f'_1, f' are parameters of the Fermi-liquid model that characterize a quasi-particle interaction. In accordance with Ref. 12 we took $f'_1 = -0.1$ and $f' = 0.25$.

We made several assumptions on the M1 strengths; first, a single-particle model assuming various strengths and, second, a magnetic giant dipole resonance model with parameters taken from Ref. 7. The best fits are shown in Figs. 3-5.

DISCUSSION AND SUMMARY

As can be seen from Fig. 3, the choice (a) is not adequate. Much better agreement between the data and model calculations is observed in Figs. 4 and 5 for the remaining two choices of S_γ^{E1}. A good agreement is achieved for the photon strength function that is represented by the approximation of Kadmenskij et al.[5], while S_γ^{M1} is assumed to be energy-independent, see Fig. 5. This accordance between the measured and calculated TSC intensities can be further improved by a slight adjustment of the measured intensities within their quoted common uncertainty of 16%. After such adjustment the agreement between the data and calculations can be characterized by a χ^2-value of 5.2 at $v = 4$. It should be stressed that the calculated expectation value of the partial radiative width $\langle \Gamma_\gamma \rangle = 72$ meV is reasonably close to the experimental value $\Gamma_\gamma(\exp) = 80 \pm 9$ meV i(see Ref. 13). If we assume that S_γ^{M1} is energy-independent the choice (b) also gives a good agreement between the calculated and measured TSC intensities, see Fig. 4. However, in this case the model calculations yield a value $\langle \Gamma_\gamma \rangle = 25$ meV which is too low to accept the considered choice for E1 photon strength function as correct.

Compared to other methods, the most important advantage of the method used is its inherent sensitivity to the photon strength of primary transitions as well as to that of secondary transitions. For the choice (c) that violates Brink's hypothesis, these strengths are substantially different both in size and their energy dependence, while in case of validity of this hypothesis the strengths are identical. For this reason the conclusion we made about the validity of the choice (c) represents a meaningful verification of the predicted violation of Brink's hypothesis.

A closer inspection of Figs. 3-5 reveals that a crucial role in testing various models is played by the integrated TSC intensities for the cascades ending at the negative-parity 1510 keV state.

Under certain simplifying assumptions one can find from Eqs. 1-3 that for $S_\gamma^{E1} \simeq S_\gamma^{M1}$ at energies $E_\gamma \simeq 3$ or 4 MeV the integrated TSC intensities will not depend on the parity π" of the final state. However, for S_γ^{E1} significantly different from S_γ^{M1} the TSC intensities for the cascades ending at states with $\pi" = \pi$ will be enhanced. By this important feature the method of two-step cascades differs substantially from previous methods that are sensitive mostly to the sum of S_γ^{E1} and S_γ^{M1}.

In the majority of cases the deviations between the modeled and measured integrated TSC intensities are much higher than experimental errors and uncertainties associated with Porter-Thomas fluctuations. The observed deviations are statistically meaningful. This demonstrates that the TSC intensities are highly sensitive to the choices of various explicit expressions for photon strength functions.

In conclusion, it is evident that the method of two-step cascades represents a promising tool. Experiments similar to that described in this paper could be repeated for a large number of nuclei with a perspective of a deeper understanding of photon strength distributions.

ACKNOWLEDGEMENTS

Research has been performed in part under contract No. DE-AC02-76CH00016 with the United States Department of Energy.

REFERENCES

1. M. A. Lone in "Neutron Capture Gamma-Ray Spectroscopy", eds. R. E. Chrien and W. R. Kane (Plenum Press, NY and London, 1979), p. 163.
2. D. M. Brink, Oxford University Thesis, 1955.
3. P. Axel, Phys. Rev. **126**, 671 (1962).
4. C. McCullagh, M. Stelts, and R. E. Chrien, Phys. Rev. **C23**, 1394 (1982).
5. S. G. Kadmenskij, V. P. Markushev, and V. I. Furman, Sov. J. Nucl. Phys. **37**, 165 (1982).
6. J. Kopecky and R. E. Chrien, Nucl. Phys. **A468**, 285 (1987).
7. J. Kopecky and M. Uhl., Phys. Rev. **C41**, 1941 (1990).
8. L. Aldea and H. Seyfarth in "Neutron Capture Gamma-Ray Spectroscopy", eds. R. E. Chrien and W. R. Kane (Plenum Press, NY and London, 1979), p. 526.
9. W. Furman, K. Niedwiedziuk, Yu. P. Popov, R. Rumi, V. Salatski, V. Tishin, and P. Winiwarter, Phys. Lett. **B44**, 464 (1973).
10. A. M. Hoogenboom, Nucl. Instr. **3**, 57 (1958).
11. S. K. Kataria, V. S. Rhamaurthy, and S. S. Kapoor, Phys. Rev. **C18**, 549 (1978).
12. A. B. Migdal, Teoriya konechnykh fermi-sistem (Mir, Moscow, 1967), p. 451.
13. S. F. Mughabghab, M. Divadeenam, and N. E. Holden, Neutron Cross Sections (Academic Press, NY, 1981), Vol. 1, Part A.

EXCITATION MODES AND DECAY CHANNELS IN COMPOUND STATES

Yu P. Popov
Laboratory of Neutron Physics, Joint Institute for Nuclear Research
141980, Dubna, USSR

ABSTRACT

Recent experimental data on the formation and decay of states at--and below--the level of neutron binding energy (compound and pre-compound states) give insight into the structural nature of the highly excited states as well as into the processes of fragmentation of simple excitation modes over compound states. The experimental data from α-decay of compound states indicate essential fragmentation of few quasiparticle components as compared with the single particle component. There is an indication on the variation of α-particle strength function on the energetic interval 5-10 keV. Single-particle excitation of compound and pre-compound states play an essential role in γ-decay of neutron resonances of heavy nuclei in the vicinity of the 4S shell.

INTRODUCTION

The study of the structure of nuclear excited states up to an energy of about 20 MeV can identify several intervals which differ from normally occurring excited states. With increasing excitation energy, simple energy states become more complex; they move to the Niels Bohr compound states and then on to the states with prevailing collective modes of excitation, i.e., the various giant resonances. From this procession we have the question: Is it a transition from the "order" of the simple low-lying state, to the "chaos" of a compound state, and even further to the "order" of a giant resonance state?

Currently, the most studied energy level for heavy nuclei is in the interval up to 2 MeV. The various theoretical models, such as the model of interacting bosons [1], the quasiparticle-phonon model[2], and others, originate from this information.

The data of neutron spectroscopy in the vicinity of the neutron binding energy ($B_n \sim 6$-8 MeV) are fairly well described to the first approximation in the frame of the compound nucleus hypothesis, and in the statistical theory that operates on the parameters averaged over a large number of neutron resonances.

The experimentalists, not satisfied with the situation, often undertake wide-ranging and sometimes successful attempts to search for "nonstatistical effects" of various types in the characteristics of neutron resonances (see, for example, Proceedings of the previous Symposia).

NEUTRON RESONANCES AS COMPOUND STATES

According to modern notions of statistical theory, the wave function of a compound state of a heavy nucleus is a random set of various components, each making a contribution of 10^{-6}, i.e. different excitation modes, are fully fragmented over compound states (the black nucleus model). However, in a real nucleus, it is not quite so. In the excitation energy range of the corresponding neutron resonances (6-8 MeV), a single-particle component of the wave function is not negligibly small--neutron strength functions systematically change from nucleus to nucleus, thus forming peculiar giant resonances of size.

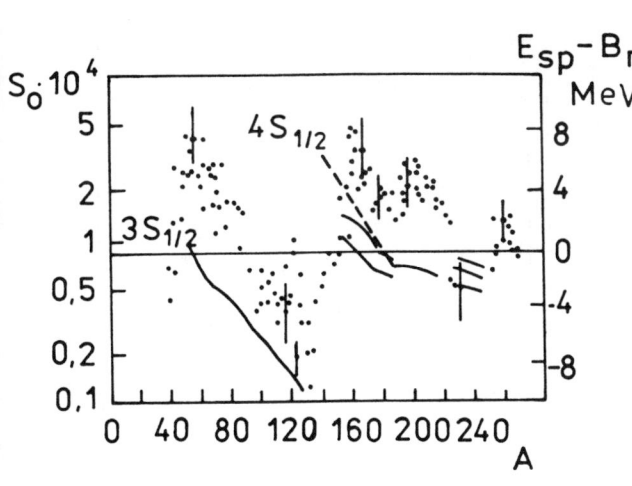

Fig. 1

The variations of neutron strength functions are in the scale of two orders of magnitude here[3]. The maxima and minima are well-correlated with the positions of single-particle shells in the nucleus with respect to the neutron binding energy (see Fig. 1). The nucleus, therefore, appears semitransparent for the neutron channel near the neutron binding energy.

Phenomenologically their interaction with neutrons is described with the optical model.

Information on the more complex components of the wave function of a compound nucleus can be derived from the analysis of α-decay of neutron resonances. But the probability of such α-decay is essentially suppressed by the Coulomb barrier of the nucleus. This results in the fact that in heavy nuclei the α-widths are in the best cases by 6-12 or more orders of magnitude smaller than the radiative widths of resonances. This explains the poor information on the α-particle strength functions of compound nuclei. Fig. 2[4] shows the dependence on atomic weight of nuclei ratios $<\Gamma_\alpha^{exp}>/<\Gamma_\alpha^{cal}>$. (Errors of points are mainly due to the fact that the average was performed over only a small number of investigated resonances.)

Fig.2 illustrates a rather satisfactory agreement between the experimental and theoretical data, which helps validate our description of the average probabilities of α-decay of compound states in the frame of the statistical model ("black" nucleus). Consequently at the neutron binding energy, the two-quasiparticle, four-quasiparticle, and the "two-quasiparticle + phonon" type wave functions (that determine α-decay of the compound state of the daughter nucleus[5]) are more strongly fragmented than in the case of single-particle neutron components (see Fig. 1).

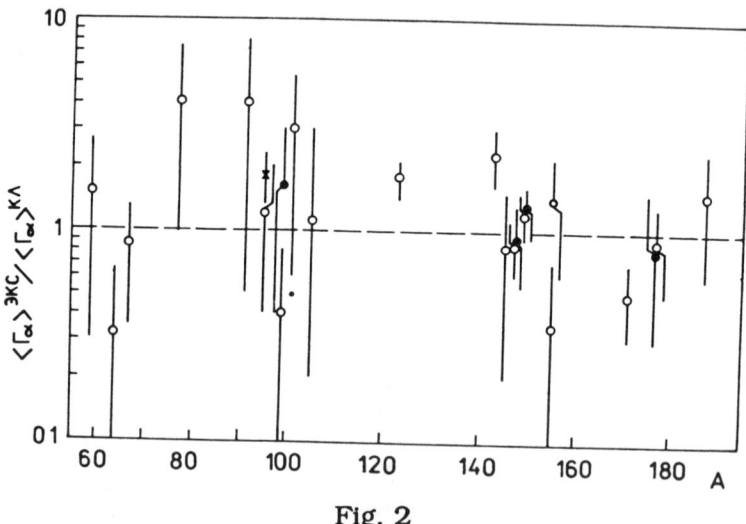

Fig. 2

However, one cannot exclude the possibility that greater accuracy of the data (by averaging over a large number of resonances) will give us an indication on incomplete fragmentation of above mentioned (four quasiparticle) components. Fig. 3 illustrates the possibility of this effect, which shows the dependence on neutron energy of $<\Gamma_\alpha/D>$ values measured for the currently most-investigated nucleus ^{147}Sm. The Γ_α widths are averaged over intervals of 200 eV (each contained 20-30 resonances)[6].

The energy dependence of $<\Gamma_\alpha/D>_J$ shows a rather obvious structure. According to the estimates obtained in the frame of the quasiparticle-phonon model of V.G. Soloviev, the average level distance between 4-quasiparticle $J^\pi=4^-$ states in the ^{147}Sm nucleus is of the order of 2-3 keV. Whether this is correct or not will be revealed in further experiments. The presence of not fully fragmented few-quasiparticle states may lead to an enhancement of α-decay of the whole group of neighbor resonances through a definite channel.

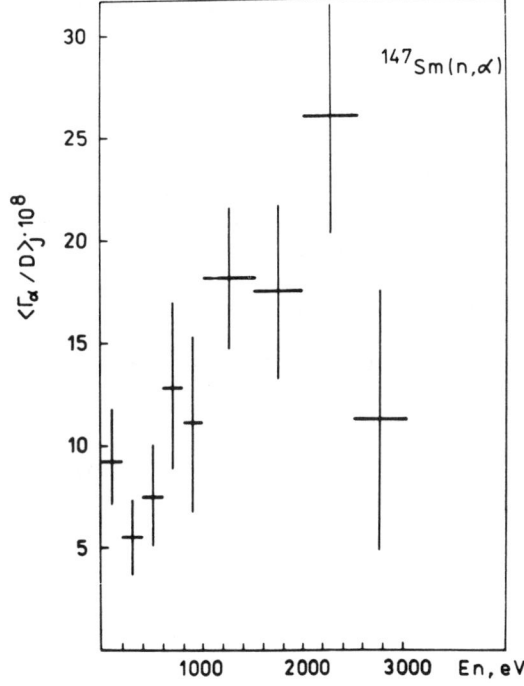

Fig. 3

THE NATURE OF THE ENHANCEMENT OF γ-TRANSITION FOLLOWING DECAY OF 4S-SHELL STATES

Usually the main regularities of the primary γ-transitions between compound states and low-lying levels in heavy nuclei (140<A<190) are successfully described in the frame of the statistical theory. However, a certain doubt concerning the application of the statistical theory arises in the study of fluctuations from resonance to resonance of the population of low-lying levels[7]. The experimentally detected fluctuations appeared to systematically exceed the analogous values calculated within the statistical theory of γ-decay for nuclei from mass region of the 4S maximum of the neutron strength function. This exceeding is not observed in the region of the strength function minimum (see Fig.1 for 90<A<140). It was suggested[7] that in γ-decay of nuclei from the 4S-shell, the observed enhancement of fluctuations are due to the enhancement of certain cascades (γ-decay channels).

This reduces the effective number of intermediate states in the cascades and as a consequence leads to increasing average population fluctuations.

Several years ago we succeeded in detecting such "channeling" of γ-cascades by investigating the compound states decay via measuring two-quanta cascades following neutron capture (the (n,2γ) reaction) with the help of two Ge(Li) detectors operating in the mode of summation over coinciding pulses[8]. This procedure provides from one side, exceptionally pure spectroscopic information on the states of energies up to 4 MeV. For example, for the compound nucleus ^{179}Hf, 239 cascades have been identified, 154 of which are located. The number of intermediate levels amounts to 48. We have studied 67±4% of the total probability of γ-decay of the compound nucleus[9].

From another side, the original data on the mechanism of γ-decay have also been obtained.

The amplitude spectrum of coinciding pulses (after background subtraction) obtained by measuring the population via two-quanta cascades of an excited level at 375 keV in ^{179}Hf is shown in Fig. 4[9] If one compares this spectrum with that expected from the statistical theory (Fig. 5), a number of essential qualitative differences can be outlined:

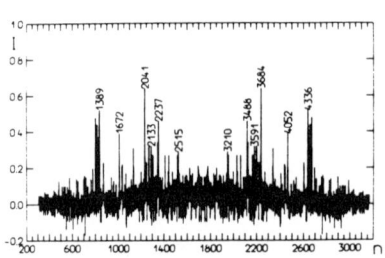 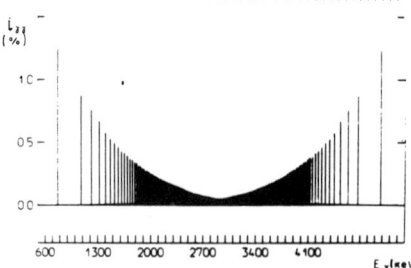

Fig. 4   Fig. 5

— in the experimental spectrum the enhanced gamma-cascades are distributed over the whole spectrum while in the calculated spectrum they concentrate on its edges;
— the high intensity cascades are often grouped,
— the two-quanta cascades from a state at $E_{\gamma 1} \sim E_{\gamma 2} \sim 0.5 B_n$ appear rather probable.

Such "visual" conclusions offer quantitative confirmation from the comparison of experimental values for the total intensity of all the two-quanta cascades populating a definite final state $f - I_{\gamma\gamma}^{exp}$ calculated on the statistical theory value of $I_{\gamma\gamma}^{cal}$. The partial radiation widths of E1 transitions were calculated within the model of giant electric dipole resonance, for the M1 and E2-transitions within the Weisskopf model. The densities of states at energies below B_n are taken in accordance with the Fermi gas model with a Strutinsky correction[10].

Table 1 presents an example of the data on ^{179}Hf. This includes $I_{\gamma\gamma}$ values along with the quantum characteristics of final states and a value of $R = I_{\gamma\gamma}^{exp}/I_{\gamma\gamma}^{cal}$ that qualitatively characterizes deviations from the statistical theory.

Table 1

Compound nucleus, I_0^π	E_f, keV	J_f^π	$K^\pi [Nn_z \Lambda]$	$I_{\gamma\gamma}^{exp}$ x)	$I_{\gamma\gamma}^{theor}$ xx)	R
^{179}Hf	214	7/2⁻	7/2⁻[514]	2.1±0.9	0.03	70±30
	375	1/2⁻	1/2⁻[510]	15.5±1.6	6.3	2.5±0.3
1/2⁺	421	3/2⁻	"	16.5±2.1	6.0	2.8±0.4
	476	5/2⁻	"	7.6±0.6	3.6	2.1±0.2
	518	5/2⁻	5/2⁻[512]	4.0±0.8	3.3	1.2±0.3
	614	1/2⁻	1/2⁻[521]	9.5±1.6	4.7	2.0±0.3
	679	3/2⁻	"	3.8±0.8	4.1	0.9±0.2
	701	5/2⁻	"	2.6±0.5	2.1	1.2±0.2
	721	3/2⁻	3/2⁻[512]	3.1±0.7	3.8	0.8±0.2
	788	5/2⁻	"	2.7±0.8	1.8	1.5±0.4
	Total			67.4±3.7	36.0	1.9±0.1

x) $I_{\gamma\gamma}^{exp}$ in % per decay; $\Delta I_{\gamma\gamma}^{exp}$ includes statistical uncertainties and random scaling errors.

xx) $I_{\gamma\gamma}^{theor}$ in % per decay.

From the analysis of the data summarized in Table 1, and of the analogous data obtained for ^{163,165}Dy, ^{167}Er, ^{175}Yb, ^{179}Hf and ^{183,187}W nuclei, one can conclude that the enhancement of the cascades (R>1) takes place if final states reveal a certain structure, i.e. they have a large single-particle component of the wave function (in particular the cascades populating the band 1/2⁻[510]).

The even-odd compound nuclei investigated lie in the mass region of 4S maximum of the neutron strength wave function. This maximum is explained by the fact that in this mass region near the neutron binding energy single quasiparticle states 1/2⁺[640] and 1/2⁺[651] are located (see Fig. 1). These states formed on neutron capture (being initial in gamma-cascades) may have essential single-particle components.

Unfortunately, until now we could study only the cascades that follow the thermal neutron capture and not those from individual resonances which would allow one to investigate the dependence of the enhancement of the cascades (R>1) on the neutron width for a given nucleus. Therefore, to clarify the role of single particle components of an initial compound state in this process, we must use a relatively reduced width of the resonance which determines the thermal cross section of a given isotope $\Gamma_n^0/\langle\Gamma_n^0\rangle$, or make use of a weighted average Γ_n, if several resonances make comparative contributions into the thermal cross section[4].

Fig. 6 presents the dependence of the value R on the relative reduced neutron width of the initial state of the cascade for the investigation even-odd deformed nuclei. The positions of the experimental points qualitatively confirm the validity of this dependence.

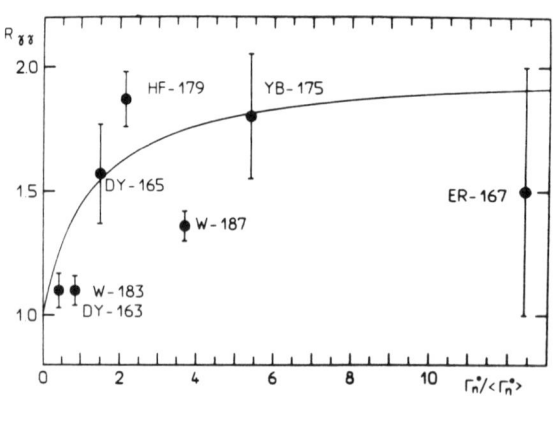

Fig. 6

The quasiparticle-phonon model being developed by a group of V.G. Soloviev[11] predicts correlation between Γ_n^0 and partial γ-widths of K-allowed E1-transitions because both widths depend on the same single-particle components of the wave functions of neutron resonances. However, because of the fact that the reduced radiation width also has other components, correlation between the neutron and radiation reduced widths may appear to be nonlinear (see Fig. 6).

Thus we have demonstrated the connection of the enhancement effect observed for certain two-quanta cascades with single-particle components of initial and final states of the cascade. What intermediate states are selected by enhanced cascades?

Some time ago in order to determine the sequence of γ-quanta in a cascade, we used the assumption that, if a transition of given energy occurs in several cascades populating different final states of the nucleus investigated, then it is a primary transition[12]. This helped to determine the positions of intermediate states in a majority of intensive gamma-cascades, and also to build the dependence of the intensity of gamma-transitions on the energy of primary transition for the main part of gamma cascades.

S.T. Boneva et al.[13] argues in favor of the assumption that the structure of intermediate states excited in most strong cascades for even-odd nuclei with Γ_n^o larger than $<\Gamma_n^o>$ contain a large part of single-particle components. This paper makes an attempt to compare the distribution of excited single-particle components according to theoretical calculations, with the experimentally obtained distributions of primary transition intensities summed over final states of cascades. Theoretical calculations are based on the quasiparticle-phonon model[5] It was supposed that it is possible to neglect more complex components than the single particle in the excitation energy range below 4 MeV.

This analysis indicated that the reason for γ-cascade "channeling" through certain intermediate states is probably due to enhancement of gamma transitions between the neutron shells 4S and 3P for investigated nuclei.

Fig. 7 illustrates the new possibility of two-step cascades method proposed in Dubna[8], namely, investigation of the strength function for soft γ-transitions between two compound states near the neutron binding energy[14]. On the figure the dependence K(E1+M1) versus $E_{\gamma 1}$ for ^{181}Hf compound nucleus is presented. About 20% of the total intensity of primary transitions was not observed for ^{181}Hf. The upper estimate of K(E1+M1) was obtained by the supposition that a "lost intensity" is uniformly distributed over an energy interval of 1.5 MeV (3 points marked by "x"). The upper curve is the dependence according to the Giant Electric Dipole Resonance (GEDR). The lower curve is the same dependence, but which takes into account the temperature- and frequency-dependent GEDR width proposed by Kadmensky et al.[15].

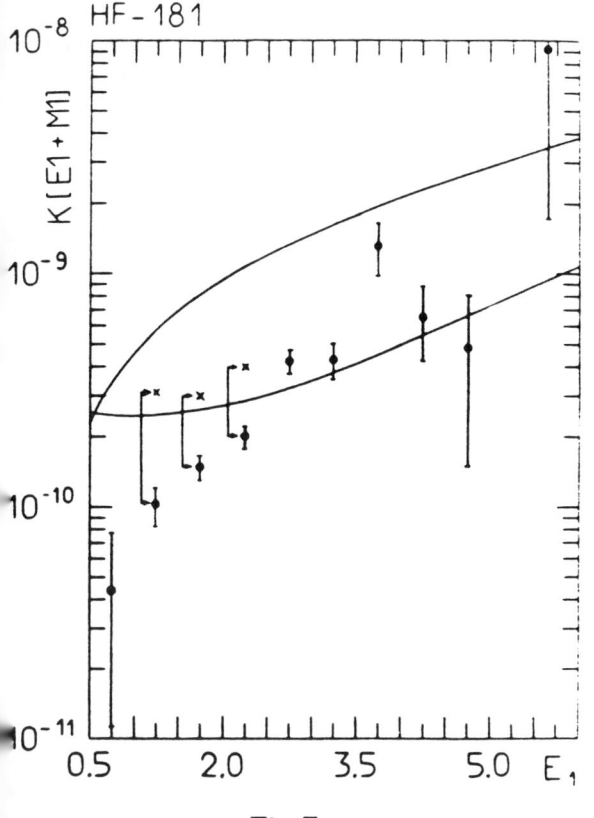

Fig 7

CONCLUSIONS

Reviewing the results of the study of two channels of the compound states decay (i.e. followed by the emission of α-particles or γ-quanta), shows novel possibilities of obtaining unique data on the structure of highly excited states of nuclei. Studies of a proton channel of decay have been initiated (see, for example, Gledenov[16] and Bowman[17]). Further progress in the study of decay through rare reactions (n,p) and (n,α) will apparently be connected with the development of methods of high-intensity neutron spectroscopy and with the application of neutron-deficient nuclei as targets. The latter is especially favorable in the study of the proton channel of decay, since in stable nuclei the proton binding energy is close to that of the neutron.

The application of the two-step gamma-quanta cascades method to the analysis of γ-decay of compound nuclei turns out to be very successful.

- Up to now a small part of compound nuclei were investigated by this method.
- An essential enhancement of experimental average intensity of E2 gamma-transitions (by 1-2 order of magnitude) was observed[9] by comparing this with calculated intensities from statistical theory.
- The new possibility has appeared to investigate soft gamma-transitions between compound states near the neutron binding energy.
- Combination of (n,γ) and (n,2γ) data give a real possibility to build a level scheme in heavy nuclei up to ~4 Mev[18].

REFERENCES

[1] A.Arima, F. Iachello, Ann. Rev. Nucl. Part. Sci., *31*, p. 75, (1981).

[2] V.G. Soloviev, "Physics of Nuclei and Particles", *9*, p. 980, (1978).

[3] S.F. Mughanghab et al. "Neutron Cross Sections: *1a*, Academic Press, N.Y., (1981); *1b*, N.Y., (1984).

[4] Yu.P. Popov, Nuclear Energy (Sofia), *25*, p. 29, (1987).

[5] V.G. Soloviev, Sov. J. Nucl. Phys., *13*, p. 48, (1971).

[6] A. Antonov, Yu. M. Gledenov, S. Marinova et al., Sov. J. Nucl Phys., *39*, p. 794, (1984).

[7] V.A. Khitrov, Yu.P. Popov, A.M. Sukhovoj, Yu.S. Yazvitsky, In: Neutron Capture Gamma-Ray Spectroscopy, R.E. Chrien, W.R. Kane (eds.) p. 655, N.Y., Plenum Press, (1979).

[8] Yu.P. Popov, A.M. Sukhovoj, V.A. Khitrov, Yu.S. Yazvitsky, Sov. J. Isv. of Acad. of Sci. of the USSR (Physics), *48*, No.5 p. 891, (1984).

[9] S.T. Boneva, V.A. Khitrov, Yu.P. Popov et al., Z. Phys. A., *330*, p. 153, (1988).

(10) A.V. Ignatiuk et al., Sov. J. Nucl. Phys, 21, p. 485, (1985).
(11) V.G. Soloviev, "Physics of Nuclei and Particles:, 3, p. 770, (1972).
(12) Yu.P. Popov, A.M. Sukhovoj, V.A. Khitrov, Yu.S Yasvitsky, Sov. J. Isv. of the Acad. of Sci. of the USSR (Physics), 49, p. 91, (1985).
(13) S.T. Boneva, E.V. Vasilieva, L.A. Malov et al., Sov. J. Nucl. Phys., 49, p. 944, (1989).
(14) V.A. Bondarenko, I.L. Kuvaga, P.T. Prokofjev et al. (A contribution to these Proceedings.)
(15) S.G. Kadmensky et al., Sov. J. Nucl. Phys, 37, p. 277, (1983).
(16) Yu.M. Gedenov et al., Z. Phys. A 308, p. 57, (1982).
(17) C.D. Bowman, In: Capture Gamma-Ray Spectroscopy, K. Abrahams and P. van Assche (eds), p. 399, IOP Publishing Ltd, Bristol, (1987).
(18) S.T. Boneva, V.A. Khitrov, Yu.V. Kholnov et al. (A contribution to these Proceedings.)

NEUTRON CAPTURE IN PERSPECTIVE

W Gelletly
SERC, Daresbury Laboratory, Daresbury, Warrington, WA4 4AD, UK

ABSTRACT

The status of Nuclear Physics is briefly reviewed, and the blanks in our knowledge are highlighted. The nature of the (n,γ) reaction and its contribution to our knowledge of nuclear properties is discussed and a recent study of the ^{188}Os (n,γ)^{189}Os reaction is cited as an example. A comparison is then made with the properties of the more fashionable, heavy ion induced, fusion-evaporation reaction which provides information about states of high spin and neutron-deficient nuclei far from stability. The recent observation of the N=Z nucleus ^{84}Mo is used as an example. Future developments in the use of the (n,γ) and (HI,xnγ) reactions and in low temperature Nuclear Physics generally are outlined.

INTRODUCTION

A complete description of atomic nuclei remains well beyond our present capabilities. Even our empirical knowledge of the properties of nuclei is tightly circumscribed by the experimental tools we have available: in consequence the predictive power of present nuclear models is extremely limited.

Fig 1 shows a version of the Segré chart. The 263 stable nuclides, the proton and neutron drip lines and those unstable nuclei which have been observed in experiment are all marked. The information we do have about nuclei is derived mainly from nuclear reactions. These reactions are diverse in character but share one characteristic, namely that they are highly specific in the nuclear states which they populate or excite, and highly specific in the properties they allow us to measure. The result is a patchwork of information stitched together from these many sources.

One can readily estimate from fig 1 that some 7000 nuclei with Z<92 lie between the drip lines. The first and most obvious gap in our knowledge is that we know something, however little, about only ~2400 of them. We know most about nuclei on or near the line of stability, but even here our knowledge is largely confined to states of low or moderate spin and low excitation energy. In neutron deficient nuclei we are largely indebted to studies of β-decay and prompt radiation from (HI,xnγ) reactions for our information. Both are highly selective: the selection rules in β-decay mean that it populates states lying close in spin to the parent state, and the (HI,xnγ) reaction predominantly populates high spin states near the yrast line. Perusal of fig 1 suggests that we have not been so successful in advancing towards the neutron drip line as we have on the neutron deficient side of the line of stability. Except in light nuclei our knowledge was again confined to the results of ß-decay studies until recently, when studies[1] of prompt γ-rays from nascent fission fragments have started to provide us with data on states of spin up to ≈14h in neutron rich nuclei.

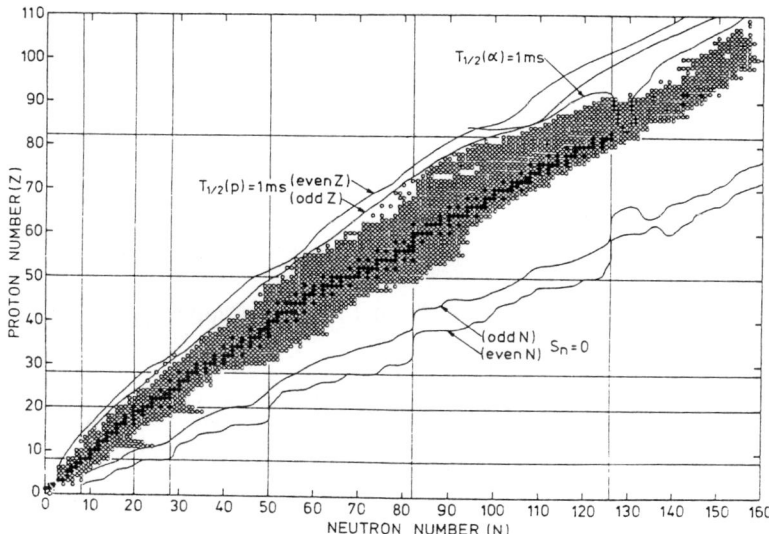

Fig 1 A version of the Segre Chart (see text)

The central theme of the present paper is that our cognizance of nuclear properties rests on this wide range of reaction studies, each reaction being limited in the information it can supply. Progress in understanding nuclei will only come from improvements in instrumentation and radical changes in the beams of nuclear projectiles available to initiate reactions. This will allow us to extend our knowledge over a wider range of Z, N, E_{exc} and rotational frequency (ω): nuclear models with a much sounder empirical base can then be constructed. To illustrate this theme we will compare and contrast studies of two nuclear reactions: the slow neutron capture reaction and the heavy ion induced fusion-evaporation reaction. Recent[2,3] studies of the ^{188}Os (n,γ) ^{189}Os and ^{58}Ni (^{28}Si, 2n)^{84}Mo reactions will be used as exemplars.

2. The Slow Neutron Capture Reaction

2.1 The (n,γ) Reaction as a Spectroscopic Tool

Slow neutron capture is the archetypal compound nucleus reaction.[4] Fig 2 provides a simple picture of how the reaction proceeds. Initially we have a stable target nucleus with an

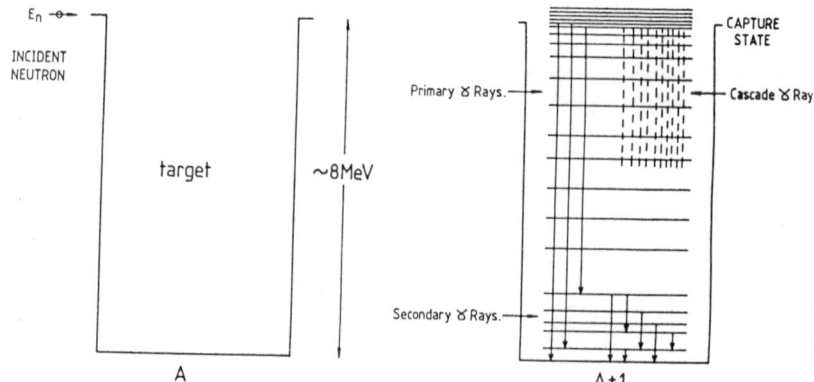

Fig 2 A simple picture of thermal Neutron Capture (see text)

incident neutron of energy $E_n \approx 1$ eV. Once captured the neutron rapidly shares its energy with other nucleons in a series of collisions and all memory of the entrance channel is lost; except for the general memory of conservation of energy and angular momentum. Typically the capture state lies in the energy range 5-10 MeV above the ground state. There are many levels below the capture state and hence many possible decay paths. The decays which are preferred are dictated by statistical considerations (the random overlaps between the complex capture state and lower states), the relative transition rates for different electromagnetic multipoles and the nuclear level density. Following these initial, primary decays from the capture state, a shower of secondary γ-rays is emitted as the levels populated decay, directly or indirectly, to the ground state.

The advantages and disadvantages of the (n,γ) reaction as a nuclear probe follow from this simple picture, and the fact that the source of neutrons used is usually a nuclear reactor. On the plus side the observation of a primary transition signals the existence of a level and defines its excitation energy. At the same time non-observance of a primary transition in thermal capture reveals nothing. More significant information can be obtained from studies (5,6) of average resonance capture (ARC). Here primary transitions are observed following capture of reactor neutrons filtered by scandium or iron. The filtered neutrons have energies of 1.95 ± 0.9 keV and 24.3 ± 2.0 keV respectively. If a sufficient number of resonances lie in this energy interval the reduced intensities of the primary transitions depend only on the transition multipolarity and the spin and parity of the level. Given sufficient sensitivity in the measurement this guarantees the observation of primary transitions to all levels within a certain range of spin. The statistical nature of the decay process also means that the reaction is generally non-selective, in the limited sense that it does not preferentially populate states of any given character. The use of reactor neutrons is also highly advantageous. Not only is the reactor an extremely stable source, but one can make use of the high flux ($\approx 10^{15}$ n cm^{-2}s^{-1}) near the core in some reactors. These two factors allow the use of instruments such as the curved crystal spectrometers[7] and beta spectrometer [8] at ILL, which have high sensitivity and resolution even although their overall efficiency is low.

Naturally the disadvantages of the use of the (n,γ) reaction are close kin to the advantages. Firstly the reaction only populates states of spin lying within a few units of angular momentum of the capture state. As observed above, failure to observe a state carries no information. The most serious limitation, however, is that it can only be applied to targets of stable or long-lived nuclei; hence its application is limited to nuclei on or near the line of stability. The low recoil velocity in neutron capture has meant few problems with Doppler shifts. At the same time, until the GRID technique (9) was developed at ILL, lifetime measurements with the (n,γ) reaction were restricted to states with half-lives measurable by direct timing.

Overall the picture which emerges is of a reaction well described by the Compound Nucleus model, which can be used as a spectroscopic tool with the specific limitations noted above.

2.2 The ^{188}Os (n,γ) ^{189}Os Reactions

Some of the virtues and defects of the (n,γ) reaction as a spectroscopic tool are revealed in a recent study[2] of the ^{188}Os(n,γ)^{189}Os reaction. It combines ARC measurements with filtered neutron beams[6] of 2 and 24 keV energy, γ-ray singles, coincidence and angular correlation measurements with a thermal beam at the Brookhaven High Flux Beam Reactor, and measurements of selected parts of the (n,e⁻) spectrum with the β-spectrometer BILL [8] at ILL.

The combined results of all the measurements lead to the partial level scheme shown in Fig 3. In this case the ARC measurements "guarantee" that all 1/2⁻ and 3/2⁻ levels in ^{189}Os have been observed, up to an empirical cut-off energy of ~ 1500 keV. New levels are proposed at 97.4 and 444.2 keV. The former decays uniquely to the 9/2[505] bandhead at 30 keV and represents a candidate for the 11/2[615] Nilsson orbit, which is expected to be at low excitation energy in this region. This assignment, coupled with the rotational parameter derived from the low-lying negative parity bands, is consistent with the suggestion from studies[10] of single particle transfer that the $J^\pi = 13/2^+$ member of the band lies at ~ 290 keV.

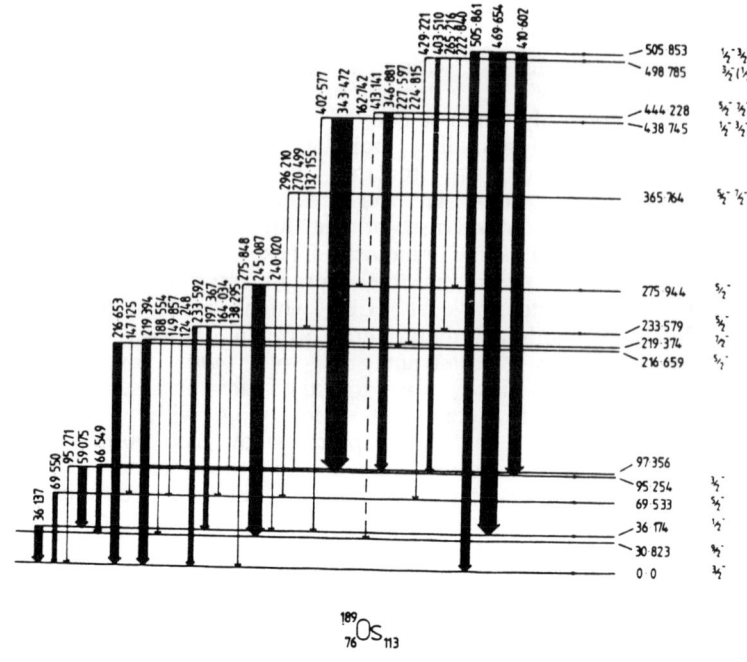

Fig 3 A partial level scheme for ^{189}Os

Fig 4 shows a portion of the (n,e⁻) spectrum with the L conversion lines of the 216 and 219 keV transitions to ground clearly visible. The levels at these energies were previously assigned[10] J^π = 7/2⁻. It is evident that the two transitions have different multipolarities. Comparison with theoretical[11] conversion coefficient ratios reveals that the data are consistent with pure E2 multipolarity for the 219 keV transition and with the 216 keV transition being a mixed M1-E2 transition with $4.1 < \delta^2 < 10.7$. Since the 216 keV level was not observed in the ARC measurements it cannot have J^π= 1/2⁻ or 3/2⁻. Hence it must have a spin and parity of 5/2⁻. This is consistent with the results of the angular correlation measurements.

This result creates a dilemma. On the one hand the transfer reaction studies[10] show a large l=3 strength to a level, or unresolved levels, near 216-219 keV excitation. This would suggest an excitation with a significant single particle component. If this strength is to the 216 keV level then it can be assigned to the 5/2[512] or 5/2[503] Nilsson states, but the former should not be strongly populated in single particle

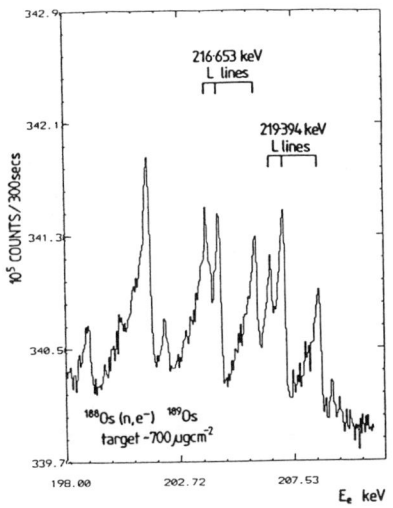

Fig 4 Part of the ^{188}Os (n,e⁻)^{189}Os Spectrum

transfer and the latter is a predominantly 'particle' state which should not be strongly populated in the (d,t) reaction. This suggests that the observed single particle strength belongs to the 219 keV level, which may be assigned to the 7/2[503] Nilsson state.

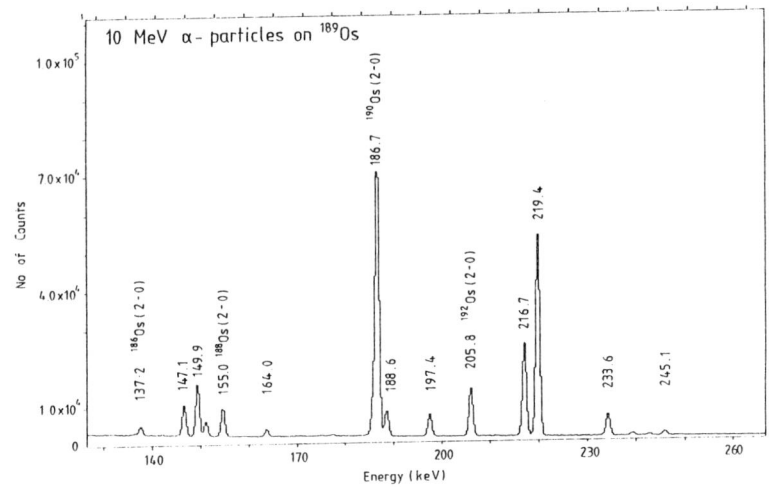

Fig 5 Part of the γ-ray spectrum from ^{189}Os bombarded by 10 MeV α particles

The information from the (n,γ) reaction cannot help us further and we must have recourse to measurements of the γ-ray spectrum resulting from the Coulomb excitation of ^{189}Os with 10 MeV alpha particles. Fig 5 shows part of such a spectrum recorded[12] at the University of Köln. The analysis is incomplete but it can be seen that both the 216 and 219 keV levels are strongly excited and the results suggest that they are both due to collective excitations.

One model which automatically includes a more complete core description is the Interacting Boson-Fermion model[13]. In the N=82-126 shell the neutrons occupy the $2p_{1/2}$, $2p_{3/2}$ and $1f_{5/2}$ shell model states. The underlying group structure is then $U^B(6) \times U^F(12)$ and symmetry schemes[14] have been deduced corresponding to SU(5), SU(3) and O(6) symmetries for the core. The ^{189}Os nucleus is thought to lie between the O(6) and SU(3) limits and an attempt[15] has been made to calculate its properties in terms of the consistent-Q formalism in odd-A nuclei (CQFOA), in which the transition between SU(3) and O(6) is characterised by one parameter χ and the boson number N.

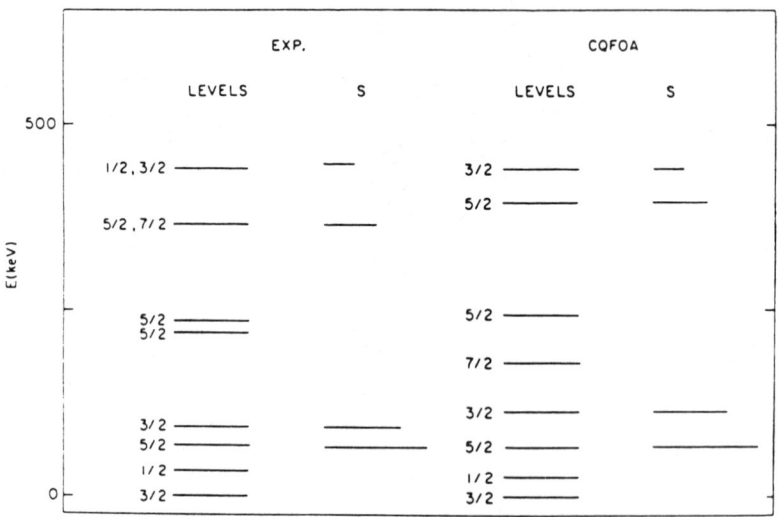

Fig 6 A comparison of part of the ^{189}Os level scheme with the results of a calculation with the CQFOA Model

Fig 6 shows a comparison of the low-lying levels in ^{189}Os with the results of a calculation with the CQFOA with the parameter $\chi = -1.5$. This value of χ was deduced from the ratio of measured (d,t) structure factors for states of the same spin. The 219 keV level has been omitted from the figure since, if it is the 7/2[503] level, it involves the $h_{9/2}$ shell model state and hence lies outside the U(6/12) basis. The 97 and 444 keV positive parity levels, the 30 keV, 9/2⁻ state, and the 5/2 state at 275 keV, which is thought to be due to a gamma vibration built on the 30 keV state have also been omitted from the comparison. The main difficulty in interpretation clearly lies in the levels near the 216-219 keV doublet. A full calculation with the IBFM code is now underway. Together with the full analysis of the Coulomb excitation data this may shed some light on the problem.

Most of the key features of (n,γ) reaction studies are revealed in this study. The ARC measurements ensure all 1/2⁻, 3/2⁻ states have been seen and the quantities one can measure give no information on the structure of the states. The stability and intensity of the reactor neutron source allowed the use of the BILL spectrometer to determine key J^π values in the level scheme. On its own the (n,γ) reaction is limited in the information it provides but combined with studies of transfer reactions and Coulomb excitation it can lead to a fuller interpretation.

Fusion Evaporation Reactions

3.1 Heavy Ion Induced Fusion Reactions

In heavy ion collisions above the Coulomb barrier many different types of reaction may occur, but close to the barrier fusion - evaporation reactions predominate. Fig 7 shows a simple picture of how a typical fusion reaction, such as the ^{92}Mo (^{40}Ca, α2p) ^{126}Ce reaction, is thought to proceed.

If the two nuclei fuse they form a highly excited (hot) compound nucleus rotating at high frequency (typically 10^{20} Hz). In simple terms the compound nucleus resembles a hot, charged, rotating liquid drop. It is highly unstable and decays rapidly with the evaporation of particles: neutrons, protons, alpha particles etc.

Neutron emission is favoured near stability but, as the relative binding energies of charged particles and neutrons change, we get increasing proton and alpha particle emission as we move

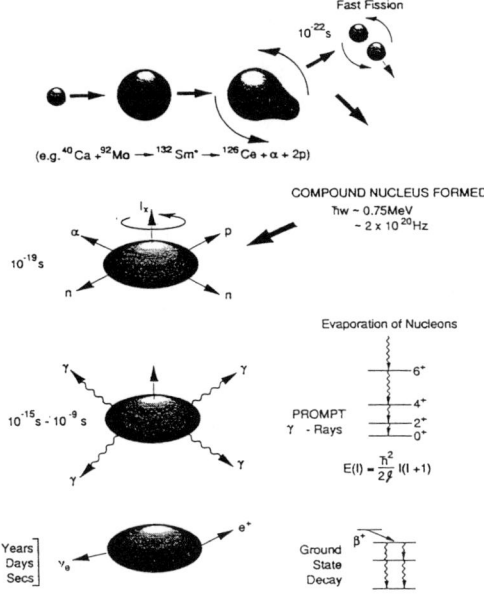

Fig 7 A Simple Picture of a fusion-evaporation Reaction

away from stability. Because of the centrifugal barrier they face the particles carry away energy but little angular momentum. Particle evaporation continues until the excitation energy lies approximately one neutron separation energy above the yrast line. We are then left with a cold nucleus rotating at high frequency. The large amount of angular momentum which remains is carried away in the emission of a cascade of 30-40 gamma rays. Initially the decays are statistical but they rapidly feed rotational bands close to the yrast line and the decays then proceed on or parallel to the yrast line, finally funnelling down into the ground state band in the case of an even-even nucleus. The decay of the ground state is governed by the Weak Interaction and so it will typically have a half-life of seconds, hours, days or years.

A number of features are important in terms of exploiting such reactions in spectroscopy. Firstly the prompt γ-rays carry information about cold nuclei at high spin. Secondly the beta decay of the ground state will provide complementary information since it will normally feed states of relatively low spin. This type of reaction is flexible in that there are many possible target - projectile combinations and we have control of the excitation energy. At the same time with the experimental sensitivity we have now it allows us to study a very limited class of states.

There are three main spectroscopic uses of these reactions. Firstly, it has proved to be the main tool for studying nuclei at high spin. Secondly, it allows us to produce and study nuclei far from stability by detecting the prompt radiation from the reaction. Thirdly we can study the subsequent β - decay of the short-lived nuclei produced.

A number of features of Fig 7 are important if the prompt radiation is to be exploited. Firstly the high multiplicity of the γ-ray cascade means that we must have γ-ray detection systems with both high efficiency and high granularity. Secondly if the compound nucleus is near stability it is mostly neutrons which are emitted. As we move away from stability, and the relative binding energies of the various kinds of particle change, the emission of protons and alphas increases rapidly. Similarly, as we increase the excitation energy the probability of charged particle emission increases. In both cases there is a large increase in the number of open channels, and it becomes imperative to have an efficient means of detecting γ-rays from a specific channel of interest. In particular to reach the nuclei farthest from stability we must be able to detect those nuclei formed with neutron emission only. The sensitivity one can achieve in detecting γ-rays in rotational cascades and in nuclei far from stability is ultimately limited by peak to total ratios in the γ-ray detectors, the detector efficiencies, the efficiency of channel selection etc. Doppler effects also limit the sensitivity in that they smear out the resolution or, if the detectors are collimated, reduce the efficiency. At the same time they allow us to measure lifetimes with the DSAM or Recoil Distance techniques.

3.2 The Observation of ^{84}Mo

The recent observation[3] of γ-rays from the N=Z nucleus ^{84}Mo illustrates how fusion-evaporation reactions may be used to study nuclei far from stability. This nucleus was produced in the ^{28}Si(^{58}Ni, 2n)^{84}Mo reaction with a cross-section close to the limit of sensitivity of our present experimental techniques.

The nuclei in the A~80 region with N=Z have a relatively low, single particle level density and there are marked single particle energy gaps at both oblate and prolate deformation. As a result the addition or removal of 1 or 2 nucleons, or changes in rotational frequency, manifest themselves in quite dramatic changes in shape. Many nuclei in the region exhibit shape coexistence. They are the heaviest nuclei we can reach with the neutrons and protons filling the same orbits: the $1g_{9/2}$, $2p_{1/2}$, $1f_{5/2}$ and $2p_{3/2}$ orbits. One may hope that the effects of the n-p interaction are exposed to observation here.

The experiment described here is one of a series carried out at Daresbury Laboratory, aimed at observing the even-even, N=Z nuclei in the A=80 region. The prompt γ-rays emitted in the ^{28}Si(^{58}Ni, 2n)^{84}Mo reaction were detected in an array of twenty, escape-suppressed Ge detectors arranged in four rings at 143°, 117°, 101° and 79° to the beam direction. They were assigned to particular nuclides by detecting them in coincidence with the evaporation residues, whose A and Z were measured with the Daresbury Recoil Separator[16]. The design of the Recoil

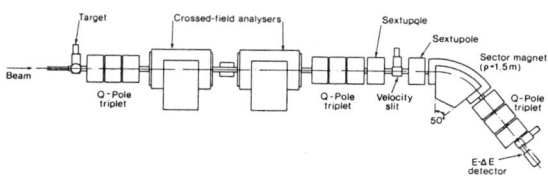

Fig 8 Schematic representation of the Daresbury Recoil Separator

separator is shown schematically in fig 8. It consists principally of two Wien filters, the first to separate the nuclear recoils from the beam particles and the second to disperse the recoils in velocity. A slit selects recoils with a ±3% range in velocity prior to momentum dispersion with a large dipole magnet. A C foil-microchannel plate detector in the focal plane gives an A/q spectrum and any mass ambiguity is resolved if the energy E is measured to a precision of 3% or less in an Ionization chamber. The ΔE signals from the first two anodes in the chamber allow a measurement of Z.

The experimental strategy adopted was common to all the experiments. The beam energy was chosen to be as close to the Coulomb Barrier as possible. This produces the nuclei of interest in pure neutron evaporation, which keeps the recoil cone small and optimises the transport efficiency. It also keeps the number of open channels small. In addition the nucleus sought has the highest Z, the highest ΔE signal in the ionization chamber, and hence it is easier to separate it in Z. The reaction used is an inverse reaction, a heavy projectile on a light target, which optimises the kinematic focusing and gives the best transmission. It also gives the highest recoil velocity, which improves the Z separation.

Fig 9 shows the intensities of three γ-rays from A=84 nuclei, produced in the ^{28}Si + ^{58}Ni reaction at 195 MeV bombarding energy, plotted as a function of ΔE. It is clear that the peaks of the intensity distributions are progressively shifted to higher ΔE. The most intense line is the 2^+-0^+ transition in ^{84}Zr and the others are assigned to ^{84}Nb and ^{84}Mo. Gamma ray spectra which are enhanced in each of these three nuclei can be created by selecting slices of this ΔE Spectrum. Careful subtraction of the different energy loss cuts allows one to create the 'pure' spectra, associated with these three nuclei, which are shown in fig 10. The lowest part of the

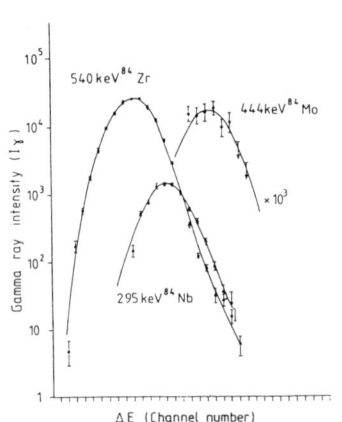

Fig 9 Iγ vs ΔE for γ-rays from the ^{28}Si+ ^{58}Ni reaction

figure shows the spectrum of ^{84}Mo γ-rays. A single γ-ray of energy 443.8 ± 0.3 keV is clearly seen, which is assigned as the 2^+-0^+ transition in ^{84}Mo on the basis of the systematic observation of γ-rays in even- even nuclei formed in fusion-evaporation reactions and the measured γ-ray angular distribution. No other γ -ray could be attributed to ^{84}Mo, which is consistent with the I(4-2)/I(2-0) intensity ratios observed in earlier[17,18] studies of ^{80}Zr and ^{76}Sr.

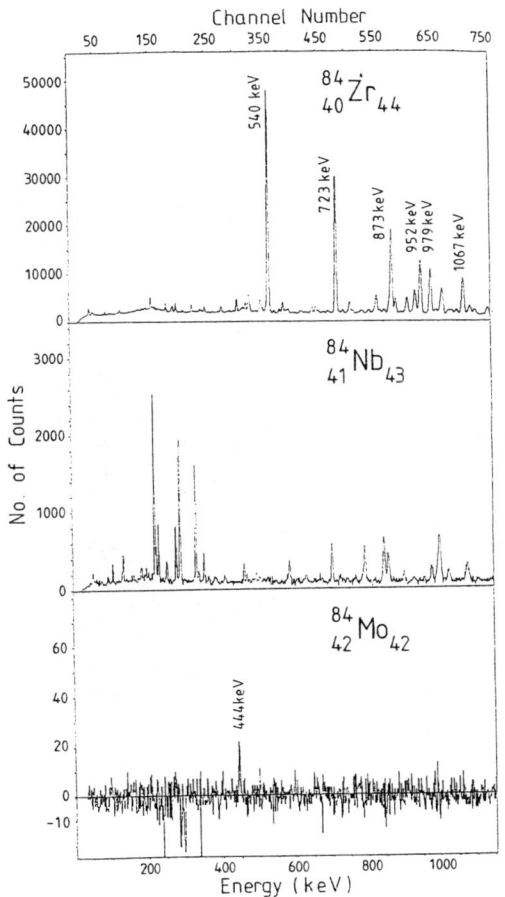

Fig 11 summarises the results of the series of experiments on N=Z, even-even nuclei. In addition to the 2+ level energies it shows the value of the quadrupole deformation,

$$\varepsilon_2 = \left[\frac{1228}{A^{7/3}E(2^+)}\right]^{1/2}$$

derived from Grodzins' formula[19] and the theoretical value derived from the potential energy surface calculations of Moller and Nix[20]. The reliability of an estimate based on Grodzins' formula can be questioned, especially in the Se and Kr isotopes where one might expect competing shapes at low excitation energy, but it indicates that the theory does quite well in predicting the

Fig 10 Pure γ-ray spectra from ^{84}Zr, ^{84}Nb and ^{84}Mo respectively

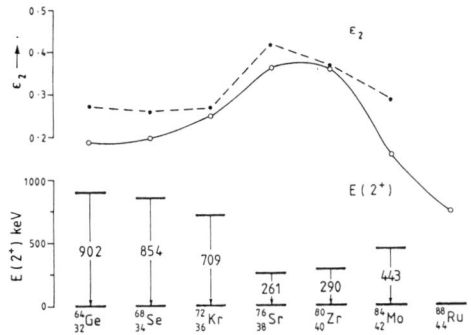

Fig 11 E(2+), ε_2 derived from Grodzins' formula (solid points) and ε_2 from Moller & Nix (open points) for even-even, N=Z nuclei

centre of this region of deformation but is poorer as we move away towards the doubly closed, spherical shell at N=Z=50.

The cross-section for the production of the 540 keV gamma ray has been measured by Lister et al[21]. If we assume that the intensities of the 540- and 444- keV transitions reflect the total decay strength to the ground states of ^{84}Zr and ^{84}Mo, and that the sum of the intensities of the 228-, 295- and 586- keV transitions represents the strength of the population of the ^{84}Nb ground state, then, after correction for the γ-ray detection efficiencies and the variation in the recoil separator transport efficiency, the observed values of Iγ summed over ΔE yield $\sigma(^{84}\text{Zr})$: $\sigma(^{84}\text{Nb})$: $\sigma(^{84}\text{Mo})$ = 35(3)mb : 1.3(3)mb : 7(3)μb. The value of $\sigma(^{84}\text{Zr})$ is taken from ref 21. This may be compared with the corresponding values of 44(1)mb : 2(1)mb : 10(5)μb determined[17] for ^{80}Sr : ^{80}Y : ^{80}Zr.

The results shown in fig 10 suggest that we are close to the present limit of observation with this technique. This particular measurement is an extreme example of the use of the fusion-evaporation reaction. It shows that it does allow us to reach nuclei far from stability, and that we have a flexibility derived from the control of the beam energy and in some cases a variety of target-projectile combinations. In the case of nuclei nearer stability observed in the same reactions[22] it allows the study of high spin states near the yrast line. At the same time its use is limited to the study of high spin states on the neutron-deficient side of stability. The main interest in its use for studies such as that described is that the results provide good tests of the predictions of models based on the properties of nuclei near stability. Unlike the case of the (n,γ) reaction it has not so far been found possible to study other reactions leading to the same states because suitable radioactive beams or targets have not been available.

4. **Experimental Sensitivities**

There is no simple way of comparing the sensitivities of the (n,γ) and (HI, xnγ) reactions, and it is not clear how meaningful such a comparison would be. A sensitivity of 1mb partial cross-section is quoted for the detection of γ-rays of energy ~200 keV with the curved crystal spectrometers[7] at ILL. Since there is a very wide variation in the cross-section for thermal neutron capture - the cross section depends on whether or not a resonance lies near the neutron separation energy - the dynamic range in (n,γ) reaction studies also varies widely. If we choose a case with a large cross-section such as the ^{167}Er (n,γ)^{168}Er reaction, where σ=650b, one can detect γ-rays with intensities ranging from ~0.5 to ~1.4 x 10^{-6} per neutron capture. This huge dynamic range plays a major role in allowing one to look at the details of the decay of levels only weakly excited in the (n,γ) reaction.

The current generation of escape-suppressed γ-ray arrays used at Daresbury allows one to detect γ-rays in a cascade associated with a fraction 10^{-4} of the main reaction channel or 10^{-2} of a weak channel. i.e. a partial cross-section of ~ 30 μb. Since the total cross-section is approximately constant, independent of the fusion reaction studied, the dynamic range is ~2-3×10^{-5}. Used in conjunction with the recoil separator the measured ^{84}Mo cross-section was 7μb as reported in section 3.2. Since the cross-section for a strong open channel in this reaction is ~ 300mb the dynamic range is again ~2×10^{-5}.

Future Measurements

Future progress in understanding nuclear properties must depend on breaking the constraints on our knowledge of nuclear properties, set by the range of Z,N, E_{exc} and ω which is accessible.

In the case of the (n,γ) reaction successive conferences in the series, the latest of which is reported in the present volume, have seen steady improvements in technique. There has been little advance in the neutron fluxes and the quality of neutron beams in the last few years, and only marginal improvements can be anticipated. Further improvements in the sensitivity of γ-ray detection have been proposed[23] in the form of a double curved crystal spectrometer. These and other ideas will lower the level of sensitivity and allow the study of the electromagnetic decay of states at higher excitation energy in more detail. Many modes of decay such as two phonon excitations or mixed symmetry states may then be exposed more readily to study.

In the case of (HI,xnγ) reactions the route to improvement is clear in the short to medium term. The Daresbury Recoil Separator presently uses a single charge state but a number of charge states are in focus simultaneously. The position sensitive detector and ionization chamber are now being modified to allow coincidences between γ-rays and recoiling ions in any of three separate charge states. This will increase the efficiency by a factor of 2.5. A new γ-ray array (EUROGAM) is under construction by a joint UK-French collaboration. This will increase the efficiency of detection of recoil-gamma coincidences by a factor of 50. Thus an improvement by a factor of ~125 overall can be expected. The cross-section limit based on the above measurement for ^{84}Mo should approach 50-100 nb.

The most important advance on the horizon is, however, the likely provision of beams of radioactive species with which one can initiate reactions. Such beams may be produced in fragmentation or in spallation. The former is already in use[24], in combination with a heavy ion storage ring, at GSI. In principle it is also possible to envisage the use of spallation and fission induced by high energy protons to produce short-lived nuclei, followed by on-line mass separation and re-acceleration of the ions to the Coulomb barrier. This would produce beams of unstable species over a wide range of Z and N which could be used to study nuclear properties. In the words of the comic book we would be able to say of nuclear physics "With one bound he was free," if we could have such beams now. It will take some time before such a facility is built somewhere in the World, as it surely will be.

References

1. W R Phillips, Proceedings of this Conference
2. A M Bruce, (1985)Ph.D Thesis (University of Manchester)
3. W Gelletly et al, Phys Letters B in press
4. N Bohr (1936) Nature 137, 344
5. L M Bollinger, (1968) Proc. Conf. Slow Neutron Capture Gamma Ray Spectroscopy ANL-7282, 523
6. R E Chrien (1981) Proc of Neutron Capture Gamma Ray Spectroscopy and Related Topics (IOP Conf. Series 62) 342
7. H R Koch et al (1980), Nucl. Inst and Methods 175, 401
8. W Mampe et al (1978) Nucl. Instr and Methods 154, 127
9. See e.g. H G Borner et al, Contemporary Topics in Nuclear Structure Physics (World Scientific, Singapore, 1988) p 27
10. D Benson et al, Phys Rev C14 (1976) 2095
11. F Rosel et al., At. and Nucl Data Tables 21 (1978) 291
12. A M Bruce et al., to be published
13. F Iachello and O Scholten, Phys Rev. Letters 43 (1979)679
14. A B Balantekin et al., (1983) Phys Rev C27 1764
15. A M Bruce et al in Symmetries in Nuclear Structure (eds R A Meyer and V Paar, Harwood Academic Publishers, London, 1986) p186
16. A N James et al., Nuclr Instr and Methods A267 (1988) 144
17. C J Lister et al., Phys Rev Letters 59 (1987) 1270
18. C J Lister et al Phys Rev C , in press
19. L Grodzins, Phys Letts 2 (1962)88; F S Stephens et al., Phys Rev. Letters 29 (1972) 438
20. P Moller and J R Nix, Nuclear Physics A361 (1981) 117
21. C J Lister et al, Z Physik A329 (1988) 413
22. C J Gross et al., to be published
23. H G Borner, private communication
24. F Bosch, Proc Int Conf Nucl Phys. Harrogate (Institute of Physics Conf Series No 86, 1986) p 537

NON-YRAST SPECTROSCOPY OF TELLURIUM NUCLEI

J.A.Cizewski, R.G.Henry, and C.S.Lee
Department of Physics and Astronomy, Rutgers University
New Brunswick, New Jersey 08903 USA

ABSTRACT

We have studied $^{118-126}$Te with Sn$(\alpha,xn\gamma)$ reactions and have extended the knowledge of these nuclei up to moderate angular momenta. In the ^{122,124,126}Te isotopes we have searched for intraband transitions in the previously proposed 4p-2h intruder band. For ^{119,121,123}Te we have found additional negative-parity states. The structure of these excitations built on an $h_{11/2}$ neutron is examined in terms of the interacting boson-fermion model.

INTRODUCTION

The tellurium nuclei with two protons outside of the Z=50 shell closure exhibit a complicated structure with signatures of collective vibrational, two-quasiparticle, and possibly intruder configurations of moderate deformation. The systematics of the low-lying states in N≥66 Te nuclei are presented in Fig.1. The yrast 2^+ and 4^+ states have a similar parabolic behavior, increasing in energy as the N=82 shell closure is approached in ^{134}Te. In contrast, the excitation energies of the lowest 6^+ states are essentially constant as a function of neutron number, a behavior characteristic of proton excitations. Therefore, while the low-lying excitations of these Te isotopes are expected to be of predominantly collective vibrational nature, two-proton configurations also play a significant role in yrast states of moderate spin. The deviation from the simple expectations of a vibrator is also evident in the energy ratios[1] of these nuclei. While the R(4/2) values are typically ≥2.0 for 66≤N≤76 tellurium

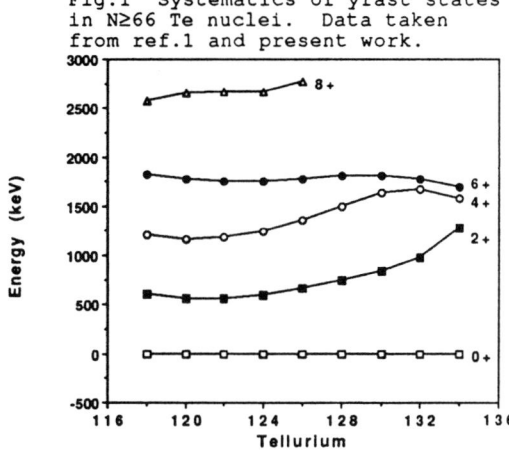

Fig.1 Systematics of yrast states in N≥66 Te nuclei. Data taken from ref.1 and present work.

isotopes, the R(6/4) values are less than the value expected for a harmonic vibrator. In addition to the moderate collective motion of vibrations, nuclei near shell closures frequently exhibit structures of a more deformed nature, from particle-hole excitations across the shell gap. For Te with Z=52, the "normal" two-valence proton structures could coexist with the more collective motions from "intruders" involving 4p-2h proton excitations across the Z=50 gap. Similar intruder configurations have been identified in Cd, In, Sn and Sb nuclei;[2] their existence in Te is still an open question. We report here on our recent measurements of ^{122,124,126}Te and ^{119,121,123}Te.

EXPERIMENTAL METHODS

In the present work γ-ray transitions in tellurium isotopes have been studied with Sn(α,xnγ) reactions at the Rutgers tandem Van de Graaff accelerator (which has since been dismantled and is operating in Australia). Gamma-gamma coincidence measurements were made with three high-purity Ge detectors, with typical efficiencies of 25%. Angular distributions were measured with a single n-type Ge detector, with and without a BGO Compton-suppression shield in an asymmetric configuration. Data were taken in 15° steps between 0°-105°. Considerable alignment was retained as the compound nucleus deexcites, enabling the extraction of angular distribution coefficients for both stretched and mixed transitions.

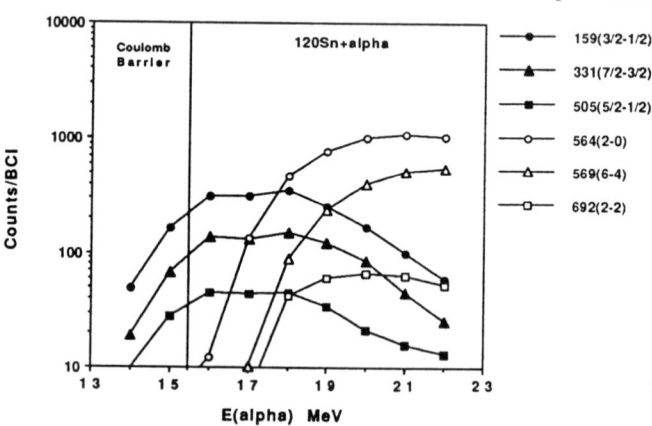

Fig.2 Excitation function for the 120Sn+alpha reaction.

A typical excitation function for the Sn+α reaction is illustrated in Fig.2, which displays our measurements for the ^{120}Sn target. The 1-n evaporation channel, which populates ^{123}Te, peaks at an energy of 17-MeV; the 2-n channel saturates at about 20 MeV. As the target mass increases, the peak of the excitation function decreases in energy. Therefore, for the heavier Sn targets, no yield was observed for the 1-n channel, because of the lack of fusion cross section at beam energies below the Coulomb barrier.

DISCUSSION

1. ROLE OF INTRUDERS IN EVEN-A ISOTOPES

In Fig.3 we present the systematics of the yrast and near-yrast excitations in ^{122,124,126}Te. These states are the ground-state band (gsb) and a quasiband, which may be related to the first excited 0^+ state, a proposed[3] 4p-2h intruder excitation. The present work has extended both of these "bands" to higher angular momenta. The quasibands in all of these nuclei have similar characteristics: they become yrast at $J \approx 8$, deexcite to the gsb, but no intraband transitions have been observed, at least below the 8^+ members.

These quasibands have been previously suggested[4] as intruder configurations, based upon 4p-2h proton excitations across the Z=50 shell gap. If this is correct, the intraband transitions must be enhanced and should dominate over transitions to the gsb. Therefore, we have made an effort to search for these intraband transitions. Though weak in intensity, when the E_γ^5 dependence is taken into account, these transitions should have strong relative E2 strengths compared to the interband transitions.

We have searched for the transitions from the 8_q^+ states in the quasibands to the 6_2^+ states; these 8^+ states are observed to decay to the 6_1^+ states. No $8_q^+ \to 6_2^+$ transitions have been observed. Rather, we have extracted upper limits for these transition intensities. For ^{122,124,126}Te

$$\frac{B(E2; \; 8_q^+ \to 6_2^+)}{B(E2; \; 8_q^+ \to 6_1^+)} \leq 4\text{-}6 \; .$$

These are considerably less than the values of 20-50 expected for the enhancement of intraband compared to interband transition rates. Therefore, the states of interest appear to have similar structure, rather than the marked differences that characterize normal and intruder configurations, and these structures are essentially identical as a function of neutron number.

326 Non-Yrast Spectroscopy of Tellurium Nuclei

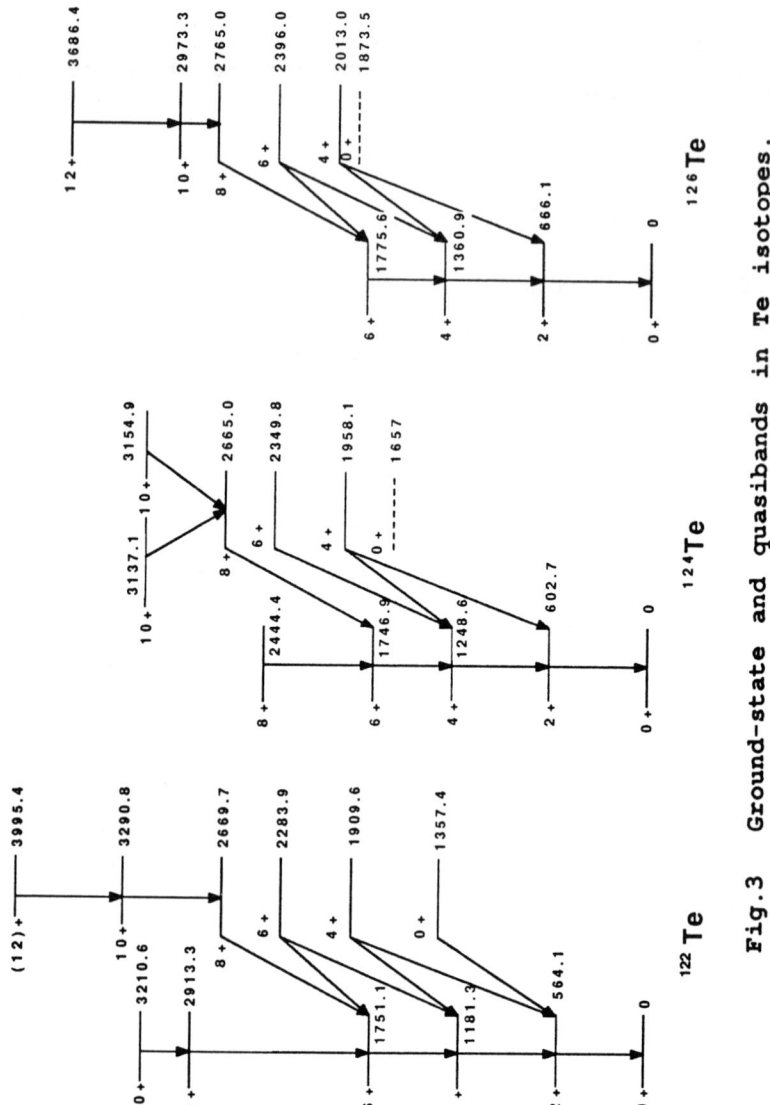

Fig.3 Ground-state and quasibands in Te isotopes.

The results could possibly be explained by strong mixing between intruder and normal 6^+ states. For two-state mixing, the matrix element is at most one-half the final energy spacing between the mixed states. For ^{122}Te and ^{124}Te two 10^+ states are observed; assuming two-state mixing, the matrix elements are at most 40 keV and 9 keV, respectively. This is considerably less than the 100-keV matrix elements[5] needed to explain the mixing between intruder and normal configurations in Cd nuclei. Either, there is a significant spin dependence to the magnitude of the mixing, causing it to decrease rapidly as the angular momentum increases (in contrast to the calculations of ref. 3, in which the purity of the intruder states increases as a function of angular momentum), or strong mixing between the 6^+ states cannot explain our results. This study of the size of mixing matrix elements may be too simple, since in ^{122}Te a third 10^+ state (at 3579.6 keV) has been identified. It is not known how this or unobserved higher excitations in the other Te isotopes could affect the size of mixing matrix elements. However, whatever model is invoked to explain the lack of observable intraband transitions has to apply equally well to all of these Te isotopes.

2. NEGATIVE-PARITY STATES IN ODD-A ISOTOPES

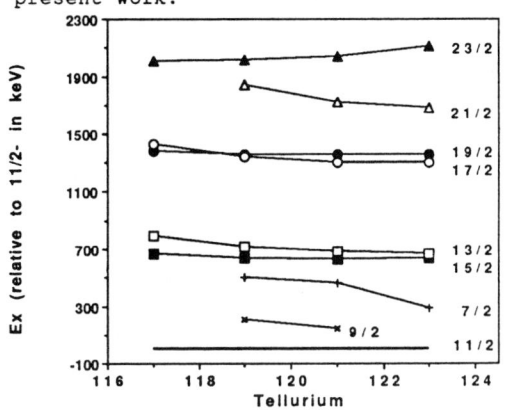

Fig. 4 Systematics of negative-parity states. Data taken from refs.6-8 and present work.

The negative-parity states displayed in Fig.4 are essentially pure excitations built on the $h_{11/2}$ neutron, since the first negative-parity states in the even cores are typically above 2.2 MeV. By studying these excitations, we hope to gain further insight into the shape of the even-mass cores, as well as to probe the role of particular two-particle configurations. The present experiments have extended the excitations to higher angular momenta in ^{123}Te and helped to confirm the non-yrast $7/2^-$ and $9/2^-$ states in ^{119,121}Te.

Our initial theoretical investigation has been in the framework of the interacting boson-fermion approximation (IBFA) model, for which we have assumed the core to be characterized by a limiting symmetry of the IBA (either vibrational SU(5) or γ-unstable O(6)). For the coupling of a single-j to a boson core, only three parameters are needed to describe the particle-core interaction. These are the quadrupole-quadrupole interaction (BFQ), the exchange interaction (BFE), and the occupation probability for the orbit j (v^2). Moreover, these parameters are not independent, but are related by [9]

$$\frac{BFQ}{BFE} = R = \sqrt{\frac{2.33 u^2 v^2}{N_v}}, \quad u^2 = 1 - v^2$$

where N_v is the number of neutron bosons in the core, and the u^2 and v^2 are the usual BCS occupation probabilities. In principle, the ratios R=BFQ/BFE are fixed by the experimentally determined occupation probabilities. While these data do exist, there is considerable fluctuation in the values as a function of neutron number and these values are accurate to no better than 25%. Therefore, we have taken a smooth interpolation of the v^2 data[8] in fitting the energy levels of ^{117,119,121,123}Te. Hence, there remains only one free parameter (either BFE or BFQ) in the comparison to experiment.

To study the range of exchange parameters BFE that reproduce the energies of the known negative-parity states, we have made a series of IBFA calculations[10]. Our results for ^{119}Te are shown in Fig.5, which displays the calculated excitation energies (normalized to that of the 15/2$^-$ state) as a function of BFE values. When these results are compared to the empirical energies, it is clear that the energy of the 9/2$^-$ state provides a severe restriction on the range of valid parameters, favoring large values, e.g., BFE=2.5-3.5. Such large values can also explain the inversion of the 17/2,19/2 and 21/2,23/2 multiplets. While the $h_{11/2} \otimes O(6)$ calculations predict a spectrum considerably expanded compared to the data, these calculations also favor large BFE values. Similar conclusions can be drawn from calculations for the other isotopes. In particular, a large BFE value reproduces the change in energy of the 7/2$^-$ state as a function of neutron number. Calculations are continuing to see the effect of these parameter choices on the relative E2 properties. In the future we shall also examine the positive-parity excitations.

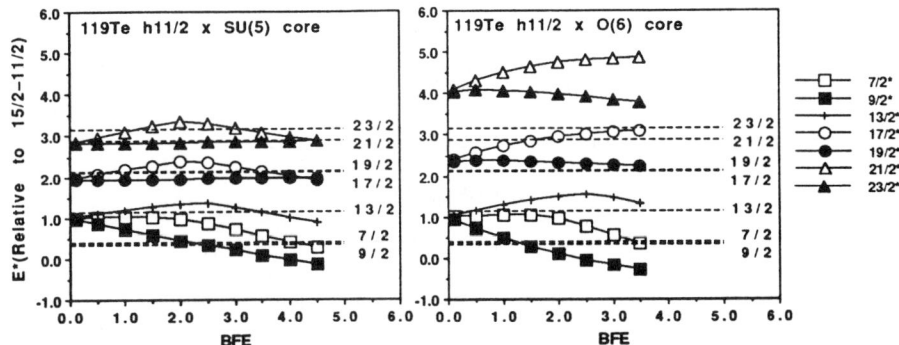

Fig.5 IBFA calculations for 119Te.

CONCLUSIONS

We have measured (α, xn) reactions on $^{116-124}$Sn targets and report here on our preliminary results for ^{122,124,126}Te and ^{119,121,123}Te. In the even isotopes we have searched for transitions within the proposed intruder band. No such transitions have been found. It is unlikely that simple mixing between intruder and normal configurations can explain the results for ^{122,124,126}Te, where not only the number of valence neutrons changes (which affects the collectivity of the normal configurations), but also the excitation energies of the states of interest change, and still the observables remain the same. In the odd-A isotopes our initial work has focused on the negative-parity states built on the $h_{11/2}$ neutron. While our IBFA calculations favor a large exchange interaction, it is difficult to reproduce simultaneously the energies of the yrast states and the non-yrast $9/2^-, 7/2^-$ excitations. It is not known if the core shape is wrong, our calculations are too simple, or if these nuclei (with only one valence proton boson), are good cases for the IBFA. However, the energy level systematics presented in Fig.4 are very simple, with essentially constant behavior as a function of neutron number. Whatever approach is found to be valid will have to apply equally well for a large range of isotopes.

ACKNOWLEDGEMENTS

We would like to thank D.Barker, R.Tanczyn, L.P.Farris, J.Szczepanski, H.Dorsett, G.Kumbartzki, H.Li, and J.W.Gan for assistance with the data acquisition and early stages of the analysis. This work is supported in part by the National Science Foundation.

REFERENCES

[1] J.A.Cizewski, Phys. Lett. **219B**, 189 (1989), and references therein.
[2] K.Heyde, et al., Phys. Rep. **102**, 291 (1983).
[3] J.Rikovska, et al., Nucl. Phys. **A505**, 145 (1989).
[4] P.Chowdhury, et al., Phys. Rev. **C25**, 813 (1982).
[5] A.Aprahamian, et al., Phys. Lett. **140B**, 22 (1984).
[6] T.Paradellis, et al., in <u>Capture Gamma-Ray Spectroscopy 1987</u>, p.547. (1988).
[7] U.Hagemann, et al., Nucl. Phys. **A329**, 157 (1979).
[8] T.Rødland, et al., Nucl. Phys. **A469**, 407 (1987).
[9] O.Scholten and A.E.L.Dieperink, in <u>Interacting Bose-Fermi Systems in Nuclei</u>, p.343 (1981).
[10] Code ODDA, O.Scholten (unpublished).

MESONIC EFFECTS IN NUCLEI NEAR ^{208}Pb DEDUCED FROM β DECAY*

E. K. Warburton

Brookhaven National Laboratory, Upton, New York 11973

ABSTRACT

The mesonic enhancement of the time-like component of the weak axial current in nuclear matter is very large and is best observed via its effect on the decay rate of $\Delta J = 0$ $(\pi_i \pi_f = -)$ β decay. Studies in the A = 16, 40, and 90 regions yield enhancements of 40–60% over the impulse approximation. The lead region is a rich source of information on these first-forbidden decays. This study is the first to extract information on mesonic enhancement from these decays. ^{206}Hg(0^+) → ^{206}Tl(0^-) → ^{206}Pb(0^+) is chosen to exemplify the approach which has been applied to 10 or so first-forbidden decays in A = 205–214 nuclei. The nuclear wave functions are evaluated via large-basis shell-model calculations. The results indicate a much larger enhancement than expected and thus the possibility of some non-nucleonic effect in addition to the mesonic enhancement considered to date.

INTRODUCTION

In 1978 Kubodera, Delorme and Rho[1] used chiral-symmetry arguments and soft-pion theorems to predict a very large ($\sim$ 40–70%) enhancement over the impulse approximation for the time-like component of the weak axial current in nuclear processes, i.e., for the matrix element of γ_5. In the same year, Guichon et al[2] made a similar prediction and showed that the expected enhancement, due to meson-exchange-current contributions to the matrix element of γ_5, is most easily observed via first-forbidden beta decay between states of the same spin since γ_5 is a rank 0 operator and the selection rule on the rank R is $|J_i - J_f| \leq R \leq J_i + J_f$. In this talk I shall endeavor to sketch the procedures and calculations which lead to the conclusion that not only is an enhancement over the impulse approximation definitely present (as was previously known[2-10]), but in A $\sim$ 208 β decays it appears to be considerably larger than the evaluations of its magnitude which have been made to date from meson-exchange alone.

The strong $\Delta J = 0$ first-forbidden decays in the lead region fall into two categories: Those for which the dominant decay is $\nu 2p_{1/2} \rightarrow \pi 2s_{1/2}$ and $\nu 1g_{9/2} \rightarrow \pi 0h_{9/2}$, respectively. The decays of the first type are shown in Fig. 1. The second type is comprised of the decays of the six A = 209–214 Pb isotopes. I shall consider here the least complicated of the decays of Fig. 1; namely, ^{206}Hg(0^+) → ^{206}Tl(0^-) → ^{206}Pb(0^+).[11-15] Analysis of the other decays of Fig. 1 involves the evaluation of rank 1 matrix elements as well as rank 0. This has been achieved successfully and the results for the other A $\sim$208 decays will be presented at a later time.

Invited talk given at the Seventh International Symposium on CAPTURE GAMMA RAY SPECTROSCOPY AND RELATED TOPICS, October 14-19 (1990).

© 1991 American Institute of Physics

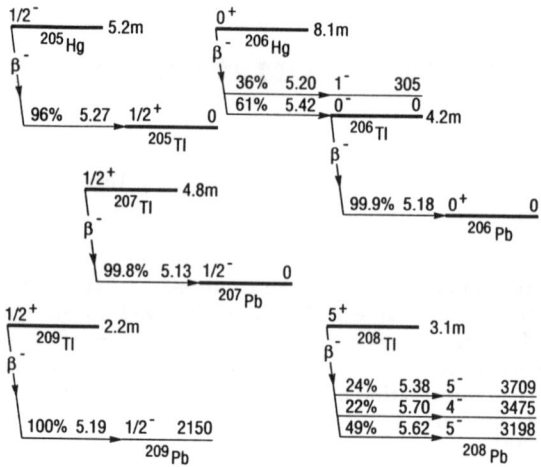

FIG. 1. Fast $\Delta J = 0,1$ first-forbidden decays in the lead region. The decays shown are those for which the dominant single-particle transition is $\nu 2p_{1/2} \to \pi 2s_{1/2}$. For each decay branch, the branching ratio (in percent) and the logft value is given.

The prediction of a sizable mesonic enhancement of the axial current was greeted with considerable interest and gave rise to a large number of experimental and theoretical studies. On the experimental side, a great deal of effort was directed to the accurate measurement of the rank 0 decay rates of first-forbidden β decays in the A = 16 region. These are reviewed by Towner[3], Warburton[5], and Millener and Warburton[6]. Recently, the decay of ^{96}Y(0^-) has also been subjected to experimental scrutiny[7] and the available data for the A ~ 40 region was examined for possible evidence of meson-exchange effects[8]. It turns out that all the experimentally known interesting decays for A < 100 are of the type $s_{1/2} \leftrightarrow p_{1/2}$ and occur near A = 16 and 96.

The activity on the theoretical side has been of two types. The more fundamental is the actual evaluation of the one-pion exchange process. The prediction of a large meson-exchange effect is of a general nature but can be expected to be modified by the nuclear structure of the initial and final states. Evaluation of nuclear structure effects on the partial decay rate due to first-forbidden beta decay as mediated by one-pion exchange is quite difficult if the same rigour is demanded as can be brought to the impulse approximation calculations via the shell model. Nevertheless, much progress has been made; mainly by Towner and his colleagues[3,4], and Kirchbach and her colleagues[9,10]. It has been found that the predicted enhancement is relatively insensitive to nuclear structure. This insensitivity is expected[3] because the effect is mainly due to the interaction of the valance nucleons with the core. Kirchbach and Reinhardt[9] have calculated meson enhancements for the matrix element of γ_5 of between 40-60% for transitions near A = 16, 96, and 206. These authors have also stressed that renormalization of the nucleon mass, as is encountered in relativistic mean-field theories, can also lead to an enhancement of the matrix element of γ_5. This effect — parameterized by using an effective nucleon mass — is expected to increase with mass number. Thus we should speak of non-nucleonic enhancement rather than the more specific meson-exchange enhancement.

The second type of theoretical activity involves as careful an evaluation of the decay rate as possible via the impulse approximation as formulated in the spherical shell model. The extent of enhancement due to effects of the nuclear medium is then ascertained by a comparison to experiment.[5-8] We now consider this approach in detail.

CALCULATIONS

There are two rank 0 operators which contribute to first-forbidden beta decay in normal order. These are the space-like operator $\vec{\sigma} \cdot \vec{r}$, and the time-like operator, γ_5. The latter is relativistic and has a non-relativistic form $\propto \vec{\sigma} \cdot \vec{p}$. We shall refer to the matrix elements of these operators as M_0^S and M_0^T, respectively. The evaluation of these matrix elements in the impulse approximation has been fully discussed elsewhere.[8] In principle, the shape factor for the rank 0 contribution to a first-forbidden beta decay has terms dependent on the beta energy, W, namely, $\propto W^{-1}$ in leading order and $\propto W$ in the next order in Z (Ref. 16). However the contribution of these terms — and all other higher-order terms — is at most a few percent of that for the energy independent terms in the cases of interest here[15-17] and we can define a rank 0 first-forbidden matrix element combination in terms of the $f_0 t$ value for the decay[8]:

$$M_1^{(0)}(expt.) = [9.15/f_0 t]^{\frac{1}{2}} \times 10^4. \tag{1}$$

Here f_0 is the Fermi integral for allowed decay, and t is the partial half-life for that portion of the decay in question which is rank 0. We can also express the prediction for $M_1^{(0)}$ as

$$M_1^{(0)} = [\epsilon_{nn} q_T M_0^T + a(Z, W_0, r_u) q_S M_0^S] \tag{2}$$

where M_0^T and M_0^S are calculated using the impulse approximation and by convention M_0^T is chosen positive, ϵ_{nn} is the non-nucleonic enhancement factor which we wish to evaluate, $a(Z, W_0, r_u)$ is the usual factor for first-forbidden decays, with W_0 being the β end-point energy and r_u the uniform charge radius corresponding to the experimental root-mean-square (r.m.s.) charge radius. ϵ_{nn} is then evaluated by equating Eqs. (1) and (2). My convention is that M_0^T and M_0^S denote matrix elements evaluated in the model space of Fig. 2, the Kuo-Herling[18] model space. There are potentially large core-excited contributions from 2p-2h excitations of the final state, see Fig. 3. Because of Pauli blocking due to the large neutron excess there are no contributions due to the initial-state admixtures which connect directly to the dominant final-state terms other than those included in the model space of Fig. 2. The final-state core excitations comprise <1% of the final-state wave functions and thus can be added perturbatively. This has been done[19] using various realistic interactions and the results are incorporated as effective "quenching" factors — the

q_T and q_S of Eq. (2). For rank 0 operators $q_T \approx -q_S$ (this is an identity for harmonic-oscillator wave functions) and $q_S \approx 0.9$ while for the rank 1 operators the quenching factors are ~ 0.5. The difference between the quenching of rank 0 and 1 is that the cental and tensor contributions are destructive for rank 0 and constructive for rank 1.[19]

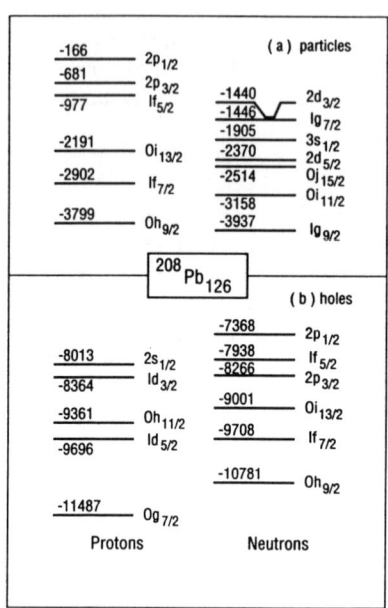

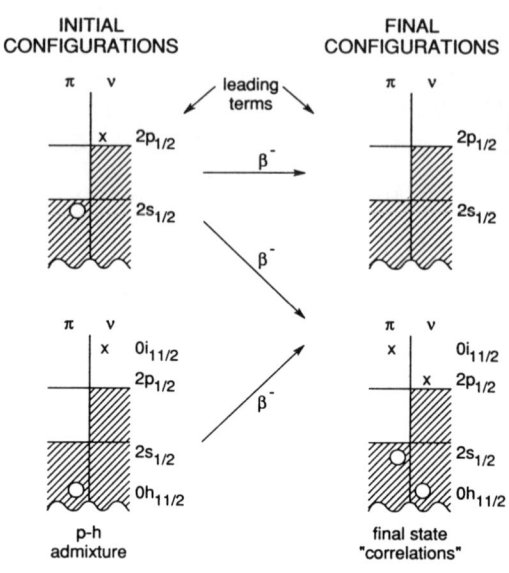

FIG. 2. The Kuo-Herling model space. For each orbit the single-particle energy (in keV) relative to ^{208}Pb is listed.

FIG. 3. Diagram indicating some of the relevant configurations entering for first-forbidden transitions in the lead region. The example shown is that of ^{206}Tl $\rightarrow$ ^{206}Pb. Arrows indicate the configurations linked by the beta decay.

It should be noted that non-nucleonic effects on M_0^S are predicted to be small, of the order of a few percent.[2–4] Application of the approach of equating Eqs. (1) and (2) has yielded an average value for ϵ_{nn} of 1.64 for four decays near A = 16 (Ref. 5) while at A = 96 a value of 1.70 is obtained.[7]

The experimental information on the A = 206 decays is as follows. The half-life and Q_β-value for ^{206}Hg decay are 489(6) s (Ref. 11) and 1309(20) keV (Ref. 20). The ground-state branching ratio is 61(12)% (Refs. 11–13). The resulting values of $f_0 t$ and $\log f_0 t$ are $2.61(55) \times 10^5$ and 5.42(9), respectively. The half-life and Q_β-value for ^{206}Tl decay are 252(1) s (Ref. 13) and 1531(4) keV (Ref. 20). The ground-state branching ratio is 99.91(2)% (Ref. 13,14). These result in values for $f_0 t$ and $\log f_0 t$ of $1.527(16) \times 10^5$ and 5.184(5), respectively. The f_0 values were calculated using the parameterization of Wilkinson and Macefield[21]. Using Eq. (1), we obtain values for $M_1^{(0)}$(expt.) of 59(6) and 77.4(4) for ^{206}Hg and ^{206}Tl decay, respectively.

For rank 0 operators we have the simple relation

$$M_0^S = \sum_j OBTD(j) M_0^S(j) \qquad (3)$$

where OBTD(j) and $M_0^S(j)$ are the one-body-transition-density and beta-decay matrix element for the single-particle-transition $j_i \to j_f$ with $j_i = j_f \equiv j$. The relationship for M_0^T is, of course, similar. The OBTD(j) are the output of the shell-model calculations to be described below. The $M_0^S(j)$ and $M_0^T(j)$ were evaluated with Woods-Saxon, Hartree-Foch, and Harmonic Oscillator wave functions. The results for Woods-Saxon and Hartree-Foch wave functions were very similar and we only describe the former. The Woods-Saxon parameters were those listed for ^{208}Pb by Street et al.[22,23]. These parameters reproduce the experimental r.m.s charge radius of ^{208}Pb, 5.502 fm, and give a difference in the r.m.s proton and neutron radii of -0.12 to -0.20 fm (depending on the occupation numbers of the orbits) which is also consistent with the known experimental evidence.[22] With these parameters M_0^S and M_0^T for both decays are $\sim 1\%$ and $\sim 4\%$ less, respectively, than calculated with harmonic-oscillator wave functions using a value of b which gives the same r.m.s charge radius. The sensitivity of the calculated matrix elements to the separation energy of the neutron in the initial state amd the proton in the final state was examined in detail and the evaluation was formulated so as to include the effect of the fragmentation of the parentages of the decaying nucleons over many states of the A-1 parent.[24]. It was a surprise to find that the results were very insensitive to the separation energies so that there appears to be negligible uncertainty from this source.

The evaluation of the OBDT(j) via shell-model calculations was done with three different interactions. (1) the surface delta interaction (SDI) of Poppelier and Glaudemans[25], (2) The Kuo-Herling hole interaction, labeled KHH[18], and (3) a large-basis interaction labeled PKH. The calculations were carried out with the shell-model code OXBASH[26]. The Poppelier-Glaudemans interaction uses a model space consisting of the nearest three particle orbits and four hole orbits with respect to the ^{208}Pb core for both neutrons and protons. The relative single-particle energies of these orbits and the two other parameters of the SDI were determined by a least-squares-fit to 74 experimental energies in A = 207-209 nuclei assuming at most 1p-1h excitations relative to the doubly-closed Z = 82, N = 126 core. This interaction gives good agreement with binding energies and other observables and gives an adequate description of the general features of the A = 206 beta decays. However, the Coulomb energy was not treated consistently in its derivation so that, e.g., it cannot be used for the ^{208}Tl decays (Fig. 1) and its single-particle energies are suspect. Thus we concentrate our attention on the results using the KHH and PKH interactions. The Kuo-Herling interaction works in the "hole" (orbits below ^{208}Pb) space of Fig. 2 using the single-particle energies listed in the figure. The version used here — labeled KHH_e — has some two-body-matrix-elements (TBME) tuned to give better agreement with binding energies of levels in ^{206}Hg, ^{206}Tl, and $^{201-206}$Pb.[24]

The PKH model space is the same 14 orbits as the Poppelier-Glaudemans interaction. It uses the KHH$_c$ TBME for the orbits below ^{208}Pb. For the particle-particle interaction it uses the Kuo-Herling TBME with the core-polarization contribution increased to compensate for the truncation.[21] The particle-hole TBME are taken from a potential fit to a G-matrix derived from a nucleon-nucleon potential.[27] The same potential was used to evaluate the q_S and q_T for each individual single-particle transition.[19] In the KHH$_c$ space the J-dimensions [D(J)] of the ^{206}Hg(0$^+$), ^{206}Tl(0$^-$), and ^{206}Pb(0+) states are 5, 4, and 6. In the PKH space all possible 1p-1h excitations across the ^{208}Pb double shell closure are included. The D(J) are now 134, 185, and 111. The results from the two calculations are combined to give a composite set of OBTD(j). This perturbative approach is justified because the contributions which are not common to the two calculations (j = 7/2, 9/2, 11/2) are due to very small components in the wave functions. The synthesis of the KHH and PKH results for the beta matrix elements are summarized in Table I.

TABLE I. Contributions to $M_1^{(0)}$ from the possible single-particle transitions. The matrix elements $q_S M_0^S$ and $q_T M_0^T$ are in fm.

orbit		^{206}Hg → ^{206}Tl		^{206}Tl → ^{206}Hg	
ν	π	$q_S M_0^S$	$q_T M_0^T$	$q_S M_0^S$	$q_T M_0^T$
$2p_{\frac{1}{2}}$	$2s_{\frac{1}{2}}$	-6.172	91.747	-7.542	82.136
$2p_{\frac{3}{2}}$	$1d_{\frac{3}{2}}$	-0.514	7.676	-0.531	7.924
$1f_{\frac{5}{2}}$	$1d_{\frac{5}{2}}$	-0.269	3.874	-0.520	7.950
$1f_{\frac{7}{2}}$	$0g_{\frac{7}{2}}$	-0.012	0.348	-0.003	0.078
$0i_{\frac{11}{2}}$	$0h_{\frac{11}{2}}$	1.274	-24.437	1.291	-24.630
$1g_{\frac{9}{2}}$	$0h_{\frac{9}{2}}$	-0.025	0.403	-0.008	0.126
total		-5.719	79.610	-5.242	73.585

The results are generally as expected. The KHH$_e$ wave function components involving the $\nu 2p_{1/2}$ and $\pi 2s_{1/2}$ orbits comprise 71%, 92%, and 52% of the ^{206}Hg, ^{206}Tl, and ^{206}Pb ground-state wave functions, respectively. Thus normalization reduces the contributions of the dominant $\nu 2p_{1/2} \to \pi 2s_{1/2}$ component to well below its $\nu 2p_{1/2} \to \pi 2s_{1/2}$ single-particle values which, for ^{206}Tl decay are -7.54 fm for $q_S M_0^S$ and 113.22 fm for $q_T M_0^T$. Because of the repulsive nature of the T = 1 particle-hole interaction in nuclear matter, the other 'pure' particle-hole term — that involving the j = 11/2 orbits — is opposite in phase to the main particle-hole term. The other four contributions involve hole-hole terms in ^{206}Tl and as such have the opposite phase to particle-hole terms. The same effect was observed in ^{96}Y decay (Ref. 7) and in a previous study of ^{206}Tl → ^{206}Pb (Refs. 15,28).

For ^{206}Hg and ^{206}Tl decays we calculate a(Z, W$_0$, r$_\nu$) values of 14.09 and 14.40, respectively. Equating Eqs. (1) and (2) we find the predictions for the two ϵ_{nn} values are 1.75(18) and 2.08(1), respectively, where the uncertainties are experimental only. The corresponding Poppelier-Glaudemans SDI values are

1.98(21) amd 2.07(1). The results for these A = 206 decays are representative of those found for all the fast A $\sim$ 208 ΔJ = 0 decays which collectively give a firm value of $\epsilon_{nn} \approx 2.0$.

SUMMARY

Let us summarize the findings. The meson-exchange contribution to the matrix element of γ_5 has been calculated by Towner and Khanna[4] for ^{16}N(0^-) decay, ϵ_{nn} = 1.45–1.60, and by Kirchbach and Reinhardt[8] for A = 16, ϵ_{nn} = 1.50, A = 96, ϵ_{nn} = 1.60, and A = 206, ϵ_{nn} = 1.40. The non-nucleonic contribution can be obtained indirectly by comparing impulse approximation calculations to experiment. This approach, which we have detailed, yields ϵ_{nn} = $\sim$1.64 for A = 16, $\sim$1.70 for A = 96, and $\sim$2.0 for A $\sim$ 206. These results for A = 16 and A = 96 are larger than the predictions but not unduly so considering the uncertainties inherent in the shell-model calculations. For A $\sim$ 208, however, the extracted value of ϵ_{nn} is considerably larger than the value of 1.40 predicted from meson-exchange alone and it can be argued that the uncertainties in the shell-model calculation are probably less than in the lighter nuclei. Thus, the present conclusions suggest the possibility of sizable effects on ϵ_{nn} from non-nucleonic effects other than those so far evaluated.

Thanks are due to N.A.F.M. Poppelier and P.W.M. Glaudemans for expert advice and criticism and to M. Kirchbach for an enlightening discussion.

---

*Work has been supported under contract DE-AC02-76CH00016 with the United States Department of Energy.

[1] K. Kubodera, J. Delorme, and M. Rho, Phys. Rev. Lett. 40, 755 (1978).

[2] P. Guichon, M. Giffon, J. Joesph, R. Laverrière, and C. Samour, Z. Phys. A 285, 183 (1978); P. Guichon, M. Giffon, and C. Samour, Phys. Letts. 74B, 15 (1978).

[3] I. S. Towner, Comments Nucl. Part. Phys. 15, 145 (1986).

[4] I. S. Towner and F. C. Khanna, Nucl. Phys. A372, 331 (1981).

[5] E. K. Warburton, in Interactions and Structures in Nuclei (R. J. Blin-Stoyle and W. D. Hamilton, Eds.), p. 81, Adam Hilger, Bristol/Philadelphia (1988).

[6] D. J. Millener and E. K. Warburton, in Nuclear Shell Models (M. Vallieres and B. H. Wildenthal, Eds.), p. 365, World-Scientific, Singapore (1985).

[7] H. Mach, E. K. Warburton, R. L. Gill, R. F. Casten, J. A. Becker, B. A. Brown, and J. A. Winger, Phys. Rev. C 41, 226 (1990).

[8] E. K. Warburton, J. A. Becker, B. A. Brown, and D. J. Millener, Ann. Phys.(N.Y.) 187, 471 (1988).

[9] M. Kirchbach and M. Reinhardt, Phys. Letts. 208, 79 (1988); The result ϵ_{nn} = 0.45 for A = 96 given in the reference has been revised to ϵ_{nn} = 0.60 (M. Kirchbach, private communication).

[10] M. Kirchbach, Proceedings of the International Symposium on Modern Developments in Nuclear Physics, Novosibirsk, USSR, edited by O. P. Sushkov,

(World-Scientific, Singapore, 1987), p. 475.

[11] G. K. Wolf, Nucl. Phys. **A116**, 387 (1968).

[12] G. Astner and G. K. Wolf, Nucl. Phys. **A147**, 481 (1970).

[13] M. P. Webb, Nuclear Data Sheets **26**, 145 (1979).

[14] H. C. Griffin and A. M. Donne, Phys. Rev. Letts. **28**, 107 (1972).

[15] W. Wiesner, D. Flothmann, H. G. Gils, R. Löhken, and H. Rebel, Nucl. Phys. **A191**, 166 (1972).

[16] H. Behrens and W. Bühring, Electron Radial Wave Functions and Nuclear beta-Decay, Clarendon, Oxford (1982).

[17] The possible effects of higher-order terms on the first-forbidden decay rates were considered following Refs. 15 and 16. One might have expected such terms — especially those involving the expansion in powers of Z — to be important in the lead region. However, after detailed consideration, their effect on the A = 206 decay rates was found to be less than 0.5%.

[18] T. T. S. Kuo and G. H. Herling, US Naval Research Laboratory Report no. 2258, unpublished (1971).

[19] E. K. Warburton, Phys. Rev. C, in press.

[20] A. H. Wapstra and G. Audi, Nucl. Phys. **A432**, 1 (1985).

[21] D. H. Wilkinson and B. E. F. Macefield, Nucl. Phys. **A232**, 1 (1974).

[22] J. Street, B. A. Brown, and P. E. Hodgson, J. Phys. G. **6**, 839 (1982).

[23] ^{206}Tl → ^{206}Pb decay in a single-particle $\nu 2p_{1/2} \to \pi 2s_{1/2}$ approximation was also considered by Behrens and Bühring (Ref. 16) using harmonic-oscillator and Woods-Saxon wave functions. The present Woods-Saxon results differ significantly from theirs because they used parameters — based on pre-1968 data — which gave, for instance, a difference in the $r.m.s$ neutron and proton radii of 0.6 fm. However, the need for a non-nucleonic contribution — which they did not discuss — is indicated by their results. The reader is directed to their book for a very enlightening discussion of first-forbidden beta decay.

[24] E. K. Warburton, to be published.

[25] N. A. F. M. Poppelier and P. W. M. Glaudemans, Z. Phys. A **329**, 275 (1988).

[26] B. A. Brown, A. Etchegoyen, W. D. M. Rae, and N. S. Godwin, OXBASH, 1984 (unpublished).

[27] A. Hosaka, K.-I. Kubo, and H. Toki, Nucl. Phys. **A244**, 76 (1985).

[28] Wiesner et al. (Ref. 15), considered ^{206}Tl → ^{206}Pb in a shell-model space including j = 1/2, 3/2, 5/2, and 13/2 orbits. The wave functions for these terms were remarkably similar to the present SDI results. Their calculation differs in the omission of the important j = 11/2 orbits (see Table I), use of pre-1969 Woods-Saxon wave functions, and — since the work was done before 1978 — neglect of the non-nucleonic contribution. Nevertheless, their results — viewed in retrospect — also indicate the need for a large non-nucleonic contribution.

SINGLE PARTICLE STATES IN THE HEAVIEST KNOWN NUCLEI

I. Ahmad, R.R. Chasman and A.M. Friedman*

Argonne National Laboratory, Argonne, Illinois 60439

S.W. Yates

University of Kentucky, Lexington, Kentucky 40506

ABSTRACT

Neutron single-particle states above the N=152 subshell have been studied by high-resolution (d,p) reaction on a ^{250}Cf target. All of the orbitals between N=152 and N=164 subshells have been identified. A tentative assignment has been made for the 1/2-[761] Nilsson state.

INTRODUCTION

Theoretical calculations[1], performed about twenty years ago, predicted long half-lives for superheavy elements with atomic number 114 and neutron number 184. Attempts to identify such elements in nature and in nuclear reactions have been unsuccessful, but neutron-deficient isotopes of elements up to 109 have been discovered[2]. The estimates of half-lives for superheavy elements are based on calculations of nuclear energies as a function of deformation. An essential ingredient in these Strutinsky type calculations is the magnitude of the shell correction, which is extremely sensitive to the single-particle energy gap at Z=114 and N=184, and to the level spacings near these gaps. For this reason any experimental measurements of energies of orbitals near these gaps is extremely important. In the case of protons, the gap at Z=114 is determined by the splitting of the $f_{7/2}$ and $f_{5/2}$ orbitals, and this was deduced from the spectroscopy of odd-proton nuclei[3,4] and used to derive the parameters of a single-particle potential as illustrated in Fig. 1. The shell correction near N=184 is largely determined by the position of the $h_{11/2}$, $k_{17/2}$ and $j_{13/2}$ spherical states[5]. It is therefore important to determine the energies of single-particle states in the heaviest nuclei.

The best way to identify orbitals above the N=152 subshell is to perform single neutron transfer reactions on a target with neutron number 152 or greater. In such a stripping reaction, the 153rd neutron will occupy orbitals above the N=152 gap. The advantage of a target with N=152 is that the population of hole state orbitals is strongly suppressed because of the large values of V^2 (V^2 is the pair occupation probability). Thus a clean (d,p)

*deceased.

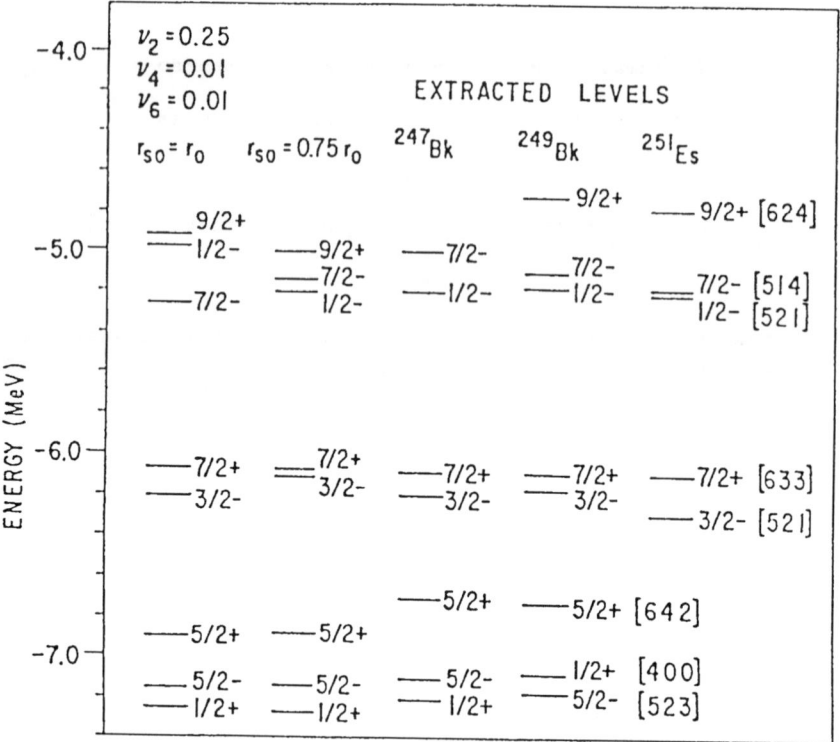

Fig. 1. Theoretical (two left columns) and extracted energies of proton single-particle states. Theoretical energies were calculated with a Woods-Saxon potential using the parameters shown in the figure. The extracted energies were obtained by removing the contribution of the pair correlation effects from the experimental level energies.

spectrum is expected. The only target with N≳152, with a reasonable half-life, is ^{248}Cm ($T_{1/2}$ = 3.5 x 10^5 y). Neutron transfer reaction[6] and neutron capture gamma[7] experiments have been performed with ^{248}Cm targets which give information on states in ^{249}Cm. However, these experiments did not have sufficient sensitivity to observe weakly populated levels. Another possible target with N≳152 is ^{250}Cf ($T_{1/2}$ = 13.1 y), which is much more radioactive than ^{248}Cm, but has the advantage that many single particle states in ^{251}Cf have been well characterized[8] in the alpha decay of ^{255}Fm.

EXPERIMENTAL PROCEDURE

The ^{250}Cf(d,p) spectra were measured with the Argonne Tandem Van de Graaf accelerator using a 12.0-MeV deuteron beam and an 8-μg/cm^2 ^{250}Cf target on a 40-μg/cm^2 carbon film which was prepared in the Argonne electromagnetic isotope separator. The emerging protons

were momentum analyzed with an Enge split-pole spectrograph and were
detected with photographic emulsion plates in the focal plane.
Spectra were recorded at 90° and 120° with respect to the beam
direction. Portions of the proton spectra showing the regions of
interest are displayed in Figs. 2 and 3. The resolution (FWHM) of
the peaks in these spectra is 7 keV and these are among the cleanest
spectra measured in the heavy-element region.

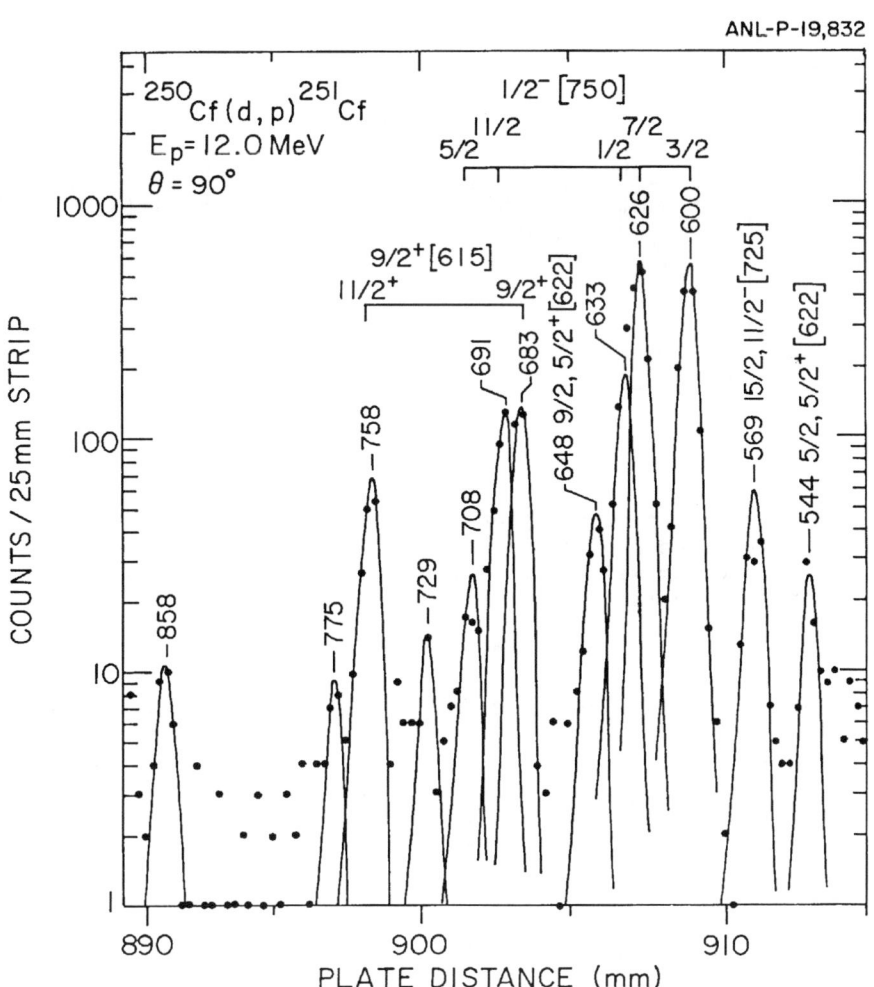

Fig. 2. Proton spectrum from the ^{250}Cf(d,p) reaction measured with
the Argonne Enge split-pole spectrograph showing the
population of the 1/2-[750] and 9/2+[615] bands. Energy
scale is ~3.5 keV per 0.25-mm strip.

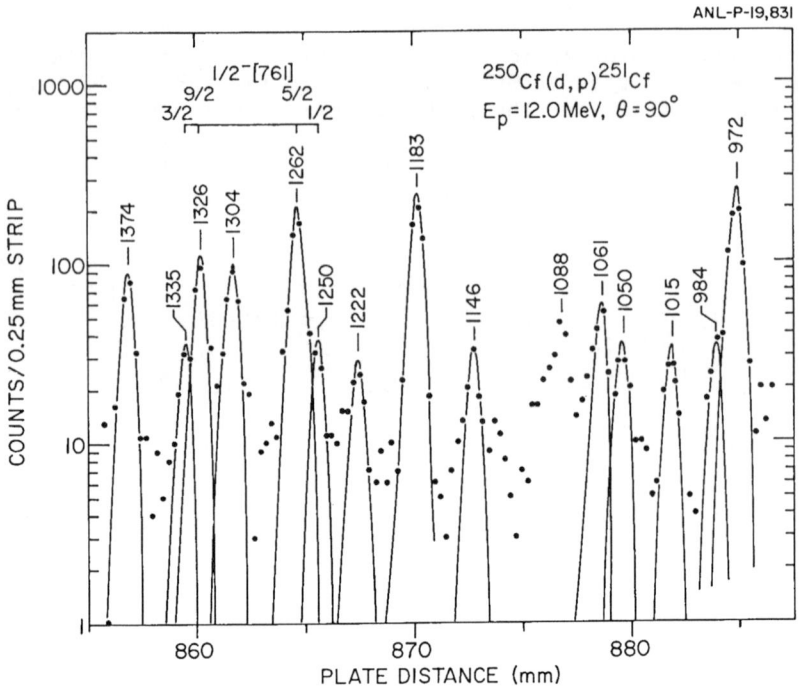

Fig. 3. A portion of the proton spectrum from the ^{250}Cf(d,p) reaction showing peaks above 1 MeV excitation. The spectrum was measured with the Argonne Enge split-pole spectrograph. Energy scale is ~3.5 keV per 0.25-mm strip.

DISCUSSION

The rotational bands 1/2+[620], 7/2+[613], 3/2+[622], 5/2+[622], 11/2-[725] and 9/2-[734] were well characterized[8] in the ^{255}Fm alpha decay. The present (d,p) reaction data (Fig. 2) fully support these assignments and enable us to compare calculated cross sections with the measured ones. All observed levels below 600 keV excitation fit extremely well as members of known bands, leaving no level unassigned. According to our calculations the remaining intense peaks below 1 MeV should be associated with the 1/2-[750] and 9/2+[615] bands. Using the calculated[9] signature of the 1/2-[750] band we have made the following assignment: 1/2 (633 keV); 3/2 (600 keV); 5/2 (708 keV); 7/2 (626 keV); 11/2 (691 keV). These level energies give a rotational constant of 4.8 keV and a decoupling parameter of -3.29, in very good agreement with the theoretical value of -3.8. The peaks at 683 and 758 keV are assigned to the 9/2 and 11/2 members of the 9/2+[615] band, respectively, which yield a reasonable rotational constant of 6.9 keV. The observed energy of the 9/2+[615] orbital in ^{251}Cf is considerably higher than the assignment made in ^{249}Cm at 209 keV[6].

The experimental level energies, after correction for pairing effects, are shown on the right hand side in Fig. 4. These are compared with the energies calculated with a momentum-dependent Woods-Saxon single particle potential using $\nu_2 = 0.25$, $\nu_4 = -0.01$ and $\nu_6 = +0.02$ (left column). The deformation parameters were chosen to give a good fit to the energies of the well characterized levels in ^{251}Cf. The agreement between the theoretical and extracted energies is good. This gives us some confidence in the predicted energies of the higher lying states. The values of C_j^2 obtained in these calculations are given in Table I. Values of C_j^2 for the lower-lying orbitals are given in Ref. 10.

Although there are many peaks in the spectrum above 1 MeV (Fig. 3), only tentative assignments have been possible for some levels. We have made a tentative assignment of I=1/2, 5/2, 9/2 and 3/2 members of the 1/2-[761] rotational band to the 1250, 1262, 1326 and 1335 keV levels. These assignments give a rotational constant of 6.9 keV, and a decoupling parameter of +3.1, which is in good agreement with the theoretical value of +3.9 for this band. We also assign the 1183 keV level to the 9/2 member of the 9/2+[604] band because it is the strongest peak in the spectrum above 1 MeV excitation.

The energy of the 1/2+[880] orbital, which is largely determined by the position of the $k_{17/2}$ spherical shell state, is calculated to be 1376 keV. The decoupling parameter of +8.3 makes this a very unusual band. In addition, this band has a very strong interaction with the 3/2+[871] band. Using appropriate Coriolis matrix elements we have made a mixing calculation which brings down the 9/2+, 13/2+, 5/2+ and 17/2+ levels below the energy of the 1/2+ level to 1218, 1224, 1267 and 1286 keV, respectively. Also, the 9/2, 13/2, 17/2, and 1/2 members of the band are expected to receive populations of 10-30 μb/sr. We do not have any unassigned peak in the spectrum below 1 MeV with the predicted cross section, indicating that the 1/2+[880] band lies above 1 MeV excitation. One important thing worth noting is the lowering of the 17/2 member of the 1/2+[880] band to ~1 MeV. Thus heavy-ion reactions, which preferentially populate high angular momentum states, could be used to identify the 1/2+[880] band.

In summary, we have identified all of the single-particle states in the subshell between N=152 and N=164 in ^{251}Cf. We have also calculated the levels above the N=164 gap and have tentatively identified the 1/2-[761] and 9/2+[604] orbitals. The position of the 1/2-[761] orbital provides information about the $j_{13/2}$ spherical shell state which lies above the N=184 neutron gap. If the excitation energy of the 1/2+[880] orbital can be determined, we will then have some experimental information on all of the spherical orbitals between N=184 and N=224. These constraints should help us to determine the feasibility of making superheavy elements.

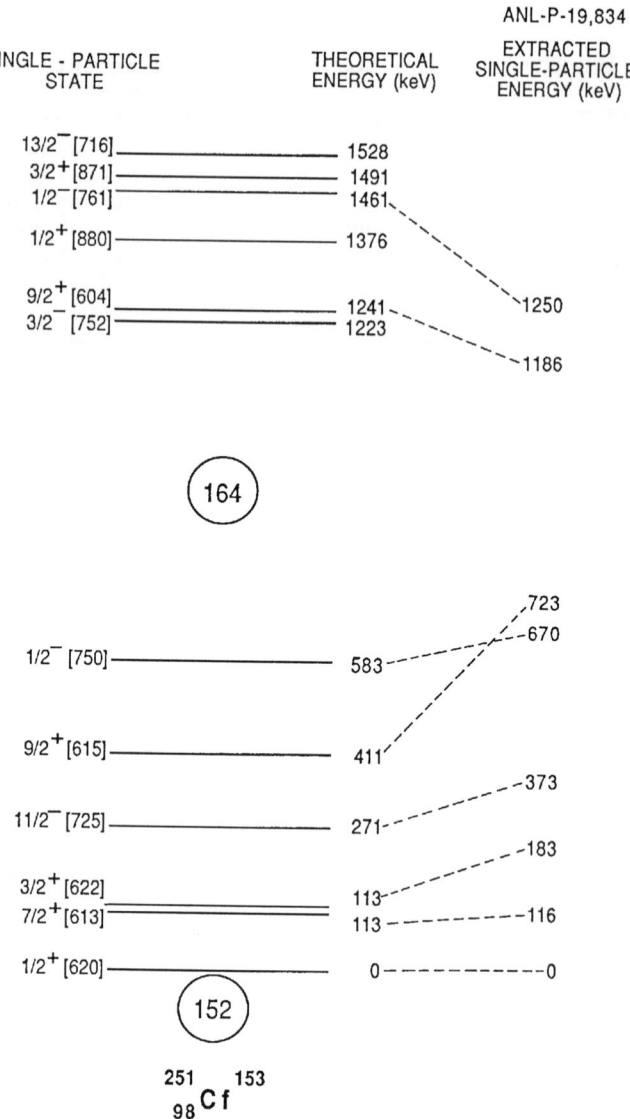

Fig. 4. Theoretical (left column) and extracted (right column) energies of neutron single-particle states. Theoretical energies were calculated with a Woods-Saxon potential, using $\nu_2=0.25$, $\nu_4=-0.01$ and $\nu_6=+0.02$. The extracted energies were obtained by removing the contribution of pair correlation effects from the experimental level energies. A constant matrix element (G=21/A MeV) was used for the pairing calculations.

Table I C_j^2 for Neutron Single-Particle States Above N=152

State	a	1/2	3/2	5/2	7/2	9/2	11/2	13/2	15/2	17/2
1/2+[620]	+0.402	0.131	0.041	0.209	0.216	0.238	0.133	0.030	-	-
7/2+[613]		0.0	0.0	0.0	0.003	0.843	0.102	0.049	0.0	0.003
3/2+[622]		0.0	0.168	0.091	0.358	0.161	0.196	0.024	0.0	0.002
11/2−[725]		0.0	0.0	0.0	0.0	0.0	0.010	0.004	0.981	0.0
9/2+[615]		0.0	0.0	0.0	0.0	0.069	0.914	0.016	0.0	0.0
1/2−[750]	−3.8	0.031	0.183	0.039	0.383	0.025	0.207	0.011	0.120	0.0
3/2−[752]		0.0	0.048	0.055	0.309	0.094	0.291	0.063	0.137	0.0
9/2+[604]		0.0	0.0	0.0	0.0	0.910	0.076	0.009	0.0	0.005
1/2+[880]	+8.3	0.006	0.001	0.002	0.004	0.029	0.001	0.209	0.0	0.743
1/2−[761]	+3.9	0.031	0.0	0.186	0.034	0.347	0.049	0.332	0.019	0.0
3/2+[871]		0.0	0.001	0.042	0.030	0.022	0.008	0.191	0.0	0.701
13/2−[716]		0.0	0.0	0.0	0.0	0.0	0.0	0.001	0.997	0.0
11/2+[606]		0.0	0.0	0.0	0.0	0.0	0.996	0.003	0.0	0.0
3/2+[611]		0.0	0.0	0.516	0.202	0.158	0.050	0.017	0.0	0.056
5/2+[862]		0.0	0.0	0.030	0.067	0.014	0.013	0.165	0.001	0.702

ACKNOWLEDGMENT

This research was supported by the US Department of Energy, Nuclear Physics Division, under contract No. W-31-109-ENG-38. The calculations were carried out on the NFMECC computers at Livermore. The authors acknowledge helpful discussions with D. G. Burke.

REFERENCES

1. S. G. Nilsson, C. F. Tsang, A. Sobiczewski, Z. Szymanski, S. Wycech, C. Gustafson, I. L. Lamm, P. Möller, and B. Nilsson, Nucl. Phys. A131, 1 (1969).
2. P. Armbruster, Ann. Rev. Nucl. Part. Sci. 35, 135 (1985).
3. I. Ahmad, A. M. Friedman, R. R. Chasman, and S. W. Yates, Phys. Rev. Lett. 39, 12 (1977).
4. R. W. Lougheed, J. F. Wild, E. K. Hulet, R. W. Hoff, and J. H. Landrum, J. Inorg. Nucl. Chem. 40, 1865 (1978).
5. R. R. Chasman, I. Ahmad, A. M. Friedman and J. R. Erskine, Rev. Mod. Phys. 49, 833 (1977).
6. T. H. Braid, R. R. Chasman, J. R. Erskine and A. M. Friedman, Phys. Rev. C4, 247 (1971).
7. R. W. Hoff, W. F. Davidson, D. D. Warner, H. G. Börner and T. von Egidy, Phys. Rev. C25, 2232 (1982).
8. I. Ahmad, F. T. Porter, M. S. Freedman, R. F. Barnes, R. K. Sjoblom, F. Wagner, Jr., J. Milsted and P. R. Fields, Phys. Rev. C3, 390 (1971).
9. P. D. Kunz, University of Colorado (unpublished).
10. R. R. Chasman, Phys. Rev. C3, 1803 (1971).

SPECTROSCOPIC STUDIES NEAR THE PROTON DRIP LINE

K. S. Toth
Oak Ridge National Laboratory, Oak Ridge, TN 37831, USA

D. M. Moltz, J. M. Nitschke, and P. A. Wilmarth
Lawrence Berkeley Laboratory, Berkeley, CA 94720, USA

J. D. Robertson
University of Kentucky, Lexington, KY 40506, USA

ABSTRACT

We have investigated nuclei close to the proton drip line by using heavy-ion fusion reactions to produce extremely neutron-deficient nuclides. Their nuclear decay properties were studied by using on-line isotope separators at Oak Ridge (UNISOR) and Berkeley (OASIS), the Oak Ridge National Laboratory velocity filter, and a fast helium-gas-jet transport system at Lawrence Berkeley Laboratory 88-Inch Cyclotron. Many isotopes, isomers, and β-delayed-proton and α-particle emitters were discovered. This contribution summarizes three topics that are part of our overall program: (a) decay rates of even-even α-particle emitters, (b) mass excesses of ^{181}Pb, ^{182}Pb, and ^{183}Pb, and (c) β-delayed proton emitters near N = 82.

DECAY RATES OF EVEN-EVEN ALPHA-PARTICLE EMITTERS

Half-lives for s-wave α transitions, i.e., decays between ground states of doubly-even nuclei, are taken to represent unhindered α-decay rates. The resultant reduced widths are regarded as standards against which widths of other types of α transitions are to be compared. Figure 1 shows s-wave α widths for nuclei with Z from 78 to 100 plotted as a function of neutron number. We utilized Rasmussen's formalism[1] to calculate the width, δ^2, which is defined as: $\delta^2 = \lambda h/P$, where λ is the decay constant, h is Planck's constant, and P is the penetrability for the α particle to tunnel through a barrier. A rather regular behavior as a function of both neutron and atomic number is observed for these reduced widths. They are largest for nuclei two or four particles beyond a closed shell (with sharp minima occurring at the closed shell), followed by a decrease as one approaches the next closure. These trends can be understood in terms of single-particle models which have shown[2] that the extremely sharp break at N = 126 is essentially a shell structure effect. There is also a dip in values at N = 152 due to the subshell at that neutron number.

As stated above, the reduced widths around N = 130 are large; in particular, the available ^{218}Ra value (labeled as "Previous Measurement" in Fig. 1) exhausted 75% of the Wigner-sum-rule limit. Suggestions have been made[3] that these large widths indicate α clustering on the nuclear surface; the distorting effect of these

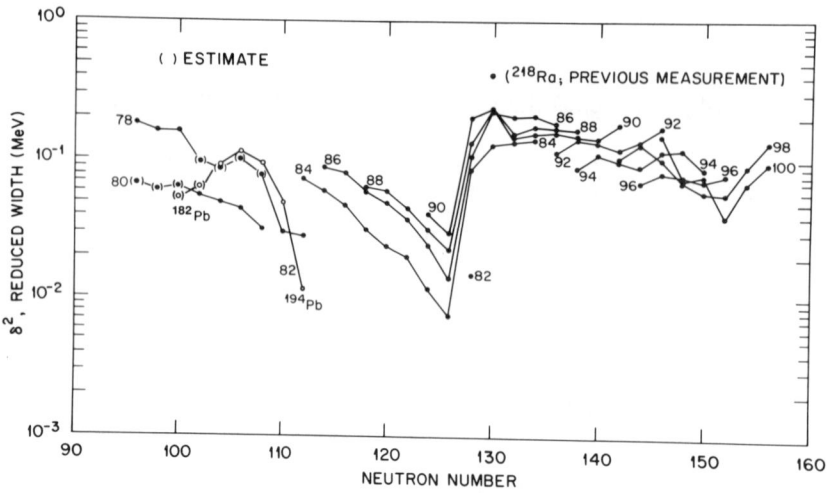

Fig. 1. Reduced widths for s-wave α transitions plotted as a function of N for isotopes with Z from 78 to 100.

clusters could account for reflection asymmetry observed in radium and thorium nuclei near N = 130. We remeasured[4] the ^{218}Ra half-life by using the ORNL velocity filter and a novel technique in which reaction products, after being separated from the incident beam, are implanted in a Si(Au) detector and their subsequent α decays are observed in the same detector. We found the half-life of ^{218}Ra to be 25.6(11) μs instead of the adopted value of 14(2) μs. It yields an α width which in Fig. 1 is indistinguishable from those of ^{216}Rn and ^{220}Th. This smooth trend of α widths from the N = 130 region to the well-deformed, prolate, Cm, Cf, and Fm nuclei weakens one argument quoted for the existence of α clusters in the heavy elements. It is the stabilizing effect of the shell closure that produces the sharp decrease in width values from N = 130 to N = 126.

However, contrary to an expected shell effect at Z = 82, the decay rates of ^{186,188,190,192}Pb (open points in Fig. 1 with N from 104 to 110) are less hindered than those of neighboring Hg isotopes. This implies that, midway between N = 82 and N = 126 the 82-proton shell is not magic.[5] We recently used the UNISOR facility to identify the α decay of ^{194}Pb for the first time and to determine the isotope's α branch.[6] The resultant width for ^{194}Pb (see Fig. 1) is much less than those of nuclei with N ≤ 110 and less than the $δ^2$ for ^{190}Pt (N = 112), indicating that perhaps the Z = 82 gap is being restored for N > 112. The reduced α widths for ^{190}Hg and ^{192}Hg should be determined to be certain about this proposal; if our suggestion is correct, the ^{192}Hg width should exceed that of ^{194}Pb.

We also show in Fig. 1 widths for ^{182}Pb and ^{184}Pb based on our data[7] and those of Ref. 8, respectively, assuming α branches of 100%. [From gross β-decay theory the calculated half-lives for ^{182}Pb and ^{184}Pb (β^+ + EC) decay yield α branching ratios > 90%.] These δ^2 values are appreciably smaller than the ^{186,188,190}Pb widths and may indicate that the Z = 82 shell reappears and retards the α decay of very proton-rich Pb isotopes.

MASS EXCESSES OF ^{181,182,183}Pb

The (55^{+40}_{-35})-ms half-life used to calculate the ^{182}Pb width was measured with the use of a rapid gas-jet-transport system at the LBL 88-Inch Cyclotron. In these experiments we identified ^{182}Pb (Ref. 7) and ^{181}Pb (Ref. 9) in ^{40}Ca bombardments of ^{147}Sm and ^{144}Sm, respectively. The α-particle spectrum accumulated in 222-MeV ^{40}Ca irradiations of ^{147}Sm is shown in Fig. 2. The (6.919 ± 0.015)-MeV peak is assigned to ^{182}Pb on the basis of excitation-function and cross-bombardment data and decay-energy systematics. Because the ^{182}Pb α-decay chain terminates at ^{146}Gd, our decay-energy measurement yields a rather precise mass excess for ^{182}Pb, i.e., -6823 ± 25 keV. The adopted[10] mass excess for ^{182}Pb is -6874 ± 28 keV.

In our investigation[7] two incident ^{40}Ca energies were used, i.e., 194 and 222 MeV, to enhance the production of ^{184}Pb [^{147}Sm(^{40}Ca,3n) product] and ^{182}Pb [^{147}Sm(^{40}Ca,5n) product], respectively. To obtain more definite data on ^{183}Pb and to provide additional cross-bombardment information we irradiated ^{147}Sm with 212-MeV ^{40}Ca ions. Figure 3 shows the accumulated α-particle spectrum. The 194-MeV spectrum was dominated by ^{184}Pb while the 222-MeV spectrum had only possible traces of ^{184}Pb and ^{183}Pb α activity (see Fig. 2). Figure 3 clearly represents an intermediate situation, namely, the intensity of ^{183}Pb is now somewhat greater than that of ^{184}Pb, and ^{182}Pb is seen with about the same intensity

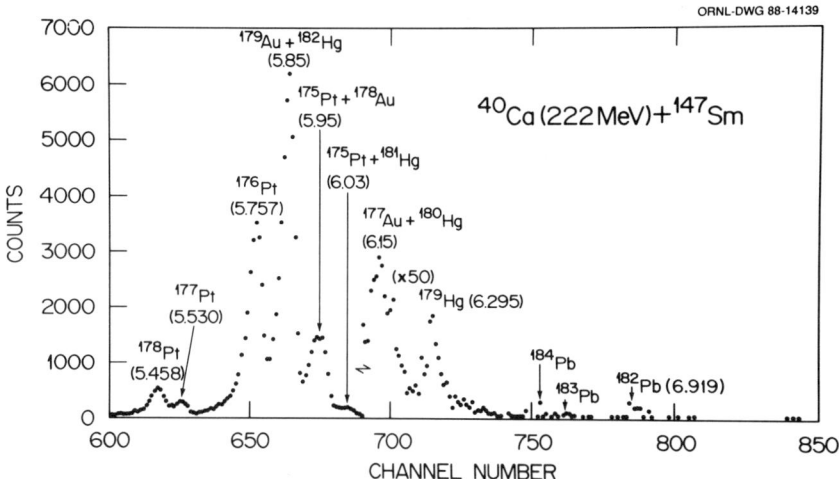

Fig. 2. Alpha-particle spectrum measured during a bombardment of ^{147}Sm with 222-MeV ^{40}Ca ions. Energies shown are in MeV.

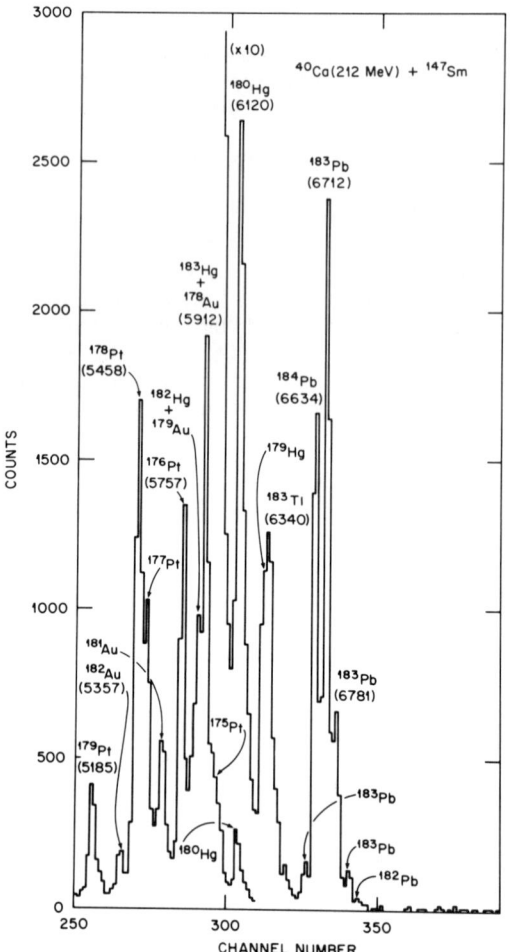

Fig. 3. Alpha-particle spectrum measured in bombardments of ^{147}Sm with 212-MeV ^{40}Ca. Energies shown are in keV.

as had been observed at 222 MeV. We assign a total of four α groups to ^{183}Pb. Previously, only two groups have been observed in ^{183}Pb α decay. A mass excess of -7720 ± 310 keV has been adopted[10] for ^{183}Pb; it is based on a Q_α value deduced from the highest-energy previously observed[8] α group of 6798 keV. Since we now have seen an α peak of 6874 keV both the adopted Q_α value and mass excess of ^{183}Pb will have to be revised.

Because the production of A < 182 lead nuclei with ^{147}Sm necessitates unfavorable reactions wherein six or more neutrons have to be evaporated (there is fission competition at each evaporation step) we used ^{144}Sm as the target in our search for ^{181}Pb. Figure 4 shows the spectrum measured at 201 MeV with the ^{144}Sm target. In

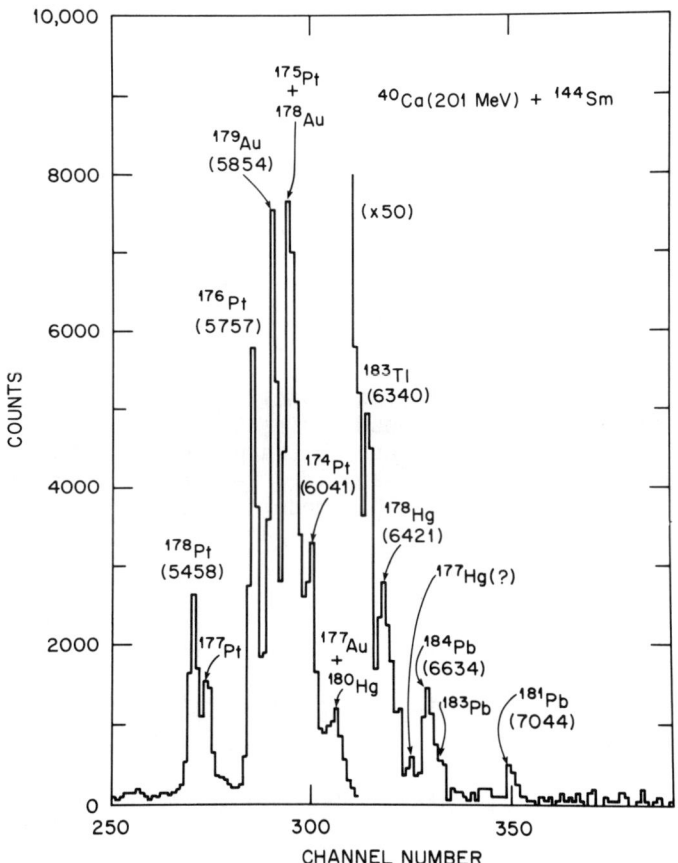

Fig. 4. Alpha-particle spectrum measured in bombardments of ^{144}Sm with 201-MeV ^{40}Ca ions. Energies shown are in keV.

addition to nuclides with $Z \leqslant 80$, one observes ^{184}Pb and ^{183}Pb produced from heavier samarium isotopes present in the target material [^{147}Sm (3.9%), ^{148}Sm (2.2%), ^{149}Sm (2.3%), etc.]. A new peak is seen at 7044 ± 15 keV. We assign it to the α decay of ^{181}Pb because its yield as a function of energy and target behaves as one would expect for the ^{144}Sm(^{40}Ca,3n) product, i.e., it is observed neither in ^{40}Ca + ^{147}Sm bombardments (Figs. 2 and 3) nor in data obtained at 212 MeV with the ^{144}Sm target.

Based on what is known about the α decays of ^{187}Pb, ^{185}Pb, and ^{183}Pb, we conclude that the 7044-keV peak is probably only the most intense of several ^{181}Pb α transitions. If one assumes that it proceeds between the ground states of ^{181}Pb and ^{177}Hg then the mass excess of ^{181}Pb can be estimated to be about −3.31 MeV based on the value of (−12950 ± 230) keV adopted[10] for ^{177}Hg. Our decay-energy measurement thus leads to the first mass determination for ^{181}Pb, a nucleus which is 26 mass units away from stability.

BETA-DELAYED PROTON EMISSION NEAR N = 82

We have investigated the decay properties of many short-lived neutron-deficient rare earth nuclei with 65 < Z < 71 by using the OASIS separator facility,[11] on-line at the Lawrence Berkeley Laboratory SuperHILAC. These isotopes were produced in fusion reactions in which targets of ^{96}Ru, ^{92}Mo, ^{94}Mo, ^{95}Mo, ^{96}Mo, and ^{93}Nb were bombarded with ^{64}Zn and ^{58}Ni projectiles. Following mass separation the radioactive products were assayed with a Si particle ΔE-E telescope, a thin plastic scintillator, and a hyperpure and two n-type Ge detectors.

Figure 5 shows a portion of the nuclidic chart which encompasses the mass region where these radioactivities are located. Some of the nuclei are at or close to the proton drip line. For many of them β-delayed-proton (and in a few instances direct-proton) emission becomes a probable mode of decay. Note that while all nuclei in Fig. 5 with N ≥ 84 are α-particle emitters, no α decay

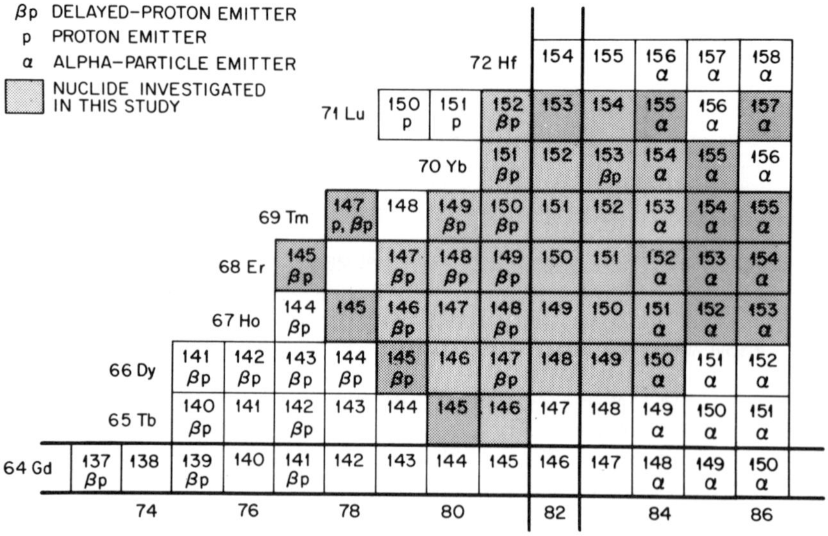

Fig. 5. Part of rare earth region encompassing investigated nuclei.

has been observed in rare earth isotopes with N < 84. This is due to the influence of the N = 82 closed shell which enhances the α-decay energies of nuclides with N = 84 and slightly higher but drastically reduces these energies for nuclides with N ≤ 83.

One thrust of our program has been an attempt to understand the pronounced peaks seen in the β-delayed-proton spectra of the even-Z N = 81 precursors $^{147}_{66}$Dy, $^{149}_{68}$Er, and $^{151}_{70}$Yb. Such structure is unusual for nuclei with A ≳ 70 because of the large level densities in the β-decay daughters at excitation energies high enough for proton emission. We have shown[12,13] that the proton peaks are associated with the β decays of the $s_{1/2}$ ground states in these N = 81 precursors and reflect regions of low densities of 1/2 and 3/2 levels in the range of 3.5 - 5.0 MeV in the N = 82 daughters. Protons emitted by the $h_{11/2}$ isomers in the same N = 81 precursors, because of energetics and angular-momentum considerations, sample levels in the daughters that have a centroid at about 7 MeV. Here the level densities are higher and the proton spectra associated with the $h_{11/2}$ states are structureless. For these three precursors one then finds peaks superposed on a statistical component which becomes progressively larger as the atomic number increases due to an increasing β-decay Q value. Gamma-ray decay studies, β-strength function measurements, and calculations of state densities and Gamow-Teller strength distributions, have led us to suggest[13] that the structure in delayed-proton spectra near N = 82 may arise from the preequilibrium decay of doorway states populated in β decay.

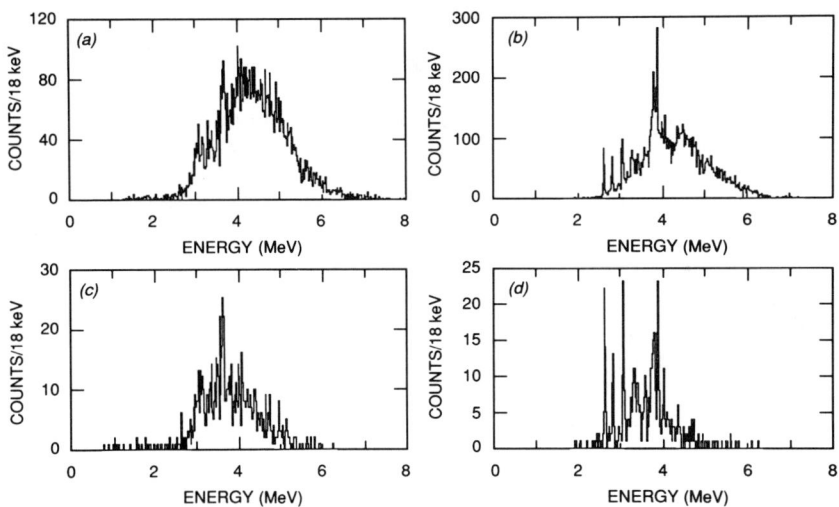

Fig. 6. Beta-delayed protons from ^{147}Er [(a)] and (c)] and ^{149}Er [(b) and (d)]; (a) and (b) show singles spectra, while (c) and (d) display protons observed in coincidence with positrons.

We have also studied the decay properties of the N = 79 precursors ^{145}Dy and ^{147}Er. Their β-decay daughters have the same atomic numbers as those of ^{147}Dy and ^{149}Er, respectively, but have 80 rather than 82 neutrons. A comparison of delayed-proton spectra from these four emitters should therefore provide information concerning the relative importance of the N = 82 and Z = 64 closures with regard to the observed proton structures.

The ^{145}Dy and ^{147}Er spectra were found to be basically structureless, though there were weak proton groups at the low-energy portions of the spectra. By requiring positron coincidences one can enhance the fraction of protons emitted from low excitation energies, i.e., from levels found[15,16] to be associated with the pronounced peaks in the N = 81 data. Figure 6 compares the singles protons [parts (a) and (b)] and protons in coincidence with positrons [parts (c) and (d)] for ^{147}Er and ^{149}Er, respectively. The structure observed in Fig. 6(a) is indeed emphasized in Fig. 6(c) though in neither spectrum are the peaks as pronounced as they are for ^{149}Er [Figs. 6(b) and 6(d)].

Thus the intense structures disappear from observed spectra almost as soon as the β-decay daughters no longer have a major closed-shell configuration. The role of the Z = 64 subshell in lowering level densities must be a minor one. This and other points are addressed in a recent thesis[14] where the results obtained at the OASIS facility for delayed-proton precursors, ranging from ^{119}Ba to ^{154}Lu, are described and discussed.

ACKNOWLEDGMENTS

We thank Y. A. Akovali, H. J. Kim, and J. W. McConnell for their contributions to this experimental program. Oak Ridge National Laboratory is operated by Martin Marietta Energy Systems, Inc. for the U.S. Department of Energy under Contract No. DE-AC05-84OR21400. Work at the Lawrence Berkeley Laboratory is supported by the Director, Office of Energy Research, Division of Nuclear Physics of the Office of High Energy and Nuclear Physics of the U.S. Department of Energy under Contract DE-AC03-76SF00098.

REFERENCES

1. J. O. Rasmussen, Phys. Rev. <u>113</u>, 1593 (1959).
2. H. J. Mang, Annu. Rev. Nucl. Sci. <u>14</u>, 1 (1964).
3. F. Iachello and A. D. Jackson, Phys. Lett. <u>108B</u>, 151 (1982).
4. K. S. Toth et al., Phys. Rev. Lett. <u>56</u>, 2360 (1986).
5. K. S. Toth et al., Phys. Rev. Lett. <u>53</u>, 1623 (1984).
6. Y. A. Ellis-Akovali, K. S. Toth, H. K. Carter, C. R. Bingham, I. C. Girit, and M. O. Kortelahti, Phys. Rev. C <u>36</u>, 1529 (1987).
7. K. S. Toth et al., Phys. Rev. C <u>35</u>, 2330 (1987).
8. U. J. Schrewe et al., Phys. Lett. <u>91B</u>, 46 (1980).
9. K. S. Toth, D. M. Moltz, and J. D. Robertson, Phys. Rev. C <u>39</u>, 1150 (1989).
10. A. H. Wapstra, G. Audi, and R. Hoekstra, At. Data Nucl. Data Tables <u>39</u>, 281 (1988).
11. J. M. Nitschke, Nucl. Instrum. Methods <u>206</u>, 341 (1983).
12. K. S. Toth et al., Phys. Lett. <u>B178</u>, 150 (1986).
13. J. M. Nitschke et al., Phys. Rev. Lett. <u>62</u>, (1989).
14. P. A. Wilmarth, Ph.D. thesis, University of California, Berkeley, 1988, Lawrence Berkeley Laboratory Report No. 26101.

Nuclear Structure of Neutron-rich Fission Fragments

W.R.Phillips

Department of Physics, Univertsity of Manchester, Manchester,M13 9PL, U.K.

Abstract

The yrast and near-yrast level structures of some very neutron-rich nuclei have been studied by observing prompt γ-rays emitted from ^{252}Cf and ^{248}Cm fission fragments. $\gamma-\gamma$ coincidence experiments have enabled levels with spins up to $\sim 14\hbar$ to be observed in strongly produced final fragments, viz those with N/Z ratios of ~ 1.55 and ~ 1.62, and with mass numbers A ~ 104 and ~ 144 respectively, near to the peaks in the mass distributions. Level schemes have been constructed from the coincidence data using, where possible, previously identified transitions between levels at low excitation energy. Several new nuclei have also been discovered, and their partial decay schemes determined, using an identification technique based on gating on transitions in complementary fragments. New nuclei ^{103,104}Zr, ^{107,108}Mo and ^{141,142}Xe, with N/Z ratios higher than given above, have been studied in this way. Spectroscopic studies in the higher of the two mass regions have revealed the extent and details of the new region near N=88 where strong octupole correlations may give rise to stable octupole deformations. Spectroscopic studies near A=100 have revealed details of how the large deformations in this region change with mass number and how they are strongly influenced by $h_{11/2}$ intruder orbitals.

1. Introduction

Fission is presently the most effective way of producing neutron- rich nuclei for spectroscopic studies. Experiments on the beta-decays of mass-separated fragments have yielded much information[1] on levels with spins close to that of the parent in many previously unexplored daughter nuclei. The advent of large arrays of high-resolution γ-ray detectors has made experiments possible[2] on discrete, prompt γ-rays emitted by final fission fragments formed following the fission process. These prompt γ-ray studies give details of partial decay schemes of yrast and near-yrast levels in neutron -rich nuclei, both for nuclei in which a few transitions were previously known and also for new nuclei about which nothing was previously known. This paper is concerned with the nuclear structure information recently obtained from experiments on prompt γ-rays in fission fragments.

2. The Extent of the Spectroscopic Studies

Many different nuclei are produced in fission and this severely hinders the examination of γ-rays emitted by any particular one of them. The mass and charge distributions of the final fragments, and where they occur on the nuclear chart, depend on the fissioning nucleus. Low energy fission of actinide nuclei, such as spontaneous fission (SF) and fission induced by low-energy light-particle bombardment, give rise to the most neutron-rich γ-

emitting fragments. Fission of systems at higher excitation energy, such as compound nuclei formed in heavy-ion fusion rections, produce less neutron-rich species. The extra neutron evaporation (both pre- and post-fission) needed to cool the fragments to the point where γ-emission occurs, causes the final fragment distribution to withdraw towards the line of stability. This is illustrated in Figure 1 which shows, for three representative fission processes, the most strongly produced isotope of each element, for isotopes produced with a yield of half or more of the strongest fragments.

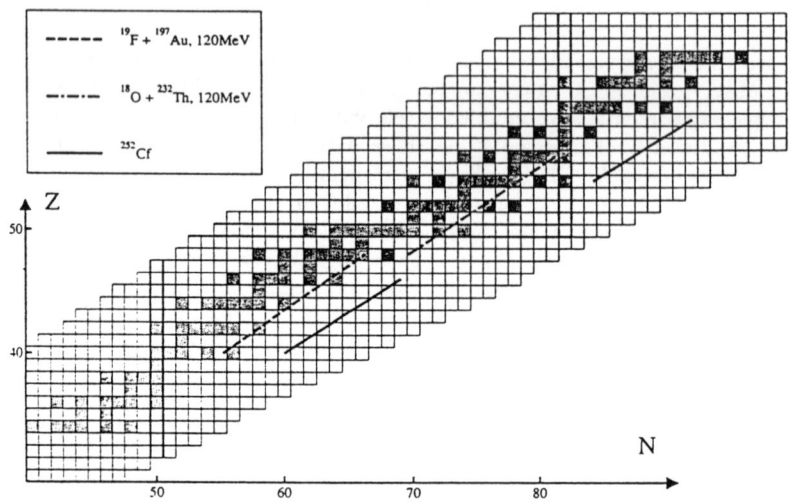

Figure 1: Yields of final fragments from three different fission processes. The lines show the most strongly produced isotopes of those elements formed with a yield of one half or more of that of the strongest species. The lines for ^{248}Cm and ^{254}Cf fall very near but to the right of the ^{252}Cf line. Hatched squares are stable isotopes.

The mass distributions from SF sources suitable for γ-ray work are double humped, and within the two humps, centred on A $\sim$104 and 144, the strongest isotopes formed as final fragments have N/Z ratios of $\sim$1.55 and 1.62 respectively. The most copiously produced fragment typically comprises $2 - 3\%$ of the total fission yield. In the immediate future the best way to study nuclei nearest the neutron-drip line, albeit still far from it, is by spontaneous or slow neutron-induced fission experiments. However, the ability to vary to some extent the most strongly produced fragments can make charged particle fusion-fission rections useful for studying specific nuclei.

The fractional yield of the nucleus being examined among the many γ-emitting fragments, and the limitations on sensitivity of existing γ-ray detector arrays, place limits on the range of neutron-rich nuclei which can be studied at the moment. Another parameter which determines the extent of the information obtained on fragments, for a given experimental sensitivity, is the average spin at which they are formed. The mechanism of production of discrete, near-yrast, levels in final fragments is somewhat similar to the way they are produced in a fusion-evaporation residue. Primary fragments are produced at scission at variable excitation energy and spin, and usually emit one or more neutrons before reaching points in the excitation energy- spin plane in the final fragments from which they emit γ-rays. However, the average spin <J> in fragments is much less than for typical heavy-ion fusion-evaporation residues, and for fusion-fission reactions can not be increased significantly by increasing the angular momentum of the fissioning nucleus. Figure 2 shows <J> for selected final fragments from three different fissioning systems[3]. The average spins from the heavy-ion reactions are only slightly higher than from SF. There is only a small change in <J> as the bombarding energy E_B is increased for fusion-fission reactions , i.e. as the average angular momenta $< l >$ of the fissioning nuclei increase. The measured average γ-ray multiplicities $< M_\gamma >$ for all pairs of fragments from several different reactions are roughly constant. Although the relationship between $< M_\gamma >$ and <J> is complex, the trend in the former should reflect the trend in the latter. The observation that $< J >$ increases only slowly with E_B is probably due to the increased pre-fission neutron emission[4] as E_B is raised, which results in primary fragments of nearly constant average temperature. Within a statistical model interpretation[5] this would lead to nearly constant spin. There is some evidence [6] that for reactions with ions of mass ≥ 40 at bombarding energies such that most of the reaction cross-section can not be attributed to fusion, M_γ, and hence <J>, increases more markedly with E_B. However, these collision processes produce γ-emitting fragments nearer to the stability line than the fissions discussed here because of increased neutron emission.

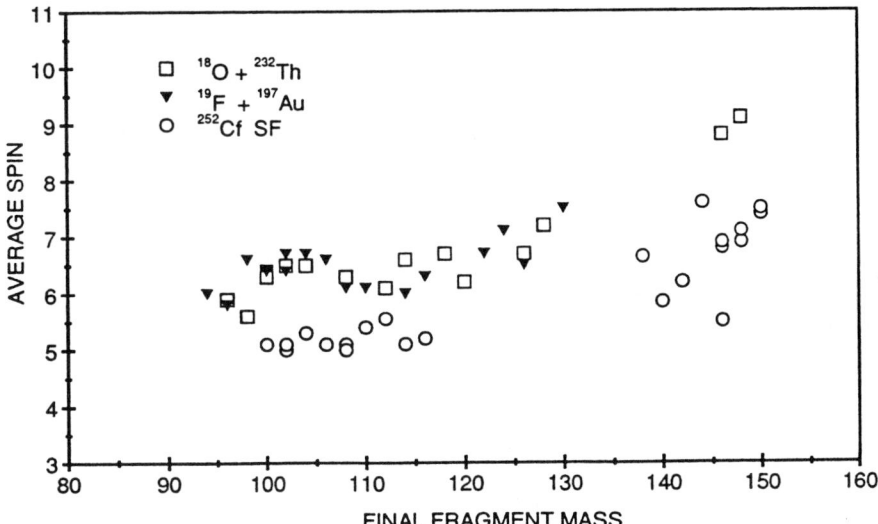

Figure 2: The average spin at which discrete levels in final fragments are populated in three diffrent fission processes: ^{252}Cf SF, the fusion-fission of a 120 MeV beam ^{19}F beam with a thick Au target, and the fusion-fission of ^{18}O with a thick ^{232}Th target at an incident energy of 120 MeV. Errors on the spins vary from 0.3 to 0.5 and have been omitted for clarity.

3. Experiment and Analysis Techniques

Yrast levels in fission fragments have recently been studied using large arrays of Compton-suppressed Ge-detectors operating in coincidence. Low energy γ-ray and X-ray detectors (LEPS), which have better energy resolution than Ge-detectors, have also been used to help clarify decay schemes and to measure internal conversion coefficients.

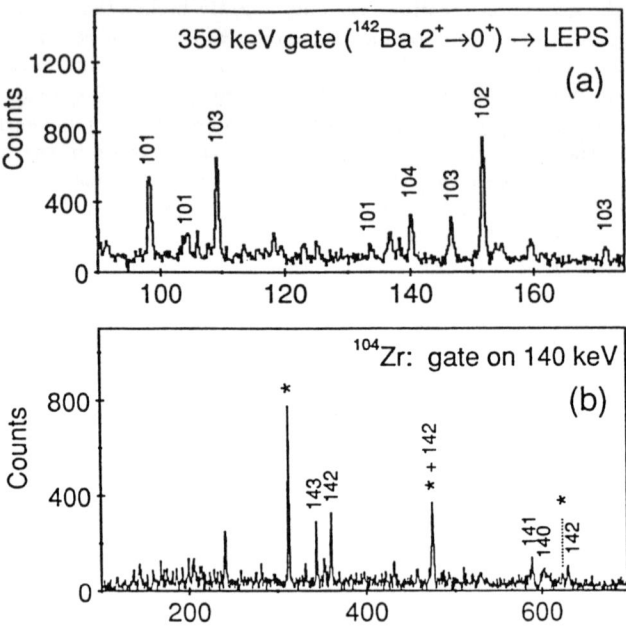

Figure 3: (a) Part of a spectrum of γ-rays observed in a LEPS detector in coincidence with the ^{142}Ba $2_1^+ - 0_1^+$ (360 keV) transition. Peaks are numbered with the masses of the corresponding complementary Zr fragments. (b) Part of a spectrum of γ-rays observed in Ge-detectors in coincidence with the peak at 140 keV shown in (a). Peaks are labelled with the complemetary Ba fragment mass, or with a star if in the ^{104}Zr ground-state band.

The starting point for constructing a decay scheme in a particular final fragment is a transition, when known, between low-lying states in that fragment. Known transitions have been determined either from the early pioneering work[7] in this field, or from beta-decay studies. Decay schemes in nuclei for which nothing was previously known can also be constructed[8] by identifying a γ-transition in a new nucleus using the fission γ-ray data themselves. These data consist of prompt coincidence events between transitions in a particular nuclear decay sequence and also coincidence events involving transitions in complementary fragments. The primary fragments are produced at a range of excitation energies such that neutrons may be evaporated from each fragment. The total number of neutrons emitted per SF event, ν, typically ranges from 1 to 5. A transition in a given nucleus will therefore appear in coincidence with a number of complementary fragments.

Hence, for example, in the SF of ^{248}Cm transitions in a range of Zr isotopes will appear in spectra gated on a particular transition in a Ba fragment. While this complicates the coincidence spectra, it also provides a means for identifying the origin of unknown transitions. By studying spectra in coincidence with transitions in a series of Ba fragments, a sequence of γ-rays in different Zr isotopes is observed, among which candidates can be found for transitions in Zr isotopes heavier than those previously known. The procedure is illustrated in Figure 3. Part (a) is a γ-ray spectrum obtained in a LEPS detector by gating on $2_1^+ - 0_1^+$ transitions observed in an array of Ge-detectors. ^{142}Ba is lighter than the strongest Ba final fragment (^{144}Ba) and thus we expect to see transitions in the higher mass Zr isotopes. Candidate transitions for previously unexplored nuclei ^{103}Zr and ^{104}Zr (corresponding to ν=3,2) are indicated on the figure along with known γ-rays in ^{101}Zr and ^{102}Zr (ν=5,4). These candidates are established as belonging to Zr nuclei because spectra obtained in coincidence with them show known transitions in a corresponding range of Ba isotopes, as well as other transitions that can be assembled into a consistent level scheme for that particular Zr isotope. Figure 3(b) shows a spectrum observed in coincidence with the candidate $2_1^+ - 0_1^+$ transition in ^{104}Zr. From the relative yields of the Ba fragments seen in this spectrum, the mean value of the Ba mass associated with this transition can be deduced. The same procedure was followed for other transitions proposed in ^{104}Zr as well as for transitions proposed in ^{103}Zr and for known transitions in the lighter Zr isotopes. In Figure 4 the resulting average Ba masses are plotted as a function of the known and proposed Zr masses. A smooth trend is seen, confirming the mass assignments for the new level sequences.

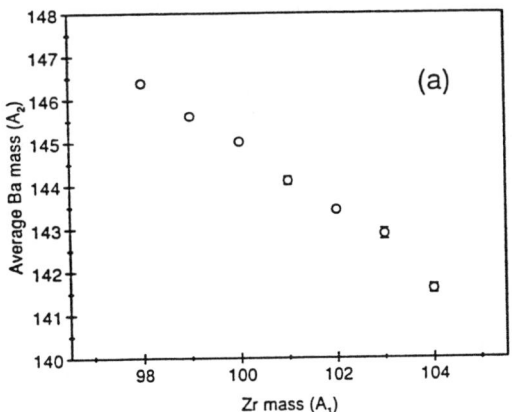

Figure 4: The average mass of Ba complementary fragments associated with different Zr final products.

Using the technique described above, transitions in new nuclei ^{103,104}Zr, ^{107,108}Mo and ^{141,142}Xe have been determined from SF experiments, and partial decay schemes in these nuclei obtained. Supporting evidence for the assignments in each case is provided by the relative yields of the appropriate isotopes. These approximate Gaussian distributions with maxima at mass numbers $A_m(Z) \cong Z(A_0 - <\nu>)/Z_0 + k$. For the SF of ^{248}Cm, A_0=248;

$Z_0 = 96$, and $<\nu> = 3.2$. The number $k \cong +1.5$ and -1.5 for heavy and light fragments respectively.

4. Spectroscopy Results

4.1 Nuclei near N=88

For neutron number N near 88, valence neutrons occupy orbitals on which there is a large influence from the octupole component of the residual nucleon-nucleon interaction. For proton number Z near 56, there are the same strong effects for protons. The resulting electric octupole (E3) correlations greatly influence the structure at low excitation energy and spin of the nuclei in this region, which chiefly comprises neutron excess nuclei amenable only to fission studies. An important question is the extent to which the strong E3 correlations produce yrast structures with a reflection-asymmetric deformation, i.e. are there level sequences with wave functions which approximate those for rotations of an octupole-deformed nuclear density distribution? For such distributions the centre of mass of the proton matter may be separated from that of the neutron matter and there could be large intrinsic static electric dipole moments. If this were so an octupole deformed even-even nucleus would exhibit a sequence of levels with $J^\pi = 0^+, 1^-, 2^+, 3^-, 4^+, \ldots$. The levels would show rotational spacings, and there would also be strong E1 transitionss between $\Delta J=1$ states, as well as strong E2 transitions between $\Delta J=2$ states. The characteristics of some octupole deformed odd-A nuclei would be near degenerate parity doublets arising from the motion of the odd nucleon in the reflection asymmetric mean field. Associated with these doublets would be rotational bands with roughly similar properties. Several nuclei with yrast structures approaching those described above have been observed[9] in the light actinides, where strong E3 correlations are also expected between valence protons and between valence neutrons. These structures occur where predicted by theory and lend support to the existence of stable octupole deformations in certain regions of nuclei, although there exist no unambiguous signatures of such deformations. It remains possible that the level structures observed could also be described, within a macroscopic picture, in terms of vibrations of octupole-soft nuclei. There is also evidence in support of octupole deformation in the N=88 region, although the examples are not so clear or numerous as in the light actinides. This is partly because of the difficulty of studying the N=88 region and partly because the spin range over which possible octupole bands remain yrast (and thus amenable to study) in these nuclei is often smaller than in the actinides; band crossings often occur at lower spins in the lighter mass region. Extensive calculations of the nuclear potential energy surfaces (PES) have been made[10] for assumed axial symmetry, using macroscopic models for the smoothly varying part and deformed Woods-Saxon potentials with the Strutinsky procedure to calculate shell corrections. These calculations predict islands of nuclei in the light actinides and near N=88, Z=56 with minima in the PES which occur at reflection-asymmetric shapes. The depths of these minima, ΔV, referred to the minima for reflection-symmetric shapes, are major factors in determining whether the nuclei have sufficiently "stiff" octupole deformation to exhibit rotational features. Figure 5 shows predicted ΔV for the ground states of several nuclei in the N=88 region. The depths of the minima are typically ≤ 0.5 MeV compared with ≤ 1.0 MeV in the light actinides, suggesting that rotational features may not be so clear in the former region as in the latter. The conclusion that several lanthanide nuclei have PES with "pear-shaped" minima is supported by fully microscopic calculations for Ba isotopes[11]. Rotation of the nuclei can modify the energy level patterns, and hence the shell corrections, in such a way as to increase ΔV and result in more stabilised octupole shapes. This occurs in the N=88 Ba, Ce, Nd and Sm nuclei for which partial yrast decay schemes have been determined, the more neutron-rich via fission experiments[12], the less neutron-rich via fusion-evaporation reactions with alpha- particle beams[13]. Figure 6 shows the level scheme obtained for ^{144}Ba (relatively large ΔV) and the way it approximates the alternating parity levels expected for reflection-asymmetric rotors. Similar structures are found in ^{146}Ce, ^{148}Nd and ^{150}Sm. The remaining even-even N=88 nucleus which may be expected to show large effects from E3 correlations is ^{142}Xe. The $2_1^+ - 0_1^+$ transition in this previously unknown nucleus was determined, as described in section 3, from ^{248}Cm SF data.

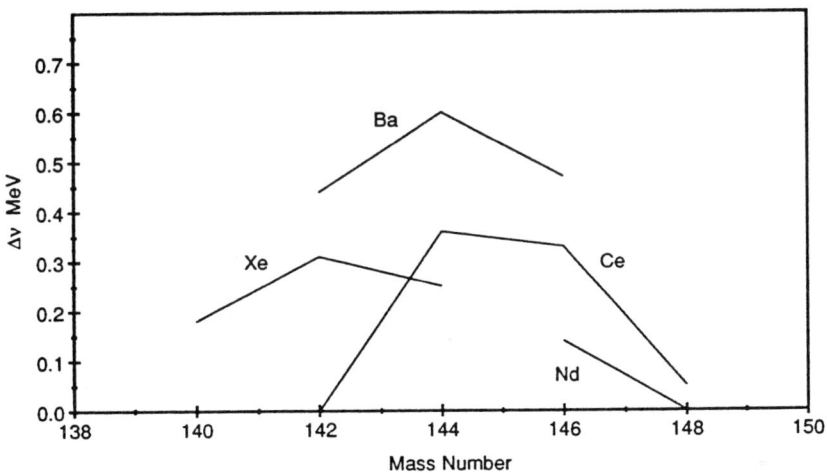

Figure 5: Predicted values of ΔV, the difference in minima of calculated ground-state PES with and without reflection asymmetry, for some even-even nuclei in the N=88 region.

Figure 6 also shows the partial decay scheme obtained for ^{142}Xe. No alternating parity sequence of levels becomes yrast below spin $\sim 10\hbar$, and no sideband is populated with greater than 20% of the intensity of the positive parity sequence. The low energy structure of ^{142}Xe is thus different to that of the other N=88 nuclei of Figure 5. This single observation does not, of course, invalidate the models used to describe the influence of E3 correlations .

The overall agreement between experiment and theory is very good, and extends to the intrinsic electric dipole moments deduced from the E1 transition rates measured in the "octupole" even-even nuclei[14]. The difficulty with ^{142}Xe is probably connected with deficiencies in the calculations as the Z=50 closed shell is approached. However, the example does illustrate how information over a wide range of nuclei is needed to complete a satisfactory model description.

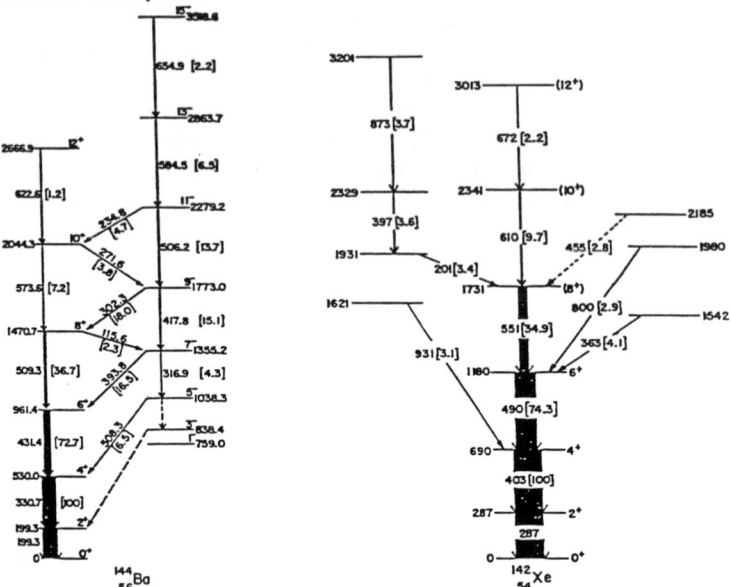

Figure 6: The partial decay schemes of ^{144}Ba and ^{142}Xe obtained from SF of ^{248}Cm.

4.2 Nuclei near A=100

The rapid onset of large deformation in the neutron-rich nuclei with Z≥ 38 as the neutron number passes through 60 is a feature of the systematics of nuclear ground states. Values of the deformation parameter β_2 approaching ∼ 0.4 are suggested by interpretations of lifetime measurements of $2_1^+ - 0_1^+$ transitions in these nuclei. These deformations, at zero spin, are as large as those met at high spin in several "superdeformed" nuclei. It is likely that over the whole nuclear chart there exists a continuous wide range of deformations to be probed in cold nuclei, both at low and at high spin.

Much theoretical work has gone into the description of deformed nuclei near A=100. Prolate ground state deformations are predicted by a number[15] of PES calculations made with axially symmetric deformed mean fields. These suggest that components of the $h_{11/2}$ unique parity neutron orbit lie below the Fermi level at N ∼ 60, and that occupancy of these components plays a major role in stabilising the large deformations. Calculations with a spherical mean field[16] attribute more importance to the isovector interaction between neutrons and protons in $g_{7/2}$ and $g_{9/2}$ orbits respectively. The distinction between the two approaches is in the degree of deformation predicted, and up to now there has been no evidence on orbit occupancies, although the 2_1^+ lifetimes suggest large deformations consistent with the first approach. The present fission work on even-even nuclei near A=100 has provided more information on deformation trends in this region; the work on odd-A nuclei has provided the first good identification of Nilsson orbitals close to the Fermi level.

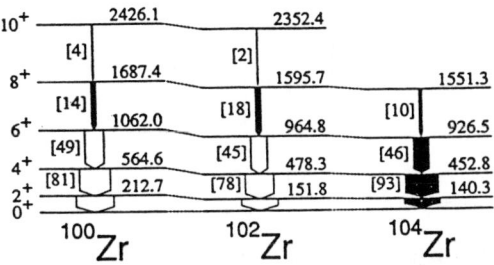

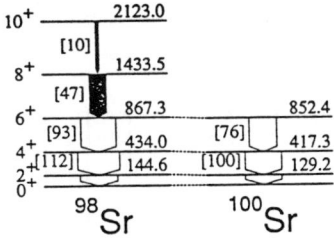

Figure 7: Ground-state bands observed in even-even Sr and Zr isotopes. New transitions are shown shaded. Relative intensities (within each isotope) of transitions are shown in square brackets.

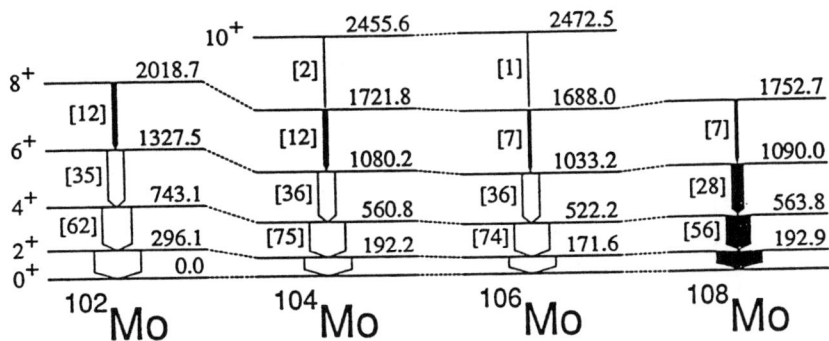

Figure 8: Ground-state bands observed in even-even Mo isotopes.

Figures 7 and 8 show the main yrast sequences observed in even-even Sr, Zr and Mo isotopes. (^{104}Zr and ^{108}Mo are new nuclei). The excitation energies of the 2_1^+ states

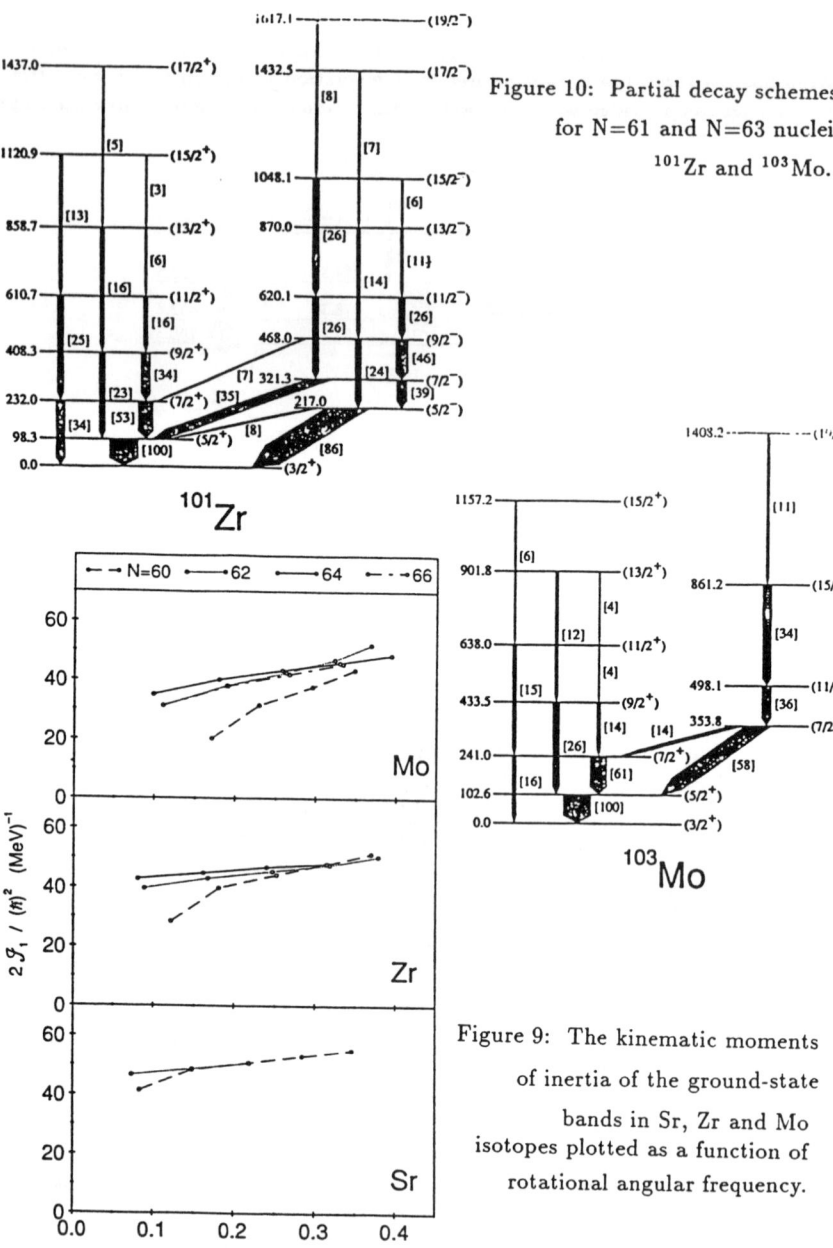

Figure 10: Partial decay schemes for N=61 and N=63 nuclei ^{101}Zr and ^{103}Mo.

Figure 9: The kinematic moments of inertia of the ground-state bands in Sr, Zr and Mo isotopes plotted as a function of rotational angular frequency.

suggest that the deformations in the Zr isotopes continue to increase up to N=64, at which point the deformation (using the same criterion) reaches a maximum in the Mo isotopes. The data of Figures 7 and 8 can be used to derive kinematic moments of inertia. These are shown in Figure 9, and, especially for the N=64 nuclei, are nearly constant over the angular frequency range shown, indicating that over this interval the angular momentum is nearly all rotational.

Partial decay schemes in the odd-proton nuclei ^{99,101}Y, ^{101,103}Nb and ^{105}Nb with Z=39 and 41, and in the odd-neutron nuclei ^{101,103}Zr, ^{103,105}Mo and ^{107}Mo with N=61, 63 and 65 have also been obtained from SF experiments. (^{105}Nb, ^{103}Zr and ^{107}Mo are new nuclei). Figure 10 shows the schemes for the N=61 and 63 nuclei which we will concentrate on here. The ground state bands in ^{101}Zr and ^{103}Mo have been assigned $K^{\pi} = 3/2^+$ and associated with the [411] Nilsson orbital[17]. The observed side-band in ^{101}Zr is most probably associated with the unique parity $5/2^-$[532] orbit arising from the spherical $h_{11/2}$ intruder level. The evidence for this comes from the decay branches of the 321 keV excited state, which predominantly decays to the $5/2^+$ member of the ground state band, there being less than 2% decay to the ground state. If the band-head of the side-band were spin 3/2, simple intensity rules predict a dominant ground state decay for the 321 keV level. The side-band is then assigned $K^{\pi}=5/2^-$, the odd parity being chosen because of the proximity of levels of the same spin in the two bands. For similar parities mixing would normally prevent such proximity.

The properties of the side-band in ^{103}Mo are also consistent with its association with $h_{11/2}$ intruder states. The trends shown on Figure 10 suggest that the moment of inertia of ^{103}Mo is somewhat lower than for ^{101}Zr. This results in a stronger Coriolis interaction in the former nucleus and the side-band in ^{103}Mo is more decoupled than in ^{101}Zr. The favoured sequence of levels is then $7/2^-, 11/2^-, 15/2^-, ...$ with only the E2 transitions observed. In the N=63 nuclei ^{103}Zr and ^{105}Mo, the ground state bands are closely similar to the ^{101}Zr side-band. In these nuclei the deformations approach the maximum observed in this region, and only a relatively small amount of signature splitting occurs, the degree of splitting in the ground state bands of ^{103}Zr and ^{105}Mo being about the same as in the ^{101}Zr side-band.

The main conclusion from the present results on the spectroscopy of fission fragments near A=100 is that the $5/2^-$[532] orbital lies very close to the Fermi level in the N=61 and 63 nuclei. The simplest picture then suggests that the $\Omega = 1/2^-, 3/2^-$ levels from the intruder $h_{11/2}$ orbitals are already occupied in these nuclei and that the $h_{11/2}$ orbitals play a major role in stabilising the large deformations observed in this region.

Acknowledgements

The work discussed in this paper has been done in collaboration with several colleagues at the Argonne Laboratory, USA and at the University of Manchester and the Daresbury Laboratory, UK. Special thanks are due to I. Ahmad, J.L.Durell, M.A.C.Hotchkis, R.V.F.Janssens, T.L.Khoo, L.R.Morss, and A.S.Mowbray.

References
1) e.g. Nucl. Spec. of Fission Fragments, IOP (UK), Conference Series, No. 51(1980)
2) W.R.Phillips et al., Phys.Rev.Lett.57(1986)3257
3) Y.Abdelrahman et al., Phys.Lett.199B(1987)504
4) D.J.Hinde et al., Nucl.Phys.A452(1986)550
5) T.Ericsson, Phil.Mag.Supp.((1960)425
6) R.Bock et al., Nucl.Phys.A388(1982)334
7) E.Cheifetz et al., Phys.Rev.C4(1971)1913
8) M.A.C.Hotchkis et al., Phys.Rev.Lett.64(1990)3123
9) M.Dahlinger et al., Nucl.Phys.A484(1988)337 and refs. therein
10) e.g.A.Sobiczewski et al., Nucl.Phys.A485(1999)16
11) J.L.Egido and L.M.Robledo, Nucl.Phys. to be published
12) Ref.2) above and W.R.Phillips et al., Phys.Lett.212B(1988)402,
A.S.Mowbray et al., Phys.Rev.C42(1990)1192
13) W.Urban et al., Phys.Lett.185B(1987)331,
W.Urban et al., Phys.Lett.200B(1988)424
14) e.g.W.R.Phillips, Proc.Int.Conf.Nucl.Phys.,Sao Paulo, Brasil (1989),Vol.2,p.363
15) e.g.A.Faessler et al., Nucl.Phys.A230(1974)302,
P.Bonche et al. , Nucl.Phys.A443(1985)39,D.Galeriu et al., J.Phys.$\bar{\text{G}}$12(1986)329
16) P.Federman and S.Pittel, Phys.Rev.C20(1979)820,
A.Etchegoyan et al., Phys.Rev.C39(1989)1130
17) e.g.T.Seo et al., Z.Phys.A320(1985)393 and refs. therein

LEVEL LIFETIMES IN THE PS REGION: SPHERICAL AND DEFORMED STRUCTURES AT A ~ 100

H. Ohm, M. Liang, M. Büscher, U. Paffrath, B. De Sutter, K. Sistemich
Institut für Kernphysik, Forschungszentrum Jülich,
D–5170 Jülich, F.R. Germany

G. Molnár
Institute of Isotopes, H–1525 Budapest, Hungary

ABSTRACT

Level lifetimes in the range between about 10 ps and 1 ns have been determined for the heavy isotopes of Zr, Nb and Mo at the fission–product separator JOSEF. The results allow the identification of spherical configurations for the lighter isotopes and the determination of the deformation of the heavier ones. It could be shown that a rapid onset of deformation takes place at N = 60 in the Nb chain since a deformation of $\beta_q = 0.40(5)$ and $0.37(8)$ is deduced for ^{101}Nb and ^{103}Nb, respectively, while spherical shell–model and particle–phonon coupled states exist in ^{99}Nb. Hence, the situation in the Nb chain is similar to that in neighbours with smaller Z. The closed shell character of ^{96}Zr is confirmed by the half–life of 127(10) ps of the 8_1^+ level at 4390 keV. This half–life is in accordance with the expectation for a $\pi(g_{9/2})^2$ configuration and an effective charge $e_{eff} = 1.65(6)e$. The energy of the state evinces a strong single particle gap at Z = 40.

INTRODUCTION

An outstanding property of the neutron–rich nuclei with masses around 100 is the unusually sudden transition from spherical to deformed shapes which takes place between N = 58 and N = 60. The levels of members of the isotopic chains around Zr with neutron numbers below 60 can typically be understood in terms of the spherical shell model whereas ground–state rotational bands and strongly enhanced B(E2) transitions are observed for N ≥ 60. Shape coexistence has been found around the border line of N = 59.

The structure of these A ~ 100 nuclei is mainly studied through nuclear spectroscopy since most of them are unstable fission products. Hence, the half–lives of individual levels and the resulting absolute probabilities of transitions between levels are particularly important experimental quantities for the determination of the properties of these isotopes. Half–lives have been obtained in some cases from Doppler–shift studies, e.g. those of the 2_1^+ states of the even–even deformed nuclei[1]. Most of the knowledge about half–lives has, however, been yielded from the study of delayed coincidences between the radiation feeding a level and that which depopulates it. Such investigations are well suited for half–lives in the ns region if Ge detectors are involved. Only exceptionally and with relatively large uncertainties half–lives as low as 0.1 ns could be assessed, see e.g. Ref. 2.

Recently a novel technique has been introduced[3,4] for the determination of half–lives in the ps region which makes use of the fast timing properties of BaF$_2$ crystals for γ radiation. This technique has been applied[5] at the fission–product separator JOSEF[6] in systematic studies of the heavy elements of Zr,

Nb and Mo. The Zr chain is of special interest because of the existence of a double subshell closure[7] at ^{96}Zr, of shape isomerism[8] at ^{100}Zr and of an especially strong deformation[1] at ^{102}Zr. The even Mo isotopes also show deformations for N ≥ 60, but these are smaller than in the corresponding Zr isotopes. There is evidence for the existence of rotational structures in the heavy Nb isotopes, see e.g. Ref. 9, but their deformation has still to be determined. As examples for the information which has been gained in these studies the cases of ^{96}Zr and the isotopes ^{99}Nb, ^{101}Nb and ^{103}Nb are discussed in the following.

EXPERIMENTAL PROCEDURE

The technique to determine the level half–lives in the ps region consists[3,4,5,10] in the measurement of the delay between the feeding of the level of interest through the ß- decay of the parent and its depopulation by γ radiation. The ß- particles have been detected with a 4 mm thick disc of NE 111 A (30 mm diameter) coupled to an XP 2020 photomultiplier tube. The γ radiation was measured with a BaF$_2$ crystal of the shape of a truncated cone (top and base diameter of 20 and 30 mm, respectively) with a XP 2020 Q tube. Dynode timing was applied.

Since the energy resolution of the BaF$_2$ detector is not sufficient to identify the individual γ rays of complex level schemes in the singles spectrum a coincidence with a Ge detector was set up. By the choice of the proper coincidence gate in the Ge spectrum, individual γ rays could be selected in the BaF$_2$ spectrum. Background has been subtracted both for the Ge and the BaF$_2$ gates.

The time resolution of the plastic–BaF$_2$ setup amounted to 150 ps for γ rays of 500 keV. Consequently, level half–lives of about 100 ps or more showed up as a delayed slope of the otherwise symmetric time distributions. An example is given in Fig. 1. The ß–γ-time distribution for the 469 keV transition in ^{99}Nb has been obtained with gates on the 469 and 546 keV transitions in the BaF$_2$ and Ge spectrum, respectively. If the gates are inverted, the distribution of the 546 keV line results, which represents a prompt case. The half–life of the level at 469 keV in ^{99}Nb has been deduced from a fit to the time distribution with a convolution of an exponential and a prompt Gaussian shape, see the curve in Fig. 1. The result of $t_{1/2}$(469 keV) = 173(4) ps agrees with that of 0.21(6)ns obtained[11] with the use of Ge detectors, but is much more precise.

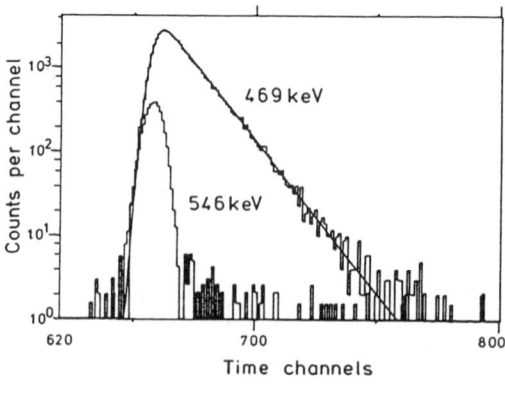

Fig. 1.
ß–γ-time distributions for the 469 and 546 keV transitions in ^{99}Nb. The time calibration is 20.9 ps/channel. The intensity of the 546 keV transition has been reduced by a factor of 13.

If the half–life of the investigated level is considerably smaller than 100 ps, then the centroid shift method is applied. An example is the determination of $t_{1/2}(3^-;{}^{96}Zr) = 46(15)$ ps which revealed[12,13] an unusually high collectivity of $B(E3;3^- \to 0^+) = 69(+34,-17)$ W.u.. For the application of the centroid shift method it is an advantage that the beam of JOSEF contains several isotopes. Thus, an intrinsic determination of the prompt timing curve is usually possible, see e.g. Ref. 14.

SPHERICAL NUCLEI

^{96}Zr: An 8^+ level has been identified[15] at 4390 keV in this nucleus with the subshell closures Z = 40 and N = 56. It is populated through an allowed ß$^-$ transition (log ft = 5.0) from the high–spin isomer of ^{96}Y and decays strongly[16] via a cascade of stretched E2 transitions into the 0_2^+ first excited state. Since the isomer of the odd–odd parent ^{96}Y is interpreted[17] as the 8^+ member of the $[\pi g_{9/2}, \nu g_{7/2}]$ multiplet and the 0_2^+ state of ^{96}Zr contains[18] a strong $\pi(g_{9/2})^2$ configuration it is reasonable to assume[13,15] that the 8^+ level at 4390 keV in ^{96}Zr is the highest spin member of the $\pi(g_{9/2})^2$ two–proton band.

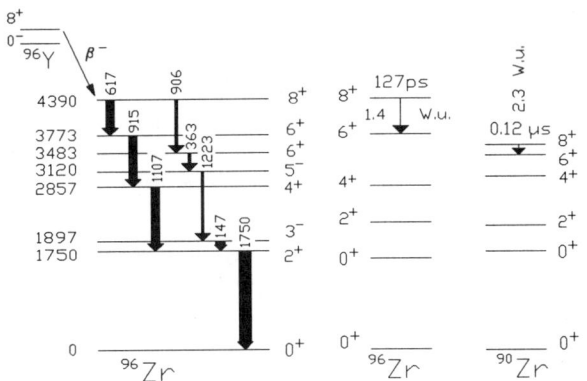

Fig. 2. Selected levels and major γ transitions from the decay of the 8^+ level in ^{96}Zr (left) and comparison of band structures in ^{96}Zr with ^{90}Zr. For more detailed information see Ref. 13.

If so, then the strength of the $8_1^+ \to 6_2^+$ transition which forms major depopulation of the 4390 keV level is expected to be of the order of one single particle unit (W.u.). Hence, the half–life of the level is expected to lie around 100 ps. The result which has been obtained at JOSEF amounts to

$$t_{1/2}(8^+;{}^{96}Zr) = 127(10) \text{ ps}$$

supporting the conjectures concerning its nature. More specifically, the deduced probability for the 617 keV transition is

$$B(E2) = 36.3 \text{ e}^2\text{fm}^4 = 1.4(1) \text{ W.u..}$$

The reduced probability for the $8^+ \to 6^+$ transition within the $(\pi g_{9/2})^2$ band calculated according to the formalism of Ref. 19 corresponds to

$$B(E2)_{th} = 13.4 \text{ e}^2 \text{ fm}^4$$

if a radial matrix element of $<g_{9/2}|r^2|g_{9/2}> = 23.4$ fm^2 is assumed[20]. Hence, an effective charge of

$$e_{\text{eff}} = 1.65(6) \text{ e}$$

is deduced which compares well with values at other magic nuclei, in particular with the value[21] of 1.72 e for ^{90}Zr.

It is of interest to compare the energy of the $\pi(g_{9/2})^2$ 8^+ level with the prediction that can be made from the known masses[22] of ^{96}Zr and its neighbours and the $\pi g_{9/2}$ excitation energies of the N = 56 isotones. Here the procedure which has been outlined in Ref. 20 for the $\pi(h_{11/2})^2$ 10$^+$ state of ^{146}Gd is followed:

$$E(8^+) = 2\,\delta_p + V_{hh} + V_{ph} + V_{pp} \qquad = 4.62 \text{ MeV}$$

where δ_p = $S_p(^{96}\text{Zr}) - S_p(^{97}\text{Nb})$ = 4.06 MeV
V_{hh} = $S_p(^{95}\text{Y}) - S_p(^{96}\text{Zr})$ = −1.85 MeV
V_{ph} = $2[E(g_{9/2},^{95}\text{Y}) - S_p(^{95}\text{Y}) - E(g_{9/2},^{97}\text{Nb}) + S_p(^{97}\text{Nb})]$ = −2.25 MeV
V_{pp} = +0.60 MeV

δ_p is the difference between the proton separation energies for the magic nucleus and its Z + 1 isotope. V_{hh} is a measure for the residual interaction in the core, V_{ph} takes into consideration the interaction between the $\pi g_{9/2}$ protons and the 0$^+$ core configuration. The particle–particle energy V_{pp} has been estimated from similar states[20] near ^{208}Pb with a correction for the different interaction volumes. The calculated value of E(8$^+$) is in reasonable agreement with the experimental energy. This also supports the assignment of $\pi(g_{9/2})^2$ and, moreover, substantiates the influence of the large Z = 40 gap.

The analogous calculation for ^{90}Zr results in E(8$^+$) = 3.70 MeV compared to the experimental value of 3.59 MeV. Hence, the difference in the energies of the 8$_1^+$ states of ^{90}Zr and ^{96}Zr can well be understood from the values of δ_p which amount to 3.20 and 4.06 MeV, respectively. It is interesting to note, that preliminary results of QRPA calculations[23] on ^{96}Zr starting from δ_p = 4.06 MeV also reproduce the 8$^+$ energy well.

<u>^{99}Nb:</u> The lowest levels of this nucleus at 0 and 365 keV are based on the $g_{9/2}$ and $p_{1/2}$ shell model configurations, respectively. Levels at about 500 keV are discussed[11,24] in terms of particle–phonon coupled states where the energy suggests the correlation to the 2$_1^+$ level at 536 keV in ^{100}Mo. A detailed interpretation was, however, hampered, by the fact that conflicting spin and parity assignments were proposed for these levels. The results of the measurements of the level half–lives which are presented in Fig. 3 remove these uncertainties[25]:

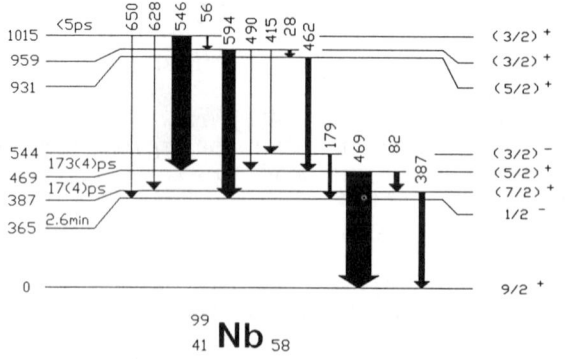

Fig. 3.
Level scheme of ^{99}Nb with those of the proposed[24,26] spin and parity assignments which are consistent with the half–life results.

Out of the alternatives[26] $I\pi = 5/2^+$ and $7/2^+$ for the 387 keV level only the choice $7/2^+$ can be accepted since otherwise the 387 keV transition would have E2 multipolarity with an unreasonable strength of 123 W.u.. Spin and parity $5/2^+$ result for the 469 keV level since the alternative[24] $7/2^+$ would require a strongly enhanced 546 keV transition from the 1015 keV level with an upper limit of the half–life of 5 ps. The half–life of the 469 keV corresponds to $B(E2; 469\text{ keV}) = 4.7(1)$ W.u..

Thus, the levels at 387 and 469 keV are candidates for being the $7/2^+$ and $5/2^+$ members of the $[\pi g_{9/2} \otimes 2^+]$ multiplet. The ordering of these levels is equal to that in the analogue ^{93}Nb with 2 neutrons beyond the N = 50 closed shell. The 469 keV transition is then of the type $[\pi g_{9/2} \otimes 2^+]_{5/2^+} \to \pi g_{9/2}$. Its reduced transition probability is, however, smaller than that of the $2_1^+ \to 0^+$ transition in the "core" ^{100}Mo with $B(E2)\downarrow = 36$ W.u.. Detailed calculations are needed to understand the cause for this difference. It is for example to be studied whether the phonons of the magic neighbour ^{98}Zr influence this transition probability[11]. In any case, the identification of the spins and parities of the levels gained from new lifetime results support the interpretation of ^{99}Nb as a nucleus with spherical character.

DEFORMATION OF HEAVY Nb ISOTOPES

Extended level schemes have recently been established for the isotopes [101]Nb through [104]Nb at the separators LOHENGRIN and JOSEF. All the

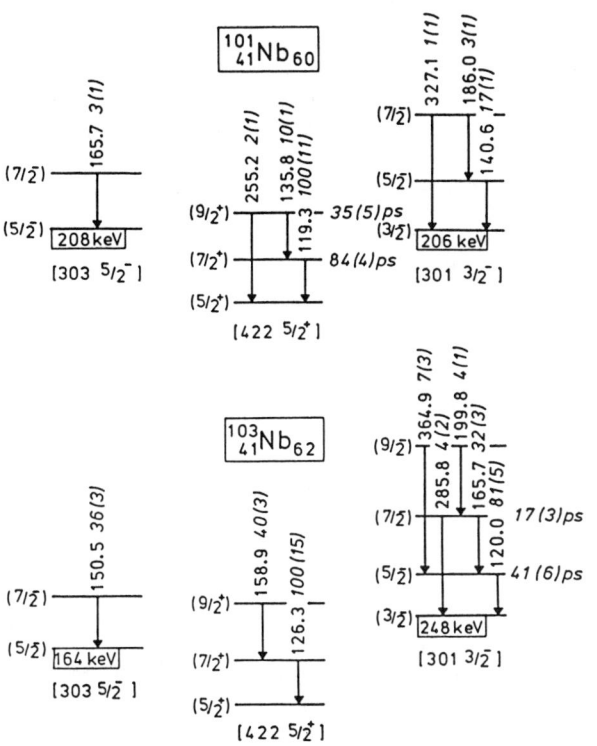

Fig. 4. Rotational bands identified in ^{101}Nb and ^{103}Nb.

low lying levels of the odd mass isotopes[9,27] can be grouped into rotational bands, see Fig. 4. The proposed assignment of the Nilsson configurations to the bands is based on the experimental knowledge about these states and their feeding from the ß⁻ parents as well as on the results of Nilsson model calculations[28].

The determination of the half–lives of the band members in connection with the knowledge about the relative intensities of the intraband transitions allows to deduce[29] the nuclear deformation. The crossover transitions in the band are expected to be of E2 nature. With the use of the geometrical model the E2 fractions of the stopover transitions can therefore be calculated from the relative intensities of the crossover and stopover transitions. Hence, the B(E2) values and the values of the deformation parameter $ß_q$[30] can be deduced.

Selected results of the present half–life studies are included in Fig. 4. The half–life of

$$t_{1/2}(119 \text{ keV}) = 84(4) \text{ ps}$$

for ¹⁰¹Nb corresponds to

$$B(E2; 119 \text{ keV}) = 241(69) \text{ W.u.}$$

and

$$ß_q = 0.40(5).$$

Here, it must be pointed out that the large uncertainties of B(E2) and $ß_q$ result from the uncertainties of the relative γ ray intensities and not from those of $t_{1/2}$. Second, it should be kept in mind that the deduced collectivity depends on the ground state spin. If the ground–state band were based on the [301 3/2⁻] Nilsson configuration (which is less probable, but not excluded[27]) then the values of B(E2; 119 keV) and $ß_q$ would be 150(43) W.u. and 0.34(5), respectively. It is interesting to note that the Nilsson model calculations[28] have predicted $t_{1/2}(119 \text{ keV}) = 92$ ps.

In the case of ¹⁰³Nb, the half–life of the first excited state at 126 keV has been determined to 73 ps, but the deformation cannot be deduced since information on the intensity of the 285 keV crossover transition is lacking. However, the deformation can be determined for the side band based on the 248 keV state since both the half–life of the 368 keV first excited member and the intensity of the crossover transition are known:

$$t_{1/2}(368 \text{ keV}) = 41(6) \text{ ps}$$

$$B(E2; 120 \text{ keV}) = 193(87) \text{ W.u.}$$

$$ß_q = 0.37(8)$$

Here again the uncertainty for the collectivity is mainly caused by that of the γ ray intensities. It is therefore important to improve the knowledge about these intensities. The preliminary result of the Nilsson model calculations is $t_{1/2}(368 \text{ keV}) = 45$ ps.

Thus, the results of the present lifetime studies evince large deformations for ¹⁰¹Nb and ¹⁰³Nb. It is remarkable that also the half–lives of the second members of the ground–state band in ¹⁰¹Nb and the 248 keV band in ¹⁰³Nb could be determined, see Fig. 4. They confirm the deduced large values of $ß_q$. Thus a sudden onset of deformation occurs also in this chain.

The values of the deformation parameters $ß_q$ for the Zr, Nb and Mo chains are compiled in Fig. 5. The deformation of ¹⁰¹Nb is larger than that of the Zr and Mo isotopes. This may be due to the effect of the odd particle. Such odd–particle effects are suggested by the systematics of the Mo isotopes where the information is now equally precise for the odd–mass and the even–mass isotopes. The deformation of ^{103}Mo and ^{105}Mo is larger than in the neighbours ^{102}Mo, ^{104}Mo and ^{106}Mo which may be caused by the influence of the odd proton.

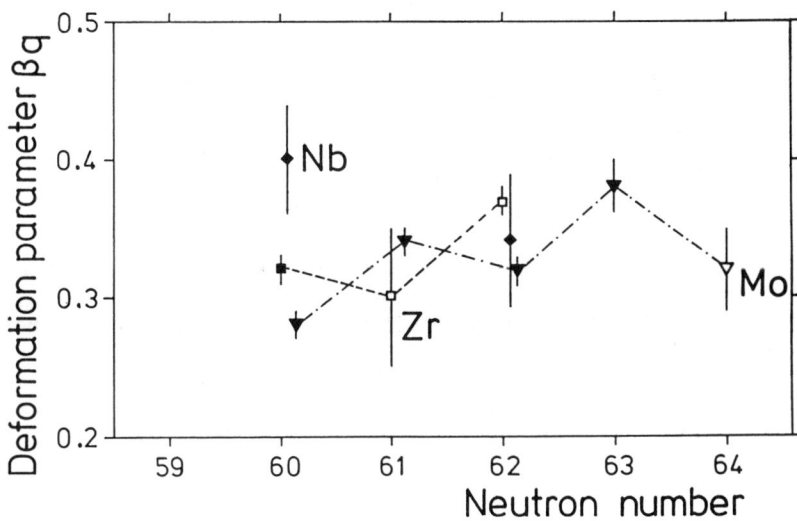

Fig. 5.
Deformation for the N $\geq$ 60 isotopes of Zr, Nb and Mo. The cases where the present lifetime studies have contributed are marked by full symbols. The data for ^{101}Nb and ^{103}Nb represent average values from the half–lives of both members of the ground–state and 248–keV band, respectively. The values of β_q of the odd nuclei depend on K for the band which has been taken as 5/2 for ^{101}Nb, ^{105}Mo and 3/2 for ^{101}Zr, ^{103}Nb, ^{103}Mo. The choice for ^{105}Mo deviates from the conjecture of Ref. 29; it is suggested by the similarity of the energy pattern of its ground state band with the side band[31] in ^{101}Zr. For ^{100}Zr the band mixing[8] has not been taken into account.

REFERENCES

1. E. Cheifetz et al., Inst. Phys. Conf. Ser. 51, 193 (1980)
2. G. Molnár et al., Z. Phys. A—Atomic Nuclei 331, 97 (1988)
3. H. Mach et al., Nucl. Instr. Meth. A280, 49 (1989)
4. M. Moszyński, H. Mach, Nucl. Instr. Meth. A277, 407 (1989)
5. H. Ohm et al., Inst. Phys. Conf. Ser. 105, 323 (1990)
6. H. Lawin et al., Nucl. Instr. Meth. 137, 103 (1976)
7. G. Molnár et al., Nucl. Phys. A500, 43 (1989)
8. H. Mach et al., Phys. Lett. B230, 21 (1989)
9. T. Seo et al., Z. Phys. A—Atomic Nuclei 315, 251 (1984)
10. R.L. Gill, Contribution to this conference
11. G. Battistuzzi et al., Z. Phys. A—Atomic Nuclei 306, 113 (1982)
12. H. Ohm et al., Phys. Lett. B241, 472 (1990)
13. G. Molnár, Contribution to this conference
14. H. Ohm et al., Z. Phys. A—Atomic Nuclei 334, 519 (1989)

15. M.L. Stolzenwald et al., Z. Phys. A–Atomic Nuclei 327, 359 (1987)
16. G. Molnár et al., Phys. Rev. C33, 1843 (1986)
17. S. Brant et al., Z. Phys. A–Atomic Nuclei 329, 301 (1988)
18. H. Mach et al., Phys. Rev. C37, 254 (1988)
19. A. Bohr, B. Mottelson, Nuclear Structure, vol. 1 (New York, 1985)
20. P. Kleinheinz et al., Z. Phys. A–Atomic Nuclei 290, 279 (1979)
21. D.H. Gloeckner, F.J.D. Serduke, Nucl. Phys. A220, 477 (1974)
22. A.H. Wapstra, G. Audi, Nucl. Phys. A432, 55 (1985)
23. O. Rosso, W. Unkelbach, KFA Jülich, private communication
24. H.A. Selič et al., Z. Phys. A–Atomic Nuclei 289, 197 (1979)
25. H. Ohm, Ann. Rep. 1989, IKP, KFA Jülich, Jül–Spez 562, 34 (1990)
26. E.R. Flynn et al., Phys. Rev. C28, 575 (1983)
27. H. Ohm et al., submitted for publication
    A.–M. Schmitt et al., Ann. Rep. 1983, IKP, KFA Jülich, Jül–Spez 225, 41 (1984)
28. T. Seo et al., Research Reports in Physics, Springer Verlag (J. Eberth et al. eds.) p. 349
29. K. Shizuma et al., Z. Phys. A–Atomic Nuclei 315, 65 (1984)
30. K.E.G. Löbner et al., ADNDT A7, 495 (1970)
31. M.A.C. Hotchkins et al., Phys. Rev. Lett. 64, 3123 (1990)

A NEW METHOD FOR PICOSECOND LIFETIME MEASUREMENTS USING ELECTRONIC TIMING: NUCLEAR STRUCTURE APPLICATIONS.

R. L. Gill
Brookhaven National Laboratory, Upton, NY 11973, USA

ABSTRACT

A technique to measure the lifetimes of nuclear states with half lives <10 ps has been developed in conjunction with the TRISTAN mass separator at the High Flux Beam Reactor at BNL. The method uses fast plastic and BaF_2 scintillators and Ge detectors in a triple coincidence (β-γ-γ) fast-slow counting system. The timing information is derived from the fast plastic-BaF_2 coincidence, while the higher resolution of the Ge detector (in slow coincidence) serves to insure that the β-γ event lies in the cascade of interest. The calibrations and corrections necessary to achieve precise results and the methods of data reduction and results from recent measurements on the A=97 mass chain are presented.

INTRODUCTION

The measurement of absolute transition rates provides very sensitive tests of nuclear models. This is particularly so for collective transitions where the E2 transition strengths provide the key signatures to their character. Near stability, absolute B(E2) values are readily obtained via Coulomb excitation and inelastic scattering techniques. On the neutron deficient side of stability, many Doppler-based techniques are employed to measure nuclear lifetimes (τ), from which absolute B(E2) values are deduced. However, for neutron rich nuclei that are far from stability, whose production and separation methods preclude the easy adaptation of Doppler techniques, electronic timing techniques must be employed. Until recently, this has limited the regime of reliable measurements to those nuclear states with τ>100 ps. This paper presents a new technique that makes it possible to measure lifetimes, with high precision, in the regime of <10 ps and with improved reliability for lifetimes >40 ps.

TIMING TECHNIQUE

The timing technique was developed for use at the TRISTAN mass separator, which supplies ion beams of neutron-rich, far from stability nuclei. The reader is referred to technical publications[1,2] where the details of the technique are presented and to numerous application-oriented publications in the few-picosecond range[3,4] and in the sub-nanosecond range[5,6]. These references are not intended be all inclusive, there are many other references, but those mentioned contain information that is germane to a discussion of the technique. The system employs a fast-slow coincidence counting system, with the β-γ timing information being derived from an NE111A plastic scintillator, coupled to an XP2020 photomultiplier, and a BaF_2 scintillator, coupled to an XP2020Q photomultiplier. The slower Ge detector provides a high resolution selection of a γ-ray in cascade with the desired β-γ event. Although this approach is rather standard, it employs a few novel features. First, is the use of a BaF_2 crystal (1.3 cm thick) which provides a fast signal[7] (especially when used with dynode timing[8]) as well as a moderate (~10%) energy resolution. Thus, in certain cases, the triple coincidence (with the Ge detector) is unnecessary and better statistics can be accumulated in shorter experiments. Secondly, a thin (3 mm) plastic scintillator is used. Since the β decay of far from stability nuclei usually involves high Q values, only a fraction of the total energy will be detected. The thickness of the scintillator was chosen so that an energy loss of about 600 keV (for $E_β≥1.5$ MeV) would be recorded, independent of the incident β energy. The importance of this feature is that it is no longer necessary to map the system response for ranges of both β and γ energies: only the γ response needs to be known. Thus, at least in principle, it is possible to calibrate the system over a wide dynamic energy range with a single calibration. Although other corrections which complicate this simple picture must be applied, it remains possible to measure lifetimes from many transitions in a single experiment. Not only does this reduce the measurement time, it also makes it possible to measure the same level lifetime via different, essentially independent, paths and provides for important internal consistency checks, as will be discussed later.

Figure 1 illustrates the timing apparatus. The ion beam is deposited onto an Al window which also serves to absorb β rays with energy below 1.5 MeV. The BaF_2 detector is covered with a Pb absorber to prevent betas from giving a signal in that detector. Two Ge detectors are positioned at 90° to the other detectors. A slit is provided for external positioning of a calibration source. When gates are set to select the desired β-γ-γ events, a semi-gaussian TAC peak with a

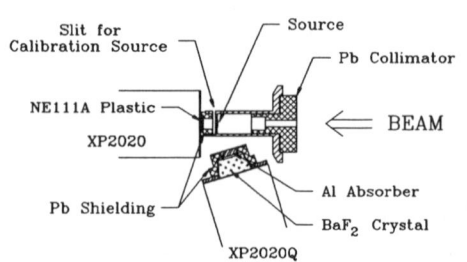

Fig. 1. Picosecond timing apparatus. The Ge detectors are not shown.

timing resolution (FWHM) of 96 ps for 1.3 MeV γ rays and 130 ps for 0.5 MeV γ rays can be achieved for prompt transitions. For τ≥200 ps, an exponential fit to the delayed side of the TAC spectrum yields the half life. For τ≥40 ps, the half life is deconvoluted from the TAC spectrum using an approximation to a prompt shape and an exponential tail. Neither of the above two cases require any detailed calibration, and provide greatly improved results in the regime of τ≥40 ps, especially where previous techniques required the use of the centroid shift analysis method. Shorter lifetimes require a measurement of the apparent shift of the centroid of the TAC peak from the position that would correspond to that of a prompt transition of the same γ energy, as shown in Fig. 2, where the corresponding prompt peak is shown as the dashed curve. Although shifts of 1-3 ps can be readily measured, detailed calibrations and corrections are required to translate these shifts into a lifetime.

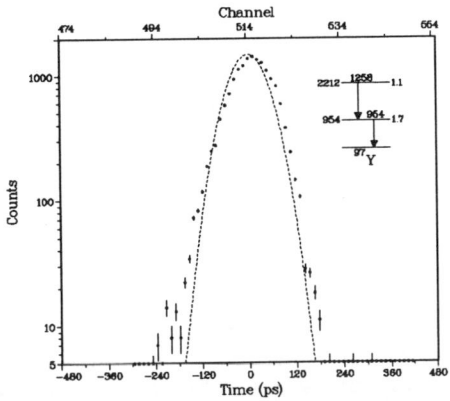

Fig. 2. TAC spectrum obtained from the 1258→954→0 cascade in ^{97}Y.

To illustrate the application of the centroid shift technique, consider the partial level scheme of ^{97}Y shown in Fig. 3. The level at 2212 keV is strongly fed by a β transition. If Ge and BaF$_2$ gates are set on the 1258-and 954-keV γ rays, respectively, the shift between the prompt TAC position and the β-γ centroid gives the sum of the mean lives $\tau_{2212}+\tau_{954}$. This is the situation shown in Fig. 2. If the gates are reversed, such that the Ge gate is at 954 keV and the BaF$_2$ gate is at 1258 keV, then the shift corresponds to just τ_{2212}. Each of these lifetimes involves an absolute measurement based on

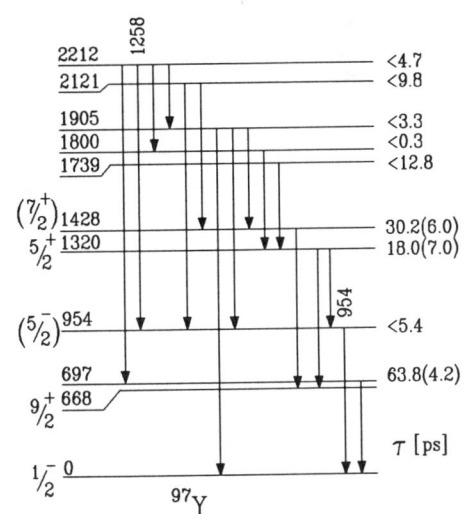

Fig. 3. Partial level scheme of ^{97}Y. The measured lifetimes are on the right.

the centroid shift from a known prompt position. If the difference of these positions is taken, the result is τ_{954}, and many uncertainties (and occasionally the corrections themselves) will cancel. By examining the level scheme, it can be seen that many paths to measure the lifetime of a particular level can be constructed. By determining the lifetime using a number of paths, each of which represents an independent measurement, an internal check of the corrections and calibrations is possible.

DATA ANALYSIS

The first step in extracting a lifetime via the centroid shift technique is to determine the energy dependence of the position of the TAC peak. A source of ^{24}Na is used to calibrate the system since it has a reasonable half life, is easily produced in the reactor, has a relatively high Q_β and high energy γ-ray cascades with no significant lifetimes. A gate is set on the β spectrum, another on the γ spectrum in the Ge detector, and a series of energy gates are selected over the range of the γ-ray spectrum from the BaF$_2$ detector. A plot of the position of the TAC peak for each BaF$_2$ energy slice for a prompt transition then results. The lower dotted curve in Fig. 4 shows the result of this calibration. This actually serves to calibrate the <u>shape</u> of the prompt curve, since the ^{24}Na source is located in a different position than the source of interest. The importance of careful geometrical positioning is evident when one considers that light requires 3 ps to traverse 1 mm. In addition, there are other factors that cause the actual prompt curve to shift. These include: electronic drifts, geometrical factors, a dependence on the energy collected in the BaF$_2$ detector, the initial γ-ray energy and a small, residual, dependence on β energy in the plastic scintillator. Nevertheless, the <u>shape</u> of the prompt curve remains the same and only the magnitude of the shift needs to be determined.

In order to determine the magnitude of the shift, transitions in the decay chain of interest that are known to be prompt (usually high energy γ rays) are used. This assumption may not always be correct, but it is usually possible to test its validity. If a number of high energy γ rays

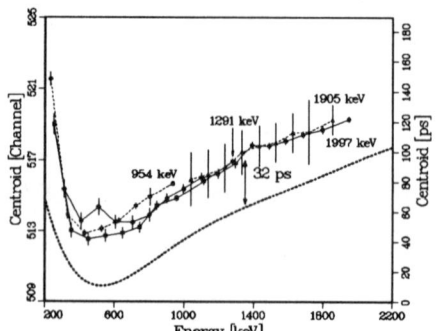

Fig. 4. Prompt curve (lower dotted line) and some ^{97}Y and ^{97}Zr centroids.

are present, select those with the smallest shift from the prompt curve. Typically, these will all have not only the smallest shifts, but will have the same shift. It is, therefore, unlikely that all these high energy γ rays will have a lifetime, and even more unlikely that they will all have the same lifetime. Thus, it is safe to assume that these transitions have no delayed component. In even-even nuclei it is often possible to find transitions with energies of 2 MeV, or higher. However, this is not always the case for odd-mass nuclei, and additional uncertainties may need to be included.

In the development of this technique, it was determined that the shift between the prompt curve and the centroids of prompt transitions depends on the transition energy itself.[1,2,3] To determine the magnitude of this transition energy (TE) correction, calibration measurements using sources of ^{88}Rb, ^{116}Ag and ^{142}Cs were made. A number of transitions in the daughters of these nuclides have well known lifetimes over a wide energy range. In addition, these sources are produced on-line at TRISTAN and, therefore, have a geometry identical to the source of interest. High energy transitions in the nuclide of interest are selected (in ^{97}Y, 1291 and 1997 keV), using the above arguments, as being prompt. The TE correction for these transitions is defined as zero. The relative shifts due to the lifetimes of the known transitions, after correction for the lifetimes, then gives the TE correction for the corresponding energy. The result of such a procedure is shown in Fig. 5 for ^{97}Y. However, for this case, due to the relatively low transition energy and the small number of transitions available, there remains the additional uncertainty that one (or both) of the so-called prompt ^{97}Y transitions may, indeed, have a lifetime of a few ps. Thus, the uncertainty associated with this correction must be increased to accommodate an undetected lifetime. The dotted line illustrates the difference

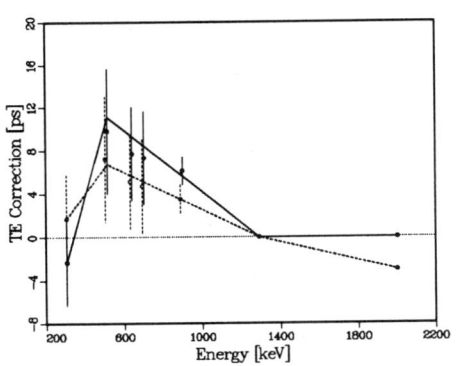

Fig. 5. Data points used for calculation of the transition energy (TE) correction.

in the TE correction if the 1997-keV transition is assumed to have an undetected lifetime of 3 ps (a lifetime larger then 3 ps would be readily observable). Thus, the uncertainty must be adjusted to allow for this possibility.

There is another correction that must be applied to the centroid shift data before a final lifetime can be extracted. As in other coincidence data, these events will have a background due to accidental coincidences, scattering and other phenomena. Thus, these contributions must be removed. However, unlike the case of γ-γ coincidences, this does not involve a simple reduction of the number of events. The background events here will arise from phenomenon which originate from differing energies or have been delayed through scattering. In any case, the background will have a different centroid from the desired events. Thus, the background must be properly shifted before it can be subtracted from the desired events. Fortunately, since the shape of the prompt curve is known, this can be easily accomplished.

Before leaving the discussion of techniques, it is worth noting that, although the method was developed to obtain lifetime information from β-γ coincidences, the technique should be readily adaptable to γ-γ coincidences in neutron capture γ-ray experiments. This can be accomplished by replacing the β detector by another BaF_2 detector. The timing resolution should remain similar to that of the β-γ system, due to the fast response of BaF_2. The primary analysis method would be the deconvolution technique, since the use of two BaF_2 detectors would necessitate the mapping of γ-ray response curves for each detector, thereby, at least in the worst case, requiring a family of curves to be determined. Thus, due to the energy dependence of the γ detector, the centroid shift technique will be more difficult to apply, and the primary region of interest would be lifetimes ≥40 ps.

VIBRATIONAL STRUCTURE IN ^{97}Y AND ^{97}Sr

The lifetimes measured in ^{97}Y are indicated in Fig. 3. From the 18.0(7) ps lifetime determined for the 1320-keV level a B(E2) value of 11(4) SPU (0.03(1) e^2b^2) was deduced. From the limit of τ<5.4 ps for the 954-keV level a B(E2) value of >7 SPU (>0.1 e^2b^2) was deduced. Both of these B(E2) values suggest the interpretation of the levels as 2^+_{phonon} excitations of a spherical band head or ground state. The B(E2) values can be compared to the vibrational transitions in ^{97}Sr and even-even ^{96}Sr, as shown in Fig. 6. The B(E2) value for ^{97}Sr was determined in the experiment described by Büscher, et al.[4] (which is high-

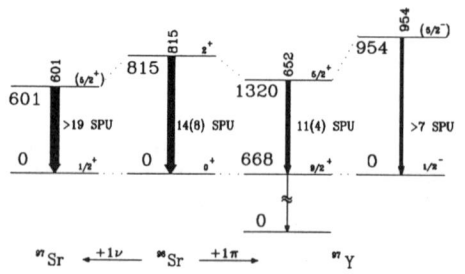

Fig. 6. Systematics of the quadrupole vibrational excitations in ^{96}Sr, ^{97}Sr and ^{97}Y.

lighted in this paper). The experimental B(E2) value for ^{96}Sr was determined by Mach et al.[9] from the lifetime of the $2^+\to 0^+$ transition (7(4) ps) using the technique described here. Thus, the data suggest the existence of vibrational bands built on the ground states in both ^{97}Y and ^{97}Sr. The B(E2) values for the vibrational transitions favor aligned states in ^{97}Y and ^{97}Sr and are of the same order of magnitude as in their even-even neighbors.

Although the focus of this paper is on ^{97}Y, it is worth noting that many lifetimes in other A=97 nuclides were also measured. In ^{97}Sr the lifetime measurements were able to confirm that these levels were part of a rotational band that was postulated earlier.[10] The deformation [β=0.34(4)] and quadrupole moment ($|Q_0|$=3.5(4) eb) of the deformed levels was deduced. Thus, the data from this experiment provided information on spherical, vibrational and deformed levels in ^{97}Sr. It was also possible to extract the coefficients for the mixing between the vibrational and deformed states. The data limited these values to upper limits of V_{mixing}≤31 keV for the spherical-deformed states and ≤5.2 keV for the vibrational-deformed states. These data can be combined with that from neighboring Sr isotopes to follow the development of the deformed intruder state, as shown in Fig. 7.

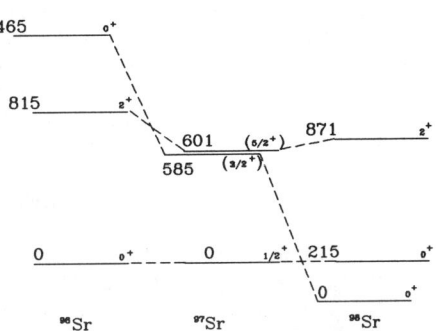

Fig. 7. Spherical and deformed states in ^{96,97,98}Sr. The relative lowering of the deformed intruder state relative to the spherical states is indicated.

CONCLUSION

Thus, a wealth of information, only part of which was discussed here, is available from a single experiment. The ability to simultaneously measure lifetimes of essentially all levels in a nucleus contributes, not only to the interpretation of level schemes, but also leads to increased confidence in the results due to numerous built-in cross checks. The wide dynamic energy range of the technique is possible due to the novel use of a thin β detector that has an energy independent response and no γ-ray sensitivity. In addition, the use of BaF$_2$ crystals yields a timing resolution that is far superior to any previously available for such measurements. This provides, even without the centroid shift method, greatly improved accuracy and precision for states with τ≥40 ps. This technique has been applied to many nuclei, in addition to those mentioned in this paper, at TRISTAN and other facilities, and is proving itself to be a valuable tool in the investigation of nuclear structure.

ACKNOWLEDGMENTS

The efforts of many persons went into the development of this technique. The author is especially grateful for the efforts of Drs. H. Mach and M. Moszyński, whose creativity and insights are largely responsible for the success of the timing technique. Research was performed under Contract No. AC02-76CH00016 with the U.S. Department of Energy.

REFERENCES

1. H. Mach, R.L. Gill and M. Moszyński, Nucl. Instr. and Meth. A280, 49 (1989).
2. M. Moszyński and H. Mach, Nucl. Instr. and Meth. A277, 407 (1989).
3. H. Mach, M. Moszyński, R.F. Casten, R.L. Gill, D.S. Brenner, J.A. Winger, W. Krips, C. Wesselborg, M. Büscher, F.K. Wohn, A. Aprahamian, D. Alburger, A. Gelberg and A. Piotrowski, Phys. Rev. Lett. 63, 143 (1989).
4. M. Büscher, R.F. Casten, R.L. Gill, R. Schuhmann, J.A. Winger, H. Mach, M. Moszyński and K. Sistemich, Phys. Rev. C41, 1115 (1990).
5. H. Mach, F.K. Wohn, M. Moszyński, R.L. Gill and R.F. Casten, Phys. Rev. C41, 1141 (1990) and Nucl. Phys. A507, 141c (1990).
6. H. Mach, M. Moszyński, R.L. Gill, G. Molnár, F.K. Wohn, J.A. Winger and J.C. Hill, Phys. Rev. C41, 350 (1990) and Phys. Rev. C42, 793 (1990).
7. M. Laval, M. Moszyński, R. Allemand, E. Cormoreche, P. Guinet, R. Odru and J. Vacher, Nucl. Instr. and Meth. 206, 169 (1983).
8. B. Bengtson and M. Moszyński, Nucl. Instr. and Meth. 204, 129 (1982).
9. H. Mach, F.K. Wohn, G. Molnár, K. Sistemich, John C. Hill, M. Moszyński, R.L. Gill, W. Krips and D.S. Brenner, Nucl. Phys. A (submitted).
10. G. Lhersonneau, B. Peiffer, K.-L. Kratz, H. Ohm and K. Sistemich, Z. Phys. A330, 347 (1988).

RELIABILITY OF SHORT LIFETIMES MEASURED BY THE DOPPLER SHIFT ATTENUATION METHOD

J. Keinonen

Accelerator Laboratory, University of Helsinki,
Hämeentie 100, SF-00550 Helsinki, Finland

ABSTRACT

Most recent development and applications of the Doppler shift attenuation method in measurements of nuclear lifetimes are discussed. Emphasis is put on the development of the Doppler shift attenuation technique performed in our laboratory. Main factors behind inconsistencies of the literature data are shortly estimated. The applications of current interest, i.e. the systematic studies of transition strengths in the sd-shell nuclei, the superdeformation of rare earth nuclei at high spin, and the studies of fp-shell nuclei with a new ultra-high resolution neutron capture γ-ray spectroscopy, are considered. The first use of molecular dynamics techniques in the simulation of the slowing-down process combined with the simulation of Doppler-broadened γ-ray lineshapes by Monte Carlo method, is discussed in the frame work of ultra-low velocity Ti recoils produced through a neutron capture reaaction.

INTRODUCTION

The accuracy and reliability of experimental electromagnetic transition strengths sufficient for meaningful comparison with theoretical values, depend essentially on the mean-lifetime values of the exited states. Known lifetime values are mainly based on Doppler-shift attenuation (DSA) studies. The reported values have, however, in many cases large uncertainties and mutual inconsistencies. Main factors behind the inconsistencies, i.e. the use of the slowing-down theory without sufficient experimental confirmation, complicated target structures in experiments and approximative methods in the DSA analysis, are shortly considered in the beginning of the paper.

The improved DSA method developed in our laboratory in connection of the proton capture reaction, has been discussed earlier.[1] In the present paper, the further development of the method is demonstrated with examples from current studies on transition strenghts in sd-shell nuclei,[2] superdeformation of rare earth nuclei at high spin,[3] and transition strenghts in fp-shell nuclei with a new ultra-high resolution neutron capture γ-ray spectroscopy.[4-6] The status of stopping power calculations in the velocity region $\beta \sim 10^{-4}$ - 10^{-2} where the measurements of the examples have been performed, is discussed first.

STATUS OF STOPPING POWER CALCULATIONS

In the DSA measurements, the velocity of recoiling nuclei at the moment of the γ-ray emission depends on the stopping power of the slowing-down material. The line shape of a γ peak in the γ-ray spectrum is the velocity distribution of the recoiling nuclei at the moment of the γ-ray emission. In the DSA analysis the reproduction of the line shape or its first moment ($F(\tau)$ value) defines the lifetime value.

The stopping power of a slowing-down material for recoiling ions is described by the following equation:

$$\left(\frac{dE}{dx}\right)_{corr} = f_n \left(\frac{dE}{dx}\right)_n + f_e \left(\frac{dE}{dx}\right)_e. \tag{1}$$

The stopping power of Ta for ^{26}Mg ions and Ti for ^{49}Ti ions is shown in fig. 1.

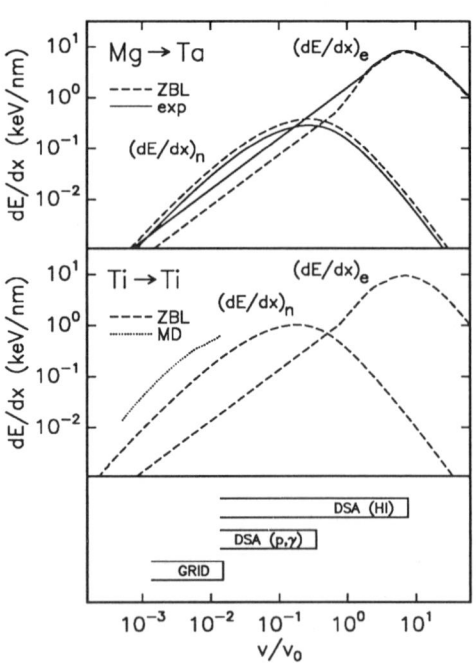

Fig. 1. Stopping power of Ta for ^{26}Mg ions and Ti for ^{49}Ti ions as a function of velocity (in units of Bohr velocity v_0). The dashed line shows the ZBL stopping power, the solid line the experimental stopping power, and the dotted line the nuclear energy loss obtained in MD calculations. The lower part of the figure illustrates velocity regions of recoiling atoms produced in heavy ion reactions (HI), proton capture reactions ((p, γ)), and thermal neutron capture reactions (GRID).

The uncorrected nuclear stopping power $(dE/dx)_n$ is derived in the Monte Carlo (MC) code used in our laboratory[1] by calculating the scattering angles of the recoiling ion directly from the classical scattering integral. The interatomic interaction is derived from the Thomas-Fermi (TF) potential (see ref. 1) or the universal interatomic potential (ZBL) deduced by Ziegler et al.[7] The correction factor f_n accounts for the stopping power reduction due to the microchanneling of the recoiling atoms in a polycrystalline backing material. This approach is based on the fact that in simulations of γ-ray line shapes no significant differences have been obtained either using the polycrystalline structure and the interatomic potential or the experimentally determined correction factors of the nuclear stopping power.[1,8] Systematic studies on the correction factor at low velocities ($v \leq v_0$, where $v_0 \sim c/137$ is the Bohr velocity) done in our laboratory have resulted in the values of f_n^{TF} = 0.7 - 0.9 (ref. 1) or f_n^{ZBL} = 0.7 - 0.8 for different light-ions in polycrystalline Ta, $f_n^{TF}(^{26}\text{Mg} \rightarrow \text{Ta}) = 0.70 \pm 0.05$.

In the description of the slowing-down process of ions at ultra-low velocities ($v \leq v_0/100$) the binary collision approximation (BCA) described above can not be used since atomic collisions can not be approximated as isolated phenomena. At these velocities, the only realistic method to simulate the atom-atom collicions is the Molecular Dynamics (MD) calculations where the equations of motion of both the recoiling and target atoms are intergrated numerically. In this method the crystalline structure of the slowing-down material and thermal motion of the atoms are in a realistic way included in the simulation. The method has recently been applied in our laboratory for the simulation of the slowing down of ultra-low velocity Ti ions in Ti,[4] to be discussed below. The energy losses obtained in the MD calculations were considerably higher than the $(dE/dx)_n$ values extrapolated to low velocities according to the BCA method.

Experimental information on the electronic stopping power $(dE/dx)_e$ has been used by Ziegler et al.[7] to construct an empirical model. The experimental information was utilised to obtain the values of the model parameters. This empirical model is now commonly used in all the cases where the electronic stopping power is needed.

For the parametrisation of the electronic energy loss, the velocity dependence was treated differently in three velocity regions. In the velocity region $v \leq v_F$ (Fermi velocity $v_F \sim v_0$ in the solids), the stopping power was assumed to be proportional to the velocity. In the velocity region $v \geq 3v_F$, the electronic energy loss was obtained by using the equivalent experimental energy loss of protons. In the medium velocity region $v_F \leq v \leq 3v_F$, the electronic stopping power and its velocity dependence were obtained by taking into account the charge-density distribution of electrons and the shielding of the nucleus. The velocity of crossover from low-velocity stopping to medium-velocity stopping was determined from the lower limits of validity of the empirical parameters. The stopping power at the crossover velocity determined then the constant of proportionality at lower velocities.

Due to the very few experimental data available at velocities $v \leq 3_F$, it is not known how well the electronic energy loss is reproduced by the model at the lower part of this velocity region. We have recently started a systematic study to test the velocity dependence of the empirical electronic stopping power in the velocity region $v \leq 2v_0$.[9] In these studies a well established lifetime value in the inverted DSA analysis is used to find out the electronic stopping power and its velocity dependence. Much higher electronic stopping power (f_e^{ZBL} up to 1.5) and different velocity dependence than predicted by the ZBL model have been obtained in these studies. At velocities $v \geq 2v_0$, the scaling factors $f_e^{ZBL} = 0.90$ - 1.10 to the ZBL stopping power have been obtained for ^{26}Mg ions in 17 (Z = 22 - 79) elemental solids, $f_e^{ZBL}(^{26}\text{Mg} \to \text{Ta}) = 1.07$.[9]

INCONSISTENCIES OF THE LITERATURE DATA

The lifetime values reported in the literature, have normally been obtained in DSA analysis where the nuclear stopping power has been taken directly from the LSS theory[10] and the large angle scattering has been described with the Blaugrund's approximation.[11] The electronic stopping power has often been taken from the LSS theory. Different corrections have, however, in many cases been applied to the theoretical values. They have been based on systematics of experimental stopping power data. The deduced lifetime values reported in the literature have only in relatively few cases been deduced by the use of experimental stopping power, e.g. ref. 12.

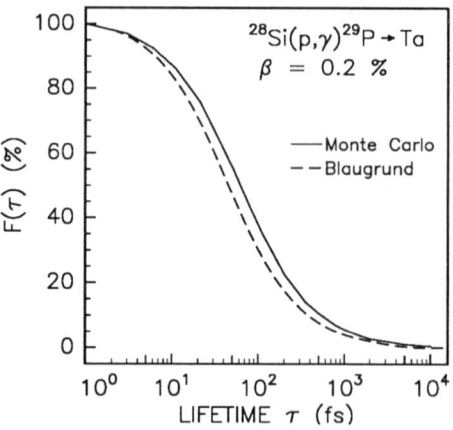

Fig. 2. Difference in lifetime values deduced in a DSA analysis by using the MC method and Blaugrund's approximation.

By the use of the large-angle scattering correction of Blaugrund[11] in the DSA analysis of low recoil-velocity data, systematically shorter lifetime values are obtained than those obtained in the MC simulations. The systematic error increases with increasing lifetime value. The attenuation factor $F(\tau) = 0.10$ for the $\beta = 0.2\%$ ^{29}P recoils in Ta, has recently been shown[13] to correspond to the lifetime values of 380 and 500 fs according to Blaugrund's approximation and the MC simulation, respectively, (fig. 2). The error due to this approximation is circumvented in the line-shape analysis of high recoil-velocity data by fitting only that part of the line shape where the effect of nuclear scattering is negligibly small.[12]

An example of the lifetime value measured several times with different methods, is the lifetime of the $E_x = 1.37$ MeV level in ^{24}Mg.[2] The comparison of the lifetime values is shown in fig. 3. In deducing the adopted value, an uncertainty of 20 % has been added in quadrature to the reported uncertainties in those cases where only a statistical error has been reported in the literature or where no information is available on the slowing-down conditions in the DSA analysis. This is done for the comparison with the values from those measurements for which the uncertainty due to the stopping power is included. Note that in the cases where the literature data include such an uncertainty, the values obtained without the experimental stopping data are still subject to systematic error. The reference value is the adopted value and contours at $\pm 2(\Delta \tau)$ are centered at this value.

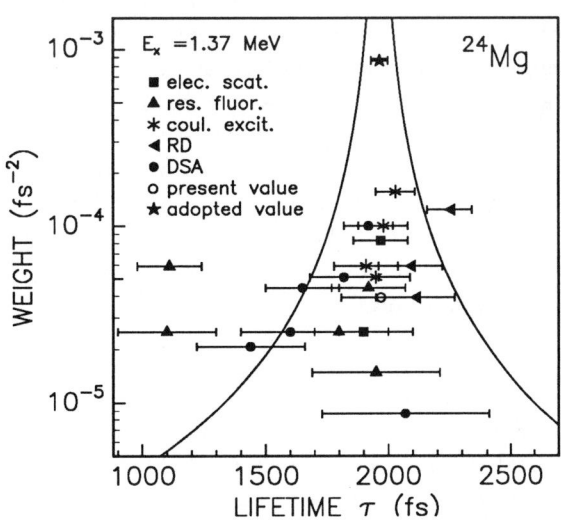

Fig. 3. Plot of the weights of lifetime measurements of the 1.37 MeV state in ^{24}Mg versus lifetime value. The values of the weight of the measurements, on a logarithmic scale, are plotted as a function of the lifetime values.[12] The weight is assumed to be $(\Delta \tau)^{-2}$, where $\Delta \tau$ is the quoted uncertainty of the lifetime measurement.

As in the most cases of the reported lifetime values, the DSA measurements of ^{24}Mg have been performed under such experimental conditions that it is very difficult to justify reasons behind disagreements between different results, e.g. the use of different evaporated targets (Na_2WO_4, NaCl, NaOH, NaBr) and backed C or ^{24}Mg foils. The use of compound or layered targets in DSA measurements complicates the DSA analysis because the uncertainty due to the stopping power increases with the increasing number of elements in a slowing-down material. In MC calculations the layered target structures and the slowing down in different layers can be taken into account in a realistic way. Approximations about the initial depths of the recoiling nuclei are not needed. Effects of the delayed feeding transitions on the lineshape can also be taken into account in MC calculations in a realistic way, see below. There are very few DSA analysis reported in the literature where the MC method has been employed. Fig. 3 illustrates also that there are inconsistencies between lifetime values obtained with other methods than DSA.

RECENT APPLICATIONS OF THE DSA METHOD IN MEASUREMENTS OF NUCLEAR LIFETIMES

Systematic studies on transition strengths in sd-shell nuclei

The initial evaluations of the results of the recent shell model calculations[14] suggest that they yield a good accounting for the level densities of positive-parity states in the first several MeV of excitation energy in mid-sd-shell nuclei and that the wave functions are able to reproduce many features of low-lying states.

Our systematic study of short lifetimes in the sd-shell nuclei has resulted in lifetime values to accuracies sufficient to permit extraction of transition strengths which can be compared meaningfully with theoretical values, e.g. refs. 2, 13 and references therein. In addition, the systematic studies with detailed and similar DSA measurements and analysis yield consistent lifetime data in different sd-shell nuclei. The reliability of the reported results obtained with implanted targets and Ta as slowing-down material, is increased by the fact that consistent results have been obtained in the DSA analysis of the proton capture and heavy-ion reaction data, measured under similar conditions; the recoil atoms are produced at velocities, where the nuclear or electronic stopping power, respectively, is dominant.

Superdeformation in heavy nuclei

Superdeformation observed in ^{152}Dy has attracted experimental and theoretical studies to find such a deformation also in other rare earth nuclei.[15] DSA measurements have been performed to prove that the observed cascades belong to superdeformed shapes. The deduced lifetime values have proved the superdeformation, shown normal collectivity, transformation from deformated state to a triaxial shape or that high-spin states have a single-particle structure.

In the DSA analysis, the MC method is the only realistic technique to take into account the delayd feeding cascades in reproduction of a γ-ray lineshape representing a transition inside a long transition cascade.[16] Fig. 4 illustrates a recent DSA analysis where complicated feeding patterns had to be included in the DSA analysis.[3] For reproducing the observed lineshapes it was necessary to assume the side-feeding intensity of each yrast state to come from a rotational cascade of about 10 transitions. For each yrast state a new side band with a variable intrinsic quadrupole moment was allowed. In this way the yrast-band lifetimes and side-band feeding times were fitted at each spin.

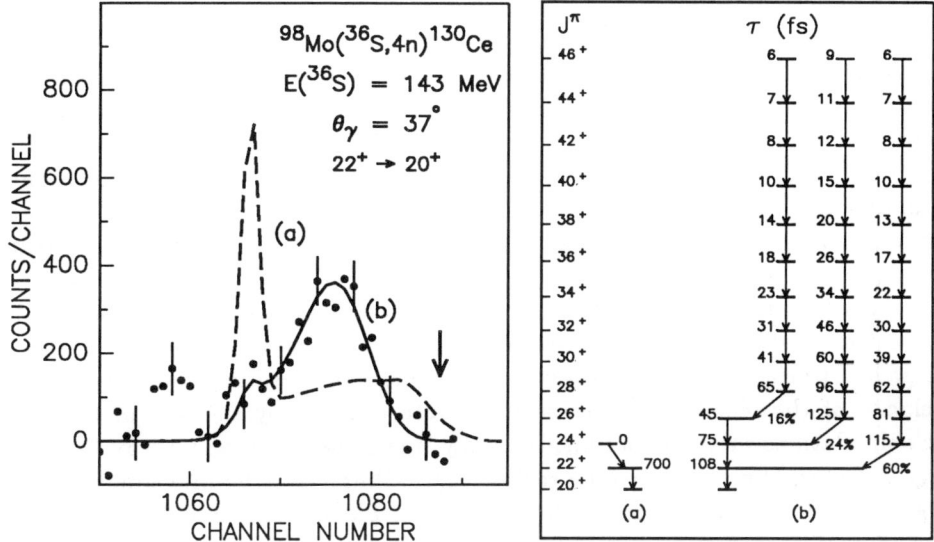

Fig. 4. Portion of the γ-ray spectrum in the DSA measurements of the 22^+ state. The dashed line is the computer simulation of the γ-ray lineshape for a prompt feeding of the 22^+ state as illustrated by the decay scheme (a) of the figure. The centroid of the lineshape corresponds to the centroid of the experimental lineshape. This effective lifetime is 700 fs. The solid line is the computer simulation of the lineshape for the feeding cascades illustrated by the decay scheme (b) of the figure. Lifetime values of the feeding cascades are shown. The side-feeding intensities of the yrast states are given as fractions of the intensity of the $22^+ \to 20^+$ transition. The lifetime value of the 22^+ state is 108 ± 20 fs. The position of the vertical arrow corresponds to the unattenuated Doppler shift, calculated from the reaction kinematics.

Analysis of recoil Doppler profiles obtained with ultra-high resolution (n, γ) spectroscopy

The analysis of the Doppler broadened line shapes obtained with a new technique, Gamma Ray Induced Doppler broadening (GRID) developed by Börner at al.[5], provides a new method to obtain lifetime values of excited nuclear states produced in thermal neutron capture. In GRID measurements, de-excitation of a capture state by a γ-ray emission after thermal neutron capture imparts a recoil velocity to the nucleus. Secondary γ-rays emitted by the recoiling nucleus before it has slowed down in matter, are thus Doppler shifted. Since the primary γ-rays are emitted isotropically the observed γ-ray lineshape is Doppler broadened. For typical recoil velocities of the order of $\beta \sim 10^{-4}$, the energy loss mechanism is dominated by elastic scattering of recoiling atoms from the screened Coulomb potential of the substrate atoms.

The GRID data for Ti atoms slowing down in Ti substrate,[5] have very recently been simulated in our laboratory by combining MD and MC techniques.[4] In the MD calculations, Ti atoms were placed in an HCP array with initial velocities to yield the target temperature. The equations of motion of the atoms were solved numerically with a time step of 0.5 fs. In the beginning of a simulation event, isotropic recoil velocity was given to one of the lattice atoms. Simulation was carried on until the excited state in the recoiling ^{49}Ti nucleus emitted the γ-ray corresponding to the lineshape of interest. The final velocity vector of the nucleus was stored for the use in the MC calculations of the γ-ray lineshape. In addition to the direct decay, the capture state can decay to the state under study via intermediate states. In these cases the recoil velocity is imparted to the atom in several steps. The effect of these cascade γ-rays to the velocity of the recoiling atom was taken into account by the MC method. In the simulations of the experimental γ-ray lineshapes, different decay branches were weighted by the branching ratios. Fig. 5 illustrates the MD simulations of the trajectories of recoiling and lattice atoms. The electronic energy loss of the recoiling ^{49}Ti atoms[7] was included in the calculations. It was, however, found to have a negligible effect on the lineshapes.

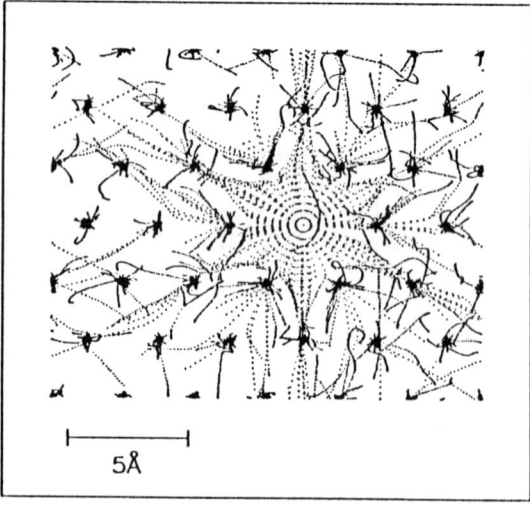

Fig. 5. Superimposed trajectories of atoms in the (001) plane from 200 simulation events. In every event the atom in the middle of the picture was given a recoil velocity of 3.21 x 10^4 m/s with a random direction in the (001) plane. Time difference between dots of the trajectories is 1 fs. The mean lifetime of the excited state of the recoiling nucleus was 15.9 fs.

The dependence of the simulated γ-ray lineshape on the interatomic potential was studied by using three potentials;[4] an universal potential (ZBL) widely used in stopping power calculations,[7] a pair potential of Born-Mayer form (BM), and a pair potential (EMT) calculated by using the effective-medium theory (EMT). The lattice sites of the atoms corresponding to their equilibrium positions, were obtained by adding an attractive potential to the three repulsive potentials. The calculated lineshapes were found to be insensitive to the exact form of the attractive potential. The total interatomic potentials used in the simulations, were obtained by adding the repulsive potential to an attractive potential. The interatomic potentials are illustrated in fig. 6.

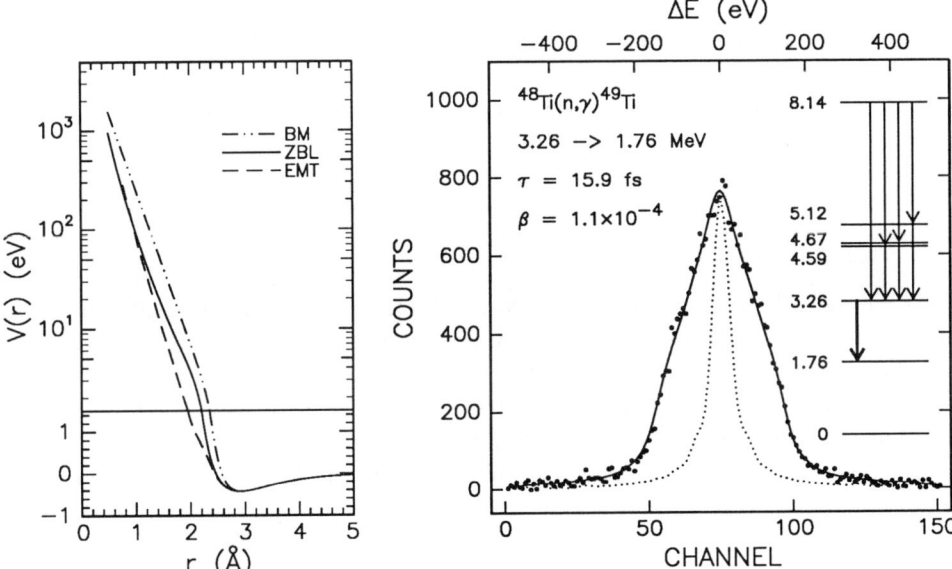

Fig. 6. Pair potentials for Ti-Ti used in the simulations.

Fig. 7. The experimental and simulated (solid line) γ-ray lineshapes for the 1499 keV transition. The dotted line is the instrumental responce function. Also shown is the decay scheme of the capture state to the 3.26 MeV state.

The dependence of the simulated γ-ray lineshapes on the interatomic potential is demonstrated by the results obtained with the ZBL, BM and EMT potentials for the 1499 keV γ-ray lineshape, the transition 3.26 → 1.76 MeV. The best fit of the experimental lineshape to a lineshape simulated with the ZBL potential, was obtained with the lifetime value of 15.9 ± 1.6 fs for the 3.26 MeV state, fig. 7. The higher repulsion between Ti atoms predicted by the BM potential than the ZBL potential, reduces the lifetime value to 9.6 ± 0.5 fs. The EMT potential is less repulsive at distances 1.0 - 2.5 than the ZBL potential. This small difference affects the lineshape and increases the lifetime value to 18.1 ± 0.7 fs. In the frame work of the MD and MC calculations, we can conclude that at the moment the uncertainty of lifetimes obtained by this new method is due to uncertainty of the interatomic potential.

CONCLUSIONS

It is demonstrated that by the use of the computer techniques available today, realistic simulations of DSA and GRID data can be performed. The improved analysis techniques and knowledge of the slowing-down theory makes it possible to obtain lifetime values whose accuracy and reliability is much better than reported in the literature during past years. On the other hand, the improved measuring and analysing techniques put more requirements on the further improvement of the slowing-down theory and interatomic potentials.

REFERENCES

1. Keinonen, in *Capture Gamma-Ray Spectroscopy, and Related Topics-1984*, Knoxville, Tennessee (American Institute of Physics, Conference Series No. 125) p. 557.
2. J. Keinonen, P. Tikkanen, A. Kuronen, Á.Z. Kiss, E. Somorjai, and B.H. Wildenthal, Nucl. Phys. A **493**, 124 (1989).
3. P. Tikkanen, J. Keinonen, A. Lampinen, R. Julin, A. Pakkanen, P. Ahonen, J. Hattula, S. Juutinen, S. Törmänen, P.J. Nolan, P. von Brentano, D. Dewald, D. Gelberg, G. Siems, and R. Wirowski, Phys. Rev. C to be published.
4. A. Kuronen, to be published.
5. H.G. Börner, J. Jolie, F. Hoyler, S. Robinson, M.S. Dewey, G. Greene, E. Kessler, and R.D. Deslattes, Phys. Lett. B **215**, 45 (1988).
6. J. Jolie, S. Ulbig, H.G. Börner, K.P. Lieb, S.J. Robinson, P. Schillesbeeckx, E.G. Kersler, M.S. Dewey, and G.L Greene, Europhys. Lett. **10**, 231 (1989).
7. J.F. Ziegler, J.P. Biersack, and V. Littmark, *The Stopping Powers and Ranges of Ions in Matter* (Pergamon, New York, 1985), Vol. 1.
8. M. Hautala, Phys. Rev. **30**, 5010 (1984).
9. K. Arstila, J. Keinonen, and P. Tikkanen, Phys. Rev. B **41**, 6117 (1990).
10. J. Lindhard, M. Scharff, and H.E. Schiftt, Mat. Fys. Medd. Dan. Vid. Selsk. 33, **14** (1963).
11. A.E. Blaugrund, Nucl. Phys. 88, 501 (1966).
12. J.S. Forster, T.K. Alexander, G.C. Ball, W.G. Davies, I.V. Mitchell, and K.B. Winterbon, Nucl. Phys. A **313**, 397 (1979).
13. P. Tikkanen, J. Keinonen, A. Kuronen, A.Ź. Kiss, E. Koltay, E. Pintye, and B.H. Wildenthal, Nucl. Phys. A, to be published.
14. B.H. Wildenthal, Prog. Part. Nucl. Phys. **11**, 5 (1984), B.A. Brown and B.H. Wildenthal, Ann. Rev. Nucl. Part. Sci. **38**, 29 (1988).
15. P.J. Nolan and P.J. Twin, Ann. Rev. Nucl. Part. Sci. **38**, 533 (1988).
16. J.C. Bacelar et al., Phys. Rev. Lett. **57**, 3019 (1986).

PRECISION GAMMA-RAY MEASUREMENTS IN ^{25}Mg FOLLOWING THERMAL NEUTRON CAPTURE IN ^{24}Mg*

S. Michaelsen, K.P. Lieb and L. Ziegeler,
II. Physikalisches Institut, Universität Göttingen,
D-3400 Göttingen, Fed. Rep. Germany

T. von Egidy,
Physik Department, TU München,
D-8046 Garching, Fed. Rep. Germany

ABSTRACT

Gamma-ray transitions in ^{25}Mg following the ^{24}Mg(n,γ) reaction have been measured by means of HP germanium and pair spectrometers and precise γ-ray and level energies have been deduced. The neutron binding energy resulted as B_n = 7330.61(14) keV.

MOTIVATION AND EXPERIMENT

In the course of a systematic study of thermal neutron capture in the sd-shell nuclei [1], and in particular in the nucleus ^{31}P [2], we have also investigated the reaction ^{24}Mg(n,γ)^{25}Mg, making use of the 5 10^{14}/(cm^{2}s) neutron flux in the H7 in-pile position of the ILL high flux reactor at Grenoble. This study was motivated by a discrepancy in the previous measurements of the neutron binding energy in ^{25}Mg [3-5] and the need for more precise level and gamma ray energies than those quoted in the compilation of Endt and Van der Leun [6].

The target consisted of either 298 mg 99.5% enriched ^{24}MgO or 290 mg ^{24}Mg$_2$P$_2$O$_7$ (and approximately 0.2 mg NaCl) and was contained in a cylinder of reactor graphite. The low energy part between 50 keV and 3 MeV and the full γ-ray spectrum (up to 8 MeV) were scanned with a 20% efficient intrinsic HP germanium detector having an energy resolution of 2.0 keV FWHM at 1.33 MeV. The high energy part of the spectrum was measured with a pair spectrometer, having 3.1 keV energy resolution at 2 MeV and 6.5 keV at 8 MeV. Details of the energy and efficiency calibration of the spectra relative to the ^{35}Cl(n,γ)^{36}Cl standard [7] are given in the paper by Michaelsen et al.[2].

RESULTS

Since the main objective of the experiment had been to extend the ^{32}P level scheme and since the thermal neutron capture cross section in ^{24}Mg is rather small (σ = 53 mb), no attempt was made to localize all possible ^{25}Mg transitions in the spectra. We rather concentrated onto the 13 transitions assigned by Spilling et al. [3]. The energies of these transitions, corrected for recoil, were obtained from the Ritz combination procedure and are listed in Table 1, together with their level placements and their intensities (normalized to 100 captured neutrons). It is seen that these transitions make up for 97(4) % of the decay of the capture state and that the γ-ray flux reaching the ground state results as 114(23) %. While the transition and level energies are more precise than the previously quoted values [6], the γ-ray branching ratios, although in agreement with [6], have larger uncertainties.

As a further result the neutron binding energy was deduced as B_n=7330.61 keV with a purely statistical error of 0.03 keV. Adding the systematic calibration error in quadrature, we find a total error of 0.14 keV. Our value is in good agreement with that measured by Hungerford and Schmidt [5], B_n=7330.83(14) keV, and by Islam et al. [4], B_n=7330.5 keV, with an error of less than 0.5 keV, but is at variance with the result of Spilling et al. [3]. Combining the present figure with that of [4,5], we propose an average value of 7330.69(10) keV.

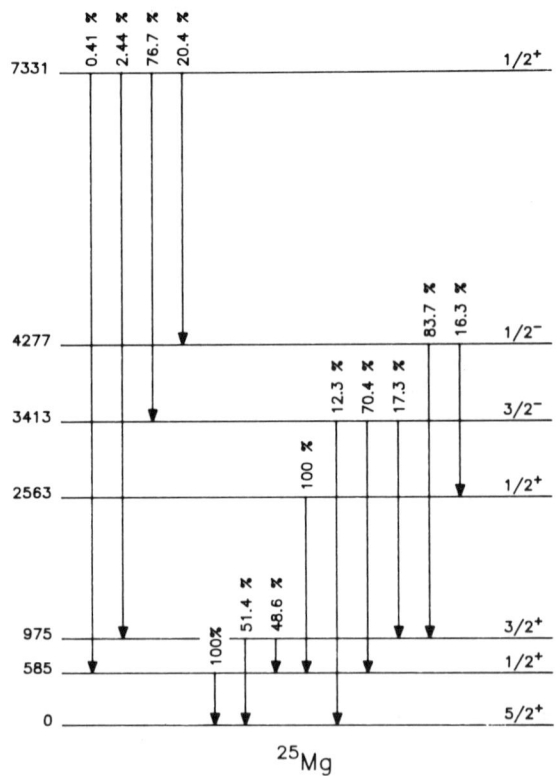

Fig. 1   Level scheme of ^{25}Mg obtained from the ^{24}Mg(n,γ) reaction. Shown are the excitation energies, branching ratios [6] and spin-parity assignments of the states.

Table 1: Placements of the γ-ray lines observed in the ^{25}Mg(n,γ)^{25}Mg reaction

E (keV) [a]	Intensity [b]	Placement		Ex (keV)
389.808(29)	18.9(48)	975	585	
585.105(28)	86 (22)	585	0	585.086(20)
974.822(32)	17.9(45)	975	0	974.853(22)
1713.10 (9)	3.0 (7)	4277	2563	
1978.30 (6)	3.3 (8)	2563	585	2563.44 (7)
2438.51 (5)	13.8(21)	3413	975	3413.45 (7)
2828.17 (5)	56 (8)	3413	585	
3053.91 (4)	19.8(10)	7331	4277	
3301.43 (5)	15.4 (8)	4277	975	4276.51 (9)
3413.16 (5)	9.8 (5)	3413	0	
3916.84 (5)	74.3(37)	7331	3413	7330.61 (14)
6354.79(10)	2.4 (1)	7331	975	
6744.04(43)	0.40(5)	7331	585	

[a] Gamma ray energies not corrected for recoil
[b] Gamma ray intensity per 100 captured neutrons.

REFERENCES

*Supported by Deutsches Bundesministerium für Forschung und Technologie under contract 06 GÖ 141.

[1] B. Krusche and K.P. Lieb, Phys. Rev. **C34**, 2103 (1986).
[2] S. Michaelsen, et al., Nucl. Phys. **A501**, 437 (1989).
[3] P. Spilling, et al., Nucl. Phys. **A102**, 209 (1967).
[4] M.A. Islam, et al., Can. J. Phys. **58**, 168 (1980).
[5] P. Hungerford and H.H. Schmidt, Nucl. Instr. Meth. **192**, 609 (1982).
[6] P. Endt and C. van der Leun, Nucl. Phys. **A310**, 1 (1978).
[7] B. Krusche, et al., Nucl. Phys. **A386**, 245 (1982).

GRID LIFETIME STUDY OF THE 1762 KEV STATE IN ^{49}Ti POPULATED VIA CASCADE FEEDING*

S.Ulbig[1], J.Jolie[2], H.G.Börner[2], M.S.Dewey[3], K.P.Lieb[1] and S.J.Robinson[2]

[1]) II. Physikalisches Institut, Universität Göttingen, D–3400 Göttingen, Fed. Rep. Germany

[2]) Institut Laue–Langevin, F–38042 Grenoble, France

[3]) National Institute of Standards and Technology, Gaithersburg, USA

ABSTRACT

One of the difficulties encountered in GRID–measurements concerns the feeding of the level under study via intermediate levels. The description for two–step cascades is derived and tested by analysing Doppler broadened lineshapes of the 1762 keV transition in ^{49}Ti after thermal neutron capture in ^{48}Ti.

MOTIVATION

The GRID method [1-3] has been successfully applied to measure lifetimes of excited nuclear states after thermal neutron capture [1,4-6] using the GAMS4 spectrometer [7] installed at the ILL as a joint ILL/NIST facility. This method exploits the fact that the emission of a γ–ray $E_{\gamma 1}$ by the excited nucleus imparts a recoil $v/c = E_{\gamma 1}/Mc^2$ to the atom containing the nucleus. If the nucleus decays further by the emission of a γ–ray $E_{\gamma 2}$, the observed energy of this γ–ray will be Doppler shifted by $\Delta E_{\gamma 2} = E_{\gamma 2} v(t)/c \, cos(\gamma 1, v(t))$ depending on the time interval t between the emissions of $E_{\gamma 2}$ and $E_{\gamma 1}$. As the directions of emission are isotropically distributed this shift leads to a measurement of a Doppler broadening[1,2]. The observed lineshape can then be used to relate the slowing down time to the lifetime τ of the excited state. The cases studied in [1,4-6] offered relative simple initial conditions, as the levels of interest were mainly primarily fed resulting in a uniquely defined recoil velocity $v/c = E_{prim}/Mc^2$.

However the case of predominant primary feeding is not the common one in (n,γ)–reactions. Usually many different transitions contribute to the feeding of a low lying state. The resulting velocity distribution then depends on the energies of all feeding transitions and the lifetimes of the intermediate levels involved in the feeding process. This is especially important for GRID studies in heavy nuclei, as reported in refs.[8-10].

The model case to investigate the effects of cascade feeding is a level with lifetime τ fed by two successive transitions $(E_{\gamma 1}, E_{\gamma 2})$ via an short lived intermediate level of known lifetime τ' and being depopulated by radiation $E_{\gamma 3}$. The initial velocity distribution for this case will be developed in the following section. The description will be tested with the 1762 keV level in ^{49}Ti which is fed to more than 90% by such a twofold cascade via the 3261 keV level [11].

THE INITIAL VELOCITY DISTRIBUTION

The description of the slowing down process follows the methods outlined in several previous papers [1-3,12]. The scattering of the recoiling atom was simulated by elastic hard–sphere collisions with energy dependent hard–sphere radii determined by a Born–Mayer potential [13]. The two feeding transitions $E_{\gamma 1}$ and $E_{\gamma 2}$ introduce recoils $v_{1,2}/c = $

$E_{\gamma1,\gamma2}/Mc^2$. The emission of $E_{\gamma2}$ happens during the slowing down process due to $E_{\gamma1}$. For a known lifetime τ' the number of decays $N(k)$ of the intermediate level between the scatters k and k+1 can be calculated. For each k one gets a distribution $D(v)$ determined by the vectorial coupling of $\vec{v}_1(k)$ and $\vec{v}_2$,

$$D(v)dv = N(k)/(2v_1(k)v_2)vdv \; for \; |v_1(k) - v_2| < v < v_1(k) + v_2.$$

The initial velocity distribution of nuclei decaying by $E_{\gamma3}$ is obtained by summing up the contributions of the individual scatters k. For computational reasons the continuous distribution is approximated by 10 discrete velocity intervals. As the population of the level to be studied may happen in between two collisions a reduced free path for the first scatter is used. For each velocity interval a mean reduced free path is taken.

EXPERIMENTS AND RESULTS

After thermal neutron capture in ^{48}Ti the 1762 keV state in ^{49}Ti is mainly fed by two γ-rays of 4.88 and 1.498 MeV via the intermediate 3261 keV state. The lifetime of this level has been determined by GRID to be $\tau' = 17.3(5) fs$ [14]. The lineshapes of the 1762 keV transition have been measured in metallic Ti and in TiC and Ti_2O_3 compound targets. In case of the composite targets the collision sequences C-Ti and O-Ti-O-Ti-O have been used in the analysis. The experimental and fitted lineshapes in the TiC target are shown in Fig.1 together with the response function of the spectrometer. The lifetimes in Ti, TiC and Ti_2O_3 were fitted with the computer code 'GRIDDLE'[15] and yielded $\tau(1762) = 55(5), 48(^4_3)$ and $53(^7_6)$ fs, respectively. The results in three different target materials are consistent and give an average value of $\tau = 52(3) fs$. They agree well with a previous DSA value of 52(12) fs [16], but are higher than the value of 36(4) fs obtained in a resonant bremsstrahlung scattering measurement [17].

Fig.1: Lineshape of the 1762 keV in the TiC-target. Shown are the experimental rocking curve, the fitted lineshape and the response function of the GAMS4 spectrometer (dashed line). The insert shows the partial decay scheme of ^{49}Ti as used in the discussion.

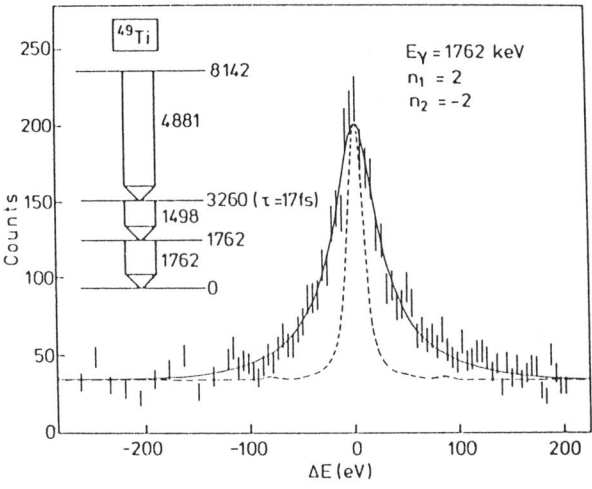

The knowledge of handling the two-step cascade feeding offers the possibility to study nuclear states with much more complex feeding patterns. The extension to cascades of higher multiplicities is straightforward[18]. GRID studies with the inclusion of cascade feeding effects are discussed in [19,20].

* Work partially supported by Deutsches BMFT under contract 06GOE141.

REFERENCES

[1] H.G. Börner, J. Jolie, S.J. Robinson, M.S. Dewey, G. Greene, E.G. Kessler and R.D. Deslattes, Phys. Lett. **B215**, 45 (1988); H.G. Börner et al., Contr. to this conference.

[2] H.G. Börner, J. Phys. **G14**, S143 (1988).

[3] J. Jolie, H.G. Börner, F. Hoyler, S.J. Robinson and M.S. Dewey, J. Phys. **G14**, S586 (1988).

[4] K.P. Lieb, H.G. Börner, M.S. Dewey, J. Jolie, S.J. Robinson, S. Ulbig, Ch. Winter, Phys. Lett. **B215**, 50 (1988).

[5] S. Ulbig, K.P. Lieb, Ch. Winter, H.G. Börner, J. Jolie, S.J. Robinson, P.A. Mando, P. Sona, N. Taccetti, M.S. Dewey, J.G.L. Booten and F. Brandolini, Nucl. Phys. **A505**, 193 (1989).

[6] S. Ulbig, K.P. Lieb, H.G. Börner, J. Jolie, S.J. Robinson, J.G.L. Booten, M.S. Dewey and Ch. Winter, XXV Zakopane School on Physics (1990), World Scientific, in press.

[7] M.S. Dewey, E.G. Kessler, G.L. Greene, R.D. Deslattes, H.G. Börner and J. Jolie, Nucl. Instr. Meth. **A284**, 151 (1989), and E.G. Kessler, G.L. Greene, M.S. Dewey, R.D. Deslattes, H.G. Börner and F. Hoyler, J. Phys. **G14**, S167 (1988).

[8] H.G. Börner, J. Jolie, S.J. Robinson and P. Schillebeeckx, Inst. Phys. Ser. No. **105**, 145 (1990).

[9] H.G. Börner, J. Jolie, S.J. Robinson, R.F. Casten and J.A. Cizewski, Phys. Rev., Rapid Comm., in press.

[10] H.G. Börner, J. Jolie, S.J. Robinson, B. Krusche, R. Piepenbring, R.F. Casten, A. Aprahamien and J.P. Draayer, subm. to Phys. Rev. Lett.

[11] T.W. Burrows, Nucl. Data Sheets **48**, 569 (1986).

[12] J. Jolie et al., Contr. to this conference.

[13] A.A. Abrahamson, Phys. Rev. **178**, 76 (1968).

[14] J. Jolie, S. Ulbig, H.G. Börner, K.P. Lieb, S.J. Robinson, P. Schillbeeckx, E.G. Kessler, M.S. Dewey and G.L. Greene, Europhys. Lett. **10**, 231 (1989).

[15] S.J. Robinson and J. Jolie, ILL internal report 90RO14T (1990).

[16] P.A. Mando, G. Poggi, P. Sona and N. Taccetti, Phys. Rev. **C23**, 2008 (1981).

[17] V.K. Rasmussen, Phys. Rev. **C13**, 631 (1976).

[18] B. Krusche et al., to be published.

[19] S. Ulbig, J. Jolie, S.J. Robinson, K.P. Lieb, H.G. Börner and P. Schillebeeckx, submitted for publication.

[20] S. Ulbig, K.P. Lieb, H.G. Börner, S.J. Robinson and J.G.L. Booten, Contribution to this conference.

LEVEL DENSITY IN ^{51}V AT SPIN 9/2-15/2 MEASURED WITH THE ^{50}V(n,γ) REACTION*

S. Michaelsen, K.P. Lieb,
II. Physikalisches Institut, Universität Göttingen,
D-3400 Göttingen, Fed. Rep. Germany

S.J. Robinson,
Institut Laue-Langevin, F-38042 Grenoble, France

ABSTRACT

Neutron capture in ^{50}V ($I^{\pi}=6^+$) was studied at the ILL high flux reactor and nearly 60 states in the spin range [9/2,15/2] were identified leading to a "complete" level scheme up to 5.5 MeV. Fermi gas model level density parameters were deduced. The energy scaling of the primary dipole spectrum and its correlation with the p-wave neutron-transfer strength are discussed.

MOTIVATION AND EXPERIMENT

The investigation of high spin states via heavy ion fusion reactions requires information on the level density and γ-ray strength function distributions at high spins which usually are being extrapolated from low spin data. However, complete neutron capture spectroscopy on targets with elevated spins may provide valuable tests of these distributions at fairly high spin. To this end we have studied, at the ILL high flux reactor, thermal neutron capture in the isotope ^{50}V, one of the very few stable nuclei with high spin in its ground state ($I^{\pi}=6^+$). S-wave neutron capture leads to $11/2^+$, $13/2^+$ compound states in ^{51}V and, after primary dipole transitions, to states within the spin window $9/2 \leq I \leq 15/2$. Because of the non-selectitvity of the (n,γ) process the complete level scheme within this spin range up to about half the binding energy should be accessible. On the basis of these data and the known low-spin states [1], it should be possible to determine the spin dependence of the level density in ^{51}V from individual level counting.

The 25 mg V_2O_3 target was enriched to 36% in the isotope ^{50}V and was exposed to the 5 10^{14}/cm^{2}s neutron flux in the H7 internal irradiation position. The γ-radiation was measured in a 20% intrinsic Ge detector and a pair spectrometer. Both detectors were energy and efficiency calibrated relative to the ^{36}Cl standard [2] by adding 15 mg of NaCl to the target. The lines in ^{52}V originating from the ^{51}V(n,γ) reaction were eliminated on the basis of measurements with ^{nat}V targets containing 99.75% ^{51}V. Out of the 720 lines attributed to ^{51}V, 330 transitions comprising 90% of the gamma ray flux, and 59 primary transitions were located in the level

*Supported by Deutsches BMFT under contract 06 Gö 141

scheme, by employing a Ritz combination procedure. Many levels previously found in the the ^{50}V(d,p) reaction to have transfer angular momentum l=1 were confirmed [3] and twice as many states in the $9/2^-$-$15/2^-$ spin range as compared to the previous (n,γ) work [4,5] were established. A full account of both nuclei will be given in a forthcoming paper [6].

LEVEL DENSITY

The density of states in ^{51}V at excitation energies E_x was parametrized with either the Bethe formula (BF) for the back-shifted Fermi gas model

(1) $\varrho(E_x) = \exp[2(a(E_x-E_1))^{1/2}]/[12\ 2^{1/2}\ \sigma\ a^{1/4}\ (E_x-E_1)^{5/4}]$

or the constant temperature Fermi gas model (CTFG)

(2) $\varrho(E_x) = (1/T)\ \exp[(E_x-E_0)/T]$

where the parameters a, E_1, T and E_0 have their usual meaning and σ denotes the spin cut-off parameter [7]. The formulae refer to the total density of states and have to be multiplied by the spin distribution function $\Sigma f(I)$ with

(3) $f(I) = \exp(-I^2/2\sigma^2) - \exp(-(I-1)^2/2\sigma^2)$

where the sum extends over the spin window $[I_1,I_2]$ of intermediate levels populated by the thermal (n,γ) process.

The level density parameters a=5.7(2) MeV^{-1}, E_1=-0.9(2) MeV and T=1.36(5) MeV, E_0=-2.1(4) MeV, respectively, were determined from a fit of (1) and (2) to the experimental density of states in the spin range [9/2,15/2] between 2.5 and 5.5 MeV. The level scheme was assumed to be complete within this spin window (see fig. 1). At the neutron binding energy B_n=11,051.11(13) keV [6], the average level spacing D_0=1.4(2) keV of s-wave resonances (I=11/2$^+$,13/2$^+$), corrected by Rohr [8] for unobserved resonances, was used. In extrapolating this value to the spin window [9/2,15/2], the spin cut-off parameter was set σ=2.5-3.1 (see below), but this quantity has only a minor influence on the other parameters. As illustrated in fig. 1 the statistical behavior of the level density in ^{51}V starts above 2.5 MeV. This is just the energy necessary to excite a nucleon across the $1f_{7/2}$-$2p1f_{5/2}$ shell gap.

We now discuss in more detail the spin distribution of the states in ^{51}V. We first note that nearly all states populated by primary (n,γ) transitions coincide with levels excited via l=1 neutron transfer in the ^{50}V(d,p) reaction and thus have negative parity. On the other hand, low spin states (I$\leq$7/2) are known from transfer reactions on even-even targets, i.e. from the reactions ^{50}Ti(^{3}He,d), ^{52}Cr(t,α), and ^{52}Cr(d,^{3}He), and from inelastic proton and neutron scattering experiments on ^{51}V [1]. Up to 5.0 MeV excitation, 106 levels are known and 73 of them have unambiguous spin-parity assignments or the spin restricted to two values. The experimental spin distribution based on these data is shown in fig. 2a. (If the spin of a state was not known exactly but only restricted to a certain range, we assumed that all allowed spins are equally probable.) For the spin 7/2 levels, only a lower limit is given, because these states are difficult to excite experimentally. Also plotted in fig. 2a is the theoretical spin distribution (4) for σ = 3.1 which fits the data very well. A more global representation of the spin distribution is the quantity $N_{all}/N_{9/2-15/2}$, i.e. the ratio between all states and the class [9/2,15/2] states. This ratio is plotted in fig. 2b as funtion of σ, giving σ=2.8(3).

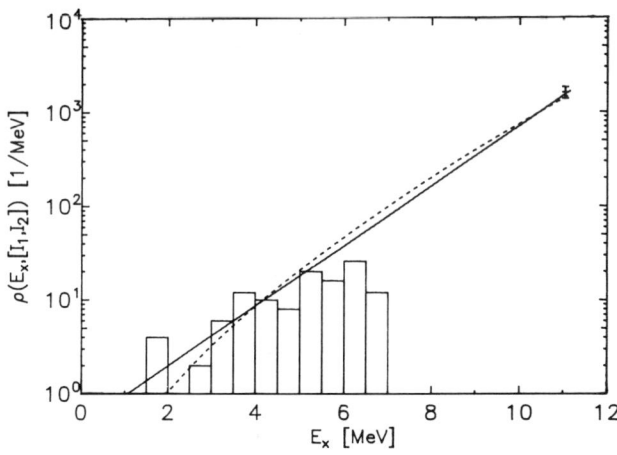

Fig. 1 Experimental level density $\varrho(E_x,[9/2,15/2])$ (histogram) obtained from the $^{50}V(n,\gamma)$ reaction and fits with the CTFG model (solid line) and the BF model (dashed line).

Recently, von Egidy et al. [9] have reviewed the level density parameters in 75 nuclei all over the Periodic Table and proposed empirical formulae for them as function of the nuclear mass A. The average spin of their studied spin windows, $\langle I \rangle$ = 2.3 ℏ, is much lower than that of the states populated in the $^{50}V(n,\gamma)$ reaction, $\langle I \rangle$ = 6 ℏ. Their "overall" level density parameters evaluated for A=51 are T≈1.31 MeV, E_0≈-2.05 MeV, a≈6.42 MeV^{-1} and E_1≈-0.51 MeV. In spite of the large difference in the spin ranges, these values agree rather well with our fitted parameters. That the single-particle level sensity parameter a is below the overall prediction, reflects the fact that ^{51}V is a semi-magic nucleus with 28 neutrons. On the basis of the experimental spin distributions in many nuclides, von Egidy et al. [9] also proposed a fit of σ,

(4) $\sigma = (0.98 \pm 0.23) A^{(0.29 \pm 0.06)}$

which in the case of ^{51}V gives σ≈3.1, in good agreement with our value. Note, however, the large interval 1.9≤σ≤4.8 allowed by eq. (4).

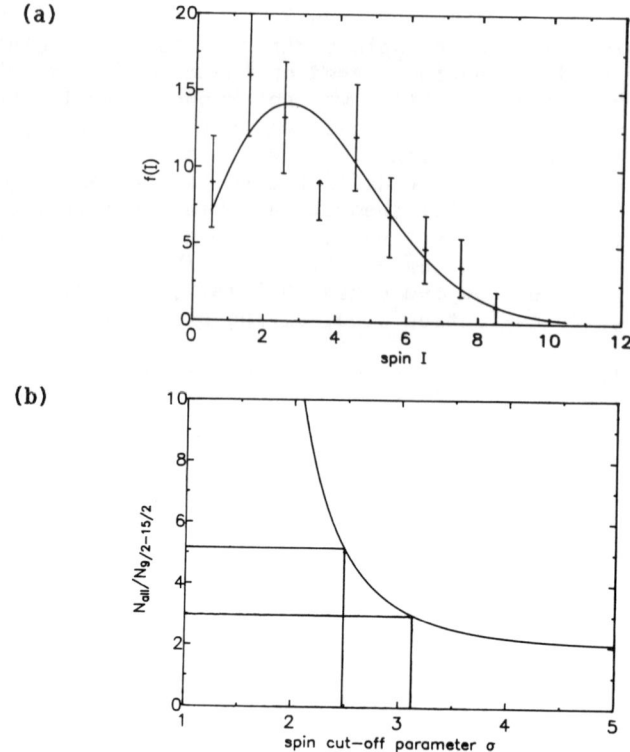

Fig. 2 (a) Experimental spin distribution for states up to 5 MeV in ^{51}V. The full line represents the calculated distribution with the spin cut-off parameter σ=3.1.
(b) Calculated and experimental ratio $R=N_{all}/N_{9/2-15/2}$ from which the spin cut-off parameter σ=2.8(3) was deduced.

PRIMARY DIPOLE SPECTRUM

Finally, we shortly discuss the distribution of the relative branching ratios $b(E\gamma)$ of the primary E1 transitions and their correlation with the p-wave spectroscopic factors $S_{l=1}$ of the ^{50}V(d,p) reaction [3]. As usual we either consider a single-particle energy scaling [$b(E\gamma) \sim E\gamma^3$] or giant dipole resonance (GDR) scaling [$b(E\gamma) \sim E\gamma^5$]. In fig. 3 we compare the experimental integrated branching ratios

(5) $\quad Y(E_x) = \sum\limits_{E\gamma \geq E_B - E_x} b(E\gamma)$

with the prediction of the statistical model

(6) $\quad Y(E_x) = c \int\limits_0^{E_x} \varrho(E')\,(E_B - E')^n\, dE', \quad n=3 \text{ or } 5$.

As it is seen in fig. 3 the suitable energy dependence is given by $E\gamma^5$. But it is also seen that between 3 and 4 MeV the increase of the experimental curve is steeper than predicted. This effect might be due to the direct capture mechanism which is known to dominate the primary dipole spectrum for nuclei in the mass region $A \approx 40-65$. Indeed, we observe a strong correlation between the p-wave (d,p) spectroscopic factors $S_{l=1}$ and the reduced strengths $b(E\gamma)/E\gamma^5$; the correlation factor is 0.92. A more detailed discussion of the primary spectrum will be presented in [6].

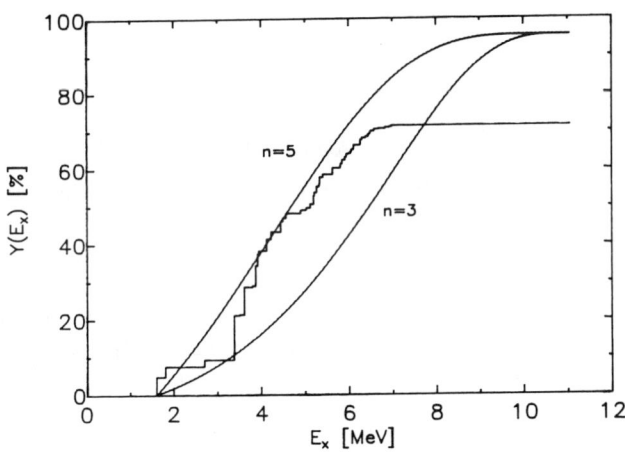

Fig. 3 Integrated intensity $Y(E_x)$ (step function) of primary E1 transitions as compared with the prediction of eq. (6) (solid lines) for the scaling exponents n = 3 and n = 5.

REFERENCES

[1] Nuclear Data Sheets **48**, 111 (1986).
[2] B. Krusche, et al., Nucl. Phys. **A386**, 245 (1982).
[3] M.E. Belote, et al., Nucl. Phys. **A94**, 673 (1967).
[4] D. Harrach, Dissertation München (1973), unpubl.
[5] A. Robertson, et al., Z. Phys. **A284**, 407 (1978).
[6] S. Michaelsen, K.P. Lieb, S.J. Robinson, B. Krusche, in prep.
[7] See B. Krusche, K.P. Lieb, Phys. Rev. **C34**, 2103 (1986).
[8] G. Rohr, private communication.
[9] T. von Egidy, et al., Nucl. Phys. **A481**, 189 (1986); **A454**, 109 (1988).

The lowest $J^\pi = 1^+$ analogue and antianalogue in ^{56}Fe

Zhendi Guo, C. Alderliesten, C. van der Leun
and P.M. Endt

*R.J. Van de Graaff Laboratorium, Rijksuniversiteit Utrecht,
P.O. Box 80.000, 3508 TA Utrecht, The Netherlands.*

Abstract

Yield curves of the ^{55}Mn(p,γ)^{56}Fe and ^{55}Mn(p,n)^{55}Fe reactions have been measured in the E_p = 1420–1470 keV region. The three components, at E_p = 1440, 1446 and 1455 keV, of the split analogue of the $J^\pi = 1^+$ 0.11 MeV ^{56}Mn level are completely resolved.

The Compton-suppressed Ge γ-ray spectra taken at these three resonances with good statistics all show strong primaries to levels at E_x = 4866 and 5023 keV, which can be regarded as components of the split antianalogue. Energies and γ-ray branching ratios of many ^{56}Fe levels have been obtained with high precision. The Q-value of the ^{55}Mn(p,γ)^{56}Fe reaction amounts to 10 184.10 ± 0.18 keV.

In previous work[1] on the ^{55}Mn(p,γ)^{56}Fe reaction it had been shown that the analogue of the 1^+ 0.11 MeV ^{56}Mn level is split into three (unresolved) components at which γ-ray spectra had been measured with a bare Ge(Li) detector with poor statistics. It was felt that much more information on these interesting states could be obtained with thinner targets, better accelerator stabilization, with Compton suppression and with better statistics.

The yield curve for the ground-state γ-ray transition (γ_0) is shown in Fig. 1. The three resonances are quite narrow (target thickness 0.7 keV) and completely resolved, and apparently there are no other γ_0 resonances in the investigated E_p interval. The yield curve for a window on the γ-ray spectrum set at $E_\gamma > 3.5$ MeV, however, shows some 45 resonances, largely unresolved. The same holds for the yield of neutrons measured with a ^{3}He detector. From current level density expressions the number of levels in ^{56}Fe at $E_x \approx 11.6$ MeV is estimated as 200 in the investigated region.

Gamma-ray spectra were measured at the three resonances for about 11 h each at a proton current of 60 μA with a Compton shielded 95 cm^3 Ge detector (CSS) at $\theta = 55°$ and a bare Ge detector at $\theta = 90°$. The three spectra are very similar.

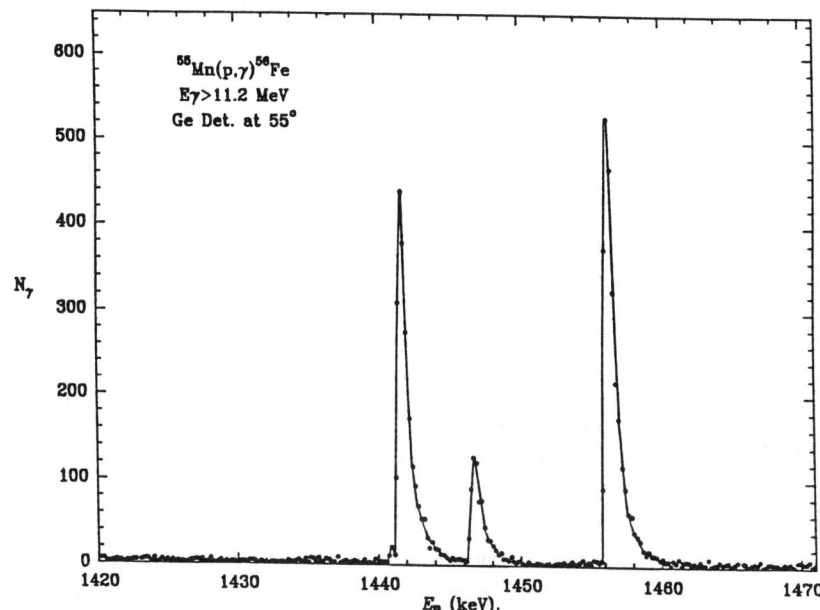

Figure 1: The yield of the ^{55}Mn(p,γ)^{56}Fe ground-state transition in the E_p = 1420–1470 keV region.

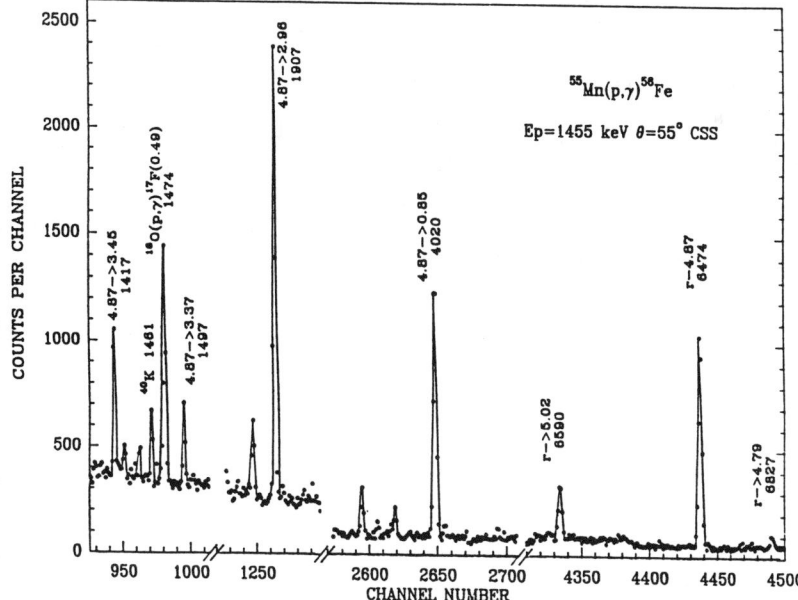

Figure 2: Selected parts of the ^{55}Mn(p,γ)^{56}Fe γ-ray spectrum at the E_p = 1455 keV measured with a Compton shielded Ge detector at $\theta = 55°$. The primaries are shown to the 4.87 and 5.02 MeV levels and the four strongest decay branches of the former.

Table I
Gamma-ray branching ratios (in %) of the E_x=4.87 and 5.02 MeV ^{56}Fe levels[a]

J_f^π:	0+	2+	2+	0+	2+	2+	1(+)	Unplaced
E_{xi} \ E_{xf}:	0	847	2658	2942	2960	3370	3449	
4866	2.3 *3*	50 *2*	2.5 *3*	<0.5	25.3 *8*	3.8 *2*	7.4 *3*	9 *2*
5023	2.8 *9*	2.1 *8*	<2	5.5 *7*	30.0 *12*	20.3 *7*	19.8 *7*	19 *2*

[a] All energies are in keV. The indices i and f relate to initial and final states, respectively. The error in units of the last decimal is indicated in italics. The branchings presented here are the averages of those measured at E_p=1440, 1446 and 1455 keV.

There are strong transitions to $E_x = 0$ and 0.85 MeV, with average branchings of 48% and 18%, respectively, weak transitions (together ≈ 13%) to many levels in the $E_x = 0.9 - 4.8$ MeV region and strong transitions to levels at $E_x = 4.87$ and 5.02 MeV (branchings of 16% and 5%, respectively). Parts of the CSS spectrum obtained at the $E_p = 1455$ keV resonance, showing the primaries to the latter two levels and the four strongest decay lines of the 4.87 MeV level, are given in Fig. 2. From the 90° spectra accurate energies (errors < 0.2 keV) of 16 bound states could be obtained, and from the 55° spectra much new information on their γ-decay.

The levels at $E_x = 4866.46 \pm 0.05$ and 5023.45 ± 0.05 keV can be regarded as the split antianalogue of the analogue resonances. Although their spin could not be measured, their γ-decay to $J^\pi = 0^+, 1^{(+)}$ and 2^+ levels is in agreement with a $J^\pi = 1^+$ assignment (see Table I).

In a separate experiment the proton energy of the strongest analogue component was measured as $E_p = 1455.18 \pm 0.07$ keV by comparison with the accurately known value[2] for the 1417 keV ^{23}Na(p,γ)^{24}Mg resonance. Combined with $E_x = 11612.93 \pm 0.17$ keV as obtained from the 90° spectrum this yields $Q = 10184.01 \pm 0.18$ keV for the ^{55}Mn(p,γ)^{56}Fe reaction, in agreement with but much more accurate than the value 10183.9 ± 0.9 keV from the most recent atomic mass table[3].

[1] Huo Junde et al., Nuclear Data Sheets **51**, 1 (1987).
[2] P.M. Endt et al., Nucl. Phys. **A510**, 209 (1990).
[3] A.H. Wapstra and G. Audi, Nucl. Phys. **A432**, 1 (1985).

GRID LIFETIME MEASUREMENTS AND SHELL MODEL CALCULATIONS IN ^{59}Ni*

S.Ulbig and K.P. Lieb
II. Physikalisches Institut, Universität Göttingen,
D-3400 Göttingen, Fed. Rep. Germany

H.G. Börner and S.J. Robinson
Institut Laue-Langevin, F-38042 Grenoble, France

J.G.L. Booten
Department of Physics and Astronomy, NL-3508 TA Utrecht, The Netherlands

ABSTRACT

The lifetimes of seven states in ^{59}Ni in the range 8×10^{-15} -$7\times 10^{-13}s$ have been determined by means of the GRID method and the $^{58}Ni(n,\gamma)$ reaction. Energies and electromagnetic transition probabilities of the negative parity levels up to 3 MeV are compared with shell model predictions within 3p0h and 4p1h model spaces.

INTRODUCTION

Measuring nuclear lifetimes by the γ-ray induced Doppler broadening method (GRID)[1] is one of the main activities in ultrahigh resolution γ-ray spectroscopy carried out at the ILL [2-5]. Gamma-rays from short-lived intermediate states populated via primary γ-rays after thermal neutron capture reactions are diffracted in the double flat-crystal spectrometer GAMS4 [6]. Their Doppler-broadened lineshapes reveal information about the slowing-down process of the atoms recoiling at $v/c \leq 10^{-4}$ and the time dependence of the deexcitation process.

EXPERIMENT

Three 25x20x2 mm^3 metallic Ni targets of natural isotopic composition (^{58}Ni:68.27%) were exposed to the $5\times 10^{14}/cm^2s$ neutron flux available at the H6 irradiation position of the ILL high flux reactor. Thermal neutron capture in ^{58}Ni has a cross-section of 4.6 b and the primary γ-ray flux is distributed over many intermediate states [7]. A standard GAMS4 arrangement as described in [2,5,6] was used.

Fig. 1 shows the Doppler-broadened profiles of the 1301 keV and 1949 keV transitions in comparison with the respective response functions of the spectrometer. Doppler-broadenings are clearly visible and lead to mean lives of $\tau(1301)$= 319 fs and $\tau(2415)$= 79 fs. The slowing-down process of the atoms was approximated by isotropic scattering from interatomic Born-Mayer potentials [8], V(r)=A exp(-r/B), with parameters A= 13271 eV and B=0.280 Å [9].

Table I lists the measured lifetimes, in comparison with the previous results of conventional DSA measurements. For some of the excited states, cascade feeding had to be considered [7,10] in addition to direct primary feeding; details are given in a forthcoming paper [11]. The errors given include the uncertainties of the feeding.

*) Work partially supported by Deutsches BMFT under contract 06GOE141.

Grid Lifetime Measurements in ^{59}Ni

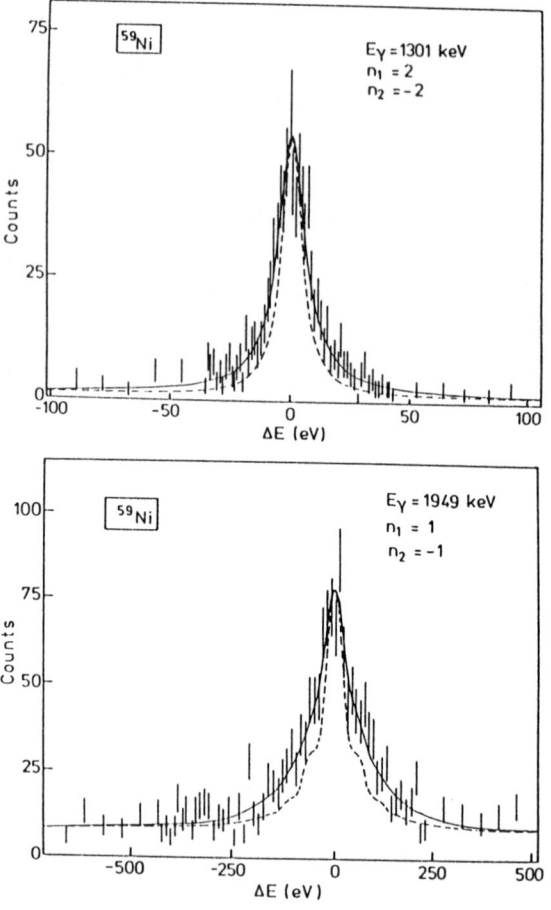

Fig.1: Lineshapes of the 1301 and 1949 keV transitions measured with GRID. Shown are the experimental rocking curves, the fitted lineshapes and the experimental response functions.

Table I: Comparison of lifetimes in ^{59}Ni obtained with conventional DSA-methods and with GRID

E_x(keV)	τ(fs)^a	τ(fs)^b	τ(fs)^c	τ(fs)^d	τ(fs)^e
878	700 (200)	700 (80)	620 (120)	900 $\binom{1300}{500}$	680 (130)
1189	370 (50)	375 (30)	440 (70)	308 $\binom{128}{88}$	381 (73)
1302	140 (20)	265 (25)	180 (20)	319 $\binom{125}{89}$	215 (55)
2415		48 (11)		79 $\binom{12}{13}$	65 (14)
		60 (30)			
2894				62 $\binom{10}{11}$	62 $\binom{10}{11}$
3182				54 (7)	54 (7)
4140				8.5 $\binom{19}{17}$	8.5 $\binom{19}{17}$

a) from ref. [12]), b) from ref. [13]), c) from ref. [14]), d) this work, e) adopted.

SHELL MODEL CALCULATIONS

In 1972, Glaudemans and coworkers [15] interpreted the ^{59}Ni level scheme and electromagnetic properties within a 3p0h model space and a modified surface delta residual interaction outside the doubly magic ^{56}Ni core. Van Hees et al. [16] studied the effects of enlarging the model space to 4p1h and using the Kuo-Brown residual interaction. The latter authors demonstrated that smaller effective E2 charges ($e_p = 2.0e, e_n = 1.0e$) and unquenched M1 matrix elements are sufficient to account for the experimental E2 and M1 transition strengths among the negative parity states.

We have performed similar shell model calculations by using the program RITS-SCHIL and the single-particle energies and two-body matrix elements given in our ^{57}Fe paper [5]. The resulting $\pi = -$ level scheme is presented in fig. 2. One notes the 1:1 correspondence of calculated levels in the 3p0h and 4p1h model spaces and to the experimental levels.

Table II compares measured and calculated B(M1) and B(E2) values. In general, our transition probabilities are very close to the ones reported in [16] for both choices of residual interactions and agree, within a factor of 2, with the experimental quantities.

Table II: a) Comparison of exp. and theor. B(M1) values calculated in 3p and 4p1h spaces. All levels have negative parities and are denoted by their spins.
b) Comparison of exp. and theor. B(E2) values calculated in 3p and 4p1h spaces with $e_p = 2e, e_n = e$.

Transition $I_i \rightarrow I_f$	B(M1) [μ_N^2] Exp.	3p	4p1h
$\frac{5}{2}1 \rightarrow \frac{3}{2}1$	0.015(2)	0.004	0.004
$\frac{1}{2}1 \rightarrow \frac{3}{2}1$	< 0.025	1.45	0.41
$\frac{3}{2}2 \rightarrow \frac{3}{2}1$	0.13(3)	0.10	0.10
$\frac{3}{2}2 \rightarrow \frac{5}{2}1$	< 0.009	0.025	0.014
$\frac{5}{2}2 \rightarrow \frac{3}{2}1$	0.07(2)	0.21	0.06
$\frac{5}{2}2 \rightarrow \frac{5}{2}1$	< 0.019	0.060	0.026
$\frac{7}{2}1 \rightarrow \frac{5}{2}1$	< 0.0013	0.020	0.008
$\frac{5}{2}3 \rightarrow \frac{5}{2}1$	< 0.083	0.000	0.001
$\frac{5}{2}3 \rightarrow \frac{3}{2}1$	< 0.007	0.003	0.000
$\frac{3}{2}3 \rightarrow \frac{5}{2}2$	< 0.319	0.71	0.25
$\frac{3}{2}3 \rightarrow \frac{1}{2}1$	< 0.024	0.045	0.005
$\frac{3}{2}3 \rightarrow \frac{5}{2}1$	< 0.028	0.004	0.001
$\frac{3}{2}3 \rightarrow \frac{3}{2}1$	< 0.053	0.19	0.01
$\frac{9}{2}1 \rightarrow \frac{7}{2}1$	0.06(2)	a.	0.095
$\frac{7}{2}2 \rightarrow \frac{5}{2}2$	< 0.129	0.035	0.044
$\frac{7}{2}2 \rightarrow \frac{5}{2}1$	< 0.033	0.008	0.013
$\frac{1}{2}2 \rightarrow \frac{1}{2}1$	0.05(2)	0.001	0.000
$\frac{1}{2}2 \rightarrow \frac{3}{2}1$	< 0.125	0.42	0.26

Transition $I_i \rightarrow I_f$	B(E2) [$e^2 fm^4$] Exp.	3p	4p1h
$\frac{5}{2}1 \rightarrow \frac{3}{2}1$	< 41	0.24	0.53
$\frac{3}{2}2 \rightarrow \frac{3}{2}1$	18 (29/13)	19	69
$\frac{5}{2}2 \rightarrow \frac{3}{2}1$	129 (89/57)	58	134
$\frac{1}{2}2 \rightarrow \frac{5}{2}1$	12 (7/5)	20	16
$\frac{7}{2}1 \rightarrow \frac{5}{2}1$	381 (109/75)	32	81
$\frac{7}{2}1 \rightarrow \frac{3}{2}1$	47 (12/9)	4	12
$\frac{5}{2}3 \rightarrow \frac{1}{2}1$	19 (12/9)	0.03	0.07
$\frac{9}{2}1 \rightarrow \frac{7}{2}1$	39 (85/33)	a.	3.6
$\frac{9}{2}1 \rightarrow \frac{5}{2}1$	158 (53/31)	a.	89
$\frac{7}{2}2 \rightarrow \frac{3}{2}1$	67 (14/10)	8	33

a.) Value not calculated.

Krusche and Lieb [17] have recently simulated the γ-ray flux pattern after thermal neutron capture in light nuclei within the statistical model. Knowing the level densities

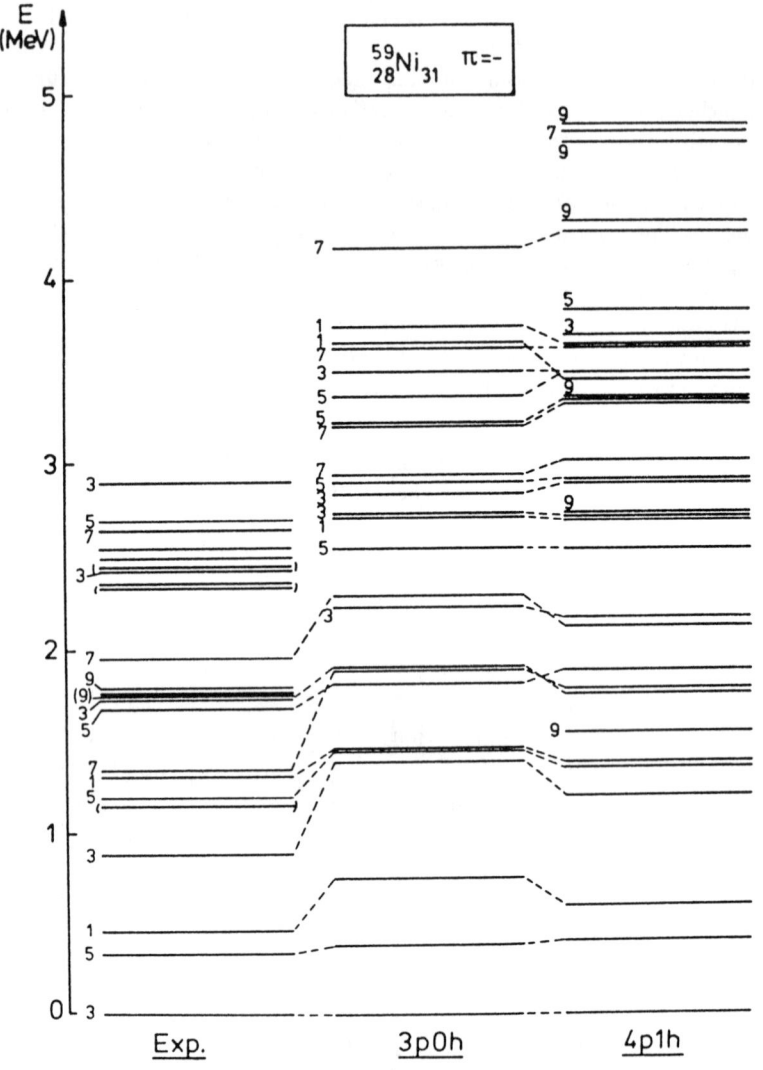

Fig.2: Comparison of experimental and calculated excitation energies of negative parity states. The results in the 3p space are given up to spin 7/2, the ones in the 4p1h space up to 9/2.

and the strength function distributions of primary E1, M1 and E2 radiations, one can estimate the average level lifetimes as function of excitation energy and spin. Fig.3 shows the result of such a calculation in ^{59}Ni; details are given in the paper by Krusche[18].

The experimental level widths span more than three orders of magnitude and are rather well reproduced by the calculation. This is astonishing for a nucleus only three valence particles away from being double magic.

Fig.3: Experimental and calculated level widths as a function of excitation energy and spin.

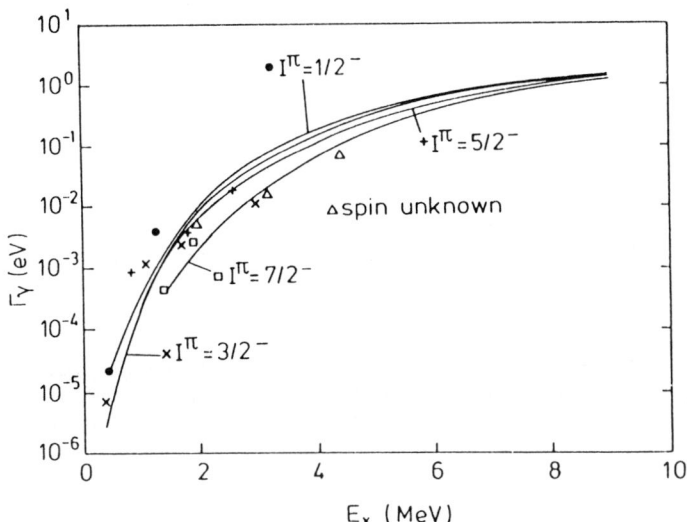

REFERENCES

[1] H.G. Börner, J. Phys. **G14**, S143 (1988).
[2] H.G. Börner et al., Phys. Lett. **B215**, 45 (1988).
[3] J. Jolie et al., Europhys. Lett. **10**, 231 (1989).
[4] K.P. Lieb et al., Phys. Lett. **B215**, 50 (1988).
[5] S. Ulbig et al., Nucl. Phys. **A505**, 193 (1989).
[6] M.S. Dewey et al., Nucl. Instr. Meth. **A284**, 151 (1989);
    E.G. Kessler et al., J. Phys. **G14**, S167 (1988).
[7] P. Anderson et al., NDS **39**, 641 (1983).
[8] J. Jolie et al., J. Phys. **G14**, S586 (1988).
[9] A.A. Abrahamson, Phys. Rev. **178**, 76 (1968).
[10] S. Ulbig et al., this conference.
[11] S. Ulbig et al., in preparation.
[12] J.D. Hutton et al., Nucl. Phys. **A206**, 403 (1973).
[13] M. Pichevar et al., Nucl. Phys. **A224**, 34 (1974).
[14] V.K. Mittal et al., J. Phys. **G9**, 91 (1983).
[15] P.W.M. Glaudemans et al., Nucl. Phys. **A198**, 609 (1972).
[16] A.G. van Hees et al., Z. Phys. **A293**, 327 (1979).
[17] B. Krusche and K.P. Lieb, Phys. Rev. **C34**, 2103 (1986);
    J. Phys. **G14**, S183 (1988).
[18] B. Krusche, to be published.

MEASUREMENT OF CONVERSION ELECTRONS FROM THE ^{75}As(n,e$^-$) REACTION
A TEST OF THE DESCRIPTION OF ^{76}As IN THE $U_\nu(6/12) \otimes U_\pi(6/12)$ SCHEME

J. Jolie[a], F. Hoyler[b], and B. Krusche[a]

a) Institut Laue-Langevin, 156X, F-38042 Grenoble Cedex, France
b) Phys. Institut, Univ. Tübingen, D-7400 Tübingen, FRG

ABSTRACT

The energy spectrum of conversion electrons following the ^{75}As(n,e$^-$) reaction was measured at the ILL. The determined E2 branchings for several transitions allow to test for the claim that the spectrum of the low-lying positive parity states of the odd-odd ncleus ^{76}As resembles the vibrational limit of the $U_\nu(6/12) \otimes U_\pi(6/12)$ supersymmetry.

At the Bill spectrometer the energy spectrum of conversion electrons following the thermal neutron capture in ^{75}As was measured with improved sensitivity up to 560 keV. Two targets made out of natural GaAs evaporated onto an Al foil were used in the experiment. The target thickness was approximately 250 μg/cm^2 for the thinner (used for electron energies below 170 keV) and 500 μg/cm^2 for the thicker target (used for the higher energies).

Table I Experimental results

E_γ (keV)[a]	I_γ(/100n)	I_{e^-}(/10^{4}n)	$\alpha_K * 10^2$	δ^2 [b]	comment
473.152(2)	3.53(14)	0.62(4)	0.175(14)		M1
402.746(4)	1.08(8)	0.30(5)	0.27(5)		M1
399.347(3)	0.71(7)	0.39(5)	0.55(9)		E2
360.383(3)	0.55(5)	0.25(4)	0.46(9)	0.81	M1+E2
357.401(4)	1.39(8)	0.83(7)	0.60(6)		E2
315.041(6)	0.20(1)	0.21(5)[c]	1.05(28)		(E2)
281.575(1)	1.79(7)	1.07(4)	0.59(4)		M1
263.894(2)	3.46(4)	2.38(7)	0.69(2)	0.04	M1+E2
237.281(4)	0.25(1)	0.15(5)	0.58(19)		E1 / M1
235.877(1)	3.63(4)	5.32(11)	1.47(3)	0.44	M1+E2
221.532(2)	1.12(2)	3.07(9)	2.74(11)	1.96	E2+M1
220.380(2)	0.81(2)	0.83(7)	1.03(8)		M1
188.062(1)	1.54(3)	2.50(8)	1.62(7)	0.03	M1+E2
178.018(2)	1.48(6)	2.76(11)	1.86(11)		M1
159.116(9)	0.08(1)	0.67(14)	8.8(19)	1.6	E2+M1
144.549(1)	2.16(15)	6.67(13)	3.10(22)		M1
137.012(6)	0.53(3)	1.86(11)	3.50(32)		M1
116.755(1)	1.82(11)	9.12(18)	5.02(35)		M1
83.280(3)	0.13(1)	1.37(22)	10.5(19)		E1 /(M1)
75.836(3)	0.23(3)	6.25(44)	27.1(41)	0.06	M1+E2

[a] Lines below E_{e^-}=170 keV are measured using the thin target.
[b] δ^2 are quoted only for E2+M2 mixing at least 1.5 σ away from the pure multipolarities.
[c] questionable line in e$^-$ - spectrum

Since in this experiment the emphasis was put on the observation of transitions connecting states of positive parity, small energy regions have been scanned several times. Calibration was performed using previously measured γ- ray energies and intensities[1]. Table I lists all transitions connecting positive parity states where both, the conversion electron and the γ- ray line, have been observed. Mixing ratios for M1+E2 transitions having an internal conversion coefficient at least 1.5σ away from the theoretical value for a pure multipolarity are also given.

The experimentally obtained M1 and E2 branching ratios for the lowest positive parity levels are a good test for the claim that ^{76}As is an example of the vibrational limit of the $U_\nu(6/12) \otimes U_\pi(6/12)$ supersymmetry[2]. Here we compare the theoretical and experimental E2 branchings, because they should be the most direct test of the quantum numbers assigned to the levels[2]. The most general transition operator of $U(6/12)^3$ can easily be extended for odd-odd nuclei and then is given by:

$$T(E2) = e_b \, (s^\dagger d + d^\dagger s)^{(2)} + e_{b'} \, (d^\dagger d)^{(2)} + e_{f_\nu}(G_\nu^{(2)}(02) + G_\nu^{(2)}(20)) + e_{f_{\nu'}} G_\nu^{(2)}(22)$$
$$+ \, e_{f_\pi}(G_\pi^{(2)}(02) + G_\pi^{(2)}(20)) + e_{f'_\pi} G_\pi^{(2)}(22)$$

This operator contains 6 parameters, which either can be chosen such that the E2 operator becomes a generator of a symmetry group, leading to selection rules, or they can be adjusted

Table II Comparison E2 branchings in ^{76}As

Experiment					Theory		
E_i(keV)	J_i	J_f	E_γ(keV)	B.R.†	B.R.	J_i	J_f
120.3	1	1	75.8	100	100	1(3)	1(1)
		1	33.5	*	10.5		1(2)
280.3	1,2	1	235.9	100	85	1(4)	1(1)
		1	160.0	< 7*	100		1(3)
308.3	2	1	263.9	47	26	2(3)	1(1)
		1	221.5	100	100		1(2)
		1	188.1	7	5		1(3)
		0	104.8	*	1.8		0(1)
		1,2	43.4	*	0.06		2(1)
401.8	1,2	1	357.4	100	100	2(2)	1(1)
		1	315.0	(14)	0.5		1(2)
		1	281.6	$0 \leq 5 \leq 9$	1.6		1(3)
		0	198.3	*	0.16		0(1)
		1,2	137.0	$0 \leq .14 \leq .72$	.05		2(1)
447.2	1,2	1	402.8	$0 \leq 103 \leq 228$	24	1(5)	1(1)
		1	360.4	100	100		1(2)
		1	326.9	!	0.15		1(3)
		1,2	182.4	*	0.016		2(1)
519.6	2,3	1	432.8	< 21*	32	2(4)	1(1)
		1	399.4	100	100		1(3)
		0	316.1	< 15*	45		0(1)
		1,2	254.8	*	2.8		2(1)
		1,2	239.3	< 24 !	25		1(4)
		2	211.3	*	15		2(3)
		1,2	72.4	*	0.01		2(2)

† limits calculated using sensitivity limit in γ - ray measurement and assuming pure E2 multipolarity - *no γ and no conversion electron observed for this transition - !only γ ray observed

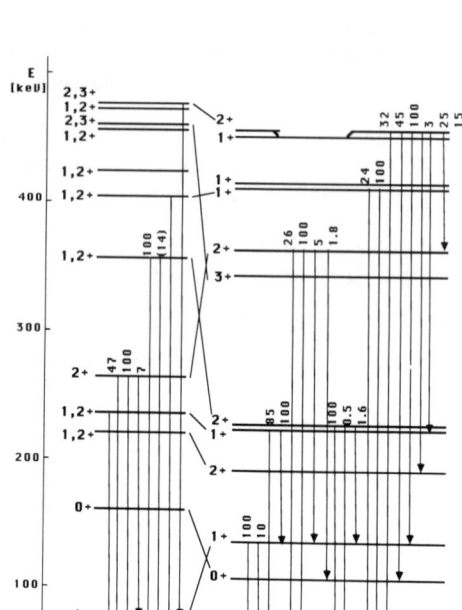

Fig. 1. Comparison of the experimental results for ^{76}As with theory

to reproduce the experimental B(E2) values and quadrupole moments in the three other nuclei of the supersymmetric quartet. As several observed transitions, namely between different $U^{BF}_{\nu\pi}(6)$ representations, do break the most obvious selection rule, we follow the second procedure. We start with the even-even core nucleus ^{76}Se. From a best fit to the B(E2) values we obtain $e_b = 0.093$ eb. and from the quadrupole moment of the first excited state we determine $e_{b'}$ to be -0.203 eb. Then we continue with these boson parameters and look at the odd-A members of the quartet. The neutron charges are determined from the B(E2) values in ^{77}Se. From a fit to the B(E2; $3/2_1, 5/2_2 \to 1/2_1$) and the B(E2; $3/2_2, 5/2_1 \to 1/2_1$) we obtain $e_{f_\nu} = 0.03$ eb. The B(E2; $3/2_1 \to 5/2_1$) and the B(E2; $3/2_2 \to 5/2_2$) yield $e_{f_{\nu'}} = 0.04$ eb. The proton charge e_{f_π} is determined from the quadrupole moment of the $3/2_1$ and the B(E2; $5/2_1 \to 3/2_1$) in ^{75}As to be $e_{f_{\pi'}} = 0.3$ eb. Finally the other B(E2) values yield $e_{f_\pi} = 0.11$ eb. Having determined all charges in the E2 operator, we calculate the B(E2) values in ^{76}As without further fitting using calculation c of ref^2. Using the experimental energies we calculated the theoretical E2 branching ratios for the lowest states. The results are given in table II for the levels with observed E2 transitions and the transitions having a theoretical branching bigger than 0.5 are compared in Figure 1 with experiment. A reasonable agreement between theory and experiment is found. For instance, the assignment of the quantum numbers to the experimental levels can be conserved with the exception of the second and third 2^+ state which should correspond to the levels at 401.8 and 308.3 keV respectively.

In conclusion we have performed a measurement of the ^{75}As(n,e$^-$) reaction. The experimental E2 branching ratios of transitions between low-lying positive parity states of ^{76}As are used to test whether the vibrational limit of the $U_\nu(6/12) \otimes U_\pi(6/12)$ supersymmetry is able to describe these branching ratios. We have shown that the E2 selection rule which forbids transitions between different U(6) representations is broken. This rules out the use of an E2 transition operator defined by symmetry constraints. On the other hand reasonable agreement between theory and experiment can be obtained with the transition operator *completely* determined from the adjacent odd-mass nuclei and the even-even core.

REFERENCES

[1] F. Hoyler, J. Jolie, G. C. Colvin, H. G. Börner, K. Schreckenbach, P. Van Isacker, P. Fettweis, H. Göktürk, J. C. Dehaes, R. F. Casten, D. D. Warner, A. M. Bruce, Nucl. Phys. **A512**, (1990) 189
[2] P. Van Isacker, and J. Jolie, Nucl. Phys. **A503**, (1989) 429
[3] P. Van Isacker, A. Frank, and H. Z. Sun, Ann. of Phys. **157** (1984) 183

MEASUREMENT OF THE NATURAL LINE SHAPE OF KRYPTON CONVERSION ELECTRONS FROM GASEOUS ^{83m}Kr

Daniel J. Decman and Wolfgang Stoeffl
University of California, Lawrence Livermore National Laboratory
Livermore, California 94551, USA

EXTENDED SUMMARY

The isomeric state in ^{83}Kr is an important calibration source for the gaseous-tritium beta-decay experiments that hope to measure the mass of the electron antineutrino [1,2]. The isomer, which has a half-life of 1.8 hours, is produced in the decay of ^{83}Rb ($t_{1/2}$ = 86 days). The isomer decays by a cascade of two electromagnetic transitions that have energies of 32 keV and 9.4 keV. The 32-keV E3 transition has a conversion coefficient of more than 2000; the K-conversion electrons from this transition have an energy of 17.8 keV, which is close to the tritium beta-decay endpoint energy of 18.6 keV. Therefore, this isomer can be used to determine the resolution function of the electron-spectrometer gaseous-source system at a very important electron energy. However the conversion electron lines are not monoenergetic calibration sources since they have a natural line shape due to the lifetime of the vacancy. More importantly though are the shakeup and shakeoff satellites that accompany the vacancy formation. Here we report the results of our study of the complex line shape of krypton conversion lines.

Our experiment measures electrons from a windowless, gaseous radioactive source with a large, toroidal magnetic field spectrometer of the Tret'yakov type. The ^{83m}Kr emanates from a thin layer of ^{83}RbCl solid in a stainless-steel container attached to our source vacuum system. We heat the container to 200°C to enhance the diffusion of the krypton gas from the solid. Electrons from decays in the gaseous source are guided into the spectrometer by five solenoidal superconducting magnets. The gas is differentially pumped at the ends of the 5-m-long source tube by eight turbomolecular pumps and is then reinjected into the middle of the tube. The gas at this position has a pressure of approximately 10^{-5} Torr and consists mostly of hydrogen. The temperature of the source tube is kept at 100 K. The radioactive gas is further prevented from entering the spectrometer tank by a 1.5-m-long pumping restriction. The magnetic spectrometer focuses the electrons onto a multisegmented, liquid-nitrogen-cooled Si(Li) detector. The position of the electron impact on this detector determines the energy. The magnetic spectrometer operates at a constant field and current setting to avoid thermal stresses and misalignment in the spectrometer hardware. We measure electron spectra by changing the electric potential of the source. A more complete description of our apparatus is presented elsewhere[1].

Figure 1 shows our spectrum of the 32-keV K-conversion line analyzed at an electron energy of 15 keV. Because the source is a low density gas the data are free of any solid state energy loss processes allowing us to make a direct measurement of the satellite spectrum. We measure with a resolution of 8.5 eV FWHM so the main Lorentzian component is clearly resolved from the shakeup satellites. The first satellite corresponds to shake up of the 4p electrons to the 5p, 6p and Rydberg states, the exponential tail on this peak represents the shake off excitations into the unbound states of the continuum. We also observe similar structures for excitations involving the 3p and 3d electrons. We fit the data using Lorentzian line shapes convoluted with our gaussian machine resolution. The shake off structures are appoximated as exponentail tails on the Lorentzian shape of the shake up satellites. The results of the fit for each of these components are shown as smooth curves in figure 1.

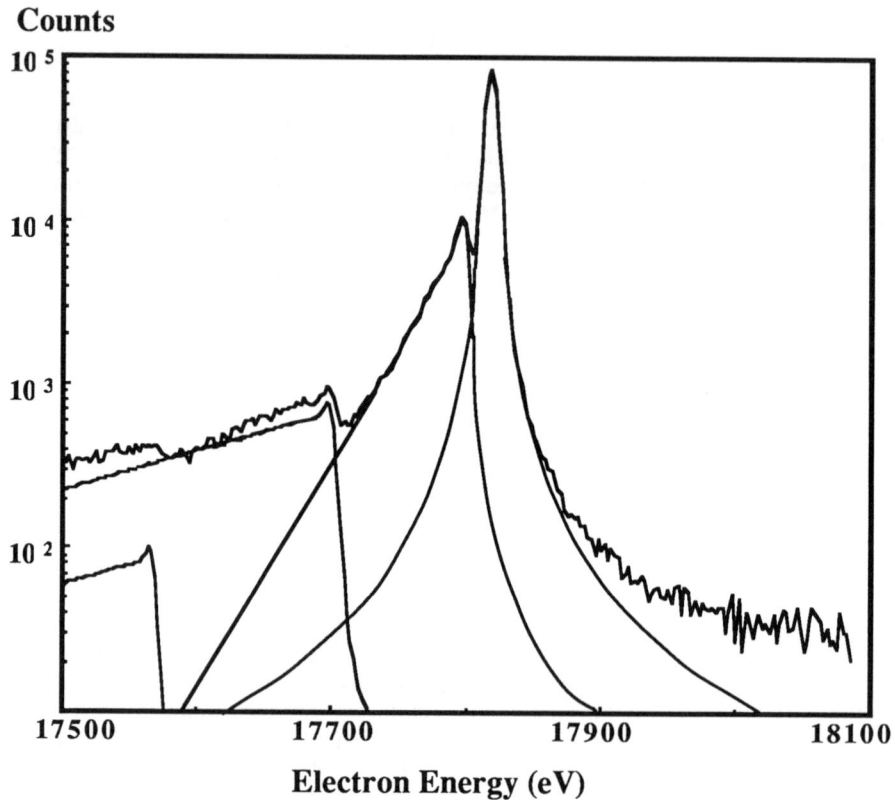

Fig. 1. The K-conversion line of the 32-keV transition of ^{83m}Kr in log scale.

The least-squares fit of our data yields a Lorentzian width of 2.9(2) eV for the krypton K orbital in good agreement with the literature value[3]. The 4p shake up peak appears at an energy 21 eV below the main peak, the intensity of this line with its exponential shakeoff tail represents about 20% of the total intensity. There is no measurable intensity between the main peak and the 4p shakeup peak. Other structures corresponding to the onset of shakeoff from the 3d and 3p shells occur at energies of 119 eV and 250 eV below the main peak. The 3d satellite has 8% of the total intensity whereas the 3p structure represents about 3%. An additional feature of our spectrum is the long tail that extends to energies of several hundred eV below the main peak.

The overall intensity of the shake off for krypton has been calculated in terms of the sudden approximation[4,5]. In this method one calculates the overlap of the atomic orbitals of the neutral atom with the orbitals of the atom with a K vacancy. The results of these calculations underestimate our measured shake-off intensity by a factor of two. Moreover there are no calculations which can reproduce the other information in our satellite spectrum, such as the shape of the shakeoff tails or the relative intensities of the shake up and shake off components of each orbital. Therefore these data should stimulate new atomic physics calculations of the complicated processes accompanying vacancy formation in atomic inner shells.

This work was performed by LLNL under the auspices of the U. S. Department of Energy under contract No. W-7405-Eng-48.

---

1. W. Stoeffl et al., in *Fifth Force - Neutrino Physics, Proceedings of the Eighth Moriond Workshop,* Les Arcs, Savoie, France, 1988, edited by O. Fackler and J. Tran Thanh Van (Editions Frontieres, Paris, 1988).
2. J. F. Wilkerson et al., Phys. Rev. Lett. **58,** 2023 (1987).
3. M. H. Chen et al., Phys. Rev. **A21**, 436 (1980).
4. T.A. Carlson, et al., Phys. Rev. **169**, 27.(1968)
5. T. Mukoyama, and K. Taniguchi , Phys. Rev **A36**, 693.(1987)

E2 AND M1 STRENGTHS AND STRONG SUB-SHELL CLOSURE EFFECTS IN NEUTRON-RICH A~100 NUCLEI*

F. K. Wohn,[a] H. Mach,[b,c,d] G. Molnár,[c,e] K. Sistemich,[c] John C. Hill,[a]
M. Moszyński,[d,f] R.L. Gill,[d] W. Krips,[d,g] D. S. Brenner,[h] and R. F. Casten[d]

[a] Ames Laboratory, Iowa State University, Ames, IA 50011
[b] Univ. of Uppsala, Studsvik Res. Lab, S–61182 Nyköping, Sweden
[c] Institut für Kernphysik, KFA Jülich, D-5170 Jülich, F.R. Germany
[d] Brookhaven National Laboratory, Upton, NY 11973
[e] Institute of Isotopes, H-1525 Budapest, Hungary
[f] Institute for Nuclear Studies, PL 05-400 Swierk-Otwock, Poland
[g] Institut für Kernphysik, Univ. Köln, D-5000 Köln, F.R. Germany
[h] Clark University, Worcester, MA 01610

ABSTRACT

E2 strengths of several $A \sim 100$ nuclei were deduced from ps level-lifetime measurements at the fission-product separator TRISTAN. The exceptionally low B(E2) values for ^{90,92,94,96}Sr reveal a close similariity between spherical Sr and Zr nuclei. For Sr and Y nuclei with $N \geq 60$, B(M1) and B(E2) values indicate that the deformation saturates just at its onset. A dramatic change in the Sr collectivity occurs at $N=60$, where the B(E2) strength abruptly increases by a factor of ~ 15, suggesting a "phase change" in the collectivity.

INTRODUCTION

For more than a decade neutron-rich $A \sim 100$ nuclei have attracted considerable interest due to their unusual features: an extremely abrupt change from spherical to highly deformed shapes, coexistence of very low-lying spherical and highly deformed shapes, very large rotational moments of inertia, and weaker than normal pairing correlations. Rotational bands in deformed $A \sim 100$ nuclei have implied axially symmetric deformations β of ~ 0.4. Discussions of these aspects of the $A \sim 100$ region are given in Ref. 1 and in references therein.

A new β-γ-γ fast-timing method[2] has made it possible to measure level lifetimes down to ~ 10 ps. Large deformations ($\beta \sim 0.4$) for Sr, Y and Zr with $N > 58$ have been directly determined[3,4,5,6] with this method. In contrast, for $N \leq 58$, sub-shell closures at $Z = 38, 40$ and $N = 56, 58$ should retard the development of collectivity for $50 < N < 60$. Level lifetime measurements for Sr nuclei with $50 < N < 60$ are needed to study the extent of this retardation effect.[7]

SATURATION OF DEFORMATION

For Sr and Y with $N \geq 60$, the B(M1) and B(E2) values indicate that the deformation saturates just at its onset.[3,4] Briefly, for ^{99}Sr and ^{99,100}Y the g-factors definitively establish the Nilsson assignments of the low-lying rotational bands built upon $\pi 5/2[422]$ and $\nu 3/2[411]$. The g-factors and the intrinsic quadrupole moment Q_o are well described by a simple picture for the structure of these nuclei, namely that the deformation of the even-even ^{98}Sr core is constant, unaffected by the presence of one or two valence nucleons.[3,4] In particular, the

---

* Supported by U.S. Department of Energy under Contracts W-7405-ENG-82, DE-AC02-76CH00016, and DE-FG02-88ER40417 and by Swedish Natural Science Research Council.

core Q_o of 3.78(6) b deduced[8] for the deformed band in ^{98}Sr is the same as the Q_o values of 3.8(5) b for ^{99}Sr and 3.9(4) b for ^{99}Y. The saturation of deformation inferred from this analysis[3,4] is confirmed by a new ^{100}Sr measurement[9] that gives a Q_o of 3.80(8) b. The "early" (i.e., before neutron midshell) saturation of deformation can be explained in terms of the valence Nilsson orbitals.[3,4,9]

SUB-SHELL CLOSURE EFFECTS

The B(E2) values[7] in Table I for ^{90,92,94,96}Sr, which fill the N=52–58 gap in the known B(E2) rates for $^{78-100}$Sr, are remarkably low. These low values establish a close similarity between spherical Sr and Zr isotopes, which along with spherical Pb nuclei exhibit the lowest B(E2) for all known nuclei past A=56.

Table I. Experimental data for 2^+ levels of Sr (Z=38) nuclei

A	E(2^+) (keV)	meanlife (ps)	B(E2,$0^+ \to 2^+$) ($e^2 b^2$)	(W.u.)	β_2(1st order)	Q_o (b)
78	278	224(27)	1.07(13)	108(13)	0.434(27)	3.28(21)
80	385	57(5)	0.84(7)	82(7)	0.377(16)	2.90(13)
82	573	12.8(5)	0.513(20)	48(2)	0.290(6)	2.27(4)
84	793	4.6(7)	0.28(4)	26(4)	0.211(5)	1.68(12)
86	1077	2.7(4)	0.106(16)	9.4(14)	0.128(10)	1.03(8)
88	1836	0.213(12)	0.092(5)	7.9(4)	0.117(3)	0.96(3)
90	832	10(3)^a	0.10(3)	8.3(27)	0.120(19)	1.00(16)
92	815	12(5)^a	0.09(4)	8(3)	0.116(24)	0.98(20)
94	837	10(4)^a	0.10(4)	8(3)	0.115(25)	0.98(21)
96	816	7(4)^a	0.17(9)	13(7)	0.15(4)	1.3(4)
98	144	4010(100)^b	1.29(3)	96.4(24)	0.409(5)	3.61(4)^d
100	129	5640(230)^c	1.44(6)	104(4)	0.426(9)	3.80(8)

^{a}present results of Mach et al.[7]
^{b}mean of 4.04(12) ns (Ref. 8) and 3.95(17) ns (Ref. 10).
^{c}new result by Lhersonneau et al.[9]
$^d Q_o$=3.78(6) b for the unmixed deformed band [Mach et al.[8]].

A strong similarity in the B(E2) values of Sr and Zr nuclei is observed from shell closure at N=50 to deformation at N=60.[7] These nuclei are more similar in their collectivity than is suggested by their 2_1^+ energies, which are nearly constant for $^{90-96}$Sr but vary significantly for $^{92-98}$Zr. These nuclei are similar due to subshell closures at Z=38($\pi 2p_{3/2}$), 40($\pi 2p_{1/2}$) and at N=56 ($\nu 2d_{5/2}$), 58($\nu 3s_{1/2}$). These sub-shells (for low-j orbits) stabilize spherical configurations,[11] thereby permitting Sr and Zr with N=50–58 to very effectively resist the normal smooth progress towards deformation as N increases above 50.

Q_o SYSTEMATICS OF SR NUCLEI

The Sr and Zr nuclei are unique in their B(E2) systematics. No other nuclei (see the compilation of Ref. 12) exhibit such an abrupt and large change in B(E2) values from "spherical" values of $\sim$8 W.u. to deformed values of $\sim$100 W.u. This change in collectivity is most dramatically seen in Fig. 1 which gives the intrinsic quadrupole moment Q_o for Sr nuclei. The Q_o given above for ^{99}Sr is included, as is the ^{97}Sr value[6] of 3.5(4)b. For ^{98}Sr, Fig. 1 includes both the "mixed" value of 3.61(4) b and the "unmixed" value of 3.78(6)b, obtained[8] by correcting for the mixing of the 0_1^+ and 215-keV 0_2^+ states in ^{98}Sr.

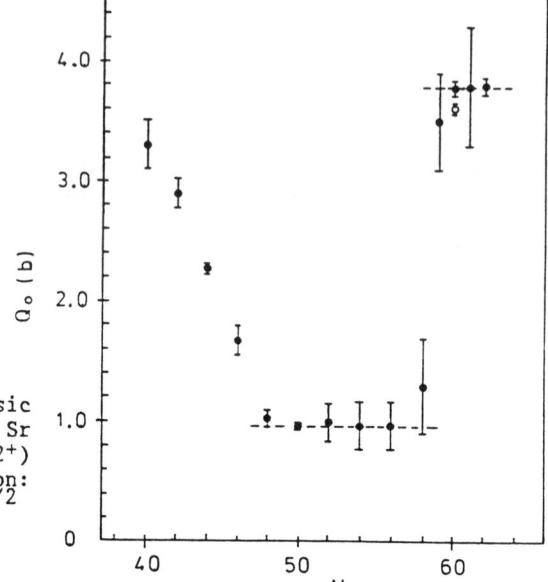

Fig.1: Systematics of intrinsic quadrupole moments Q_o for the Sr nuclei obtained from $B(E2, 0^+ \to 2^+)$ using the rotational expression: $Q_o = [(16\pi/5)B(E2, 0^+ \to 2^+)/e^2]^{1/2}$

Figure 1 reveals two plateaus: one for N=50–58, with mean Q_o of 0.96(3) b, and one for N=59–62, with mean Q_o of 3.79(4) b. This change in the Sr collectivity, which corresponds to a factor of ~15 increase in B(E2), is so remarkably large and abrupt that it suggests a "phase transition" in the Sr collectivity. The term "phase transition" is used to emphasize that the abruptness contrasts sharply with the rather smooth "evolution" (such as is observed in Fig. 1 for neutron-deficient Sr nuclei) of nuclear collectivity that occurs for all other known shape-transition regions.

REFERENCES

1. Nuclear Structure of the Zirconium Region, edited by J. Eberth, R. A. Meyer, and K. Sistemich (Springer-Verlag, Berlin, 1988).
2. H. Mach, et al., Nucl. Instr. Methods A280, 49 (1989); see also invited paper by R.L. Gill, this symposium.
3. F. K. Wohn, et al., Nucl. Phys. A507, 141c (1990).
4. H. Mach et al., Phys. Rev. C 41, 1141 (1990).
5. G. Lhersonneau et al., Z. Phys. A 332, 243 (1989).
6. M. Buescher et al., Phys. Rev. C 41, 1115 (1990).
7. H. Mach et al., Nucl. Phys. A (in press).
8. H. Mach et al., Phys. Lett. B 230, 21 (1989).
9. G. Lhersonneau et al., Z. Phys. A (in press).
10. H. Ohm et al., Z. Phys. A 327, (1987).
11. G. Molnár et al., Nucl. Phys. A500, 43 (1989).
12. S. Raman et al., At. Data Nucl. Data Tables 42, 1 (1989).

THE LEVEL SCHEME OF ^{110}Ag FROM THE (n,γ) REACTION

I.A.Kondurov[1], P.A.Sushkov[1], M.Bogdanović[2], J.Simić[2], H.Seyfarth[3], H.A.Baader[3], D.Breitig, R.Koch[3], G.Barreau[4], H.G.Börner[4], R.Brissot[4], S.Kerr[4], T.D.Mac Mahon[5], T.Mitsunari[5], H.Faust[4], K.Schreckenbach[4]

1. Leningrad Nuclear Physics Institute, Gatchina
2. The Boris Kidrič Institute of Nuclear Sciences, Belgrade
3. Institut für Kernphysik, KFA, Jülich
4. Institut Laue-Langevin, Grenoble
5. University of London Reactor Centre, Silwood Park, Ascot

A thorough investigation of the excited states of the doubly odd $^{110}_{47}$Ag$_{63}$ nucleus produced in the ^{109}Ag(n,γ) reaction has been performed. Some of the results were already published in Refs. 1,2.

Secondary gamma rays in the energy range 45 - 1200 keV were measured with the bent crystal spectrometer at Risö. Also new measurements using the bent crystal spectrometers GAMS1 and Gams2/3 at Alltogether ILL Grenoble have been made in the energy range from 35 - 1800 keV. About 1070 transitions were observed. Energies were colibrated using the 657.7622(21) keV line from the decay[3] of ^{110}Ag to ^{110}Cd. The low-energy spectrum from 10 - 100 keV has been measured with a Si(Li) detector at LNPI in Gatchina. Here we used for the energy calibration energies of X - rays from Ag and In and the Ag gamma-ray energies for Ag as deduced from the GAMS experiments.

The conversion electron spectrum from ^{110}Ag was studied with the spectrometer BILL at ILL Grenoble. In the energy range from 20 to 650 keV about 580 electron lines were identified.

The spectra of the primary gamma rays from the thermal neutron capture were measured with the pair spectrometers at the ILL Grenoble and IKP Jülich, respectively, in the energy range from 1380 to 8600 keV. The energy and intensity calibration were made with Al (n,gamma) reaction lines.

Prompt γγ coincidences were studied with the two Ge(Li) detectors. Coincidence events were registrated in a matrix with 1024 x

2048 channels covering the energy range from 15 - 240 and 40 - 640 keV, respectively. At IKP in Jülich γγ coincidence measurements were performed in a matrix of 4096 x 4096 channels, in the energy range from 50 - 300 and 80 - 2000 keV, respectively. The delayed coincidence and singles experiments to investigate the population of the 118.7 keV isomeric state ($T_{1/2}$ = 36.8(7) ns) performed in Belgrade with Ge(Li)-NaI(Tl) detectors resulted in about 600 gamma lines in the energy interval from 100 - 1200 keV. Into the level scheme were included the γγ coincidence data made at LNPI in order to investigate the population of the isomeric 117.55 keV state ($T_{1/2}$ = 250.4 days, 6^+).

From all these extensive experimental data and also including (d,p) reaction data[4] the level scheme has been constructed up to 1.2 MeV excitation energy with 84 levels (10 of them are new).

The level energies with spin and parity assignments are listed in Table 1.

Some of the proton-neutron multiplets are identified. For the ground state 1^+, followed by the 6^+ state at 117.5 keV, 5^+ at 174.6 keV, 4^+ at 255.0 keV, 2^+ at 198.7 keV, the configuration is ($\pi 7/2^+$, $\nu 5/2^+$). The 1.11 keV level is probably the 2^- state of the proton-neutron doublet ($\pi p_{1/2}^-$, $\nu d_{5/2}^+$)$^{2^-,3^-}$. We have tried to identify the higher lying proton-neutron multiplets.

Table 1

Eex (keV)	I^π	Eex (keV)	I^π
0	1^+	640.259(5)	$2^+,3^+,4^+$
1.113(2)	2^-	653.848(2)	$1^-,2^-$
117.532(5)	6^+	663.405(4)	$1^+,2^+,3^+$
118.710(2)	3^+	664.865(2)	$1^-,2^-$
174.547(5)	$5^+,6^+$	683.085(7)	$1^+,2^+,3^+$
191.613(4)	$2^+,3^+$	689.109(3)	$1^+,2^+,3^+$
198.676(2)	2^+	689.162(2)	$2^-,3^-$
236.833(1)	$1^-,2^-,3^-$	698.485(2)	$1^+,2^+$

Eex (keV)	I^π	Eex (keV)	I^π
237.027(2)	$1^-,2^-$	706.062(2)	$1^+,2^+$
254.992(3)	$4^+,5^+$	725.738(2)	$1^-,2^-$
267.195(2)	$1^+,2^+$	741.596(2)	$2^+,3^+$
267.230(4)	$1^+,2^+,3^+$	748.506(3)	$1^+,2^+$
271.451(2)	$2^+,3^+,(4)$	750.825(2)	$1^+,2^+$
304.503(2)	$(1)^+,2^+,3^+$	753.273(2)	$2^+,3^+$
338.884(3)	$1^-,2^-$	759.550(4)	$1^-,2^-$
360.587(2)	$1^+,2^+$	766.963(3)	$2^+,(1)^+$
379.365(7)	$1^+,2^+,3^+$	773.584(3)	$1^+,2^+$
381.161(1)	$1^-,2^-$	785.569(2)	$2^+,(1)^+$
424.661(3)	$1^-,2^-$	789.628(2)	$1^-,2^-,3^-$
432.328(3)	2^-	811.329(2)	$1^+,2^+$
446.529(2)	$3^-,4^-$	818.790(5)	$1^-,2^-,3^-$
468.807(2)	$2^+,3^+$	818.886(4)	$1^+,2^+,3^+$
471.199(3)	$1^+,2^+,3^+$	820.640(8)	$1^-,2^-$
485.728(2)	$2^+,3^+$	854.332(5)	$1^-,2^-$
496.833(3)	$1^-,2^-$	880.444(3)	$(1,2,3)$
525.612(1)	$1^-,2^-$	881.422(4)	$1^-,2^-$
527.461(4)	$2^+,3^+$	910.815(2)	$2^+,3^+$
536.130(2)	$0^+,1^+,2^+$	925.035(5)	$(1,2,3)$
539.525(4)	$1^-,2^-,3^-$	953.096(3)	$1^+,2^+$
539.760(4)	$1^+,2^+,3^+$	954.314(2)	$1^+,2^+$
549.328(2)	$1^+,2^+$	979.724(5)	$(1,2,3,4)^+$
551.309(2)	$2^-,3^-$	985.576(3)	$1,2$
557.014(2)	$2^-,3^-$	994.969(3)	$1^+,2^+$
579.241(2)	$2^-,3^-$	1012.956(3)	$1^+,2^+$
589.722(4)	$1^-,2^-$	1034.821(3)	$1^+,2^+$
592.856(2)	$2^+,3^+$	1036.792(4)	$2,3$
594.968(3)	$1^-,2^-$	1097.429(2)	$1^+,2^+$
605.149(2)	$2^-,3^-$	1106.637(3)	$1^+,2^+$
613.582(2)	$2^-,3^-$	1164.354(8)	$1,2,3$
615.071(1)	$1^-,2^-$	1169.006(3)	$1^+,2^+,3^+$

The Level Scheme of ^{110}Ag from the (n,γ) Reaction

Eex (keV)	I$^\pi$	Eex (keV)	I$^\pi$
633.384(1)	$0^-,1^-,2$	1175.735(4)	1,2,3
		1178.251(6)	2,3

1. M.Bogdanović, S.Koički, J.Simić, B.Lalović
   Fizika 11 (1979) 157

2. I.A.Kondurov, P.A.Sushkov, H.A.Baader, D.Breitig, H.R.Koch,
   M.Bogdanović, G.Barreau, H.G.Börner, R.Brissot, K.Schreckenbach,
   T.D.Mac Mahon, T.Mitsunari, H.Seyfarth
   ANSSSR, LNPI, Preprint No 1087, july 1985

3. R.C.Greenwood, R.G.Helmer and R.J.Gehrke
   Nucl. Instr. Methods 159 (1979) 465

4. C.E.Brient, R.J.Riley, H.Seitz, S.Sen
   Phys. Rev. C6 (1972) 1837

INTERACTING BOSON-FERMION-FERMION DESCRIPTION OF ODD-ODD INDIUM NUCLEI

Zs. Dombrádi and T. Fényes
Institute of Nuclear Research, 4001 Debrecen, Hungary
S. Brant and V. Paar
Prirodoslovno-matematički fakultet, University of Zagreb,
41000 Zagreb, Croatia, Yugoslavia

ABSTRACT

The low-lying states of the odd-odd $^{104-116}_{49}$In nuclei have been classified into proton-neutron multiplets on the basis of the measured and calculated level energies and electromagnetic properties. The systematic trends and the observed anomalies were interpreted on the basis of the interacting boson-fermion-fermion model (IBFFM). It was found that the collective degree of freedom plays a significant role in the observed phenomena.

INTRODUCTION

In the last few years a large amount of experimental data was obtained on the odd-odd indium nuclei, making possible to advance the theoretical interpretation of most of the low-lying states. The systematic analysis of the data shows the deviations from the quasiparticle predictions and provides evidence for description in terms of IBFFM.

IDENTIFICATION OF THE STATES

In order to determine the wave functions of the low-lying states in odd-odd In nuclei the calculations of the level energies and electromagnetic properties have been performed in the frame of IBFFM. The model parameters used were close to the parameters determined from the neighbouring odd tin and indium nuclei. The boson core was approximated with the SU(5) limit of IBM and the d-boson energy was taken form the corresponding even-even tin isotope.

The spectroscopic information used for the identification was obtained from (n,γ), (p,nγ), (α,nγ) reactions and β^+ decay.[1-7] In the heavy isotopes also the one-nucleon transfer reactions supplied the information on the multiplet structure of the states. In these nuclei we tested the applicability of the model and the method of identification.

After fine tuning of the quasiparticle energies the position of the states could be predicted with about 100 keV uncertainty. The close lying states of the same spin were distinguished on the basis of their decay properties. Even some states with spins not determined unambiguously could be assigned to multiplets on the basis of the decay properties, as the strong ($\approx$1 W.u.) M1 transitions connecting the $\Delta J = 1$ members of the multiplets were observed in most cases. The weaker M1 transitions helped to determine the mixing of the states. The transition rates were described with about 0.05 W.u. reliability. We could make new assignments for 38 states in the $^{104-110}$In nuclei on the basis of electromagnetic properties.

SYSTEMATICS OF THE MULTIPLET STATES

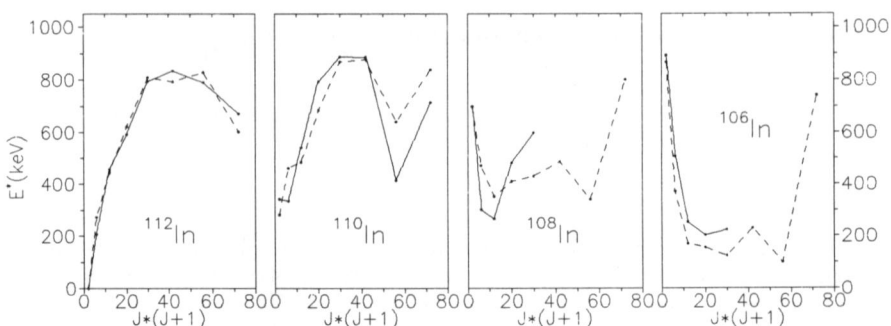

Fig.1. Splitting of the $\pi g_{9/2}^{-1} \nu g_{7/2}$ multiplet in $^{112-106}$In as a function of $J*(J+1)$. Solid line: experimental, dotted: calculated data.

The proton-neutron multiplet-like components were the most significant ones in the wave functions of the low-lying states, typically with 50–60% weight. 35–45% was fragmented over the one d-boson states and an additional 4–8% over the two d-boson states. The wave functions were based mainly on a single multiplet, suggesting that even a diagonal quasiparticle description should give a reasonable result. In spite of this, two striking deviations form the quasiparticle predictions were found (see Fig.1.).

First, the parabolic like splitting of the $\pi g_{9/2}^{-1} \nu g_{7/2}$ multiplet is not inverted through smoothing out of the splitting, as expected from the quasiparticle model, but through the kinking of the ends of the parabola. The $\pi g_{9/2}^{-1} \nu d_{5/2}$ multiplet begins its inversion in a similar way.

Secondly, the splitting of the $\pi g_{9/2}^{-1} \nu g_{7/2}$ multiplet in $^{112-116}$In does not show the odd J even J staggering characteristic for a short range interaction. The phenomenon was observed also in the case of $\pi g_{9/2}^{-1} \nu d_{5/2}$ and $\pi g_{9/2}^{-1} \nu h_{11/2}$ multiplets.

THE ROLE OF COLLECTIVITY

The W-like splitting was predicted in the O(6) limit of the IBFFM by Brant and Paar [8] as a consequence of the boson-neutron exchange interaction. In the SU(5) limit the exchange interaction acts only on the d-boson states, but the proton-boson dynamical interaction makes it effective also for the low-lying states by admixing more and more d-boson components to them. The change of the shape of the multiplet splitting from the quasiparticle prediction ($\Gamma_p = 0$ MeV) to the typical IBFFM shape ($\Gamma_p = 3$ MeV) is shown in fig.2.

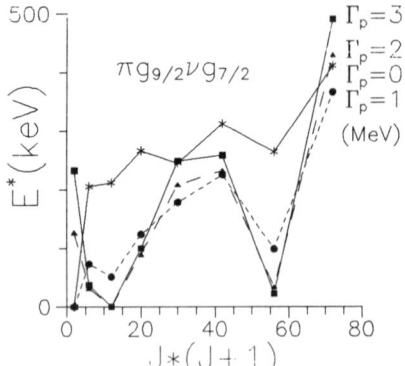

Fig.2. Calculated splitting of the $\pi g_{9/2}^{-1} \nu g_{7/2}$ multiplet as a function of $J*(J+1)$. ($V_n^2 = 0.5, \Lambda_n = 2.5$ MeV)

The presence of the ≈40% d-boson component in the wave function explains the other deviation, too. The stronger the dynamical boson-fermion interaction, the larger is the weight of the d-boson configurations in the wave function. The admixture of the d-boson components means adding multiplet components with different spin values with appropriate weight, that is the boson-fermion coupling averages the energy of the neighbouring states, like a smoothing procedure. This is seen in Fig.3 as a function of the particle-vibration coupling strengths. The energy difference of the 1^+ and 2^+ states strongly drops, and the staggering, characteristic for the short range interaction ($\Gamma_p = 0$ MeV), is also smoothed out. In addition to the splitting shown in Fig.3, a quadrupole core polarization interaction is also present, if $\Gamma_n \neq 0$.

The presence of the above mentioned two anomalies shows that although the skeleton structure of the odd-odd In nuclei can be described in quasiparticle approximation with appropriate effective interaction, the collective degree of freedom must be taken into account explicitly in order to understand the fine details.

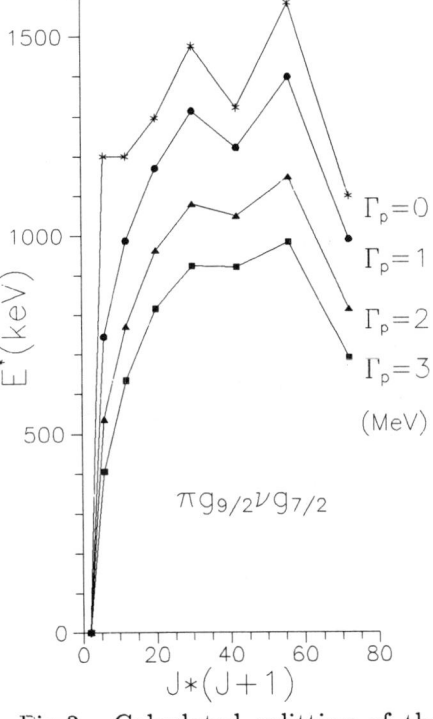

Fig.3. Calculated splitting of the $\pi g_{9/2}^{-1} \nu g_{7/2}$ multiplet as a function of $J * (J + 1)$. ($V_n^2 = 1, \Gamma_n = 0$ MeV)

REFERENCES

1. V.L. Alexeev, B.A. Emelianov, D.M. Kaminker, Yu.L. Khazov, I.A. Kondurov, Yu.E. Loginov, V.I. Rumiantsev, S.L. Sakharov and A.I. Smirnov, Nucl. Phys. A262, 19 (1976)
2. J. Tímár, T. Fényes, T. Kibédi, A. Passoja, M. Luontama, W. Trzaska and V. Paar, Nucl. Phys. A455, 477 (1986)
3. T. Kibédi, Zs. Dombrádi, T. Fényes, A. Krasznahorkay, J. Tímár, Z. Gácsi, A. Passoja, V. Paar, D. Vretenar, Phys. Rev. C37, 2391 (1988)
4. A. Krasznahorkay, Zs. Dombrádi, J. Tímár, Z. Gácsi, T. Kibédi, A. Passoja, R. Julin, J. Kumpulainen, S. Brant, V. Paar, Nucl. Phys. A503, 113 (1989)
5. A. Krasznahorkay, Zs. Dombrádi, J. Tímár, T. Fényes, J. Gulyás, J. Kumpulainen, E. Verho, Nucl. Phys. A499, 453 (1989)
6. J. Gulyás, Zs. Dombrádi, T. Fényes, J. Tímár, A. Passoja, J. Kumpulainen, R. Julin, Nucl. Phys. A506, 196 (1990)
7. R. Barden, R. Kirchner, O. Klepper, A. Płochocki, G.-E. Rathke, E. Roeckl, K. Rykaczewski, D. Schardt and J. Zylicz, Z. Phys. A 329, 319 (1988)
8. S. Brant and V. Paar, Z. Phys. A 329, 151 (1988)

NEW FEATURES IN THE SYSTEMATICS OF LOW-SPIN STATES IN EVEN $^{106\text{-}120}$Cd

J. Kumpulainen, R. Julin, J. Kantele, A. Passoja*, W. H. Trzaska**,
E. Verho and J. Väärämäki
Department of Physics, University of Jyväskylä, SF-40100 Jyväskylä, Finland

ABSTRACT: Low-spin states in even $^{106\text{-}112}$Cd and ^{116}Cd were investigated by in-beam and off-beam γ-ray and conversion-electron spectroscopy. New spin assignments and decay branching ratios for the levels in ^{106}Cd, ^{108}Cd, ^{110}Cd and ^{112}Cd were obtained. From the new systematical data for the even $^{106\text{-}120}$Cd, it is inferred that two sets of low-lying 0^+ states cross between ^{114}Cd and ^{116}Cd. One of these sets appears to have features of both intruder and two-quadrupole-phonon states.

No consistent picture has emerged from the various mixing calculations for the low-lying states of even Cd isotopes[1-4]. Attempts to establish the systematics of the suggested intruder states have failed mainly because of the scarce experimental information, especially on the neutron deficient Cd isotopes[2,5].

We used the (p,p'), (p,2n) and (α,2n) reactions and the odd-odd In decay to populate levels in the even ^{106,108,110,112}Cd. Proton-γ- and γγ-coincidence, γ-ray angular-distribution and excitation-function measurements were carried out, as well as conversion-electron spectroscopy in the (p,p') reaction and In decay. A few complementary pγ-coincidence experiments for ^{116}Cd were performed via the (p,p') reaction.

In ^{106}Cd, the γ-ray angular-distribution and excitation-function measurements reveal that the 1795 keV level has $I^\pi = 0^+$, not 4^+ as reported[6]. The 2035 keV level, tentatively assigned[7] as 0^+ was not observed. A new 0^+ state at 2144 keV, connected via an E0 transition to the ground state, was found. At 2254 keV we found a new level with a tentative spin assignment of 3^+, and the I^π of the 2371 keV level is revised as 2^+.

In ^{108}Cd, we firmly establish $I^\pi = 0^+$ for the 1721 keV level based on the γ-ray angular-distribution and excitation-function data. The 0^+ assignment[5] for the 1913 keV level is confirmed by observing the E0 transition from this state to the ground state. We do not observe the 2021 keV level reported in Ref. 5. On the basis of new γ-ray data and conversion coefficients, the 2163 keV level is assigned as 2^+ replacing the earlier[5] $I^\pi = 3^-$.

In ^{110}Cd, on the basis of the γ-ray angular-distribution and excitation-function results, we confirm the $I^\pi = 0^+$ assignment[8] of the 1731 keV level. The 1810 keV level reported in Ref. 8 was not observed.

In ^{112}Cd the main result is the observed band built on the 0_2^+ state. We found at 1870 keV a new level with $I^\pi = 4^+$, in addition to the 0^+ level at 1871 keV.

With the present results included, the even-mass Cd data are now sufficient for relating low-lying levels of similar characteristics in $^{106\text{-}120}$Cd (Fig. 1).

The most conspicuous feature of the level systematics is the behaviour of the 0_A^+, 0_B^+ and 2_3^+ states. The 0_2^+ (0_A^+) states in the even $^{106\text{-}114}$Cd are characterized by enhanced E2($0_A^+ \to 2_1^+$) transitions, which is reflected in the small value of $X_{211} = B(E0;0_2^+ \to 0_1^+)/B(E2;0_2^+ \to 2_1^+)$ for these states (Table I). In our earlier work[9] we have measured the B(E2;$0_2^+ \to 2_1^+$) in ^{112}Cd and ^{114}Cd to be 51 W.u. and 41 W.u., respectively. In ^{116}Cd we identify the 0_3^+ state as the 0_A^+ state, since this state and not the 0_2^+ state is populated in Coulomb excitation (B(E2;$0_3^+ \to 2_1^+$) = 30 W.u. [10]). This interpretation is supported by a recent lifetime measurement[11]. On the basis of the excitation energies of[12-14], it is now straightforward to identify the 0_3^+ states in ^{118}Cd and ^{120}Cd as the 0_A^+ states. This

---

* Present Address: Department of Physics, University of Joensuu, Joensuu, Finland
** Present Address: Cyclotron Institute, Texas A&M University, College Station, U.S.A.

© 1991 American Institute of Physics

interpretation, i.e. the crossing of the 0^+_A and 0^+_B states between the Cd isotopes 114 and 116, is supported by a recent (t,p) study[3]. However, according to our data, no corresponding crossing occurs on the neutron-deficient side, contrary to the suggestion of Ref. 3.

A characteristic feature of the 0^+_B (0^+_3) states in the even $106-114$Cd isotopes is the large value for the ratio R = $B(E2;0^+_3 \rightarrow 2^+_2)/ B(E2;0^+_3 \rightarrow 2^+_1)$ (Table I), apparently due to the hindrance of the $E2(0^+_B \rightarrow 2^+_1)$ transition (for $112,114$Cd, see [9]). For the 0^+_B (0^+_2) states in the even $116-120$Cd, this ratio is not known, since it is difficult to observe the low-energy $E2(0^+_B \rightarrow 2^+_2)$ transitions. The lifetime-measurement [11] of the 0^+_B (0^+_2) state in 118Cd indicates the hindrance of the $E2(0^+_B \rightarrow 2^+_1)$ transition, although one would expect even longer lifetime.

In the simple quadrupole-vibrator model, the 0^+_2 (0^+_A), 2^+_2 and 4^+_1 states in the even $106-114$Cd isotopes form a two-phonon triplet. The observed strong E2 branches from these states to the 2^+_1 state are in agreement with this interpretation. Very recently, Fahlander et al.[15] have

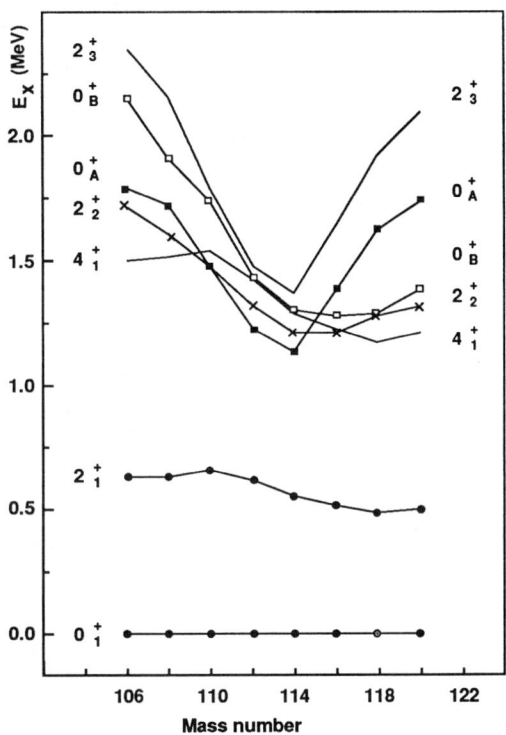

Fig. 1. Partial level systematics of even $106-120$Cd.

shown that all of the observed levels and the measured B(E2) values below 2.6 MeV in 114Cd can be accounted for in view of the simple vibrator. The 0^+_3 (0^+_B) and 2^+_3 states play the role of the three-phonon states which in $112,114$Cd are pushed down to form the well-known quintuplet of states. The application of the vibrator picture to the heavier even Cd isotopes would lead to the conclusion that the three-phonon 0^+ state lies lower than the two-phonon 0^+ state. This kind of phenomenon has been explained as being due to γ softness of the nucleus. This interpretation conflicts with [12] where 118Cd is introduced as an almost perfect harmonic vibrator up to the three-phonon excitations.

Table I.   Ratios of transition rates in $106-120$Cd.

	106	108	110	112	114
X_{211} (10^{-3})	< 9	< 12	< 8	26(4) [a]	26(5) [a]
R	230	1100	170	8500	40000 [b]

a From 9. b From 15.

The intruder states in the even Cd isotopes should involve proton 2p-4h excitations, i.e. 6 valence quasiprotons. The total neutron-proton interaction in these states should be similar to that in the ground state band of the Ru (Z = 50 - 6) and Ba (Z = 50 + 6) isotopes. The new systematics reveals that the energy difference between the third 2^+ states (2^+_3) and the 0^+_A states (0^+_2 in $106-114$Cd, 0^+_3 in $116-120$Cd) is close to the excitation energy of the 2^+_1 states in Ru and Ba isotopes (Fig.2), indicating that the 0^+_A and 2^+_3 states represent the two lowest members of the proton 2p-4h intruder band. Moreover, we point out that it is clearly the 0^+_A (0^+_2) state (and not the 0^+_B (0^+_3) state) in ^{110}Cd and ^{112}Cd which is populated in the two-proton transfer i.e. in the (^{3}He,n) reaction[16]. Only in ^{110,112,114}Cd candidates for the 4$^+$ and 6$^+$ members of the band have been found (Fig. 2).

The above considerations lead to a paradoxical conclusion: the intruder-like behaving 0^+_A state is the state which in the phonon picture plays the role of the two-phonon 0$^+$ state.

In Ref. 2 the intruder 0$^+$ state has been associated with the 0$^+$ state having a large R-value, i.e. 0^+_B state in our notation. This interpretation is clearly different from ours.

Many of the E2 and E0 decay properties of even Cd nuclei have been reproduced by introducing a strong mixing of the vibrational and intruder states [1,2,4]. However, the similarities in Fig. 2 indicate that the 2^+_3 - 0^+_A energy differences in the Cd isotopes are not seriously affected by the mixing. Moreover, the selective population in the (^{3}He,n) [16] and (t,p) [3] reactions do not support the idea of strong mixing.

In our view the concepts of intruder and phonon states in even Cd nuclei represent different approaches, which may not fit into the same frame.

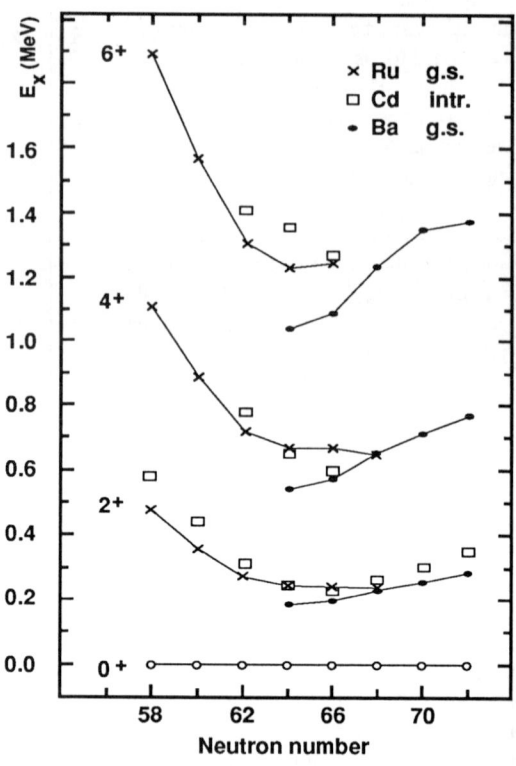

Fig. 2. The states of the intruder band in $^{106-120}$Cd and the ground-state bands in even Ru and Ba isotopes. The intruder 0$^+$ band-head is normalized to 0 MeV.

References

1. K. Heyde et al., Phys. Rev. C **25**, 3160 (1982)
2. A. Aprahamian et al., Phys. Lett. **140B**, 22 (1984)
3. J. M. O'Donnell, A. Kotwal and H. T. Fortune, Phys. Rev. C **38**, 2047 (1988)
4. D. Kusnezov et al., Helv. Phys. Acta **60**, 456 (1987)
5. B. Roussiere et al., Nucl. Phys. **A419**, 61 (1984)
6. S. Flanagan et al., J. Phys. **G2**, 589 (1976)
7. B. Roussiere et al.,CERN 81-09, 465 (1981)
8. P. De Gelder, E. Jacobs, and D. De Frenne, Nucl. Data Sheets **38**, 545 (1983)
9. R. Julin et al., Z. Phys. **A296**, 315 (1980)
10. F. K. McGowan et al., Nucl. Phys. **66**, 97 (1965)
11. H. Mach et al., Phys. Rev. Lett. **63**, 143 (1989)
12. A. Aprahamian et al., Phys. Rev. Lett. **59**, 535 (1987)
13. A. Aprahamian et al., BAPS **29**, 1041 (1984)
14. A. Aprahamian, Ph. D. thesis, Clark University, 1985
15. C. Fahlander et al., Nucl. Phys. **A485**, 327 (1988)
16. H. W. Fielding et al., Nucl. Phys. **A281**, 389 (1977)

LEVELS IN ^{129}Te POPULATED IN THE ^{128}Te(n,γ) REACTION

Craig A. Stone*
San Jose State University, San Jose, CA

B. E. Zimmerman, C. E. Ford, P. F. Mantica, Jr., and W. B. Walters
University of Maryland, College Park, MD 20742 USA

ABSTRACT

In this paper, new data are reported for the ^{128}Te(n,γ) ^{129}Te reaction. These data are combined with data from the radioactive decay of both the 4.4-h 7/2+ and 17-min 19/2- isomers of ^{129}Sb to provide considerable detail for the low-energy structure of ^{129}Te.

INTRODUCTION

The study of the odd-N Te nuclides has proven a fruitful region in which to test calculations in which collectivity is not large. The even-even Te core structures have been studied in some detail and the structures described in extensive interacting boson model-2 (IBM-2) calculations.[1] In those calculations, particle-hole intruder levels were found to play an important role in the low-energy structure near mid-shell at N = 66(^{118}Te).[2] The particle-hole intruder structures are well known in the adjacent Sb and I nuclides and rise quickly in energy as N moves toward either the N = 50 or N = 82 closed shells.[3] Consequently, the influence of particle-hole structures diminishes significantly in the even even Te nuclides as N approaches 82. Thus, away from mid shell, Te nuclides have structures that enable the testing of weak and intermediate coupling models in some detail.

Study of the structure of ^{129}Te is particularly attractive, as there are two isomers in ^{129}Sb, the ground-state 7/2+ isomer with a 4.4-hr half life [4,5] and an isomeric 19/2- 17-min state that can populate a wide range of levels with spins above 5/2. Studies of the decay of the high spin isomer has been previously reported and we have taken new data for the decay of the low-spin isomer using mass separated sources produced at the on-line mass separator TRISTAN located at the High Flux Beam Reactor (HFBR) at Brookhaven National Laboratory.[6] This study of the ^{128}Te(n,γ) reaction was carried out to provide complementary data for the low-spin levels that are not populated in the decay studies. Previously reported data for this reaction was focussed on the levels in the 2-4 MeV range that are populated in transfer reactions and showed a strong correlation between transfer strength and direct population from the capture state at 6082.4 keV.[7,8,9]

* Previously at Center for Analytical Chemistry, National Institute for Standards and Technology, Gaithersburg, MD 20799.

EXPERIMENTAL WORK

These data were collected at the reactor at the National Institute of Standards and Technology using a large shielded Ge(Li) detector. The sample consisted of 150 mg of ^{128}Te enriched to over 95% placed in the vertical beam developed for trace-element measurements.[10] Samples of urea and iron were used for energy and efficiency calibrations and Cl and F impurities in the Te source and the radioactive daughter provided for *in situ* calibrations.

RESULTS

The principal goal of this work was the identification of the $1/2^+$ and $3/2^+$ levels below 1 MeV that arise from the coupling of the single-particle $d_{3/2}$ ground state and $s_{1/2}$ level at 180 keV to the ^{128}Te core 2^+ level at 743 keV. In Figure 1 is shown a partial level scheme containing the most intense gamma rays and the depopulation of the low-energy levels. A level near 874-keV with an $\ell = 2$ value in transfer reactions had not been observed in earlier decay studies. In the (n,γ) study, the precise energies and depopulation pattern of this level has been identified and with these energies and intensities, evidence for this level also found in our new study of ^{129}Sb decay. The absence of direct population in the decay of $7/2^+$ ^{129}Sb indicates $3/2^+$ spin and parity. The rather weak 936.94 and much stronger 756.58-keV γ rays are the only two new and unplaced gamma rays below 1200 keV with a 180.42-keV energy difference. We tentatively suggest that these γ-rays depopulate a level at 936 keV that is the $1/2^+$ level arising from the coupling of the $d_{3/2}$ ground state with the 2^+ core. However, the branching ratio of about 5/100 for depopulation to the ground and 180-keV levels is opposite to that observed for the $1/2_2^+$ level in ^{131}Te and calculated in PTQM calculations.[11] As the 756-keV transition is the largest unplaced transition in this energy region, it could indicate that the $1/2^+$ level is at 756 keV.

This work has been supported by the U. S. Dept. of Energy through Grant DE-FG05-88ER-40418 with the University of Maryland.

---

[1] J. Rikovska *et al.*, Nucl. Phys. **A505**, 145 (1989).
[2] P. F. Mantica *et al.*, Phys. Rev. C **42**, (October 1990).
[3] K. Heyde *et al.*, Phys. Rep **102**, 291 (1983).
[4] S. Ohya, T. Tamura, and S. Kageyama, J. Phys. Soc. Japan **29**, 1435 (1970).
[5] P. G. Calway and H. D. Sharma, Nucl. Phys. **A156**, 338 (1970).
[6] C. A. Stone and W. B. Walters, Z. Phys. A **328**, 257 (1987).
[7] J. Honzátko *et al.*, Z. Phys. A **299**, 183 (1981).
[8] M. A. M. Shahabuddin, J. A. Kuehner, and A. A. Pilt, Phys. Rev. C **23**, 64 (1981).
[9] S. Galés *et al.*, Nucl. Phys. **A381**, 173 (1982).
[10] D. L Anderson *et al.*, J. Radioanal. Chem. **63**, 97 (1981).
[11] R. Meyer, V. Paar, E. Henry, R. Chrien, and B. Koene, Private Comm. (1989).

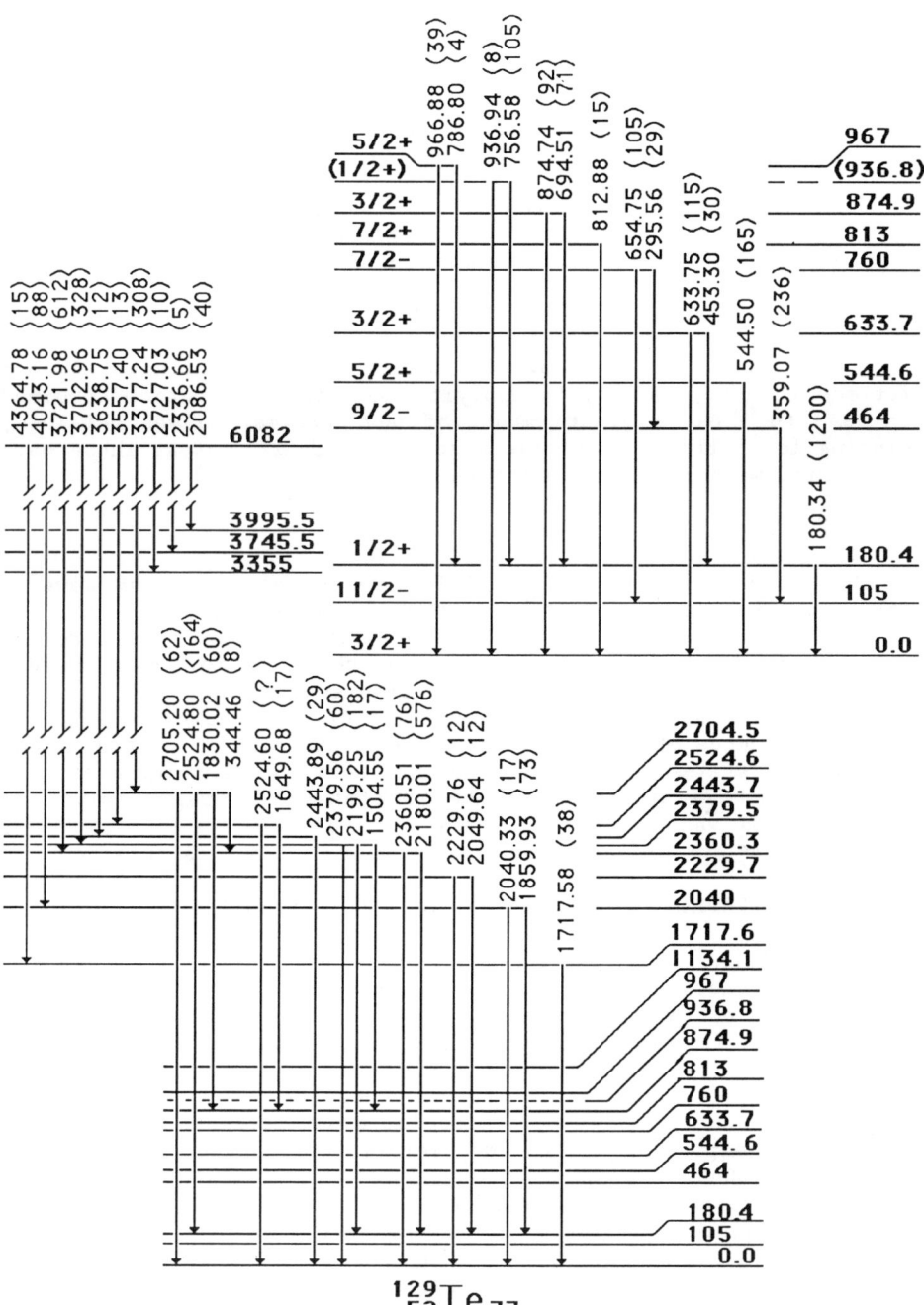

FIGURE 1. LEVELS IN ¹²⁹TE POPULATED IN THE ¹²⁸Te(n,γ)¹²⁹Te REACTION.

DOPPLER SHIFT LIFETIME MEASUREMENTS AT LOW RECOIL VELOCITIES AND
MIXED-SYMMETRY STATES IN ^{134}Ba

T. Belgya, B. Fazekas, G. Molnár, Á. Veres
Institute of Isotopes of the Hungarian Academy of Sciences, Budapest,
H-1525, Hungary

R. A. Gatenby, E. M. Baum, E. L. Johnson, S. W. Yates
Department of Chemistry, University of Kentucky,
Lexington, KY 40506-0055, USA

ABSTRACT

It is shown that the low recoil energy DSAM provides reliable lifetimes in heavy nuclei. By applying the (n,n'γ) reaction combined with DSAM to ^{134}Ba, a great amount of new information was obtained, including level lifetimes of two mixed-symmetry 2^+ states. These data are compared with IBM-2 calculations.

INTRODUCTION

The value of nuclear level lifetimes is clear. Among several lifetime determination methods, the DSAM is the most generally used to extract lifetimes in the 10^{-15}-10^{-12} s range, but it has been observed that stopping theory fails at low recoil energy-i.e., for heavy nuclei. We present evidence that inelastic neutron scattering (INS) combined with DSAM can be used to extract level lifetimes in heavy nuclei.
The influence of the Majorana force on certain, so-called mixed-symmetry (MS) states is obvious, and our incomplete knowledge of this stimulates the investigation of the MS states. The report of 2^+ MS states in ^{134}Ba was the first in the case of O(6) nuclei[1]. Now our latest results[2] are presented and compared with IBM-2 calculations.

LIFETIME MEASUREMENT AT LOW RECOIL VELOCITIES

INS combined with DSAM is described by Belgya et. al.[3] for accelerator-produced neutrons and by Elenkov et. al.[4] for reactor fast neutrons. When nuclear excitations are produced with INS, the initial recoil velocity is very small and v/c is in the range of 0.05%-0.1%. In the literature, it can often be found[5] that the knowledge of the stopping processes is uncertain for $\epsilon<2$. Therefore, the first question is how well known lifetimes determined by other methods which are not suffer from the problems of the deceleration processes can be reproduced by INS-DSAM. In table I, the known lifetimes obtained from resonance fluorescence of four different heavy nuclei are compared with our INS determined lifetimes. It can be seen that the agreement is generally good, except for those values marked with stars. In these cases, possible feeding can account for the differences. Further study of the low recoil energy DSAM is in progress.

Table I. Comparison of INS with (γ,γ') lifetimes[6,7]

Nucleus	E_{lev}, keV	τ_{INS}, fs	$\tau_{(\gamma,\gamma')}$, fs
^{90}Zr	3308.3	140±20	107±20
	3842.4	15±5	20±3
^{138}Ba	2218.0	164±45	166±13
^{142}Nd	2384.4	240±40	200±40
	2845.9	48±4	57±10
	3045.2	140±20	470±200
	3424.2	6.7±2.4*	2.0±0.3
	4093.5	9.4±4.3	6.0±1.1
	4253.8	35±25	33±12
^{144}Sm	2423.3	52±6	43±6
	2799.7	99±9	140±27
	3225.3	10.7±1.6*	3.0±0.3
	3890.1	7.7±3.8	3.1±0.5
	3906.8	31±9	26±8

Table II. Comparison of decay properties of ^{134}Ba with IBM-2 results. The level energies and the transition rates with parentheses were obtained from INS measurements. The % denotes F^2/F^2_{max} values in percent.

E_i (keV)	$J_i \to J_f$	E_x Ref. 10	%	B(E2) ($e^L b^L$)	B(M1) (μ_n^{2L})	B(E2) Ref. 10	B(M1)
605	$2_1\ 0_1$	575	98	0.137(3)	-	0.111	-
1168	$2_2\ 2_1$	1175	87	0.215(30)	3E-04(1)	0.139	0.087
	$2_2\ 0_1$			0.001(1)	-	-	-
1401	$4_1\ 2_1$	1275	99	0.216(24)	-	0.151	-
2029	$2_3\ 2_2$	1595	54	0.015(9)	0.008(2)	0.001	0.003
	$2_3\ 2_1$			0.004(2)	0.062(8)	0.014	0.301
	$2_3\ 0_1$			0.002(1)	-	0.001	-
2088	$2_4\ 2_2$	2119	49	<0.003	0.068(9)	0.013	0.080
	$2_4\ 2_2$			0.096(24)	0.011(9)	0.013	0.080
	$2_4\ 2_1$			<0.001	0.137(12)	0.001	-
	$2_4\ 0_1$			0.006(1)	-	-	-
2160	$0_3\ 2_1$	1840	77	0.060(13)	-	0.005	-
2335	$3_3\ 2_1$	2742	42	0.001(1)	0.034(12)	-	0.078
2337	$0_4\ 2_1$	2399	62	0.037(9)	-	0.001	-
2371	$2_5\ 3_1$	2405	55	0.040(28)	0.015(6)	0.056	0.122
	$2_5\ 2_1$			0.006(2)	0.001(1)	-	0.001
	$2_5\ 2_1$			0.002(1)	0.008(4)	-	0.001
	$2_5\ 0_1$			2E-04(1)	-	-	-
2488	$0_6\ 2_2$	3070	48	0.083(39)	-	-	-
	$0_6\ 2_1$			0.004(2)	-	0.001	-
2506	$4_5\ 4_1$	2728	37	0.006(13)	0.118(93)	-	0.008
	$4_5\ 4_1$			0.100(93)	0.038(47)	-	0.008
	$4_5\ 2_1$			0.005(4)	-	-	-

MIXED-SYMMETRY STATES IN ^{134}Ba

IBM-2 calculations have concentrated on the general description of the Ba chain[8] or on the behavior of MS states in ^{134}Ba.[9,10] The calculations have been repeated, but none of them can describe well the MS energy region. Since van Egmond's calculation[10] was performed so that the Majorana term described well some MS states, it is used for a comparison in table II to a few selected levels where our measured transition rates are well determined. The three lowest levels were also included where the lifetimes were taken from other sources[11]. The calculation can describe well the B(E2)'s for the lowest three symmetric states, at least within a factor of two, but it fails for the MS states. Moreover there are sizeable discrepancies when the level energies are compared. These facts lead us to the conclusion that it is necessary to make a more realistic calculation, and this calculation is still in progress. The 2_3^+ and 2_4^+ states have a reduced M1 transition probability sum of 0.2 μ_N^2. That value is in the expected range for O(6) nuclei. Therefore, the proposed shared MS properties[1] can be similarly interpreted as in the case of ^{56}Fe.[12] However, in this case, the lower 2_3^+ level is more symmetric. It is also important to note that lifetimes were determined for three 0^+ levels. According to the calculated F^2/F^2_{max} values, one or more of them may be the first example of a MS 0^+ state.

We wish to thank M. T. McEllistrem, J. L. Weil for fruitfull discussion. This work was supported by U. S. National Science Foundation Grants No. PHY-9001465 and INT-8917990 and by the Hungarian Academy of Sciences.

REFERENCES

1. G. Molnár, R. A. Gatenby, S. W. Yates, Phys. Rev. C37, 898 (1988).
2. T. Belgya, B. Fazekas, G. Molnár, Á. Veres, R. A. Gatenby, J. R. Vanhoy, E. M. Baum, E. L. Johnson, S. W. Yates, 3rd Int. Spring Seminar on Nucl. Exc., Ischia Italy, May 21-25, 1990 (in press).
3. T. Belgya, G. Molnár, B. Fazekas, Á. Veres, R. A. Gatenby, S. W. Yates, Nucl. Phys., A500, 77 (1989).
4. D. Elenkov, D. Lefterov, G. Toumbev, Nucl. Instr. Methods, 228, 62 (1984).
5. J. Lindhard, M. Scharff, H. E. Schiott, Mat. Fys. Medd. Dan. Vid. Selsk., 33 No. 14 (1963).
6. C. M. Lederer and V. S. Shirley, "Table of Isotopes 7th Edition", (John Wiley & Sons, Inc., New York, 1978).
7. F. R. Metzger Phys. Rev. C17, 939(1978), C18, 1603 (1978).
8. G. Puddu, O. Scholten, T. Otsuka, Nucl. Phys. A348, 109 (1980).
9. H. Harter, P. O. Lipas, R. Nojarov, Th. Taigel, Amand Faessler, Phys. Lett. B205, 174 (1988).
10. A. van Egmond and K. Allaart, Nucl. Phys. A436, 458 (1985).
11. Yu. V. Sergeenkov, V. M. Sigalov, Nucl. Data Sheets 34, 475 (1981).
12. S. A. A. Eid, W. D. Hamilton, J. P. Elliott, Phys. Lett. B166, 267 (1986).

TESTS OF OCTUPOLE BAND STRUCTURES*

P.D. Cottle, S.M. Aziz, J.W. Holcomb, T.D. Johnson, K.W. Kemper,
M.L. Owens, E.L. Reber, K.A. Stuckey, S.L. Tabor and P.C. Womble
Physics Department, Florida State University
Tallahassee, Florida 32306 USA

J.D. Brown, E.R. Jacobsen and Y.Y. Sharon
Physics Department, Princeton University
Princeton, New Jersey 08544 USA

S.G. Buccino and F.E. Durham
Physics Department, Tulane University
New Orleans, Louisiana 70118 USA

ABSTRACT

Experimental investigations of possible octupole band structures in ^{74}Se and ^{144}Nd are reported. These investigations used the techniques of γ-ray spectroscopy and proton inelastic scattering.

The existence of octupole bands, bands of negative parity states resulting from the coupling of positive parity yrast states to an octupole vibrational state, in even-even nuclei in several regions of the Periodic Table has made possible the study of the effects of angular momentum on octupole phonons. On the basis of early experimental results on octupole bands in the rare earth and actinide regions, Vogel[1] pointed out that the quasiparticle structures on which octupole phonons are built are unstable at high angular momenta. As a result, octupole bands "align" and become instead two quasiparticle bands above a certain angular momentum. One implication of Vogel's work is that in a nucleus for which $E(6_1^+) > 1.5$ MeV the octupole phonon will be completely unstable against rotation, and no octupole band structure will exist. Here we briefly discuss two cases, ^{74}Se and ^{144}Nd, for which $E(6_1^+)$ is greater than 1.5 MeV, and demonstrate that ^{144}Nd does not appear to have any octupole band structure, but that such a structure in ^{74}Se appears to exhibit remarkable stability.

We have studied a 5_1^- state in the transitional nucleus ^{144}Nd which has been identified[2] as a possible member of an octupole band structure by means of proton inelastic scattering of 35 MeV protons using the Princeton University cyclotron and the QDDD spectrograph. Proton scattering is well suited for such an investigation because of its selectivity for two quasiparticle (2qp) structures which can also form high spin negative parity states. A coupled channels analysis of the 5_1^- state was performed in order to find if it has an octupole band origin. If this state indeed results from the coupling of the 2_1^+ quadrupole phonon state to the 3_1^- octupole phonon state, which we may call an octupole coupled (OC)

438 Tests of Octupole Band Structures

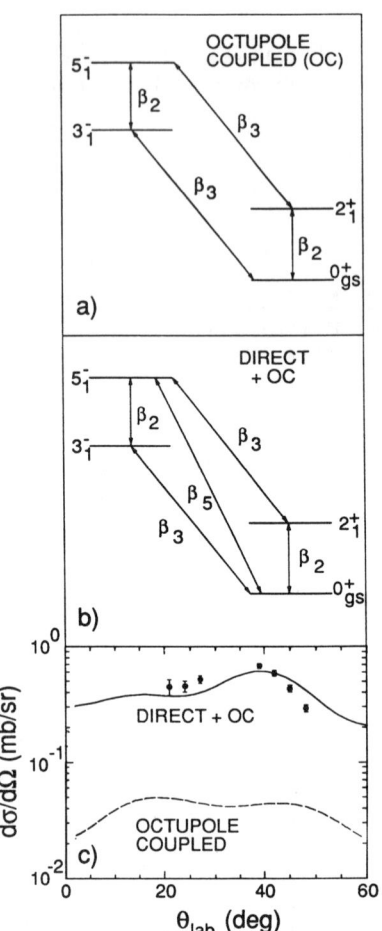

Fig. 1. Channel coupling schemes for (a) octupole coupled and (b) direct + OC excitation of the 5^- state. Results are compared to the data in (c).

configuration, then the population of the state would take place primarily through the two-step processes illustrated in Fig. 1a. However, if a large 2qp component exists in the 5_1^- state, it would be populated directly, as shown in Fig. 1b (Direct+OC). A coupled channels calculation performed using the OC scheme and β_2 and β_3 parameters extracted from the 2_1^+ and 3_1^- states underpredicted the data by a factor of ten. However, the inclusion of the direct route allowed us to reproduce the data well with a reasonable value for β_5. From this, we concluded[3] that the 5_1^- state could not be described as an octupole coupled or octupole band state, and that the octupole phonon in ^{144}Nd is completely unstable against angular momentum, as Vogel's condition would predict.

The similarities between the energy structures of the yrast and $K^\pi=3^-$ band in the deformed nucleus ^{74}Se suggest that the 3^- band is an octupole band. However, measurements of the lifetimes of states in the two bands provide a more stringent test of this hypothesis. If the negative parity band is not an octupole band, the E2 transitions linking the members of the band to the 3^- state would be weaker (in terms of the transition quadrupole moment Q_t) than those in the yrast band. Measurements of lifetimes of states in these bands were performed[4] at the Florida State University Superconducting Linear Accelerator Facility using an array of Compton suppressed γ-ray detectors. The Doppler Shift Attenuation Method was used with two different heavy-ion induced reactions. The Q_t values for the yrast and negative parity bands are shown as a function of J_{yrast} in Fig. 2 (J_{yrast} = J for the yrast band and J-3 for the negative parity band). Values for J_{yrast}=2 and 4 are taken from Ref. 5. The negative parity band transitions are either as strong or stronger than the yrast

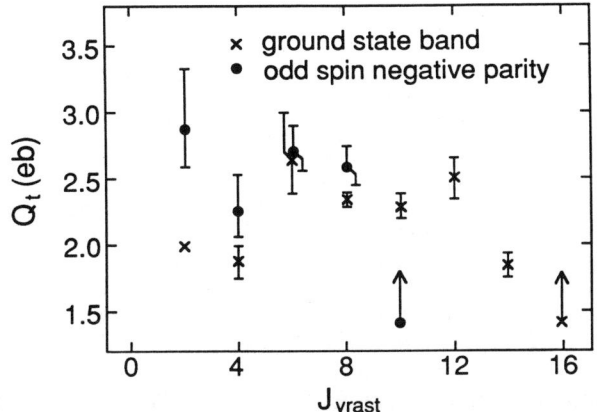

Fig. 2. Transition quadrupole moments for the ground state and odd spin negative parity bands in ^{74}Se.

transitions for all states measured. These results are consistent with an octupole band interpretation for the negative parity band up to J=13.

The apparent stability of the octupole band structure in ^{74}Se seems to conflict with Vogel's condition. Further experimental and theoretical study would be useful in order to clarify this situation.

* Work supported by the National Science Foundation and the State of Florida.

REFERENCES

[1] P. Vogel, Phys. Lett. 60B, 431 (1976).
[2] P.D. Cottle, S.M. Aziz, J.D. Fox, K.W. Kemper, and S.L. Tabor, Phys. Rev. C40, 2028 (1989).
[3] P.D. Cottle, S.M. Aziz, K.W. Kemper, M.L. Owens, E.L. Reber, J.D. Brown, E.R. Jacobsen, and Y.Y. Sharon, Phys. Rev. C42, 762 (1990).
[4] P.D. Cottle, J.W. Holcomb, T.D. Johnson, K.A. Stuckey, S.L. Tabor, P.C. Womble, S.G. Buccino, and F.E. Durham (in press, Phys. Rev. C).
[5] J. Adam, M. Honusek, A. Spalek, D.N. Daynikov, A.D. Efimov, M.F. Kudojarov, I.Kh. Lemberg, A.A. Pasternak, O.K. Vorov, and U.Y. Zhovliev, Z. Phys. A 332, 143 (1989).

Detailed Spectroscopy of ^{144}Nd

S.J. Robinson, H.G. Börner, S. Judge, J.Jolie and P. Schillebeeckx
Institut Laue Langevin
156X, 38042 Grenoble Cedex
FRANCE

The nucleus ^{144}Nd lies only two neutrons away from the N=82 closed shell and thus it might be expected to exhibit strong two-particle character in its structure. This has indeed been demonstrated in particle-core model (PCM) calculations [1,2,3] in which the 10 valence protons are treated as a system of phonon excitations and the two valence neutrons are treated explicitly. Despite this, attempts have also been made to describe its low lying structure in terms of purely collective models, in particular IBM-2 [4,5,6], and led to the proposal of the 2+ level at 2.072 MeV as one of the first examples of a vibrational mixed symmetry state [5]. This interpretation is not in agreement with recent calculations [2,3], in which IBM-2 wave functions are mapped onto PCM wave functions, which suggest that the mixed symmetry strength is fragmented between several 2+ levels. Thus ^{144}Nd provides an ideal opportunity to study the interaction of collective and single particle degrees of freedom.

In order to provide more information for further theoretical study an extensive experimental investigation of the ^{144}Nd level scheme has been undertaken using the neutron capture reaction. In all cases targets enriched to 91.1% ^{143}Nd were used.

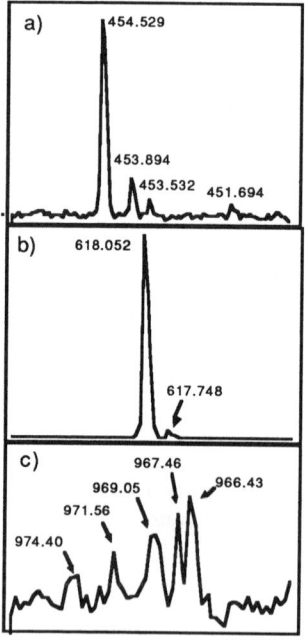

1) The GAMS bent crystal spectrometers at the ILL were used to study the gamma spectrum in the range 30 to 2000 keV, in which over 350 transitions have been identified. This is to be compared with less than 150 in previous Ge detector studies [4]. As well as identifying many new weak transitions some stronger lines, previously reported as singlets [4], have revealed a more complex character. Fig. 1 shows examples in the energy regions around 450 keV, 618 keV and 967 kev.

2) In two separate experiments gamma rays were studied in the energy range 1.5 to 8 MeV using the ILL pair spectrometer PN4. For part of the second run an NaCl target was added in order to provide precise energy and efficiency calibrations [7]. In addition the first target was also studied using an n-type Ge detector in the range 100 keV to 4 MeV and during the second run the pair spectrometer was run in parallel in Compton suppression mode in the range 400 keV to 3 MeV. A portion of the primary spectrum is shown in Fig.2. In total, from the GAMS and PN4 measurements, over 1000 gamma transitions have been identified with the ^{143}Nd(n,γ)^{144}Nd reaction.

Fig 1. Selected regions of the ^{143}Nd(n,γ) spectrum observed using GAMS2 in a) 2nd order, b) 5th order and c) 2nd order of reflection. Peak energies are indicated in keV.

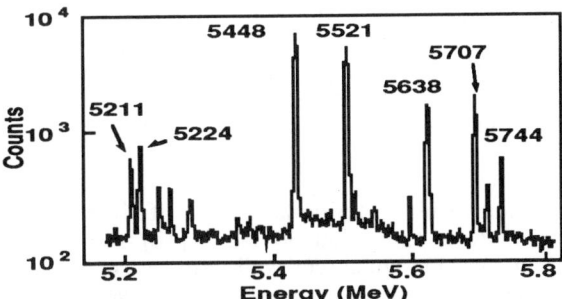

Fig.2. A portion of the primary gamma-ray spectrum from the ^{143}Nd(n,γ) reaction. The energies of some prominent peaks are marked in keV. This measurement was characterised by an excellent resolution of 3.6 keV at 5 MeV and 4.9 keV at 8 MeV.

3) In a separate experiment several of the stronger lines in the spectrum were used to perform GRID measurements and the lifetimes, or limits thereon, of 10 excited levels were determined. Due to the limited knowledge of the feeding patterns of these levels a statistical feeding model [8], based on a Constant Temperature Fermi gas model of the intermediate level densities, was employed. In addition very precise transition energies were obtained. Table 1 gives the gamma energies, the levels which they depopulate and the lifetimes extracted (where appropriate). These energies were used to place the GAMS measurements on an absolute energy scale.

Table 1. Level energies, gamma energies and lifetime values extracted from a GRID measurement of the ^{143}Nd(n,γ) reaction.

Level Energy (keV)[a]	J^π	Gamma Energy (keV)[b]	Lifetime[c]
696.5151(2)	2+	696.5133(2)	
1314.5671(3)	4+	618.0506(1)	
1510.6578(4)	3-	196.0905(5)	> 1 ps
		814.1403(3)	1.1(2) ps
1560.9389(8)	2+	864.4210(8)	550(50) fs
1791.3918(3)	6+	476.8238(2)	> 500 fs
2072.7756(37)	2+	1376.2468(38)	$145(^{+40}_{-30})$ fs
2093.1449(5)	5-	301.7530(4)	> 600 fs
		778.5747(11)	> 350 fs
2218.1592(9)	(5)+	426.7667(9)	> 700 fs
2295.2647(15)	4+	980.6919(19)	$250(^{+200}_{-100})$ fs

a. Taken from least squares fit to all placed transitions
b. Energies are not recoil corrected. Only statistical errors given.
c. No systematic error due to uncertainties in the feeding pattern is included

4) The electron spectrometer BILL was used to measure conversion electrons and these, together with previous measurements [9], will be used to determine the multipolarities of many of the stronger transitions in the level scheme.

Construction of the level scheme

By using the precise energies obtained from GAMS4 it was possible to determine the energies of several excited states to a very high degree of precision

(Table 1). These can then be used as a basis on which to construct the rest of the level scheme using the Ritz combination principal. In this way excitation energies in the 2 to 4 MeV range can be determined with a precision of only a few eV. To date a level scheme involving 32 levels, in the range 0 to 3.5 MeV, and 162 transitions has been constructed. The neutron binding energy has been determined to be 7816.84(8) keV.

Results

The most important result so far concerns the lifetime of the 2073 keV 2^+_3 level. The determined lifetime of approximately 145 fs [10] is in disagreement with the value determined by Metzger [11] but agrees well with a recent (p,p') study [12]. The revised B(M1) and B(E2) values for the decays from this level are shown in Table 2. These show good agreement with the values produced by recent PCM calculations [2] which show the fragmentation of the mixed symmetry strength amongst several levels. A discussion of the relevance of these values to nuclear structure interpretations can be found in another contribution to this conference [13].

Table 2. Comparison of experimentally determined and calculated absolute transition rates for the 2^+_3 level in ^{144}Nd.

	Expt.	Theory[a]
$B(E2; 2^+_3 - 0^+_1)$	0.004(1)	0.010
$B(E2; 2^+_3 - 2^+_1)$	0.007(3)	0.009
$B(M1; 2^+_3 - 2^+_1)$	0.10(4)	0.17

a. Taken from ref. 2

References
[1] K. Heyde and P.J. Brussard, Nucl. Phys. **A104** (1967) 81.
    G. Vanden Berghe, Z. Phys. **A272** (1975) 245.
[2] J. Copnell, Ph.D. thesis, University of Sussex (1988).
    J. Copnell, S.J. Robinson, J. Jolie and K. Heyde, Phys. Lett. **B222** (1989) 1, and to be published
[3] R.A. Meyer, O. Scholten, S. Brandt and V. Paar, Phys. Rev. **C41** (1990) 2386.
[4] D.M. Snelling and W.D. Hamilton, J. Phys. G **9** (1983) 763.
[5] W.D. Hamilton, A. Irbäck and J.P. Elliott, Phys. Rev. Lett. **53** (1984) 2469.
[6] S. Robinson et al in: Proc. 6th Inter. Symp. on Capture Gamma-Ray Spectroscopy, IOP Conference Series No. 88 (1987) 506.
    A.B. Balantekin and B.R. Barrett, Phys. Rev. **C35** (1987) 878.
    W.J. Vermeer, C.S. Lim and R.H. Speer, Phys. Rev. **C38** (1988) 2982.
[7] B. Krusche et al, Nucl. Phys. **A386** (1982) 245.
[8] B. Krusche, 'Statiistical Description of the Gamma Ray Cascades after Thermal Neutron Capture', contribution to this conference.
[9] M. Harder, M.Sc. thesis, University of Sussex (1986)
[10] H.G. Börner et al, 'Gamma Ray Induced Doppler Broadening ', contribution to this conference.
[11] F.R. Metzger, Phys. Rev. **187** (1969) 1700.
[12] P.D. Cottle et al, 'Proton Inelastic Scattering on the Transitional Nucleus ^{144}Nd, preprint (1990)
[13] S.J. Robinson, J. Jolie and J. Copnell, 'Mixed Symmetry States in Nuclei Near Closed Shells', contribution to this conference.

Extraction of Spins and Mixing Ratios from Directional Correlation Data

S.J. Robinson
Institut Laue Langevin
156X
38042 Grenoble Cedex
FRANCE

Introduction

In order to perform meaningful interpretations of nuclear model calculations it is generally agreed that the most stringent test is the reproduction of experimentally determined transition rates. When the transition parameters permit the presence of radiation of more than one multipole order the partitioning of the total intensity among two of these is expressed in terms of the multipole mixing ratio δ. In order to experimentally determine this partitioning the absolute value of δ is needed. The sign of δ also gives the relative sign of the transition matrix elements for the multipoles under consideration, also a stringent test for any model. A much more basic test is the reproduction of the spins and parities of the excited states. Thus it is essential to have reliable information on both J^π and δ values.

Directional Correlation Formalism

One of the common experimental techniques used to determine this information is that of directional correlation. Here the angular dependence of the coincidence counting rate for two cascading transitons (J_1-J_2-J_3) is determined. The angular distribution probability, $W(\theta)$, is then given (for dipole/quadrupole mixing) by:

$$W(\theta) = \sum_{\lambda=0,2,4} a_\lambda P_\lambda(\cos\theta) \tag{1}$$

where:
$$a_\lambda = a_0 A_{\lambda\lambda} \tag{2}$$

$$A_{\lambda\lambda} = B_\lambda(\gamma_1) A_\lambda(\gamma_2) \tag{3}$$

$$B_\lambda(\gamma_1) = \frac{F_\lambda(1\,1\,J_1 J_2) - 2\delta(\gamma_1) F_\lambda(1\,2\,J_1 J_2) + \delta(\gamma_1)^2 F_\lambda(2\,2\,J_1 J_2)}{1 + \delta(\gamma_1)^2} \tag{4}$$

$$A_\lambda(\gamma_2) = \frac{F_\lambda(1\,1\,J_3 J_2) + 2\delta(\gamma_2) F_\lambda(1\,2\,J_3 J_2) + \delta(\gamma_2)^2 F_\lambda(2\,2\,J_3 J_2)}{1 + \delta(\gamma_2)^2} \tag{5}$$

Here the $P_\lambda(\cos\theta)$ are the appropriate Legendre polynomials and the F_λ are geometrical factors which are defined and tabulated by Frauenfelder and Steffen [1].

In general the a_λ of eq. (3a) can include other factors such as:
i) U_λ which describe the deorientation effects of any, unobserved, intermediate transitions in the cascade;
ii) Q_λ, which describe the effects of finite source and detector geometries and;
iii) G_λ, which describe the attenuation of the correlation due to the interaction of any, relatively long lived, intermediate state with extra-nuclear fields.

The question now is how to extract, from these values, information concerning the spins of levels and d-values of transitions involved in the cascade, in a *statistically meaningful* way. Since, in the vast majority of cases, J_2, J_3 and $\delta(\gamma_2)$ are already known (eg the pure E2 $2^+{}_1 \to 0^+{}_1$ transition in an even-even nucleus) this means trying to extract J_1 and $\delta(\gamma_1)$.

It has recently been shown [2] that one of the common methods used to extract this information, based on a linear least squares fit of the a_λ coefficients, can produce unpredictable (in the statistical sense) results, due to misinterpretation of the results. In particular three mistakes are usually made.

i) Since the information to be extracted is contained in both a2 and a4 a *joint* 1σ confidence region must be defined. This cannot be constructed simply from the separate 1σ limits on a2 and a4.

ii) The fact that a2 and a4 may be strongly *correlated* (because they are extracted from the *same* data set) is usually ignored.

iii) No constraint that a2 and a4 correspond to a *meaningful solution* for the spins and δ values involved is placed on the fitting procedure.

As shown in ref 2. the correct results can be extracted if the results of any linear fit of the correlation coefficients are correcly treated. However the simplest method is based on a direct non-linear least squares fit for δ, as follows.

Data Analysis

For a chosen cascade spin sequence $\tan^{-1}\delta$ is varied from -90° to 90°. At each point the statistic

$$S^2 = \sum_i \left[\left(W(\theta_i)_{exp} - W(\theta_i)_{th} \right) \Big/ \Delta W(\theta_i)_{exp} \right]^2 \tag{5}$$

(where the index i runs over the experimentally measured points) is minimised with respect to the normalisation factor a_0 and the whole curve is plotted versus $\tan^{-1}\delta$. If the spins involved allow only one multipole for the transition of interest the only one value is calculated. From standard χ^2 tables a confidence level is chosen for the acceptance or rejection of solutions and if any part of any S^2 curve (or single value - see ref. 2 for discussion of degrees of freedom) falls below this value it is accepted as a possible solution. The mixing ratio, where relevant, is taken as the value corresponding to the minimum in the S^2 curve.

We now come to the question of the extraction of 1σ error limits. It is common in non-linear fitting to take the analytical error estimates from the covariance matrix of the last iteration in the fitting procedure. However this may be incorrect for two reasons. Firstly, if the Marquardt method has been used (as is common), then the gradient factor may be large and the diagonal elements of of the covariance matrix may be correspondingly small. In this case the correct procedure would be to recalculate the matrix, first setting the gradiant factor to zero. These values can then be used as a first estimate for the error limits but, because the S^2 curve is not a true parabola, the correct limits should be searched for explicitly.

It has been shown [2,3] that, if internal errors are to be used (ie if the dominant errors are due to counting statistics, or if systematic errrors are well understood) then the correct 1σ errors limits should be taken from the points in the S^2 curve corresponding to

$$S^2_{lim} = S^2_{min} + 1 \quad (6)$$

The principle of this procedure is shown in Fig. 1.

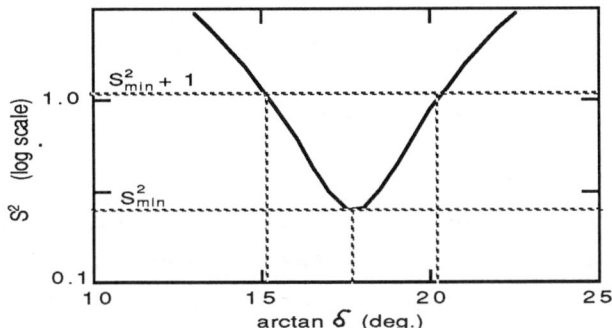

Fig1. The extraction of internal error limits from the S^2 curve.

If the dominant errors are likely to be systematic and can only be estimated then the external error estimates should be used. These are taken at the points around the minimum corresponding to

$$S^2_{lim} = S^2_{min}\left[1 + \frac{F_{68.3\%}(1, N-2)}{N-2}\right] \quad (7)$$

where $F_{68.3\%}(1, N-2)$ is the relevant value of the statistical F distribution with 1 and N-2 degrees of freedom (see ref. 3 for a useful table of these values). Notice that, for large N, this factor comes close to unity and so, assuming a parabolic form for the S^2 curve, this corresponds closely to the common practice of multiplying the internal errors by the square root of the chi-square per degree of freedom. For small numbers of degrees of freedom this practice can result in a large deviation from the true value.

Errors in the Input Paramenters
Some of the parameters used to determine $W(\theta)_{th}$ in eq (5) may themselves be taken from experiment, and so have errors associated with them. Examples might be the mixing ratio of the gating transition, the magnetic moment which determines the G_λ parameters.or incertainties in detector dimensions for the Q_λ These errors must naturally be taken into account and this can be done by including them in the weighting factors used in eq. (5). However it must be remembered that the same parameter will give errors in both a_2 and a_4 which will be perfectly correlated [4] and this must be taken into account also.

References
[1] H. Frauenfelder and R.M. Steffen, in: The Electromagnetic Interaction in Nuclear Physics, ed. W.D. Hamilton (North Holland, Amsterdam, 1975) p. 505.
[2] S.J. Robinson, Nucl. Instr. and Meth. **A292** (1990) 386.
[3] D.W.O. Rogers, Nucl. Instr. and Meth.**127** (1975) 253.
[4] H. Marshak and C.H. Spiegelman, Nucl. Instr. and Meth. **A234** (1985) 455.

EXPERIMENTAL EVIDENCE FOR OCTUPOLE DEFORMATION FROM OBSERVATION OF PARITY DOUBLETS IN ^{152}Eu, ^{154}Eu, ^{156}Eu NEUTRON CAPTURE STUDIES

M.K.Balodis, P.T.Prokofjev, A.V.Afanasjev
Physics Institute, Latvian Academy of Sciences,
229021 Riga-Salaspils, Latvia

ABSTRACT

Experimental results of ^{156}Eu from triple neutron capture are reported, those of ^{154}Eu and ^{152}Eu from (n,γ) reactions are shortly reviewed. Comparatively intense E1 transitions between levels of bands with $\Delta K=0$ and opposite parities allow to propose existence of octupole deformation in ^{156}Eu, ^{154}Eu, and ^{152}Eu, which results in observation of parity doublets.

INTRODUCTION

During the last decade, an idea of octupole deformation was used to analyze nuclear structure in the regions of Ra-Pa and Ba-Sm (see, e.g., refs.[1,2]). Having much information on ^{152,154,156}Eu from neutron capture studies, it is of interest to continue the discussion started in publications of Sheline and Sood for ^{154}Eu and ^{152}Eu (refs.[3,4]).

^{156}Eu results, being reported here in some detail, are published[5] in a near future. A short review is given for earlier published[6,7] results of ^{154}Eu and ^{152}Eu.

We propose 4 parity doublet (PD) bands in ^{156}Eu, in addition to 15 resp. 8 PDs in ^{154,152}Eu. Existence of parity doublets is one of experimental criteria for octupole deformation.

^{156}Eu: EXPERIMENTAL RESULTS AND LEVEL SCHEME

Using enriched ^{153}Eu targets (99.1%), low-energy conversion electrons (18-1100 keV) and gamma-rays (25-1600 keV) were measured from thermal neutron capture. Spectrometers BILL, GAMS1, GAMS2/3 were used [8,9] at Institut Laue-Langevin, Grenoble. Due to thermal neutron fluxes of 3×10^{14} and 5.5×10^{14} cm^{-2}s^{-1} respectively, and high capture cross sections of 390, 1500, and 4040 barns for three successive captures, about 1000+300+100 transitions of three europium nuclei were observed in the reaction chain ^{153}Eu(n,γ)^{154}Eu(n,γ)^{155}Eu(n,γ)^{156}Eu.

Comparing several measurement series, isotopic identification was made for γ-transitions and electron lines in ^{154}Eu, ^{155}Eu, and ^{156}Eu. Results obtained with ^{154}Sm targets [10] were used since almost 80 transitions of ^{156}Eu were detected via chain of neutron captures and β-decay ^{154}Sm(n,γ)^{155}Sm $\xrightarrow{\beta^-}$ ^{155}Eu(n,γ)^{156}Eu. The result of analysis is 95 transitions from 22 to 800 keV in ^{156}Eu, and 70 multipolarities are determined for them.

We have developed the ^{156}Eu level scheme (fig. 1) containing 31 levels with energies up to 515 keV, confirming 6 levels known from the ^{156}Sm β-decay measurements [11]. We observe the K=3 band head with the structure p5/2$^+$[413] - n11/2$^-$[505] known from the (t,p) reaction work [12]. 13 rotational bands are interpreted due to 3 proton and 4 neutron orbits: p5/2$^+$[413], p5/2$^-$[532], p3/2$^+$[411], and n5/2$^+$[642], n3/2$^-$[521], n5/2$^-$[523], n11/2$^-$[505].

Our scheme is in agreement with the ground state spin measurements [13,14] indicating J=0 since the ground state band is as expected for K=0. Four well-developed low-energy bands have K=0 or K=1, and their counterparts in Gallagher-Moszkowski doublets have K=4 or K=5.

In the first approximation, a simple rotational mo-

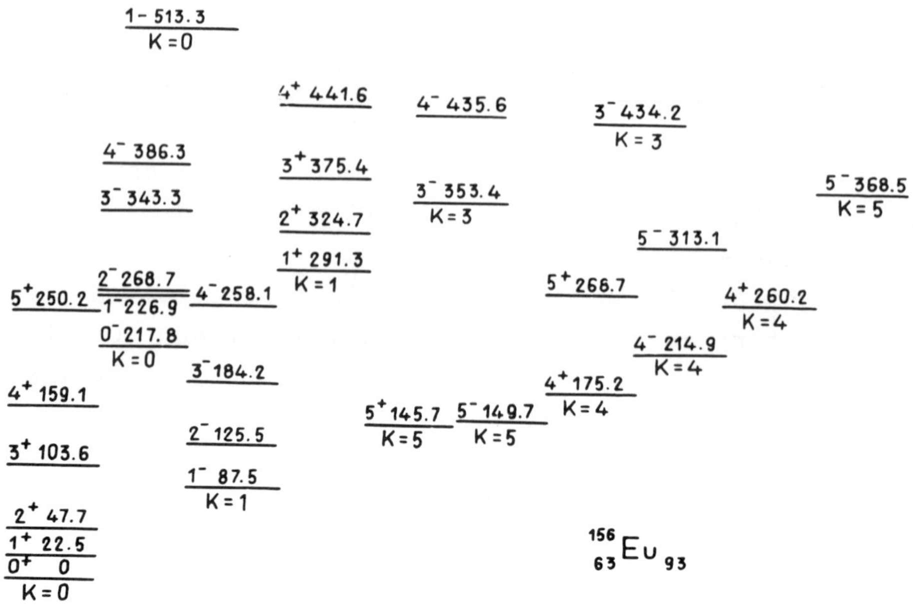

448 ^{152}Eu, ^{154}Eu, ^{156}Eu Neutron Capture Studies

```
                1⁻ 513.3
                ─────────
                 K = 0

                              4⁺ 441.6     4⁻ 435.6        3⁻ 434.2
                                                           ────────
                                                            K = 3
       4⁻ 386.3                                                              5⁻ 368.5
                              3⁺ 375.4                                       ────────
       3⁻ 343.3                             3⁻ 353.4                          K = 5
                              2⁺ 324.7      ────────
                                             K = 3
                                                           5⁻ 313.1
                              1⁺ 291.3
                              ────────
               2⁻ 268.7        K = 1
       5⁺ 250.2 ─────── 4⁻ 258.1            5⁺ 266.7                         4⁺ 260.2
                1⁻ 226.9                                                     ────────
                0⁻ 217.8                                                      K = 4
                ────────                                   4⁻ 214.9
                 K = 0   3⁻ 184.2                          ────────
       4⁺ 159.1                                             K = 4
                                            4⁺ 175.2
                         2⁻ 125.5   5⁺ 145.7  5⁻ 149.7     ────────
       3⁺ 103.6                      K = 5    K = 5         K = 4
                         1⁻ 87.5
                         ────────
        2⁺ 47.7           K = 1
        1⁺ 22.5                                                        156
        0⁺ 0                                                            63Eu₉₃
        ────
         K = 0
```

Fig. 1. Level scheme of ^{156}Eu.

del well describes level energies in ^{156}Eu. A few interband transitions of M1 or E2 character are understood due to Coriolis and residual interaction. We will see that a number of intense E1 transitions can be understood in octupole deformation approach.

Level scheme is estimated to be complete for all spins up to 240 keV, and, for J=0-4, to 350 keV.

^{152}Eu AND ^{154}Eu: A SHORT REVIEW

Cross sections of ^{153}Eu and ^{151}Eu being 390 and 9200 barns, rich low-energy gamma-ray and conversion electron spectra were obtained. Transfer reaction measurements and high-energy neutron capture results, including ARC data for ^{154}Eu, helped to establish levels[6,7].

99 energy levels were observed in ^{154}Eu connected with about 530 transitions ~380 multipolarities being determined for them. 90 levels were assigned to 32 rotational bands interpreted using 3 proton and 9 neutron or-

bits. The scheme of ^{152}Eu contained 95 levels with a total of nearly 900 identified transitions and more than 130 multipolarities. 70 levels were interpreted as members of 33 bands basing on 3 proton and 8 neutron orbits.

OCTUPOLE DEFORMATION, PROPERTIES OF PARITY DOUBLETS, E1 TRANSITIONS IN ^{152}Eu, ^{154}Eu, ^{156}Eu

^{154}Eu and ^{156}Eu, in the first approximation also ^{152}Eu, can be successfully interpreted due to rotational model. Including Coriolis and residual interactions, model calculation results[3,15] well agree with experimental energy relations and transition probabilities for levels of the same parity. Similar calculations result in E1 transition probabilities for neighbour odd nuclei ^{153},^{155}Eu being 4-5 orders smaller than corresponding experimental values. From halflife data on a numbers of nuclei between A=150 and A=190, E1 transitions were found to have a tendency to be ~2 orders of magnitude faster for $\Delta K=0$ than for $\Delta K=1$, and nonadiabatic rotational model was used for explanation (see, e.g., ref.[16]).

For understanding of E1 transitions, identification of parity doublet (PD) bands connected with octupole deformation of nuclear core can be considered as more general model approach.

In order to study the effects of octupole deformation, nuclei of Eu and a few neighbour elements present favourable situation. The number of neutron states of positive and negative parity is nearly the same, and we have two low-lying proton states with different parities and the same Ω_p value. An indication for octupole deformation of Eu is given by odd-even staggering of differential nuclear radii (see, e.g., ref.[17], and references therein).

Both in odd and odd-odd nuclei, simple selection

rules for PD's should be taken into account:

$$K' = K, \quad \pi' = -\pi. \tag{1}$$

For parity doublet of odd-odd nucleus, a transition should proceed between configurations

$$\Omega_p[Nn_z\Lambda]_p \pm \Omega_n[N\pm1\,n'_z\Lambda']_n \leftrightarrow \Omega_p[Nn_z\Lambda]_p \pm \Omega_n[Nn_z\Lambda]_n \tag{2}$$

or $\Omega_p[N\pm1\,n'_z\Lambda']_p \pm \Omega_n[Nn_z\Lambda]_n \leftrightarrow \Omega_p[Nn_z\Lambda]_p \pm \Omega_n[Nn_z\Lambda]_n.$

From model point of view, parity quadruplets can be discussed. An attempt to find experimental evidence for this idea is reported elsewhere[18].

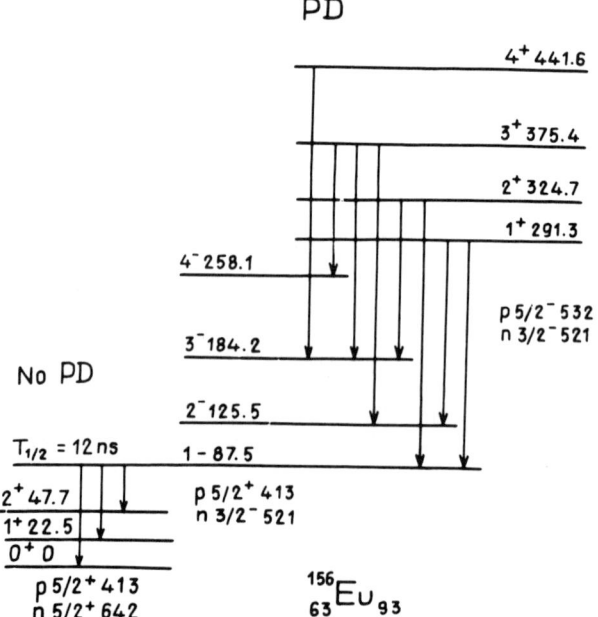

Fig.2. Partial level scheme of ^{156}Eu.

Our aim is to search for new PD bands in ^{156}Eu, and to give comments concerning earlier proposed PD's in ^{152}Eu and ^{154}Eu.

The following parity doublets in ^{156}Eu can be proposed (figs. 1,2): the 0^+ ground state band and the 0^- 217.8 keV band, (p5/2$^+$413 or p5/2$^-$532) − n5/2$^+$642, the 1^-87.5 and 1^+291.3 keV bands, (p5/2$^+$413 or p5/2$^-$532) − n3/2$^-$521, the 4^+175.2 and 4^-214.9 keV bands, (p5/2$^+$413 or p5/2$^-$532) + n3/2$^-$521, and the 5^+145.7 and 5^-149.7 keV bands, (p5/2$^+$413 or p5/2$^-$532) + n5/2$^+$642.

All available information is in agreement with these assignments of PD bands.

For the 1^+291.3 keV band head, the halflife limit $T_{1/2}<0.2$ ns gives limits $F_w<10^3$ for the E1 transitions depopulating this level. In the depopulation of J=2÷4 levels of the 1^+ rotational band, E1 transitions have relatively large intensities.

All known levels of the 0^- band, and the levels of the 4^- band, decay with E1 transitions mainly; this is a good argument for the $0^{\pm}$ and $4^{\pm}$ parity doublets.

For the decay of the 5^-149.7 keV band head, the 4 keV E1 transition to the other member of PD is the only reasonable decay possibility. Since $F_w \sim 10^2$-10^3 for E1 transitions inside PD's, we estimate $T_{1/2} \approx 10^{-7}$ ns for the 5^- band head. "Normal" E1 transitions outside PD's depopulate the 1^-87.5 keV band head to the levels of the 0^+ ground state band, having $F_w \sim 10^4$.

For a comparison, we present normal E1 transitions in ^{154}Eu, ^{152}Eu on figs. 3 and 4.

From publications of Sheline et al 15 parity doublets in ^{154}Eu and 8 doublets in ^{152}Eu should result in relatively fast E1 transitions.

Most confident PD's of ^{154}Eu and ^{152}Eu, respectively, are pairs of bands where transitions between band heads have $T_{1/2}$ resp. F_w values: 1^-82.8 keV and 1^+71.9 keV bands ($T_{1/2}$=20 ns, F_w=6.3x10^3), 4^-141.8 and 4^+89.8 keV bands ($T_{1/2}$=2.5 ns, F_w=4.5x10^3). These two PD's, as well as most of PD's proposed in ^{154}Eu, ^{152}Eu are due to neutron change.

There is only one PD observed in each of these two nuclei due to proton change: $3^{\pm}$, (p5/2$^+$413 or p5/2$^-$532)--n11/2$^-$505, whex the 3^- band head is the ground state in both cases.

Due to larger mixing effects in ^{154,152}Eu, several of measured $T_{1/2}$ values result in medium F_w values bet-

452 ¹⁵²Eu, ¹⁵⁴Eu, ¹⁵⁶Eu Neutron Capture Studies

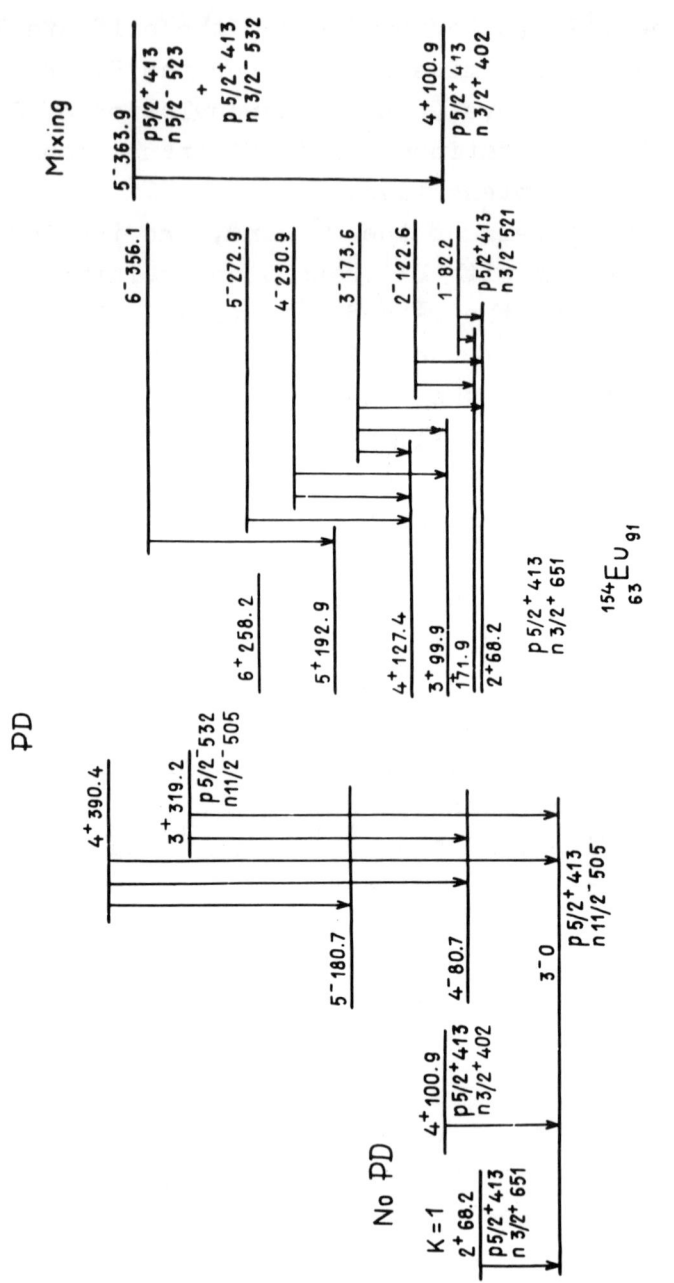

Fig. 3. Partial level scheme of ¹⁵⁴Eu.

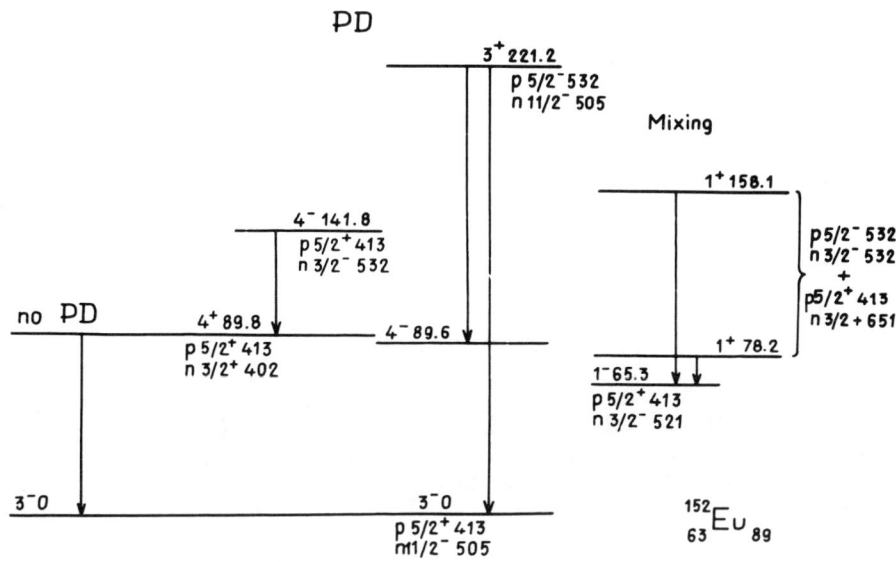

Fig.4. Partial level scheme of ^{152}Eu.

ween those F_W of "normal" and "enhanced" E1 transitions. We give on fig. 4 one of mixing cases proposed in ref.[4], but a case on fig. 3 is correlated with a Coriolis interaction between the K=5 an K=4 states.

CONCLUSIONS

Octupole deformation model argumented with data on odd-odd Eu nuclei helps to understand E1 transition probabilities for parity doublet bands with $\Delta K=0$ and corresponding change of proton or neutron orbit. For ^{156}Eu, all four parity doublets are connected with change of proton orbit. A larger role of neutron orbit change in ^{154}Eu can be correlated with a neutron orbit density larger than in ^{156}Eu.

REFERENCES

1. G.A.Leander, R.K.Sheline, Nucl.Phys. A413, 375 (1984)
2. G.A.Leander et al., Phys.Lett. 152B, 284 (1985)
3. R.K.Sheline, Phys.Lett. B219, 222 (1989)
4. R.K.Sheline and P.C.Sood, Progr.Theor.Phys. 81, 1057 (1989)
5. M.K.Balodis et al., Nucl.Phys. A, accepted for publication
6. M.K.Balodis et al., Nucl.Phys. A472, 445 (1987)
7. T.von Egidy et al., Z.Phys. A286, 341 (1978)
8. W.Mampe et al., Nucl.Instr.Meth. 154, 127 (1978)
9. H.R.Koch et al., Nucl.Instr.Meth. 175, 401 (1980)
10. K.Schreckenbach et al., Nucl.Phys. A376, 149 (1982)
11. P.G.Hansen et al., Nucl.Phys. 82, 614 (1966)
12. R.G.Lanier et al., Nucl.Phys. A413, 236 (1984)
13. C.Ekstrom and L.Robertson, Z.Phys. A302, 101 (1981)
14. G.O.Alkhazov et al., Izv. AN SSSR, ser. fiz. 49, 24 (1985)
15. T.V.Guseva et al., Izv. AN SSSR, ser. fiz. 51, 856 (1987)
16. W.Andrejtscheff, K.D.Schiling and P.Manfrass, At. Data Nucl. Data Tables 16, 515 (1975)
17. S.A.Ahmad et al., Nucl.Phys. A483, 244 (1988)
18. A.V.Afanasjev et al., these Proceedings

NUCLEAR STRUCTURE INVESTIGATIONS AND LIFETIME MEASUREMENT IN ^{156}Gd

J. Klora[a,b], H.G. Börner[b], T. von Egidy[a], H. Hiller[a],
S. Judge[b], B. Krusche[b], V. A. Libman[c], H. Lindner[a],
L. L. Litvinsky[c], U. Mayerhofer[a], A. V. Murzin[c], S. J. Robinson[b]

a) Physik-Department, Technische Univ. München, D-8046 Garching, FRG

b) Institut Laue-Langevin, 156X, F-38042 Grenoble Cedex, France

c) Institut for Nuclear Research, 252028, Kiev, Prospect Nauki 47, USSR

ABSTRACT

The level structure of ^{156}Gd has been studied using the neutron capture reaction at the ILL in Grenoble with the Gams 2/3 spectrometer and the pair spectrometer PN4. The conversion electron spectrum has been measured with BILL. Lifetimes have been measured using the GRID-Method. ARC measurements have been carried out at the Institut for Nuclear Research in Kiev. The direct reactions ^{155}Gd (d,p) ^{156}Gd and ^{155}Gd(d,t) ^{156}Gd have been investigated at the Munich Tandem Accelerator.

INTRODUCTION

The strongly deformed even-even nucleus ^{156}Gd has long been an excellent testing ground for nuclear models[1]. Even more interest was drawn to this nucleus when a new M1-Mode was discovered and investigated by Richter et al.[2] during studies of the (γ,γ'), (e,e') and (p,p') reaction. This was the motivation for us for new studies of this nucleus, using several complementary techniques, listed below.

1. SECONDARY GAMMA RAYS

Secondary gamma rays have been measured with the bent-crystal spectrometer GAMS 2/3 at the ILL. Three targets were constructed from 90% enriched ^{155}Gd oxide. Two were 2 mg/cm^2 thick and one 20 mg/cm^2 thick. About 700 transitions from 80 keV to 2500 keV were measured up to the 7 th order of diffraction. The enery resolution was about 300 eV at 1000 keV in 5 th order. To have a good basis for the Ritz Combination principle, eight selected transitions were measured with the double flat crystal spectrometer GAMS4 with an energy resolution of about 2 eV at 1000 keV ($\delta\epsilon/\epsilon = 2*10^{-6}$). A preliminary level scheme has been constructed: so far, 220 transitions have been placed in 14 bands with 52 levels[3].

2. PRIMARY GAMMA RAYS

In order to get information about the primary gamma rays in the ^{155}Gd (n,γ) ^{156}Gd reaction, the energy spectrum has been measured from 1200 keV to 8500 keV with the pair spectrometer PN4 at the ILL. In this energy region, 600 lines were found. The FWHM of the detector was about 5 keV at 8 MeV.

3. CONVERSION ELECTRONS

The energy spectrum of conversion electrons following the thermal neutron capture in ^{155}Gd was measured from 80 to 8600 keV with the electron spectrometer BILL at the ILL. Due to the high neutron capture cross section for ^{155}Gd the target material was rapidly used up in the intense neutron flux of $3*10^{14}$ n/cm^2/s. This problem was overcome by changing the target every two days. A total of 16 targets was used, each target consisting of a thin deposit (between 200 μg/cm^2 and 450 μg/cm^2) of 90% enriched ^{155}Gd with 1.3% of ^{157}Gd on a Ni foil. About 3000 conversion electron transitions have been observed. The energy resolution was about 400 eV at 1000 keV.

4. ARC MEASUREMENTS

Primary transitions following capture of 2 keV neutrons were observed at the Institut for Nuclear Research in Kiev. Three pair spectrometers, with a FWHM of about 10 keV at 8 MeV, were used. The target consisted of 60 g of 91.9 % enriched ^{155}Gd. The neutron beam was filtered by ^{45}Sc, ^{60}Ni and ^{54}Fe, resulting in a beam of 10^8 n/sec/cm^2 with an average energy of 2 keV and a width of 1.6 keV. As the average level spacing of ^{155}Gd is 1.8 eV , the averaging takes place over about 900 compound-nucleus states. The effect of the different level densities of states with spin and parity of $I^\pi=1^-$ and $I^\pi=2^-$ could be seen. ARC-measurements with other averaging energies (3.5 keV, 134 keV and one out of 24.5 keV, 45 keV, 59 keV) are planned.

5. PARTICLE TRANSFER REACTIONS

The direct reaction ^{155}Gd (d,p) ^{156}Gd and ^{157}Gd (d,t) ^{156}Gd have been measured at the Munich Tandem Accelerator with 22 MeV deuterons and a 50 μg/cm^2 90 % enriched ^{155}Gd target. The (d,p) reaction was observed at the angles 15,25,40 and 60 degrees in an enery range from 0 keV to 3000 keV, and the (d,t) reaction at angles 15,30 and 50 degrees in an energy range from 0 keV to 3500 keV. In order to obtain spectroscopic factors, DWBA calculations were carried out. A resolution (FWHM) of about 5 keV was obtained.

6. LIFETIME MEASUREMENTS

Lifetimes of selected levels have been measured with the GAMS4 spectrometer at the ILL using the GRID Methode (Gamma-ray Induced Doppler broadening) [4]. Preliminary lifetime results of some of the measured levels are listed in Table I (only statistical errors given). The calculation was based on a statistical distribution for the feeding paths [5] (taking into account the known experimental feeding e.g. primary and strong secondary transitions).

Table I Lifetimes

Band	I^π	E_{level}(keV)	E_γ(keV)	Lifetime (fs)
Gamma	2^+	1154.1512 (5)	1065.1820 (2)	1210 ($^{170}_{140}$)
		1154.1512 (5)	1154.1513 (2)	2700 ($^{1500}_{700}$)
Gamma	4^+	1355.4248 (8)	1067.2325 (2)	1000 ($^{320}_{200}$)
$K=1_1$	1^-	1242.476 (7)	1242.486 (10)	166 ($^{14}_{13}$)
$K=1_1$	3^-	1276.1372 (11)	1187.1679 (2)	160 ($^{17}_{15}$)
$K=0_1$	1^-	1366.455 (5)	1277.488 (10)	48 ($^{11}_{10}$)
$K=0_1$	3^-	1538.845 (7)	1449.904 (11)	35 ($^{10}_{9}$)
$K=0_4$	2^+	1771.088 (6)	1682.172 (18)	500 ($^{180}_{110}$)
	1^-	1946.382 (14)	1857.414 (28)	50 ($^{20}_{17}$)

REFERENCES

P. Van Isacker, K. Heyde, M. Waroquier, G. Wenes, Nucl. Phys. **A380**, (1982) 383

D. Bohle, A. Richter, W. Steffen, A. E. L. Dieperink, N. Loiudice, F. Palumbo, O. Scholten, Phys. Lett. **137B**, 27 (1984)

A. Bäcklin, G. Hedin, B. Fogelberg, M. Saraceno, R. C. Greenwood, C. W. Reich, H. R. Koch, H. A. Baader, H. D. Breitig, O. W. B. Schult, K. Schreckenbach, T. von Egidy, W. Mampe, Nucl. Phys. **A380**, (1982) 189

H. Börner, et al. Phys. Lett. **B215**, 45 (1988) and contribution to this conference

B. Krusche, to be published and contribution to this conference

This work was supported by the BMFT, Bonn.

INVESTIGATION OF GAMMA TRANSITIONS POPULATING THE 1094 keV ISOMERIC STATE IN THE ^{167}Er(n,γ) REACTION

M. Bogdanović, J. Simić

The Boris Kidrič Institute of Nuclear Sciences, Belgrade
Yugoslavia

Time differential γγ coincidence measurements have been made in order to investigate the population of the 1094.04 keV ($T_{1/2}$=112 ns) isomeric level in ^{168}Er following the ^{167}Er(n,γ) reaction. With these data the levels in the levels scheme obtained in the erlier study[1,2,3] have been confirmed and some new levels have been proposed.

The population of the 1094.04 keV isomeric level was investigated using γγ coincidence spectrometer consisting of a Ge(Li) detector with 16 cm^3 volume and NaI(Tl) scintilation detector to measure singles and coincidence spectra in the energy interval 0.1 - 6.7 MeV. The target of Er_2O_3 powder was exposed to an external horizontal filtred neutron beam of the RA reactor at Vinča, Belgrade. The thermal neutron flux was 4.2 x 10^7 neutrons cm^{-2} sec^{-1}. The ratio of coincidence to single γ-ray intensity, normalized to 100 for a transition populating directly the isomeric state, gives the parameter P associated with each transition in the level scheme. Taking the values from about zero (depending on chance events) up to about 100, P indicates the position of the transition in the level scheme. Combining new results with previous data[1] we could propose about 30 new levels beyond 2.2 MeV excitation energy.

In Table 1 are given the energy levels found from the P values for the secondary γ transitions. In the first column are listed energies of transitions associated with large values of parameter P given in the second column. In the third an fourth columns are shown the positions of the corresponding transitions in the level scheme.

Table 1

gamma transition (keV)	P (%)	E_i(keV) (new level)	E_f(keV)
374.683(4)	100	2434.66	2059.99
407.984(6)	100	2526.76	2118.85
408.457(8)	100	2663.31	2254.80
436.673(5)	143(90)	1629.68	1193.02
471.875(6)	100	2125.43	1653.55
535.64(2)	100	1629.68	1094.04
537.76(6)	100	2663.31	2125.43
538.684(31)	100	2144.63	1605.85
688.79(3)	63(18)	1881.86	1193.02
938.198(26)	80(24)	2249.61	1311.46
969.509(27)	110(27)	2874.63	1905.11
1009.678(21)	103(25)	2663.31	1653.55
1173.56(2)	92(18)	2267.62	1193.02
1176.42(5)	108(28)	2270.42	1094.04
1284.08(8)	97(50)	2378.13	1094.04
1432.75(7)	92(23)	2526.76	1094.04
1433.75(7)		2527.79	1094.04
1470.41(17)	80(23)	2663.31	1193.02
1569.31(11)	97(24)	2663.31	1094.04
1780.52(8)	94(16)	2874.63	1094.04

In Table 2 are given the energy levels found from the P values from the $\gamma\gamma$ coincidence measurements with the primary transitions in ^{168}Er.

Table 2

E_i(keV) (new level)	$E_{\gamma p}$(keV)	P (%)
2302.58(19)	5468.75(19)	34(17)
2601.44(8)	5169.94(18)	11(5)

E_i(keV) (new level)	$E_{\gamma p}$(keV)	P (%)
2651.58(9)	5119.46(24)	21(8)
2656.85(8)	5114.64(35)	40(9)
2683.55(15)	5087.59(24)	35(13)
2689.27(15)	5082.35(31)	119(30)
2727.50(15)	5043.48(23)	20(10)
2746.25(10)	5024.79(30)	14(7)
2810.75(15)	4960.45(34)	14(8)
2844.20(15)	4928.10(95)	large
2933.20(21)	4838.13(24)	17(4)
2969.80(12)	4801.68(24)	15(5)
2984.35(25)	4787.32(23)	30(10)

1. W.F.Davidson, D.D.Warner, R.F.Casten, K.Schreckenbach, H.G.Börner, J.Simić, M.Stojanović, M.Bogdanović, S.Koički, W.Gelletly, G.B.Orr, M.L.Stelts
   J.Phys. G: Nucl. Phys. 7(1981) 455 and 843

2. W.F.Davidson, W.R.Dixon, R.S.Storey
   Can.J.Phys. 62 (1984) 1538

3. U.Mayerhofer, T.Von Egidy, G.Hlawatsch, J.Klora, H.Linder
   Jnst. Phys. Conf. Ser. No 88, p. S137

STUDY OF ODD-ODD ^{196}Au VIA THE (p,d) REACTION

G. Rotbard, G. Berrier, M. Vergnes, J.M. Maison, S. Fortier,
J. Vernotte, J. Kalifa, L. Rosier
Institut de Physique Nucléaire, F-91406 Orsay, Cedex, FRANCE

P. Van Isacker
Daresbury Laboratory, Daresbury, Warrington WA4 4AD, U-K

ABSTRACT

^{196}Au has been studied with an energy resolution of 12 keV via the ^{197}Au (p,d) ^{196}Au reaction, using a 25 MeV proton beam from the Orsay MP tandem and the split pole spectrometer. The angular distributions, measured at 12 angles, permit to clearly characterize the ℓ = 1,3 and 6 transfers. Data concerning the energies, ℓ values and absolute spectroscopic factors have been obtained ; a comparison is made, for negative parity levels, between these results and theoretical energies, spins and strengths calculated in the framework of the extended supersymmetry. A preliminary level scheme is proposed for ^{196}Au and a few tentative spin attributions are suggested.

INTRODUCTION

The supersymmetry (Susy) concept has been applied in Nuclear Physics with the goal to treat - in a common framework - both the even-even and the odd-A nuclei, described respectively by a boson core - SU(3), O(6) or U(5) - and by a fermion coupled to the core. The success of this approach has encouraged theorists[1] to further extend the model (extended supersymmetry : E. Susy) to also treat the odd-odd nuclei (core plus two fermions). The Susy model has been particularly successful in the Pt region, ^{196}Pt being considered[2] as the "best O(6) nucleus", and the odd-A nuclei described by the group $U_\pi(6/4)$ for the odd-Z and by the group $U_\nu(6/12)$ for the odd-N. One of the "quartets" of nuclei expected[1,3] to be well described by the E. Susy model ($U_\nu(6/12) \times U_\pi(6/4)$ with an O(6) core) consists of the nuclei ^{194}Pt, ^{195}Pt, ^{195}Au and ^{196}Au. The level scheme of the odd-odd ^{196}Au being rather badly known[4], it was first necessary to get as much data as possible.

EXPERIMENTAL RESULTS

^{196}Au has been studied via the ^{197}Au(p,d)^{196}Au reaction at 25 MeV. The overall energy resolution, with the split pole spectrometer and a thin gold target, was 12 keV (F.W.H.M). The angular distributions, measured at 12 angles between 5° and 60°, permit to clearly differentiate and characterize the ℓ=1, 3 and 6 transfers (see Fig. 2).

Below 600 keV, we have observed 25 levels : 9 with ℓ=1, 4 with ℓ=3, 5 with ℓ=1+3, 5 with ℓ=6 and 1 with ℓ=6+(1), for which absolute spectroscopic factors, C²S, have been extracted by comparison with our previous Pt results[5]. It is striking that

Study of Odd-Odd ^{196}Au via the (p,d) Reaction

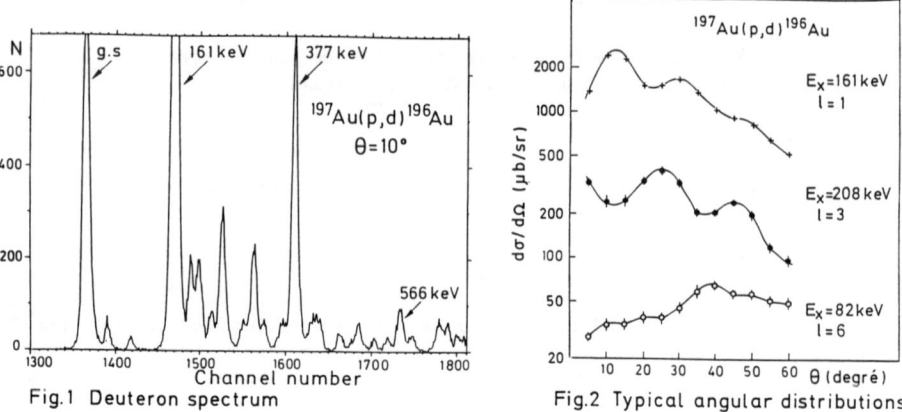

Fig.1 Deuteron spectrum

Fig.2 Typical angular distributions

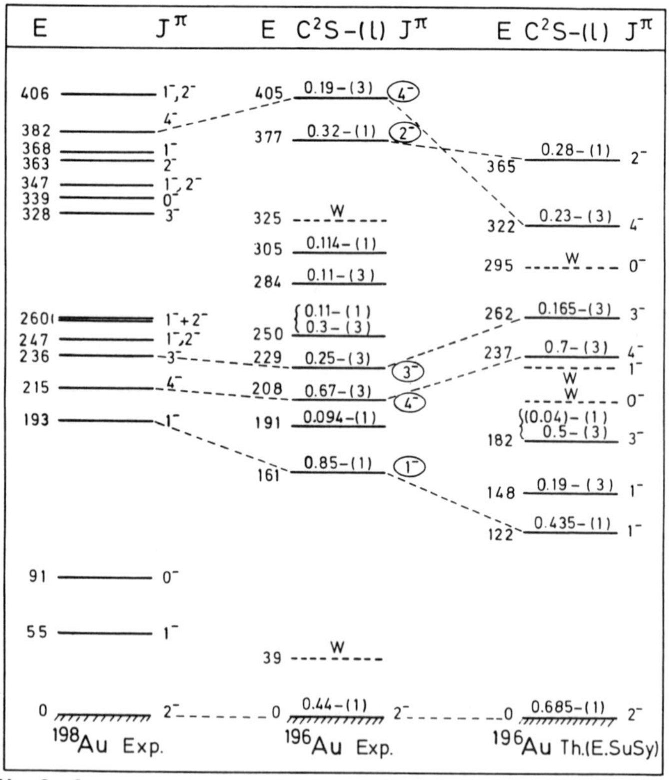

Fig.3 Comparison of the present results with the known data for ^{198}Au and with E.SuSy calculations (see text).

a very recent study[6] of the (d,t) and (^{3}He,α) reactions, at only a few angles but with a very good resolution (6.7 keV F.W.H.M), has shown only 2 additional (weakly populated) levels in the same energy range.

COMPARISON WITH THE E. SUSY MODEL AND WITH ^{198}Au LEVELS

The theoretical energy spectrum of ^{196}Au is obtained from a fit to levels in ^{194}Pt, ^{195}Pt and ^{195}Au and a subsequent adjustment of some parameters guided by the ^{198}Au calculation [7]. Preliminary values of the spectroscopic factors (ℓ = 1 or 3 in the model) for the reaction studied have been computed, using occupation probabilities chosen in ref. 8 for the Susy calculation of the ^{196}Pt $\to$ ^{195}Pt p.u reaction. The summed strengths (Σ C^{2}S) compare well with experiment :

For ℓ=1 : E. Susy : 2.25, Exp. : 2.14
For ℓ=3 : E. Susy : 2.06, Exp. : 1.84

To be able to make a more detailed comparison (see Fig. 3 for levels with E < 420 keV), we have selected - among the experimental and theoretical levels - only those having large spectroscopic factors (C^{2}S > 0.09), such levels carrying about 90 % of the strength. Weakly populated levels are shown as dotted lines.

For some of the strongly populated levels of ^{196}Au we feel that it is possible (using similarities in energies and in the nature (ℓ=1 or 3) and strength, of the transfer) to establish tentative but reasonable correspondences - shown as dashed lines - between experimental and theoretical levels of ^{196}Au. Additional support comes in several cases from similarities in energies between levels in ^{196}Au and ^{198}Au (ref. 7). On such basis, we suggest (circles in Fig. 3) : J^π = 1$^-$ for the 161 keV level of ^{196}Au, J^π = 2$^-$ for the 377 keV level, J^π = 3$^-$ for the 229 keV level and J^π = 4$^-$ for the 208 keV and 405 keV levels. It is clearly difficult, with only the present data, to seriously test the E. Susy. However, the total experimental strengths (ℓ=1 and 3) agree reasonably with theory and Fig. 3 shows even some agreement for individual levels.

REFERENCES

1. P. Van Isacker, J. Jolie, K. Heyde and A. Frank, Phys. Rev. Letters 54 (1985) 653.
   P. Van Isacker, J. Math. Phys. 28 (4), April 1987.
2. J.A. Cizewski, R.F. Casten, G.J. Smith, M.L. Stelts, W.R. Kane, H.G. Börner and W.F. Davidson, Phys. Rev. Letters 40 (1978) 167 ; Nucl. Phys. A323 (1979) 349.
3. J. Jolie, private communication.
4. J. Halperin, Nuclear Data Sheets 28 (1979) 485.
5. G. Berrier-Ronsin, M. Vergnes, G. Rotbard, J. Vernotte, J. Kalifa, R. Seltz and H.L. Sharma, Phys. Rev. C17 (1978) 529.
6. J. Jolie, U. Mayerhofer, T. Von Egidy, H. Hiller, J. Klora, H. Lindner and H. Trieb, submitted to Phys. Rev. Letters.
7. D.D. Warner, R.F. Casten and A. Frank, Phys. Letters B180 (1986) 207.
8. M. Vergnes, G. Berrier-Ronsin and R. Bijker, Phys. Rev. C28 (1983) 360 ; M. Vergnes, G. Berrier-Ronsin and G. Rotbard, Phys. Rev. C36 (1987) 1218.

THE LEVEL SCHEME OF ^{198}Au STUDIED WITH (d,p), (n,γ) AND (n,e$^-$) REACTIONS AND THE LEVEL SCHEME OF ^{196}Au STUDIED WITH (d,t) AND (^{3}He,α) REACTIONS

U.Mayerhofer, T.von Egidy, H.Lindner, H.Hiller, J.Klora, H.Trieb, A.Walter, Physics Department, Technical University Munich, 8046 Garching, Germany

The description of odd-odd nuclei with the extension of supersymmetry was recently proposed in theoretical publications[1]. This scheme starts with the descripiton of an even-even nucleus with the IBM. Supersymmetry describes simultaneously odd isotope and isotone neighbours of an even-even nucleus within the same mathematical framework. Extended supersymmetry contains then in addition to these three nuclei as a fourth member the corresponding odd-odd nucleus.

With the high resolution Munich Q3D spectrograph the ^{197}Au(d,p)^{198}Au reaction was measured with a resolution of 5 keV FWHM. At the ILL with the crystall spectrometer GAMS1 and GAMS2/3 the (n,γ) ^{198}Au reaction was studied to obtain high precision gamma energies. The conversion electrons of ^{198}Au were measured with the electron spectrometer BILL at the ILL to receive information about the multipolarities of these gamma transitions. The results of these experiments[2] allow to extend the existing level scheme of ^{198}Au up to an excitation energy of 1.56 MeV. The level scheme consists now of 125 levels, most of them with known spin and parity. The experimental results are compared with the calculations in the framework of the IBFFM[3] and the extended supersymmetry[4]. In figure 3 the experimental levelscheme up tp 800 keV excitation energy is compared with the IBFFM energy spectrum, in figure 1 a) for the negativ parity levels the experimental transfer intensities are compared with the intensities calculated with the IBFFM. Two regions in the chart of the nuclides were proposed to test the extended supersymmetry: the Ir-Au and the W-Os-Pt regions. The best test proposed is the quartett with the odd-odd nucleus ^{196}Au. Because there existed almost no experimental information on ^{196}Au, the level scheme of ^{196}Au was studied with the ^{197}Au(d,t) (see fig. 2) and ^{197}Au(^{3}He,α) reactions at the Munich Tandem Accelerator. A resolution of 6 keV FWHM was obtained. Up to an excitation energy of 950 keV 49 levels could be established with information on spins. These data are compared with the recently predicted results of the extended supersymmetry theory[5]. The experimental transfer intensities for levels of ^{196}Au with negativ parity are compared with the intensities calculated with the SuSy in figure 1 b).
This work is supported by the BMFT, Bonn.

REFERENCES

[1] P.Van Isacker et.al. Phys.Rev.Lett.54 (1985) 653
[2] U. Mayerhofer, to be published
[3] A.B. Balantekin, V. Paar Phys.Rev.C34(1986)
[4] J. Jolie, contributed to this conference
[5] J. Jolie, et.al, submitted to Phys.Rev.C

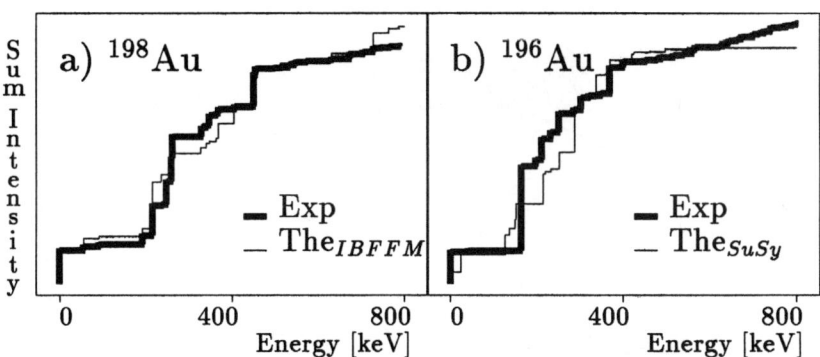

Figure 1: Experimental transfer data for the negativ parity states ($E_{exc} < 800 keV$) in comparison with the calculated data. a) for ^{198}Au and IBFFM, b) for ^{196}Au and SuSy. Sumintensity in arbitrary units.

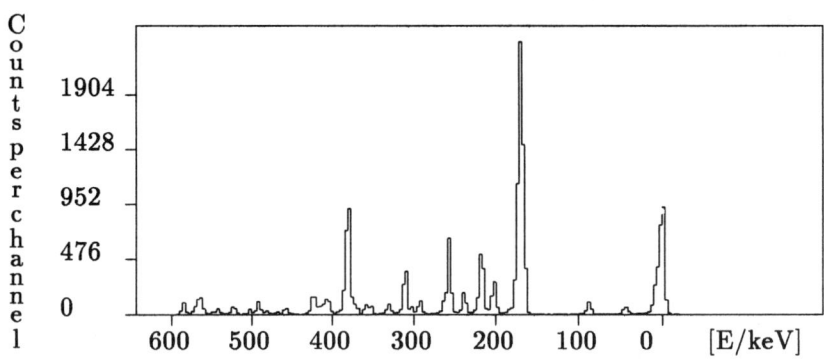

Figure 2: Energy spectrum of ^{197}Au(d,t)^{196}Au, $\Theta_{Lab} = 30°$, FWHM = 6 keV

466 The Level Scheme of ^{198}Au and ^{196}Au

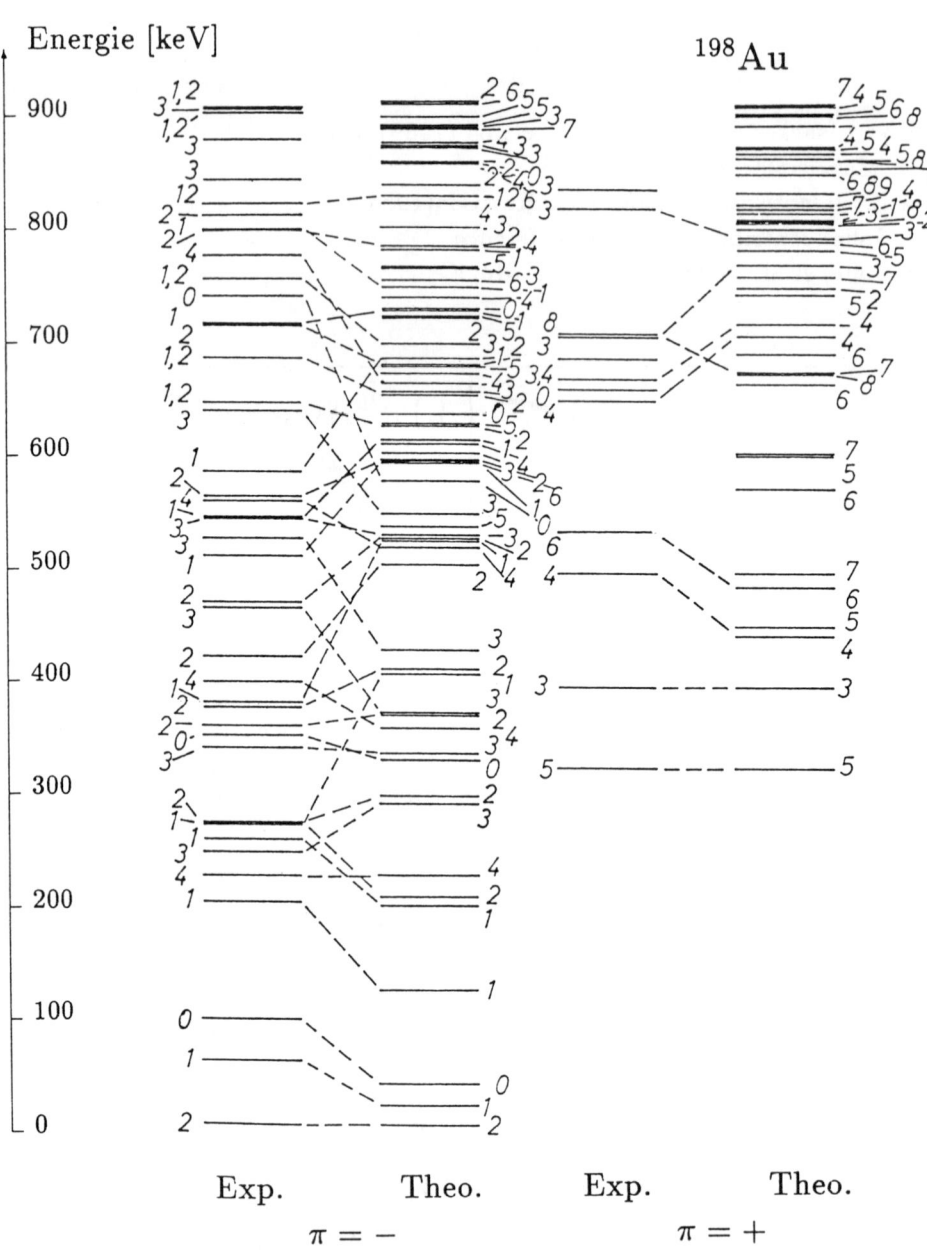

Figure 3: IBFFM energy spectrum for ^{198}Au in comparison with the experimental levels.

HIGH RESOLUTION SPECTROSCOPY USING TRANSFER REACTIONS

H. Lindner, T. von Egidy, H. Hiller, J. Klora, U. Mayerhofer,
H. Trieb, A. Walter
Physik-Department, Technische Universität München, D-8046 Garching,
Germany

ABSTRACT

With a recently developped focal plane detector for the Q3D-spectrograph at the Munich Tandem Accelerator the Q-values of the reactions $^{205}Tl(d,t)^{204}Tl$, $^{206}Pb(d,t)^{205}Pb$ and $^{205}Tl(^{3}He,d)^{206}Pb$ were measured. With this data the Q-value of the $^{205}Tl(ec)^{206}Pb$ reaction was determined to be 51.4(6) keV.
In order to obtain more complete experimental data on ^{104}Rh the $^{103}Rh(d,p)^{104}Rh$, $^{106}Pd(d,\alpha)^{104}Rh$ and $^{105}Pd(d,^{3}He)^{104}Rh$ reactions were measured. The odd-odd nucleus ^{104}Rh has many low lying energy levels because of the neutron-proton coupling. In comparison with $^{103}Rh(n,\gamma)^{104}Rh$ data new energy levels were found and the angular momentum transfer was determined.

INTRODUCTION

The low lying energy levels of nuclei are frequently well known and identified as single particle or collective excitations. At higher excitation energies (0.5 - 3 MeV) there is a lack of information from both theory and experiment. But especially in this energy region there are many questions of great importance for the understanding of nuclei (mixing of single particle and collective excitations, pairing at higher excitation energies, coupling of neutrons and protons, new M1-modes, energy level densities of nuclei, chaotic or regular behaviour).
The combination of the non selective (n,γ)-reaction with selective transfer reactions yields up to a certain excitation energy a complete set of energy levels with spin, parity and structure information. At the Munich Tandem Accelerator transfer reactions and nuclear spectroscopy investigations are performed with the high resolution Q3D-magnetic spectrograph and a new focal plane detector.

THE DETECTOR

For the Q3D-spectrograph at the Munich Tandem Accelerator a new focal plane detector was recently built. In the gas-filled detector avalanches around the anode wires parallel to the focal plane cause by electrical influence signals on several adjacent cathode strips. Each of these strip signals is amplified, shaped and converted into a digital word by an individual fast ADC. In the new read-out system an ASIC (Application Specific Integrated Circuit) determines within 1 μsec the multiplicity and the number of the first strip seeing the avalanche. With these informations a transputer - based unit calculates the center of the avalanche by a Gaussian fit, considering the non-central cathode strip signals by a two-step Newton-approach. The whole operation lasts about 20 μsec.[1,2,3]

MEASUREMENT OF Q-VALUES

In the discussion about the solar neutrino problem the reaction $^{205}Tl(\nu_e,e^-)^{205m}Pb$ is important. The resulting ^{205m}Pb has a half live of $1.48 \cdot 10^7$ a. There seems to be a chance to determine the integral solar neutrino flux over the last 10^7 a from the ^{205}Pb-^{205}Tl-ratio in geological formations. A precise ^{205}Pb-^{205}Tl mass difference is relevant for the evaluation of this process.[4]

In order to determine the Q-value between ^{205}Pb and ^{205}Tl, the Q-values of the reactions $^{205}Tl(d,t)^{204}Tl$, $^{206}Pb(d,t)^{205}Pb$ and $^{205}Tl(^3He,d)^{206}Pb$ have been measured and found to be -1288.7(6) keV, -1831.2(5) keV and 1761.7(14) keV, respectively. The energies of these reactions were calibrated with the $^{158}Gd(d,t)^{157}Gd$, $^{172}Yb(d,t)^{171}Yb$, $^{184}W(d,t)^{183}W$ and $^{187}Re(^3He,d)^{188}Os$ reactions. The decay energy of ^{205}Pb was deduced to be 51.3(6) keV.[5]

THE NUCLEUS ^{104}Rh

The odd-odd nucleus ^{104}Rh with its unpaired neutron and proton offers a great variety of n-p-couplings. This leads to many low lying energy levels and demands a high resolution apparatus. In order to extend the level scheme up to higher excitation energies, the $^{103}Rh(d,p)^{104}Rh$ reaction and as an attempt the $^{106}Pd(d,\alpha)^{104}Rh$ and $^{105}Pd(d,^3He)^{104}Rh$ reactions were measured. In comparison with $^{103}Rh(n,\gamma)^{104}Rh$ data [6] new energy levels were found. We measured 91 energy levels up to 1785 keV excitation energy. The transfered angular momentum in the (d,p) reaction was determined for all levels.

CONCLUSIONS

In this paper we want to point out that the transfer reactions in the nuclear spectroscopy are a powerful tool to investigate nuclear structure and to establish more complete level schemes. Thereby a very precise detector arrangement is necessary.

This work is supported by the BMFT, Bonn.

REFERENCES

1. H. Lindner, H. Angerer and G. Hlawatsch, Nucl. Instr. and Meth. A 273 (1988) 444
2. H. Hiller, Diploma Thesis, TU München 1989
3. H. Trieb, Diploma Thesis, TU München 1990
4. M.S. Freedman, C.M. Stevens, E.P. Horwitz, L.H. Fuchs, J.L. Lerner, L.S. Goodman, W.J. Childs and J. Hessler, Science 193 (1976) 1117
5. H. Lindner, H. Trieb, T. von Egidy, H. Hiller, J. Klora, U. Mayerhofer and A. Walter, Nucl. Instr. and Meth. A, 1990 in print
6. I.A. Kondurov, T.D. Mc Mahon et al., private communication

HIGH RESOLUTION STUDY OF THE ^{222}Ra EXOTIC DECAY

M. Hussonnois, J.F. Le Du, L. Brillard
Institut de Physique Nucléaire, B.P. 1, 91406 Orsay-Cédex, France

J. Dalmasso, G. Ardisson
Laboratoire de Radiochimie, Université de Nice, 06034 Nice, France

Abstract

The ^{14}C-decay of ^{222}Ra has been reinvestigated using a strong ^{230}U source and the magnetic superconducting spectrometer SOLENO in view of looking for a possible feeding of the first ^{208}Pb excited state. 210 ^{14}C-events were recorded in a single peak of (30.930±0.090) MeV energy; the ^{14}C branching ratio values b = (2.31±0.31)×10^{-10} and < 1×10^{-12} to the respective ^{208}Pb ground and I$^\pi$ = 3$^-$ octupole state were deduced.

Introduction

The recent discovery in the ^{223}Ra ^{14}C-decay of favored transitions to the ^{209}Pb first excited states[1] has revived the analogy between this exotic decay and α-emission, in particular concerning the preformation mechanism[2].

Nevertheless, the calculation of the hindrance factors (H.F.) for the ^{14}C emissions from ^{223}Ra decay supposes that ^{14}C transition of even Ra nuclei to ground states of Pb daughter nuclei have HF=1 value and assumes as ^{14}C ground-state partial half-lives the total half-lives, still only measured. Hence, it seemed us essential to measure the ^{222}Ra ^{14}C spectrum then to test the possible influence of the ^{222}Ra octupole deformation in the feeding of the I$^\pi$=3$^-$ state of the ^{208}Pb doubly closed shell nucleus.

Experimental Details

The ^{222}Ra ^{14}C-decay has been reinvestigated by means of the decay chain ^{230}U→^{226}Th→^{222}Ra. With this end in view, a strong ^{230}Pa source of about 1.8 GBq activity has been prepared by proton irradiation (Ep=34 Mev) of a thick metallic Th target at the CERI isochronous cyclotron. A multistep radiochemical separation was performed based on organic ion exchanger chromatography[3], to isolate ^{230}Pa; after a one month growing time the highly purified 85.1 MBq ^{230}U source was eluted and electroplated on a Pt backing.

The superconducting magnetic spectrometer SOLENO, at Orsay, has been used to focus the ^{14}C^{6+} ions on a good resolution Si(Au). detector (F.W.H.M. = 20 keV on 6.01 MeV α-line) of 400 mm^2 area. Amplified pulses, in the range 100 keV to 40 MeV, were simultaneously recorded and analysed with two 4096 channels multichannel analysers.

Measurements and Discussion

Three runs were performed with two ^{230}U of 22 and 85.1 MBq. The SOLENO spectrometer was calibrated in energy by means a 23 MBq activity ^{227}Ac source, in equilibrium with daughters (^{227}Th-^{223}Ra), wich was counted during 4.1 days. A total of 43 ^{14}C events was shared into 3 peaks, as previously observed[1], with relative intensities 11%, 85% and 4%, assigned to the respective feeding of the ground, first and second excited states of ^{209}Pb. The calibration line was fitted using the calculated ^{14}C kinetic energies, from Audi and Wapstra Tables[4] as well as the know ^{209}Pb level scheme. A first experiment, for 2.8 days was performed with the 22,2 MBq ^{230}U source and the same amplifier gain, in which 18 ^{14}C events of (30.930±0.090) MeV energy were recorded, in good agreement with the expected energy (30.969±0.008) MeV. The 85.1 MBq ^{230}U intense source was used, in a second experiment, ajusting conveniently the current setting of SOLENO, in an attempt to observe the ^{14}C feeding to the ^{208}Pb first excited state at 2.614 MeV ($I^\pi = 3^-$). Two successive runs, for 6 and 9 days, were performed with two different amplifier gain in wich 104 and 106 events were respectively recorded. The summed spectrum (Fig 2) exhibits that all ^{14}C events but one are grouped in a 30.93 MeV energy single peak of F.W.H.M = 200 keV.

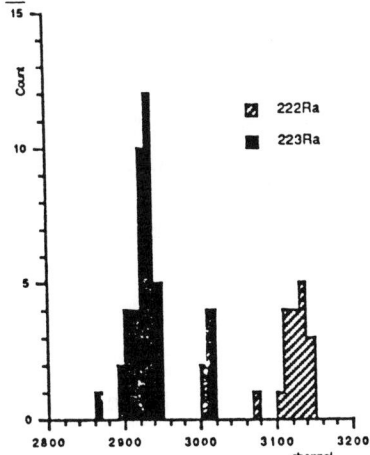

Fig. 1: ^{14}C spectrum from ^{222}Ra and ^{223}Ra

The energy value 28.83±0.1 MeV of the single event observed seems too far to the expexted value 28.519+0.008 MeV assigned to the ^{14}C transition to the $I^\pi = 3^-$ state of ^{208}Pb.

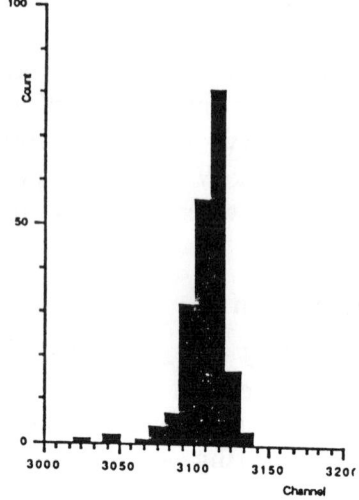

Fig. 2 : ^{14}C sum spectrum from ^{222}Ra

The branching ratio, $b = \lambda_{14C}/\lambda_\alpha$, was deduced from the total number of ^{14}C events number N_{14C}, the total number of α-particules emitted by ^{222}Ra N_α, the effective solid angle $\Delta\Omega$ and the fraction ρ of the $q=6^+$ charge state in the charge distribution according to

$$b = \frac{4\pi . N_{14C}}{N_\alpha . \Delta\Omega . \rho}$$

The value $b=(2.31\pm0.31)\times10^{-10}$ deduced here is lower than the values $(3.7\pm0.5)\times10^{-10}$ measured by Price et al[5], as well as that of Hourani et al[6], i.e. $(3.1\pm1.0)\times10^{-10}$, although it overlaps the last value at the confidence level 1 σ. The upper limit of 1×10^{-12} (HF<200) deduced for the ^{14}C branching ratio to the ^{208}Pb first excited state is not in disagreement with the HF=10^5 expected value[7].

References

1. L. Brillard, A.G. Elayi, E. Hourani, M. Hussonnois, J.F. Le DU, L.H. Rosier and L. Stab, C.R. Acad. Sci. Paris, 309, ser. II, 1105(1989).
2. M. Hussonnois, J.F. Le Du, L. Brillard and G. Ardisson, J. Phys. G: Nucl. Part. Phys., 16, 177 (1990).
   ibid., Phys. Rev., 42 C, R495 (1990).
3. M. Hussonnois et al., to be published.
4. A.H. Wapstra and G. Audi, Nucl. Phys., A 432, 1 (1985).
5. P.B. Price, J.D. Stevenson, S.W. Barwick and H.L. Ravn, Phys. Rev. Lett. 54, 297 (1985).
6. E. Hourani, M. Hussonnois, L. Stab, S. Gales and J.P. Shapira, Phys. Lett. B, 160, 375 (1985).
7. S. Landowne and C.H. Dasso, Phys. Rev., C 33, 387 (1986).

HIGH-SPIN STATES OF ^{238}Pu USING A HEAVY-ION ONE-NEUTRON TRANSFER REACTION

M.A. Stoyer, J.O. Rasmussen, A.A. Shihab-Eldin,
Lawrence Berkeley Laboratory, University of California,
Berkeley, CA 94720, USA

D. Cline, K. Helmer, A.E. Kavka, W.J. Kernan, B. Kotlinski,
E. Vogt, C.Y. Wu,
Nuclear Structure Research Laboratory, University of
Rochester, Rochester, NY 14627, USA

C. Bingham, M.W. Guidry, X.L. Han, R.W. Kincaid, X.T. Liu,
H. Schechter,
Department of Physics and Astronomy, University of
Tennessee, Knoxville, TN 37996-1200, USA

M.L. Halbert and D. Hensley
Oak Ridge National Laboratory, Oak Ridge, TN 37830

ABSTRACT

The heavy-ion one-neutron pickup reaction ^{239}Pu(^{90}Zr,^{91}Zr)^{238}Pu and the rotationally inelastic excitation reaction ^{239}Pu(^{90}Zr,^{90}Zr$'$)^{239}Pu$'$ (E_{lab} = 500 MeV) were used to study the rotational level schemes of ^{238}Pu and ^{239}Pu, respectively. Spectroscopic information on ^{238}Pu is presented and discussed.

INTRODUCTION

Coulomb excitation[1] and (α,xn) reactions[2] have yielded much information on high-spin states in actinide nuclei. Because (HI,xn) reactions populate states with higher excitation energy, such reactions are not suitable in the actinide region due to fission competition. Recently, the one-neutron transfer reaction has been used[3] to study the ground rotational band in ^{234}U. The heavy-ion one-neutron pickup reaction ^{239}Pu(^{90}Zr,^{91}Zr)^{238}Pu (E_{lab} = 500 MeV) has been used to study the rotational states in ^{238}Pu in this work.

EXPERIMENTAL DETAILS

This experiment was performed at the Holifield Heavy Ion Research Facility at Oak Ridge National Laboratory in the Spin Spectrometer setup. Inside the Spin Spectrometer, position-sensitive parallel-plate avalanche counters were used to detect backscattered Zr-like ions. Scattering kinematics enabled a reconstruction of the event to determine the scattering angle of the Pu-like fragment for the Doppler corrections. Eighteen NaI(Tl) detectors of the Spin Spectrometer were replaced with BGO or NaI Compton suppressed intrinsic Ge detectors. Particle-γ coincidence events were recorded for subsequent analysis. A thin ^{239}Pu target ($\sim$ 400μg/cm^2) was used to minimize target energy-loss effects in the scattering kinematics.

© 1991 American Institute of Physics

RESULTS

The brevity of this article precludes a complete presentation of the results; however, the main features will be discussed. Figure 1 shows a Ge γ-ray spectrum gated on three previously known ^{238}Pu transitions, and the ground rotational band is identified as high as the 26$^+$ level. In addition, four "mystery" transitions are observed in Figure 1. An examination of the γ-γ coincidence data reveals that they are most likely an excited side-band in ^{238}Pu.

The probability for neutron transfer as a function of the distance of closest approach follows a simple WKB barrier penetration model for the range of distances investigated in this experiment. The kinematic and dynamic moments-of-inertia are gradually increasing with increasing spin; there is no evidence for a sharp backbend.

The (E, M) plot (total energy, total multiplicity of the reaction) gated on one-neutron transfer exhibits one peak near the Yrast-line, indicating the reaction is most likely dominated by one-neutron pickup of the odd-neutron in the $\frac{1}{2}^+$[631] orbital. Figure 2 presents the rotational population pattern for ^{238}Pu and compares it with theory. The rotational population pattern has a minimum in the population probability at spin 14$\hbar$. Theoretical calculations using a modified Alder-Winther-deBoer method[4], exact details to be published elsewhere[5], also indicate a dip in the probability around spin 12 − 14$\hbar$.

ACKNOWLEDGEMENTS

This work was supported in part by the Director, Office of Energy Research, Division of Nuclear Physics of the Office of High Energy and Nuclear Physics of the U.S. Department of Energy under Contract DE-AC03-76SF00098. We are most appreciative of the essential support of the technical staff at the Holifield Heavy Ion Research Facility, Oak Ridge National Laboratory.

REFERENCES

[1] T. Czosynka, D. Cline, L. Hasselgren, C.Y. Wu, R.M. Diamond, H. Kluge, C. Roulet, E.K. Hulet, R.W. Lougheed and C. Baktash **Nucl. Phys. A458** (1986) 123.

[2] K. Hardt, P. Schüler, C. Günther, J. Recht and K.P. Blume **Z. Phys. A314** (1983) 83.

[3] C.Y. Wu, X.T. Liu, S.P. Sorensen, R.W. Kincaid, M.W. Guidry, D. Cline, W.J. Kernan, E. Vogt, T. Czosynka, A.E. Kavka, M.A. Stoyer, J.O. Rasmussen and M.L. Halbert **Phys. Lett. 188B** (1987) 25.

[4] L.F. Canto, R.J. Donangelo, J.O. Rasmussen, P. Ring and M.A. Stoyer in press **Phys. Lett. B** (1990).

[5] J.O. Rasmussen, M.A. Stoyer, L.F. Canto, R. Donangelo and P. Ring to be published **Z. Phys.** (1990).

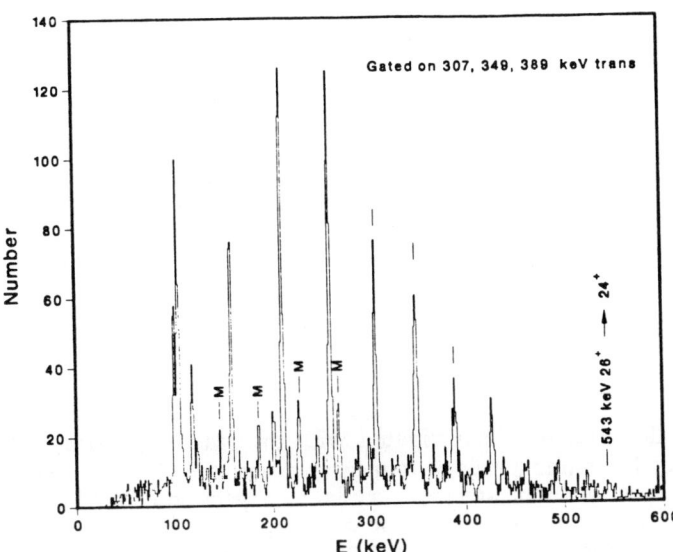

Figure 1: Ge γ-ray spectrum gated on the 307 keV $12^+ \to 10^+$, 349 keV $14^+ \to 12^+$ and 389 keV $16^+ \to 14^+$ ^{238}Pu transitions (indicated by dashes). The "mystery" transitions are marked with "M's", and the highest transition in the ground band of ^{238}Pu is noted.

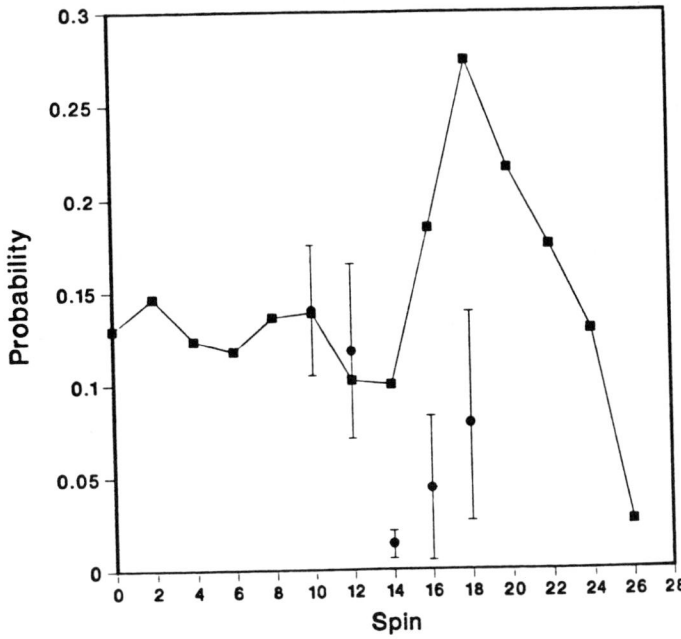

Figure 2: Probability for populating a given spin in ^{238}Pu in the one-neutron pickup reaction (•) compared with a modified Alder-Winther- deBoer calculation (solid blocks).

NUCLEAR STRUCTURE OF ^{241}Pu FROM (n,γ) AND (n,e) REACTION MEASUREMENTS*

D.H. White**, R.W. Hoff
Lawrence Livermore National Laboratory
Livermore, CA 94550, USA

and

H.G. Boerner, G. Colvin, F. Hoyler, AND K. Schreckenbach
Institut Laue-Langevin
Grenoble, France

ABSTRACT

The nuclide ^{241}Pu was investigated with the reactions ^{240}Pu(n,γ)^{241}Pu and ^{240}Pu(n,e)^{241}Pu. The gamma rays were measured with the GAMS 1 and GAMS 2/3 curved crystal spectrometers, and the 3-crystal pair spectrometer at the Institut Laue-Langevin (ILL) High-Flux Reactor. The conversion electrons were measured with the BILL magnetic spectrometer. Gamma-rays attributed to de-exitation of ^{241}Pu were identified against an intense background of fission gammas by comparison of growth patterns. A total of 45 primary and 151 secondary gamma rays are tentatively identified with de-exitation of ^{241}Pu, of which 22 primary and 70 secondary have been placed on a level scheme consisting of 22 levels below 1 MeV. The conversion electrons have aided in the assignment of multipolarity to the transitions. Neutron binding energy was determined to be 5241.57± 0.20 keV.

INTRODUCTION

Secondary γ rays following thermal neutron capture in ^{240}Pu were measured using the curved-crystal spectrometers GAMS1 and GAMS2/3 at the Institut Laue-Langevin (ILL) in Grenoble. Two targets were used. One run scanned upward in energy, and the other scanned downward. Each target consisted of around 4 mg ^{240}PuO$_2$ (enriched to 99.46%) in the form of a wafer of height 4.65 cm and width 0.57 cm. The targets were situated in a flux of 5.5 x 10^{14} neutrons/cm^2-s. The GAMS1 spectrometer was used to study γ rays in the energy range 35-500 keV and the GAMS2/3 spectrometer was used to cover the range 150-1500 keV. Comparison of the two runs allowed us to identify the Pu-241 peaks among the intense background of fission (and multiple capture) gamma-rays using growth systematics.

Primary gamma rays were measured in the energy range 3600-5100 keV using the three-crystal pair spectrometer, consisting of a Ge(Li) detector, flanked by two NaI(Tl) counters for coincident detection of the 511 keV annihilation radiation. This spectrum was calibrated with reference to the γ-rays resulting from capture in the carbon and aluminum present in the target.

---

*Work performed in part under the auspices of the US Department of Energy by the Lawrence Livermore National Laboratory, contract #W-7405-ENG-48.

**Present address: Western Oregon State College, Monmouth, OR 97361

The conversion electrons following thermal neutron capture in ^{240}Pu have been studied with the BILL magnetic spectrometer at ILL. A thin target (32 µg/cm^2) was used for the energy range 18-100 keV, and a thicker target (207 µg/cm^2) was used for the energy range 100-1500 keV.

Multipolarities of strong transitions were determined by L-subshell ratios, and relative intensity calibration between (n,γ) and (n,e) measurements was determined by comparison with K,L, and M conversion coefficients. These comparisons also permitted correction for target self-absorption, a feature increasingly important in both the gamma-ray and electron spectrum, below 150 keV.

CONCLUSION

We have constructed a level scheme consisting of 22 levels below 1000 keV, using direct population of high-energy primary γ-rays, Ritz combinations of secondary γ-rays, and consideration of previous work on (α,γ), (β,γ), (n,γ), and charged-particle reactions. A total of 45 primary and 151 secondary γ-rays have been fit to the level scheme using the least squares program LEVEL (See Fig. 1). The resulting neutron binding energy was determined to be 5241.57±0.20 keV.

478 Nuclear Structure of ^{241}Pu

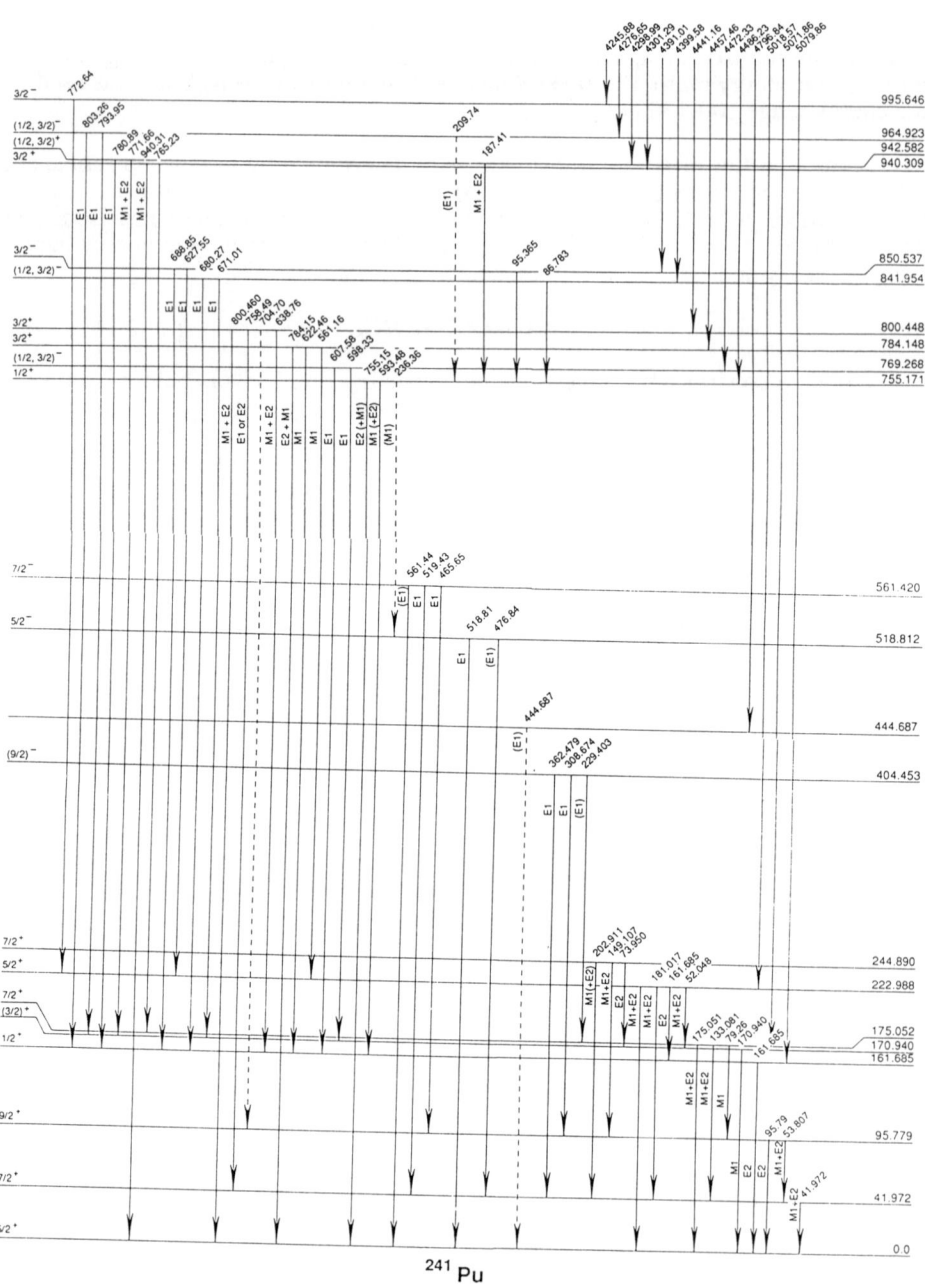

Figure 1.

POLARIZATION MEASUREMENTS IN NUCLEAR RESONANCE FLUORESCENCE EXPERIMENTS

R. D. Heil, A. Degener, H. Friedrichs, A. Jung, U. Kneissl, S. Lindenstruth,
J. Margraf, H. H. Pitz, H. Schacht, B. Schlitt, U. Seemann, R. Stock, C. Wesselborg
Institut für Kernphysik, Universität Giessen, D-6300 Giessen

P. v. Brentano, A. Zilges
Institut für Kernphysik, Universität zu Köln, D-5000 Köln

G. Müller, M. Schumacher
2. Physikalisches Institut, Universität Göttingen, D-3400 Göttingen

ABSTRACT

Systematic photon scattering experiments have been performed over the recent years at the Giessen Electron Linac and Stuttgart Dynamitron in order to investigate low lying dipole transitions in deformed nuclei. With the high resolution and sensitivity of these Nuclear Resonance Fluorescence (NRF) experiments detailed dipole strength distributions in numerous isotopes could be extracted. By using linearly polarized bremsstrahlung or by measuring the linear polarization of the scattered photons the parities of numerous J = 1 states in ^{48}Ti, ^{70,72}Ge, the rare earth nuclei ^{142,150}Nd, ^{162}Dy and in ^{232}Th could be determined in an unambigous, model-independent way.

INTRODUCTION

Recently, there have been experimental indications for the occurence of enhanced electric dipole transitions[1], that had been proposed theoretically[2] due to the excitation of α-cluster modes and/or reflection asymmetric deformations. Since angular distribution measurements in NRF cannot distinguish between E1 and M1 excitations parity assignments are of crucial importance for the interpretation of the observed dipole excitations. This is of particular importance for the identification of the M1 excitations of the so called "scissors mode", discovered in high resolution (e,e') experiments by Bohle et al.[3]. This "weakly collective" isovector mode is closely related to the orbital motion of the protons with respect to the neutrons[4]. It seems to be a rather general phenomenon in deformed nuclei, since it has been found in photon and electron scattering experiments in rare earth, in actinide, and in medium heavy fp shell nuclei[5].

The systematics of the total M1 strength in the nuclei of the rare earth region (^{150}Nd, the Gd, Dy, Er and Yb isotopes) found in the NRF experiments exhibits a pronounced drop near A=165 (see P.v.Brentano, this conference). Up to now the parities of the states in this compilation were inferred from electron scattering form factors or from the energy systematics of the M1 mode by summing up the strength of K=1 states in the energy range of the scissors mode ($E^* = 66 \cdot \delta \cdot A^{-1/3}$). The first method is limited to strong and well isolated excitations while in the second case for $\Delta K=1$ both E1 and M1 transitions are possible. Therefore, direct and unambigous parity assignments are highly desirable.

POLARIZATION EXPERIMENTS

Model independent parity assignments can be achieved in NRF experiments either by using linearly polarized photons in the entrance channel ($\vec{\gamma}, \gamma'$) or by measuring the linear polarization of the scattered photons ($\gamma, \vec{\gamma}'$). With the first technique, which has been applied successfully in numerous photon scattering experiments at the Giessen accelerator using partially linearly polarized off-axis bremsstrahlung[6], the azimuthal asymmetry $\varepsilon(\theta) = (N_\perp - N_\parallel)/(N_\perp + N_\parallel) = P_\gamma \cdot \Sigma(\theta)$ is measured. $N_{\perp,\parallel}$ denote the countrates in planes perpendicular and parallel to the polarization plane defined by the incoming photon beam and its electric field vector. P_γ is the degree of polarization of the photon beam. The analyzing power $\Sigma(\theta)$ has its maxima of +1 for E1 and −1 for M1 and E2 transitions under a scattering angle of $\theta = 90°$ for spin cascades 0-1-0 and 0-2-0. Fig.1 shows as an example the E1 and M1 strength distributions of ^{48}Ti, that could be deduced from the azimuthal asymmetries ε in the $(\vec{\gamma}, \gamma')$ reaction with E_{BS}= 11 MeV at the Giessen facility[7]. The observed low energy M1 strength is in agreement with recent Darmstadt (e,e') data. A pronounced concentration of E1 strength between 6-8 MeV can be stated.

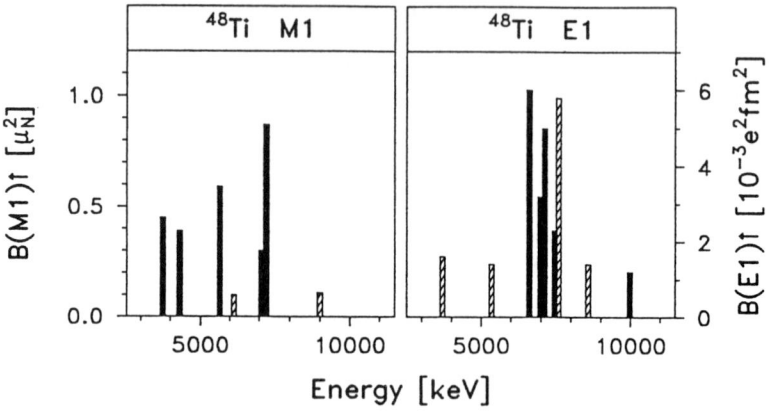

Fig.1. M1 and E1 strength distributions observed in ^{48}Ti$(\vec{\gamma}, \gamma')$ at the Giessen polarized bremsstrahlung beam. Transitions with a tentative parity assignment are shown as shaded bars.

However, this method is restricted to higher excitation energies preferable in lighter nuclei because of experimental limitations. Thus for the low energies (2-3.5 MeV) of the E1 and M1 transitions of interest here, we alternatively measure the linear polarization of the scattered photons using Compton polarimeters. The azimuthal asymmetry $\varepsilon(\theta)$ of photons with polarisation $P_\gamma(\theta)$ measured with a polarimeter of polarization sensitivity Q amounts to $\varepsilon(\theta) = P_\gamma(\theta) \cdot Q(E_\gamma)$. Utilizing the polarization sensitivity of the Compton scattering cross section, which is larger for scattering directions perpendicular to the electric field vector $\vec{E}$ of the incoming photons, one experimentally observes an asymmetry $\varepsilon_{exp} = (N_\perp - N_\parallel)/(N_\perp + N_\parallel)$ with $N_{\perp,\parallel}$ number of Compton scattered photons perpendicular and parallel to the (γ, γ') plane. The polarization $P_\gamma(\theta)$ of the photons is maximal for $\theta = 90°$ and amounts to +1 for M1 and E2 transitions and −1 for E1 transitions.

In our experiments we simultaneously used two different Compton polarimeters consisting either of a sectored Ge detector or a classical 5 detector set up. With a new, improved fourfold sectored Ge(HP) polarimeter the positive parities of two weaker J=1, K=1 states in ^{150}Nd ($E^* = 3096$ keV, B(M1)↑ = 0.13 μ_k^2 ; $E^* = 3103$ keV, B(M1)↑ = 0.13 μ_k^2) in the energy range of the M1 mode could be established[8] in addition to our previous results[9,10]. The summed M1 strength in ^{150}Nd amounts to B(M1)↑= 1.4μ_k^2. Fig.2 shows the asymmetry ε_{exp} measured in a recent ^{162}Dy $(\gamma, \vec{\gamma}')$ experiment. The asymmetries for the unpolarized ^{27}Al transitions used in the mixed target vanish while 5 transitions in ^{162}Dy show a positive asymmetry and thus the corresponding levels have positive parity. The three states at 3 MeV that probably belong to the orbital M1 mode have a summed M1 strength of $B(M1) \uparrow= 2.5\mu_k^2$ according to the M1 strength systematics in Ref.5.

But the most surprising result is the negative parity of the J=1 level at $E^* = 2520$ keV, $(B(E1)\uparrow = 5 \cdot 10^{-3}e^2 fm^2)$. A similar state with negative parity has been found in ^{150}Nd at $E^* = 2414$ keV, $(B(E1)\uparrow = 3 \cdot 10^{-3}e^2 fm^2)$. Both excitations might be due to the octupole or α-cluster mode using reasonable values for the cluster admixture amplitude ($\eta \approx 3\%$). Furthermore, both levels show a K-mixing[11]. The structure of the M1/E1 excitations around 2.5 MeV that also occur in other rare earth nuclei is still uncertain and will be investigated in further (γ, γ') experiments and the model-independent parity determinations obtained from the linear polarization measurements will help towards solving this problem.

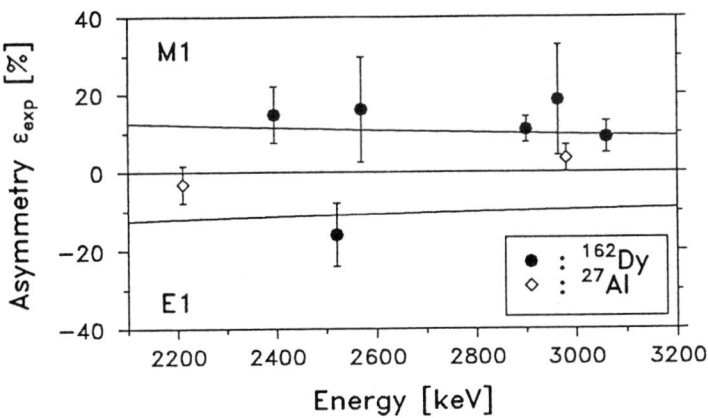

Fig.2. Experimental asymmetries ε_{exp} for ^{162}Dy determined with the fourfold sectored Ge(HP) Compton polarimeter at the Stuttgart Dynamitron. The experimental data are compared with the calculated polarization sensitivity (lines) of the detector for pure dipole transitions.

REFERENCES

1.) M.Gai et al., Phys. Rev. Lett. **50**, 239 (1983).
2.) F.Iachello, Phys. Lett. **B160**, 1 (1985).
3.) D.Bohle et al., Phys. Lett. **B148**, 260 (1984).
4.) A.Richter, Nucl. Phys. **A507**, 99c (1990).
5.) U.Kneissl, Prog. Part. Nucl. Phys. **24**, 41 (1990).
6.) U.E.P.Berg and U.Kneissl, Ann. Rev. Nucl. Part. Sci. **37**, 33 (1987).
7.) A.Degener et al., Nucl. Phys. **A513**, 29 (1990).
8.) H.Friedrichs et al., to be published
9.) B.Kasten et al., Phys. Rev. Lett. **63**, 609 (1989).
10.) R.D.Heil et al., Nucl. Phys. **A506**, 223 (1990).
11.) A.Zilges et al., Phys. Rev. C, in press (1990).

POTENTIALITIES OF A MULTIDETECTOR SYSTEM TO STUDY CASCADE γ DECAY OF A COMPOUND STATE OF COMPLEX NUCLEI

V.A.Khitrov, Yu.P.Popov, A.M.Sukhovoj.
Joint Institute for Nuclear Research, Dubna 141980, USSR.

ABSTRACT

Decades of investigation of the compound-state γ-decay of complex nuclei (A $\geq$100) have not given sufficient information about this process. Traditional ("rough") studies of γ-spectra using best modern detectors cannot provide full information.

So one should look for a new approach to the experimental study of this process.

It seems that the most effective way is to distinguish cascades of two, three and more γ-transitions with total energy being equal to the neutron binding energy B_n or to the corresponding difference between B_n and excitation energy of some low-lying states.

The Laboratory of Neutron Physics, JINR, has been measuring the cascades with energies $E_1+E_2=B_n$ since 1982 [1]. For the present time they have been studied in 15 nuclei (in ^{137}Ba, ^{181}Hf and ^{187}W - jointly with IP AS LatvSSR).

The main information obtained is a decay-scheme of levels excited by cascades with a threshold intensity $i_{\gamma\gamma} \geq 3*10^{-4}$ per decay up to an excitation energy not lower than 3-4 Mev. Very important are also the two-quanta cascade intensities averaged over transition energy (and in some cases[3] - over excitation energy of a nucleus). (First estimates of radiative strength functions of soft primary transitions were evaluated too). These averaged intensities could be easily compared with those calculated in the frame of any of the models of excited states density and partial widths [2,3]. The experiment reveals a large enough portion of the total intensity of all the primary transitions in compound-state decay as illustrated in the table.

Sometimes the sum $i_{\gamma\gamma}+i_\gamma$ (see the table) amounts to 80% of the total radiative width of a compound-state. This fact means that the problem of the complete experimental investigation of γ-decay of several nuclei is near to its practical solution.

Table I. $\Sigma i_{\gamma\gamma}$ is the sum of cascade itensities (% of the total radiative width Γ_γ of a compound-state) N_f - the number of final levels excited by cascades; Σi_γ - the intensity sum of primary transitions to levels N_f.

Compound-nucleus	^{144}Nd	^{146}Nd	^{168}Er	^{174}Yb	^{178}Hf	^{180}Hf
$\Sigma i_{\gamma\gamma}$	50(4)	36(1)	29(2)	21(1)	17(1)	10(1)
N_f	5	4	6	3	8	9
Σi_γ	11	11	1	3	1	1

Compound-nucleus	^{163}Dy	^{165}Dy	^{175}Yb	^{177}Yb	^{179}Hf	^{181}Hf	^{183}W	^{187}W
$\Sigma i_{\gamma\gamma}$	28(1)	54(7)	73(10)	59(5)	67(4)	52(4)	37(2)	44(2)
N_f	7	7	9	3	10	5	5	5
Σi_γ	2	13	10	1	11	28	23	14

The cascade intensity $i_{\gamma\gamma} = i_{\lambda g} * i_{gf} / \Sigma i_{gf}$ is expressed as a product of the primary transition intensity $i_{\lambda g}$ between the compound-state λ and intermediate level g and the branching coefficient of the decay of the level g to final levels f.
It is obvious, that:
a) knowing the energy $E_{\lambda g}$ of a cascade quantum one could find the intensity $i_{\lambda g}$ of a considerable number of primary transitions from the single detector spectrum;
b) detectors of higher (than we had) efficiency could permit one to increase both N_f and $\Sigma i_{\gamma\gamma}$.
As it is seen from the table, even-even and some of even-odd compound-nuclei have a relatively small sum $i_\gamma + i_{\gamma\gamma}$. This is obviously due to the fact that three-, four- and more quanta cascades are competing with direct primary transitions to several low-lying levels and two-quanta cascades to the same levels.
Extrapolation of the experience we have acquired in the study of two-quanta cascades shows that it is highly probable to distinguish cascades from triple coincidences. So a cascade 6522+589+454 keV in ^{146}Nd is registered[4] in total absorption peaks at 6522 and

1043=(589+454) kev by 10% efficiency detectors at a rate of $\simeq 2$ events per 100 hours.

Taking into account that this cascade corresponds to $\simeq 3.5\%$ of decay events and that 30% efficiency detectors could be used in the experiment one could expect to have during the measuring time of 100 hrs the count statistics of up to 2000 events in the total absorption peak of a cascade $E_1+E_2+E_3=B_n$. This level of statistics was achieved in the registration of two-quanta cascades with 10% detectors in individual cases only. The background at registration of three-quanta cascades with total energy $\Sigma E_\gamma = B_n$ cannot be worse than that at registration of two-quanta cascades.

By selecting the last (most intensive) transition from a three-quanta cascade the procedure of studying γ-decay is reduced to the wrought up method of investigation of two-quanta cascades.

So nowadays the real possibility appeared to fully investigate the process of the compoumd-state cascade γ-decay of any nuclei. For that one needs a system of four (or six) detectors of high enough efficiency (not less than 30%) allowing the distinguishing of coincidences of two, three (and desirably four) γ-quanta.

The mathematical support for the experiment to analyse coincidences has been perfected in its main part at LNP, JINR.

REFERENCES

1. A.A.Bogdzel et al., JINR preprint P15-82-706, Dubna, 1982.
2. S.T.Boneva et al.,Z.Phys., A330, 153 (1988).
3. S.T.Boneva et al., JINR preprint E3-90-45, Dubna, 1990.
4. S.T.Boneva et al.,Izv.Akad.Nauk SSSR, Ser.Fiz., 53,2092 (1989)

ON THE CONSTRUCTION OF A COMPLEX GAMMA-DECAY SCHEME ON THE BASIS OF THE SPECTROSCOPIC DATA FROM REACTIONS (n,2γ) and (n,γ).

S.T.Boneva, V.A.Khitrov, Yu.V.Kholnov, V.D.Kulik, Le Hong Khiem, Pham Dinh Khang, Yu.P.Popov, A.M.Sukhovoj, E.V.Vasilieva.
Joint Institute for Nuclear Research, Dubna 141980, USSR.

ABSTRACT

The extensive and complite level scheme (~ 100 levels and 500 transitions) one may determined by the combination of the experimental data from (n,2γ) and (n,γ) reaction.

Experimental investigation of two-step cascades between compound-states of complex nuclei (A ≥140) and a group of their low-lying levels ("reaction (n,2γ)") permits one to reveal an essential portion of intermediate levels with an excitation energy up to 4 MeV. But such experiments are possible only with intensive cascades ($i_{\gamma\gamma} \geq 3\text{-}5*10^{-4}$ per decay). Much more spectroscopic information on the decay-scheme could be obtained by a joint analysis of the spectroscopic data from reactions (n,2γ) and (n,γ).

Then the data processing is to have the following stages:

1. To make more precise the cascade transition energies observed in the reaction (n,2γ) by using precise energies from the reaction (n,γ).

At that for the decomposition of the (n,γ)-spectrum into primary and secondary transitions one has to use the data on cascade transitions from the reaction (n,2γ) as it is more informative than those on γ-transitions from the (n,γ)-reaction.

2. To determine with higher precision the intermediate level energies. The comparison of the sums of energies ($E_\gamma + E_f$), where E_γ is the energy of a cascade transition from inermediate level E^* to final state E_f, gives information about possible doublets of intermediate levels.

The E_γ transitions to E^* levels excited by them were placed in a decay-scheme by processing of the spectroscopic data from the reaction (n,2γ). A mean error of transition energy determination is ≃1.5 keV for the reaction (n,2γ) and ≃0.1 keV - for the (n,γ) reaction.

3. To place in the scheme obtained all the transitions (from the reaction (n,γ)) between an intermediate level $E^* \geq 2$ MeV and final levels $E_f \geq 1$ MeV.

It is due to the fact that only cascades to final

states $E_f \leq 1$ MeV are possible to be observed in the experiment on the investigation of the $(n,2\gamma)$-reaction by means of 10% efficiency detectors.

A ratio of primary transition intensities i_γ from the reaction (n,γ) to cascade intensities $i_{\gamma\gamma}$ from the reaction $(n,2\gamma)$ provides information also on: a) transition doublets in the list of (n,γ)-data; b) an upper limit of the total intensity Σi_γ of all the possible transitions depopulating any of intermediate levels with the energy $E^* \geq 2$ MeV observed in the reaction (n,γ).

4. To determine the order of transitions in an individual cascade not placed in the decay-scheme in the study of the reaction $(n,2\gamma)$. This is done by using data from the reaction (n,γ), if one of the placing versions appears more probable.

The quanta sequence in a cascade is determined by the method reported in [1] independently of other methods of the decay-scheme construction. This method employs the fact, that the primary transitions in two-step cascades which excite the same intermediate level E^* but different final $E_f \leq 1$ MeV levels have one and the same energy. Secondary transitions from this intermediate level have the difference in their energies equal to the difference in final level energies. If the intermediate level E^* is depopulated by only one transition to levels $E_f \leq 1$ MeV, then the method[1] does not work. Then a transition will find its place in a decay-scheme if the intermediate level E^* is depopulated, as the data from the (n,γ)-reaction indicate, by several secondary transitions to final states $E_f \geq 1$ MeV.

5. To place using the Ritz combination principle the transitions that were impossible to find place for in the level scheme obtained in the $(n,2\gamma)$-reaction.

6. To avoid wrong placing of γ-transitions in a decay-scheme by comparing sums of transition intensities populating and depopulating the same level. If a continuous parameter q is introduced ($0 \leq q \leq 1$) for any transitions from the reaction (n,γ), then the condition to perform this comparison may be written in the form:

$$\Sigma i_\gamma^{nj} q^{nj} \leq \Sigma i_\gamma^{jk} q^{jk} \qquad (1)$$

Here, n, j and k are the arbitrary levels with energies $E_n^* > E_j^* > E_k$ and i_γ is the intesity of an individual transition measured in the (n,γ)-reaction. System of inequalities (1) allows one to exclude any random transition between two levels n and j only if this transition is not placed in a decay-scheme by analysis of the data from the reaction $(n,2\gamma)$. System (1) of 100-200 inequalities is solved by the Monte-Carlo method.

7. To estimate the reliability of the decay-scheme constructed.

We have realized the algorithm described above for a ^{187}W decay-scheme, for example. It contains ≃520 transitions populating more than 100 levels up to the exitation energy of 3.4 MeV. About 420 transitions were placed in a decay-scheme with a high degree of reliability.

REFERENCE
1. Yu.P.Popov et al.,Izv.Akad.Nauk SSSR, Ser.Fiz., 48, 891 (1984)

DEPENDENCE OF TWO-STEP CASCADE INTENSITIES ON THE EXCITATION ENERGY OF INTERMEDIATE LEVELS FOR THREE NUCLEI

S.T. Boneva, V.A. Khitrov, A.M. Sukhovoj, A.V. Vojnov

Joint Institute for Nuclear Research, Dubna 141980, USSR.

ABSTRACT

The energy dependence of two-step cascade intensities is obtained for compound- nuclei ^{146}Nd, ^{174}Yb, and ^{183}W. Some excitation energy intervals are revealed in which the experimentally obtained cascade intensities are inconsistent with model calculations.

The method of amplitude summation of coinciding pulses from two Ge(Li)-detectors applied to the analysis of the compound-state γ-decay process allowed systematic study of excited states in the energy range below the neutron binding energy B_n.

The main result was that in a majority of cases it is impossible to reach satisfactory agreement between the predicted by statistical theory and observed experimentally values of cascade intensities.

To compare experimental and theoretical data one has to decompose (separate) the observed cascade distributions into primary and secondary transitions. However, nowadays no experiment tells the order of transitions. This order determination with the following decay-scheme construction could be done by data processing provided some conditions are met. The main conditions to be satisfied are given in[1]. The decomposition is performed with respect to such parameters of strong cascades as transition intensities and intermediate-level energies from a decay-scheme we constructed in[2-4].

The sufficient statistical accuracy of the measurement of ^{145}Nd(n,2γ), ^{173}Yb(n,2γ) and ^{182}W(n,2γ) reactions permits the determination within acceptable errors of the energy dependence of cascade intensities from compound to 1-3 low-lying states in energy intervals of 500 keV each.

The experimental histograms are compared with the curves calculated using two different level density models[5,6] and the conventional idea of the energy dependence of radiative widths of cascade transitions[7].

As it is seen from Figs.1-4, essential divergences between experimental and calculated cascade intensities is observed at some energies. Therefore it seems necessary both to perform further study by the method reported here and continue development of the model

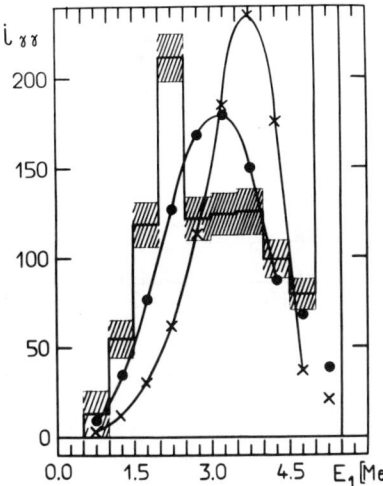

Fig.1. The distribution of two-step cascade intensities to the ground state of ^{183}W (per 10^4 capture events) as a function of the primary transition energy E_1 (MeV).

The histogram shows the experimental data. Statistical errors are shaded. ●-calculated in the frame of the level density model ^{6}, x-of the model reported in ^{5}

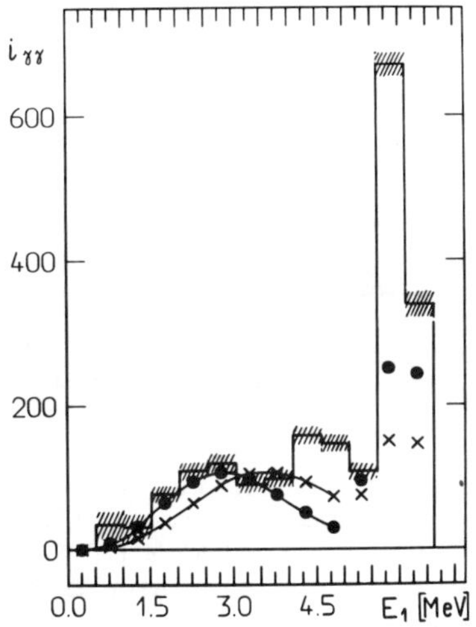

Fig.2. The intensity of cascades to the first excited state of ^{146}Nd. The notations are analogous to those in fig.1.

description of the compound-state decay. A usually used (e.g.^{7}) simple model with constant radiative strength functions and a smooth energy dependence of level densities does not provide sufficiently good description of γ-decay below the neutron binding energy.
Ge(Li)-detectors of higher efficiency will enable the investigation of the cascade γ-decay for all primary transitions and excitation energies.

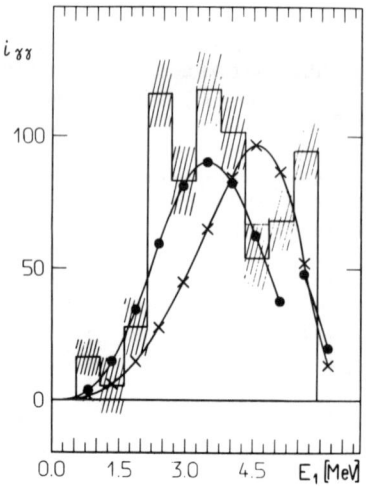

Fig.3. The intensity distribution of cascades to the first excited state of ^{174}Yb. The notations are analogous to those in Fig.1.

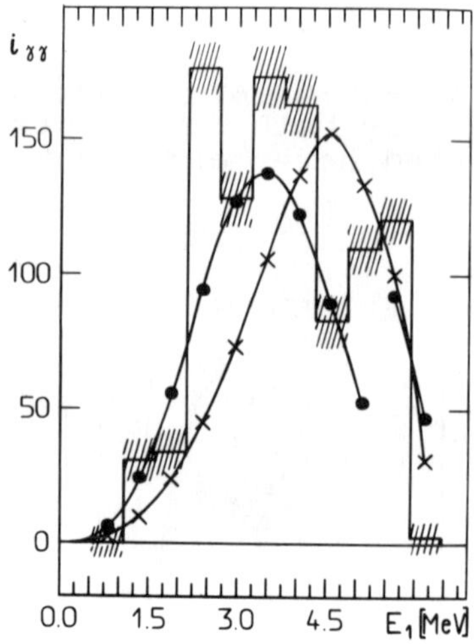

Fig.4. The intensity of cascades to the second excited state of ^{174}Yb. The notations are analogous to those in fig.1

REFERENCES

1. S.T.Boneva et al., JINR preprint E3-90-45, Dubna, 1990.
2. S.T.Boneva et al.,Izv.Akad.Nauk SSSR, Ser.Fiz., 53,2401 (1989)
3. S.T.Boneva et al.,Izv.Akad.Nauk SSSR, Ser.Fiz., 53,2092 (1989)
4. S.T.Boneva et al.,Izv.Akad.Nauk SSSR, Ser.Fiz., 53, 7 (1989).
5. W.Dilg et al., Nucl.Phys., A217, 269 (1974).
6. V.A.Ignatiuk et al., Yad.Fiz., 21, 485 (1975).
7. S.T.Boneva et al., Z.Phys., A330, 153 (1988).

ON ESTIMATES OF RADIATIVE STRENGTH FUNCTIONS OF SOFT PRIMARY TRANSITIONS IN ^{181}Hf.

Bondarenko V.A., Kuvaga I.L., Prokofjev P.T.,
Rezvaya G.L., Simonova L.I.
Institute of Physics, Salaspils 229021, Latvia
Khitrov V.A., Yu.V.Kholnov, Le Hong Khiem,
Pham Dinh Khang, Yu.P.Popov, Sukhovoj A.M.
Joint Institute of Nuclear Research, Dubna 141980, USSR.

ABSTRACT

Compound state de-excitation of ^{181}Hf by means of two-step cascades was studied. More than 80% of total intensity of primary radiation was extracted. The dependence of the two-step cascade intensity on the primary transition energy ($E_1 \geq 0.5$ MeV) was established. The lower estimate of the radiative strength function of soft primary E1 plus M1 transitions was obtained.

The intermediate energy region ($B_n > E_m \geq 1-3$ MeV) may be effectively investigated by measuring the two-step cascades of γ-quanta emitted in the radiative neutron capture process. Experimental intensity distributions of these cascades have the simplest shape[1]: a pair of narrow peaks of equal width and height, provided the background is effectively reduced. Unfortunately, these distributions are generaly a superposition of numerous cascades with nearly equal energies of primary and secondary transitions. But hapily their number and intensity are different.

Nevertheless, as it was shown in [2] the experimental superposition may be decomposed into components depending on the energy of primary (secondary) transitions only. This decomposition is possible under general assumption that cascades by hard primary transitions form a small enough number of intensive cascades while cascades by hard secondary transitions usually form a continuous distribution.

The sum of absolute intensities of all the two-step cascades to final 5 low-lying levels in ^{181}Hf is shown in Fig.1. This intensity distribution was obtained under the assumption that main part of the cascades with intensity above $\cong 5 \cdot 10^{-4}$ quanta per compound state decay could be identified and placed into the level scheme up to 2200 keV excitation energy. The experimental histogram is compared with the calculated one using two models[3,4] of the level density below the neutron binding energy.

The intensity of the individual cascade $i_{\gamma\gamma}$ is determined by the intensity of its primary transition i_1 and by the intensities of all the secondary transitions i_2 as follows :

$$i_{\gamma\gamma} = i_1 \cdot i_2 / \Sigma\, i_2 \qquad (1)$$

By taking sum (1) over the final states j of two-step cascades and over the excitation energy interval ΔE, one obtains the following expresion :

$$\Sigma\, i_{\gamma\gamma} = i_1 <m> \varkappa , \qquad (2)$$

where $<m>$ is the averaged number of the levels in the interval ΔE_M populated by primary gamma-quanta and $\varkappa \leq 1$ is some coefficient which depends on the completeness of secondary transitions observed in experiment.

If intensites i_1 of all primary transitions are determined in an independent experiment and put into expresion (2), then $\varkappa$ may be chosen to be equal to 1. This possibility exists because the single γ-radiation spectrum measured in the (n,γ) reaction forms a basis for the construction of a complete level scheme of γ-decay in a wide range of excitation energies by using schemes obtained in the study of the $(n,2\gamma)$ reaction [5].

The radiative strength function can be obtained from (2) if one knows the total radiative width Γ_γ of the compound-state and the level density D_j in the whole range of excited states :

$$K(E1+M1) \geq \Sigma\, i_{\gamma\gamma}\, \Gamma_\gamma / E_\gamma^3\, D_j\, A^{2/3} \qquad (3)$$

The corresponding dependence is presented in Fig.2. Calculation in the frame of the statistical theory gives a broad maximum at 2 MeV for the product of primary transition widths and number of excited states. In our experiment only 20% of the total intensity of primary transitions from the compound state of ^{181}Hf was not observed.

It is possible to obtain the "upper" estimate K(E1+M1) under the condition that the "lost" intensity is uniformly distributed over the energy interval of 1.5 MeV. These estimates are also shown in Fig.2 (3 points marked by "x").

The comparison of the values of the radiative strength function obtained in experiment with those predicted by the theory was done with the result presented as a function of the primary transition energy E_1 (MeV) in Fig.2 for two cases of extrapolation to a Lorentzian curve of :
a) the cross section dependence of the Giant Electric-Dipole Resonance (GEDR) in the A=181 region (upper curve),
b) the same dependence but taking into account the temperature- and frequency-dependent GEDR width proposed

in Ref.[6] (lower curve).

Agreement between experiment and theory is obviously better in case b).

Enhancement of K(E1+M1) at two values of intervals of E_1 could be caused by a single particle component, as it was pointed out in [7]. Then this enhancement must be related to the value Γ_n^o of some resonance which governs the thermal capture cross section.

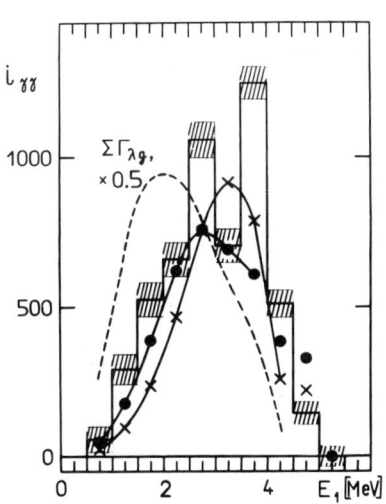 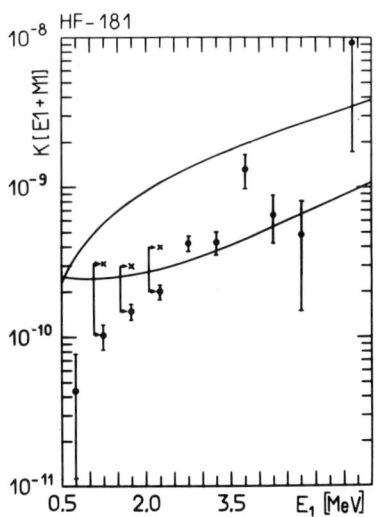

Fig.1.   Fig.2.

REFERENCES

1. A.M.Sukhovoj,V.A.Khitrov,Sov.J.PTE,N 5, .27(1984).
2. S.T.Boneva et al., JINR, E3-90-45, Dubna(1990).
3. W.Dilg et al., Nucl.Phys.,A217, 269(1973).
4. A.V.Ignatiuk et al.,Yad.Fiz., 21, 485(1985).
5. S.T.Boneva et al.,JINR, E3-90-307, Dubna(1990).
6. S.G.Kadmensky et al.,Yad.Fyz., 37, 277(1983).
7. S.T.Boneva et al., Z.Phys., A330, 153(1988).

CASCADE γ DECAY OF THE ^{187}W COMPOUND STATE

S.T.Boneva, V.A.Khitrov, L.A.Malov,
Yu.P.Popov, A.M.Sukhovoj, E.V.Vasilieva.
Laboratory of Neutron Physics, JINR,
Dubna 141980, USSR.
M.R.Beitins, P.T.Prokofjev, G.L.Rezvaya, L.I.Simonova.
Institute of Physics, Salaspils 229021, Latvia

ABSTRACT

In the decay scheme of ^{187}W below 3.4 MeV the excitation energies and decay modes of 100 levels are established by joint analysys of (n,2γ) and (n,γ) data. The comparison of the experimental with calculated cascade intensities shows that the compound-state decay of ^{187}W is more due to the excitation of simple (few-quasiparticle) structure levels than in the case of the ^{183}W compound-state decay.

Using the Summation Amplitude of Coinciding Pulses (SACP) method we can establish reliably the energy values and decay-modes of several tens of levels at an excitation energy $E \leq 3-4$ MeV[1]. By this method we investigated γ-transition cascades from the ^{186}W(n,2γ)[2] reaction and identified all intensive γ-transitions taking place at thermal neutron capture in the (n,γ) reaction[3]. Joint analysis of the experimental data from the (n,γ) and (n,2γ) reaction is described in detail in[4]. A new decay-scheme of the nucleus ^{187}W has about 460 γ-transitions and contains about 100 excited states. The product $I_\gamma E_\gamma$ for γ-transitions placed in the level scheme equals 64 B_n (% MeV). Figure 1 gives the cumulative sum for observed excited states of ^{187}W in the energy range 1-3 MeV. For comparision with this cumulative sum we present the expected dependence calculated a) from the model developed by Ignatyuk A.V.[5] (curve 1) and b) from the Fermi-Gas model with Back Shift[6] (FGBS) for the moment of inertia equal to the half of the solid state one with parameters $\alpha=18.57$ MeV^{-1}, $\delta=-0.4$ MeV (curve 2) The parameters are obtained under assumption that only states with $I^\pi=1/2^-$ and $3/2^-$ are present in the level scheme constructed by us. If we assume that all states of both parities are found, then model parameters[6] should be replaced by the following values : $\alpha=17.45$ MeV^{-1} and $\delta=-0.9$ MeV (curve 3). Despite of the fact that the experimental density of excited states defines the parameters of the model and is in agreement with the model in the energy interval $E_M \leq 1.6-1.7$ MeV, where "loss" of level is small, no agreement is achieved between the calculated and experimental cascade intensities for transitions from the compound-state to five final levels. This is clearly seen from Table.1 which shows the experimental sums of intensities of all the cascades to final states $I^\pi K[Nn_z\Lambda]$ and the sums calculated for different excited state densities given in Fig.1.

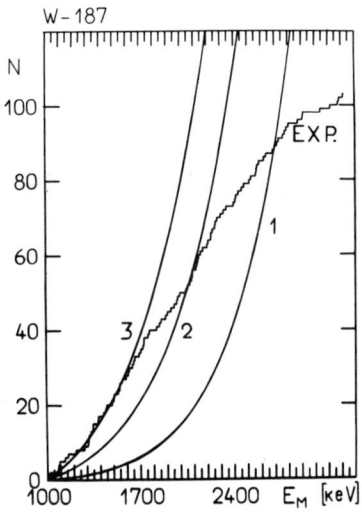 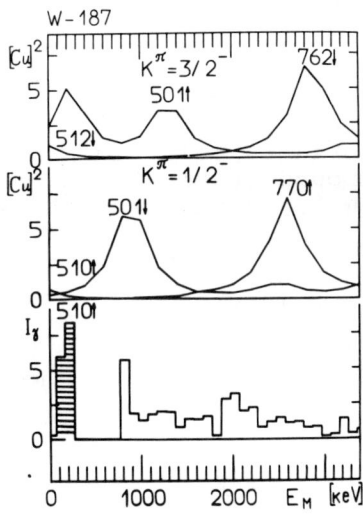

Fig.1. Plot of the experimentally obtained number of levels $N(E_M)$ together with the model predictions for spins 1/2 and 3/3 only.

Fig.2. Primary transition intensities and one-quasiparticle state $(Cu)^2$ - strength distribution as a function of the excitation energy (keV).

Excess of cascade intensities over values calculated by the model for even-odd compound-nuclei with a relatively large single particle component in the structure of the compound-state was established earlier [7]. This received natural explanation : the experimentally observed enhancement of part of primary transition partial widths depends on the correlation with the resonance values Γ_n^0 which form the compound state. Analysis made in [8] shows that within errors of the quasiparticle-phonon nuclear model to a group of few intermediate levels excited by strong primary transitions corresponds an energy range where the calculation by this model predicts strength localization of one-quasiparticle states with $K^{\pi}=1/2^-$ and $3/2^-$. The E1(M1) transition matrix element depends on the distribution of one-, three-, quasiparticle ⊕ phonon,...states, and on the Bogolubov canonical transformation parameters u and v, etc. In Fig.2, the comparison between the primary transition sum intensities I_γ and one-quasiparticle state $(Cu)^2$-strength distribution shows that there is really observed the correlation of positions of maxima (except for the [512]! state). So, we can expect that transitions between states with a relatively simple structure play an important role in the cascade decay of the ^{187}W compound-state, differently from its neighbour nucleus ^{183}W, which does not exhibit such effects.

Table 1.

Final cascade level (keV)	$I^{\pi}K[Nn_z\Lambda]$	Cascade intensities Experimental	Calcul. Ignatiuk[5]	Models FGBS $\delta=-0.4$	FGBS $\delta=-0.9$
0	3/2⁻ 3/2[512]	12.6(13)	10.5	9.2	8.6
77	5/2⁻ 3/2[512]	5.6(5)	4.6	4.0	3.8
146	1/2⁻ 1/2[510]	9.5(10)	7.1	6.4	5.9
201+204	7/2⁻ 3/2[512] +3/2⁻ 3/2[510]	10.7(6)	7.0	6.0	5.5
303	5/2⁻ 1/2[510]	5.1(10)	2.9	2.5	2.3
sums		43.5(21)	32.1	28.1	26.1

REFERENCES

1. S.T.Boneva et al., Izv.Akad.Nauk SSSR, Ser.Fiz.52, 2082,(1988).
2. V.A.Bondarenko et al., JINR Communication P6-89-10 (Dubna,1989).
3. L.I.Simonova et al. Preprint LAFI-149(Salaspils,1989).
4. S.T.Boneva et al.,JINR Preprint E3-90-307(Dubna,1989).
5. A.V.Ignatiuk et al. Yad.Fiz. 21,485,(1985).
6. W.Dilg et al. Nucl.Phys.,A217, 269,(1973).
7. S.T.Boneva et al., Z.Phys., A330, 153 (1988).
8. S.T.Boneva et al.,Jamada Conference XXIII NuclearWeak Process and Nuclear Structure, Osaka,Japan,June 12-15, 1989,p.372.
9. S.T.Boneva et al.,Izv.Acad.Nauk SSSR, Ser.fiz.53,7 (1989).

Superdeformation in Nuclei

SUPERDEFORMATION IN THE MERCURY REGION

R. R. Chasman
Physics Division, Argonne National Laboratory, Argonne, IL 60439

ABSTRACT

A brief history of superdeformation is presented. Calculations of superdeformed well depths, single particle energy level diagrams at I=0, orbital alignments as a function of rotational frequency, and dynamical moments of inertia are given. The sensitivity of calculations to the details of the spin-orbit interaction are examined.

INTRODUCTION

Superdeformed nuclides (axis ratio >1.5:1) were first proposed[1] some twenty five years ago as the explanation of the phenomenon[2] of fission isomerism seen in the actinides at I=0. The fission isomers are nuclear states associated with very elongated shapes (axis ratio of ≃2:1). Although the calculations using the Strutinsky method[1] give superdeformed minima in the energy surfaces of many of the actinides, these calculations do not seem to explain some of the spectroscopic information obtained in the study of the fission isomers; most notably the magnetic moment measurements[3] in ^{239}Pu. This failure of the calculations has given rise to much theoretical speculation about the nature of the effective potential at large deformations; however the difficulty of obtaining sufficient spectroscopic information on superdeformed actinides has made it impossible to resolve this problem. A detailed, useful treatment of superdeformation in the actinides can be found in the review[4] of Bjornholm and Lynn.

In the past few years, heavy ion reactions have been used to produce[5] several superdeformed nuclides at high spin near ^{152}Dy. Calculations using the cranked Strutinsky method suggested[6,7,8,9] the existence of superdeformed states in several nuclides of this region at high spin. These nuclides decay from superdeformed to less deformed shapes at high spins (I= ≃25) and little is known about the energy differences of rotational bandheads in this region. It is interesting to note that the same large gap in the neutron single particle spectrum (N≃86) that gives rise to these superdeformed nuclides at high spin also explains[10] many features of the mass yields observed in the fission of the actinides. This large shell gap gives a minimum in the energy surface at high spins because the moments of inertia are much larger for superdeformed states than they are for normally deformed states and the rotational energy is lower. In the fission case, the superdeformed shapes are favored because of the coulomb repulsion of the nascent fission fragments.

We have carried out calculations on nuclides in the A=190 region, using the cranked Strutinsky method, and predicted[11] that

© 1991 American Institute of Physics

there are many nuclides in the Hg region in which superdeformed states are experimentally accessible at moderate spins (I≈40). Based on these predictions, an experimental study[12] was made and superdeformation was first discovered in ^{191}Hg in this mass region. We also found in our calculations that this region differs substantially from the A=150 region in that the superdeformed minimum persists to low spin. There are other[13,14,15] calculations that show superdeformed minima for nuclides in this region at I=0; but the stability of superdeformed states at I=0 remains an important open theoretical and experimental question and is an area of active research. As of now, superdeformed states with I=0 have not been seen in this region. Our calculations suggest that the superdeformed states are at least 4 MeV higher in energy than the ground state in the A=190 region, at I=0. This is roughly twice as much excitation energy as the fission isomers seen in the Uranium isotopes and implies an order of magnitude increase in E2 decay rates from the superdeformed state if one assumes the same transition matrix element.

THE MERCURY REGION

In the Strutinsky method, the total energy is the sum of a smooth contribution to the nuclear energy, given by a liquid drop model; and a fluctuating part the "shell correction", that measures the difference between the level density of the nucleus of interest and some smoothed level density. When the level density is low at the fermi level, we get extra stabilization. In Fig. 1, we show the proton and neutron levels appropriate to the superdeformed Hg region. One can see that the shell corrections are particularly favorable for Z=80 and for 112<N<120. In our calculations, we use the Myers-Swiatecki[16] parameterization of the liquid drop model and a Woods-Saxon single particle potential for the level density calculation. This parameterization[9] of the Woods-Saxon potential is based on single particle energy level spacings obtained from a many-body analysis of near yrast high spin states in the ^{148}Gd region.

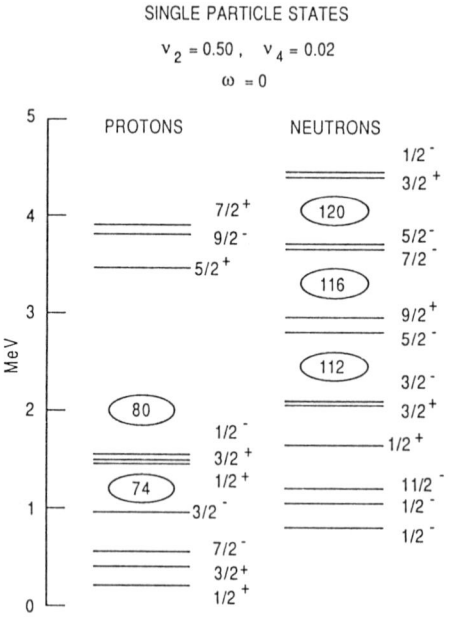

Fig. 1. Proton and neutron single particle levels for superdeformed Hg region.

The energy is calculated on a grid in deformation space. The ground state shape is that shape with lowest total energy

Excited states are predicted at all local minima in the energy surface. Angular momentum is introduced by adding a cranking term ωJ_x [17] to the single-particle Hamiltonian and the calculation is extended thereby to finite angular momenta. It is worth noting that super-deformed states come down in energy relative to less deformed ones as the angular momentum increases, because they have larger moments of inertia and hence less rotational energy.

There are some caveats that apply to the Strutinsky method. Ground state masses are calculated with an accuracy of ≃0.8 MeV. Also, there are calculations in the A=146 region showing a reflection asymmetric minimum about 1.5 MeV deeper than the reflection symmetric minimum, but experimentally the nuclides do not show reflection asymmetry in their ground states. This suggests that a well depth of at least 2 MeV is needed before one has a reasonable prediction of shape stability. On more general theoretical grounds, one expects that there will be strong mixing of configurations with different shapes when the barriers between the shapes are small. When one goes beyond the Strutinsky method, e.g. with a generator-coordinate calculation, one finds just this sort of shape mixing. A second caveat is that one may not have looked at the region of deformation space where the true minima are found; i.e. by neglecting octupole,

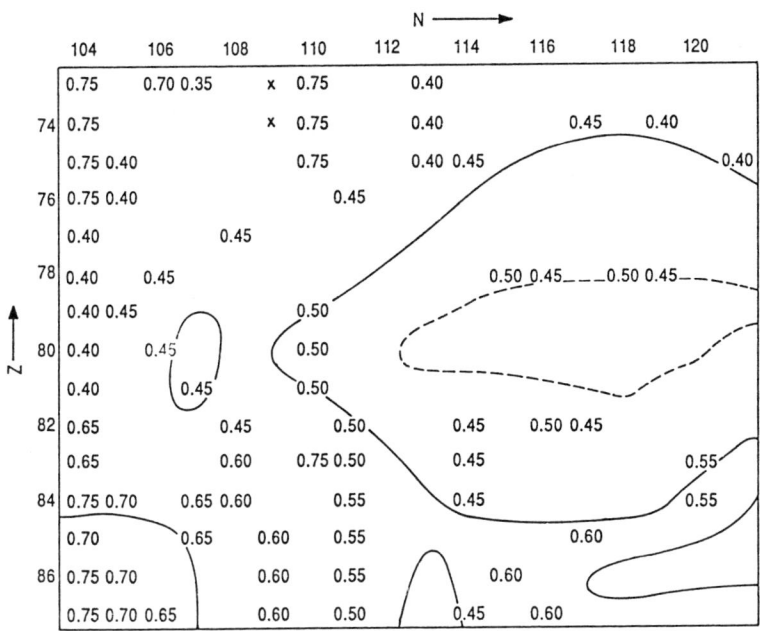

Fig. 2. Depth of the superdeformed minimum for nuclides near A=190 at I=40. The continuous line is the 2 MeV contour; the dashed line is the 3.5 MeV contour. The numbers inside the figure are quadrupole deformations ν_2 of the superdeformed minimum. In each row,, all nuclides to the right of a written value of ν_2 have the same value.

hexadecapole 2^5-pole or 2^6-pole degrees of freedom, one might get a totally erroneous picture of nuclear structure, even if the Strutinsky method were exact.

With these qualifications in mind, we turn to our[11] calcula-tions of superdeformation in the Hg region. These calculations include only quadrupole and hexadecapole deformations. In Fig. 2, we display the calculated superdeformed well depth at I=40 in the A=190 region. The first thing to note is that there is a large region of nuclides with a well developed superdeformed minimum at a quadrupole deformation of $\nu_2=0.5$, corresponding to an axis ratio of 5:3. We chose a 2 MeV well depth as a minimum requirement for the reasons cited above. In Fig. 3, we display the excitation energy of the superdeformed state relative to yrast at I=40. We have assumed that the lower the excitation energy of the superdeformed state, the easier it is to produce. Taking both of these figures into consideration, we are led to a prediction of the most experimentally accessible superdeformed states being in ^{191}Hg and ^{192}Hg. These predictions are borne out by the experimental observation[12,18,19] of superdeformation in these nuclides. Subsequently, superdeformed states have been found in ^{190}Hg [20], ^{193}Hg [21], ^{194}Hg [22], ^{193}Tl [23], ^{194}Tl [24], ^{194}Pb [25] and ^{196}Pb [26]. All of these nuclides are expected to have superdeformed minima (cf. Fig. 2). In Fig. 4, we

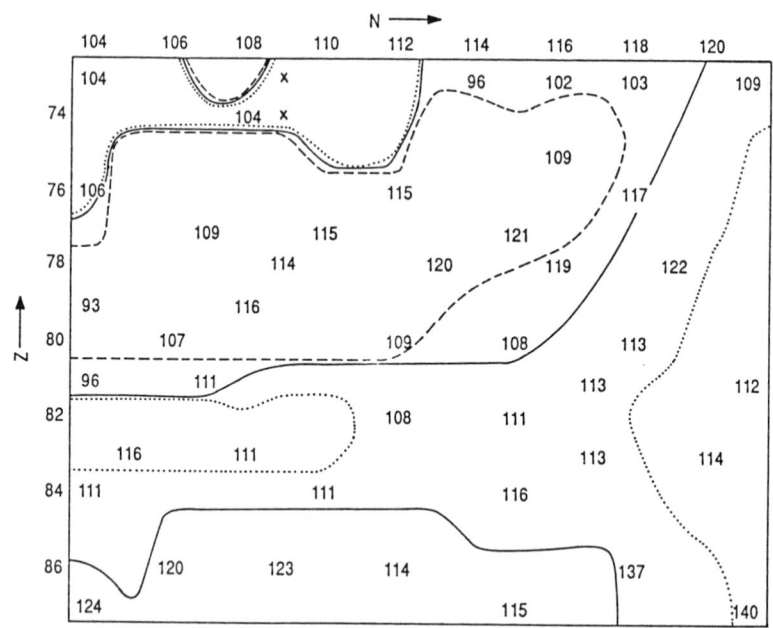

Fig. 3. Excitation energy (E*) of superdeformed state at I=40, relative to yrast. The dashed line is the 0.5 MeV contour; the continuous line, 2 MeV; the dotted line, 4.5 MeV. The numbers inside the figure are moments of inertia calculated without pairing.

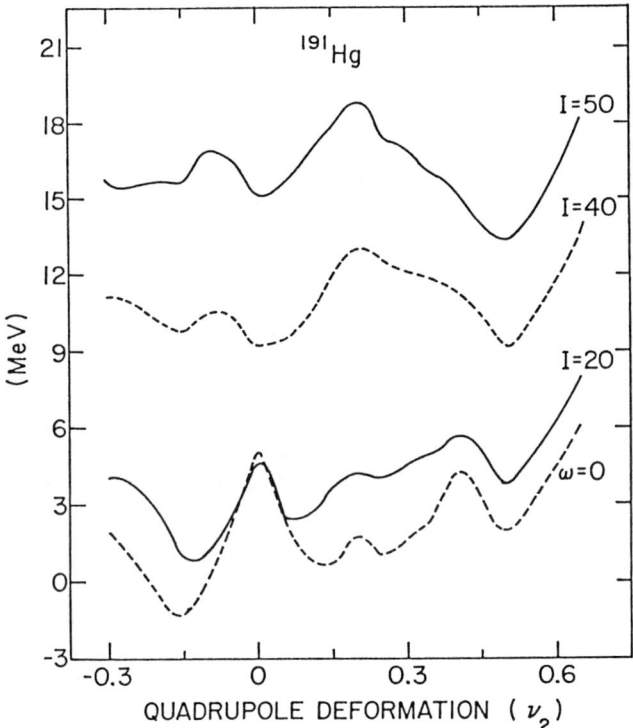

Fig. 4. A one dimensional cut of the Strutinsky energy surface of ^{191}Hg, as a function of angular momentum.

display the energy surface obtained with the Strutinsky method for ^{191}Hg at different values of the angular momentum. We have made a one dimensional cut of the energy surface so as to connect the minima. We get qualitatively similar results for the other known super-deformed nuclides in this region.

However, there is a serious problem! Our calculations indicate that the superdeformed minimum in ^{194}Hg and ^{196}Hg is roughly 4 MeV above yrast even at I=40. Conventionally, one assumes that the superdeformed band is populated in that range of angular momentum states in which it is near yrast and fission competition is not yet serious. The population of superdeformed bands in Pb is not consistent with our calculation if one makes this assumption. This suggests that either our single particle scheme is not quite right or that highly excited superdeformed states can be populated in heavy ion reactions.

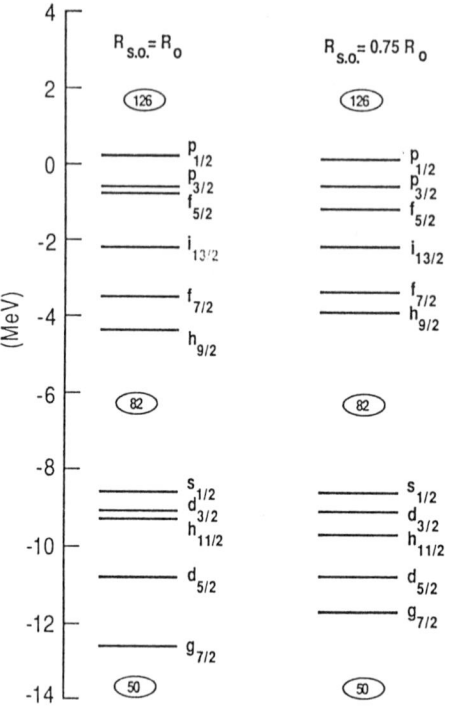

Fig. 5. Comparison of proton spherical single particle levels for $R_{s.o.}=R_0$ and $R_{s.o.}=0.75R_0$.

Here we explore the possibility that the single particle potential is not quite right. Many years ago Rost[27] pointed out that one could improve the agreement of Woods-Saxon calculations with experimentally determined proton energy level spacings in the heavy elements by using a spin-orbit radius parameter that is 3/4 the value used for the central part of the potential. As many of these actinide levels are at or even below the fermi level in the super-deformed A=190 region, it is worthwhile to explore the effects of this parameterization on superdeformation in the Hg region. We note that this parameterization gives results similar to our earlier work for the ground state deformations in the A=180-190 region. Whether or not this choice is valid, these calculations provide an estimate of the sensitivity of the model to reasonable changes in the parameters. We have carried out such a calculation. In Fig. 5, we compare the spherical single particle energy levels obtained with $R_{s.o.}=R_0$, and the spherical levels obtained using $R_{s.o.}=0.75R_0$. The level orderings and spacings are quite similar apart from an increased splitting of the $h_{11/2}$-$h_{9/2}$ pair and a decreased splitting of the $f_{7/2}$-$f_{5/2}$ pair with the Rost parameterization. It should be noted that there is evidence for a decreased $f_{7/2}$-$f_{5/2}$ splitting from the (d,p) studies of Pb and also from the observation[28] of the 1/2-[521] orbital in the A=250 region. In Fig. 6, we display the well depths associated with the superdeformed minima at I=40, using the Rost spin-orbit radius. Comparing with Fig 2, one immediately sees that there are many more superdeformed nuclides with well depths greater than 2 MeV. Also, the deepest wells are on the order of 5 MeV and are ≃2 MeV deeper than we found with our usual parameterization, for the nuclides in the vicinity of ^{199}Po. We emphasize that the deep

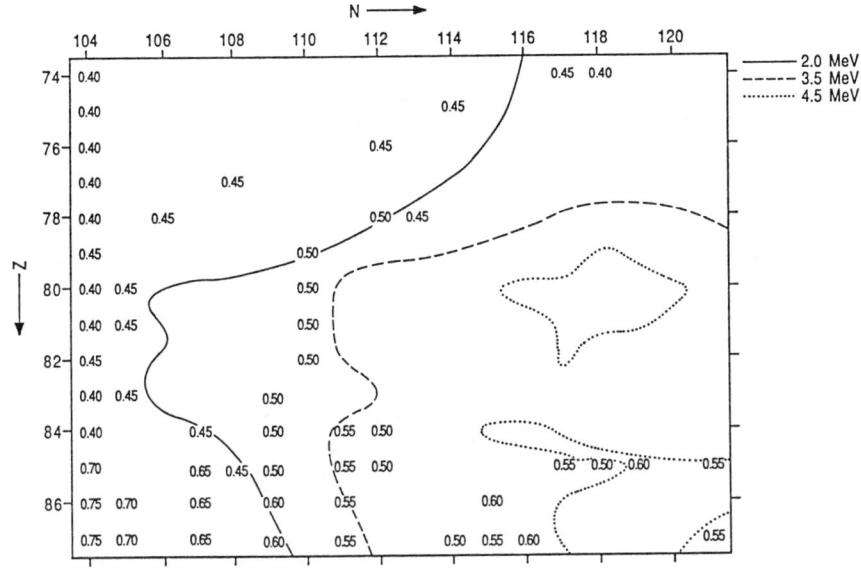

Fig. 6. Depth of superdeformed minimum at I=40 obtained with the Rost spin-orbit parameterization. The continuous line is a 2 MeV contour; the dashed line, the 3.5 MeV contour; the dotted line, a 4.5 MeV contour. See caption for Fig. 2.

superdeformed well depths are retained in the Hg-Tl-Pb isotopes in which superdeformed bands are known. Unfortunately, the excitation energy of the superdeformed minima are predicted to be as high or higher in this calculation than the values shown in Fig. 3. This calculation does not provide an easy explanation of the population of the superdeformed bands in ^{194}Pb and ^{196}Pb. To put things more strongly, there is a maximum in E* at I=40, for the superdeformed band for neutron numbers 110<N<118, when Z=81 or Z=82. With our original parameterization, this maximum always occurs for Z=82. This means that the superdeformed states in Pb should be the hardest to produce according to this criterion. If the Rost parameterization is the correct one, we might argue that the decay out of the superdeformed minimum would be inhibited by the existence of a such deep wells and perhaps one might observe transitions in the superdeformed rotational bands of nuclides near ^{199}Po. Both the Rost and the conventional spin-orbit interactions predict superdeformed minima in this mass region. However, there are large differences; i.e. the results are sensitive to the details of the potential. It is then important to fix the potential as accurately as possible.

OTHER SURPRISES

There have been several other surprises in the experimental study of superdeformed bands in the Hg region. The first has been how relatively easy it is to produce these bands in heavy ion reactions. The second is the large regular increase in the dynamic (second derivative) moment of inertia with rotational frequency. The third is the observation of of almost exactly the same transition energies in superdeformed bands of different nuclides in the Hg region. This feature was first noted in superdeformed bands of the A=150 region.

We consider the dynamic moments of inertia. In the high spin superdeformed rotational bands of the Dy region one finds very little variation in the moment of inertia as a function of angular momentum. In the Hg region, there is in some instances a ≃40% increase in the moment of inertia in the superdeformed bands. In the A=150 region, the superdeformed bands terminate at I≃25, while they continue to rather low spin in the Hg region. One might assume that pairing plays a rather different role in the two regions and that a pairing attenuation is related to the increase of the moment of inertia in the Hg region. The role of pairing is quite paradoxical in superdeformed nuclides. In order that there be a superdeformed minimum in a nucleus, the level density must be small near the Fermi level and pairing correlations must be small. On the other hand, a superdeformed shape has a large moment of inertia and small energy differences between members of a rotational band. This means that a small pairing correlation energy that decreases with increasing angular momentum could have a dramatic effect[29,30] on the moments of inertia of superdeformed states without really affecting their structure.

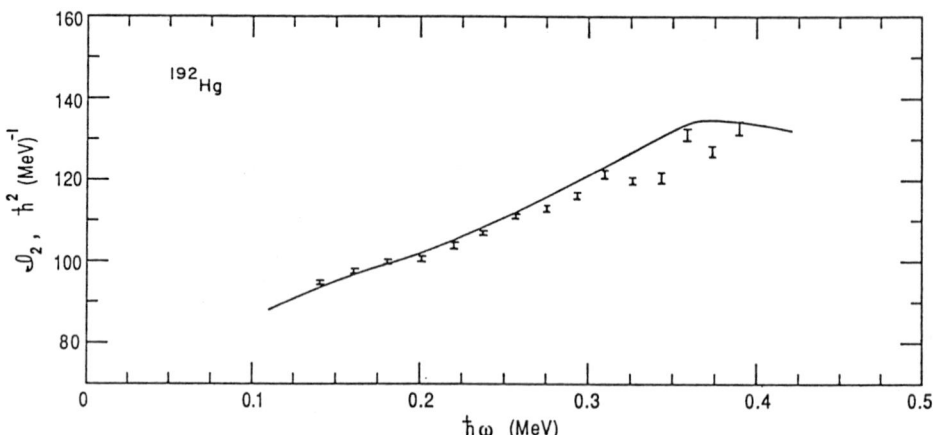

Fig. 7. Dynamic moment of inertia of ^{192}Hg as a function of rotational frequency. The continuous line is the calculation.

Because the level density is low for superdeformed shapes, a quasiparticle treatment of pairing forces is not appropriate in this case. If one uses a gap parameter to get a finite pairing correlation energy, this gives excessive pairing correlation and too low a moment of inertia at low spin. The approach that we take[31] is to calculate the pairing correlation energy accurately at $\omega=0$ using the method of correlated quasiparticles[32] which goes beyond the quasiparticle (BCS) approximation and is particularly appropriate when pairing effects are small. Motivated by the observed smooth rise in the moment of inertia in this region, we assume that the pairing correlation energy is smoothly attenuated as a function of increasing angular momentum. The functional form that we choose is an exponential attenuation, i.e. we set

$$E_{pair}(\nu_2, \nu_4, \nu_6, \omega) = E_{pair}(\nu_2, \nu_4, \nu_6, \omega=0)\, e^{-(\omega/\omega_0)^2}$$

at each point in the deformation space grid. We use a single parameter ω_0 for all nuclides in the A=190 region. Comparing with the experimental data, we choose $\omega_0=0.25$ MeV (h)-1 and get a reasonable agreement with observed moments of inertia. This means that the pair correlation energy is reduced to ≃40% of its ground state value at I=30. In Fig. 7, we compare the calculated and observed moments of inertia for ^{192}Hg, where the agreement is quite good and for ^{193}Hg in Fig. 8 where the agreement is not as good. It appears that one can account for the increase in the moments of inertia quite reasonably in this way. In doing this calculation, we included 2^6-pole deformations in addition to quadrupole and hexadecapole deformation. This shifted the superdeformed minima somewhat; giving much smaller values of ν_4 than we obtained without

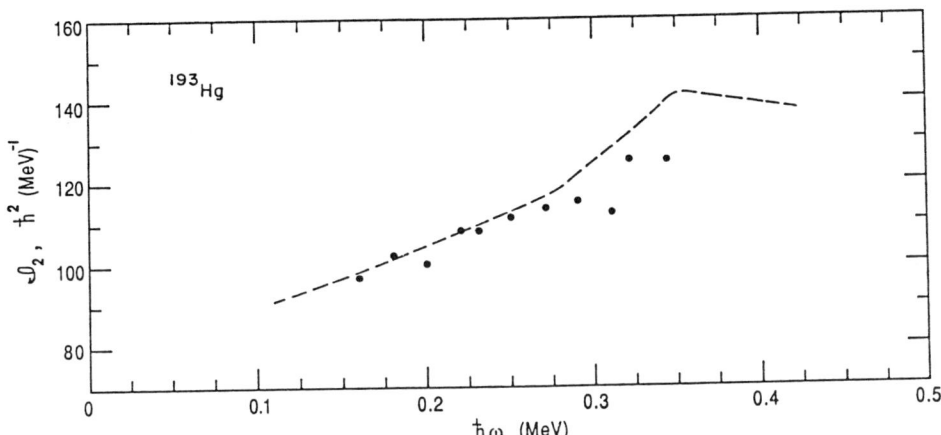

Fig. 8. Dynamic moment of inertia of ^{193}Hg. See caption for Fig. 6.

the inclusion of 2^6-pole deformations. It also lowered the excitation of the superdeformed state by several hundred keV in some instances; providing a nice example of the need to use an extended deformation space in Strutinsky calculations.

The problem of almost identical transition energies in different nuclides is not easy to understand. This was first seen[33,34] in the superdeformation bands of the A=150 region with almost identical transition energies in N=86 isotones. For obvious reasons, such bands are sometimes called twinned bands. There have been many more such bands seen in the Hg region. The differences in the transition energies are just a few keV, which is considerably less than one expects from the rigid rotor estimate of the moment of inertia. A calculation accurate to a few keV goes far beyond my assessment of the accuracy of the cranked Strutinsky method. Stephens has emphasized[35] that the incremental alignment in the twinned bands is either integer or half integer relative to ^{192}Hg. We have examined the calculated single particle neutron alignments for the orbitals near N=112. These single particle alignments as a function of rotational frequency are shown in Fig. 9. We do not seem to find orbitals with constant alignments, or with half integer or integer alignment. One expects to see such alignments with a pseudo-SU(3)[36,37] coupling scheme, which has been suggested as an explanation of this phenomenon. However, this is known to be only an approximate symmetry in deformed nuclides. This remains an open and fascinating problem!

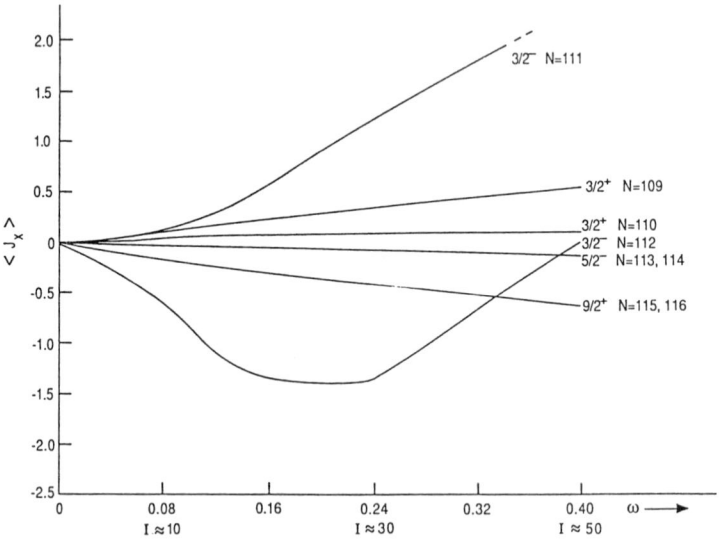

Fig. 9. Single particle alignments near N=112 at ν_2=0.50, ν_4=0.02. The neutron numbers and Ω^π values are for ω=0.

SUMMARY

To summarize, we find that the cranked Strutinsky method, used with some care, has led us to the prediction of a new region of superdeformed nuclides that are accessible in heavy ion reactions. The experimental study of this region has raised some interesting questions: (1) why are the transition energies in superdeformed bands so similar in different nuclides; (2) why do we populate superdeformed bands, when the calculations suggest that they are highly excited; (3) why is there a large increase in the second moment of inertia with increasing spin. We believe that the last problem is semi-quantitatively understood; the second question may be answered within the context of the cranked Strutinsky method; and that one must go beyond the Strutinsky method and beyond such approximate symmetry concepts as pseudo-SU(3) to deal with the first. There is a reciprocal interaction between theory and experiment that is making nuclear structure the stimulating area of research that it is today. It is just this interaction that leads us to suggest experiments to clarify the role of E* in the production of superdeformed states. This means the study of nuclides with Z=84 and Z=85, with N near 115.

This work supported by the U.S. Department of Energy, Nuclear Physics Division, under contract W-31-109-ENG-38. The calculations were carried out on the NERSC computers at Livermore.

REFERENCES

1. V. M. Strutinsky, Arkiv Fys. 36, 629(1966); Nucl. Phys. A95, 420 (1967).
2. S. M. Polikanov et al., JETP (USSR) 24, 1464 (1962).
3. H. Backe et al., Phys. Rev. Lett. 42, 490 (1979).
4. S. Bjornholm and J. E. Lynn, Rev. Mod. Phys. 52, 725 (1980).
5. P. J. Twin et al., Phys. Rev. Lett. 57, 811 (1986).
6. K. Neergaard and V. Paskevich, Phys. Lett B59, 218 (1975).
7. J. Dudek et al., Phys. Lett B112, 1 (1982).
8. J. Dudek and W. Nazarewicz, Phys. Rev. C 31, 298 (1985).
9. R. R. Chasman, Phys. Lett B187, 219 (1987).
10. B. D. Wilkins et al., Phys. Rev. C 14, 1832 (1976).
11. R. R. Chasman, Phys. Lett. B219, 227 (1989).
12. E. F. Moore et al., Phys. Rev. Lett. 63, 360 (1989).
13. C. F. Tsang and S. G. Nilsson, Nucl. Phys. A140, 275 (1970).
14. M. Girod, J. P. Delaroche and J. F. Berger, Phys. Rev. C 38, 1519 (1988).
15. P. Bonche et. al, Nucl Phys. A500, 308 (1989).
16. W. D. Myers and W. J. Swiatecki, Arkiv for Fysik 36, 343 (1966).
17. M. Brack and B. K. Jennings, Nucl. Phys. A258, 264 (1976).
18. D. Ye et al., Phys. Rev. C 41, 13 (1990).
19. J. A. Becker et al., Phys. Rev. C 41, 9 (1990).
20. M. W. Drigert et al. (to be published).
21. E. A. Henry et al., Z. Phys. A335, 361 (1990).
22. M. A. Riley, Nucl. Phys. A512, 178 (1990).

23. P. B. Fernandez et al. (to be published).
24. F. Azaiez et al. (to be published).
25. M. J. Brinkman et al., Z. Phys. A336, 115 (1990).
26. K. Theine et al., Z. Phys. A336, 113 (1990).
27. E. Rost, Phys. Lett. 26B, 184 (1968).
28. I. Ahmad et al., Phys. Rev. Lett. 39, 12 (1977).
29. W. Nazarewicz, Symposium on Nuclei at High Spin, Copenhagen, October 1989.
30. T. Bengtsson, Symposium on Nuclei at High Spin, Copenhagen, October 1989.
31. R. R. Chasman, Phys. Lett 242B, 317 (1990).
32. R. R. Chasman, Phys. Rev. C 5, 29 (1972).
33. P. Twin, Symposium on Nuclei at High Spin, Copenhagen, October 1989.
34. T. Byrski et al., Phys. Rev. Lett. 64, 1650 (1990).
35. F. S. Stephens et al., Phys. Rev. Lett. 65, 301 (1990).
36. K. T. Hecht and A. Adler, Nucl. Phys. A137, 129 (1969).
37. A. Bohr, I. Hamamoto and B. R. Mottelson, Phys. Scr. 26, 267 (1982).

SHAPE ISOMERS: MEAN-FIELD DESCRIPTION AND BEYOND[+]

P. Bonche*, S.J. Krieger, M.S. Weiss
Department of Physics,
Lawrence Livermore National Laboratory, Livermore, CA 94550, USA

J. Dobaczewski
Institut of Theoretical Physics, Hoża 69, PL00681 Warsaw, Poland

H. Flocard
Division de Physique Théorique[†], Institut de Physique Nucléaire,
91404 Orsay Cedex, France

P.H. Heenen[‡]
Physique Nucléaire Théorique, Université Libre de Bruxelles,
CP229, 1050 Bruxelles, Belgium

J. Meyer
Institut de Physique Nucléaire, IN2P3-CNRS, Université Lyon 1,
69622 Villeurbanne Cedex, France

INTRODUCTION

Nuclear Hartree-Fock (HF) + BCS calculations have led to predictions of shape isomerism in isotopes of Pt, Hg and Os nuclei[1]. These have been confirmed through the observation of superdeformed rotational bands in $^{190,\cdots,194}$Hg. Encouraged by these measurements and similar observations in ^{194}Pb, we have extended these calculations to a wide range of contiguous nuclei. These HF results, for ^{192,194}Pt, $^{190,\cdots,198}$Hg and ^{194}Pb, have been employed in a Generator Coordinate Method (GCM) calculation utilizing the quadrupole deformation as the generating variable[2]. The resulting spectra confirm the conclusions drawn from the HF results and agree with those experiments which have been performed.

Adding a phenomenological assumption for the moments of inertia of our GCM states, we can construct the radiative transitions within and out of the superdeformed band. The results are in good agreement with the observed de-population of the superdeformed band built upon the shape isomer both in minimum angular momentum and in rapidity of de-population. Inferences for the existence of shape isomers will be drawn.

---

* Permanent address: SPT-CEN Saclay, 91191 Gif sur Yvette Cedex, France.
† Laboratoire associé au CNRS
‡ Directeur de Recherches FNRS.

MEAN FIELD CALCULATIONS

Our description is based upon the Hartree Fock (HF) mean field theory which can be derived from a variational principle where the energy

$$E = \langle \Psi | \hat{H} | \Psi \rangle / \langle \Psi | \Psi \rangle \qquad (1)$$

is minimized for a given set of trial many-body wave-functions $|\Psi\rangle$. When all possible wave-functions are allowed, the exact many body Schödinger equation is recovered. To get the HF approximation, the variations are restricted to the Hilbert space built on Slater determinants only.

From an effective two-body force V_{ij} which includes among other terms a spin-orbit force, a pairing interaction and the coulomb repulsion between the protons, the HF approximation constructs the mean one-body potential which averages the interactions of a given nucleon with all the others. The full many-body problem is reduced to the solution of the one-body HF equations,

$$h \varphi^\lambda = e^\lambda \varphi^\lambda \quad . \qquad (2)$$

The HF hamiltomian, $h = t + U$, is the sum of a kinetic energy term and a mean field,

$$U(\mathbf{r}) = \int d^3 r' \, v(\mathbf{r} - \mathbf{r}') \rho(\mathbf{r}') \quad , \qquad (3)$$

which depends upon the solution of the HF equations through the density ρ. Due to this dependence the HF equations are non-linear. In addition, the mean field may be non local, depending upon the choice of the effective interaction. The solution of these equations yields both the single particle spectrum $\{e^\lambda, \varphi^\lambda\}$, and the total energy of the nucleus, $E = \sum_\lambda e^\lambda - \frac{1}{2} \sum_{\lambda\mu} V_{\lambda\mu}$, where the sum runs over occupied states only. The density matrix, $\rho_{\lambda\mu}$, built from the single particle spectrum, is used to calculate expectation values of one-body observables: $\langle \hat{A} \rangle = A_{\lambda\mu} \rho_{\mu\lambda}$. Finally, the many-body wave function representing the nucleus is a Slater determinant constructed on the occupied single particle orbitals φ_λ.

The HF energy being variational is guaranteed to be an extremum. However it may be only a local minimum. As the HF equations are non linear, they are solved iteratively from a suitable "first guess" of the converged solution. Depending on the starting point, it is a matter of experience that besides the absolute minimum, it is often possible to generate secondary local minima corresponding for instance to collective excitations of the nucleus.

To explore a given collective degree of freedom, such as the quadrupole deformation, it is convenient to generate a sequence of shapes evolving continuously from the ground state to an eventual secondary local minimum. For that purpose the two-body Hamiltonian $\hat{H}$ is replaced in the variational principal by the operator sometimes called the routhian $\hat{H} - \lambda \hat{Q}$, where $\hat{Q}$ is a collective operator which generates the different shapes and λ the corresponding Lagrange mutiplier. In general we choose for $\hat{Q}$ the quadrupole operator $\hat{Q} = (3z^2 - \mathbf{r}^2)$. The resulting new HF equations read as:

$$(h - \lambda \hat{Q})\varphi^\lambda = e^\lambda \varphi^\lambda \quad . \tag{4}$$

The value of λ is adjusted so that the expectation value of the collective operator $\hat{Q}$ is equal to a desired value $q = \langle \hat{Q} \rangle$. For each choice of q, one obtains a Slater determinant which minimizes the energy. In particular, the energy is minimal with respect to any other unconstrained degrees of freedom, at least locally. This implies that no parametrization of higher multipoles of the nuclear shape has to be done.

In the calculations presented below, we have used Skyrme-type two-body effective interactions[3]. This particular choice of interaction leads to a local mean field. In addition, we include in the Hamiltonian a term which describes the pairing correlations since thay are known to be crucial for a description of nuclear collective motion. We choose the standard constant G pairing matrix element ansatz.

We solve the HF equations in coordinate space according to the discretization procedure described in Ref. 4. The collective space is mapped through doubly contrained HF calculations defined by the quadrupole deformation q and the asymmetry angle γ. These two parameters are adapted to the description of any axial and non-axial ellipsoidal nuclear shape.

In fig. 1, we have drawn the quadrupole deformation energy curve for ^{194}Hg with the Skyrme interaction SkM*[5]. Only the axial degree of freedom is investigated so that $\gamma = 0°$ for the prolate deformations ($q > 0$) and $\gamma = -180°$ for oblate ones ($q < 0$). The curve exhibits three local minima. The absolute minimum is oblate and slightly deformed ($q \approx -10$ b). On the prolate side ($q \approx 8$ b) there is a second minimum whose excitation energy is about 1.0 MeV. At a much larger deformation ($q \approx 44$ b), there is a third minimum 5.0 MeV above the absolute minimum and separated from the main well by a barrier of 2.2 MeV (see Ref. 6).

These predictions have been verified through the observation of superdeformed (SD) bands extending to low angular momentum in the Hg nuclei[7,8]. They are qualitatively consistent with the model calculation of Pashkevitch in 1969 (Ref 9), with more recents calculations[10,11] and with the mean field calculations of the Gogny group[12].

Fig. 2a,e displays a more complete survey of the Hg isotopes. The deformation energy curves (solid lines) are obtained from our HF+BCS calculations using the axial quadrupole constraint described above and the same SkM* interaction[5]. The ground state configuration is associated with the absolute minimum. For each five isotopes, a secondary minimum occurs at a rather large deformation corresponding to a mass quadrupole moment of approximately 44 b. This deformation corresponds to a typical value of the β_2 parameter of 0.5, or equivalently to a 1.6 to 1 axis ratio of the ellipsoidal shape of the nuclear density. On the basis of these static calculations, one can identify these nuclei whose second minimum coresponds to an isomeric state which will ultimately decay electromagnetically.

The indirect confirmation of this assumption[7,8] through the experimental observation of the SD decay has led us to make an extensive study of the mass region ranging from Gadolinium nuclei up to Radium. The choice of this mass region is twofold. First these nuclei are heavy enough so that shape isomerism will occur at low spin: no additional centrifugal force is required to stabilize them as in the case of those Dysprosium isotopes which exhibit SD rotational bands which terminate at angular momentum above $20\hbar$. On the other hand, they must not be too heavy otherwise spontaneous fission will take over and we would enter the known region of fission isomers. The occurence of a second well deformed minimum for many nuclei in this mass region is a theoretical common pattern: fig. 3 presents our results for the region which we call the isthmus of shape isomerim. It may be used as an indication of candidate nuclei for the experimental observation of shape isomerism.

BEYOND THE MEAN FIELD

The generator coordinate method (GCM) is the next term of the Hartree-Fock theory which incorporates collective and single-particle dynamics into a coherent quantum mechanical formulation. Given a family of N-body wave functions $|\Phi(q)\rangle$, depending on a collective continuous variable q, the GCM determines approximate eigenstates of the Hamiltonian $\hat{H}$ having the form:

$$|\Psi_k\rangle = \int dq\, f_k(q)\, |\Phi(q)\rangle \ . \tag{5}$$

The weight functions f_k are found by requiring that the mean energy (1) calculated with $|\Psi_k\rangle$ is stationary with respect to arbitrary variations δf_k. This prescription leads to the Hill-Wheeler (HW) integral equation[13]

$$\int \left(\mathcal{H}(q,q') - E_k\, \mathcal{J}(q,q') \right) f_k(q')\, dq' = 0 \ , \tag{6}$$

in which the Hamiltonian ($\mathcal{H}$) and the overlap kernel ($\mathcal{J}$) are:

$$\mathcal{H}(q,q') = \langle \Phi(q)|\hat{H}|\Phi(q')\rangle \ , \qquad \mathcal{J}(q,q') = \langle \Phi(q)|\Phi(q')\rangle \ . \tag{7}$$

The HW equation is solved by using the generating states $|\Phi(q)\rangle$ defined for a finite set of deformations q_i[6], and discretizing the integrals in (5) and (6). One obtains a discrete approximation to the GCM, where kernels become matrices, and the integral equation (6) a matrix equation.

All the numerical problems related to the solution of the HW equations have been addressed to in Ref. 6. The same two-body Hamiltonian is used in the HF+BCS calculations and in the construction of the kernel (7). Since HF+BCS wave functions are not eigenstates of the particle number operator, the GCM states may not have the correct particle number. To restore it, one introduces additional constraints on the neutron and proton particle number operators in the GCM Hamiltonian kernel, with appropriate Lagrange multipliers.

Let us note that the GCM incorporates automatically the effects of ingredients of the Bohr collective Hamiltonian[14,4] method such as the zero point motion energy and the effective collective masses. Moreover, it does not rely on assumptions such as the gaussian overlap approximation nor on the validity of quadratic expansions of the kernels[15] which are often used to derive microscopically these ingredients[12].

The short horizontal bars in fig. 2a,e show the results of the diagonalization of the HW equations for five Hg isotopes ($N = 110 - 118$). The ordinate is the total energy of the states (where the GCM ground state is taken as zero) and the abscissa of the center of the bar the average of the quadrupole moment for that particular state. The energy shifts of the two HF minima relative to their corresponding GCM states are nearly equal. This means that the excitation energy of the shape isomer as predicted from static HF calculations is not significantly modified by the collective correlations induced by configuration mixing.

The modifications of the deformation energy curves as a function of N strongly affect the shape isomeric state. In fig. 2a,e the vertical dotted lines point to the mass quadrupole moment of the second well minima. One notes that the mean value $\langle \hat{Q} \rangle$ of the best superdeformed candidate amongst the GCM eigenstates (vertical arrow on the Q axis) is close to that of the HF+BCS minima at large N and slowly moves toward lower Q as neutrons are removed. This phenomenon can be understood by the changes as a function of N in the thickness and height of the barrier separating the first and second well. The smaller of either of these quantities, the more likely the GCM will mix HF states of comparable energy in the first and second well, shifting the shape isomer toward smaller deformations.

Assuming that the moment of inertia of the shape isomer is given by the rigid body value, our calculations reproduce the band head moment of inertia inferred from the recent measurements on $^{190-194}$Hg. Moreover, our results are compatible with the observation of ^{192}Hg and ^{194}Hg having the same moment of inertia while that measured in ^{190}Hg is slightly smaller[16]. Concomitantly, we predict that the moment of inertia for the SD band in ^{196}Hg and ^{198}Hg should be the same as observed in ^{192}Hg and ^{194}Hg.

DEPOPULATION OF THE SUPERDEFORMED BANDS

From this study, it is possible to make predictions on the de-excitation rates in and out of the superdeformed band. A necessary ingredient for that purpose is the set of matrix elements of the charge quadrupole operator. It can be calculated within the GCM, assuming that the rotational states can be described as a product of a collective rotor times an intrinsic state.

In addition, we introduce phenomenological estimates of the moments of inertia of the bands in order to allow an extrapolation of the GCM results to non-zero spin states. For the SD band, we choose the rigid body moment of inertia[17,18]. We then assume that the collective GCM states at smaller values of the quadrupole moment are band heads for rotational bands which function as doorways at the appropriate values of the angular momentum. From the choice of the moment of inertia for each band, one constructs a rotational spectrum based on GCM states in the first well. The moment of inertia $\mathcal{J}_k$ of a given band k is chosen phenomenologically[2] according to:

$$\mathcal{J}_k = \sqrt{a^2 + b^2(Q_k/Q_{SD})^2} \quad , \tag{8}$$

where $Q_k = \langle \Psi_k | \hat{Q} | \Psi_k \rangle$ is the quadrupole moment of the corresponding band head. The coefficients $a = 12.5\hbar^2/\text{MeV}$ and $b = 100\hbar^2/\text{MeV}$ have been selected to reproduce the rigid body moment of inertia at large Q and to have the order of magnitude suggested by experiment[19] at small Q. The last ingredient of the model is the energy $E_k^{0^+}$ of the first state of each band. It has been determined from the corresponding GCM intrinsic state energy E_k by subtracting the rotational energy:

$$E_k^{0^+} = E_k - \frac{\hbar^2}{2\mathcal{J}_k} \langle \hat{J}^2 \rangle \quad . \tag{9}$$

In this formula $\langle \hat{J}^2 \rangle$ denotes the HF+BCS expectation value total angular momentum for quadrupole moment value of the band head.

From the calculated spectrum and using standard formulae for transition rates[15], we can follow the population of the SD band as a function of angular momentum, assuming it to be unity at $30\hbar$. Our results are shown on the right side of fig. 4 for the five Hg isotopes. A nearly complete depopulation of the SD bands occurs in two to three transitions and begins at about 8 to $14\hbar$. Both results appear qualitatively consistent with experiments in the Hg isotopes. The left side of fig. 4 displays the spectrum of depopulating γ-rays expected from our calculated transitions. The pattern changes systematically as a function of neutron number, with the predicted spectrum hardening with increasing N. If this is the correct mechanism for the emptying of the SD rotational band, then perhaps these γ-rays could be observed. However, the actual γ-ray spectrum may be more complicated since we expect the states in the first well to dilute their strength within states of the non-collective continuum. The resulting spreading will render more difficult the identification of the depopulating γ-rays.

The mechanism responsible for the sudden depopulation of the SD bands has its origin in the structure of the transition rates which are proportional to the square of the quadrupole transition matrix element and to the fifth power of the energy of the transition. The former dependence favors in band transitions whereas the second is responsible for the depopulation as soon as the energy of an out of band transition is large enough to overcome the smallness of the off-diagonal matrix elements of $\hat{Q}$.

It is relevant to be concerned that other than quadrupole transitions will alter our conclusions. However we believe that direct out of band $E1$ transitions will not be important as most of the observed dipole strength lies at too high an excitation energy to affect this process in the Hg isotopes. The situation might be quite different for the SD bands observed in the Dysprosium region since they have been observed to disappear at a much higher angular momentum and absolute energy.

CONCLUSION

Microscopic mean field theory calculations have been performed for nuclei ranging from Gadolinium to Radon and a large number of these isotopes are predicted to exhibit shape isomerism. Where experiments have been performed within this isthmus of isomerism through observation of superdeformed bands built upon these predicted shape isomers, they are completely consistent with the measurements. We have compared the energy and position in collective space of the GCM eigenstates with those obtained in static HF+BCS calculation. We conclude from this study that the excitation energy of the SD band head or shape isomer predicted from HF+BCS methods is almost unaltered by the introduction of quadrupole correlations. The next step in this theory, GCM, yields subtle variation in the moment of inertia of the observed SD rotational bands in the Hg isotopes which are consistent with experiment and correctly reproduces the observed rapid de-population and limiting angular momentum. This explanation of the sudden de-population of the SD band makes it unlikely that the SD band head can be reached through the direct feeding of high lying rotational states at high angular momentum. Alternative experiments involving a direct feeding of SD states at low spin will have to be implemented.

REFERENCES

+ This work was supported in part by the NATO grant RG 85/0195, in part by the U.S. Department of Energy under Engineering Contract NW 7405-ENG-48, in part by SDIO/IST administered by NRL, in part by DOE-DOD(OM)MOU and in part by the Polish Ministry of National Education under Contract CPBP 01.09.
1. P. Bonche, S.J. Krieger, P. Quentin, M.S. Weiss, J. Meyer, M. Meyer, N. Redon, H. Flocard and P.-H. Heenen, Nucl. Phys. A500, 308 (1989).
2. P. Bonche, J. Dobaczewski, H. Flocard, P.-H. Heenen, S.J. Krieger, J. Meyer and M.S. Weiss, Nucl. Phys. A, in press.
3. D. Vautherin and D.M. Brink, Phys. Rev. C5 626 (1972).
4. P. Bonche, H. Flocard, P.-H. Heenen, S.J. Krieger and M.S.Weiss, Nucl. Phys. A443 39, (1985).
5. J. Bartel, P. Quentin, M. Brack, C. Guet and H.-B. Håkanson, Nucl. Phys. A385 269 (1982).
6. P. Bonche, J. Dobaczewski, H. Flocard, P.-H. Heenen and J. Meyer, Nucl. Phys. A510 466 (1990).
7. E.F. Moore et al, Phys. Rev. Lett. 63 360 (1989).
8. E.A. Henry, invited talk to this symposium.
9. R.R. Pashkevitch, preprint JINR (Dubna) P4-4383 (1969) unpublished.
10. See for instance S. Åberg, Phys. scripta, 25 23 (1982).
11. R.R. Chasman, Phys. Lett. B219 227 (1989), and invited talk to this symposium.
12. M. Girod, J.P. Delaroche, D. Gogny and J.F. Berger, Phys. Rev. Lett. 62 2452 (1989); J.P. Delaroche, M. Girod, J. Libert and I. Deloncle, Phys. Lett. B232 145 (1989).
13. D.L. Hill and J.A. Wheeler, Phys. Rev. 89 1102 (1953).
14. Å. Bohr and B. Mottelson, Nuclear Structure, volume 2, W. Benjamin, New York (1975).
15. P. Ring and P. Schuck, The Nuclear Many-Body Problem, Springer Verlag, Berlin (1980).
16. M.J. Brinkman et al, Z. Phys. A336 115 (1990); J.A. Becker et al, Proc. Conf. on nuclear structure in the nineties, Oak Ridge, Tennessee, ed. N. Jlhnson, Nucl. Phys. to be published.
17. H.J. Specht, J. Weber, E. Konecny and D.Heunemann, Phys. Lett. B41 43 (1972).
18. V. Metag, D. Habs and H.J. Specht, Phys. Reports 65 1 (1980).
19. B. Singh, Nucl. Data Sheets 56 75 (1989).

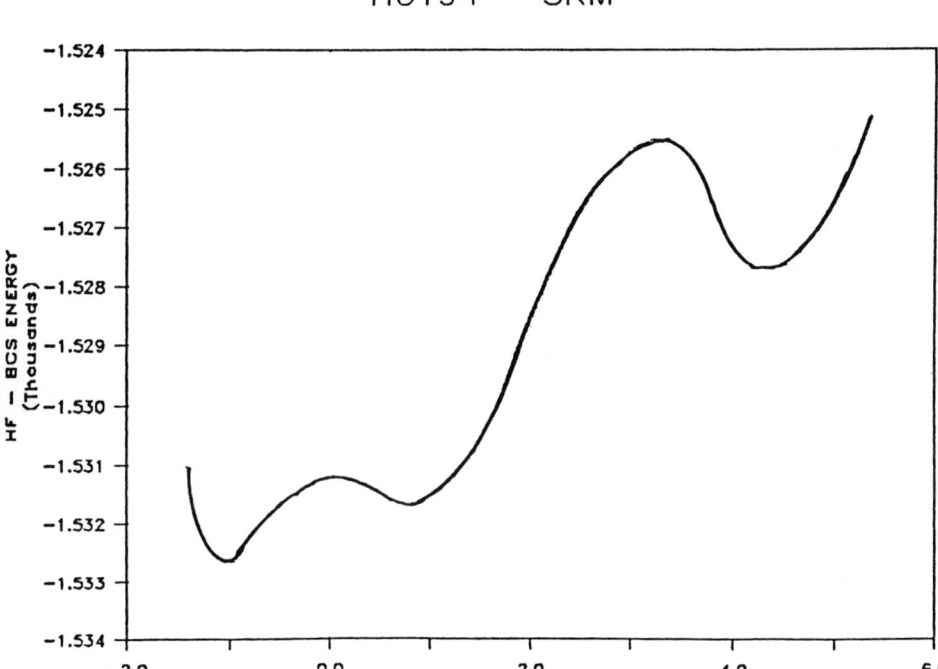

Fig. 1. HF+BCS deformation energy curve of ^{194}Hg as a function of the quadrupole moment (b) with the SkM* interaction.

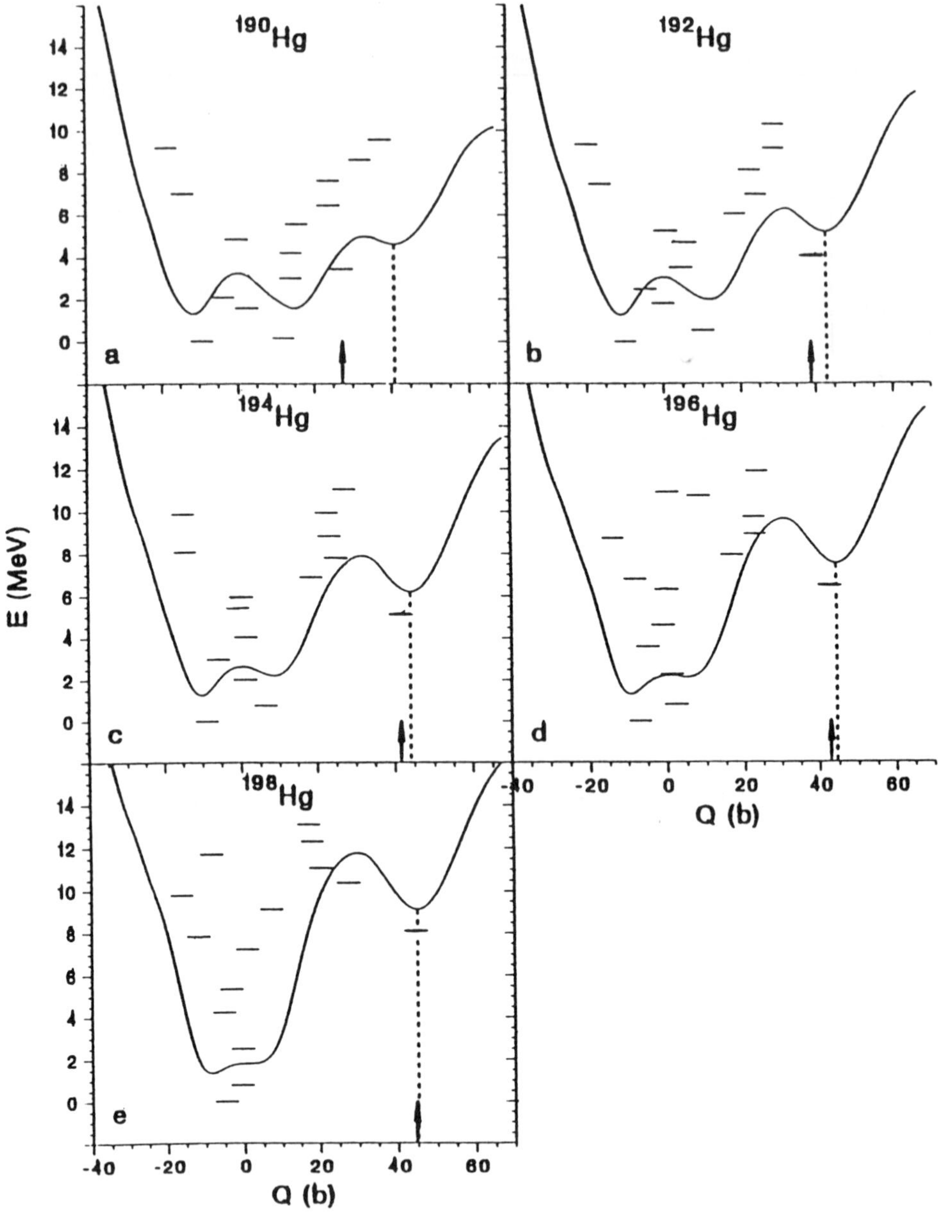

520 Shape Isomers

Fig. 2. HF+BCS deformation energy curves as a function of the quadrupole moment (b). The ordinate of the horizontal bars are the energies of the GCM states. The mid-segment abscissae give the GCM mean values of the quadrupole monent. The dashed vertical lines point to the HF+BCS SD minimum and the arrows to the GCM SD state. The interaction is SkM*.

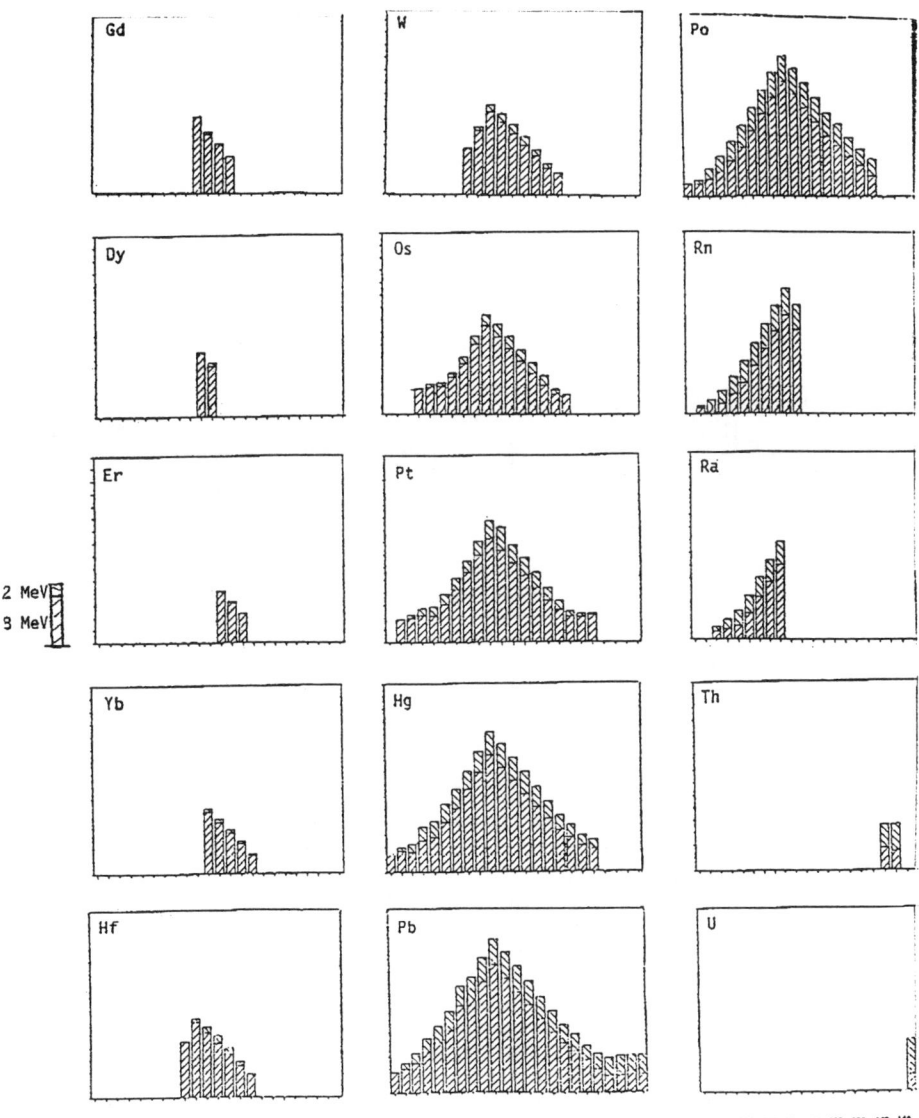

Fig. 3. Isthmus of shape ismerism: excitation energies and well depths of secondary minima for various isotopes of even nuclei ranging from Gadolinium to Radium. Some isotopes of Thorium and Uranium are also shown for comparison. The lower (upper) part of each bar diagram corresponds to the excitation energy (well depth).

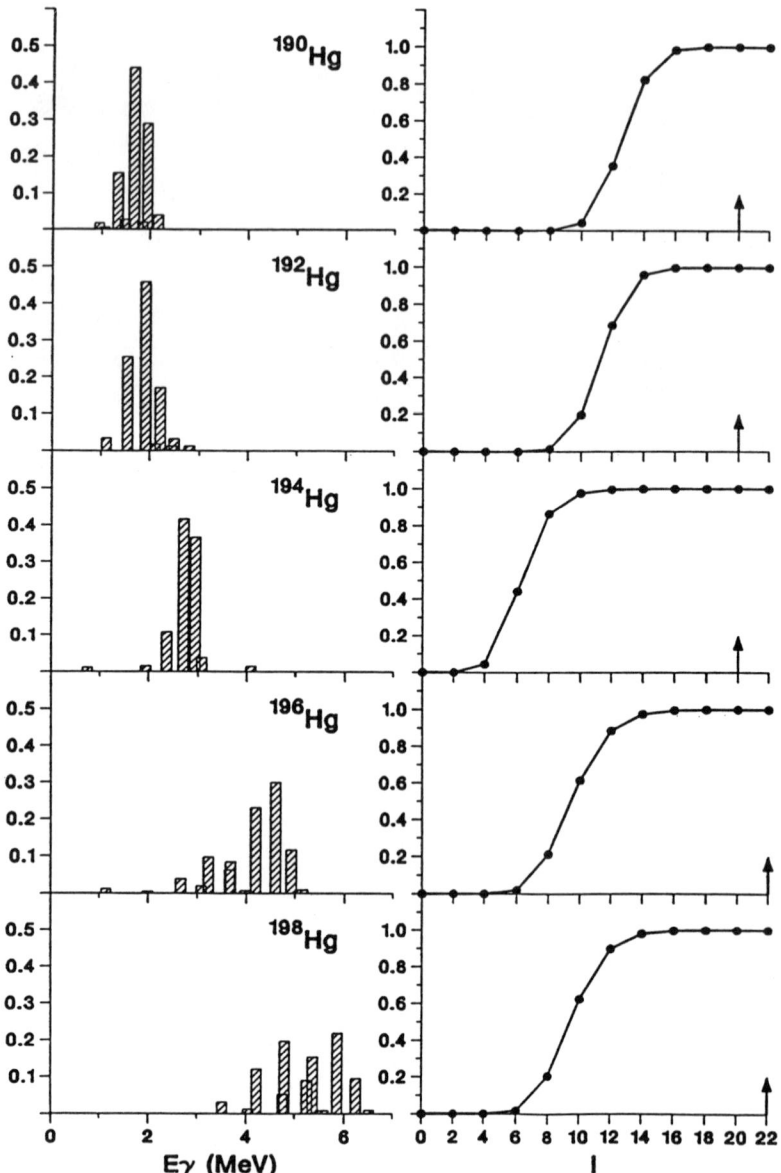

Fig. 4. Decay of the SD band for five Hg isotopes. Left side: relative intensities of the depopulating E2 gamma rays as a function of their energies. Right side: evolution of the population of the SD band as a function of angular momentum. The vertical arrow indicates the spin below which the SD band is excited with respect to at least one positive parity band.

Superdeformation in the Hg-Tl-Pb Region

E. A. Henry, J. A. Becker, S. W. Yates*, T. F. Wang, and A. Kuhnert
Lawrence Livermore National Laboratory
Livermore, California 94550

M. J. Brinkman and J. A. Cizewski
Rutgers University, New Brunswick, New Jersey 08903

M. A. Deleplanque, R. M. Diamond, F. S. Stephens
F. Azaiez, W. Korten, and J. E. Draper#
Lawrence Livermore Laboratory
Berkeley, California 94720

ABSTRACT

Superdeformation in the Hg-Tl-Pb region is discussed, with concentration on the experimental results. At least twenty-five superdeformed bands are known in this region, providing much new data to test theoretical calculations.

I. Introduction

The phenomenon of superdeformation has been recently identified in the region of the neutron deficient mercury, thallium, and lead nuclei. Superdeformation was first identified in the actinide nuclei nearly thirty years ago when Polikanov et al.[1] observed delayed fission from ^{242}Am. Experiments at the time showed that the fission isomers were neither of the ordinary "high spin" type, nor caused by small transition energies. The idea of shape isomerism was advanced to explain these nuclear states. In this view, the shape isomer is a state of the nucleus at an extreme deformation. Consequently, decay to normal nuclear states with little or no deformation is inhibited by the dramatic shape change required, and isomerism results. With the incorporation of shell effects into the liquid drop description of nuclear potential energies, deformed minima arose naturally in calculations and provided a theoretical basis for shape isomerism[2]. Theory predicted that secondary minima exist in the

*Permanent Address: University of Kentucky, Lexington, KY 40506
#Permanent Address: University of California, Davis, CA 95616

potential energy surface of nuclei in many regions of the nuclear chart, but additional examples of shape isomerism were found only in the actinides.

This situation changed dramatically in 1986 when Twin et al.[3] observed a cascade of seventeen gamma rays in ^{152}Dy data obtained with a heavy-ion reaction. They showed that these gamma rays were mutually coincident with each being about 47 keV from its neighbors. They recognized that this band was the signature of a rapidly rotating nucleus with a large prolate deformation. They established that the axis ratio is 2:1 for ^{152}Dy, as it is in the actinide nuclei. Soon high spin superdeformation was identified in other nuclei near A = 150, as well as A = 135. Recently superdeformation has been identified in the A = 142 region, and in the Hg-Tl-Pb region where the axis ratio is about 1.65:1. The rapid accumulation of experimental data has been accompanied by intense theoretical effort. Though significant understanding of nuclei at a large deformation is emerging, many unanswered questions remain. A partial list of properties necessary to characterize nuclei at large deformation includes: extent of the region in Z and N, excitation energy of the superdeformed (SD) states with respect to the ground state, spins and parities of the SD band members, level lifetimes, band depopulation, moments of inertia, the population mechanism, and the systematics of and relationships between the SD bands. In this paper, we review the experimental results for some of these properties for nuclei in the Hg-Tl-Pb region.

II. A Typical Superdeformed Band in the Hg-Tl-Pb Region

Superdeformed bands have been identified in data from heavy-ion reactions accumulated with large multidetector arrays such as HERA located at the 88-Inch Cyclotron facility at the Lawrence Berkeley Laboratory. These arrays are particularly effective for the study of events with a high gamma-ray multiplicity. The SD band in ^{192}Hg is typical of those found in the region (Fig. 1). The band consists of eighteen levels connected by seventeen gamma rays with energies ranging from about 200 to 800 keV. The energy spacing between gamma rays is about 30 keV near 800 keV, and increases to about 40 keV near 200 keV. The transition intensity gradually increases as the gamma-ray energy decreases, and reaches a maximum near the middle of the sequence. This maximum intensity is sustained down to the one or two lowest energy gamma rays. Then the intensity drops precipitously, and the gamma-ray sequence ends. These gamma-ray energy and intensity patterns are followed by almost all SD bands in the region. The ^{192}Hg SD band is atypical in one respect. It is populated relatively strongly, with about 2% of reactions that make ^{192}Hg resulting in population of the SD band. Many SD bands in the region are populated much more weakly, some as much as a factor of ten less.

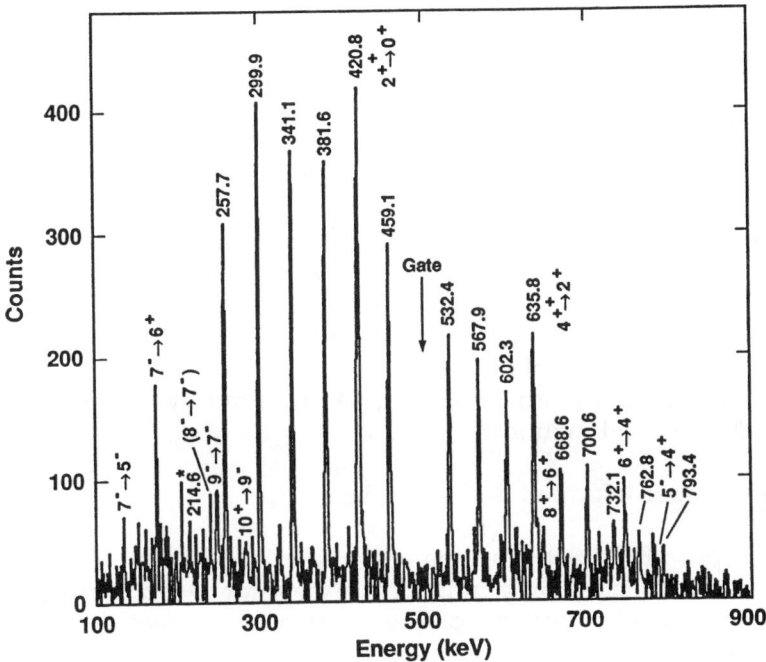

Fig. 1 Spectrum of the ^{192}Hg SD band gamma rays gated by the 496-keV gamma ray. Low-lying yrast transitions in ^{192}Hg are labeled by spin and parity, and ^{193}Hg gamma rays are indicated by an asterisk. (From ref. 5).

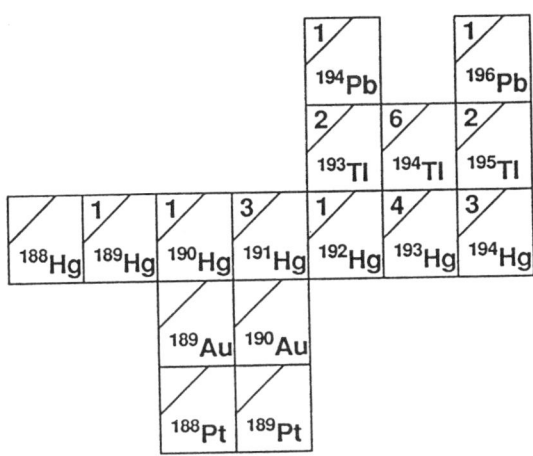

Fig. 2 Nuclei in the Hg-Tl-Pb region where SD bands have been identified. The number in the upper left corner indicates the number of SD bands known in each nucleus.

III. Extent of the Region of Superdeformation

At least twenty-five SD bands have been identified in eleven nuclei in the Hg-Tl-Pb region[4-15] (see Fig. 2). Many of these nuclei exhibit multiple bands. Much of the current research activity around the world is concentrated on defining the limits of this region of superdeformation. Searches have been made in ^{188}Hg (ref. 16), ^{190}Au (ref. 17), and ^{189}Pt (ref. 18) but no SD bands have been identified in these nuclei. All the more neutron rich nuclei in this region that have been studied in sufficient detail with heavy-ion reactions have been shown to possess SD bands. No results for nuclei with Z > 82 have been reported, although studies of Bi, Po, and Ra nuclei are planned or underway. Many nuclei that are predicted to be superdeformed can not be produced by currently available target-projectile combinations.

IV. Spin of Superdeformed Band Members

The spin of the SD band members is an important property that so far has been established only indirectly, except for several actinide nuclei. However, the regular pattern of the gamma-ray sequence immediately suggests that the SD bands are rotational in nature. The gamma rays of several strongly populated SD bands in the Hg-Tl-Pb region have been shown to be quadrupole transitions. Thus the level energies for these bands can be given by the rotational formula:

$$E_I = (\hbar^2/2\mathscr{I})[I(I+1) - K^2] + E_0. \qquad (1)$$

As can be seen from the spectra, the spin of a given level is also a smooth function of the gamma-ray energy. Following Harris[19], we have expanded the level spin as a function of rotational frequency:

$$I^* = 2\alpha\omega + 4/3\beta\omega^3 + \cdots, \qquad (2)$$

where $I^* = \sqrt{I(I+1)}$, I is the spin, and α and β are the inertial parameters, and $\hbar\omega = E_\gamma/2$ is the rotational frequency[20]. A series of such equations is generated, one for each level, with spin values increasing by $2\hbar$ with each transition. A least squares fit to this set of coupled equations yields I_f, the spin of the lowest level, as well as α and β. This method to determine spin of SD band members is effective in the Hg-Tl-Pb region because the gamma-ray sequence extends to low energies, where a change of one unit of angular momentum has a large affect on the level energy and the gamma-ray energy. This is in contrast to the A = 150 region, where gamma-ray sequences end at higher energy and the method cannot

Table 1. Results of least squares fits to eqn. 2 for sixteen superdeformed bands. The bands are identified by the nuclide and the energy of the lowest-energy band transition. Band energies were first fit with I_f, α, and β as free parameters. I_f was then fixed to the nearest integer or half integer (given here) and α and β fit again. The results of the final fits are tabulated here.

Nuclide	$E\gamma$	α $(\times 10^{-2} \hbar^2/\text{keV})$	β $(\times 10^{-8} \hbar^2/\text{keV}^3)$	I_f
^{190}Hg	360.0	4.129(3)	8.51(6)	14
^{191}Hg	350.6	4.444(5)	6.48(10)	29/2
^{192}Hg	214.6	4.366(2)	9.03(8)	8
^{194}Hg	254.3	4.430(1)	8.49(4)	10
^{194}Pb	169.7	4.398(4)	8.33(16)	6
^{196}Pb	215.8	4.363(5)	6.13(14)	8
^{191}Hg	292.0	4.713(3)	5.21(8)	25/2
^{191}Hg	311.8	4.709(4)	6.14(9)	27/2
^{193}Hg	293.6	4.663(7)	6.45(18)	25/2
^{193}Hg	234.5	4.662(8)	6.37(25)	19/2
^{194}Hg	201.2	4.677(2)	6.23(8)	8
^{194}Hg	262.5	4.682(4)	6.38(11)	11
^{193}Tl	228.1	4.777(5)	5.07(2)	19/2
^{193}Tl	248.3	4.773(4)	5.92(11)	21/2
^{194}Tl	268.0	4.988(5)	4.08(14)	12
^{194}Tl	240.5	4.767(8)	5.30(19)	10

distinguish between similar spin values for the lowest band member.

As a result of our fitting procedure, we know that many of the SD bands in the Hg-Tl-Pb region are depopulated completely near spin 10 (See Table 1). In ^{194}Pb the SD band is observed down to spin 6. Since the spins of the SD band members can be known with confidence now, important quantities such as the kinematic moment of inertia and the spin alignment can be determined. In cases where the population of low-lying yrast states can be observed, an approximate constraint on the number of gamma-rays present in the cascade from the SD band is established. In addition, the fitting procedure provides values for inertial parameters that can be used to compare different SD bands.

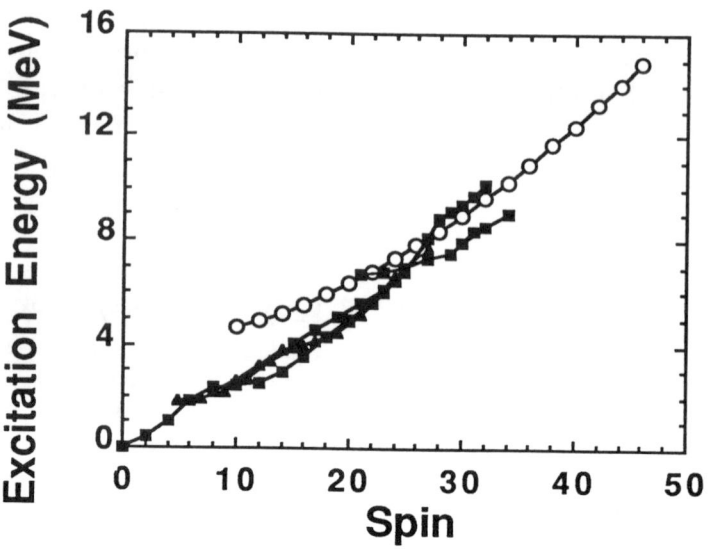

Fig. 3 ^{194}Hg normal levels (solid symbols) and SD band levels (open circles) plotted as a function of spin. The SD band becomes yrast at about spin 34. As a result, the SD spin-0 bandhead (unobserved) has an excitation energy of about 4 MeV.

V. Excitation Energy of Superdeformed Bands

An important, and as yet unresolved, question about SD bands is their excitation energies relative the the ground state. The precise way to determine a band's excitation energy is by identifying the gamma rays that connect the levels in the SD band to the normal levels in the nucleus. These connecting transitions have not been identified so far (except possibly in ^{191}Hg, an observation as yet unconfirmed). To gain insight into the excitation energy of SD bands in these nuclei, we have made the usual assumption that the SD band becomes yrast at the transition which exhibits half the maximum band intensity. In ^{194}Hg that transition connects levels with spins 36 and 34. In order to improve our knowledge of the yrast sequence of normal states in ^{194}Hg, we have extended the known level scheme up to approximately spin 34. A reasonable positioning of the spin 34 SD band member relative to the normal states of about the same spin results in the spin 10 SD band member being about 2.2 MeV above the lowest 8+ level (see Fig. 3). The spin-0 SD bandhead is then found by extrapolation to be about 4 MeV above the ground state. The three-dimensional microscopic Hartree-Fock calculations of Bonche et al.[21], which yield a SD minimum that is 4 to 5 MeV above the ground-state minimum, are in agreement for our estimate of the excitation energy for the SD bandhead in ^{194}Hg.

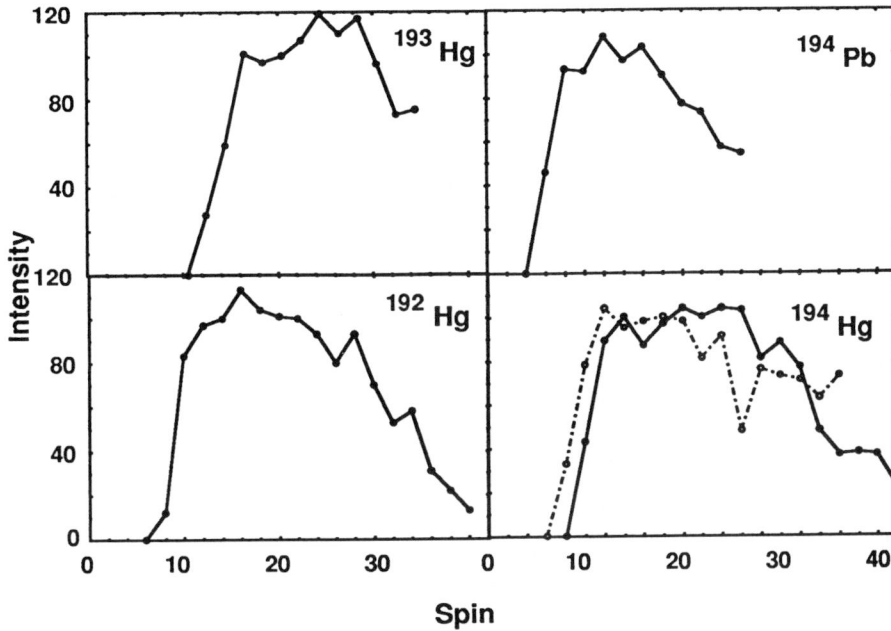

Fig. 4 Intensity remaining in the SD band as a function of level spin for some bands in ^{192}Hg, ^{193}Hg, ^{194}Hg, and ^{194}Pb. For ^{194}Hg, the intensity plots are for the yrast SD band (solid curve), and one of the excited bands (dashed curve).

VI. Depopulation of Superdeformed Bands

The intensities of the transitions in the all SD bands in the Hg-Tl-Pb region follow the same general pattern of an increase as the gamma-ray energy decreases until a maximum is attained about midway through the band, followed by a rapid decrease in intensity for the one or two lowest energy transitions. The SD bands are completely depopulated before the bandhead is reached. This observation is understood conceptually as competition between in-band and out-of-band transitions. However, detailed understanding of band depopulation is only beginning to emerge and probably involves the density of states, mixing of SD and normal states, a decrease of transition rates within the SD band, the influence of octupole and higher multipole degrees of freedom, and other phenomena.

Fig. 4 summarizes the depopulation of the SD bands as a function of level spin for ^{192}Hg, ^{193}Hg, ^{194}Hg (two bands) and ^{194}Pb. The intensity of all of these bands has decreased to below detectable limits by the time the gamma-ray cascade has reached levels in the spin range of 6 to 12. The SD band in ^{194}Pb is observed down to the spin 6 level, the lowest spin

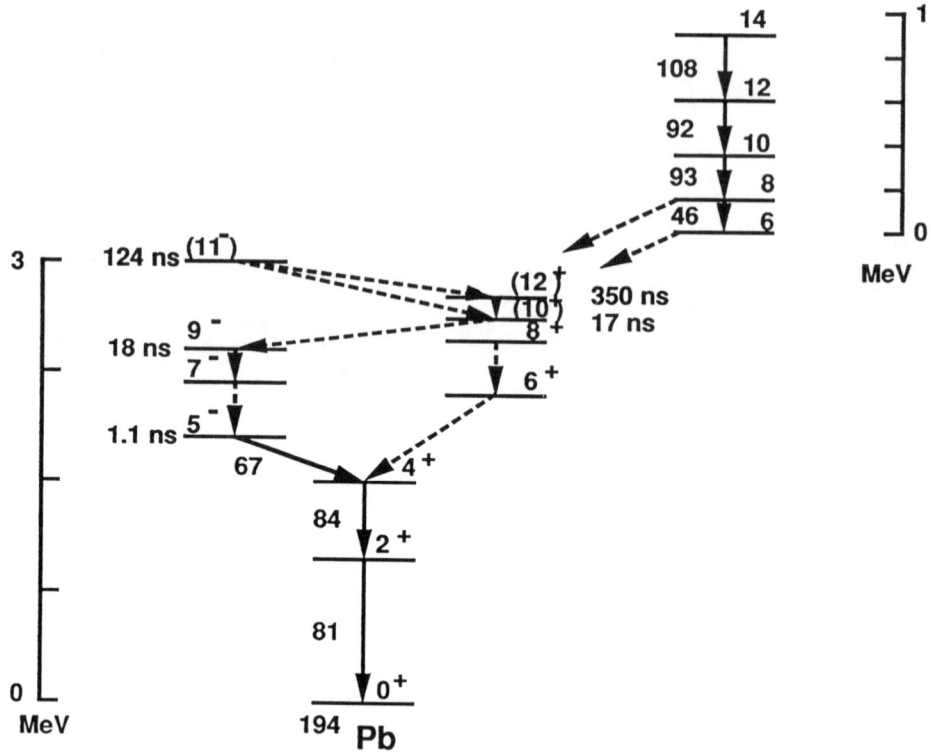

Fig. 5 Level scheme of ^{194}Pb, including the SD band at about 5 MeV of excitation. The transition intensities at the right of the gamma-ray arrows are normalized to 100 for the maximum intensity of the SD band. Intensity values are preliminary.

observed in an SD band outside the actinide region. On the basis of energy alone, the quadrupole transition rates for the lowest-energy gamma rays observed in these bands differ by about an order of magnitude or more, if all other factors are equal. Particularly interesting is ^{194}Hg where the most strongly populated (yrast) SD band is depopulated at spin 10, while a more weakly populated (excited) band is depopulated at spin 8. Theoretical calculations of Bonche et al.[22] are beginning to reproduce general band depopulation patterns such as the ones exhibited by these nuclei. However, detailed understanding of band depopulation will be necessary in order to predict with confidence the existence of shape isomers in this region.

In some even-A nuclei with few low-lying isomers, the feeding of normal yrast states can be measured, even if the connecting transitions can not be observed. Fig. 5 depicts a portion of the ^{194}Pb level scheme including the SD band. The maximum intensity of the band is taken as 100 units. The intensity of the lower SD band transitions and the normal states in coincidence with the band are indicated. Those transitions for which the intensity is not given cannot be observed and an upper limit on their intensity is about 20 units. The intensity of the 4^+--2^+ and 2^+--0^+ transitions is about 80 units, indicating that up to 20 units of the intensity may go through the 9^-, 10^+, 11^-, or 12^+ isomers and could not be detected in thin target experiments. Most of the intensity that is observed goes through the 5^---4^+ transition. If the SD band in ^{194}Pb has positive parity as would be expected, this observation indicates that E1 decay from the SD bands in this region may populate low-lying negative-parity levels directly. The importance of the octupole degree of freedom in the decay of the SD bands could be assessed once the E1 transition rates are measured. Cullen et al.[13] have suggested that octupole deformation may play a role in the decay of one SD band to another in ^{193}Hg.

This work is supported in part by the U. S. Department of Energy under Contract No. W7405-ENG-48 (LLNL), the SDIO/IST (LLNL), in part by the U. S. Department on Energy under Contract No. DE-AC03-76SF00098 (LBL), and in part by the National Science Foundation (Rutgers).

[1] S. M. Polikanov, et al., Zh. Eksp. Teor. Fiz. 42, 1464 (1962) [Sov. Phys,. JETP 15, 1016 (1962)].
[2] V. M. Strutinsky, Nucl. Phys. A95, 420 (1967), and Nucl. Phys. A122, 1 (1968).
[3] P. J. Twin, et al., Phys. Rev. Lett. 57, 811 (1986).
[4] E. F. Moore, et al., Phys. Rev. Lett. 63, 360 (1989).
[5] J. A. Becker, et al., Phys. Rev. C42, R9 (1990).
[6] D. Ye, et al., Phys. Rev. C41, R13 (1990).
[7] C. W. Beausang, et al., Z. Phys. A335, 325 (1990).
[8] E. A. Henry, et al., Z. Phys. A335, 361 (1990).
[9] M. Riley, et al., Nucl. Phys. A512, 178 (1990).
[10] M. P. Carpenter, et al., Phys. Lett. B 240, 44 (1990).
[11] M. J. Brinkman, et al., Z. Phys. A336, 115 (1990).
[12] K. Theine, et al., Z. Phys. A336, 113 (1990).
[13] D. M. Cullen, et al., Phys. Rev. Lett. 65, 1547 (1990).
[14] F. Azaiez, et al., submitted for publication (1990).
[15] F. Azaiez, et al., to be published.
[16] R. V. F. Janssens, private communication (1990).

[17] M. A. Deleplanque, private communication (1990).
[18] R. M. Diamond, private communication (1990).
[19] S. M. Harris, Phys. Rev. B 509, 138 (1965).
[20] J. Becker, et al., in Proceedings of Nuclear Structure in the Nineties II, Oak Ridge, Tennessee, 1990; ed. N. R. Johnson, Nucl. Phys. (to be published).
[21] P. Bonche, et al., Nucl. Phys. A500, 309 (1989).
[22] P. Bonche, et al., presented at the Conference on Nuclear Structure in the Nineties, Oak Ridge, April 23-27 (1990).

GLOBAL SYSTEMATICS OF SUPERDEFORMATION

R.K. Sheline and P.C. Sood
Florida State University, Tallahassee, FL 32306
and
Ingemar Ragnarsson, Lund Institute of Technology
Department of Mathematical Physics, P.O. Box 118, S-221 00 Lund, Sweden

ABSTRACT

The systematics of superdeformation ($\epsilon \approx 0.6$) is studied throughout the mass regions from A~8 to A~242, including the recently observed superdeformed nuclei with A~152 and ~192. Emphasis is placed on the light nuclei where calculated potential energy surfaces suggest new regions of superdeformation.

INTRODUCTION

We begin by looking at the general features of shell structure, the potential surfaces which derive from them and the differences between the different mass regions. We then proceed to take a more in-depth view of the systematics of superdeformation in the light nuclei, including the calculation of a number of potential energy surfaces, followed by a somewhat more cursory view of the intermediate and heaviest mass regions. A particular effort is made to suggest nuclei in which superdeformation might be experimentally observed.

GENERAL FEATURES OF NUCLEAR SHELL STRUCTURE AND THE POTENTIAL ENERGY FUNCTIONS RESULTING FROM THE NILSSON-STRUTINSKY FORMALISM

1. Shell Energies in the Pure Harmonic Oscillator

Fig. 1[1] shows the inhomogeneous nature of the level density of the harmonic oscillator (without spin-orbit coupling or other modifications) when these levels are plotted as a function of the quadrupole deformation coordinate, ϵ. Gaps (closed shells) occur for certain particle numbers when the axis ratios are small whole numbers (arrows at the bottom of Fig. 1). The degeneracy of the levels also has a profound effect on the calculated shell landscape. Those levels with large degeneracies are shown in Fig. 1 as bold. Specifically, levels with $n_z = 0$ and 1 are bold, where [$Nn_z\Lambda\Sigma$] are the usual asymptotic Nilsson quantum numbers. If we now calculate the shell energy as a function of quadrupole deformation, ϵ, using the Strutinsky formalism, in which the smoothed sum of the energy levels is subtracted from the actual sum of the energy levels, we get the shell energy landscape shown in Fig. 2.[2,3] Those levels with $n_z = 0$ produce the ridges in the shell energy (Fig. 2) shown with double dashed lines; levels with $n_z = 1$ give rise

© 1991 American Institute of Physics

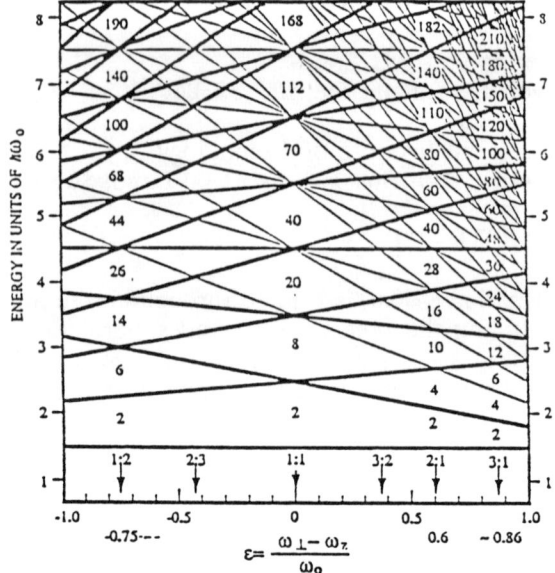

Fig. 1. Single particle energy levels of the spherical oscillator as a function of quadrupole deformation, ϵ. For certain axis ratios shell stability is observed at the number of particles indicated in the gaps. Levels with large degeneracies with $n_z=0$ or 1 shown bold.

to ridges with single dashed lines. In between adjacent ridges with $n_z = 0$, where the principal quantum numbers, N, differ by 1, lie valleys of stability. It is easy to note in Fig. 2 that these valleys of stability go "north east" from oblate through spherical to prolate with increasing particle number as discussed in more detail in refs 2 and 3 and more recently in ref. 4.

Thus we can expect local minima and special stability connected with a sequence of shapes which proceed from oblate to spherical to normal deformed to superdeformed to hyperdeformed with axis ratios, z:x, and deformation (ϵ) of $\approx 1{:}2(-0.75)$, $1{:}1(0.0)$, normal deformed (0.25-0.30), $\approx 2{:}1(\approx 0.6)$ and $\approx 3{:}1(\approx 0.86)$ for certain numbers of particles (nucleons). These particle numbers are shown in Fig. 2 at the appropriate minima. In this research we are especially interested in superdeformed nuclei with particle numbers 4, 10, 16, 28, 40, 60, 80, 110 and 140. It should, however, be noted that there is a systematic variation or alternation in the values of ϵ for the superdeformed shapes in Fig. 2. Specifically $\epsilon > 0.6$ for the particle numbers 4, 16, 40, 80, and 140, and < 0.6 for 10, 28, 60 and 110. This alternation effect plays an important role in the variation in deformation in superdeformed nuclei.

2. Shell Energies for the Nilsson Model

The main shell energy features of Fig. 2 are qualitatively unchanged when spin-orbit and l^2 terms are added to the harmonic oscillator Hamiltonian. The shell energy landscape for the resulting Nilsson model[2,3] shows quite clearly the well known fact that, for spherical nuclei with nucleon number $N < 20$, the closed shells are the same while for $N > 20$ the effect of adding the intruder orbital

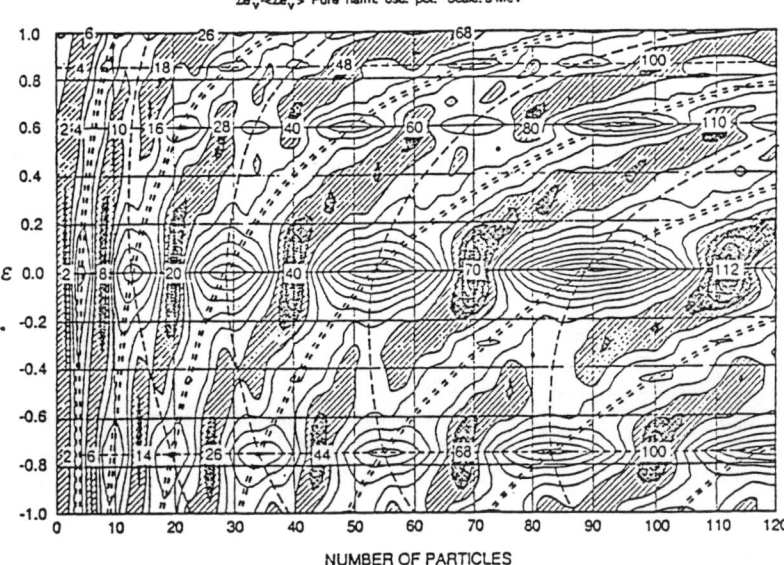

Fig. 2. A contour diagram of the shell effects of the spherical harmonic oscillator (Fig. 1) as a function of the number of particles and the quadrupole deformation, ϵ. Spacings between contour lines are 3 MeV. The shaded areas have $E_{shell}<0$. The double dashed and single dashed lines result when $n_z=0$ and 1 respectively.

increases the closed shell nucleon numbers in a systematic way in the Nilsson model. In a similar way the closed shell particle numbers for superdeformed nuclei are the same for the lightest nuclei but shift upward and become somewhat more diffuse for nuclei above N=20. In particular, 28 → 28-32, 40 → 40-44, 60 → 60-66, 80 → 80-86, 110 → 110-116 and 140 → 140-150. In addition, shell effects are smaller in the Nilsson model. Similar shell effects are found for other axis ratios. This is shown in Fig. 3 in which cuts are made through the shell energy landscapes of the simple harmonic oscillator (Fig. 2) and the Nilsson model for certain specific values of ϵ where the ratio of oscillator frequencies $\omega_\perp:\omega_z$ (corresponding to axis ratios z:x) are small whole numbers. The values of ϵ chosen are 0 (spherical) in 3a, 0.6 (superdeformed prolate) in 3b and -0.75 (oblate) in 3c. The dashed curves show the shell effects for the unmodified harmonic oscillator, and the solid curves, the Nilsson modified oscillator. It is obvious that shell effects are smaller for the Nilsson model with shell closures moving to higher and less well defined nucleon numbers particularly for deformed shapes.

3. General Features of Potential Energy Surfaces from the Nilsson-Strutinsky Formalism.

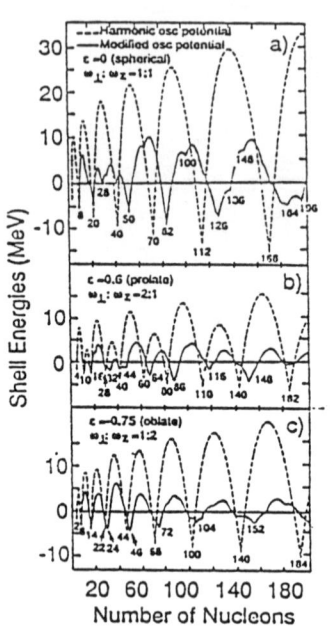

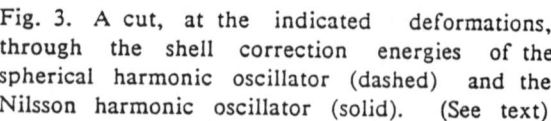

Fig. 3. A cut, at the indicated deformations, through the shell correction energies of the spherical harmonic oscillator (dashed) and the Nilsson harmonic oscillator (solid). (See text)

The total potential energy nuclear surface is calculated by adding the Nilsson shell energies described above together with pairing effects to the liquid drop macroscopic energy of Myers & Swiatecki[5] with their original parameters.

In the case of superdeformed nuclei the changing nature of the liquid drop with increasing proton and neutron number has a profound effect on the total potential energy surface. This is shown schematically in Fig. 4[6] for light, intermediate and heavy nuclei. In the lightest nuclei at the top of Fig. 4 the shell effects (shown cross-hatched) dominate the liquid drop potential (dashed). Therefore both the antishell effects near $\epsilon=0$ and the normal shell effects near $\epsilon\sim0.6$ produce a final potential (solid line) with a superdeformed ground state. In the intermediate case (middle of Fig. 5) the liquid drop potential dominates and superdeformation is only observed at high spin and high energies where superdeformation is stabilized (shown with the potential line labelled 60⁺). In the heaviest nuclei the combined effect of the Coulomb potential and the fissionability of the liquid drop causes the potential to reach a maximum at $\epsilon\approx0.6$. When the shell effect is added to this liquid drop potential, a second minimum, about 2-3 MeV above the deformed ground state, is produced in the total potential, implying a coexisting spectroscopy (bottom of Fig. 4). This second minimum is responsible for the fission isomers.

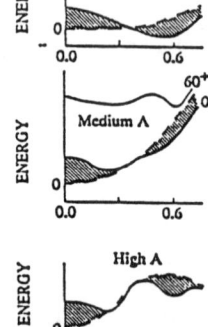
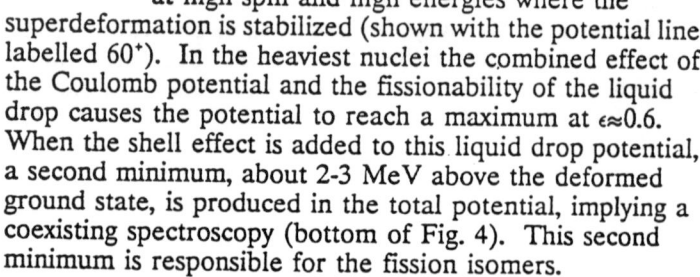

Fig. 4. Schematic drawing of the interactions of the liquid drop and shell effects for light (top), medium (middle) and heavy (bottom) nuclei as a function of quadrupole deformation, ϵ.

GLOBAL SYSTEMATICS OF SUPERDEFORMATION

Fig. 5 is a schematic representation of the nuclear periodic table with neutrons plotted against protons and the line of ß-stability shown dashed. Horizontal and vertical lines represent proton and neutron numbers with large shell effects for superdeformation ($\epsilon \approx 0.6$). When these lines cross we have nuclei or regions of nuclei in which both the neutron and proton shell effects reinforce this prediction of superdeformation. Individual nuclei and groups of nuclei at or near these crossing points are identified in Fig. 5.

As we shall see, the experimental spectra and detailed calculations for the nuclei in Fig. 5 agree well with each other and indicate that superdeformation occurs systematically all the way from ^{8}Be to the fission isomers with $A \approx 240$.

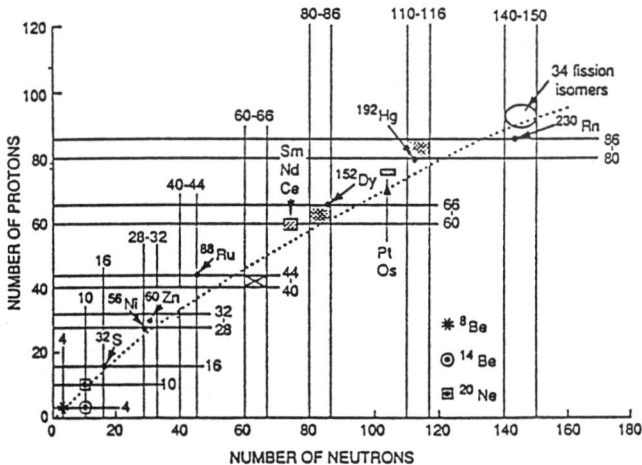

Fig. 5. Schematic representation of the nuclear periodic table. The line of ß-stability is shown dashed. Horizontal and vertical lines represent proton (horizontal) and neutron (vertical) numbers and regions of nuclei in which superdeformation ($\epsilon \approx 0.6$) is predicted. Those nuclei or regions of nuclei discussed in this paper are specifically identified.

1. Superdeformation in the Ground States of ^{8}Be, ^{14}Be, and ^{20}Ne

Knowing that 4 and 10 protons or neutrons correspond to a superdeformed system, we could expect that ^{8}Be$_4$ and ^{14}Be$_{10}$ should both have superdeformed structures for both protons and neutrons while in all other Be isotopes it is only the protons that tend toward superdeformation. With these expectations in mind, calculations have been made for the potential energy surfaces of ^{8}Be, ^{10}Be, ^{12}Be

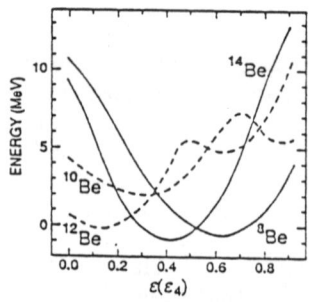

Fig. 6. Calculated potential energy surfaces of ^{8}Be, ^{10}Be, ^{12}Be and ^{14}Be ground states.

and ^{14}Be using the microscopic-macroscopic method. Note especially the comparatively large l·s coupling strength in the lower shells, $\kappa = 0.12$ for $N=1$ and $\kappa = 0.105$ for $N=2$. The results are shown in Fig. 6. ^{8}Be and ^{14}Be are clearly more strongly deformed than ^{10}Be and ^{12}Be in this calculation.

It is well known experimentally that ^{8}Be has a large deformation. On the other hand, nothing is known about the excited level structure of ^{14}Be. However the mass excess for ^{14}Be projected from the systematics in the 1983 Mass Table[7] was 41020±300 keV. This compares with the recently measured value 40100±130 keV. The difference, which is assumed to be the stabilization effect due to the shape change, is 920 keV. This tentatively suggests that ^{14}Be is considerably more deformed than ^{12}Be.

We can also ask what we should expect the quadrupole deformation of ^{20}Ne to be. With neutron and proton superdeformed shells, but with both expected to have $\epsilon < 0.6$, we might expect the deformation to be less than in ^{14}Be. In fact the calculated deformation is $\epsilon = 0.34$. Although the deformation in ^{20}Ne is relatively small, the same $N = Z = 10$ single-particle gap covers the whole deformation range from $\epsilon = 0$ to $\epsilon \approx 0.9$ with the gap being largest for $\epsilon \approx 0.55$. Therefore it appears appropriate to refer to the ^{20}Ne ground state as caused by the special shell structure for the 2:1 deformation especially in view of the alternation effect. Experimentally the ground state rotational band in ^{20}Ne is observed from 0^+ to band termination at 8^+, implying a strongly deformed system.

In summary, then, we have suggested both from the experimental spectroscopy, the mass-energy surface and the theoretical calculation of the potential energy surfaces that ^{8}Be, ^{14}Be and ^{20}Ne have <u>superdeformed ground states</u>. Finally it should also be noted that the stabilization and its implied shape change in ^{14}Be has been explained in terms of a neutron halo effect resulting from the large neutron excess and the near zero neutron binding.

2. Superdeformation in ^{32}S, ^{56}Ni, ^{60}Zn and ^{88}Ru

In this section we mostly consider $N=Z$ nuclei in which the superdeformed band is no longer the ground state but lies relatively low in energy. In ^{32}S and ^{56}Ni the 0^+ band head of the superdeformed band is tentatively identified. As we progress to ^{60}Zn and ^{88}Ru, these $N=Z$ nuclei are increasingly far from the line of stability; detailed experimental data are almost totally lacking.

One of the best candidates for additional experimental study is ^{32}S. However, no state of ^{32}S with definite spin more than 4 has been observed.[8]

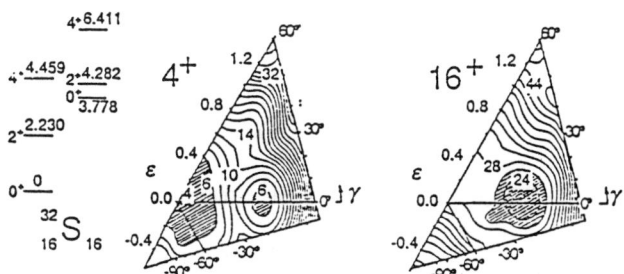

Fig. 7. The calculated 4^+ and 16^+ potential energy surfaces for ^{32}S in the ϵ, γ space. To the left the low energy spectroscopy of ^{32}S is presented. The 4.459 and 6.411 MeV 4^+ states may correspond to the 4 and 6 MeV minima in the 4^+ potential energy surface.

Calculations of potential energy surfaces for the 4^+ and 16^+ states of ^{32}S using the Nilsson-Strutinsky formalism are shown in Fig. 7 together with the experimental band structure which may correspond to the calculated minima. As expected from the alternation effect, a secondary minimum is observed in the calculated 4^+ energy surface at $\epsilon = 0.63$ (slightly greater than $\epsilon = 0.6$) but the ϵ of this minimum decreases with increasing spin until it reaches 0.56 ($\gamma = 3°$) at spin 16^+ (Fig. 7), by which spin the ground state band has disappeared. The superdeformed band becomes Yrast by spin 12^+. The energies of the experimental band structure (Fig. 7) are in excellent agreement with calculation.

The next superdeformed bands might be expected in the $N = Z$ nuclei ^{56}Ni, ^{60}Zn and ^{64}Ge with 28-32 protons and neutrons. A candidate for a superdeformed band is present in the ^{56}Ni literature at 5004 keV with a 0^+-2^+ energy difference of 351 keV. There is at present no evidence to connect the 0^+, 2^+ and 4^+ members of this possible band. Although little significant experimental work has been done on ^{60}Zn and ^{64}Ge, a study of ^{60}Zn is now underway at Nordball.[9] Finally, we might consider the $N = Z$ nuclei ^{80}Zr, ^{84}Mo and ^{88}Ru. These nuclei are quite far from stability and represent a severe experimental challenge. Calculations of the potential energy structures of ^{60}Zn and ^{88}Ru are shown in Fig. 8 for $I = 22^+$ and 36^+ respectively. As expected, the superdeformed minimum of ^{60}Zn at $\epsilon \approx 0.42$ is considerably smaller than that of ^{88}Ru ($\epsilon \approx 0.56$) because of the alternation effect. We point out that the calculations strongly

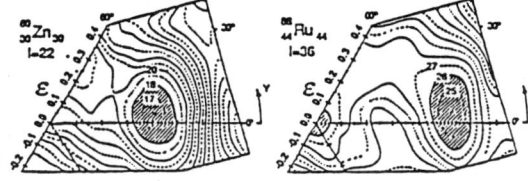

Fig. 8. The calculated 22^+ potential energy surface of ^{60}Zn and the 36^+ potential energy surface of ^{88}Ru in the ϵ, γ space. Solid contour lines differ by 2 MeV.

suggest that superdeformation is favored if we stay above the 440 1/2 and 550 1/2 orbits. This in turn implies that particle numbers 30 and 44 (^{60}Zn and ^{88}Ru) are better candidates for superdeformation than particle numbers 28 and 40 (^{56}Ni and ^{80}Zr). It should be noted that there are many other possible superdeformed nuclei not considered here--nuclei in which N≠Z and in odd-A nuclei.

3. Superdeformation in the Regions around ^{152}Dy and ^{192}Hg.

Fig. 5 suggests that the next superdeformed region should involve 60-66 protons and 80-86 neutrons. It is interesting to note in passing that this region of superdeformation had been predicted for more than 20 years.[10,11,12] Ragnarsson et al.[13] first noted that the ≈2:1 shape was stabilized at high spin.

These theoretical predictions have been verified using the new multidetector Ge-balls. Nine superdeformed nuclei were known in the ^{152}Dy region as of early 1990. They include ^{146,148,149,150}Gd, ^{150,151}Tb and ^{151,152,153}Dy.[14]

It should also be noted that a region of deformation with $\epsilon = 0.3$-0.4 has been studied in the Ce and Nd and Sm nuclei noted in Fig. 5 as a rectangular box. A summary[15] of this region has recently been given.

The next major region of superdeformation in Fig. 5 involves 80-86 protons and 110-116 neutrons. Very recently the first example (^{191}Hg) of superdeformation in this region was reported.[16] Since that time ^{189,190,191,192,193}Hg, ^{193}Tl and ^{194}Pb have been shown to be superdeformed.[17] These nuclei seem to be somewhat less deformed ($\epsilon \approx 0.50$) than superdeformed nuclei in the A=150 region. The Os and Pt nuclei with slightly smaller mass (see Fig. 5) have also been shown to have some of the characteristics of superdeformed nuclei. It seems probable that these nuclei like the Ce, Nd and Sm nuclei just discussed have an intermediate deformation ($\epsilon \approx 0.4$).

4. Superdeformation in the Fission Isomers

The discovery of the fission isomers (with half lives 24-30 orders of magnitude shorter than normal spontaneous fission) by Polikanov et al[18] ultimately required an interpretation[19] in terms of a second minimum in the potential energy surface. These fission isomers with major to minor axis ratios of 2:1 and $\epsilon = 0.6$ were the prototypes of superdeformation. In contrast to the ^{152}Dy and ^{192}Hg regions in which the second minimum is only stabilized at quite high energy and spin, a stable second minimum is achieved at the lowest spins. Thus, the fission isomers are more akin to nuclei such as ^{32}S or ^{56}Ni which are proposed in this research to have low lying K=0$^+$ bands corresponding to $\epsilon \approx 0.6$.

Fission isomerism is known[24] in 34 nuclei with proton numbers from 90-97 and neutron numbers from 141-151 (Fig. 5). They lie in the region where there are strong neutron shell effects but not large proton shell effects. In all other cases in Fig. 5 superdeformation occurs at or near the concurrence of strong 2:1 shell effects for both neutrons and protons. Although the experiment would be difficult, we should study a nucleus like ^{230}Ra (Fig. 5) where both proton and neutron shell effects might lead to superdeformation. Calculations[19] suggest that the superdeformed state would lie only about 1 MeV above the ground state. Fission from such a low lying superdeformed minimum should be severely impeded by the much larger barrier to fission to the point that it cannot compete with gamma decay to the ground state minimum.

5. $0^+ \rightarrow 2^+$ Energies in Superdeformed Rotational Bands

The $0^+ \rightarrow 2^+$ energies of the superdeformed rotational bands in the even-even nuclei ^{8}Be, ^{20}Ne, ^{32}S, ^{56}Ni, ^{152}Dy, ^{194}Hg and ^{238}U are plotted vs. mass number using a log-log format in Fig. 9. These energies are seen to alternate around the straight line which is the theoretical value with slope -5/3 expected for a rigid 2:1 structure. The $0^+ \rightarrow 2^+$ values for ^{152}Dy and ^{194}Hg are estimates which have been extrapolated from inertial values at much higher spins. We note the alternation in these $0^+ \rightarrow 2^+$ energies ($3\hbar^2/\Im$) described in Section II.

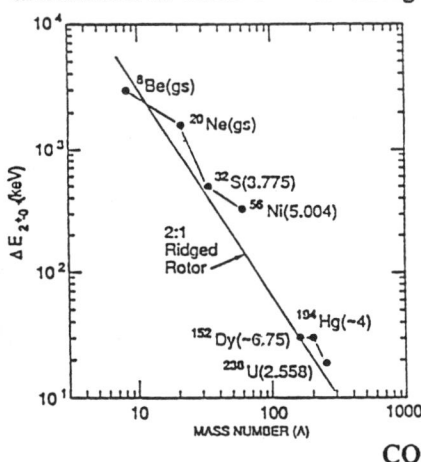

Fig. 9. A log-log plot of the rotational 0^+-2^+ energies of superdeformed nuclei against their mass numbers. Individual superdeformed nuclei are labeled with the nucleus and the energy of excitation of the band head in MeV. The $0^+ \rightarrow 2^+$ energies of a rigid rotor with $\epsilon=0.6$ are defined by the straight solid line with slope -5/3. The alternation effect described in the text is immediately obvious.

CONCLUSIONS

We have correlated superdeformation from the lightest to the heaviest nuclei. The recent efforts, particularly in the regions around ^{152}Dy and ^{192}Hg, fit well with both the lighter nuclei and the fission isomers. Superdeformation in the light nuclei ^{8}Be, ^{14}Be, ^{32}S, ^{60}Zn and ^{88}Ru has been emphasized by the calculation of their potential energy surfaces. Observation of superdeformation in ^{230}Rn might give evidence for a very low lying second minimum.

We are grateful to Aage Bohr, Ben Mottelson and Sven Gösta Nilsson whose interest and encouragement whetted our interest in this subject beginning already many years ago. We thank Sven Åberg for the unpublished calculated potential energy surfaces of ^{32}S, Pieter Endt who sent us the current level structure of ^{32}S and Gene Henry for the estimation of the superdeformed band head energy for ^{192}Hg. Financial support from the National Science Foundation under grant number PHY89-06613 at Florida State University, and from the Swedish Natural Science Research Council, is gratefully acknowledged.

REFERENCES

1. A. Bohr and B.R. Mottelson, Nuclear Structure. Vol II (W.A. Benjamin, Inc., New York, 1975).
2. R.K. Sheline, Proceedings of the Third International Conference on Nuclei far from Stability, May 19-26, Cargese, Corsica (France) CERN 76-13.
3. I. Ragnarsson, S.G. Nilsson and R.K. Sheline, Phys. Reports 45, 1 (1978).
4. J. Dudek, W. Nazarewicz, Z. Szymanski and G.A. Leander, Phys. Rev. Lett. 59, 1405 (1987).
5. W.D. Myers and W.J. Swiatecki, Ark Fys 36, 343 (1967).
6. A.K. Jain, R.K. Sheline, P.C. Sood and K. Jain, Revs. Mod Phys. 62, 393 (1990).
7. A.H. Wapstra and G. Audi, Nucl. Phys. A432, 1 (1985).
8. P.M. Endt and C. van der Leun, Nucl. Phys. A310, 1 (1978), and private communication from P.M. Endt, June 1990.
9. J. Nyberg, private communication to I. Ragnarsson, May 1990.
10. V.M. Strutinsky, Nucl. Phys. A95, 420 (1967).
11. V.M. Strutinsky, Nucl. Phys. A122, 1 (1968).
12. K. Neergard, V.V. Pashkevich and S. Frauendorf, Nucl. Phys. A262, 61 (1976).
13. I. Ragnarsson et al., Nucl. Phys. A347, 287 (1980).
14. D. Habs, Nucl. Phys. A502, 105c (1989).
15. P. Nolan, Proceedings of the Conference on High-Spin Nuclear Structure and Novel Nuclear Shapes, April 13-15, 1988, Argonne Nat. Lab.
16. E.F. Moore et al., Phys. Rev. Lett. 63, 360 (1989).
17. J.A. Becker et al., Phys. Rev. C41, R9 (1990); D. Ye et al., Phys. Rev. C41, R13 (1990); see also Vol 1 of the Proceedings, Nuclear Structure in the Nineties April 23-27, 1990 Oak Ridge Nat. Lab, and M.A. Riley et al., Nucl. Phys. A512, 178 (1990).
18. S.M. Polikanov, V.A. Druin, V.A. Karnaukhov, V.L. Mikheev, A.A. Pleve, N.K. Skobelov, V.G. Subbottin, G.M. Terakopyan and V.A. Fomichev, Sov. Phys. JETP 15, 1016 (1962).
19. I. Ragnarsson and R.K. Sheline, Physica Scripta 29, 385 (1984).

INTERACTING BOSON MODEL FOR SUPERDEFORMATION

M.Honma and T.Otsuka

Department of Physics, University of Tokyo, Hongo, Bunkyo-ku, Tokyo, 113, Japan

ABSTRACT

The existence of the small "spherical core" and the wide "valence shell" in the shell structure of the superdeformed nucleus are studied by a Nilsson calculation with a large deformation. Collective S and D pair dominance in the Nilsson wave function is also discussed. A simple model for the electromagnetic transition based on a boson picture is given, which can reproduce the observed sudden decay of the superdeformed band.

INTRODUCTION

The superdeformation is characterized by the axially symmetric deformation with axis ratio 2:1:1 (or 3:2:2, etc.), which is one extremity of a quadrupole deformation. The aim of this paper is to give one picture of such a limit from the viewpoint of IBM. In section 1, we study basic characters, such as the shell structure and the wave function of a significantly deformed nucleus by using a Nilsson model. In section 2, one schematic model for the electromagnetic transition is given. One of the advantages of a boson approach is the exact treatment of angular momentum, in other words, exact angular momentum projection can be carried out easily for boson systems. Thus, the description of the drastic change of some physical quantities may be possible.

§1. THE BASIC STUDY FOR BOSON DESCRIPTION

In order to investigate the relation between the superdeformation and IBM, we carried out a Nilsson calculation for ^{152}Dy proton orbits (Z=66) with a deformation parameter $\delta=0.5$ (axis ratio 2:1:1). The Nilsson orbit can be expanded in the spherical harmonic oscillator basis. The occupation probability in the spherical basis is shown in the lower row of Table I. It is also calculated for the case of the normal deformation ($\delta=0.25$) and listed in the upper row. For $\delta=0.25$, the orbits below the magic number Z=50 are completely occupied and there is only 2.1 nucleons in the orbits above the Z=82 magic number. This implies the stability of the Z=50 spherical core and it is sufficient to take one major shell above the core as an active valence shell. In the case of $\delta=0.5$, the spherical orbits are mixed significantly because of the strong quadrupole field. But the orbits below the Z=28 magic number is still fully occupied, indicating the existence of the small spherical core. In fact, by treating nucleons below Z=28 as the inert core, the calculated intrinsic quadrupole moment decreases only by 6%. The rest of the nucleons distribute over a large number of shells above the core. The number of nucleons in the orbits above Z=82 magic number is 13.8, requiring very wide valence

Table I. Occupation probabilities in spherical basis

orbit	$0s_{1/2}$~$0f_{7/2}$	28	$1p_{3/2}$~$0g_{9/2}$	50	$0g_{7/2}$~$0h_{11/2}$	82	$0h_{9/2}$~$0i_{13/2}$
$\delta=0.25$	100%		98~94%		16~66%		0~6%
$\delta=0.5$	94~100%		60~93%		17~30%		6~16%

space. Thus, the bosons which describe the superdeformation (super bosons) carry the collectivity of more than one major shell, in contrast to the normal ("usual") boson defined within one major shell. Since the number of bosons is given by the half of the number of valence nucleons in IBM, the number of the super boson becomes (66-28)/2=19, while the number of the normal boson is (66-50)/2=8. In general, the number of the super boson is about three times larger than that of the normal boson.

We next consider the structure of the wave function of valence nucleons. The Nilsson wave function is written as $|\psi> \propto (A^\dagger)^N |c>$, where $|c>$ denotes the spherical inert core, N stands for half of the number of valence nucleons, and $A^\dagger$ creates the Cooper pair in the Nilsson potential. The A pair can be decomposed as $A^\dagger = x_0 S^\dagger + x_2 D^\dagger + x_4 G^\dagger + \cdots$, where $S^\dagger, D^\dagger, G^\dagger$, etc. denotes the collective nucleon pairs with $J^\pi = 0^+, 2^+, 4^+$, etc. The probability of each pair shows that the S and D pairs dominate (77%) the A pair and the validity of the truncation is improved further (89%) if the G pair is included.

§2. A SCHEMATIC MODEL FOR THE ELECTROMAGNETIC TRANSITION

Based on the results obtained above, we give a simple model for the electromagnetic transition from the superdeformed band. Experimentally, the superdeformed band is composed of 10~20 γ-ray cascades with nearly constant energy difference. At a certain spin (about 10 in ^{194}Hg), the intra super-band transition suddenly becomes weak and terminates within the spin change $\Delta I \sim 4$. We shall consider this phenomenon below.

We adopt the sd-model for simplicity. The super bosons are denoted as s', d', while the normal bosons s, d. The number of the super (normal) boson is N (N_1) and in general $N \sim 3N_1$. A new boson σ with $J^\pi = 0^+$ is introduced to account for nucleons between the usual spherical closed shell and the closed shell for the superdeformed band as introduced above. Consequently the number of σ is $N-N_1 \equiv N_0$. We choose $(N, N_0, N_1) = (42, 27, 15)$ for ^{194}Hg. By definition, super bosons and normal bosons have finite overlaps: $[s'^\dagger, s] \equiv u_0, [d'^\dagger, d] \equiv u_2, [s'^\dagger, \sigma] \equiv w$, which are treated as input parameters.

We adopt two different types of one-body E2 transition operators. Although E1 transition may be sizable for superdeformed-to-normal transition in the realistic case, E1 transition has the similar spin dependence of the transition rate and its effect can be included approximately in the input parameters. The first type operator $T^{(1)}(E2)$ is defined by the linear combination of all possible boson one-body operators of rank two, $s\tilde{d}^\dagger, \sigma\tilde{d}^\dagger, [d\tilde{d}^\dagger]^{(2)}$, etc. The strength parameters are determined to reproduce the intrinsic quadrupole moment (about 18±3eb for ^{191}Hg) of the superdeformed state and turned out to be about 0.1eb. The second type operator $T^{(2)}(E2)$ is defined by $T^{(2)}_\mu(E2) = t^{(2)}_{ji} [b_j^\dagger \tilde{b}_i]^{(2)}_\mu$, where $b_i' = s'$ or d' and b_{jm} denotes a non-collective boson which corresponds to the two-quasi particle state. We consider b-bosons with $J^\pi = 2^+, 4^+$. The strength parameter $t^{(2)}_{ji}$ is varied up to 0.5eb, while the results are not changed appreciably. $T^{(1)}(E2)$ contributes to both the super-to-super and the super-to-normal transition, while $T^{(2)}(E2)$ causes only the super-to-normal (two-quasi particle state) transition.

The superdeformed state is obtained by the angular momentum projection from the intrinsic state which is the condensation of the intrinsic super bosons $b_0 = x_0's' + x_2'd_0'$ and expressed as $|SD;JM> = P^J_{M0}\sqrt{1/N!}(b_0^\dagger)^N|0>$ where P^J_{M0} denotes the angular momentum projection operator. We consider the vibrational normal state projected from the intrinsic state $\sqrt{1/N_0!N_1!(N_0-N_s-N)!}(\sigma^\dagger)^{N_\sigma}(s^\dagger)^{N_s}(d_0^\dagger)^{N_d}|0)$, where all possible combination (N_s, N_d) is included under the condition of $N_s + N_d = N - N_0$. These states cover almost the whole space spanned by s and d bosons. In addition, the two-quasi particle state $|n;(jJ_1)JM\rangle$ is constructed by coupling the N-1 boson normal state $|n;N-1,J_1M_1\rangle$ and b_{jm}-boson to total spin J. The N-boson normal state $|n;N,JM\rangle$ is given by $|n;JM\rangle = P_{M0}^J \sqrt{1/N_0!N_1!} \, (\sigma^\dagger)^{N_0}(b_0^\dagger)^{N_1}|0\rangle$, where b_0 denotes the the intrinsic normal boson $b_0=x_0s+x_2d_0$ and amplitudes x_0', x_2', x_0, x_2 are determined from those of collective nucleon pairs in the Nilsson+BCS calculation.

The energy level of the superdeformed state is estimated by a simple rotational band: $E(I)=E_0+I(I+1)/2\mathfrak{J}$. We adopt $E_0=9$MeV and $\mathfrak{J}=100\hbar^2/$MeV for ^{194}Hg. Because there is no experimental data, the value of E_0 is only an assumption. Since rather many normal states are included, it is no use treating each state separately. We parametrize the average energy level of the normal-state by a rotational formula with $E_0=2.5$MeV and $\mathfrak{J}=75\hbar^2/$MeV.

The calculated result is illustrated in Fig.1. The γ-ray flow means the probability that the transition intra super-band through a given spin occurs after the superdeformed band is excited at the maximum spin. By choosing input parameters properly, this model can reproduce the sudden decay of the superdeformed band. As long as we use one-body operators, the super-to-normal transition matrix elements can be divided into two parts approximately. The first part contains the strength parameters, C-G coefficients, etc. and its spin dependence is rather weak. The second part is the overlap between superdeformed and normal states with N-1 bosons. As is shown in Fig.2, this overlap has strong spin dependence and enhances the super-to-normal transition matrix element rapidly as spin goes down, leading to the sudden decay of the superdeformed band.

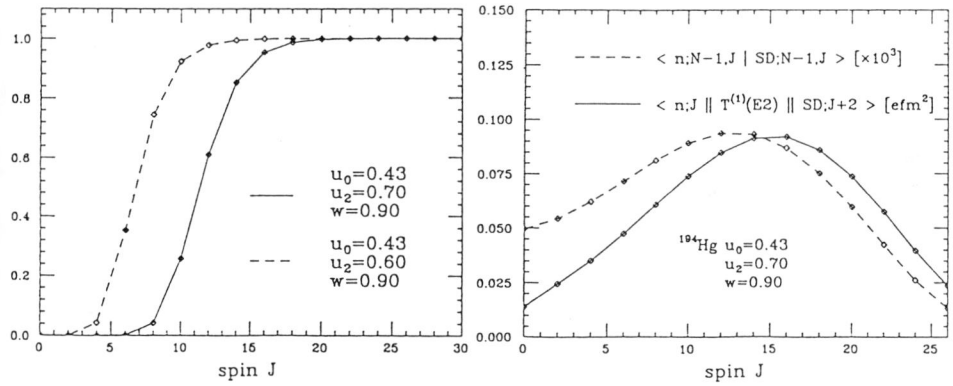

Fig.1. γ-ray flow in superdeformed band   Fig.2. Overlap and matrix element

SUMMARY

In summary, the superdeformed state is characterized by smaller spherical core and wider valence shell, which requires a large boson number in comparison with the normal case. Collective S and D pair dominate the wave function of the superdeformed state, similar to the normal case. Our model can reproduce the observed sudden decay of the superdeformed band. The strong spin dependence appears through the exact angular momentum projection.

Analysis of High-Multiplicity Gamma-Ray Events

M. J. Brinkman, J. A. Cizewski
Department of Physics and Astronomy, Rutgers University
New Brunswick, New Jersey 08903, USA

D. R. Manatt, J. A. Becker, E. A. Henry, N. Roy
Lawrence Livermore National Laboratory
Livermore, California 94550, USA

R. M. Diamond, F. S. Stephens, M. A. Deleplanque, C. W. Beausang, and J. E. Draper
Nuclear Science Division, Lawrence Berkeley Laboratory
Berkeley, California 94720, USA

ABSTRACT

A significant improvement in the measured peak-to-background ratio of highly-correlated gamma-ray cascades can be achieved through the use of high-multiplicity gamma-ray coincident events. We studied this effect using data from the ^{176}Yb(^{22}Ne,6n)^{192}Hg reaction at a beam energy of 122 MeV using HERA at the LBL 88-Inch cyclotron facility. An enriched set of all four- and higher-fold coincidences was culled from these data.

The use of high-multiplicity (i.e. four- and higher-fold) gamma-ray events provides unique opportunities to study nuclear structure. One of the primary advantages is that analysis of this type allows one to study highly correlated gamm-ray cascades without the necessity of background subtraction! This ability has already been exploited in earlier work[1].

The data presented here were collected following the bombardment of three, stacked ^{176}Yb foils (~450, 450, and 700 μg/cm² thick) by a ^{22}Ne beam with an energy of 122 MeV. The gamma rays were measured using the High Energy-Resolution Array (HERA) located at the LBL 88-Inch Cyclotron facility. HERA was designed to optimize the detection of high-multiplicty events; it is composed of 21 Compton-suppressed germanium (CSGe) detectors and was augmented by a 4π "inner ball" of 40 bismuth germanate (BGO) detectors which were put in place just prior to this run. All events in which two CSGe detectors fired in coincidence with eight or more BGO detectors were recorded on tape as were all events where three or more CSGe detectors fired in coincidence. In this run ~600 million events were recorded, and ^{192}Hg made up ~75% of the total reaction cross section. From the data recorded on tape an enriched data set that contained all of the events in which four or more CSGe detectors fired in coincidence (~10.7 million such events) was created and stored on disk.

We created four spectra to study the effects of different gating conditions on the measured peak-to-background ratio of high multiplicity events. The first was a total projection of all the gamma-ray energies. We then generated a list of seventeen energy gates that correspond to known[2,3] energies of the superdeformed band found in ^{192}Hg. This band was chosen as a candidate for study because it is highly correlated (being composed of 17 transitions) and is populated at only ~2% of the reaction channel's cross section

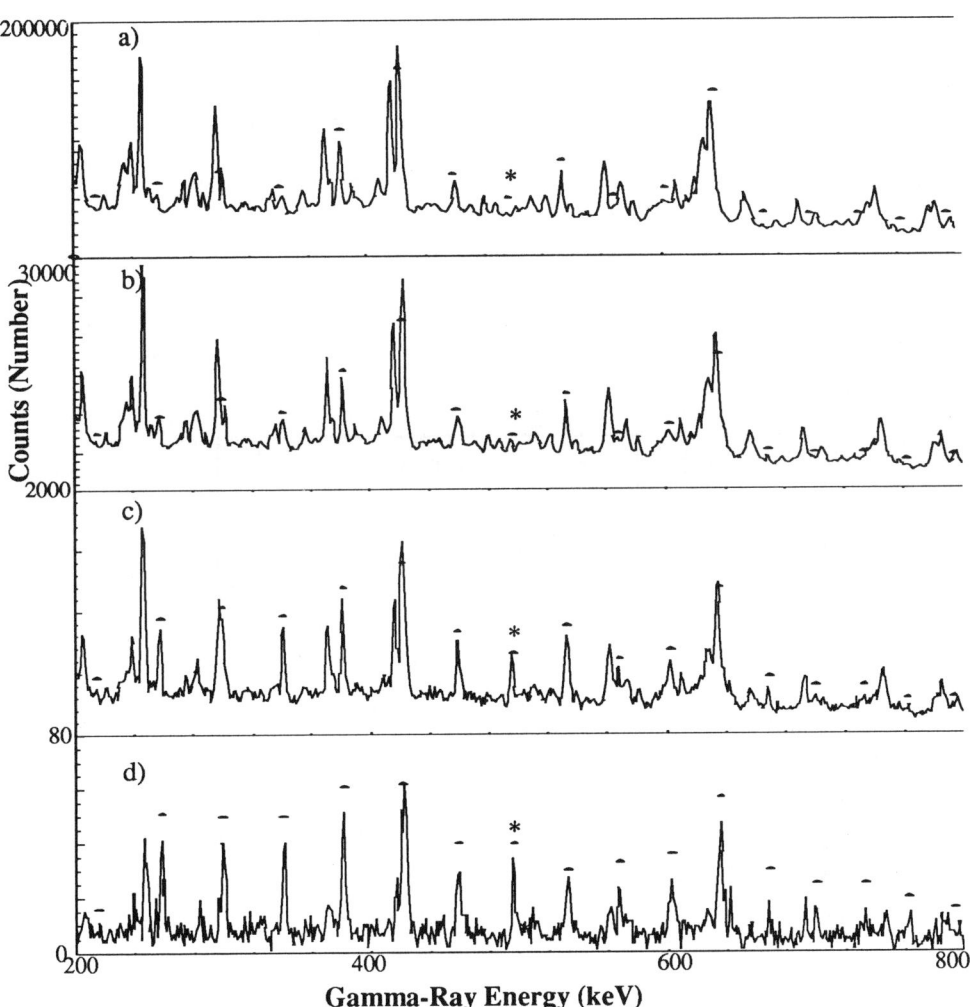

Figure 1: The ^{192}Hg superdeformed band, composed of 17 transitions at energies of 214.6, 257.7 299.9, 341.1, 381.1, 420.8, 459.1, 496.2, 532.4, 567.9, 602.3, 635.8, 668.6, 700.6, 732.1, 762.8, and 793.4, are denoted in each of the spectra by the black triangles. The spectra show the same quadruples data sorted with different gating conditions. The spectra are a) a total projection of all the quadruples, b) a sum of all spectra coincident with at least one band member, c) a sum of all spectra coincident with two or more distinct band members, and d) a sum of all spectra coincident with three or more distinct band members. In spectrum d) all of the transitions that are not members of the superdeformed band are known[4] low-lying states in ^{192}Hg that the superdeformed band depopulates through. The increase in the peak-to-background ratio can be best seen in the 496.2-keV transition (denoted by *) since it is a relatively "clean" peak.

(thus letting us check the sensitivity of the technique). In each of the three remaining sorts we increased by one the number of gamma rays in the quadruple that had to lie within distinct energy gates. In Figure 1 we show the spectra that resulted from these sorts.

In these spectra we can measure the changes in the peak-to-background ratio under different gating requirements. For the case of the 496.2-keV transition (see Table 1) the peak-to-background ratio improves by an average of 5.0 (6) for each increase in the number of gates set. This results in an overall improvement by a factor of ~125 from the total projection (where the superdeformed band can barely be seen) to the triple-gated spectrum (where the superdeformed band is the principle structure in the spectrum). Thus, even for weakly populated cascades, triple-gated quadruples sorting provides a means of studying their member transitions without the necessity of background subtraction.

Number of Gates	Peak-to-Background Ratio	Improvement Factor
0	0.0352 (2)	—
1	0.200 (2)	5.7 (3)
2	1.00 (4)	5.0 (2)
3	4.5 (9)	5 (2)

Table 1: The peak-to-background ratio for the 496.2 keV transition measured from sorts using differing number of gates. This transition was chosen since it is relatively "clean". The values are different for other members of the superdeformed band, but for all but the most "contaminated" transitions the improvement factor going from two to three gates is ~ 5.

This work was supported in part by the NSF (Rutgers) and in part by the U.S. Department of Energy under contract numbers W-7405-ENG-48 (LLNL) and DE-AC03-76F00098 (LBL).

[1] M. J. Brinkman et al., "Workshop on the Interface Between Nuclear Structure and Heavy Ion Reaction Dynamics", Notre Dame, 1990.
[2] J. A. Becker et al., Phys. Rev. C **41** (1990) R9.
[3] D. Ye et al., Phys. Rev. C **41** (1990) R13.
[4] H. Hubel, A. P. Byrne, S. Ogaza, A. E. Stuchbery, and G. Dracoulis, Nucl. Phys. A **453**, (1986), 316.

PATTERN RECOGNITION IN GAMMA-GAMMA COINCIDENCE DATA SETS

D. R. Manatt, F. L. Barnes, J. A. Becker
J. V. Candy, E. A. Henry
Lawrence Livermore National Laboratory
Livermore, CA 94550, USA

M. J. Brinkman
Rutgers University
New Brunswick, NJ 08903, USA

ABSTRACT

Considerable amounts of tedious labor are required to manually scan high-resolution 1D slices of two dimensional γ-γ coincident matrices for relevant and exciting structures. This is particularly true when the interesting structures are of weak intensity. We are working on automated search methods for the detection of rotational band structures in the full 2D space using pattern recognition techniques. For nominal sized data sets (1024x1024), however, these techniques only become computationally feasible through the use of Fourier Transform methods. Furthermore the presentation of data matrices as images rather than series of 1D spectra has been shown to be useful. In this paper we will present the data manipulation techniques we have developed.

INTRODUCTION

By examining a two-dimensional data set as a multi-dimensional entity, features that are obscured by interfering lines in the one-dimensional gated slices can be characterized. Fast fourier transform methods allow the use of generalized correlation techniques to detect and characterize patterns in the full two-dimensional space.

We use these correlation techniques to search a γ-γ coincident matrix for a fixed pattern of gaussian peaks in two dimensions, which is representative of the decay of a high spin rotational band. These methods extend our detection capability to include groups of very low intensity peaks.

PATTERN LOCATION

The correlation technique provides only the probable location of the centroid of patterns matching a given mask pattern in our data set, but with the proper choice of mask pattern, additional visually useful clues can be extracted. If the mask is chosen to have about half the extent of the pattern in the data set, multiple centroids will be indicated. If, in addition, the mask is configured with a fixed spacing and an odd number of rows and columns, the centroid of the mask corresponds to a point that falls on the middle row and middle column of the mask pattern, thus the indicated centroids will lie on many of the corresponding peaks in the data set.

ROTATIONAL BANDS

The correlation mask is determined by the pattern that one is searching for, in this case the gamma energies characteristic of high spin rotational bands. These energies are given by:

$$E_\gamma = \frac{\hbar^2}{2J} (4I - 2) \tag{1}$$

Thus theory tells us that the band members are evenly spaced. Furthermore, in a γ-γ coincidence matrix the peaks will be found in a square matrix with the diagonal absent. Figure 1 shows a typical 5x5 correlation mask with a spacing of 40 keV, which is characteristic of the rotational structures of interest.

ANALYSIS METHOD

To perform this analysis we first compress the data matrix by by a factor of four and apply an uncorrelated background removal.[1] A correlation mask is then built with the expected gamma-ray energy spacing. The cross correlation of the mask and the γ-γ coincidence matrix is calculated and the correlation spectrum is examined for peaks indicating a pattern match. The peaks thus detected

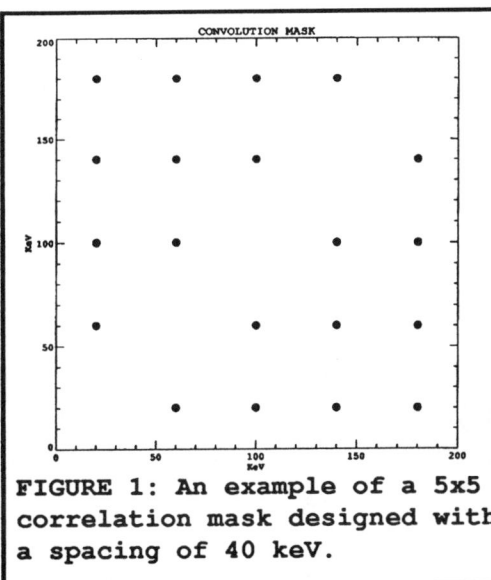

FIGURE 1: An example of a 5x5 correlation mask designed with a spacing of 40 keV.

are then characterized in the background subtracted data matrix.

To increase the sensitivity of the correlation process, additional transformation can be performed on the data matrix. The most common examples are upper limit thresholding to reduce the contribution of large ground state peaks to the correlation spectrum, and lower limit thresholding to eliminate artifacts resulting from the background removal process.

SUMMARY

Our preliminary results with this method indicate that the use of two-dimensional correlation techniques increases our ability to detect and characterize weak rotational band structures. We are continuing to develop this and other multi-dimensional analysis methods to characterize the shapes and the intensities of rotational band transitions.

* Work performed under the auspices of the U.S. Department of Energy by the Lawrence Livermore National Laboratory under contract number W-7405-Eng-48 and the U.S. National Science Foundation (Rutgers).

[1] O. Andersen et al., Phys. Rev. Let. **43** 687 (1979)

Capture Reaction Mechanisms, Proton Capture, and Resonances

DIRECT NEUTRON CAPTURE AND RELATED MECHANISMS

J. E. Lynn
Los Alamos National Laboratory, New Mexico 87545, USA
S. Raman
Oak Ridge National Laboratory, Tennessee 37831, USA

ABSTRACT

We consider the evidence for the role of direct and related mechanisms in neutron capture at low and medium energies. Firstly, we compare the experimental data on the thermal neutron cross sections for E1 transitions in light nuclei with careful estimates of direct capture. Over the full range of light nuclei with small cross sections direct capture is found to be the predominant mechanism, in some cases being remarkably accurate, but in a few showing evidence for collective effects. When resonance effects become substantial there is evidence for an important contribution from the closely related valence mechanism, but full agreement with the data in such cases appears to require the introduction of a more generalised valence model. The possibility of direct and valence mechanisms playing a role in M1 capture is studied, and it is concluded that in light nuclei at relatively low gamma ray energies, it does indeed play some role. In heavier nuclei it appears that the evidence, especially from the correlations between E1 and M1 transitions to the same final states, favours the hypothesis that the main transition strength is governed by the M1 giant resonance.

I. INTRODUCTION

After more than half a century of study we still do not have a fully quantitative, or even universally applicable, theory of nucleon radiative capture. What we have is a number of models that are applicable in different mass number or energy regions, and a certain amount of basic theory that explains in principle why some of these models work in certain situations. Fortunately, high quality experimental work and data continue to be produced in this area and this is leading to continued quantitative development and application of capture models. The present authors, together with a number of colleagues, have been attempting to build up a systematic study of the relative importance of the simple direct process[1] in thermal neutron capture. We have found that for all the nuclides so far studied up to mass number 50, direct neutron capture appears indeed to be the dominant capture mechanism for electric dipole transitions, and in many cases gives a remarkably accurate account of the cross sections[2,3,4,5]; the present status of this topic is reviewed in Section II.

In the cases so far studied the cross sections have been low (total thermal capture cross section < 1 b). Such low cross sections are typical of truly off-resonant energy regions, in which direct (or potential) capture theory is applicable. Higher thermal capture cross sections are indicative of substantial resonance contributions, in which the question of relative importance of a simple valence model, closely related to potential capture, with respect to more complex compound nucleus processes must be addressed. One example of interplay, at similar levels of amplitude, of these three effects is found in thermal neutron capture of ^{43}Ca, and this is briefly described in Section III.A and B. The simple valence model of resonance capture[5] has been found to be successful for a number of nuclides (see, for example, refs.[6,7]) and some further remarks are made on this topic in Section III.C.

The success of the direct capture model in explaining E1 thermal neutron capture by light nuclei invites the speculation that it could also be a significant model for magnetic dipole transitions. We survey the evidence for this in Section IV. Valence M1 transitions from resonance states are also possible and some evidence for this process is discussed in Section V. Finally, the possibilities and evidence for some more complex but definitive mechanisms are recounted in Section VI.

II. DIRECT E1 THERMAL NEUTRON CAPTURE

Although low energy neutron capture can be regarded as the classical compound nucleus process (the original experimental evidence for large neutron capture cross sections motivated Bohr in the creation of the compound nucleus concept) it has also been known for a long time that far from resonances a direct form of radiative capture, not involving the compound nucleus, can exist. In this, the neutron in its initial scattering orbital can simply fall into a bound single-particle state of the zero-order potential well representation of the neutron-target interaction. What makes this form of direct capture particularly amenable to calculation is that the major component of the radial factor of the electric dipole matrix element arises outside the potential radius where the relevant wave-functions are well-known, being dependent only on the neutron scattering length (for the initial state), the binding energy of the final state and its single-particle spectroscopic factor θ_f^2. Indeed, these factors are so dominant that many thermal neutron partial E1 cross sections can be quite accurately reproduced by a formalism that treats the potential with a sharp cut-off at the well radius and ignores all internal contributions to the matrix element. This version of the theory is known as channel capture, in distinction to the fuller version, called potential capture, that attempts to represent the wave functions more realistically (using a Woods-Saxon form of optical model for the initial state and a similar Eckart potential for the final state) and includes the internal contribution (from the wave functions projected on the internal continuation of the channels) to the matrix element.

The lightest nuclei for which E1 primary transitions are prominent in slow neutron capture are those from mass number A = 6-15, i.e. those in which final states of the $1p_{3/2}$ and $1p_{1/2}$ shell are available. Because in these nuclei (other than Li) a significant component of the $2s_{1/2}$ single particle state is just bound, the slow neutron scattering length is large compared with the potential radius, and hence there are large cancellations in the E1 radial matrix element giving rise to small cross sections despite the large spectroscopic factors for the single-particle p-wave content of the final states. Because of the large cancellations the cross sections are difficult to calculate accurately. Nevertheless, there is good agreement between experiment and theory, and the residual discrepancies indicate a very small admixture of more complex compound nucleus effects in the capture process[3].

In the next group of nuclei, A=19-22, the single-particle admixture of the final states is governed by the $2p_{3/2}$ shell, which is largely unbound in these nuclei with some spreading into weakly bound states. The E1 direct cross sections are therefore expected to be quite small because of small spectroscopic factors and the low energies of the transitions, but nevertheless we find that the experimental data[8,9] are in quite good agreement with the direct capture theory. The comparison is shown in Table I.

Table I. Primary E1 transitions from $^{19}F(n,\gamma)$, $^{20,22}Ne(n,\gamma)$ reactions

E_f (keV)	J^{π}	θ_f^2	E_γ (keV)	σ_{dir} (mb)	σ_{exp} (mb)	σ_{CN} (mb)
(A). $^{19}F(n,\gamma)$; a = 5.38 fm							
984	1⁻	0.0047	5618	0.096	0.14±0.01	0.004
1309	2⁻	0.0034	5292	0.080	0.23±0.02	0.052
1844	2⁻	0.0040	4757	0.103	0.19±0.02	0.013
4371		0.0006	2230	0.020	0.04±0.01	
4592		0.0040	2009	0.110	0.05±0.01	
4892		<0.0048	1709	<0.130	0.03±0.01	
5047	(2⁻)	0.0052	1554	0.130	<0.01	(0.130)
5224		0.018	1337	0.41	<0.01	
5319		0.011	1283	0.15	0.09±0.02	
5463		<0.023	1138	<0.3	<0.01	
5555	1⁻	0.01	1046	0.12	0.18±0.02	0.004
5810		0.10	791	0.1	<0.01	
5936	2⁻	0.086	665	1.22	1.44±0.09	0.009
6018	2⁻	0.136	584	1.75	3.48±0.20	0.29
(B). $^{20}Ne(n,\gamma)$; a = 4.43 fm							
4725	3/2⁻	0.475	2036	38.0	27±3	0.9
5694		0.1	1067	5.1	6.7	0.1
(C). $^{22}Ne(n,\gamma)$; a = 3.70 fm							
3221	3/2⁻	0.243	1980	23.5	25±4	0.02
3836		0.045	1365	3.2	3.6	0.01

The quantities σ_{CN} in the last column of Table I are inferred compound nucleus cross sections for the transition, deduced from the deviation between the direct capture theory estimate and experiment. While the largest cross section in the fluorine reaction is for the softest primary ray transition of 584 keV energy and is in strong qualitative agreement with the theory, the value of σ_{CN}/E_γ^3 is some two orders of magnitude greater than would be expected from Cameron's semi-empirical rule for partial radiation width magnitudes[10]. This cross section is now being remeasured so that the necessity of taking account of other mechanisms (e.g. collective coupling[4]) can be assessed.

In the next heavier group of nuclides(A = 24-38), the $2p_{1/2}$ single particle state becomes more deeply bound. In these conditions, we have found that agreement between the measured cross sections and theory is still generally good for both the isotopes of magnesium[4] and of sulphur[2]. The residual differences between theory and experiment, interpreted as "compound-nucleus" cross sections[3,4], are within the region expected from Cameron's semi-empirical relation[10].

In the calcium isotopes the $2p_{3/2}$ state is admixed among final nucleus states of low excitation, and the $2p_{1/2}$ single particle state is to be found at excitations of 3 to 4 MeV. Unlike the Be,C group the $s_{1/2}$ state is not bound to any significant fraction, and the potential scattering length is expected to be lower than the potential well radius, although individual measured scattering lengths are somewhat larger than the expected value owing to local level fluctuations. In this group of nuclei some of the most precise examples of agreement of measured E1 cross sections with the direct capture theory are to be found[5]. The cross sections for ^{40,42}Ca have recently been remeasured[11], and also new assessments of the spectroscopic factors have been made. The new comparison between theory and experiment is shown in Table II. It is apparent that the discrepancy between theory and experiment is very small, even for the weaker transitions to states at higher excitation energy, and the residual "compound nucleus" cross section (calculated neglecting experimental errors) is approximately two orders of magnitude lower than the measured cross sections. Agreement between theory and experiment for the target ^{44}Ca is not so good, the cross sections being systematically lower than those calculated from the model. This nucleus shows evidence of vibrational character, however, and it is plausible that collective coupling effects play a role in reducing systematically the direct cross sections[4]. Some other nuclei, especially in the Mg group, also show some evidence of systematic deviation from the rigid spherical direct model. Such target nuclei also have characteristic collective effects in their low-lying spectra.

Table II. Primary E1 transitions from ^{40,42}Ca(n,γ) reactions

E_f (keV)	J^π	θ_f^2	E_γ (keV)	σ_{dir} (mb)	σ_{exp} (mb)	σ_{CN} (mb)
(A). ^{40}Ca(n,γ); a = 4.64 fm							
1943	3/2⁻	0.535	6420	158	164±11	0.06
2462	3/2⁻	0.152	5900	47	25±2	3.4
3614	1/2⁻	0.065	4749	6.7	12.7±1.0	0.95
3731	3/2⁻	0.015	4632	3.7	0.4±0.1	1.7
3944	1/2⁻	0.435	4419	42.5	70±7	3.4
4603	3/2⁻	0.03	3760	6.1	9.9±0.9	0.46
4753	1/2⁻	0.135	3610	11.8	30±2	4.2
5072	1/2⁻	0.01	3291	0.8	1.5±0.2	0.1
5077	1/2⁻	0.01	3286	0.8	0.5±0.1	0.03
5120	3/2⁻	0.005	3242	0.9	1.1±0.2	0.01
5370	3/2⁻	0.005	2993	0.9	0.9±0.2	0.0
5450	1/2⁻	0.02	2913	1.4	1.5±0.2	0.0
5468	3/2⁻	0.022	2895	3.9	1.9±0.2	0.4
5670	3/2⁻	0.008	2693	2.9	1.5±0.2	0.2
5703	1/2⁻	0.045	2660	6.8	3.0±0.5	0.8
5889	1/2⁻	0.015	2474	2.1	1.9±0.2	0.01
(B). ^{42}Ca(n,γ); a = 3.31 fm							
593	3/2⁻	0.075	7339	55	42±3	0.9
2046	3/2⁻	0.68	5886	413	375±25	0.9
2103	3/2⁻	0.012	5830	7.5	6.7±0.7	0.02
2611	1/2⁻	0.135	5322	29	32±3	0.07
2878	1/2⁻	0.095	5055	20	20±2	0.0
2943	3/2⁻	0.05	4989	26	29±2	0.2
3286	3/2⁻	0.032	4647	16	21±2	0.3
3316	1/2,3/2⁻	0.04ᵃ	4617	4	4.6±0.5	0.02
3572	3/2⁻	0.04	4360	18	25±3	0.6
3869	1/2,3/2⁻	0.05ᵃ	4063	4	2.7±0.4	0.1
4207	1/2⁻	0.21	3726	67	67±5	0.0
4249	1/2,3/2⁻	0.055ᵃ	3684	8.5	7.9±1.1	0.01
4901	1/2,3/2⁻	0.09ᵃ	3032	11.5	11.5±0.8	0.0
5037	1/2,3/2⁻	0.095ᵃ	2896	11.7	13.8±1.0	0.1

ᵃ Assumption J = 1/2

III. E1 VALENCE CAPTURE

A. Valence capture in the thermal neutron region

The thermal neutron total capture cross sections in the above cases are all small (<1 b) indicating that the influence of resonance wings is rather small. In some other cases of target nuclei with mass number less than 50, nearby resonances quite clearly have a major influence on the cross section, and their more complex capture mechanisms have to be considered. Of these, the most elementary is the simple valence mechanism.

The formulation of thermal neutron direct capture theory within an optical model framework[1] gives rise to an imaginary term that is interpreted as the average resonance-resonance interference cross section of a simple valence transition from the single-particle component of the resonance to an equivalent component of the final state. From this term an estimate of the average valence radiation width $<\Gamma_{\gamma,\text{val})}>$ may be obtained. Alternatively, the valence width may be calculated directly within the framework of R-matrix theory[12,13]. The valence contribution appears in the thermal neutron cross section amplitude as a resonance type term proportional to $\Gamma_{\lambda(n)}^{1/2}\Gamma_{\lambda\gamma,\text{val})}^{1/2}/(E_\lambda-E)$. The individual valence radiation width for resonance λ has the important property of being proportional to the neutron width $\Gamma_{\lambda(n)}$. The valence term in the capture amplitude is constructive or destructive with respect to the potential capture amplitude, the resonance level E_λ being above or below thermal neutron energy E, respectively.

An interesting and important case of thermal neutron capture, in which the resonance influence is clearly dominant but not overwhelming, is that of ^{43}Ca. The total capture cross section is 6.2b, which is at least an order of magnitude greater than the cases considered in Section 2. The coherent scattering cross section is well known[5], but the total scattering cross section is not, nor are the resonance parameters (especially of the nearest resonance at 1.48 keV) known accurately, so the resonance contribution to the spin-dependent scattering lengths at thermal cannot be accurately assessed. However, using the currently available values of the 1.48 keV resonance parameters[14], the assumption that the total angular momentum $J = 3$ and the computed value of the potential scattering length from Moldauer's optical model parameters[15], we expect that $a_{J=3}$ is -3.7 fm with an uncertainty of ±50%. The value of $a_{J=4}$ is constrained, on choosing $a_{J=3}$, by the measured value of the coherent scattering length. Comparison of the primary E1 cross sections with direct capture calculations shows most consistency with a value of $a_{J=3} \approx -5.7$ fm, which is within the range of uncertainty of the value expected from the resonance parameters. The corresponding value of $a_{J=4}$ is 1.8 fm, and the ensuing discussion is confined to this choice of the scattering properties.

The calculated values of the direct capture cross section to the final states of ^{44}Ca are shown in Table III, in comparison with experiment.

In many cases the final state spin is assigned on the basis of the closest match to the experimental datum. Even with this selection it is apparent that for a relatively small number of transitions there is no correlation between experiment and theory. (Part of this can be explained by the fact that the ℓ_n assignments are not known for many states, especially at higher excitation, and some of these may have large $1f_{7/2}$ admixture). The overall magnitude of the theoretical cross sections matches the measured strength, however, the total direct capture cross section for the transitions listed in Table 3 being 3.1 b. These are the transitions to states up to excitation 6.15 MeV and exhaust most of the $2p_{3/2}$ strength. There is also expected to be a contribution to the direct capture cross section from the $2p_{1/2}$ state. This is presumably to be found at higher excitation energies, to which the primary E1 transition strengths are unknown. We can make an estimate of the cross section to a single $p_{1/2}$ state 3.3 MeV below the neutron separation energy; it is 0.6b, giving a total direct cross section of 3.7b. These figures represent a very substantial fraction of the measured total capture cross section of 6.2b, indicating that even in this quasi-compound nucleus situation direct capture plays a significant role.

Table III. Experimental data and calculated direct capture cross sections in the ^{43}Ca(n_{th},γ) reaction

E_f (MeV)	J_f^π	θ_f^2	E_γ (MeV)	$\sigma_{\gamma,dir}$ [a] $J_f=2$	$J_f=3$	$J_f=4$	$J_f=5$	$\sigma_{\gamma,expt}$ (mb)
1.157	2$^+$	0.08	9.974	179				6.8
2.283	4$^+$	0.007	8.848			12.3		22.9
2.656	2$^+$	<0.02	8.474	28.5				4.3
3.044	4$^+$	*	8.086			~0		41.5
3.301	2$^+$	*	7.829	~0				37.2
3.357	2-4$^-$	~0	7.773	~0	~0	~0	~0	188.5
3.776	2	~0	7.354	~0				30.4
3.923	3$^+$,4,5	0.32 [b]	7.208		71.2	46.2	14.9	96.1
4.196	2$^+$	0.03	6.935	42.5				54.6
4.480	2$^+$	0.06	6.651	80.0				26.0
4.584	2$^+$,3,4	~0	6.547	~0	~0	~0		146.3
4.651	2$^+$	0.45	6.480	539				142.6
4.690	1-4	~0	6.441	~0	~0	~0		24.2
4.914	?	0.19 [b]	6.217	217	172	112	37.8	~0
4.992	?	0.08 [b]	6.139	88.9	70.5	46.1	15.5	~0
5.006	4$^+$	0.40	6.125			229		230.6
5.130	2$^+$,3$^+$	0.96 [b]	6.001	206	164			210.2
5.231	2-5$^+$	4.32 [b]	5.901	901	716	469	160	432.1
5.289	2-5$^+$	2.16 [b]	5.842	445	353	231	79.2	72.5
5.342	2$^+$	0.45	5.789	454	361	237	81.3	21.7
5.375	2-5$^+$	0.56 [b]	5.756	113	89.6	58.8	20.2	52.7
5.459	2-4$^+$	2.64 [b]	5.673	519	413	271		20.5
5.549	2-4$^+$	3.20 [b]	5.582	615	489	321		61.4
5.733	2-5$^+$	6.00 [b]	5.398	1095	872	574	210	232.5
5.867	2$^+$-5	1.28 [b]	5.264	225	179	118	42	73.8
6.040	2-5$^+$	0.64 [b]	5.092	107	85	56	20	24.8
6.146	2-5$^+$	3.68 [b]	4.984	594	473	313	112	69.4

[a] Single entries are under the known spin of the final state. Multiple entries for unknown or range of spin assignments. Bold entries indicate favoured spin on basis of theory-expt.
[b] Quantity shown is $(2J_f+1)\theta_f^2$

This is in some contrast to the resonance itself. The radiation width has not been directly measured, but if it is inferred from the thermal capture cross section and the 1.48 keV resonance parameters, it is found to be approximately 1.9 eV, but could be less than 1 eV if the interference between potential and valence capture is taken into account. The radiation width for the valence mechanism, evaluated as in ref.[13], is 0.13 eV for the final states listed in Table III and there will be an extra contribution of 0.028 eV from states centred from a $2p_{1/2}$ state centred about 7.8 MeV excitation. The smallness of the valence contribution compared with the apparent total radiation width of 1.9 eV is in contrast with the much larger ratio of direct to total at thermal and emphasises the importance of the constructive interference between resonance valence and potential capture at off-resonance energies.

The difference between the estimated direct and the measured capture cross sections of Table III can be used to estimate the magnitude of an assumed "compound nucleus" contribution to the capture. This turns out to be on average $\sigma_{\gamma,cn}/E_\gamma^3 = 3.10^{-8}$ b.MeV^{-3}, and leads to an estimate for the average radiation width for a single transition $\langle \Gamma_{\gamma,cn}/E_\gamma^3 \rangle = 3.10^{-8} E_\lambda$ eV.MeV^{-3} = $4 \cdot 5.10^{-5}$ eV.MeV^{-3}. This is almost four times larger than the Cameron semi-empirical estimate[10] of compound nucleus radiation widths, but is about equal to the estimate obtained from the Brink-Axel giant resonance model[16,17]. The latter model does not, however, explain the general spectrum shape; the calculated radiation width for gamma rays of energy below 5 MeV is only 0.13 eV, accounting for only 0.4b of the measured cross section of 3.9b for these transitions.

B. Generalised valence capture

The simple valence mechanism discussed above is special in its relationship to the entrance channel, which correlates the radiative width through this process to the neutron width of the resonance and gives systematic interference with the potential capture cross section in off-resonance situations. It is possible, however, to describe other aspects of the capture mechanism through single particle transitions of this type, but with the initial single particle motion not being that from the entrance channel. In this generalised valence model[12] the single particle motion is in the field of an "inert" core, which can be one of many excited states of the target, or of the residual nucleus that results from removing a proton from the compound nucleus.

In a resonance state λ there will be an intensity admixture $c_{\lambda,xa}^2$ of a single particle state a in the field of a non-changimg core state x. If x represents the ground state of the target the radiative width to a final state $x \otimes b$ can be calculated by the R-matrix treatment of ref.[11]. For excited states x' of the core, the single particle state and associated radiative width in resonances at low neutron energy can be computed as a discrete bound state with binding energy approximately equal to the excitation of x'. The single particle radiative width is obtained directly and multiplied by $c_{\lambda,x'b}^2$. While the configuration $x' \otimes b$ is spread over the actual final states it may be assumed that this spreading is sufficiently confined that the summed radiation width to these states is quite accurately represented by the single particle radiation width.

The single particle neutron states that are likely to be significant in the low energy s-wave neutron resonances of ^{43}Ca are the $3s_{1/2}$, $2d_{5/2}$ and $1g_{9/2}$. From the optical model parameters[15] it is found that the R-matrix representation of the first two will lie a little above the neutron separation energy and the third will be a weakly bound discrete state. The states that are likely to be dominant in the final state configuration are $2p_{3/2}$, $1f_{7/2}$ and, to much smaller extent, $2p_{1/2}$ and $1f_{5/2}$. The $1f_{7/2}$ sub-shell will already be nearly half-filled in the core states of ^{43}Ca, so we ought, as an approximate procedure, to reduce the computed radiation widths to configurations involving the $1f_{7/2}$ state by a factor of about one half. The median energy of the $2p_{3/2}$ state appears experimentally at 5.8 MeV below the neutron separation energy. With the well depth that reproduces this value the $1f_{7/2}$ state would be bound by about 10 MeV. We find from the spectroscopic factors of the (d,p) data that there are very strong $\ell_n=3$ states near 8 MeV binding.

With single particle states of this character, with an assumed Lorentzian spreading of half-width 2 MeV, and discrete core states of ^{43}Ca that comprise the ground state, the 0.37 MeV state, the 0.59 MeV state and a simple level density representation of core states above that, we can compute roughly the neutron contribution to the valence width and its spectral form. We find a total radiation width for high energy gamma rays of the right magnitude, about 0.4 eV, and a component for low energy gamma rays (<5 MeV) of about 0.2 eV. This is last is a factor of about 3 too low, but we note that we have not allowed for the greater strength of $1f_{7/2}$ state associated with more highly excited cores, nor included the valence proton contributions which are expected to be centred at lower gamma ray energies. We conclude that the generalised valence model is capable of a better representation of the electric dipole capture process, at least for light nuclei, than either of the other two commonly used models in resonance level studies.

C. E1 valence mechanism in resonances

A great deal of experimental work has been done to establish evidence for the existence of the simple valence mechanism in neutron resonance levels. Most of this work has centred on C, Fe, Ni and Mo isotopes. It is summarised here.

The lightest nucleus for which E1 valence capture has been investigated in detail is the ^{13}C target. This has a substantial p-wave resonance at 153 keV. Using the R-matrix description of the $2p_{3/2}$ component of the resonance[13] it is found that the valence model reproduces the partial radiation widths of the three primary E1 transitions to states in ^{14}C with substantial $2s_{1/2}$ and $1d_{5/2}$ character to within experimental error. This is quite remarkable inasmuch as the initial state does not stem from a size resonance in the low energy p-wave strength function, and the conditions might therefore be said not to be too favourable for valence capture.

In the much higher mass number region, A=50-60, the conditions for valence capture are expected to be optimal. Here there is a peak in s-wave neutron strength function due to the location of the $3s_{1/2}$ single particle state close to binding, and the low lying final states are often dominated by the $2p_{3/2,1/2}$ single particle states. There are several strong E1 transitions from the strong s-wave resonance at 7.68 keV in the cross section of ^{54}Fe (Γ_n=1043 eV) to a host of final states with very significant p-wave spectroscopic factors. Raman et al[18] measured the relative intensities of these and found that they were in good agreement with the ratios of valence calculations (using the optical model method), but that the absolute total of the calculated valence widths exceeded the measured total radiation width by about 30%. We have recalculated the valence widths using the R-matrix method and find agreement with the total width and the renormalised partial widths. This resonance appears to provide an almost perfect example of valence capture.

However, in the radiative cross section of the neighbouring even nucleus, ^{56}Fe, anomalies appear. From the strong 27.7 keV resonance (Γ_n = 1420 eV) the transitions to the ground and first excited states are of about equal strength[19], yet the valence theory indicates that the ground transition should be about an order of magnitude weaker than the experimental value. The valence theory reproduces accurately the width to the first excited state. A similar phenomenon is observed for thermal neutron capture, which is dominated not so much by potential capture as by the strong bound level lying between a few and about 10 keV below the neutron separation energy. The valence plus potential capture model gives a cross section of 0.037 b for the ground transition, which is to be compared with the experimental value, 0.65 b. We speculate that the ground state could contain a major component of $|2^+\rangle \otimes 1f_{5/2}$, where $|2^+\rangle$ is the first vibrational state of the ^{56}Fe core, and the strong transitions to this state are from the $|2^+\rangle \otimes 2d_{5/2,3/2}$ components of the s-wave resonances.

In the mass number 90-100 region, the p-wave strength function peaks, and valence transitions to $3s_{1/2}$ and $2d_{3/2,5/2}$ components of the final states become possible. Studies of p-wave resonances in the cross sections of several Mo isotopes reveal evidence of such transitions. They are most apparent in the cross section of ^{98}Mo, in which the lowest three J^{π} = 1/2$^-$ resonance levels have partial radiation widths to a sequence of ten even parity final states that are reproduced to within a factor of two or three by the valence model, using the measured data on neutron widths and spectroscopic factors, the agreement being better the larger the reduced neutron width of the resonance[6]. The J^{π} = 3/2$^-$ resonances show a roughly similar order of magnitude of model calculation and data for the final states with large spectroscopic factors, but there is little correlation for individual transitions. For ^{100}Mo, order of magnitude agreement is found for the single measured J^{π} = 1/2$^-$ resonance, but not for the J^{π} = 3/2$^-$ initial states[7]. In the cross section of ^{92}Mo the correlation between theory and experiment is rather weak for the individual transitions from the J^{π} = 1/2$^-$ resonances, but the resonance averaged calculated partial radiation width agrees with the average experimental width for final states with large spectroscopic factors[20]. There is not even this degree of agreement for the $J^\pi = 3/2^-$ resonances. It is still suggestive that the valence radiation widths often approach the order of magnitude of the measured partial widths, however, and again we might speculate on the role of a generalised valence model; in the molybdenum case a major individual component might be $|2^+\rangle \otimes 2f_{7/2} \to |2^+\rangle \otimes 2d_{5/2}$, $|2^+\rangle$ again indicating the first vibrational state of ^{92}Mo. The single particle radiation width of this transition is approximately twice that of the $|2^+\rangle \otimes 3p_{3/2} \to |2^+\rangle \otimes 2d_{5/2}$ component. The former transition could certainly feature significantly in the $J^\pi = 3/2^-$ resonances, but is not allowed from the $J^\pi = 1/2^-$ resonances and would thus explain the qualitative difference between the two initial spins.

IV. MAGNETIC DIPOLE DIRECT CAPTURE

Because of the success of the simple direct capture theory in explaining a great deal of the data on cross sections of primary E1 transitions following thermal neutron capture by a wide range of light nuclei, it is natural to ask if a similar theory may be valid for M1 capture. In this case, the chances of a successful theory are slimmer for two reasons. The first is that in the E1 case the matrix elements are concerned with the electrostatic dipole moment, whereas in the second case the M1 operator depends on currents that may be much more poorly described by the simple wave functions of such a model. The second reason is that the radial component of the E1 matrix element is more strongly weighted to the channel region, where the wave functions are well established by the energies of the initial and final states, than is the radial M1 element. In fact in the M1 case it is often stated that there can be no direct capture that is analogous to E1 direct capture because the radial wave functions are necessarily orthogonal in a simple potential well. However, because of the complexity of the nucleus, we usually find it necessary to describe the radial wave functions of the initial and final states with different potentials (just as we usually do for E1 capture).

The simple form for the magnetic dipole operator for a neutron (magnetic moment μ_n in nuclear magnetons) impinging on the potential field provided by a target nucleus with magnetic moment μ_I is

$$H'_{M1,M} = \frac{e\hbar}{2mc}\left(\frac{3}{4\pi}\right)^{1/2}\left(\frac{\mu_I I_M}{I} + \frac{\mu_n \sigma_{n,M}}{\sigma}\right) \quad (1)$$

Here, e is the protonic charge, c the velocity of light, I is the target nucleus spin operator and σ the Pauli spin operator for the neutron. From this expression the reduced matrix element for the spin factor can be computed, while the radial matrix element is simply the integral of the product of the initial state X_λ radial wave function and the final state, Φ_μ, radial wave function.

It is well-known[21] that the use of this simple impulse operator in the cross section expression for thermal neutron capture by the proton accounts for the major part of the cross section (0.3326b), with only about 10% remaining to be explained by meson exchange effects. Application to the capture cross section of ^{3}He leads to controversial results[22], with some authors claiming that meson exchange currents account for the major part of the cross section. With the elementary approach we are testing here, using Eckart potential forms with parameters adjusted to give the thermal neutron scattering length (a_{J-1} = 3.0±0.1 fm) for the initial state and the binding energy (20.58 MeV) for the final state we find $\sigma_{\gamma,\text{dir}} \approx 80$ μb, with a spread of about 20% due to reasonable changes in the Eckart parameters and about 15% due to the experimental uncertainty in the scattering length. This is to be compared with the most recent experimental result[23], 54±6 μb, and indicates that the simple direct capture M1 model may be of some validity in the $1s_{1/2}$ shell nuclei.

Significant M1 transitions are not to be expected in slow neutron capture by the 1p shell nuclei, until the upper end of the group is approached. Here, starting with the ^{10}Be compound nucleus, a significant fraction of the $2s_{1/2}$ single particle level becomes just bound. The low gamma ray energies of the transitions to the states sharing in this component of the $2s_{1/2}$ orbital make these viable candidates for the direct capture mechanism. The thermal neutron cross section for the transition to the J_f^π = 2⁻ state bound by 0.547 MeV is 13.5 μb. The direct capture model gives $37\Theta_f^2$ μb, and the spectroscopic factor Θ_f^2 is expected[3] to be in the range 0.2 - 0.3. Similarly, in the ^{12}C(n,γ) reaction the cross section to the -1.857 MeV state is 5.6±0.5 μb, and the model cross section is about $25\Theta_f^2$ μb. In the ^{13}C(n,γ) reaction the cross section to the J_f^π = 0⁻ state at -1.274 MeV is 67 μb, and the model cross section is $68\Theta_f^2$, while the experimental and theoretical cross sections to the J_f^π = 1⁻ level at -2.08 MeV are 34 μb and $24.5\Theta_f^2$ μb, respectively. All these comparisons signify a qualitatively important role for the direct capture mechanism in this region of mass and energy.

Beyond the 1p shell the $2s_{1/2}$ single particle state becomes moderately to strongly bound. Magnetic dipole transitions to moderately bound levels with substantial components of this single particle state are found in ^{19}F and ^{20,22}Ne thermal neutron capture. In the ^{19}F case the direct capture model is found to under-represent the cross sections to the two strongest $s_{1/2}$ states (at binding energies of 3.075 and 3.113 MeV) by a factor of about 4. But that to the more strongly bound state at 5.545 MeV is under-calculated by three orders of magnitude. It is also to be remarked that transitions to states with $1d_{3/2}$ content, which are inaccessible on the direct capture model, have cross sections at least as great as those to the $2s_{1/2}$ states. In the ^{20}Ne(n,γ) reaction, there is one state with substantial $s_{1/2}$ content, the 2.794 MeV state (E_γ = 3.966 MeV) with Θ_f^2 = 0.85, with a cross section of 0.14 mb. The calculated direct cross section is one order of magnitude greater. In the ^{22}Ne(n,γ) reaction, the state at 1.02 MeV (E_γ = 4.18 MeV, $\Theta_f^2 \approx 0.5$) has a cross section of 0.9 mb, and the calculated cross section is 0.6 mb. These comparisons indicate that the simple direct process has a substantial but not predominant role to play in this regime of M1 capture.

In the higher mass group of nuclides around magnesium the direct capture theory fails to explain the observed M1 primary transitions by up to two orders of magnitude, and this ratio rapidly worsens in general for heavier nuclei.

V. MAGNETIC DIPOLE VALENCE CAPTURE

Magnetic dipole capture transitions from resonances are surprisingly strong compared with electric dipole gamma rays of similar energy, the former often being found, on average, to be less than one order of magnitude weaker than the latter. In addition, suggestive correlations have been found in some cases between transitions to the same final states from s-wave and p-wave resonances. Again, one can speculate that a valence mechanism, analogous to that of E1 capture, may play a role.

Following the study of thermal neutron capture it follows that the most likely candidates for valence capture will occur in very light nuclei. The lithium nuclei are the lightest nuclei with suitable resonances. The fast neutron capture cross section of ^{7}Li through the p-wave resonance at 255 keV ($J^\pi = 3^+$) has attracted special interest because of its astrophysical importance, and has recently been remeasured[24]. ^{8}Li has its ground state ($I^\pi=2^+$) at 2.033 MeV below the neutron separation energy and one excited state ($I^\pi=1^+$) bound by 1.058 MeV. The thermal neutron capture cross section to these two states is reproduced to within the accuracy of the spectroscopic factors by the E1 direct capture model[25]. The s-wave scattering length can be deduced from the total cross section and angular distribution data to well above 0.5 MeV and the E1 capture cross section estimated through this energy range, enabling the M1 radiation width in the resonance to be deduced. The best estimate of this quantity is $\Gamma_\gamma = 0.08 \pm 0.02$ eV. The valence radiation width can be calculated by the R-matrix method[13], substituting the operator of eq.1, to give $\Gamma_{\gamma,\mathrm{val}} = 0.093$ eV. This good agreement between the model and observation is in accord with the evidence from thermal neutron capture that this simple mechanism is predominant for very light nuclei and relatively low energy transitions.

More data are required to test the M1 valence model at slightly higher mass numbers. As it is, little more information is available until we reach ^{35}Cl. From the bound s-wave resonance dominating the thermal neutron cross section of this nucleus there are strong M1 transitions to the ground state and several even parity excited states of the compound nucleus[26]. The valence model fails to explain the radiative widths for these transitions by some 3 to 4 orders of magnitude. Similarly, for the p-wave resonances at 9.48 keV and 11.2 keV in the cross section of ^{54}Fe and at 1.157 keV in that of ^{56}Fe there is three orders of magnitude discrepancy[27,28]. This trend is borne out by studies of the M1 transitions from s-wave resonances of the Mo isotopes[6,7,20].

VI. CORRELATIONS BETWEEN M1 AND E1 TRANSITION STRENGTHS

In some of the above cases (especially the Cl and Fe isotopes) strong correlations have been observed between M1 radiative widths and widths or cross sections to the same final states reached from another resonance or energy region by E1 gamma rays. Such correlation can also imply a correlation between the M1 radiative width and the neutron width, as well as the (d,p) spectroscopic factor. These suggest initially a simple valence explanation, although, as we have seen such an explanation is not borne out quantitatively.

Similar correlations between thermal neutron M1 radiative cross sections and (d,p) spectroscopic factors to s-wave single particle components of final states have been noted previously and explained by a semi-direct mechanism[29]. This can be extended readily to resonance transitions. Let us denote a residual nucleus (core) state by χ_x and a single particle orbiting in the average field of this core by ϕ_a. Then we can assume that a compound nucleus resonance state can be expanded as

$$X_\lambda = \sum_{xa} c_{\lambda,xa} \chi_x \phi_a \tag{2}$$

The final state X_μ has a similar expansion

$$X_\mu = \sum_{x'a'} c_{\mu,x'a'} \chi_{x'} \phi_{a'} \tag{3}$$

The core states for these two expansions do not need to be actual states of the possible residual nuclei but can be conceptual states provided that these form a complete set.

Suppose that we have in the former expansion a significant core state $\chi_{x=(M1)x'}$ that has the form $D_{M1}\chi_{x'}$, where D_{M1} is proportional to the (full) magnetic dipole operator, i.e. x is the magnetic dipole giant resonance state[30] built on x'. Then the partial radiation width of compound state λ to state μ due to this component and a common single particle state p alone will be

$$\Gamma_{\lambda\gamma\mu,GR} = c_{\lambda,(M1)x'a}^2 c_{\mu,x'a}^2 \Gamma_{\gamma(M1GR)} \tag{4}$$

where the giant dipole radiation width is

$$\Gamma_{\gamma(M1GR)} = (16\pi k_\gamma^3/9) |\langle \chi_{(M1)x'} \| H_{M1} \| \chi_{x'} \rangle|^2 / (2J_\mu+1) \tag{5}$$

Clearly, eq.4 is maximised for some combination of x', (M1)x' and a for which the expansion coefficients $c_{\lambda,(M1)x'a}$ and $c_{\mu,x'a}$ are large. If the latter is large then the former can be large when the energy of the transition from λ to μ is about equal to the magnetic dipole giant resonance energy. The partial radiation width is further intensified if the M1 giant resonance is narrow, and this is most likely to be the case if x' is a residual nucleus ground state (x'≡0 for the target nucleus).

The electric dipole transition from some opposite parity resonance λ' to the same final state μ can be large by the generalised valence mechanism if the expansion coefficient $c_{\lambda',x'_a'}$ for a suitable single particle state a', connecting to a by the E1 operator, is large. The commonality of c_{μ,x'_a} thus gives rise to correlations of the M1 and E1 partial radiation widths from λ and λ' respectively, although the mechanisms for the two kinds of transition can be totally different. If $x' \equiv 0$, c_{μ,x'_a}^2 is the spectroscopic factor for the single particle state a, and the correlation with (d,p) is thereby established; if $x'(\equiv 0)a'$ is the entrance channel for the neutron the connection with the simple valence model for resonance λ' is apparent.

In ^{35}Cl the available neutron and proton orbits for forming the collective 1^+ spin excitation[30] are the $1d_{5/2}$ to $1d_{3/2}$ states. The excitation energy can be estimated to be 11.5 MeV and the M1 giant resonance radiation width to be 137 eV. Using the latter figure, a giant resonance half-width of 1 MeV and an s-wave resonance spacing of about 40 keV we estimate an expansion coefficient $c_{\lambda(M1)x'_a}^2$ of about 0.01 and a partial M1 radiation width of about $0.6 c_{\lambda x'_a}^2$ eV if the giant resonance is centred at 11.5 MeV and about ten times greater than this if it is centred at the gamma ray energy. There is some degree of correlation of the thermal neutron capture M1 transition strengths with the spectroscopic factors to the even parity final states, so if we use the latter to obtain the final estimates of the partial widths we find agreement with the assumption that the M1 giant resonance is centred close to the gamma ray energy, i.e. about 3 MeV lower than the collective spin excitation estimate. That correlation with (d,p) together with the observed correlations between thermal neutron M1 transitions and the 398 eV p-wave resonance E1 transitions[26] also implies that the latter have a simple valence origin, and this is borne out by comparison with calculations of the latter.

The ^{56}Fe(n,γ) reaction provides another interesting example of E1-M1 correlations. In this case the final state spins are predominantly of odd parity, several of them having considerable values of spectroscopic factors to $p_{1/2}$ and $p_{3/2}$ single particle orbits. We have already considered the E1 transitions from thermal neutron capture in Section III, showing that for some there is substantial agreement with direct capture theory and for others surprisingly little. The M1 transitions to the same final states from the 1157 eV resonance, a p-wave resonance with $J^\pi=1/2^-$, show a considerable degree of correlation with the thermal neutron transitions[89]. In ^{56}Fe the M1 giant resonance is formed principally from the $1f_{7/2} \rightarrow 1f_{5/2}$ spin-orbit change, and its energy is expected to be about 14 MeV. With this value (and a half-width of about 1 MeV) the M1 partial radiation width will be about $50 c_{\lambda x'_a}^2$ meV. This is considerably smaller than some of the observed widths from the 1157 eV resonance, suggesting again that the M1 giant resonance energy is considerably lower than the above estimate. There is some, but not perfect, correlation of the M1 radiation widths with the final state (d,p) strength, suggesting that the target core is an important, but not the only significant, configuration for the spin excitation.

VII. CONCLUSION

There is now a very substantial body of evidence to show that the simple direct or potential capture process is the predominant mechanism for electric dipole transitions at off-resonance energies in light nuclei (A ≤ 50). Closely related to the potential capture process is the valence process, which explains many E1 capture transitions from cross section resonance states. At resonance energies, however, many E1 transitions are clearly not of simple valence character, but the magnitude of the widths for such transitions are qualitatively consistent with the valence process even though the required correlation with the individual resonance neutron width is absent. This suggests that a generalised valence process may be operative, a nucleon changing its single particle orbital in the field of an excited state of the core.

It is not expected that M1 capture mechanisms can be dominated to such an extent by direct or valence processes. Nevertheless, for very light nuclei (A ≤ 15) and comparatively low energy gamma ray energies, such processes do seem to carry some validity. It is clear, however, that for heavier nuclei the simple single particle transition mechanism fails by some two to four orders of magnitude to explain both thermal and resonance capture data. For these, and especially to explain some of the remarkable correlation effects involving both M1 and E1 transitions from initial states of different parity, the extension to resonances of the semi-direct mechanism[29], in which the core state undergoes a M1 giant resonance transition while the extra nucleon remains in a relatively low-lying single particle orbital, appears to be capable of explaining the data, but there is an implication that the M1 giant resonance lies at considerably lower excitation energies than estimates from theory of correlated nucleon motions suggest[30].

ACKNOWLEDGMENT

This work was sponsored by the U.S.A. Department of Energy under Contract W-7405-eng-36 with the University of California (Los Alamos) and Contract DE-AC05-84OR21400 with Martin Marietta Energy Systems, Inc. (Oak Ridge).

REFERENCES

1. A.M.Lane and J.E.Lynn, Nucl.Phys.**17**,563(1960), **17**,586(1960)
2. S.Raman, R.F.Carlton, J.C.Wells, E.T.Jurney and J.E.Lynn, Phys.Rev.C**32**,18(1985)
3. J.E.Lynn, S.Kahane and S.Raman, Phys.Rev.C**35**,26(1987)
4. S.Raman, S.Kahane and J.E.Lynn, Nuclear Data for Science and Technology (Proc.Int.Conf.,1988, Mito, Japan) p.645 (Saikon, Tokyo) and references therein
5. S.Raman, S.Kahane, R.M.Moon, J.A.Fernandez-Baca, J.L.Zaretsky, J.E.Lynn and J.W.Richardson, Phys.Rev.C**39**,1297(1988)
6. R.E.Chrien, G.W.Cole, G.G.Slaughter and J.A.Harvey, Phys.Rev.C**13**,578(1976)

7. H.Weigmann, S.Raman, J.A.Harvey, R.L.Macklin and G.G.Slaughter, Phys.Rev.C**20**,115(1979)
8. T.J.Kennett, W.V.Prestwich and J.S.Tsai, Can.J.Phys.**65**, 1111 (1987)
9. W.V.Prestwich, T.J.Kennett and J.S.Tsai, Z.Phys.**A325**,321(1986)
10. A.G.W.Cameron, Can.J.Phys.**37**,322(1959)
11. E.T.Jurney, S.Raman and W.Starner, private communication
12. J.E.Lynn, *The Theory of Neutron Resonance Reactions*, pp.306,326 (Oxford University Press, 1968)
13. S.Raman, M.Igashira, Y.Dozono, H.Kitazawa, M.Mizaumoto and J.E.Lynn, Phys.Rev.C**41**,458(1990)
14. S.F.Mughabghab, M.Divadeenam and N.E.Holden, *Neutron Cross Sections* vol.1 (Academic, 1981)
15. P.A.Moldauer, Nucl.Phys.**47**,65(1963)
16. D.M.Brink, D.Phil.Thesis (Oxford University,1955)
17. P.Axel, Phys.Rev.**126**,671(1962)
18. S.Raman, G.G.Slaughter, J.C.Wells and B.J.Allen, Phys.Rev.C**22**, 328(1980)
19. M.J.Kenny, Aus.J.Phys.**24**,805(1971)
20. O.A.Wasson and G.G.Slaughter, Phys.Rev.C**8**,297(1973)
21. J.M.Blatt and V.F.Weisskopf, *Theoretical Nuclear Physics*, p.604 (Wiley, New York, 1952)
22. J.Carlson, D.O.Riska, R.Schiavilla and R.B.Wiringa, Phys.Rev.C**42**,830(1990) and references therein
23. F.L.H.Wolfs, S.J.Friedman, J.E.Nelson, M.S.Dewey and G.L.Greene, Phys.Rev.Lett.**63**,2721(1989)
24. M.Wiescher, R.Steininger and F.Käppeler, Astrophys.J.**344**, 464(1989)
25. J.E.Lynn, E.T.Jurney and S.Raman, to be submitted for publication
26. R.E.Chrien and J.Kopecky, Phys.Rev.Lett.**39**,911(1977)
27. S.Raman, G.G.Slaughter, J.C.Wells and B.J.Allen, Phys.Rev.C**22**,328(1980)
28. R.E.Chrien, M.R.Bhat and O.A.Wasson, Phys.Rev.C**1**,973(1970)
29. C.F.Clement, A.M.Lane and J.Kopecky, Phys.Lett.**71B**,10(1977)
30. A.Bohr and B.Mottelson, *Nuclear Structure*, vol.2, p.636 (Benjamin, Reading, Mass., 1975)

GAUSSIAN DISTRIBUTION FOR SPACINGS OF SIMPLE NEUTRON RESONANCES.

G. ROHR
Commission of the European Communities, Joint Research Centre,
Central Bureau for Nuclear Measurements, Geel, Belgium.

ABSTRACT

The nearest level spacing distribution is studied for simple excited (2p-1h) neutron resonances. According to a comparison of level densities for neutron resonances and for calculated values of doorway states such resonances are available at neutron separation energy for A<38 and closed shell nuclei. A Gaussian-like distribution for ^{28}Si, ^{32}S, ^{40}Ca, ^{52}Cr, ^{54}Fe, ^{58}Fe and ^{96}Zr is obtained. The standard deviation of this distribution determines the lifetime of phonons and the latter agrees quite well with the lifetime of resonances based on the average neutron width. These values, using the same amount of resonance data for both lifetimes, are more precise in the phonon picture and favour the lattice model for a nucleus.

I. INTRODUCTION

In contrast to the Bohr assumption of statistical equilibrium there is evidence in the literature that the compound nuclear reaction process can be very simple. Calculations of the level density of doorway states (2p-1h) at neutron separation energy, based on the independent particle model including residual interaction, result in spacings which agree reasonably well with those observed in medium light (A<38) and closed shell nuclei [1,2,3]. In this case the resonances can be considered as created in a single two-body collision process which is the lowest mode of nucleon-nucleon interaction and the spacing distribution belonging to it is fundamental. Beyond A=38 more complicated states (3p-2h, 4p-3h etc.) become energetically possible, indicated by steps in the level density systematics [3] and the additional states will obscure this fundamental spacing distribution. There is no possibility to reconstruct it since in general there is no possibility to distinguish experimentally between resonances of different seniority.
Therefore only a level spacing distribution based on doorway resonances characterized by a defined spin, parity and s(seniority)=3 will reflect the properties of nucleon-nucleon interaction in nuclei.

II. SELECTION OF SIMPLE RESONANCES

The limitation to resonances for A<38 and closed shell nuclei considerably reduces the number of states available for the adjacent level spacing fluctuation studies. Therefore only a narrow distribution of these values is expected to become detectable. In this contribution examples for a Gaussian-like distribution using simple s-wave resonances are discussed. As will be shown, their spacings are almost equidistant and for a small number of resonances this distribution is characterized by having no spacing smaller than $D_{ave}/2$ [4].

a) Examples of Gaussian-like spacing distributions.

In the following, s-wave neutron resonance data of $^{40}Ca, ^{54}Fe, ^{58}Fe$ and ^{96}Zr are studied. The data have been taken from the literature and are used up to a maximum energy that prevents inclusion of 3p-2h states. In this small energy range no correction for the energy dependence of the level spacings has been applied since this effect is assumed to be small compared with the inherent spacings fluctuation.

In fig. 1 the s-wave resonance parameters of $^{40}Ca+n$ up to 750 keV are plotted versus their resonance energy. Resonance energies and their corresponding reduced neutron widths have been taken from ref.[5]. The plot contains no resonances of small spacing, indicating a strong mutual level repulsion effect. This property is seen even more clearly in fig. 2, which represents a plot of the 15 adjacent level spacings in a histogramme. For comparison fig. 2 includes the Wigner distribution. The experimental spacing distribution is narrower and is represented by the Gaussian distribution using a standard deviation calculated from the experimental spacings. The level spacings are distributed in an interval $0.56 \leq D/<D> \leq 1.76$; the probability that a spacing inside this interval is distributed according to a Wigner distribution is 0.693 and consequently the probability for fifteen spacings is $P_{Wig} = (0.693)^{15} = 0.41\%$. ^{40}Ca is

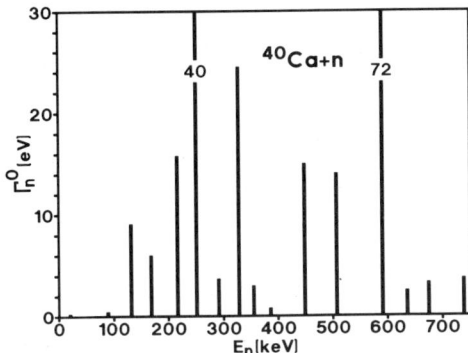
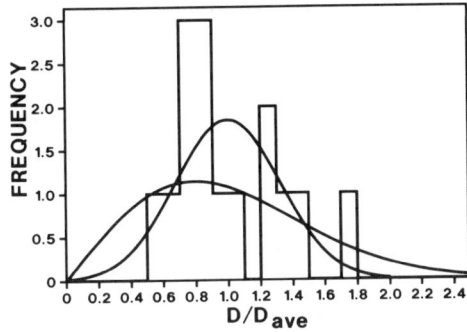

Fig. 1: s-wave resonance energies and reduced neutron widths for $^{40}Ca+n$

Fig. 2: Histogram of the level spacing distribution for $^{40}Ca+n$ resonances compared with Wigner and Gaussian functions

the largest natural nucleus which has the same neutron and proton number. Since it has a double closed shell (N=P=20) the resonances at neutron separation energy are doorway resonances.

Fig. 3 contains the s-wave resonance parameters of ten resonances in ^{54}Fe up to an energy of 250keV[5,6]. The adjacent level spacings are distributed in the interval $0.55 \leq D/<D> \leq 1.71$; the probability that the resonances are distributed according to a Wigner distribution is 3.4.%. Also ^{54}Fe has a closed neutron (N=28) shell and the resonances are expected to be of 2p-1h type.

A similar spectrum to ^{54}Fe is observed for ^{58}Fe and the resonance parameters of nine resonaces up to 310keV[5,7,8] are represented in fig 4. The spacings are distributed in the interval of $0.62 \leq D/<D> \leq 1.69$ and the probability that they obey a Wigner distribution is $P_{wig}=1.5\%$. ^{58}Fe has no shell closure but has a neutron separation energy which is 2.72 Mev lower than that for ^{54}Fe. The low excitation energy results in an average level spacing which is even larger than that for ^{54}Fe. Therefore both Fe isotopes should have resonances of the same type at neutron separation energy.

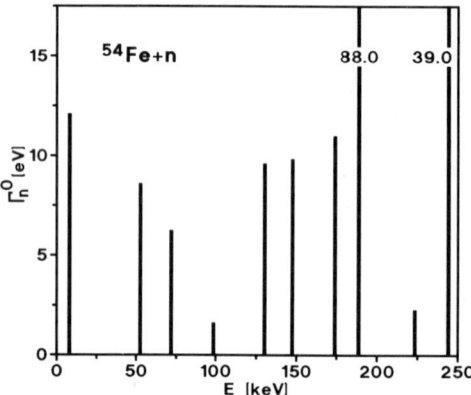
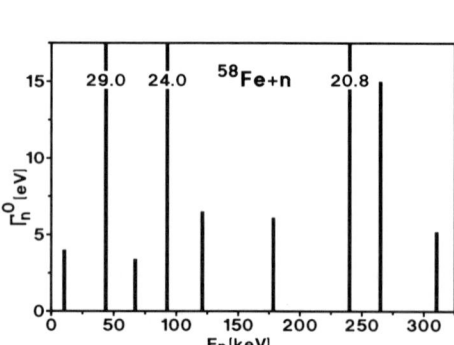

Fig. 3 : s-wave resonance energies and reduced neutron widths for ^{54}Fe +n

Fig.4 : s-wave resonance energies and reduced neutron widths for ^{58}Fe+n

A very regular neutron spectrum is seen in fig. 5 which represents the eight ^{96}Zr resonances up to an energy of 100 keV taken from the literature[5,9]. Seven spacings are distributed in the interval $0.62 \leq D/<D> \leq 1.69$ and the distribution has a probability of being represented by a Wigner distribution of $P_{wig} = 2\%$. ^{96}Zr is a closed sub-shell (N=40) nucleus and has a neutron separation energy which is 0.9 Mev lower than that of ^{94}Zr. Both effects suggest that the ^{96}Zr resonances are of the 2p-1h type.

The plots of figs. 1,3,4,5 contain no resonances of small spacing indicating a strong mutual level repulsion effect. This property is seen even more clearly in fig. 6, which represents a plot of all 39 adjacent level spacings from the four above mentioned nuclides in one histogram. For comparison this figure includes the Wigner distribution. The level spacings are distributed in an interval $0.55 \leq D/<D> \leq 1.76$; the probability that a spacing inside this interval is distributed according to a Wigner distribution is 0.705 and the probability for thirty nine spacings can be estimated to be $P_{wig} = (0.705)^{39} = 1.2 \cdot 10^{-4}$ %. The result excludes the Wigner distribution and supports a Gaussian-like distribution. However the histogramme is not symmetric around the average level spacing and may indicate larger spacings (gaps) compared with the spacing defined at the peak of the distribution.

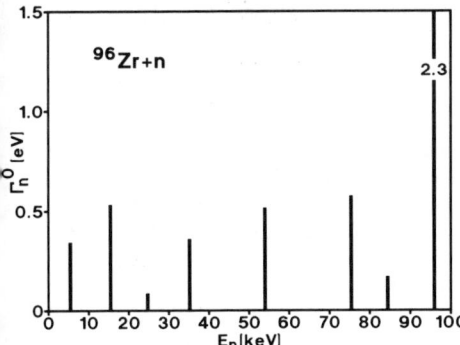

Fig. 5: s-wave resonance energies and reduced neutron widths for ^{96}Zr+n

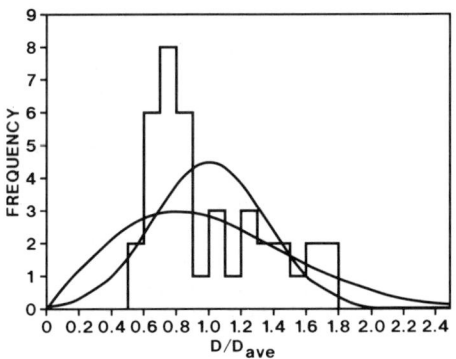

Fig. 6: Distribution of the nearest level spacings for ^{40}Ca+n, ^{54}Fe, ^{58}Fe and ^{96}Zr resonances (39 spacings)

b) Examples with discontinuities in the level spacing

The Gaussian-like nearest spacing distribution would fail if there were gaps indicating a lack of two or more resonances. One example for such a non-continuous neutron spectrum is shown in fig. 7. In this figure the level distribution of ^{52}Cr resonances, plotted against the neutron energy, is not as linear as might be expected[10]. Two gaps are to be seen, at 200 and 670 keV where at least two resonances for each gap are missing. Therefore in fig. 8 the nearest level spacing distribution for 18 values is shown, where the gap values and those which are nearest to them (in total four) are excluded. The remaining spacings are limited to the interval $0.55 \leq D/<D> \leq 1.71$ for which the probability to be elements of a Wigner distribution is $P_{wig}=0.011$%. This distribution is symmetric compared to fig. 6 because the spacings of the gaps have been excluded. Since ^{52}Cr is a closed shell (N=28) nucleus the resonances are doorway states.

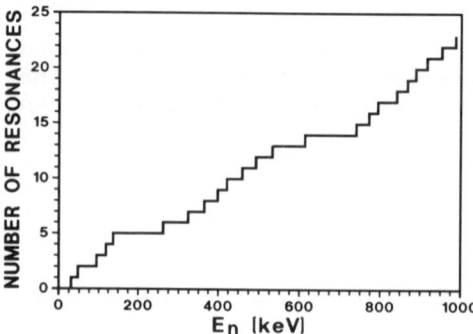
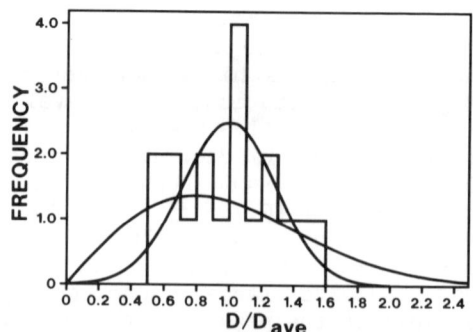

Fig. 7: Level distribution for ^{52}Cr+n resonances with l=0

Fig. 8: Distribution of the nearest level spacings for ^{52}Cr+n resonances (18 spacings)

Similar to the steps in the level density observed as a function of the atomic number (level density systematics) there should also be steps expected in the level density as a function of the excitation energy at the threshold energy of 3p-2h states. In order to observe these discontinuities in the level spacing a small spreading of the additional states near threshold is needed. If in addition the fragmentation of the doorway state were small, there might be a chance to distinguish between the doorway states and 3p-2h states[11]. Two examples are discussed in ref. [11]: ^{28}Si and ^{32}S.

In total seven spacings between the levels assigned as 2p-1h in the interval $0.78 \leq D/<D> \leq 1.2$ are obtained. The probability that these are elements of a Wigner distribution is P_{wig} = 0.016 %. The distribution belonging to it is shown in fig. 9.

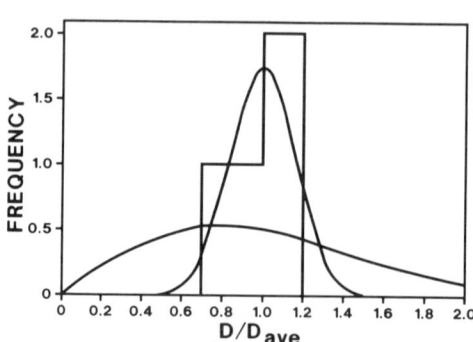

Fig. 9: Distribution of the nearest level spacings for ^{28}Si+n and ^{32}S+n resonances (7 spacings)

It can be shown that the resonance spacing of almost all nuclides, where simple resonances are to be expected from the level density systematics and where resonance data are available, agree with a Gaussian-like distribution. However, for closed neutron shell nuclei above the N=28 shell the resonance data are expected to be incomplete: the reduced average neutron width decreases with A and with a constant detection limit s-wave resonances are missed.

III. DISCUSSION OF THE RESULTS

The Gaussian-like spacing distribution of resonances represents a spectrum of almost equidistant levels which can be explained with the lattice vibration model [12]. The neutron resonances in this model can be interpreted as consisting of a large number of lattice vibration quanta, the so-called phonons. The average lifetime of resonances will then correspond with the lifetime of phonons which, based on the fluctuations of the phonon energy can be calculated as $\tau_q = \hbar/\Delta\varepsilon_q$ where $\Delta\varepsilon_q$ is the standard deviation of the Gaussian-like distribution. These lifetimes (full dots) of six nuclides are shown in fig. 10 together with the lifetimes of resonances calculated from the average neutron widths (open squares) $\tau_{p\text{-}h}=\hbar/<\Gamma_n>$.

A good agreement is observed, especially for nuclides with ten or more resonances and the lifetime is seen to increase with atomic number. The lattice picture provides a more accurate value for the average lifetime of neutron resonances based on the same amount of data. These results favour the validity of a lattice nuclear model which characterises the dynamics of nucleons in a nucleus.

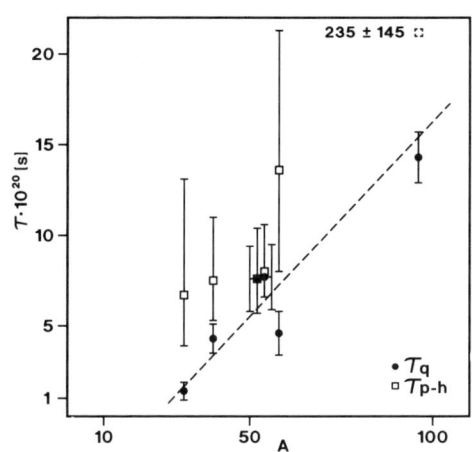

Fig. 10: The lifetime of phonons (full dots) and the average lifetime of resonances (open squares) for six nuclides discussed in this contribution.

REFERENCES

1. G.L. Payne, Phys. Rev. <u>174</u>, 1227 (1968)
2. K.N. Müller and G. Rohr, Nucl.Phys.<u>A164</u>, 97 (1971)
3. G. Rohr, Z.Phys. <u>A318</u>, 299 (1984)
4. G. Rohr, Nuclear Data for Science and Technology (1988 Mito), Editor S. Igarasi, Saikon Publishing Co. Ltd., 707 (1988)
5. S.F. Mughabghab, M. Divadeenam and N.E. Holden, Neutron Cross Sections Vol. 1 Academic Press (Part A 1981, Part B 1984)
6. E.M.R.Cornelis, C.R.Jungmann, L.Mewissen and F.Poortmans, Neutron Cross Sections for Technology(1979 Knoxville) Editors J.L. Fowler, C.H.Johnson and C.D. Bowman U.S. Government Printing Office:Washington:1980. p. 159
7. E.Beer,Ly Di Hong,and F.Käppeler, Nucl.Sci.Eng. <u>67</u>, 184 (1978)
8. J.B. Garg,S. Jain and J.A. Harvey, Phys. Rev. <u>C18</u>, 1141 (1978)
9. C.Coceva,P.Giacobbe and M. Magnani, Neutron Cross Sections for Technology (1979 Knoxville) Editors J.L.Fowler, C.H.Johnson and C.D. Bowman, U.S.Government, Printing Office:Washington:1980. p. 319
10. G. Rohr, R. Shelley, A. Brusegan, F. Poortmans and L.Mewissen, Phys. Rev. <u>C39</u>, 426 (1989)
11. G. Rohr, Proceedings of the 6th Inernational Conference on Capture Gamma-Ray Spectroscopy, Leuven, Belgium 1987, editors K. Abrahams and P. Van Assche (Institute of Physics, Bristol, 1988) p. 643
12. D. Mattuck, A Guide to Feyman Diagrams in the Many-Body Problem, Mc. Graw-Hill Publishing Company Limited (1967)

THE REAL OPTICAL– AND SHELL–MODEL POTENTIALS[+]

R. D. Lawson, S. Chiba[*], P. T. Guenther and A. B. Smith
Argonne National Laboratory, Argonne, Illinois 60439

ABSTRACT

From fits to neutron scattering data over a wide range of nuclei it is shown that r_v, the reduced radius of the real optical–model potential, decreases with increasing A. The value of the isovector part of the real potential is discussed and a simple argument is given for its magnitude. The dispersion relationship and the method of moments are used to extrapolate the scattering potential to the bound–state regime. The possibility of deducing the spin–orbit strength from the observed single–particle binding energies is discussed.

* * * * * * *

Over the past few years the elastic scattering of neutrons with incident energies of 1.5 – 10 MeV has been studied extensively in this laboratory using ^{51}V, ^{58}Ni, ^{59}Co, ^{89}Y, Zr, ^{93}Nb, In and ^{209}Bi targets. (Where the mass numbers are not shown, elemental samples were used.) The data obtained were analyzed using the spherical optical model (SOM) assuming Woods–Saxon and derivative Woods–Saxon shapes for the real and imaginary potentials, respectively, and a Thomas term for the spin–orbit interaction. The parameters describing the real Woods–Saxon well for 8 MeV incident neutrons[1] are shown in Table I. From this table it is clear that the reduced radius, r_v, ($R = r_v \cdot A^{1/3}$) is decreasing as A increases, and if we parameterize this decrease by assuming that

$$r_v = r_0 + r_1/A^{1/3}, \quad (1)$$

the values shown in column 5 of Table I are obtained when

$$r_0 = 1.148 \text{ fm} \qquad (2)$$
$$r_1 = 0.442 \text{ fm}.$$

These values are similar to those found by Moldauer[2] from considerations of lower–energy neutron processes.

[+] Work supported by the U. S. Department of Energy, Office of Energy Research contract W–31–109–Eng–38.
[*] Permanent address, Japan Atomic Energy Research Institute, Tokai, Ibaraki, Japan.

Table I. Values of r_v, a_v and J_v for the real SOM Woods Saxon potential used in describing scattering of 8 MeV neutrons from the targets of column 1. Columns 5 and 6 (Systematics) show the values obtained using Eqs. 1, 2 and 5.

Target	SOM Fits			Systematics	
	r_v^b	a_v^b	J_v^c	r_v^b	J_v^c
^{51}V	1.268	0.615	440.4	1.267	451.3
^{58}Ni	1.254	0.646	462.9	1.262	461.7
^{59}Co	1.262	0.636	454.8	1.262	448.6
^{89}Y	1.240	0.703	424.5	1.247	423.9
^{90}Zr	1.259	0.667	431.9	1.247	426.6
^{93}Nb	1.250	0.700	426.3	1.246	423.7
^{115}In	1.234	0.640	400.1	1.239	410.1
^{209}Bi	1.220	0.700	385.5	1.222	380.9

[b] Radii and diffusenesses in fm.
[c] Volume–integrals–per–nucleon in MeV fm^3.

Turning to the diffusenesses, given in column 3 of Table I, there is some slight evidence that this quantity increases with increasing A. However, the improvement in χ^2 over that obtained when the average value, $a_v = 0.663$ fm, is used is not significant.

Finally, the values of J_v, the volume integral per nucleon of the real potential, show a very distinct decrease with increasing A. Since the SOM potential is expected to have an isovector component[3], on might try fitting these values assuming

$$J_v = J_0(1 - \xi \cdot (N-Z)/A), \tag{3}$$

where J_0 is a constant and ξ is the relative strength of the isovector potential. An excellent fit to the data is obtained with $J_0 = 486.8$ MeV fm^3 and $\xi = 1.04$. These parameters are quite similar to the 8 MeV values obtained by Holmqvist and Wiedling[4] ($J_0 = 480$ MeV fm^3, $\xi = 0.98$), and to the 11.1 MeV conclusions of Ferrer et al.[5] ($J_0 = 495$ MeV fm^3, $\xi = 0.95$). On the other hand, the nucleon–nucleon scattering data[6] indicates that ξ should be ≈ 0.48, and the (p,n) results are consitent[7] with a value of 0.4. The global SOM fits of Walter and Guss[8] and of Rapaport[9] give $\xi = 0.32$ and $= 0.42$, respectively, at 8 MeV. Thus, in order to fit the rapid A–variation of J_v one needs a value of ξ much larger than deduced from other considerations.

The volume–integral–per–nucleon of the Woods–Saxon potential is given by the expression[10]

$$J_v = \frac{4\pi}{3} \cdot r_v^3 \cdot V_0 \cdot \left[1 + \left[\frac{\pi \cdot a_v}{R}\right]^2\right], \quad (4)$$

where $R = r_v \cdot A^{1/3}$. However, according to Eq. 1, r_v decreases with increasing A. Thus part of the rapid decrease in J_v with mass number observed experimentally can be attributed to the decrease in r_v. To take this into account we parameterize J_v by the expression

$$J_v = K_0 \cdot (1 - \xi \cdot (N-Z)/A) \cdot (r_0 + r_1/A^{1/3})^3. \quad (5)$$

When r_0 and r_1 have the values given by Eq. 2, one finds $K_0 = 234.2$ MeV and $\xi = 0.53$. This parameterization yields a value of ξ similar to those obtained by other considerations and, moreover, the predicted J_v, shown in the last column of Table I, are in good agreement with experiment.

That a value of ξ in the neighborhood of $0.3 - 0.5$ should be observed can be deduced from very simple considerations. Let us assume that the SOM potential arises from the interaction of the incident nucleon with the constituent nucleons of the target. From experiment, we know that the range of nucleon–nucleon interaction is small compared to nuclear dimensions. Therefore, in first approximation, we assume that this interaction is a delta function, and from this we estimate the isospin dependence of the SOM. Since the exclusion principle does not get in the way when one of the Z protons in the nucleus interacts with an incident neutron, there are Z possible pairs. On the other hand, the number of target neutrons with which the projectile can interact is only N/2; i.e., the Pauli Principle demands that if the two neutrons sit on top of each other spatially, their spins must be oppositely oriented. Thus the SOM for the neutron will be given by

$$\begin{aligned} V &= U_0 \cdot Z + U_0 \cdot N/2 \\ &= \tfrac{3}{4} \cdot U_0 \cdot (N+Z) - \tfrac{1}{4} \cdot U_0(N-Z) \\ &= \tfrac{3}{4} \cdot U_0 \cdot A \cdot (1 - \tfrac{1}{3} \cdot (N-Z)/A), \end{aligned} \quad (6)$$

where $A = N + Z$ is the total number of target nucleons and U_o is the product of the potential strength and the probability that two nucleons sit on top of each other. A similar argument can be made for an incident proton, the only difference being that $N \leftrightarrow Z$ so that the coefficient $-1/3$ becomes $+1/3$. Furthermore, if the nucleon wave function in the nucleus is uniformly spread over the nuclear volume, the normalization factor will be proportional to $A^{-1/2}$. Thus the product $U_o A$ in Eq. 6 is expected, and indeed found, to have a very weak A dependence. Therefore, the coefficient of the isovector part of the SOM, that is ξ, should be about $1/3$. Thus the observed magnitude of ξ is a direct consequence of the short-range nature and nearly equal strengths of (n,p) and (n,n) free-nucleon interactions.

We now examine what can be said about the shell-model potential by the use of the scattering data. There is a well known dispersion relationship linking real and imaginary optical potentials[11]

$$V(r,E) = V_{HF}(r,E) + \frac{P}{\pi} \cdot \int_{-\infty}^{+\infty} \frac{W(r,E')}{E-E'} \cdot dE', \tag{7}$$

where $V(r,E)$ is the total real OM interaction, $V_{HF}(r,E)$ is its Hartree-Fock component, P is the principal-value integral, and $W(r,E)$ is the absorptive potential. The same dispersion relationship holds for the radial moments of the potentials, so that

$$<r(E)^q>_V = <r(E)^q>_{HF} + \frac{P}{\pi} \cdot \int_{-\infty}^{+\infty} \frac{<r(E')^q>_W}{E-E'} \cdot dE', \tag{8}$$

where, for example,

$$<r(E)^q>_W = \frac{4\pi}{A} \cdot \int W(r,E) \cdot r^q \cdot dr. \tag{9}$$

In general the radial moments of the imaginary potential can be fairly well parameterized by an expression of the form

$$<r(E)^q>_W = \frac{C_q (E - E_F)^2}{(E - E_F)^2 + D_q^2}, \tag{10}$$

where C_q and D_q are constants to be fitted to the various moments and E_F is the Fermi energy. In addition, one expects $V_{HF}(r,E)$ to be a smooth function of the energy, and it is reasonable to approximate its moments by

$$<r(E)^q>_{HF} = A_q + B_q \cdot E. \tag{11}$$

When Eqs. 10 and 11 are substituted into Eq. 8, an analytic expression can be obtained for the various moments, $\langle r(E)^q \rangle_v$, of the total real potential,

$$\langle r(E)^q \rangle_v = A_q + B_q \cdot E + \frac{C_q \cdot D_q \cdot (E - E_F)}{(E - E_F)^2 + D_q^2}. \tag{12}$$

To be specific, we shall apply the foregoing considerations to the case of ^{51}V. For this nucleus high quality data are available[12] over the energy range 1.8 − 11.1 MeV. These data were fitted using the SOM defined earlier. The radii, strengths and diffusenesses found from this fitting were used to calculate the radial moments of the real and imaginary SOM potentials. Mahaux and Sartor[13] have studied which moments are most stable and have concluded that the best results are obtained when q, in Eq. 8, takes the values 0.8, 2 and 4. For these values, the calculated $\langle r(E)^q \rangle_w$ were parameterized by Eq. 10, and in this way values of C_q and D_q were obtained. Finally, the parameters A_q and B_q were deduced using Eq. 12 to fit the moments $\langle r(E)^q \rangle_v$ obtained from the real potential.

If one now assumes that Eq. 12 holds for all values of E and , further, takes V(r,E) to be a Woods–Saxon potential, the three moments of the interaction can be used to determine the strength and geometry, V_o, r_v and a_v, for energies outside the 1.8 − 11.1 MeV region. This provides a way of estimating the bound–state potential to be used in shell–model calculations.

The variations in the strength and geometry of the real Woods–Saxon well for ^{51}V are shown in Fig. 1. For this nucleus there are six shell–model single particle states for which reasonable estimates of the binding energy can be made, and these are listed in Table II. Using these experimental values, one can obtain from Eq. 12 the predicted V_o, r_v and a_v which should reproduce the binding energies. In column 3 of Table II we list the values actually obtained when these Woods–Saxon parameters are used. These binding energies, which are based upon the assumption that the spin–orbit interaction has the Thomas form, and the same strength as used in the analysis of the scattering data[12],

$$V_{so} = 8.36 \text{ MeV}$$
$$r_{so} = 1.00 \text{ fm}$$
$$a_{so} = 0.65 \text{ fm}, \tag{13}$$

are somewhat less than experiment for the particle states, whereas two of the three hole states are predicted to be overbound. The rms deviation between predicted and observed values is 650 keV.

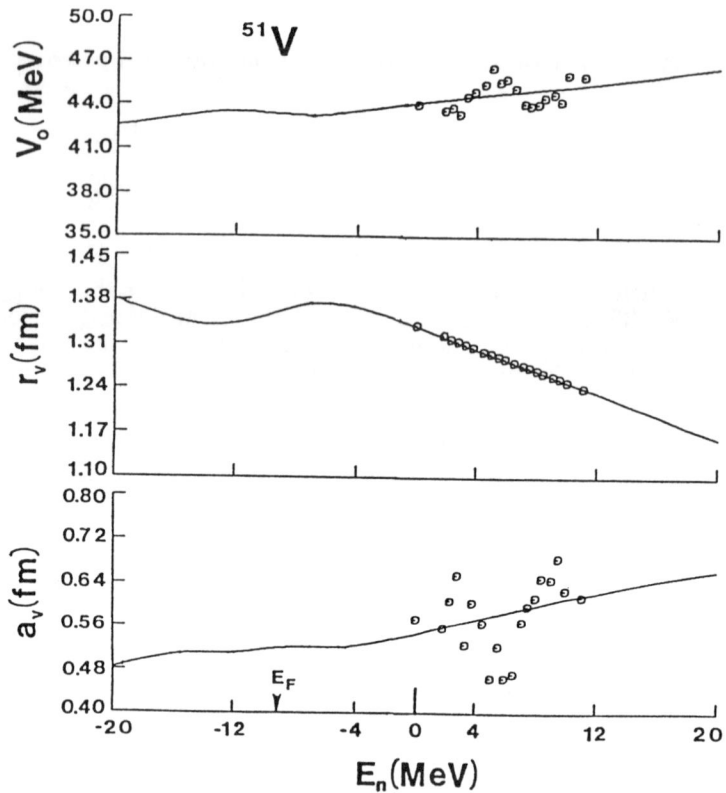

Fig. 1. The solid lines are the predicted variation with energy of V_o, r_v and a_v for a Woods–Saxon potential. These result when the q = 0.8, 2 and 4 moments of the real potential, obtained from the best four–parameter fit to the experimental data, are approximated by Eq. 12. The "O" symbols at positive energies are the values obtained from the fitting of the experimental scattering data. The Fermi energy, $E_F = -9.68$ MeV, is also shown.

The fit to the elastic–scattering and total cross section data is not very sensitive to the spin–orbit potential. Thus, in the absence of low–energy polarization data for ^{51}V, the strength of this interaction, Eq. 13, could be, and was, somewhat arbitrarily chosen. On the other hand, the binding energies of all but the $s_{1/2}$ level are quite sensitive to this force. Therefore, for nuclei with none or only sketchy polarization data, the spin–orbit strength can be estimated by optimizing the single–particle binding energies. If this is done the optimum value of V_{so}, assuming that r_{so} and a_{so} are fixed at the values of Eq. 13, is 6.6 MeV, which is quite close to the global values.[8,9] The rms deviation between theory (shown in column 4 of Table II) and experiment is reduced to 340 keV.

Table II. Single–particle and single–hole binding energies in ^{51}V. The second column gives the experimental estimate, and the third and fourth the predicted results when V_0, r_v and a_v have the values given by Eq. 12 at the experimental binding energies. The spin–orbit interaction used in column 3 is given by Eq. 13, whereas in column 4 $V_{so} = 6.6$ MeV. In column 5 results are listed when r_v and a_v are given by Eq. 14, and V_0 is adjusted so as to reproduce the J_v values (Eq. 12 with $q = 2$) obtained from the analysis of the scattering data.

State	Binding Energy (MeV)			
	Exp.	$V_{so}=8.36$	$V_{so}=6.6$	$V_{so}=6.6$
$p_{1/2}$	−5.5	−5.38	−5.65	−5.94
$f_{5/2}$	−6.0	−5.26	−6.14	−5.98
$p_{3/2}$	−7.3	−7.18	−6.99	−7.23
$f_{7/2}^{-1}$	−11.1	−12.26	−11.48	−10.23
$d_{3/2}^{-1}$	−15.1	−14.55	−15.35	−14.16
$s_{1/2}^{-1}$	−15.2	−15.79	−15.79	−14.89

The possibility of determining the spin–orbit interaction from the single–particle binding energies depends on the sensitivity of these quantities to the parameters characterizing the Woods–Saxon well. To investigate this sensitivity let us assume that r_v and a_v are constant for negative energies and have the values given by the scattering analysis when $E = 0$. For ^{51}V this implies that

$$r_v = 1.34 \text{ fm}$$
$$a_v = 0.574 \text{ fm}. \tag{14}$$

Generally, the scattering data lead to a rather smooth energy variation of J_v and J_w, the volume integrals per nucleon of the real and imaginary SOM potentials. Therefore, we will use the negative energy values predicted by Eq. 12 with $q = 2$, together with Eq. 14, to determine the depth of the bound–state potential. Under these conditions the binding energies shown in the last column of Table II are obtained when $V_{so} = 6.6$ MeV. As can be seen from Fig. 1, the dispersion relation approach using q = 0.8, 2 and 4 moments of the SOM potential gives $V_o \approx 43.3$ MeV and $r_v \approx 1.37$ fm for the $p_{1/2}$, $f_{5/2}$ and $p_{3/2}$ particle states, whereas when Eq. 14 is used the constraint that J_v be the same for the two approaches implies that V_o be about 2 MeV larger. Despite these differences in V_o and r_v, the calculated binding energies for the particle states are quite similar for the two approaches. On the other hand, when Eq. 14 is used, the $f_{7/2}$, $d_{3/2}$ and $s_{1/2}$ hole states are all less tightly bound by about 1 MeV. From Fig. 1 it is clear that for these states the radii used in the two different approaches are almost identical. However, a_v given by Eq. 14 is more than 10% larger than the values shown in Fig. 1, and as a consequence V_o will be smaller than shown in the figure. Thus binding energies are quite sensitive to details of the Woods–Saxon potential. Consequently, before determination of the spin–orbit strength can be made using the bound–state data, one must check whether or not the parameters describing the Woods Saxon well depend significantly on which moments of the potential are used in the extrapolation to negative energies.

REFERENCES

1. S. Chiba, P. T. Guenther, R. D. Lawson and A. B. Smith, Argonne National Laboratory Report, ANL/NDM–116 (unpublished); Phys. Rev. C in press.
2. P. A. Moldauer, Nucl Phys. 47, 65 (1963).
3. A. M. Lane, Phys. Rev. Lett. 8, 171 (1962).
4. B. Holmqvist and T. Wiedling, Nucl. Phys. A188, 24 (1972).
5. J. C. Ferrer, J. D. Carlson and J. Rapaport, Nucl. Phys. A275, 325 (1977).
6. G. W. Greenlees, W. Makofske and G. J. Pyle, Phys. Rev. C1, 1145 (1970).
7. C. J. Batty, E. Friedman and G. W. Greenlees, Nucl. Phys. A127, 368 (1969).
8. R. L. Walter and P. P. Guss, Nucl. Data for Basic and Applied Science, edited by P. G. Young et al., (Gordon and Breach, New York, 1986) vol. 2, p. 1079.
9. J. Rapaport, Phys. Reports 87, 25 (1982).
10. L. R. B. Elton, Nucl. Phys. 5, 173 (1958).
11. G. R. Satchler Direct Nuclear Reactions (Clarendon Press, Oxford, 1983).
12. R. D. Lawson, P. T. Guenther and A. B. Smith, Nucl. Phys. A493, 267 (1989).
13. C. Mahaux and R. Sartor, Advances in Nuclear Physics, edited by J. W. Negele and E. Vogt, (Plenum Press, New York, to be published).

TENSOR POLARIZED DEUTERON CAPTURE REACTIONS AND D-STATE EFFECTS IN LIGHT NUCLEI

H. R. Weller, G. Feldman, M. J. Balbes, L. H. Kramer, J. Z. Williams
Duke University and Triangle Universities Nuclear Laboratory (TUNL), Durham,
North Carolina 27706

D. R. Tilley
North Carolina State University, Raleigh, North Carolina 27695 and TUNL

ABSTRACT

The usefulness of tensor polarized deuteron capture reaction studies arises from the fact that the observables of these reactions are particularly sensitive to various tensor force effects in nuclei. Since these effects are more transparent in the very light nuclei, our studies have concentrated on these systems. I will report on results which utilize tensor polarized deuterons to study ^{3}He, ^{4}He, ^{5}Li, ^{5}He and ^{8}Be.

Angular momentum algebra shows us how the tensor analyzing powers are related to tensor force effects for polarized-deuteron capture reactions.[1] If we use L-S coupling, where S is the channel spin ($\vec{S} = \vec{i} + \vec{I}$, where $\vec{i}$ is the incident particle spin and $\vec{I}$ is the target spin), then we find that all tensor analyzing powers vanish unless S, S' and 2 triangulate. Two examples will serve to illustrate how this leads to a sensitivity to tensor force effects. First, consider the ^{1}H(d,γ)^{3}He reaction. As shown in Fig. 1, this reaction in the vicinity of E_d = 10-20 MeV is dominated by E1 radiation which is mostly ΔS = 0. If the ground state of ^{3}He is viewed as a d + p configuration, it will have L = 0 and S = 1/2 to form the 1/2$^+$ ground state. So the continuum will have S = 1/2. Then, with S and S' (two labels for the continuum S value) equal to 1/2, we cannot satisfy

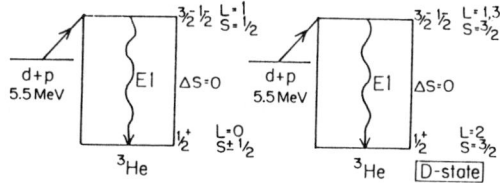

D-state presence leads to S=3/2 strength so (S,S',2) is satisfied

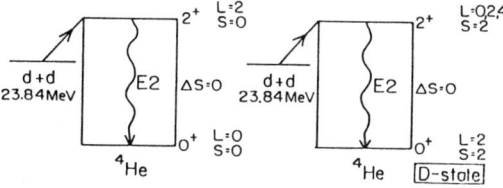

D-state presence leads to S=2 strength so (S,S',2) is satisfied

Fig. 1 Transition matrix elements for ^{1}H(d,γ)^{3}He (dominant E1) and ^{2}H(d,γ)^{4}He (dominant E2) produce finite tensor analyzing powers when D-state is present in the ground states of ^{3}He and/or ^{4}He.

the triangle rule and the tensor analyzing powers are all zero. If, however, there is an L = 2 admixture in the ground state, then S_{gnd} = 3/2 and S_{cont} = 3/2 (for ΔS = 0 - E1). Then, with S = 1/2 and S' = 3/2, for example, we can satisfy the triangle rule and have tensor analyzing power. So the finite tensor analyzing powers in the $^1H(\vec{d},\gamma)^3He$ reaction should be sensitive to the D-state strength in the ground state of ^3He. We will come back to this example later.

In our second example, we consider the $^2H(d,\gamma)^4He$ reaction. In this case isospin forbids E1 and the reaction in the 10-20 MeV region should be predominantly E2. Also, the ℓ + S = even symmetry requirement, due to the identical Boson entrance channel, results in 1⁻ states only with S = 1. Since the d + d ground state of ^4He will be predominantly L=0–S=0 (0⁺), ΔS = 0 E1 is forbidden. Also, for E2 radiation, if S = S' = 0, the triangle rule leads to zero tensor analyzing powers. However, if we have L=2–S=2 (0⁺) ground state strength we can have S = 0 and S' = 2 interference and satisfy the triangle rule. So we expect the tensor analyzing powers in the $^2H(d,\gamma)^4He$ reaction to be sensitive to the D-state of ^4He. Of course other tensor force effects, such as deuteron D-state and/or tensor coupling in the continuum itself can, in principle, give rise to tensor analyzing powers and must be considered in a detailed analysis of the data.[2]

^3He

The experimental and theoretical situation regarding the study of the D-state of ^3He using the $^1H(\vec{d},\gamma)^3He$ reaction is summarized as of 1988 in Ref. 3. The results of precision measurements of $A_{yy}(90°)$ at E_d = 29.2 MeV were compared to a full three-body Faddeev calculation which treated both the bound and continuum states using a RSC interaction. The conclusion was that the D-state of ^3He was ~7.6%.[4]

The above calculation[5] had a very surprising result: it indicated that 89% of A_{yy} was due to capture into the part of the D-state of ^3He which has $\ell = \lambda = 1$, where ℓ denotes the relative angular momentum between the "interacting" pair of particles, and λ denotes the angular momentum of the free particle relative to the center of mass of ^3He. There is now a new-full Faddeev calculation underway, with the preliminary results already available.[6] This calculation uses a separable non-local interaction which is adjusted to fit the deuteron's properties. It uses the same Hamiltonian to describe the bound and the scattering states, and it obeys three-body unitarity. The prediction of this calculation for E1 only and including just the S and D parts of the two-body force are shown in Fig. 2. The T_{20} data shown here were obtained in a collaborative experiment which we performed at McMaster University.[7] The agreement, when rescattering effects are included, is excellent. The D-state probability of the deuteron used here was 5.5%, corresponding to a calculated ^3He D-state probability of 7.6%.

The inclusion of the P-wave part of the two-body force as well as the addition of E2 strength in these calculations is presently underway. If the claims of Ref. 5 are correct and 89% of A_{yy} is due to capture to these P-wave parts, then the excellent agreement shown above will surely be destroyed. The answer to this should be available in the next few months.[6]

Fig. 2 The T_{20} data at E_d = 19.7 MeV from Ref. 7 along with the <u>full</u> Faddeev calculation of Lehman and Fonseca.[6] The deuteron D-state was taken to be 5.5%, leading to a D-state probability of 7.6% in ^{3}He.

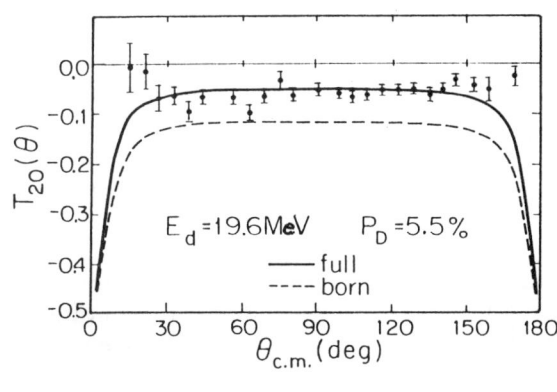

^{4}He

A review of the study of ^{4}He using the ^{2}H($\vec{d}$,γ)^{4}He reaction as of 1988 is given in Ref. 3. The latest TUNL results are scheduled for publication in the October (1990) issue of Phys. Rev. C.

The low-energy results on the tensor analyzing powers can be discussed using Fig. 3. The measured values of A_{yy} and T_{20} at 130°, for incident deuteron energies from 0.8 to 15 MeV, are shown here along with the results of the multi-channel resonating group model calculation of Ref. 8. This calculation predicts a two-deuteron D-state probability of 2.2% for the ground state of ^{4}He. The calculation shown here used a rather crude effective two-body force and treated all fragment (d,^{3}H, ^{3}He) as having only S-states. The neglect of D-states, especially for the deuteron, is expected to be a serious over-simplification, especially at very low energies.[8,9]

The very low energy ($E_d \lesssim 1$ MeV) behavior of the ^{2}H(d,γ)^{4}He reaction remains something of a puzzle. A transition matrix element analysis of the σ(θ), A_y(θ) and A_{yy}(θ) data for E_d < 0.8 MeV (the 0.8 MeV beam stopped in the target) has been performed. The fits to the data are shown

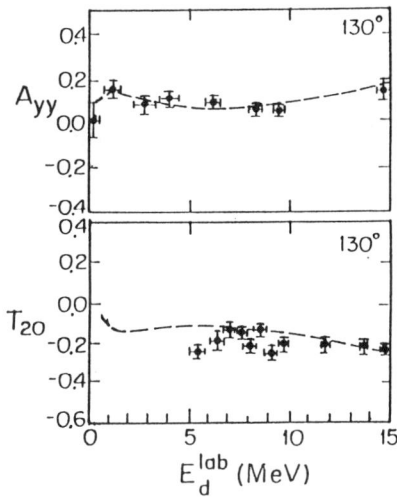

Fig. 3 The A_{yy} and T_{20} data at θ = 130° for the ^{2}H(d,γ)^{4}He reaction below E_d = 15 MeV and the results of the MCRGM calculation which predicts a d+d D-state of 2.2% in ^{4}He.

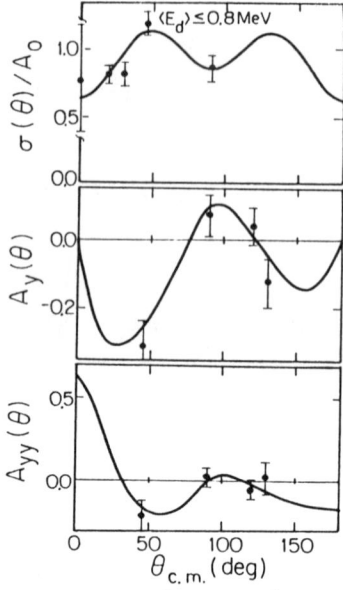

Fig. 4 The ^{2}H(d,γ)^{4}He data for $E_d \leq 800$ keV and the results of fitting these data in terms of T-matrix elements. (See Table.)

in Fig. 4. The results are summarized in Table 1. The results indicate that around 38% of the yield at this low energy is due to E1 and M2 radiations. The fits also indicate that the s- and d-wave E2 capture strengths ($^5s_2$ and $^5d_2$) are important, giving rise to 45% of the total capture strength. On the contrary, the MCRGM calculations predicts that these E2 strengths comprise only 1.5% of the total capture strength, with E1 + M2 radiation giving about 66% of the strength. This result is, however, particularly untrustworthy due to the neglect of D-states in the fragments and inaccurate threshold predictions which produce unrealistic p-wave strengths at these low energies. A new calculation is in progress; it is designed to remedy these deficiencies by using a realistic two-body force and including D-states in all fragments.[8]

Table 1 Allowed transitions, contributions to total cross section (%) and phases in ^{2}H(d,γ)^{4}He. Results of the simultaneous fit to data, $\langle E_d \rangle = 0.8$ MeV, $\chi^2/\nu = 1.6$.

Partial Wave (Continuum State)	γ Multipolarity	Contribution (%)	Phase (deg)	Final ^{4}He State
$^1d_2$	E2	16.4 ± 1.4	0	$^1S_0$
$^5s_2$	E2	24.0 ± 0.6	51. ± 8.	$^5D_0, ^1S_0$
$^5d_2$	E2	21.0 ± 1.0	-103. ± 5.	$^5D_0$
$^3p_1$	E1	16.1 ± 1.3	-169. ± 6.	$^5D_0, ^1S_0$
$^3p_2$	M2	22.5 ± 1.1	167. ± 5.	$^5D_0, ^1S_0$
$^5d_1$	M1	not used	not used	$^5D_0$

The interest in the "very" low energy behavior of the ^{2}H(d,γ)^{4}He reaction is based on the fact that s-wave capture to the D-state of ^{4}He can greatly enhance the low energy cross section or, equivalently, the Astrophysical S-factor.[3] Fig. 5

summarizes the experimental results for the cross section at these low energies (plotted here as the Astrophysical S-factor), and shows the prediction of the MCRGM model. The large cross sections, which have led to a revised S-factor (by a factor of 35)[10] are described by this model. However, a thorough understanding of the details of this reaction at these very low energies will require high quality polarization data in this energy region below $E_d = 1$ MeV.

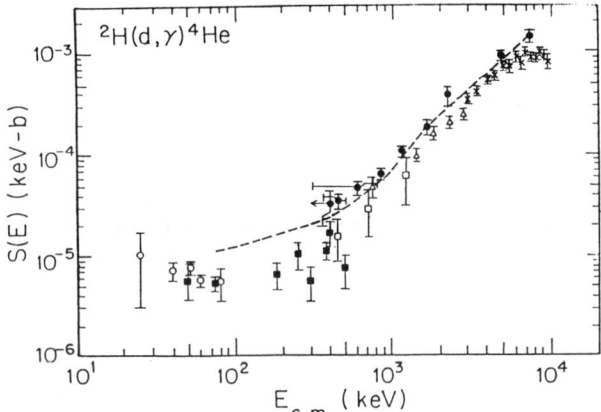

Fig. 5 The world data (references are given in Ref. 3) on the astrophysical S-factor for the $^2H(d,\gamma)^4He$ reaction and the results of the MCRGM calculation.

^{5}He and ^{5}Li

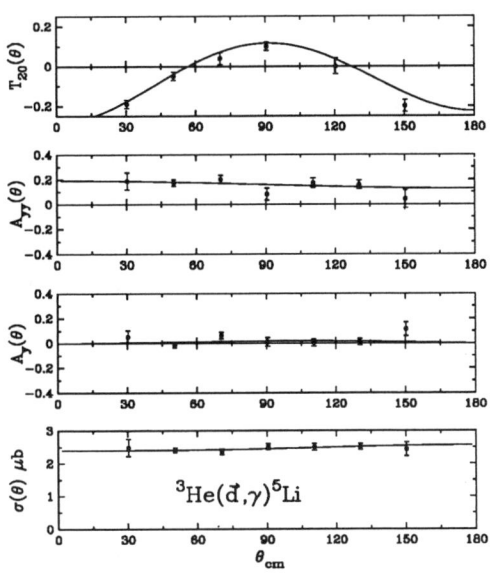

Fig. 6 Data for $E_d \leq 800$ keV for the $^3He(d,\gamma)^5Li$ reaction and the fits obtained from the T-matrix element analysis (see text).

Our work on the $^3H(d,\gamma)^5He$ reaction consisted of measuring $\sigma(\theta)$, $A_y(\theta)$ and $A_{yy}(\theta)$ with a 400 keV beam which was stopped in a tritiated titanium target. The results[11] were fitted and indicated that this "fusion resonance" region is dominated by $S = 3/2$ E1 capture with 2.6% $S = 1/2$ E1 and 2.5% M1 admixtures.

We have recently obtained data on the $^3He(d,\gamma)^5Li$ reaction[12] in the fusion resonance region which is centered at $E_d = 450$ keV (lab).[13] A beam of 800 keV polarized deuterons was stopped in a gaseous ^3He target. Data have been obtained for $\sigma(\theta)$, $A_y(\theta)$, $A_{yy}(\theta)$ and $T_{20}(\theta)$ and are shown in Fig. 6. The fits shown were obtained by searching on four T-matrix elements' amplitudes, and their three relative phases. The terms included were S = 3/2 and S = 1/2 - E1, and S = 3/2 and S = 1/2 - M1. Two solutions were found. One of these corresponds to 91.5% S = 3/2 - E1, 2.7% S = 1/2 - E1, 5.3% S = 3/2 - M1 and 0.6% S = 1/2 - M1. The second solution, which is indistinguishable with regard to the χ^2 value, consists of 42% S = 3/2 - E1, 52% S = 1/2 - E1, 5.1% S = 3/2 - M1 and 1.2% S = 1/2 - M1.

A preliminary multi-channel resonating group model (MCRGM) calculation[8] has been performed. The results emphasize the importance of the tensor force in this reaction. To date, three channels have been included: $[p + {}^4He]^{S=1/2}$, $[d + {}^3He]^{S=1/2}$, and $[d + {}^3He]^{S=3/2}$. The ground state of ^5Li was found to be dominated by the $[p + {}^4He]^{S=1/2}$ configuration. Even so, the $^4S_{3/2}$ (E1) transition matrix element dominates the capture reaction channel. This results primarily from the effects of the tenor force which couples the S = 3/2 and S = 1/2 continuum channels, thereby resulting in the S = 1/2 decay of the S = 3/2 fusion resonance. The S = 1/2 continuum strength couples directly to the nucleon channels and is thereby spread over a large energy region, according to this model.

This calculation is being refined by adjusting the interactions in an attempt to reproduce the $^3He(d,d)^3He$ and the $^4He(p,p)^4He$ experimentally determined phase shifts. Spin-orbit and tensor force strengths will then be adjusted in an attempt to describe the $^3He(d,p)$ data[13] in addition to the $^3He(d,\gamma)$ data of the present work.

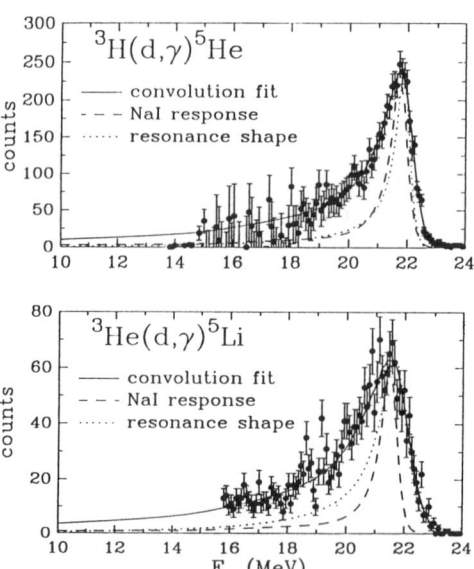

Fig. 7 The $^3H(d,\gamma)^5He$ and $^3H(d,\gamma)^5He$ spectra and the fits obtained from convoluting the NaI response with a Breit-Wigner resonance shape.

In the course of studying the $^3H(d,\gamma)^5He$ and the $^3He(d,\gamma)^5Li$ reactions, we realized that our gamma-ray spectra made it possible for us to determine the ground-state widths of ^5He and ^5Li. To do this, the gamma-ray spectra were fit with a convolution of the NaI lineshape response function and a Breit-Wigner single-level expression. The energy dependent penetrabilities and shift functions were included and found to be critical. The results are shown in Fig. 7. The widths which we obtained were Γ = 1.36 ± 0.19 MeV for ^5He and Γ = 2.44 ± 0.21 MeV for ^5Li, results which are quite different from previously tabulated results.[14] This difference is due mainly to the effect of the energy dependent shift function. Our widths produce a ratio of reduced widths of 1.0 ± 0.16, as expected from the charge-symmetry property of the nuclear force. Previous widths were inconsistent with this result.

^{8}Be

My final example is our study of ^{8}Be via the ^{6}Li(d,γ)^{8}Be reaction. The spin-1 on spin-1 nature of this reaction makes it similar to ^{2}H(d,γ)^{4}He from the angular momentum viewpoint. The idea here is to measure tensor analyzing powers with the hope of seeing a sensitivity to D-state effects in ^{8}Be; the simplest type term being a d + ^{6}Li cluster structure having L = 2 and S = 2 in the ground state of ^{8}Be.

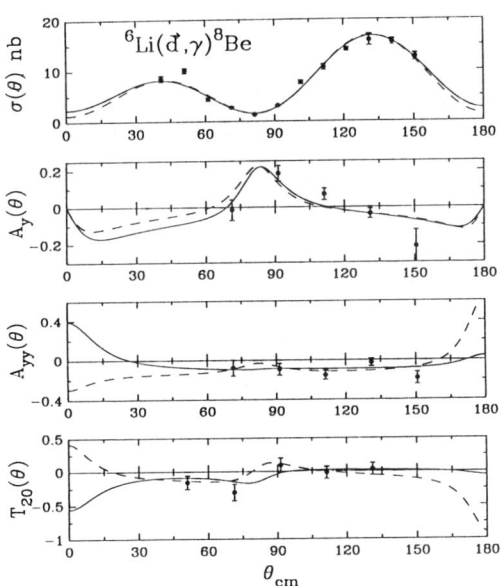

Fig. 8 The ^{6}Li(d,γ)^{8}Be data obtained at E_d = 9 MeV. The solid and dashed lines correspond to two different T-matrix element fits. The solid curves correspond to 21% E1 solution, the dashed to 10.9% E1.

The data we have obtained at E_d = 9 MeV are shown in Fig. 8. The fits were obtained by searching on an selected set of T-matrix elements. The asymmetry in $\sigma(\theta)$ indicates the presence of non-E2 radiation. We have included five matrix elements in the fits of Fig. 8: S = 0 - L = 2 (E2), S = 2 - L = 0 (E2), S = 2 - L = 2 (E2), S = 0 - L = 1 (E1), and S = 1 - L = 1 (E1). Two solutions were obtained. The solid lines correspond to 21% E1 admixture and to 6.5% S = 2 - L = 0 (E2) with almost no S = 2 - L = 2 (E2). It is the S = 2 strength which we expect to be related to D-state strength in ^{8}Be since the ΔS = 0 E2 operator should lead us to the L = 2 - S = 2 ground state component. The second solution (dashed) has 12.84% E1, 13.2% S = 2 - L = 2 (E2) and 2.3% S = 2 - L = 0 (E2).

This solution indicates that 15.5% of the capture strength is leading us to the D-state of ^{8}Be. It is interesting to note that these A_{yy} data have about the same magnitude but are of the opposite sign as those obtained for the ^{2}H(d,γ)^{4}He reaction. It will be very interesting to understand this in a detailed manner.

MCRGM calculations are in progress for this reaction. ^{4}He + ^{4}He, d + ^{6}Li, n + ^{7}Be and p + ^{7}Li configurations are being included. The n and p channels are critical for producing the E1 strength. Preliminary results indicate a d + ^{6}Li ground state strength with L = S = 2 of about 5%. This is about the amount of strength indicated by a direct capture calculation, which considers only E2 radiation and treats the d and ^{6}Li nuclei as point particles.

I hope that these examples have convinced you of the utility of studying tensor polarized deuteron capture reactions and especially of their sensitivity to tensor force effects in nuclei. At present, the theory for A = 3 is in good shape and on the threshold of becoming excellent. Beyond this, our most useful and powerful tool these days is the MCRGM model. It can help us to relate and to understand many features of our data. But even in the A = 4 system, it is becoming apparent that effective two-body models are inadequate. Perhaps future calculations using numerical techniques such as the Green's function Monte Carlo method will help remedy this problem.

Work supported by the U.S. Department of Energy, Office of High Energy and Nuclear Physics, under Contract Nos. DE-AC05-76ER01067 and DE-FG05-88ER40441.

REFERENCES

1. R. G. Seyler and H. R. Weller, Phys. Rev. C20, 453 (1979).
2. S. Mellema, T. R. Wang and W. Haeberli, Phys. Lett. 166B, 282 (1986); and Phys. Rev. C34, 2043 (1986).
3. H. R. Weller and D. R. Lehman, Ann. Rev. Nucl. Part. Sci. 38, 563 (1988).
4. I. Sick, in *The Three-Body Force in the Three-Nucleon System*. (Heidelberg: Springer-Verlag (1986), 260), p. 42.
5. J. Jourdan *et al.*, Nucl. Phys. A453, 220 (1986).
6. D. Lehman and A. Fonseca, private communication (1990).
7. M. Vetterli *et al.*, Phys. Rev. Lett. 54, 1129 (1985).
8. B. Wachter, T. Mertelmeier and H. Hofmann, Phys. Lett. B200, 246 (1988); and H. Hofmann, provate communication (1990).
9. A. Arriaga, A. M. Eiró, F. D. Santos and J. E. Ribeiro, Phys. Rev. C37, 2312 (1988).
10. H. J. Assenbaum and K. Langanke, Phys. Rev. C36, 17 (1987).
11. J. C. Riley, H. R. Weller and D. R. Tilley, Phys. Rev. C40, 1517 (1989).
12. F. E. Cecil *et al.*, Nucl. Instr. and Meth. in Phys. Res. B 10/11, 411 (1985).
13. W. Möller and F. Besenbacher, Nucl. Instr. and Meth. 168, 111 (1980).
14. F. Ajzenberg-Selove, Nucl. Phys. A490, 1-225 (1988).

ISOVECTOR GIANT QUADRUPOLE RESONANCE IN ($\vec{p},\gamma$)

G. Feldman, L.H. Kramer and H.R. Weller
Duke University, Durham, NC 27706, USA
and Triangle Universities Nuclear Laboratory, Durham, NC 27706, USA

E. Hayward and W.R. Dodge
National Institute of Standards and Technology, Gaithersburg, MD 20899, USA

ABSTRACT

The ^{30}Si$(\vec{p},\gamma)^{31}$P reaction has been studied in the energy range E_p = 20-36 MeV. A transition matrix element analysis of $\sigma(\theta)$ and $\sigma(\theta)A_y(\theta)$ at E_p = 25.5 MeV reveals substantial E2 strength (σ_{E2}/σ_{tot} ~ 26%) in the (p,γ_1) channel, in excess of a direct E2 capture estimate (~7%). The energy dependence of $A_y(90°)$ for γ_1 shows a resonance structure which can be reproduced by a direct-semidirect calculation including an E2 resonance (E_{IVGQR} = 38.6 MeV, Γ_{IVGQR} = 5.0 MeV, S_{IVGQR} = 50%) at the expected peak of the isovector giant quadrupole resonance built on the first excited state of ^{31}P. Angular distributions of $\sigma(\theta)$ and $\sigma(\theta)A_y(\theta)$ have also been obtained for the ^{89}Y$(\vec{p},\gamma)^{90}$Zr reaction at E_p = 22.5 MeV. A preliminary analysis reveals large E2 contributions (~28%) for the γ_0 transition, as well as possible E3 strength.

INTRODUCTION

Giant multipole resonances (other than E1) have been studied for almost two decades. During this period, the isoscalar component of the giant quadrupole resonance has been extensively explored via hadron scattering, and a considerable data set exists which has led to a reasonable understanding of the systematics of this isoscalar collective E2 mode. The isovector component of the E2 resonance, however, has been rather poorly documented, due to the difficulties of unfolding the E2 intensity from electron scattering data, which are also plagued by large radiative backgrounds.

Polarized proton capture reactions offer several unique advantages in the search for resonant E2 strength. First of all, real photons are detected, which restricts the contributing multipoles to relatively low order ($\lambda \leq 3$). Secondly, E1/E2 interference effects can give rise to noticeable signatures in the measured observables, such as cross section asymmetry or vector analyzing power. There is evidence[1] based on the direct-semidirect model to suggest that the 90° analyzing power is especially sensitive to the presence of resonant E2 capture strength.

We have begun a program of $(\vec{p},\gamma)$ studies in three nuclei (^{31}P, ^{90}Zr, ^{209}Bi) in an effort to identify the isovector giant quadrupole resonance (IVGQR) and to understand its systematics. In the first part of this work, we have measured the cross section $\sigma(\theta)$ and the vector analyzing power $A_y(\theta)$ at 9 points in the angular range θ_γ = 37.5°–145° for the ^{30}Si$(\vec{p},\gamma)^{31}$P reaction using polarized protons at E_p = 25.5 MeV. We have also measured the vector analyzing power at θ_γ = 90° at 11 energies over the proton energy range E_p = 20-36 MeV. In the second part of this work, which is currently underway, we have measured $\sigma(\theta)$ and $A_y(\theta)$ at 6 angles in the range $\theta_\gamma = 38.8°$–$130°$ for the ^{89}Y($\vec{p},\gamma)^{90}$Zr reaction at $E_p = 22.5$ MeV.

EXPERIMENT

These experiments have been performed at the 88" Cyclotron of the Lawrence Berkeley Laboratory. Polarized proton beams were incident on a 10 mg/cm² self-supporting ^{30}Si target enriched to 95% and a 26 mg/cm² rolled ^{89}Y metal foil. Capture γ rays were detected in a new anticoincidence-shielded 25.4 cm x 27.9 cm NaI(Tl) detector. The detector was surrounded by 10 cm of Pb and 20 cm of boric acid bricks, as well as 25 cm of paraffin between the target and the detector. The pulsed structure of the cyclotron beam enabled us to use time-of-flight techniques to eliminate neutron-induced backgrounds from the γ-ray spectra.

The beam polarization was determined by ^{4}He($\vec{p}$,p)^{4}He elastic scattering, for which the analyzing powers in this energy region are well known.[2] During the angular distribution measurements, the polarization was monitored during the course of the experiment by measuring the up-down asymmetry of protons scattered by the ^{30}Si or ^{89}Y target using two solid-state detectors located symmetrically on either side of the beam axis. During the excitation function, however, the polarization was checked at each energy using the ^{4}He($\vec{p}$,p)^{4}He reaction.

ANALYSIS AND RESULTS: ^{30}Si($\vec{p},\gamma)^{31}$P

A typical γ-ray spectrum for the ^{30}Si($\vec{p},\gamma)^{31}$P reaction measured at $\theta_\gamma = 50°$ is given in Fig. 1. Several strong γ-ray transitions are evident in this spectrum. The spectrum was fitted with NaI lineshapes located at the positions of the first 7 strong single-particle states in ^{31}P (up to $E_x = 5.02$ MeV), as given by (^{3}He,d) proton stripping results.[3] The figure shows the individual fits to the first two transitions γ_0 and γ_1 explicitly (dashed curves), along with the total fit to the spectrum (solid curve). Transition strengths were obtained by integrating the response function down to zero energy -- a linear extrapolation to zero was used below the point at which the NaI response is known.

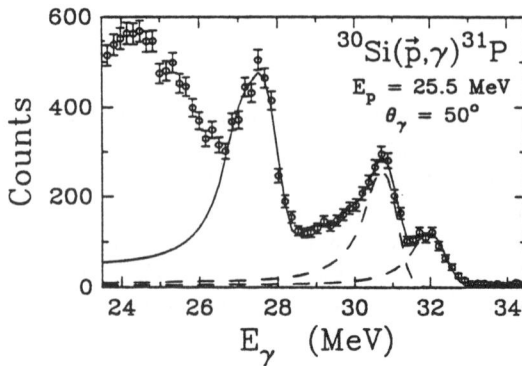

Fig. 1: Measured γ-ray spectrum for the ^{30}Si($\vec{p},\gamma)^{31}$P reaction at $E_p = 25.5$ MeV, showing the total lineshape fit (solid curve) to the data, along with the γ_0 and γ_1 transitions (dashed curves).

The angular distributions of the relative cross section $\sigma(\theta)/A_0$ and the product of analyzing power and relative cross section $\sigma(\theta)A_y(\theta)/A_0$ for γ_0 and γ_1 are plotted in Fig. 2. Both cross sections are strongly forward-peaked, implying significant mixing of opposite parity radiation (M1 or E2) with the dominant electric dipole (E1) radiation. The analyzing power for γ_1 is large, especially at 90°, but for γ_0 it is small at all angles. The analyzing power for γ_1 also deviates substantially from the typical $\sin(2\theta)$ angular dependence expected for pure E1 radiation. The angular distributions of $\sigma(\theta)/A_0$ and $\sigma(\theta)A_y(\theta)/A_0$ were fitted with Legendre polynomials and associated Legendre polynomials, respectively, up to order $k = 4$. The least-squares fits to the data are shown by the solid curves in Fig. 2. Finite values for the odd coefficients indicate the interference of multipoles of opposite parity, and non-zero values of the Legendre coefficients of order $k = 3,4$ suggest that the interfering amplitudes originate primarily from E2 radiation (rather than M1).

To obtain a model-independent estimate of the contributing amplitudes, we have performed a transition matrix element (TME) analysis of these angular distributions, assuming only E1 and E2 terms. An intermediate continuum state I is formed by coupling the projectile angular momentum $j = \ell + s$ to the target spin J_i — the final state J_f is reached by the emission of Eλ radiation. The complex reduced transition matrix element T can then be written in terms of a real amplitude R and a phase ϕ:

$$T = \langle J_f \| E\lambda \| (\ell s)j \times J_i = I \rangle \equiv R\, e^{i\phi}. \tag{1}$$

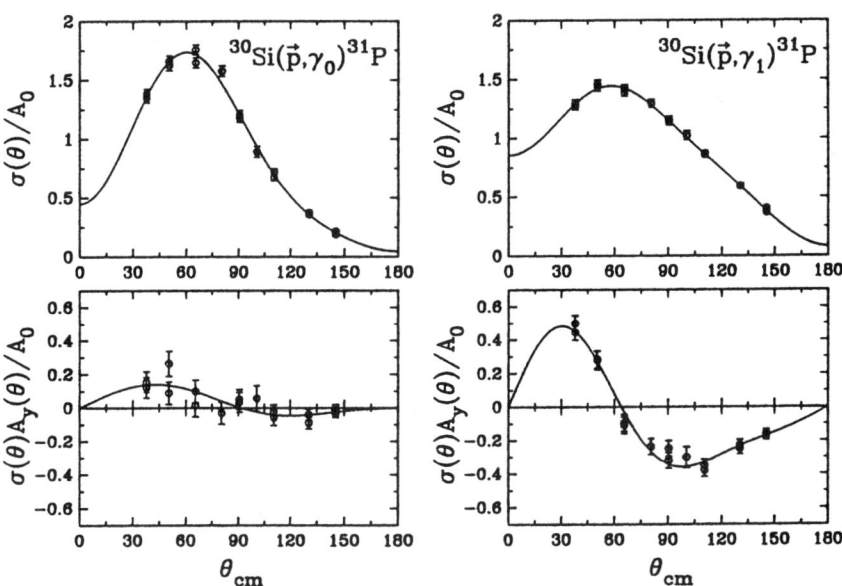

Fig. 2: Angular distributions of cross section and analyzing power for the γ_0 (left panel) and γ_1 (right panel) transitions from the $^{30}\text{Si}(\vec{p},\gamma)^{31}\text{P}$ reaction at $E_p = 25.5$ MeV. The solid curves are Legendre polynomial fits.

For E1 capture to the 3/2+ first excited state of ^{31}P at E_x = 1.27 MeV (primarily a $1d_{3/2}$ single-particle state), there are 3 possible partial waves ($^2p_{1/2}$, $^4p_{3/2}$, $^6f_{5/2}$) that can contribute. The notation $^{2I+1}\ell_j$ refers to the orbital angular momentum ℓ and the total angular momentum j of the projectile, and the spin I of the intermediate state. For E2 capture, there are 4 possible partial waves ($^2s_{1/2}$, $^4d_{3/2}$, $^6d_{5/2}$, $^8g_{7/2}$). Transitions to the 1/2+ ground state (primarily a $2s_{1/2}$ single-particle state) are much simpler ($^2p_{1/2}$ and $^4p_{3/2}$ for E1; $^4d_{3/2}$ and $^6d_{5/2}$ for E2).

The relationships between the transition matrix elements and the angular distribution coefficients a_k and b_k in jj coupling have been previously deduced.[4] Using these relationships, the amplitudes and phases of the TME's were treated as free parameters and were varied to fit the cross section and analyzing power data *simultaneously*. The quality of the fits was identical to that obtained when fitting in terms of Legendre polynomials.

The results of the TME analysis are given in Table I. Each column corresponds to a possible solution; all of the solutions for each final state give equivalently good fits. The quantities listed in the table represent the percentage of the total cross section contributed by each partial wave. For the γ_0 case, two solutions are possible that give similar χ^2 values, but the E2 strength fraction (10%) is independent of the solution.

The γ_1 case gives multiple solutions, two of which are listed in Table I to show the range of E2 strengths obtained. Despite the uncertainty (20-32%) in the E2 fraction it is clear that substantial E2 contributions arise in (p,γ_1) at this energy, more than twice as much as in (p,γ_0). Direct-semidirect model calculations indicate that direct E2 capture can account for only ~7% of the cross section for both γ_0 and γ_1 at this energy, suggesting the presence of substantial excess E2 strength in the γ_1 channel.

Table I: Results of the TME analysis for capture to the ground state and first excited state of ^{31}P. Each column corresponds to a different solution. The quantities represent the percentage contribution of each partial wave to the total cross section. The total strength of each multipole (E1 or E2) is given at the bottom of the column.

Partial Wave	^{30}Si($\vec{p},\gamma_0$)^{31}P		^{30}Si($\vec{p},\gamma_1$)^{31}P	
$^2p_{1/2}$ (E1)	2	72	34	7
$^4p_{3/2}$ (E1)	88	18	7	10
$^6f_{5/2}$ (E1)	---	---	39	51
Σ (E1)	90	90	80	68
$^2s_{1/2}$ (E2)	---	---	1	8
$^4d_{3/2}$ (E2)	1	8	3	10
$^6d_{5/2}$ (E2)	9	2	3	0
$^8g_{7/2}$ (E2)	---	---	13	14
Σ (E2)	10	10	20	32

The energy dependence of $A_y(90°)$ for the ground-state and first-excited-state transitions is shown in Fig. 3. The 90° analyzing power for γ_1 is sizable and increases monotonically over most of the measured energy range, reaching a peak at $E_p = 34$ MeV. No such peaking behavior is evident in the γ_0 case, however, and the analyzing power is nearly zero at all energies.

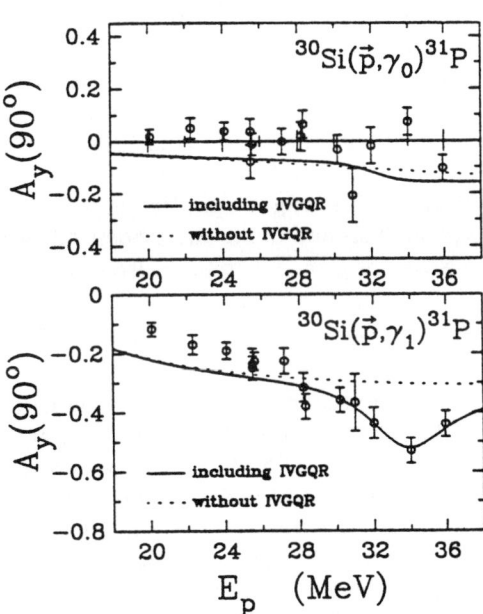

Fig. 3: Energy dependence of the 90° analyzing power for the γ_0 (top panel) and γ_1 (bottom panel) transitions from the ^{30}Si($\vec{p},\gamma$)^{31}P reaction. The curves are direct-semidirect calculations that either include the IVGQR (solid curves) or neglect it (dashed curves). The parameters for the IVGQR were obtained by fitting the γ_1 excitation function and are discussed in the text.

We have used the direct-semidirect (DSD) model to calculate $A_y(90°)$ as a function of energy. In these calculations, we have included the giant dipole resonance as well as isoscalar and isovector components of the giant quadrupole resonance. For the isovector interaction, a complex particle-vibration coupling form factor was used, with a real volume term of strength $V_1 = 70$ MeV and an imaginary surface term of strength $W_1 = 35$ MeV. The isoscalar quadrupole coupling interaction was treated as a real surface-peaked form of strength $U_0 = 50$ MeV. Optical model parameters were obtained from the global fits of Becchetti and Greenlees.[5] The bound-state potential well depth was adjusted to obtain the correct binding energy. Based on previous work,[6] a two-component GDR (due to isospin splitting) was used with the following parameters: $E_< = 17.3$ MeV, $\Gamma_< = 3.4$ MeV, $S_< = 54\%$ and $E_> = 20.6$ MeV, $\Gamma_> = 4.0$ MeV, $S_> = 46\%$. The isoscalar quadrupole resonance was included with $E_{ISGQR} = 19.7$ MeV, $\Gamma_{ISGQR} = 7.1$ MeV and $S_{ISGQR} = 32\%$.[7] All resonance energies presented here and below are given in terms of the energy *above* the state upon which the resonance is built. The resonance parameters of the IVGQR were treated as free parameters in order to fit the energy dependence of $A_y(90°)$ for the γ_1 transition. No such fitting procedure was employed for the γ_0 transition.

The solid curves in Fig. 3 are the results of the above DSD calculations, including the IVGQR with energy $E_{IVGQR} = 38.6 \pm 1.0$ MeV and width $\Gamma_{IVGQR} =$

5.0±1.2 MeV and exhausting (50±15)% of the isovector E2 energy-weighted sum rule. The dashed curves are the same calculations but without the IVGQR. The γ_0 transition shows no particular distinction between the two calculations. While the energy dependence is nearly flat, the calculated magnitude of $A_y(90°)$ for γ_0 is small but non-zero, in marginal agreement with the data. The peaking behavior of $A_y(90°)$ for the γ_1 transition, however, is reproduced by the DSD calculation including the IVGQR, in contrast with the smooth energy variation of the curve with no E2 resonance. Clearly, in the γ_1 case, the need for a localized (collective) enhancement of E2 strength in this higher-energy region is apparent.

The location of the E2 resonance built on the first excited state of ^{31}P is entirely consistent with the expected location[8,9] of the IVGQR in a nucleus of mass A = 31: $E_{IVGQR} \sim 120\text{-}130\ A^{-1/3} \sim 38\text{-}41$ MeV. Thus, the present results indicate that the collective E2 strength identified in the γ_1 channel, as seen in the model-independent TME analysis and in the energy dependence of the 90° analyzing power, arises from the isovector giant quadrupole resonance built on the first excited state of ^{31}P.

ANALYSIS AND RESULTS: $^{89}Y(\vec{p},\gamma)^{90}Zr$

In the second part of our systematic study of the IVGQR, we have obtained angular distributions of cross section and analyzing power for the $^{89}Y(\vec{p},\gamma)^{90}Zr$ reaction at E_p = 22.5 MeV. The γ-ray spectrum (θ_γ = 65°) depicted in Fig. 4 shows the lineshape fit to the data (solid curve), as well as the first 4 single-particle states (dashed curves) of ^{90}Zr. The ground-state transition (γ_0) is clearly resolved from the strong cluster of 3 states near $E_x \sim 2.3$ MeV. The angular distributions for γ_0, shown in Fig. 5, reveal large asymmetries about θ_γ = 90° in both $\sigma(\theta)$ and $\sigma(\theta)A_y(\theta)$, indicative of E1/E2 interference effects. Legendre polynomial fits give non-zero Legendre coefficients of order k = 3,4, thus confirming the presence of E2 radiation.

We have performed a TME analysis of the γ_0 transition only, since the 3 states in the group near $E_x \sim 2.3$ MeV could not be individually resolved. There are two possible partial waves ($^3s_{1/2}$ and $^3d_{3/2}$) that can account for E1 transitions to the 0+ ground state of ^{90}Zr, and there are two possible E2 transitions ($^5p_{3/2}$ and $^5f_{5/2}$). Two

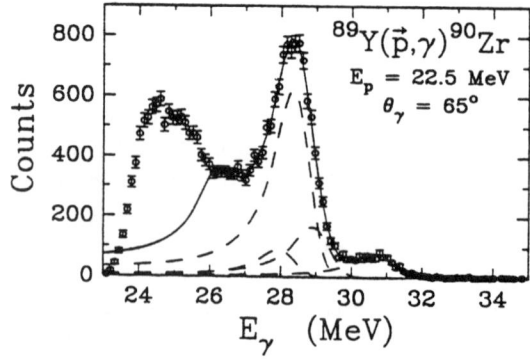

Fig. 4: Measured γ-ray spectrum for the $^{89}Y(\vec{p},\gamma)^{90}Zr$ reaction at E_p = 22.5 MeV, showing the total lineshape fit (solid curve) to the data, along with transitions to the ground state and the first 3 excited states (dashed curves).

solutions were found, as shown in Table II, both of which give an E2 contribution of 28%. These fits are shown as the dashed curves in Fig. 5. The direct E2 capture strength given by a DSD calculation is 6%, indicating a substantial concentration of non-direct E2 strength in the reaction at this energy.

The quality of the TME fits to these angular distributions can be improved significantly by adding two E3 transition matrix elements ($^7d_{5/2}$ and $^7g_{7/2}$) and then fitting the data with Legendre functions up to order k = 5. This new fit ($\chi^2 \approx 7.5$), given by the solid curves in Fig. 5, represents a factor of 3 improvement over the previous fit using only E1 and E2 amplitudes ($\chi^2 \approx 25$). The non-zero a_5 coefficient ($a_5 = 0.2\pm0.1$) further supports the necessity of including E3 amplitudes. Several solutions are possible in this case (two of which are listed in Table II), but the E3 strength falls in the range 2-6%, while the E2 contribution is 22-32%. The resulting a_k coefficients from the TME fits (up to k = 5) are all in good agreement with an earlier (p,γ) measurement[10] of this reaction with unpolarized protons in which direct E3 radiation was included in the analysis.

CONCLUSION

In summary, we have studied the ^{30}Si($\vec{p}$,γ)^{31}P reaction over the proton energy range E_p = 20-36 MeV, corresponding to excitation energies of E_x = 26.6-42.1 MeV in ^{31}P. A transition matrix element analysis of the angular distributions of $\sigma(\theta)$ and $\sigma(\theta)A_y(\theta)$ at E_p = 25.5 MeV reveals that E2 radiation accounts for 10% of the total cross section for γ_0 and ~26% for γ_1. For γ_1, this constitutes a considerable excess of E2 strength above a direct E2 capture estimate (~7%), thus providing strong evidence for collective E2 strength in the γ_1 channel at this energy. Somewhat surprisingly, the γ_0 channel does not show a similar result, and direct E2 capture alone can account for the observed E2 strength in the cross section.

The energy dependence of $A_y(90°)$ reveals a strong signal for the presence of substantial collective E2 strength in ^{31}P located near E_p = 34 MeV in the γ_1 channel. In this case, the 90° analyzing power data for γ_1 are well described by including an E2 resonance built on the first excited state of ^{31}P with parameters E_{IVGQR} = 38.6±1.0 MeV, Γ_{IVGQR} = 5.0±1.2 MeV and S_{IVGQR} = (50±15)%. This resonance energy is in good agreement with the expected position of the IVGQR in ^{31}P. By contrast, the γ_0 channel does not show much sensitivity to the presence of collective E2 strength in this energy region, and these data can be equally well described by a DSD calculation with or without an IVGQR.

We have also measured angular distributions of $\sigma(\theta)$ and $\sigma(\theta)A_y(\theta)$ for the ^{89}Y($\vec{p}$,γ)^{90}Zr reaction at E_p = 22.5 MeV. A transition matrix element analysis of these data gives ~28% of the capture strength as E2 radiation, compared to 6% for direct E2 capture. The possibility of small ($\leq$ 6%) E3 contributions was discussed.

We plan to continue our investigation of the ^{89}Y($\vec{p}$,γ)^{90}Zr reaction by mapping out the energy dependence of $A_y(90°)$ over the proton energy range E_p = 12-30 MeV. Subsequently, we will conclude our systematic study of the IVGQR by investigating the ^{208}Pb($\vec{p}$,γ)^{209}Bi reaction. As before, we will obtain a complete angular distribution near the peak of the giant E2 resonance and also determine the energy dependence of the 90° analyzing power in order to extract resonance parameters.

Isovector Giant Quadrupole Resonance in $(\vec{p},\gamma)$

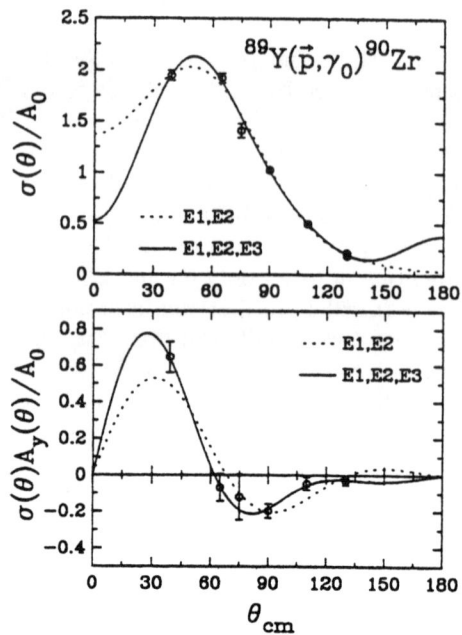

Fig. 5: Angular distributions of the cross section and analyzing power for the ground-state transition from the $^{89}Y(\vec{p},\gamma)^{90}Zr$ reaction at $E_p = 22.5$ MeV. The curves are transition matrix element fits, including either E1, E2 amplitudes (dashed) or E1, E2 and E3 amplitudes (solid).

Table II: Results of the TME analysis for capture to the ground state of ^{90}Zr. Each column corresponds to a different solution. The quantities represent the percentage contribution of each partial wave to the total cross section. The total strength of each multipole (E1, E2 or E3) is given at the bottom of the column.

Partial Wave	$^{89}Y(\vec{p},\gamma_0)^{90}Zr$ [E1 and E2]		$^{89}Y(\vec{p},\gamma_0)^{90}Zr$ [E1, E2, E3]	
$^3s_{1/2}$ (E1)	62	5	23	3
$^3d_{3/2}$ (E1)	10	67	43	69
Σ (E1)	72	72	66	72
$^5p_{3/2}$ (E2)	25	1	6	8
$^5f_{5/2}$ (E2)	3	27	26	14
Σ (E2)	28	28	32	22
$^7d_{5/2}$ (E3)	---	---	1	3
$^7g_{7/2}$ (E3)	---	---	1	3
Σ (E3)	---	---	2	6

In the present work, we have used the $(\vec{p},\gamma)$ reaction to explore a region of excitation energy previously unstudied in polarized capture reactions. We have demonstrated the utility of the $(\vec{p},\gamma)$ reaction as a viable probe of the isovector giant quadrupole resonance in this high excitation energy regime. The restriction to low-order multipoles in the capture reaction and the sensitivity of the polarization observables to resonant E2 strength are the primary advantages that have been exploited in this investigation.

ACKNOWLEDGEMENTS

We are grateful to R.-M. Larimer, E.B. Norman, D.J. Clark, R.J. McDonald, S.E. Kuhn, W. Rathbun and the technical personnel of the 88" Cyclotron at the Lawrence Berkeley Laboratory for their assistance. This work was supported in part by the U.S. Department of Energy, under Contract No. DE-AC05-76ER01067.

REFERENCES

1. F. Saporetti and R. Guidotti, Nucl. Phys. **A390**, 207 (1982).
2. A. D. Bacher et al., Phys. Rev. **C5**, 1147 (1972).
3. P. M. Endt and C. van der Leun, Nucl. Phys. **A310**, 1 (1978).
4. R. G. Seyler and H. R. Weller, Phys. Rev. **C20**, 453 (1979).
5. F. B. Becchetti, Jr. and G. W. Greenlees, Phys. Rev. **182**, 1190 (1969).
6. C. P. Cameron et al., Phys. Rev. **C22**, 397 (1980).
7. F. E. Bertrand, Ann. Rev. Nucl. Sci. **26**, 457 (1976).
8. A. Bohr and B. Mottelson, Nuclear Structure Vol. II, (Benjamin, N.Y., 1975).
9. F. E. Bertrand, Nucl. Phys. **A354**, 129c (1981).
10. F. S. Dietrich et al., Phys. Rev. Lett. **38**, 156 (1977).

RESONANCES IN THE DIRECT RADIATIVE CAPTURE OF A LIGHT NUCLEUS BY A LIGHT NUCLEUS

A. Mondragón and E. Hernández
Instituto de Física, UNAM, Apdo. Postal 20-364, 01000 México, D.F. MEXICO

ABSTRACT

The $E1$ radiative capture of a nucleus A by a nucleus B to form bound or unbound states of a nucleus C is described as a direct process in which the nucleus-nucleus interaction in the entrance channel produces resonances, that is, it gives rise to the formation of short lived quasi-molecular states of C with a dominant [A,B] structure, which then decay by photon emission to form the final, ground or excited state of C; A and B being s-shell clusters. Explicit analytical expressions for the partial transition amplitudes are derived in the framework of the resonating group method. The resonant structure of the cross-section is made apparent expanding the partial transition amplitude in singularities in the complex k-plane, k being the wave number of the relative motion of the clusters A and B. Explicit expressions for the elastic and radiative widths are given in terms of integrals containing Gamow radial wave functions of the relative motion of the two clusters. Some features of our treatment of the problem are: we use totally antisymmetric wave functions, the centre of mass motion is correctly treated, bound state and continuum wave functions are treated in a unified way, the radial wave functions used have the correct asymptotic behaviour.

The differential cross section for direct radiative capture from the continuum to a bound state of the A-B system with emission of electric dipole radiation may be written in terms of partial transition amplitudes [1] as

$$Q_{Jl}(E) = i^\ell \exp[i(\sigma_\ell - \sigma_o)] \sum_{\ell'=|J-1|}^{J+1} (\Psi_{J_f L_f} \| M_{E1} \| \Psi_{J\ell'\ell}) \qquad (1)$$

The reduced matrix element in the right hand side of (1) may be computed from the identity

$$(\Psi_{J_f L_f} \| M_{E1} \| \Psi_{J\ell'\ell}) = \frac{1}{2\pi} \lim_{\Lambda \to 1} \frac{\partial}{\partial \Lambda} \Lambda \int d\Omega_k \Big\{ Y^*_{1M}(\hat{K}_\gamma)$$
$$\times \Big(\Psi_{J_f L_f} \| \sum_{i=1}^C \frac{e}{2}[1 + \tau_3(i)] e^{i\Lambda(\vec{K}_\gamma \cdot (\vec{r}_i - \vec{R}_{CM}))} \| \Psi_{J\ell'\ell} \Big) \Big\}. \qquad (2)$$

A matrix element of this type was considered by Kanada, Liu and Tang [3,4] in the framework of the resonating group method. We made the computation of $Q_{Jl}(E)$ following essentially the procedure described in these references but we avoided the long-wavelenght limit approximation. In the single channel approximation, the initial and final state wave functions are written as $\mathcal{A}\{\phi_A \phi_B [\frac{1}{kr} f_{J\ell'\ell}(r) \mathcal{Y}^{M_S}_{J\ell;S,T}] Z_{CM}\}$, the function $\mathcal{Y}^{M_S}_{J\ell;ST}$ is a spin-isospin-angle function appropriate for spin S, isospin T and the required values of orbital angular momentum ℓ and total angular momentum J with Z-component M_S. The functions ϕ_A and ϕ_B represent the internal spatial structures of the A and B nuclear clusters, respectively [4], and, $Z(\vec{R}_{CM})$ is a normalized gaussian function which describes the center of mass motion of the entire system. A straightforward calculation gives

$$Q_{J\ell}(E) = i^{\ell+1} \exp[i(\sigma_\ell - \sigma_0)] \sum_{\ell'=|J-1|}^{J+1} \mathbf{A}_{\ell'} \left\{ \left[Z_A \mathcal{I}_{0\ell'}\left(\frac{B}{C}K_\gamma\right) - Z_B \mathcal{I}_{0\ell'}\left(\frac{A}{C}K_\gamma\right) \right] + \right.$$

$$\left. + (2L_f+1)^{-1/2} \sum_{x=1}^{\min(A-1,B-1)} T_x \sum_{n=1}^{2} \left(C_x^{(n)} \mathcal{I}_{x\ell'} - \tilde{C}_x^{(n)} \tilde{\mathcal{I}}_{x\ell'} \right) \right\} \quad (3)$$

where

$$\mathbf{A}_{\ell'} = (-1)^{M_f - M_i + \ell + 1} \sqrt{(2J+1)(2J_f+1)} C(\ell' 1 L_f; 000)$$

$$\sqrt{(2\ell'+1)(2S+1)} \begin{Bmatrix} L_f & S & J_f \\ \ell' & S & J \\ 1 & 0 & 1 \end{Bmatrix} \left(\frac{mK_\gamma}{\hbar^2 k}\right)^{1/2} e2\sqrt{8\pi} \sqrt{C!} C_f^* \quad (4)$$

The direct radial integral is

$$\mathcal{I}_{0\ell'}\left(\frac{B}{C}K_\gamma\right) = \int_0^\infty f_{J_f L_f}(r) \left[\Theta_{E1}\left(\frac{B}{C}K_\gamma r\right) - 2\tilde{\mu}_o K_\gamma^2 j_1(K_\gamma r) \right] f_{J\ell'\ell}(r) dr \quad (5)$$

the radial integral in the term coming from the exchange of x nucleons is

$$\mathcal{I}_{x\ell'} = \int_0^\infty \int_0^\infty f_{J_f L_f}(r) \left[\Theta_{E1}^{(x)}(\tilde{\alpha}_{nx} K_\gamma r, \tilde{\beta}_{nx} K_\gamma r') k_{L_f,\ell}^{(x)}(r,r') \right] f_{J\ell'\ell}(r') dr dr' \quad (6)$$

Explicit expressions for the constants $C_f, T_x, \tilde{\alpha}_{nx}, \tilde{\beta}_{nx}$ and $\tilde{\mu}_o$ in terms of the size parameter $\alpha_{A,B}$, charge $Z_{A,B}$ and mass number A, B of the two clusters, as well as the exchange kernel $k_{L_f,\ell}(r,r')$, may be found in the work of Kanada, Liu and Tang [3,4]. Explicit expressions for the radial part of the direct and exchange dipole operators $\Theta_{E1}(\rho)$ and $\Theta_{E1}^{(x)}(\rho,\rho')$ may be found in Mondragón and Hernández [2]. The coefficients $C_x^{(n)}$ are given by

$$C_x^{(1)}(A,B) = \sum_{i=0}^{x-1}(x-i)\binom{\min(Z_A,Z_B)}{x-i}\binom{\min(N_A,N_B)}{i} \quad (7)$$

$$C_x^{(2)}(A,B) = \sum_{i=1}^{x}(Z_A - x + i)\binom{\min(Z_A,Z_B)}{x-i}\binom{\min(N_A,N_B)}{i} \quad (8)$$

$\tilde{C}_x^{(1)}$ is equal to $C_x^{(1)}$, and $\tilde{C}_x^{(2)}$ is obtained from $C_x^{(2)}$ exchanging A and B in (8).

The operators $\Theta_{E1}(\rho)$ and $\Theta_{E1}^{(x)}(\rho,\rho')$ are entire functions of the energy and the wave number k, the radial function of the final state, $f_{J_f L_f}(r)$, is independent of k. Hence, the singularities of the transition amplitude $Q_{J\ell}(E)$ in the complex k-plane are those of the wave function of the relative motion in the entrance channel, [2]. In order to make apparent the resonant behaviour of $Q_{J\ell}(E)$ we make an expansion of the integrals (5) and (6) in singularities in the complex variable k [5]. The continuum wave function of the initial state $f_{J\ell'\ell}(kr)$ is expressed in terms of the complete Green's function of the relative motion in the entrance channel $G_{J\ell'\ell}^{(+)}(k;rr')$ written in its spectral representation. After some manipulation, the term in the radial integral containing the singularities may be cast in the form

$$I^{(1)}_{x\ell'}(k) = C_\ell(k) \int_0^\infty < f_{J_f L_f} |\Theta^{(x)}_{E1}(G^{(+)}_{J\ell'\ell''}(k')$$
$$- G^{(-)}_{J\ell'\ell''}(k'))V^{(J)}_{\ell''\ell} |M_{i\eta,\ell+1/2} > \frac{1}{(k+i\epsilon)^2 - k'^2} dk' \qquad (9)$$

where $M_{i\eta,\ell+1/2}(i2kr)$ is the regular Whittaker function and $C_\ell(k)$ is the Gamow factor. The integration contour in (9) is deformed into the lower half of the complex k-plane. When the deformed contour crosses over resonance poles but avoids other singularities of the analytically continued Green's function, the theorem of the residue yields the desired result which allows us to write

$$Q_{J\ell}(E) = (2J+1)i^{\ell+1} \exp[i(\sigma_\ell - \sigma_0)] \exp[i(\phi_{\nu J\Lambda} + \chi_{\nu J\Lambda})]$$
$$\times \frac{\Gamma^{1/2}_{\gamma,\nu J\Lambda}\Gamma^{1/2}_{el.,\nu J\Lambda}}{(E - E_{\nu J\Lambda}) + i\frac{1}{2}\Gamma_{\nu J\Lambda}} N_\ell(k_\nu, k) + Q^{(B)}_{J\ell}(E). \qquad (10)$$

The elastic and radiative widths are proportional to the absolute value squared of the transition amplitude from a state in the continuum to the resonance state and the transition amplitude from the resonance state to the final state, respectively. Hence,

$$\exp[i\phi_{\nu J\Lambda}]\Gamma^{1/2}_{el.,\nu J(\Lambda)\ell} = \left[\frac{4m}{\hbar^2 k}\right]^{1/2} (2i)^{-\ell-1} C_\ell(k) \times$$
$$\sum_{\substack{\ell''=|J-1|\\ \ell''\neq J}}^{J+1} \langle u^{(\Lambda)}_{\nu J\ell''}(k_{\nu J})|V^{(J)}_{\ell''\ell}|M_{i\eta,\ell+\frac{1}{2}}(2ik)\rangle \qquad (11)$$

and

$$\exp[i\chi_{\nu J\Lambda}]\Gamma^{1/2}_{\gamma,\nu J\Lambda} = \sum_{\ell'=|J-1|}^{J+1} A_{\ell'}\left\{\left[Z_A \mathcal{I}^{(G)}_{0\ell'}\left(\frac{B}{C}K_\gamma\right) - Z_B \mathcal{I}^{(G)}_{0\ell'}\left(\frac{A}{C}K_\gamma\right)\right] + \right.$$
$$(2L_f+1)^{-1/2} \sum_{x=1}^{\min(A-1,B-1)} T_x \sum_{n=1}^{2} \left(C^{(n)}_x \mathcal{I}^{(G)}_{x\ell'}(\tilde{\alpha}_{nx}k_\gamma) - \tilde{C}^{(n)}_x \mathcal{I}_{x\ell'}(\tilde{\beta}_{nx}K_\gamma)\right)\right\}$$
$$(12)$$

The integrals $\mathcal{I}^{(G)}_{x\ell'}$ are of the same form as those in (5) and (6) with the Gamow or resonance function $u^{(\Lambda)}_{\nu J\ell}$ belonging to the resonance energy $E_{\nu J(\Lambda)\ell}$ in place of $f_{J_f L_f}(r)$ in (11) and in place of $f_{J\ell'\ell}(r)$ in (12). The details of the expansion in resonances of the transition amplitude as well as the rules for normalization of the Gamow functions may be found in Hernández and Mondragón [2,5].

REFERENCES

1. J.M. Blatt and L.C. Biedenharn, *Rev. Mod. Phys.* **24**, 258 (1952).
2. A. Mondragón and E. Hernández, *Phys. Rev.* **C41**, 1975 (1990).
3. H. Kanada, Q.K.K. Liu and Y.C. Tang, *Phys. Rev.* **C22**, 813 (1980).
4. Q.K.K. Liu, H. Kanada and Y.C. Tang, *Phys. Rev.* **C23**, 645 (1981).
5. A. Mondragón, E. Hernández and J.M. Velázquez-Arcos, To be published in *Ann. Phys.* (Leipzig), **48**, 1991.

CALCULATIONS OF CAPTURE CROSS SECTIONS AND GAMMA-RAY SPECTRA WITH DIFFERENT STRENGTH FUNCTION MODELS

J.Kopecky
Netherlands Energy Research Foundation ECN
P.O. Box 1, 1755 ZG Petten, The Netherlands

M. Uhl
Institut für Radiumforschung und Kernphysik, Universität Wien
A1090 Wien, Boltzmanngasse 3, Austria

ABSTRACT

We studied the impact of different models for the E1 and M1 gamma-ray strength function on the results of neutron capture cross sections calculations for spherical and deformed target nuclei with mass numbers between 55 and 197. Assuming an M1 strength based on a standard Lorentzian we found in many cases strong support for an E1 strength in terms of a generalized Lorentzian with an energy dependent width and and a non-zero limit as the energy tends to zero. These features are founded in microscopic models.[1,2] For some of the investigated deformed nuclei, namely ^{151,153}Eu, ^{156}Gd and ^{175}Lu, an enhancement of the width is required in order to reproduce the experimental data.

INTRODUCTION

Direct experimental information on gamma-ray strength functions mainly stems from photonuclear reactions and from resonance capture gamma spectroscopy; it is available only for energies exceeding a few MeV. Compound nucleus model calculations of the average capture cross sections and related quantities, on the other hand, critically depend on the properties the E1 and M1 strength functions at low energies where direct experimental information is lacking. Therefore such model calculations allow an investigation of the low energy behaviour of these strength functions, provided that the compound nucleus mechanism dominates the capture process.

Of particular interest is the low energy behaviour of the E1 strength function f_{E1} resulting from microscopic models proposed by Kadmenskij et al.,[1] and by Sirotkin.[2] In contrast to many others which yield a vanishing E1 strength as the energy ε_γ tends to zero these models predict a non-zero limit depending on the excitation energy of the final state and therefore imply a partial breakdown of Brinks hypothesis.[3] A representation of f_{E1} in terms of a generalized Lorentzian exhibiting the essential features of the model of Kadmenskij et al.[1] was proposed by Kopecky and Chrien.[4] In a recent investigation[5] we supported the energy dependent non-zero limit of f_{E1} by means of statistical model calculations for some selected spherical nuclei. Here we report on preliminary results for an extended data basis with emphasis on problems encountered with deformed nuclei.

STRENGTH FUNCTION MODELS EMPLOYED

We searched for evidence for a representation of the **E1 strength function** in terms of a *generalized Lorentzian* with an energy dependent width and a non-zero $\varepsilon_\gamma \to 0$ limit[4]:

$$f_{E1}(\varepsilon_\gamma, T) = \frac{8.68 \times 10^{-8}}{\text{mbMeV}^2} \sum_{i=1}^{N} \sigma_{0i} \Gamma_{0i} \left[\frac{\varepsilon_\gamma \Gamma_i(\varepsilon_\gamma, T)}{(\varepsilon_\gamma^2 - E_{0i}^2)^2 + \varepsilon_\gamma^2 \Gamma_i(\varepsilon_\gamma, T)^2} + \frac{0.7 \Gamma_{0i} 4\pi^2 T^2}{E_{0i}^5} \right], \quad (1)$$

where N = 1 or 2, mainly depending on the mass region under consideration. The temperature of the final state $T = \sqrt{(U - \Delta)/a}$ is given by the excitation energy U, the Fermi gas level density parameter a and the pairing correction Δ. For the energy dependence of the width we adopted the result of the theory of Fermi liquids[6]

$$\Gamma_i(\varepsilon_\gamma,T) = \beta_i(\varepsilon_\gamma^2 + 4\pi^2 T^2), \qquad (2)$$

with the normalization constant $\beta_i = \Gamma_{0i}/E_{0i}^2$ as proposed by Kadmenskij et al.[1] For the parameters of the Lorentzians ($\sigma_{0i}, E_{0i}, \Gamma_{0i}$) we used the results of fits to photonuclear data compiled by Dietrich and Berman.[7] In some comparisons (see Ref. 5) we use for f_{E1} also a **Lorentzian with energy dependent width** which results from Eq. (1) by omitting the second term and a **standard Lorentzian** in which in addition $\Gamma_i(\varepsilon_\gamma,T)$ is replaced by Γ_{0i}. For some of the considered nuclei the contribution of a *pygmy resonance* of standard Lorentzian shape was incoherently added to the E1 strength function; the parameters were chosen so as to improve the reproduction of primary resonance capture transitions and/or gamma-ray spectra.

For the **M1 strength function** we used either a single standard Lorentzian with global parameters for E_0 and Γ_0 and σ_0 adjusted to reproduce resonance data or f_{E1}/f_{M1} systematics[8] or the **adjusted single particle model** i.e. an energy independent strength function adjusted to experimental data or systematics; more details are given in Ref. 5.

Simple models were adopted for the **strength functions of quadrupole and octupole radiation** : a standard Lorentzian with global parameters[9,10] for E2- and the single particle model[11] for M2-, E3-, and M3 radiation.

RESULTS AND DISCUSSION

All calculations were performed for two level density models, the model of Kataria et al.[12] (KRK model) and the backshifted Fermi gas model[13] (BSFG model), both with parameters relying on experimental data (see Ref. 5 for more details). The neutron transmission coefficients were derived from optical potentials given in the literature; for deformed nuclei reaction channel coupling was accounted for. Results of these calculations have **not** been subject to any normalization.

Previously[5] we obtained for ^{93}Nb, ^{105}Pd, ^{143}Nd and ^{197}Au a satisfactory **simultaneous** reproduction of photonuclear data and primary transitions on the one hand and total s-wave radiation width, capture cross sections and gamma-ray spectra on the other hand, in terms of an E1-strength function with a non-zero $\varepsilon_\gamma \to 0$ limit as e.g. exhibited by Eq.(1) and a standard Lorentzian for f_{M1}. The above results of model calculations do not sensitively differentiate between E1 and M1 contributions. A comparable reproduction of the data was also obtained with a Lorentzian with energy dependent width for f_{E1} and the adjusted single particle model for f_{M1}; however, the latter is not supported by M1 transitions between low lying levels[5] and is at variance with a finite energy weighted sum rule. For nuclei with mass numbers between 50 and 60 it proved more difficult to support our preferred strength function models for E1 and M1 radiation.[14] Among the reasons are strong nonstatistical effects related to the 3s maximum of the neutron strength function and less reliable information on primary resonance capture transitions.

When extending these investigations to neutron capture in deformed nuclei we found contradictory results. For ^{182,183,184,186}W and ^{175,177}Re a reasonably good reproduction of the experimental data was achieved with a generalized Lorentzian for f_{E1} and a standard one for f_{M1}. On the other hand, the same assumptions on f_{E1} and f_{M1} underpredict the s-wave radiation width, the capture cross sections and the gamma-ray production spectrum for ^{151,153}Eu, ^{156}Gd and ^{175}Lu ; for these nuclei a standard Lorentzian for f_{E1} seems to do a better job. As representative examples we show in Fig. 1 the gamma-ray spectra for neutron capture in ^{156}Gd and ^{182}W, respectively, calculated with level densities according to the KRK model and three different assumptions on f_{E1} : a generalized Lorentzian (full curve), a standard Lorentzian (dashed curve) and a Lorentzian with energy dependent width; apart from those derived from a standard Lorentzian the results for ^{182}W account for a pygmy resonance. Test calculations showed that the generalized Lorentzian and the formula by Kadmenskij et al.[1] give very similar results. The failure of the generalized Lorentzian as given by Eq.(1) shows up also in comparisons with primary resonance capture transitions for ^{156}Gd[16] and ^{175}Lu[17], although the uncertainty in absolute calibration of f_{E1} can be partly responsible for that.

An improvement of the reproduction of *all* experimental data can be achieved by increasing the normalization of the energy dependent width given in Eq. (2). According to the theory of Fermi liquids[2,6,18] the normalization constant β_i depends on the density of quasiparticle states and thus on energy and shell structure. In order to guarantee also the reproduction of the the photonuclear data we assumed the following linear energy dependence:

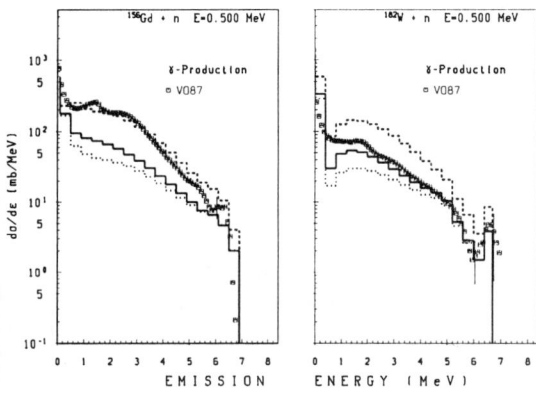

Fig.1 Capture cross sections and γ-spectra for ^{156}Gd and ^{182}W (see text). The data were measured by Voignier et al..[15]

$$\beta_i(\varepsilon_\gamma) = \frac{\Gamma_{0i}}{E_{0i}^2}\left(k_0 + \frac{\varepsilon_\gamma - \varepsilon_\gamma^0}{E_{0i} - \varepsilon_\gamma^0}(1 - k_0)\right).$$

The constants k_0 and ε_γ^0 can be used to reproduce the primary transitions following resonance capture; $k_0 = 1$ reproduces our original value of β_i. In the spirit of the Kopecky-Chrien approach the second terms in Eq. (1) must be multiplied by $(k_0 E_{0i} - \varepsilon_\gamma^0)/(E_{0i} - \varepsilon_\gamma^0)$.

The experimental data for ^{151,153}Eu, ^{156}Gd and ^{175}Lu can reasonably well be fitted with values of k_0 between 2 and 4 and ε_γ^0 around 5 MeV while for the considered W- and Re targets k_0 is close to 1. For a better understanding of the mass and energy dependence of the enhancement of the widths of the Lorentzians further investigations are required.

REFERENCES

1. S. G. Kadmenskij, V. P. Markushev, and V. I. Furmann, Sov. J. Nucl. Phys. **37**,165(1983).
2. V. K. Sirotkin, Sov. J. Nucl. Phys. **43**, 362(1986).
3. D. M. Brink, Ph. D. thesis, Oxford University, 1955.
4. J. Kopecky and R. E. Chrien, Nucl. Phys. **A468**, 285(1987).
5. J. Kopecky and M. Uhl, Phys. Rev. **C41**, 1941(1990).
6. D. Pines and P. Noziers, The Theory of Quantum Liquids, (Benjamin, New York 1966).
7. S. S. Dietrich and B. L. Berman, At. Nucl. Data Tables **38**, 199(1989).
8. J. Kopecky and M. Uhl, Proc. IAEA Specialists' Meeting on the Measurement, Calculation, and Evaluation of Photon Production Cross Sections, Febr. 5-7,1990, Smolenice, CSFR, in press.
9. J. Speth and A. van der Woude, Rep. Prog. Phys. **44**, 719(1981).
10. W. V. Prestwitch, M. A. Islam, and T. J. Kennet, Z. Phys. **A315**, 103(1984).
11. M. Blatt and V. F. Weisskopf, Theoretical Nuclear Physics (Wiley, New York, 1952).
12. S. K. Kataria, V. S. Rhamamurty, and S. S. Kapoor, Phys. Rev. **C18**, 545(1978).
13. W. Dilg, W. Schantl, H. Vonach, and M. Uhl, Nucl. Phys. **A217**, 269(1973).
14. M. Uhl and J. Kopecky, in Ref. 8
15. J. Voignier, S. Joly, and G. Grenier, Nucl. Sc. Eng. **96**, 343(1987).
16. R. E. Chrien, private communication (unpublished data).
17. R. W. Hoff, R. F. Casten, M. Bergoffen, and D. D. Warner, Nucl.Phys. **A437**, 285(1985).
18. V. K. Sirotkin and Yu. V. Adamchuk, Sov. J. Nucl. Phys. **26**, 262(1977).

STATISTICAL DESCRIPTION OF THE γ-RAY CASCADES AFTER THERMAL NEUTRON CAPTURE

B.Krusche

Institut Laue-Langevin,156X Centre de Tri, 38042 Grenoble Cedex,France *

ABSTRACT

Statistical model calculations of the γ-ray cascades after thermal neutron capture have been carried out. They are based on the average level density, nearest neighbor spacing distributions and radiative decay widths given by average γ-ray strengths functions and their fluctuations. Different parametrisations of the level density and different γ-ray strengths functions have been tested for an ensemble of 169 nuclei in the mass range $24 \leq A \leq 244$. For this purpose the average radiative widths of s-wave resonances were calculated and compared to measured values. The calculations were also used to predict feeding patterns and time development of the γ-cascades which are used as input for the GRID lifetime method.

INTRODUCTION

The study of statistical proprties of nuclear reactions has regained considerable interest during the last few years. The general picture which emerged is that one can characterize the physical observables by their average behavior which reflects the underlying dynamics of the system and rather universal fluctuation properties [1]. In the case of the thermal (n,γ)-reaction these observables are the positions of the excited states described by an average density and nearest neighbor spacing fluctuations and the radiative decay widths given by γ-ray strengths functions and Porter-Thomas fluctuations.

It was already shown [2] for some nuclei that a statistical simulation based on the constant temperature Fermi gas model (CTF) level density and giant resonance γ-ray strength functions reproduces many features of the γ-ray cascades. On the other hand even an extensive analysis of level schemes [3] did not allow to decide if the CTF-model or the Bethe formula is better suited to describe the level density at energies below ≈10 MeV. Furthermore the average radiative s-wave widths for odd-odd nuclei with $A \leq 80$ were nicely reproduced [2] from an E1-strengths function that did not include the spreading of the GDR. However such a simple strengths function was found to disagree with experimental results [4] for some heavier nuclei. One aim of the present work was therefore to study in more detail this average properties of the nuclei.

Apart from the interest in the basis of the statistical model a reliable description of the γ-ray cascades is e.g. very important for the application of the life time measuring method GRID [5] to nuclear states which are not predominantly fed by a primary transition. Here the strengths functions enter much more directly than into spectral shapes, energy and intensity distributions because the time development of the cascades is involved which depends on the absolute values of the decay widths while most other properties are primarily sensitive to branching ratios.

INPUT TO THE STATISTICAL MODEL CALCULATIONS

1) The level density can be parametrized [2,3] by the CTF-model:

$$\rho_C(E_x,J^\pi) = f(\sigma,J) \times \frac{1}{T} \times \exp[(E_x-E_o)/T] \tag{1}$$

*present address: II Physikalisches Institut, Justus Liebig Universität, 6300 Gießen, W.-Germany or the Bethe formula:

$$\rho_B(E_x, J^\pi) = f(\sigma, J) \times \frac{\exp(2\sqrt{a(E_x - E_1)})}{12\sqrt{2}\sigma a^{1/4}(E_x - E_1)^{5/4}} \qquad (2)$$

with:

$$f(\sigma, J) = \frac{1}{2}(exp(-J^2/2\sigma^2) - exp(-(J+1)^2/2\sigma^2)) \qquad (3)$$

The level density parameters T, E_o, respectively a, E_1 and σ were taken from von Egidy et al.[3] for 75 nuclei with $24 \leq A \leq 244$. For another 94 nuclei in the same mass range they were fitted to the known level schemes and the resonance spacings at the capture state [6] as described in ref.[2].

2) Radiative strength
The average E1-widths derived from the Axel-Brink hypothesis in the framework of giant dipole resonances are given by:

$$\Gamma_a(E1) = 8.68 \times 10^{-2} E_\gamma^2 D_\lambda \frac{\sigma_D \Gamma_D^2 E_\gamma^2}{(E_D^2 - E_\gamma^2)^2 + E_\gamma^2 \Gamma_D^2} \qquad (4)$$

Here $\Gamma_a(E1)[eV]$ is the partial radiative width of a transition with energy $E_\gamma[MeV]$, $D_\lambda[MeV]$ the average level spacing at the initial state and $E_D[MeV]$, $\Gamma_D[MeV]$, $\sigma_D[mb]$ are the excitation energy, width and peak cross section of the Lorentz curve describing the photoabsorption cross section.

In order to incorporate the spreading of the GDR and to achieve a correct low energy behavior, Kopecky and Chrien [4] suggested a modified strengths function:

$$\Gamma_b(E1) = 8.68 \times 10^{-2} E_\gamma^2 D_\lambda \sigma_D \Gamma_D [\frac{E_\gamma^2 \Gamma(E_\gamma)}{(E_D^2 - E_\gamma^2)^2 + E_\gamma^2 \Gamma^2(E_\gamma)} + \frac{0.7 E_\gamma \Gamma_D \times 4\pi^2 T^2}{E_D^5}] \qquad (5)$$

with

$$\Gamma(E_\gamma) = \Gamma_D(E_\gamma^2 + 4\pi^2 T^2)/E_D^2 \qquad (6)$$

where

$$T = \frac{d}{dE_f} log\rho(E_f) \qquad (7)$$

stands for the nuclear temperature at the final state energy. The generalisation for deformed nuclei, where the GDR is described by two Lorentz curves is obvious. The Lorentz parameters were taken from the tables of Dietrich and Bermann [7].
The semi empirical approximation [8]:

$$\Gamma(M1) = 9 \times 10^{-13} E_\gamma^5 A^{4/3} D_\lambda \qquad (8)$$

was used for the M1 transitions.
For E2 transitions again the Axel-Brink hypothesis was employed:

$$\Gamma(E2) = 5.22 \times 10^{-8} E_\gamma^2 D_\lambda [\sigma_{GQR}^{T=0} + \sigma_{GQR}^{T=1}] \qquad (9)$$

The parmeters for the Lorentz curves of the photoabsorption cross section

$$\sigma_{GQR}(E_\gamma) = \frac{\sigma_Q \Gamma_Q^2 E_\gamma^2}{(E_\gamma^2 - E_Q^2)^2 + E_\gamma^2 \Gamma_Q^2} \qquad (10)$$

for the isoscalar and isovector part were approximated from the compilation of Bertrand [9].

AVERAGE RADIATIVE s-WAVE WIDTHS

The average radiative s-wave widths for 169 nuclei were calculated for the simple and the more elaborate form of the E1-strengths function and for the Bethe and CTF formula. The M1 and E2 contribution was included in all four calculations using eqs. 8,9. The ratio of calculated to measured widths [10] for $\Gamma_b(E1)$ and the CTF (1a) and Bethe (1b) formula is shown in figure 1 as function of the s-wave resonance spacing D_o. (The behavior for $\Gamma_a(E1)$ as function of D_o is very similar but the ratios are shifted upwards).

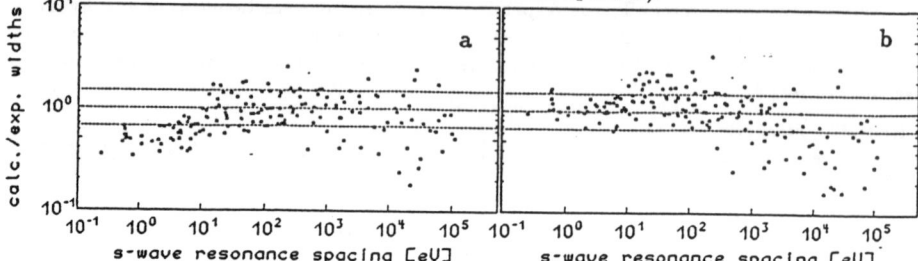

Fig 1: Ratio of calculated and measured average radiative s-wave widths for different level density formula. The dashed lines indicate where agreement is better than a factor of 1.5.

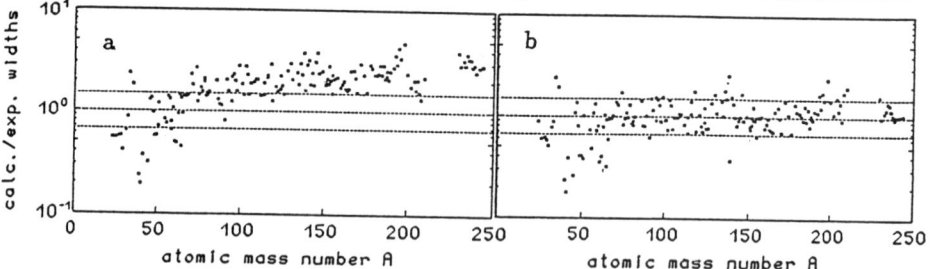

Fig 2: Ratio of calculated and measured widths as function of the mass number A. The E1-contribution was calculated from eq.4 for (2a) and from eq.5 for (2b).

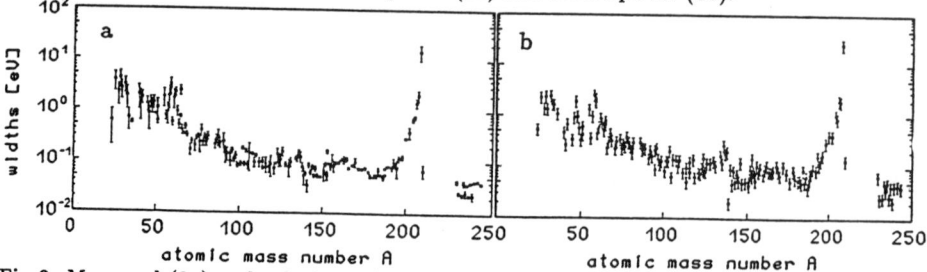

Fig 3: Measured (3a) and calculated (3b) radiative s-wave widths.

The agreement is similar or even better for the CTF model for nuclei with a level spacing larger than 10 eV, but this calculation underestimates the widths for nuclei with higher level density, where the Bethe formula gives better results. This effect can be understood in the following way. The level density parameters are fitted to the low energy parts of measured level schemes and the level spacing at the capture state. It is then clear from the shape of the two parametrisations that the Bethe fit will predict more levels in the high energy range (say between $2/3\ E_B$ and E_B) than the CTF fit. This discrepancy becomes more important the higher the density at the binding energy. Therefore the results seem to indicate, that the Bethe formula is a more natural description at high level density as one would expect from earlier work on level densities [11]. In the following we have used the Bethe formula for nuclei with a level density parameter a $\geq$ 15 MeV^{-1} and the CTF fit for the rest.

The calculations for $\Gamma_a(E1)$ and $\Gamma_b(E1)$ are displayed in figure 2a,b. It is evident from this figure that wheras the agreement is comparable for the lighter nuclei, the more elaborate γ-ray strengths function $\Gamma_b(E1)$ gives a much better agreement over the whole mass range. Apart from a group of nuclei around mass 50 which are strongly effected by direct capture [2,3], agreement whithin almost a factor of 1.5 is achieved. The measured s-wave widths [10] and the calculation with $\Gamma_b(E1)$ are shown in figure 3. It should be noted that the strong shell effects e.g. around mass 200 are very nicely reproduced. Shell effects enter into the statistical calculations only via the level density parameters. The Behavior of the widths around ^{208}Pb is dominated by the term $D_\lambda = \rho^{-1}(E_B)$ in equs. 5,8,9. The strong increase of the widths reflects the increase of D$^\lambda$ towards the double shell closure which is not balanced by the corresponding decrease of the final state density. Also the reproduction of the broad peak between mass 150 and 170 contradicts the usual explanation that this peak is mainly due to direct capture components. Such components are not included in this calculation.

LIFE TIMES OF LOW LYING STATES

The good agreement for nuclei with very different masses, giant resonance properties, level densities and nuclear structure is the main justification to use the same model also for levels below the capture state and thus to predict the development of the entire statistical cascades which may be used for the analysis of GRID measurements. We have already shown [2] that properties like spectral shapes etc. are very well reproduced by such calculations. However a much more crucial test is the prediction of lifetimes for lower lying levels. Here the GRID measurements open up new possibilities because in some cases lifetimes of levels populated in the (n,γ)-reaction can be determined over a fairly large range of excitation energy. One example is the nucleus ^{59}Ni where lifetimes up to 4 MeV excitation energy have been measured [12]. The statistical simulation reproduces this lifetimes within a factor of $\approx$ 2 and comparably well as a full scale shell model calculation [12].

CONCLUSION

It was shown that not only features like spectral shapes etc. [2] but also the _average_ lifetimes of nuclear states can be described by a statistical model calculation over a wide range of different nuclei. In the energy range of the capture states an agreement within about a factor of 1.5 is achieved and we have some evidence that reasonable agreement exists down to relatively low excitation energies. A more detailed analysis for states at excitation energies a few MeV below the binding energy will be published elsewhere [13]. Best results are obtained if the CTF level density formula is used only for nuclei with fairly low level density. Clear evidence was found that the spreading of the E1 giant dipole resonance must be included.

ACKNOWLEDGEMENTS
Many stimulating discussions with K.P.Lieb, T.von Egidy, J. Kopecky, H.G.Börner, J. Jolie, S. Robinson and S.Ulbig are gratefully acknowledged.

REFERENCES
[1] H.A.Weidenmüller, Comments on Nucl. Part. Phys.16 (1986)189
[2] B.Krusche et al., Phys.Rev. C34(1986)2103, B.Krusche et al.,J.Phys.G14 suppl.(1988)183
[3] T.von Egidy et al.,Nucl.Phys.A481(1988)189
[4] J.Kopecky et al., Nucl.Phys.A468(1987)285
[5] H.G.Börner et al., Contr. to this Conference
[6] G.Rohr, privat communication
[7] S.S.Dietrich et al., ADNDT 38(1988)199
[8] J.Kopecky, Inst.Phys.Conf.Ser. No 62, Chapter 2 (1981)423
[9] F.E.Bertrand, Ann.Rev.Nucl.Sci.26(1976)457
[10] S.F.Mughabghab et al., Neutron Cross Sections (Academic, N.Y. 1981) Vol 1, Pts A,B
[11] A.Gilbert et al.,Can. J. Phys.43(1965)1446
[12] S.Ulbig et al., in preparation
[13] B.Krusche, in preparation

Calculated Neutron Radiative Capture Cross Section Using Modifying Exciton Model

Z.H.Lu   C.Z.MO   Y.K.HO

(Zhengzhou University, China)   (Fudan University, China)

The fast neutron radiative capture cross sections were successfully treated by the direct semidirect theory[1]. Recently, the preequilibrium exciton model has been extended to the reaction process including γ-ray emission, but the calculated γ-ray spectra were lower than the experimental results[2]. The exciton-phonon couplying model has been treated the neutron radiative capture reaction by including a phonon to take account of the collective effect of the gaint dipole resonance[3]. The calculated (n,γ) cross sections are still lower than the experimental results at low energy region.

In this paper, the direct potential capture, as an one-exciton componente, is included in the exciton model and the exciton-phonon couplying model to calculate the (n,γ) cross sections for ^{40}Ca. The apparent improvement has been achieved in the (n,γ) cross sections especially in low-energy region. The neutron radiative capture reaction mechanism is represented schematically in Fig 1, where n, a and γ denote the exciton, phonon and photon respectively. The incident neutron is considered as an one-exciton state undergoing pontential scattering. It may further excites a three-exciton state or excites an one-exciton-phonon state. Both pure exciton states and exciton-phonon couplying states can decay with emission of γ-ray. We only consider primay process and neglect the cascade low energy γ-ray emissing process. The (n,γ) cross sections of the exciton-phonon couplying system :

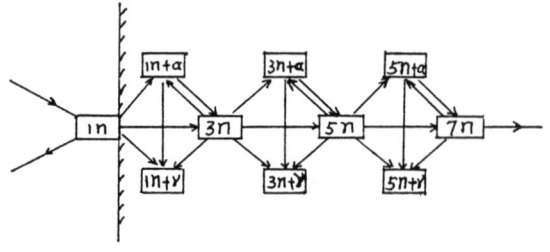

Figure 1, The reaction mechanism of the exciton-phonon couplying model including direct capture. a=1 and 0 denote the exciton-phonon couplying states and pure exciton states respectively.

$$\sigma_\gamma = \sigma_c \sum_n \left\{ W_r(n,0) \int_0^T P(n)Q(0)dt + \left[W_r(n,1) + W_{pho\rightarrow}(n) \right] \int_0^T P(n)Q(1)dt \right\} \quad (1)$$

σ_c is the neutron absorbing cross section, $W_r(n,0)$ and $W_r(n,1)$ [4] denote the probability of emitting γ-ray by the n-exciton and n-exciton-phonon states respectively, $P(n)Q(0)$ and $P(n)Q(1)$ are the occupying probability of the n-excitons and n-exciton-phonon respectively, which can be get from the main equation of the exciton-phonon couplying system, T is the equilibrium time of the system.

The (n,γ) cross section of the one-exciton state :

$$\sigma_\gamma(1,0) = \sigma_c W_r(1,0) \int_0^T P(1)Q(0)dt$$

$$= \sigma_c W_r(1,0) \tau(1) \quad (2)$$

where the life of one-exciton state $\tau(1)$ can be expressed as

$$\tau(1) = \int_0^T P(1)Q(0)dt \quad (3)$$

observably in our model, the cross section for one-exciton state which is considered as an initial condition in the exciton chain can be determined by setting its magnitude to be equal to the potential capture cross section σ^d [5]

$$\sigma^d = \sigma_\gamma(1,0)$$
$$= \sigma_c W_r(1,0) \tau(1)$$

$$\tau(1) = \frac{\sigma^d}{\sigma_c W_r(1,0)} \quad (4)$$

σ^d and $W_r(1,0)$ may be calculated. The (n,γ) cross section of any exciton-phonon states in the exciton chain can be get by the main equation and the initial condition. In the calculation, except the exciton-phonon trasitional matrix

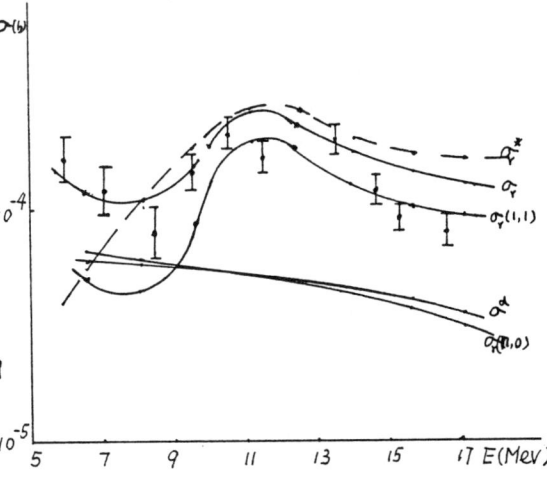

Figure 2, (n,γ) reaction excitation curve for ^{40}Ca, E is the incident neutron, σ_γ^x is total cross section of the original exciton-phonon model, σ_γ, $\sigma_\gamma(1,1)$ and $\sigma_\gamma(1,0)$ is the cross sections of total, one-exciton-phonon state and pure exciton states in our model respctively. σ^d is the direct capture cross section.

element,we didn't make use of any independent adjuslable paremater,all paremaers are from the free preequilibrium exciton model. The exciton trasitional matrix element $|M|^2 = 190\ A\ E^{-1} Mev^2$ is used. The exciton-phonon trasitional matrix element $|M|^2 = K_c\ A^{-3}$, K_c is the adjustable paremater,k_c=800 was chosen.

The (n,γ) excitation curve for ^{40}Ca are represented schematically in Fig 2,where the dashed line denote the result of the original exciton-phonon couplying model,the solid lines denote the results of our model.

We can get the following conclusions from Fig 2 and above the discussion :

1. The apparent improvement has been achieved in the calculated (n,γ) cross sections for ^{40}Ca especially in the low energy region.

2. The gaint dipole resonance mainly originate from one-exciton-phonon state. The conclution consistes with that of the direct semidirect theory which considers that the gaint dipole resonance is caused by the doorway phonon.

3. The (n,γ) excitation curves are calculated to indicate the clear physical picture in our model.

References

1. R.E.Chrien   Neutron Radiative Capture   Vol 3,1984.
   Y.K.HO and M.A.Lone   Nucl.Phys. A406,1983.
2. J.M.Akkermans and H.Gruppelaar   Phys.Lett. Vol 157B,No 2,3, 1985.
3. Ma Zhongru et al   Chinese Journal of Nuclear Physics Vol 3,No 3, 1983.
4. E.Betak and J.Dobes   Phys.Lett. Vol 84B,No 4,1979.
5. B.J.Allen and A,R.Musrgrove   Advance in Nucl.Phys. Vol 10,Plenum Press,New York,1979.

POLARIZED NEUTRON-PROTON CAPTURE FOR NEUTRON ENERGIES FROM 19 TO 50 MEV

G.Fink
Ohio University Accelerator Laboratory
Athens, Ohio 45701 U.S.A.

P.Doll, S.Hauber, M.Haupenthal, H.O.Klages, H.Schieler
Kernforschungszentrum Karlsruhe, Institut für Kernphysik I,
Postfach 3640, D-7500 Karlsruhe 1, Germany

F.Smend and G.D.Wicke
University Göttingen, Germany

Abstract

A continuous energy beam of polarized neutrons has been used to determine the differential cross section and the analyzing power for neutron-proton capture at laboratory angles of 55°, 90° and 125° in the energy range from 19 to 50 MeV. Capture gamma rays in the energy range E_γ = 11-27 MeV and recoil deuterons were detected in coincidence, using an organic scintillator NE213 as hydrogen target. 10^5 times larger background due to elastic neutron-proton or carbon scattering was removed by pulse-shape discrimination technique in the NaI detectors and by measuring the time of flight for reaction products and pulse height information in the target. Multipe scattering effects in the target were studied. The differential cross sections and analyzing powers are in agreement with extensive theoretical calculations using modern nucleon-nucleon potentials and including meson exchange currents. Discrepancies between former experiments and theoretical calculations in both observables cannot be confirmed.

Recent sudies of the $^2H(\gamma,n)H$ reaction [1,2] and the inverse process have indicated that besides meson exchange currents (MEC) corrections even the basic reaction mechanism is not fully understood already at low photon energies. Therefore we performed a neutron-proton capture experiment using polarized neutrons [3] between 18 and 50 MeV. Our set-up consists of three 16x16x24 cm^3 NaI detectors shielded with lead, polyethylene and boron loaded polyethylene. We used a scintillating target (NE213) allowing simultaneous measurement of neutron-proton and neutron-carbon capture.

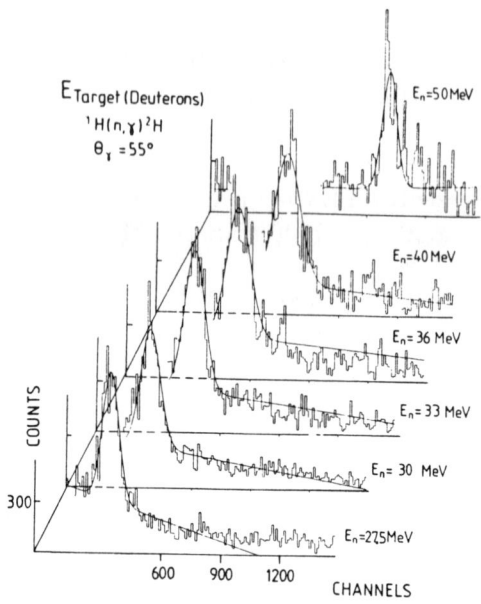

Fig. 1. Pulse height spectra of recoil deuterons for various incident neutron energies

Fig. 1 shows the pulse height resolution in the target for recoiling deuterons arising from neutron-proton capture. The pulse height resolution turned out to be sufficient ($\sim 16\%$) to separate recoil events from multiple scattering events in the target. The evaluation of the photon yields was performed in the NaI spectra, supported by pulse-shape discrimination technique in the NaI crystal [4] and the time-of-flight measurement between the capture target and the NaI-detector. The relative efficiency of the NaI detectors at 55°, 90° and 125° in the laboratory system was determined through the angular distribution of the decay of the 15.1 MeV state in ^{12}C.

Fig. 2 shows the angular distributions of the differential cross sections for various neutron energies. The absolute cross sections are based on total capture and disintegration cross sections related through detailed balance. The solid lines in Fig.2 represent Legendre fits to the data and the dashed curves are theoretical calculations by P.Wilhelm et al [5]. The cross sections for 0° and 180° were taken for the Legendre fit in the range known in the literature. The angular distribution shows only a smooth variation of the forward-backward asymmetry as a function of neutron energy due to the small admixture of higher multipole transitions. Analyzing powers at the three scattering angles were measured simultaneously. Our results compared to data from ref. 1,2 for $\Theta = 90°$ seem to indicate slightly larger analyzing powers in the present experiment, after considering multiple scattering effects in the target.

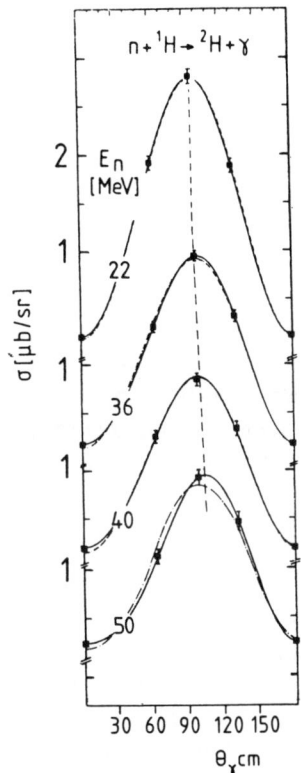

Fig. 2. Angular distributions of diffential cross sections for $^1H(n,\gamma)^2H$ at various neutron energies.

1. J.P.Soderstrum, L.D.Knutson, Phys. Rev. C35, 1246 (1987).
2. R.J.Holt, K. Stephenson, J.R.Specht, Phys. Rev. Lett. 50, 5 (1983).
3. G.Fink, P.Doll, S.Hauber, M.Haupenthal, H.O.Klages, H.Schieler, F.Smend, G.D.Wicke, Report KfK 4405, Kernforschungszentrum Karlsruhe (1987) p.29 and to be published
4. P.Doll, G.Fink, R.W.Finlay, S.Hauber, M.Haupenthal, H.O.Klages, H.Schieler, F.Smend, G.D.Wicke, Nucl. Instr.and Meth.285, 464 (1989).
5. P.Wilhelm, W.Leidemann, H.Arenhövel, private communication (1989)

SYSTEMATICS OF 3.5 to 100 MeV $^2H(\gamma,n)$ DATA

A. Wolf, S. Kahane and Y. Birenbaum

*Physics Department, Nuclear Research Center-Negev,
P.O.B. 9001, Beer-Sheva, Israel 84190*

ABSTRACT

A study of the systematics of low energy deuteron photodisintegration data was undertaken, utilizing data from twelve publications in the 3.5 to 100 MeV energy range. While the absolute $D(\gamma,n)$ cross section seems to follow a regular pattern, the differential cross section data below 30 MeV is spread out in magnitude. More, high accuracy data is required before a clear pattern for that quantity may emerge.

INTRODUCTION

The need for analytical expressions for the total and differential $D(\gamma,n)$ cross sections was emphasized in the past[1],[2]. Such expressions are important for comparison between theoretical calculations and experimental data, and for calculations in which the two-body cross section is needed as input.

De Pascale et al.[1] treated experimental data between 10 and 120 MeV, and obviously could not consider the low energy data published in the last 5 years. The phenomenological formulae recently proposed by Rossi et al.[2] for the total cross section and for the angular distribution coefficients of the $D(\gamma,n)$ reaction, were based upon experimental data between 20 and 400 MeV, and provide reasonable results for photon energies above 30 MeV. However, at lower energies (*i.e.* in the 5-25 MeV range) the agreement with experimental data is not satisfactory. In the present contribution we used the same approach, utilizing recently published low energy data, together with older data up to 100 MeV, to obtain analytical relations, which provides a better description of the experimental situation at low and medium energies ($E_\gamma \leq 100 MeV$). The new low energy data is important for establishing reliable relations over a larger energy range.

DATA ANALYSIS

Both photodisintegration and neutron capture data were considered, employing detailed balance when necessary. Nine different data sets[4] - [11] in the

3.5 MeV to 100 MeV energy range, were included in the fit of the total cross section to a function of the form:

$$\sigma_T(E) = \sum_{i=1}^{2} B_i e^{-c_i E} \qquad (1)$$

each data point was weighed by its quoted experimental error. Differential cross section data was examined by looking at the neutron angular distribution coefficients, obtained from a fit of the experimental data to a sum of legendre polynomials of the form:

$$(d\sigma/d\theta)_{c.m.} = A_0 + \sum_{\nu=1}^{3} A_\nu P_\nu(\cos\theta); \qquad (2)$$

Whenever necessary, these coeficients were deduced from the published data. In the cases where absolute values for A_0 were not experimentaly available, the prediction of the analytical expression for σ_T obtained in the present work were used. Six data sets taken at photon energies between 3.5 and 100 MeV, were considered for the angular distribution coefficients [4],[7],[12],[13], including our recently published[14], and unpublished results between 5.97 and 11.4 MeV.

RESULTS AND DISCUSSION

The experimental data for σ_T, A_1, A_2 and A_3, considered in the present investigation, are presented in fig. 1(a) to 1(d) respectively, together with prediction of De Pascale(dotted line) [1] and Rossi(solid line) [2]. In the case of σ_T (fig. 1(a)), the extrapolation of the relation suggested by Rossi et al. clearly underestimates the data below 20 MeV. The agreement between the experimental data above 10 MeV, and the relation given by De Pascale et al.[1], seems to be better, but the extrapolation to lower energies tends to predict too low values. A similar situation exists in the case of A_2 (fig. 1(b)), except that here the experimental data is **overestimated** by Rossi's expression. The experimental A_2 values were fitted in the present work to a function similar to that for σ_T. In the case of A_1 and A_3 (figures 1(c) and 1(d) respectively), the experimental data below 30 MeV does not follow a clear trend, and due to the spread in values, any fit to the data in this energy region is meaningless.

Figures 1(f) and 1(e) show the fits obtained in the present work to σ_T and to A_2 respectively. The expansion coefficients obtained for σ_T and for A_2 are given in Table I. In both cases the agreement between experiment and theory is seen to be better than in figures 1(a) and 1(b), in particular in the 6 - 20 MeV energy region.

In conclusion, new experimental data was used to obtain reliable analytical relations for the low energy $D(\gamma,n)$ data. The total cross section could be assigned a simple expression, valid down to 5 MeV. As for the angular distribution coefficients, only the A_2 data was found to vary relatively smoothly with energy, while for A_1 and A_3 more high accuracy data is needed before the trend followed by these coefficients can be established.

Table I. The coefficients arising from the fits of σ_T and A_2 to eq. (1). B_i are in $\mu b/sr$ and c_i are in MeV^{-1}.

	B_1	B_2	c_1	c_2
σ_T	3816.5	537.5	0.135	0.024
A_2	−217.8	−137.4	0.197	0.063

References

[1] M.P.De Pascale, et al., Phys. Lett. 119B 30 (1982)

[2] P. Rossi et al., Phys. Rev. C40 2412 (1989)

[3] J. .Ahrens et al. Phys. Lett. 52B 49 (1974)

[4] D. M. Skopik, Y. M. Shin, M. C. Phenneger and J. J. Murpy,II Phy. Rev. C9 531 (1974)

[5] T. Stiehler et al. Phys. Lett. 151B 185 (1985)

[6] Y. Birenbaum, S. Kahane and R. Moreh, Phys. Rev. C32 1825 (1985)

[7] E. De Sabtis et al. Phys. Rev. Lett. 54 1639 (1985)

[8] C. Dupont et al. Nucl. Phys. A445 13 (1985)

[9] R. Bernabei et al. Phys. Rev. Lett. 57 1542 (1986)

[10] P. Michel, K. Moeller, J. Moesner and G. Shmidt, J. Phys. G 15 1025 (1989)

[11] P. Levi Sandri et al. Phys. Rev. C39 1701 (1989)

[12] M. De Pascale et al. Phys. Rev. C32 1830 (1985)

[13] K. E. Stephenson, R. J. Holt, R. D. McKeown and, J. R. Specht, Phys. Rev. C35 2023 (1987)

[14] Y. Birenbaum, Z. Berant, A. Wolf, S. Kahane and R. Moreh, Phys. Rev. Lett. 61 810 (1988);Erratum Phys. Rev. Lett. 62 115 (1989)

Fig. 1

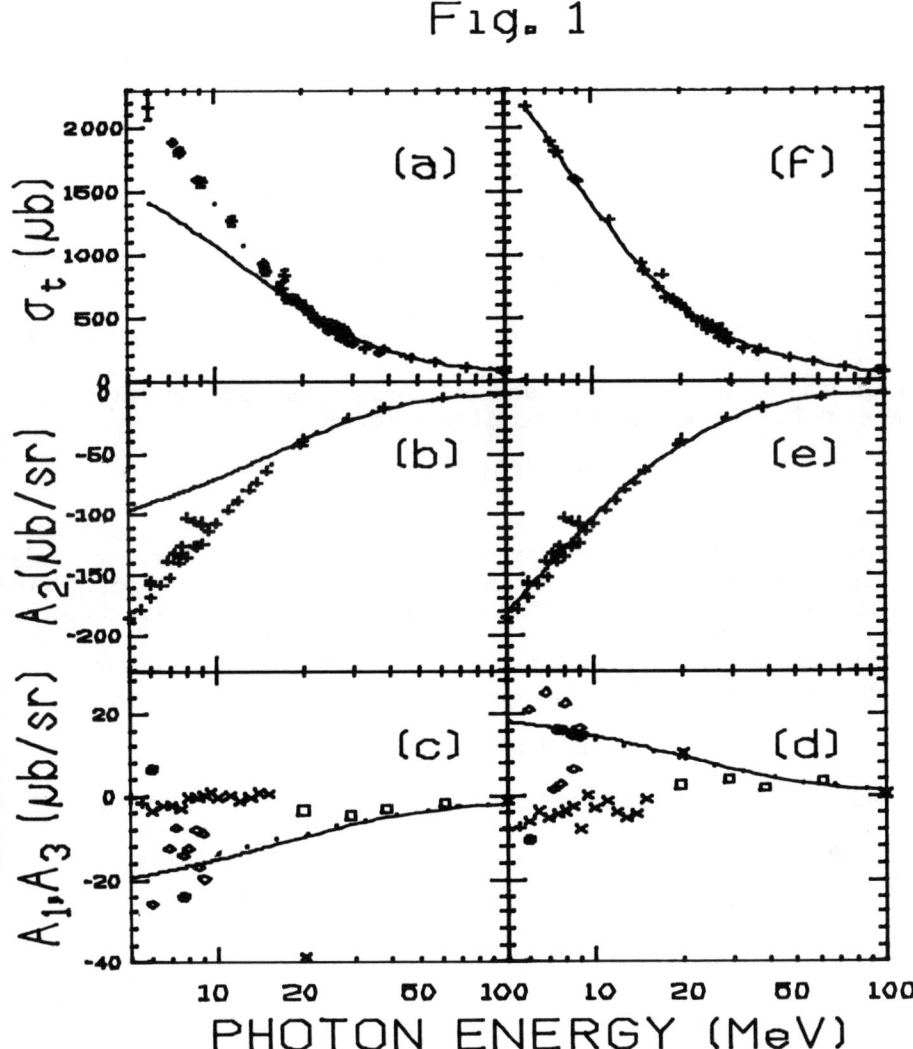

CAPTURE GAMMA RAYS FROM BROAD NEUTRON RESONANCES IN THE PROTON-ODD NUCLEI ^{14}N, ^{19}F AND ^{27}Al

M. Igashira, H. Kitazawa, S. Kitamura, H. Anze, and M. Horiguchi
Tokyo Institute of Technology, O-okayama, Meguro-ku, Tokyo 152, Japan

ABSTRACT

Capture gamma rays were measured from broad neutron resonances in the proton-odd nuclei ^{14}N, ^{19}F and ^{27}Al. Strong primary M1 transitions to low-lying states were observed in all residual nuclei, in addition to some primary E1 transitions to low-lying states in ^{15}N and ^{20}F. Partial radiative widths were extracted for all these primary transitions. The valence-capture-model calculation based on the Lane-Mughabghab formula was made for both E1 and M1 transitions, however the results never explain the observed E1 and M1 transitions.

INTRODUCTION

We have studied capture gamma rays from broad neutron resonances in 1p and 2s-1d shell nuclei.[1-4] For nuclei with even protons, electric dipole transitions were successfully explained[2-4] by the Lane-Mughabghab valence capture model[5] and a particle-vibrator or particle-rotator coupling model. However, we lack definite information on nuclei with odd protons. Therefore, measurement was performed of capture gamma rays from broad neutron resonances in ^{14}N, ^{19}F, and ^{27}Al.

EXPERIMENTS

Employing an anti-Compton NaI(Tl) detector and a time-of-flight technique, we have observed neutron capture gamma rays from the 644-keV s-wave resonance(1/2+, Γ=43 keV) in ^{14}N, from the 27-keV p-wave resonance(2−, Γ=0.36 keV), 49-keV p-wave resonance(1−, Γ=1.8 keV), and 97-keV p-wave resonance(1−, Γ=16.4 keV) in ^{19}F, and from the 35-keV s-wave resonance(2+, Γ=3.0 keV) and 142-keV s-wave resonance(3+, Γ=17.5 keV) in ^{27}Al. Pulsed neutrons were produced from the ^{7}Li(p,n)^{7}Be reaction by bombarding a natural lithium target with a 1.5-ns bunched proton beam from the 3-MV Pelletron accelerator at the Tokyo Institute of Technology. The average proton-beam current was 7-10 µA for a pulse repetition rate of 2 MHz. Neutrons incident on a capture sample were monitored by two ^{6}Li-glass scintillation detectors. The capture sample was a disk of about 6 cm in diameter and 0.010-0.055 atoms/b in thickness, and was placed at the distance of 15.6 cm from the neutron source. The gamma-ray detector was located at the distance of about 80 cm from the sample, and its axis made an angle of 125° with respect to the proton beam direction. A gold capture sample was used as a standard.

RESULTS AND DISCUSSION

Strong primary M1 transitions to low-lying states were observed in all residual nuclei, and some primary E1 transitions to low-lying states in 15N and 20F. Tables 1-3 show the extracted partial radiative widths for all these primary transitions. As seen from Tables 1-3, the radiative width for M1 transition ranges from 0.1 eV to 3.7 eV, particularly the transition to the second excited state in 15N has a strength of 0.8 Weisskopf unit. Also, the radiative width for the E1 transition to the ground state in 15N is 9.0 eV, and this value is twice larger than the old one[6] obtained from the 14C(p,γ)15N reaction.

Table 1. Experimental and theoretical partial radiative widths for 644-keV s-wave neutron resonance(1/2+) in 14N

Final states MeV (J^π)	Partial radiative widths (eV)		
	Experiment	F and G[6]	Valence
Gnd:0.00(1/2−)	9.0±1.2	4.2±0.7	0.33
2nd:5.30(1/2+)	3.7±0.4	—	—
3rd:6.32(3/2−)	1.3±0.5	—	—

Table 2. Experimental and theoretical partial radiative widths for 27-keV p-wave resonance(2−), 49-keV p-wave resonance(1−), and 97-keV p-wave resonance(1−) in 19F

Final states MeV (J^π)	Partial radiative widths (eV)		
	27-keV res.(2−)	49-keV res.(1−)	97-keV res.(1−)
Gnd:0.00(2+)	0.01±0.004 (0.000)	0.02±0.008 (0.000, 0.24)	0.38±0.05 (0.00, 0.026)
1st:0.66(3+)	0.06±0.01 (0.009)	—	—
3rd:0.98(1−) 4th:1.06(1+)	—	<0.24±0.03 (0.054, 3.5)	0.32±0.05 (0.002, 0.004)
5th:1.31(2−)	0.36±0.04 (0.00)	—	0.48±0.07 (0.00, 0.00)
7th:1.84(2−) 8th:1.97(3−?) 9th:2.04(2+)	0.51±0.07 (0.000)	0.45±0.04 (0.00, 4.7)	0.29±0.06 (0.00, 0.54)
15th:3.49(1+) 16th:3.53(0+) 17th:3.59	0.05±0.01 (0.002)	0.14±0.02 (2.0, 2.1)	0.73±0.09 (0.074, 0.28)

Note: Theoretical values are shown in parentheses. The left and right values in the parentheses correspond to the results with j=1/2 and j=3/2, respectively.

Table 3. Experimental and theoretical partial radiative widths for 35-keV s-wave resonance(2+) and 142-keV s-wave resonance(3+) in ^{27}Al

Final states MeV (J^π)	Partial radiative widths (eV)			
	35-keV res.(2+)		142-keV res.(3+)	
	Experiment	Valence	Experiment	Valence
Gnd:0.00(3+) ⎤ 1st:0.03(2+) ⎦	1.89±0.19	0.16	3.41±0.53	0.41
2nd:0.97(0+) ⎤ 3rd:1.01(3+) ⎦	0.094±0.021	0.19	—	—
5th:1.62(1+) ⎤ 6th:1.62(2+) ⎦	0.42±0.06	0.22	—	—

The valence-capture-model calculation based on the Lane-Mughabghab formula was made for both E1 and M1 transitions. For the M1 transitions in ^{28}Al, the tensor part introduced by Bohr and Mottelson[7] in a velocity-dependent field was included in the M1 operator to take the l-forbidden M1 transition($\Delta l = 2$) into account[3], where the free parameter k was set to a value of 0.78 that was obtained by an analysis of the 188-keV s-wave resonance in ^{28}Si.[3] Calculated results are compared with observed ones in Tables 1-3. The calculations never explain the observed E1 and M1 transitions. For E1 transitions, taking into consideration that the gamma-ray transitions subsequent to neutron capture by the 153-keV p-wave resonance in ^{13}C were successfully explained by a valence capture model[8], it is probable that the unpaired protons in these nuclei play an important role in radiative neutron capture.

REFERENCES

1. M. Shimizu, M. Igashira, K. Terazu, and H. Kitazawa, Nucl. Phys. A452, 205 (1986).
2. H. Kitazawa, M. Ohgo, T. Uchiyama, and M. Igashira, Nucl. Phys. A464, 61 (1987).
3. H. Kitazawa and M. Igashira, J. Phys. G 14, S215 (1988).
4. T. Uchiyama, M. Igashira, and H. Kitazawa, Phys. Rev. C 41, 862 (1990).
5. A. M. Lane and S. F. Mughabghab, Phys. Rev. C 10, 412 (1974).
6. A. J. Ferguson and H. E. Gove, Can. J. Phys. 37, 660 (1959).
7. A. Bohr and B. R. Mottelson, Nuclear Structure Vol. 1 (Benjamin, N. Y., 1969), p. 336.
8. S. Raman, M. Igashira, Y. Dozono, H. Kitazawa, M. Mizumoto, and J. E. Lynn, Phys. Rev. C 41, 458 (1990).

ELECTROMAGNETIC TRANSITIONS FROM BROAD s-WAVE NEUTRON RESONANCE IN THE sd-SHELL NUCLEI ^{24}Mg, ^{28}Si, AND ^{32}S

H. Kitazawa, M. Igashira, Y. Achiha, Y. Lee,
N. Mukai, K. Muto, and T. Oda
Tokyo Institute of Technology
2-12-1 O-okayama, Meguro-ku, Tokyo 152, Japan

ABSTRACT

Observation was performed for electromagnetic transitions from the s-wave neutron resonances at 658 keV in ^{24}Mg, 180 keV in ^{28}Si, and 103 keV in ^{32}S. As a result, we found strong E1 and M1 transitions from these resonances to low-lying states with a considerable spectroscopic factor. The observed radiative widths for E1 transition are in substantial agreement with the predictions of the Lane-Mughabghab valence-capture model. Moreover, a brief discussion is devoted to M1 transitions in a core-particle coupling scheme.

INTRODUCTION

From the uncertainty principle, broad resonance states are expected to have a simple configuration. In fact, strong E1 transitions, which are successfully explained by a valence model or by a core-particle coupling model, have been observed for the radiative capture in resonances with large reduced neutron width.[1,2]

On the other hand, available experimental data on the total radiative width of broad neutron resonances have been found to frequently include some systematic error. It was due to the difficulty in resolving capture gamma-ray events and the background produced by the neutrons scattered in the capture sample. The problem is in particular serious for s-wave neutron resonance data.

From this viewpoint, we have examined electromagnetic transitions from the s-wave neutron resonances at 658 keV in ^{24}Mg, 180 keV in ^{28}Si, and 103 keV in ^{32}S.

EXPERIMENT

Experiment was made with the 3.2-MV Pelletron accelerator of the Tokyo Institute of Technology, employing a time-of-flight technique. Pulsed neutrons were produced from the ^{7}Li(p,n)^{7}Be reaction by bombarding a Li-evaporated copper disk with the 1-ns bunched proton beam from the accelerator. The average beam current was typically 7 µA at 2-MHz pulse repetition rate. Capture gamma rays were measured with an anti-Compton NaI(Tl) detector,[3] which was located at an angle of 125° with respect to the proton beam direction, 80 cm distant from the sample. In the present experiment, moreover, we gave special consideration to measurements of gamma rays from strong resonances: the ^{6}Li-powder was inserted into the collimator of the gamma-ray detector to remove many neutrons scattered from the sample.

© 1991 American Institute of Physics

RESULTS AND DISCUSSION

Strong gamma rays have been observed for transitions from the s-wave resonances in ^{24}Mg, ^{28}Si, and ^{32}S to low-lying states with a considerable spectroscopic factor. Tables 1-3 show the partial radiative widths derived from these transitions. The ^{28}Si data were remeasured in the present experiment, since a mistake was found for the data processing in our previous work.[4]

Table I Partial radiative widths of the 658 keV resonance in ^{24}Mg

Ex (MeV)	radiative widths (eV)		
	measured	valence	configuration-mixing
0.59(1/2+)	1.24±0.39		(1.24)
0.98(3/2+)	0.65±0.12		(0.65)
2.80(3/2+)	0.47±0.09		0.27
3.41(3/2−)	0.31±0.09	0.22	(0.31)
4.28(1/2−)	0.11±0.04	0.05	0.12

Table II Partial radiative widths of the 180 keV resonance in ^{28}Si

Ex (MeV)	radiative widths (eV)		
	measured	valence	configuration-mixing
0.00(1/2+)	1.29±0.26		(1.29)
1.27(3/2+)	1.20±0.27		(1.20)
4.93(3/2−)	0.76±0.28	0.64	(0.76)
6.38(1/2−)	0.35±0.15	0.10	0.15

Table III Partial radiative widths of the 103 keV resonance in ^{32}S

Ex (MeV)	radiative widths (eV)		
	measured	valence	configuration-mixing
0.00(3/2+)	0.85±0.11		(0.85)
0.84(1/2+)	0.87±0.13		(0.87)
3.22(3/2−)	1.17±0.12	0.88	(1.17)
5.72(1/2−)	0.23±0.07	0.15	0.21

The observed widths for E1 transition were compared with the calculations based on the Lane-Mughabghab valence capture model. In the model calculation, the optical potential parameters were taken from the work of Moldauer.[5] However, the strength of the central potential part was determined so that the s-wave neutron resonated to the nucleus. The neutron widths for the s-wave resonances in ^{24}Mg and ^{32}S were taken from the compilation of Mughabghab et al.[6] The width for ^{28}Si was newly determined by Kitazawa et al.,[7] using a Breit-Wigner multilevel fitting for resonances in ^{28}Si, since it was found that the value given in the compilation was wrong. The spectroscopic factors for ^{25}Mg and ^{32}S were taken from the work of Endt,[8] and those for ^{29}Si taken from the work of Peterson et al.[9] As is seen from the tables, the observed values for E1 transition are in substantial agreement with the theoretical predictions. Probably, the success of the model calculation is due to the fact that the E1-transition matrix element cancels out in the interior region of the nucleus.

In addition, it is conceivable that the spin-flip core excitation in the resonance state contributes largely to the observed M1 transitions. The excitation may be of isovector type, since isoscaler states are strongly suppressed in electromagnetic transitions. Therefore, we assumed a configuration-mixed wave function, $\Psi = a(0+ \otimes 1/2+) + b(1+ \otimes 1/2+) + c(1+ \otimes 3/2+)$, for the resonance state, neglecting the sign of each term. The probability amplitudes, a, b, c, were derived from the observed E1 and M1 transitions. Here, the reduced strength $B\downarrow$ (M1) was taken to be 1.0 μ_0^2 for ^{24}Mg and ^{28}Si, and 1.8 μ_0^2 for ^{32}S. In the tables, the values in parentheses were used to determine these amplitudes. The assumed wave functions seems to give a reasonable explanation to observed transitions. More detailed shell-model calculations using the Wildenthal interaction in the (sd)^n full space predict large components of $|(T=1, J=1) \otimes s1/2\rangle$ and $|(T=1, J=1) \otimes d3/2\rangle$ near the s-wave resonance states in ^{24}Mg and ^{29}Si.

REFERENCES

1. H. Kitazawa, M. Ohgo, T. Uchiyama, and M. Igashira, Nucl. Phys. A464, 61 (1987)
2. T. Uchiyama, M. Igashira, and H. Kitazawa, Phys. Rev. C41, 862 (1990).
3. M. Igashira, H. Kitazawa, and N. Yamamuro, Nucl. Instrum. Methods A245, 432 (1986).
4. H. Kitazawa and M. Igashira, J. Phys. G14, S215 (1988).
5. P. A. Moldauer, Nucl. Phys. 47, 65 (1963).
6. S. F. Mughabghab, M. Divadeenam, and N. E. Holden, Neutron Cross Sections (Academic, N. Y., 1981), Vol. 1., p. 12-2.
7. H. Kitazawa, Y. Harima, and T. Fukahori, Proc. Int. Conf. on nuclear data for science and technology, Mito, 1988, ed. S. Igarasi (Saikon, Tokyo, 1988), p.473.
8. P. M. Endt, At. Data Nucl. Data Table 19, 49 (1977).
9. R. J. Peterson, C. A. Fields, R. S. Raymond, J. R. Thieke, and J. L. Ullman, Nucl. Phys. A408, 221 (1983).

THE MEASUREMENT OF THE $^{12}C(n,\gamma)^{13}C$ REACTION IN THE PYGMY RESONANCE REGION *

Huang Zhengde Zhu Lihua Ho Long Shi Xiamin Ding Dazhao
Institute Of Atomic Energy, P.O.Box 275(15), Bejing,102413,CHINA

ABSTRACT

The excitation function of $^{12}C(n,\gamma_0)^{13}C$ reaction has been measured in the pygmy resonanace region. The present results are in agreement with the results of ref.8 and DSD calculation of ref.7. The angular distributions of $^{12}C(n,\gamma_0)$ reaction were fitted by Legendre polynomials. It is concluded that the E1 transitions dominate in the $^{12}C(n,\gamma_0)^{13}C$ reaction.

INTRODUCTION

An interesting phenomenon has been observed in studies of the nuclear photoeffect in light nuclei having one or two nucleons outside closed (4n) shell. Photon absorption in light 4n nuclei is concentrated in a few states which form the giant dipole resonance and exhaust a large fraction of the dipole sum rule. The addition of an extra nucleon profoundly affects the character of the absorption. The giant dipole resonance is observed to be accompanied by a so called 'pygmy' resonance at lower energy. For the photo-nuclear reaction $^{13}C(\gamma,n)^{12}C$ many experimental measurements and theoretical calculations have been reported [1-4]. For photon energies above 9 MeV, the measured ground state cross section displays the so called 'pygmy' resonance (10-18 MeV) [2], with isospin T=1/2 [5]. The giant resonance region from 17 to 30 MeV excitation energy contains two distinct peaks. It is one of the best examples of the isospin splitting of the giant dipole resonance. The lower peak is T=1/2 and the upper is T=3/2 [6] . For inverse reaction $^{12}C(n,\gamma)^{13}C$ recently only two laboratories reported their results. The results of TLU and LANL[7] agree with the results of TUNL[8] in the region of the giant dipole resonance while the discrepancy in the 'pygmy' resonance region is in a factor of 1.7.

EXPERIMENT

We have measured the excitation function and angular distributions for $^{12}C(n,\gamma_0)^{13}C$ reaction at neutron energies from 7 to 14 MeV which equivelent to excitation energy in 11 to 18 MeV at the tandem accelerator in CIAE. A large plastic-NaI anticoincidence sheilded spectrometer was used to detect gamma-rays. The absolute efficiency of the spectrometer system has been determined by $^{12}C(p,\gamma_0)^{13}N$ reaction at 15.07 MeV resonance, while the variation of efficiency with energy was obtained by taking into account of all energy dependent effects. Neutrons were produced by D(d,n) reaction. The neutron production target consisted of a 2.8 cm long gas cell with 5.27 mg/cm² havor entrance window. The gas pressure is 6.0 atm. The neutron energy spread varies from 580 kev at E_n = 7.0 MeV to 240 kev at E_n = 14.0 MeV. Deuteron beam was chopped and bunched to produce 1 ns beam burst at a repeatition rate of 4 MHz. The time of flight techniqure has been used to separate the scattered neutrons and gamma-rays and eliminate events not associated with target interaction prompt gamma-rays. The time resolution of the spectrometer system is 2.04 ns. The ^{12}C target was a 4 times 4 cm

---

* The project supported by National Natural Science Foundation of China

right-cicular shape cylinder of natural carbon having mass of 90.20 g.

RESULTS AND CONCLUSIONS

The 90° differential cross section is shown in fig.1 with pervious measurement results [7,8]. The agreement between present results and results of ref.8 is reasenable at measured energy region. The present data quite well agreee with DSD model calculation of ref.7 which successfully fitted the $^{12}C(p, \gamma_0)^{13}N$ data while failed to reproduce their $^{12}C(n, \gamma_0)^{13}C$ data below E_n = 15.1 MeV. The peak position of pygmy resonance at 13.0 MeV could be obtained from our data in fig.1 by eye guiding.

The angular distribution of the $^{12}C(n, \gamma_0)^{13}C$ reaction cross section were measured at five angles from 55° to 125° at 3 neutron energies, 9, 11 and 14.2 Mev. The angular distribution measurement at 14.2 MeV was performed at Cockraft-Walton accelerator and the result has been published[9]. The angular distribution of cross section can be expressed in terms of Legendre polynomials

$$\sigma(E, \theta) = A_o(E) [1 + \sum_{k=1}^{2L_{max}} a_k p_k(\cos\theta)] \quad (1)$$

The fitting results of angular distribution by equation (1) indicate that a_2 is dominate while a_1, a_3 and a_4 are very small. a_2 = -0.682 ± 0.178, -0.573 ± 0.092 and -0.385 ± 0.064 for neutron energies E_n = 9 MeV (E_x = 13.25 MeV), 11 MeV (15.1MeV) and 14.2 MeV (18.1MeV) respectively. The results are in agreement with the results of the inverse reaction [1]. It could be concluded that the reaction $^{12}C(n, \gamma_0)^{13}C$ in this energy region is dominated by E1 radiative transition. There are no obvious interference from M1 and/or E2. Using channel spin angular momentum coupling scheme the coefficients of Legendre polynomials can be written as a function of transition amplitude and phase.

$$A_o = S_{1/2}^2 + 2 D_{3/2}^2$$

$$A_o a_2 = -2 S_{1/2} D_{3/2}^2 \cos(\phi_d - \phi_s) - D_{3/2}^2 \quad (2)$$

by using $s = S_{1/2} / \sqrt{A_o}$ and $d = \sqrt{2} D_{3/2} / \sqrt{A_o}$, equation (2) can be rewritten as

$$1 = s^2 + d^2 \quad (3)$$

$$a_2 = -d^2/2 - \sqrt{2} s d \cos(\phi_d - \phi_s)$$

The relation between the transition amplitude ratio d/s and phase difference ($\phi_d - \phi_s$) was shown in fig.2. The ref.10 shows that the remarkable feature is the constancy of the relative d wave to s wave amplitudes across the entire region of GDR. The preferred solution can be obtained from ref.11. The average ratio value d/s of the $^{12}C(n, \gamma_0)^{13}C$ reaction was 1.91 ±0.34 between excitation energies from 16 to 22 MeV.

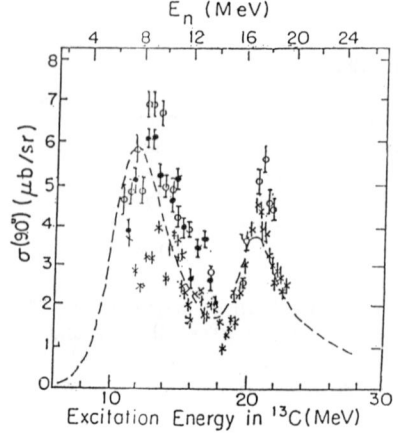

Fig.1 The present data for the $^{12}C(n,\gamma_0)^{12}C$ 90° cross section are shown here along with previously published results of ref.7,8. The direct-semidirect model (DSD) calculation is shown as the dashed line.
● present results,
○ TUNL [8]
× TLU, LANL and
---- DSD calculation [7]

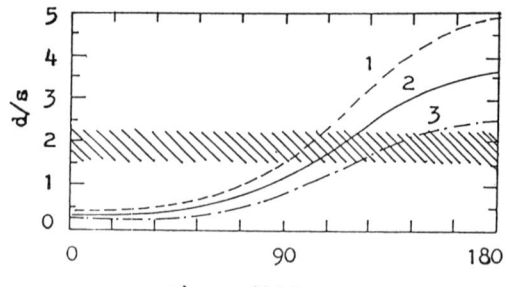

Fig.2 relation between amplitude ratio d/s and phase difference ($\phi_d - \phi_s$).
1. E_n = 9.0 MeV
2. E_n = 11 MeV
3. E_n = 14 MeV

If we assume that it will be held down to E_x = 11 MeV from fig.2 one gets phase difference ($\phi_d - \phi_s$) = 101° ±10°, 112° ±10° and 133° ±15° at E_n = 9 MeV (E_x = 13.25 MeV), 11 MeV (15.1 MeV) and 14.2 MeV (18.1 MeV) respectively.

The authors would like to thank the operators of the Tandem Laboratory and Computer Center in CIAE for their cooperation during the course of this word.

REFERENCES

(1) J. G. Woodworth et al., Nucl. phys. A327,53 (1979).
(2) J. G. Woodworth et al., Can. J. phys. 55,1704 (1977).
(3) J. W. Jury et al., Phys. Rev. C19,1684 (1979).
(4) M. Marangoni et al., Nucl. Phys. A277,239 (1977).
(5) D. F. Measday et al., Nucl. Phys. 61,269 (1965).
(6) E. Hayward, Photonuclear reactions, U. S. Natl. Bur. Stand. Monograph No.118 Washington, DC, 1970.
(7) I. Bergqvist et al., Nucl. Phys. A456,426 (1987).
(8) R. A. August, H. R. Weller and D. R. Tilly, Phys. Rev. C35,393 (1987).
(9) Huang Zhengde, Liu Jishi, Cao Zhong, Wang Huizhu, Ding dazhao
    Chinese Jour. of Nucl. Phys. Vol.11,55(1989).
(10) H. R. Weller and N. R. Roberson, Rev. Mod. Phys. 52,699 (1980).
(11) R. A. August, Thesis, TUNL, Duke University, (1984).

GAMMA-RAY INTENSITIES FOLLOWING THERMAL NEUTRON CAPTURE IN ^{117}Sn.

V.R. Skoy and E.I. Sharapov
Laboratory of Neutron Physics, Joint Institute for Nuclear Research, Dubna, Head Post Office, Box 79, USSR

ABSTRACT

Results of γ-ray measurements on an enriched ^{117}Sn sample in a neutron beam from the pulsed reactor IBR-2 are reported. The total capture cross section and the intensities of primary and secondary γ-transitions following thermal neutron capture have been deduced.

A capture state with $J^\pi = 1^+$ (negative resonance $E_0 = -29$ eV) prevails in thermal neutron capture in the ^{117}Sn. The total capture cross section was known from [1,2] to be $\sigma_\gamma = 2.3 \pm 0.5$ b and $\sigma_\gamma = 1.32 \pm 0.18$ b, respectively. We have made an attempt to reduce this uncertainty (see below). High-energy γ-lines E_γ are formed by transitions from capture state to final states E_f, the latter being known from other reactions. The positive parity of the E_f-state indicates an M1 transition $E_{\gamma 1}$ to occur. However, the intensities I_γ of these transitions, but for $E_{\gamma 0} = 9325$ keV, were unknown to us.

Our measurements were made at the pulsed reactor IBR-2. A curved neutron guide was used to minimize the background of fast neutrons and γ-rays. The neutron flux as a function of wavelength λ is shown in Fig. 1 as obtained by the time-of-flight method. The structure of the flux is due to a Pb-filter in the beam.

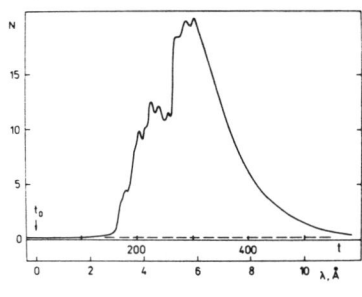

Fig. 1

For cold neutrons the capture cross section σ_γ dominates over the constant incoherent scattering cross section ($\sigma_{inc} = 1$ b). We made use of this observation to obtain the σ_γ cross section from the total cross section σ_t, which is the sum $\sigma_t = (\lambda/\lambda_{therm})\sigma_\gamma + \sigma_{inc}$. The value $\sigma_\gamma = 1.07 \pm 0.05$ b was deduced from our data in Fig. 2 on the total cross section measurements.

The γ-spectra were obtained by using a Ge(Li)-dectector placed 10 cm from the 117Sn metal sample. A pulse-height analyzer and the time window circuit reported in [3] were used. The background was low at high energies but started to increase at energies below 7.6 MeV. The principal background lines were those from capture in Al, Fe and Pb. As an example part of the (n,γ) spectrum is given in Fig. 3. It was accumulated for 52 hours.

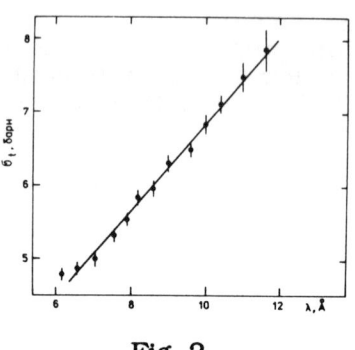

Fig. 2

To acquire these data we used a sample (91% of ^{117}Sn) 7.0 10^{21} at/sm^2 thick. The energy calibration of the detector was made by using Ni and Fe capture lines. The intensities of the γ-ray lines were obtained from absolute intensities of the Cl(n,γ) and Fe(n,γ) lines given in [4]. The measurements with NaCl and Fe-metal samples were made before and after every ^{117}Sn run.

Table 1 Intensities of high-energy M1-transitions the ^{117}Sn(n,γ)^{118}Sn reaction.

E_f	0	1230	1760	2056 2044	2329	2400	2911
J_f^π	0^+	2^+	0^+	$0^+(2^+)$	$1^+(2^+)$	(2^+)	(2^+)
E_γ	9325	8095	7565	7269 7281	6996	6925	6414
I_γ	3.±0.5	0.36±0.1	≤0.1	0.6±0.15	0.4±0.1	≤0.1	1.3±0.2

The results are summarized in Table 1.

The average value of the M1 radiation width is $\langle\Gamma_{\gamma i}(M1)\rangle$=0.8±0.4 meV, where the Porter-Thomas fluctuations are included.

The intensities of some secondary low-energy γ-lines were measured also. They are given in Table 2. They correspond to transitions from levels in ^{118}Sn to the first exited state and to the ground state.

<u>Table 2</u> Intensities of secondary transitions following thermal neutron capture in ^{117}Sn.

E_γ	1099	1230	1447	1506	2044
I_γ	10.3±1.2	48.0±8.0	1.35±0.15	2.21±0.24	2.5±0.3

The γ-ray data obtained in this work for the ^{117}Sn target contribute to γ-ray spectroscopy. They are of particular importance for the analysis of parity violation in the direct E_γ=9325 keV transition [5] and integral γ-spectra [6].

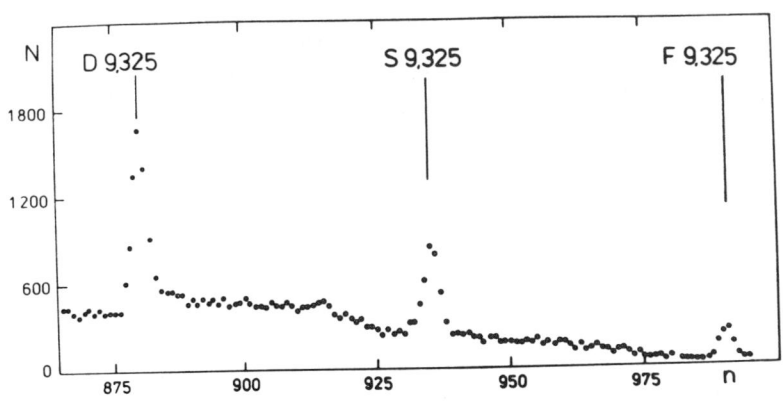

Fig. 3

REFERENCE

1. S.P. Mughabghab et al., Neutron Cross Sections, (N.Y. Academic Press, 1981), <u>v.1,</u> part 1.
2. V.P. Alfimenkov et al., Yad.Fiz. <u>v.39</u>, p.1057 (1984)
3. V.A. Vagov et al., JINR Report p.10-86-562, Dubna (1986)
4. M.A. Lone, R.A. Leavitt, D.A. Harrison, Nuclear Data Tables, <u>v.26</u>, p.511 (1981).
5. G.V. Daniljan et al., Pisma ZhETF, <u>v.24,</u> p.380 (1986)
6. V.E. Bunakov et al., Yad Fiz. <u>v.49</u>, p.988 (1989)

STUDY OF THE $^{91}Zr(n,\alpha)^{88}Sr$ AND $^{187}Os(n,\alpha)^{184}W$ REACTIONS ON RESONANCE NEUTRONS

J.Andrzejewski, Yu.M.Gledenov, M.P.Mitrikov,
Yu.P.Popov, P.V.Sedyshev, I.Chadraabal, L.Ho Bom

ABSTRACT

This paper reports on the results of the study of α-decay of highly excited states of ^{92}Zr and ^{188}Os. The experiment was performed by the time-of-flight method and utilized neutrons from the IBR-30 booster of the Laboratory of Neutron Physics, JINR.

For ^{91}Zr the α-widths of three neutron resonances at $E_0=2.474$ keV, $E_0=1.531$ keV and $E_0=0.449$ keV: (165+46), (12+7), (31+10) meV, respectively were measured. These results were compared with those calculated in terms of the cluster model.

Measurements of the $^{187}Os(n,\alpha)^{184}W$ reaction yielded α-widths equal to $(6.8\pm1.0)\times10^{-9}$ eV and $(2.0\pm0.3)\times10^{-9}$ eV for two resonances at $E_0=9.47$ eV and $E_0=12.7$ eV, respectively. The α-width averaged over the neutron energy interval from 38 to 302 eV was found to be $(10\times4)\times10^{-9}$ eV. Experimental α-particle strength functions are compared with those calculated within the model of "the black nicleus".

The investigation of α-decay of neutron resonances, being carried out for many years, shows, that the experimental values of α-widths for nuclei in the interval $60<A<180$[1,2] are well reproduced by the calculation in the cluster model of α-decay[3]. However, on the basis of the data[4,5] from the measurements of the ^{238}U cross-section on thermal neutrons the values of α-width exceding the calculated ones by six orders of magnitude were obtaned. What is it due to? Is it due to a new mechanism of α-decay of heavy nuclie? Then it would be interesting to measure the $^{187}Os(n,\alpha)^{184}W$ in order to test high excited states decay models.

The ^{91}Zr nucleus is one of few nuclie with atomic weight A<100 for which one can observe α-decay of neutron resonances. Besides, the data from the $^{91}Zr(n,\alpha)^{88}Sr$ reaction are of interest from the viewpoint of the estimation of helium accumulation in the reactor construction materials[6].

The measurements of both isotopes were carried out on neutrons beam of the IBR-30 pulsed booster of the Laboratory of Neutron Physics, JINR. The neutron spectrometry was performed by the time-of-flight method.

A multisection proportional chamber[7] (MPC) was used as a detector in the investigation of the $^{91}Zr(n,\alpha)$

reaction. It consists of six double proportional chambers. A mixture of Ar+20% CO_2 was used as a working gas.
The experimental time-of-flight spectra selected by an amplitude window were written on the magnetic disk of the SM-3 minicomputer. The information was registered simultaneously by all sections.
The measurements of the $^{187}Os(n,\alpha)$ reaction were carried out with a two-section ionization chamber with a grid (DIC). A working gas is a mixture of Ar+3%CO_2. Two-dimensional data on the energy of detected particles and on the time-of-flight captured neutrons were recorded on a magnetic tape using a measuring module. The data were selected and processed on a PDP-11/70 computer.
The measurements with ^{147}Sm and ^{143}Nd were made for the caliblation and normalization. Table 1 provides information on the targets used.

Table 1.

Target-nucleus	Kind of combination	Enriched %	Thickness mg/cm	Area cm	Detector
^{91}Zr	ZrO_2	91.1	5.00	625x4, 625x8	MPC
^{147}Sm	Sm_2O_3	95.3	5.00	625x2	MPC
^{187}Os	metal	99.0	2.00	625, 625x2	DIC
^{143}Nd	Nd_2O_3	83.2	0.52	625	DIC

Experimental data on the $^{187}Os(n,\alpha)$ reaction allowed one to observe the α-decay of two resonances at E_0=9.47 eV and E_0=12.7 eV. Moreover, a bump in the interval 38<E_n<302 eV, related to a group of levels ^{188}Os was observed. The α-widths of both resonances were determined. The average α-width was calculated in the given neutron energy interval. These data make it possible to obtain the values of the α-particle strength function. The results of the investigation are given in Table 2.

Table 2.

E eV	J^π	Γ_α neV	$<\Gamma_\alpha>$ neV	S_α MeV^{-1}x10^{-2}
9.47	1	6.8+1.0	4.4+2.6	6.0+3.5
12.70	1	2.0+0.3		
38-302	1		10+4	13+5

Due to the fact that α-decay of ^{188}Os levels: 0^- 0^+, 2^+, 4^+...is forbidden by selection rules, transitions from $J^\pi=1^-$ are realized. The values obtained for the strength functions of α-particles may be compared with $S_\alpha^{b.n.}=4.8\times10^{-2}$ MeV, the α-particle strength function for the ^{187}Os nucleus, calculated in the frame of the "black nucleus" model. It is seen that the experimental values slightly exceed the calculated ones.

Table 3 summarizes the results of investigations of the ^{91}Zr(n,α) reaction. The values of α-widths for resonanses at $E_0=1.531$ keV and $E_0=2.474$ keV were determined. Upper limits of α-widths of resonances at 892.7 eV, 292.6 eV, 181.1 eV were estimated.

Table 3.

E_0 eV	J^π	Γ_α μeV
181.1	3^-	< 0.16
292.6	$2^+;3^-$	< 0.43
449.4	$1^-;3^-$	31+10
892.7	3^-	< 42
1531.4	2^+	12±7
2474.2	2^+	165±46

The experimental data make it possible to obtain average α-widths over the energy interval up to 2500 eV: $\langle\Gamma_\alpha\rangle=44+25$ μeV. The calculation within the cluster model gives $\langle\Gamma_\alpha\rangle^{c.m.}=21$ μeV. The experimental value agrees with the calculated one within the accuracy of the experiment.

References
1. Antonov A.D. et al. - Sov.Nucl.Phys.,1978, v.27,p.18.
2. Pikelner L.B. et al. - Usp.Fiz.Nauk.,1982,v.137,p.39.
3. Kadmensky S.G., Furman V.I. In: Alpha Decay and Related Nuclear Reactions. Moscow, Energoatomizdat, 1985, p.194.
4. Asghar M. et al. - Nucl. Phys., 1976, v.A259, p.429.
5. Wagemans C. et al. - Nucl. Phys., 1981, v.A362, p.1.
6. Balabanov N.P., Gledenov Yu.M. JINR Comm., P3-81-276, Dubna, 1981.
7. Antonov A. et al. - Nucl.Technology, 1982,v.59,p.526.
8. Popov Yu.P. et al. - Sov.Nucl.Phys.,1971, v.13,p.913.

Asymmetries in nucleon radiative capture caused by nuclear structure

R. Guidotti and F. Saporetti
Dipartimento di Fisica dell' Universita' di Bologna
and INFN , Sezione di Bologna
Via Irnerio 46 , Bologna (Italy)

G. Maino and A. Ventura
ENEA , Viale Ercolani 8 , Bologna (Italy)

Abstract

The form factor describing the particle-vibration coupling for the excitation of the M1 giant resonance in radiative capture of fast neutrons is derived microscopically for ^{208}Pb and compared with the phenomenologic form, which assumes a Fermi nuclear density. The a_1 coefficient of the Legendre polynomial expansion of the differential cross section, related to the E1-M1 interference process, is calculated on the basis of the direct-semidirect (DSD) model. The effects caused by the nuclear structure on the asymmetries with respect to $\theta = 90°$ of the γ − ray angular distributions are investigated.

Previous study of the coupling between the incident nucleon and the collective degrees of freedom of the target nucleus in (N, γ) reactions has shown the need of microscopic calculation of the form factors describing the radial dependence of the effective interaction responsible for the emission of E1 and E2 photons in the semidirect reaction channel ([1,2]) .

The inadequacy of the phenomenologic approach to the form factors , based on an almost uniform Fermi distribution of nuclear mass, and neglecting the nuclear shell configuration , reveals itself even more clearly in M1 collective capture, a process strongly influenced by the nuclear structure , since it is due to spin-flip transitions of nucleons in external j-subshells allowed by the exclusion principle. In support of the above statements , Fig. 1 shows the radial form factors $h_{E1}(r)$ and $h_{M1}(r)$ evaluated in the frame of the phenomenologic and the shell models for the $^{208}Pb(N, \gamma)$ reaction ([1,3]) : as far as the M1 capture is concerned, the phenomenologic approach yields an interaction that receives contributions not only from the nuclear surface , as suggested by the microscopic interpretation of the M1 excitation process, but also from the nuclear bulk.

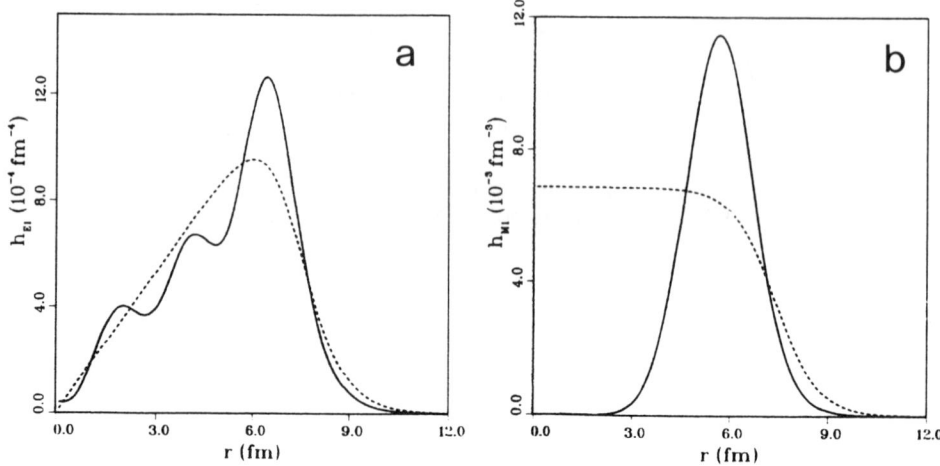

Fig.1: Radial form factors: solid lines: microscopic model; dashed lines phenomenologic model; (1a): E1; (1b): M1.

As is known, the effects of excitation of the M1 giant resonance in capture reactions are to be expected only on the asymmetries with respect to $\theta = 90°$ in the angular distributions of emitted photons. These effects derive from the E1-M1 interference in the a_1 coefficient of the Legendre polynomial expansion of the differential cross section:

$$\frac{d\sigma}{d\Omega} = \frac{\sigma}{4\pi}(1 + \sum_{n=1}^{\infty} a_n P_n(\cos(\theta))) \qquad (1)$$

It is worth while to point out that, in the energy region of the M1 resonance, the semidirect M1 channel seems to be the principal mechanism of the emission asymmetry, since the statistical (compound nucleus) emission appears to be almost symmetric with respect to $\theta = 90°$ and the E1-E2 interference is effective at higher excitation energy [3]

In order to appreciate the extent to which the microscopic features of M1 capture affect the asymmetries, we have computed, in the frame of the direct-semidirect model, the contribution of the E1-M1 interference to the a_1 coefficient in the energy region of the M1 giant resonance for neutron capture to the $2g_{9/2}$ and $1i_{11/2}$ levels of ^{209}Pb.

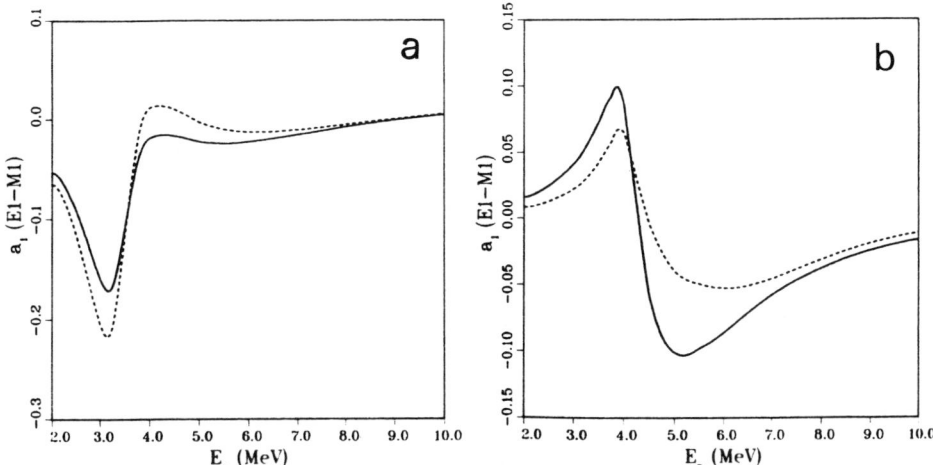

Fig.2: E1-M1 interference contribution to the a_1 coefficient: solid curves: microscopic model; dashed curves: phenomenologic model; (1a): (n, γ_0) ; (1b): (n, γ_1)

As shown in Fig. 2, the results obtained by using the microscopic and phenomenologic form factors, h_{E1}, h_{E2} and h_{M1}, exhibit smaller differences than the form factors themselves.

In any case, the microscopic effects on observable quantities, e. g. distribution asymmetries, might turn out to be more or less significant, depending on the degree of overlapping of the initial and final wave functions of the nucleon with the form factor in the interaction region, and might change their strength according to the target nucleus, and, for a given target, according to the level where the incident nucleon is captured, as shown in Fig. 2 for the levels of ^{209}Pb considered in this work.

References

(1) R. Guidotti, F. Saporetti, G. Maino and A. Ventura, Nucl. Phys. **A 480** (1988) 253 .
(2) R. Guidotti, F. Saporetti, G. Maino and A. Ventura, Nuovo Cimento **A 101** (1989) 111 .
(3) R. Guidotti, F. Saporetti, G. Maino and A. Ventura, Nucl. Phys. **A** (in press).

Statistical Properties
of Nuclear Levels

PRECISION AND COMPLETENESS

C. van der Leun
R.J. Van de Graaff Laboratorium, Rijksuniversiteit Utrecht,
P.O. Box 80.000, 3508 TA Utrecht, The Netherlands.

Abstract

One of the most attractive aspects of modern nuclear spectroscopy, and especially of capture γ-ray spectroscopy, is the completeness of the resulting nuclear level schemes.

Experimentally, completeness is closely related to high precision (E_γ, I_γ, E_p), high resolution (Ge detector, proton beam) and low background (Compton suppression spectrometer). Therefore recent progress in these fields is discussed here in some detail.

Completeness also implies a huge amount of data. Consequently new ways of handling the data are of interest. As an example we discuss the possiblity of statistical spin assignments.

INTRODUCTION

This contribution is a sequel to my paper[1] presented at the Leuven conference under a slightly more elaborate titel. It concentrates on recent progress made in the precision and the completeness of the data from capture γ-ray experiments.

Related theoretical aspects, in particular chaos in nuclear level schemes and level densities, will be discussed during the present conference by J.F. Shriner[2] and V. Paar[3], respectively.

PRECISION

In proton and neutron capture experiments the final precision is basically determined by the precision with which γ-ray spectra can be measured and analyzed. Due to the broad use of high resolution Ge-detectors and low- background

Compton-suppression spectrometers, capture reactions are presently in the frontline of nuclear precision measurements. Here I would like to mention shortly a few topics on which progress has been made in recent years i.e. (a) *γ-ray energies* and (b) *γ-ray intensities*. A third topic is (c) *proton energies*, often a limiting "external" factor for the precision of proton-capture data.

(a) Gamma-ray energies.
The bulk of the available γ-ray energy data is based on two practically independent standards[4]. (i) The "gold standard", or the wavelength of the 412 keV ^{198}Au line determined relative to optical standards with double flat-crystal spectrometers with a precision of better than 1 ppm[5,6] and (ii) the "mass-doublet standard" based on mass-spectrometer determinations of nuclear mass diferences[7]. Comparison of the two standards suggested[8,9] that the two standards deviate by 6 - 10 ppm. The 1986-evaluation of fundamental constants[10] implies a 7 ppm reduction of the gold standard[11] and this brings the two standards in line even within the recently appreciably reduced uncertainties.

(b) Gamma-ray intensities.
Precision analysis of γ-ray spectra also requires precision detector efficiencies. This is a problem especially for high-energy γ-rays. The "two-line method" is used in thermal-neutron and proton capture reactions. Ideally one uses a two-step cascade of a high- and low-energy γ-transition ($γ_1$ and $γ_2$, respectively) which each have 100% γ-ray branchings to one level. This implies an intensity ratio $R = I(γ_1)/I(γ_2) \equiv 1$ and thus connects the (usually well-known) low-energy efficiency to that for the high-energy line. This ideal case of $R = 1$ is rare, and one thus selects cascades approaching this ideal as closely as possible. In this respect the (p,γ) reactions on light nuclides, with usually a broad choice of sharp resonances, have an advantage over the thermal-neutron capture reactions. The selected resonances have been studied in detail with a Compton suppression spectrometer (CSS). They are listed in table I together with the measured intensity ratios; the latter have uncertainties ranging from 0.1% to 1.2%. These ratios are lower than the previously reported ratios, because the present decay schemes are more complete, and the indirect feeding of the lower-lying levels escaped detection in earlier studies. The present ratios R are in general an order of magnitude more precise than the older values. One previously recommended cascade, the 12.67 → 4.24 → 0 MeV cascade in ^{24}Mg, had to be rejected for precision calibrations since it turns out to be contaminated.

It has also been demonstrated that the two-line method can be extended to γ-ray energies above 12 MeV via the ^{11}B(p,γ)^{12}C reaction. The γ-decay scheme of ^{12}C is in some respects ideal for the two-line method, since R is very close to unity ($\Delta R < 0.001$) for the r→ 4.44 → 0 MeV cascade. In this case, however, the γ-ray angular distributions rather than R limit the precision.

The present measurements of the ratios R with a precision of the order of 0.5%

Table I Measured intensity ratios $R = R(\gamma_1)/I(\gamma_2)$.

E_p (keV)	E_{γ_1}-E_{γ_2} (MeV)	Present work[a] R	ΔR_0	ΔR_u	Previous work R
^{11}B$(p,\gamma)^{12}$C:					
675	12.14-4.44	1.000			
1388	12.79-4.44	1.000			
2626	13.92-4.44	1.000			
^{23}Na$(p,\gamma)^{24}$Mg:					
1318	11.58-1.37	0.960 ± 0.002	0.002	0.0005	1.00 ± 0.02
1417	8.92-2.75	0.985 ± 0.0011	0.0008	0.0008	1.00 ± 0.02
^{27}Al$(p,\gamma)^{28}$Si:					
767	7.70-2.84	0.981 ± 0.002	0.0009	0.002	0.99 ± 0.02
992	10.76-1.78	0.806 ± 0.010	0.010	0.00002	0.808 ± 0.016
1317	6.57-4.54	1.017 ± 0.006	0.006[e]	0.0016	1.11 ± 0.06

[a] ΔR_0 is the uncertainty due to uncertainties in the observed intensities, ΔR_u is the uncertainty related to the upper limits for the intensities of unobserved transitions. The stated total uncertainty in R equals $\sqrt{(\Delta R_0^2 + \Delta R_u^2)}$.

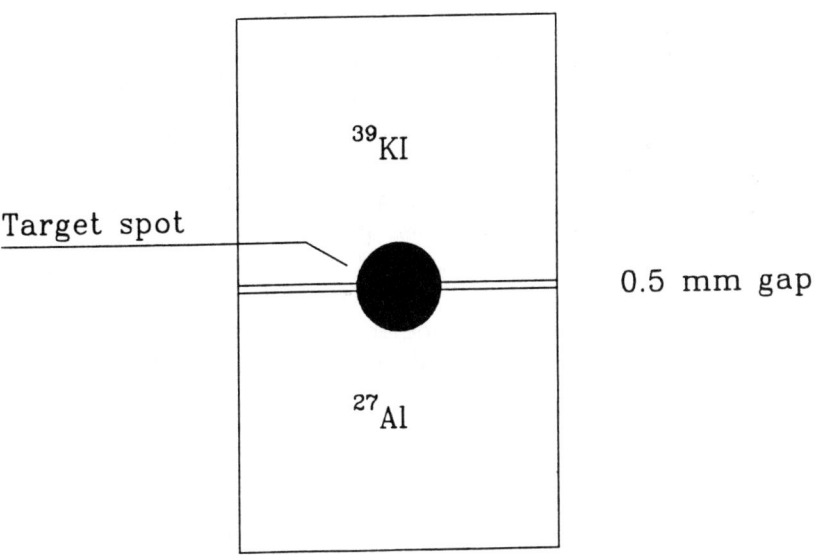

Fig. 1. Combined KI-Al target for the simultaneous measurement of proton energies of ^{39}K$(p,\gamma)^{40}$Ca resonances relative to neighbouring ^{27}Al$(p,\gamma)^{28}$Si resonances.

do not imply that γ-ray efficiency calibrations of such accuracy are only a matter of accumulating sufficient spectral statistics. At the 1% level several other effects should be considered, like those due to angular distributions, summing effects and differences in peak shapes.

For more details the reader is referred to ref.[12]

(c) Proton energies.
An accurate determination of the Q-value of a (p,γ) reaction requires not only precise γ-ray energy measurements, but also determination of precise proton energies E_p of the resonances studied. One usually sandwiches yield curve measurements of the resonance studied with those of a nearby calibration resonance, most often a ^{27}Al(p,γ) resonance. This requires repeated target changes, sometimes causing unreproducible shifts.

Kikstra[13] circumvented this problem in a very simple way by using a combined K-Al target to calibrate his ^{39}K(p,γ) resonances against neighbouring well-known ^{27}Al(p,γ) resonances; see the self-explanatory fig. 1.

It is obvious that high precision in this context is closely related to high-resolution proton-beams, as was nicely demonstrated by Zhendi Guo in her poster contribution.[14]

A suggestion on E_p calibrations may be in place here. Almost all E_p-values published are directly or indirectly connected to the well known ^{27}Al(p,γ)^{28}Si calibration resonance at E_p = 992 keV. The most recent precision value for this resonance energy[15], E_p= 991.835 ± 0.027 keV, has an uncertainty which is basically more than 20 years old.[16] I am looking forward to a measurement with modern techniques of a considerably more precise calibration energy. My suggestions in planning such an experiment would be (i) to select another resonance, since the 992 keV resonance has a large, precision-limiting natural width (Γ = 100 ± 15 eV[17]), and (ii) to choose another target material in order to rule out the error due to the influence of a possible oxide layer on the Al targets[18] ; ^{26}MgO and ^{30}SiO$_2$ might be good candidates.

Finally, I would like to mention a few experiments that became feasible or worthwhile, due to the improved precision discussed above.

- According to the conserved vector-current hypothesis (CVC) the ft-values for the superallowed $0^+ \to 0^+ \beta$-transitions should all be identical after small corrections have been applied. Precision ft-values have been obtained by combining the results of (n,γ) and (p,γ) reactions on the same target, e.g. ^{25}Mg(n,γ)^{26}Mg and ^{25}Mg(p,γ)^{26}Al^m for the ^{26}Al^m(β^+)^{26}Mg decay. The resulting ft-values for superallowed ^{26}Al^m(β^+)^{26}Mg and ^{42}Sc(β^+)^{42}Ca decay[19,2] agree within about 10 ± 10 parts in 10^4 with that of the "standard" $0^+ \to$

0^+ ^{14}O decay. For a detailed discussion see Raman's contribution to this conference[21].

- Gamma-ray resonant absorption experiments via cross-excitation, initiated in the seventies[22,23], still heavily rely on precision γ-ray energy measurements.

- For the light nuclei an almost unbroken chain of (p,γ) and (n,γ) Q-values forms the backbone of Wapstra's forthcoming atomic mass evaluation[7].

COMPLETENESS

A nuclear level scheme is considered as "complete" up to the energy E, when *all* the levels with excitation energies $E_x < E$ have been observed and their spins, parities, isospins (and preferably also lifetimes and decay modes) have been determined experimentally.

The most obvious goal of complete and precise experiments is an adequate and thorough comparison of the data with results from nuclear structure calculations. A few examples will be discussed here. Other related fundamental problems like chaos, level density calculations, analogue states and tests of CVC-theory will be discussed by others[2,3,14,21] during this conference.

During the Leuven conference I mainly discussed[1] the data on ^{26}Al obtained by P.M. Endt and colleagues in a study of the ^{25}Mg(p,γ)^{26}Al reaction, which was in progress at that time. The data have now been published[25,26,27]

To me, the most remarkable result of these comprehensive experiments is the fact that the in proton capture experiments traditional gap between bound states and resonance levels has completely disappeared. This indeed is a break-through in completeness. Impressive is the one-to-one correspondence between the lowest 45 levels of ^{26}Mg and the lowest 45 $T = 1$ levels of ^{26}Al. A one-to-one correspondence is also found between the lowest 30 $T = 1$ plus the lowest 40 $T = 0$ even-parity states of ^{26}Al and the extensive calculations of the even-parity states of the sd-shell nuclides by Wildenthal and co-workers. It should be mentioned that the comparison extends up to the 14th $J = 2$, 17th $J = 3$, 13th $J = 4$ and 12th $J = 5$ states(!). Comparison of calculated and many measured γ-ray transition strengths was also fruitful, especially for the E2 transitions.

Later experiments on other nuclides[13,14], however, made us realise that ^{26}Al is not exactly a "typical" case. The low-lying $T = 1$ states of the odd-odd self conjugated nuclides like ^{26}Al make the application of the RUL's[28,29] for spin- and isospin assignments very fruitful. The forbidden and retarded character of some dipole transitions is not only distinctive, but also makes the E2 transitions stand out more clearly. In addition, the modest level density of ^{26}Al and the relatively low Q-value of 6.3 MeV make ^{26}Al a lucky case. All this, however, is not the end of the game. After all, there are many more odd-odd selfconjugated nuclides. And for all the others the progress in precision measurements discussed above will be of some help. The precise γ-ray energies lead to better excitation energies, and the smaller the uncertainties in E_x, the smaller the risk that observed γ-lines are misplaced and that doublets go undetected. Precise intensities have a similar effect, they facilitate the identification of energetically unresolved doublets by observing the different decay of the components. For the detection of doublet resonances, stable and energetically sharp beams are equally important. The main problem of non-selfconjugated nuclides, the less effective role of the RUL's in spin assingments, may be overcome by a new, statistical method of spin assignments. This is the subject of the final section.

SPIN ASSIGNMENTS

The high resolution of the Ge detectors combined with the low background of Compton-suppression spectrometers, allows us to unravel the γ-decay of nuclear levels in hitherto unknown detail. During the Leuven conference I tried to convince you that it is possible to assign spins to nuclear levels on the basis of a statistical treatment of the huge amounts of information on the γ-decay of the levels studied[1].

In particular it was shown that the spins of many *resonance* levels can be very simply assigned by taking the (weighted) average of the spins of the final levels to which they decay, $<J_f>_w$, provided that at least ten γ-decay branches to levels with known spins have been observed. A χ^2-like test of the distribution of the final spins improves the method, especially for low spins. Fig. 2 is an example of the many similar figures that convinced us of the feasibility of the method. It is a plot of $<J_f>$ as a function of $n = 1,2,3...$, when the n strongest decay branches from the ^{26}Al resonance at $E_x = 7.46$ MeV are taken into account; $<J_f>$ converges rapidly to $J = 3$, which in fact is the spin of the $E_x = 7.46$ MeV level of ^{26}Al. With increasing n the value of $<J_f>$ becomes so stable that only a miracle could move $<J_f>$ to one of the neighbouring values $J = 2$ or $J = 4$. For the more than one hundred resonances considered so far, such a miracle did not occur.

Here we concentrate on similar spin assignments to *bound* states. The fact that the levels discussed previously[1] are (p,γ) resonance states and thus unbound, is not relevant in the discussion of statistical spin assignments. In principle, the method can thus be used just as well for J-assignments to bound states. The problem is, however, that the number of bound states for which a sufficient number ($n \geq 10$) of γ-decay branches is known, is very limited. For the 17 levels of light nuclides where this condition is fulfilled (all in ^{26}Al), the statistical spin assignments confirm the spins determined on classical arguments, except for the $E_x = 5.51$ MeV

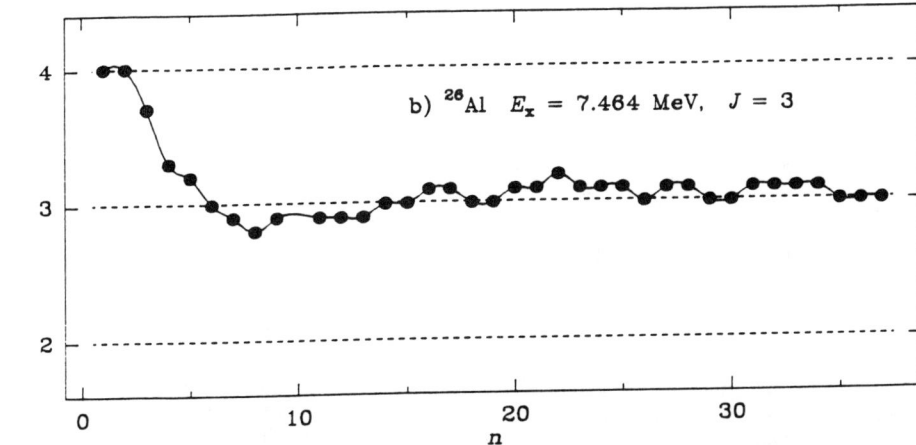

Fig. 2.   Average final spin $<J_f>$ of the ^{26}Al levels to which the $E_x = 7.464$ MeV level of ^{26}Al γ-decays as a function of n, i.e. the number of decay branches taken into account.

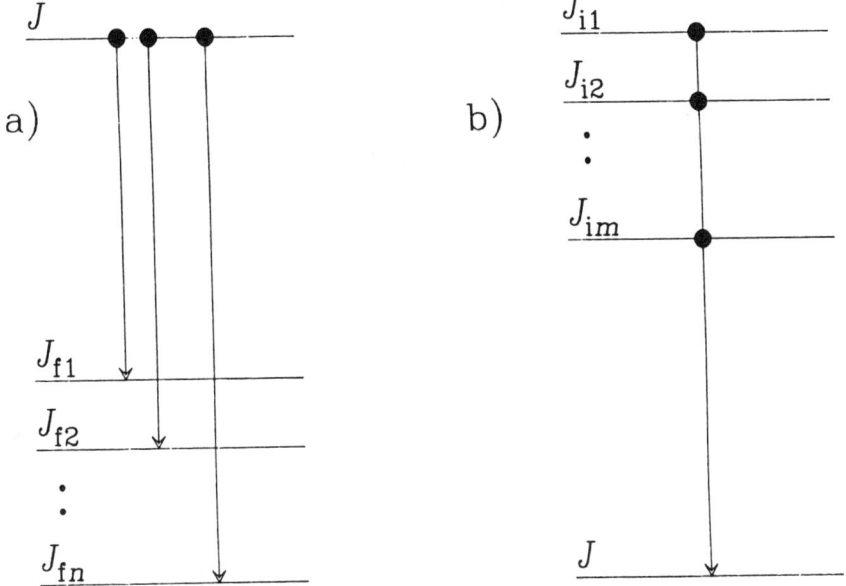

Fig. 3.   Under well-defined conditions the spin J of a nuclear level can be inferred from a) the spins J_{fk} ($k = 1,2,....n$) of the final states to which the level decays, and from b) the spins V_{ik} ($k = 1,2,....n$) of the states that decay to the level.

Table II Comparison of traditional and statistical spin assignments to bound states of ^{26}Al.

E_x(MeV)	N^a	$J^\pi;T^b$	$J_{statist}$	E_x(MeV)	n^a	$J^\pi;T^b$	$J_{statist}$
0	31	5+	5	4.71	33	4+;1	4
0.23	22	0+	1(0)	4.77	31	4+	4
0.42	60	3+	3	4.940	19	1−	2(1)
1.06	34	1+	1	4.941	12	5+	5
1.76	49	2+	2	4.95	30	3+	3
1.85	34	1+	1	5.007	19	2−	3(2)
2.069	43	4+	4	5.010	14	1+	1
2.070	42	2+;1	2	5.13	28	4+;1	4
2.072	31	1+	1	5.14	41	2+;1	2
2.37	56	3+	3	5.20	13	0+;1	0
2.55	53	3+	3	5.25	31	4+	4
2.66	42	2+	2	5.40	24	4−	4
2.74	35	1+	1(2)	5.43	13	1−	1(2)
2.91	49	2+	2	5.457	30	3−	3
3.07	45	3+	3	5.495	27	2+	2
3.16	54	2+;1	2	5.51	20	4+	4
3.40	20	5+	5	5.55	31	2+;1	2
3.51	12	6+	6	5.59	13	1−	1
3.60	46	3+	3	5.60	22	3−	3
3.67	34	4+	4	5.67	17	1+	1
3.68	47	3+	3	5.68	11	4−	5(4,6)
3.72	29	1+	1	5.69	34	3−	3
3.751	46	2+	2(3)	5.73	29	4+;1	4
3.754	13	0+	0	5.85	21	2+	2
3.96	51	3+	3	5.88	11	3+	3(2)
4.19	56	3+;1	3	5.916	25	2−	(2,3)
4.21	37	4+	4	5.924	24	4+;1	4
4.35	44	3+	3	5.95	12	1−	1
4.43	33	2−	3(2)	6.03	13	1(+)	1
4.55	41	2+;1	2	6.08	13	5+	5
4.60	52	3+;1	3	6.28	14	3+	3(2)
4.62	32	2−	2	6.36	21	3+;1	3

[a] Number of feeding branches taken into account.
[b] Literature.

level, where the present χ^2 analysis gives $J = 3$ instead of the previous value $J = 4$. This may be ascribed to the fact that nuclear structure effects of a non-statistical nature are more important in the lower part of the spectrum than in the resonance region. For the bound states, however, another procedure is more fruitful in which not the final spins J_f of the decay γ-transitions (see fig. 3a) but the initial spins J_i of the feeding γ-transitions (see fig. 3b) are statistically treated in the way discussed above. As an example we consider the bound states of ^{26}Al that are excited in the decay of 10 or more higher states (resonances and bound states); they are listed in table II. For none of the 64 levels, including the E_x = 5.51 MeV level (!), the statistical spin assignment conflicts with the previous assignment. Similar tables have been prepared for the bound states with $n \geq 10$ of the other nuclides. No disagreements have been found.

Summarizing it may be stated that the spins of almost 200 levels of the nuclides ^{23}Na, ^{26}Al, ^{40}Ca and ^{47}V each with at least 10 decay or feeding γ-ray transitions to or from states with known spin, have been assigned via a χ^2 analysis of the distributions of the spins of the final and initial states, respectively. The method is essentially based on the combined evidence of *many* γ-ray transitions instead of on specific properties (like strength, angular distribution, polarization) of one or a few transitions.

For the cases studied so far, the results agree very well with the spins assigned along traditional lines. The confidence level of the new spin assignments is estimated at >99%. The 11 levels for which the present method yields more information than the previous assignments are listed in table III. A publication with all the details is in preparation. Sofar the method has only been applied to light and medium-weight nuclides. Tests for heavier nuclides are in progress.

Table III   Statistical spin assignments to ^{47}V, ^{26}Al and ^{40}Ca levels

Nuclide	E_p(keV)	E_x(MeV)	N^a	$J^\pi ; T^b$	$J_{statist}$
^{47}V		2.77	10	1/2$^-$ (3/2$^-$)	1/2
		3.77	12	1/2 (3/2$^-$, 5/2$^-$)	1/2
		4.2718	10	1/2 (3/2$^-$)	1/2
		4.35	12		1/2
	1011	6.16	16	5/2$^{(+)}$ (3/2$^+$)	5/2
	1045	6.19	26	3/2 (5/2$^-$)	3/2
	1127	6.27	26	3/2$^{(-)}$ (5/2)	3/2
	1209	6.350	33		3/2
	1232	6.374	13	1/2 (3/2)	1/2
^{26}Al	516	6.802	12	1$^+$ (1$^-$, 2$^-$); 1	1
^{40}Ca	2371	10.64	9	(3$^-$–5$^-$)	4

^{a}Number of branches taken into account.
^{b}Ref.[30] for ^{47}V, refs.[25-27] for ^{26}Al and ref.[12] for ^{40}Ca.

REFERENCES

1. C. van der Leun, Inst. Phys. Conf. Ser. **88**, S109 (1988).
2. J.F. Shriner et al., "Chaos in Nuclear Level Schemes," these Proceedings.
3. V. Paar et al., "New Formula for Spin-Dependent Level Density," these Proceedings.
4. R.G. Helmer, P.H.M. Van Assche and C. van der Leun, At. Data Nucl. Data Tables **24**, 39 (1979).
5. E.G. Kessler, R.D. Deslattes, A. Henins and W.C. Sauder, Phys. Rev. Lett. **40**, 171 (1978).
6. G.L. Greene, E.G. Kessler and R. Deslattes, Phys. Rev. Lett. **56**, 819 (1986).
7. A.H. Wapstra, Proc. 2nd Int. Conf. on Capture Gamma-ray Spectroscopy (R.C.N., Petten, The Netherlands, 1975) p. 686.
8. P.F.A. Alkemade, C. Alderliesten, P. de Wit and C. van der Leun, Nucl. Instr. **197**, 383 (1982).
9. P. de Wit, F. Stecher-Rasmussen and C. van der Leun, Proc. 4th Int. Conf. on Capture Gamma-ray Spectroscopy Grenoble (1981).
10. E.R. Cohen and B.N. Taylor, CODATA Bulletin **63** (1986).
11. R.G. Helmer (Idaho Falls, USA), private communication (1990).
12. F. Zijderhand, F.P. Jansen, C. Alderliesten and C. van der Leun, Nucl.Instr. **A28**, 490 (1990).
13. S.W. Kikstra et al., Nucl. Phys. **A512**, 425 (1990).
14. Z. Guo et al., "The Lowest $J^\pi = 1^+$ Analogue and Antianalogue in ^{56}Fe," these Proceedings.
15. A.H. Wapstra, Nucl. Instr., to be published.
16. M.L. Roush, L.A. West and J.B. Marion, Nucl. Phys. **A147**, 235 (1970).
17. P.M. Endt and C. van der Leun, Nucl. Phys. **A310**, 1 (1978).
18. R.O. Bondelid and J.W. Butler, Phys. Rev. **130**, 1078 (1963).
19. S.W. Kikstra et al., Nucl. Phys, to be published (1991).
20. S.W. Kikstra et al., Nucl. Phys. **A496**, 429 (1989).
21. S. Raman, "Selected Topics in Thermal Neutron Capture," these Proceedings.
22. R.J. Sparks, H. Lancman and C. van der Leun, Nucl. Phys. **A259**, 13 (1976).
23. Ph.B. Smith, Phys. Rev. **C13**, 2071 (1976).
24. A.H. Wapstra (Amsterdam, The Netherlands), private communication (1990).
25. P.M. Endt, P. de Wit and C. Alderliesten, Nucl. Phys. **A459**, 61 (1986).
26. P.M. Endt, P. de Wit and C. Alderliesten, Nucl. Phys. **A476**, 333 (1988).
27. P.M. Endt, P. de Wit, C. Alderliesten and B.H. Wildenthal, Nucl. Phys. **A487**, 221 (1988).
28. P.M. Endt and C. van der Leun, At. Data Nucl. Data Tables **13**, 67 (1974).
29. P.M. Endt, At. Data Nucl. Data Tables **23**, 3 (1979).
30. H.P.L. de Esch and C. van der Leun, Nucl. Phys. **A454**, 1 (1986).
31. S.W. Kikstra and C. van der Leun, Nucl. Phys., to be published (1991).

CHAOS IN NUCLEAR LEVEL SCHEMES

J. F. Shriner, Jr.
Tennessee Technological University, Cookeville, TN 38505

G. E. Mitchell
North Carolina State University, Raleigh, NC 27695 and
Triangle Universities Nuclear Laboratory, Durham, NC 27706

T. von Egidy
Technische Universität München, D-8046 Garching, Germany

ABSTRACT

It has been conjectured that the fluctuation properties of energy levels of quantum systems can serve as a signature for the presence of quantum chaos. We have examined a large collection of low-lying nuclear energy levels and find that the fluctuation behavior shows a strong mass dependence. Effects due to spin and to deformation are also observed. Possible reasons for this behavior, including the effects of broken symmetries, are discussed.

INTRODUCTION

For classical systems chaotic behavior is a well-defined concept. However, that is not yet true for quantum systems. One criterion which has been applied for quantum chaos is that the fluctuation properties of the energy levels show behavior which is consistent with the Gaussian orthogonal ensemble (GOE) of random matrix theory (RMT).[1] This idea was first proposed by Bohigas, Giannoni, and Schmit[2] when they studied the quantized Sinai's billiard. A variety of systems have now been studied numerically (see, e.g.,[3-6]) in which the quantum analog of a classically chaotic system shows GOE fluctuations and the quantum analog of a classically integrable system displays Poisson fluctuations. Systems have also been studied for which this generalization does not hold.[7-9]

Although the relation of quantum fluctuation properties to the underlying chaoticity or regularity is not completely understood, the evidence strongly suggests a connection. For this reason, it is of interest to study the fluctuation properties of a variety of experimental systems in an attempt to shed light on their behavior. Study of experimental systems also may help clarify the theoretical situation. The GOE is known[10-11] to describe the fluctuation properties of nuclear resonances; in this paper we examine the behavior of nuclear energy levels at lower excitation energies. This work represents an extension of previous, less extensive studies along the same lines.[12-14]

ANALYSIS PROCEDURE

When comparing data to the predictions of RMT, it is necessary to consider only levels which have the same symmetries (i.e., the same quantum numbers). For nuclear physics applications this requirement generally means the same total angular momentum J and the same parity π, although additional quantum numbers are required in some cases. We shall use the term "sequence" to mean a group of levels from one nuclide with the same quantum numbers.

A meaningful analysis of the fluctuation properties of a set of energy levels requires data of extremely high quality. The sequence being studied must be both complete (few or no missing levels) and pure (few or no misassigned levels) over the energy range considered, since failure to meet either of these conditions can severely affect the fluctuation behavior. We have chosen for examination data from 90 nuclides whose level schemes appear to be complete over an energy region starting at the ground state and over some range of angular momenta; the masses range from A = 24 to A = 244. In choosing these nuclides, we required that at least the first ten levels of the level scheme have definite spin and parity assignments. The upper limit of the selected levels frequently is determined by the absence of a well-established spin and/or parity for one level. In a few cases where spin/parity assignments were ambiguous, model information has been used. A large fraction of the nuclides have been investigated by the neutron capture reaction, as the non-selectivity and sensitivity of this reaction guarantees a high degree of completeness.

Testing the completeness of a level scheme in a given energy and spin range is difficult. The level schemes included here have been checked by considering the experimental reactions and the quality of the measurements, by comparing adjacent nuclides, and by considering theoretical predictions of level schemes. For instance, in ^{233}Th there is a one-to-one correspondence between observed and predicted levels below 800 keV; this provides an argument for completeness. We estimate that in this analysis less than 5% of the levels are missing or have erroneous spin or parity assignments for the level schemes included.

Several different statistics are generally employed in studying energy-level fluctuations. These include the nearest-neighbor spacing (NNS) distribution, the Dyson-Mehta Δ_3 statistic,[15] and the linear correlation coefficient between adjacent spacings. We have performed Monte Carlo studies which indicate that the NNS distribution is the most reliable of these three statistics for the small sample sizes considered here. We find that spacing distributions can distinguish between GOE and Poisson behavior with sequences of as few as 5 levels, provided that several such sequences are combined. Requiring a minimum of five levels in a sequence yields 168 sequences in 60 different nuclides (the other 30 nuclides originally selected have no sequences meeting this criterion). Note that data from ^{26}Al, probably the most complete nuclear level scheme available,[16-17] are not included here because the majority of these levels are much closer to the resonance region than to the ground state. Since in the present analysis we are interested in the properties of low-lying levels, we chose to omit the ^{26}Al data set.

Nearest-neighbor spacings S_i for a sequence of levels are generated by calculating the energy differences between adjacent levels of that sequence. It is simplest to express the spacing distribution as a function of the dimensionless parameter $x=S/D$, where D is the average spacing. For spacings which obey GOE statistics, the NNS probability distribution function $P(x)$ is very nearly a Wigner distribution:[18]

$$P(x) = \frac{\pi}{2} x e^{-\pi \frac{x^2}{4}} \qquad (1)$$

If the spacings are from a Poisson distribution, then

$$P(x) = e^{-x} \qquad (2)$$

An interpolation formula between these two extremes has been proposed by Brody:[19]

$$P(x;\omega) = \alpha (\omega+1) x^\omega e^{-\alpha x^{\omega+1}};$$
$$\alpha = \left[\Gamma\left(\frac{\omega+2}{\omega+1}\right)\right]^{\omega+1} \qquad (3)$$

In this expression $\omega = 0$ corresponds to a Poisson distribution, while $\omega = 1$ yields a GOE distribution. The value of ω provides a convenient method of characterizing NNS distributions; however, it must be emphasized that this is an empirical formula, and no physical significance is known for the parameter ω. However, it has recently been observed by Cheon[20] that this formula provides a surprisingly good description of the transition from Poisson to GOE.

To compare the different sequences to each other, each set of energy levels must be converted to a set of normalized spacings $\{x_i\}$, where $x_i = S_i/D$ and D is the average level spacing for that sequence. However, D itself is a function of energy and decreases as the excitation energy increases; knowing D is equivalent to knowing the level density ρ, since $\rho(E)=1/D(E)$. We have normalized the spacings by fitting each sequence to a constant temperature level density.[21] The NNS distribution is then determined for each sequence; errors are estimated using the bootstrap method.[22]

ANALYSIS

The sample sizes for most of these sequences are too small to allow definitive conclusions about the nature of the spacing distribution for a single sequence. Thus, it is necessary to combine data from several sequences. The values of P(x) and ∫P(x) for a group of sequences are determined by computing a weighted average of the appropriate values from the individual sequences. Thus, at any value of x, the probability (or cumulative probability) for the group of sequences is just the weighted average of the probability (or cumulative probability) for the individual sequences. (In practice, one needs to average only at values of x where P(x) or ∫P(x) changes in one of the individual sequences.) Uncertainties for these averages are calculated using standard techniques for a weighted average. The histogram for P(x) is then normalized to area one. The value of ω characterizing a group of sequences is determined by fitting the probabilities determined in this way to (3).

We have examined the data in several different ways. First the data are divided into six groups based on the mass of the nuclide: $A \leq 50$, $50 < A \leq 100$, $100 < A \leq 150$, $150 < A \leq 180$, $180 < A \leq 210$, and $230 \leq A$ (there are no nuclides in this analysis with A between 210 and 230). To search for possible effects of nuclear shape on the NNS distribution, the rare earth ($150 < A \leq 180$) and heavy actinide ($230 \leq A$) nuclides have been combined in the group labelled "deformed"; the other four mass regions are combined to form the "spherical" nuclides.

To look for effects depending on proton or neutron number, nuclides are grouped according to whether they are even-even, odd-odd, or odd mass. Further, within the even-even nuclides, states are subdivided into groupings which include 2^+ and 4^+ states, 0^+ and 3^+ states, and states which are not 2^+ or 4^+. This last subdivision is to look for spin effects and follows suggestions by Abul-Magd and Weidenmüller[12] of such effects in a smaller data set and by Paar and Vorkapić[23] of such effects in interacting boson model calculations.

The results for the six different mass regions are shown in Fig. 1. The distribution for the lightest masses appears very close to GOE behavior, while the heaviest masses show a distribution which is only slightly different from a Poisson. Visually there is an obvious trend from GOE to Poisson as the mass increases. The values of ω are 0.72 ± 0.16, 0.88 ± 0.41, 0.55 ± 0.11, 0.33 ± 0.07, 0.43 ± 0.17, and 0.24 ± 0.11, respectively; thus, ω also shows a trend from GOE to Poisson behavior as the mass increases, although it is not monotonic since the decrease in ω appears to be interrupted for masses in the ranges 50-100 and 180-210. In both cases, however, the uncertainties in ω are too large to draw definitive conclusions. The levels with masses 50-100 constitute the smallest sample of these six groupings, and the error on ω is significantly larger than in the other five cases. For levels with masses 180-210, there may be a real effect, since these are spherical nuclides surrounded by two groups of deformed nuclides; such shape effects are indeed apparent in the data, as we discuss below.

Spin effects are more difficult to search for in this data set, because the number of levels of a given spin and parity is

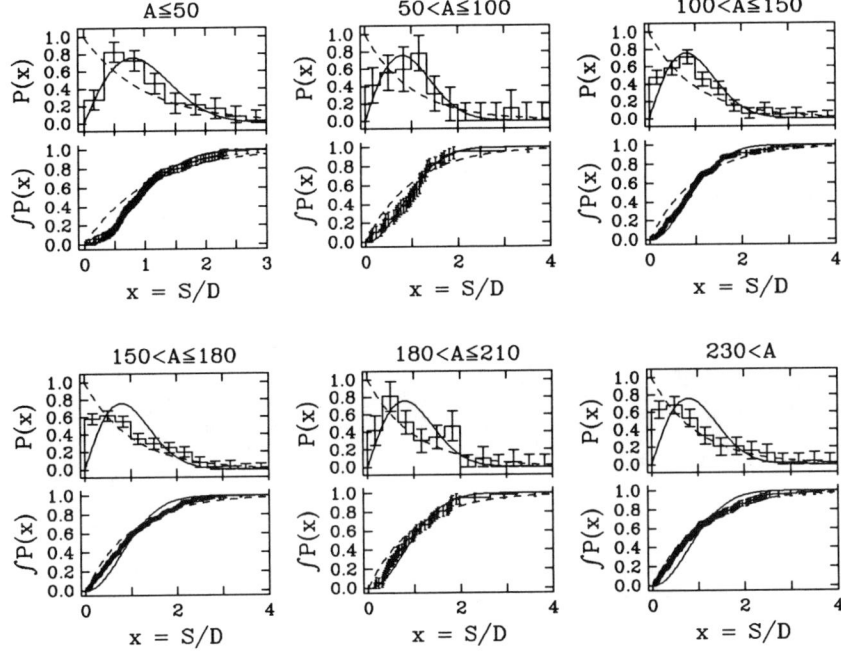

Figure 1. NNS distributions and cumulative distributions for all levels in six different mass regions.

generally small and the uncertainties are large. However, as shown in Fig. 2, the 2^+ and 4^+ states in deformed nuclides are close to Poisson ($\omega = 0.16 \pm 0.13$), while the 0^+ and 3^+ states in the deformed nuclides are near the other extreme ($\omega = 0.74 \pm 0.52$), albeit with limited statistical accuracy. Other spins or 2^+ and 4^+ states from spherical nuclides generally show values of ω which are intermediate between these two groups of states.

Fig. 3 demonstrates that the nuclear shape also has an effect on NNS distributions in even-even nuclides. The 2^+ and 4^+ states behave quite differently for deformed and spherical nuclides ($\omega = 0.16 \pm 0.13$ vs. $\omega = 0.52 \pm 0.15$), while the remainder of the states show NNS distributions which have little if any shape dependence ($\omega = 0.51 \pm 0.21$ for deformed vs. $\omega = 0.57 \pm 0.16$ for spherical).

DISCUSSION

In attempting to understand these results, it is interesting to consider what could cause a transition from GOE to Poisson fluctuations. In light of the discussion in the Introduction, one might expect this to be indicative of the system moving from

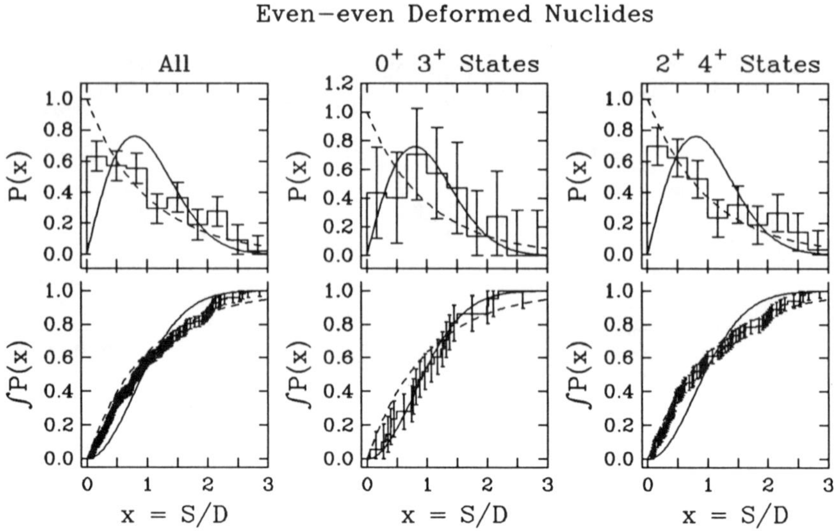

Figure 2. NNS distributions and cumulative distributions for levels in even-even deformed nuclides.

chaotic behavior to regular behavior. However, it is well known (see, e.g., Mehta[24]) that the NNS distribution also would move from GOE toward Poisson if the spacings result from a superposition of unrelated sequences. Therefore, if there were a conserved quantum number which had not been considered when the sequences were identified, GOE behavior would not be expected, even if that were the true underlying behavior of the system.

Considering the role of other quantum numbers leads quickly to the question of what a broken symmetry does to the fluctuation properties. This general problem has been studied by Dyson[25] and by Pandey.[26] A summary was given by Pandey: "...even a small breaking of a fundamental or model symmetry is shown to yield fluctuation patterns which would be found in the complete absence of the symmetry." Therefore, if any unidentified symmetries are sufficiently broken, there would be little or no difference in the fluctuation properties from the case in which the symmetry was unbroken.

In nuclei two additional quantum numbers which come to mind are isospin and the K quantum number (used in describing rotational nuclei). The only experimental data available for a broken symmetry test of this type are from nuclear levels in ^{26}Al, where not only J and π but also the isospin T have been identified for $\approx$160 states by Endt et al.[16-17] There the data are qualitatively consistent[27-28] with the predictions of Dyson and Pandey, although even for a pure isospin sequence the behavior is

not GOE. A detailed
examination of how isospin
symmetry-breaking affects
the fluctuation properties
has been performed by Guhr
and Weidenmüller,[29] who
constructed a random-
matrix model for the
process. Within the
framework of this model,
they were able to provide
an explanation of the ^{26}Al
data and extract a value
for the Coulomb matrix
element which was in
agreement with previous
results.

For the K quantum
number there are not
suitable data at present
for an experimental test
of how K symmetry-breaking
affects the fluctuation
properties. However, K
symmetry-breaking has been
studied by Paar and
Vorkapić[30] within the
framework of the
interacting-boson model.
Study of a level scheme of
^{164}Er calculated with this
model shows approximate K-
independence for the fluctuation properties.[31]

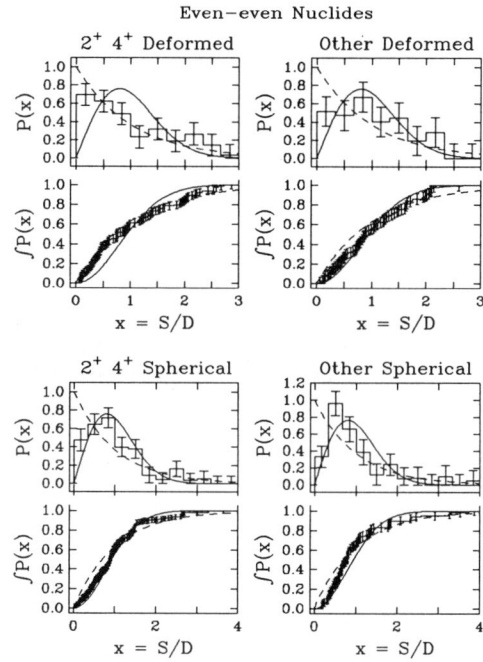

Figure 3. NNS distributions and cumulative distributions for several groups of levels in even-even nuclides.

Do these theoretical studies offer any insights into the
present results, in which the only symmetries considered are
angular momentum and parity? When the data are examined as a
function of mass, the two regions where nuclei are strongly
deformed display the NNS distributions which are nearest to
Poisson behavior. Of course, K is expected to be most relevant in
these two regions. Thus, one might speculate that the fact that
more regular behavior is observed in these two regions may simply
reflect the importance of K, and that the failure to separate
sequences by K (in addition to J and π) has biased these results.
However, if K (or some other quantum number) explains the observed
mass dependence, then the results for different spins in the
deformed nuclides indicate different amounts of symmetry breaking
for different spin groups. Current experimental data, except for
^{26}Al, are insufficient to consider the effects of symmetry
breaking on the fluctuation properties in any single nuclide.

On the other hand, the lightest nuclides show behavior which
is close to GOE. These nuclides are generally well described by

the shell model, and it has long been known that shell-model
states generally agree with the GOE (see, e.g., the discussion in
Brody et al.[1]). Therefore, this behavior does not seem very
surprising. As we pointed out earlier, even in the spherical
nuclides the fluctuation properties appear to be moving toward
regularity at heavier masses. It is not clear whether this
represents the increasing importance of another quantum number or
some other effect.

SUMMARY AND CONCLUSIONS

Fluctuation properties have been analyzed for a large
collection of low-lying nuclear energy levels. Emphasis has been
placed on nearest-neighbor spacing distributions; 988 spacings
from 60 different nuclides have been included in the analysis. A
strong mass dependence is observed, with the lightest nuclides
showing behavior close to GOE and the heaviest nuclides displaying
Poisson behavior. There is a spin dependence, in that 2^+ and 4^+
states in deformed nuclides behave differently from 0^+ and 3^+
states in the same nuclides. Evidence also has been presented
that the nuclear shape has an effect on the fluctuation
properties.

Whether the empirically determined behavior of the
fluctuation properties is a manifestation of regular behavior for
low-lying states in heavy nuclei, a reflection of the omission of
other relevant symmetries, or an indicator of some other effect,
it is clear that more extensive data in a single nuclide are
required. Then one could search for a possible phase transition
(if the data are indeed regular at low energies and chaotic at
high energies) or study the effects of symmetry breaking
explicitly (if neglected symmetries explain the results). For
light nuclides, level schemes as extensive at that obtained for
^{26}Al are in principle attainable by studying (p,γ) and (p,p)
resonance reactions with different targets. Additional data on
nuclides such as ^{22}Na and ^{30}P would be most valuable, and such
experiments are being planned. To gain sufficient data in a
single heavy nuclide is difficult and will probably require the
next generation of γ-ray detectors (Gammasphere or Euroball); to
experimentally study K symmetry-breaking in this manner would seem
to require such data.

Since the experimental data needed to study fluctuation
properties are so difficult to obtain in sufficient quantity and
quality, it would be extremely valuable to have an alternative
approach in which more limited experimental data could be used in
tests for the presence of chaos. One method which has been
applied in molecular physics[32] involves the Fourier transform of
the intensity autocorrelation function; we are beginning to study
this technique to see if it can be applied to nuclear data. In
any event, the fluctuation properties of low-lying states in
nuclei show very interesting trends which are not yet understood
and which deserve further experimental and theoretical study.

ACKNOWLEDGMENTS

This work was supported in part by the U. S. Department of Energy, Office of High Energy and Nuclear Physics, under Contracts No. DE-FG05-87ER40353 and DE-FG05-88ER40441 and by the Bundesministerium für Forschung und Technologie, Bonn, Germany. We wish to thank T. Guhr and H. A. Weidenmüller for valuable discussions on theoretical issues and for their computer code to calculate GOE spectra and R. C. Spirko and B. L. Winn for their assistance with the analysis.

REFERENCES

1. T. A. Brody et al., Rev. Mod. Phys. 53, 385 (1981).
2. O. Bohigas, M. J. Giannoni, and C. Schmit, Phys. Rev. Lett. 52, 1 (1984).
3. T. H. Seligman, J. J. M. Verbaarschot, and M. R. Zirnbauer, J. Phys. A 18, 2751 (1985).
4. D. Delande and J. C. Gay, Phys. Rev. Lett. 57, 2006 (1986); 57, 2877 (1986).
5. D. Wintgen and H. Marxer, Phys. Rev. Lett. 60, 971 (1988).
6. D. C. Meredith, S. E. Koonin, and M. R. Zirnbauer, Phys. Rev. A 37, 3499 (1988).
7. M. V. Berry and M. Tabor, Proc. R. Soc. London A 356, 375 (1977).
8. G. Casati, B. Chirikov, and I. Guarneri, Phys. Rev. Lett. 54, 1350 (1985).
9. T. Cheon and T. D. Cohen, Phys. Rev. Lett. 62, 2769 (1989).
10. R. U. Haq, A. Pandey, and O. Bohigas, Phys. Rev. Lett. 48, 1086 (1982).
11. O. Bohigas, R. U. Haq, and A. Pandey, Phys. Rev. Lett. 54, 1645 (1985).
12. A. Y. Abul-Magd and H. A. Weidenmüller, Phys. Lett. 162B, 223 (1985).
13. T. von Egidy, A. N. Behkami, and H. H. Schmidt, Nucl. Phys. A454, 109 (1986).
14. T. von Egidy, H. H. Schmidt, and A. N. Behkami, Nucl. Phys. A481, 189 (1988).
15. F. J. Dyson and M. L. Mehta, J. Math. Phys. 4, 701 (1963).
16. P. M. Endt, P. deWit, and C. Alderliesten, Nucl. Phys. A459, 61 (1986).
17. P. M. Endt, P. deWit, and C. Alderliesten, Nucl. Phys. A476, 333 (1988).
18. E. P. Wigner, Oak Ridge National Laboratory Report ORNL 2309, 1957, p. 59.
19. T. A. Brody, Lett. Nuovo Cimento 7, 482 (1973).
20. T. Cheon, Phys. Rev. Lett. 65, 529 (1990).
21. A. Gilbert and A. G. W. Cameron, Can. J. Phys. 43, 1446 (1965).
22. B. Efron, SIAM Review 21, 460 (1979).
23. V. Paar and D. Vorkapić, Phys. Lett. 205 B, 7 (1988).

24. M. L. Mehta, *Random Matrices and the Statistical Theory of Energy Levels* (Academic, New York, 1967).
25. F. J. Dyson, J. Math. Phys. 3, 1191 (1962).
26. A. Pandey, Ann. Phys. (NY) 134, 110 (1981).
27. G. E. Mitchell, E. G. Bilpuch, P. M. Endt, and J. F. Shriner, Jr., Phys. Rev. Lett. 61, 1473 (1988).
28. J. F. Shriner, Jr., E. G. Bilpuch, P. M. Endt, and G. E. Mitchell, Z. Phys. A 335, 393 (1990).
29. T. Guhr and H. A. Weidenmüller, Ann. Phys. (NY) 199, 412 (1990).
30. V. Paar and D. Vorkapić, Phys. Rev. C 41, 2397 (1990).
31. V. Paar and D. Vorkapić, to be published.
32. L. Leviander, M. Lombardi, R. Jost, and J. P. Pique, Phys. Rev. Lett. 56, 2449 (1986)

SCARS AND THE ORDER TO CHAOS TRANSITION

Hans Frisk

Nuclear Science Division, Lawrence Berkeley Laboratory
Berkeley, California 94720

ABSTRACT

A numerical investigation of the localization of eigenfunctions around periodic orbits is presented. The classical motion in the potenial shows a generic order to chaos transition when the deformation is increased. Some eigenfunctions are found to be remarkably orded even if the underlying classical motion is strongly chaotic.

INTRODUCTION

The periodic orbits constitute the skeleton for the organization of phase space in Hamiltonian systems. However, the importance of the periodic orbits in semiclassical mechanics was recognized rather late[1,2]. The profound result was that the density of states can be divided into a Thomas-Fermi part and an oscillatory part expressed as a sum over all periodic orbits. If the smooth oscillations of the level density are of interest only the shortest periodic orbits are necessary to include in the summation. The level density is obviously related to the shell structure and the truncated periodic orbit sum is very powerful in nuclear structure calculations[3,4].

In the most orded and predictable Hamiltonian systems the number of constants of motion equals the degrees of freedom and generally infinitely many periodic orbits of a given type exist. These systems are called integrable. An example is a two dimensional infinite well of circular shape where the particles motion inside the well is free and it is perfectly reflected at the boundary. This circular billiard is obviously integrable since both energy and angular momentum are conserved. The shortest periodic orbits are the diameters, next shortest are the equilateral triangles etc., and all orientations of the periodic orbit are possible[2]. However, if a step towards chaos is taken and the boundary is slightly deformed, e.g. to an octupole shape, generally only two isolated periodic orbits remain from each type, one is stable the other is unstable. If the deformation is even further increased bifurcations take place and the short stable periodic orbits become unstable. Finally chaos is seen everywhere in phase space.

Both stable and unstable isolated periodic orbits give rise to imprints in the eigenfunctions[5]. This is in contrast to the integrable case where generally no particular orientation of the periodic orbit is prefered. That the

© 1991 American Institute of Physics

unstable isolated periodic orbits show up in the eigenfunctions was a surprise when it was observed in computer explorations of the stadium billiard[6] and the term scars was coined for these imprints.

In the rest of this contribution I will present a numerical investigation of a generic order to chaos transition with the focus on how the the eigenstates localize around the shortest periodic orbits. At least in the actinides the wavelength is sufficiently small to make the concept of localization meaningful in nuclear physics. For convenience I have chosen a two dimensional system which is relevant since the planar motion is very important for the nuclear structure[3,4].

THE CLASSICAL MOTION

The integrable system we start from is the circle billiard described above. Let us now deform the boundary $(u,v)=(\cos\varphi,\sin\varphi)$ in the following way

$$u = A(\cos\varphi + \sigma\cos 2\varphi)$$
$$v = A(\sin\varphi + \sigma\sin 2\varphi) \quad (1)$$

by increasing the parameter σ. The parameter range is $0.0 \leq \sigma \leq 0.5$ and A is fixed by requiring area conservation. An appropriate choice of deformation parameter instead of σ is $R_>/R_<$, in the following denoted μ, where $R_>$ ($R_<$) is the largest (shortest) distance from the center, $(A^3\sigma,0)$, to the boundary. For the circle $\mu=1.0$ and in the interval $1.0 \leq \mu < 1.139$ the boundary is convex everywhere. At $\mu=1.139$ the curvature at $\varphi=\pi$ vanishes and a concave part exists for larger deformations but at $\mu=1.581$ it is reduced to a cusp at $\varphi=\pi$. Classical results show that the order to chaos transition mainly takes place between $\mu=1.04$ and $\mu=1.27$ and no stable periodic orbit is seen above $\mu=1.14$[7]. For not too large deformations the classical motion in this billiard and an octupole shaped one is qualitatively the same.

The localization of the probability density around the remnants of the diameter, triangle and square orbits in the circle is considered. Three of these periodic orbits are unstable for $\mu > 1.0$ and are shown in fig. 1, taken from ref. 5, at $\mu=1.1$. They are in the following denoted 2U, 3U and 4U depending on how many collisions with the boundary that occur during one period. The other three periodic orbits considered, see the $\mu=1.3$ billiard in fig. 1, are denoted 2SU, 3SU and 4SU. They are stable for small deformations but bifurcate in the interval $\mu=1.067$-1.095 and become very unstable for large deformations. The lenghts of the SU (U) orbits decrease (increase) with increasing deformation and the 2SU orbit has the strongest μ-dependence which is reflected in the quantum levels, see fig. 2.

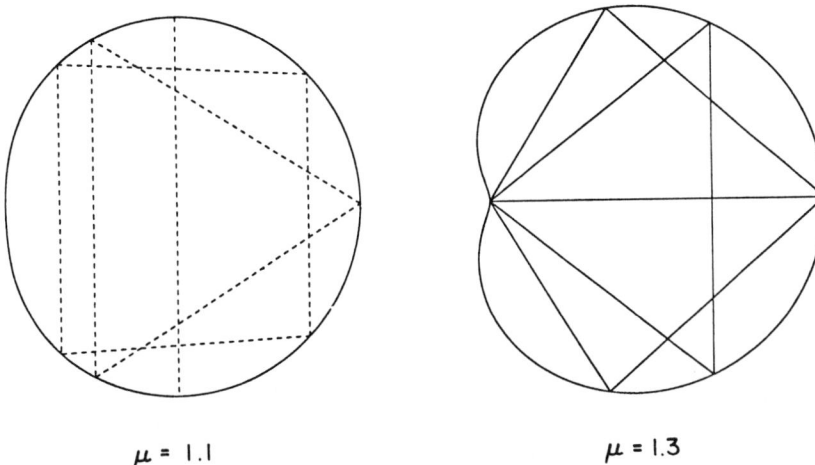

Fig. 1. Two members of the family of billiards given by (1). The SU orbits (see text) are shown at μ=1.3 and the U orbits at μ=1.1.

QUANTUM MECHANICAL RESULTS

Details how to solve the quantised billiard can be found in refs. 5 and 8. The deformation dependence for positive parity (solid curves) and negative parity (dashed curves) levels are shown in fig. 2 in the energy region E=50-300$\hbar^2/2mR_o^2$ where R_o denotes the circle radius. In the actinide region the neutron Fermi level lies around 150 $\hbar^2/2mR_o^2$. Notice the upsloping positive parity levels for $\mu \lesssim 1.25$. Their eigenfunctions have a large imprint of the 2SU orbit for which the length decreases most strongly with increasing μ. In the interval 1.0< μ <1.095 the 2SU orbit is stable and it is expected that eigenfunctions separated in wave number by $\Delta k \approx 2\pi/L_{2SU}(\mu)$ have a strong imprint of this orbit. Here L_{2SU} denotes the length of the 2SU orbit and the wave number, k, is related to the energy by $E=\hbar^2 k^2/2m$. This can be understood by sending out a gaussian wave packet along the periodic orbit[6]. Due to the stability infinitely many recurrences of the wave packet will occur and the overlap with the original wave packet will be periodic in the covered distance.

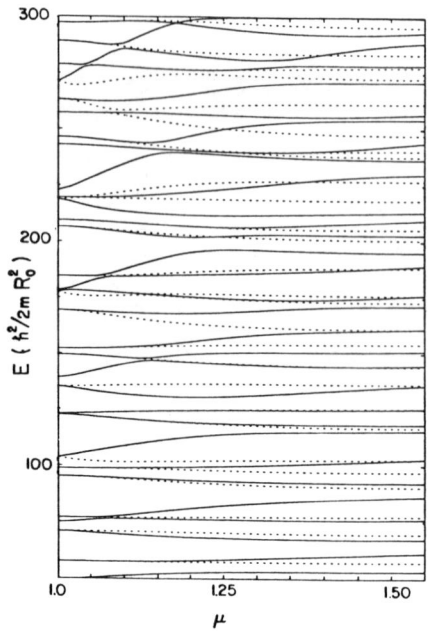

Fig. 2. Positive parity (solid curves) and negative parity (dashed curves) eigenenergies as function of deformation.

This periodicity is reflected in the wave number domain by strong imprints of the periodic orbit in equidistant spaced eigenstates. The upsloping levels continue to some extent into the chaotic domain and therefore scars of the 2SU orbit will be seen. For $\mu > 1.25$ the slopes are generally small and the scar effect from the 2SU orbit is reduced but fairly constant. The localization around the other SU orbits show a similar deformation dependence. Contour plots of the probability density for two scarred eigenfunctions are shown in fig. 3. The distance between the contours is $0.5R_o^{-2}$ which should be compared to the value R_o^{-2}/π for the mean probability density.

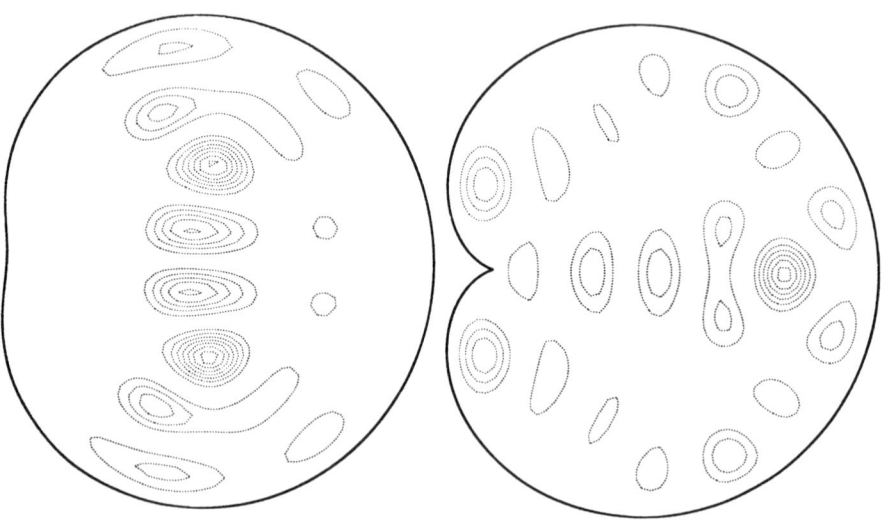

Fig. 3. Two examples of eigenfunctions showing scars of the 2U orbit (left) and 2SU orbit (right).

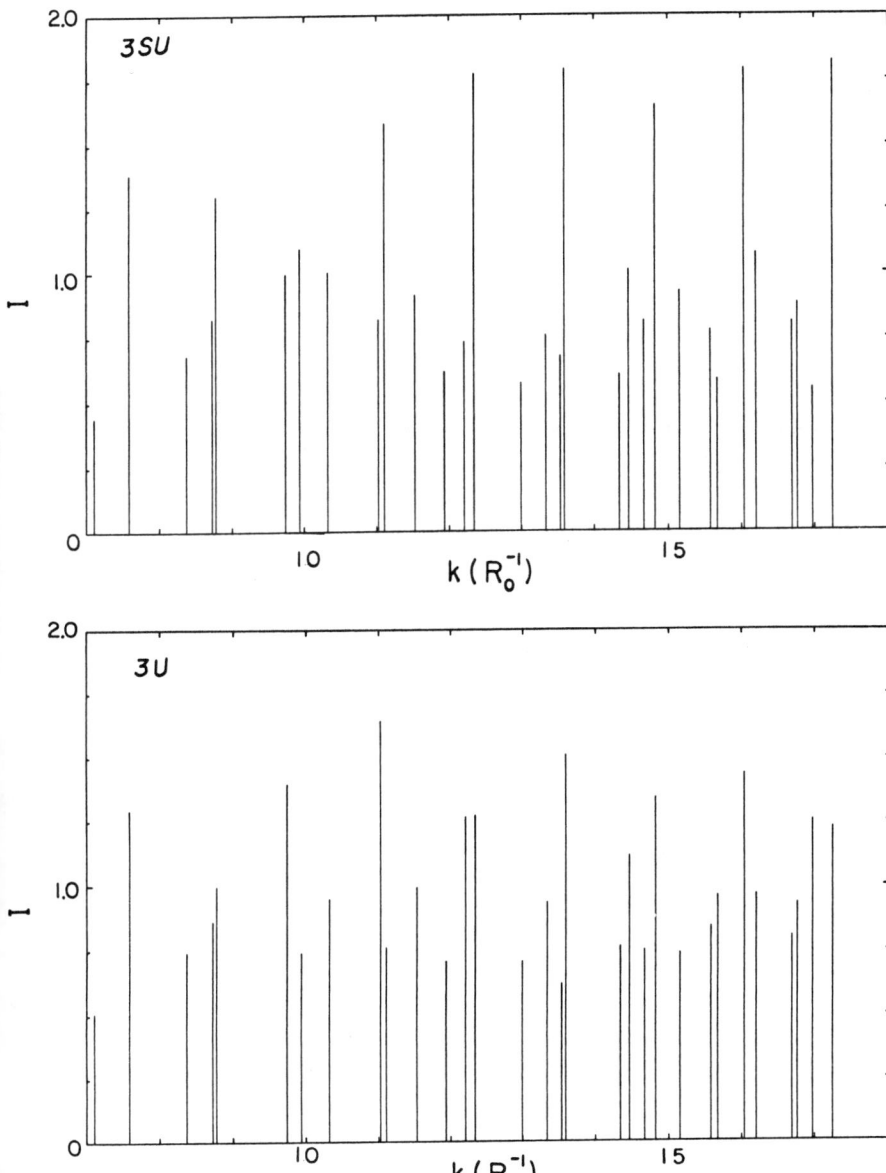

Fig. 4. The quantity defined in (2) in the same wave number region as in fig. 2 for the 3SU orbit (top) and the 3U orbit at $\mu=1.06$

In fig. 3 (left) the negative parity state with $E=163.4\hbar^2/2mR_o^2$ at $\mu=1.2$ is displayed. I apologize for the lack of symmetry around the $v=0$ line due to the plotting routine. The 2U orbit is only slightly unstable at $\mu=1.2$ and the concentration around it is not surprising. Much more surprising is the positive parity state with $E=151.1\hbar^2/2mR_o^2$ at $\mu=1.55$. The 2SU orbit is strongly unstable at this deformation but despite that a strong localization around it occurs.

To get a measure of the scars, the probability density for each eigenfunction is integrated along each periodic orbit

$$I = \frac{\pi R_o^2}{L} \oint |\Psi(u,v)|^2 \, ds \qquad (2)$$

where $ds=\sqrt{du^2 + dv^2}$. The prefactor provides the mean value of I to be 1 except for the 2SU orbit and postive parity states for which the mean value is 2 due to the symmetry around the $v=0$ line. Negative parity states have of course $I=0$ for the 2SU orbit. In fig. 4 I values for the 3U and 3SU orbits at $\mu=1.06$ and positive parity states are shown. The 3SU orbit is stable at this deformation and for $k > 11R_o^{-1}$ almost equally separated eigenstates have a strong concentration around this orbit. For the unstable 3U orbit the scar strength appears more irregular and often two eigenstates share the strength. Note that the 3SU (3U) orbit has a shape opposite (similar) to the boundary, see fig. 1. Thus, a nucleon located in an eigenstate connected with the 3SU orbit polarize the nucleus towards spherical shape. A calculation of the r.m.s. values for the oscillations of I around the mean[5] confirms the qualitative statements made in connection with fig. 2. The localization effect for the SU orbits is maximal around $\mu=1.04$ and decrease up to $\mu \approx 1.25$, i.e. where the order to chaos transition mainly takes place. For larger deformations the scar effect from the SU orbits is remarkably constant even if the orbits are strongly unstable. The scar effect from the 3U and 4U orbits have a rather weak μ-dependence while an interesting symmetry effect occurs for the 2U orbit. In the limit $\mu=1.0$ positive parity states with odd angular momentum have $I_{2U}=0$ while $I_{2U}=I_{2SU}$ for positive parity states with even angular momentum. This symmetry effect gives rise to strong oscillations of I which survives up to the largest deformations.

To conclude, the isolated periodic orbits in non-integrable systems show up in the eigenfunctions and some eigenfunctions can be remarkably orded even if the underlying classical motion is strongly chaotic. This is good news for us who want to observe nuclear structure.

My stay at LBL has been made possible by a grant from the Swedish Natural Science Research Council

REFERENCES

1. M. C. Gutzwiller, J. Math. Phys. 11, 1791 (1971).
2. R. Balian and C. Bloch, Ann. of Phys. 69, 76 (1972).
3. V. M. Strutinsky et. al., Z. Phys. A283, 269 (1977).
4. H. Frisk, Nucl. Phys. A511, 309 (1990).
5. H. Frisk, NORDITA preprint 90/46 and submitted to Phys. Scripta.
6. E. J. Heller, Quantum Chaos and Statistical Nuclear Physics (Springer-Verlag 1986), p. 162.
7. M. Robnik, J. Phys. A16, 3971 (1983).
8. M. Robnik, J. Phys. A17, 1049 (1984).

NEW FORMULA FOR SPIN-DEPENDENT LEVEL DENSITY

V. Paar and S. Brant
Prirodoslovno - matematicki fakultet, University of Zagreb
41000 Zagreb, Croatia, Yugoslavia and Lawerence Livermore
National Laboratory, Livermore, CA 94550

D.K. Sunko
Prirodoslovno - matematicki fakultet, University of Zagreb
41000 Zagreb, Croatia, Yugoslavia

M.G. Mustafa and R.G. Lanier
University of California, Lawrence Livermore National
Laboratory
Livermore, CA 94550 USA

ABSTRACT

The first realistic calculations of spin-dependent level densities using the Gaussian polynomial generating function method (GFM) are performed for ^{114}Cd and ^{244}Am. Contrary to the results of previous combinatorial calculations, the curvature of the logarithmic energy dependence of the calculated total level density is consistent with the Bethe formula. The GFM results support the assumption that the spin-dependent level density can be factorized analogous to the Bethe formula, but would require a sizeable modification to the spin distribution function. A new algebraic formula for the spin distribution function is proposed and applied to ^{114}Cd and ^{244}Am.

INTRODUCTION

The calculation of nuclear level densities in large shell model spaces is an important and long-standing problem in nuclear physics because the cumbersome nuclear reaction calculations depend crucially on the level densities. Several approaches to the microscopic calculation of nuclear level densities have been used.[1-8] In particular, much attention has been paid to the thermodynamic approach with the saddle point approximation for the grand partition function and the combinatorial approach (for excellent reviews see Refs. 2,3). On a practical level, in the nuclear reaction calculations the central role is still played by the pioneering work by Bethe,[1] with an approximate algebraic formula. The Bethe formula was originally derived in a schematic equidistant spacing model, for noninteracting Fermions with non-degenerate equidistant single-particle levels[1]. However, the range of applicability was extended beyond the scope of the equidistant model by the realization that both the shell effects and the effect of the residual pairing force can be, approximately, simulated by introducing the ground-state back-shift in the Bethe formula[2].

A particular sensitive aspect of the Bethe formula is the spin dependent function. Namely, the Wigner-type form employed in Bethe formula corresponds to the first term in a series expansion, which is an approximation applicable to the low-spin states only[9-11]. Without resorting to the low-spin limit, Lang has obtained a more complex spin-dependent level density formula[10] which, in an approximate way, takes into account higher order contributions in the equidistant model. In the low-spin limit the Lang formula reduces to the Bethe formula. Lang formula relates the spin cut-off parameter σ to the yrast spin. However, the total level density in Lang formula is not factorized from the spin dependent function. Another difficulty associated with the Lang formula lies in the problem of fitting the level density and spin cut-off parameters. For these reasons, the Bethe formula, in spite of its shortcoming as a low-spin limit, has been almost universally used in the reaction calculations.

For a long time, attention has been paid to the exact solution of the level density problem by using a combinatorial approach. Hillman and Grover have developed a method[4] which is similar in spirit to the investigations carried out by Motz and Feinberg[12] as long as 1930's. In the Hillman-Grover method all the possible configurations are obtained by means of a straightforward method of enumeration and classification, with the use of realistic single-particle level spacings and with an approximate treatment of the pairing interaction employing the BCS approximation.

GAUSSIAN POLYNOMIAL GENERATING FUNCTION METHOD (GFM) FOR LEVEL DENSITY

A new method[13] has been recently introduced for an exact solution of the combinatorial problem of the nuclear level density. This method is based on the use of Gaussian polynomials as generating functions. The basic generating function of GFM is

$$G(n,q) = q^{-n(r-n)/2} \begin{bmatrix} r \\ n \end{bmatrix}_q \qquad (1)$$

where

$$r = 2j+1 \quad \text{(for fermions)}$$
$$r = 2j+n \quad \text{(for boson)} \qquad (2)$$

and the square bracket denotes a Gaussian polynomial[14] in q. The multiplicity of the states with a given total angular momentum projection M for n particles of spin j is equal to the coefficient of q^M in the generating function (1). An efficient recursion algorithm is available to calculate these coefficients. The extension of the generating function (1) to the multilevel case is straightforward[13]: The generating function of the M state multiplicities of a configuration of z levels j_i with n_i particles per level j_i is the product of the generating functions (1) of the individual levels. Furthermore, the GFM enables an exact treatment of the pairing force within the levels: The generating function of a configuration with a given seniority is[13]

$$V(s = n,j) = G(n,j) - G(n - 2,j) \qquad (3)$$

since any angular momentum state containing pairs coupled to zero must also be present in the (n-2)-particle configuration. With an extension of (3) to the multilevel case[13], the GFM is equivalent to an exact diagonalization of the Hamiltonian[15]

$$H = \sum_i n_i \varepsilon_i + G \sum_i s_+^{(i)} s_-^{(i)} \qquad (4)$$

with a schematic pairing force that acts only within the levels. Here, n_i denote the number operators of the single-particle levels with energy ε_i and $s_+^{(i)}, s_-^{(i)}$ are the quasispin operators.

The advantage of GFM is the use of a powerful mathematical formalism for Gaussian polynomials which enables us to produce the exact energy spectrum of (3) in very large configuration spaces without the need to resort to the BCS and associated approximations. This provides an exact treatment of the energies of promoted pairs, which is a difficult problem of the Hillman-Grover method. However, we assume that the part of the pairing force which is not included in GFM can be accounted for by renormalization of the pairing strength G.

In Fig.1 we present the result of the GFM calculation for the total level density of ^{114}Cd, using the Seeger's single-particle energies from Ref.4. In analogy to the Hillman-Grover method[4], for each nucleus we find the value of the effective pairing strength G necessary to cause the calculated level densities for the appropriate spins to pass through the experimental level density given by the neutron resonance data[16]. The corresponding renormalized pairing strength for ^{114}Cd is G = 0.46. Assuming that the long-range components of the

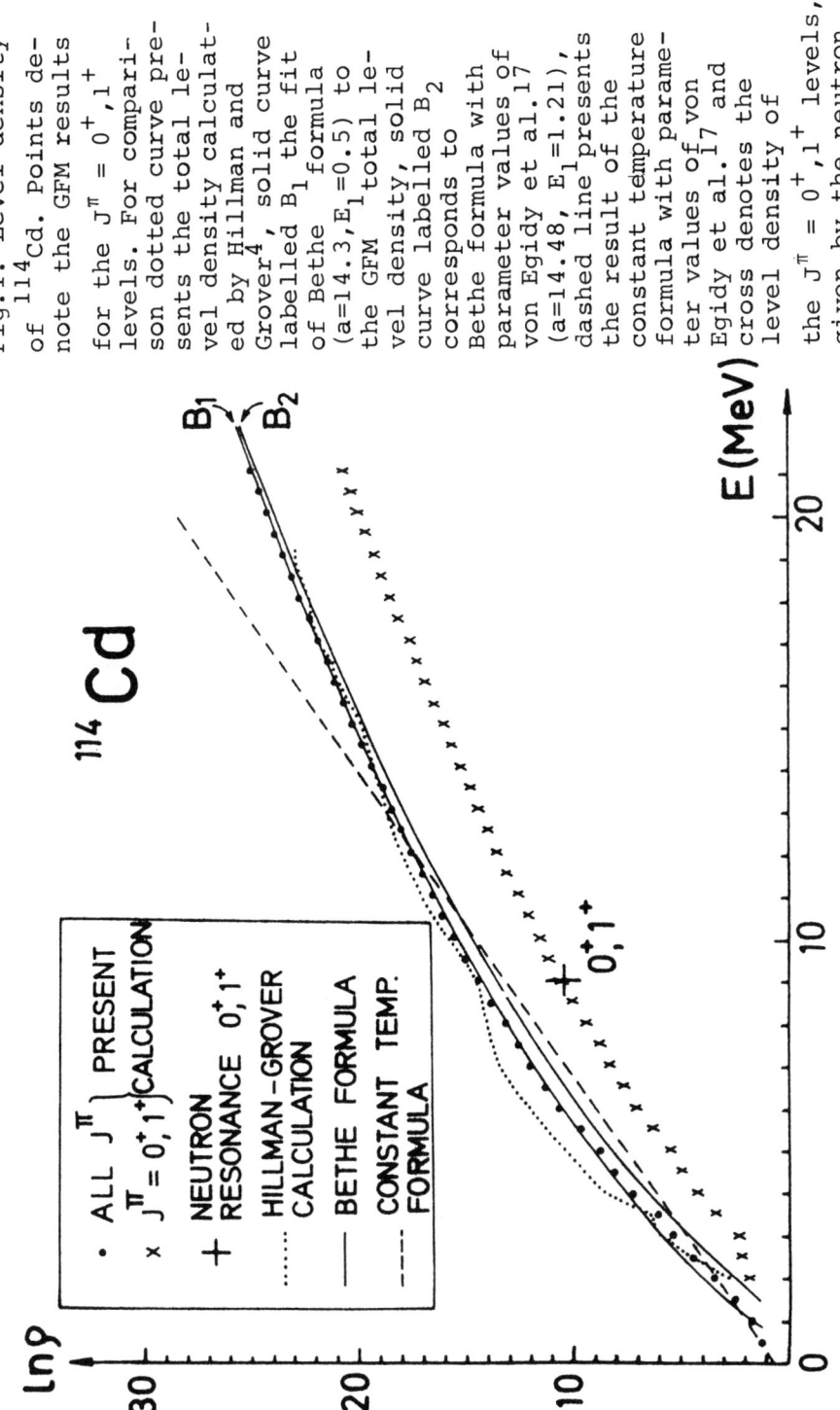

Fig.1. Level density of ^{114}Cd. Points denote the GFM results for the $J^\pi = 0^+, 1^+$ levels. For comparison dotted curve presents the total level density calculated by Hillman and Grover[4], solid curve labelled B_1 the fit of Bethe formula (a=14.3, E_1=0.5) to the GFM total level density, solid curve labelled B_2 corresponds to Bethe formula with parameter values of von Egidy et al.[17] (a=14.48, E_1=1.21), dashed line presents the result of the constant temperature formula with parameter values of von Egidy et al.[17] and cross denotes the level density of the $J^\pi = 0^+, 1^+$ levels, given by the neutron resonance data.

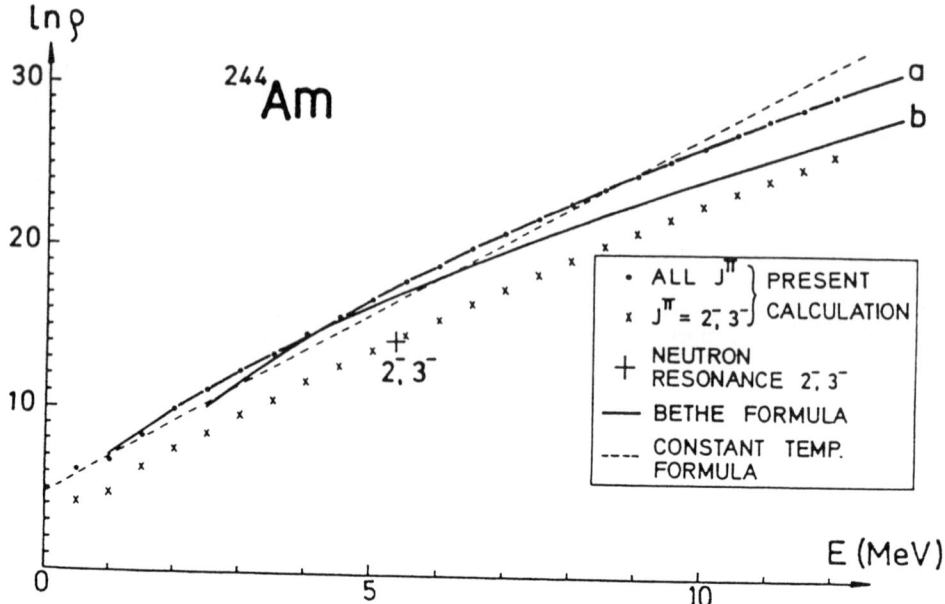

Fig.2. Level density for ^{244}Am. Points denote the GFM results for the total level density and x´s the GFM results for the $J^\pi = 2^-, 3^-$ levels. Solid line labelled a presents the fit of Bethe formula (a=31.95, E_1=0.38) to the GFM total level density, solid curve labelled b corresponds to Bethe formula with parameter values from von Egidy et al.[17] and dashed line the result of constant temperature formula with the parameter values from von Egidy et al[17].

residual force (which are not included in the calculation of level densities) would largely scatter the levels calculated for the pairing residual force alone, their effect is simulated by averaging the level density over an energy interval of a suitable with: The calculated level density $\rho(E)$ is given by the number of levels in the energy interval (E-0.5MeV, E+0.5MeV). The solid curve with a label B_1 in Fig.1 presents the total level density given by Bethe formula for the values of the level density parameter a and the ground state back shift E_1 obtained by a least squares fit to the combinatorial GFM level densities above 6 MeV excitation energy. These values (a = 14.3, E_1 = 0.5) are in agreement with the estimate[3] a = A/8 $\doteq$ 14.25. (For definition of Bethe formula we use the standard form from Ref.17.) The solid curve with a label B_2 presents the result of Bethe formula for the parametrization of von Egidy et al.[17]

As seen, the logarithmic behaviour of the GFM level density above the low-energy region agrees very well with the Bethe formula. This is in contrast to the results of the Hillman-Grover calculation[4] (shown, for comparison, by a dotted line in Fig.1); the level density calculated by using the Hillman-Grover method differs significantly from that amenable to the analytical solution of Bethe formula. A big difference between the two combinatorial calculations, employing the GFM and the Hillman-Grover method for the same set of Seeger´s single particle levels is indeed a challenge. This difference might be due to basically different methods of treating the pairing interaction and/or due to possible truncations in the Hillman-Grover calculation. This point requires further consideration.

In Fig. 2 we present the result of GFM calculation for the level density of ^{244}Am, using the Seeger´s single-particle energies. Fitting the renormalized pairing strength to the neutron resonance data from Ref. 17 we obtain G = 0.337. The calculated total level density above 4 MeV of excitation energy is in good agreement with the Bethe formula for the parameters a = 31.95, E_1 = 0.38 (solid curve labelled a).

Above the low-energy region both in ^{114}Cd and ^{244}Am the total level densities calculated by using GFM are in accordance with Bethe formula for the value of level density parameter close to the estimate a = A/8. In the low-energy region there appear fluctuations caused by the absence of the long-range components of the residual force in the GFM calculation. In this region the constant temperature formula[2,17] can give somewhat better fit to the GFM results, then given by the Bethe formula. Such a pattern is in agreement with previous phenomenological investigations[17,18]. We note that a caution is necessary in connection to the comparison with total level densities deduced from experiment. Namely, as pointed out in Ref.4, the experimental values are usually deduced using an assumed spin distribution which is not in agreement with combinatorial calculations.

Let us now turn our attention to the spin distribution calculated in GFM. It is convenient to introduce the transformation

$$y_H = \ln \left[N(J)_{E,E+\Delta E} / (J + \tfrac{1}{2}) \right] \qquad (5)$$

where $N(J)_{E,E+\Delta E}$ denotes the number of levels of spin J in the energy interval (E, E+ ΔE). In Fig.3 we present the plot of y_H versus $(J + \tfrac{1}{2})^2$ for several energy intervals in the GFM energy spectrum of ^{244}Am. These results are compared with the predictions of the approximate algebraic relations. For this purpose, we express the number of levels in terms of the spin dependent level density

$$N(J)_{E,E+\Delta E} \simeq \rho(E,J) \cdot \Delta E \qquad (6)$$

For Bethe formula the spin-dependent level density has the factorized form[1]

$$\rho(E,J) = \rho_B(E) \cdot f(J) \qquad (7)$$

where $\rho_B(E)$ is the Bethe formula for the total level density and $f(J)$ is the Bethe formula for spin distribution function

$$f(J) = \frac{2J+1}{2\sigma^2} \exp\{-(J+\tfrac{1}{2})^2/(2\sigma^2)\} \qquad (8)$$

where σ is the energy dependent spin cutoff parameter. For Bethe formula there is

$$y_H = C - \frac{1}{2\sigma^2}(J+\tfrac{1}{2})^2 \qquad (9)$$

$$C = \ln\{\rho_B(E) \cdot \Delta E/\sigma^2\} . \qquad (10)$$

Adjusting σ to the GFM spin distributions for low J, we obtain the linear dependence shown by dashed lines in Fig.3. As seen, the GFM spin distribution exhibits at low spins linearity for $y_H\{(J+\tfrac{1}{2})^2\}$ and with increasing spin the deviations from linearity increase monotonically. Similar pattern of the combinatorial approach has been observed previously using the Hillman-Grover method[4,11].

NEW FORMULA FOR SPIN DISTRIBUTION

The GFM calculations are in accordance with Bethe formula for the total level density, but they display a significant deviation from Bethe formula for the spin distribution function, as illustrated in the previous section. This provides a microscopic hint that an approximate algebraic formula inherent to GFM preserves the factorization relation (7), only the expression (8) for f(J) should be modified. A guideline for a guess of a form of modified spin-distribution function is provided by the approximate solution of all-configurations problem[19] as well as by the mathematical formalism of the generating function method[13]. The Gaussian polynomial may be interpreted as generating a certain class of restricted partitions[20]. Rigorous derivation of the approximate algebraic formula for spin distribution without resorting to the low-spin limit would contain, through the coefficients associated with the expansion of

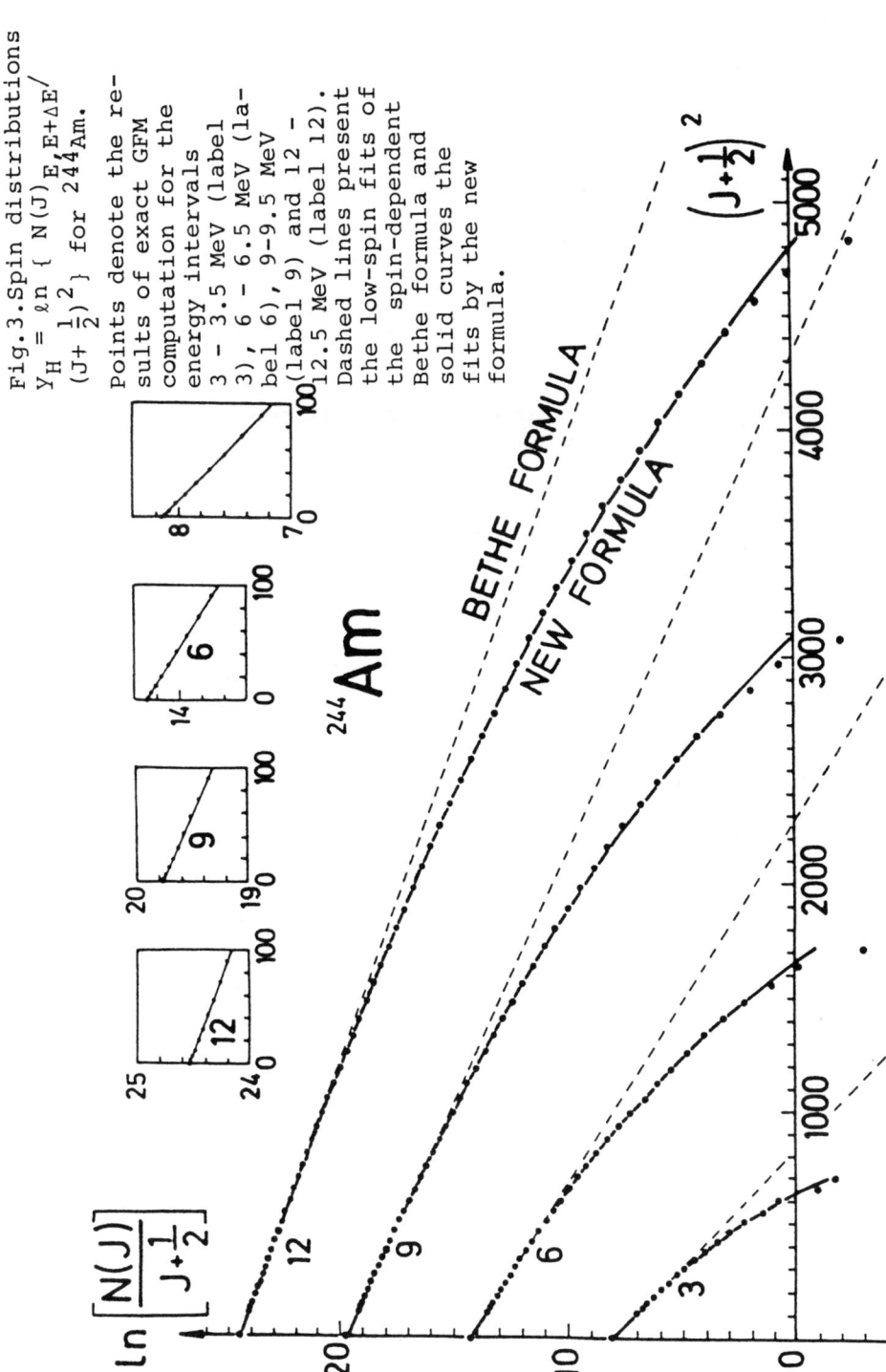

Fig.3. Spin distributions $Y_H = \ln\{N(J)_{E,E+\Delta E}/(J+\frac{1}{2})^2\}$ for ^{244}Am. Points denote the results of exact GFM computation for the energy intervals 3 – 3.5 MeV (label 3), 6 – 6.5 MeV (label 6), 9-9.5 MeV (label 9) and 12 – 12.5 MeV (label 12). Dashed lines present the low-spin fits of the spin-dependent Bethe formula and solid curves the fits by the new formula.

Gaussian polynomial[13] some general statements about the divisibility of numbers. This is, however, a notoriously difficult problem of the number theory which, to our knowledge, is not solved in the mathematical literature. However, the functional form appearing in the Hardy-Ramanujan expressions[19] is sinh, which is also consistent with the mathematical forms appearing in the GFM formalism. Following this hint and having in mind the GFM calculations for spin distribution in fig.3, we introduce the new formula for the spin distribution function

$$f(J) = \frac{2J+1}{2\sigma^2} e^{-\frac{1}{2\sigma^2} \zeta \sinh\{(J+\frac{1}{2})^2/\zeta\}} \quad (11)$$

It is convenient to express the new parameter ζ in terms of another parameter η, defined by

$$\zeta = \frac{2\eta^3}{\sqrt{3}\,\sigma} \quad (12)$$

which will be referred to as the spin shell-correction parameter. A Taylor series expansion of the logarithmic form for Eq. (11) is

$$\ln\{f(J) \cdot \sigma^2/(J+\tfrac{1}{2})\} = -\frac{1}{2\sigma^2}(J+\tfrac{1}{2})^2 - \frac{1}{2(2\eta^2)^3}(J+\tfrac{1}{2})^6 - \ldots \quad (13)$$

The first term in the series expansion reproduces the Bethe formula (8) for the spin distribution function. Inserting the new spin distribution function (11) into (7) and (6) we obtain

$$y_H = C - \frac{1}{2\sigma^2} \zeta \sinh\{(J+\tfrac{1}{2})^2/\zeta\} \,. \quad (14)$$

In each energy interval (E, E + ΔE) for E = 1 MeV, 1.5 MeV, 2 MeV,... and ΔE = 0.5 MeV we have fitted the energy-dependent parameters C, σ and η in the expression (14) to the corresponding GFM results for y_H. The fitting procedure is simplified here as follows: In the first step we consider the low-spin limit of Eq. (14), i.e. the expression (9), corresponding to Bethe formula, is used. Thus, the parameters C and σ are fitted to the GFM results for each pair $\{y_H(J=0), y_H(J=J_\kappa)\}$, with $J_\kappa = 1,2,3,\ldots$ Thus we obtain

$$\sigma^{(\kappa)} = \{\tfrac{1}{2} J_\kappa(J_\kappa+1)/(y_H(0) - y_H(J_\kappa))\}^{1/2} \quad (15)$$

$$C^{(\kappa)} = y_H(0) + \{y_H(0) - y_H(J_\kappa)\}/\{4J_\kappa(J_\kappa+1)\} \,. \quad (16)$$

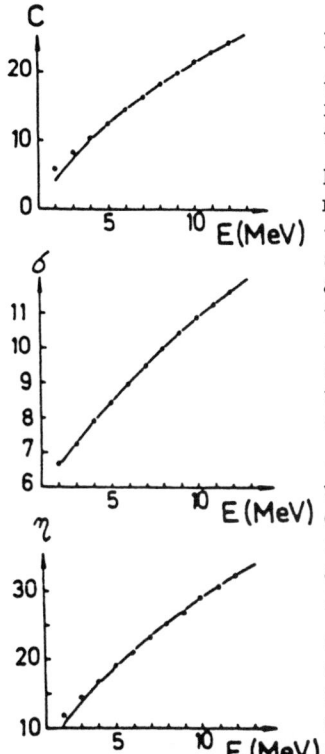

Fig.4. Energy dependence of the parameters C,σ,η in new formula for ^{244}Am. Solid curves present fits of approximate algebraic formulas for energy dependence.

In the second step, for each pair $\{\sigma^{(\kappa)}, C^{(\kappa)}\}$ we fit the spin shell-correction parameter η (i.e. ζ) in Eq.(14) to the GFM values of $y_H(J = J_{max}-4)$. Here J_{max} denotes the yrast angular momentum in the energy interval $(E, E+ \Delta E)$. Since the number of states with spins J_{max} and $J_{max}-1$ is small ($\lesssim 50$) and thus not appropriate for the statistical consideration underlying an approximate algebraic expression, these states are omitted from the fitting procedure. In the third step, among the sets of parameters $\{C^{(\kappa)}, \sigma^{(\kappa)}, \eta^{(\kappa)}\}$ we select the one which gives the distribution (14) closest to the GFM distribution for $\{y_H\}$, in the sense of smallest quadratic deviation. The values of C, σ and η for ^{244}Am, obtained in this way, are presented in Fig.4 in dependence on the excitation energy. Above the low-energy region rather good fit of these values can be obtained by functions $C = \ln\{\rho_B(E)\}, \eta = \ln\{\rho_B(E)\}$ where $\rho_B(E)$ has the same form as Bethe formula for total level density, with parameters analogous to (a, E_1) being fitted to the GFM results for C and η, respectively. On the other hand, the parameter σ can be fitted by the expression $\sigma = \alpha(E-\beta)^{1/4}$, which is the same form as derived for σ in the Fermi gas model[2]. It should be noted that the GFM values of σ which correspond to the calculated GFM spin distributions are higher than the values commonly used[2,17]. Similar effect has been observed in the previous combinatorial calculations[4].

From the GFM spin distribution follows a simple approximate prescription, as a rule of thumb, for the yrast angular momentum in each energy interval $(E, E+ \Delta E)$: J_{max} is the angular momentum which most closely satisfies the relation

$$y_H(J_{max}) = -2 .\qquad(17)$$

In Fig.5 we illustrate deviations of fits by three approximate algebraic expressions from the combinatorial GFM results for y_H for ^{244}Am. Solid curves labelled (a)

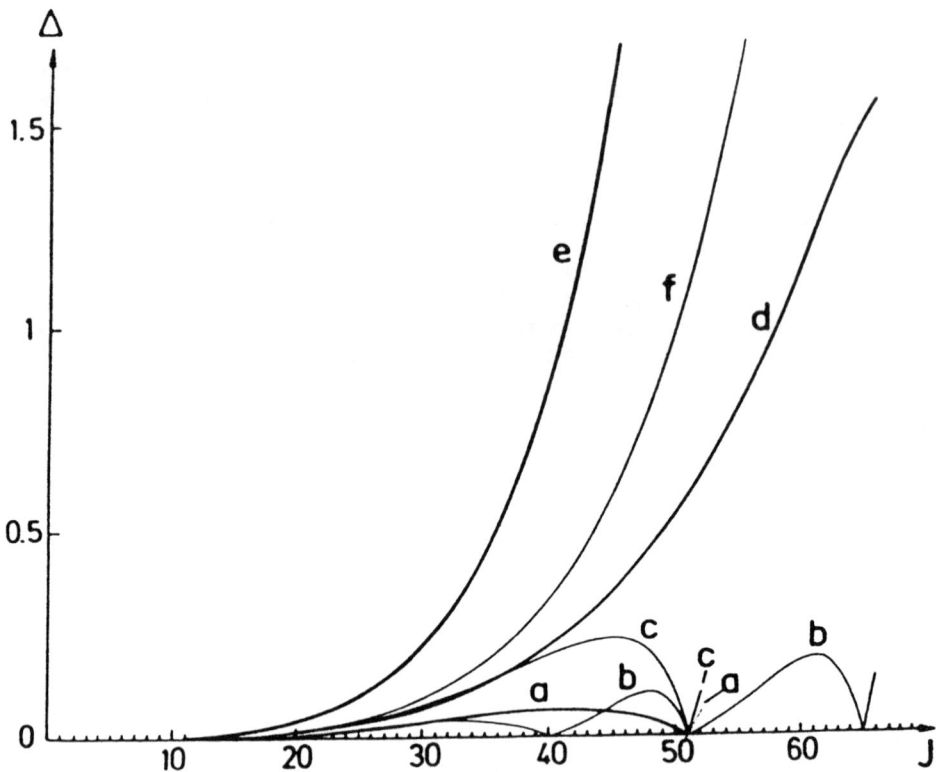

Fig.5. Deviations of the approximate algebraic formulas from the combinatorial GFM spin distributions for ^{244}Am. A measure of deviation is

$$\Delta y = \sum_{J=0}^{J_{max}-3} \left| y_H^F(J) - y_H^{GFM}(J) \right|$$

where y_H^F corresponds to each one of the approximate algebraic formulas (Bethe-, Lang- or the new formula) and y_H^{GFM} to the combinatorial GFM result. Solid curves present deviations for: new formula in the energy intervals 9-9.5 MeV (curve a) and 12-12.5 MeV (curve b); Lang formula in the energy intervals 9-9.5 MeV (curve c) and 12-12.5 MeV (curve d); Bethe formula in the energy intervals 9-9.5 MeV (curve e) and 12-12.5 MeV (curve f).

and (b) present deviations of the new formula (14) from GFM in the energy intervals 9 - 9.5 MeV and 12 - 12.5 MeV, respectively. For comparison, the corresponding deviations are presented for Bethe and Lang formulas. In the first case we use the expression (9) and in the second the expression obtained by inserting the Lang formula from Ref. 10 into Eqs.(6) and (5). Solid curves labelled (e) and (f) present the results of Bethe formula and those labelled (c) and (d) represent the Lang formula. In the case of the Lang formula, the parameters a and σ were determined from a fit to the 9-9.5 MeV interval, using the same fitting procedure as for the new formula. But for the 12-12.5 MeV interval, we kept the value of a fixed and determined σ anew, as in Ref. 11. As seen, the quality of fit in the higher energy region has significantly worsened. We note that a similar limitation of the Lang formula was shown in a previous investigation by Gilat.[11]

CONCLUSIONS

The first realistic calculations of level densities and spin distributions using the Gaussian polynomial generating function method are performed for ^{114}Cd and ^{244}Am. Contrary to the results of previous combinatorial calculations using the Hillman-Grover method, the GFM total level density can be well fitted by the Bethe formula. In this way, we obtain the microscopic values of the level density parameter, which are very close to the estimate $a = A/8$. The GFM calculations provide evidence that the factorization formula for spin-dependent level density has a microscopic foundation, but the Bethe spin distribution formula must be modified. We propose a new formula for the spin distribution function, with a sinh functional dependence in the exponent. The spin cutoff parameter σ and the spin shell-correction parameter η in the new formula are determined by fitting the new formula to the GFM spin distribution. The energy dependence of these microscopically calculated parameters can be described by using approximate algebraic expressions similar to those associated with the Bethe formula and the Fermi gas model.

The authors thank R.A. Meyer for encouragement, N. Sarapa for useful discussion, and N. and D. Paar for computational assistance.

REFERENCES

1. H. Bethe, Rev. Mod. Phys. $\underline{9}$, 69 (1937).
2. T. Ericson, Advan. Phys. $\underline{9}$, 425 (1960).
3. J.R. Huizenga and L.G. Moretto, Ann. Rev. Nucl. Sci. $\underline{22}$, 427 (1972).
4. M. Hillman and J.R. Grover, Phys. Rev. $\underline{185}$, 1303 (1969).
5. J.B. French and V.K.B. Kota, Ann. Rev. Nucl. Part. Sci. $\underline{32}$, 35 (1982).
6. C. Jacquemin and S.K. Kataria, Z. Phys. $\underline{A\ 324}$, 261 (1986).

7. Z. Pluhar and H.A. Weidenmüller, Phys. Rev. C 38, 1046 (1988).
8. B. Lauritzen, P. Arve and G. F. Bertsch, Phys. Rev. Lett. 61, 2835 (1988).
9. C. Bloch, Phys. Rev. 93, 1094 (1954).
10. D. W. Lang, Nucl.Phys. 77, 545 (1966).
11. J.Gilat, Phys. Rev. C 1, 1432 (1970).
12. L. Motz and E. Feinberg, Phys. Rev. 54, 1055 (1938).
13. D. K. Sunko and D. Svrtan, Phys. Rev. C 31, 1929 (1985); D.K.Sunko, Phys. Rev. C 33, 1811 (1986), C 35, 1936 (1987).
14. I.G. Macdonald, Symmetric Functions and Hall Polynomials (Oxford University Press, Oxford, 1979).
15. P. Ring and P. Schuck, The Nuclear Many Body Problem (Springer Verlag, New York, 1980), p. 449.
16. W. Dilg, W. Schantl, H. Vonach and M. Uhl, Nucl. Phys. A 217, 269 (1973).
17. T. von Egidy, H.H.Shmidt and A. N. Behkami, Nucl. Phys. A 454 , 109 (1986); Nucl. Phys. A 481, 189 (1988).
18. A. Gilbert and A. G. W. Cameron, Can. J. Phys. 43, 1446 (1965).
19. G. S. Hardy and S. Ramanujan, Proc. London Math. Soc. 17, 75 (1918).
20. G. Andrews, Theory of partitions (Addison Wesley, New York, 1976).

LEVEL DENSITIES, EXPECTATION VALUES AND STRENGTHS WITH INTERACTIONS: CONVOLUTION FORMS AND APPLICATIONS

V. K. B. Kota

Physical Research Laboratory, Ahmedabad 380009, India

ABSTRACT

Recently a theory is given [Ref. 4] for nuclear state densities generated by a Hamiltonian H which takes into account the two-body interaction V. It is based on the well known Central Limit Theorem which acts, even in presence of interactions, in many particle spectroscopic spaces, a physically significant group (tensorial) decomposition of the interaction, the recognition that distant configurations (defined, say, by oscillator energy $\hbar\omega$) interact weakly and they should be ignored or treated by special methods and the spreading widths of the strongly interacting configurations are nearly constant. The result being that the density is a convolution of the densities produced by the (modified) non-interacting particle part h of H and the interaction V respectively. The essential role of V is to Gaussian spread the h-density. The convolution form for state densities extends to expectation value and strength densities and they define smoothed expectation values and strengths respectively. Moreover the various h-densities themselves involve large number of convolutions and hence they can be constructed with ease. The above theory, together with the measured neutron resonance densities in rare-earth and actinide nuclei, is used to determine significant parameters of the nuclear Hamiltonian.

1. INTRODUCTION

Smoothed forms of state densities and their decompositions by conserved symmetries (say J) which give rise to level densities (a level being a set of degenerate states), by weakly broken symmetries and by configurations (partitioning of active particles among the single particle (s.p.) orbits) are important because sometimes they are mesurable, comparison with theory is then of obvious consequence and because they enter into calculations of reaction cross sections, transition strength densities, in astrophysical problems and also, as recently shown, in determining bounds[1] on the amount of breaking of fundamental symmetries (time reversal, parity, etc) in nucleon-nucleon force etc. Moreover, the smoothed forms of level densities, as they determine a smoothed form of the partition function, leads to smoothed forms for expectation values of operators (good examples being occupancies $< n_\alpha >^E$ and spin cut-off factors $< J_z^2 >^E$) and strengths (squares of matrix elements connecting two distinct eigenstates).

Extending the spectral distribution theory of French[2,3] to indefinitely large spectroscopic spaces, recently a theory for state densities[4] and expectation value densities[5] (expectation value multiplied by state density) is given and the later defines smoothed forms for expectation values; see Fig. 1 ahead. An important element of the theory is that here interactions enter via convolutions and the non-interacting part (NIP) itself is expressible in terms of convolutions. The theory is described briefly in Section 2 and tests of the convolution forms for various NIP densities is given in Section 3. Applications of the theory to level densities in neutron resonance domain of heavy nuclei, which will tell us something significant about the interaction will form Section 4. Convolution forms for strength densities are briefly mentioned in Section 5 along with some concluding remarks.

2. STATE AND EXPECTATION VALUE DENSITIES

Let us begin with the decomposition of the m-particle spectroscopic space generated by distributing the particles in spherical j-orbits (which generate N s.p. states) $j_1, j_2,...$ First the space is decomposed according to distant subspaces, that interact weakly, denoted by S (oscillator excitation $S\hbar\omega$ or a generalization of it). The S-subspaces are further decomposed into 'unitary' orbit (U.O.) configurations (unitary orbit is a set of spherical orbits) and further into spherical orbit configurations. The spherical and unitary orbits are denoted by $\alpha, \beta,...$ and $\mathbf{a}, \mathbf{b}...$ and the corresponding configurations by $\mathbf{m}$ and $[\mathbf{m}]$ respectively. Note that $\mathbf{m} = m_1, m_2,...m_\alpha,...$ where m_α is the number of particles in orbit α and similarly $[\mathbf{m}]$ is defined. Thus,

$$m \to \sum S \quad ; \quad S \to \sum [\mathbf{m}] \quad ; \quad [\mathbf{m}] \to \sum \mathbf{m} \qquad (1)$$

Similarly, given a (1+2)-body Hamiltonian $H = h + V$, the interaction V can be decomposed into two parts, the part $V_{t=0}$ that preserves S and the other ($V_{t\neq 0}$) that mixes S. The $V_{t=0}$ part is further decomposed into a part ($V^{[0]}$) that generates (as h) the spherical orbit configuration centroid energies and a remaining part (V) whose main function, due to action of Central Limit Theorem (CLT)[6], is to produce local Gaussian spreadings. The $V^{[0]}$ part is further decomposed into tensors (denoted by ν where ν takes values $\nu = 0, 1,$ and 2; see Ref. 2) with respect to $U(N)$ where N is defined in the beginning. With the observation that the spherical configuration variances (belonging to a given S) have nearly the same variance but slowly varying with S and neglecting the $V_{t\neq 0}$ and $V^{\nu=2[0]}$ parts (the smallness of the later is well verified in a large array of numerical calculations[4] and a schematic model describing the effects due to $V_{t\neq 0}$ part is

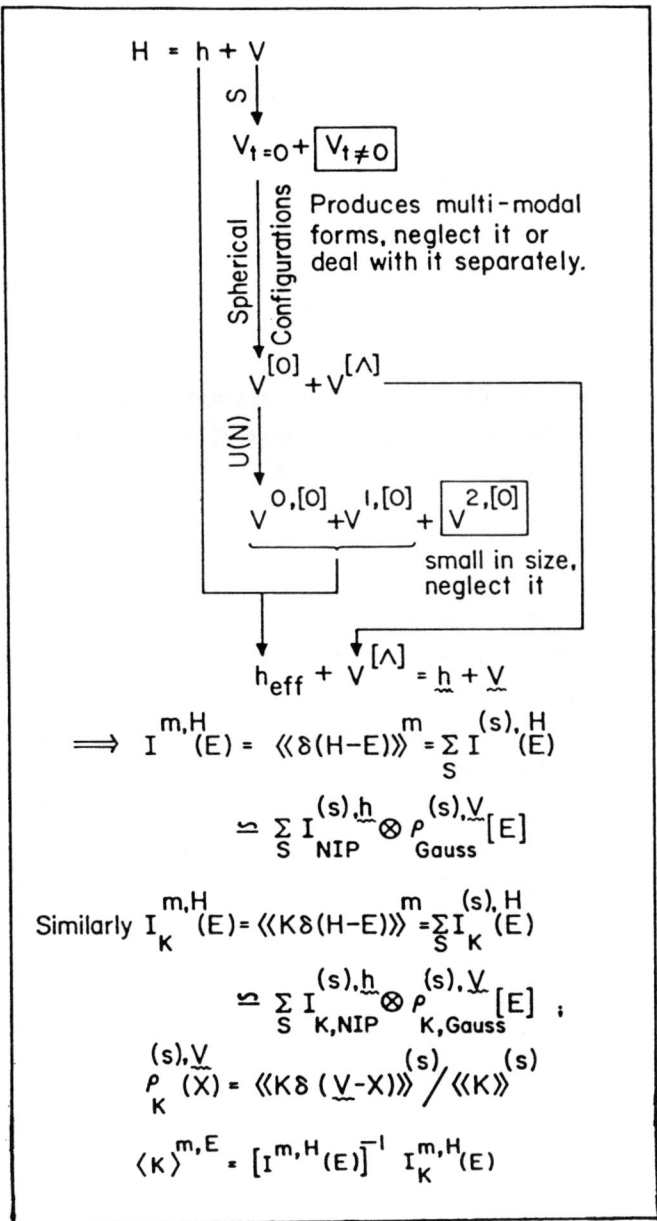

Fig. 1 The hierarchy of (unitary) decompositions of the Hamiltonian and the resulting state $(I(E))$ and expectation values $(I_K(E))$ densities. Note that $\underset{\sim}{V}$ in the figure stands for V in the text.

available[4]), as shown in Fig. 1, one obtains the remarkable result that fixed-S state density generated by H is a convolution of densities produced by h (with $V^{0,[0]}$ and $V^{1,[0]}$ added to the primary h; see Fig. 1) and **V** respectively and the spreading due to **V** is a Gaussian[2,6];

$$I^{m,H}(E) = \sum_S I^{S,h}_{NIP} \otimes \rho^{S,V}_{Gaussian}[E] \quad . \tag{2}$$

The symbol $\otimes$ in (2) denotes a convolution; $f \otimes g[x] = \int_{-\infty}^{+\infty} f(y)g(x-y)dy$.

The convolution form (2) for state densities extends to expectation value densities. For positive definite operators K (which have non-negative eigenvalues, examples are n_α, J_z^2, O^+O etc.) once can define expectation value densities $I^{m,H}_K(E)$ as the product of expectation value $< K >^E$ and the state density $I^{m,H}(E)$. The transcription of (2) for I^H_K is,

$$I^{m,H}_K(E) = \sum_S I^{S,h}_K \otimes \rho^{S,V}_K[E] \quad ; \quad < K >^E = I^{m,H}_K(E)/I^{m,H}(E) \tag{3}$$

Fig. 1 gives definitions of various quantities appearing in (3). The fact that the spreading function ρ^V_K, in general, is a Gaussian, the near constancy of spherical configuration variances (for fixed S) of $\rho^{m,V}_K$ and the near equality of the average variance of $\rho^{m,v}_K$ densities with that of $\rho^{m,V}_K$ that are essential for the validity of (3) are well verified in numerical calculations with n_α and J_z^2 operators[5] (see also the second reference in ref. 3 and ref. 7). It is worth remarking that the $\rho^V_{K:Gaussian}$ is defined by $K-V$ and $K-V^2$ correlations. Forms analagous but slightly different from (3) can be derived by first decomposing K, as it is done for V, and then dealing with the expectation value densities of the separate parts of K by parametric differentiation of I^H (i.e. dealing with I_α of $H_\alpha = H + \alpha K$). A final remark about (3) is that it follows from the marginal density of the bivariate strength density (see Sect. 5 ahead and ref. 1) which has a bivariate convolution form that is valid for a wide class of operators O (note that $K = O^+O$).

Technical feasibility of the theory given by (2) and (3) is due to the following: (1) it is easy to write down, via combinatorials and particle-hole theorems, the unitary decompositions of V (2) the m-particle variances of the spreading Gaussians ρ^V, ρ^V_K are calculable using the trace propagation methods[2,4,8] and (3) the NIP densities I^h and I^h_K, with K that are commonly encountered, involve convolutions of easily calculable functions. It is useful to remark that a proton-neutron version of the theory can be written down by simply replacing $m \to (m_p, m_n)$ everywhere and $U(N) \to U(N_p) \times U(N_n)$. The S and parity, as well as finer $V^{[0]}$, decompositions follow from an appropriate choice of U.O. Before turning to applications of the theory, convolution forms for NIP (h)-densities are given together with some tests.

3. CONVOLUTION FORMS FOR NIP DENSITIES

The NIP densities I^h and I^h_K can be constructed in huge spaces by exploiting the CLT. For simplicity we restrict K to be n_α or J^2_z and they suffice for the applications given in Sections 4, 5. Decomposing the m-particle space into U.O. configurations [m] (decomposing with respect to **m** gives rise to the exact counting method which is well known[9]), one immediately has

$$I^{m,h}(E) = \sum_{[m]} I^{[m]h}(E) \quad ; \quad I^{[m]h} = I^{m_a,h_a}_{N_a} \otimes I^{m_b,h_b}_{N_b} \otimes \ldots \quad (4)$$

where the number of s.p. states (N_a, N_b ...) in a given U.O. are shown explicitly. The result follows from the fact that h is a sum of h_a where h_a acts in U.O. a and the different parts of h commute. As cumulants (K_r) add under convolutions, using the lower order cumulants an Edgeworth representation[3,4] of $I^{[m],h}$ can be constructed and hence $I^{m,h}$. The m-particle cumulants for NIP densities defined over a single U.O. follow from standard methods[2]. Figure 2 shows that (4) gives a good theory for constructing I^h. Appropriate choice of U. O. automatically produce S and parity decomposition and for (p,n) systems the density follows by convoluting the separate proton and neutron densities. Going further, it is seen easily that the convolution form (4) extends to n_α and J^2_z expectation value densities. For example, the occupancy densities can be written as

$$I^{m,h}_{n_\alpha}(E) = \ll n_\alpha \delta(h-E) \gg^m = \sum_{[m]} I^{[m],h}_{n_\alpha}(E);$$

$$I^{[m],h}_{n_\alpha} = I^{m_a,h_a}_{N_a} \otimes I^{m_b,h_b}_{N_b} \otimes \ldots \otimes I^{m_t,h_t}_{N_t;n_\alpha} \otimes \ldots \quad (5)$$

In (5) it is assumed that the spherical orbit α belongs to the U.O. t. The important point here is that the moments $<n_\alpha(h^t)^p>$ (and hence cumulants) of $I^{h_t}_{n_\alpha}$ derive from I^{h_t} by parametric differentiation; $\mathcal{L}t_{\lambda \to 0} \partial/\partial\lambda < (h_t + \lambda n_\alpha)^p > = p < n_\alpha(h_t)^{p-1}>$. Thus an Edgeworth representation of $I^{[m],h}_{n_\alpha}$ can be constructed and hence $I^{m,h}_{n_\alpha}$. Spin cut-off densities $I^{m,h}_{J^2_z}$ follow from (5) by replacing $n_\alpha \to J^2_z(\alpha)$ and recognizing that (i) $<J^2_z> = \sum_\alpha <J^2_z(\alpha)>$ where $J^2_z(\alpha)$ acts only in the spherical orbit α and (ii) double parametric derivatives of h_t-moments produce the moments of $I^{h_t}_{J^2_z(\alpha)}$. In Figs. 3a,b results for occupancies and spin cut-off factors, calculated using the convolution form (5) and its extension for spin cut-off, are shown. The Edgeworth representation for $I^{[m],h}_{n_\alpha}$ and $I^{[m],h}_{J^2_z}$ is seen to give a good description of the exact results.

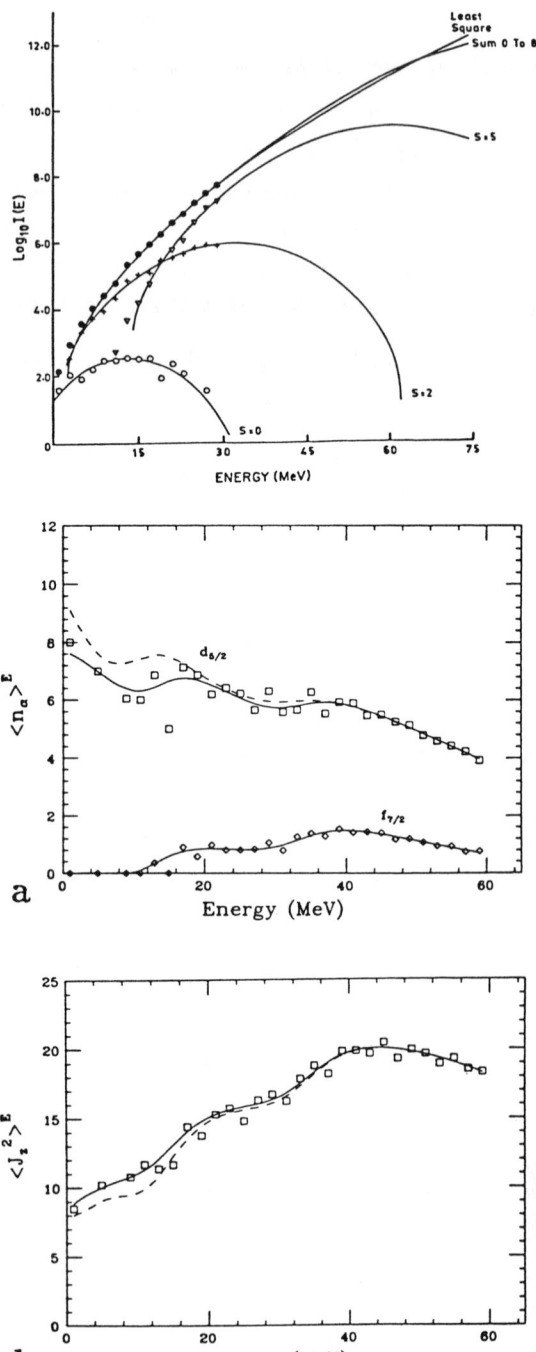

Fig.2 State densities for a system of 36 non-interacting identical particles. The major shells $(s, p, sd, fp,...)$ are chosen as unitary orbits and the s.p. energies derive from Seeger energies[9]. The total density and its S-decomposition (continuous curves) are in good agreement with the results from exact counting (stars, circles, diamonds etc.). The S-space densities use lower order cumulant corrections. The total density is fitted very well to the standard $\exp 2\sqrt{aE}$ form. The calculations are by J.F. Smith[10].

Fig.3(a) Occupancies $<n_\alpha>^E$ for a system of 24 non-interacting nucleons. The orbits $(s, p, sd, f_{7/2}, (p_{3/2}\ f_{5/2}\ p_{1/2}))$ are chosen as unitary orbits and the s.p. energies derive from Seeger energies[9]. Configurations, with positive parity, up to $4\hbar\omega$ excitation are included. Exact results with 2 MeV bin size (boxes and diamonds) are compared, for $d_{5/2}$ and $f_{7/2}$ occupancies, with the results obtained via occupancy densities (continuous curve) with Edgeworth representation for $I_K^{[m],h}$. The dashed curve represents results where $I_K^{[m],h}$ is constructed via the polynomials[11] defined by $I^{[m],h}$. The results obtained with the Edgeworth representation are in good agreement with exact results. (b) same as Fig. 3 but for spin cut-off factors $<J_z^2>^E$.

4. APPLICATIONS TO ^{168}Er AND ^{234}U

Level densities in neutron resonance domain provide high quality data for testing the theory outlined in Sect. 2,3. Results of some preliminary calculations are described below taking ^{168}Er and ^{234}U as examples; see also ref. 10). The state densities are calculated using (2) and they are decomposed according to angular momentum, thereby producing level densities, using spin cut-off factors and the later are calculated using (3). The proton-neutron formalism is used throughout. In order to produce numerical values for level densities, a set of s.p. orbits and an interaction are to be selected. Usually a few orbits around the Fermi surface will suffice if one is restricting to 6-8 MeV excitation in heavy nuclei (as is the case with the resonance domain). For ^{168}Er the spectroscopic space is chosen to be $(^2d_{5/2}\ ^1h_{11/2}\ ^3d_{3/2}\ ^3s_{1/2})^{m_p=10}$ and $(^1h_{9/2}\ ^1i_{13/2}\ ^3p_{1/2}\ ^2f_{5/2}\ ^3p_{1/2})^{m_n=10}$ for protons and neutrons respectively and similarly for ^{234}U they are respectively $(^1h_{9/2}\ ^1i_{13/2}\ ^2f_{7/2})^{m_p=10}$ and $(^2g_{9/2}\ ^1j_{15/2}\ ^1i_{11/2}\ ^3d_{5/2})^{m_n=16}$. The s.p. energies are taken to be the zero-deformed Nilsson energies[12]. The whole space is assumed to belong to a single S and hence there is only one spreading (interaction) variance. Choosing surface Delta Interaction with a strength G(MeV), the spreading variances (σ_V^2) are 183.6 G^2 and 294.2 G^2 for ^{168}Er and ^{234}U respectively. As a reference energy (to have an association between calculated and observed levels) a high enough level in experimental spectrum, below which the spectrum is complete (for all J^π), is chosen. Recent compilation by von Egidy et al.[13] and nuclear data sheets are used for this purpose. With E_0 and J_0 the energy and J-value of the reference energy and $N(E_0)$ the number of states up to and including E_0, we have $[N(E_0) - (2J_0 + 1)/2] = \int_{-\infty}^{E_0} I(E)dE$. For ^{168}Er and ^{234}U $(E_0, J_0^\pi, N(E_0))$ are (1.943 MeV, 12^+, 291) and (1.354 MeV, 6^+, 303) respectively. Starting with some value of G, the reference energy is fixed and then the resonance level density, in the observed spin range, is calculated. Spin cut-off factors are calculated using (3) with the approximation $\rho_{J_z^2}^V \simeq \rho^V$, (i.e.) the $J_z^2 - V$ and $J_z^2 - V^2$ correlations are neglected. Finally from the observed resonance densities, the value of G (more appropriately the interaction variance) is determined. The resonance energy, spins and density[14] and the deduced value of G are (7.771 MeV, (3, 4)$^+$, 250000 $\pm$ 25000 MeV^{-1}, 0.183 MeV) and (6.843 MeV, (2, 3)$^+$, 1818182 $\pm$ 165289 MeV^{-1}, 0.125 MeV) respectively for ^{166}Er and ^{234}U. It is remarkable that the values of G are consistent with the value of $G \sim A/25$ - $A/28$ derived from the spectroscopy of low-lying levels; see also ref. 1. Fig. 4a shows that $I(E, J)$ is extremely sensitive to the value of G (or spreading width σ_V); a similar behavior, in a slightly different context, is found in ref. 15. Thus, with spherical orbits, interactions play an important role in producing resonance densities. As shown in Fig. 4b exponential form of $I(E)$ is still maintained after convoluting I^h with ρ^V. It is worthwhile pointing out that complete data say up to 3 MeV excitation, when available, can be used to fix the value of G from low-energy data and then predict the resonance densities; at present this cannot be done reliably. Finally, more complete calculations that incorporate: $J^2 - V^2$ correlations, the renormalizations of h (surface delta interactions do not renormalize), larger spectroscopic spaces, more realistic interactions (with different pp, nn and pn strengths), the $V_{t\neq 0}$ effects, etc. are in progress.

5. STRENGTH DENSITIES AND CONCLUSIONS

Given a transition operator O, the bivariate strength density S^H is,

$$S_O^{m,m',H}(E,E') = I^{m',H}(E') |<m',E'|O|mE>|^2 I^{m,H}(E). \qquad (6)$$

Theory given by (2) leads to a bivariate convolution form for S_O^H (for O that behave as H); $S_O^H \to S_O^h \otimes \rho_{O:G}^V$ where the bivariate Gaussian $\rho_{O:G}^V$ is normalized to unit integral and the correlation coefficient is $\sim <O^+VOV>/<O^+O><VV>$. Strength densities enter into the problem of deriving bounds on the amount of breaking of time reversal invariance (TRI) in nucleon-nucleon interaction. With α the ratio of TRNI (time reversal non-invariance) to TRI, analysis of energy level fluctuations (from neutron resonance data) determine bound on a quantity (Λ) that is α^2 times a strength density. The theory for I^H an S^H is used to reduce the Λ-bound to α-bound. The final result is that $\alpha \lesssim 2 \times 10^{-3}$, a value which by some arguments is at the boundary of fundamental interest; see ref. 1 for details. Similar analysis for parity[16] and isospin[17] are possible.

In conclusion, new convolution forms are derived for state, expectation value and strength densities including interactions. They will lead to new ways of dealing with nuclear thermodynamics. They are shown to determine significant parameters of the nuclear Hamiltonian.

The author acknowledges collaborations with J. B. French, A. Pandey, J. F. Smith and S. Tomsovic.

REFERENCES

1. J.B. French, V.K.B. Kota, A. Pandey and S. Tomsovic, Ann. Phys. (N.Y.)**181**, 235 (1988).
2. F.S. Chang, J.B. French and T.H. Thio, Ann. Phys. (N.Y.) **66**, 137 (1971).
3. J.B. French and V.K.B. Kota, Ann. Rev. Nucl. Part. Sci. **32**, 35 (1982); V.K.B. Kota and K. Kar, Pramana J. Phys. **32**, 647 (1989).

4. J.B. French and V.K.B. Kota, Phys. Rev. Lett. **51**, 2183 (1983); Univ. of Rochester Report UR-1116 (1989).
5. J.B. French, V.K.B. Kota and J.F. Smith, Univ. of Rochester Report UR-1122 (1989).
6. K.K. Mon and J.B. French, Ann. Phys. (N.Y.) **95**, 90 (1975).
7. S. Sarkar and K. Kar, Phys. Rev. **C40**, 1826 (1989).
8. S.S.M. Wong, Statistical Nuclear Spectroscopy, (Oxford Univ. Press, N.Y., 1986).
9. M. Hillman and J.G. Grover, Phys. Rev. **185**, 1303 (1969).
10. J.F. Smith, PhD thesis (Univ. of Rochester, 1987), unpublished.
11. J.P. Draayer, J.B. French and S.S.M. Wong, Ann. Phys. (N.Y.) **106**, 472 (1977).
12. A. Bohr and B.R. Mottelson, Nuclear Structure (vol.2)(Benjamin, Reading, 1975).
13. T. Von Egidy, A.N. Behkami and H.H. Schmidt, Nucl. Phys. **A454**, 109 (1986); **A481**, 189 (1988).
14. S.F. Meghabghab, Neutron Cross Sections (vol 1, part B) (Academic Press, N.Y., 1984).
15. K. Sato and S. Yoshida, Z. Phys. **A333**, 141 (1989).
16. J.D. Bowman et al., (TRIPLE collaboration), to be published.
17. G.E. Mitchell et al., Phys. Rev. Lett. **61**, 1473 (1988); T. Guhr and H. Weidenmüller, Ann. Phys. (N.Y.) **199**, 412 (1990).

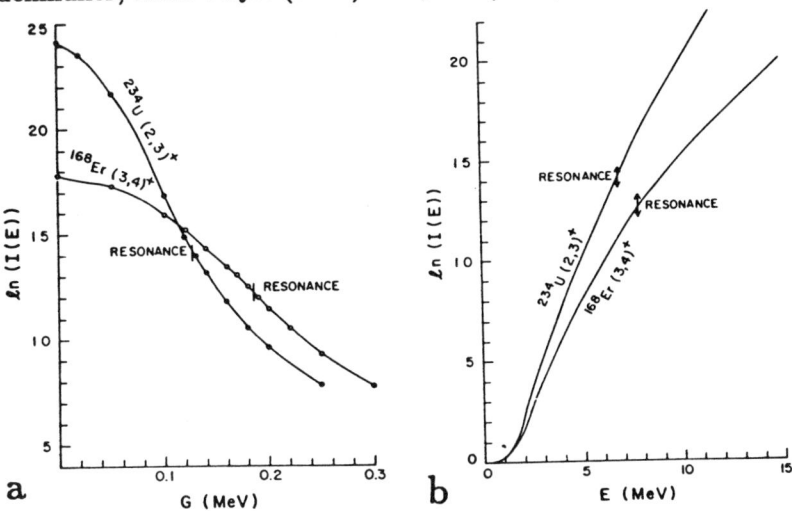

Fig. 4 (a) Level densities at neutron resonance energy for ^{168}Er and ^{234}U vs G, (b) Level density $I(E)$ vs energy E for ^{168}Er and ^{234}U.

High-Energy Neutrons, Gamma-Ray Production

NEUTRON-INDUCED GAMMA-RAY PRODUCTION

R.O. Nelson, D.M. Drake, R.C. Haight, C.M. Laymon, S.A. Wender
and P.G. Young,
Los Alamos National Laboratory
Los Alamos, NM 87545, USA

M. Drosg, A. Pavlik and H. Vonach
Institut für Radiumforschung und Kernphysik
University of Vienna, Vienna, Austria A-1090

D.C. Larson
Oak Ridge National Laboratory
Oak Ridge, TN 37831, USA

ABSTRACT

High resolution Ge detectors coupled with the WNR high-intensity, high-energy, pulsed neutron source at LAMPF recently have been used to measure a variety of reactions including (n,xnγ) for $1 \leq x \leq 11$, (n,nαγ), (n,npγ), etc. The reactions are identified by the known gamma-ray energies of prompt transitions between the low-lying states in the final nuclei. With our spallation neutron source cross section data are obtained at all neutron energies from a few MeV to over 200 MeV. Applications of the data range from assisting the interpretation of the planned Mars Observer mission to map the elemental composition of the martian surface, to providing data for nuclear model verification and understanding reaction mechanisms. For example, a study of the Pb(n,xnγ) reactions, for $2 \leq x \leq 11$, that populate the first excited states of the even Pb isotopes is underway. These data will be used to test preequilibrium and other reaction models.

INTRODUCTION

Recently a program to perform measurements of (n,xγ) cross sections for individual gamma-ray lines in the incident neutron energy range from a few MeV up to a few hundred MeV has been initiated at the Weapons Neutron Research (WNR) target area of the Los Alamos Meson Physics Facility (LAMPF). Results of a feasibility study on a natural Fe sample were reported in 1989[1], and preliminary results of (n,xnγ) measurements on Pb isotopes have been presented this year[2]. A series of similar measurements on structural materials (e.g. Fe, Cr, Ni, etc.) was performed at the Oak Ridge Electron Linear Accelerator (ORELA) facility by Larson, et al.[3] The ORELA data were measured up to 40 MeV incident neutron energy. The WNR neutron source produces intense neutron fluxes from below 1 MeV to over 200 MeV. This energy range and the high intensity available make the WNR a unique facility for fast-neutron research.

The good energy resolution of Ge detectors allows identification of individual gamma rays from final state nuclei produced in neutron induced reactions. When used in conjunction with the WNR neutron source, this technique can provide excitation function data for "exotic" reactions that involve the emission of several particles, such as (n,xnγ), which are difficult to measure by other means.

The data obtained in these measurements support an ongoing effort at the Los Alamos National Laboratory to extend the capabilities of nuclear models for improved reliability of calculations in the 20-100 MeV incident nucleon energy range[4]. Initial calculations using the current implementation of the GNASH computer code[5] are presented along with the data in this paper.

Our measurements also have application to the Mars Observer mission to map the elemental composition of the surface of Mars using a Ge gamma-ray spectrometer. In order to accurately interpret the Mars Observer data a good knowledge of neutron-induced nonelastic reaction cross sections is needed.

Following a brief description of the experiments and analysis, new results will be presented for ^{207}Pb and ^{56}Fe, and the applications of the data will be discussed.

EXPERIMENT

The WNR neutron source has been described in detail elsewhere[6]. The pulsed proton beam typically has a macropulse repetition rate of 40 Hz. The macropulses are usually 725 μs long and are comprised of narrow (<0.5 ns) micropulses with ~10^8 protons/micropulse. The spacing between micropulses may be varied, but is typically 1.8 μs. The neutron production target is a small tungsten cylinder. Extensive shielding and collimation produce a well-defined neutron beam of uniform intensity at the sample. The earlier experiments which will be described were performed on an 18 m flight path at 15° with respect to the incident proton beam. Currently the measurements are carried out on a 41 m flight path at 30° with respect to the incident proton beam. The neutron flux is monitored during experiments with a fission ionization chamber[7] containing separate foils of ^{235}U and ^{238}U. The fluence measurement is based on either the ^{235}U(n,f) or ^{238}U(n,f) cross section, both of which are well known below 20 MeV. From 20 to 200 MeV the ^{235,238}U(n,f) cross sections have been measured relative to the H(n,p) reaction at our laboratory with an accuracy of approximately 10%.[8] The flux on the 18 m, 15° flight path is shown in fig. 1.

Two Ge detectors positioned at 90° and 125° provide some limited angular distribution information. The efficiencies of the detectors (relative to a 7.6 cm diameter x 7.6 cm long NaI detector) are in the range from 12 to 30%. The

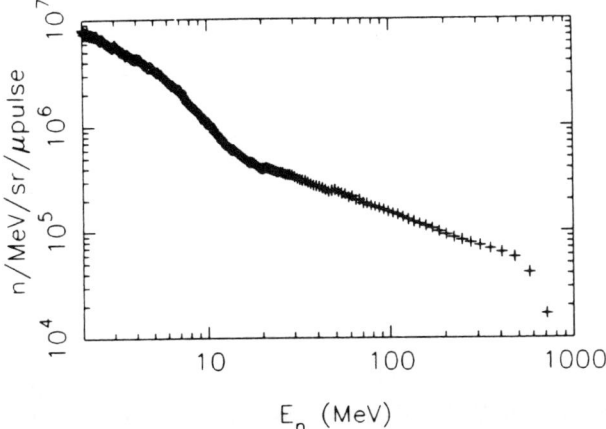

Fig. 1. The neutron flux measured by the ^{238}U(n,f) reaction on the 18 m, 15 degree flight path. The flux is in units of neutrons/MeV/steradian/micropulse. The beam solid angle is 33 μsr and typical beam conditions give an average of 16 k micropulses/s.

energy resolution obtained varied from 1.8 to 3.0 keV full width at half maximum at 846 keV. The absolute detector efficiencies were determined *in situ* by measuring the count rate from calibrated radioactive sources.

To reduce the background from neutrons scattered from the sample into the detectors, a 2 cm thick ^{6}LiD absorber was mounted in front of the detectors. The Ge detectors were shielded and collimated with a 5 cm thick layer of tungsten powder sealed in an annulus constructed from an inner tube of Plexiglass and an outer tube of 1 mm thick steel.

The data are stored in 2-dimensional (2D) arrays, with 4096 channels of pulse height (PH) versus 512 channels of time-of-flight (TOF). The data are also stored event by event.

DATA ANALYSIS

The event-by-event data allow us to take advantage of the full resolution available from our data acquisition electronics by re-sorting the data into larger 2D arrays if necessary. Usually the 2D data acquired during the runs are sufficient.

The analysis is performed in four steps. First the data are binned into selected neutron energy bins depending upon the neutron energy resolution desired and on the statistics available. During the binning process we correct for any timing variation with pulse height and for the time-dependent dead time of the time-to-digital converter. If desired, the data may be binned in pulse height. Second, the yields of the peaks of interest are extracted by fitting an appropriate (usually linear) background to the regions adjacent to the peaks and obtaining the net sums in the peaks. Third, the neutron fluence is determined from the fission yield in neutron energy bins identical to the Ge detector data. Finally, the fluence, sample thickness, detector efficiency and solid angle, dead time and yields are combined to calculate the absolute cross sections. Corrections are made at this point for multiple scattering and attenuation effects. To handle the large amounts of data obtained in these experiments the analysis procedure is automated as much as possible.

RESULTS AND DISCUSSION

Preliminary data on the ^{204,206,207,208}Pb(n,xγ) reactions were obtained in 1989 on the 18 m flight path. Reactions on ^{208}Pb were observed up to (n,11n) which has a threshold of 78 MeV. This year more extensive data were acquired on ^{207}Pb on the 41 m flight path. The major goal of this experiment is to determine the cross sections for the (n,xn) reactions, where $2 \leq x \leq 11$, populating the first excited states of the even Pb isotopes from ^{206}Pb to ^{198}Pb. These cross sections may serve as a benchmark for testing nuclear reaction models in the medium energy regime.

A spectrum taken on the 41 m flight path with a sample isotopically enriched to 92.78% in ^{207}Pb is shown in fig. 2. The reactions populating the first

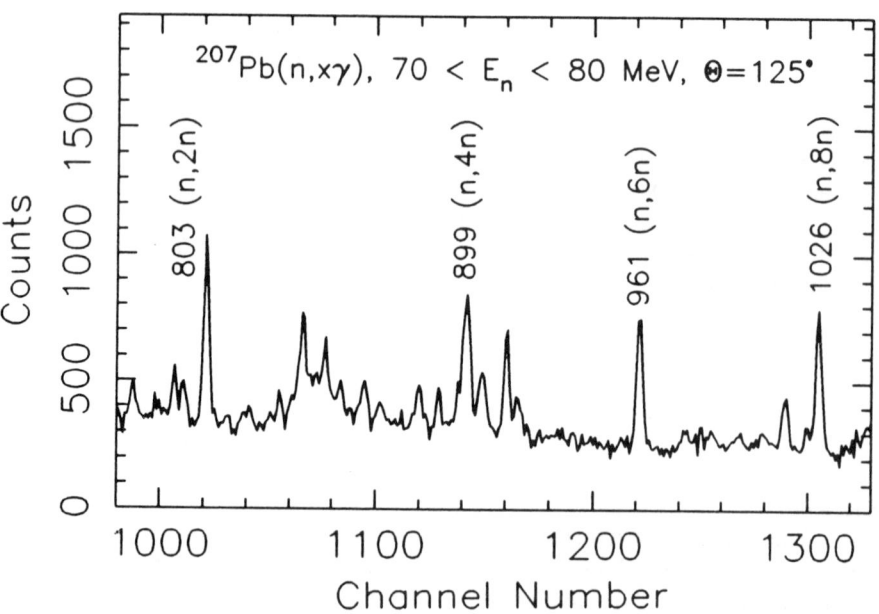

Fig. 2. A pulse height spectrum for ^{207}Pb(n,xγ). This spectrum is for incident neutron energies from 70 to 80 MeV. The gamma-ray energies and the associated reactions are listed above the peaks.

excited states of the even Pb isotopes and the gamma-ray energies are indicated on the figure. This spectrum was projected from our 2D array of TOF versus PH for incident neutrons in the energy range from 70 to 80 MeV. The data were acquired in 40 hours of running.

If we assume the a_4 coefficient in the usual Legendre polynomial expansion of the angular distribution is small, the angle-integrated reaction cross section may be approximated by $4\pi\sigma(125°)$. This reaction cross section for the ^{207}Pb(n,2nγ) reaction populating the first excited state of ^{206}Pb is shown as a function of energy from 3 to 200 MeV in fig. 3.

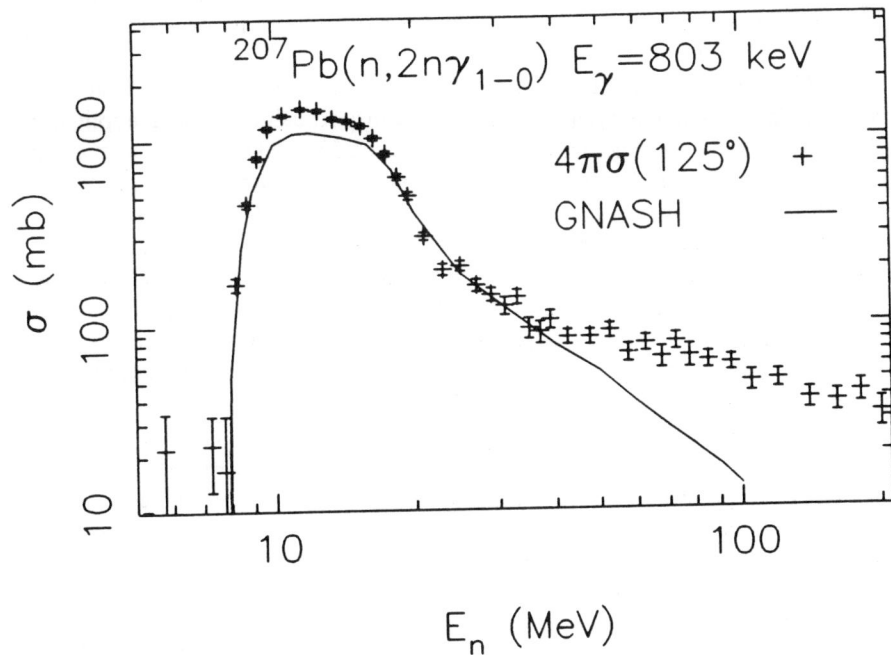

Fig. 3. The excitation function for ^{207}Pb(n,2nγ) and the GNASH model calculation. The contribution of the 2.200 MeV isomer has been subtracted (see text).

At an excitation energy of 2.2 MeV in ^{206}Pb there is an isomeric state with a lifetime of 126 μs. Because of its long lifetime the decay of this state appears as a time-random background on the 1.8 μs time scale of our measurement. In order to make a comparison with the GNASH model calculation we have subtracted the calculated contribution from the decay of this state from the calculated reaction cross section. Correspondingly, a background, constant in time, has been subtracted from the data.

A detailed description of the initial GNASH model calculations for Pb and ^{56}Fe is given by Young et al.[9] These preliminary results indicate that the Pb data will provide a good test of the different level density parameterizations used in the code. In addition, the observation of individual gamma-ray transitions provides a more stringent test of the code because different levels (with different spins) contributing to a transition may be investigated.

The GNASH model calculation predicts the energy dependence well up to 40 MeV, although it somewhat underestimates the cross section in the 10 to 20 MeV range. The compound-nucleus spin population used in the code is most likely inadequate above 40 MeV where non-equilibrium reactions are occurring. We plan to modify GNASH to improve the high energy predictions.

The lighter even-mass Pb isotopes all possess isomeric states with lifetimes in the range from 40 to 500 ns, and thus require a more sophisticated correction procedure. In each case at least one of these isomeric states is strongly populated by the (n,xn) reaction.

Gamma-ray production data on Fe are of interest both for the Mars Observer mission and as input for radiation transport calculations in applications where Fe is used as a structural material or as shielding. Data have been acquired on an isotopically enriched ^{56}Fe sample on both flight paths. Absolute cross sections calculated from data taken on the 18 m flight path are shown in fig. 4 for the ^{56}Fe(n,n'γ) reaction for the first excited state of ^{56}Fe, and in fig. 5 for the ^{56}Fe(n,2nγ) reaction for the second excited state of ^{55}Fe with the decay to the ground state. These data are in good agreement with the results of Larson[3], and extend the energy range from 40 to 200 MeV. The GNASH model calculations are in good agreement up to 25 MeV. As mentioned above we plan to extend the model to improve its higher energy predictions.

One goal of the Mars Observer mission is to map the elemental composition of the surface of Mars using a Ge gamma-ray spectrometer. The majority of the gamma-rays from the martian surface are expected to be produced by neutron-induced reactions. The neutrons are produced in much the same manner as in our source, by the spallation of cosmic-ray protons in the martian crust. Analysis of the Mars Observer data to obtain elemental abundances requires a knowledge of the energy and composition of the incident cosmic-ray flux, the neutron spectrum produced at the planet's surface, and the cross sections for gamma-ray production as a function of neutron energy up to energies as high as a few hundred MeV. Elements which have been measured to determine cross sections for the stronger transitions include: B, C, N, O, Mg, Al, Si, S, Ca, Ti, Fe, Cr, and Mn.

Identification of the nuclear reactions that contribute to the observed yield of individual lines, and improving the data base on gamma-ray production will make more accurate interpretation of the Mars Observer data possible. Our higher energy data should also provide benchmarks for improving the predictive capabilities of nuclear models in the medium energy range, as well as serving as a database for improving radiation transport calculations.

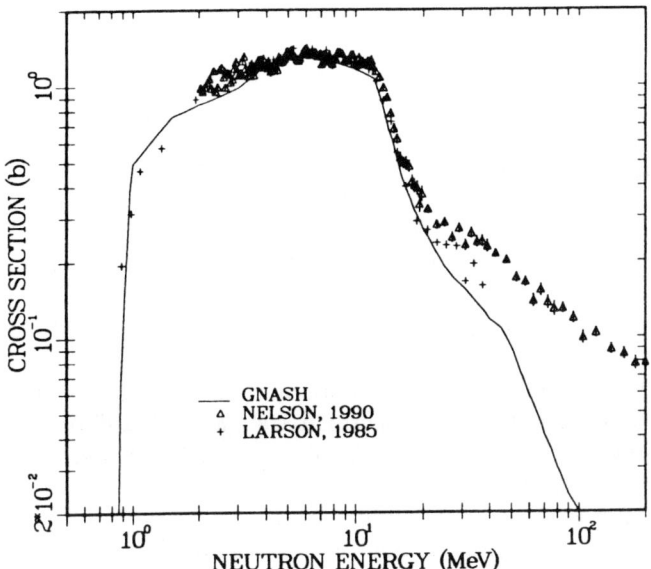

Fig. 4. The excitation function for ^{56}Fe(n,n'γ) populating the 846 keV first excited state of ^{56}Fe. The present data are compared with the data of Larson[3] and a GNASH model calculation.

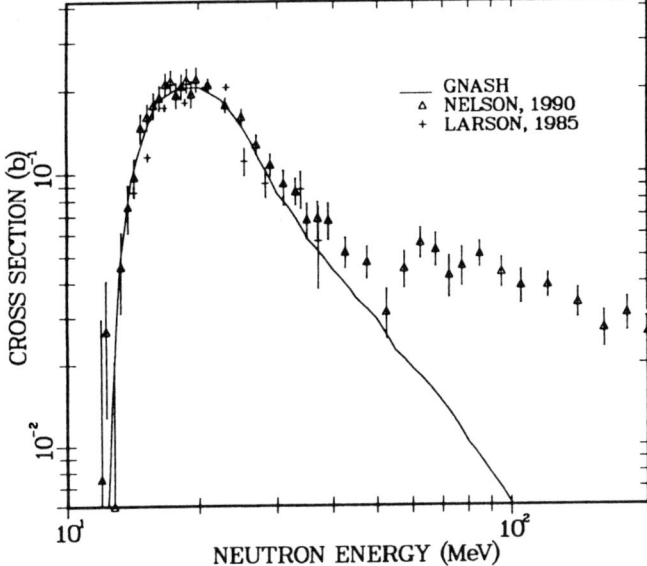

Fig. 5. The excitation function for the 931 keV gamma ray from the ^{56}Fe(n,2nγ) reaction populating the second excited state of ^{55}Fe. The present data are compared with the data of Larson[3] and a GNASH model calculation.

REFERENCES

1. R.O. Nelson, S.A. Wender, and G.L. Morgan, Bull. Am. Phys. Soc. 34, 1223 (1989).
2. R.C. Haight, et al., Bull. Am. Phys. Soc. 35, 1038 (1990).
3. D.C. Larson in: Nuclear Data for Basic and Applied Science, eds. P.G. Young, et al. (Gordon and Breach, 1985) Vol. 1, p. 71.
4. E. Arthur, P. Young, and C. Kalbach in: Proc. Int. Conf. on Nucl. Data for Science and Tech., Mito, Japan (1988), ed. S. Igarasi (Saikon Publishing Co., Ltd., Tokyo, 1988)
   E.D. Arthur, et al., International Atomic Energy Agency report IAEA-TECDOC-483, 278 (1988).
5. E.D. Arthur, The GNASH Preequilibrium-Statistical Nuclear Model Code, Los Alamos National Laboratory Report LA-UR-88-1753 (1988).
6. P.W. Lisowski, C.D. Bowman, S.A. Wender, and G.J. Russell, Nucl. Sci. and Eng. to be published.
7. S.A. Wender, et al., A Fission Ionization Detector for Neutron Flux Measurements at a Spallation Source, Los Alamos National Laboratory Report LA-UR-90-3399 (1990).
8. A. Gavron and P.W. Lisowski, private communication (1990).
9. P.G. Young, et al., Calculation of $(n,x\gamma)$ Cross Sections Between Threshold and 100 MeV for Fe and Pb Isotopes: Comparisons with Experimental Data, Los Alamos National Laboratory Report LA-UR-90-2129 (1990).

HARD PHOTON PRODUCTION IN NUCLEAR REACTIONS AT LOW AND INTERMEDIATE ENERGIES

Yu-Kun Ho*

Center of Theoretical Physics, China Centre of Advanced Science and Technology
(World Laboratory), Beijing, China
Nuclear Science Department, Fudan University, Shanghai, China

ABSTRACT

This paper presents a general formalism for hard photon production mechanisms in nuclear reactions at intermediate and low energies with projectiles ranging from nucleon to heavy-ion. The mechanisms can be divided into three categories, namely the direct bremsstrahlung radiations, doorway mechanisms and statistical process. For each mechanism, its physical significance, experimental characteristics identifing the mechanism, calculations and comparison with experiments, as well as problems left over are outlined.

1. INTRODUCTION

1.1 Hard photon production in nuclear reaction has attracted attention, and is generally considered as a sensitive and unambiguous probe to study the reaction dynamics, especially of the first phase. This is because: 1) the electromagnetic interaction, which operates the photon emission process, is well known; 2) there is no notable final state interaction, i.e. the emitted photons are not as seriously affected by the residual nuclei as baryons; 3) hard photons are mostly produced in the early stage of reaction, the nonstatistical process. There is not definite energy limit refering to hard photon, it depends on the kinds of reaction. In low energy (n, γ) reactions (thermal, resonance or MeV neutrons), hard photon refers to an energy range around the neutron binding energy. Whereas it extends to a region of $10 - 100 MeV$ for heavy-ion (HI) reactions at intermediate energies. In last thirty years, a geart interest has been given to study the (n,γ) reaction mechanisms These investigations have played an important role in developing the advanced nuclear physics. Recently, high-energy photons become one of the most attractive subjects in HI reactions. It is interested to indicate that there exist pronounced similarities between the mechanisms for photon productions in low and intermediate energies. This paper will review those connections and present an unified formalism discribing the mechanisms.

1.2 characteristics of experimental results

One of the most distinct features for nucleon and light-ion induced capture reactions is that the gamma-ray spectra display a coexistence of two components: the statistical component resulting from cascade of complicated configurations, and nonstatistical contribution coming mainly from transitions of simple particle or phonon configurations in the early stage of reaction and showing correlations with other channels, such as the spectroscopic tactors[1-3].

As for the main characteristics of photon emission deduced in HI reactions at intermediate energies, the reader is refered to the recent review papers[4-6]. We brief the distinct features in the following.

-Photons with energies up to three-to-four times of the incident energy per nucleon were observed. No distinct upper limit was established. Photon production cross section ($E_\gamma > 30 MeV$) is in the order of mb.

-The gamma-ray spectrum consists generally of three parts: below $10 MeV$, it shows a exponential shape with inverse slope $T \approx 1 - 2 MeV$; within 10 to $25 MeV$, the contribution from GDR is prominent; above $20 - 30 MeV$, it shows again a distinct exponential behaviors with inverse slope $E_0 \approx 10 MeV$. The magnitude of E_0 depends on the incident energy, TKEL (total kinetic energy loss) in exclusive measurements, angle of observation, as well as weakly on the target-projectile masses.

-The shapes of angular distribution are almost independent of the projectile-target masses at same beam velocities. The source velocities close to the half beam velocities, i.e. the nucleon-nucleon c.m. velocity in the first collisions. The shape of the c.m. angular distribution is almost isotropic with small amount of dipole component, which increases with beam energy. Those features close to that of a nucleon-nucleon collision.

-The total photon production cross section can be factorized as

$$\sigma_\gamma = \sigma_R N_{np} P_\gamma \tag{1}$$

where, σ_R is the total reaction cross section, N_{np} is the average number of $n-p$ collision, P_γ is the photon production probability in a single $n-p$ collision.

1.3 A veriety of model mechanisms have been proposed to explain the features of photon production in reactions at low and intermediate energies. These mechanisms can be classified into three categories according to the characteristic time scales, which measure the complexities of reaction processes. Direct bremsstrahlung radiations take place in the very initial phase of reaction, $10^{-22} - 10^{-23} sec$. The direct bremsstrahlung radiations can be elaborated to include: coherent nucleus-nucleus potential bremsstrahlung; incoherent nucleon-nucleus potential bremsstrahlung, and incoherent nucleon-nucleon bremsstrahlung, relying on the beam energy. The incident energy decides the extent of excitation of intrinsic degrees of freedom due to dissipation of relative motion. The typical time scale for doorway mechanism to operate is $10^{-19} - 10^{-20} sec$. In this process, the intermediate states (doorway states, e.g. GDR) are populated in a prompt time scale by projectile bombardement, following its decay with emission of gamma-ray. The projectile or substantial part of it can be regarded as a single entity at lower incident energies, or as incoherent mixing of the constituent nucleons at higher energies. The cascade radiations from reaction fragments reaching thermal equlibrium would last for $10^{-16} - 10^{-18} sec$. This is the typical life time of compound nuclei, called statistical process. These radiations came from transitions of a large amount of compicated configurations.

In the following, we will outline each mechanism, including its physical significance and representation, characteristics identifing the mechanism, calculations and comparison with experimental data, as well as problems left over.

2. DIRECT BREMSSTRAHLUNG RADIATIONS

2.1 Conherent nucleus-nucleus potential bremsstrahlung.

In this process, the projectile or its substantial part as a single entity is scattered in its counterpart potential field and captured into a low-lying orbit with an emission of photon. The potential field, including both the nuclear and Coulomb interactions, could be a complex well. The imaginary part, simulating the absorption or deceleration of projectiles, is proportional to the friction coefficient[7]. The most distinguished feature of this process is the structures, exhibited in the energy spectra or excitation functions, which show correlations with other reaction channels, such as the spectroscopic factors. Since the effective charge of projectile-target system for an electric transition of multipole order L is

$$e_f = \mu^L \left(\frac{Z_1}{M_1^L} + (-1)^L \frac{Z_2}{M_2^L} \right) \qquad (2)$$

μ is the reduced mass, M_1, Z_1, M_2, and Z_2 are the mass and charge of the projectile and target, respectively. Thus for HI reaction under the conditions $M_1 \approx M_2$ and $Z_1 \approx Z_2$, this process can only radiate quadrupole photons.

The total Hamiltonian of the projectile-target system can be written as

$$H = \frac{P_R^2}{2\mu} + U(R) + H_2(\alpha) + H_3(x_i) + H_{12}(R,\alpha) + H_{13}(R,x_i) + H_{23}(\alpha,x_i) \qquad (3)$$

where, $P_R^2/2M$ and $U(R)$ are the kinetic and potential energy of relative motion, $H_2(\alpha)$ and $H_3(x_i)$ design the collective and particle Hamiltonian, $H_{12}(R,\alpha)$, $H_{13}(R,x_i)$ and $H_{23}(\alpha,x_i)$ designate the relative-collective, relative-particle and collective-particle coupling, respectively. For collective potential radiation, only the first two terms of eq.(3) play role.

The following approaches may be used to calculate the collective bremsstrahlung cross sections. The potential capture model, which was originally developed to study neutron potential capture, can be extended to study collective radiation by treating the projectile or its substantial part as an entity[8]. In the TDHF model the collective radiation is attributed to the coherent space-time radiation caused by one-body current[9]. In addition, the friction model[10-12] also attracted much attention.

As for application of this mechanism, perhaps the most simple and successful example is the nucleon potential capture model, which works still quite well at relatively high energies for transitions between unbound states, such as the noted $^{11}B(p,\gamma_{19})$ reactions[13,14]. Correlations are also found in low energy light-ion induced reactions[15], as well as a part of HI reaction experiments[16,17]. In addition, HI reaction near Coulomb threshould is tested to decay preferably to final states with large prolate deformation.

In quantitative comparison with experiments, collective bremsstrahlung model is able to reproduce the data at relatively low energies for nucleon or light-ion induced reactions. This model is an order of magnitude too small to explain the data for HI reactions at higher energies of $20 - 80 MeV^9$.

2.2 Incoherent nucleon-nucleus potential bremsstrahlung

In the incident energy range of $(E - V_c)/A \approx 6 - 20 MeV$, and mostly for light-ion projectiles, the direct bremsstrahlung radiations are appeared as incoherent nucleon-nucleus potential bremsstrahlung, where the constituent nucleons of projectile, moving in the target field, undergo incoherent potential transitions with emission of photons. In this energy region, the projectile becomes rather loose and the particle degree of freedom of the projectile begins to be excited by a dissipation of the relative motion $(H_{13}(R, x_i)$ in eq.(3)). V_c is the Coulumb barrier.

Niita and his collaborators[9], based upon the TDHF model, studied the electromagnetic transitions between occupied and unoccupied time dependent single-particle states during the collision. By applying to $^{16}O + ^{16}O$ collision at $80MeV/n$, it was found that this term may account for only 10 per cent of the measured data. It is also available to calculate this contribution based upon the preequilibrium model.

2.3 Incoherent nucleon-nucleon collision bremsstrahlung

As the incident energies exceed $20-30 MeV$, the direct radiations are dominanted by the nucleon-nucleon collision bremsstrahlung due to the drastic dissipation of relative motion in the participated zone ($H_{13}(R, x_i)$ in eq.(3)). In this case, the intrinsic degrees of freedom are fully excited. The criterions for this model to work may be attributed to that: the mean free-path of a moving nucleon in nucleus is less than the nuclear dimension; and its de Broglie wave length is less than the mean space occupied by a nucleon in nucleus. Since proton-proton collision can only radiate dipole photons (see eq. (2)), high energy photons are chiefly produced by neutron-proton collisions in the early stage of reactions.

This model contains two major ingredients in calculations. One is the elementary photon production cross section in a proton-neutron collision in nuclear medium ($n-p-\gamma$). A variety of the $n-p-\gamma$ models were used in literatures, such as the semiclassical model with nuclear-medium correction[18-19]; quantum approach including the charged pion and meson exchange current contributions[9,20-23], as well as the phenomenological expression[24].

The second problem is to study the collision history. The extensively adopted is to simulate the BUU (Boltzmann-Uehling-Uhlenback) equation[9,18,20,23] to find the time evolution of Wigner function in phase-space $f(\vec{r},\vec{k},t)$, and then to calculate the photon production cross sections. There also exist a number of simplified models to find the collision history. In the fermi gas model[22], both the target and projectile are treated as Fermi spheres in configuration and momentum space. Only the first collisions are taken into account. Remington et al.[25] treat the incident nucleons as excitons, their decay behaviors are described by a Boltzmann master equation.

The nucleon-nucleon collision model is quite successful in reproducing the experimental data ($E_\gamma > 20 MeV$) for HI reaction at $E/A = 20-60 MeV/n$ or (p,γ) reactions up to $140 MeV$[4-6]. In addition, this kind of study may also provide knowledges on particle distribution in phase-space in the early stage of reactions before and after photon emissions.

Some problems are left over for further study. Pronounced discrepancy between the predicted and measured data has been found at higher incident energies[4], e.g. $E/A \geq 80 MeV/n$. It is speculated that other unknown mechanism is playing role in this case. The violation of the sum-rule for the $n-p-\gamma$ model, where the radiations are incoherently summed up, is also needed for clarification[26]

3. DOORWAY MECHANISM

2.1 Doorway mechanism with coherent nucleus-phonon interaction.

This process mostly takes place at relatively lower incident energies, say, $(E - U_c)/A < 6 - 7 MeV$. The projectile or substantional part of it, interacting as a single entity with the target via $H_{12}(R, \alpha)$ in eq.(3), is captured into a low-lying orbit with an excitation of intermediate phonon state (e.g. GDR or GQR) in a prompt time scale ($10^{-19} - 10^{-20} s$). Then the intermediate state may decay with an emission of γ-ray. The simplest example of this process is the semidirect capture model in nucleon or light-ion induced reactions27. Both the nuclear and Coulumb interactions can contribute to the excitation of intermediate states. In contrast to nuclear interaction, the Coulumb field can excite both the $IVGR$ and $ISGR$ states with equal efficiency.

To calculate contributions from this model, one may use the extended semidirect model. The other approach is to calculate the nucleus-phonon interaction matrix elements induced by one body current based upon the TDHF formalism.

As a comparison with experiments, the semidirect model is proving to work very well, at least, for nucleon and light-ion induced capture reactions[15,16,27].

3.2 Doorway mechanism with incoherent nucleon-phonon interaction

At a little higher incident energies, say, $20 MeV > (E-U_c)/A > 6-7 MeV$, the light ion projectile becomes loose due to the dissipation of the relative motion ($H_{13}(R, x_i)$ in eq.(3)). The intermediate phonon state is excited by individual constituent nucleon-phonon interaction at a prompt time scale ($H_{23}(x_i, \alpha)$ in eq.(3)), following a decay of the phonon state with a emmision of photon.

Reffo et al.[28], based upon the preequilibrium formalism, applied this model to calculate the photon production cross sections in ^3He and $\alpha +^{154} Sm$ reactions at $27 MeV$. They were able to claim that a improvement was achieved, specifically for E_γ in the range of $10-25 MeV$. One may also count the incoherent excitation contribution by calculate the nucleon-phonon interaction matrix elements between occupied and unoccupied single-particle states in TDHF framework.

4. COMPOUND NUCLEUS PROCESS (CN)

In this process, photons are emitted by fragments or residual nuclei, reaching thermal equilibrium with typical life time of $10^{-16} - 10^{-18} sec.$, in the exit channels. This is Ho/Hard Photon production...

a stochastic multistep emission process (incoherent cascade), and can be handled with the statistical model.

A crucial question in the statistical model is to calculate the photon strength function, which is related to the photon absorption cross section by the reciprocal theorem. According to the latest viewpoint[16,29,30], the photon absorption in the energy range concerned consists of two parts. One is the GDR contribution, which mainly accounts for photons of $5 - 20 MeV$. The second is the quasideutron process[30], which play an important role at photon energies of $20 - 60 MeV$

The CN contributions may also be calculated with the fireball model[21], where the incoherent multi-step $n - p - \gamma$ collisions are taken into account in the equilibrated hot-spot (participated zone).

In comparison with experiments, CN model is proved being able to well account for the low energy data ($E/A < 6 - 7 MeV$). In the energy region of $6 - 20 MeV$, CN model also may reproduce the data, provided both the GDR and quasideutron components are taken into account in the photon strength function[4,5,6,16,29]. At still higher energies, say, $E/A \geq 44 MeV/n$, CN model severally underestimate the measured high-energy photons[5]. Also, this model can not explain the dipole component observed in photon angular distributions at E/A above $30 MeV/n$.

Many problems are left over for further clarification. It has long been argued that the whole process of hot nuclei, formed in HI reactions at intermediate energies, can not be divided into two distinct stages: formation and decay[32]. Also, the relationship between CN model and $n - p - \gamma$ model is an interesting subject. Provided the CN model takes into account both the GDR and quasideutron mechanisms, whereas the $n - p - \gamma$ model is extended to including multi-step collisions, then it seems that these two models would overlap each other.

References

* Mailing address: Nuclear Science Dep[artment, Fudan University, Shanghai, China
1. B.J. Allen and A.R.de L. Musgrave, Adv. Nucl. Phys. 10(1978)129
2. Y.K. Ho, J.F. Liu and C. Coceva, Nucl. Data Sci. Tech. (MITO, Japan, 1988) p.703
3. H.R. Weller and N.R. Roberson, Rev. Mod. Phys. 52(1980)699
4. M.Kwato Njock et al., Nucl. Phys. A482(1988)489c
5. V. Metag, Nucl. Phys. A488(1988)483
6. H. Nifenecker et al., Nucl. Phys. A495(1989)3c; Progress in Particle and Nuclear Physics (1989)
7. D.H.E. Gross Nucl. Phys. A240(1975)472
8. Y.K. Ho, J.F. Liu and Z.S. Yuan, Phys. Rev. C40(1989)2541
9. K. Niita et al., Nucl. Phys. A482(1988)525c
10. N. Alamanos et al., Phys. Lett. B173(1986)392
11. D. Vasak, Phys. Lett. B176(1986)276; J. Phys. G: Nucl. Phys. 11(1985)1309
12. T. Stahl, Z. Phys. A327(1987)311
13. H.R. Weller et al., Phys. Rev. C25(1986)2921
14. Y.K. Ho and Z.S. Yuan, to be published
15. S.L. Blatt et al., Phys. rev. C30(1984)423
16. K. Snover, Ann. Rev. Nucl. Part. Sci. 36(1986)545
17. A.M. Nathan, A.M. Sandorf and T.J. Bowles, Phys. Rev. C24(1981)932
18. W. Bauer et al., Phys. rev. C34(1986)2127
19. W. Cassing et al., Phys. Lett. B181(1986)217
20. W. Bauer, Phys. Rev. C40(1989)715
21. D. Neuhauser and S.E. Koolin, Nucl. Phys. A462(1987)163
22. K. Nakayama and G.F. Bertsch, Phys. Rev. C40(1989)685; C39(1989)1475
23. T.S. Biro et al., Nucl. Phys. A475(1987)579
24. J. Bandrup and R. Vandenbosch, Nucl. Phys. A490(1988)418; A487(1988)397
25. B.A. Remington, Phys. Rev. Lett. 57(1986)2909; Phys. Rev. C35(1987)1720
26. M. Durand and J. Knoll, Nucl. Phys. A496(1989)539
27. I. Bergqvist, In Neutron Radiative Capture, Edited by R.E. Chrien (Pergamon, New York, 1984) p.33
28. G. Reffo, M. Blann and B.A. Remington, Phys.Rev. C38(1988)1190
29. N. Herrmann et al., Phys. Rev. Lett. 60(1988)1630
30. J.S. Levinger, Phys. Lett. 82B(1979)181
31. H.G. Clerc, AIP Conf. Proc. No.125, Edited by S. Raman (AIP, New York 1986) p.636
32. J. Galin, Nucl. Phys. A488(1988)297

RADIATIVE CAPTURE AND PREEQUILIBRIUM γ EMISSION

P. Obložinský
Institute of Physics, Slovak Academy of Sciences
842 28 Bratislava, Czechoslovakia

ABSTRACT

We review theory of preequilibrium γ emission and discuss its applications in radiative capture as induced by nucleons in the energy range $\approx$ 10 MeV-200 MeV. Two mechanisms of γ emission during preequilibrium cascade can be distinguished. Single-particle radiative transitions dominate the spectral energy range of giant resonances. This is discussed in terms of the preequilibrium exciton and hybrid models as well as the quantum-mechanical multistep compound model. Emission of hard photons is dominated by two-particle quasi-deuteron radiative mechanism as discussed in terms of the preequilibrium hybrid model.

INTRODUCTION

It is the purpose of the present paper to discuss radiative capture of fast nucleons within the framework of preequilibrium decay. Preequilibrium reaction models[1,2] are in use for more than twenty years and interpretation of nucleon spectra in nuclear reactions induced by fast neutrons has always been their primary interest. Nevertheless, γ ray emission has for long been neglected as a domain for applying the ideas of preequilibrium decay. This is being changed, and in recent several years remarkable progress has been achieved in understanding preequilibrium γ emission. It is natural that new developments are usually first tried on nucleon radiative capture since the entrance channel is very simple, intranuclear cascade is well known and computational problems are relatively simple.

PREEQUILIBRIUM DECAY AND RADIATIVE MECHANISMS

Basic concept of preequilibrium reaction models is illustrated in Fig. 1. Nuclear states are characterized by the number of excited particles and holes (excitons), initial state in nucleon induced reactions is 1p0h, and excited system develops via residual two-body interactions as a chain of n = p+h = 1,3,5,... exciton states towards equilibrium. At each stage of this chain an emission of particles and/or γ rays is possible. State densities are often sufficiently described by the simple equidistant spacing model as[3]

$$\rho_n(E, J) = \omega_n(E) R_n(J) = \frac{g^{n+1} E^n}{p! h! (n-1)!} \frac{2J+1}{2\sqrt{2\pi}\, \sigma_n^3} e^{-(J+1/2)^2/2\sigma_n^2}, \qquad (1)$$

where E, J stand for energy and spin, g is the single-particle state density, and σ_n is the spin cut-off factor.

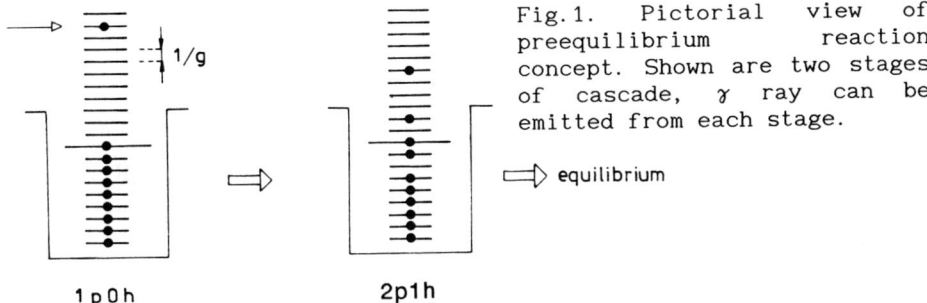

Fig. 1. Pictorial view of preequilibrium reaction concept. Shown are two stages of cascade, γ ray can be emitted from each stage.

From preequilibrium models most popular are phenomenological ones, exciton and hybrid, which differ in statistical assumptions and in treatment of preequilibrium cascade. More recently, quantum-mechanical statistical multistep compound and multistep direct models have been developed by Feshbach et al.[3] So far, γ ray emission has been incorporated into the exciton, hybrid as well as multistep compound models. Radiative capture in each of these models can be described as the product of three terms, formation cross section of the initial composite nucleus (σ_a or σ^{in}), γ emission rate (λ or width Γ) relative to total emission plus intranuclear transitions rates, and depletion factor D. Summation is made over all preequilibrium reaction states up to the equilibrium exciton number $\bar{n}$. In the exciton model one has[4]

$$\frac{d\sigma}{d\varepsilon_\gamma} = \sum_J \sigma_a(E,J) \sum_{\substack{n=1 \\ \Delta n=+2}}^{\bar{n}} \frac{\sum_S \lambda_n^\gamma(EJ \to^{\varepsilon_\gamma} US)}{\lambda_{nJ}} D_{nJ} . \quad (2)$$

In the hybrid model the capture cross section reads[5]

$$\frac{d\sigma}{d\varepsilon_\gamma} = \sigma_a \sum_{x=\nu}^{\pi} \sum_{\substack{n=1 \\ \Delta n=+2}}^{\bar{n}} n^p x \int_0^E \frac{\omega_{n-1}(E-\varepsilon)g d\varepsilon}{\omega_n(E)} \frac{\lambda_\gamma(\varepsilon,\varepsilon_\gamma)}{\lambda_{tot}(\varepsilon)} D_n , \quad (3)$$

where ε is the excitation energy of the selected particle, and neutrons ν are explicitly distinguished from protons π. In the statistical multistep compound model one considers only bound nuclear states and the cross section reads[6]

$$\frac{d\sigma}{d\varepsilon_\gamma} = \sum_J \sigma_J^{in} \sum_{N=1}^{r-1} \frac{<d\Gamma_{nJ}/d\varepsilon_\gamma>}{<\Gamma_{nJ}>} D_{nJ} , \quad (4)$$

where the nuclear stage N is related to the number of excitons as

$2N+1 = n$ (note that $n=1$ state is unbound and therefore excluded), and r refers the the equilibrium stage.

Preequilibrium models are statistical in nature, therefore, γ emission rates should be evaluated from the principle of detailed balance. γ emission mechanism is thus reverse of that of photoabsorption, and it is instructive to examine this latter process. Shown in Fig. 2 is the photoabsorption cross section as a function of γ ray energy up to the pion threshold as calculated for ^{53}Cr. Three γ ray energy ranges can be distinguished, low energy ($\lesssim 10$ MeV), medium energy ($\approx 10 - 30$ MeV, GDR region) and high energy ($\gtrsim 30$ MeV, hard photons).

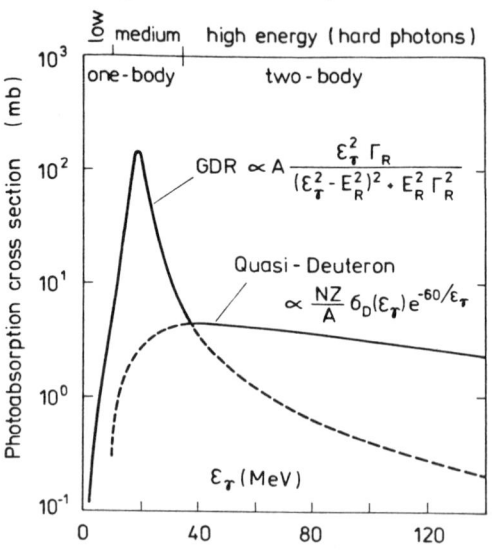

Fig. 2. Photoabsorption cross section as a function of γ ray energy calculated for ^{53}Cr. Shown are three energy ranges and the interactions involved.

Photoabsorption in the medium and low energy ranges is sufficiently characterized by the giant dipole resonance (Lorentzian) lineshape and the mechanism of photoabsorption can be identified with single-particle transition. In other words, coordinates of only one nucleon are changed. Since low spectral energy range is dominated by equilibrium statistical processes, preequilibrium γ rays of single-particle origin should be predominantly found in the giant resonance region.

In the range of hard photons, γ ray is increasingly absorbed by a neutron-proton pair (quasi-deuteron) in a nucleus. This is because of the wavelength of the incident γ ray is comparable to the intranucleonic distance in the nucleus. The quasi-deuteron photoabsorption mechanism has been proposed by Levinger[7] and is now considered as well established. An incident photon interacts with n-p pair rather than n-n or p-p pairs because they have no dipole moment. The photoabsorption cross section is proportional to the number of n-p pairs and to the photoabsorption of o free deuteron, reduced by a Pauli-blocking factor that is of importance in a real nucleus. Preequilibrium emission of hard photons should thus have a two-body character.[8] This observation is supported also by alternative description of hard photon emission via incoherent n-p bremsstrahlung.[9]

GIANT DIPOLE RESONANCE REGION

A single-particle radiative transition can be viewed as an electromagnetic interaction of a single nucleon with the nuclear core, the exciton number changing as $\Delta n = 0, -2$, see Fig. 3. To treat this process via detailed balance one should invoke the Brink-Axel hypothesis implying that photoabsorption cross section for any n-exciton state can be identified with that for the ground state. Application of detailed balance to preequilibrium γ emission is not trivial and it was first done properly by Akkermans and Gruppelaar[10] who improved an earlier result of Ref. 11. It is important to realize[12] that neutrons do radiate thanks to their dipole effective charge, $-Z/A$, and their radiation does not differ essentially with that of protons which posses N/A effective charge (in units of e). Another important point concerns parity constraint that the dipole operator imposes on single-particle radiative transitions. It was shown that, within the approximation of equal contribution of positive and negative parities to nuclear state densities, parity effects can be neglected.[13] Still another point concerns angular-momentum conservation and full treatment of angular-momentum coupling. This was studied in the exciton and the multistep compound models and the corresponding effects were found to be small.[4,6]

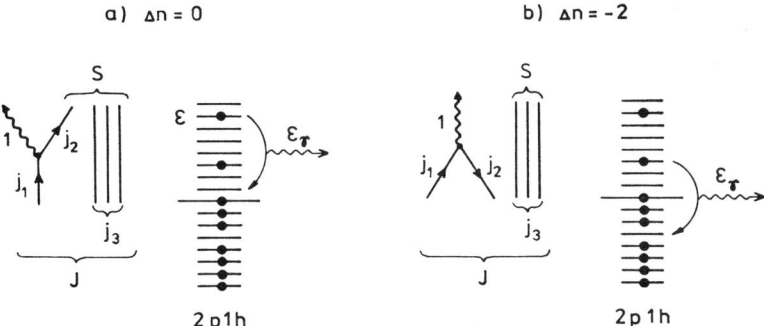

Fig. 3. Schematical view of single-particle radiative transitions in the 2p1h state. The exciton number can be changed as $\Delta n = 0, -2$. Initial and final spin are J and S, single-particle spins are j_1, j_2, and nuclear core has spin j_3.

We first explain how preequilibrium γ emission rates are calculated in the phenomenological preequilibrium models. Photoabsorption microscopically means either creation of a particle-hole pair, $n \to n+2$, or simple transition of one exciton, $n \to n$. Branching ratios for these possibilities are assumed to be given by available phase space, and in the exciton model one has[10]

$$b_n^{n+2} = \frac{g^2 \varepsilon_\gamma^2}{gn + g^2\varepsilon_\gamma^2} \quad , \quad b_n^n = \frac{gn}{gn + g^2\varepsilon_\gamma^2} \quad . \tag{5}$$

Detailed balance then leads to γ emission rate per Mev per sec

$$\lambda_n^\gamma(\varepsilon_\gamma) = \frac{\varepsilon_\gamma^2 \sigma(\varepsilon_\gamma) \; \omega_{n-2}(E-\varepsilon_\gamma) b_{n-2}^n + \omega_n(E-\varepsilon_\gamma) b_n^n}{\pi^2 \hbar^3 c^2 \qquad \qquad \omega_n(E)} \quad , \tag{6}$$

where photoabsorption cross-section is

$$\sigma(\varepsilon_\gamma) = \sigma_{GDR}(\varepsilon_\gamma) = \sum_{R=1}^{2} \sigma_R \frac{\varepsilon_\gamma^2 \Gamma_R^2}{(\varepsilon_\gamma^2 - E_R^2)^2 + E_R^2 \Gamma_R^2} \quad , \tag{7}$$

the notation being selfexplanatory. When considering also angular momentum coupling, the formalism becomes more complicated. For example, instead of the branching b_n^n one has

$$b_{nS}^{nJ} = \frac{gnx_{nS}^{nJ}}{gnx_{nS}^{nJ} + g^2\varepsilon_\gamma^2 x_{nS}^{n+2J}} \tag{8}$$

with the angular momentum coupling term

$$x_{nS}^{nJ} = \frac{3(2J+1)}{R_n(S)} \sum_{j_1 j_2 j_3} (2j_1+1)R_1(j_1)(2j_2+1)R_1(j_2)R_{n-1}(j_3) \begin{pmatrix} j_2 & 1 & j_1 \\ \frac{1}{2} & 0 & -\frac{1}{2} \end{pmatrix}^2 \begin{Bmatrix} j_2 & j_3 & S \\ J & 1 & j_1 \end{Bmatrix}^2 \quad , \tag{9}$$

where the notation is explained in fig. 3 and R_n is the spin part of the level density, see Eq. (1).

In the hybrid model one needs the γ emission rate for a particle or hole with the excitation energy ε. Using the idea of branching ratios for photoabsorption the γ emission rate is[5]

$$\lambda_\gamma(\varepsilon,\varepsilon_\gamma) = \frac{\varepsilon_\gamma^2 \sigma(\varepsilon_\gamma)}{\pi^2 \hbar^3 c^2} \frac{g}{gn + g^2\varepsilon_\gamma^2} \qquad \text{for } \Delta n = 0 \tag{10a}$$

and

$$\lambda_\gamma(\varepsilon,\varepsilon_\gamma) = \frac{\varepsilon_\gamma^2 \sigma(\varepsilon_\gamma)}{\pi^2 \hbar^3 c^2} \frac{g}{g(n-2)+g^2\varepsilon_\gamma} \frac{\omega_{n-2}(E-\varepsilon_\gamma)}{\omega_{n-1}(E-\varepsilon)} \qquad \text{for } \Delta n = -2 \quad . \tag{10b}$$

This rate was subject of debate in view of two different solutions (Ref.5,12), and it was proved to be correct.[13,14]

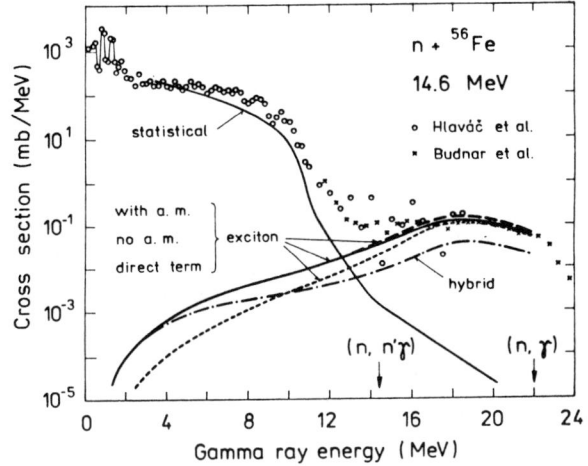

Fig.4. Experimental and calculated γ ray spectra for ^{56}Fe+n at 14.6 MeV incident energy. Shown are spectra calculated by the statistical model, preequilibrium exciton model (with angular momentum, without angular momentum, and the direct n=1 term), and by the preequilibrium hybrid model.

The above models are illustrated on ^{56}Fe(n,γ) at 14.6 MeV. Shown in Fig.4 is total γ ray spectrum with the capture component seen above 14 MeV. Standard statistical model calculations do well in describing strong (n,n'γ) component and completely fail to explain the (n,γ) part in the GDR region. Preequilibrium models do substantially better in this latter region. For the exciton model three curves are shown. It is seen that spectra with and without angular momentum coupling differ only slightly. Remarkable contribution comes from the direct term, n=1. The dominance of this term in the GDR region is supported also by more systematic calculations.[15,16] The spectrum calculated in the preequilibrium hybrid model has similar shape, dominant direct term, but it is lower in the GDR range by about a factor of 4-5. The difference between the two models has been studied in detail[17] and it was suggested that a careful two-component calculation (neutrons and protons distinguished) in the exciton model is consistent with the hybrid model results. This point together with objections to a purely statistical treatment of the direct n=1 term call for cautious approach as regards the physics involved. On the other hand, it seems clear that the above concept provides very useful tool in nuclear technology.

The quantum-mechanical multistep models as developed by Feshabch et al.[3] treat compound and direct processes separately. The multistep compound model considers only bound states and thus explicitly excludes the above n=1 term. The contribution from unbound states should then be added as multistep direct component. The γ emission width in the multistep compound model should be averaged and it is given as[6]

$$\langle d\Gamma_{nJ}/d\varepsilon_\gamma \rangle = \frac{\varepsilon_\gamma^2 \, \sigma(\varepsilon_\gamma)}{3\pi^2 \hbar^2 c^2} \frac{\rho_{n'}(U,S)}{\rho_n(E,J)} \beta_{n'S}^{nJ} , \qquad (10)$$

where n'= n,n-2. The branching ratios βare no more normalized to unity as is the case in the exciton model, since they are now restricted to photoabsorption leading to bound states,

$$\beta_{nS}^{nJ} + \beta_{n'S}^{n+2J} < 1 . \qquad (11)$$

As shown in Fig.5 on the case of ^{181}Ta(n,γ) at 14 MeV incident energy the multistep compound model can explain only some 50% of γ rays observed in the GDR region. It is to be expected that the rest is due to the multistep direct radiative capture. Such formulation, though, is still missing.

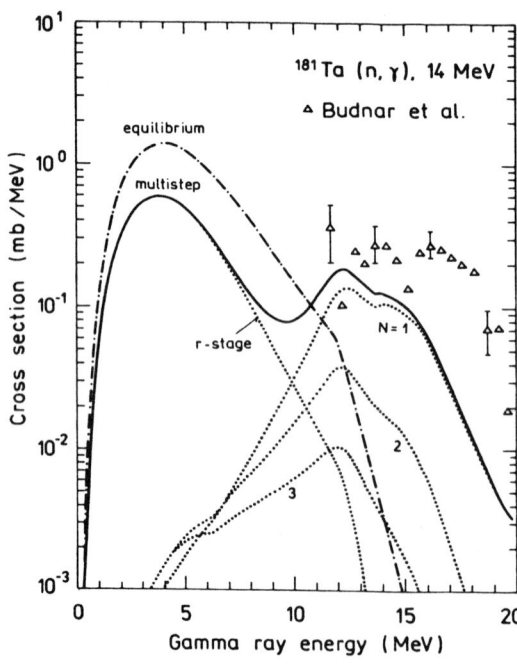

Fig.5. Radiative capture on ^{181}Ta for 14 MeV neutrons. Shown are contributions of first three preequilibrium multistep compound stages (N=1,2,3, corresponding to 2p1h, 3p2h and 4p3h) as well as the equilibrium r-stage.

HARD PHOTONS

There is growing evidence that hard photons come from incoherent neutron-proton collisions in the initial stages of nuclear reactions. This two-particle radiative mechanism is usually identified with bremsstrahlung. We shall discuss purely preequilibrium approach based on the hybrid model and the quasi-deuteron radiative mechanism. This mechanism is illustrated in Fig.6. In the emission the dominant process is $\Delta n = +2$ that for the initial 1p0h state leads to 2p1h state. In the absorption the dominant is creation of neutron plus proton particle-hole pairs, that is 4-exciton state. The hard photon emission rate thus, to a good approximation, is[8]

$$\lambda_\gamma(\varepsilon,\varepsilon_\gamma) = \frac{\varepsilon_\gamma^2\,\sigma_{QD}(\varepsilon_\gamma)}{\pi^2\hbar^3 c^2}\,\frac{\omega_3(\varepsilon-\varepsilon_\gamma)}{\omega_4(\varepsilon_\gamma)}, \quad (12)$$

where

$$\sigma_{QD}(\varepsilon_\gamma) = 6\,\frac{NZ}{A}\,\sigma_D(\varepsilon_\gamma)\,e^{-60/\varepsilon_\gamma},\quad \sigma_D(\varepsilon_\gamma) = 61.2\,\frac{(\varepsilon_\gamma - 2.22)^{3/2}}{\varepsilon_\gamma^3}. \quad (13)$$

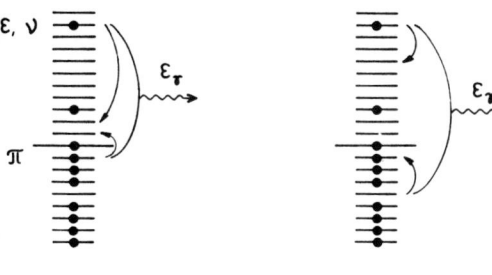

a) Δn = +2   b) Δn = 0

Fig. 6. Quasi-deuteron radiative mechanism. Selected neutron ν of energy ε interacts with proton π. Shown are transitions changing the exciton number by Δn = +2 and 0. Two other possibilities, Δn = -2, -4, are very weak.

Shown in Fig. 7 is the case of Au+p (72 MeV). It appears that hard photons are emitted from the moving n-p pair in the very early stage of reaction. There is a considerable Doppler shift that explains the angular distribution. The agreement with data is excellent.

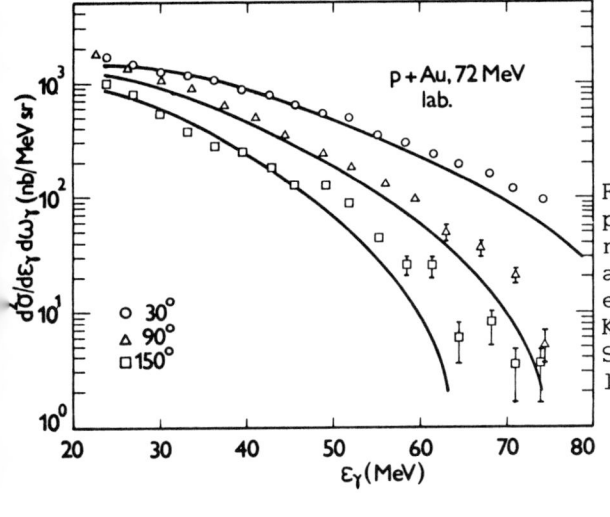

Fig. 7. Calculated hard photon spectra for the reaction p+Au at 72 MeV are compared with experimental data of Kwato Njock et al.[20] Spectra are given in the laboratory frame.

CONCLUSIONS

Preequilibrium γ ray emission concept seems to be a promising tool for describing radiative capture of fast nucleons in the GDR spectral energy range as well as in the spectral energy range of hard photons. An important unsolved problem is incorporation of γ emission into the quantum-mechanical statistical multistep direct model.

REFERENCES

1. M. Blann, Ann. Rev. Nucl. Sci. 25, 123 (1975).
2. H. Gruppelaar, P. Nagel, and P.E. Hodgson, Riv. Nuovo Cimento 9, 1 (1986).
3. H. Feshbach, A. Kerman, and S. Koonin, Ann. Phys. (NY) 125, 429 (1980).
4. P. Obložinský, Phys. Rev. C35, 407 (1987).
5. P. Obložinský, Phys.Lett. B215, 597 (1988).
6. P. Obložinský and M.B. Chadwick, Phys. Rev. C42 (1990) in press.
7. J.S. Levinger, Phys. Rev. 84, 43 (1951);
   J.S. Levinger, Phys. Lett. 82B, 181 (1979).
8. P. Obložinský, Phys. Rev. C40, 1591 (1989).
9. B.A. Remington, M. Blann, and G. Bertsch, Phys. Rev. C35, 1720 (1987).
10. J.M. Akkermans and H. Gruppelaar, Phys. Lett. B157, 95 (1985).
11. E. Běták and J. Dobeš, Phys. Lett. B84, 368 (1979).
12. G. Reffo, M. Blann, and B.A. Remington, Phys. Rev. C38, 1190 (1988); 38, 1188(E) (1989).
13. P. Obložinský, Phys.Rev. C41, 401 (1990).
14. G. Reffo, M. Blann, and B.A. Remington, Phys. Rev. C41, 403 (1990).
15. F. Cvelbar and E. Běták, Zeit. Phys. A332, 163 (1989).
16. F. Cvelbar, E. Běták and J. Merhar, J. Phys. G, in press.
17. A. Musacchio Lasa, Diploma work (Comenius University, Bratislava 1989) unpublished.
18. S. Hlaváč and P. Obložinský, International Atomic Energy Agency Report No. INDC(CSR)-5/GI, Vienna 1983.
19. M. Budnar et al., International Atomic Energy Agency Report No. INDC(YUG)-6/L, Vienna 1979.
20. M. Kwato Njock et al., Phys. Lett. B207, 269 (1988).

Neutron-Proton Bremsstrahlung Studies using the White Neutron Source at the LAMPF/WNR

S. A. Wender, R.O. Nelson, M.E. Schillaci
Los Alamos National Laboratory
Los Alamos, NM 87545

and

M. Blann
Lawrence Livermore National laboratory
Livermore, CA 94550

Abstract

Nucleon-nucleon bremsstrahlung is a few-body radiative process that provides insight into several areas of nuclear physics. It is one of the simplest systems for studying the off-shell behavior of the nucleon-nucleon potential. The physics involved in neutron-proton bremsstrahlung (NPB) is significantly different from that of proton-proton bremsstrahlung (PPB). In particular, NPB cross sections are much larger than PPB cross sections because NPB allows E1 radiation, and the contribution to the cross section from the meson exchange currents has been calculated to be as large as the contributions from external radiation.

To date there have been essentially four NPB experiments. These measurements have covered only a small part of the available phase space. A major experimental problem in performing these measurements has been the lack of a suitable intense, high-energy neutron beam.

We are planning a measurement of the NPB cross section using the white neutron source at the WNR target area at the LAMPF accelerator. We plan to implement the experiment in three phases. In the first stage, we shall measure inclusive hard-photon production using a multi-element gamma-ray telescope that is insensitive to neutrons. In the second phase, we shall measure the bremsstrahlung gamma-rays in coincidence with recoil protons. In the last phase, we shall detect the scattered neutrons in coincidence with the recoil protons and gamma rays.

Introduction

The simplest interaction between two nucleons is nucleon-nucleon elastic scattering. The next level of complexity is inelastic NN scattering involving the emission of electromagnetic radiation. The electromagnetic interaction is a nice way to probe the strong interaction because it is "understood" and "weak". When the two outgoing nucleons emerge from their interaction with small relative velocities and couple together, it is called a capture reaction. The radiative processes that occur between the kinematic regions of elastic scattering where the two particles scatter with $90°$ between them and the highly inelastic capture reaction ($0°$ between the two outgoing particles) are called nucleon-nucleon bremsstrahlung (NNB).

Figure 1 shows the diagrams for NNB below the pion threshold. The diagrams in Fig. 1a-d are call external diagrams and involve emission of photons either before or after the strong interaction, from either one of the particle legs. Figure 1e is an internal diagram that involves meson exchange. Figure 1F is the diagram for rescattering. Above the pion threshold, additional diagrams that involve coupling to the delta resonance are necessary.

Diagrams for Nucleon–Nucleon Bremsstrahlung

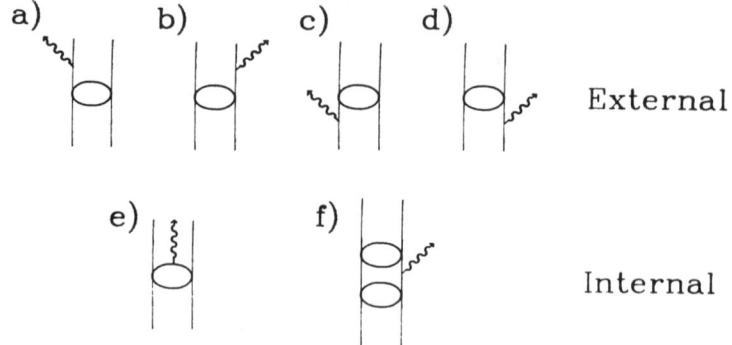

Figure 1. Diagrams involved in nucleon-nucleon bremsstrahlung. Diagrams a-d are external diagrams, e) meson exchange diagram and f) rescattering diagram.

In the language of the Direct Semi-Direct capture model, the external diagrams are analogous to the direct matrix elements, and the internal diagrams are analogous to the semi-direct amplitudes.

The physics of NNB depends strongly on the types of nucleons involved in the collision. Because the center of mass and center of charge are coincident in the case of proton-proton bremsstrahlung (PPB), dipole radiation is not allowed. The leading term in PPB is therefore E2. Because PPB involves nucleons of the same charge, to first order only neutral mesons are exchanged that do not radiate. Therefore, the meson exchange term is quite small in PPB. These two differences between PPB and neutron-proton bremsstrahlung (NPB) explain why the cross section for NPB is much greater than for PPB.

The major goal in the study of NPB is to understand the nucleon-nucleon interaction by looking at its off-shell behavior with particular emphasis on the meson exchange terms. In addition, NPB data may also shed light on the source of high-energy photons that have been observed in heavy-ion reactions. Electroweak scientists are also very interested in NPB data because NPB processes are thought to produce significant backgrounds in several neutrino experiments.

Nucleon-Nucleon Physics

Figure 2 shows a calculation of the NPB cross section by V. R. Brown[1]. In Fig. 2a, Hamada-Johnston and Bryan-Scott potentials are used to calculate the NPB cross section for scattered neutrons and protons at +/- 30° for an incident neutron energy of 200 MeV as a function of gamma-ray angle. As seen from this plot, it will be very difficult for any NPB experiment in this region of phase space to distinguish between different potentials even with excellent data. In Fig. 2b, the contributions to the cross section for the meson exchange term and the external radiation diagrams are plotted separately for the Bryan-Scott potential. As seen from the figure, the contribution of the exchange term depends on the gamma emission angle. In this calculation, the exchange term is smaller than the external term for gamma-rays emitted on the same side as the protons (positive angles) and larger than the external radiation when gamma emission occurs on the neutron side, especially at back angles.

Because this experiment will span the incident neutron energy range from 50 to 400 MeV, we shall be able to study the heavy meson contribution to the exchange term and investigate the contribution of the Δ resonance above the pion threshold.

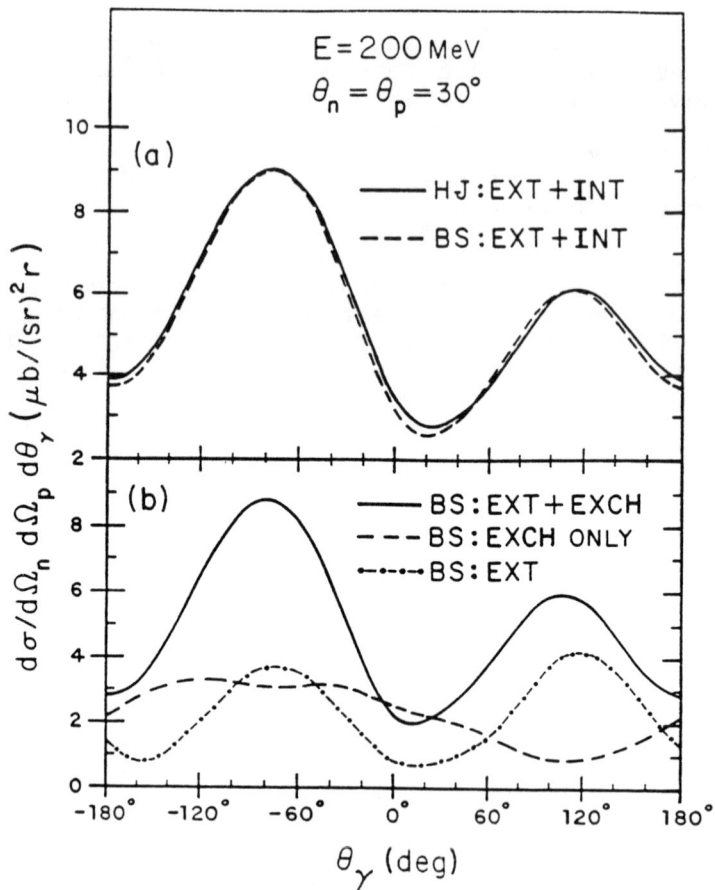

Figure 2. Calculated NPB cross section using the Hamada-Johnston and Bryan-Scott potentials.

Heavy-Ion Physics

Since the observation of hard photon emission following heavy-ion reactions[2], there has been considerable recent interest in NPB cross sections. Understanding the origin of these gamma rays would give information on the early stages of the collision, where the nucleons in the two heavy ions accelerate and radiate. The interpretation of these gamma rays should be much less ambiguous than pion emission because gamma rays are not strongly absorbed. Because the NPB process allows E1 radiation, it is much larger than PPB. The dominant gamma-ray production process is therefore thought to be from NPB processes.

The gamma-ray spectra obtained[2] following the ^{14}N + ^{12}C reaction at 40, 30 and 20 MeV/nucleon are shown in Fig. 3. As seen in the figure, the gamma ray energies extend far beyond the bombarding energy per nucleon. The question is whether these high-energy gamma rays can be explained in terms of the elemental NPB process or is some collective process required. The classical NPB calculation, multiplied by 2 to roughly account for meson exchange currents, and the quantum mechanical calculations of the NPB cross section tend to agree with the data; however, these calculations have not been tested by comparing their results with measurements of the elemental NPB cross section.

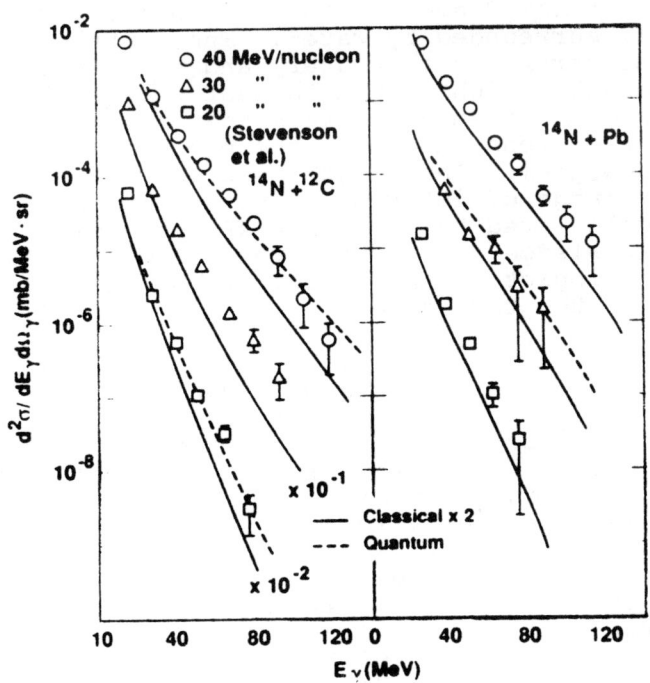

Figure 3. Inclusive gamma-ray production data from ^{14}N on ^{12}C and Pb. The solid lines are calculations using semi-classical NPB cross sections. The dashed lines are calculations using quantum mechanical expressions for the NPB cross sections.

Large Cerenkov Detector Backgrounds

An experiment to measure the Weinberg angle, θ_W, has been proposed as a test of the Standard Model[3]. The experiment involves measuring the ratio, R, of the neutrino-electron scattering cross sections for different neutrinos following the decay of positive pions.

$$\pi^+ \longrightarrow \mu^+ + \nu_\mu \qquad 26 \text{ nsec}$$
$$ \hookrightarrow e^+ + \nu_e + \bar{\nu}_\mu \quad 2.2 \text{ }\mu\text{sec}$$

$$R = \frac{\sigma(\nu_\mu, e)}{\sigma(\nu_e, e) + \sigma(\bar{\nu}_\mu, e)}.$$

The pions are produce by stopping the 800 MeV proton beam from the LAMPF accelerator in a heavily shielded target cell surrounded by water. The neutrinos scatter from the electrons in the water, and the electrons produce Cerenkov light. The largest source of error in this ratio comes from gamma rays that convert into energetic electrons and mimic the electron-neutrino scattering signal. The largest source of gamma rays in the energy range of interest in this experiment (10-50 MeV) are calculated to come from NPB processes. An accurate knowledge of the NPB cross section will have significant impact on the design of this detector and the shielding of the target cell.

Past experiments

There have been very few experiments to measure the NPB cross section. Figure 5 is a plot of the off-shell parameter, ΔM^2, as a function of gamma-ray energy divided by the projectile energy for all of the NPB experiments. The off-shell parameter is the square of the difference between the off shell mass and the proton mass, in units of the proton mass for the external diagrams. The curve labelled ref 4 is the experiment of Brady et al.[4] at 208 MeV, the curve labelled ref 5 is the measurement by Edgington et al.[5] at 130 MeV. Both of these experiments integrated over the gamma-ray energy. A more recent experiment by DuPont et al.[6] at 76 MeV is shown labeled as ref 6. This experiment was similar to the preceeding experiments in that it detected the proton and neutron but used a segmented detector. Also plotted in Fig. 5 is a similar curve for the region of phase space that would be obtained using the white neutron source at LAMPF with 400 MeV incident neutrons and detected outgoing nucleons at +/-20 degrees. As can be seen from the plot, this experiment probes regions significantly farther off shell than any previous experiment.

In these three past experiments the scattered neutron and the recoil proton were measured directly and the gamma-ray energy was deduced. A recent experiment by Nifenecker and Pinston[7] at SATURNE measured photons directly. In this experiment the inclusive gamma-ray spectrum for neutrons of 180 +/- 80 MeV incident on a liquid hydrogen target was measured using a "smart" gamma-ray detector. The detector consisted of a charged-particle veto, a BaF_2 active radiator, two ΔE detectors and a large NaI calorimeter. Because the two ΔE detectors identified the converted electrons, this technique is very insensitive to neutrons.

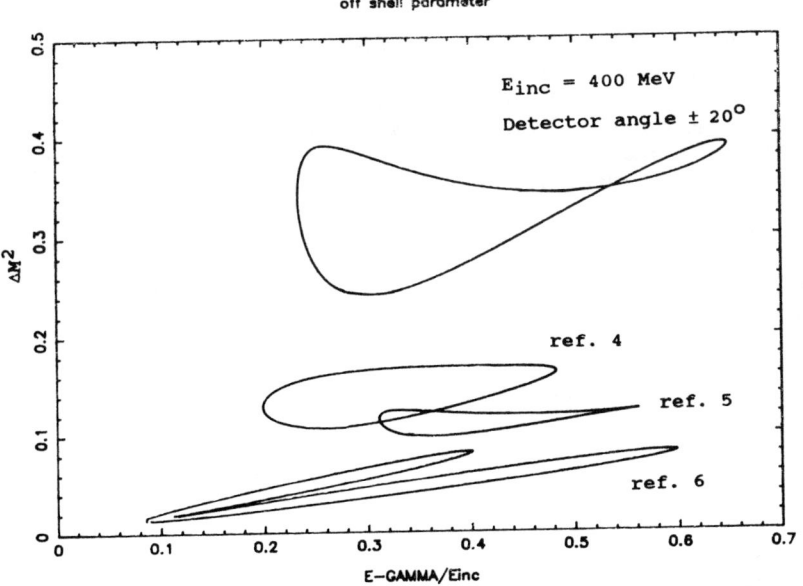

Figure 4. Plot of the off-shell parameter ΔM^2 as a function of the gamma-ray energy for the three previous NPB experiments. The uppermost curve represents the proposed experiment at LAMPF for incident neutrons at 400 MeV and particles detected at +/- 20°.

LAMPF/WNR Experiment

The proposed experiment will be performed at the continuous energy, white neutron source at the WNR target area at LAMPF. This source has high intensity and a neutron spectrum that extends above 400 MeV. This source has been previously described in detail.[8] Figure 5 shows the experimental setup. The NPB measurements will proceed in three phases. The first phase will be the measurement of the inclusive gamma-ray spectrum over the incident neutron energy range from approximately 50 MeV to 400 MeV, and for gamma rays from 20 to 200 MeV. The target will consist of a liquid hydrogen cell approximately 6 cm thick and 15 cm in diameter. The gamma-ray detector will be a multi-element gamma-ray telescope similar to that used by Nifenecker and Pinston[7] in their inclusive experiment. The gamma-ray detector will consist of a charged-particle veto, a 1 cm thick BGO active radiator, two 5 mm thick plastic ΔE detectors, and a 40.5 x 40.5 x 40.5 NaI calorimeter. The calorimeter is segmented into 16 optically isolated 10 x 10 x 40.5 cm NaI sections. The active radiator consists of 10 pieces each 7 cm wide and 20 cm long. The efficiency of the detector was calculated to be approximately 30 percent, and the resolution is approximately 15 percent for gamma-ray energies of 150 MeV. The detectors will be located at +/- 90° on either side of the neutron flight path. We estimate the count rate due to NPB events to be approximately 112 events/day in a 10 MeV neutron bin and a 10 MeV gamma-ray bin. Because we shall acquire 35 pulse-height spectra in the incident neutron energy range from 50 to 400 MeV simultaneously, the total count rate should be approximately 24,000 NPB events/day. This first phase of the experiment will not be able to distinguish gamma rays from bremsstrahlung processes from gamma rays following capture or π^0 decay. The data, however, will be

NPB EXPERIMENTAL APPARATUS

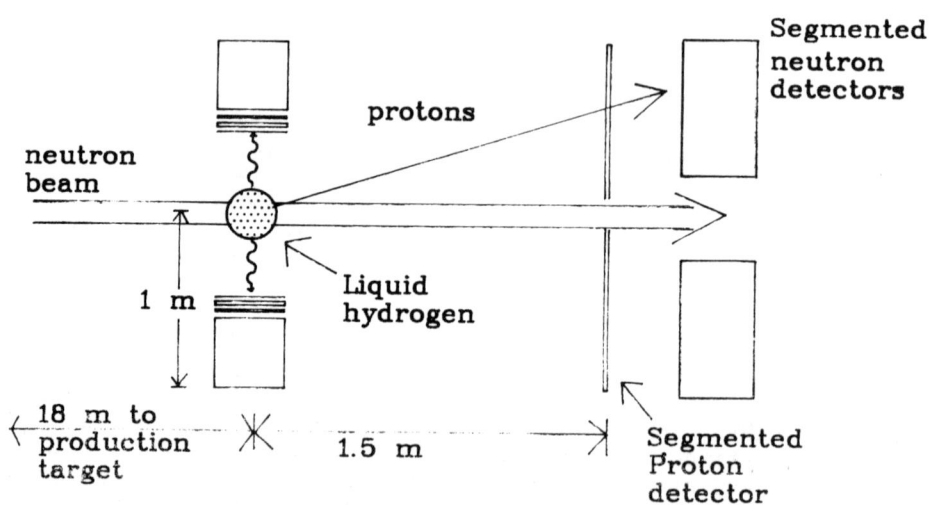

Figure 5. The experimental apparatus for the neutron-proton bremsstrahlung experiment.

immediately useful to test the heavy-ion calculations and contribute to background information for the LCD experiment.

In the second phase, we shall measure the gamma rays in coincidence with the recoil protons. The proton detector will be a 200 x 60 x 1 cm thick segmented plastic array. It will be segmented into 10 cm squares and located 1.5 m downstream from the liquid hydrogen target. By measuring the recoil proton in coincidence with a gamma ray, it is possible to uniquely identify the radiation as coming from a bremsstrahlung process instead of from a capture reaction or from neutral pion decay.

Figure 6 shows the kinematics for the NPB reaction for incident neutron energies of 100, 200 and 400 MeV. In the example shown in the plot we assume that the proton is detected at 10° on the opposite side from the gamma ray. The band represents the uncertainty in the incident neutron energy which comes from the time resolution of the gamma-ray detector. As we move along the kinematic locus, the neutron angle and energy are also varying. If we set a window on a particular gamma-ray energy, we should observe two proton groups.

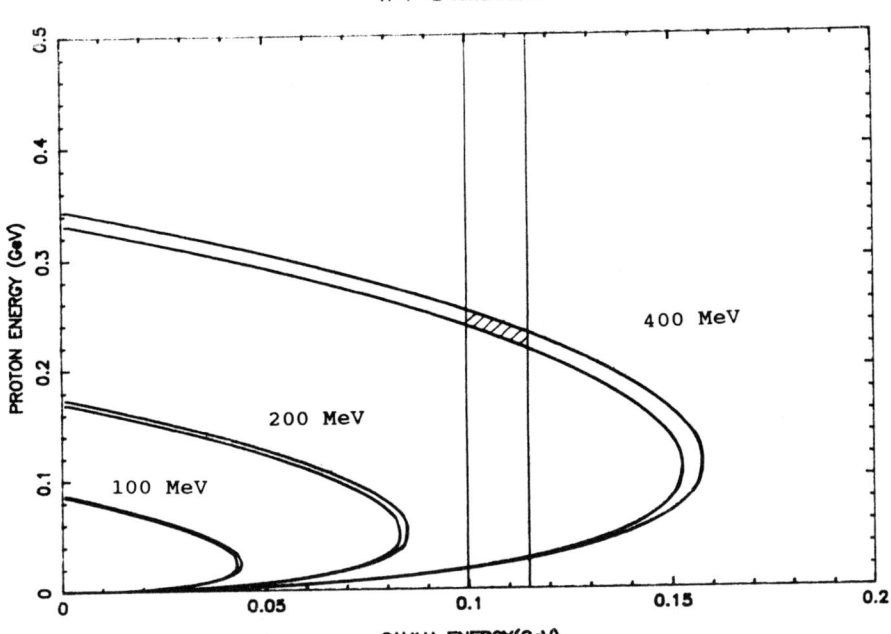

Figure 6. An example of the NPB kinematics for protons detected at 10°, gamma-ray detector at 90°, for incident neutrons of 100, 200 and 400 MeV.

In the final phase, we shall detect the scattered neutron. This triple coincidence experiment should allow the measurement of the NPB cross section over a wide region of phase space. Three variables (energy and two angles) for each outgoing particle will be measured, as well as the energy of the incoming neutron. Thus, the kinematics are highly overdetermined, which should allow excellent background discrimination.

We expect to mount this experiment over the next several months, with phases 1-3 taking place during the next three summer running periods.

References

1. V. R. Brown in "Lectures from the Workshop on Nucleon-Nucleon Bremsstrahlung" Los Alamos National Laboratory, January 1990, (LA-11877-C) Compiled by B. F. Gibson, M. E. Schillaci and S. A. Wender.

2. B. A. Remington, M. Blann and G. Bertsch, Phys. Rev. C35 (1987) 1720.

3. "A Proposal for a Precision Test of the Standard Model by Neutrino-Electron Scattering (Large Cerenkov Detector)" (LA-11300-P), April 1988; D. H. White, spokesman.

4. F. P. Brady, J. C. Young and C. Badrinathan, Phys. Rev. Lett. 20 (1968) 750.

5. J. A. Edginton, V. J. Howard, I. M. Blair, B. E. Bonner, F. P. Brady and M. W. McNaughton, Nucl. Phys. A218 (1974) 151.

6. C. Dupont, C. Deom, P. Leleux, P. Lipmik, P. Macq, A. Ninane, J. Pestieau, Sindano Wa Kitwanga and P. Wauters, Nucl. Phys. A481 (1988) 424.

7. J. A. Pinston and H. Nifenecker, in "Lectures from the Workshop on Nucleon-Nucleon Bremsstrahlung" Los Alamos National Laboratory, January 1990, (LA-11877-C) Compiled by B. F. Gibson, M. E. Schillaci and S. A. Wender.

8. P. W. Lisowski, C. D. Bowman, S. A. Wender and G. J. Russell, Nucl. Sci. and Eng., to be published.

HIGH RESOLUTION MEASUREMENT OF RADIATIVE PROTON CAPTURE AND PAIR PRODUCTION AT INTERMEDIATE ENERGIES

B. Höistad, E. Nilsson, J. Thun, S. Dahlgren, and S. Isaksson.
Dept. of Radiation Sciences, Uppsala University, S-75121 Uppsala, Sweden

G.S. Adams and C. Landberg
Physics Dept., Rensselaer Polytechnic Institute, Troy, New York 12180

ABSTRACT

A new type of compact pair spectrometer is developed for studies of radiative capture reactions and internal pair production at intermediate energies. The first results from capture reactions on ^{11}B are presented together with theoretical predictions based on a continuum shell model calculation.

INTRODUCTION

The radiative capture is one of the most fundamental processes in nuclear physics. At low energies this process is well understood, but our knowledge is rather poor when the energy is greater than about 100 MeV. At these energies, the Born amplitudes predict cross sections which are generally too small at large angles [1,2]. This is probably a consequence of the large momentum transfer involved in these exclusive reactions. The common approach to solve this problem has been to include two-body amplitudes of meson exchange types. These attempts have however been met with rather limited success. A more promising avenue might be to use a continuum shell model approach, which treats the nuclear structure and the reaction mechanism in a unified way. The success in our attempt to understand this process is certainly contingent upon the availability of experimental data, which so far are very scarce in the intermediate energy region. A research program for experimental studies of the exclusive radiative proton capture on light nuclei at intermediate energies is therefore initiated at the The Svedberg Laboratory, Uppsala Sweden. We intend to make comprehensive measurements of the (p,γ) reaction in the energy region between 100 and 200 MeV, including angular distributions of the differential cross section as well as of the analyzing power. The gamma detection should be made with very good energy resolution to be able to isolate the different transitions to excited states in the residual nuclei. Moreover, the recorded spectra should be free from background in order to uncover the minute cross sections (fractions of nb/sr) at large angles.

Our experimental program requires the development of a high resolution detector for intermediate energy photons. The most crucial parameter is the energy resolution which must be excellent in order to match the

Invited talk, presented by B. Höistad to the 7th Int. Symposium on CAPTURE GAMMA-RAY SPECTROSCOPY and related topics, Asilomar, 14-19 October 1990.

level spacing for excited states in light nuclei. The only way to accomplish this, using known technology, is to build a pair spectrometer.

THE PAIR SPECTROMETER

In a pair spectrometer, the high energy gamma is converted to an electron-positron pair whose momenta are determined in a magnetic spectrometer. This method, however, leads to low detection efficiency since only very thin converters of merely a few percents conversion probability can be used in order to keep the unknown energy loss of the electron and positron small in the converter. This fact has necessitated the employment of magnets with very large gap, in order to obtain a reasonable total acceptance. An obvious improvement of the existing pair spectrometers would therefore be to design a magnet with two directional focussing properties so that the whole available field volume is used by the e^+e^- pair. This should reduce the size of the magnet considerably without loss of acceptance. A smaller movable magnet can also cover the photon detection in a large angular interval.

The small acceptance of the pair spectrometer is in fact not a serious problem as long as this can be compensated for by using a high beam current. It is, however, then necessary to have a detector and magnet arrangement which is insensitive to the background radiation. In particular, it is not possible to use a position sensitive detector behind the converter which is exposed to the whole particle flux from the beam-target collision. The conversion point must therefore be determined indirectly from the raytracing procedure.

We have solved the focussing problem in the pair spectrometer by using a clamshell type configuration for the pole pieces. This pole shape generates a field in the median plane of the form $B_o(y) = const. y^{-n}$. In the ideal case with sharp field boundaries and $n=1$, this field causes the particles to move in a plane tilted with respect to the median plane.

A comprehensive description of the pair spectrometer and its performance is given in ref. 3. A schematic picture of the pair spectrometer and its detector system is shown in fig. 1. The detectors generating the signals for the electron and positron are placed in two identical groups which are positioned symmetrically above and below the scattering center. Each detector group consists of three driftchamber packages (x and z) and three planes of trigger scintillators, where each scintillator plane consists of four separate detector blocks. An event trigger is generated by a simultaneous hit in all three scintillator planes above and below the scattering center. The trajectories from the e^+e^- pair leaving the magnet are determined from the three (x,z) coordinates given by the driftchambers. Knowledge of the magnetic field allows these trajectories to be traced back to a common source point on the converter, which is placed in the fringe field. The energy and direction of incoming photons emerging from the target is thereby determined.

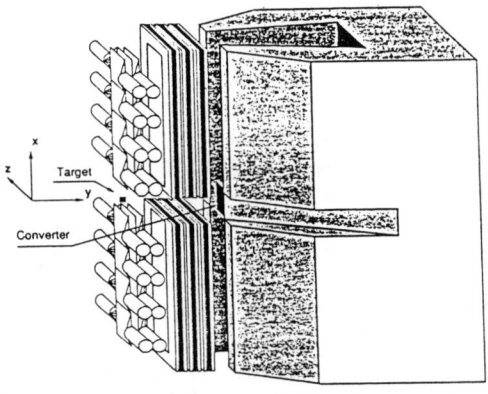
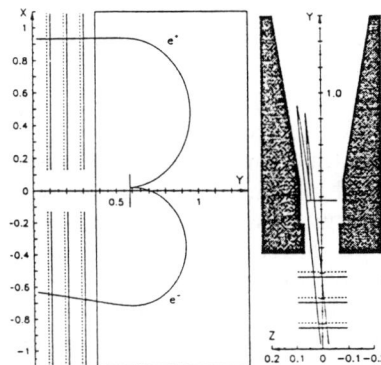

Fig. 1. Schematic overview of the pair spectrometer and its detector system.

Fig. 2. Illustration of the particle motion in the PACMAN field

The pole faces determining the field configuration have been designed to give a field form resulting in maximum acceptance, considering useable converter area and transmission through the magnet. A profile of the pole face is shown in fig. 2. The generated field form has a slightly steeper fall off than the usual clamshell type which has a y^{-1} dependence. This additional focussing is introduced in order to compensate for the unavoidable defocussing effects in the fringe field. We show, also in fig. 2, the projection of a typical electron and positron trajectory on the xy-plane and on the yz-plane. The focussing property of the field is clearly demonstrated by this figure.

Our pair spectrometer, which is named PACMAN (Photon And Charged Meson ANalyser), has a weight of about 45 tons, and is movable on air pads. The nominal scattering angle range spans from 35 to 140 degrees, but more forward angles can be reached with special arrangements. The maximum allowed momentum for a single trajectory is roughly 200 MeV/c, which means that gamma energies up to about 300 MeV can be analyzed without significant loss of acceptance. The scattering center is placed 60 cm from the front end of the pole edge. A 0.1 mm gold converter, with the size 12 x 20.6 cm^2, is placed in the fringe field 58.5 cm from the scattering center.

One experimental difficulty with the present detector and magnet configuration is that the detectors must be situated very close to the target. This is necessary in order to detect very asymmetrically produced pairs. It is therefore extremely important to have maximal shielding between the target and the detectors. With a layer of about 20 radiation lengths of tungsten above and below the target used as shielding we can conservatively run with a beam current of about 50 nA (100% duty factor) on a 50 mg/cm^2 boron target with acceptable background rates in the trigger.

The quantity which specifies the overall acceptance in the pair-spectrometer is the fraction of the photons created in the target that are possible to detect as an electron and a positron in the lower and upper set of detectors. This acceptance $A(E_\gamma)$ depends on the solid angle for the converter, the conversion probability, the geometrical detection acceptance for the trajectories, and the fraction of events which are lost due to bremsstrahlung in the converter. Note that the acceptance varies with the photon energy as well as with the setting of the magnet current.

In order to calculate $A(E_\gamma)$ an extensive Monte Carlo simulation of the whole experimental set up has been performed. The simulation of the pair production process and the subsequent interactions in the converter, i.e. energy loss from ionization, multiple Coulomb scattering and bremsstrahlung, was done by using the EGS code. This simulation code provided the input to a tracking program, from which the geometrical detection acceptance was found by tracking the generated e^+e^- pairs through the magnetic field and requiring that the relevant detectors should be crossed. The geometrical acceptance is mainly determined by the available momentum range of the magnet for the electron and positron, but also other phenomena like the shape of the momentum distribution for the e^+e^- pair, and effects from multiple Coulomb scattering in the helium gas inside PACMAN and in various mylar foils are important. Fig. 3 shows the total acceptance $A(E_\gamma)$ for one magnet current over a wide range of photon energies. This acceptance curve has also been verified by experimental (p,γ) data. The useable momentum bite of the spectrometer is roughly given by the FWHM of the acceptance curve, which gives $\Delta E_\gamma = 40$ MeV with a maximum acceptance of 2.5×10^{-5} at $E_\gamma = 95$ MeV.

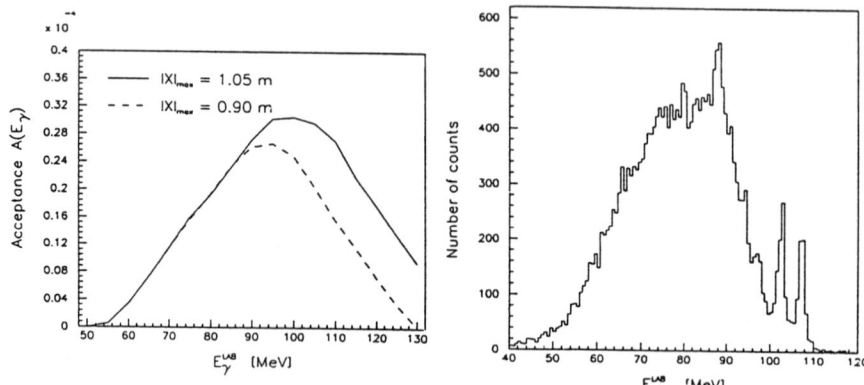

Fig. 3. The total acceptance $A(E_\gamma)$ as a function of photon energy for two different cuts on the maximal accepted x-coordinate.

Fig. 4. A gamma spectrum for the $^{11}B(p,\gamma)$ ^{12}C reaction analyzed with a fast on-line method.

A gamma spectrum from the $^{11}B\,(p,\gamma)\,^{12}C$ reaction analyzed with a fast on-line method is shown in fig. 4. The gross shape of this distribution reflects the acceptance function of PACMAN. Examples of gamma spectra in its final form are shown in fig. 5. The energy resolution in these spectra is about 0.8 MeV. A detailed analysis of the different contributions to this energy width shows that the intrinsic resolution of PACMAN is about 0.6 MeV.

The measurements of internal pair production, (p,e^+e^-), which involves an internal conversion of a virtual photon inside the target nucleus can also be studied with PACMAN. The only necessary modification in the experimental set up is to remove the converter. The momentum determination of the e^+e^- pair is made by raytracing the measured outgoing trajectories from PACMAN back to the target point. The opening angle of the electron and positron and their energies determine the invariant mass of the decaying system, i.e. the photon mass, which has a lower limit of $2m_e$ and an upper limit set by the reaction kinematics. In the (p,e^+e^-) case the acceptance becomes strongly dependent on the mass of the virtual photon. Detailed calculations of this acceptance are in progress.

It should be pointed out that although the cross section for the (p,e^+e^-) reaction is much smaller than for the (p,γ) reaction, the count rate from the (p,e^+e^-) reaction can be quite large, since no external conversion is required. This could in fact impose some difficulties in distinguishing between the two reactions when PACMAN is set up to study the (p,γ) reaction. Using a very thin converter of 1% conversion probability, the contribution from the internal conversion is not negligible. Owing to the placement of the converter in the fringe field, it is however possible in the raytrace analysis to determine whether the e^+e^- pair originates from the target or the converter.

PRELIMINARY RESULT FROM THE $^{11}B\,(p,\gamma)\,^{12}C$ REACTION

This first experiment using PACMAN was devoted to an investigation of the (p,γ) reaction on ^{11}B at 98 MeV. This reaction was chosen from experimental point of view as being the most suitable testground, since the level spacing for the low lying states in the residual nucleus is large and the break up channels start at reasonably high excitation energies ($^8Be+\alpha$ at 7.4 MeV and $^{11}B+p$ at 16 MeV). It should thus be possible to isolate well defined peaks free from background due to the continuum. Spectra of the gamma energies from the $^{11}B\,(p,\gamma)\,^{12}C$ reaction, recorded at laboratory angles 37.5° and 120°, are shown in fig. 5. The first spectrum represents a case with relatively large cross sections, while the second one is recorded in a region where the cross section is only a few nb/sr. The first four peaks in these spectra correspond to individual states in the residual nucleus ^{12}C, while other peaks can correspond to groups of states. A brief inspection of these spectra reveals a clear nuclear structure dependence in this reaction. We note that the cross section to the 4.4 MeV state is very dominating at large momentum transfer (120° corresponds to 461 MeV/c). We also note, from fig. 4, the ocurrence of peaks at high excitation energies.

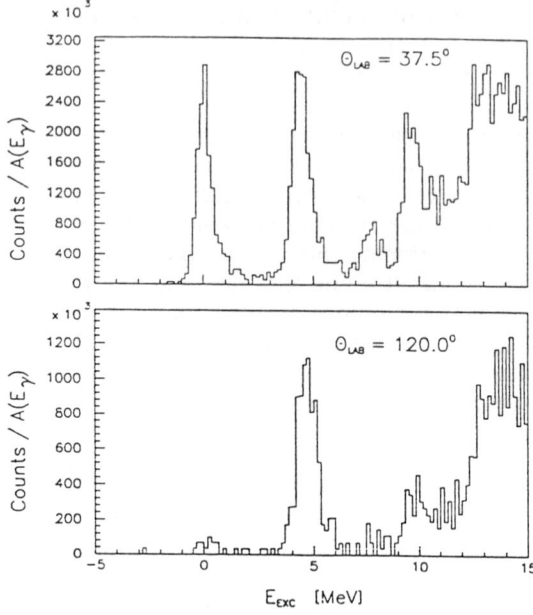

Fig. 5. Excitation energy spectra from the $^{11}B(p,\gamma)^{12}C$ reaction at 98 MeV.

Angular distributions of the differential cross section have been extracted for the transitions to the first four states in ^{12}C, i.e. the ground state (0+), the 2+ state at 4.4 MeV, the 0+ state at 7.6 MeV and the 3- state at 9.6 MeV. These distributions are shown in fig.6. The main features of these distributions are similar. It is only for large momentum transfer one can see some differences. The two 0+ states, i.e. the ground state and the 7.6 MeV state, have a more or less identical shape. They fall off steeper at large momentum transfer than the angular distributions for the transitions to the 4.4 and 9.6 MeV states. The implications of the measured angular distribution in terms of reaction dynamics and nuclear structure are certainly not appearant from a brief inspection of the data or by e.g. any simple model calculation based on a DWBA analysis. Such attempts, although sometimes successful, to interprete earlier data from ground state transitions for different nuclei and energies have not given any reliable conclusive answer. Too many problems in the theoretical analysis, like distortion effects, ortogonality, single particle wave functions, exchange currents, effects of the Δ resonance, applicability of Siegert's theorem, gauge invariance etc, have obscured the true character of this reaction. The experience from all these attempts has made it clear that this reaction can not be treated in a simple way, but progress is going to require a realistic treatment of all aspects involved in this reaction. Such a program is undertaken by S. Cotanch and T. Bright at NCSU, who are calculating the cross sections for the transitions in ^{11}B presented in this paper. Their analysis consists of a coupled channel calculation based on the continuum shell model calculation by Buck and Hill[4]. Their calculation, which is completely microscopic with no adjustable parameters, includes a realistic model space, which leads to a set

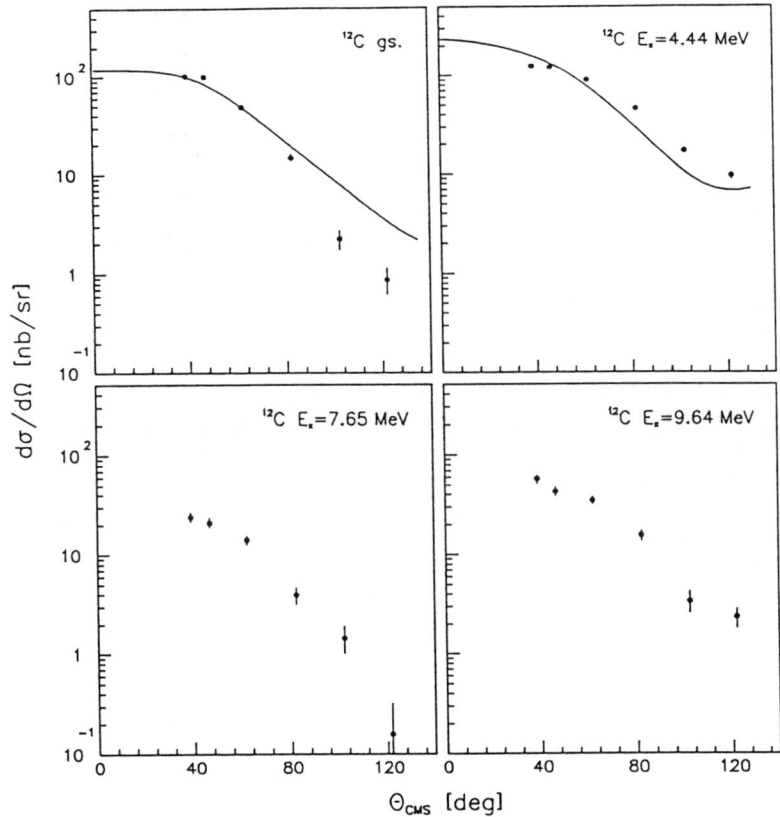

Fig. 6. Angular distributions of the differential cross section for transitions in the $^{11}B(p,\gamma)^{12}C$ to different bound states in ^{12}C.

of inhomogeneous, integro-differential equations. The photonuclear amplitude includes a multipole decomposition of E1 to E11 and M1 to M9, and exchange currents from pions, vector mesons and isobars will be included. Preliminary results [5] from calculations for the transition to the ground state and first excited state in ^{12}C are shown as the solid line in fig. 6. These calculations do not yet include exchange currents or virtual $\Delta(1232)$ amplitudes and hence are not expected to be accurate at high momentum transfer. Nevertheless there is an excellent agreement between theory and experiment for the transition to the 2^+ state over the full angular range. The agreement between the data and the calculated transition to the ground state is equally impressive at small angles, but the calculation overestimates in this case the experimental cross section at large angles. Further calculations, which includes exchange currents, will shed light on the slight discrepancy at the back angles.

PRELIMINARY RESULT FROM THE ^{11}B(p,e$^+$e$^-$)^{12}C REACTION

The rare (p, e^+e^-) reaction, which involves an internal conversion of a virtual gamma inside the nucleus, has large similarities with the (p, γ) reaction, but it also exhibits some distinct differences. In the simple one-body process the virtual photon is emitted directly from the incident proton. The fact that the emitted photon is virtual, implies that a longitudinal part appears in the transition matrix, as opposed to the case with real photon emission, which only contains a transverse part. It is however difficult to say anything quantitatively about the sensitivity to the longitudinal part in the phase space covered by PACMAN before detailed calculations are done. Such calculations are under way by a group at ISV[6].

The massive photon decays in an electron and a positron, which are emitted with an opening angle related to the magnitude of the mass and the energy of the photon. For a photon of total energy 100 MeV, the minimum opening angle (degrees) in the lab is approximately numerically the same as the photon mass (MeV). The large opening angle of PACMAN allows the virtual gamma emission to be studied for an invariant mass of the e$^+$e$^-$ pair up to about 10 MeV. Fortunately, the cross section for the (p, e^+e^-) reaction is peaked strongly towards low values of the photon mass. For example, approximately 50% of the cross section is expected to be found below a mass of 10 MeV for a photon with 150 MeV total energy[6].

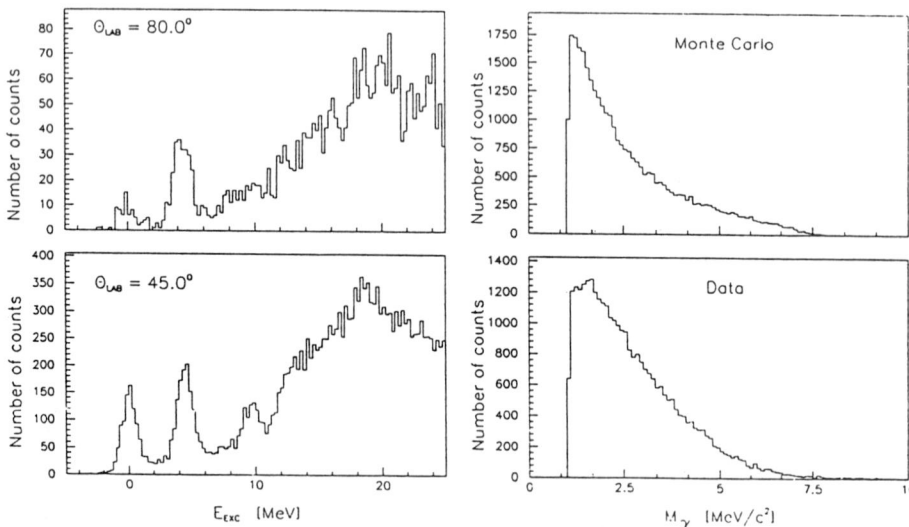

Fig. 7. Excitation energy spectra from the ^{11}B(p, e$^+$e$^-$)^{12}C reaction at 98 MeV.

Fig. 8. Invariant mass distribution for e$^-$e$^-$ pairs from real data and from Monte Carlo.

We have measured the $^{11}B(p,e^+e^-)^{12}C$ reaction at 45 and 80 degrees, which data are shown in fig. 7. The similarity between the (p,γ) data is almost total in this way of presentation. In order to find any differences, if there are any, it is thus necessary to select data far away from the (p,γ) kinematics region, i.e. at large masses of the photon or at selected azimuthal angles for the e^+e^- pair. Such analysis is in progress.
Another experimental result from the (p,e^+e^-) measurement is the invariant mass distribution of the virtual photon. This distribution can easily be calculated in our case since the momenta of both the e^+ and e^- are measured as well as the opening angle. The mass resolution obtained by PACMAN is about 1 MeV. Using the whole sample of (p,e^+e^-) data available at 45° we get the result shown in fig. 8. The shape of this distribution is very much affected by the acceptance of PACMAN for different invariant masses of the e^+e^- pair, and the work with unfolding the acceptance in the spectrum is in progress. At present we can show, in fig. 8, how a mass spectrum from Monte Carlo generated e^+e^- pairs, based upon a mass distribution calculated for the $^2H(p,e^+e^-)^3He$ reaction [6], should look like when detected by PACMAN. Comparing the experimentally detected data set and the Monte Carlo generated set we find one peculiarity. In the real data there is a weak indication of a peak around 1.7 MeV which does not seem to be reproduced by the present Monte Carlo simulation. It is however extremely unlikely that a peak should occur in the invariant mass distribution of the e^+e^- pair. This fact calls for further detailed analysis of the present data and possibly also further experimental measurements. It would certainly be most interesting to establish the existence of some new phenomena pertinent to the e^+e^- system, in particular in the light of the structure found in the energy spectra from electron and positron emission in heavy ion collisions [7].

REFERENCES

1. R.S. Turkey et al., Phys. Lett. 157B (1985) 19,
   G.S. Adams et al., Phys. Rev. C 38 (1988) 2771.
2. S. Boffi et al., Nucl. Phys. A359 (1981) 91,
   M.Leitch et al., Phys. Rev. C 31 (1985) 1633.
3. B. Höistad et al., TSL/ISV-90-0034 internal report, to be published in Nuclear Instruments and Methods.
4. Buck and Hill, Nucl. Phys. A95, 271 (1967).
5. S. Cotanch and T. Bright, North Carolina State University , private communication.
6. U. Tengblad, ISV, Uppsala, private communication.
7. P. Salabura et al. Phys. Lett. 245B (1990) 153, and references in this paper.

A Study of Reaction Mechanisms for Gamma Production in Fast-Nucleon Induced Reactions

A. Höring, H. A. Weidenmüller

Max Planck Institut für Kernphysik, D-6900 Heidelberg, West Germany

F. S. Dietrich

Lawrence Livermore National Laboratory, Livermore, CA 94550

M. Herman, and G. Reffo

Ente Nazionale Energie Alternative, 40138 Bologna, Italy

Abstract

Using the example of 14-MeV neutrons incident on ^{93}Nb, we have studied the contributions to the gamma spectrum from the multistep-compound reaction mechanism and from direct and semidirect capture.

Introduction

The energy averaged cross-section of a nuclear reaction can be split in two parts. One part depends on the energy-averaged S matrixelement and therefore describes the fast part of the process, the direct reactions. The other, given in terms of the fluctuating part of the S-matrix, describes the slow part of the reaction and is refered to as the compound nucleus cross-section.

$$<\sigma_{ab}> = \pi \lambda_a^2 (|\delta_{ab} - <S_{ab}>|^2 + <|S_{ab}^{fl}|^2>)$$

Both terms include pre-equilibrium contributions. The second term contains the so-called multistep compound (MSC) process in which all particles occupy bound states or bound states imbedded in the continuum. The inclusion of gamma emission in the MSC via the decay of giant dipole resonances (GDR) is studied in the next section. Direct and semidirect capture to the continuum which are part of the direct cross section are then discussed in the following section.

The Inclusion of the GDR into the MSC and its Contribution to the Gamma Cross-Section

The MSC as described by Nishioka et al.[1] allows for particle emission in each stage of the reaction distinguished by its complexity which is described by the exciton number (number of particles plus number of holes). The following representation of the S matrix is used.

$$S_{ab} = \delta_{ab} - 2i\pi \sum_{M\mu,N\nu} W_{a,M\mu}(D^{-1})_{M\mu,N\nu} W_{N\nu,b}$$

$$D_{M\mu,N\nu} = E\delta_{MN}\delta_{\mu\nu} - H_{M\mu,N\nu} + i\pi \sum_c W_{M\mu,c} W_{N\nu,c}$$

where the phases have been left out and M, N are labels for the exciton number defining a given class, μ, ν are running indices within the classes, and W couples bound states to channel states. Each exciton class is described in the Hamiltonian by a Gaussian Orthogonal Ensemble (GOE) and the part of the Hamiltonian which connects two neighbouring classes consists of Gaussian distributed matrix elements. The energy average is replaced by the ensemble average; both are equal by ergodicity.

To describe the gamma decay of a giant dipole resonance (GDR) in the framework of MSC we describe the GDR as a coherent linear superposition of one particle one hole states and use the Brink-Axel hypothesis, which states that the GDR cannot only be built on the ground state but also on every excited state. In order to construct a basis that distinguishes between states that are GDR's built on some low lying states from states that do not contain a GDR, we build a GDR on each uncorrelated exciton-model state by application of the appropriate particle-hole operator. Using energy conservation, angular momentum conservation and the fact that the GDR is described as a two exciton state we find that each H_{MM} is now described by four instead of one GOE, three of which can directly decay via the GDR channel (distinguished by the three different possibilities of the angular momentum of the final state) and one that can make gamma decay only via coupling to the GDR states of the next higher exciton class. This is how the GDR gets its spreading width.

The Ensemble average is performed as in reference 1 and yields

$$\overline{\sigma_{a\gamma}^{fl}} = \sum_{M,N,m,n} T_{Mm}^a \Pi_{Mm,Nn} T_{Nn}^\gamma$$

The additional indices m and n label the four different subclasses of each exciton class, $T^a = \sum_{Mm} T_{Mm}^a$ are the optical model transmission coefficients and the probability transport matrix $\Pi_{Mm,Nn}$ is given by:

$$(\Pi^{-1})_{Mm,Nn} = \delta_{MN}\delta_{mn} 2\pi\rho_{Mm}(\Gamma_{Mm}^\uparrow + \Gamma_{Mm}^\downarrow) + (1-\delta_{MN}\delta_{mn}) 2\pi\rho_{Mm} v_{Mm,Nn} 2\pi\rho_{Nn}$$

with $\Gamma_{Mm}^\uparrow$ and $\Gamma_{Mm}^\downarrow$ being the decay and spreading width of class Mm, respectively, ρ_{Mm} is the level density of class Mm, and $v_{Mm,Nn}$ the average matrix element between the classes Mm and Nn. γ stands symbolically for the E1 gamma channel. The transmission coefficient for the gammas T^γ can be related to its strength function and through microreversibility directly to the GDR absorpion cross-section. Experimental data are used for the parameters of the latter.

$$T_\gamma = \frac{S_\gamma}{2\pi} = \frac{2}{3\pi}\sigma_{abs}(E_\gamma^{E1})\frac{(E_\gamma^{E1})^2}{(\hbar c)^2}$$

Figure 1 shows the MSC contribution to the gamma spectrum using the example of 14-MeV neutrons incident on ^{93}Nb.

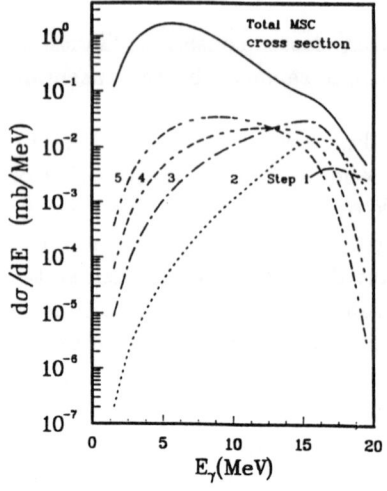
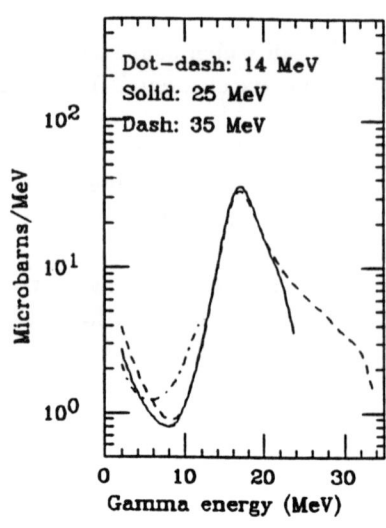

Fig. 1 $^{93}Nb(n,\gamma)^{94}Nb$, MSC contribution

Fig. 2 $^{93}Nb(n,\gamma)^{94}Nb$, DSD contribution

Direct and Semi-Direct Capture

We supplement the MSC by a direct-semidirect (DSD) calculation to describe the initial stages of the reaction. Such a calculation has already been performed by Rigaud et al.[3] for capture to bound final states in ^{94}Nb, with satisfactory results. We have extended the direct-semidirect calculation to higher-lying (unbound) final states, using the method of integration in the complex plane described by Weller et al.[2] to treat the radial integrals for the direct term, which extend far beyond the nuclear surface. Whereas in the calculation to bound states the spectroscopy of the final single-particle state is described by a Woods-Saxon potential, in the continuum case it is contained in the optical potential for the outgoing channel. The results of DSD calculations to the unbound final states are shown in Figure 2 for neutrons incident on ^{93}Nb at energies of 14, 25, and 35 MeV. DSD capture to unbound states is not very large at 14 MeV because the gamma energies lie below the GDR; however, the mechanism becomes more important as the incident energy increases.

References

(1) H. Nishioka, J.J.M. Verbaarschot, H. A. Weidenmüller, and S. Yoshida, Ann. Phys. 172, 67 (1986)
(2) H. R. Weller et al., Phys. Rev. C25, 2921 (1982)
(3) F. Rigaud et al., Nuc. Phys. 173, 551 (1971)

Work by LLNL under USDOE Contract No. W-7405-ENG-48, and by ENEA Contracts 37481(30/12/88) and 27566.

Fundamental Physics
with Neutrons

PARITY AND TIME REVERSAL SYMMETRY VIOLATION IN NEUTRON-NUCLEUS SCATTERING

C. R. Gould, J. E. Bush, C. M. Frankle, D. G. Haase, G. E. Mitchell
*North Carolina State University, Raleigh, NC 27695 and
Triangle Universities Nuclear Laboratory, Durham, NC 27706*

C. D. Bowman, J. D. Bowman, J. Knudson,
S. Penttilä, S. J. Seestrom, J. J. Szymanski, S. H. Yoo, V. W. Yuan
Los Alamos National Laboratory, Los Alamos, NM 87545

N. R. Roberson and X. Zhu
*Duke University, Durham, NC 27706 and
Triangle Universities Nuclear Laboratory, Durham, NC 27706*

P. P. J. Delheij
TRIUMF, Vancouver, British Columbia

H. Postma
University of Technology, Delft, The Netherlands

ABSTRACT

The TRIPLE collaboration has begun a multi year program of study of symmetry violation using the intense pulsed polarized epithermal neutron beam available at LANSCE (Los Alamos Neutron Scattering Center). The parity violation experiments consist of measurement of the helicity dependence of neutron transmission at p-wave resonances in compound nuclei. A recent success has been the first determination of the spreading width of the weak interaction in ^{239}U.

INTRODUCTION

Coherent neutron scattering provides a unique way of studying spin dependent forces in nuclear physics. In combination with recent work on symmetry breaking in quantum chaotic systems, it has emerged as a powerful new technique for the study of parity (P) and time reversal (T) symmetry violation in nuclei. In this talk we will review the basic ideas of symmetry tests in neutron transmission measurements, primarily with reference to the TRIPLE collaboration's recent work[1] on P violation in p-wave resonances in ^{239}U. The goals of the TRIPLE program are to map the A dependence of the spreading width of the weak interaction, and to exploit compound nuclear enhancement mechanisms to perform sensitive new tests of T violation in both the weak and strong interaction regimes.

Work on P violation prior to 1987 has been summarized by Krupchitsky[2], and issues relating to T violation were reviewed in a 1987 TUNL-LANL workshop proceedings[3]. Some of the theoretical background was recently discussed by Weidenmüller[4] and previous summaries of the TRIPLE program have been given by Bowman[5] and Seestrom[6] and in references therein.

POLARIZED NEUTRON TRANSMISSION

The operators of interest in coherent neutron transmission are the neutron momentum k, the neutron spin σ, and the nuclear spin I. The forward scattering amplitude f will contain terms of the form $\sigma \cdot c$ where c is a vector along k, I or $I \times k$. The simplest terms are $\sigma \cdot I$, the familiar central spin-spin interaction, $\sigma \cdot k$ the P-violating helicity dependence, and $\sigma \cdot (I \times k)$, the three-fold (TC) correlation which is both P odd and T odd. There are more complicated combinations, for example the five-fold correlation (FC), $\sigma \cdot (I \times k)(I \cdot k)$, the simplest T-odd P-even term. In the expression for the forward-scattering amplitude each combination is associated with a pair of statistical tensors describing the orientation of the beam and target[7]. A complete list of the allowed vector combinations is given in the appendix along with expressions for the helicity-dependent cross sections.

By coherent we mean that the forward-scattered neutron wave can interfere with the incident wave. This coherence leads to two classes of phenomena: neutron spin rotation about c, and a difference in the total cross sections for neutrons with spins parallel and anti-parallel to c. The former is due to the real part of the spin-dependent forward-scattering amplitude, f_r, which changes the phase of the transmitted wave, and the latter is due to the imaginary part f_i which changes the amplitude of the transmitted wave.

Both phenomena were observed in 1980 in P-violating thermal neutron helicity-dependent transmission at Grenoble[8] The effects, while small, were about 1000 times larger than expected considering the magnitude of the Fermi coupling constant G_F. They clearly showed the importance of taking into account the structure of the compound nuclear system. At higher energies (1-1000eV), only transmission asymmetry measurements are feasible, but the enhancements, as first pointed out by Sushkov and Flambaum[9], can be even another factor of 1000 larger.

The experiments which stimulated much of the present interest in symmetry violation studies with polarized neutrons were carried out at Dubna in 1982, and consisted of transmission asymmetry measurements at energies corresponding to p-wave neutron resonances in heavy nuclei. Several large analyzing powers were observed[10], including a 10% effect for the 0.73eV resonance in ^{139}La. This effect is so large that we now routinely use it as a polarization calibration for the neutron beams at LANSCE.

The origin of this enhancement is discussed in many of the references cited earlier. It arises from the known properties of the strong interaction and the long lifetimes of the CN states, and is not due to any special feature of the weak interaction *per se*. We emphasize that despite these large enhancements, the experiments are time consuming and difficult because in general the p-wave resonances themselves are only weakly excited. Information in the literature is sparse and in many nuclei, holmium for example, a p-wave resonance has yet to be seen.

SPREADING WIDTH OF THE WEAK INTERACTION

The experiments all measure a matrix element $M = <\phi_s \mid H_W \mid \phi_p>$ of the weak interaction Hamiltonian, H_W, between compound nuclear (CN) states, ϕ_s and ϕ_p of the same spin but opposite parity. Typical values for M are found to be of order $10^{-3} eV$. The states are so complicated, with $\sim 10^6$ single particle components, that for some time it was not considered possible to interpret these values in anything but a qualitative fashion. But recent work by French et al.[11] has shown that the powerful methods of statistical nuclear theory can be used to make quantitative statements about symmetry breaking in these highly complex systems, and that the symmetry breaking can in turn be related directly to the underlying properties of the effective nucleon-nucleon interaction.

The quantity of interest is the spreading width Γ_{PV}, related to the mean square of the parity violating matrix elements, $<M^2>$ and the mean spacing between the levels, $<D>$ via

$$\Gamma_{PV} = 2\pi <M^2>/<D>$$

The spreading width is expected to be independent of mass number or excitation energy; in contrast to the mean spacings or mean square matrix elements from which it is defined. The first determination of Γ_{PV} was obtained from our recent study[1] of seventeen p-wave resonances in ^{239}U where a value $\sim 10^{-7} eV$ was obtained.

Spreading widths arise in many contexts in nuclear physics. Isospin symmetry breaking is for example characterized by the spreading width of the electromagnetic interaction, found by Harney et al.[12] to be of order 10keV. Intuitively, spreading widths set a time scale $\tau \sim \hbar/\Gamma$ for the symmetry breaking to occur. Our value of Γ_{PV} indicates CN states would be completely mixed in parity after $\sim 10^{-9}s$ if they did not decay rapidly via n or γ emission.

The quantitative link to symmetry breaking in the underlying effective N-N interaction is given[11] by the dimensionless order parameter Λ_{PV}

$$\Gamma_{PV} \equiv 2\pi \Lambda_{PV} D = 2\pi 10^5 (eV) \alpha_{PV}^2$$

where α_{PV} is the ratio of the P-odd to P-even effective N-N interactions. Our work on ^{239}U gave $\alpha_{PV} \sim 4 \times 10^{-7}$, in qualitative agreement with the free N-N value of $G_F m_\pi^2 / G_S \sim 2 \times 10^{-7}$. The way in which α_{PV} is related to the free parity-violating N-N interaction is a major theoretical issue yet to be explored. But it seems clear that it must reflect the strengths of the coupling constants in the weak nucleon nucleon potential. The measurement of these coefficients has been the object of much beautiful experimental work over the last decade (see Adelberger and Haxton[13] for a recent review).

The requirements for determining Γ_{PV} are P-odd cross sections σ_{PV} strong enough to be seen and a sufficient sample of p-wave resonances to be able to compute a variance of the analyzing powers. From the expression in the appendix, we see $\sigma_{PV} \sim \sqrt{S_0 S_1 D}$ where $S_{0,1}$ are the s- and p-wave strength functions and D is the mean spacing. (We use $M \sim \sqrt{D}$ from the fact that Γ_{PV} is constant, and $\Gamma_{s,p}^n \sim S_{0,1} D$ from the definitions of the strength functions.) This argues for nuclei with large spacings, but in this case there will be only a few p-wave resonances in the energy range accessible for study— up to $\sim 300eV$ for our present experiments at LANSCE. As the facility is upgraded we will be able to study nuclei with smaller spacings, but for the moment we are concentrating on nuclei with spacings in the 20-50eV range, with sample sizes limited to at most 10-20 resonances per nucleus.

EXPERIMENTAL METHODS AND RESULTS

The experiments are carried out on a 56m flight path at LANSCE and take advantage of the intense pulses of 800-MeV protons delivered by the Proton Storage Ring (PSR). The H^- beam from LAMPF is first converted to H^0 with 100% efficiency in a high-field stripper magnet and then stripped to H^+ with $\sim 90\%$ efficiency in a thin carbon foil. Up to 2800 turns are injected and accumulated before the beam pulse is extracted and transported to the neutron-production target. The pulse is roughly triangular in shape with a width at the base of 260ns. The present beam intensity of 75μA at 20Hz corresponds to a peak current pulse of 30A. Each proton produces 15-20 neutrons which are moderated in water to give a $1/E$ neutron flux. We have been using a dynamically pumped LMN cryostat to polarize the neutrons. The neutron polarization was ∼ 40% in 1989, but has dropped to ∼ 20% this year. We believe this decrease is due in part to radiation damage in the crystals. We will be bringing up a new system in 1991 consisting of ammonia pumped in a 5T magnetic field at a temperature of 1K. The polarization is expected to be > 90% and the beam area will be $50 cm^2$, together giving a figure of merit $P^2 I$ at least ten times larger than our present system.

The neutron spins are rapidly reversed every 10s in a magnetic spin flipper system located after the cryostat. The spins can also be reversed by changing the microwave pumping frequency. This is a slow process and we do this only once a day as a check on systematic errors. The ^{238}U data were taken with the sample at room temperature. This year we have been using a liquid nitrogen chiller to reduce Doppler broadening, which dominates the energy resolution below about 100eV. Most of our data so far have been taken with an array of seven ^6Li glass scintillators operating in current mode. A transient digitizer is used to sample the neutron yield as a function of time. The efficiency of ^6Li falls as $1/v$ and we have also investigated using ^{10}B loaded liquid scintillators to improve the detection efficiency for neutrons up to 1keV in energy. The large hydrogen cross section gives a high efficiency for scattering and the neutrons, once thermalized, are readily captured by the ^{10}B.

Figure 1 shows the analyzing powers, P_i and the likelihood function obtained for the seventeen p-waves studied in ^{238}U. The biggest effect was obtained for the 63-eV resonance ($2.5 \pm 0.4\%$); four other resonances showed 2σ effects (only one would be expected from random fluctuations). Weak ^{235}U s waves showed no effects giving us confidence that systematic errors are small or absent. The most compact 68% confidence interval in the likelihood function gives $M = 0.58^{+0.50}_{-0.25}$meV. The dotted line shows the likelihood function for the 63-eV resonance alone and shows the importance of the small or null values in bounding M from above. An upper limit is necessary in integrating the likelihood functions. We use 10meV, but the final value for M is relatively insensitive to this choice.

The ^{232}Th data from this year's running have shown more large effects than ^{238}U, and preliminary analysis indicates out of 21 resonances at least four with > 4σ effects. Resonances at 38 and 64eV show 5σ effects of order 10%, comparable to the 0.73eV resonance in ^{139}La, previously the largest known effect. The most pronounced asymmetry is for a resonance at 128eV, shown in figure 2. This is an 8σ effect, and indicates the importance of having good neutron intensity and resolution up to high energies.

We saw no effects in a later run with natural indium, but the sensitivity was low due to low neutron polarization from the LMN cryostat.

SUMMARY

Coherent neutron scattering has emerged as a powerful new technique for the study of parity and time reversal violation in nuclei due to the large enhancements present in CN scattering processes. Symmetry violation is characterized by the spreading width of the underlying effective N-N interaction and we have made the first measurement of the spreading width of the weak interaction in a heavy nucleus, ^{238}U. The value of 10^{-7}eV is qualitatively consistent with the expected magnitude of weak interaction effects $\sim 10^{-7}$. More data for investigating the A-dependence of the spreading width, and theoretical study of the relation between the free and effective weak N-N interactions are clearly needed. We look forward to continued progress in this new field.

ACKNOWLEDGEMENTS

This work was supported in part by the U. S. Department of Energy, under Contracts. No. DE-AC05-76ER01067 and DE-FG05-88ER40441. One of the authors (CRG) acknowledges financial support from Associated Western Universities for a sabbatical fellowship at Los Alamos National Laboratory.

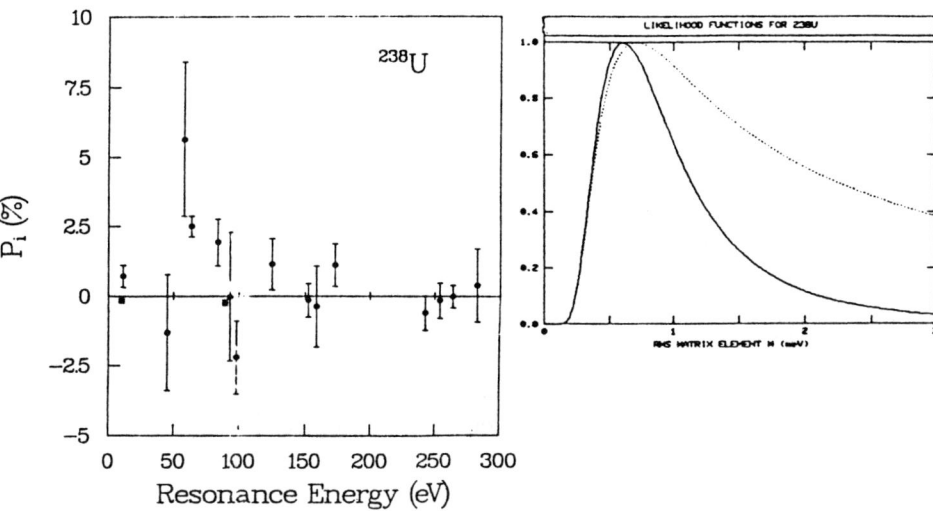

Fig 1. Analyzing powers and likelihood function for ^{239}U

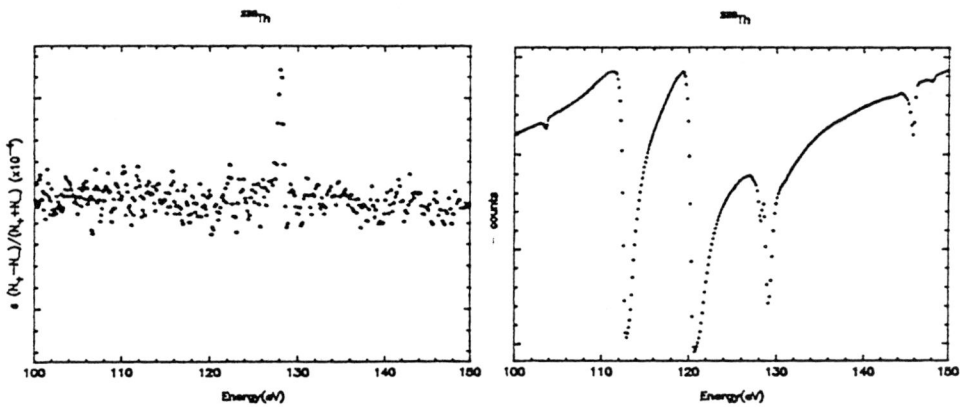

Fig 2. Asymmetry and transmission yield near the 128eV resonance in ^{232}Th

APPENDIX

The vector combinations which contribute to the spin dependent part of the neutron forward scattering amplitude are listed in the table. Angular momentum coupling limits the number of k's to ≤ 2 for s- and p- wave neutrons. Additionally the number of times I can appear is limited to $2I$. Higher order terms are obtained simply by multiplying the $I \geq 3/2$ terms by $I \cdot k$.

The helicity dependence of the resonance cross section is parametrized by an analyzing power P defined by $\sigma^\pm = \sigma_p(1 \pm P)$. Experiments measure the transmission count rate asymmetry $(N^+ - N^-)/(N^+ + N^-) = -\sigma_{PV}(nx)f_n$ where $\sigma_{PV} = P\sigma_p$, nx is the target thickness and f_n is the neutron polarization. At neutron energy $E = E_p$ in two level s-p interference[7,14,15] for $I = 0$ targets:

$$\sigma_{PV} = \frac{8\pi}{k^2} \frac{<\phi_s | H_W | \phi_p>}{(E_s - E_p)} \frac{\sqrt{\Gamma_s^n \Gamma_p^n}}{\Gamma_p}$$

and

$$\sigma_p = \frac{4\pi}{k^2} \frac{\Gamma_p^n}{\Gamma_p}$$

The s-wave width is to be evaluated at $E = E_p$. The expression for the analyzing power of the ith p wave P_i can be generalized[1] to take into account interference with many s waves ϕ_{sj}:

$$P_i = \sum_j A_{ij} <\phi_{sj} | H_W | \phi_{pi}>$$

with

$$A_{ij} = \frac{2}{(E_{sj} - E_{pi})} \sqrt{\frac{\Gamma_{sj}^n}{\Gamma_{pi}^n}}$$

The matrix elements are considered to be Gaussian distributed random variables with mean zero and variance $<M^2>$. The variance of a sum is the sum of the variances so $Var(P_i) = K_i^2 <M^2>$, where $K_i^2 = \sum_j A_{ij}^2$. In practice we define a new observable $Q_i = P_i/K_i$ whose variance is the mean square matrix element $<M^2>$. The spins of the p-wave resonances are in general not known. We extract $<M^2>$ from a likelihood analysis that takes into account the fact that only the $J = 1/2$ resonances can show parity mixing. For the 17 resonances in ^{238}U the likelihood function is (probability 1/3 for $J = 1/2$, probability 2/3 for $J = 3/2$)

$$L(M) = \prod_{i=1}^{17} \left\{ \frac{1/3}{\sqrt{2\pi(\sigma_{Qi}^2 + M^2)}} \exp \frac{-Q_i^2}{2(\sigma_{Qi}^2 + M^2)} + \frac{2/3}{\sqrt{2\pi\sigma_{Qi}^2}} \exp \frac{-Q_i^2}{2\sigma_{Qi}^2} \right\}$$

Target Spin	Vectors	Odd
$I \geq 0$	1	
	$\sigma \cdot k$	P
$I \geq 1/2$	$\sigma \cdot I$	
	$(\sigma \cdot k)(I \cdot k)$	
	$\sigma \cdot (I \times k)$	PT
$I \geq 1$	$(\sigma \cdot I)(I \cdot k)$	P
	$(\sigma \cdot k)(I \cdot k)^2$	P
	$\sigma \cdot (I \times k)(I \cdot k)$	T
$I \geq 3/2$	$(\sigma \cdot I)(I \cdot k)^2$	
	$(\sigma \cdot k)(I \cdot k)^3$	
	$\sigma \cdot (I \times k)(I \cdot k)^2$	PT

REFERENCES

1 J. D. Bowman et al. Phys. Rev. Lett. **65**, 1192 (1990).
2 P. A. Krupchitsky, *Fundamental Research with Polarized Slow Neutrons*, (Springer-Verlag, Berlin, 1987).
3 *Tests of Time Reversal Invariance in Neutron Physics*, ed. N. R. Roberson et al. (World Scientific, Singapore, 1987)
4 H. A. Weidenmüller, *Fundamental Symmetries in Nuclei and Particles*, ed. H. Henrikson and P. Vogel (World Scientific, Singapore, 1989),p. 30
5 J. D. Bowman et al., op. cit. p. 1
6 S. J. Seestrom et al., Proceedings of Workshop on Polarized Beams and Polarized Targets, McCormicks Creek (APS, New York, 1989),p. 1
7 C. R. Gould, D. G. Haase, N. R. Roberson, H. Postma, and J. D. Bowman, Int. J. Mod. Phys. **A5**, 2181 (1990).
8 M. Forte, B. R. Heckel, N. F. Ramsey, K. Green, G. L. Greene, J. Byrne, and J. M. Pendlebury, Phys. Rev. Lett. **45**, 2088 (1980).
9 S. P. Sushkov and V. V. Flambaum, JETP Pis'ma **32**, 377 (1980).
10 V. P. Alfimenkov, S. B. Borzakov, Vo Van Thuan, Yu. D. Mareev, L. B. Pikelner, A. S. Khrykin, and E. I. Sharapov, Nucl. Phys. **A398**, 93 (1983).
11 J. B. French, V. K. B. Kota, A. Pandey and S Tomsovic, Ann. Phys. (N. Y.) **181**, 198 (1988); **181**, 235 (1988)
12 H. L. Harney et al., Rev. Mod. Phys. **58**, 607 (1986)
13 E. G. Adelberger and W. C. Haxton, Ann. Rev. Nuc. Sci. **35**, 501 (1985)
14 V. E. Bunakov and V. P. Gudkov, Nucl. Phys. **A401**, 93 (1983).
15 J. R. Vanhoy, E. G. Bilpuch, J. F. Shriner Jr., and G. E. Mitchell, Z. Phys. A–Atoms and Nuclei **331**, 1 (1988).

THE MEASUREMENTS OF PARITY VIOLATION IN RESONANT NEUTRON-CAPTURE REACTIONS

E.I. Sharapov[1], S. A. Wender[2], H. Postma[3], S. J. Seestrom[2], C.R. Gould[4], A. Wasson[5], Yu. P. Popov[1], and C. D. Bowman[2]

[1] *Joint Institute for Nuclear Research, Dubna, USSR*
[2] *Los Alamos National Laboratory, Los Alamos, NM USA 87545*
[3] *University of Technology, Delft, The Netherlands*
[4] *North Carolina State University, Raleigh, NC USA 27695*
[5] *National Institute of Science and Technology, Washington, DC*

ABSTRACT

The study of parity violation in total (n,γ) cross sections on ^{139}La and ^{117}Sn targets was performed at the LANSCE pulsed neutron source using longitudinally polarized neutrons and a BaF$_2$ detector. The effect of parity nonconservation in the ^{139}La(n,γ) reaction for the resonance at $E_n=0.73$ eV was confirmed. New results for p-wave resonances in the ^{117}Sn(n,γ) reaction were obtained. A comparison between the capture and transmission techniques is presented.

INTRODUCTION

Parity nonconservation (PNC) in neutron p-wave resonances has been studied by using the transmission technique for several years[1-4]. Definite cases of parity violation have been found for compound states in ^{81}Br, ^{111}Cd, ^{117}Sn, ^{139}La and ^{238}U. The transmission method measures the total cross section and seems to be well established. In this method the neutron detector is located directly in the beam and the count rate for each helicity is measured as a function of incident neutron energy. The p-wave resonance appears as a weak dip on the high counting rate level which plays the role of a background. Large target masses of approximately 1 kg are necessary. Only the capture part of the cross section is responsible for the PNC effect; the potential scattering does not contribute and leads only to the loss of neutron intensity.

The measurement of the helicity dependence of neutron capture cross sections promises to be a more direct way to observe PNC effects in p-wave resonances. This method is not sensitive to potential scattering and smaller targets may be used. This is especially important when enriched isotopes are required. The γ-ray detectors are located outside the neutron beam and should have greater than 50% efficiency. The background from scattered neutrons must also be minimized.

The (n,γ) technique was successfully used by Masuda et al.[5] to measure parity violation in the ^{139}La resonance at $E_n = 0.73 eV$. An unpublished communication from IAE Moscow[6] treats the case of the ^{111}Cd p-wave resonances by the (n,γ) method. The present report discusses the first measurements to investigate the possibility of using polarized (n,γ) reactions to study parity violation at the Los Alamos Neutron Scattering Center (LANSCE) pulsed neutron source.

EXPERIMENTAL APPARATUS

Measurements were made using the time-of-flight technique. The 800 MeV, 125 nsec wide (FWHM), beam from the proton storage ring (PSR) at the Los Alamos Meson Physics Facility (LAMPF) produced neutrons by spallation reactions in a tungsten target. Beam spills occurred at a rate of 20 hz. The water-moderated neutron flux is described by the expression[7]:

$$F(E_n) = \frac{2 \times 10^8}{E_n(eV) L^2(m)} (cm^2 * eV * sec)^{-1}$$

This equation assumes an average proton current of 100 μamps and a sample viewing 100 cm² of the moderator. The validity of this expression has been confirmed in experiments by Koehler[8].

Neutrons are polarized by transmission through a polarized proton filter. We use an LMN polarizer and a spin-flipper that have been described in detail in Ref. 9. The polarization of the neutron beam was rather low in this experiment, $f_n = 17\%$ instead of the normal value of 40%; we believe this is due to deterioration of the LMN crystals. Positive and negative neutron helicity states were alternated every 10 seconds using the spin flipper. Data were also obtained reversing the spin direction by changing the microwave pumping frequency. The asymmetries quoted in this paper are the weighted average of the results for the two microwave frequencies.

Polarized neutrons reached the metallic samples of ^{139}La (6.8×10^{22} and 3.1×10^{22} atoms/cm²) and of ^{117}Sn (92% enrichment, 2.9×10^{22} atoms/cm²) located approximately 22.5 m from the water moderator. The beam was collimated to a spot 5 cm in diameter at the sample location using collimators made from boron-loaded polyethylene. A longitudinal magnetic guide field ($B = 16G$) was provided by a solenoid that extended from the spin-flipper to the sample position.

The γ-ray detector consisted of two 15x15 cm^2, 15 cm thick cubic BaF$_2$ crystals located 10 cm from the center of the beam on either side of the flight path. ^{6}Li-metal disks (2.5 cm thick, 15 cm in diameter) were placed between each detector and the sample. The detector was shielded with 10 cm lead and borated polyethylene plates. The solid angle for the two detectors was approximately 25% of 4π. The background of high-energy neutrons and γ- rays paralyzed the detector for 150 μsec after the initial beam burst.

The anode outputs of the photomultiplier tubes (PMT) were connected to two fast discriminators. One discriminator determined the lower γ-ray energy level; the other discriminator determined the upper γ-ray energy level. The logical sum of these discriminators provided a γ-ray energy window; these signals for each detector were summed and connected to a multiscaler. The width of the discriminator output pulses was 20 nsec. The lower level discrimination was set to approximately 1 MeV. The upper level was set to 5 MeV for ^{139}La and 9 MeV for ^{117}Sn.

The data acquisition system, which was developed for the TRIPLE transmission measurements[4], used the XSYS language on a DEC μVAXII workstation. Typical time-of-flight spectra are shown in Fig. 1 for 0.8 cm thick ^{139}La (the channel width is 1 μsec) and in Fig. 2 for 0.8 cm thick ^{117}Sn (the channel width is 0.5 μsec). The continuum spectra outside the peaks are due to s-wave capture and to background. The high-energy strong s-wave resonances were distorted because of the large instantaneous current in the photomultipliers.

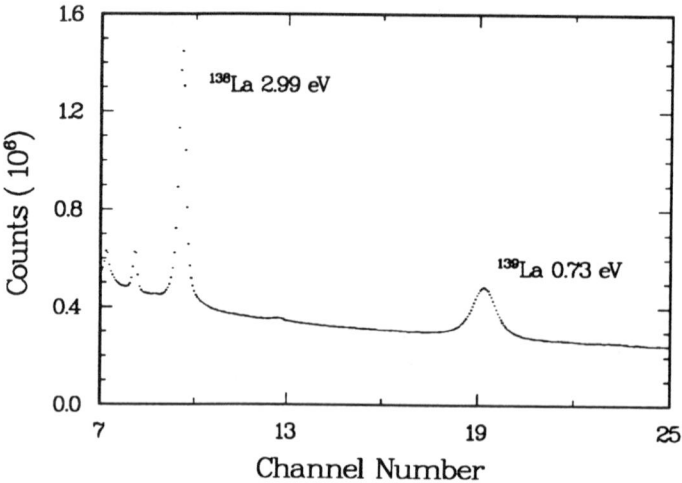

Fig. 1 Time of flight spectrum for La.

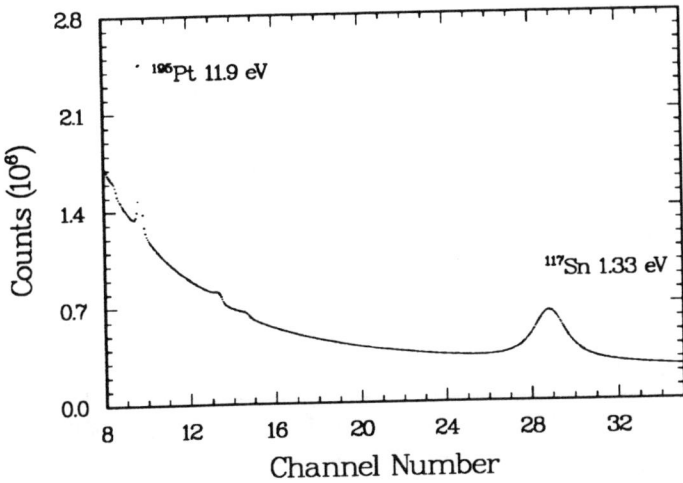

Fig. 2   Time of flight spectrum for ^{117}Sn.

RESULTS

The data were analyzed to obtain the longitudinal asymmetry ϵ_γ for the p-wave resonance peaks. ϵ_γ is defined as: $\epsilon_\gamma = (1/f_n)(N_+ - N_-)/(N_+ + N_-)$ where f_n is the polarization of the neutron beam and N is the number of counts for each spin state. The number of counts $N_+ + N_-$ was obtained by fitting a Lorenztian peak shape to the summed spectrum. The energy and width determined in this way were used to fit the difference spectrum to determine $N_+ - N_-$. Plots of the difference spectra and fits for ^{139}La and ^{117}Sn are shown in Fig. 3 and Fig. 4.

The γ-ray counts N are not linearly proportional to capture cross sections σ_{CAP} for thick samples. The effects of self-attenuation and the capture of scattered neutrons in the target should be taken into account. We therefore measured ^{139}La data for two different sample thicknesses that are shown in Fig. 5 together with the data of Ref. 5. The relative beam polarization was obtained by NMR measurements; the absolute beam polarization was obtained by normalizing our two La measurements to the curve defined by the data of Ref. 5. For the thin La sample this corresponds to a beam polarization of $f_n = (16.5 \pm 1)\%$. The results of this preliminary analysis are shown in Table I.

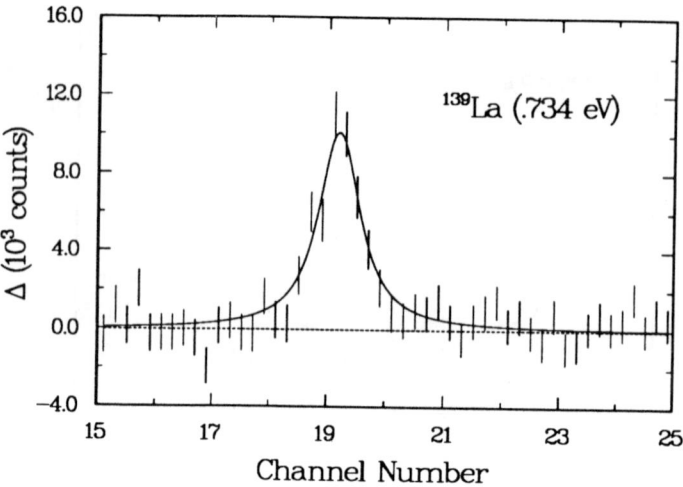

Fig. 3 Difference spectrum in region of 0.734 eV resonance in ^{139}La.

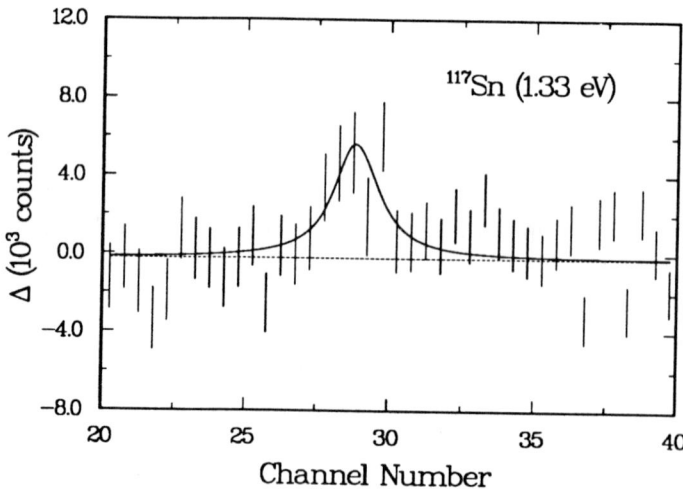

Fig. 4 Difference spectrum in region of 1.33 eV resonance in ^{117}Sn.

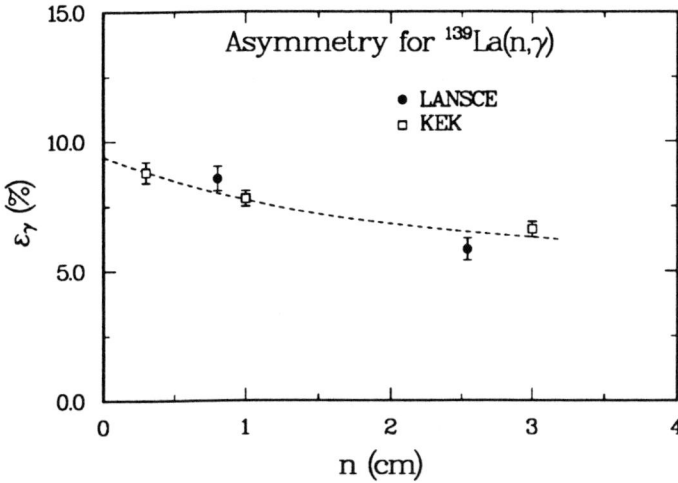

Fig. 5 Comparison of ϵ_γ from KEK and LANSCE.

We made no correction for sample thickness effects for the 1.33 eV ^{117}Sn resonance. It will certainly be less than in the case of ^{139}La because of the two times better ratio of $\sigma_{CAP}/\sigma_{POT}$ for ^{117}Sn. The ^{117}Sn result is two times larger than the value from transmission measurements in Ref. 1 but not in disagreement with the result of Ref. 2. A weak neutron resonance was found at $E_n = 26.4 eV$ in addition to the known resonance[10] at $E_n = 34.0 eV$ in ^{117}Sn. Upper limits of the longitudinal asymmetry for these resonances are shown in Table I.

Table I: Parity-Violating Asymmetries Measured in Neutron Capture

Sample	n (10^{22} atom/cm^2)	E_n (eV)	ϵ_γ (%)
La	3.1	0.73	8.6 ± 0.5
La	6.8		5.8 ± 0.4
^{117}Sn	2.9	1.33	1.1 ± 0.2
^{117}Sn	2.9	26.4	−1.3 ± 1.6
^{117}Sn	2.9	34.0	0.45 ± 0.49

COMPARISON OF CAPTURE AND TRANSMISSION

Capture experiments were the first to show parity-violating effects in slow neutron reactions but are in general more difficult to analyze than transmission experiments because of the need for additional spectroscopic information relating to the decay channels[11]. An important exception arises for 4π gamma ray detection where the two methods are equivalent. The p-wave neutron width is negligible compared to the gamma ray width and both experiments are essentially measuring the helicity dependence of the total resonance cross section. Capture measurements can have a sensitivity advantage because there is no contribution from potential scattering.

Both experiments measure a parity-violating cross section σ_{PV} as defined, for example, in the appendix of Ref. 12. The capture asymmetry $\epsilon_\gamma = \sigma_{PV}/\sigma_{CAP}$ will nearly always be larger than the transmission asymmetry $\epsilon_n = -n\sigma_{PV} \equiv 2\sigma_{PV}/\sigma_{TOT}$ for an $n = 2$ mean free path target (the optimum thickness for transmission). This ratio particularly favors capture if the p wave is weak ($\sigma_{CAP} \sim \sigma_p \ll \sigma_{TOT}$). Capture is also favored for thin targets; in capture the error in σ_{PV} scales as $1/\sqrt{n}$ compared to $1/n$ for transmission. Large quantities of isotopic material will usually be impossible to obtain. The actual errors will depend of course on the relative efficiencies of the detectors, the neutron flux striking the samples, and on the backgrounds in each experiment.

Transmission experiments overcome these limitations by working with thick samples close to the neutron source and by placing a large detector at a long flight path. This resolves neutron resonances up to high energies, which is important in surveys of parity-violating asymmetries. This is a problem for capture experiments in which small samples are usually placed close in to the neutron source with detectors at the same location. Capture experiments also must analyze multiple scattering effects which dilute the asymmetry[5]. Most of these effects are due to neutrons scattered by s-wave amplitudes which do not flip the spin, so the corrections can be reliably made. But it is a complication that is not present in the transmission work.

CONCLUSION

An experiment was performed at the LANSCE pulsed neutron source to measure the PNC effect in neutron capture. Although the experiment was not optimized, the capture γ-ray method has proved to be a viable and promising technique for parity violation study at LANSCE/LAMPF. It is particularly advantageous for weak resonances and in cases of enriched isotopes in small quantities and is complementary to the transmission method. This technique may prove to be a very useful tool for future time-reversal experiments.

There are many improvements that should be made before the next set of experiments are performed. The background from fast neutrons must be reduced with additional shielding. The effect of detector paralysis at short flight times could be reduced by segmenting the detector which would in addition allow multiplicity and angular distribution measurements to be performed.

ACKNOWLEDGEMENTS

This work was supported in part by the U. S. Department of Energy, under Contracts. No. DE-AC05-76ER01067 and DE-FG05-88ER40441. One of the authors (CRG) acknowledges financial support from Associated Western Universities for a sabbatical fellowship at Los Alamos National Laboratory.

REFERENCES

1. V. P. Alfimenkov et al., Nucl. Phys. **A398**, 93 (1983).
2. S. A. Biryukov et al., Yad. Fiz. **45**, 1511(1987).
3. Y. Masuda et al., Nucl. Phys. **A478**, 737c (1988).
4. J. D. Bowman et al., Phys. Rev. Lett. **65**, 1192 (1990).
5. Y. Masuda et al., Nucl. Phys. **A504**, 269 (1989).
6. B. V. Muradijan, Report to the JINR (Dubna). Workshop on Nuclear Physics with pulsed neutrons, March 1990.
7. G. J. Russell, Private communication.
8. P. E. Koehler, Nucl. Instr. Meth. **A292**, 541 (1990).
9. C. D. Bowman et al., Hyperfine Interactions **43**, 119 (1988).
10. S. F. Mughabghab, Neutron-Cross Sections, Academic Press New York, 1984.
11. J. R. Vanhoy, E. G. Bilpuch, J. F. Shriner, Jr. and G. E. Mitchell, Z. Phys. A–Atoms and Nuclei **331**, 1 (1988).
12. C. R. Gould et al., in these Proceedings.

RADIATIVE CAPTURE IN FEW-NUCLEON SYSTEMS AND EXCHANGE CURRENTS

K. Abrahams, M.W. Konijnenberg and R. Wervelman

FOM/ECN Nuclear Structure Group,

Netherlands Energy Research Foundation ECN,

P.O. Box 1, 1755 ZG Petten, the Netherlands

October 10, 1990

Abstract

Low-energy radiative neutron capture by the target nuclei ^{1}H, ^{2}H, and ^{3}He depends on exchange currents. For radiative capture of thermal neutrons in the hydrogen isotopes, contributions of meson-exchange currents (MEC) range from 10-50%, but these do not explain the photon circular polarization in polarized neutron capture by protons. In order to confirm previous data a feasibility study on the ^{1}H$(\vec{n}_{th}, \vec{\gamma})$ reaction has been performed. Experiments on polarized deuterium in a ZrD$_2$ target have been performed in order to unravel the channel-spin admixtures for the ^{2}H(n, γ) reaction, and to enable a comparison with model calculations. For ^{3}He a radiative thermal neutron capture cross section of 55(3) μb has been measured for single-photon emission, and 30(80) μb for ^{3}He$(n, \gamma\gamma)$ double-photon emission.

Accurate cross-section data for radiative capture of thermal neutrons in ^{3}He are relevant to a solution of the solar-neutrino problem. Shell-model calculations, which include meson-exchange currents, were performed for the ^{3}He(n, γ) reaction and for the so called hep process: ^{3}He + p $\rightarrow$ ^{4}He + e$^+$ + ν_e. A theoretical result of 48(17) μb for the thermal neutron capture cross section, which agrees with the experiment, and a much more accurate estimate for the cross-section ratio of thermal neutron capture- and hep-process were obtained. Together with the present experiments this ratio would imply a hep-neutrino flux on earth equal to 5.7(8)$\cdot 10^4$ cm^{-2}s^{-1}. This flux corresponds to about 10 (respectively 3) percent of the measured (respectively calculated) count rate of the ^{37}Cl detector.

1 Introduction

In a nucleus one might observe besides nucleons also mesons and isobaric nucleon excitations (deltas). Far below their threshold (ranging from 138 MeV to 300 MeV) the electro-magnetic current due to these particles is only a few percent of the corresponding nucleonic current, as the nucleons are only bound by about 10 MeV. Static and dynamic properties of nuclei may therefore be subject to corrections, which below the nucleon binding energy are much larger than the relativistic corrections, (the latter corrections are maximally one percent, as the nucleon mass is about 1000 MeV). Meson-exchange currents show up in M1 transition probabilities in few-nucleon systems[1], but also in static and dynamic properties, such as magnetic dipole moments of 1p-shell nuclei[2], and Gamow-Teller matrix elements[3]. At low energy it is not possible to distinguish the mesonic currents from the main nucleonic currents in the E1-transition probability due to Siegert's theorem[4].

Even for M1 transitions following thermal neutron capture it is inherently impossible to separate the nucleonic current from the exchange current in a model-independent way. Although each current has a well defined contribution, the data might be described by several different models. Apparently the wavelength of the gamma is too large (compared with the dimension of the nucleus) to enable a distinction between model dependent effects such as D-state admixture and exchange currents (see ref. 5). An advantage of few-nucleon systems is that calculations can be made on a microscopic basis, which allows a treatment of the small exchange currents. Polarisation data produce cross-section ratios for the two channel-spins and relative signs of the amplitudes.

Like most amplitudes in nuclear physics, the non-nucleonic contributions to the M1 matrix elements have a gaussian distribution around zero, for MECs with a range of percents of nucleons-only amplitudes. These small effects are often cancelling each other by interference, and they are drowned in the chaotic currents of nucleons in a nucleus. It will be therefore very hard to detect among this noise also quark-exchange currents[6], for which the amplitudes are at least one order of magnitude smaller. For the ^{1}H(n,γ) process, however, both experiment and theory are simple and the possibility to observe such QECs is of interest.

2 Non-nucleonic effects in the reaction ^{1}H(n,γ)

As it is the simplest radiative capture reaction, the cross section for thermal neutron capture by protons can be very well calculated. A calculation in nucleons only impulse approximation (NOIA), yields 302(4) mb in an average over all viable models[9], which is about 10% lower than the value 334.2(5) mb from experimental work[7]. Inclusion of mesonic and isobaric currents brings the calculated value close to the experimental value, but a discrepancy of 1.2(2)% with the NOIA amplitude remains. It should be noted that only a singlet capture state has been considered[9]. In principle the 2.2 MeV deuteron capture state is a coherent mixture of the 0^+ singlet state and the 1^+ triplet state. In fact both states contribute in the scattering process, with as scattering amplitudes: $a_s = 5.42$ fm and $a_t = -23.75$ fm (see ref. 7). In ref. 8 the circular polarisation of gamma-radiation after polarized neutron capture has been written as:

$$R = -\frac{(\mu_n + \mu_p) - \frac{3}{2}P_D(\mu_n + \mu_p - \frac{1}{2})}{(\mu_n - \mu_p)(1 - P_D)} \cdot \frac{1 - 1/ka_t - \frac{1}{4}k(r_{0s} + r_{0t})}{1 - 1/ka_s - \frac{1}{4}k(r_{0s} + r_{0t})}$$

Inserting all known values for the magnetic moments and effective ranges r_0, this leads to $R = -0.75\%$ (for $P_D = 0$-10%). This theoretical value does not fit the experimental value $R = -0.29(9)\%$, which has been obtained by capture of polarised neutrons in a para-hydrogen target[8]. Because the discrepancy between measured and calculated values can not be explained by admixing E2 radiation[27], nor by an extremely large D- state contribution[9], it is tempting to try an explanation involving meson-exchange currents[28]. The requirement that the Pauli principle must be fulfilled by the six constituent quarks even introduces the possibility of observing quark-exchange currents[6]. However, before stating anything regarding MECs or QECs from the discrepancy in the R-value, it will be necessary to reproduce the NaI work of ref. 8 with Ge- detectors, in order to reduce errors due to background.

An explorative measurement on samples of ZrH_2 and TiH_2 at the Petten polarised thermal neutron beam showed that the depolarisation of the neutrons by spin-flip scattering due to hydrogen ortho-para transitions is small because the distance between the atoms is longer than the wave-length of the thermal neutrons. This is not the case for most other hydrogen containing targets, which therefore should be specially kept in a para state in order to reduce depolarisation due spin-flip neutron scattering. At least one year of measurements would be needed in order to reach the statistical significance level of ref. 8. As one would like to reach an improvement in the accuracy with about a factor three about ten years of measurement would be needed at the Petten set-up. Because such an undertaking is not likely to be feasible, it would be better to perform such an experiment at a setup with a polarised beam of ten times higher flux at the ILL (or at some other high flux reactor).

3 Non-nucleonic effects in the reaction $^2H(n,\gamma)$

Once the nucleon-nucleon interaction is known, it is possible to calculate the cross section of reactions involving three nucleon systems accurately with Faddeev equations[10]. Radiative capture of thermal neutrons and deuterons may proceed through two reaction channels, as the deuteron spin 1 may combine with the neutron spin either to spin 1/2 or to spin 3/2, but the value of the relative admixture of the two channel-spins is badly known.

Circular polarisation measurements have defined the sign of the matrix elements of the transition[11], and could be used to show that MECs do play a role in the capture process. All viable models disagree with the measured cross section as well as with the measured polarisation value if exchange currents are omitted. In the present experiment an attempt was made to measure the absolute value of the capture channel-spin admixture to a high precision by means of a polarised deuterium sample. A compressed block of ZrD_2 has been used, and in this sample material the deuterons could be polarized by cooling to 27(2) mK in an 8 T magnetic field (the temperature was obtained from gamma-ray thermometry with the 6294 keV Zr line). From temperature and magnetic field, a value of only 4.5(4)% for the polarisation of the deuterium nuclei followed.

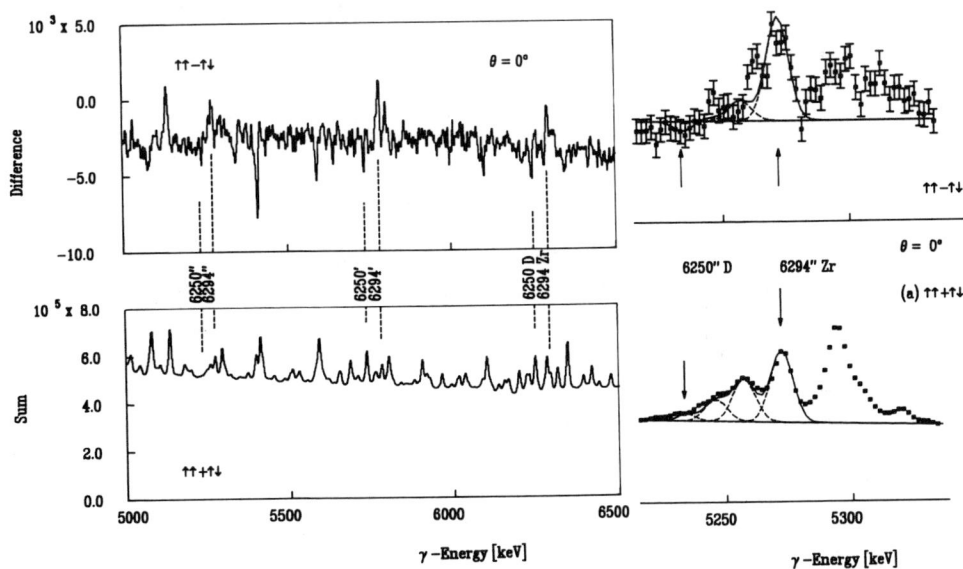

Figure 1: Gamma-spectra from polarized capture in ZrD_2.

In figure 1 the sum- and difference-spectra (1a and 1b, respectively) are shown of measurements with the polarisation vectors of deuteron and neutron parallel and anti-parallel. It is clear that the polarisation was too small, and the background too high to allow a statistically significant result after measuring for 28 days. This feasibility study suggests that it will be better to polarize by frozen spin method, with a target that is free of Zr.

In figure 2 the result of all $^2H(n_{th},\gamma)$ measurements is shown. MECs are responsible for about 50% of the cross section and two solutions for the channel-spin admixture are indicated as wedges. Old results on muon induced radiative

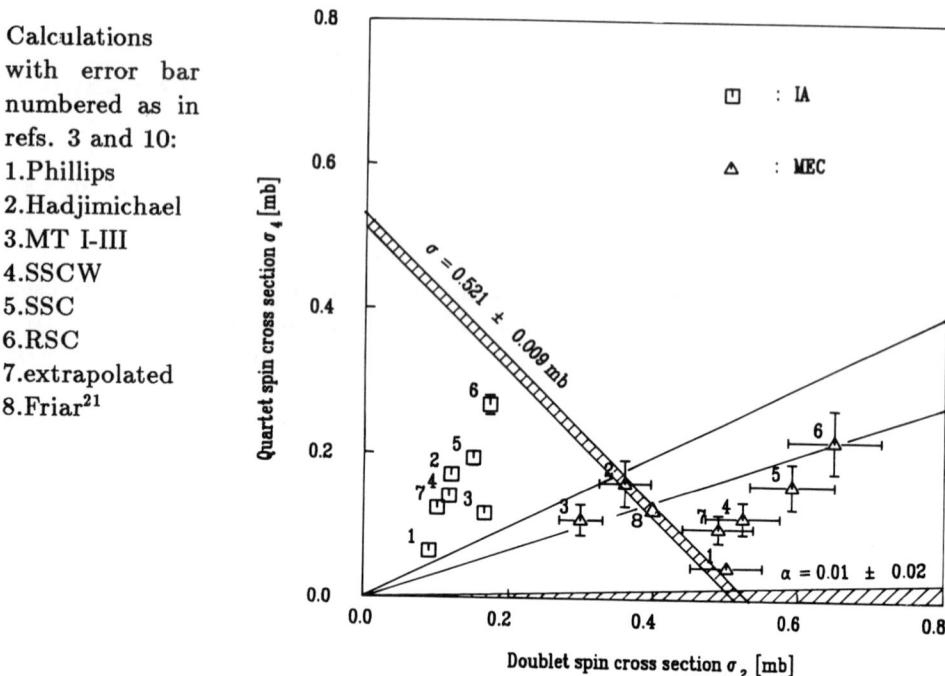

Figure 2: Cross sections for the ^{2}H(n,γ) reaction for spin-channel 1/2 or 3/2, either with exchange currents ($\triangle$) or without ($\square$). A diagonal band shows the total capture cross section[7], and wedges indicate standard deviation boundaries for the measured channel-spin admixture[11,12].

capture of protons by deuterons, imply that the smaller of the two solutions (given as shadowed wedge) is valid[12]. This solution, which is in agreement with near orthogonality of a pure s-wave capture state and a pure s-wave ground state, is nevertheless excluded by all models. Therefore it would be interesting to repeat the old radiative pd fusion experiments.

4 The reaction ^{3}He(n_{th}, γ) and the hep process

In first order NOIA a simple rotation in isospin-space will directly relate the ^{3}He(n,γ) reaction to the so called hep process: ^{3}He $+$ p $\rightarrow$ ^{4}He $+$ e^+ $+$ ν_e (Q= 18.8 MeV), which is the source of high-energy solar neutrinos. These solar hep neutrinos should be detectable with a new generation of detectors[13]. In contrast to radiochemical detectors (for instance the chlorine detector in the well known Davis experiment), which have no energy resolution, detection of individual recoil electrons will enable a measurement of the energy spectrum of solar-neutrinos. It would be especially interesting to measure the possible variation of this flux during a solar cycle, as has been done for ^{8}B neutrinos[26].

At solar temperatures the proton kinetic energy (of the order of a few keV) is far below the ^{3}He Coulomb barrier, and the hep cross-section will be extremely small. A measurement of this extremely small hep cross section is not feasible, but calculations are possible[13-15]. Confidence in such calculations will increase if these also predict the correct ^{3}He(n_{th},γ) cross section.

Although the Q-value for radiative neutron capture is 20.6 MeV, the cross-section for capture of thermal neutrons is very small, as orbital symmetries of wave functions forbid a one-body M1 ground-state transition. Meson-exchange currents follow different selection rules and therefore a higher fraction of exchange currents can be expected than for radiative capture in deuterium[15]. Other calculations have shown however that contributions of different exchange currents may cancel each other[16].

In order to measure the radiative capture cross section for ^{3}He, a gaseous target was placed in a beam with flux 10^8 n/cm^{2}s. Two 5"x5" NaI(Tl) spectrometers recorded the spectra given in fig. 3. With a stack of gold foils, flux, average density, and energy spectrum were calibrated upstream as well as downstream of the target, in order to account for the influence of the dominating ^{3}He(n,p) reaction in the target cell (see also ref. 16). To derive the Na(Tl)-detector efficiency up to 20 MeV, the Electron Gamma Shower Monte Carlo code (EGS) was used, calibrated with radioactive sources and the 17.6 MeV line from the ^{7}Li(p,γ) reaction. This resulted in a cross section of 55(3) μb, which is twice as accurate as the recent result of ref. 18. The three main error sources -calibration of flux, EGS-efficiency, and statistics - have about equal weight, so it will be hard to improve on the value. The accepted[7] value 31(9) μb is definetely excluded.

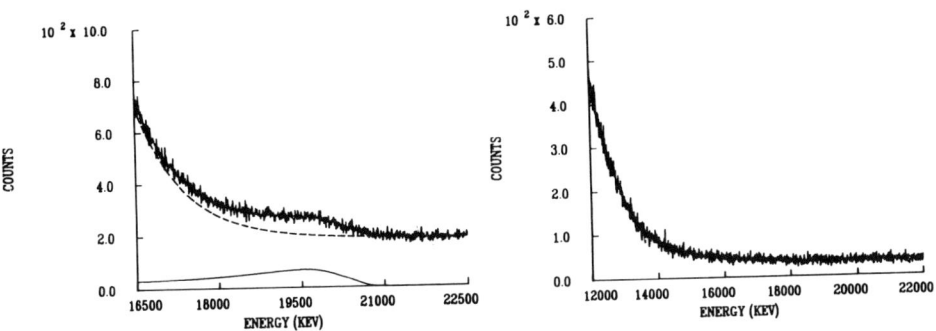

Figure 3: Gamma-spectra with and without ^{3}He in the holder.

Magnetic dipole and the Gamow-Teller transition amplitudes were calculated in a complete $(0+2)\hbar\omega$ model space, with the Utrecht shell-model program RITSSCHIL[22] (see tables 1 and 2). Both the ^{3}He and ^{4}He ground state wave functions were deduced from Sussex relative harmonic-oscillator matrix elements, which are derived from measured nucleon-nucleon phase shifts[23].

	[f] ($\lambda\mu$) L S	A^2
^{3}He	[3] (00) L=0 S=$\frac{1}{2}$	.948
	[3] (20) L=0 S=$\frac{1}{2}$	.001
	[21] (20) L=0 S=$\frac{1}{2}$	.001
	[21] (20) L=2 S=$\frac{3}{2}$	.049
^{4}He	[4] (00) L=0 S=0	.912
	[4] (20) L=0 S=0	.048
	[22] (20) L=0 S=0	.001
	[22] (20) L=2 S=2	.038

Table 1: The dominant components in the ^{3}He and ^{4}He ground state wave functions.

matrix elements in fm$^{3/2}$	$M_\gamma^{T=0}$	$M_\gamma^{T=1}$	$M_\beta/g_A C_0$
one-body	-.12	-.51	-1.78
π seagull		-.40	-.20
π current		.53	
π isobars		.16	-.64
π-ρ current	.02		
π-ω current		.01	
ρ seagull	.00	-.14	
ρ current		-.07	
ρ isobars		-.05	
ω seagull	-.01	-.02	
TNI π current		-.02	
TNI ρ current		.01	
sum	-.11	-.50	-2.62
	-.61		

Table 2: Contributions to the matrix elements M_γ and M_β from various transition operators. With 'TNI' the translationally non-invariant currents are denoted.

The cross section of radiative capture of thermal neutrons is given by:

$$\sigma_{n\gamma} = \frac{2J_i + 1}{(2J+1)(2j+1)} \left(\frac{e^2}{\hbar c}\right) \left(\frac{E_\gamma}{Mc^2}\right)^2 \sqrt{\frac{E_\gamma}{2E_n}} \sqrt{\frac{E_\gamma M}{\hbar^2}} \frac{1}{3} |M_\gamma|^2$$

With the matrix-elements of table 2, the value 47(18) μb is obtained. Uncertainties in the wave function of the triplet capture state produce an error as large as 20% in the matrix elements. These errors cancel to a large extend in the ratio of the magnetic dipole and Gamow-Teller cross-sections, and by combining expressions from ref. 24 one may write:

$$\frac{\sigma_{pe^+\nu_e}}{\sigma_{n\gamma}} = \frac{\tilde{g}_A^2}{e^2/\hbar c} \frac{f^+(Z,W_0)}{E_\gamma^3} \frac{3m_e M^2 c^6}{4\pi^3} \sqrt{\frac{E_n}{E_p}} \frac{|M_\beta|^2}{|M_\gamma|^2}$$

With the matrix-elements from table 2, the dimensionless coupling constant $\tilde{g}_A = 3.84 \cdot 10^{-12}$, the reduced Fermi function $f^+(Z = 2, W_0 = 37.7 \cdot m_e c^2) = 2.4 \cdot 10^6$, and $\sigma_{n\gamma} = 55(3)$ μb the astrophysical S-factor of the hep-process can be calculated. With as input the standard solar model[25,14] this S-factor predicts $5.7(8) \cdot 10^4$ cm^{-2}s^{-1} for the flux of hep-neutrinos on earth, leading to about 3% in terms of theoretically derived count-rate in the chlorine detector.

In fig. 4 it is shown how the magnetic dipole and the Gamow-Teller matrix elements depend on D-state admixtures in target and product nucleus. Any difference between the two patterns would imply that the D-state admixture in ground state wave functions is an observable, which is contradiction with theory[5].

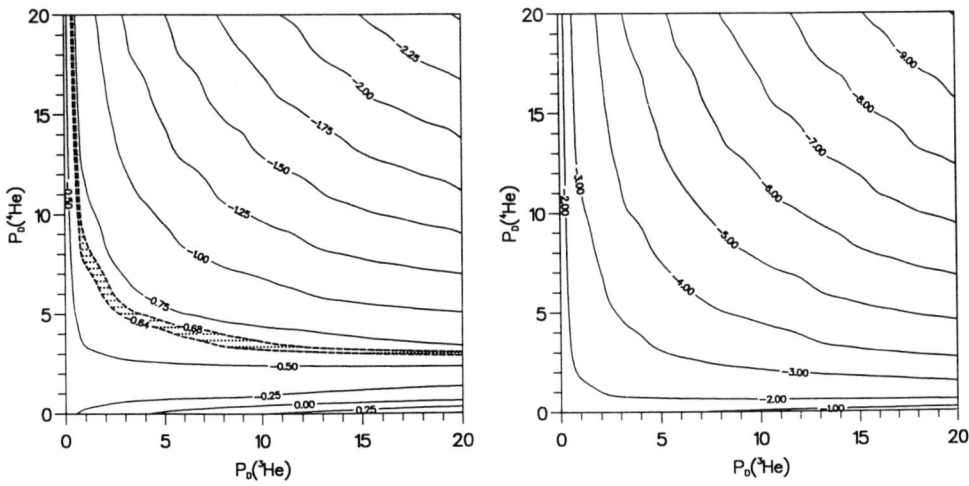

Figure 4: M_γ (left) and M_β (right) as a function of $P_D(^3\text{He})$ and $P_D(^4\text{He})$.

5 Double-photon emission in the radiative capture of thermal neutrons by ^{3}He

Thermal neutron capture in ^{3}He involves two spin-channels, the 1^+ radiative capture channel, and the 0^+ channel through which the ^{3}He(n,p) reaction proceeds. Due to spin-selection rules the 0^+ channel radiative-capture contribution can only proceed either by monopole radiation or by double-photon emission. In ref. 17 the cross section for the double-photon process was measured to be equal to 17(6) μb. This value is much too high in comparison with a value of about 1.2 μb, as calculated by Lee et al.[19] under the assumption that double-photon emission proceeds by two E1 transitions. Recently strong double-photon emission from ^{16}O, ^{40}Ca and ^{90}Zr has been reported, for which, due to nuclear structure effects the 2E1-component was quenched but not the 2M1 component[20,29]; so it seemed worth while to remeasure the double-photon emission from ^{4}He with a better precision. This was especially needed as an EGS calculation showed that the detector efficiency for the softer double-photon spectrum has been underestimated in ref. 17 by a factor two. As a consequence the cross section result from ref. 17 should therefore be enlarged by a factor two.

Measurements were done in the same geometry as the single-photon measurement, with the spectrometers at a $90°$ angle between the two NaI detectors in the plane perpendicular to the neutron beam. A multi-parameter setup, with a 64x64 channel Nuclear Data analyzer was used to measure the coincidence spectrum. In contrast with the previous work of ref. 17 no anti-coincidence scintillator active shield against cosmic radiation has been applied. A pile-up rejection circuit eliminated most of the random counts and the rest was corrected for by separate measurements with a time delay in the fast- coincidence signal. No coincidence signal appeared however for a sum energy around 20 MeV. Analyses of the sum spectrum at 20 MeV (about 20 coincidences a day) yielded a cross-section value equal to 30(80) μb, which neither supports nor contradicts ref. 17.

Regarding the uncertainty in the 2γ-detection efficiency in the Suffert and Berthollet experiment, it is urgent to perform a new measurement of the doubly radiative thermal neutron capture cross section of ^{3}He at a well collimated beam of a high flux reactor. In addition, the angular correlation between the two photons needs to be measured, in order to investigate the interference of 2E1 and 2M1 radiation.

6 Conclusions

Higher neutron fluxes may open doors towards the solution of problems related to quark-exchange currents, especially for the ^{1}H(n_{th}, γ) reaction. For three-nucleon systems more precise radiative capture measurements are clearly needed, in order to answer rather fundamental questions regarding meson-exchange currents. Exchange-currents are highly relevant in four-nucleon systems because they enhance the hep solar-neutrino flux. It is necessary to reconfirm double-photon emission in ^{4}He.

References

[1] K. Abrahams et al., Proceedings of the 6th Symp. on Capt. Gamma-Ray Spectroscopy, (Leuven 1987) 373
[2] J.L.G. Booten et al., accepted for publication in Phys. Rev. C
[3] M. Chemtob and M. Rho, Nucl. Phys. **A163**, 1 (1971)
[4] A.J.F. Siegert, Phys. Rev. **52**, 787 (1937)
[5] J.L. Friar, Phys. Rev. **C22**, 796 (1980)
[6] Y. Yamaugi, R. Yamamoto and M. Wakamatsu, Nucl. Phys. **A443** 628 (1985)
[7] S.F. Mughabghab, M. Divadeenam, and N.E. Holden, *Neutron Cross Sections*, Vol. 1 Part A, Acad. Press (N.Y. 1981)
[8] V.A. Vesna et al., Nucl. Phys. **A352**, 181 (1981)
[9] J.F. Mathiot, Phys. Rep. **173**, 63 (1989)
[10] J. Torre and B. Goulard, Phys. Rev. **C28**, 529 (1983)
[11] M.W. Konijnenberg et al., Phys. Lett. **B205**, 215 (1988)
[12] A.C. Phillips, Phys. Rev. **170**, 952 (1968)
[13] J.N. Bahcall and R.K. Ulrich, Rev. Mod. Phys. **60**, 297 (1988)
[14] P.E. Tegnér and C. Bargholtz, Astrophys.J. **272**, 311 (1983)
[15] J. Carlson et al., Phys. Rev. **C42**, 830 (1990)
[16] R. Wervelman et al., ECN preprint RX-090-061 (1990)
[17] M. Suffert and R. Berthollet, Nucl. Phys. **A318**, 54 (1979)
[18] F.L.H. Wolfs et al., Phys. Rev. Lett. **63**, 2721 (1989)
[19] H.C. Lee et al., Phys. Rev. Lett. **65B**, 201 (1976)
[20] J. Schirmer et al., Phys. Rev. Lett. **53**, 1897 (1984)
[21] J.L. Friar, to be published, private communication (1990)
[22] D. Zwarts, Comp. Phys. Comm. **35** 365 (1985)
[23] J.P. Elliot et al., Nucl. Phys. **A121** 241 (1968)
[24] P.J. Brussaard and P.W.M. Glaudemans, *Shell model applications in nuclear spectroscopy*, North Holland (Amsterdam 1977)
[25] J.N. Bahcall et al., Rev. Mod. Phys. **54**, 767 (1982)
[26] R. Davis et al., Ann. Rev. Nucl. Part. Sci. **39** 467 (1989)
[27] N. Austern, Phys. Rev. **85** 147 (1951)
[28] A.P. Burichenko and I.B. Kriplovich, Nucl. Phys. **A515** 13 (1990)
[29] J. Kramp et al., Nucl. Phys. A474, 412 (1987)

CURRENT RESULTS AND FUTURE PROSPECTS FOR A NEUTRON LIFETIME DETERMINATION USING TRAPPED PROTONS

Maynard S. Dewey
National Institute of Standards and Technology, Gaithersburg, MD 20899

ABSTRACT

The availability of an accurate value for τ_n has important implications for tests of the standard $V - A$ theory of semi-leptonic weak processes, cosmology, and astrophysics. The lack of agreement between values of g_A/g_V derived from recent lifetime measurements and values obtained from recent β-decay asymmetry experiments hints at a problem with the current theoretical framework. In this neutron beam measurement the lifetime is determined by counting decay protons stored in a Penning trap whose magnetic axis coincides with the neutron beam axis. The recent result, $\tau_n = 893.6 \pm 5.3$ s, is based upon data accumulated during one reactor cycle at the Institut Laue-Langevin in Grenoble, France. In the next two years a final accuracy approaching 1 s will be sought. This improvement will be a consequence of long periods of running time available at NIST, several improvements to the apparatus, and an aggressive program aimed at development and intercomparison of advanced methods for the absolute determination of thermal and cold neutron fluxes.

INTRODUCTION

Self consistency among experimental values for the neutron lifetime τ_n, the various angular and polarization correlation coefficients in free neutron β-decay, and ft-values of pure Fermi $0^+ \to 0^+$ superallowed β-transitions, provides one of the best tests of the standard $V - A$ theory of semi-leptonic weak processes.[1] The availability of an accurate value for τ_n also has important implications for cosmology[2] and astrophysics.[3] Experimentally the determination of τ_n is a significant challenge with results from the last 18 years in some degree of disarray; see Table I. Within the last year several new results have been reported which have had, in addition to much higher precision, greater immunity to systematic errors. For neutron decay, the accumulated data are sufficiently accurate to establish useful limits on departures from the Standard Model.[4–7]

According to the standard $V - A$ model the neutron decay rate is given by

$$\tau_n^{-1} = \frac{\ln 2}{K} f^R g_V^2 (1 + 3\lambda^2) , \qquad (1)$$

where $\lambda = g_A/g_V$, $K = 2\pi^3 (ln2) \hbar^7 / m_e^5 c^4$, and f_R the suitably corrected phase-space factor $= 1.71465 \pm 0.00015$.[1] The constants g_V and g_A are real when time

Table I: Direct determinations of the neutron lifetime.

τ_n (sec)	σ_{τ_n}	Reference
1108	216	Robson, 1951
1039	130	Spivak et al., 1956
1099	164	D'Angelo, 1959
1013	26	Sosnovski et al., 1959
935	14	Christensen et al., 1967
918	14	Christensen et al., 1972
907[a]	70	Paul and Trinks, 1978
875[a]	95	Morozov et al., 1980
937	18	Byrne et al., 1980
899[a]	11	Kosvintsev et al., 1986
876	21	Last et al., 1988
886	30	Schreckenbach et al., 1989
877[a]	10	Paul et al., 1989
887.6[a]	3	Mampe et al., 1989
893.6	5.3	This work, 1989
885[b]	9	Spivak, 1990
888.4[a]	2.9	Serebrov et al., 1990
883.2[a]	2.9	Morozov et al., 1990

[a]These results were obtained using bottled ultra-cold neutrons.
[b]This value is a re-evaluation of the Sosnovski et al., 1959 result.

invariance holds. The vector coupling constant g_V can be derived from nuclear superallowed $0^+ \to 0^+$ decays, if conservation of the weak vector current (CVC) is assumed. The most recent evaluation gives a value $(ft)_{0^+ \to 0^+} = K/2g_V^2 = 3073.3 \pm 3.5$ s.[8] The β-decay correlation coefficient that has been measured most accurately is the asymmetry parameter A. According to $V - A$ it is related to λ through the relation

$$A \approx A_o = -2\frac{\lambda(\lambda+1)}{1+3\lambda^2} . \qquad (2)$$

A differs from A_o at the percent level due to energy dependent recoil order corrections.[9] In Table II measured values of A have been combined with g_V from nuclear superallowed $0^+ \to 0^+$ decays to derive indirect lifetime values. The quoted errors reflect only the uncertainty reported for λ; the uncertainty on g_V is added in quadrature to the combined result. If our assumptions about the Standard Model are valid, then the indirect and direct lifetimes should be equal. Combining direct values $\tau_n = 918 \pm 14$ s,[10] 899 ± 11 s,[11] 876 ± 21 s,[12] 886 ± 30 s,[13] 877 ± 10 s,[14] 887.6 ± 3.0 s,[15] 893.6 ± 5.3 s,[16] 888.4 ± 2.9 s,[17] 883.2 ± 2.9 s,[18] and 885 ± 9 s[19] gives $\tau_n = 887.3 \pm 1.5$ s with $\chi^2 = 10.9$ for nine degrees of freedom. Combining indirect values $\tau_n = 905\pm 19\,\text{s}$,[20] $899\pm 20\,\text{s}$,[21] $901\pm 15\,\text{s}$,[22] $895.1\pm 5.9\,\text{s}$,[23] and $904.1\pm 4.3\,\text{s}$[24] gives $\tau_n = 901.0 \pm 3.5\,\text{s}$ with $\chi^2 = 1.57$ for four degrees of freedom. The difference between these two results is $13.7 \pm 3.8\,\text{s}$ which deviates from zero by $3.6\,\sigma$. We note that the result which individually provides the greatest weight to this experimental difference is that of Erozolimskii et al.[24]

Table II: Indirect determinations of the neutron lifetime.

τ_n (sec)	σ_{τ_n}	Measured coefficient	Reference
905	19	A	Krohn and Ringo, 1975
899	20	a	Stratowa et al., 1978
901	15	A	Erozolimskii et al., 1979
895.1	5.9	A	Bopp et al., 1985
904.1	4.3	A	Erozolimskii et al., 1990

The extension to the Standard Model that is usually made is a manifestly left-right symmetric model $SU(2)_L \times SU(2)_R \times U(1)$.[4-7] In such a model, the mass eigenstate gauge-bosons W_1 and W_2 are linear combinations of the weak eigenstates W_L and W_R which couple to left and right handed currents respectively. According to Gaponov et al.[7] three new parameters η, ζ, and λ_N are required in this model. Their meaning is as follows: $\eta = (M_1/M_2)^2$ denotes a squared mass ratio of the W_1 and W_2 bosons; ζ is the mixing angle of these bosons; and λ_N reflects a relative renormalization of the axial nucleon current. In the Standard Model, these parameters tend to $\eta = 0$, $\zeta = 0$, and $\lambda_N = \lambda$. Results from β-decay measurements can be combined with similar results from other leptonic and semi-leptonic decays to place even tighter restrictions on allowable values for these parameters. At the present time only the β-decay measurements are inconsistent with the Standard Model, therefore additional measurements of A (or other correlation coefficients) with an accuracy < 1% and of τ_n with an accuracy < 5 s are crucial.

PRESENT RESULTS

There are two distinct strategies for the direct determination of τ_n. The usual procedure[10,12,13,19] is an "in-beam" method whereby the neutron decay rate $\dot{N}_n$ is measured in a well-defined volume through which a neutron beam passes. If N_n is the mean neutron number in that volume, τ_n is determined by application of the differential equation $\dot{N}_n = -N_n/\tau_n$. This method requires absolute measurements of the event rate, beam volume, and time-averaged neutron density within that volume. In neutron "bottle" experiments[11,14,15,17,18] an isolated ensemble of $N(0)$ neutrons is confined for a time t and τ_n is determined by application of the exponential relation $N(t) = N(0)e^{-t/\tau_n}$. Such methods have advanced dramatically in recent years; nevertheless, they require very careful assessment of loss mechanisms.

The present experiment is of the "in-beam" variety where the beam volume is defined by the boundaries of an electromagnetic Penning trap.[25] The trap retains all protons produced by neutron decay in the cold neutron beam which traverses it parallel to the magnetic axis. These protons are subsequently ejected from the trap and counted with near unit efficiency. Simultaneously the mean neutron density is determined by counting with known efficiency the α-particles from the $^{10}B(n,\alpha)^7Li$ reaction. In an earlier version of this technique[26] the magnetic field was oriented normal to the neutron beam.

In the parallel configuration any dependence on the spatial distribution and velocity distribution of the neutrons within the neutron beam is eliminated[25] and τ_n is given by

$$\tau_n = \left(\frac{N_\alpha}{\epsilon_o}\right)\left(\frac{\epsilon_p}{N_p}\right)\left(\frac{L}{v_o}\right). \qquad (3)$$

Here N_p and N_α are the numbers of protons and α-particles respectively recorded in an arbitrary counting period, L is the length of neutron beam which decays in the trap, and ϵ_p is the efficiency of the proton detector. $\epsilon_o = (\Omega/4\pi)\rho_s\sigma_o N_A/A$ is the efficiency for counting a thermal neutron of velocity $v_o = 2200\,\mathrm{ms}^{-1}$, where $(\Omega/4\pi)$ is the relative solid angle for α-particle collection, ρ_s is the surface density in g/cm^2 of ^{10}B atoms ($A = 10.0129$), σ_o is the thermal cross-section and N_A is Avogadro's number.

A schematic of the apparatus used in the experiment is shown in Fig. 1. The Penning trap consists of two potential barriers $\approx 1\,\mathrm{kV}$ high, superimposed on a coaxial 5 Tesla uniform magnetic field produced by a superconducting solenoid. Trapped protons of energy $< 0.751\,\mathrm{keV}$ move in cyclotron orbits of radii $< 1\,\mathrm{mm}$ about the local magnetic field lines. Their guiding centers move along the magnetic field lines between the confining or "mirror" electrode and the "gate" electrode through which the protons are periodically released. Trapping times between 5 and 100 ms were studied to confirm that there was no diffusion loss of trapped protons. Final data were all taken with a 10 ms trapping time.

Due to the presence of energy and electric field dependent end effects, the mean trap length is difficult to determine precisely. The trap is therefore constructed from 16 segments, each of which is fabricated from fused quartz to optical tolerances, coated with a conducting layer, and electrically connected to its own high voltage switch. Each segment is nominally 22 mm long and its length is known to better than $\pm 10\,\mu\mathrm{m}$. By employing a range of trap lengths, it is possible to eliminate the end effects entirely and determine the differential decay rate dN_p/dL, i.e. the number of trapped protons per unit length of beam, which replaces N_p/L in Eq. 3. An experimental plot of N_p against L is shown in Fig. 2 which reveals no deviation from linearity.

To release the trapped protons the "gate" electrode facing the detector is lowered to ground potential, and the protons exit with their guiding centers moving

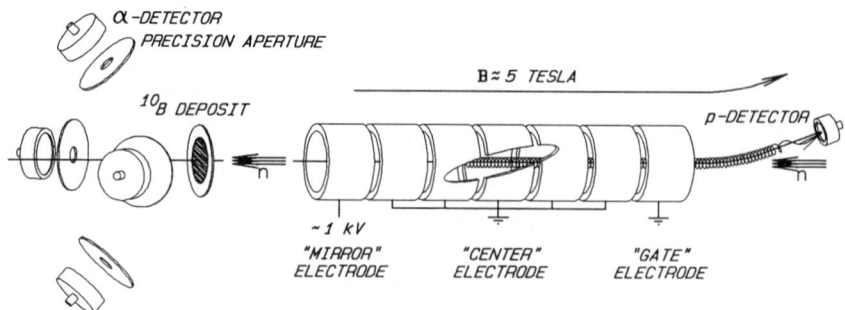

Figure 1: Schematic outline of the experimental method showing the gate electrode in the trap-open configuration. The actual proton trap has 16 independent electrode segments.

adiabatically along the magnetic field lines. These field lines bend by 9.5° in the region beyond the trap. Besides bending, the magnetic field decreases slightly in this region to avoid creating a magnetic mirror for low energy protons. Finally the protons are accelerated into a silicon surface barrier detector, which is held at a high negative potential. The resulting signal is transmitted to ground through an optical fiber. Typical acceleration voltages range between 20 kV and 40 kV, allowing easy discrimination of protons from counter noise. The background is further suppressed by a factor > 100 by gating the counter only during the appropriate "trap open" interval.[26] Following proton collection, the "mirror" electrode is grounded for 10 μs to release any trapped decay electrons of energy < 1 keV. This is the source of the electrode timing correction listed in Table III.

The distribution of observed proton events versus energy and time of arrival is shown in Fig. 3 where the signal to noise ratio is typically about 500. The timing spectrum was used in the final data analysis because the background and deadtime corrections are more straightforward in this case. A timing spectrum event signals the arrival of one or more protons in the trap or a background pulse. In the energy spectrum, it is sometimes difficult to distinguish multi-proton events from true background.

After traversing the proton trap the neutron beam passes through a 0.34 mm thick single crystal of silicon which supports a 94% ^{10}B enriched deposit. Those α-particles from the reaction ^{10}B$(n,\alpha)^{7}$Li that are emitted in the backward direction are inhibited by the silicon wafer from reaching the proton trap where they could generate background by ionization. Those emitted in the forward direction suffer negligible scattering or energy loss when emerging from the deposit and are recorded by four surface barrier detectors. The total collection solid angle Ω was determined by four precision apertures. It was measured in two ways; by mechanical contact metrology and by calibration with α-sources of known absolute activity. These methods agreed to within 0.1% yielding the result $\Omega/4\pi = 0.004196 \pm 0.1\%$.

The ^{10}B$(n,\alpha)^{7}$Li cross-section $\sigma(v)$ is known to deviate from pure $1/v$ behavior

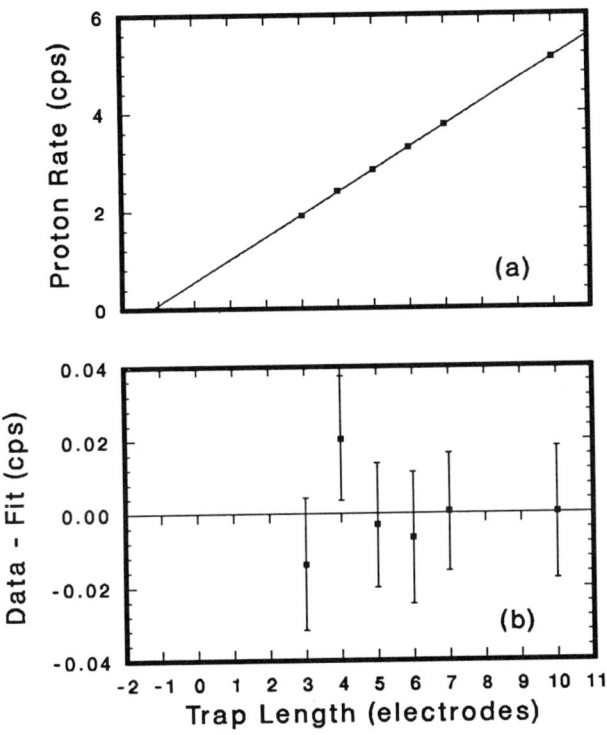

Figure 2: A plot of proton count rate vs. trap length (a) and residuals (b). The slope of this line is the activity per unit length of the neutron beam.

(as reflected by the Westcott g-factor) by $\lesssim 0.03\%$[27] and has the thermal value $3839.5 \pm 0.16\%$ barn.[28] The ^{10}B surface density ρ_s was determined on the basis of isotope dilution mass spectrometry (IDMS) on selected samples from a range of ^{10}B and ^{6}LiF deposits, whose counting rates were compared in a neutron beam at the BR1 reactor, SCK/CEN, Mol, Belgium.[29] The result for the ^{10}B foil number 2-H4, which was used in the present experiment, was $\rho_s = 11.934 \pm 0.037 \,\mu g/cm^2$.

The data used to determine the present result were derived from 5 complete runs carried out at the cold beam position PN7 at the Institut Laue-Langevin, Grenoble, France. Several different accelerator voltages were used in conjunction with several surface barrier detectors having a variety of measured gold window thicknesses. The data are summarized in Table IV, together with calculated values of ϵ_p which depart slightly from unity because of Rutherford scattering in the gold window or, with less probability, in the silicon substrate. Most of the data were obtained with detectors having $20 \,\mu g/cm^2$ of gold so that backscattering was minimized. The other detectors were used primarily as a check on the systematic variation of the backscattering. Corrections with uncertainties for important systematic effects are listed in Table III. The largest correction is present because the boron deposit was not uniformly distributed on the foil (there were more target

Table III: Final results and corrections.

Correction (sec)	Error (sec)	Source of correction
0.0	2.8	^{10}B foil mass per unit area
0.0	1.4	^{10}B cross section
0.1	1.0	n-detector solid angle for finite n-beam
0.0[a]	1.0	p-backscattering from Au surface of detector
2.8	0.7	Absorption of neutrons by ^{10}B foil
0.0[a]	0.5	p-backscattering from Si in detector
-3.6	0.5	Finite radius n-beam on nonuniform ^{10}B deposit
-0.9	0.5	Electrode timing
-0.7	0.5	Incoherent n-scattering in Si backing of ^{10}B foil
0.0	0.5	Trap length
0.8	0.2	Absorption of neutrons by Si backing
-1.5	3.7	Total correction for systematic effects
895.1	3.8	Uncorrected statistical result
893.6	5.3	**Final 1-σ result**

[a]Corrections for these effects were applied to individual runs.

atoms available in the center).[29] This distribution was sampled by a neutron beam roughly 16 mm in diameter at the foil. As it happens, this adjustment is nearly compensated for by one made for absorption of neutrons in the boron foil (there were fewer neutrons deeper in the target). The final result is $\tau_n = 893.6 \pm 5.3$ s where the stated error is the quadratic sum of systematic and statistical errors.

We note the discrepancy between our result and the earlier result of Byrne et al.[26] We now recognize that incorrect values for the calibration and uniformity of the neutron counting foils were used in the 1980 experiment.[29] Other defects associated with end effect uncertainties and proton loss by magnetic mirror trapping have been eliminated in the present work as well.

FUTURE PROSPECTS

In late 1990, NIST's Cold Neutron Research Facility (CNRF) will begin operation. The lifetime apparatus has been reassembled for installation on the end of a neutron guide, and it is anticipated that data collection will begin when the neutrons become available. An anticipated error budget reflecting several improvements to the apparatus and an aggressive program of flux determination (Phase 2) is outlined in Table V. The two largest uncertainties in the current

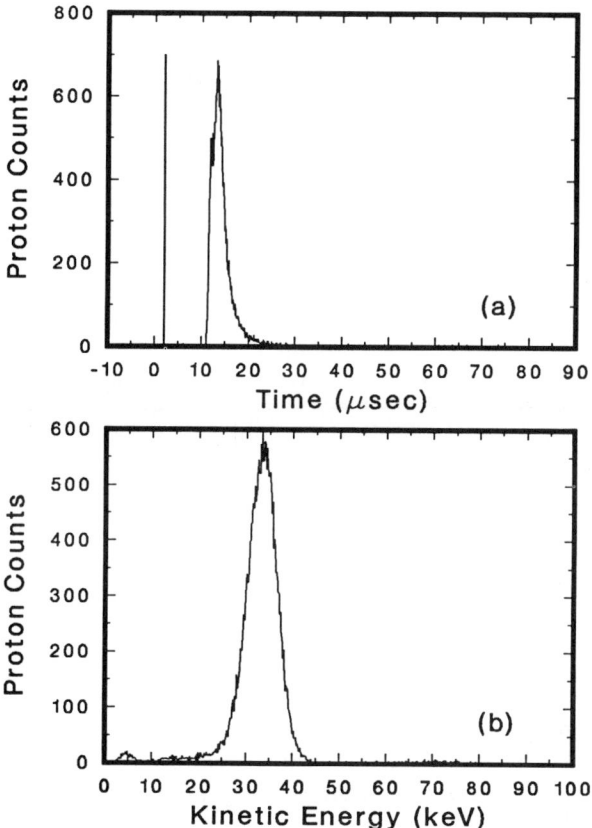

Figure 3: (a) A histogram of proton counts vs. arrival time at the detector. Time $t = 0$ corresponds to the signal which triggers the "opening" of the Penning trap. The sharp peak shortly after $t = 0$ is an artifact. (b) A histogram of proton counts vs. energy. The decay signal is seen with very high signal to noise ratio.

result (Phase 1) are statistics on decay detection and uncertainty in the efficiency for counting thermal neutrons, ϵ_o; consequently the largest gains will be made in these areas.

A. Statistics

Statistical errors will be reduced by running for longer periods of time. This will require that the apparatus be less prone to breakdown problems. A redesigned trap is being constructed that will alleviate high voltage instability. In it, the spacers that separate the electrodes are recessed more deeply from the inner bore than before, the corners of the electrodes are rounded, and gold has replaced chromium as the coating material. We anticipate that it will be possible to run for much longer continuous periods of time than was possible at ILL. With the

Table IV: Results for several detector voltages and gold thicknesses. ϵ_p corrects for elastic Rutherford scattering of protons from the detector surface. The quoted errors are the statistical errors on each individual dN_p/dL determination. The statistically combined result needs to be adjusted for several small effects.

Detector volts (kV)	Au thickness ($\mu g/cm^2$)	ϵ_p	$\left(\frac{N_a}{\epsilon_o}\right)\left(\frac{\epsilon_p}{v_o}\right)(dN_p/dL)^{-1}$ (sec)
24.5	38.6	0.9743	886.0 ± 11.4
29.5	18.3	0.9924	894.3 ± 5.7
32.0	18.3	0.9936	892.7 ± 4.7
34.5	18.3	0.9945	903.1 ± 8.4
32.0	58.9	0.9762	938.0 ± 21.5

ILL neutron flux, it was possible to get a 0.5% statistical error in 12 hours of running. NIST fluxes are expected to be comparable to, though slightly less than, ILL fluxes.

B. Flux Determination

Uncertainties in the ^{10}B foil mass per unit area, ^{10}B cross section, and α-detector solid angle will not contribute to uncertainty in future measurements because the α-detector that monitors flux will be calibrated against an absolute neutron flux monitor. The anticipated uncertainty in this scheme is ≈ 1 s. The basis of calibration will be shifted to "black" neutron detector methods that do not require knowledge of the mass of the boron deposit. The calibration against a "black" detector must be carried out on a monochromatic neutron beam. There are currently two black detectors under development, one based on ^{6}Li calorimetry[30] and the other based on ^{10}B $\alpha - \gamma$ coincidence methods.[31] The two methods both aim for errors $\lesssim 0.1\%$. Intercomparisons between these methods will be important in establishing the validity of measurement techniques that are venturing into unprecedented levels of accuracy.

In initial tests of the calorimetry method carried out at the NIST thermal column in May 1989, results were compared with counts from a fission chamber flux monitor and found to agree well within the statistical accuracy of the measurements (1%). Work is currently progressing on the design of an improved calorimeter based on a liquid helium reservoir.

A newly constructed $\alpha - \gamma$ coincidence apparatus features two separate gamma detectors in opposing positions above and below the beam to reduce sensitivity to angular correlation and positioning effects. It also features a second alpha detector in an optimum position for independent calibration of the system by use of standard alpha sources.

A third possibility for intercomparison also exists. In a recent preprint Spivak discusses remeasuring τ_n to an accuracy of ± 3 s using an improved version of his apparatus.[32] As part of their work, he and his co-workers have refined their technique for assaying irradiated gold foils, and they argue that the neutron density in their cold neutron beam can now be measured to an accuracy of 0.22%. They report that the largest uncertainty in the method is due to beta absorption in the gold foil.

C. Further Refinements

Long-term availability of a neutron beam will make it possible to study the systematics of Rutherford backscattering of the protons so as to reduce uncertainties associated with it. The use of chemically etched, rather than mechanically polished silicon wafers as a substrate for ^{10}B should reduce incoherent scattering of neutrons by a factor of 10. A measurement of the effect of changes in neutron detector efficiency due to beam profile will be made on the beam used to measure the lifetime with a precision x-y translation stage designed to carry the α-detector. Variable apertures will be used between the detector and beam to define the geometry precisely.

ACKNOWLEDGEMENTS

This collaboration includes David Gilliam, Geoffrey Greene, George Lamaze, W. Michael Snow and the author at the National Institute of Standards and Technology; James Byrne, Peter Dawber, and Jeffrey Spain at the University of Sussex; Andreas P. Williams at the Institut Laue-Langevin; Roger Scott at the Scottish Universities Research and Reactor Centre; and J. Pauwels, R. Eykens, and A. Lamberty at the Central Bureau for Nuclear Measurements. Primary responsibility for the calorimetry experiment lies with Timothy Chupp and Jonathan Richardson at Harvard University; and Hamish Robertson and John Wilkerson at Los Alamos National Laboratory.

We wish to thank Emeritus Professor J.M. Robson for his support and interest in this work, Dr. Klaus Schreckenbach for the considerable assistance and technical support provided by him and other members of the staff at the Institut Laue-Langevin, J. Van Gestel for his outstanding contribution in the preparation of the ^{10}B and ^{6}LiF reference deposits, and Dr. Nalin Parikh of the University of North Carolina for the measurements of the gold window thicknesses. This work is supported in part by DOE interagency agreement No. DE–AI05–87ER40340 and by the SERC of the U.K. We also acknowledge a travel grant from NATO Scientific Affairs Division.

Table V: Error Budget for Neutron Lifetime Experiment. Phase 1 refers to results already obtained; Phase 2 refers to results anticipated from use of the newly redesigned apparatus at NIST's Cold Neutron Research Facility.

Source of Error	Phase 1 (sec)	Phase 2 (sec)
Statistics on decay detection	3.8	<1.0
^{10}B-foil mean mass per unit area (IDMS)	2.8	NA
^{10}B cross section	1.4	NA
^{10}B-foil central shape correction	0.4	NA
^{10}B-foil central shape × beam profile	0.4	< 0.1
^{10}B-foil absorption	0.7	0.1
Si-backing absorption	0.2	0.1
Si-backing incoherent scattering	0.5	0.2
α-detector solid angle Ω	1.0	NA
α-detector $\delta\Omega$ due to beam profile	0.3	0.1
p backscattering from Au	1.0	0.2
Near normal forward scattering of protons	0.1	0.1
Si-p backscattering	0.5	0.2
Trap length 20°C	0.5	0.2
Timing of mirror electrode	0.5	0.3
Trappable background	< 0.1	< 0.1
Trap thermal contraction	< 0.1	< 0.1
Si-backing Bragg Scattering	< 0.1	< 0.1
$\Omega(\vec{B})$ modification of trajectory	< 0.1	< 0.1
Neutron trajectory in trap	< 0.1	< 0.1
Trap alignment	< 0.1	< 0.1
Diffusion loss of protons	< 0.1	< 0.1
Neutron beam centroid misalignment	< 0.1	< 0.1
Trap end effect and beam divergence	< 0.1	< 0.1
Calibration of neutron monitor	NA	≈ 1.0
Total Error	5.3	≈ 1.5

NA Not applicable

REFERENCES

1. D. H. Wilkinson, Nucl. Phys. **A377**, 474 (1982).
2. D. N. Schramm and L. Kawano, Nucl. Instrum. Methods **A284**, 84 (1989).
3. J. N. Bahcall et al., Rev. Mod. Phys. **54**, 767 (1982).
4. A. S. Carnoy, J. Deutsch, and B. R. Holstein, Phys. Rev. D **38**, 1636 (1988).
5. B. G. Erozolimskii, Nucl. Instrum. Methods **A284**, 89 (1989).
6. D. Dubbers, W. Mampe, and J. Döhner, Europhys. Lett. **11**, 195 (1990).
7. Y. V. Gaponov, N. B. Shul'gina, and P. E. Spivak, Neutron beta-decay and right-handed current problem, Preprint IAE–5032/2, 1990, I. V. Kurchatov Institute of Atomic Energy, Moscow.
8. J. C. Hardy, I. S. Towner, V. T. Koslowsky, E. Hagberg, and H. Schmeing, Nucl. Phys. **A509**, 429 (1990).
9. B. R. Holstein, Rev. Mod. Phys. **46**, 789 (1974).
10. C. J. Christensen et al., Phys. Rev. D **5**, 1628 (1972).
11. Y. Y. Kosvintsev, V. I. Morozov, and G. I. Terekhov, JETP Lett. **44**, 571 (1986).
12. J. Last, M. Arnold, J. Döhner, D. Dubbers, and S. J. Freedmann, Phys. Rev. Lett. **60**, 995 (1988).
13. K. Schreckenbach, G. Azuelos, P. Grivot, R. Kossakowski, and P. Liaud, Nucl. Instrum. Methods **A284**, 120 (1989).
14. F. Anton, W. Paul, W. Mampe, L. Paul, and S. Paul, Nucl. Instrum. Methods **A284**, 101 (1989).
15. W. Mampe, P. Ageron, J. C. Bates, J. M. Pendlebury, and A. Steyerl, Phys. Rev. Lett. **63**, 593 (1989).
16. J. Byrne et al., Phys. Rev. Lett. **65**, 289 (1990).
17. V. P. Alfimenkov et al., Results of neutron lifetime measurements with gravitational UCN trap, Preprint 1629, 1990, Leningrad Nuclear Physics Institute.
18. Morozov et al., 1990, submitted for publication.
19. P. E. Spivak, Sov. Phys.-JETP **67**, 1735 (1988), A further error in the proton count was found and the neutron lifetime should be decreased down to $\tau_n = 885$ s.
20. V. Krohn and G. Ringo, Phys. Lett. B **B55**, 175 (1975).
21. C. Stratowa, R. Dobrozemsky, and P. Weinzierl, Phys. Rev. D **18**, 3970 (1978).
22. B. G. Erozolimskii, A. I. Frank, Y. A. Mostovoy, S. S. Arzumanov, and L. R. Voytzik, Sov. J. Nucl. Phys. **30**, 356 (1979).

23. P. Bopp et al., Phys. Rev. Lett. **56**, 919 (1986).
24. B. G. Erozolimskii, I. A. Kuznetsov, I. V. Stepanenko, I. A. Kuida, and J. A. Mostovoi, New measurements of the electron-neutron spin asymmetry in the neutron beta-decay, submitted for publication, 1990.
25. J. Byrne et al., Nucl. Instrum. Methods **A284**, 116 (1989).
26. J. Byrne et al., Phys. Lett. **92B**, 274 (1980).
27. B. A. Magurno, R. R. Kinsey, and F. M. Scheffel, *Guidebook for the ENDF/B-V Nuclear Data Files*, BNL–NCS–31451, 1982.
28. R. Peele and H. Condé, Neutron standard data, in *Proceedings of the International Conference on Nuclear Data for Science and Technology*, edited by S. Igarasi, page 1005, Mito, Japan, 1988, Japan Atomic Energy Research Institute.
29. J. Pauwels et al., The preparation and characterisation of ^{6}LiF and ^{10}B reference deposits for the measurement of the neutron lifetime, Talk at the 15th World Conference Meeting of the International Nuclear Target Development Society, 1990.
30. R. G. H. Robertson and P. E. Koehler, Nucl. Instrum. Methods **A251**, 307 (1986).
31. D. M. Gilliam, G. L. Greene, and G. P. Lamaze, Nucl. Instrum. Methods **A284**, 220 (1989).
32. P. E. Spivak, On the project of measuring the neutron lifetime within an error of ±3s in the experiment on cold neutron beam transport in the field-free space, Preprint IAE-4844/2, 1989, I. V. Kurchatov Institute of Atomic Energy, Moscow.

RESULTS OF NEUTRON LIFETIME MEASUREMENTS WITH GRAVITATIONAL UCN TRAP

V.P.Gudkov, A.G.Kharitonov, V.V.Nesvizhevsky, A.P.Serebrov,
S.O.Sumbaev, R.R.Taldaev, V.E.Varlamov and A.V.Vasilyev
*Leningrad Nuclear Physics Institute, Gatchina,
Leningrad district, 188350, USSR*

V.P.Alfimenkov, V.I.Lushikov, V.N.Shvetsov, A.V.Strelkov,
*Joint Institute for Nuclear Research, Dubna,
Moscow district, 141980, USSR*

ABSTRACT

In the present experiment the neutron lifetime has been measured with ultra-cold neutrons (UCN) confined in a gravitational trap. The measured neutron lifetime is equal to (888.4 ± 2.9)s, $\lambda = -(1.2677 \pm 0.0025)$. This result is in agreement with the results of other experiments: 887.6 ± 3.0)s [1] and (893.6 ± 5.3)s [2]. However parameters λ_τ, λ_A extracted from the neutron lifetime experiments and from the β-decay asymmetry experiments differ (3.2σ). The reasons of the discrepancy are discussed.

INTRODUCTION

The precise measurements of the neutron lifetime and of β-decay asymmetry give the additional information for check-up of the Standard Model of the electroweak interaction and seach for different deviations. As is known the vector coupling constant G_V is accurately determined from studies of "super-allowed Fermi ($0^+ - 0^+$) transitions":

$$(ft)^{00}(1-\delta_c)(1+\delta_R^{00}) = \frac{\pi^3 \hbar^7 \ln 2}{m_e^5 e^4 G_V^2 (1+\Delta_\beta)} \quad (1)$$

Where f is the phase space factor, t is half life, δ_c is the nuclear structure correction, δ_R is the outer radiative correction, Δ_β is the inner radiative correction. Because of small β-decay energy the role of the energy-dependent form-factors is negligible and the neutron β-decay is almost completely described by single parameter - ratio of the weak-coupling constants, $G_A/G_V = \lambda$. There exist following relations for the neutron lifetime τ and the β-decay asymmetry A:

$$(f\tau)^n(1+\delta_R^n) = \frac{2\pi^3 \hbar^7}{m_e^5 c^4 G_V^2 (1+\Delta_\beta)(1+3\lambda^2)} \quad (2)$$

$$A = -2 \frac{\lambda^2 + \lambda}{1+3\lambda^2} \quad (3)$$

Thus the experimental check-up task is the independent determination of λ from the neutron lifetime (λ_τ) and from the β-decay asymmetry (λ_A).

EXPERIMENT

In the present experiment the neutron lifetime has been measured with ultra-cold neutrons (UCN) confined in a gravitational trap. A low loss rate due to the neutron interaction with the wall has been achieved by keeping the trap at a temperature of 10-15 K and by choosing weakly absorbing materials for the trap walls.

A scheme of our method is shown in fig.1. The setup is the cryostat with a gravitational trap for UCN storage, being used at the same time as a gravity spectrometer. UCN confinement in the storage regime is achieved by a sphere rotating

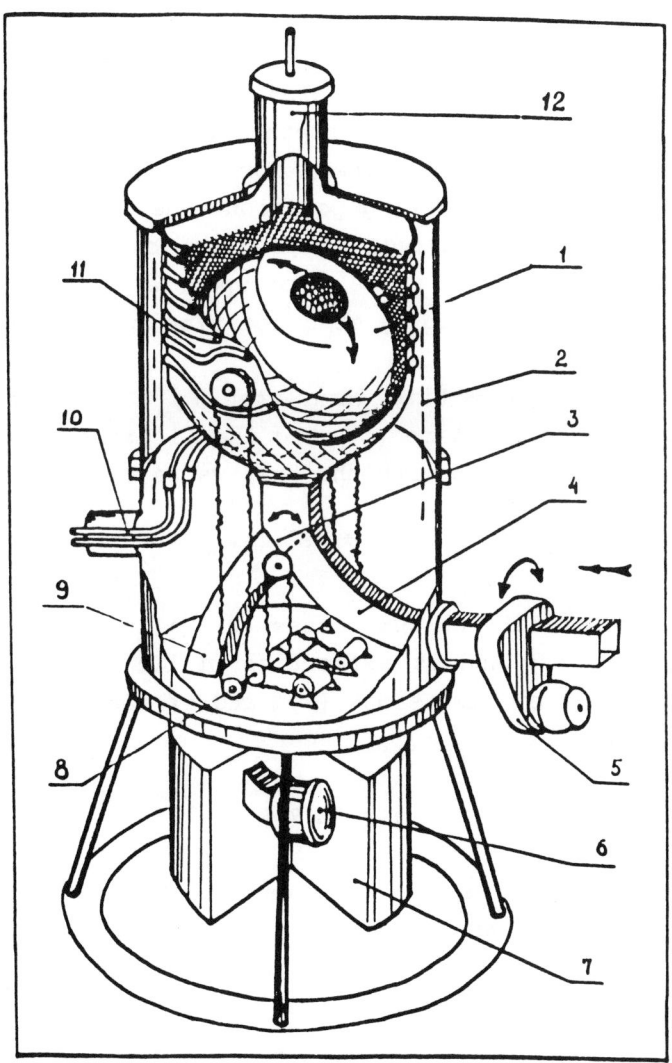

Fig.1. Setup for the measurement of the neutron lifetime with a gravitational trap: (1) UCN storage trap; (2) nitrogen screen; (3) distribution valve; (4),(9) neutron guide; (5) injection valve; (6) UCN detector; (7) detector shield; (8) (8) system for rotation; (10) cryopipes; (11) cryostat; (12) system for frozen cover.

around a horizontal axis from a position with the hole pointing down to a position with the hole upwards. In this case low energy neutrons become enclosed in the trap due to the gravitation field. After the storage of UCN during the given time the neutrons are detected, for this purpose the sphere is rotated into the starting position. The change between regimes (filling or emptying) is provided by a distribution --valve and a injection-valve. The spectral measurements are carried out due to step-by-step rotation of the sphere from the top position.

The UCN storage time in a trap (τ_{st}) is determined by the free neutron lifetime (τ_n) and the lifetime of the losses due to the interaction with the trap walls (τ_{los})

$$\tau_{st}^{-1} = \tau_n^{-1} + \tau_{los}^{-1} \qquad (4)$$

The probability of UCN losses (τ_{los}^{-1}) is in proportion to the loss factor η and is a function of UCN energy (E), of a shape and the trap dimensions (R), and of the critical energy of the trap walls (E_{lim}).

$$\tau_{st}^{-1} = \tau_n^{-1} + \eta \, \gamma \, (E,R,E_{lim}) \qquad (5)$$

In the equation (5) τ_{st}^{-1} is a linear function of the argument γ, which is calculated for the given E,R,E_{lim}. The neutron lifetime (τ_n^{-1}) may be obtained by the extrapolation to the zero value of the argument γ. The loss probability has to be much less than the probability of β-decay to reach the precise measurements of the neutron lifetime.

In our experiment an aluminium sphere of 75 cm diameter covered inside with beryllium (3000-5000 A) was used. At low temperature (15 K) pure oxygen (impurity concentration less than 10^{-4}) was frozen on the walls. The thickness of the oxygen coating was 3-7 μm. The measured probability of UCN losses accounted for 3% of the neutron β-decay probability, and

τ_{los} = 8-10 hours. In this case the 10% accuracy of the factor η makes it possible to obtain the 0.3% accuracy of the neutron lifetime. However the range of γ-function variation due to energy dependence was not sufficient, therefore the additional measurements with the flat cylinder trap (diameter - 75 cm, distance between the flat walls - 15 cm) were carried out.

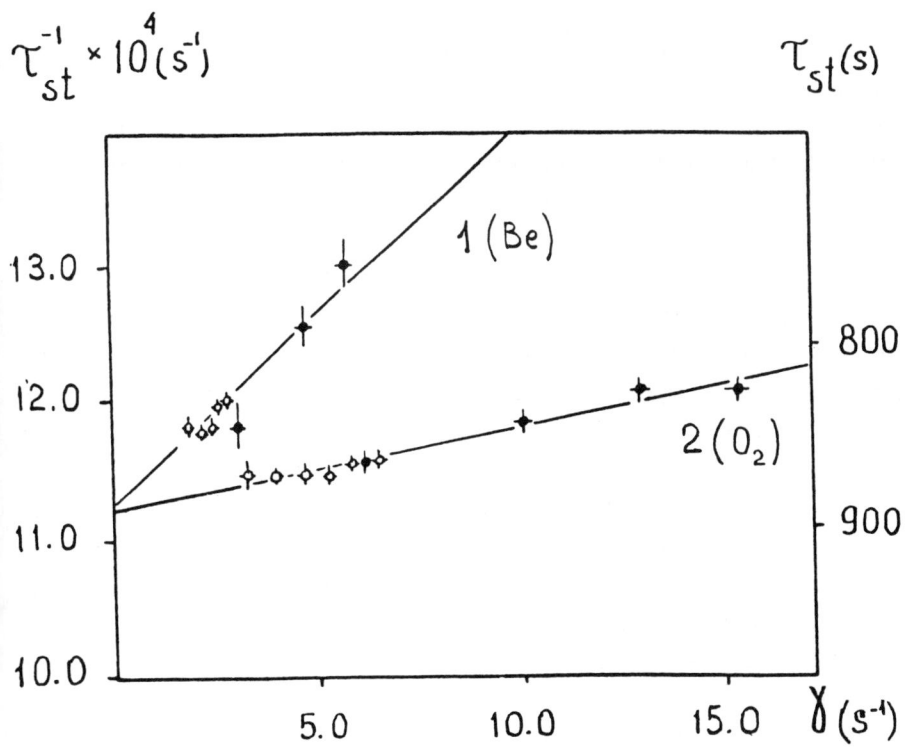

Fig.2. Results of the UCN storage time measurements (τ_{st}^{-1}) as function of the parameter γ. 1(Be) - extrapolation to the neutron lifetime (τ_n) from the beryllium traps data. 2(O_2) - extrapolation to the neutron lifetime (τ_n) from the oxygen traps data.

o - results of measurements with the sphere traps.
● - results of measurements with the cylinder traps.

The resuls of measurements are shown in fig.2 as a linear function $\tau_{st}^{-1}(\gamma)$. Besides the measurements with the oxygen traps, the measurements with the beryllium traps were performed to demonstrate the conditions for UCN storage before the freezing of oxygen. As an additional analysis gave the vertical errors shown in fig.2 define from the statistic accuracy and correspond to the error variance of the separate measurements. The horizontal errors define from the accuracy of the energy interval for UCN, they equal to approximately 1% of the argument γ and do not effect the accuracy of the final result. The possible systematic error concerned with the accuracy of the critical energy (E_{lim}) was found to be very small (0,1 s) due to the compensation of the uncertainty for the two traps with the different shape.

The result from the extrapolation the following neutron lifetime values were obtained: τ_n= (885.0 ± 7.7)s for the traps with the beryllium coating, τ_n= (889.0 ± 3.1)s for the traps with the oxygen coating. The values of the loss factor η accounted for: $(28.1 \pm 4.0) \cdot 10^{-6}$ for the beryllium trap and $(6.1 \pm 0.6) \cdot 10^{-6}$ for the oxygen trap. The validity of extrapolation is confirmed by χ^2 value 0.81 calculated from oxyden data. The final result obtained in our experiment for the neutron lifetime is (888.4 ± 2.9)s, the corresponding value of λ is -(1.2677 ± 0.0025).

Evaluating a prospects of the neutron lifetime accuracy it is worth while to mention the chance of accuracy increasing by the factor of 1.5 or 3 times due to the continuation of the measurements or the use of the more intensive UCN beam. However the investigations to decrease the losses of UCN in the storage process are of a great interest.

CONCLUSION

The result obtained is in agreement with results of the other experiments: (887.6 ± 3.0)s [1] and (893.6 ± 5.3)s [2]. The neutron lifetime data [1-7] including the present result give the following values of the neutron lifetime and of the parathe parameter λ_τ: τ_n= (888.5 ± 1.8) s, λ_τ= -(1.2676 ± 0.0016). At the same time the β-decay asymmetry data [8-11] give the average value λ_A= -(1.2570 ± 0.0028). The disagreement λ_A, λ_τ account for 0.0106 ± 0.0033 or 3.2σ.

Though the given disagreement is statistical significant, it is impossible to exclude both the randon deviation and the experimental systematic errors, which are not revealed yet.

The mentioned discrepancy may be decreased if the effect of nuclear matter on the value of the inner radiative correction Δ_β to be taken into account. As was shown in [12], the part of the radiative correction, which is depended on the structure of strong interaction, is different for decay of free neutron and nucleus. This causes the change of Δ_β for the difference ($\Delta_\beta - \Delta_\lambda$) in equation (1) and the decreasing of λ_τ by 0.3%. The discrepancy remained accounts for 2.1σ. However another variants of interpretation are of the interest.

If it is supposed, that the discrepancy may be explained by the right-handed current, then using the results of the work [13] one can obtain the connection between the experimental values A, τ and the model parameters η, ζ:

$$0.23 \eta^2 + 2.2 \eta \zeta + 1.1 \zeta^2 = A + 2 \frac{\lambda_\tau^2 + \lambda_\tau}{1 + 3\lambda_\tau^2} \qquad (6)$$

Where η - the squared mass ratio of left and right bosons, ζ - mixing angle. The right part of equation (6) is determined by the experimental results, it accounts for (2.62 ± 1.22).10^{-3} after including correction Δ_λ. The region of restrictions at 90% c.l. is shown in fig.3, where the region of restrictions from μ-decay experiment is also presented [14]. Though these regions some what overlap one another, the crossing regions are in disagreement with the restriction of mixing angle, which follows from the experimental check of uni-

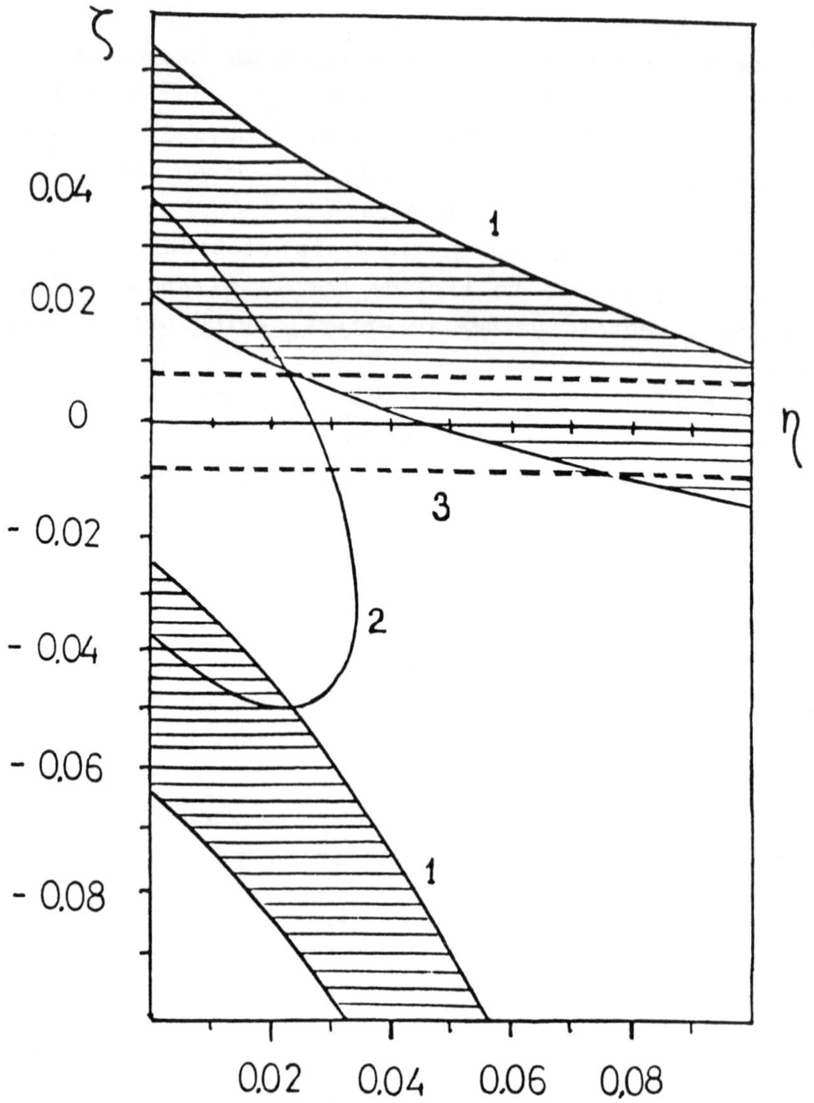

Fig.3. Restrictions on the boson mixing angle ζ and the squared mass ratio η provided by: 1 - the combination neutron-decay asymmetry, neutron lifetime and superallowed-pure-Fermi - decay probabilities; 2 - the muon-decay asymmetry; 3 - the unitarity (extended region, see text).

tarity of Kobayashi-Maskawa matrix - $|\zeta| \leq 4.10^{-3}$ [15-17]. The
The disagreement remains even if it is supposed that the correction Δ_λ is involved completely in the uncertainty of unitarity. Region of restrictions thus extended is shown in fig. 3. At last, analysis of the contribution of right-handed currents into the difference of mass K_L, K_S-mesons [18] and into nonleptonic decays of K-mesons [19] gives the more strong restrictions: $\eta \leq 3.10^{-3}$, $|\zeta| \leq 4.10^{-3}$. Thus the explanation of λ_A, λ_τ-discreapancy by right-handed currents, as it was supposed in [20,21], encounters the evident difficulties.

We have analysed another variant of interpretation of λ_A, λ_τ-discreapancy due to nonzero Fierz term about 1-2%. As is known, Fierz term is related with scalar and tensor interactions. However, the tensor interaction, besides its unattractiveness from theoretical point of view, is rather strongly restricted experimentally [22]. For Fierz term with scalar interaction there exists a restriction - $b_F \leq 0.004$ [15], which closes appreciable contribution of b_F in λ_A, λ_τ-discreapancy. Moreover, procceeding from the restrictions for the mass of charged Higgs bosons $M_H \geq 19$ GeV [23], yet less value of b_F ($b_F \leq 3.10^{-8}$) should be expected.

At the present time there are no reasons to consider that the discussed λ_A, λ_τ-discreapancy may be explained from theory position. The further experimental investigation of neutron β-decay is necessary to clear up this question.

R e f e r e n c e s

1. W.Mampe et al., NIM. A284, 1, 111 (1989).
2. J.Byrne et al., Phys.Rev. Lett. 65, 289 (1990).
3. V.J.Morozov, Nucl.Instr. and Meth. A284, 108 (1989).
4. P.E.Spivak, JETR. 94,1 (1988).
5. J.Last et al., Phys.Rev.Lett. 60, 995 (1988).
6. K.Schreckenbach K. et al., NIM. A284, 1, 120 (1989).
7. F.Anton et al., NIM. A284, 1, 101 (1989).

8. P.Bopp , D.Dubbers et al., Phys.Rev.Lett. 56, 9, 919-922 (1986).
9. B.G.Erozolimskii, I.A.Kuznetszov, I.A.Kuida, Yu.A.Mostovoy, I.V.Stepanenko, Preprint LNPI-1589 (1989); Sov.J.Nucl.Phys. in press.
10. V.Krohn, G.Ringo, Phys.Lett. B55, 175-177 (1975).
11. B.G.Erozolimskii, A.I.Frank, Yu.A.Mostovoy, S.S.Arzumanov, L.R.Voytzik, Sov.J.Nucl.Phys. 30, 692 (1979).
12. V.P.Gudkov, JETR Lett., in press.
13. B.R.Holstein, S.B.Treiman, Phys.Rev. D16, 2369 (1977).
14. A.Jodidio et al., Phys.Rev. D34, 1967 (1986).
15. J.C.Hardy et al., Nucl.Phys. A509, 429 (1990).
16. P.R.Wolfenstein, Phys.Rev. D29, 2130 (1984).
17. J.Deutsch in: Fundamental Symmetries and Nuclear Structure eds J.N.Ginocchio, S.P.Rosen (World Scientific, 1989), p.36.
18. G.Beall, M.Bender, A.Soni, Phys.Rev.Lett.,48, 848 (1982).
19. J.F.Donoghue, B.R.Holstein, Phys.Lett. B113, 382 (1982).
20. Yu.V.Gaponov, N.B.Shulgina, Sov.J.Nucl.Phys. 49, 1359 (1989).
21. Yu.V.Gaponov, N.B.Shulgina, P.E.Spivak, Preprint IAE-5032/2.(1990).
22. H.Paul, Nucl.Phys. A154, 160 (1970).
23. "Review of Particle Properties", Phys.Lett. B204 (1988).

PARITY VIOLATION IN RESONANT NEUTRON REACTIONS

A. Müller*, H. L. Harney, E. D. Davis**
Max-Planck-Institut für Kernphysik
6900 Heidelberg, FRG

ABSTRACT

A formalism is developed that incorporates parity violation into the statistical theory of nuclear reactions. It allows the calculation of average square values of observables sensitive to parity breaking starting from well known average parameters (transmission coefficients and level densities) as input. For the case of low-energy neutron resonance scattering, a systematic study of the expected size of the effect is presented. An enhanced sensitivity of these reactions to parity violation is borne out.

1. INTRODUCTION

It is now a well established, but still surprising phenomenon, that neutron induced compound nuclear reactions show a strong "enhancement" of parity violation: in nucleon-*nucleon* scattering the following value of the P-forbidden difference of cross-sections has been found ($\pm$ denotes the helicity of the incoming particle): $\frac{\sigma_+ - \sigma_-}{\sigma_+ + \sigma_-} = -(1.71 \pm 0.4)10^{-7}$. Here, longitudinally polarized protons have been scattered on hydrogen[1]. The corresponding quantities in nucleon-*nucleus*-scattering however, may be orders of magnitude larger, e.g. $\frac{\sigma_+ - \sigma_-}{\sigma_+ + \sigma_-} = (7.3 \pm 0.5)10^{-2}$. This value[2] has been found in the scattering of longitudinally polarized neutrons on ^{139}La. Similar experiments with other targets have shown, that this "enhancement" is not a special feature of a special place in the nuclear table (for a survey see the monograph of Krupchitsky[3]) and very recently a quite extensive study on ^{238}U has shown, that it is not a special feature of a special value of excitation energy, neither[4]. Before giving a pictorial explanation and an exact model of the mechanisms, a list of very closely related observables of P-violation in neutron induced reactions is presented.

2. OBSERVABLES

There are three neutron-optical effects sensitive to P-violation, depending on what polarization state we have in the entrance channel.

First, consider longitudinally polarized neutrons impinging on an unpolarized target. Parity violation here reveals itself in a different absorption for the

---

* Supported by Studienstiftung des Deutschen Volkes
** Present address: Department of Physics, University of Arizona, Tucson, AZ 85721

two different helicities, and we consider the quantity

$$\Delta = \frac{N_+ - N_-}{N_+ + N_-} \quad \begin{cases} N : \text{forward counting rate} \\ \pm : \text{entrance helicity} \end{cases} \qquad (1)$$

The corresponding pseudoscalar is $\vec{s_i} \circ \vec{p_i}$ (where i denotes the entrance channel) and an optical analogy is circular dichroism[5].

Next, consider transversally polarized neutrons impinging on an unpolarized target. Transverse polarization can be written as a linear superposition of the two longitudinal polarizations with equal weights. Parity violation here reveals itself in a different wave vector of the two helicities inside the target, or, alternatively, in a different index of refraction. This difference leads to a phase shift between the two longitudinal components, which in turn, means a rotation of the resulting transverse polarization[6]. Thus, we consider the rotation angle

$$\Phi[rad] = kl Re(n_- - n_+) \begin{cases} n : \text{index of refraction} \\ \pm : \text{entrance helicity} \\ l : \text{target thickness} \\ k : \text{wavenumber} \end{cases} \qquad (2)$$

The corresponding pseudoscalar is again $\vec{s_i} \circ \vec{p_i}$ and an optical analogy is birefringence or optical activity[5].

Last, consider unpolarized neutrons impinging on an unpolarized target. This means a statistical mixture of the two longitudinal polarizations with equal weights. As in the first case, we have parity violation if we have different absorption for the two different helicities, but now leading to unequal weights in the exit channel (they are scattered out of the beam with different probabilities). This means, in turn, that there will be a net polarization in the transmitted beam

$$P = \frac{N_+ - N_-}{N_+ + N_-} \quad \begin{cases} N : \text{forward counting rate} \\ \pm : \text{exit helicity} \end{cases} \qquad (3)$$

The corresponding pseudoscalar $\vec{s_f} \circ \vec{p_f}$ here pertains to the exit channel and the optical analogy is a circular dichroic medium, now acting as a polarizer.

By virtue of the optical theorem, all these observables can be expressed by the difference of forward scattering amplitudes:

$$\left.\begin{array}{c}\Phi \\ \Delta, P\end{array}\right\} = -2\pi \frac{L}{k} \begin{cases} Re\Delta f(0°) \\ Im\Delta f(0°) \end{cases}, \qquad (4)$$

The raw effect is proportional to the target thickness L (now to be understood as area density $L = \rho l$, where ρ is the number density of the target), but the counting statistics goes exponentially down with it. Balancing the two influences gives an optimal target thickness of the order of the mean free path $L_{opt} = 2L_{mean} = 1/\sigma^{tot}$. The difference of forward scattering amplitudes, in turn, is given by[6]

$$\Delta f = \frac{2i}{k} \sum_{\substack{\ell=1 \\ J}} \Delta_{I,\ell-\frac{1}{2},J} \frac{2J+1}{2(2I+1)} S^{J}_{\ell,\ell-1} \quad . \tag{5}$$

Here, k is the wave number and $\Delta(I,j,J)$ is one if I,j,J fulfill the triangle rule, otherwise zero. Note, that the element of S is truly P-violating, because it connects partial waves, which have a angular momentum different by one unit and thus, because of $\pi = (-1)^{\ell}$, also different parity.

3. MODEL AND MECHANISMS

3.1. Enhancement Mechanisms: Consider the following pictorial representation of slow neutron capture, first without P-violation, see Fig. 1a.:

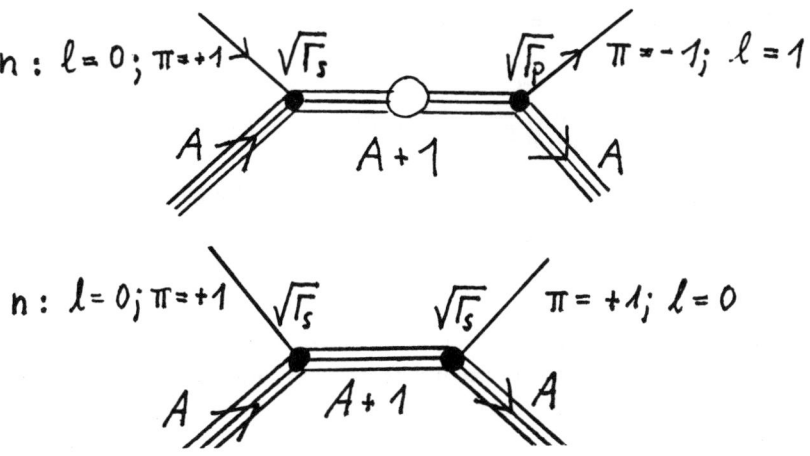

Fig.1. Compound nucleus scattering without (1a., lower half) and with P-violation (1b., upper half).

The incoming neutron combines with the target nucleus of mass number A to the compound system, which then evolves till decay, here back to the neutron channel. For slow neutrons the process is dominated by the s-wave ($\ell = 0$) and the corresponding coupling amplitude $\sqrt{\Gamma_s}$. With parity violation we have the following picture, see Fig. 1b.: Formation of the compound system as before; transition between states of opposite parity; decay back to an exit channel with parity opposite to that of the entrance channel. The channels with parity opposite to the s-wave are those with an odd value of ℓ, for slow neutrons they are dominated by the p-wave ($\ell = 1$) and the corresponding coupling amplitude $\sqrt{\Gamma_p}$. Comparing the two processes, we see that the ratio $\Delta f/f$ will contain a factor $\sqrt{\Gamma_p/\Gamma_s} \simeq kR$ which is *small*, again due to the low energy of the neutrons.

What other influences do we expect? Think in terms of the "golden rule", which gives the transition probability

$$(\Delta f)^2 \propto \text{transitionrate} \cdot \text{time} \sim \hbar^{-1}\frac{1}{D}(H^w)^2 \cdot \frac{\hbar}{\Gamma} \sim \frac{H^w}{D} \cdot \frac{H^w}{\Gamma} \quad . \tag{6}$$

Here, we have estimated the "time" by the lifetime $\frac{\hbar}{\Gamma}$ of the resonant states, where Γ is the average width of the decaying resonances; and we have written the "density of final states" (occurring in golden rule expressions) as $\frac{1}{D}$, where D is the average distance of the resonances. The factors showing up in the last line are *large* factors (in comparison to a nucleon nucleon system) : the first one due to the high level density, the second one due to the long lifetime of the compound nucleus.

Summing up, we expect the following factors:

$$\frac{\Delta f}{f} \propto \sqrt{\frac{\Gamma_p}{\Gamma_s}} \quad \text{"kinematic suppression"}$$
$$\Delta f \propto \frac{H^w}{D} \quad \text{"dynamical enhancement"} \tag{7}$$
$$\Delta f \propto \frac{H^w}{\Gamma} \quad \text{"lifetime enhancement"} \quad .$$

Once more put differently, the first factor reflects the angular momentum barrier for the p-wave neutrons; the second factor reminds the common perturbation theory result, which is proportional to the ratio 'perturbation/level distance'; the third factor can be visualized as follows: imagine two channels with a weak "cross-talking"; the longer this holds on, the more signal will be accumulated in the "wrong" channel.

3.2. Statistical Model: Trying to make a quantitative model, we meet the following difficulty: all the quantities mentioned above, i.e. level distances, partial and total widthes and also H^w in reality fluctuate from resonance to resonance in an unpredictable manner. The obvious solution is to consider averages, what in the foregoing heuristic argumentations already has been done for D and Γ by hand. A well-founded averaging procedure is obtained by the following construction. Bearing in mind, that Δf is given the S-matrix, which in turn is given by the Hamiltonian, we take the latter as starting point for both the incorporation of statistics and parity violation:

$$H = \begin{pmatrix} H^0_+ & \emptyset \\ \emptyset & H^0_- \end{pmatrix} + \begin{pmatrix} \emptyset & H^w_{+-} \\ H^0_{-+} & \emptyset \end{pmatrix} \tag{8}$$

These are matrices in the space of compound nucleus resonances, their dimension thus beeing prohibitively large for any numerical treatment. The first part represents the unperturbed strong interaction, connecting resonances of the same parity; the second part represents the perturbing weak interaction, connecting resonances of opposite parity. In order to mimic the fluctuations, all matrix elements are assumed to be uncorrelated Gaussian variables, characterized by the variances $\overline{(H^0_\pm)^2_{\mu\nu}}$ and $\overline{(H^w)^2_{\mu\nu}} = w^2$. There are two gratifying features of this ansatz[8]: first, it can be treated analytically, second, the final results - to which we come in the next section - contain only averaged quantities as average level spacings D and transmission coefficients $T_c \approx 2\pi \Gamma_c/D$.

4. RESULTS

4.1. Equations: In order to provide a reference frame for the new results with P-violation present, I first give the basic result for the case without P-violation[8]. Up to kinematical and geometrical factors, the cross-sections are given by the average modulus square of the S-matrix elements, and it turns out, that they can be cast in a familiar form:

$$\overline{|S^0_{ab}|^2} = \frac{T_a T_b}{T_{sum}} \cdot K(T_a, T_b, ...) \tag{9}$$

Here, T_a and T_b are the transmission coefficients of initial and final channel, respectively, T_{sum} is the sum of all transmission coefficients. This is very close to a Hauser-Feshbach-expression, with a modifying function K, which contains all the physical subtleties and all the mathematical effort, but which is numerically not very different from 1 and henceforth does not matter for a qualitative understanding. Note, that the total absorption cross-section in channel a is essentially the sum of this expression over b.

Similarly, the perturbation expansion in H^w up to lowest non-vanishing order gives the average modulus square of the P-violating S-Matrix by

$$\overline{|S^{PV}_{ab}|^2} = \left(\frac{w}{D}\right)^2 \frac{T_a T_b}{T^2_{sum}} \cdot K'(T_a, T_b, ...) \tag{10}$$

What shows up, is the ratio 'perturbation/level distance', as expected from perturbation theory; the function K' again contains all the mathematical effort and numerical details, but does not very much differ from unity.

Finally, inserting in the formulae connecting S via Δf with the observables (Sec 2.) we obtain

$$\Phi_{RMS} = P_{RMS} = \Delta_{RMS} = \left|\frac{w}{D}\right|\sqrt{\frac{T_{np}}{T_{ns}}}T_{sum}^{-1}K'' \quad . \tag{11}$$

The first factor is a remainder of dynamical enhancement, the second of kinematic suppression, the third of lifetime enhancement; K'' contains K, K' and other numerical factors not very different from 1.

4.2. Graphs: In order to get numbers or graphs we have to supply the following input parameters:

- Transmission coefficients and level densities; these can be extracted from extensive tabulations[9]

- Parametrization of the weak interaction; we rewrite

$$2\pi\left(\frac{w}{D}\right)^2 = \frac{\Gamma_{weak}^{\downarrow}}{D} = \frac{\Gamma_{weak}^{\downarrow}}{\Gamma_{strong}^{\downarrow}}\cdot\frac{\Gamma_{strong}^{\downarrow}}{D} \sim 10^{-14}\cdot 2\pi 10^5\frac{[eV]}{D} \quad . \tag{12}$$

The first equation is the definition of $\Gamma_{weak}^{\downarrow}$, the "weak spreading width". The second equation relates it to the "strong spreading width" which is a "constant" in the sense of very weak dependence on mass number and exitation energy; the ratio of spreading widths is the squared strength ratio of effective weak to effective strong interaction in the shell model and hence approximately 10^{-14}. These considerations can be found in early work of T.E.O. Ericson[10], and the fact of "constancy" is backed by experimental material in the case of isospin breaking[11]. The third equation gives a numerical value obtained by B. French and coworkers[12].

Now we have toghether all input data for the graphs in Fig. 2., which show two general tendencies: a maximal effect around 100eV up to 1keV which stems from the interplay of the increasing p-wave transmission and decreasing lifetime; values of this maximal effect around $10^{-4} - 10^{-3}$. Hence we have an enhancement of 3 or 4 orders of magnitude, but now *on average*, not on the selected positions of p-wave resonances. A more detailed discussion of these curves has already been published[6].

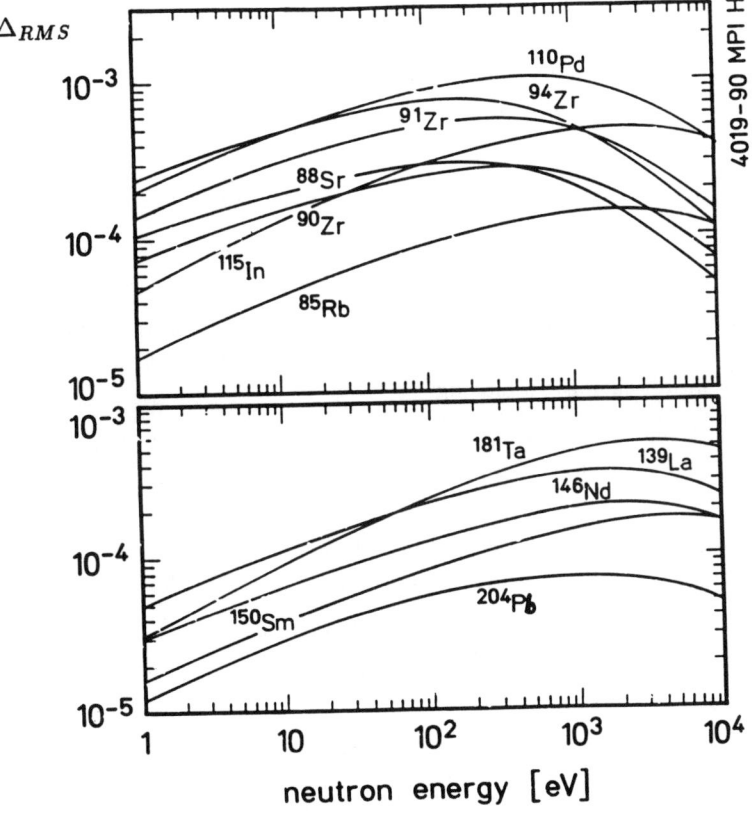

Fig.2. Root mean square observables as a function of energy.

5. SUMMARY AND OUTLOOK

1) A systematic approach was developed up to numerical predictions for a whole series of nuclei. "Systematic" here means "statistical", i.e. we use only average input parameters ($D's$ and $T's$) and we parametrize the outcome by an average stength parameter, namely $\Gamma^{\downarrow}_{weak}$.

2) We consider it to be an interesting phenomenon in itself, that tiny effects (weak interaction!) survive (even enhanced!) upon averaging over a so strongly fluctuating system as it is the compound nucleus.

3) "Back of the envelope estimate:" very recently[4], Δ has been measured in a whole series of p-wave resonances in ^{238}U, showing values between 1% and 5 %. These "on-resonance-measurements" are a very specific choice of energy points. The GOE, however, treats them all equivalently. In the language of statistics, a specific choice introduces a "bias" in the ensemble. There is work in progress[13] to introduce this biasing in a formalism, and the general outcome

is the following: due to the lifetime argument, the effect is "concentrated" in the resonances; the ratio of the RMS-effect on resonances to that in between is roughly $\sqrt{D/\Gamma}$. Taking the target thickness and the weak spreading width from Ref. 4. we have the following relationship

$$\Delta'_{RMS} \sim \Delta_{RMS} \cdot \left(2\pi \frac{\Gamma}{D}\right)^{-1/2} \sim 4 \cdot 10^{-3} \cdot 10 \sim 4\% \quad , \tag{13}$$

where the primed average refers to the on resonance biased ensemble.

We think, that this is in satisfactory coincidence with the experimental values and take it as a preliminary confirmation of both the underlying GOE-Model and its modification by biasing.

REFERENCES

1. W. Häberli et al., Proc. 6th Int. Symp. Polar. Phenom. in Nucl. Phys., edited by M. Kondo et al (Komiyama, Tokyo, 1986) 996

2. V. P. Alfimenkov et al., JETP Lett. **35**/1, 51-54 (1982)

3. "Fundamental Research with Polarized Slow Neutrons", P.A.Krupchitsky (Springer, Berlin, 1987)

4. J.D. Bowman et al., Phys. Rev. Lett. **65**, 1192 (1990)

5. "Encyclopaedic Dictionary of Physics" Edt. J. Thewlis (Pergamon Press, Oxford, 1967)

6. A. Müller, E.D. Davis, H.L. Harney, Phys. Rev. Lett. **65**, 1329 (1990)

7. "Mecanique Quantique", C.Cohen-Tannoudji, B.Diu, F.Laloe (Hermann, Paris, 1973)

8. J.J.M.Verbaarschot, H.A.Weidenmüller, M.R. Zirnbauer, Phys. Rep. **129**, 367 (1985)

9. "Neutron Cross Sections", Bd. 1A and 1B, S.F.Mughabghab et al. (Ac. Press, New York, 1981 and 1984)

10. T.E.O.Ericson, Phys. Lett. **23**, 97 (1966)

11. H.L.Harney, A.Richter, H.A.Weidenmüller, Rev. Mod. Phys. **58**, 607 (1987)

12. J.B.French, A.Pandey, J.Smith in: "Tests of Time Reversal Invariance in Neutron Physics", Eds. N.R.Roberson, C.R.Gould, J.D.Bowman, conf. Chapel Hill, 1987 (World Scientific, Singapore, 1987)

13. V. Bunakov, A. Müller, manuscript in preparation

NEUTRON-ANTINEUTRON OSCILLATION EXPERIMENT: RESULTS AND PERSPECTIVES

T.Bitter, F.Eisert, P.El-Muzeini, M.Kessler, U.Kinkel, E.Klemt, W.Lippert, R.Werner
Physikalisches Institut, University of Heidelberg,
D-6900 Heidelberg, FRG

D.Dubbers, K.Gobrecht
Institut Max von Laue-Paul Langevin (ILL),
F-38042 Grenoble, France

M.Baldo-Ceolin, F.Bobisut, D.Gibin, A.Guglielmi, M.Laveder, F.Mattioli, M.Mezzetto, G.Puglierin, A.Sconza, M.Vascon, L.Visentin
Dipartimento di Fisica "G.Galilei", University of Padova and INFN Sezione di Padova,
I-35131 Padova, Italy

P.Benetti, E.Calligarich, R.Dolfini, M.Genoni, A.Gigli Berzolari, F.Mauri, A.Piazzoli, A.Rappoldi, G.L.Raselli, D.Scannicchio
Dipartimento di Fisica Nucleare e Teorica, University of Pavia
and INFN Sezione di Pavia, I-27100 Pavia, Italy

ABSTRACT

Data taking of the experiment on free neutron oscillation is in progress at ILL reactor in Grenoble. The already achieved oscillation time limit is $\tau_{N\bar{N}} \geq 0.7 \cdot 10^8$ s.

INTRODUCTION

Small violations of the barion number conservation law ($\Delta B \neq 0$) are predicted in the framework of most new physics beyond the standard electroweak model (S.M.). The recent experimental confirmation of S.M. and the lack of any evidence for proton decay ($\Delta B = 1$), give new effort to search for $\Delta B = 2$ transitions. Much earlier A.Sacharov (ref.1a) suggested in 1967 the possibility of non conserving barion number interactions, to explain the observed matter-antimatter asimmetry in the Universe; in 1985 Kuzmin, Rubakov and Shaposnikov (ref.1b) related the barion violating processes to the early universe electroweak phase transition. Marshak and Mohapatra (ref.2) pointed out in 1980 that a crucial test for the unified gauge theories is the neutron-antineutron oscillation in the range 10^6-10^{10} s.
The nuclear physics complications (ref.3) suggest that an experiment on free neutron oscillation is the only clean way to check this limit.
The present experiment (ref.4) gives for $\tau_{n\bar{n}}$ the lower limit 10^7 s, one order of magnitude that the previous ones (ref.5), at the moment the experiment after running $1.5 \cdot 10^7$ s with $1.3 \cdot 10^{11}$ neutrons per second and an antineutron detection efficiency of 0.6 allowes to the experimental limit $\tau_{N\bar{N}} \geq 0.7 \cdot 10^8$ s.

© 1991 American Institute of Physics

THE N-N EXPERIMENT

The experiment is carried out at the 57 MW reactor of the Institut Max von Laue - Paul Langevin (ILL) in Grenoble. As shown in figure the beam of could neutrons

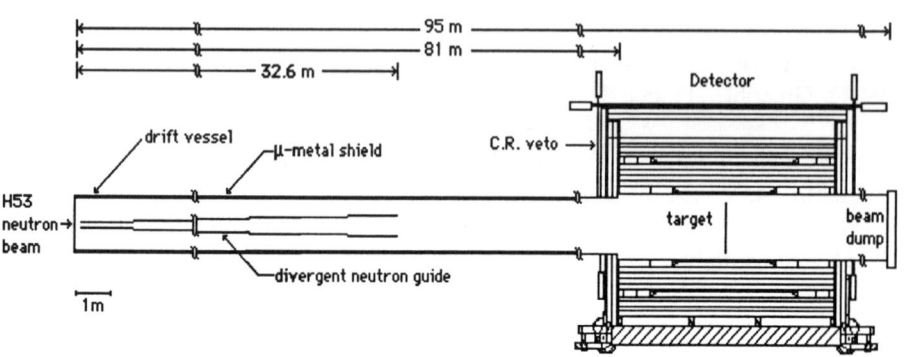

$(1.3 \cdot 10^{11}$ n/s with a mean wavelenght $\lambda \approx 6.5$ Å) moderated in liquid deuterium at 27°K, travels a system of totally reflecting neutron guides 60 m long. The quasi-free neutron flight is obtained in a cilindrical vacuum ($2 \cdot 10^{-3}$ Pa) vessel 81 m long. The earth magnetic field is reduced to B<10 nT, by a passive μ metal shield installed inside the drift pipe and by an external active shield. The effective neutron observation time turns out to be $(<t^2>)^{1/2} = 0.105$ s. The 120 μm thick carbonium target (110 cm diameter) is surrounded by a 2 mm ^{6}LiF layer in order to absorb the scattered neutrons.

The antineutron annihilation detector covers 94% of the solid angle and consists of limited streamer tube (0.9x0.9x500 cm³) planes (LST) and scintillation counters. The inner vertex detector is made of ten LST planes supported by Al honeycomb plates with a mean density $\rho = 0.3$ g/cm³. The outer calorimeter consists of 12 LST planes with a total of 4 radiation lenghts, allowing the detection of γ's produced by π^0 and providing an extimation of the total energy associated to the recorded events. The vertex reconstruction is better than 3 cm in each spatial coordinate.

The scintillator planes are used as trigger and also as a measurement of the time of flight of the charged particles (the time resolution is 700 ps), giving the track direction of flight. The trigger is vetoed against cosmic rays by an array of 235 plastic scintillation counters covering a total area of 115 m².

A 10 cm lead shield (canned into inox boxes), placed between apparatus and cosmic ray veto counters, prevents the auto-veto induced by particles generated in the target.

The experiment is going to run during 1991 to achieve the final sensitivity $\tau_{N\bar{N}} \geq 10^8$ s at 90% C.L.

REFERENCES

1. a) A.Sacharov, JEPT Lett. 6 (1967) 236
   b) V.Kuzmin, V.Rubakov and M.E. Shaposnikov, Phys. Lett. 155B (1985) 36
2. R.N.Mohapatra and R.E.Marshak, Phys.Rev.Lett.44 (1980) 1316
3. P.K. Kabir, Phys.Rev.Lett. 51 (1983) 231;
   P.K. Kabir and J.Noble, University of Virginia preprint
4. M.Baldo-Ceolin et al., Phys.Lett. B236 (1990) 95
5. G.Fidecaro et al., Phys.Lett. B156 (1985) 122;
   G.Bressi et al., Z.Phys. C43 (1989) 175

STUDY OF PARITY VIOLATION IN THE ^{6}Li$(n,\alpha)^3$H REACTION WITH POLARIZED NEUTRONS

J.Andrzejewski, A.D.Antonov, Yu.M.Gledenov, M.P.Mitrikov, Yu.P.Popov.
Laboratory of Neutron Physics, Joint Institute for Nuclear Research, Dubna, USSR.
I.S.Okunev, B.G.Peskov, E.V.Shul'gina, V.A.Vesna.
B.P.Konstantinov Leningrad Institute of Nuclear Physics, Gatchina, USSR.

Measurements were carried out of the P-odd asymmetry of the W~1+α_{pN} ($\vec{\sigma}_n,\vec{k}_t$) type at $\vec{\sigma}_n \uparrow\uparrow \vec{k}_t$, where $\vec{\sigma}_n$ is the spin of the neutron, $\vec{k}_t$, $\vec{k}_n$ are the unit vectors in the direction of the triton emission and neutron incidence, respectively. The experimental value for the coefficient of P-odd asymmetry with respect to a zero experiment is

$$\alpha_{pN}^{exp} = -(6.44 \pm 5.50) \times 10^{-8}.$$

In the recent years much attention is being paid to experimental and theoretical studies[1-4] of P-violation effects in few-nucleon systems of the type nd-tγ, pp-pp, np-d, ^{18}F* ^{18}F+γ etc. They are aimed at identification of contributions of a charged and neutral weak current and estimation of constants of the weak interaction, f_π, $h_{\rho,\omega}^{\Delta t}$. By studying P-odd asymmetry in polarized thermal neutron capture reactions, e.g. ^{6}Li$(n,\alpha)^3$H, ^{10}B$(n,\alpha)^7$Li having large capture cross-sections, one achieves good precision of measuring. The VVR-M reactor built at the B.P.Konstantinov Leningrad Institute of Nuclear Physics of the Academy of Sciences of the USSR to generate high flux of cold neutrons allowed the improvement of the precision of the earlier measurements of P-violation effects by more than one order of magnitude.

The measurements were performed on a beam of polarized neutrons with an integral flux of ~2×10^{10} n/s, 80% polarization and average wavelength $\lambda=4$ A. A multisection proportional chamber with wire electrodes working in the ionization current mode was used. Twenty four double chambers were positioned one after another along the direction of the beam travel. Half of a chamber was to detect tritons emitted in the direction of the neutron pulse and the other half of it to detect those emitted in the opposite direction. All first halves (or all second halves) were electrically connected to one amplifier. There was used as a target a ^{6}LiF layer sprayed on an aluminium foil of the thickness d=20 μm. Two targets were used for each double chamber. They were positioned with their backs close in to form an aluminium backing 40 μm thick in which the tritons emitted were fully absorbed so that one target worked for one half chamber only. The target thickness for the first group of 12 chambers (nearest to the reactor) was $d=400$ μg/cm^2, for the second group of 12 chambers $d=600$ μg/cm^2. To absorb α-particles and form a necessary solid angle of triton emission an aluminium foil 10 μm thick was pasted on every target, mean cosinus $(\vec{\sigma}_n;\vec{K}_t)=0.8$. Neutron absorption in the chamber was 90%. Thanks to the geometry chosen a suppression factor of left-right hand asymmetry $(\vec{\sigma}_n[\vec{K}_t\times\vec{k}_n])$ was achieved to be not worse than 10^{-4}, what allowed the exclusion of the possible contribution of this asymmetry into the effect to a level of 10^{-8}. The polarization of the beam was reversed with the help of an adiabatic flipper[5]. For the details of reactor fluctuations compensation and measuring procedure the reader is offered to address papers[6,7].

The result obtained was:

$$\alpha_{pN}^{exp} = -(4.65\pm5.28)\times10^{-8} .$$

In order to perform a zero (background) experiment the targets were additionally covered with an aluminium foils each 20 μm thick to fully absorb the tritium component. With a ratio of background to reaction intensity taken into account the result was:

$$\alpha_{background} = (1.80\pm1.54)\times10^{-8} .$$

When this value is subtracted from the value obtained in the main experiment one obtains

$$\alpha_{pN}^{exp} = -(6.44\pm5.50)\times10^{-8}$$

to essential improvement of the precision of earlier measurements[6]

$$\alpha_{pN} = (0.7\pm8.0)\times10^{-8} .$$

Of the existing models of the P-violation effect in nuclei most advantageous in the case considered is the model accounting for the distinct cluster structure of the ^{6}Li nucleus.

Having taken this structure into account it is possible according to[8] to reduce the problem of the neutron-nucleus interaction to the three-particle problem. In[8] a theoretical estimate of P-odd asymmetry was obtained to be

$$\alpha^{theor} = (0.45f_\pi - 0.06h_\rho) \sim 3\times10^{-7}$$

with the "best" available values for the constants of

weak interaction from[4] used; $h_\rho^{o.b.v.} = -11.4 \times 10^{-7}$, $f_\pi^{b.v.} = 4.6 \times 10^{-7}$.
This estimate appears to essentially exceed the value measured in our experiment. This fact evidences in favour of a very small value of the neutral current, provided the model used in[8] is correct. A similar result was obtained in the measurement of γ-quanta circular polarization ($E_\gamma = 1091$ keV) in ^{18}F [9].

The experimental value for the constant of weak neutral current $f_\pi = (0.4^{+1.4}_{-0.4}) \times 10^{-7}$ was obtained to be essentially smaller the theoretical estimate[4] ($f_\pi^{b.v.} = 4.6 \times 10^{-7}$).

REFERENCES

1. Dubovik V.M., Zenkin V. Ann.Phys., 1986, v.172, p.100.
2. Adelberger E.G. Ann. Rev. Sci., 1985, v.35, p.501.
3. Kaiser M., Meisser G. NPA 499, 1989, p.699.
4. Despllanques B. et al. Ann. Phys (1980), 124, p.449.
5. Altarev J.S. et al. Pisma ZhETF, 1986, v.44, p.269-272.
6. Vesna V.A. et al. Pisma ZhETF, 1983, v.38, p.265.
7. Antonov A.D. et al. Sov.Nucl.Phys., 1988, v.48, p.305.
8. Nesterov M.M., Okunev I.S. Pisma ZhETF, 1988, v.48, p.573.
9. McDonald A.B. et al. Phys.Rev.C, 1987, v.35, p.1119.

P-EVEN EFFECTS IN A DIRECT γ TRANSITION FOLLOWING NEUTRON CAPTURE IN 113Cd AROUND E_P=7 eV RESONANCE

V.P. Alfimenkov, S.B. Borzakov, Yu.D. Mareev, L.B. Pikelner,
V.R. Skoy, A.S. Khrykin, and E.I. Sharapov
Laboratory of Neutron physics, Joint Institute for
Nuclear Research, Dubna, Head Post Office, Box 79, USSR

ABSTRACT

The polarization left-right aymmetry and angular fore-aft asymmetry of γ yeilds were investigated for the direct transition E_γ=9.04 MeV in ^{114}Cd near the p-wave resonance E_p=7. eV. The experiments were performed at the pulsed reactor IBR-30 using an LMN-polarizer of neutrons and NaJ (Tl)-detectors.

Analysis of experiments on parity violation in compound nuclear states requires spectroscopic information on nuclear level parameters, in particular, on the mixing of γn1/2, γn3/2 neutron width amplitudes. This mixing is characterized by parameters $x=\gamma n1/2/(\Gamma_n)^{1/2}$ and $y=\gamma n3/2/(\Gamma_n)^{1/2}$, with $x^2+y^2=1$. One of the ways to obtain this information is to study polarization and angular correlations in a capture reaction. Interfering E1 and M1 γ-rays are emitted in the case of simultaneous occurrencs of s- and p-waves at a given neutron energy E. Interference gives rise to the A1 and B1 terms (which are proportional to the fore-aft $\mathcal{E}_{f-a}$ and left-right $\mathcal{E}_{l-r}$ asymmetries) in the expression for the differential cross section of the ^{113}Cd(n,γ)^{114}Cd reaction:

$$\sigma(\theta,\phi,E)=A_0(E)+A_1(E)\cos\theta+P_nB_1(E)\sin\theta\cos\phi+A_2(E)P_2(\cos\phi).$$

Here θ is the polar angle of γ-ray emission relative to the neutron beam direction, Pn the neutron polarization vector, φ -the azimuthal angle, that is between vectors Pn and [Kn kγ]. The terms $A_1(E)$ and $B_1(E)$ have the form of fA(x,y)Refp(E) and fB(x,y)Imfp(E) respectively, where fp(E) is the Breit-Wigner amplitude of the p-wave resonance and fA, fB known functions of (x,y). Explicit expressions for them can be found in [1-4]. Such measurements were made first for the E_p=1.33 eV p-wave resonance of the compound nucleus ^{118}Sn in [3].

Now we have measured the Ep=7.0 eV resonance in the ^{113}Cd(n,γ)^{114}Cd reaction at the IBR-30 pulsed reactor. The same polarization setup with the LMN - crystal was used. The direct transition to the ground state of ^{114}Cd with the energy E_γ=9.04 MeV was studied with three NaJ(Tl) crystals each 200 mm in diameter and 200 thick. The samples were enriched Cadmium (95% of ^{113}Cd) of masses from 100 to 400 g. Flight paths were 35 m (polarized beam) and 52 m (unpolarized beam). Geometries for both cases are schematically shown in Fig 1. The

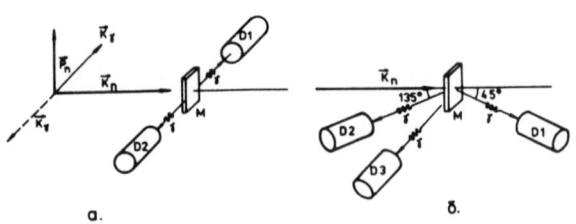

Fig. 1

fore-aft asymetry was measured for two pairs of angles 45° - 135° and 55° - 125°, the left-right asymmetry - for θ=90° only. A Ta- filter with black" resonances at 10.39 eV and 4.3 eV was permanently in the beam. This allowed measurement of the background. The time-of-flight spectrum of an integral γ-ray yeild is shown in Fig. 2 as obtained with a threshold of 1.5 MeV The fore-aft asymmetry $\mathcal{E}_{f-a}$ and the left-right asymmetry $\mathcal{E}_{l-r}$ for the 9.04 MeV ground state transition are shown in figures 3 and 4 respectively. They

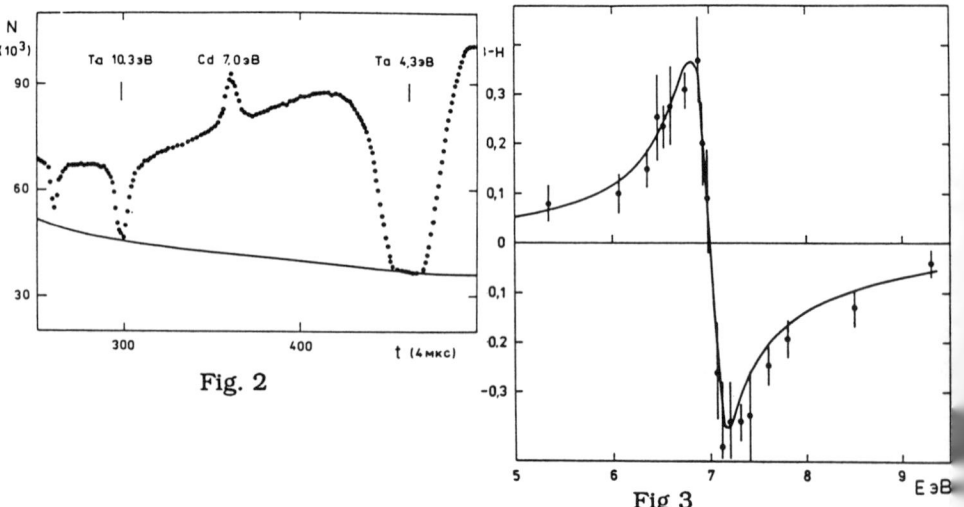

Fig. 2

Fig 3

show the expected energy behavior. The maximum asymmetries are near to 30%. The data of Fig. 3 were fitted using the explicit formulae from Ref. 4 taking into accound the Doppler effect. The resonance parameters were E_0=7.0 eV, $g\Gamma_n$=(3.1 ±0.3) 10^{-7} eV, and Γ=0.16±0.02 eV from Ref. 5, and the p-wave to s-wave cross section ratio $\sigma p(\theta,E)/\sigma s(\theta,E)$ was obtained in this work. The fit is shown by the solid line. Two pairs of (x,y) - parameters, (x,y)1 =(0.97, +0.22) and (x,y)2=(0.10,-0.99), describe the data equally well. Then we introduced these parameters into the forumlae for the ε1-r asymmetry and obtained in Fig.4 curve 1 for (x,y)1 and curve 2 for (x,y)2. Both curves are in disagreement with the experimental points. We note, however, the points of ε1-r can be fitted well independent of the data in Fig. 3, but with a quite different pair of (x,y)-parameters.

Therefore, mixing parameters were not obtained. The same result was arrived at in Ref.4 for the Ep=1.33 eV resonance in the $^{117}Sn(n,g)^{118}Sn$ reaction. The energy dependence of ε_{f-a} and ε_{1-r} assymmetries definitely demonstrates their being due to interference between the parity conserving s- and p-wave neutron capture amplitudes. However, they are not described consistently in the frame of the existing calculations based on the resonance model.

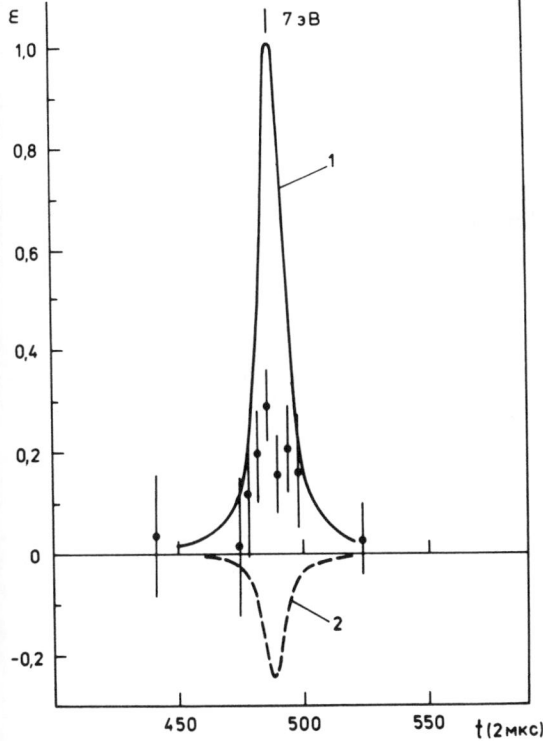

Fig. 4

REFERENCES

1. O.P. Shushkov, V.V. Flambaum; Nucl. Phys. A435, 352(1985)
2. V.N. Efimov, JINR Reports P4-88-848, P4-88-528, Dubna (1988)
3. V.P. Alfimenkov et al. In: JINR Rapid Communications, No. 10-85, p. 19, Dubna (1985).
4. V.P. Alfimenkov et al. JINR Rep P3-90-270, Dubna, (1990), Yad. Fiz. to be published.
5. V.P. Alfimenkov et al. JINR Rep P3-87-117, Dubna (1987).

NEW MEASUREMENTS OF THE ELECTRON-NEUTRON SPIN ASYMMETRY IN THE NEUTRON BETA DECAY

B.G.Erozolimskii, I.A.Kuznetsov, I.V.Stepanenko

Leningrad Nuclear Physics Institute, Gatchina, Leningrad district, 188350, USSR

I.A.Kuida, Ju.A.Mostovoi

Kurchatov Atomic Energy Institute, Moscow, USSR

ABSTRACT

The A-coefficient, which characterizes the angular correlation between the electron escape and the decaying neutron spin directions, is measured with an intence polarized neutron beam from the new vertical channel of the LNPI reactor. The result is

$$A = -0,1131 \pm 0,0014 \quad (A_0 = -0,1116 \pm 0,0014),$$

implying $\lambda = g_A/g_V = -1,2544 \pm 0,0036$ for the ratio of the weak coupling constants.

INTRODUCTION

Last years there are several high class researches carried out, the purpose of which was to measure the neutron beta-decay parameters with extreemly improved accuracy. Theese are the investigations [1-8] in which the neutron lifetime τ_n was determined, and the recent measurement of the angular correlation between the electron momentum and the neutron spin directions, completed at ILL (Grenoble, France) [9].

A new measurement of the correlation constant A was performed at LNPI (Gatchina, USSR) on a vertical polarized neutron beam with $3 \cdot 10^{10}$ 1/s intensity from a liquid hydrogen source, placed in the centre of the reactor core [10].

EXPERIMENT

The decay events, taking place in the neutron beam inside the experimental chamber (fig.1) were recorded with the help of an electron detector (photomultiplier with a plastic scintilator) and a proton detector (photomultiplier with a thin CsI(Tl) layer) working in coincidences with each other. The magnetic guide field, directed along the chamber axis is provided by a system of coils. Each 2-3 seconds the beam polarization was reversed with the help of an adiabatic spin-flipper, which is placed under the chamber, one meter from its centre.

The main principle of the measurement, developed in the previous work [11] is the creating of such experimental conditions, which ensure an independence of the proton recoil registration efficiency from the antineutrino escape direction for neutron decay events, recorded by the electron detector.

Realization of such condition releases from the necessity of taking into account the spin asymmetry of the antineutrino escape.

For this purpose, the neutron beam is separated from the electron detector by a diaphragm on the cylindric electrode (6) and and thus only such electrons can reach the detector, which were created in a confined part of the neutron beam (shaded on the fig.1).

The electric field between the electrodes (5) and (6) (2,5 kV) turns all the recoil protons, created in the shaded region of the beam, toward the spherical grid (7) and, as a result, they are phocùsed on the scintillation layer of the proton detector (1) independently of their initial energy and escape direction.

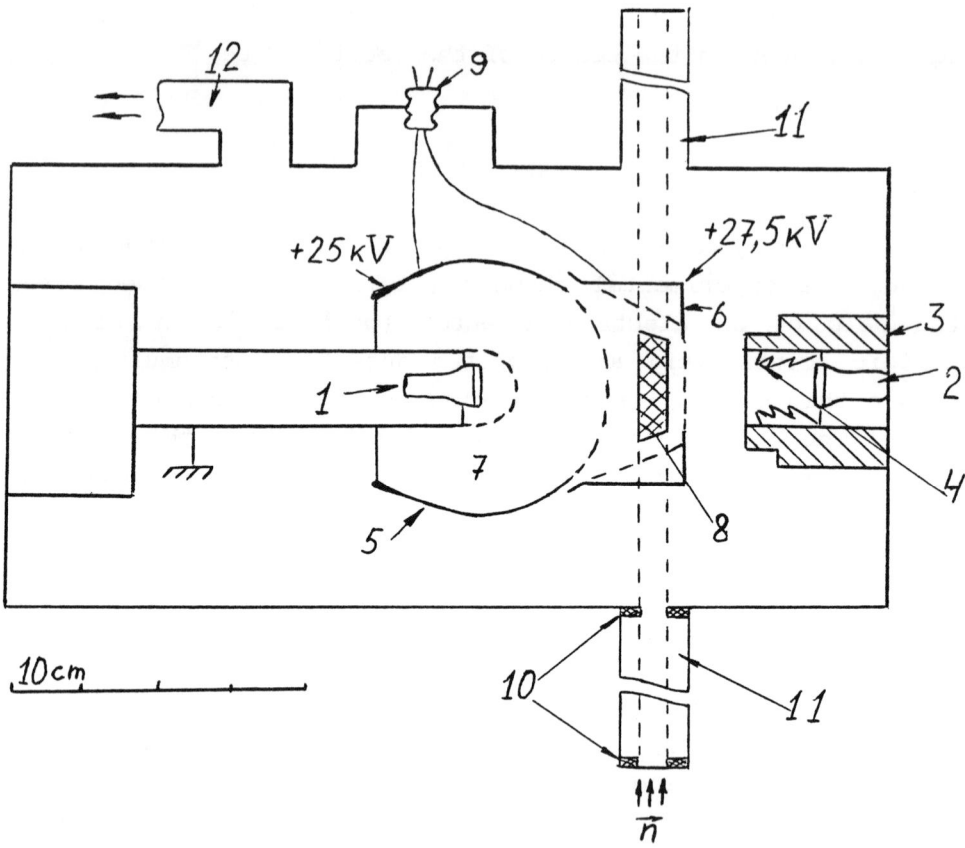

Fig.1 Experimental setup. 1- proton detector; 2- electron detector; 3- lead; 4- collimator; 5- spherical electrod; 6- cylindrical electrod; 7- spherical grid; 8- decay region of the neutron beam; 9- high voltage insulator; 10- ^{6}LiF diafragm; 11- input and output tubes; 12- vacuum pumping.

The most part of the efforts made in the experiment [11] were devoted to the proof of reliable executing of the pointed out condition.

The counting rate of the electron detector in this experiment was 1000 1/s and the main part of it was the radiative background due to gamma-rays and electrons in the chamber. This background was essentially reduced with the aid of a lead shield and a lucite collimator before the detector.

The counting rate of the proton detector was about 3000 1/s, including about 2000 1/s due to decay protons, which come from the whole volume of the neutron beam inside the electrode (6) and the remaining 1000 1/s due to gamma-rays, which cause scintillations in the glass of the photomultiplier and in the lightguide. This background component was appreciably suppressed due to special electronic device, which separates the signals with different shapes.

Such counting rates of the detectors lead to 1,5 accidental coincidences per sec, and the number of the neutron decay events was 3-3,5 1/s.

The determination of the accidental coincidences counting rate was carried out with the help of delayed coincidence technic, the way it was in the former experiment [11].

The desired asymmetry is then calculated from the expression:

$$X = \frac{(N_- - N_{del}) - (N_- - N_{del})}{(N_- - N_{del}) + (N_- - N_{del})}, \quad (1)$$

where N_- and N_- are the readings of the counting devices, which register the summs of the decay events and of the accidental coincidences with different polarization signs, and N_{del} and N_{del} mean the same for the accidental background (delayed coincidences) and are stored in separate counters.

The angular correlation constant A can be found from the

well known expression:

$$X = A P \frac{V}{C} \cos \theta_{e\sigma}, \qquad (2)$$

where P is the neutron beam polarization, averaged over the whole beam crossection and for the both spinflipper states; $\frac{V}{C}$ is averaged over the whole electron energy spectrum; the geometry factor $\cos\theta_{e\sigma}$ is calculated for the whole registration solid angle.

The polarization of the neutron beam was carried out with the help of a vertical four-slit polarizating neutronguide with Fe-Co mirror walls and a solenoid for creating an axial magnetic field [10].

It turned out that the beam polarization is very inhomogenios in such device, particularly in the plane, perpendicular to the neutronguide slits. A method of measuring the mean polarization value was created, bazed on the use of a short section of a polarizing neutronguide with the analyzing capacity, which is determined by means of absolute Stern-Gerlach method. Then the calibrated in such way analyzer-neutronguide is placed perpendicular to the polarizer slits and thus we obtain a mean polarization value for all of the neutron reflecting angles in one cros-section of the beam [12]. Fourty three such measurements were carried out and finally a weighted mean value of P was calculated. It is worth while to mention, that the time, the neutrons spend in the experimental device, is proportional to 1/V, and thus a "thin" neutron detector (fission chamber with ^{235}U) was used in polarization measurements.

The experimental error of P determination was estimated, as ±0,7%.

The calculation of $\frac{V}{C}$ needs the knowledge of the real electron spectrum detected. Systematic calibrations of the energy scale of the electron detector were carried out with the help of standart ^{113}Sn and ^{137}Cs sources.

The electron energies of these sources were preliminari compared with the conversion lines in ^{198}Au electron spectrum.

The calibrated values of the lower and upper thresholds of the electron energy interval and the shape of the detector response were used for the $\frac{V}{C}$ calculation.

The electron detection geometry, needed for $\cos\theta_{e\sigma}$ calculation, is defined by the location of the measurement system elements and the compressed ^{6}LiF powder diaphragms along the neutron beam path (fig.1). A special experiment for the investigation of the electron scattering in the measuring chamber were performed with the help of an ^{198}Au source. This effect can cause some change in the effective $\cos\theta_{e\sigma}$ value. It was estimated, that the quota of the electrons, fitting the detector after scattering in the chamber makes up 0,2±0,2%.

RESULTS

About $7 \cdot 10^6$ neutron decay events were registered. From time to time the sign of the magnetic field in the guide and in the experimental setup was changed in order to avoid all of the apparatus asymmetries and some possible effects due to the influence of the spinflipper on the detecting systems. It must be said, nevertheless, that no differencies between X_+ and X_- were found through the whole experiment.

The whole statistics was divided in groups corresponding to various values of $\frac{V}{C}$, derived as a result of calibration measurements and the quantities $\frac{X_i}{|V/C|_i}$ were calculated. The weighted mean value $\frac{X}{V/C}$ was then found, and it is $\left|\frac{X}{V/C}\right|$ = =0,08660 ± 0,00061.

The values of other coefficients in the expression (2) for finding the A constant are: $\cos\theta_{e\sigma}$ = 0,9705 ± 0,004 (the effect of the electron scattering is taken into account, and

its uncertainty is included in the error magnitude)
$P = 0,7867 \pm 0,0070$.

Thus from (2) it can be derived, that

$$A = -0,1134 \pm 0,0015$$

Besides this main part of the experiment there were two preliminary measurements, which gave following results [13]:

$$A = -0,1140 \pm 0,0057$$

and $\quad A = -0,1096 \pm 0,0039$.

The weighted mean value of A turns out to be:

$$A = -0,1131 \pm 0,0014$$

and it is the final result of this experiment. Detailed description of this work is given in [14].

As it is well known, a small correction arises from the fact, that the nuclei have final masses and dimensions. This correction can be calculated in the first order of momentum transfer approximation following the procedure, discribed in [15] and taking in account, that the main part of the experiment was carried out with the electron energy interval 250 keV - 780 keV.

The corrected value of the constant A is:

$$A_o = -0,1116 \pm 0,0014,$$

and it leeds to the ratio of the fundamental coupling constants

$$\lambda = g_A/g_V = -1,2544 \pm 0,0036.$$

CONCLUSION

A_o and λ values, derived in this work are, in consent with previous results of measurements, acomplished in the middle of 70th in Argonn (USA) ($A_o = -0,1116 \pm 0,006$ [16]) and in the Kurchatov Institute (USSR) ($A_o = -0,1126 \pm 0,005$ [11]), but agree worse with the recent result of the ILL group (Grenoble, France) (A_o =-0,1146 ± 0,0019 [9]).

The weighted mean value of the fundamental ratio λ from theese four measurements of the constant A is

$$\lambda_A = -1,2570 \pm 0,0029$$

The mean λ value, derived from the lifetime measurements τ_n, carried out in [1-8] (together with the $(f\tau)_{0^+-0^+}$ for allowed 0^+-0^+ transiions, calculated in [17]) is

$$\lambda_\tau = -1,2676 \pm 0,0016$$

As a result there are some kind of discrepancy, since

$$\lambda_\tau - \lambda_A = 0,0106 \pm 0,0033$$

and it is worthwhile to think about possible reasons for it.

As it was pointed out not once, the experimental situation in the neutron beta-decay can be analized from the point of view of the right-hand current hypotesis (for instance [18,19]). Recently [20] it has been shown, that all of the experimental data (including the data of the µ-e transition) can be put in agreement with the aid of such assumption. At the same time, it can be mentioned, that the sign of the difference $\lambda_\tau - \lambda_A$ excludes the possibility of the presense of the scalar or tensor type terms in the weak hamiltonian, if we suppose, that Cs = - Cs'and Cт = -Cт' (the case with equal signs is closedby the Fierz terms data) - it is the asumption which was discussed in various articles (look for instance [18]).

It must be emphasised however, that an essential rise of the experimental accuracy is still needed, and we think, that for the present it is to early to make categorical dramatic statements.

References

1. P.E.Spivak, JETP. 94, 1 (1988).
2. V.J.Morozov, Nucl.Instr.and Meth. A284, 108 (1989).
3. J.Dohner, H.Abele, I.Reichert, H.Borel, D.Dubbers, S.J.Freedman, J.Last,- ibid. p.123.
4. K.Schreckenbach, G.Azuelos, P.Grivot, R.Kossakowski, P.Liaud, ibid. p.120.
5. F.Anton, W.Paul, W.Mampe, L.Paul, S.Paul, - ibid. p.101.
6. W.Mampe, P.Ageron, J.C.Bates, J.M.Pendelbury, A.Steyerl, ibid. p.111.
7. J.Byrne, P.G.Damber, J.A.Spain, M.S.Dewey, D.M.Gilliam, G.L.Greene, G.P.Lamaze, A.P.Williams, J.Pauwels, R.Eykens, Van J.Gestel, A.Lamberty, R.Scott. - ibid. p.116.
8. A.G.Kharitonov, V.V.Nesvizhevsky, A.P.Serebrov, R.R.Taldaev, V.E.Varlamov, V.P.Alfimenkov, V.I.Lushchikov, V.V.Shvetsov, A.V.Strelkov, - ibid. p.98.
9. P.Bopp, D.Dubbers, L.Hornig, E.Klemt, J.Last, H.Schutze, S.J.Freedman, O.Scharpf. Phys.Rev.Lett. 56, 919 (1986).
10. I.S.Altarev, N.V.Borovikova, A.P.Bulkin et al. JETP Lett. 44, 269 (1986).
11. B.G.Erozolimskii, A.I.Frank, Yu.A.Mostovoy, S.S.Arzumanov, L.P.Voytzik. Sov.J.Nucl.Phys. 30, 692 (1979).
12. B.G.Erozolimskii, I.A.Kuznetsov, N.F.Maslov, I.V.Stepanenko. Preprint LNPI-1574 (1990).
13. B.G.Erozolimskii, I.A.Kuznetszov, I.A.Kuida, Yu.A.Mostovoy, I.V.Stepanenko. Preprint LNPI-1589 (1990).
14. B.G.Erozolimskii, I.A.Kuznetsov, I.A.Kuida, Yu.A.Mostovoy, I.V.Stepanenko. Sov.J.Nucl.Phys. in press.
15. D.H.Wilkinson. Nucl.Phys. A377,474, (1982).
16. V.E.Krohn, G.R.Ringo. Phys.Lett. 55B, 175 (1975).
17. A.Sirlin. Phys.Rev. D35, 3423 (1987).
18. B.G.Erozolimskii. Nucl.Instr.and Meth. A284, 89 (1989).

19. D.Dubbers, W.Mampe, J.Dohner. Europhys. Lett. 11, 195 (1990).
20. Yu.V.Gaponov, N.B.Shulgina, .P.E.Spivak. Preprint IAE-5032/2, (1990).

Nuclear Astrophysics

NUCLEAR AND ASTRONOMICAL CONSTRAINTS ON THE SITE FOR R-PROCESS NUCLEOSYNTHESIS

G. J. Mathews
Lawrence Livermore National Laboratory, Livermore CA 94550

G. Bazan
University of Illinois, Urbana, IL 61801

J. J. Cowan
University of Oklahoma, Norman, OK 73019

ABSTRACT

New measurements and theoretical studies of nuclear properties, together with new astronomical data on the growth of heavy-element abundancess during the early history of the Galaxy, now provide a clearer picture of where in nature the elements heavier than iron are produced by rapid (r-process) neutron capture reactions. The nuclear data suggest that the r-process involves a high-temperature and high-neutron-density beta-flow equilibrium environment. The astronomical data, when compared with simple galactic chemical evolution models, suggests that the r-process is associated with Type II supernovae and that the neutron source must be manufactured by the star independent of the initial heavy- element abundance. Low-mass type II supernovae are proposed as the most important contributors to the r-process.

INTRODUCTION

It has been known for over three decades that the solar-system abundance of elements heavier than iron must have been produced by neutron-capture reactions.[1-3] In the r-process neutron captures occur on time scales much shorter than typical beta-decay lifetimes so that sequential neutron captures lead to the production of extremely neutron-rich isotopes. Later, these isotopes beta decay back to the regime of stable atomic nuclei. In the s-process neutron captures occur much more slowly than typical beta-decay times so that the capture flow is along stable nuclei. These two processes are schematically indicated on Figure 1.

The r-process is probably the most poorly understood of the stellar nucleosynthesis mechanisms. One reason for this lack of understanding has been that the extremely neutron-rich isotopes formed during the r-process have been virtually impossible to study in terrestrial laboratories. Another reason is that abundances of these elements are so rare that they have been difficult to observe astronomically. In this paper we review some important recent progress in both of the areas of nuclear properties and astronomical abundances which contribute substantially toward unraveling the mystery as to which of the many proposed sites for r-process nucleo- synthesis are most appropriate.

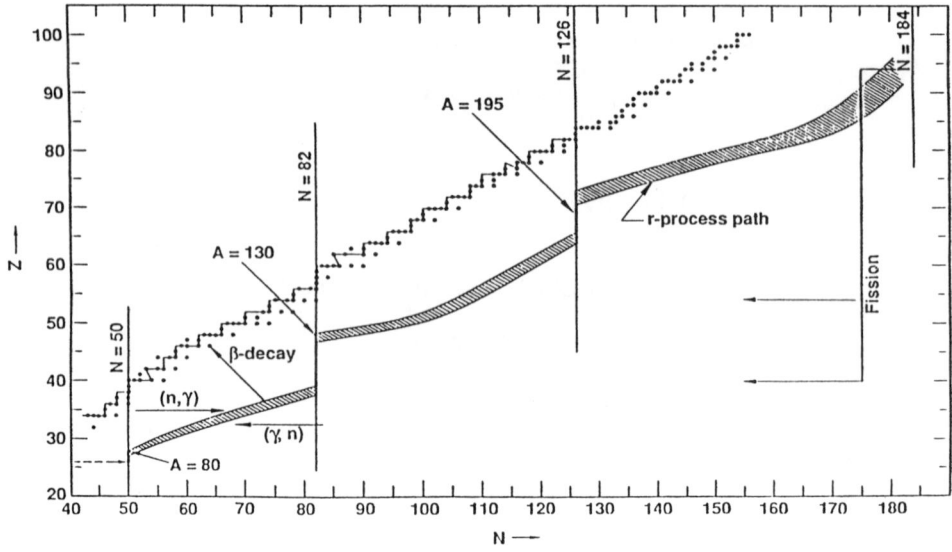

Figure 1. A schematic view of the isotopes produced during the r-process.

One important development is that, for the first time, measurements[4-8] have been made of some of the crucial beta-decay rates for the highly unstable nuclei produced during the r-process. There are also reliable neutron-capture cross sections for s-process elements[9]. Such measurements allow one to place new constraints[7,8] upon the neutron densities and temperatures appropriate to the r-process. The other development is the accumulation of recent astronomical stellar abundance determinations[10-14] for elements which are largely produced by either the r-process or s-process. Such observations show the growth of heavy-element abundances during the early history of the Galaxy and provide indications of the kind of stellar objects which could contribute to heavy-element nucleosyntheis.

THE R-PROCESS

At least ten possible sites for r-process nucleosynthesis have been proposed over the past forty years. It was initially suggested (along with the first discussions of heavy-element nucleosynthesis[1]) that the r-process could be associated with supernovae. A number of models were developed[15-19] to describe how heavy nuclei might be produced in the regions just outside neutronized supernova cores. Unfortunately, none of the models were able to produce and eject r-process abundance yields of the correct magnitude and relative abundances to explain the solar-system abundances in a realistic and self-consistent manner.

Because of the difficulties with supernova-core models, a number of other possibilities have been studied with varying degrees of success. All of the proposed r-process sites can be divided into two catagories: Those in which the r-process elements are produced as primary products directly in a star (e.g. from a photodissociated neutron-rich nucleon gas); and those in which the r-process elements are secondary, i.e. requiring the presence of preexisting heavy "seed" nuclei which undergo neutron capture reactions to produce the r-process elements. In the former catagory are included high-temperature, neutron-rich environments such as shock[20] or jet[21-24] ejection of material from neutronized supernova cores as mentioned above, or other scenarios such as may be encountered in inhomogeneous big-bang cosmologies[25,26], or ejecta from binary neutron-star tidal disruption[27-29] or coalescence.[30] In the latter catagory are included numerous less neutron-rich and lower temperature environments such as, nova outbursts[31], shock-induced explosive helium[32,33] or carbon[34] burning, the core helium flash in low-mass stars[35], super-heated proton-rich helium burning in a precollapse 11 M_A star[36], neutron-star accretion disks[37], or neutrino inelastic scattering in the outer envelopes of core-bounce supernovae[38,39]. Each of these proposed mechanisms has its advantages and disadvantages. Ultimately it will only be possible to distinguish among them through a careful compilation of nuclear and astronomical data. It is the intention of this paper to summarize briefly the available nuclear and astronomical constraints. This will provide some insight as to which of the proposed r-process sites is most likely to contribute to the observed abundances.

SOLAR-SYSTEM ABUNDANCES AND NUCLEAR PROPERTIES

Figure 2 shows the solar-system r-process abundances obtained[40] by subtracting the contribution from slow-neutron capture (s-process nucleosynthesis) from solar-system abundance tables[41-43]. The total mass fraction for r-process elements is only $\sim 10^{-7}$, i.e. r-process elements account for only $\sim 10^4 M_A$ of the $\sim 10^{11}$ M_A of material in the Galaxy. The amount of r-process material produced per supernova over the history of the Galaxy (for a supernova rate[44] of $\sim 10^{-2}$ yr^{-1} and an age when the solar system formed of $\sim 10^{10}$yr) is then only $\sim 10^4/(10^{-2} \times 10^{10}) \sim 10^{-4}$ M_A Thus, the r-process either occurs in a small subset of supernovae, or it occurs in a small region of supernovae - smaller than the numerical zoning of many supernova models.

Another striking feature in figure 2 is the presence of sharp peaks in the r-process abundances near atomic mass numbers A = 130 and A = 195 (and to a lesser extent A=80). Such narrow peaks suggest that most of the solar-system r-process material was produced in a single well defined environment rather than some combination of possibilities. Otherwise, averaging over environments with different temperatures and neutron densities might broaden the abundance peaks. The peaks are caused by different

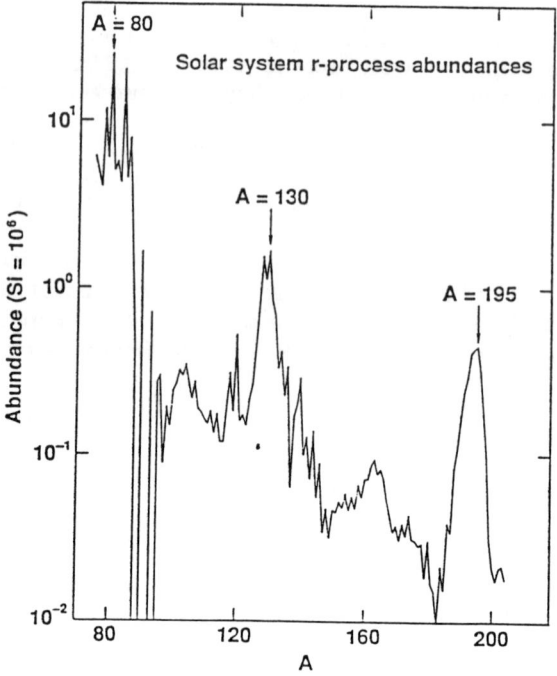

Figure 2. An example of solar-system r-process abundances from the analysis of Ref. 40.

nuclear properties associated[1,2] with the neutron closed shells at N = 82 and 126 (see Fig. 1). At high density and temperature (n,q) ~ 1 (q,n) equilibrium is established. The peaks are then caused[1-3] by decreased beta-decay rates for nuclei near neutron closed shells (waiting points). This environment is referred to as a "classical" r-process. Its numerical formulation is well summarized in Ref. 2. For the lower-neutron-density secondary r-process sites, the waiting-point peaks are caused[45] by diminishing neutron capture rates near neutron closed shells.

The prospect for understanding the conditions for the r-process was recently improved by measurements of beta-decay lifetimes for the waiting-point neutron-closed-shell nuclei, ^{130}Cd (Ref. 4) and ^{80}Zn (Refs. 5,6). It has been shown[4] that the r-process abundances in the neighborhood of these peaks are in good agreement[7,8] with that expected from both (n,q) and beta-flow equilibrium (approximate equality of abundance times beta-decay rate).

This is argues that the r-process occurs in a classical (n,q) equilibrium environment. It is possible that a secondary r-process could reproduce these data. However, we have studied a secondary steady-flow r-process[45,46] in the vicinity of the A = 80 and 130 peaks. Using the same beta-decay input data as in Ref.'s 7 and 8, the abundances can be fit by adjusting the neutron capture cross sections within their uncertainty and maintaining the neutron density above 10^{22} cm^{-3} to avoid over abundance of the doubly-closed-shell nucleus, ^{132}Sn. A definitive understanding of which environment best characterizes the r-process will require better neutron capture rates away from stability. In the future it may be possible to derive[47] sufficiently accurate neutron capture rates from a combination of measured level structure and beta-delayed neutron emission spectra. For the present time, however, the classical (n,q) environment seems most likely because of the remarkable agreement with observed abundances.

From the measured decay rates for waiting point nuclei one can deduce[48] a critical neutron density at which the (n,q) rate is equal to the beta rate. Using a direct-radiative-capture[48] estimate of 30 mb for the neutron capture cross section, s, and 280 ms for the ^{130}Cd mean beta-decay lifetime,[8] t_b, gives $n_n \sim 1/<sv>t_b \sim 10^{20}$ cm^{-3}. For a classical r-process, this corresponds to the minimum neutron density necessary to maintain the waiting-point condition.[45] For a secondary r-process this represents an estimate of the neutron density required for neutron capture to reach the waiting point before beta decay. It is difficult[49] for most secondary r-process models to obtain neutron densities in excess of this critical value, supporting the notion of a classical primary r-process.

The critical neutron density also determines the freezeout conditions for the r-process. Figure 3 shows allowed values of neutron density verses temperature[7,8] and the conditions necessary[45] for the waiting-point approximation to be valid. Neutron source reactions are quite temperature sensitive. Therefore, the r-process probably begins at a higher temperature and significantly higher neutron density than the critical neutron density as depicted in Figure 3. As the environment cools and the neutron density decreases the final observed abundances result from the conditions just before the neutron density drops below the critical value. This explains why the r-process peaks are so sharp. While the neutron density may have been higher during the r-process, the final observed abundances only reflect the condidtions during the short time just before the neutron density droped below the critical value.

The fact that the the r-process appears to last long enough for beta-decay flow to equilibrate allows for an estimate of the minimum time duration for the r-process. Beta-flow equilibrium requires that the r-process continue for at least about twice the mean beta-decay lifetime of the slowest waiting-point nucleus (^{80}Zn, t_b = 540 ms[5,6]). Thus, the r-process probably lasts for at least ~ 1 sec. This is comparable to the time for the r-process to flow from iron to actinide nuclei (~ 2 sec[7,8]), and to the time required (~ 1 sec) to shift the s-process peaks to r-process peaks in a secondary r-process.[33]

However, the time scale could be shorter if the neutron density is a rapidly decreasing function of temperature and time. The

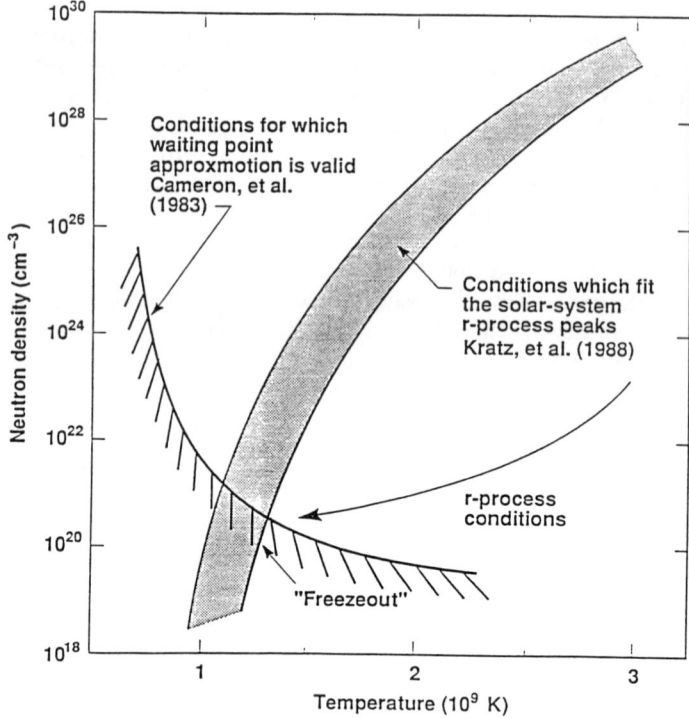

Figure 3. Neutron densities and temperatures consistent with observed r-process peaks and measured beta-decay rates from Refs. 7 and 8. Also shown are the conditions for which the classical waiting-point approximation is valid from Ref. 45. The uncertainty in the allowed conditions is due to the uncertainty in the nuclear mass predictions away from stability. The thick arrow is a schematic illustration of the temperatures and densities produced in an r-process environment before freezeout at $n_n \sim 10^{20}$ and $T \sim 10^9 K$.

r-process path may be further from stability where shorter beta-decay lifetimes imply a shorter timescale for equilibration.

The new nuclear data have provided valuable clues as to the nature of the r-process site. Although much more data for nuclei along the r-process path are needed it appears most likely that the r-process occurs in a classical primary environment. We also have a clearer picture of the neutron density ($\sim 10^{20}$cm^{-3}), temperature ($\sim 10^9$K), and time scale (~ 1 sec) for the r-process.

ASTRONOMICAL DATA AND GALACTIC EVOLUTION

Recently, numerous measurements have been made[10-13] of heavy-element abundances on the surface of metal-poor stars. Such data display the enrichment of r-process abundances during the history of the Galaxy. Abundance correlations of nearly pure r-process elements (like Eu) can now be constructed with respect to iron "seed" material over several orders of magnitude in abundance.

One of the first exciting discoveries[14] was that the the r-process abundances relative to iron in metal poor stars look much like the solar-system r-process distribution. This important result confirms the notion that the r-process is a unique event which always produces an abundance distribution similar to that observed in the solar-system.

The correlation of heavy-element abundances with iron are also useful for two other reasons. For one, the astrophysical sites for the production of iron are reasonably well understood[50-51] to be primary. The other reason is that the abundance of iron over the history of the Galaxy has been constructed[52] from studies of iron abundance as a function of stellar age. Therefore, correlations of r-process elements with respect to iron also give information of the time history of the r-process.

We know the nucleosynthesis mechanism for iron from the fact that the light curve from all types of supernovae (Ia, Ib, and II) are powered by the decay of radioactive ^{56}Ni into ^{56}Fe. This fact was established[53] by the observation of the growth of the Co absorbtion line from ^{56}Ni decay in the spectra of supernovae. There are theoretical reasons to expect that Type Ia supernovae (and perhaps Ib's as well) are produced by the ignition of a thermonuclear runaway which produces ~ 0.5 M$_A$ of ^{56}Fe per event. The production of ^{56}Fe in Type II supernovae was calibrated by the explosion of SN 1987A in the large magellanic cloud. The light curve indicates[52] a yield of 0.07 M$_A$ of Fe which is in good agreement with theoretical predictions.[53] Thus, Fe is a primary element. This makes correlations of iron with r-process elements useful. Primry r-process elements will grow at a similar rate to Fe during the history of the Galaxy. Secondary r-process elements will be delayed until iron has grown to a significant fraction of its present abundance.

We have constructed models[55] for the evolution of the iron and r-process abundances based upon various rates[56] of star formation and the relative probability of forming stars of different masses and estimates[57] of stellar lifetimes. Our models for iron production from supernovae are constrained by observed supernova rates[44] and the iron abundance as a function of stellar age[52].

Figure 4 shows the observed correlation[10-12] of the log$_{10}$ of the Eu/Fe ratio relative to its solar-system value (designated by [Eu/Fe]) as a function of [Fe/H] compared with a number of r-process scenarios. Eu is a good indicator of the r-process since it is about 90% of pure r-process origin in the solar system.[40]

The fact that the observed Eu/Fe ratio is nearly constant for most of this figure suggests that the r-process elements have

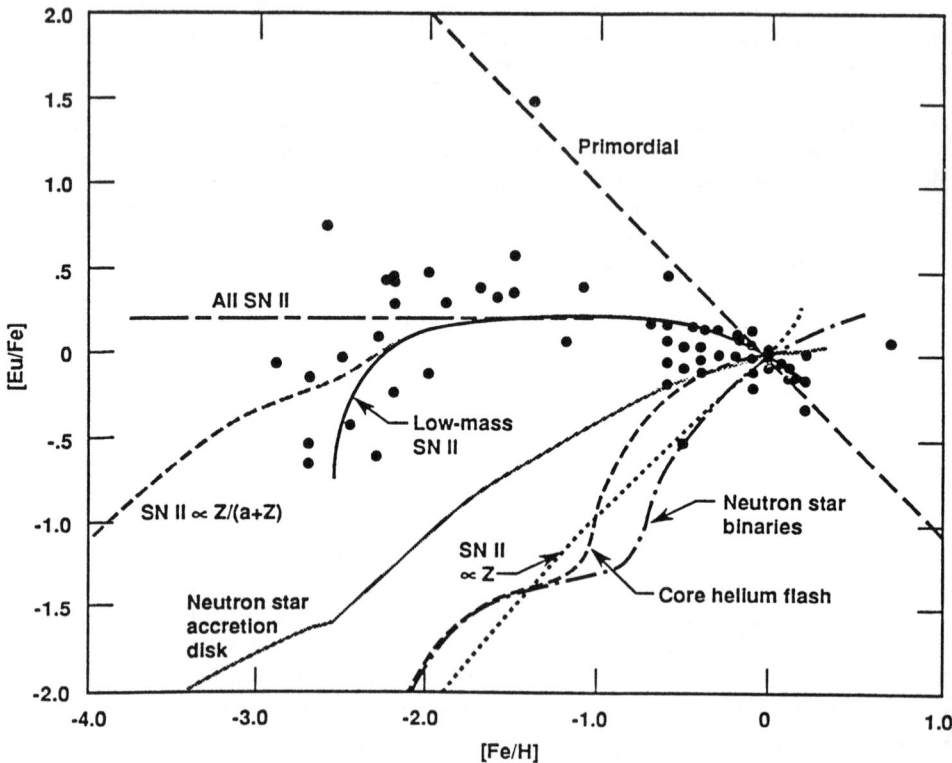

Figure 4. Correlation of [Eu/Fe] with [Fe/H]. The points are data from Refs. 10-12. The curve labeled "Primordial" corresponds to an r-process occuring before the galactic production of iron. The curve labeled "All SN II" corresponds to a primary r-process occuring in all Type II supernovae. Similarly the curve labeled "Low-mass SN II" is for a primary r-process occuring in 10 to 11 M_A stars. The curve labeled "SN II & Z" corresponds to a secondary r-process with yields proportional to the star's initial metallicity, and the curve labeled "SN II & Z/(a+Z)" is for a secondary r-

process which depends upon the initial metallicity but has a primary neutron source. The other three curves are for a core-helium-flash r-process in low-mass stars, an r-process occuring in neutron-star binary coalescence, and neutron-star accretion disks as labeled. For the core-helium-flash and accretion-disk models the initial rise in [Eu/Fe] at [Fe/H] ~ -2.0 is due to the s-process contribution to Eu. The r-process contribution does not appear until [Fe/H] ~ -1.0 in these models.

increased along with iron over the history of the Galaxy. This rules out a pregalactic primordial source for the bulk of the r-process which appears as the dashed line labeled "Primordial" in Figure 4. Nevertheless, the possibility of some primordial r-process (< 10^{-3} of the solar abundance) remains an intriguing possibility[25,26] as a signature of baryon-number inhomogeneities during the big bang.

The dashed line on Figure 4 labeled "All SN II" shows a calculation in which initially both the r-process elements and Fe are produced as primary elements from Type II supernovae. The decrease in the [Eu/Fe] ratio as [Fe/H] increases from -2.0 to +1.0 is a consequence of the fact that iron is also produced by Type I supernovae which eject their iron later.

There may be a decrease[12] in [Eu/Fe] for [Fe/H] < -2.0 (see however Ref. 13). This drop suggests that the r-process begins a little bit later than the processes which form iron. Since Fe is first produced in massive stars, the fall off in [Eu/Fe] suggests that r-process elements are produced in lower-mass stars which explode later. The solid line in Figure 4 labeled "Low-mass SN II" is from a model in which the r-process occurs in Type II supernovae with initial masses of 10-11 M_A.

The fact that low-mass stars may fit the abundances best is consistent with results from computer models[53,58] for core-bounce supernovae. Low-mass supernovae explode and eject neutronized core material more easily. This conclusion is supported by SN 1987A, which had a high-mass progenitor, and ejected only a small amount[54] of ^{56}Ni. Thus, very little neutron-rich core material (which lies below the ^{56}Ni) could have been ejected either.

Nevertheless, this arguement is model dependent. For example, if the production of Fe in SNe II is skewed toward more massive stars, this would decrease the Eu/Fe ratio early on without constraining the r-process to low-mass supernovae. Also we note that the precise mass range for the r-process supernovae is dependent upon stellar age estimates. In some models low mass stars evolve more quickly relative to high-mass stars than our estimates. This could imply a lower mass range for r-process supernovae.

The dot-dashed line on Figure 4 labeled "neutron star binaries" shows a calculation of the prodiction of [Eu/Fe] from the merger[27-29] of two neutron stars. The r-process is delayed by the time required for gravitational radiation to bring the system to merger. We estimate this time from the minimum orbital separation which allows two massive stars to evolve to a neutron star. When either star is in the red-giant phase the other star must be far enough away that the helium envelope of the red giant is not lost by Roche-lobe overflow. Otherwise, the stars would evolve[59] to carbon-oxygen white dwarfs rather than neutron stars.

From this we deduce that the minimum orbital separation of a neutron-star binary when formed is ~ 2.5 solar radii. This is comparable to the separation of the binary-pulsar candidate, PSR-1913+16, which is at a separation distance of ~1.5 solar radii. The time for gravitational radiation damping of this orbit is ~ 10^9 yr which is too long to reproduce the observations. However, if the binary is formed in a high eccentricity orbit, the orbital decay lifetime could be an order of magnitude less. The [Eu/Fe] curve will then better follow the data. Also, the time scale for damping of a black-hole plus neutron-star system would be more rapid than for a neutron-star binary. By allowing[60] for these two effects this scenario can give a reasonable fit to the data.

The fact that the [Eu/Fe] remains very close to but slightly in excess of the solar ratio as the iron abundance decreases by two orders of magnitude argues against a secondary r-process in which the yield is proportional to the preexisting heavy-element abundance. The dotted line labeled "SN II & Z" on Figure 4 shows a model for such an r-process in Type II supernovae. Independent of model parameters, the Eu/Fe ratio is always below the solar-system ratio. However, even a secondary r-process may produce yields which are independent of the amount of preexisting seed material as long as there is enough seed material available. The r-process abundance yields depend upon the product of the neutron density and the abundance of heavy seed material. However, the neutron abundance depends upon the competition between neutron production and absorbtion by capture reactions. Because the heavy nuclei have much larger neutron capture cross sections, the average absorption cross section is dominated by heavy-elements once the abundance of iron and other heavy "seed" nuclei has increased above a certain threshold. Increased r-process yields from increased seed material can therefore be cancelled by the decreased neutron density from neutron absorbtion and final r-process yields only depend upon the neutron source abundance.

For a secondary r-process to be metallicity independent it is therefore necessary for the neutron source to be independent of metallicity. Examples of such sources are the $^{13}C(a,n)^{16}O$ reaction which utilizes ^{13}C that has been manufactured in the star by mixing hydrogen with ^{12}C produced in the helium burning zone,[36] or neutrino-induced neutron emission from carbon or helium in the outer envelopes of core-bounce supernovae.[38,39]

One can estimate from the neutron capture cross sections that the initial seed abundance must be [Fe/H] ~ -3 before the r-process yields are metallicity independent. The dashed line on figure 4 labeled "SN II & Z/(a+Z)" shows a good fit to the observed [Eu/Fe] ratio based upon a secondary r-process in which neutron capture on primary and secondary nuclei has been included.

A core-helium-flash model[35] for the r-process in low-mass stars (~ 1-3 M_A) is shown by a dashed line on Figure 4. The time delay for such stars to contribute is too slow compared to the growth of iron. Figure 4 also shows a model[37] in which r-process material is ejected from neutron-star accretion disks. If the accretion occurs over a time scale comparable to the galactic age, then the rate of ejection of r-process material becomes proportional to the total number of neutron star remnants. This scenario does not well fit the Eu/Fe ratio although it may be possible to shorten the accretion time and improve the agreement. In the limit of a short accretion time this scenario becomes indistinguishable from the z/(a+z) model.

CONCLUSION

There are still large uncertainties in the available constraints on the astrophysical site for r-process nucleosynthesis. If I had to bet on what the site is, I would guess that the r-process occurs in low-mass Type II supernovae. This is the only site which provides both a classical r-process and is consistent with the stellar anundance ratios.

ACKNOWLEDGMENTS

Work performed under the auspices of the U. S. Department of Energy by Lawrence Livermore National Laboratory under contract number W-7405-ENG-48 and supported in part by NSF grants AST 89-17442 at the University of Illinois and AST 85-21705 at the University of Oklahoma.

REFERENCES

1. G. R. Burbidge, E. M. Burbidge, W. A. Fowler, and F. Hoyle, Rev. Mod. Phys., 29, 547 (1970).
2. P. A. Seeger, W. A. Fowler, and D. D. Clayton, Ap. J. Suppl., 11, 121 (1965).
3. G. J. Mathews and R. A. Ward, Rep. Prog. Phys., 48, 1371 (1985).
   G. J. Mathews and J. J. Cowan, Nature, 345, 491 (1990).
   J. J. Cowan, F.-K. Thielemann, and J. W. Truran, Phys. Rep. (1990) in press
4. K.-L. Kratz, H. Gablmann, W. Hillebrandt, B. Pfeiffer, H. L. Raven, and F.-K. Thielemann, Z. Phys., A325, 483 (1986).
5. E. Lund, K. Aleklett, B. Fogelberg, and A. Sangariyavanish, Phys. Scripta, 34, 614 (1986).
6. R. L. Gill, R. F. Casten, D. D. Warner, A. Plotrowski, Phys. Rev. Lett., 56, 1874 (1986).
7. K.-L. Kratz, F.-K. Thielemann, W. Hillebrandt, P. Moller, V. Harms, A. Wohr, and J. W. Truran, in "Proceedings of the Sixth International Conference on Capture Gamma-Ray Spectroscopy", Leuven (1987).
8. K.-L. Kratz, Rev. Mod. Astron., 1, 184 (1988).
9. F. Kappeler, H. Beer, and K. Wisshak, Rep. Prog. Phys., 52, 945 (1989).
10. H. R. Butcher, Ap. J., 199, 710 (1975).
11. R. E. Luck and H. E. Bond, Ap. J., 292, 559 (1985).
12. K. K. Gilroy, C. Sneden, C. A. Pilachowski, and J. J. Cowan, Ap. J., 327, 298 (1988).
13. P. Magain, Astron. Astrophys., 209, 211. (1989).
14. C. Sneden and M. Parthasarathy, Astrophys. J., 267, 757 (1983).

15. A. G. W. Cameron, M. D. Delano, and J. W. Truran, "Proc. Int. Conf. on the Properties of Nuclei Far from the Region of Beta Stability," Leysin, CERN Rep. 70-30, P. 75 (1970).
16. T. Kodama and K. Takahashi, Phys. Lett., 43B, 167 (1973).
17. D. N. Schramm, Astrophys. J., 185, 293 (1973).
18. K. Sato, Prog. Theor. Phys., 51, 726 (1974).
19. W. Hillebrandt, K. Takahashi, and T. Kodama, Astron. Astrophys., 52, 63 (1976).
20. S. A. Colgate, Astrophys. J.,163, 221 (1971).
21. J. M. Leblanc and J. R. Wilson, Ap. J., 161, 541 (1970).
22. D. N. Schramm and Z. Barkat, Astrophys. J., 173, 195 (1972).
23. D. L. Meier, R. I. Epstein, W. D. Arnett, and D. N. Schramm, Astrophys. J., 204, 869 (1976).
24. E. Symbalisty, D. N. Schramm, and J. R. Wilson, Astrophys. J. Lett., 291, L11 (1985).
25. J. H. Applegate, C. J. Hogan, and R. J. Scherrer, Astrophys. J.,
26. T. Kajino, G. J. Mathews, and G. M. Fuller, Submitted to Astrophys. J. (1990).
27. J. M. Lattimer, and D. N. Schramm, Astrophys. J. Lett., 192, L145 (1974).
28. _____, Astrophys. J., 210, 549 (1976).
29. E. Symbalisty and D. N. Schramm, Astrophys. Lett., 22, 143 (1982).
30. C. R. Evans and G. J. Mathews, in "Origin and Distribution of the Elements", G. J. Mathews, ed., (World Scientific, Singapore) pp. 619-624.
31. F. Hoyle and D. D. Clayton, Astrophys. J., 191, 705 (1974).
32. W. Hillebrandt and F. K. Thielemann, Astron. Astrophys., 58, 357 (1977). H. V. Klapdor, T. Oda, J. Metzinger, W. Hillebrandt, and F.-K. Thielemann, Z. Phys., A299, 213 (1981). F.-K. Thielemann, J. Metzinger, and H. V. Klapdor, Astron. Astrophys., 123, 162 (1983).
33. J. R. Truran, J. J. Cowan, and A. G. W. Cameron, Ap. J. Lett., 222, L63 (1978).
34. T. Lee, D. N. Schramm, J. P. Wefel, and J. B. Blake, Ap. J., 232, 854 (1979).
35. J. J. Cowan, A. G. W. Cameron, and J. R. Truran, Ap. J., 252, 348 (1982).
36. S. E. Woosley and T. A. Weaver in "Nucleosynthesis and Its Implications for Nuclear and Particle Physics," J. Audouze and N. Mathieu, eds., NATO ASI Ser. C, Vol 163., (D. Reidel, Dordrecht) pp. 145-166 (1986).
37. C. J. Hogan and J. H. Applegate, Nature, 320, 236 (1987).
38. R. I. Epstein, S. A. Colgate, and W. C. Haxton, preprint (1988).
39. S. E. Woosley, D. Hartmann, R. D. Hoffman, and W. C. Haxton, submitted to Astrophys. J. (1989).
40. W. M. Howard, G. J. Mathews, K. Takahashi, and R. A. Ward, Astrophys. J., 309, 633 (1986).
41. A. G. W. Cameron, in "Essays in Nuclear Astrophysics," C. A. Barnes, et al., eds. (Cambridge University Press; Cambridge) pp. 23-43 (1982).
42. E. Anders and M. Ebihara, Geochim. Cosmochim. Acta, 46, 2362 (1982).
43. E. Anders and N. Grevesse,Geochim. Cosmochim. Acta, 53, 197 (1989).

44. S. Van den Bergh, R. D. McClure, and R. Evans, Astrophys. J., 323, 44 (1987).
45. A. G. W. Cameron, J. J. Cowan, and J. W. Truran, Ap. Space Sci., 91, 235 (1983).
46. G. J. Mathews and F.-K. Thielemann, unpublished.
47. K.-L. Kratz, W. Ziegert, W. Hillebrandt, and F.-K. Thielemann, Ap. J., 125, 381 (1983).
48. G. J. Mathews, A. Mengoni, F.-K. Thielemann, and W. A. Fowler, Ap. J., 270, 740 (1983).
49. J. J. Cowan, A. J. W. Cameron, and J. R. Truran, Astrophys. J., 294, 656 (1985).
50. F. Matteucci in "Origin and Distribution of the Elements", G. J. Mathews, ed., (World Scientific, Singapore) pp. 186-198 (1988).
51. L. Greggio and A. Renzini, Astron. Astrophys., 118, 217 (1983).
52. B. A. Twarog, Astrophys. J., 242, 242 (1980).
53. S. E. Woosley and T. A. Weaver, Ann. Rev. Astron. Astrophys., 24, 205 (1986).
54. P. Pinto and S. E. Woosley, Astrophys. J. (1987).
55. G. J. Mathews, C. R. Alcock, and G. M. Fuller, Astrophys. J. (1989) in press.
56. G. E. Miller and J. M. Scalo, Astrophys. J. Suppl., 41, 513 (1979).
57. R. J. Talbot and W. D. Arnett, Astrophys. J., 170, 409 (1971).
58. W. Hillebrandt, Astron. and Astrophys., 110, L3 (1982).
59. I. Iben and A. V. Tutokov, Astrophys. J., 313, 727 (1987).
60. G. J. Mathews, G. Bazan, and J. J. Cowan, Astrophys. J. (1991) in press.

CAPTURE REACTIONS ON ^{14}C IN NONSTANDARD BIG BANG NUCLEOSYNTHESIS

Michael Wiescher, Joachim Görres
University of Notre Dame, Notre Dame, IN. 46556

Friedrich Karl Thielemann
Harvard University, Cambridge, MA. 02138

ABSTRACT

Several capture reactions on ^{14}C, important in a nonstandard big bang scenario have been studied, the experimental techniques and results are presented here. Network calculations for big bang conditions have been performed and show that the production of heavy elements in the early universe may be possible.

INTRODUCTION

The standard big bang scenario assumes a homogeneous, isotropic density expansion in a radiation dominated scenario. As the temperature declines below 1 MeV the photodisintegration of ^{2}H slows down and nucleosynthesis starts with a sequence of proton, neutron and light particle capture reactions to build up ^{3}He, ^{4}He and in small amounts also ^{7}Be, ^{7}Li [1]. The mass A=5 and A=8 gaps among stable nuclei inhibit the formation of heavier nuclei beyond A=8 at standard big bang conditions, therefore only appreciable amounts of ^{2}H, ^{3}He, ^{4}He, and ^{7}Li are produced.

A first order quark–hadron phase transition at T≈100 MeV will prevent such a homogeneous evolution [2] leading to proton rich high density regions, and low density regions with a high neutron abundance [3]. Because of the different proton and neutron abundances the nuclear reaction sequences differ considerably in the two regions. Detailed calculations [4] have shown that, with the exception of ^{7}Li the observed primordial abundances can also be reproduced in such an inhomogeneous scenario assuming certain conditions for the density and volume ratio between the two regions [5].

The most fascinating aspect of the inhomogeneous big bang scenario, however, is the possibility to bridge the mass A=8 gap in the neutron rich zones and to produce heavier elements A≥12. The predominant reaction sequence leading to the production of ^{12}C is

$$^7Li(n,\gamma)^8Li(\alpha,n)^{11}B(n,\gamma)^{12}B(\beta^-)^{12}C$$

with the possible alternative branchings

$$^7Li(\alpha,\gamma)^{11}B(n,\gamma)^{12}B(\beta^-)^{12}C \quad [5]$$

and

$$^7Li(t,n)^9Be(t,n)^{11}B(n,\gamma)^{12}B(\beta^-)^{12}C \quad [6].$$

Neutron capture reactions, ^{12}C(n,γ)^{13}C(n,γ)^{14}C, will then result in the production of ^{14}C.

The nucleus ^{14}C has a half–life of 5730±40 y, and is therefore stable in the time scales of the big bang nucleosynthesis. The production of heavier elements, A≥20, therefore depends mainly on capture reactions on ^{14}C. It was suggested [7] that the production of heavier elements A≥20 is initiated by the reaction sequence

$$^{14}C(\alpha,\gamma)^{18}O(n,\gamma)^{19}O(\beta^-)^{19}F(n,\gamma)^{20}F(\beta^-)^{20}Ne \; .$$

The reaction rate of ^{14}C(α,γ) has recently been updated [8]. It is expected that at the big bang temperature conditions the rate is mainly determined by the influence of the E = 1.14 MeV resonance [9].

Because of the high neutron abundance, however, the neutron capture ^{14}C(n,γ)^{15}C may compete strongly with the alpha capture reaction leading to an alternative sequence,

$$^{14}C(n,\gamma)^{15}C(\beta^-)^{15}N(n,\gamma)^{16}N(\beta^-)^{16}O(n,\gamma)^{17}O(n,\gamma)^{18}O(n,\gamma)^{19}O \; .$$

Isotopes with masses A≥20 will be produced by subsequent β–decay and neutron capture reactions, depending on the neutron abundance and the density.

Sufficient proton abundance in the neutron rich zones will allow an alternative production branch of ^{15}N by the proton capture reaction ^{14}C(p,γ)^{15}N, depending on the rate of the latter reaction.

A third alternative reaction sequence for depleting ^{14}C was pointed out recently [10]. A small abundance of deuterium will allow deuteron capture on ^{14}C, ^{14}C(d,n)^{15}N is suggested to be the dominant reaction branch yielding an alternative way for producing ^{15}N.

We investigated experimentally the cross section of the α–capture ^{14}C(α,γ)^{18}O and the p–capture reaction ^{14}C(p,γ)^{15}N, in addition we have calculated the cross section of the n–capture ^{14}C(n,γ)^{15}C to determine the rates of these three competing reactions and to investigate their influence on the reaction flow in this mass range.

EXPERIMENTAL PROCEDURES

a. The ^{14}C(α,γ)^{18}O Reaction

The ^{14}C(α,γ)^{18}O reaction was investigated in the energy range 1.1 to 2.3 MeV at the 3 MV Pelletron at the California Institute of Technology. A 30 μg/cm^2, 97% enriched, electrodeposited ^{14}C target was bombarded with a 10–20 μA alpha beam. The excitation curves for all transitions have been measured with a high resolution 35% Ge–detector positioned at an angle of 55^0 with respect to the beam direction.

Figure 1 shows the obtained yield curves for the secondary transition from the first excited state at 1.98 MeV to the ground state for the two observed resonances at 1.140 MeV and 1.790 MeV. Also observed are the resonances at 2.098 and 2.330 MeV. The decay branchings and resonance strengths agree well with the results of previous work [9]. The obtained resonance energies and resonance strengths are listed in table I.

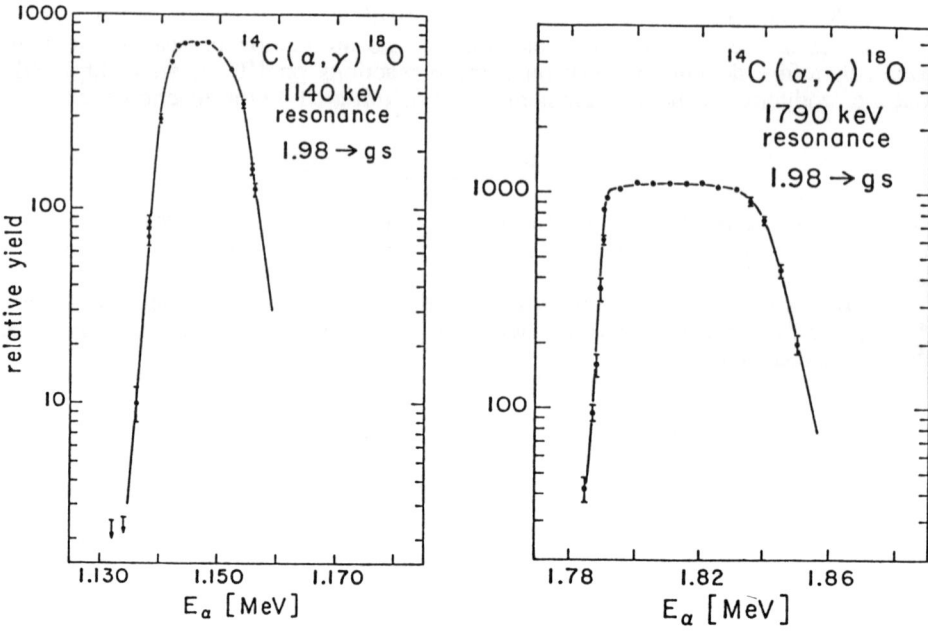

FIG. 1. Yield curve for the two resonances at 1.14 MeV and 1.79 MeV in ^{14}C(α,γ)^{18}O.

Table I Resonance strengths in the ^{14}C(α,γ)^{18}O reaction

E_i^{cm} (MeV)	J^π	$\omega\gamma$ (eV)
1.140	4^+	0.46±0.08
1.790	1^-	1.00±0.17
2.330	1^-	3.70±0.73

The analysis of the off-resonant spectra indicate nonresonant transitions to the groundstate and to the first excited state at 1.98 MeV in ^{18}O. No other transitions to higher excited states have been observed. Figure 2 shows the excitation curve for these two transitions; the open circles indicate the nonresonant γ–yield, the black dots indicate the resonant γ–yield with constant γ–energy. The dashed line represents the thick target yield curves for the two resonances. The nonresonant yield is determined by the direct capture and by strong interference patterns between the two p–wave resonances at 1.790 MeV, and 2.330 MeV, as well as between the resonances and the p–wave contributions of the direct capture.

The solid line in figure 2 indicates the fit of the data integrating the calculated nonresonant cross sections over the target thickness ΔE to reproduce the γ-yield. Both fits result in a consistent set of parameters for the resonance strengths (see table I), the resonance widths and the direct capture cross section. Table II lists the resulting cross sections for the different direct capture transitions obtained at an energy of 1.725 MeV

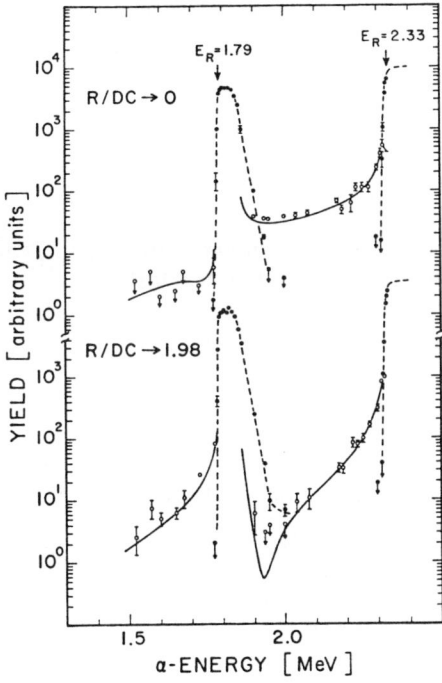

FIG. 2. Excitation curve for the resonant and nonresonant transitions to the ground state and the first excited state in ^{18}O

Table II Nonresonant transition in ^{14}C$(\alpha,\gamma)^{18}$O with the experimental cross section at 1.725 MeV

E_f (MeV)	J^π	σ (nbarn)
0.000	0^+	6.00
1.980	2^+	1.44
3.630	0^+	≤ 1.30
4.460	1^-	≤ 3.10
5.260	2^+	≤ 2.00

The reaction rate is determined by both, the resonant and the nonresonant component. The resonant rate depends mainly on the resonance strengths $\omega\gamma$ (eV) of the observed resonances, in particular the strong resonance at 1.140 MeV and can be expressed by [11],

$$N_A <\sigma v>_r = 2.81 \cdot 10^4 \, T_9^{-3/2} \sum \omega\gamma_i \exp(-11.605 \cdot E_i/T_9),$$

with the resonance strengths listed in table I.

The nonresonant reaction contribution can well be expressed in terms of the astrophysical S-factor, defined by $S(E)=\sigma(E)\cdot E \cdot \exp(2\pi\eta)$ [11], and can be fitted by a polynomial function,

$$S(E) = 2.18 - 2.60 \cdot E + 0.82 \cdot E^2 \; [\text{MeV barn}].$$

Taking into account only the nonresonant contributions with energies below the 1.79 MeV resonance yields a nonresonant reaction rate of

$$N_A <\sigma v>_{nr} = 2.68 \cdot 10^{10} \cdot T_9^{-2/3} \cdot (1 + 1.3 \cdot 10^{-2} T_9^{1/3} - 0.686 T_9^{2/3} -$$
$$-6.1 \cdot 10^{-2} T_9 + 0.164 T_9^{4/3} + 3.7 \cdot 10^{-2} T_9^{5/3}) \cdot \exp(-35.051/T_9^{1/3} - 0.174 T_9^2).$$

While the resonant contributions to the reaction rate agree fairly well with previous results, the experimentally determined nonresonant rate deviates considerably from the theoretically predicted [8] nonresonant rate. These calculations are based on the prediction of strong alpha direct capture transitions to the alpha cluster states at 3.63, 4.46 and 5.26 MeV. The measured upper limits of the cross section for the direct capture to these levels is more than one order of magnitude smaller than the predicted values [8].

b. The $^{14}C(p,\gamma)^{15}N$ Reaction

The $^{14}C(p,\gamma)^{15}N$ reaction was investigated in the energy range 0.25 to 0.75 MeV at the 3 MV Pelletron at the California Institute of Technology. The ^{14}C target was bombarded with a 30–60 µA proton beam. The excitation curves for all transition to the ground state and several excited states have been measured with a 35% Ge detector positioned at an angle of 55⁰ with respect to the beam direction. Figure 3 shows the obtained yield curve for the ground state transition and the secondary transition from the excited state at 5.27 MeV to the ground state. Several resonances have been observed with resonance energies of 0.634, 0.527, 0.519, 0.351 and 0.261 MeV. The resonance strengths were obtained from the measured thick target yield and agree well with the results of previous studies [12]. Nonresonant reaction contributions are the E1 direct proton capture transitions to the ground state as well as to excited states at 5.27, 5.30, 6.32 and 9.05 MeV in ^{15}N. Additional contributions may result from the low energy tails of broad resonances at higher energies, 1.413 and 1.617 MeV [13]. Interference effects between these resonances and the direct processes also may effect the cross section at lower energies. To study the different components angular distribution measurements have been performed at different energies in the nonresonant part of the excitation curve. The analysis showed that the direct capture process clearly dominates the cross section in the investigated energy range, the measured cross section for transitions to excited states agrees well with the theoretically predicted cross section for the direct capture. Only for the ground state transition an additional contribution from the low energy tail of the higher resonances is required to explain the observed cross section.

The reaction rate is determined by both, a resonant and a nonresonant component. The resonant rate depends mainly on the resonance strengths $\omega\gamma$ (eV) of the observed resonances, in particular the strong resonance at 0.351 MeV and can be expressed by:

$$N_A \langle \sigma v \rangle_r = 1.708 \cdot 10^5 \, T_9^{-3/2} \sum_i \omega\gamma_i \exp(-11.605 \cdot E_i/T_9),$$

with the resonance strengths listed in table III.

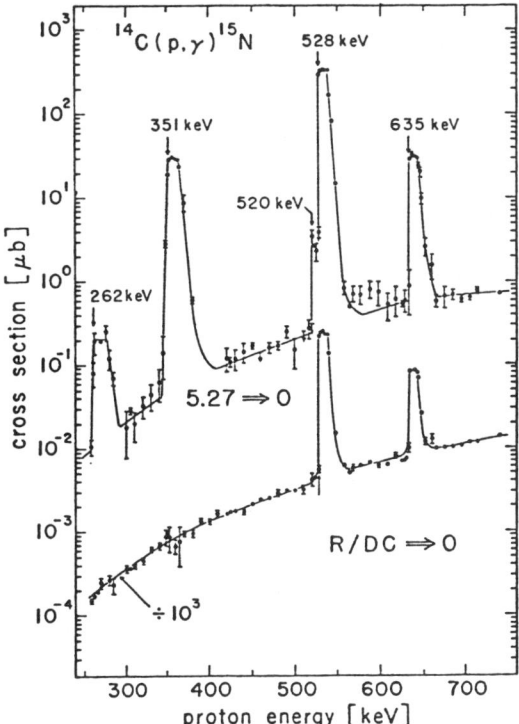

FIG. 3. Excitation curves for two transitions in $^{14}C(p,\gamma)^{15}N$.

The nonresonant reaction contribution can well be expressed in terms of the astrophysical S–factor can be fitted by the polynomial function,

$$S(E) = 7.50 \cdot 10^{-3} + 2.00 \cdot 10^{-3} \cdot E + 5.31 \cdot 10^{-3} \cdot E^2 \; [\text{MeV barn}].$$

Table III Resonance strengths in the $^{14}C(p,\gamma)^{15}N$ reaction

E_i^{cm} (MeV)	J^π	$\omega\gamma$ (eV)
0.244	$5/2^-$	0.29±0.05
0.328	$5/2^+$	37.0±6.0
0.484	$9/2^+$	2.8±0.5
0.492	$3/2^-$	0.95±0.15
0.592	$3/2^+$	0.29±0.05

This yields in a nonresonant reaction rate of

$$N_A \langle \sigma v \rangle_{nr} = 1.09 \cdot 10^8 T_9^{-2/3} \cdot (1 + 3.04 \cdot 10^{-2} T_9^{1/3} + 0.105 T_9^{2/3} + 2.24 \cdot 10^{-2} T_9 + 0.109 T_9^{4/3} + 5.94 \cdot 10^{-2} T_9^{5/3}) \cdot \exp(-13.71/T_9^{1/3} - (T_9/4.694)^2).$$

For higher energies the actual S–factor will deviate from the parametrization. However, the total reaction rate is dominated by the resonant contributions at the corresponding temperatures $T_9 \geq 2$. Therefore this deviation is of no significant consequence.

The resonant term agrees well with the results of the former tabulations [11], however, the nonresonant contribution is more than an order of magnitude larger than quoted before.

REACTION BRANCHING AT ^{14}C

The neutron capture cross section on ^{14}C is extremely small. Measurements with thermal neutrons resulted only in an upper limit of $\sigma < 1\mu$barn [14]. The Q–value of the reaction Q=1.218 MeV is small; no excited states in ^{15}C exist in that excitation range [13]. The reaction cross section is therefore only determined by the direct capture to the ground state and the first excited state in ^{15}C at 0.74 MeV. The two levels have a predominant single particle s,d–shell structure of [15], which is listed in table IV.

Table IV Parameter for the configuration of the final states in the $^{14}C(n,\gamma)^{15}C$ direct capture transitions

E	J^π	configuration	C^2S_{theo}	C^2S_{exp}
gs	$1/2^+$	$(p_{1/2}^2, s_{1/2})$	0.97	0.88
0.74	$5/2^+$	$(p_{1/2}^2, d_{5/2})$	0.94	0.69

Therefore, E1 direct capture to these states is only possible for p–wave neutrons, p→s and p→d, respectively. For the transition to the first excited state an E2 direct capture of s–wave neutrons, s→d has to be taken into consideration.

We have calculated the cross section for the direct radiative capture using first order perturbation theory [16]. The specific nuclear amplitude of the transition is taken into account by the spectroscopic single particle factor C^2S of the final state. The resulting total cross section of the direct capture is then,

$$\sigma_{DC} = \sigma_0(E1) \cdot C^2S_0 + (\sigma_1(E1)+\sigma_1(E2)) \cdot C^2S_1 .$$

The spectroscopic factors C^2S_0, C^2S_1 for the ground state and the first excited state in ^{15}C have been measured by the direct neutron transfer reaction $^{14}C(d,p)^{15}C$ [17]. These values are in good agreement with the results of shell model calculations [15], both are listed in table IV. The experimental values have been used for the direct capture calculations.

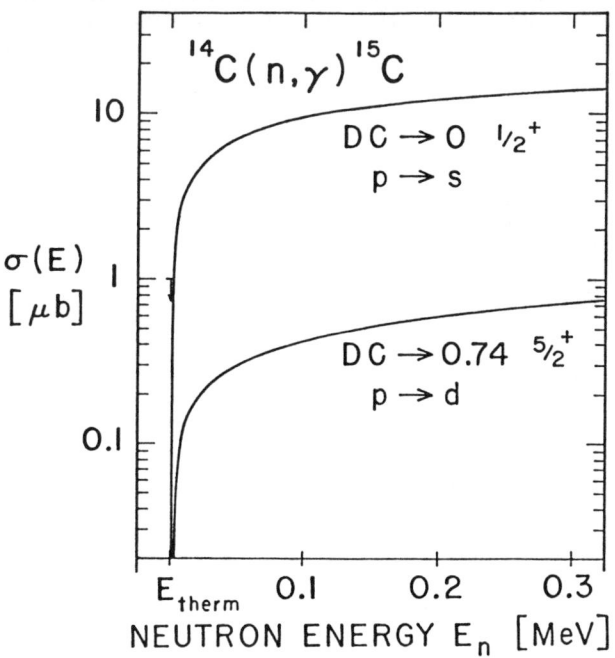

FIG. 4. The calculated neutron direct capture cross section on ^{14}C for the E1–transitions to the ground state and first excited state in ^{15}C.

Figure 4 shows the cross section for the two direct capture E1–transitions to the ground state and the first excited state. The ground state p→s E1–transition is considerably stronger than the p→d E1–transition to the first excited state. For the s→d E2–transition to the first excited state we calculate a thermal cross section of $\sigma(E2) = 0.92$ nb which is the same order of magnitude as the thermal cross sections for the E1–transitions (5.6 nb). For increasing neutron energy, however, the cross section σ(E2) decreases with 1/v. At higher energies, therefore, the E2 contribution to the direct capture is negligible.

The reaction rate for the direct capture of s–wave neutrons is constant, $\langle \sigma v \rangle = \sigma \cdot v$, and can therefore be directly extrapolated from the thermal neutron capture cross section [18]. In the specific case here, however, the influence of the orbital momentum barrier for p–wave neutrons has to be taken into account. The resulting reaction rate is proportional to the temperature [19]. Expressed in terms of the cross section σ(E) for p–wave neutrons, the rate is given by,

$$N_A \langle \sigma v \rangle = 1.12 \cdot 10^8 \cdot \sigma(E) \cdot E^{-1/2} \cdot T_9 \quad [\text{cm}^3/\text{mole s}]$$

and is shown in figure 5.

Also shown in figure 5 are the total rate for the competing reactions $^{14}C(p,\gamma)^{15}N$, and $^{14}C(\alpha,\gamma)^{18}O$ as discussed above. At temperatures below $T_9=0.8$ the the neutron capture rate dominates, while at higher temperatures the proton capture

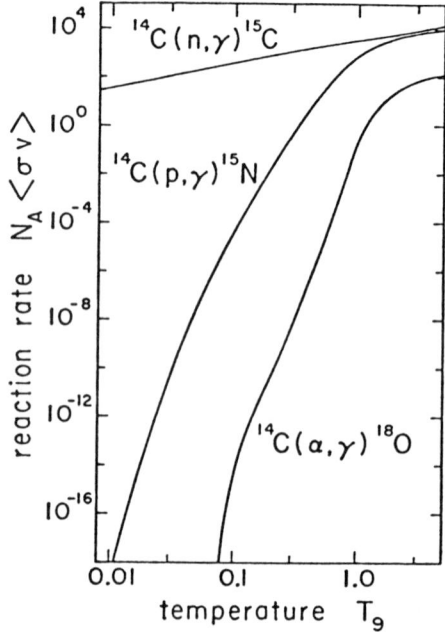

FIG. 5. Reaction rates for $^{14}C(\alpha,\gamma)^{18}O$, $^{14}C(p,\gamma)^{15}N$ and $^{14}C(n,\gamma)^{15}C$ at big bang temperature conditions.

rate is becoming larger. The alpha capture rate is small in the entire temperature range.

The destruction of ^{14}C is determined by its decay rates

$$\lambda = \rho \cdot \frac{X}{A} \cdot N_A \langle \sigma v \rangle$$

due to the various discussed capture reactions (p,γ), (n,γ) and (α,γ), with X, A as mass fraction and atomic mass of protons, neutrons and helium, respectively. In big bang nucleosynthesis, mass fractions and density vary as a function of temperature, which decreases as a function of time. Calculations of the tempertaure dependent decay rates [10] indicate a strong depletion of ^{14}C by neutron capture in the early big bang phase, while proton and possibly deuteron capture determine the depletion in the later burning phase. However, the influence of inverse reactions have also to be taken into account. Because of the low Q–value of the ^{14}C(n,γ)^{15}C reaction, photodisintegration of ^{15}C is very likely in such a high temperature, low density environment. To determine the net–flow towards heavier elements, network calculations [20] have been performed for big bang temperature and density conditions. The results indicate that the net–flow via ^{14}C(n,γ)^{15}C is always smaller than via the competing ^{14}C(p,γ)^{15}N and ^{14}C(d,n)^{15}N reaction links leading to the formation of ^{15}N. The calculations also show, that subsequent ^{15}N(p,α)^{12}C competes with the neutron capture reaction ^{15}N(n,γ)^{16}N. This generates a CNO–cycle in which part of the material is being stored, leading to a substantial enrichment of mass A=14 nuclei in the early universe. However, a sufficient mass flow towards heavier elements is suggested by the calculations and may allow considerable production of isotopes with masses A>20 at the conditions of an inhomogeneous big bang.

References:

1. R.V. Wagoner, Ap.J.Suppl. 18, 247 (1969)
2. E. Witten, Phys.Rev. D30, 272 (1984)
3. J.H. Applegate, C.J. Hogan, Phys.Rev. D31, 3037 (1985)
4. C.R. Alcock, G.M. Fuller, G.J. Mathews, Ap.J. 320, 439 (1987)
5. R.A. Malaney, W.A. Fowler, Ap.J. 333, 14 (1988)
6. R.N. Boyd, T. Kajino, Ap.J.Lett. 336, L55 (1989)
7. J.H. Applegate, C.J. Hogan, R.J. Scherrer, Ap.J. 329, 572 (1988)
8. C. Funck, K.H. Langanke, Ap.J., 344, 46 (1989)
9. M. Gai, R. Keddy, D.A. Bromley, J.W. Olness, E.K. Warburton, Phys.Rev. C36, 1256 (1987)
10. L.H. Kawano, W.A. Fowler, R.A. Malaney, preprint OAP–704 (1990)
11. W.A. Fowler, G.E. Caughlan, B.A. Zimmerman, Ann.Rev. Astron.Astrophys. 13, 69 (1975)
12. R.P. Beukens, PhD Thesis, University of Toronto (1976)
13. F. Ajzenberg–Selove, Nucl.Phys. A449, 1 (1986)
14. L. Yaffe, W.H. Stevens, Phys.Rev. 79, 893 (1950)
15. B.S. Reehal, B.H. Wildenthal, Part.Nucl. 6, 137 (1973)
16. C. Rolfs, C., Nucl.Phys. A217, 29 (1973)
17. J.D. Goss, P.L. Jolivette, C.P. Browne, S.E. Darden, H.R. Weller, R.A. Blue, Phys.Rev. C12, 1730 (1975)
18. W.A. Fowler, G.E. Caughlan, B.A. Zimmerman, Ann.Rev. Astron.Astrophys. 5, 525 (1967)
19. M. Wiescher, J. Görres, F.K. Thielemann, Ap.J. 363, in print
20. F.K. Thielemann, J.H. Applegate, J.J. Cowan, M. Wiescher, in preparation

^{176}Lu – AN s-PROCESS THERMOMETER

N. Klay[*], F. Käppeler, H. Beer, and G. Schatz
Kernforschungszentrum Karlsruhe, Karlsruhe, Federal Republic of Germany

ABSTRACT

The use of ^{176}Lu as a cosmic clock has been in question since the possibility of a thermal coupling between the long-lived ground state ($t_{1/2} = 41 \cdot 10^9$ yr) and the beta unstable isomer with 3.68 hr half-life was considered. This coupling requires induced transitions to higher lying states because direct transitions are highly forbidden by selection rules. First experimental evidence for the enhancement of the overall decay rate by that mechanism was provided via photoactivation measurements, but a quantitative discussion became possible only after the level scheme of ^{176}Lu had been carefully established in two recent investigations. These studies identified an $I^\pi = 5^-$ state at 838.6 keV to be the most efficient mediating level for the coupling of ground state and isomer. As a consequence, the half-life of ^{176}Lu is drastically reduced above $1.5 \cdot 10^8$ K, i.e. at conditions typical for the production of ^{176}Lu in the s-process during the helium burning phase of stellar evolution. Hence, any interpretation of ^{176}Lu as a cosmic clock is strongly dependent on the conditions at the stellar site, a problem that cannot be solved reliably enough with present stellar models. Instead, the strong temperature dependence of the ^{176}Lu half-life makes it a sensitive tool for constraining the s-process temperature to the range between $2.4 \cdot 10^8$ to $3.6 \cdot 10^8$ K.

INTRODUCTION

For many years, ^{176}Lu was considered as a promising chronometer with respect to galactic history[1,2]. The fact that this isotope can be attributed to the slow neutron capture process (s-process) allows for a rather precise determination of the original ^{176}Lu abundance, which could be compared with the present abundance in the solar system for deriving an age of the s-process elements.

The s-process flow in the Yb-Lu-Hf region is sketched in Fig. 1; it follows the solid line due to subsequent neutron captures and beta decays. At A = 176, abundance contributions from the r-process (indicated by arrows) are accumulated at ^{176}Yb, leaving the s-process as the only production mechanism for the isobars ^{176}Lu and ^{176}Hf (apart from minor p-process contributions). The original s-process abundance of ^{176}Lu can reliably be determined by means of the classical approach[3], and the only complication in interpreting ^{176}Lu as an s-process chronometer seemed to be the existence of an isomeric state, which causes part of the s-process flow to bypass the long-lived ground state. This effect is considered in Fig. 1 by showing ground state and isomer separately. As indicated by the strength of the lines, neutron capture on ^{175}Lu leads more frequently to the isomer, leaving only a small probability for the long-lived ground state.

---

[*] present address: Asea Brown Boveri, Forschungszentrum, CH-5405 Baden.

N. Klay et al. 851

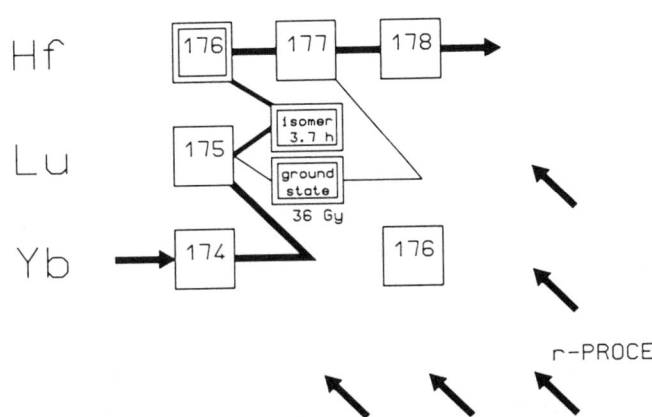

Fig. 1 The s-process neutron capture chain between Yb and Hf. Note, that ^{176}Lu and ^{176}Hf are shielded against the r-process.

Since the isomer decays exclusively to ^{176}Hf with a half-life of only 3.68 h, the original s-process abundance of ^{176}Hf is determined by the fraction of neutron captures feeding the isomer. The idea that ground state and isomer could be treated like two different nuclei appeared plausible, because direct transitions are completely inhibited by selection rules. However, this picture was questioned by an obvious discrepancy: The partial capture cross section for populating the isomer in ^{176}Lu was found to be much larger[4,5,6,7] than in a previous measurement[8], leading to a strongly reduced ^{176}Lu production that was even not sufficient to account for the abundance observed today[9]. On the other hand, Beer et al.[10] found from a comparison with the ^{176}Hf abundance that the s-process flow through ^{176g}Lu was considerably more efficient than expected from this result.

Therefore, additional feeding mechanisms for the long-lived ground state had to be identified. The most important one was, that higher lying states may be excited in the hot photon bath at the stellar s-process site[11,12], which then decay back either to the isomer or to the ground state. In this way, the initial population of these two states can be changed, and in the extreme case even thermal equilibrium can be achieved. Then, the s-process abundances of ^{176}Lu and ^{176}Hf are no longer determined by their cross sections alone, but by the competition between their stellar beta decay and neutron capture rates, resulting in an abundance ratio that is sensitive to temperature and neutron density during the s-process.

Experimental evidence for electromagnetic transitions between ground state and isomer was obtained by photoactivation of Lu samples in intense gamma-ray fields[13,14,15]. Since the isomer could be produced only with ^{60}Co and ^{24}Na sources but not with ^{137}Cs, the mediating levels are expected at excitation energies above the 662 keV gamma-ray energy of ^{137}Cs. However, quantitative information on the stellar transition rates could not be inferred from these measurements, since the relevant properties of the mediating levels, i.e. their excitation energies and quantum numbers remained unknown.

An improved theoretical approach was presented by Gardner et al.[16], who complemented the experimentally determined levels in ^{176}Lu by postulated states deduced from systematic trends in neighboring nuclei. With this 'complete' level scheme, a pronounced temperature dependence of the ^{176}Lu half-life could be obtained without violating the K-selection rule. This result emphasized the importance of an experimental extension of the level scheme, in order to arrive eventually at a quantitative solution of the ^{176}Lu riddle.

EXPERIMENTS

We have reinvestigated the level scheme of ^{176}Lu by detailed experimental studies at the high flux reactor of the ILL Grenoble. Gamma transitions in ^{176}Lu were investigated with the bent crystal spectrometers GAMS[17,18], which exhibit excellent resolution up to ~1 MeV. Complementary information on the conversion electron spectrum for determination of multipolarities was obtained with the magnetic spectrometer BILL[19] at the ILL that yields an energy resolution compatible with the gamma spectra. An additional measurement of $\gamma-\gamma$ coincidences with two germanium detectors served for the reliable assignment of the gamma transitions within the level scheme. Furthermore, information on the lifetimes of relevant levels was obtained from a measurement of delayed coincidences, and in one case from the Doppler broadened line shape in the gamma spectra.

The measurements in Grenoble were supplemented by a study of the ^{175}Lu(d,p)^{176}Lu reaction with the Q3D spectrograph[20] at the tandem accelerator of the University and Technical University Munich; the results obtained in this experiment were important for determining the energy of the $I^\pi = 1^-$ isomer as well as for the assignment of levels at high excitation energies.

A detailed presentation of the experiments is given in Ref.(21).

THE LEVEL SCHEME OF ^{176}Lu

In total, 509 gamma transitions could be identified in ^{176}Lu, and multipolarities were determined for 228 of these transitions. Based on this information, on the results from the $\gamma-\gamma$ coincidence measurement and from the ^{175}Lu(d,p)^{176}Lu study, as well as on previous investigations[22-27], the level scheme was constructed in three steps:

(i) By means of the Ritz combination principle, two separate level schemes were established, built on the ground state and on the isomer, respectively. In both schemes, a precision of $\Delta E/E = 5.7 \cdot 10^{-5}$ is achieved for the level energies, but the absolute energies of the second scheme remain uncertain due to the ~2 keV uncertainty of the isomer energy.

(ii) That energy could be considerably improved by calibrating the spectrum of the (d,p) measurement with the well established GAMS data. Eventually, the excitation energy of the isomer was precisely defined to 122.855 ± 0.009 keV by several mediating transitions between the two schemes that were identified in the GAMS spectra. Accordingly, the neutron separation energy of ^{176}Lu could be revised to $S_n = 6287.91 \pm 0.15$ keV.

(iii) The unified level scheme was checked for consistency with all existing data; the final scheme consists of 97 energy levels connected by 270 gamma transitions. About 30 Nilsson configurations and corresponding rotational bands were identified. The comparison with model calculations indicates that the level scheme comprises *all* excited states with spins $1 < I < 8$ up to 900 keV. In particular, it contains transitions that connect the ground state with the isomer via mediating levels at higher excitation energy. This result is in agreement with a recent investigation by Lesko et al.[28]

THE s-PROCESS BRANCHING AT ^{176}Lu

With this improved level scheme, the effect of temperature on the s-process flow through ^{176}Lu can now be treated quantitatively; it is expressed by means of a branching factor f_n,

$$(\sigma N)_{^{176}Lu} = f_n [(\sigma N)_{^{176}Lu} + (\sigma N)_{^{176}Hf}]. \quad (1)$$

Here, σN denotes the product of stellar neutron capture cross section, σ, and the respective abundance produced by the s-process. As long as only the ground state decay of an unstable isotope is involved, the branching factor

$$f_n = \lambda_n/(\lambda_n + \lambda_\beta) \quad (2)$$

can be expressed for this one-level system by the ratio of the rates for neutron capture, λ_n, and for beta decay, λ_β. If necessary, thermally populated excited states must also be included; in case of ^{176}Lu, it is sufficient to consider beta decays and neutron captures from the ground state (g) and the isomer (m). The many higher lying states may act as mediators for redistributing the initial populations of ground state and isomer, but they themselves are so weakly populated that their contributions to λ_n and λ_β can be neglected. Then, f_n becomes

$$f_n = \frac{n_g \lambda_n(g) + n_m \lambda_\beta(m)}{n_g [\lambda_n(g) + \lambda_\beta(g)] + n_m [\lambda_n(m) + \lambda_\beta(m)]}. \quad (3)$$

According to Eq. (3), the population probabilities of ground state and isomer are decisive for the determination of the branching factor f_n. The general time-dependent solution of this problem[29] was applied to ^{176}Lu by Beer et al.[12], who used Weisskopf estimates with constant retardation factors for the transition rates. Since the level scheme was incomplete at that time, only qualitative results could be obtained. With the now available level scheme, the calculation can be restricted to the relevant energy levels, so that the branching factor can be expressed by an analytic approximation.

Fig. 2 shows the possible transitions in a three-level system consisting of ground state, isomer, and a mediating level i. The evolution of this system to thermal equilibrium is followed first without considering the *external* production and destruction rates λ_{po}, λ_{od}, λ_{pm}, and λ_{md}. Changes of the initial population probabilities are to be expected if the time constant for approaching thermal equilibrium becomes comparable to the life times for beta decay or neutron capture.

This time dependence of the population probabilities, n, can be described by a system of differential equations:

$$\frac{dn_o}{dt} = -\rho B_{oi} n_o + (A_{io} + \rho B_{io}) n_i \quad (4a)$$

$$\frac{dn_i}{dt} = \rho B_{oi} n_o - (A_{io} + \rho B_{io}) n_i - (A_{im} + \rho B_{im}) n_i + \rho B_{mi} n_m \quad (4b)$$

$$\frac{dn_m}{dt} = (A_{im} + \rho B_{im}) n_i - \rho B_{mi} n_m. \quad (4c)$$

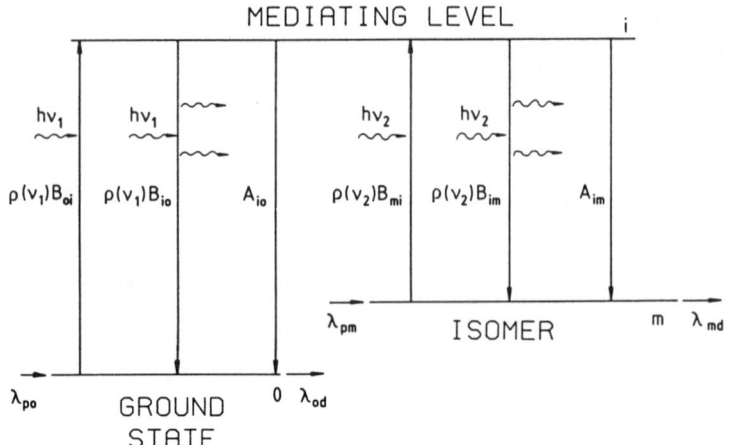

Fig.2 Possible transitions between ground state and isomer in the presence of a mediating state; direct transitions are forbidden by selection rules.

The transition probabilities between two levels 1,2 can be expressed by the Einstein coefficients for emission (A_{21}, B_{21}) and for absorption (B_{12}), as well as by the radiation density (ρ) that follows from the Planck formula. In thermal equilibrium, the Einstein coefficients are related by:

$$\rho B_{12} = A_{21} \frac{g_2}{g_1} \frac{1}{\exp[(E_2-E_1)/kT] - 1}, \quad (5a)$$

$$B_{21} = B_{12}\, g_1/g_2, \quad (5b)$$

where $g_{1,2} = 2I_{1,2}+1$ are the statistical factors according to the level spins I. These relations are not restricted to thermal equilibrium, but hold in general as long as the radiation density in the stellar plasma corresponds to a Planck distribution. Since rather high photon energies $h\nu = \Delta E \gg kT$ are involved in the thermal coupling of ground state and isomer in ^{176}Lu, stimulated emission can be neglected compared to spontaneous emission, and Eq. (5a) simplifies to

$$\rho B_{12} = A_{21} \frac{g_2}{g_1} \exp[-(E_2-E_1)/kT]. \quad (5c)$$

Non-trivial solutions of Eqs.(4a – 4c) are

$$\lambda_{o1} = -(A_{io} + A_{im}), \text{ and} \quad (6a)$$

$$\lambda_{o2} = -\frac{\rho B_{mi} A_{io} + \rho B_{oi} A_{im}}{A_{io} + A_{im}} = -\frac{A_{io}\, A_{im}}{A_{io} + A_{im}} \frac{g_i}{g_m} \exp[-(E_i - E_m)/kT]. \quad (6b)$$

The first solution corresponds to the relatively quick relaxation of the system, if the population of the mediating state exceeds its equilibrium value; the related mean life time is given by $\tau_1 = (A_{io} + A_{im})^{-1}$.

The second solution describes how the populations of the levels o and m approach thermal equilibrium. Defining the branching ratio of the transitions to the isomer as $V \equiv A_{im}/(A_{io} + A_{im})$, this time constant can be expressed in terms of measurable quantities:

$$\lambda_{02} = -\frac{V(1-V)}{\tau_1}\frac{2I_1+1}{2I_m+1}\exp[-(E_1-E_m)/kT]\ . \tag{7}$$

The larger this rate λ_{02}, the more efficient is the coupling between ground state and isomer. If more mediating states exist, the corresponding rates can be estimated accordingly. From Eq. (7) one finds, that the mediating level with the lowest excitation energy is most important for linking the isomer and the ground state, at least as long as its half-life is sufficiently short.

In the level scheme of ^{176}Lu, the lowest established mediating level occurs at 838.6 keV; this level and the related transitions are shown in Fig. 3. Obviously, this mediating state does not decay directly to the isomer but by a cascade that involves other states as well. Therefore, the above approximation can not be applied directly. It has been demonstrated[30], that the extension of Eqs. (4) to a four-level scheme reduces again to the three-level situation, provided that the half-life of the additional state is sufficiently short. With the results obtained in the lifetime measurements[21] it could be shown that this condition is satisfied for all states involved in Fig. 3. Consequently, the thermal coupling of ground state and isomer under typical s-process conditions is exclusively determined by the properties of the 838.6 keV state. Limits for the time constant, by which the closed system approaches equilibrium are obtained by means of the estimates for the life time of that level; for a thermal energy of kT = 25 keV one obtains:

$$1.3\cdot 10^{-4} < |\lambda_0| < 7\cdot 10^{-3}\ s^{-1} \qquad \text{at kT = 25 keV.}$$

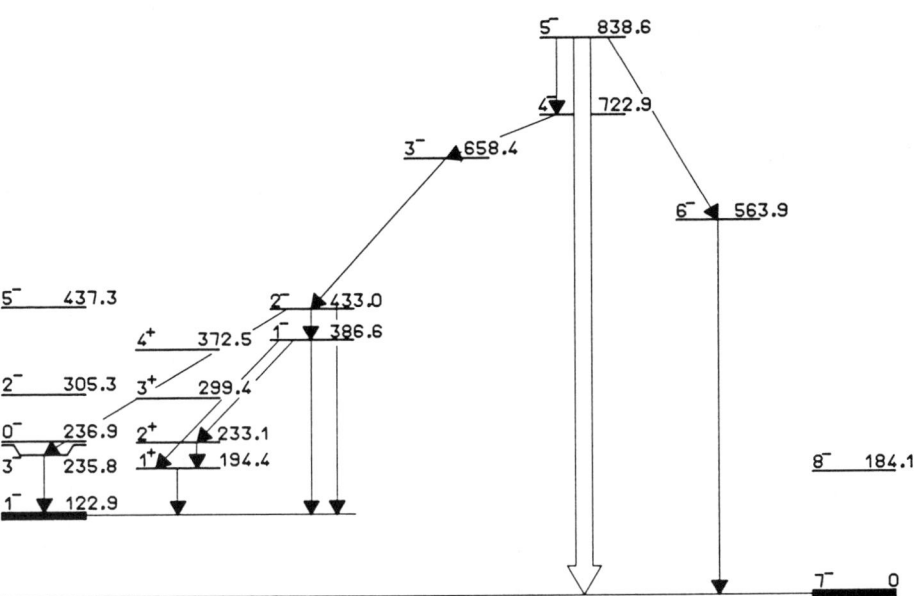

Fig. 3 The relevant part of the level scheme of ^{176}Lu for the discussion of thermally induced transitions between ground state and isomer. The coupling is dominated by the 5$^-$ state at 838.6 keV. If K-forbidden transitions contribute as well, the 437.3 keV state may also be important.

This time constant, that describes the effect of internal electromagnetic transitions, is now to be compared with the external production and destruction processes by beta decay and neutron capture.

THE BRANCHING FACTOR AT EQUILIBRIUM

The branching factor f_n is determined by the beta decay rate of the isomer
$$\lambda_\beta = \ln 2/t_{1/2} = 5.2 \cdot 10^{-5} \, s^{-1},$$
and by the neutron capture rate of the ground state
$$\lambda_n = n_n \, v_T \, \sigma = (1.4 \pm 0.5) \cdot 10^{-7} \, s^{-1} \approx 4 \, yr^{-1},$$
weighted with the population probabilities n_m and n_g (Eq. 3). In the above relations, v_T is the mean thermal velocity at temperature T, n_n the neutron density[30], and σ the neutron capture cross section[10].

If the system were closed, it would approach equilibrium with a time constant that, in any event, is not smaller than $|\lambda_0|$. At kT = 25 keV, this time scale is much shorter than the inverse neutron capture rate of the ground state, but comparable to the half-life of the isomer. This means that the population probabilities will be affected by temperature. Since the beta decays from the isomer are not negligible, the system is not closed, and *thermal* equilibrium must not be attained necessarily. Instead, another equilibrium will be achieved with a time constant large compared to $|\lambda_0|^{-1}$, provided that the external conditions are stable. This equilibrium is determined by the interplay between internal transitions and the external production and destruction processes.

The general time-dependence can be derived by including the respective terms for the external processes in Eqs. (4); solving these equations for the ratio of the population probabilities yields[30]

$$\left(\frac{n_m}{n_0}\right)_{equilib} = \frac{V \rho B_{ol} + (1-B) \lambda_n}{(1-V) \rho B_{ml} + B \lambda_\beta}. \quad (8)$$

The relative production rate $B = \lambda_{po}/(\lambda_{po} + \lambda_{pm})$ corresponds to the partial capture cross section of ^{175}Lu to the isomer in ^{176}Lu, which has recently been remeasured with improved accuracy (B = 0.11 ± 0.04)[7].

The branching factor can now be expressed in terms of neutron density, of temperature, and of the properties of the mediating state at 838 keV:

$$f_n^{-1}(n_n, T) = 1 + \frac{\lambda_\beta^{(m)}}{\lambda_n} \frac{n_m}{n_0} =$$

$$= 1 + \frac{\lambda_\beta^{(m)}}{\sigma v_T n_n} \frac{V(1-V)\tau_i^{-1} \frac{9i}{9o} \exp[-E_i/kT] + (1-B)\sigma v_T n_n}{V(1-V)\tau_i^{-1} \frac{9i}{9m} \exp[-(E_i - E_m)/kT] + B \lambda_\beta^{(m)}}. \quad (9)$$

With the neutron density limits $2.3 \cdot 10^8 < n_n < 4.5 \cdot 10^8 \, cm^{-3}$ of Ref. (31), this relation yields the shaded band in Fig. 4 (solid lines), that exhibits 3 different regions:

(i) At low temperatures, f_n is only determined by the ratio B, i.e. by the ^{175}Lu cross sections; thermal effects are insignificant.

(ii) Between 200 and 300 million degrees, thermally induced transitions cause drastic changes in the population probabilities of ground state and isomer, leading to a strong feeding of the ground state. It is this effect that yields more ^{176}Lu than would be created in a "cool" environment. In this range,

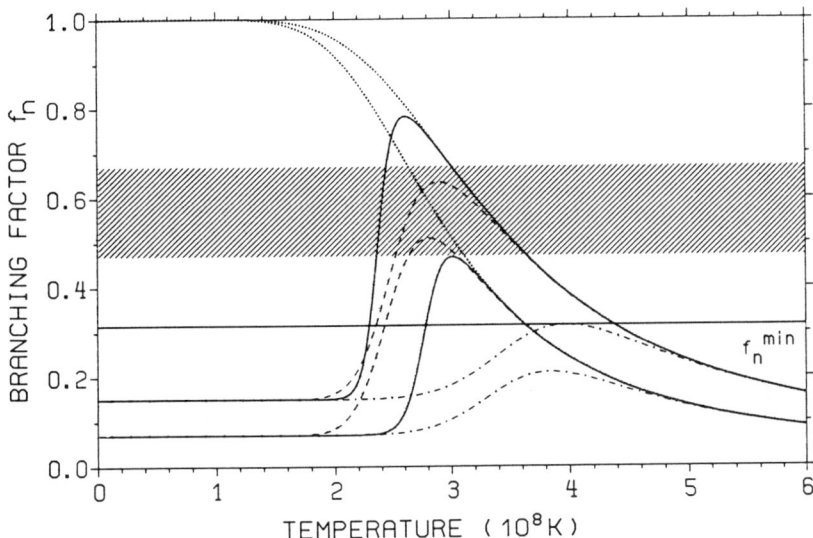

Fig. 4 The branching factor f_n as a function of temperature. The transition between the low and high temperature regime occurs at 200 to 300 million degrees (see text).

internal transitions, beta decays, and neutron captures are equally important.
(iii) For temperatures higher than 300 million degrees, the relative populations of ground state and isomer are in thermal equilibrium, since now the internal transition rates are much larger than those for beta decay or neutron capture.

The influence of K-forbidden transitions is illustrated in Fig. 4 by dashed-dotted lines (if Weisskopf estimates are used for the respective transition rates) and by dashed lines (if ~600 times faster transition rates are estimated via the E2/M1 mixing ratios in ^{176}Lu). This effect appears to be weaker, or, at most, comparable to that of the allowed transitions mediated by the 838.6 keV state. The dotted lines indicate the shape of f_n for the – unrealistic – case of complete thermal equilibrium.

The result for $f_n(T)$ is compared in Fig. 4 with observational constraints, with a clear lower limit f_n^{min} derived from Eq. (1) by using the ^{176}Lu/^{176}Hf abundance ratio at the formation of the solar system, and with a realistic branching factor $f_n = 0.57 \pm 0.10$ determined from s-process systematics (hatched band). One finds, that thermal effects are, indeed, *required* to meet the observations, and that the overlap between $f_n(T)$ with the hatched band yields a possible range for the s-process temperature: $2.4 \cdot 10^8 < T_s < 3.6 \cdot 10^8$ K.

THE TEMPERATURE DEPENDENCE OF THE BETA DECAY RATE

After termination of the neutron exposure, there is no more production of ^{176}Lu, and destruction occurs only by beta decay. Therefore, the populations of ground state and isomer exhibit a continuous decrease that can be described again by Eqs. (4) with beta decay as the only external process. The resulting expression for $\lambda_0(T)$ is plotted in Fig. 5: At low temperatures, λ_0 equals the

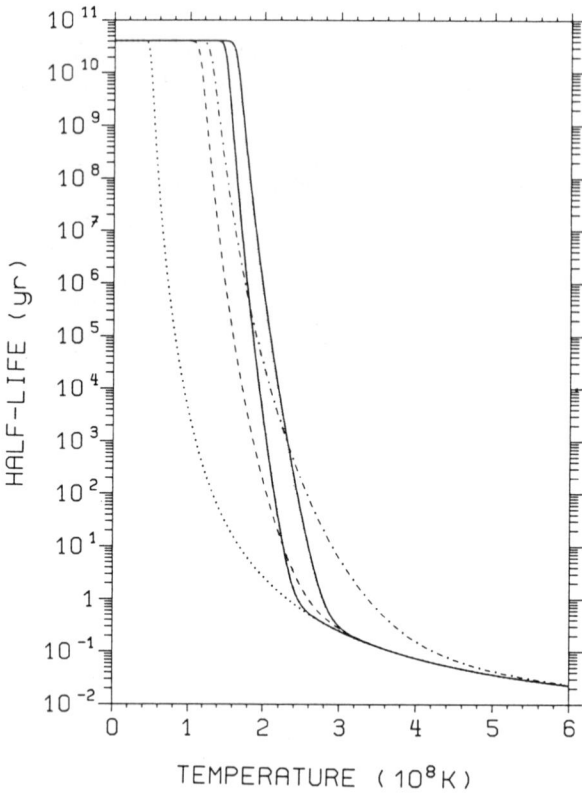

Fig. 5 The half-life of ^{176}Lu as a function of temperature.

beta decay rate of the ground state, and at very high temperatures, it reduces to the solution for thermal equilibrium (dotted line).

For the astrophysical discussion it is now very important that thermal equilibrium is reached only at a certain temperature. The way, how equilibrium is approached, depends again on the properties of the mediating state. The shaded band in Fig. 5 is obtained with the experimental limits for the life time of the 838.6 keV level, whereas the dashed-dotted and the dashed lines indicate the solutions for K-forbidden transitions as discussed in detail for f_n.

Note the extremely sharp decrease of $t_{1/2}$ over 10 orders of magnitude! Presumably, this pronounced temperature dependence makes ^{176}Lu a sensitive probe for details of the stellar site of the s-process.

CONCLUSIONS

The thermal enhancement of the beta decay rate of ^{176}Lu was found to be due to the coupling of ground state and isomer by allowed transitions via a $I^\pi = 5^-$, $K = 4$ state at 838.6 keV; the effect of K-forbidden transitions was estimated to be less important or, at most, comparable.

As a consequence of the thermal coupling between ground state and isomer, the s-process production of ^{176}Lu becomes a sensitive function of temperature. Competition between beta decays and neutron captures results in an s-process branching that can, for the first time, be described quantitatively. It is found that the abundance of ^{176}Lu is determined by the neutron densities and temperatures at the s-process site, and remains, therefore, too uncertain for using ^{176}Lu as a chronometer for the age of the s-process. Instead, it can rather be considered as a suitable s-process thermometer, yielding a temperature range between $2.4 \cdot 10^8$ and $3.6 \cdot 10^8$ K, in good agreement with similar estimates from other s-process branchings and with current stellar models[31].

REFERENCES

1. J. Audouze, W. A. Fowler, and D. N. Schramm, Nature *238*, 8 (1972).
2. M. Arnould, Astron. Astrophys. *22*, 311 (1973).
3. F. Käppeler, H. Beer, and K. Wisshak, Rep. Prog. Phys. *52*, 945 (1989).
4. B. J. Allen, G. C. Lowenthal, and J. R. de Laeter, J. Phys. *G7*, 1271 (1981).
5. B. J. Allen, G. C. Lowenthal, J. W. Boldeman, and J. R. de Laeter, in *Neutron Capture Gamma-Ray Spectroscopy and Related Topics*, eds. T. von Egidy, F. Gönnenwein, and B. Maier (Institute of Physics, Bristol, 1982) p.573.
6. F. Stecher-Rasmussen, K. Abrahams, J. Kopecky, J. Lindner, and P. Polak, P., in *Capture Gamma-Ray Spectroscopy 1987*, eds K. Abrahams and P. Van Assche (Institute of Physics, Bristol, 1988) p.754.
7. W. R. Zhao and F. Käppeler, in *Astrophysical Ages and Dating Methods*, eds. J. Audouze, M. Casse', and E. Vangioni-Flam (Editions Frontières, Gif sur Yvette, 1990) p.357.
8. H. Beer and F. Käppeler, Phys. Rev. *C21*, 534 (1980).
9. E. Anders and N. Grevesse, Geochim. Cosmochim. Acta *53*, 197 (1989).
10. H. Beer, G. Walter, R. L. Macklin, P. J. Patchett, Phys. Rev. *C30*, 464 (1984).
11. R. A. Ward, 1980 (private communication).
12. H. Beer, F. Käppeler, K. Wisshak, and R. A. Ward, Ap. J. Suppl. *46*, 295 (1981).
13. A. Veres and I. Pavlicsek, Acta Phys. Acad. Sci. Hung. *28*, 419 (1970).
14. E. B. Norman and S. Kellogg, Ap. J. *291*, 834 (1985).
15. L. Lakosi, I. Pavlicsek, and A. Veres, in *Capture Gamma-Ray Spectroscopy 1987*, eds. K. Abrahams, P. Van Assche (Institute of Physics, Bristol, 1988) p.S745.
16. D. G. Gardner, M. A. Gardner, and R. W. Hoff, in *Capture Gamma-Ray Spectroscopy 1987*, eds. K. Abrahams and P. Van Assche (Institute of Physics, Bristol, 1988) p.S315; J. Phys. G: Nucl. Phys., *14 Suppl.*, S315.
17. H. R. Koch, H. G. Börner, J. A. Pinston, W. F. Davidson, J. Faudou, R. Roussille, and O. W. B. Schult, Nucl. Instr. and Meth. *175*, 401 (1980).
18. H. G. Börner, J. Jolie, F. Hoyler, S. Robinson, M. S. Dewey, G. L. Greene, E. Kessler, and R. D. Deslattes, Phys. Lett. *B215*, 45 (1988).
19. W. Mampe, K. Schreckenbach, P. Jeuch, B. P. K. Maier, F. Braumandl, J. Larysz, and T. von Egidy, Nucl. Instr. and Meth. *154*, 129 (1978).
20. M. Löffler, H. J. Scheerer, and H. K. Vonach, Nucl. Instr. Meth. *111*, 1 (1973).
21. N. Klay et al., submitted to Phys. Rev. (1990).
22. M. M. Minor, R. K. Sheline, E. B. Shera, and E. T. Jurney, Phys. Rev. *187*, 1516 (1969).
23. G. L. Struble and R. K. Sheline, Sov. J. Nucl. Phys. *5*, 862 (1967).
24. M. K. Balodis, J. J. Tambergs, K. J. Alksnis, P. T. Prokofjev, W. G. Vonach, H. K. Vonach, H. R. Koch, U. Gruber, B. P. K. Maier, and O. W. B. Schult, Nucl. Phys. *A194*, 305 (1972).
25. R. A. Dewberry, R. K. Sheline, R. G. Lanier, L. G. Mann, and G. L. Struble, Phys. Rev. *C24*, 1628 (1981).
26. M. Délèze, A. Bruder, S. Drissi, J. Kern, and G. L. Struble, Phys. Rev. *C36*, 1826 (1987).
27. R. W. Hoff, R. F. Casten, M. Bergoffen, and D. D. Warner, Nucl. Phys. *A437*, 285 (1985).
28. K. T. Lesko, E. B. Norman, R.-M. Larimer, and B. Sur, this conference, and submitted to Phys. Rev. (1990).
29. R. A. Ward and W. A. Fowler, Ap. J., *238*, 266 (1980).
30. N. Klay, F. Käppeler, H. Beer, G. Schatz, submitted to Phys. Rev. (1990).
31. F. Käppeler, R. Gallino, M. Busso, G. Picchio, and C. M. Raiteri, Ap. J., *354*, 630 (1990).

Changes in Nuclear Decay Rates as a Result of High Stellar Temperature: ^{148}Pm and ^{176}Lu.

Kevin T. Lesko
Nuclear Science Division Lawrence Berkeley Laboratory
1 Cyclotron Road Berkeley, CA 94720

Nucleosynthesis of the heavier elements (A>60) is thought to proceed principally through two processes: slow neutron capture (s process) and rapid neutron capture (r process). The time scales of the successive neutron captures distinguishes these two processes. In the s process the typical time between captures is long enough that beta-unstable nuclei will almost always have time to decay before the next capture occurs. As a result, the s process follows the line of beta stability. In contrast, during the r process the neutron flux is so high that many neutron captures can occur before the subsequent beta decays occur. This is illustrated in the Figs. 1a,b, where we have highlighted the regions surrounding ^{148}Pm and ^{176}Lu.

s-Process Neutron Density

Branch-point nuclei have relatively long beta decay half-lives, and, consequently, both neutron capture and beta decay compete favorably. The branching ratio (BR) between these two paths can be expressed in terms of the effective decay rates λ_β and λ_n:

$$BR = \frac{\lambda_\beta}{\lambda_\beta + \lambda_n}, \quad (1)$$

where

$$\lambda_\beta = \frac{\ln(2)}{t_\beta} \quad (2)$$

and

$$\lambda_n = N_n <\sigma(n,\gamma)v>. \quad (3)$$

t_β is the beta-decay half-life, N_n is the neutron number density, and the last term on the right hand side of Eqn. 3 is the thermal average of neutron-capture cross section times the neutron velocity.

If equilibrium is reached during the s process we would expect $\sigma_{n\gamma}N(Z,A-1) = \sigma_{n\gamma}N(Z,A)$, where $\sigma_{n\gamma}$ is the neutron capture cross section and $N(Z,A)$ is the isotopic abundance of a particular species. In fact, away from the neutron magic numbers this is essentially what is observed. Coupling this equation with measured neutron-capture cross sections and meteoritic abundances one can deduce the s process neutron density. Beer et al.[1] have done this for several "branch point" nuclei and deduce neutron densities ranging between 0.8-3.5 x 10^8/cm^3. Evidence for the existence of s-process branching at A=147, 148 has been presented by

Fig. 1a. A section of the chart of the nuclides, highlighting the vicinity of ^{148}Pm. The possible branch points at ^{147}Pm and ^{148}Pm are shown. The path of the s-process is indicated by the solid lines. Stable nuclei are indicated by shaded triangles. The presence of stable isobars on either side of each of these nuclei shield them from both r- and p- process contributions.

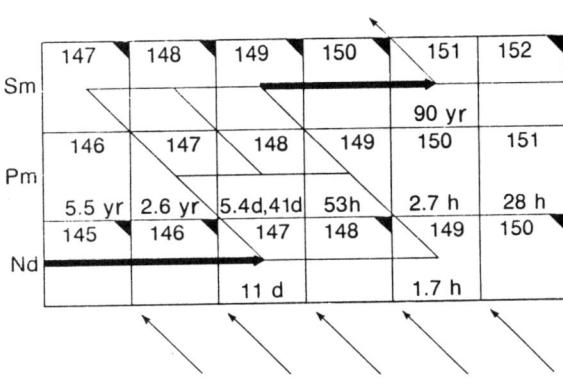

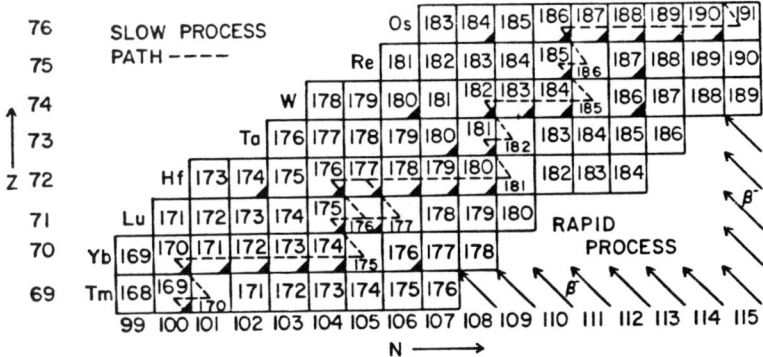

Fig. 1b. The s-process path in the vicinity of ^{176}Lu. The stability of ^{176}Yb and ^{176}Hf guarantees that ^{176}Lu can only be produced in the s process.

Winters et al.[2]. From their measurements of the neutron capture cross sections and reported isotopic abundances for ^{148}Sm and ^{150}Sm, they found $\sigma_{n\gamma}N(^{148}Sm)/\sigma_{n\gamma}N(^{150}Sm) = 0.91 \pm 0.03$. In Fig. 1a we indicated the possible branch points at ^{147}Pm and ^{148}Pm. The calculation of the branching at ^{148}Pm is complicated by the existence of a long lived isomer state at 137 keV ($J^\pi = 6^-$, $t_{1/2} = 41.3$ d).

^{148}Pm is synthesized in the s-process via the ^{147}Pm(n,γ) reaction. Because of the $J^\pi = 7/2^+$ nature of the target and the dominance of s-wave neutron capture at the energies appropriate for the s-process, this reaction will populate $J^\pi = 3^+$ and 4^+ levels in ^{148}Pm at energies just above the neutron binding energy. It has been estimated that the subsequent γ-cascades will produce roughly equal amounts of ^{148}Pm^g and ^{148}Pm^m. Furthermore, it has been calculated that the neutron capture cross section is larger for the isomer than for the ground state, (2500 mb v.s. 1500 mb). However, at s-process temperatures there exist several processes which could serve to equilibrate the ground and isomeric states. In particular if there exists a nuclear level which decays to both the ground state and the isomer, as is illustrated in Fig. 2, then photoexcitation, inelastic scattering, Coulomb excitation and positron-annihilation excitation could all serve as means to equilibrate the ground state and isomer. In thermal equilibrium one can calculate the relative populations of the isomer and the ground state and find that at a temperature of 30 keV only 4% of the nuclei would exist in the isomeric state.

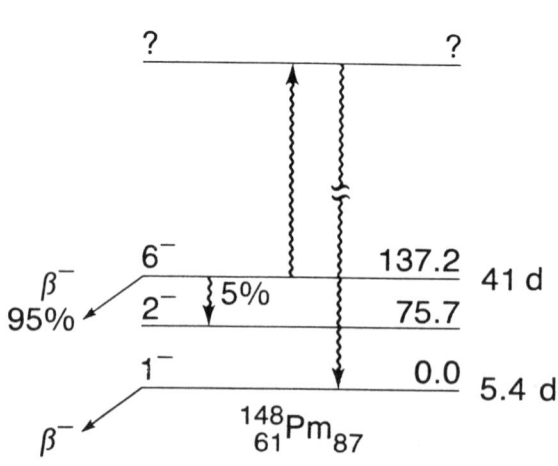

Fig. 2. The previously known level scheme of ^{148}Pm is shown along with a proposed level which would permit the equilibration of the ground state and the isomeric state.

One can deduce the neutron density in the s-process if one knows the ratio of the population of the isomer and ground state. Assuming that the ground state and isomer are in thermal equilibrium results in a neutron density for the s-process of $N_n = 3.1 \times 10^8/\text{cm}^3$. If one assumes that the isomer and ground state are populated equally, then the deduced density is: $N_n = 1.0 \times 10^8/\text{cm}^3$ Ref (3). This ambiguity can be resolved by determining the detailed nuclear structure of ^{148}Pm.

s Process Chronometer

As shown in Fig. 1b), ^{176}Lu can be produced only via the slow neutron capture process. Due to the the long half-life of the ground state, ^{176}Lu^g, it was suggested that ^{176}Lu would be a candidate for a s-process chronometer[4,5]. However, there exists a much shorter lived isomer at 122.9 keV ($t_{1/2} = 3.7$ hr). As Fig. 3 shows, the large spin difference between these two levels prevents decays from the isomer to the ground state;

rather the isomer β decays to ^{176}Hf. The presence of this isomer could affect the decay of ^{176}Lu in astrophysical environments, providing a method of communication exists between the two levels. An example of

Fig. 3. A partial level scheme of ^{176}Lu, showing the positions and decays of the ground state and isomer at 122.9 keV. The equilibration of the these two levels could be achieved by way of a level of intermediate spin, as illustrated in the figure.

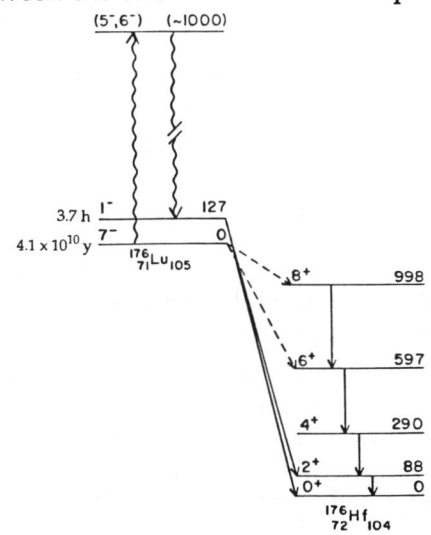

this communication is illustrated in Fig. 3 where an additional level of intermediate spin is populated and subsequently decays to both the ground state and the isomer. The time scale for obtaining equilibration between the isomer and g.s. is determined by the rate of excitation of the mediating level, its spin, parity and excitation energy, and its decay properties, as well as the half-lives of the g.s. and isomer.

The presence of such an equilibration path would severely compromise the usefulness of ^{176}Lu as an s-process chronometer due to the effective decay constant, λ_{eff}, being temperature sensitive. The presence of such a mediating level lying between 662 keV and 1332 keV excitation energy can be inferred from the photoexcitation work of Norman et al.[6]. Our aim in this experiment was to determine the level scheme of ^{176}Lu up to ~ 1 MeV to search for levels which could serve as a mediating level between the ground state and isomer.

Experiments and Analysis

The question of whether or not the isomer and ground states are in equilibrium for these two systems greatly affect the conclusions deduced from these systems. We have addressed this question using the same technique, by determining the level schemes using coincident γ ray data. We collected γ-ray coincidences using HERA, an array of 21 Compton-suppressed germanium detectors.

To form states in ^{148}Pm, we used the ^{148}Nd(p,n)^{148}Pm reaction. Possible equilibrating levels would have spin values in between those of the isomer and the ground state. The (p,n) reaction is well suited to the formation of such relatively low-spin states. An 8.0 MeV proton beam, which was chosen to maximize the (p,n) yield without introducing significant yields from other reactions, was provided by the LBL 88-Inch Cyclotron.

A coincidence event was defined as any two (or more) of the germanium detectors reporting an unvetoed event within a hardwire timing requirement of ± 50 ns. In addition we collected more detailed timing information (TAC) which consisted of a start signal generated by one specific detector and a stop signal from any of the other 19 detectors.

These data were then sorted into a two-dimensional array for off-line gating and analysis. Gates were placed on ~150 of the strongest transitions in ^{148}Pm and coincidence relationships were established from the background subtracted spectra. From the coincidence relationships tentative levels schemes were created. The levels in these schemes were then checked for self-consistency with parallel and sequential decays and for γ-ray decay intensities. These levels agree with previously reported particle transfer data. The lower energy part of our proposed level scheme[7] is presented in Fig. 4. We were unable to establish many transitions which fed the high spin isomer. Rather, we were able to establish many levels which feed the low spin ground state and several levels which populated both the isomer and ground state and are presumably of intermediate spin. No attempt was made to produce quantitative branching ratios for all these decays, nor was any attempt made to determine the spins and parities of the transitions from the intrinsic angular distribution data.

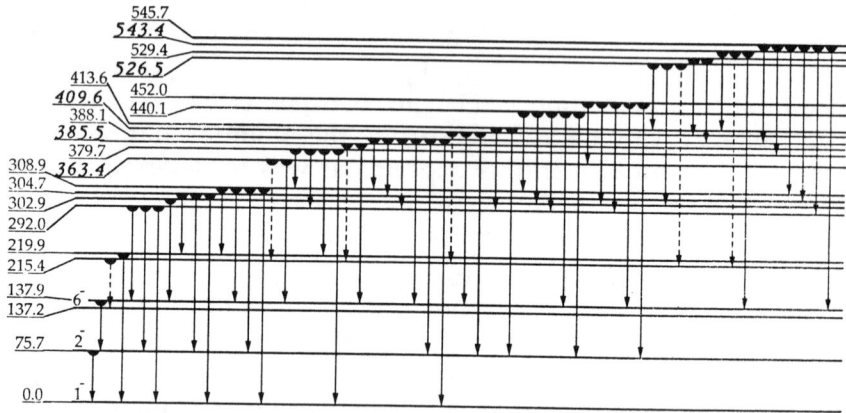

Fig. 4. The level scheme of ^{148}Pm from 0 to 561.2 keV of excitation energy deduced from this work. Those levels which decay to both the ground state and to the isomer are labelled in italic type. Transitions which directly or indirectly feed the isomer are shown with dashed lines.

We used the ^{176}Yb(p,n)^{176}Lu reaction to populate levels in ^{176}Lu. Again an 8-MeV proton beam was provided by the LBL 88-Inch Cyclotron to maximize the yield of ^{176}Lu. The target was a 2 mg/cm^2 metallic foil enriched to 97.04% ^{176}Yb. In this experiment the TAC signals were generated by designating a subset of ten of the detectors as start detectors and the stop signal was generated by a coincident event in any of the other ten detectors.

Gates were placed on ~400 of the strongest transitions in ^{176}Lu and coincidence relationships were established in the background subtracted gated spectra. Using these data, the procedures and tests listed above, the previously established level scheme, and ^{176}Lu levels established with particle transfer experiments, we constructed the level scheme[8]. We have emphasized the transitions which feed and decay from the 838.5 keV level in Fig. 5.

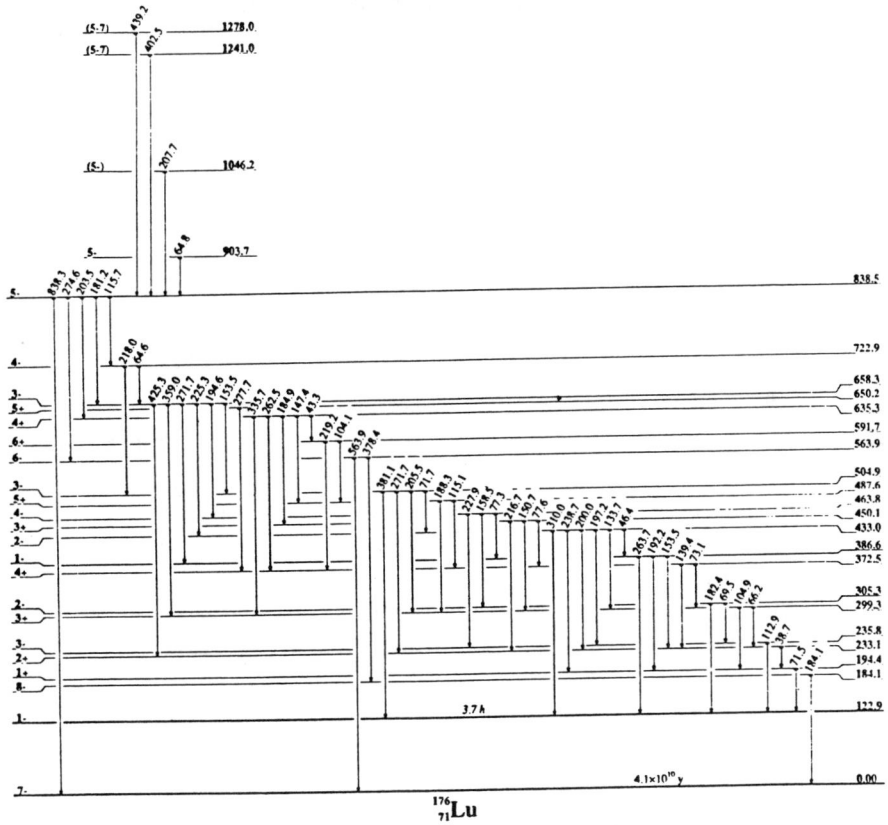

Fig. 5. A subset of the proposed level scheme of ^{176}Lu emphasizing the transitions originating from and populating the level at 838.5 keV.

There are at least six levels in ^{148}Pm which decay to both the ground state and the isomer with substantial rate: those at 363.4, 385.5, 409.6, 526.5,

543.4, and 611.2 (not shown) keV. In order to determine the spontaneous decay lifetimes of these levels, τ_{sp}, we generated TAC spectra which were gated on all of the feeding and decay transitions, separately, and confirmed that all the involved transitions had the same time relationships. From the TAC spectra, we established an upper limit on τ_{sp} of ≤10 ns, which is consistent with the observed resolution of the TAC spectra.

As can be seen in Fig. 5 the level at 838.5 keV in ^{176}Lu decays with significant decay strengths to both the ground state and to the isomer. From the transitions shown in the figure we can infer the (J,π) of this level as being their 5- or 6-. This assignment agrees with the assignments of 5- suggested in the literature. To corroborate the decays into and from the level at 838.5 keV, we generated the TAC spectra for all combinations of start detectors feeding and stop detectors decaying from this level. These TAC spectra showed the same time relationships between all combinations of the feeding and exiting γ rays which increases our confidence in their placement. From the TAC spectra, we established an upper limit on τ_{sp} of ≤10 ns.

Discussion

The relation between the photoexcitation rate of a level I to a level I* and the spontaneous decay rate of the level I* is:

$$\frac{\tau(I \Rightarrow I^*)}{\tau_{sp}(I^* \Rightarrow I)} = \frac{2J_I + 1}{2J_{I^*} + 1} \left\{ \exp(\frac{\Delta E}{kT}) - 1 \right\} \quad (4)$$

where τ_{sp} is the spontaneous decay rate, J_I and J_{I^*} are the spins of the states I and I* and ΔE is the energy difference between these two states. If we assume that the spin of the equilibration state is intermediate between the isomer and the ground state, (eg. J_{I^*} = 3 for ^{148}Pm) and that τ_{sp} = 10 ns, then we can calculate the relationship between the photoexcitation time scale, $\tau(I \Rightarrow I^*)$, and the temperature needed to obtain equilibration. This relationship is presented in Fig. 6. The two curves are for two separate levels which can serve as equilibration paths. In order for equilibration between ^{148}Pm^{g,m} to occur, this photoexcitation timescale must be substantially less the lifetime of the ground state. We can then deduce that the g. s. and isomer are in thermal equilibrium for temperatures ≥ 0.9 × 10^8 K. This temperature is somewhat below estimates of s-process temperatures. The effective beta-decay rate of a nucleus, λ_{eff}, is then:

$$\lambda_{eff} = \frac{\sum_i g_i \lambda_i \exp(\frac{-E_i}{kT})}{\sum_i g_i \exp(\frac{-E_i}{kT})} \quad (5)$$

where $g_i = (2J_i+1)$, λ_i is the beta decay rate of state i, E_i is the excitation energy of state i, and the summation is extended over all states that are in thermal equilibrium. Based upon the β-decay rates observed in neighboring nuclei, none of the low-lying levels in ^{148}Pm nor in ^{176}Lu are expected to have a drastically larger β-decay rate than that of the ground state. Thus, in our calculations, we included only the effects of the isomer and ground state. The results of our calculations for the temperature range of $T_8 = 0$ to 5 are listed in Table I.' Once thermal equilibration between ^{148}Pm^{g,m} is achieved, the effective half-life of the nucleus slowly increases with increasing temperature as the fractional population of isomer grows.

TABLE I
Effective half-life of ^{148}Pm and ^{176}Lu versus temperature.

Temperature (10^8 K)	$t_{1/2}(^{148}$Pm) (days)	$t_{1/2}(^{176}$Lu) (years)
0	23.3 a	4.08×10^{10}
1	5.4	3.29×10^3
2	5.4	2.63
3	5.5	0.245
3.5	5.6	0.125
4	5.7	0.0750
5	6.2	0.0370

a) Assuming $N_m=N_g$

Based upon the results of our in-beam γ ray study, we conclude that ^{148}Pm^{g,m} are guaranteed to be in thermal equilibrium during the s-process. Therefore, from the measured $\sigma_{n\gamma}$ cross sections and measured isotopic abundances, the s-process neutron density deduced from the ^{148}Pm branch point is $N_n = 3 \times 10^8/\text{cm}^3$. This is consistent with neutron densities inferred from other s-process branch points.

For ^{176}Lu the level at 838.9 keV serves as the mediating level and its inferred spin we can calculate the photoexcitation rate as a function of temperature using the same expression as above. From our TAC data we could place an upper limit on the spontaneous decay lifetimes of the potentially mediating levels of $\tau_{sp} \leq 10$ ns, which is consistent with the observed resolution. The single particle Weisskopf estimate for the rates of these decays are substantially faster than this limit. Using the theoretical estimates for τ_{sp} we show three curves in Fig. 7 corresponding to the population of the 838.5, 722.9, and 563.9 keV levels from the ground state. We see in Fig. 7, that for temperatures greater than 3×10^8 K, the isomer and ground state will be in thermal equilibrium using the 838.5

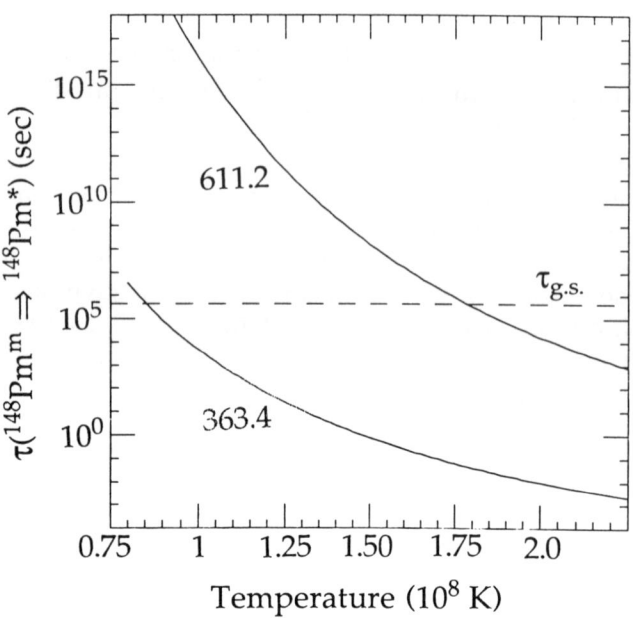

Fig. 6. The relationship between the temperature and the photoexcitation time scale deduced from Eqn 4. The 363.4 keV and the 611.2 keV levels in ^{148}Pm are the lowest and highest energy levels we found that can serve as equilibration paths. Choosing the photoexcitation time to be equal to the g. s. lifetime (horizontal dashed line), we deduce that ^{148}Pm^g and ^{148}Pm^m will be in equilibrium for T $> 0.9 \times 10^8$ K.

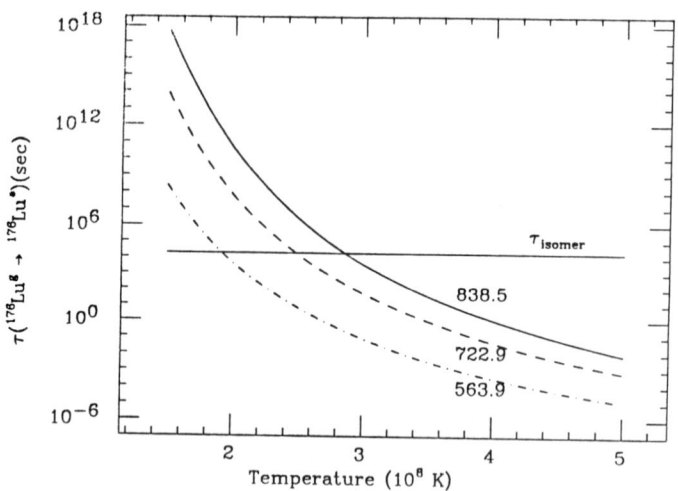

Fig. 7. The population time of the mediating level at 838.5 keV in ^{176}Lu as a function of temperature for the ground state (solid curve). Assuming that the populating time is short in comparison to the meanlife (horizontal solid line) of the isomer establishes the temperatures where

keV level as the mediating level. From Fig. 7, we might also expect that the band head at 722.9 keV would also serve as a mediating level. Calculating the single particle transition strength for this level we find that the direct decay from the 722.9 to the ground state would be only ~3% of the 838.5 decay strength. Consequently, it is possible that we would not directly observe this decay with our coincident gamma-ray technique. However, even a 1% branch to the ground state would be adequate to equilibrate the ground state and isomer via this level. Evaluating the Eqn. 4 assuming the moderating level is at 722.9 keV rather than 838.5 keV yields an estimate of the equilibration temperature of 2.6×10^8 K. A more careful examination of the level scheme yields several other levels which could act as mediating levels, the lowest one being at 563.9 keV. This level results in equilibration being reached at 2×10^8 K. In addition to photo-excitation, the processes of Coulomb excitation, inelastic neutron scattering, positron annihilation excitation also will contribute to the equilibration of the two levels and will reduce the temperature where the two levels achieve equilibrium.

The resulting effective half-life of ^{176}Lu is an extremely sensitive function of the temperature. The effective beta-decay rate, λ_{eff}, for the nucleus is Table I. This vividly illustrates how a relatively small change in s-process temperature can result in a major change in the decay constant for ^{176}Lu. This strong temperature sensitivity effectively rules out ^{176}Lu for use as an cosmochronometer and complicates efforts to use 176Lu as an s-process thermometer.

This work supported by the Director, Office of Energy Research, Office of High Energy and Nuclear Physics, Division of Nuclear Physics, of the U.S. Department of Energy under Contract No. DE-AC03-76SF00098 and by the National Science Foundation's Science and Technology Center Research Program under Cooperative Agreement No. AST-8809616.

References:
1) H. Beer, G.Walter, R.L.Macklin, and P.J.Patchett, Phys. Rev. C30, 464 (1984)
2) R.R. Winters, F. Käppeler, K. Wisshak, A. Mengoni, and G. Reffo. Astrophys. J.300, 41, (1986).
3) F. Käppeler, Proc. 5th Internat. Sym. on Capture γ-Ray Spectro. Related Topics, Knoxville, TN, ed. S. Raman (AIP, New York, 1985), p. 715.
4) J. Audouze, W.A. Fowler, and D.N. Schramm, Nature 238, 8, (1972).
5) M. Arnould Astron. Astrophys. 22, 311 (1973).
6) E.B. Norman, T. Bertram, S.E. Kellogg, S. Gil and P. Wong, Astrophys.. J. 291, 834 (1985).
7) K.T. Lesko, E.B. Norman, R.-M. Larimer, J.C. Bacelar, and E.M. Beck, Phys Rev C39, 619 (1989)
8) K.T. Lesko, E.B. Norman, R.-M. Larimer, B. Sur, and C.B Beausang, submitted to Phys. Rev. C.

SECOND GENERATION NEW THEORETICAL DATA OF BETA DECAY FAR FROM STABILITY AND OF DOUBLE BETA DECAY AND IMPLICATIONS FOR ASTROPHYSICS

H.V. Klapdor-Kleingrothaus
Max-Planck-Institut für Kernphysik, Heidelberg, Germany

Abstract

Beta decay data of nuclei far from stability are one of the most important nuclear physics inputs for the understanding of the element synthesis in the universe and determination of the age of the universe from cosmochronometers and by the latter have implications for cosmology. Results of new second-generation microscopic calculations of β^- half-lives, will be presented. The impact on the above astrophysical questions will be discussed. New results for $\beta\beta$-matrix elements for all potential $\beta\beta$-emitters with $A > 70$ are given. It results that neutrinoless double beta decay remains the most powerful tool for investigating the electron neutrino mass. Perspectives of a second-generation $\beta\beta$-experiment using enriched ^{76}Ge (Heidelberg-Moscow cooperation) are briefly presented.

1 BETA DECAY FAR FROM STABILITY AND CONSEQUENCES FOR ASTROPHYSICS AND COSMOLOGY

1.1 General

The recognition of the close connection between the laws of microphysics, astrophysics and cosmology is one of the most important discoveries of this century (see, e.g. [1,2]). The weak interaction in nuclei and more specifically the distribution of nuclear beta strength is of considerable importance. Its understanding is not only decisive for reliable calculation of nuclear double beta decay matrix elements, which are prerequisite for deduction of the neutrino mass from $\beta\beta$-decay experiments (see later sections). To give another example, the nuclear beta strength distribution determines also the efficiency of solar neutrino detectors, such as the gallium detector [3]. Turning to macrophysics beta decay of nuclei far from stability is the most important nuclear physics input in calculations of element synthesis in the universe and for determining the age of the universe from the so-called cosmochronometers (see Fig. 1) [4]. The age of the universe is, on the other hand, closely connected by cosmological models with the cosmological constant Λ of general relativity and the equivalent vacuum energy density of the universe, which again is intimately connected with the GUTs [4,5].

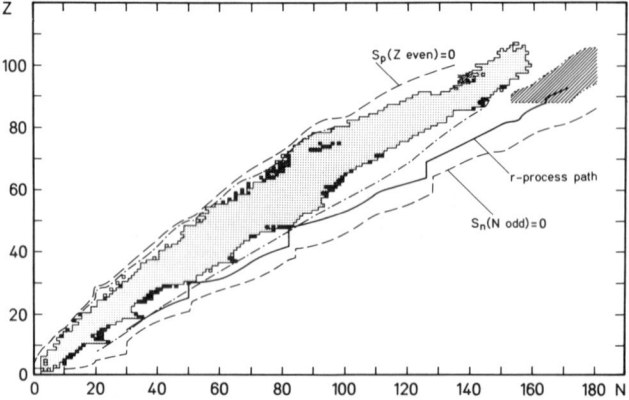

Fig. 1. The range of stable and unstable nuclei realized or realizable in nature (in a Z versus N plane). For explanation see text.

The present situation of our knowledge of beta decay properties far from stability and of the needs for astrophysical applications are summarized in Fig. 1. Shown is in a Z vs. N plane the range of stable and unstable nuclides realized, or realizable in nature. Border lines of the latter are given by the neutron and proton drip lines (denoted by $S_n = 0$ and $S_p = 0$, respectively). The hatched "peninsula" corresponds to nuclei whose ß-decay properties are known experimentally. The dashed "island" right to the upper end of the peninsula denotes the range of nuclides for which ß-delayed fission is expected. The black points denote nuclides which have been discovered after completion of the ß half-life calculations of Klapdor, Metzinger and Oda (KMO) published in 1984 [6]. The dash-dotted line denoted the border line which may be optimistically expected to be reached experimentally in future with the new SIS accelerator in Darmstadt. The nuclides of interest for the r process are those between the line of ß stability and the line denoted by "r-process path." This latter line indicates the element distribution built up for short time during supernova explosions. By the successive ß decay of this distribution the r process is thought to build up the observed cosmic r-element distribution.

ß-decay half-lives, ß-delayed neutron emission rates and for very heavy nuclides beta-delayed fission, therefore, have decisive influence on the r-process. They determine its time scale and yield severe constraints on possible astrophysical sites for this process (see, e.g., [4, 7, 8]).

Fig. 1 shows that experimental progress in investigating new neutron-rich isotopes is rather slow on the scale of astrophysical needs. Therefore, although in two points, with ^{80}Zn and ^{130}Cd, probable r-process waiting point nuclei could be reached [9-11], most nuclides participating in the r process for the foreseeable future are not accessible in terrestrial laboratories. Consequently, for studies of cosmic element synthesis theoretical predictions are required and remain indispensable. The improved understanding of nuclear beta decay reached in recent years [4, 12] which had allowed for the set of predictions published in the 1984 At. Data Nucl. Data Tables [6] and which allows for an improved second-generation set discussed in section 1.2, is, therefore, of great importance, also for other fields of physics.

There exist at present three calculations for all neutron-rich nuclides far from stability out to the neutron drip line

(1) the gross-theory calculations by Takahashi, Yamada and Kondoh [13], recently modified [14]
(2) microscopic calculations using a Tamm-Dancoff approximation by Klapdor, Metzinger and Oda [6] (KMO)
(3) second-generation microscopic calculations using the proton-neutron quasiparticle RPA by Staudt, Bender, Muto and Klapdor [15] (SBMK).

A discussion also of the first two approaches is given in [4] and [16]. In this paper we shall concentrate on the second-generation calculations.

1.2 Second-generation microscopic calculations
1.2.1. Half-Lives
This approach [15] uses the proton-neutron quasiparticle RPA with a GT residual interaction. The QRPA takes in addition to a BCS treatment (which has been realized in some simplified way in [6]) spin-isospin g.s. correlations into account. For our application an extension of the QRPA formalism discussed in [17] had to be made for GT decay of odd-odd nuclei, for which formulae did not exist in the literature. This has been published recently [18].

In nuclei with an odd number of protons and/or neutrons there occur two types of transitions: one is a phonon transition, where the odd quasiparticle(s) act(s) as a spectator. The other is a transition of the odd particle(s), which is (are) described as a phonon-correlated one-quasiparticle state or two-quasiparticle state. In that case we treat phonon correlations in first order perturbation [18].

In deriving the QRPA formulae for ß⁻-decay we can neglect the nonseparable particle-particle force, which plays, however, a decisive role in calculations of 2ν double beta decay (see below). We further use instead of a realistic NN interaction used in ßß-decay (G matrix of the Paris or some other potential (see [19-22]), a schematic GT interaction as residual interaction in the RPA. These assumptions simplify the QRPA matrix equation to an algebraic equation, because the particle-hole interaction has separable matrix elements.

Some results of the new calculations are shown in Figs. 2,3. The mean deviation of the ratio $t_{1/2}^{cal}/t_{1/2}^{exp}$ from 1 is smaller than in the other existing calculations. The description of light nuclei (Fig. 3) is seen to be able to compete with that given be the large-scale shell-model calculations by Wildenthal et al. [23].

The dependence of the calculated half-lives on the parameters involved: GT interaction strength, deformation of the Nilsson potential, strength of the pairing force and ß-decay Q values (mass formulae) has been investigated in [15], where also the results for about 6000 nuclides are given.

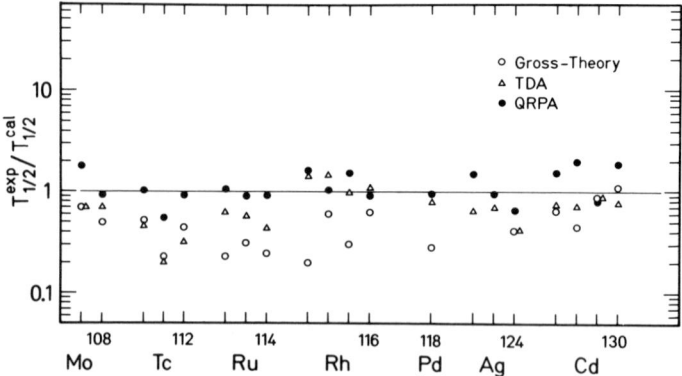

Fig. 2. Comparison of experimental half-lives with the predictions of the revised gross theory [14], the TDA model [6] and the QRPA [15].

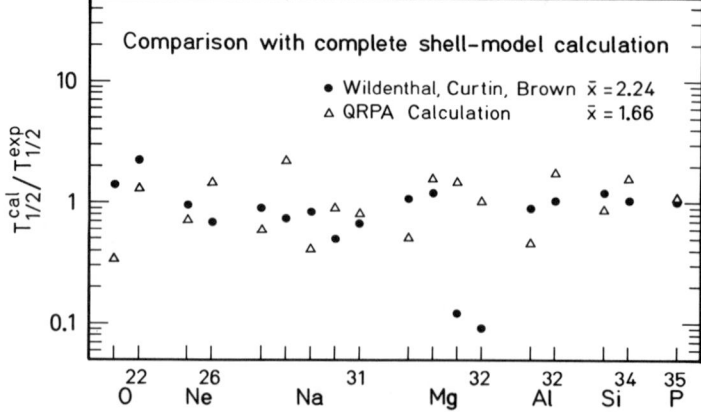

Fig. 3. Comparison of experimental ß⁻-decay half-lives of s-d shell nuclei with the second-generation microscopic approach and the calculations of Wildenthal et al. [23] (from [15]).

Concluding, the second-generation calculations yield a set of ß⁻-decay half-life predictions improved over those from other approaches. This is in particular the case for light nuclei.
Fig. 4 shows that this approach also seems to work remarkably well for ß⁺ decay. In general on the ß⁺ side the particle-particle force has to be included. Calculations of this type for all ß⁺-emitters of the nuclidic charge (~ 2000 nuclides) have been performed recently [25].

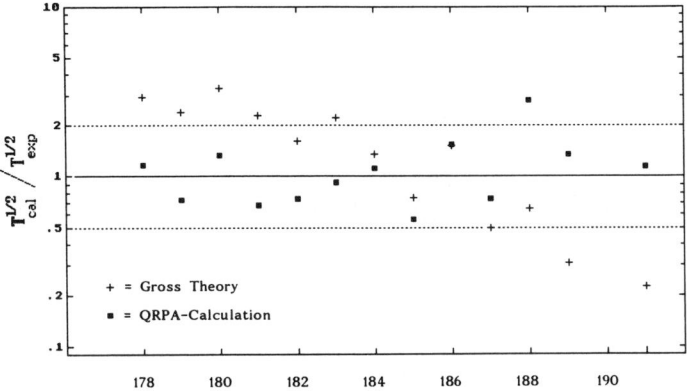

Fig. 4. Comparison of measured ß⁺-decay half-lives with calculations by the second-generation microscopic approach (from [25]).

1.2.2. ß-delayed fission
Berlovich and Novikov were the first to point out the possible existence of a large area of neutron-rich nuclides, where fission may follow the ß-decay to an excited state [26]. In the following years this island of ß⁻-delayed fission (ßdf) was confirmed by subsequent studies [8,27,28]. We performed earlier [8] a systematic calculation of ßdf (and ßdn) using matrix elements calculated in a TDA approach and discussed in detail the consequences for the r-process and age of the galaxy. However, Hoff [29] pointed out somewhat later that the values obtained by [8] seem to be too large compared to thermonuclear explosion yields. We therefore have performed a new calculation of ßdf [30] in the region of n-rich nuclei with $75 < Z < 100$ and $140 < N < 184$ using the strength functions of [15], and applying essentially the same model as used by [8].
The calculation of ß-delayed fission probabilities involves essentially two inputs. One is the ß-strength distribution which can reliably be predicted within the framework of the pn QRPA. The second - maybe more delicate - input which is required are nuclear masses, especially fission barrier heights of nuclei far off stability. The predictions of fission barrier heights by Howard and Möller [31] are applied in the present work. Q_β-values and neutron separation energies S_n and S_{2n} are chosen according to the mass formulae of Groote et al. [32] and Hilf et al. [33], respectively. The latter was shown to give reliable results for the half-lives of neutron-rich isotopes far away from the line of ß-stability [15]. We apply these two sets of masses for the sake of consistency, since the ß-strength distributions of [15] are computed with the deformations given by these two mass formulae.
The results of our calculations are displayed in fig. 5.

874   Beta Decay Far from Stability

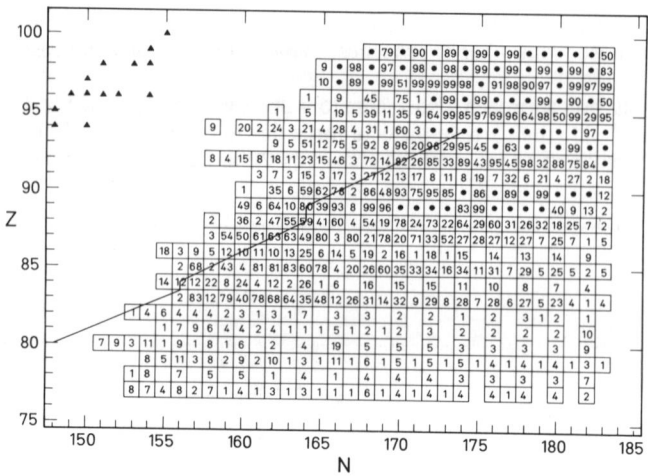

Fig. 5. Calculated rates of ß-delayed fission [30] using the mass formula of [33].

for the masses from [33]. The ß-delayed fission rates P_f are given in % for the fissioning daughter nucleus. Only P_f-values greater than 1 % are considered.
It its obvious that delayed fission has to be considered in the decay back to stability either in a nuclear explosion or in the very heavy region of the r-process.
In the region around $Z = 94$ we predict P_f-values of 100%, independently of the underlying mass formula and in agreement with [8]. This area acts as a sink to the r-process and prevents the synthesis of superheavy elements in nature.
The recently observed discrepancy between calculated and measured abundances in nuclear explosions with a ^{238}U target, traced back by Hoff [29] to overestimated ß-delayed fission rates by [8], seems to be removed in the present work [30,34].
Preliminary calculations of fission barrier heights of neutron-rich heavy elements within the Yukawa-plus-exponential [35] model seem to indicate higher fission barriers than in the work of [31]. Another difference is that the ground-state shapes may differ considerably. Both features may affect the probabilities of ß-delayed fission [30]. Despite some promising features of the new mass formulae, there are still large uncertainties in the predictions of nuclear masses of neutron-rich nuclides far from stability (see for example [15,36]). In consequence, the new estimations of fission barriers by [35] must be considered as uncertain as those of [31]. This leads to the conclusion that, at present, there seems to be no alternative to the extensive work of [31]. But one should keep in mind that new developments in this field are in progress and that new results on fission barriers might affect the fission rates discussed here and presented in [30].
We find as the most important conclusion for astrophysics, that the area of 100% ß-delayed fission remains, even if we increase the fission barriers by about 3 MeV, to an extent, that the r-process still will be stopped by ßdf (see [30]).

1.3. Beta decay far from stability and the astrophysical r process
Since recent years have proved the KMO ß-decay half-life predictions to be rather reliable (see the disc. in [16]), the essential conclusions we have drawn earlier on the astrophysical r process concerning its dependence on ß half-lives remain valid (corresponding investigations using the second-generation half-life calculations have not been made up to now but are being planned).

Also the recent measurement of ß half-lives of the (probable) r-process waiting point nuclei ^{80}Zn and ^{130}Cd do not lead to new conclusions, as claimed in [41] since they just were a confirmation of the calculations within 10% (see [16]). They rather confirm earlier conclusions.
That the cosmic r abundances in their peaks are in good agreement with that expected from an (n,γ)-(γ,n) equilibrium has been shown already by [7,8]. Also the conclusion [37] that the neutron density necessary to allow for this equilibrium is $n_n > 2 * 10^{17}$ cm^{-3} was made already ten years ago in [7].
Concerning the site of the r process, in the end of the 70ies and in the beginning 80ies the 'classical' r-process site close to the forming neutron star was ruled out on the basis of many arguments (see e.g. [38,39,7]). One of the most promising suggested new sites [40] was explosive helium burning. It is the most quantitatively investigated r-process site as far as its dependence on ß-decay properties is concerned. Its crucial dependence on the ß-decay input data had been investigated first by [7,8] who found that this scenario did not work when using the gross-theory half-lives, but yielded a good reproduction of the observed cosmic element distributions, when using the KMO microscopic half-lives. Concerning ß decay the calculated element distribution is mainly determined by the half-lives and P_n values, only for very heavy elements, essentially the cosmochronometers (see below) ß-delayed fission becomes important.

A point still under discussion in the explosive helium burning is that the number of neutrons produced in the ^{22}Ne(α,n) reaction is not much beyond the requirement for an (n,γ)-(γ,n) equilibrium and might eventually be marginal. As alternative the neutron source ^{13}C (α,n) has been discussed.
An interesting recent discussion of r-process sites was given by Mathews and Cowan([41] and these proceedings). They showed that correlations of heavy element abundances measured in recent years for the surfaces of metal-poor stars with respect to iron 'seed' material rule out most of the r-process models or sites invented in the past (although the correlations might not yet be considered as extremely striking). They rule out in particular a primordial r-process in an inhomogeneous big bang which may contribute only at a level of 10^{-3} of the solar abundance [37]. They conclude that 'the apparent decrease in [Eu/Fe] for [Fe/H] < -2.5 suggests, that the r-process may be a primary process in low-mass type II supernovae or a secondary event with a neutron source that does not require pre-existing seed material'. An example for the latter would be explosive He-burning fed by ^{13}C (α,n) (curve denoted SN$_{II}$ - Z/(s + Z) in their Fig. 4). Other investigations (Woosley [42]) favour a primary process.

1.4. The r process and the age of the universe

The production ratios of the actinide cosmochronometers (^{232}Th, ^{235,238}U, ^{244}Pu) in the r process allow one by comparison to ratios observed at the time of condensation of the solar system (in meteorites) to get information on the age of the galaxy. These production ratios are sensitive, in addition to half-lives and P_n values, to ß-delayed fission. Earlier work [43,44] on these chronometers ignored completely the effect of ß-delayed fission and led to ages of the universe systematically too short, compared to independent other sources. Inclusion of ß-delayed fission rates calculated from the KMO ß-strength functions allowed one for the first time to reach consistency of actinide chronometer ages with ages from the most important other independent sources, the globular clusters and the Hubble constant [4,8]. The conclusions from the cosmochronometers are independent of the final site of the r process. The arguments for this were already given by [8].

A calculation of the cosmochronometer production using our new half-lives [6], P_n-values [45] and ßdf-values [30] has still do be done. One important conclusion from the new calculated ßdf-values was given in section 1.2.2: The area of 100% ß-delayed fission already introduced in [8] remains to an extent that no superheavy elements can be formed in nature by the r-process.

1.5. Age of the universe and cosmology

Particular interest in improving the precision of the cosmochronometers comes from cosmology. The age of the universe is one of the boundary conditions for cosmological models and thus of importance to determine the value of the cosmological constant, which is of extreme importance for the structure of grand unified theories (see, e.g. [1,2,46]), and the amount of non-baryonic dark matter in the universe. This has been stressed by Blome and Priester [47] and by Klapdor and Grotz [5,2]. We refer to the latter and will not go into details here.

2 DOUBLE BETA DECAY, NEUTRINO MASS AND NUCLEAR STRUCTURE

The neutrino mass is one of the key quantities for the structure of grand unified theories (GUTs, SUSYs, SUGRAs) [2,48]. It plays nowadays a role for such theories as important as the role it played for the understanding of the weak interaction at the time of Pauli and Fermi.

2.1. Calculation of double beta decay rates

2.2.1 Two-Neutrino Double Beta Decay

The key observation for the solution of the problem of earlier discrepancies between calculated and experimental 2νßß half-lives (see the discussion in [19]) was the recognition [49] of the strong effect of g.s. correlations on the calculated rates. One observed a strong suppression of the calculated 2ν rates when including beyond spin-isospin correlations quadrupole-quadrupole correlations. By angular momentum recoupling this can be seen to correspond to an increase of the 1^+ (spin-isospin) correlations.

Once this dependence on the g.s. correlations was observed, it was natural to look for further ways to increase the spin-isospin correlations.

The next step along this line was made by including particle-particle forces in QRPA calculations [50,51,20], which had been ignored before. They lead to a further increase of g.s. correlations and further suppression of 2νßß rates as effect of destructive interference of forward- and backward-going amplitudes X,Y in the second ßß-decay step, which can be considered as ß$^+$ decay from the final state to the intermediate nucleus (for details see [19]).

Fig. 6 shows the sensitive dependence of the calculated matrix element $M_{2\nu}$ on the strength of the pp force (the renormalization factor g_{pp} should be not too far from 1). A precise determination of g_{pp} is obviously extremely important.

We have determined g_{pp} by fitting observed ß$^+$ decay rates for about 30 nuclei with masses $60 < A < 170$, excluding semimagic nuclei [19,21]. Using these values we calculated the rates for all potential ßß-emitters [52]. Fig. 7 shows a comparison to experiment. The result is that only lower limits for $T^{2\nu}_{1/2}$ can be given now, the upper limit in all cases being infinity. The given average value is obtained by averaging the rates over the 1σ uncertainty interval of g_{pp}. The progress of solving the earlier problem of a systematic discrepancy between experimental and theoretical 2ν rates is seen to be paid by a large loss of precision in the predictions in 2ν calculations.

2.2.2 Neutrinoless double beta decay

A reliable m mass can be extracted from a measured rate only if the nuclear matrix element $M^{0\nu} = (M^{0\nu}_{GT} - M^{0\nu}_F)$ can be reliably calculated (see [53,19], since the rate $\Omega \sim |M^{0\nu}|^2 m_\nu^2$.

The situation for calculating $0\nu\,\beta\beta$ rates fortunately is much better than in the 2ν case. The matrix element $M^{0\nu}$ is, in contrast to $M^{2\nu}$, only weakly dependent on g_{pp} [55]. The reason is the neutrino potential H occurring in $M^{0\nu}$, but not in $M^{2\nu}$. As a consequence states of various multipolarities in the intermediate nucleus are excited in the 0ν case (not only 1^+ states, as in the 2ν case), because the neutrino is virtual here.

The total matrix element $M^{0\nu}$ remains large, because the g.s. correlations affect essentially only the 1^+ component [21]. The physical reason for this is that the g.s. correlations corresponding to the higher multipoles are almost not affected by the pp force. This reflects a well-known feature of the nucleon-nucleon (proton-neutron) interaction: The pp interaction is strongly attractive in the $J^\pi = 1^+$ (the 0^+ and the highest multipole) channel, in other channels less attractive or slightly repulsive. We thus understand that and why $0\nu\,\beta\beta$ matrix elements can be predicted reliably.

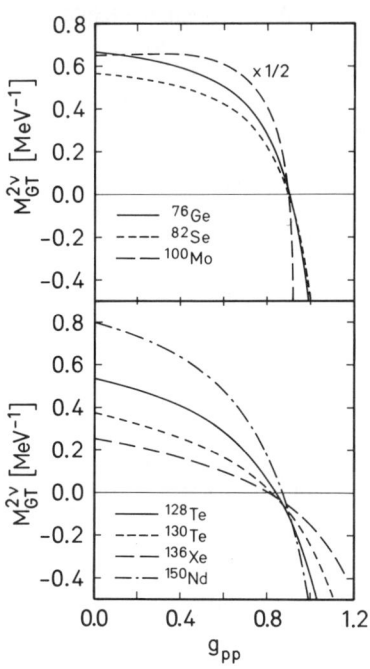

Fig. 6. Nuclear matrix elements of the $0^+ \rightarrow 0^+\ 2\nu\,\beta\beta$ decay calculated as function of the particle-particle interaction strength in the proton-neutron quasiparticle RPA. We have calculated with a realistic interaction (G matrix from Paris potential) the $M^{0\nu}$ for all potential $\beta\beta$-emitters [52] and particularly for the seven nuclei of main experimental interest. For the latter we have extracted limits for a (Majorana) neutrino mass from experimental half-life limits [21]. The results are shown in Table 1. Where we can compare to the Tübingen calculation, also using a realistic interaction (G matrix from Bonn potential) we find good agreement. Good agreement we find also with the older calculations by Haxton et al. [56]-when their unfortunate scaling procedure (which they invented because of their strong discrepancy on the 2ν sector) is not applied. The results remain stable against differ-

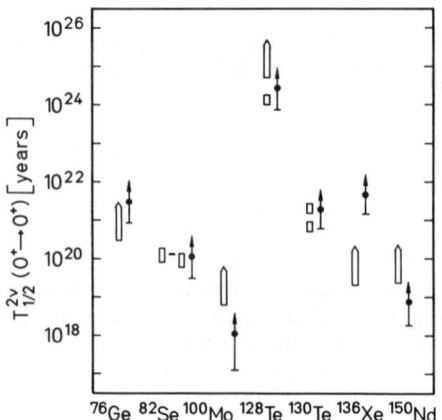

Fig. 7. Comparison of experimental and calculated half-lives of $0^+ \to 0^+$ 2ν ßß decay. The experimental values are given by open squares or open arrows for lower limits. The theoretical half-lives, calculated in the pn QRPA model, are presented by arrows. The arrow starts at the lower limit and the circle on it denotes the average over the 1σ interval around g_{pp} (see text) (from [20,21]).

Table 1 Present status of (effective) neutrino masses deduced from double beta decay experiments, for the nuclear matrix element of various nuclear structure calculations. Contributions of right-handed currents are neglected. Experimental $0^+ \to 0^+$ 0ν ßß decay half-life limits are given in the last line.

Refs.	^{48}Ca	^{76}Ge	^{82}Se	^{128}Te	^{130}Te	^{100}Mo	^{136}Xe	^{150}Nd
HS84 GK86	<40	<1.8	<7.3	<0.90	<10	--	--	--
no pp force	<0.7	<2.9	<0.43	<5.4	<8.7	<13.3	--	<3.2
TF87	<2.0	<7.4	<1.4	<19	--	--	--	--
MBK89	<2.1	<7.4	<1.2	<18	<54	<36	--	<3.8
EVZ88+	<7-17	<32-64	<2.2-3	<29-39	<66	<61-80	--	--
$T^{0\nu}_{1/2}[y]$	>2x10^{21}	>5x10^{23}	>1.1x10^{22}	>5x10^{24}	>1.5x10^{21}	>4.4x10^{20}	>1.7x10^{21}	>2.3x10^{21}

+Not realistic (zero range) interaction

ent choices of the renormalized effective interactions, e.g. Bonn-A and Reid potentials instead of the Paris potential [22]. A discrepancy occurs only when a not realistic interaction is used (see [54]).

Concluding, 0ν ßß decay remains the most powerful tool for probing the mass of the electron neutrino.

2.3 Future experimental possibilities. The Heidelberg-Moscow experiment using enriched ^{76}Ge

Double beta decay yields at present the sharpest limit on the Majorana mass of the neutrino (see Table 1). Among the direct (non-geochemical) experiments the most stringent limit is given by 0ν ßß decay of ^{76}Ge: $m_\nu < 1.4$ eV, corresponding to the claimed present half-life limit of $T^{0\nu}_{1/2} \geq 1.2 * 10^{24}$ years [57,58].
In next generation ßß decay experiments ^{76}Ge will play also the most important role. Use of enriched ^{76}Ge (86%) [59] instead of the presently used detectors from natural Ge (containing 7.8% of ^{76}Ge) could explore the half-life of 0ν ßß decay up to ~10^{25} years

and correspondingly the neutrino mass down to ~10^{-1} eV, probing a class of left-right symmetric GUT models, with a right-handed Majorana mass term of about 1 TeV, based on SU(2)L * SU(2)R * U(1).

The Heidelberg-Moscow ßß experiment [59] makes use of 16.9 kg of ^{76}Ge metal enriched to 86%, corresponding to 14.5 kg of the isotope ^{76}Ge. The full amount of ^{76}Ge has been transferred from Moscow to Heidelberg. Up to now one enriched detector of ~1 kg and 1 crystal of 3.4 kg have been produced. The detector is running in the GRAN SASSO Underground Laboratory in Italy since end of July 1990. The results of the first 108 days of measuring are $T_{1/2} > 5.66 * 10^{23}$ (90% confidence limit) or $> 1.19 * 10^{24}$ years (68% confidence limit). The corresponding limit for the neutrino mass are < 2.0 and < 1.4. The background level which characterizes the quality of the setup, is 0.45 ± 0.14 events/kg year keV in an 80 keV interval around the hypothetical 2041 keV 0νßß-line (10 events). The prospects of the experiment

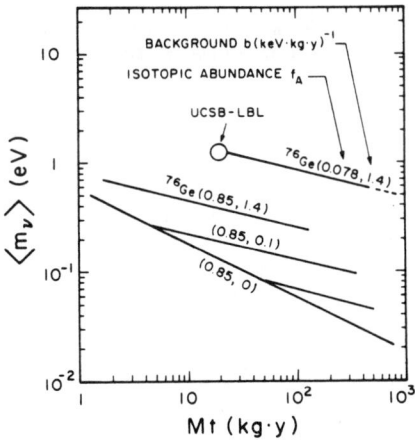

Fig. 8. The ranges of neutrino mass which can be probed with Ge-detectors of different enrichment in ^{76}Ge, as function of the product detector mass times measuring time (kg y).

are illustrated in Fig. 8.

3 Conclusion

The strong cancellation of 2ν ßß decay, which is reminiscent of a broken symmetry requiring $M^{2\nu} = 0$, seems to be understood from nuclear structure. According to recent results, the crossing of the matrix element through zero is not an artifact of QRPA but occurs also in the shell model.

The matrix elements of 0ν ßß decay can be calculated reliably making second-generation double beta experiments highly desirable.

Further experimental and theoretical investigation of properties of nuclei far from stability - after the reached degree of sophistication in the calculations of ß matrix elements, particularly of mass formulae - is of importance for exciting astrophysical questions such as element synthesis by the r process and determination of the age of the universe from cosmochronometers, which have impact also on cosmology.

References
1 E.W. Kolb, M.S. Turner, 'The Early Universe', Addison-Wesley, Redwood City, 1990
2 K. Grotz, H.V. Klapdor, 'The Weak Interaction In Nuclear, Particle and Astrophysics', Adam Hilger, Bristol, N.Y., 1990
3 K. Grotz, H.V. Klapdor and J. Metzinger, Phys. Rev. C33 (1986) 1263.
4 H.V. Klapdor, Prog. Part. Nucl. Phys. 10 (1983) 131.
5 H.V. Klapdor and K. Grotz, Astrophys. J. 301 (1986) L39.
6 H.V. Klapdor, J. Metzinger and T. Oda, Z. Phys. A309 (1982) 91 and At. Data Nucl. Data Tables 31 (1984) 81.
7 H.V. Klapdor et al., Z. Phys. A199 (1981) and CERN Report 81-09 (1981) 341.
8 F.K. Thielemann, J. Metzinger and H.V. Klapdor, Z. Phys. A309 (1983) 301 and Astron. Astrophys. 123 (1983) 162.
9 K.L. Kratz, H. Gabelmann, W. Hillebrandt, B. Pfeiffer, H.L. Raven and F.K. Thielemann, Z. Phys. A325 (1986) 489.
10 E. Lund, K. Aleklett, B. Fogelberg and A. Sangariyavanish, Phys. Scripta 34 (1986) 614.
11 P.L. Gill, R.F. Casten, D.D. Warner and A. Piotrowski, Phys. Rev. Lett. 56 (1986) 1874.
12 H.V. Klapdor, Fortschr. d. Physik 33 (1985) 1.
13 K. Takahashi, M. Yamada and Z. Kondoh, At. Data Nucl. Data Tables 12 (1973) 101.
14 T. Tachibana, S. Ohsugi and M. Yamada, AIP Conf. Proc. 164 "Nuclei far from Stability," p. 614 (1988) and M. Yamada, Proc. Int. Conf. on Nuclear Data for Science and Technology, May 30-June 3, 1988, Mito, Japan, p. 841.
15 A. Staudt, E. Bender, K. Muto and H.V. Klapdor, Z. Phys. A334 (1989) 47, At. Data Nucl. Data Tables 44 (1990) 79, and E. Bender, K. Muto and H.V. Klapdor, Phys. Lett. 208B (1988) 53.
16 H.V. Klapdor, Invited talk presented at Internat. Conf. on Selected Topics in Nuclear Structure, Dubna, USSR, June 20-24, 1989.
17 J.A. Halbleib and R.A. Sorensen, Nucl. Phys. A98 (1967) 542.
18 K. Muto, E. Bender and H.V. Klapdor, Z. Phys. A333 (1989) 125.
19 K. Muto, H.V. Klapdor, in: "Neutrinos," Springer, Heidelberg, New York, 1988, ed. H.V. Klapdor, p. 183.
20 K. Muto, H.V. Klapdor, Phys. Lett. 201B (1988) 420.
21 K. Muto, E. Bender, H.V. Klapdor, Z. Phys. A334 (1989) 177 and 187
22 A. Staudt, T.T.S. Kuo, H.V. Klapdor-Kleingrothaus Phys. Lett. 242B (1990) 17
23 B.H. Wildenthal, M.S. Curtin and B.A. Brown, Phys. Rev. C28 (1983) 1343.
24 R.C. Greenwood et al., Proc. Int. Conf. on Nucl. Data for Science and Technology, May 30-June 3, 1988, Mito, Japan.
25 M. Hirsch, A. Staudt, K. Muto, H.V. Klapdor-Kleingrothaus, subm. to Phys. Lett. (1990).
26 E.Ye Berlovich and Yu.N. Novikov, Phys.Lett. B29 (1969) 155
27 C.O. Wene, Astron. & Astrophys. 44 (1975) 233
28 C.O. Wene and S.A.E. Johansson, CERN-Report 76-13(1976) 584
29 R. Hoff, in: Proc. Int. Symp. on Weak and Electromagnetic Interactions in Nuclei, ed. H.V. Klapdor, Springer, Heidelberg, (1986) p. 207
30 A. Staudt and H.V. Klapdor-Kleingrothaus, submitted for publication
31 W.M. Howard and P. Möller, At. Data Nucl. Data Tables 25 (1980) 219
32 H. von Groote, E.R. Hilf and K. Takahashi, At. Data Nucl. Data Tables 17 (1976) 418
33 E.R. Hilf, H.V. Groote and K. Takahashi, CERN-Report 76-13, (1976) 142
34 A. Staudt, M. Hirsch, K. Muto, H.V. Klapdor-Kleingrothaus, Phys.Rev.Lett. 65 (1990) 1543

35 W.M. Howard, P. Möller, G. Mathews and B. Meyer, Symp. on The Origin and Distribution of the Elements at the 194th American Chemical Society National Meeting, Aug. 31 - Sept. 4, 1987, New Orleans, USA
36 M. Graefenstedt et. al., Z.Phys. A 336 (1990) 247
37 G.J. Mathews and J.J. Cowan, Int. Symp. on Heavy Ion Physics and Nuclear Astrophysical Problems, Tokyo, 1988, World Scientific, 1989, p. 143.
38 W. Hillebrandt, Space Sci. Rev. 21 (1978) 639.
39 W. Hillebrandt, Proc. 4th EPS Gen. Conf. (1979), p. 255 and Phys. Bl. 38 (1982) 189.
40 W. Hillebrandt and F.K. Thielemann, Mitt. Astron. Ges. 43 (1977) 234.
41 G.J. Mathews, J.J. Cowan, Nature 345 (1990) 491
42 S.E. Woosley, 1990, priv. comm.
43 W.A. Fowler, in Proc. of the Welch Foundation Conferences on Chemical Research, Vol. 21, Cosmochemistry, ed. W.D. Milligan, p. 61.
44 W.A. Fowler and C.C. Meisl, in Cosmogonical Processes, eds. W.D. Arnett, C.J. Hansen, J.W. Truran and S. Tsuruta (Singapore, VNU Press), p. 83.
45 A. Staudt, M. Hirsch, H.V. Klapdor-Kleingrothaus, to be publ.
46 L. Abbott, Sci. American, May 1988, p. 82.
47 H.J. Blome and W. Priester, Naturwissenschaften 71 (1984) 456, 515, 528.
48 P. Langacker, in: "Neutrinos," Springer, Heidelberg, New York, 1988, ed. H.V. Klapdor, p. 71.
49 K. Grotz, H.V. Klapdor, Nucl. Phys. A460 (1986) 395, Phys. Lett. 142B (1984) 32, 153B (1985) 1, and 157B (1985) 242.
50 P. Vogel, M.R. Zirnbauer, Phys. Rev. Lett. 57 (1986) 314.
51 O. Civitarese, A. Faessler, T. Tomoda, Phys. Lett. 194B (1987) 11.
52 A. Staudt, K. Muto, H.V. Klapdor-Kleingrothaus, Europhys.Lett. 13 (1990) 31
53 M. Doi, T. Kotani, E. Takasugi, Progr. Theor. Phys. Suppl. 83 (1985) 1.
54 J. Engel, P. Vogel, M.R. Zirnbauer, Phys. Rev. C37 (1988) 731.
55 T. Tomoda, A. Faessler, Phys. Lett. 199B (1987) 475.
56 W.E. Haxton, G.J. Stephenson, Progr. Part. Nucl. Phys. 12 (1984) 409.
57 D. Caldwell, Proc. 14th Europhys. Symp. on Nucl. Phys., Bratislava, 22. - 26. Okt. 1990
58 I. V. Kirpichnikov, Proc. 14th Europhys. Symp. on Nucl. Phys., Bratislava, 22. - 26. Okt. 1990
59 Heidelberg-Moscow-Cooperation: H.V. Klapdor, A. Piepke, G. Heusser, A. Buchner, A. Müller, U. Schmidt-Rohr, H. Strecker (MPI); S.T. Belyaev, A. Balish, A. Gurov, A. Demehin, I. Kondratenko, V.I. Lebedev (Kurchatov Institute), Proc. Internat. Sympos. on Weak and Electromagn. Interactions in Nuclei, Montreal, 1989 (WEIN '89).

SOME LIMITATIONS OF DETAILED BALANCE FOR INVERSE REACTION CALCULATIONS IN THE ASTROPHYSICAL p PROCESS*

D.G. Gardner and M.A. Gardner
University of California, Lawrence Livermore National Laboratory
Livermore, CA 94551, USA

ABSTRACT

p-Process modeling of some rare but stable proton-rich nuclei requires knowledge of a variety of neutron, charged particle, and photonuclear reaction rates at temperatures of 2 to 3 x 10^9 °K. Detailed balance is usually invoked to obtain the stellar photonuclear rates, in spite of a number of well-known constraints. In this work we attempt to calculate directly the stellar rates for (γ,n) and (γ,α) reactions on ^{151}Eu. These are compared with stellar rates obtained from detailed balance, using the same input parameters for the stellar (n,γ) and (α,γ) reactions on ^{150}Eu and ^{147}Pm, respectively. The two methods yielded somewhat different results, which will be discussed along with some sensitivity studies.

INTRODUCTION

The origins of the rare, proton-rich stable "p-nuclei" have been the subject of numerous investigations[1-5] in which proton capture or photonuclear reactions or both have been proposed to be the important reactions involved in the p-process, the stellar mechanism for producing such nuclei. The stellar model utilizing the latter reaction type is sometimes called the γ-process.[4] Such modeling requires stellar photonuclear rates which are usually obtained by detailed balance.[6] The approach is sound if the forward and reverse reactions involved proceed statistically such that, for the case A+a=[C]=B+b, the compound system [C] has the same distributions in energy, angular momentum, and parity. However, there are a number of well-recognized constraints and in this paper we will try to determine the magnitude of the error introduced by some of these unfulfilled assumptions. We are able to attempt to analyze the validity of detailed balance for a variety of situations because our STAPLUS(Hauser-Feshbach plus spin-dependent preequilibrium) code[7] allows photons in the entrance channel and also permits the spin-parity distributions in the first compound nucleus to be supplied as input.

CALCULATIONAL METHOD

The calculations presented here are for the target nucleus, ^{151}Eu; the separation energies from ^{151}Eu are: S_n = 7.937 MeV, S_p = 4.891 MeV and S_α = -1.964 MeV. We have chosen europium for this effort because more experimental information exists here than for the light Gd-Sm nuclei, which are of more astrophysical interest. We wrote a code called SUNBURN to calculate photonuclear transitions from target ^{151}Eu states available by the Boltzmann distribution at a given temperature, from some lower excitation energy (such as the ground state) up to an arbitrary maximum target excitation, the neutron separation energy for ^{151}Eu. In most of this work, this range has been divided up into 0.1 MeV energy bins. The photonuclear transitions

---

*Work performed under the auspices of the U.S. Department of Energy by the Lawrence Livermore National Laboratory under contract No. W-7405-Eng-48.

proceed from these target bins to states above the specific reaction threshold up to another arbitrary maximum energy, which we have chosen to be 15 MeV. This is slightly above the (γ,2n) threshold. Actually, we compute the photon absorption cross section, and multiply this by the branching fraction for the reaction of interest. This absorption cross section is computed from the user's choice of multipole strength functions, $f_{xl}(E\gamma)$, and so is used with each target-state spin, $J(E_t)$, to produce the spin and parity distribution in the ^{151}Eu compound nucleus:

$$\sigma_{\gamma ab}^{J\pi}(E_\gamma) = 3.85 \times 10^3 \left(\frac{2J+1}{2J(E_t)+1}\right) \sum_{xl} E_\gamma^{2l-1} f_{xl}(E_\gamma) \text{ barn}. \tag{1}$$

Multiplying this by the appropriate energy-, spin-, and parity-dependent branching fraction greatly simplifies the programming, and allows easy problem modifications.

STELLAR PHOTO–REACTION RATE
Assume Boltzmann–type thermal equilibrium

$$\lambda_{\gamma x} = \text{CON}/(2J_0 + 1) * \underbrace{\int_{E_l}^{E_s} dE_t * \rho(E_t)*BF(E_t)*(2J(E_t) + 1)}_{Q/(2J_0 + 1)} \int_{E_x - E_t \geq 0}^{E_h - E_t} dE_\gamma * \text{PLNK}(E_\gamma) * \sigma_{\gamma ab}(E_\gamma) * \left(\frac{\sigma_{\gamma x}(E_\gamma + E_t)}{\sigma_{\gamma ab}(E_\gamma)}\right) \sec^{-1}$$

$Q = \sum \rho(E_t)*BF(E_t)*(2J(E_t) + 1)$

$\rho(E_t)$ = target level density MeV^{-1}

CON = 2.9979e+10 * 1.e-24 cm^3 sec^{-1} b^{-1}

$J(E_t)$ = most probable J at target E_t

$BF(E_t)$ = Boltzmann factor at E_t

$PLNK(E_\gamma)$ = Planck spectrum photons cm^{-3} MeV^{-1}

$\sigma_{\gamma ab}$ = photon absorption cross section b

$\sigma_{\gamma x}(E)$ = (γ,x) cross section at $E = E_\gamma + E_t$

E_t is target energy where photon transitions start

E_l is lowest target energy (MeV)

E_h is highest target energy (MeV)

E_x is (γ,x) threshold (MeV) E_s = neutron separation energy (MeV)

Fig. 1. Equation for computing stellar photonuclear reaction rates that is incorporated in the computer code SUNBURN.

The stellar photonuclear rate expression is given in Fig. 1. All calculations were performed on an energy grid with 0.1 MeV steps. The excitation functions $\sigma_{\gamma ab}(E_\gamma)$ and $\sigma_{\gamma x}(E)$ were calculated with the Hauser-Feshbach plus precompound evaporation code STAPLUS,[7] and interpolated onto the energy grid using a piece-wise cubic Hermite polynomial routine;[8] the integrations were done using the trapezoidal rule. The integration accuracy was checked at $T_9=2$ by integrating the product of the level density and the Boltzmann factor, from 0 to 8 MeV, and comparing the answer with that obtained in closed form. The agreement was better than 1%, which may not be sufficiently accurate. We hope to try a Gaussian quadrature method in the future.

The neutron transmission coefficients were obtained using the coupled channel code ECIS[9] and a deformed optical model potential.[10] The proton[11] and alpha particle[12] potentials were both spherical. The alpha particles presented the greatest problem when computing transmission coefficients. None of our spherical or deformed potential codes were accurate for transmission coefficients below about 10^{-15} in magnitude. We intend to modify our codes to incorporate double-precision routines that compute Coulomb wave functions with greatly increased accuracy. In the present work, we graphically extended both $\sigma(\alpha,\gamma)$ and $\sigma(\gamma,\alpha)$ by the barrier penetration method from $E_\alpha=6$ down to about 4.5 MeV. This will not affect our results at $T_9=3$, but might at $T_9=2$. The multipole strength functions we favor for dipole and isoscalar E2 transitions have been described elsewhere.[13] Our unadjusted strength functions produce photoneutron reactions on ^{151}Eu and ^{153}Eu that compare well with the experimental data.[14] Our unadjusted calculations agree very well with the ENDF/B evaluations for (n,2n) and (n,3n) reactions and with the neutron capture data[10] for the (n,γ) reactions on targets of both ^{151}Eu and ^{153}Eu. Testing the choice of alpha particle optical potentials is quite important; using the Igo volume potential[15] yields a (γ,α) excitation function that is between 3 and 6 times the magnitude generated using the McFadden surface derivative potential[12]. We compared our calculations with data[16] for the (α,n) and (α,γ) reactions on ^{139}La and found the results to be quite satisfactory.

In order to compare our calculated stellar photonuclear reaction rates or rate factors with those computed via detailed balance, we follow the usual treatment[6] and calculate the capture cross sections using the discrete target levels (14 to 32) weighted in their true sequence by the Maxwell-Boltzmann and (2J+1) factors and the partition function. In the continuum we use a representative capture cross section derived from the above weighting sequence. In standard notation, the detailed balance expression for the stellar photonuclear reaction rate factor $N_A\langle\bar{\sigma}_{\gamma x}\upsilon\rangle^*$ is related to the stellar rate factor for the (x,γ) capture reaction, $N_A\langle\bar{\sigma}_{x\gamma}\upsilon\rangle^*$, by

$$N_A\langle\bar{\sigma}_{\gamma x}\upsilon\rangle^* = \left(\frac{g_I g_x}{g_L}\right)\left(\frac{Q_I}{Q_L}\right)\left(\frac{A_I A_x}{A_L}\right)^{3/2}\left(\frac{kT}{2\pi\hbar^2 N_A}\right)^{3/2} N_A\langle\bar{\sigma}_{x\gamma}\upsilon\rangle^* \exp[-(S_x+E_x)/kT]. \quad (2)$$

We have changed the symbol for the partition function from G to Q, and that for the Q-value to S_x+E_x to conform with Fig. 1. Here, S_x is the separation energy of particle X from the compound nucleus (say, ^{151}Eu) and E_x is that particle's incident energy in the center of mass system. Note that the stellar photonuclear rate factor is equal to the actual stellar rate for nuclei at thermal equilibrium exposed to a Planck photon distribution, i.e., the rate $\lambda_{\gamma x}$ sec^{-1} in Fig. 1 is numerically equal to the rate factor $N_A\langle\bar{\sigma}_{\gamma x}\upsilon\rangle^*$.

We may see that the two rate factors in Eq. 2 are linearly proportional; if Igo's transmission coefficients had been used instead of McFadden's, the stellar photonuclear reaction rate would have been a factor of 3 to 6 times faster than the rate we calculate. Of particular interest is the effect produced by low-energy charged particles trying to penetrate the Coulomb barrier in the (x,γ) reaction. For example, alpha particles on ^{147}Pm, producing the compound system ^{151}Eu, are extremely insensitive to the photon widths, $\Gamma_\gamma(E_\gamma)$, and therefore to either the photon transmission coefficients or the level density in ^{151}Eu. Either of the two may be divided by one or more orders of magnitude without changing the $\sigma(\alpha,\gamma)$ cross section, because of the very small size of $\Gamma_\alpha(E_\alpha)$. Below the neutron separation energy in ^{151}Eu, the $\sigma(\alpha,\gamma)$ cross section is only the alpha particle reaction cross section. There is essentially no information concerning photons contained within $N_A<\bar{\sigma}_{\alpha\gamma}\upsilon>^*$, and yet Eq. 2 predicts a stellar (γ,α) reaction rate! The explanation, of course, is that Eq. 2 predicts one number, for a given temperature, that is the convolution of the Planck spectrum with some photonuclear excitation function, properly weighted and summed over all available target states. There are an infinite number of multipole strength-function combinations that will yield the same numerical rate, even if concurrent and complimentary variations in the ^{151}Eu level density are ignored.

One final comment should be made concerning Eq. 2. Assume for a moment that only dipole transitions are important for photonuclear reactions. These will produce a very narrow compound nucleus spin and parity distribution, as compared with that produced by incident particles, particularly the large number of partial waves brought in by alpha particles. To make the comparison of the rates calculated by the equation in Fig. 1 with that derived from detailed balance, we have utilized the feature of the STAPLUS code to read as input the normalized spin and parity distributions that a particle would produce in the compound nucleus and to distribute the photon absorption cross section accordingly, before allowing the photonuclear reactions to proceed.

RESULTS

In Fig. 2a, we show the calculated excitation function for ^{147}Pm$(\alpha,\gamma)^{151}$Eu, with photon transmission coefficients from our systematics, and with these decreased by a factor of 10. Below the neutron separation energy in ^{151}Eu there is no visible difference between the two calculations. The same effect is found when the ^{151}Eu level density is decreased by the same amount. As mentioned in the previous section, the alpha particle is primarily exploring the Coulomb barrier, while the exponentially decreasing tail of the nuclear wave function interacts only slightly with the nucleus.

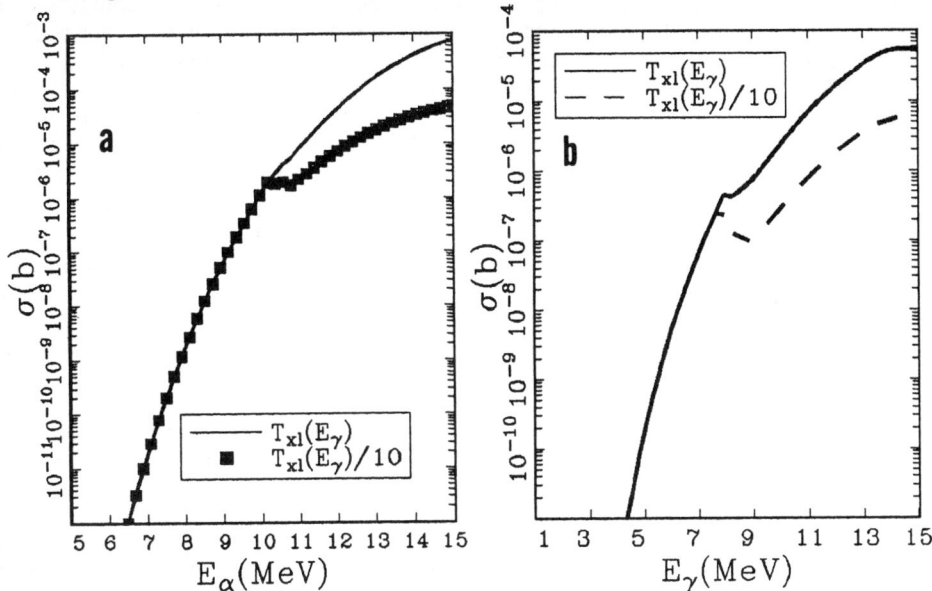

Figs. 2a,b. Change in the ^{147}Pm$(\alpha,\gamma)^{151}$Eu and ^{151}Eu$(\gamma,\alpha)^{147}$Pm cross sections due to 1/10 decrease in photon transmission coefficients.

Suppressing spin and parity,

$$\sigma_{\alpha\gamma}(E_\alpha) \approx \pi \lambdabar^2 \left(\frac{2\pi\Gamma_{\alpha 0}}{D}\right)\left(\frac{\Gamma_\gamma}{\Gamma_\gamma+\Gamma_\alpha+\Gamma_n}\right). \qquad (3)$$

Recall that at excitation energies in ^{151}Eu below the neutron separation energy, $\sigma_{\alpha\gamma}(E_\alpha)$ is merely the alpha particle reaction cross section, $\sigma_{\alpha Rx}(E_\alpha)$. The symbol $\Gamma_{\alpha 0}$ represents the transition from the ground state of ^{147}Pm to the continuum in ^{151}Eu, and we have observed that $\Gamma_{\alpha 0}$ is usually $\geq 90\%$ of the total width, Γ_α. Because Γ_γ is many orders of magnitude larger than Γ_α, very large changes in Γ_γ produce no noticeable effect. In Fig. 2b, the reverse reaction is shown. Following the above argument concerning the relative sizes of the widths Γ_α and Γ_γ, it is easy to show why changing Γ_γ when Γ_n is zero produces no effect, but at higher excitation energies in ^{151}Eu, accessed by photons, a decrease in their transmission coefficients by a factor of 1/10 results in essentially the same decrease in $\sigma(\gamma,\alpha)$. For temperatures in the range of $T_9=2$ or 3, the Planck photon distribution convolutes primarily with the low-energy portion of the (γ,α) excitation function, and detailed balance calculations should be impacted only slightly. At higher temperatures a correct representation of the multipole strength becomes steadily more important, until such high temperatures are reached that multiple-particle emission occurs, such as $(\gamma,2n)$ or $(\gamma,\alpha n)$, at which point the detailed balance approach is no longer useful. The approach of calculating directly the photonuclear reactions and their stellar rates, does not suffer from these restrictions, and clearly is useful in signaling the energy threshold where some of the restrictions of detailed balance begin to appear and hence the temperature range where detailed balance becomes less accurate.

Perhaps a more interesting illustration of the fact that detailed balance provides only a minimal constraint on the multipole strength functions, $f_{xl}(E_\gamma)$, involves a comparison of the results using a giant dipole resonance shape for E1 strengths with the energy-independent f_{E1} from the Weisskopf single-particle approach. We applied a hindrance factor of 1×10^{-3} to the Weisskopf f_{E1} and normalized the $f_{E1}+f_{M1}$ such that $2\pi\Gamma_\gamma/D = 3.95$ at the neutron separation energy in ^{151}Eu - the same value our dipole systematics produced. Figs. 3a and 3b show the anticipated effects on the (γ,α) reaction of the two vastly different functional forms for f_{E1}, even when both Γ_γ's were set equal at S_n. The effect on $\sigma(\alpha,\gamma)$ is barely discernable. This complete dominance by the Coulomb barrier may also be observed in the calculation of $N_A\langle\bar\sigma_{\alpha\gamma}v\rangle^*$, the forward rate factor in the detailed balance equation shown in Eq. 2. For ^{147}Pm, we used 14 discrete levels up to 0.807 MeV. Figure 4a presents a selection of ^{147}Pm$(\alpha,\gamma)^{151}$Eu excitation functions. For target spins of 0.5 to 5.5, and for level energies from zero to 0.807 MeV, the excitation functions overlap so well that a single curve was used to represent the collection in the convolution with the Maxwell-Boltzmann function. The order of the spin sequence in the $(2J+1)$ weighting was immaterial. This was not true in the (n,γ) case, as may be seen in Fig. 4b. Because of the Boltzmann factor weighting of the $(2J+1)$ term, the order of the spin sequence with excitation energy should not be neglected; the Maxwell-Boltzmann averaged capture cross section from each of the 27 levels in ^{150}Eu varied significantly.

The insensitivity of the stellar (γ,α) rate, below the (γ,n) threshold, to variations in the multipole strength-function modeling is not observed in the stellar (γ,n) rate because Γ_γ cannot be neglected as compared with Γ_n until the ^{151}Eu nucleus is several MeV above the (γ,n) threshold. Figure 5 displays the effect produced in the calculated ^{151}Eu (γ,n) ^{150}Eu reaction by a 1/10 decrease in the photon transmission

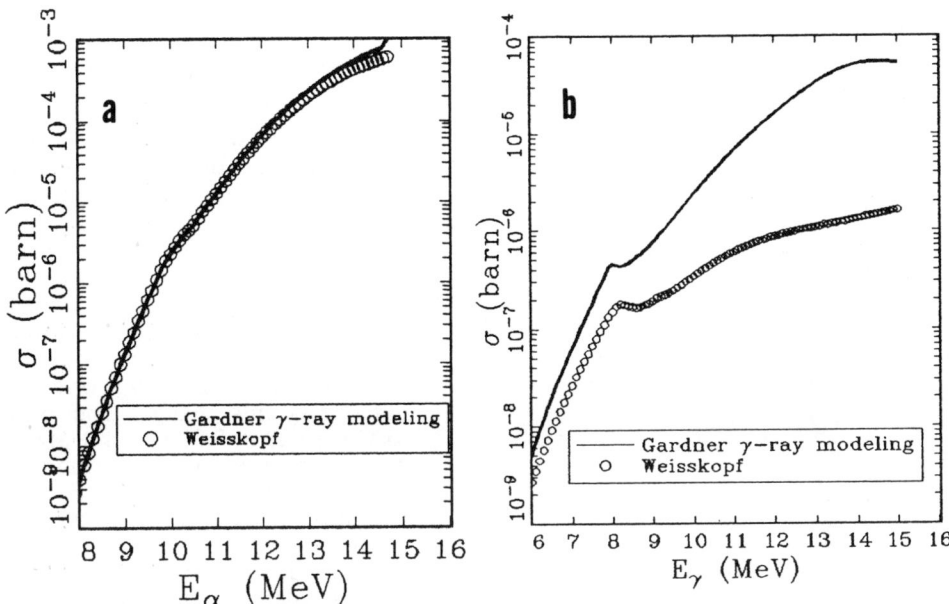

Figs. 3a,b. Change in the ^{147}Pm$(\alpha,\gamma)^{151}$Eu and ^{151}Eu$(\gamma,\alpha)^{147}$Pm cross sections due to different dipole strength functions.

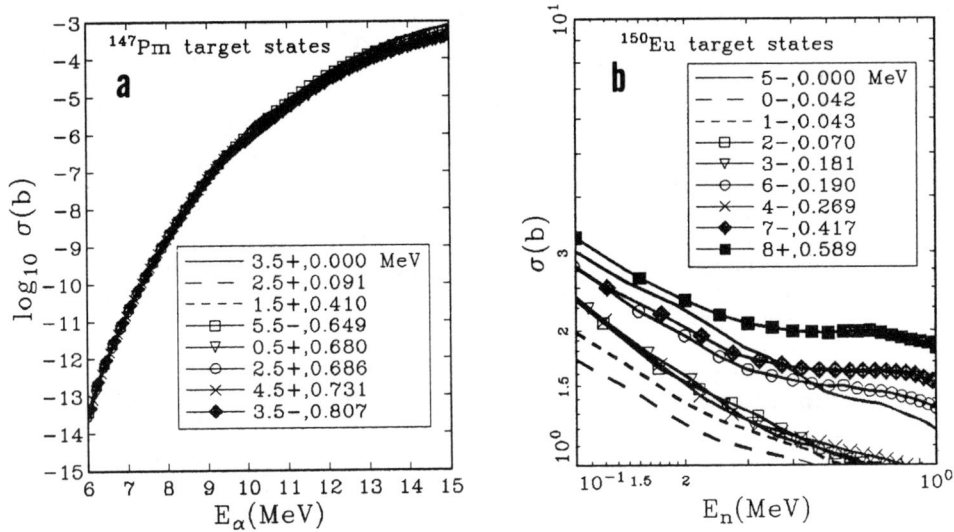

Figs. 4a,b. ^{147}Pm$(\alpha,\gamma)^{151}$Eu and ^{150}Eu$(n,\gamma)^{151}$Eu excitation functions for a selection of discrete level target states.

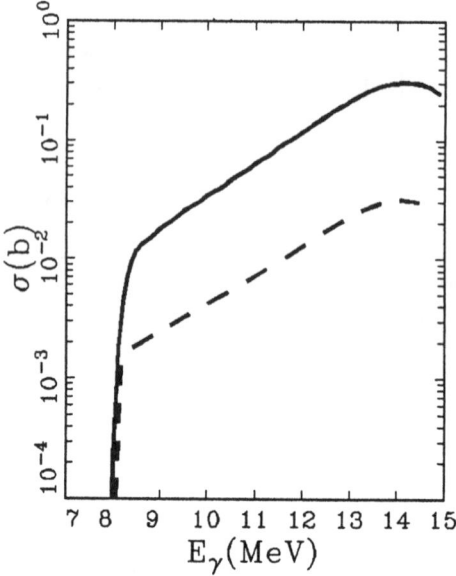

Fig. 5. Change in ^{151}Eu(γ,n)^{150}Eu cross section due to 1/10 decrease (--) in multipole strength functions.

coefficients; the decrease in $\sigma(\gamma,n)$ is also about 1/10, while the (n,γ) capture cross section decreases only about 40%. The stellar photoneutron rates show a factor-of-two decrease for T$_9$=2 to 3. Because the Planck distribution decreased by about 10^5 at T$_9$=2, for ^{151}Eu excitation energies of 8 to 10 MeV, we are not sure that our integration of the equation in Fig. 1 in steps of 0.1 MeV is sufficiently accurate. From 8.5 to 9.5 MeV of excitation, for example, the branching fraction $\sigma(\gamma,n)/\sigma_{\gamma ab}$ only changes from 0.59 to 0.77, but is 0.014 at 8.0 MeV. Not allowing γ-ray competition in the first compound nucleus increases the stellar photoneutron rate by almost a factor of 5. We plan to study other integration methods in the near future.

In Figs. 6a and 6b, we show our calculated stellar rates for the (γ,n) and (γ,α) reactions on ^{151}Eu, for temperatures of T$_9$ = 2 to 3. The curves show the rates vs target excitation energy, after which the total integrated rate is given. The Gamov peak values for the (α,γ) reaction at temperatures of T$_9$=2.0, 2.5, and 3.0 are, respectively, 5.5, 6.9, and 8 MeV; for the corresponding E$_\alpha$ energies add 2 MeV. These peak values are reasonable for the convolution of the Maxwell-Boltzmann function with the (α,γ) excitation function, whereas for the stellar (γ,α) rate functions shown in Fig. 6a the skewing of the curves' peak values to much lower excitation energies emphasizes the dramatic effect due to the shape of the Planck photon distribution. Hence, very low energy alpha particles contribute greatly to the stellar photonuclear rates. In Fig. 6b, our arbitrary cut off of ^{151}Eu target states at 8 MeV might appear to be too low an energy. However, an examination of the curve values near 8 MeV indicates little of the total integral was lost. Also, the behavior of the curves at low energies shows that discrete level information for the target nucleus is essentially not needed due to the moderation of the Boltzmann factor by the rapidly increasing level density function. Should one wish to compute stellar photonuclear rates at even higher temperatures, rate curves such as these, along with the calculated values of the neutron widths, should help decide the highest energy in ^{151}Eu where states are sufficiently long lived to be considered available target states.

We now wish to compare our stellar photonuclear reaction rates calculated directly with those obtained via detailed balance. First we must discuss the spin-parity distribution in the first compound nucleus, which must be the same for the forward and reverse reactions if detailed balance is to obtain. Photons produce a narrow distribution regardless of photon energy; an extreme example would be dipole transitions from a spin-zero target--only J=1 states could be produced. Neutrons incident on the (Z,N-1) nucleus can produce a much broader distribution, even for relatively low incident energies. Alpha particles can bring in many partial waves and

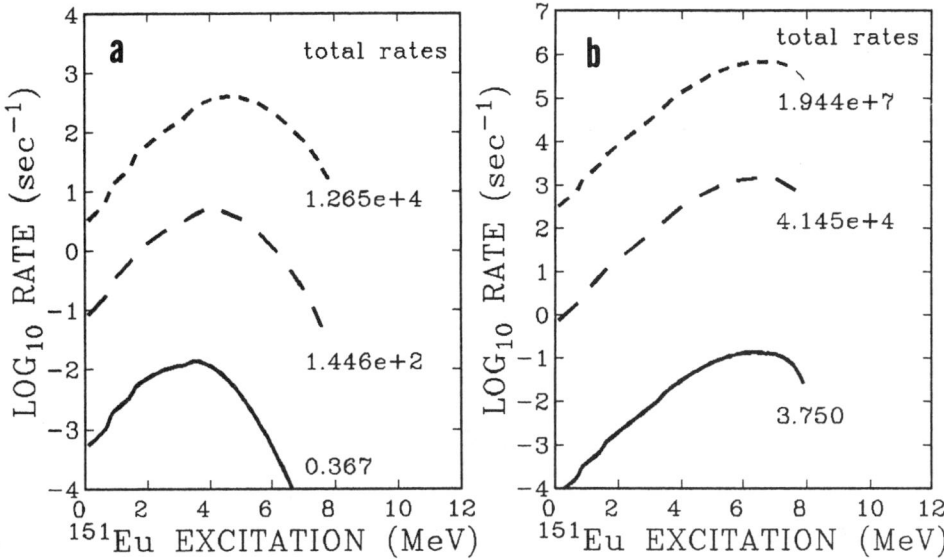

Figs. 6a,b. Stellar photonuclear rates vs target excitation energy for the $(\gamma,\alpha)^{147}$Pm and $(\gamma,n)^{150}$Eu reactions, respectively, plus total rates.

produce very broad distributions. As noted above, our STAPLUS code allows us to supply a spin and parity distribution for the first compound nucleus, normalized to the actual (photon) reaction cross section. We then repeated our (γ,n) and (γ,α) excitation function calculations, changing the initial spin distribution to that which would have been produced by a neutron or an alpha particle of the appropriate center-of-mass energy, for ^{151}Eu excitation energies of 15 MeV down to the reaction threshold. The results appear in Figs. 7a and 7b, where the ratio of the directly calculated stellar rates to those obtained via detailed balance are given for the two types of photonuclear calculations--the normal calculation using the photon-produced spin distribution and the modified calculation using the particle-produced distribution as input. For the first case, both the (γ,α) and (γ,n) rates are about twice those inferred from detailed balance. When the particle-produced distributions were used, the (γ,α) ratios differed not greatly from unity, whereas the (γ,n) ratios decreased about 20% or so.

From our results so far, it cannot be said unequivocally that the direct calculation of stellar photonuclear rates is a more accurate procedure than the traditional detailed balance approach. However, because both the (γ,α) and (γ,n) stellar rates differ from the detailed balance results by about the same factor, and the reactions are so vastly different, there is some suggestion that the direct calculational approach is more reliable. Recall the function $J(E_t)$ in the equation in Fig. 1. For ^{151}Eu this function ranges from 1.5 at low target excitation energies to 5.5 for target energies near 8 MeV. As such, it tends to resemble the neutron distribution (although the higher spins are not represented), but it does not emulate the alpha particle distribution. Therefore, when the latter distributions are inserted into the (γ,α) calculations, Fig. 7a shows that the results reproduce those from detailed balance. Because the neutron distributions are more accurately reproduced by the $J(E_t)$ function, in the direct stellar rate calculation, their insertion as input produces only the small change shown in Fig. 7b. This suggests to us that the detailed balance requirement of identical spin-parity distributions for the forward and reverse reactions is not being fulfilled--certainly not in our calculations and probably not in nature.

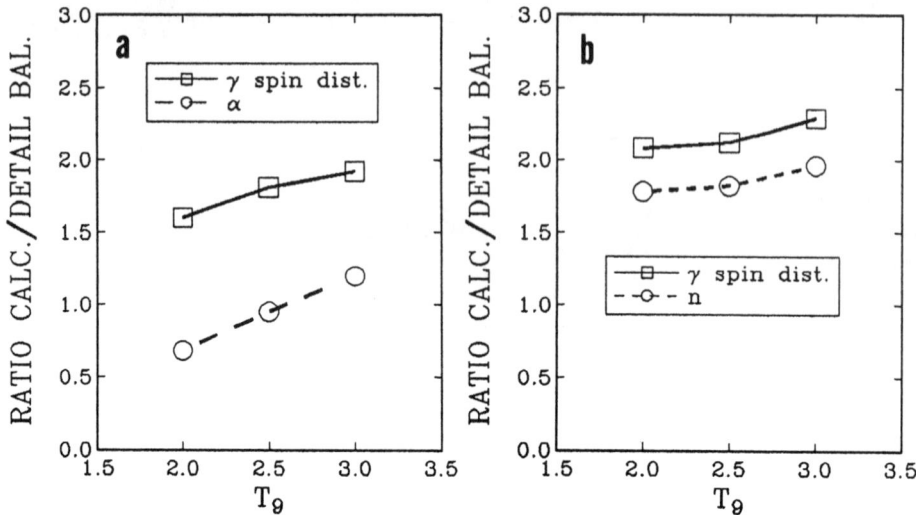

Figs. 7a,b. Stellar rates for ^{151}Eu$(\gamma,\alpha)^{147}$Pm and ^{151}Eu$(\gamma,n)^{150}$Eu reactions obtained from direct calculations ratioed to those inferred from detailed balance.

CONCLUSIONS

Using ^{151}Eu as the target isotope and with realistic nuclear parameters, we have compared the direct calculations of the stellar photo-reaction rates with those inferred from detailed balance. Because interest in the astrophysical p-process initiated this study, the temperature range was confined to $T_9 = 2$ to 3. A number of conclusions may be drawn:

1. The directly computed stellar photonuclear rates for the (γ,α) and (γ,n) reactions were about 2 to 3 times the rates inferred from detailed balance. It is argued, but not proved, that the direct calculational approach is the more accurate.
2. The direct computational approach is not limited to a single emitted particle; multiple particle or no particle emission may be treated as easily, i.e. $(\gamma,2n)$, $(\gamma,\alpha n)$, a ground state populating an isomer, an isomeric state depopulated back to the ground state, etc.
3. The direct calculation is much less dependent on knowledge of the low-energy nuclear level structure than the usual approach.
4. The photonuclear excitation functions themselves provide useful information concerning the temperature at which the inverse reaction may begin to deviate from the restrictions imposed by detailed balance. Further, when they are used to produce stellar photonuclear rates as a function of the target excitation energy, one obtains additional information on the best choice of the energy range of the integrals shown in Fig. 1.
5. The p-process nuclei are stable while their predecessors often are not. Photonuclear reactions on stable nuclei should be easier to compute than capture reactions on unstable nuclei, simply because more experimental data are available to supply input to the nuclear reaction calculations.

ACKNOWLEDGMENTS

The authors are indebted to S.E. Woosley and R.A. Ward for valuable conversations that greatly increased our understanding of the astrophysical aspects of this work, and to W.M. Howard who first introduced us to the p-process and some of the problems involved in the relevant rate factor calculations. Special thanks are given to Sue Frumenti for the preparation of this manuscript.

REFERENCES

1. E.M. Burbidge, G.R. Burbidge, W.A. Fowler, and F. Hoyle, Rev. Mod. Phys. 29, 547 (1957).
2. J. Audouze and J.W. Truran, Astrophys. J. 202, 204 (1975).
3. M. Arnould, Astron. Astrophys. 46, 117 (1976).
4. S.E. Woosley and W.M. Howard, Ap. J. Suppl. 36, 285 (1978).
5. M. Rayet, N. Prantzos, and M. Arnould, Astron. Astrophys. 227, 271 (1990).
6. J.A. Holmes, S.E. Woosley, W.A. Fowler, and B.A. Zimmerman, Atomic Data and Nuclear Data Tables 18, 305 (1976) and references therein.
7. Our version of the STAPRE code originally written by M. Uhl and B. Strohmaier, Report IRK 76/01 (1976, updated 1978).
8. F.N. Fritsch, Lawerence Livermore National Laboratory Report, UCID-30194 (1982).
9. J. Raynal, Report IAEA-SMR-9/8 (1972).
10. R.L. Macklin and P.G. Young, Nucl. Sci. Eng. 95, 189 (1987).
11. C.M. Perey and F.G. Perey, Atomic Data and Nuclear Data Tables 17, 6 (1976).
12. L. McFadden and G.R. Satchler, Nucl. Phys. 84, 177 (1966).
13. D.G. Gardner, Chapter III, Neutron Radiative Capture, Vol. 3, ed. R. Chrien (Pergamon Press, Oxford, 1984), p. 91-101; one free parameter, E_x, changed to 11 MeV (1984).
14. T.J. Boal, E.G. Muirhead, and D.J.S. Findlay, Nucl. Phys. A406, 257 (1983).
15. J.R. Huizenga and G.J. Igo, Argonne National Laboratory Report ANL-6373 (1961).
16. E.V. Verdieck and J.M. Miller, Phys. Rev. 153, 1253 (1967).

CROSS-SECTION MEASUREMENTS ON RADIOACTIVE SAMPLES

P.E. Koehler and H.A. O'Brien
Physics Division, Los Alamos National Laboratory,
Los Alamos, New Mexico 87545, USA

ABSTRACT

We have developed a system over the past few years at the Manuel Lujan, Jr. Neutron Scattering Center (LANSCE) for making (n,p) and (n,α) measurements on (mainly) radioactive nuclei. Measurements have included ^{7}Be(n,p)^{7}Li, ^{14}N(n,p)^{14}C, ^{17}O(n,α)^{14}C, ^{22}Na(n,p)^{22}Ne, ^{22}Na(n,α)^{19}F, ^{35}Cl(n,p)^{35}S and ^{36}Cl(n,p)^{36}S. The major basic physics motivation for these measurements has been the nuclear astrophysics to be learned. Currently, we are assembling a 4π detector of barium fluoride (BaF$_2$) for making (n,γ) measurements on radioactive nuclei with relatively short half lives. Once operational, this new detector should allow us to expand our measurements to many more nuclei, and to a broader range of nuclear physics and nuclear astrophysics issues addressed. Results of recent measurements are given and future plans are discussed.

INTRODUCTION

Measurements of A^*(n,p), (n,α) and (n,γ) cross sections, where A^* is a radioactive nucleus, have applications in nuclear physics, nuclear astrophysics, and in applied programs. The major basic physics motivation for our measurements has been the nuclear astrophysics to be learned. For example, the ^{7}Be(n,p)^{7}Li reaction[1] is important in calculations of the big-bang nucleosynthesis of ^{7}Li, cross sections for the ^{22}Na(n,p)^{22}Ne and ^{22}Na(n,α)^{19}F reactions[2] may aid in the explanation of the neon-E anomaly in meteorites[3] or in the possibility of observing nucleosynthetic ^{22}Na with future gamma-ray telescopes[4], and the ^{36}Cl(n,p)^{36}S reaction may be important in the nucleosynthesis of the very rare stable isotope ^{36}S in both explosive nucleosynthesis[5] and in the s-process[6]. There are a few other (n,p) and (n,α) reactions on radioactive nuclei of importance to nucleosynthesis calculations[5], and even some stable targets of importance[7]. In addition, there are a number of (n,γ) cross sections on radioactive targets which if measured would provide an almost model independent understanding of the dynamics of the s-process of nucleosynthesis[8]. This is important because tests of current stellar models of the dynamical s-process environment are still obscured by the present uncertainties of the nuclear input data[9].

The measurement of fission cross sections on radioactive targets is another area of interest. We have recently completed (n,f) measurements on ^{250}Cf, ^{254}Es and ^{247}Cm (ref. 10). Finally, cross sections on radioactive targets are important in the study of the feasibility of using an accelerator driven neutron source to transmute nuclear waste[11].

Most previous measurements[12-15] on radioactive samples have been limited to nuclei with very long half lives due to potentially large backgrounds associated with the sample activity. The advent of the pulsed spallation neutron source at LANSCE[16] has opened up the possibility of making cross-section measurements for neutron-induced reactions on nuclei with short half lives. In this paper, we will give some examples of recent measurements of this type and briefly discuss the nuclear astrophysics to be learned. We will then discuss our plans for additional measurements, outlining the techniques involved and the nuclear astrophysics to be learned.

GENERAL EXPERIMENTAL TECHNIQUE

Cross section measurements on short-lived radioactive samples require a large peak neutron intensity and a properly designed detector so that the detected rate for the reaction of interest is larger than the background rate associated with the decay of the sample under study. At LANSCE, neutrons are produced by spallation when the 800 MeV proton beam from LAMPF strikes a tungsten target. The neutrons are then slowed down in a water moderator and, on our beam line, are collimated by layers of iron, brass, borated polyethylene, concrete, and lead to define a beam spot 0.5 cm in diameter at a position of 7 m from the moderator. The flux at 1 eV has been measured[17] to be 2.3×10^6 neutrons/(s eV cm^2) for a proton current of 57 µA, and hence, at the 100 µA design intensity of LANSCE, the flux is expected to be about 4×10^6 neutrons/(s eV cm^2). The flux follows an approximately 1/E shape in the neutron energy range from 0.2 eV to 100 keV. This high neutron intensity allows measurements to be made with very small sample sizes in the approximately 100 ng to a few hundred µg range. In addition, because LANSCE runs at the low repetition rate of 20 Hz, the peak neutron intensity is very high, allowing us to make measurements on relatively short-lived radioactive samples. The small sample size means that the necessary radioactive samples are easier to produce and that only relatively modest activities must be handled. However, even these small samples can still be difficult to obtain and can present some rather large background problems.

The small beam spot size of 0.5 cm was chosen so that a large solid angle could be covered with a relatively small detector. This is important in (n,p) and (n,α) measurements where, to minimize the background, solid state detectors of minimum thickness, and hence relatively small area (typically 50-450 mm^2), have usually been employed. Having a compact detector is important for (n,γ) measurements because the best scintillator for this use, BaF$_2$, is relatively expensive. Because the requirements for the samples as well as the detectors differ, we will discuss (n,p) and (n,γ) measurements separately below.

A*(n,p) AND A*(n,α) MEASUREMENTS

For A*(n,p) and (n,α) measurements, there are three major requirements that a target must satisfy. First, the target must be thin enough to allow the charged particles to emerge without too much straggling. Typical Q-values for reactions of interest range from about 0.5 to a few MeV. Hence acceptable target thicknesses range from a few tens of µg/cm^2 to as high as about 1 mg/cm^2. Because of our small beam spot size, this is usually the hardest requirement to satisfy, and often results in the need for a fairly high specific activity. Second, impurities with positive Q-values for (n,p) and/or (n,α) reactions and having comparatively large cross sections must be kept to a minimum. In practice the most troublesome impurities are ^{10}B and ^{6}Li. Hence, targets made from chemical compounds are almost always acceptable provided the target is not too thick and the ^{6}Li and ^{10}B contamination is low. Third, the target backing should be as thin as practical, and should have a low neutron scattering cross section. In most cases this requirement is easy to satisfy by using a backing of aluminum foil as thick as 25 µm.

For A*(n,p) and A*(n,α) measurements the sample-related backgrounds can be reduced to manageable levels by choosing a charged-particle detector of minimal thickness. The detection efficiency for radioactive decay emissions from the sample can thus be reduced to order 10^{-6} of the proton or alpha detection efficiency. For most of our measurements, we have detected the charged particles with silicon surface-barrier detectors[1]. With these detectors it has been routine to measure cross sections from thermal energy to at least 10 keV. In favorable cases, the upper energy range has been extended to about 100 keV.

To extend our capabilities to isotopes with relatively small cross sections, or to samples available at only low specific activity (or low enrichment), or to samples only available as gasses, we have been developing a compensated ionization chamber[18]. By mounting the chamber closer to the moderator where the collimated beam is larger, we could in principle gain a factor of 50-100 in counting rate over our solid state detector setup. In practice, the high energy part of the flux from the LANSCE source causes a large background at short time-of-flight for even a compensated ionization chamber, and this has so far limited (n,p) and (n,α) measurements with this ionization chamber to below approximately 1 keV. On the other hand, we have made measurements of the $^{35}Cl(n,p)^{35}S$ cross section (E_p=0.598 MeV) with this chamber up to 100 keV (ref. 18), by inserting about 5 cm of lead in the neutron beam. This reduces the neutron flux to about 15% of its unattenuated intensity, so that the counting rate gain is only about 5-10 over our solid state detector setup. We are currently investigating ways of improving the performance of this ionization chamber.

$A^*(n,\gamma)$ MEASUREMENTS

For $A^*(n,\gamma)$ measurements, potentially everything on the target (including the isotope under study) is a source of background caused by neutrons scattered from the target later capturing in the BaF_2 detector. Hence the ideal target would be monoisotopic and self supporting. First, this means that chemical compounds are usually not acceptable. Second, the specific activity must be high. Finally, the target backing foil should be as thin as possible. Calculations indicate that, for example, carbon foils of about 30 μg/cm^2 or thinner will be needed in most cases.

Pileup of the low-energy decay γ-rays from the radioactive sample is a much more difficult background to overcome for (n,γ) measurements than in the case of (n,p) measurements. To overcome this potentially large background one can make use of the fact that the γ-ray decay energy, E_d, is almost always much lower in energy than the total energy, E_c, of the neutron capture cascade. Hence, a detector which registers all of the energy from the capture cascade, and which has a very short output pulse width ,τ, can effectively overcome this background. Of course, the size of the pileup is a very strong function of the ratio, E_d/E_c, and of τ, and one can always think of very difficult cases for which measurements are still not possible. Our calculations[19] indicate that measurements on many interesting samples with half lives as short as a few months can be made.

We feel that BaF_2 is the best choice as a detector for these measurements. The fast component of BaF_2 (τ≈20 ns for 15 cm crystals[20]) provides for effective pileup rejection. Monte Carlo calculations we have made using the computer code CYLTRAN[21] indicate that a thickness of 15 cm of BaF_2 is adequate to make an approximately 100% efficient detector in essential agreement with the calculations of Wisshak et al.[22] In our application, the small neutron beam allows a very compact detector. A 30 cm cube of BaF_2 with a 4 cm beam hole is sufficient to make a detector which is approximately 100% efficient. We are currently constructing such a detector composed of eight 15 cm cubes. A drawing of a side view of our beam line with the BaF$_2$ detector installed is shown in fig. 1 while an isometric drawing of the detector is shown in fig. 2.

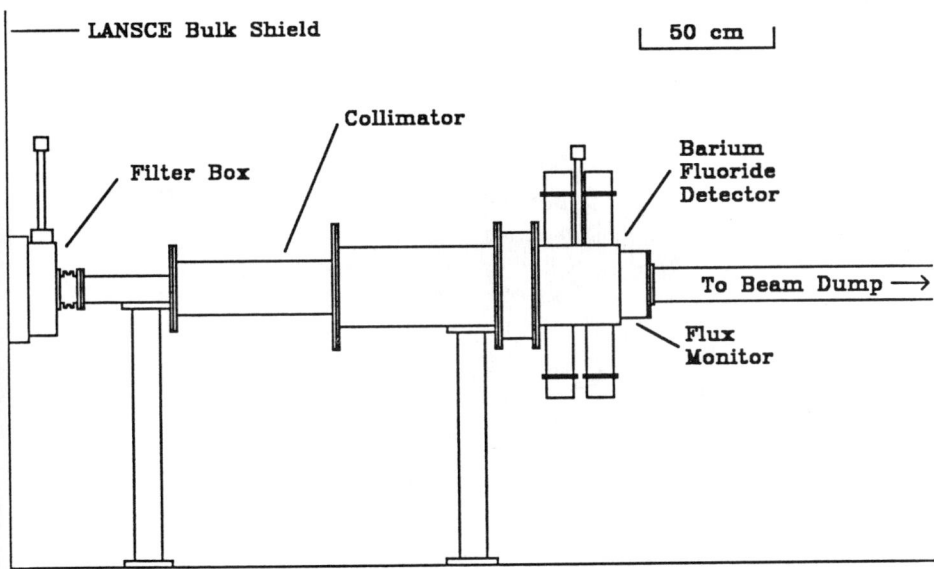

Fig. 1 Side view of collimator and BaF$_2$ detector on flight path 4 in experimental room 1 at LANSCE. The center of the BaF$_2$ detector is 7 m from the neutron source/moderator. The LANSCE bulk shield ends at about 4.7 m from the moderator. Inside the bulk shield is additional collimation as well as a mercury shutter for "turning off" the neutron beam. Outside the bulk shield, in experimental room 1, the collimator is stepped so that there are no straight unobstructed paths which a background neutron can take to the region of the detector. Just outside the bulk shield, a filter box allows us to remotely insert black filters (e.g. U, Co, etc.) into the beam to study backgrounds as well as to calibrate the time-of-flight to energy conversion. Not shown is the substantial shielding both parallel and perpendicular to the collimator, the "house" of iron, borated polyethylene and lead shielding around the detector, and the beam dump (at about 14 m from the moderator).

Initial tests during the previous (1989) LAMPF run cycle had indicated that the background was very large, but that a large percentage originated from "streaming paths" along the outside of our collimator, from a lack of adequate shielding perpendicular to our collimator, and from poor shielding of the adjacent LANSCE beam line. During the cycle break all three of these problems were addressed. Our collimator (which had a constant outside diameter) outside the LANSCE bulk shield was replaced by the stepped collimator shown in fig. 1. All available space between the bulk shield and the detector was filled with layers of iron, polyethylene, and borated polyethylene perpendicular to the beam direction. Outside of this shielding there are other layers of these materials parallel to the beam. Finally, the adjacent

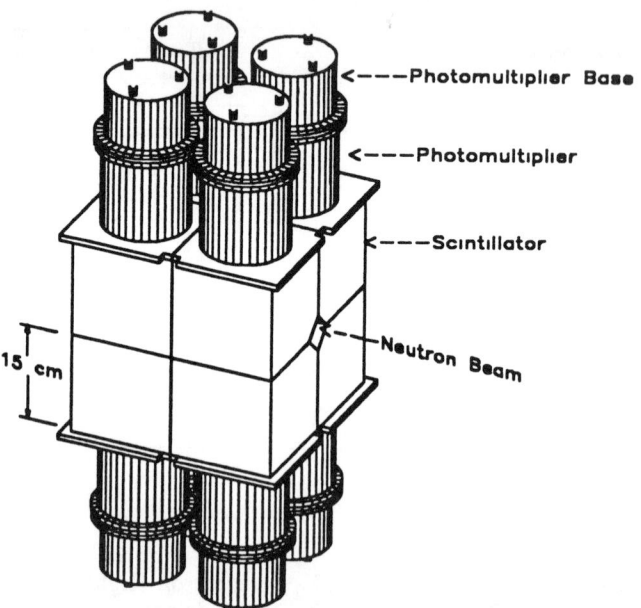

Fig. 2. Isometric drawing of the BaF$_2$ detector for A*(n,γ) measurements at LANSCE. The detector is a 30 cm cube of BaF$_2$ composed of eight 15 cm cubes each coupled to its own photomultiplier and base. Each of the eight cubes is beveled on two edges so that when the detector is assembled there is a 4 cm wide hole through the center of the detector to allow the beam to pass through, and another 4 cm hole for a target ladder.

beam line was completely realligned and reshielded. All of this new shielding and collimation followed the best recipe resulting from the extensive Monte Carlo calculations of Russell[23]. Initial tests during the current run cycle indicate that the background from neutrons capturing in the iron shielding of both our beam line as well as adjacent LANSCE flight paths is still very large. We are currently investigating ways of reducing this background.

SOME EXAMPLES OF RESULTS TO DATE

A first example of an important A*+n reaction rate is ^{7}Be(n,p)^{7}Li. This reaction is important to the primordial nucleosynthesis[24-26] of ^{7}Li. Our measurements[1] have substantially reduced (by a factor of almost ten at thermal energy) the uncertainty in the reaction rate. Furthermore, the rate based on our data is only 60% to 80% of the old rate[27] in the temperature range of interest in big-bang calculations. This difference can lead to as much as a 20% increase in the amount of ^{7}Li calculated to be produced in the big bang.

The ^{22}Na(n,p)^{22}Ne reaction may play a role in the nucleosynthesis of ^{22}Na and/or ^{22}Ne in explosive environments. An understanding of the nucleosynthesis of these isotopes is important because the origin of the Neon-E anomaly in meteorites[3] is not well understood, and because ^{22}Na has been suggested as a candidate for observation by gamma-ray telescopes[4]. The theoretical rate[28] is about a factor of ten lower than the rate determined from our data[2] one at very low temperatures. However, due to a resonance at E_n=170 eV, the two rates cross at 0.05 GK, and the theoretical rate is about a factor of six low at the highest temperatures measured (T=0.3 GK). If this difference between the experimental and theoretical rates persists to higher temperatures, it may result in a significant increase in the calculated production of ^{22}Na in explosive environments.

The ^{36}Cl(n,p)^{36}S reaction was calculated by Howard et al.[5] to be important in the nucleosynthesis of rare stable nucleus ^{36}S in explosive carbon burning. Using a theoretical estimate for the rate for this reaction, they claculated that ^{36}S was much overproduced in this environment. Alternatively, Beer and Penzhorn[6] calculate that most or all of the ^{36}S may have been produced in the s process, in which case the ^{36}Cl(n,p)^{36}S reaction is also important. The ^{35}Cl(n,p)^{35}S reaction may also be important in these environments. Our measurements were made with few hundred microgram targets of ^{35}Cl and ^{36}Cl. Because ^{35}Cl is stable, and the half life for ^{36}Cl is very long, a high peak neutron intensity is not essential to the measurements, but the relatively high average neutron intensity available from LANSCE is still important to measuring these comparatively small cross sections within a reasonable time. Our preliminary data for ^{36}Cl have been reported in ref. 19, while the data for ^{35}Cl were reported in ref. 18. The data for both reactions reveal several resonances for energies greater than about 400 eV. It remains to be seen how our measurements will affect the results of future nucleosynthesis calculations.

As a final example, our ^{14}N(n,p)^{14}C measurements[7] indicate that this cross section has very close to a 1/v shape from thermal energy to approximately 30 keV. The reaction rate[29] based upon an interpolation between direct measurements at thermal energy[30] and for E_n>150 keV (ref. 31), and upon measurements made using the inverse reaction[32] for E_n>29 keV, is in good agreement with our data. These results are in disagreement with a recent direct measurement[33] which is a factor of 2 to 3 lower. With the larger reaction rate indicated by our data and previous data, it appears that, in contrast to the claim in ref. 33, ^{14}N acts as an effective neutron poison in many scenarios in which the ^{13}C(α,n)^{16}O reaction might otherwise be a viable s-process neutron source.

FUTURE PLANS

There are a few more (n,p) and (n,α) cross sections of interest to nuclear astrophysics which should be measurable at LANSCE. Potential samples include ^{25}Mg, ^{26}Al, ^{37}Ar and ^{41}Ca. These measurements could have a significant impact on our understanding of nucleosynthesis both in explosive environments and during the s process.

Potentially, the largest number of future measurements to be made are of $A^*(n,\gamma)$ cross sections which are mainly of interest for a better understanding of the dynamics of s-process nucleosynthesis. Our planned measurements on radioactive samples, coupled with the new program of very precise measurements on stable isotopes at Karlsruhe[34], and new calculational approaches[9,35] should lead to a much better understanding of the s-process, including its dynamics.

In addition to reducing the background in the BaF_2 detector, the greatest challenge for the future measurements appears to be obtaining suitable targets[36]. There are substantial difficulties both in obtaining the radioactive isotopes and in manufacturing the samples. We have attempted measurements on ^{26}Al, ^{54}Mn, ^{55}Fe, and ^{57}Co. In all these cases, the low specific activity and/or impurities (^6Li and/or ^{10}B) in the targets prevented successful measurements. To overcome the problem of the lack of availability of high specific activity radioisotopes we have installed a target irradiation station at the M.D. Anderson Hospital Cyclotron Facility. The target can be remotely inserted into the beam line for irradiation during periods between patient treatments. It is anticipated that some high specific activity targets will be made by (p,n) reactions on (enriched) stable isotopes. Alternatively, the isotope may be made by some less expensive means (e.g. a reactor or the IPF at LAMPF[37]) and later enriched using an isotope separator. Such targets have already been made by using an "on-line" isotope separator[38]. In the end, both techniques may be fairly expensive.

Also, most of our radioactive targets have been made by deposition from a water solution[36]. This technique works well when not much material is needed. However, when the specific activity and/or the cross section is low, and hence a fairly thick target is needed, this technique yields very non-uniform results. To obtain more uniform and robust targets, we are currently assembling a vacuum evaporation apparatus because we were unable to find one that was available for making non-actinide radioactive targets.

CONCLUSIONS

With the advent of intense pulsed spallation neutron sources such as LANSCE cross section measurements on small quantities of short-lived radioisotopes has become possible. These measurements have applications in nuclear physics, nuclear astrophysics, and in applied programs. We have given some examples of how these data can be important to the study of the nucleosynthesis of the elements in different astrophysical environments.

We are indebted to R.M. Mortenson and W.A. Teasdale for valuable technical assitance.

REFERENCES

1. P.E. Koehler et al., Phys. Rev. C **37**, 917 (1988).
2. P.E. Koehler and H.A. O'Brien, Phys. Rev. C **38**, 2019 (1988).
3. D.C. Black, Geochim. Cosmoschim. Acta **36**, 377 (1972).
4. D.D. Clayton, Astrophys. J **198**, 151 (1975).
5. W.M. Howard et al., Ap. J. **175**, 201 (1972).
6. H. Beer and R.-D. Penzhorn, Astron. Astrophys. **174**, 323 (1987).
7. P.E. Koehler and H.A. O'Brien, Phys Rev. C **39**, 1655 (1989).

8. F. Kappeler et al., Ap. J. **257**, 821 (1982).
9. F. Kappeler, R. Gallino, M. Busso, G. Picchio, and C.M. Raiteri, Astrophys. J **354**, 630 (1990).
10. M.S. Moore, these proceedings
11. C.D. Bowman, P.W. Lisowski, and E.D. Arthur, In Proc. Second Inter. Symp. on Advanced Nucl. En. Research, (Japan Atomic Energy Research Institute, 1990) p. 149.
12. A. Emsallem et al., Nucl. Phys. **A368**, 108 (1981); H. Weigmann et al., Nucl. Phys. **A368**, 117 (1981).
13. Yu.M. Popov et al., Z. Phys. A **322**, 685 (1985).
14. H.P. Trautvetter et al., Z Phys. A **323**, 1, (1986).
15. Yu.M. Gledenov, Yu.P. Popov, J. Rigol,and V.I. Salatsky, Z. Phys. **A322**, 685 (1985).
16. R.N. Silver, Physica **137B**, 359 (1986).
17. P.E. Koehler, Nucl. Instr. and Meth. **A292**, 541 (1990).
18. P.E. Koehler and H.A. O'Brien, in Reports to the DOE Nuclear Data Committee, May 1990, Brookhaven National Laboratory Report BNL-NCS-44362, p. 79; and P.E. Koehler and H.A. O'Brien, Los Alamos National Laboratory Report LA-11699-PR, p. 28.
19. P.E. Koehler, H.A. O'Brien and C.D. Bowman, American Institute of Physics Conf. Proc. 170, eds. N. Gehrels and G.H. Share, (Am. Inst. Phys., New York, 1988) p. 143.
20. K. Wisshak and F. Kappeler, Nucl. Instr. Meth. **227**, 91 (1984).
21. J.A. Halblieb, Sr. and W.H. Vandevender, "CYLTRAN", Sandia National Laboratories Report, SAND 74-0030 (1974); H.H. Hsu et al., IEEE Trans. Nucl. Sci. **NS-31**, 390 (1984).
22. K. Wisshak, F. Kappeler and G. Schatz, Nucl. Instr. Meth **221**, 385 (1984).
23. G.J. Russell, Proc. 10th Meeting of the International Collaboration on Advanced Neutron Sources, American Institute of Physics Conf. Series, No. 97 (1989) p. 809.
24. G. Beadet and H. Reeves, Astron. Astrophys. **134**, 240 (1984).
25. J. Yang et al., Ap. J. **281**, 493 (1984).
26. P. Delbourgo-Salvador, C. Gry, G. Malinie and J. Audouze, Astron. Astrophy. **150**, 53 (1985).
27. N.E. Bahcall and W.A. Fowler, Ap. J. **157**, 659 (1969).
28. S.E. Woosley et al., At. Data Nucl. Data Tables **22**, 371 (1978).
29. W.A. Fowler, G.R. Caughlan and B.A. Zimmerman, Ann. Rev. Astron. Astrophys. **13**, 69 (1975).
30. F. Ajzenberg-Selove, Nucl. Phys. **A268**, 1 (1976).
31. C.H. Johnson and H.H. Barschall, Phys. Rev. **80**, 818 (1950).
32. J.H. Gibbons and R.L. Macklin, Phys. Rev. **114**, 571 (1959).
33. K. Brehm et al., Z. Phys. A **330**, 167 (1988).
34. K. Wisshak et al., Nucl. Instr. and Meth. **A292**, 595 (1990).
35. W.M. Howard et al., Ap. J. **309**, 633 (1986).
36. P.E. Koehler, H.A. O'Brien, and J.C. Gursky, submitted to Nucl. Instr. and Meth. (1990).
37. H.A. O'Brien, Nucl. Instr. and Meth. **B40/41**, 1126 (1989).
38. R.A. Naumann et al., Nucl. Instr. and Meth. **B26**, 59 (1987).

CROSS SECTION OF THE ^{7}Li(n, γ)^{8}Li REACTION AT STELLAR ENERGY

Y. Nagai, M. Igashira, N. Mukai, K. Takeda, F. Uesawa, T. Ohsaki,
T. Ando and H. Kitazawa
Tokyo Inst. Tech., Oh-okayama, Meguro-ku, Tokyo 152, Japan
S. Kubono and T. Fukuda
University of Tokyo, Tanashi, Tokyo 188, Japan

ABSTRACT

The reaction ^{7}Li(n, γ)^{8}Li is a trigger reaction for the nucleosynthsis of intermediate-mass nuclei in inhomogeneous big-bang models[1]. The reaction rate was measured at neutron energy of 30 keV by detecting prompt γ-ray from the reaction. The obtained value is two times larger than that measured recently and is consistent with the estimated value from the thermal neutron capture cross section by usig 1/v law.

INTRODUCTION

In inhomogeneous big-bang models, where two regions of high-density proton-rich and low-density neutron-rich could be formed and in the latter region the nucleosynthesis of intermediate-mass nuclei could occur, the ^{7}Li(n, γ)^{8}Li reaction is important to bridge an unstable A=8 nucleus to a stable A=11 one. Recently the reaction cross section was reinvestigated and the value was about two times smaller than both previously measured one[2]. In order to understand the discrepancy we have measured the cross section with a different method.

EXPERIMENTAL PROCEDURE

The experiment was carried out by using pulsed neutrons and detecting prompt γ-ray from the reaction. Pulsed neutron beam was produced by bombarding ^{7}Li target with the 1.5 ns pulsed proton beam from the 3.2 MV Pelletron accelerator atthe Tokyo Institute of Technology. The proton energy was adjusted to 1.898 MeV in order to produce neutrons with the most probable energy of 30 keV. γ-rays from the captured state were measured by an anti-Compton NaI(Tl) detector, which consists of a central 7.6 cmϕ * 15.2 cm NaI(Tl) detector surrounded by annular NaI(Tl) detector with the size of 25.4cmϕ * 28cm^3. A metalic Li sample with size of 60 mmϕ * 26 mm was packed by thin vinyl film and was placed on the proton beam axis, 8.6 cm away from the Li target. The absolute capture cross section of ^{7}Li was obtained by comparing the prompt γ-ray yield of Li with that of ^{197}Au.

EXPERIMENTAL RESULT

The background subtracted γ-ray spectrum from the ^{7}Li(n, γ)^{8}Li reaction is shown in fig. 1. The 2.06 MeV γ-ray peak corresponds to the γ-raytransition from the captured state of ^{8}Li to the ground state. The γ-ray intensity was obtained by a stripping method using the response fuction of the NaI(Tl) detector. The partial capture cross section of ^{7}Li was obtained as 35.4(60) μb for the γ-transition from the captured state of ^{8}Li to the ground state. Here the abundance of ^{6}Li in the Li sample, measured by the mass separator as 92.48 % of ^{7}Li and 7.52 % of ^{6}Li, was corrected. As the γ-ray from the captured state to the first excited state was not clearly detected, the partialcross section correspondig to the γ-ray transition was estimated by using the γ-ray branching ratio from the captured state of thermal neutrons to the first excited state[4]. Thus total capture cross section was obtained as 39.3(60) μb, which is two times larger than the previously measured value and is consistent with the estimated value from the thermal neutron capture cross section of ^{7}Li by using a 1/v law.

CONCLUSION

The neutron capture rate of the important reaction to produce ^{8}Li was measured by the ^{7}Li(n, γ)^{8}Li reaction using most probable neutron energy of 30 keV. Direct prompt γ-ray detection method gave the total capture rate of 39.3(60) μb.

As the value is two times larger than the previous one, it enhances the nucleosynthesis of intermediate-mass nuclei in the early universe.

We would like to thank Dr. T. Kajino for his useful discussions.

REFERENCE

1. R. A. Malaney and W. A. Fowler, Astrophys. J. 333, 14 (1988)
2. M. Wiescher, R. Steininger and F. Käppeler, Astrophys. J. 344, 464 (1989)
3. M. Igashira, H. Kitazawa and N. Yamamuro, Nucl. Instr. Meth. A245, 432 (1986)
4. F. Ajzenberg-Selove, Nucl. Phys. A490, 1 (1988)

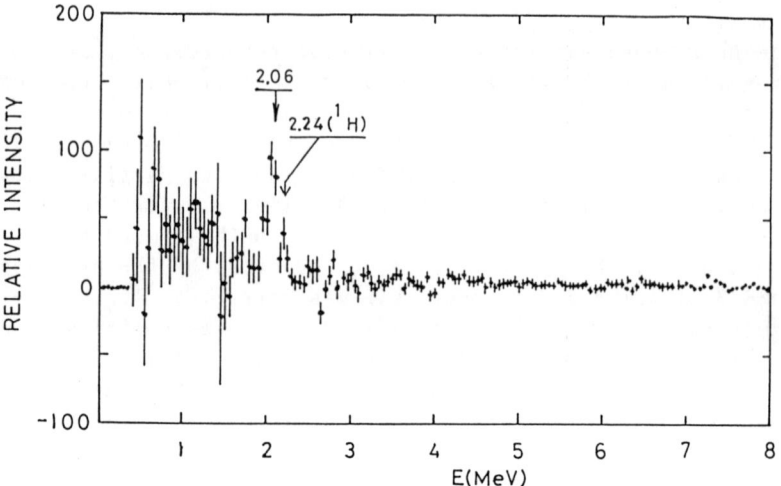

Fig. 1.
Background subtracted γ-ray spectrum from the ^{7}Li(n, γ)^{8}Li reaction

CAPTURE RATE OF THE $^{12}C(n,\gamma)^{13}C$ REACTION IN INHOMOGENEOUS BIG-BANG MODELS AND STELLAR EVOLUTION

Y. Nagai, M. Igashira, K. Takeda, N. Mukai, F. Uesawa, S. Motoyama and H. Kitazawa
Tokyo Inst. Tech., Oh-okayama, Meguro-ku, Tokyo 152, Japan
T. Fukuda
Univ. Tokyo, Tanashi, Tokyo 188, Japan

ABSTRACT

The cross section of the $^{12}C(n,\gamma)^{13}C$ reaction was measured at neutron energy of 30 keV by detecting prompt γ-rays from the reaction. The cross section was obtaine as 16.8 ± 2.1 μb, which favors the nucleosynthesis of intermediate-mass nuclei in inhomogeneous big-bang models and in neutrino induced nucleosynthesis.

INTRODUCTION

It has been discussed extensively that baryon inhomogenity caused by a quark-hadron phase transition could affect dramatically on primordial nucleosynthesis[1]. The inhomogeneity could form both high-density proton-rich and low-density neutron-rich regions. Consequently a sufficient production of intermediate mass nuclei has been expected by a primodial r-process in low-density neutron-rich regions[2]. This should be compared with the homogeneous standard big-bang model, which predicts an extremely low production of intermediate-mass nuclei[3].

These intermediate-mass nuclei can be synthesized by the following reaction sequence[4]:

$^1H(n,\gamma)^2H(n,\gamma)^3H(d,n)^4He(t,\gamma)^7Li(n,\gamma)^8Li$
$^8Li(\alpha,n)^{11}B(n,\gamma)^{12}B(\beta^-)^{12}C(n,\gamma)^{13}C(n,\gamma)^{14}C$ etc...

Since the cross section for the $^{12}C(n,\gamma)^{13}C$ reaction is not known, it is quite important to measure it for any quantitative discussion on the production rate of heavy elements. On the other hand quite recently strong interest has been aroused concerning a neutrino-induced nucleosynthesis[5].
In this process neutrinos emitted from a core-collapsed supernovae excite nuclei to particle unbound states. The neutrons, protons and α-particles, thus emitted, can be used to produce heavy elements. In particular if this neutrino process occurs in the helium shell of massive stars, where many carbons exist, and if the cross section of the $^{12}C(n,\gamma)^{13}C$ reaction is small nucleosynthesis by the rapid neutron process would occur.

Though the capture rate for the ^{12}C(n,γ)^{13}C reaction at stellar energy(kT=30keV) plays a crucial role as discussed above, the value is poorly known. An experimental value of 0.2±0.4 mb has been reported.

In the present work the neutron capture rate of ^{12}C was measured by detecting prompt γ-rays from the reaction.

EXPERIMENTAL PROCEDURE

The cross section of the ^{12}C(n,γ)^{13}C reaction measured by using pulsed neutrons and a γ-ray detector. The experimental details are discussed in ref. 7. The incident neutron energy spectrum measured with a ^{6}Li-glass scintillation detector using a time-of-flight(TOF) technique and a most probable neutoron energy of 30 keV was obtained.

Capture γ-rays were detected by an anti-Compton NaI(Tl) detector. The γ-ray events were stored ina minicomputer as two-dimensional data of TOF vs pulse height. Here four digital gates were set to obtain the prompt γ-ray spectrum as well as the background spectrum. A natural carbon disk of 5.4 cm in diameter and 1.5 cm in thickness (reactor grade) was used. The absolute capture cross section of ^{12}C was determined by comparing the prompt γ-ray yield of ^{12}C with that of ^{197}Au.

EXPERIMENTAL RESULTS

The γ-ray spectrum from the ^{12}C(n,γ)^{13}C reaction is shown in fig. 1. The γ-rays from the captured state of ^{13}C to the ground, first excited (3089keV), second excited (3684 keV) and third excited (3854keV) states and also from these excited to the ground state can be seen. The intensity of each peak in fig. 1 was extracted by a stripping method. The cross sections were derived individually by using the γ-ray intensities from the captured state. By adding these values the total capture cross section is obtained as 16.8(21)μb.

CONCLUSION

In the present work the cross section of the ^{12}C(n,γ)^{13}C reaction was measured by detecting prompt γ-rays and obtained as 16.8 ±2.1 μb. The result favors the nucleosynthesis of intermediate-mass nuclei in inhomogeneous big-bang models.

As for neutrino-induced nucleosynthesis, the present result indicates that neutrons emitted in the helium shell of massive stars could be used for the nucleosynthesis of heavy elements by a rapid neutron process.

We would like to thank to Dr. T. Kajino and Prof. H. Ohtsubo for useful discussions.

REFERENCES

1. J. H. Applegate, C. J. Hogan and R. J. Scherrer, Phys. Rev. D35, 1151 (1987)
2. J. H. Applegate, C. J. Hogan and R. J. Scherrer, Astrophys. J. 329, 529 (1988)
3. A. M. Boesgaard and G. Steigman, Ann. Rev. Astron. Astrophys. 23, 319 (1985)
4. R. A. Malaney and W. A. Fowler, Astrophs. J. 333, 14 (1988)
5. S. E. Wooseley, D. H. Hartmann, R. D. Hoffman and W. C. Haxton, Astrophys. J. 365, 272 (1990)
6. B. J. Allen and R. L. Macklin, Phys. Rev. C3, 1737 (1971)
7. Y. Nagai et al, in these proceedings

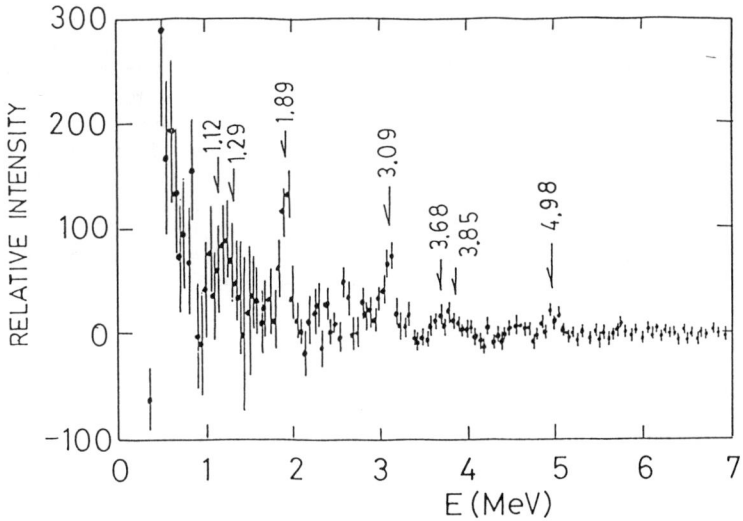

Fig. 1.
Background subtracted γ-ray spectrum from the $^{12}C(n,\gamma)^{13}C$ reaction

THE 30keV AVERAGED NEUTRON CAPTURE CROSS SECTIONS OF ^{56}Fe AND ^{60}Ni

F. Corvi, G. Fioni*, A. Mauri+ and T. Babeliowsky

Commission of the European Communities, Joint Research Centre
Central Bureau for Nuclear Measurements, Geel, Belgium

ABSTRACT

High resolution neutron capture measurements of ^{56}Fe and ^{60}Ni were performed with total energy detectors in the energy range 1-200 keV. The data have been used to derive the Maxwellian-averaged cross sections for thermal energies around kT=30 keV, corresponding to the mean temperature of the s-process of stellar nucleosynthesis.

INTRODUCTION

Accurate values of the neutron capture cross section in the keV region, as well as stellar abundances, are needed for most nuclei with atomic weight A≥56 lying in the valley of beta stability in order to study the details of the s-process of stellar nucleosynthesis. In fact, the actual quantity needed is the cross section averaged over a Maxwellian neutron energy distribution centred around kT=30 keV. Compilations including more than 250 cross sections have already been published[1] though not all items have the required precision. It should however be noted that a large fraction of these data were obtained with the so called total energy detectors i.e. those hydrogen free liquid scintillators to which a pulse-height weighting technique was applied in order to make their efficiency proportional only to the total γ-energy emitted. A better knowledge of response functions and efficiencies of these detectors, and therefore of the related weighting, has now been achieved thanks to a recently proposed experimental method [2,3]. Such advances have resulted in an improved accuracy of (n,γ) determination particularly for light and medium nuclei thus removing a 20% systematic error in the area of the 1.15 keV ^{56}Fe resonance. In view of this situation, it was decided to revisit the keV neutron capture field: to begin with, we chose to investigate ^{56}Fe, which is generally recognized as the seed nucleus for the starting of the s-process, and ^{60}Ni.

EXPERIMENTAL

The measurements were performed with the time-of-flight technique at the 150 MeV Geel pulsed electron linac using a flight distance of 58m. In the case of ^{56}Fe, the raw data base obtained several years ago [4] was retrieved by applying the correct weighting function determined in ref. 3. A concise account of the experimental conditions can be found in ref. 4: essentially, an iron oxide disk enriched to 99.93% ^{56}Fe, of thickness 0.015 Fe at/b, was viewed by two cylindrical C_6D_6 liquid scintillators placed at 90° with respect to the neutron beam. The rela-

---
* Fellow of the Commission of the European Communities
+ Seconded from ENEA, Italy

tive neutron flux was measured with a ^{6}Li-glass 0.5 mm thick below 100 keV and with a multi-plate ^{235}U fission chamber above.

Neutron capture measurements of ^{60}Ni were recently performed with the linac upgraded with respect to the ^{56}Fe case, featuring a 1 ns pulse width and 7 kWatt average beam power at a repetition frequency of 800 Hertz. The sample, on loan from ORNL, consisted of a metallic nickel disk of 8 cm diameter and thickness N = 0.018 at/b, enriched to 99% ^{60}Ni. An entirely new detection setup was used, consisting of four C_6D_6 scintillators viewing the sample at an angle $\theta=125°$ with respect to the neutron direction. The angle was chosen in order to minimize possible systematic errors due to the anisotropy of γ-rays emitted from resonances with spin larger than J=1/2. The relative neutron flux was measured as a function of energy with an ionization chamber containing three back-to-back deposits of ^{10}B 40µg/cm^2 thick. The dependence on energy of the prompt neutron scattering sensitivity relative to gold capture $\varepsilon_n/\varepsilon_c$ was measured for both geometries by comparing the counting rate from a graphite disk to that from a gold sample. Values of $\varepsilon_n/\varepsilon_c$ in the range 10^{-3} to 10^{-4} were found.

Some tests were performed to study the dependence of the weighting function on sample thickness, a point which was not sufficiently clarified in refs. 2,3. These tests are based on our traditional benchmark, i.e. measurements of capture in the 1.15 keV ^{56}Fe resonance normalized to gold capture. The results, uncorrected for γ-ray self-absorption, are plotted in Fig. 1 for both geometries (circles): one may notice that, for the $\theta=90°$ case, the measured neutron width stays approximately constant and agrees with the transmission value Γ_n = 61.7±0.9 meV (dotted line) up to a thickness N≃8.6·10^{-3} at/b, corresponding to 1 mm metallic sample. Then it decreases steadily.

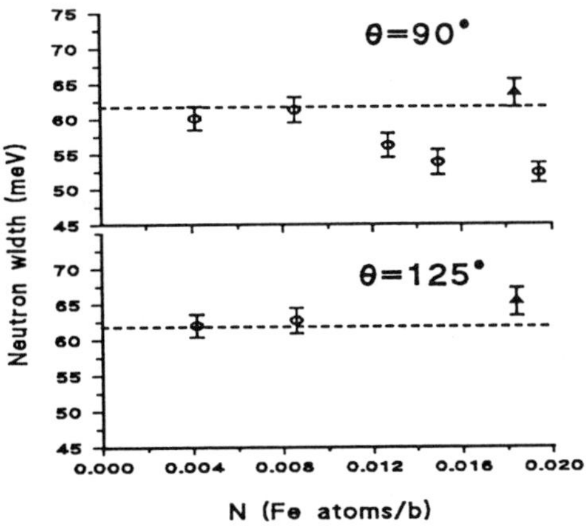

Fig. 1: The neutron width of the 1.15 keV resonance vs sample thickness

Samples thicker than 1 mm were used for both ^{56}Fe and ^{60}Ni but in these cases the difficulty of gold calibration was avoided by normalizing the data to the capture area of the lowest energy resonance of each isotope (E_0=1.15 and 2.25 keV, respectively) whose parameters had been independently determined. In this case, the question left open is whether the weighting function used is still able to correctly deal with the degraded γ-ray spectral shapes emerging from a thick sample. This point was investigated using a mixed sample of Fe and Au: the neutron width obtained by normalizing to capture in the gold contained in the sample (triangles) was only 5% larger than the thin-sample value. This shift is considered acceptable since the differences in spectrum shapes amongst the various ^{56}Fe resonances are much less than that between ^{56}Fe and Au spectra. The

tests for θ=125° reflect the same situation of the θ=90° case except that less thicknesses were measured. These data constitute a sort of validation of the ^{60}Ni measurements since the weighting function used was only determined for the θ=90°geometry.
The ^{60}Ni data were normalized to the value $g\Gamma_n\Gamma_\gamma/\Gamma$ =60.1±1.6 meV of the kernel of the E_0=2.25 keV resonance obtained by us from a gold normalization of a 0.5 mm thick ^{60}Ni sample. The ^{56}Fe data were normalized to the ORNL transmission value.

RESULTS

The data were analysed with the single level Breit-Wigner area analysis code TACASI[6] and/or with the R-matrix shape fitting code FANAC[7] up to 200 keV neutron energy for a total of 57 ^{56}Fe and 102 ^{60}Ni resonances. The Maxwellian-averaged cross sections calculated according to the prescription of ref.8 are given in Table I for the two isotopes and for a series of kT values. The given errors stem from the quadratic combinations of the uncertainties of the following quantities: parameters of the calibrating resonance, measured area of the calibrating resonance, relative neutron flux, areas of the individual resonances (including statistics, background corrections, analysis).

Table I : Maxwellian-averaged (n,γ) cross sections of ^{56}Fe and ^{60}Ni for various kT values

kT (keV)	$<\sigma_\gamma v>/v_T$		kT (keV)	$<\sigma_\gamma v>/v_T$	
	^{56}Fe	^{60}Ni		^{56}Fe	^{60}Ni
12	13.1 ± 0.5	47.4 ± 3.5	30	13.5 ± 0.5	26.7 ± 1.4
20	13.8 ± 0.5	34.8 ± 2.1	40	12.7 ± 0.5	22.3 ± 1.1
25	13.8 ± 0.5	30.1 ± 1.7	52	11.5 ± 0.4	18.9 ± 0.9

REFERENCES

1. Z.H. Bao and F. Käppeler, At.Data Nucl. Data Tables 36, 411 (1987)
2. F. Corvi, A. Prevignano, H. Liskien and P.B. Smith, Nucl. Instr. Meth. in Phys. Rev. A265, 475 (1988)
3. F. Corvi, G. Fioni, F. Gasperini and P.B. Smith, Nucl. Sci. Eng. to be published
4. F. Corvi, A. Brusegan, R. Buyl, G. Rohr, R. Shelley and T. van der Veen, Nuclear Data for Science and Technology,Reidel, p.131 (1983)
5. F. Perey, Nuclear Data for Basic and Applied Science, Gordon and Breach, New York, (1986) p. 1523
6. F.H. Fröhner, Report GA-6906 (1966)
7. F.H. Fröhner, Report KfK-2145 (1976)
8. B.J. Allen, R.L. Mäcklin, and J.H. Gibbons, Adv. Nucl. Phys. 4, 205 (1971)

BETA DECAY OF SOME FP SHELL NUCLEI FOR PRESUPERNOVA STARS

K.Kar and S.Sarkar
Saha Institute of Nuclear Physics
92,Acharya Prafulla Chandra Road
Calcutta 700009,India
and
A.Ray
Tata Institute of Fundamental Research
Homi Bhabha Road,Bombay 400005,India

Abstract : We describe a method for the calculation of beta decay rates for some fp shell nuclei which have important bearings on the presupernova evolution of massive stars(M > $10M_\odot$).

Electron capture plays an important role during the late stages of stellar evolution and the gravitational collapse of type II supernovae(SN)[1]. The capture rates on sd shell nuclei can be obtained from microscopic shell model calculations of Gamow-Teller (GT) strengths.But for heavier nuclei particularly for A> 60 , complete shell model calculations are prohibitively difficult.In this mass range beta decays of some nuclei having high Q-value can compete with the electron captures and it is important to take them into account as this process acts against the deleptonisation of the matter and reduces the entropy . Earlier work on the electron capture and beta decay rates for the fp shell used GT strength estimates based mainly on zero-order -shell model and used experimental energy levels and log ft values for the computation of the rates[2]. These were restricted to nuclei with A≤ 60 .

In this work we describe a method which calculates the sum rule strengths using neutron and proton occupancies obtained by spectral averaging methods and integrates the giant resonance strength times the f-factor of the relevant final states over the experimental Q-value . This method is well-suited for the finite temperature situation and sums over the contributions of the relevant excited states of the mother nucleus for typical preSN temperatures . The beta decay rate is written as

$$\lambda = \frac{\ln 2}{(g_v)^2}(6250\text{sec})^{-1}\sum_i (2J_i+1)\exp(-E_i/kT)\int_0^Q |M(E')|^2 f(Q-E')\,dE' \qquad -(1)$$

where the summation is over the initial states of energy E_i and spin J_i . In eq(1) we approximate the summation over the final states by an integration. $|M(E')|^2$ has contributions from Gamow-Teller(GT) and Fermi transition and is given by

$$|M(E')|^2 = |M_F(E')|^2 + |M_{GT}(E')|^2 = g_v^2\,\rho(E')\,B_F(E') + g_A^2\,\rho(E')\,B_{GT}(E') \qquad -(2)$$

with $\rho(E')$ the density of states at final energy E' and $B_F(E')$ ($B_{GT}(E')$) are squared Fermi(GT) matrix element . We include an empirical quenching factor Z_n in the GT strength needed to explain the consistent quenching of observed experimental strengths . Both $|M_F(E')|^2$ and $|M_{GT}(E')|^2$ are assumed to be Gaussians which seems to be a reasonable approximation[3] . The Fermi resonance is sharp and the area below it is taken as the sum rule (N−Z) , where N/Z are the neutron/proton number of the mother nucleus . The GT sum rule strength S_{β^-} is calculated using the expression

$$S_{\beta^-} = 3\sum_{nljj'} |C_{nl}^{jj'}|^2 (1 - <n_{nlj}^P>)<n_{nlj}^N> \qquad -(3)$$

where $C_{nl}^{jj'} = [2(2j+1)(2j'+1)]^{1/2} W(l1/2jl;j'1/2)$ and $<n_{nlj}^P>$ and $<n_{nlj}^N>$ are the fractional occupancies in the proton and neutron orbit (nlj). Spectral distribution theory shows that the expression of eq.(3) reproduces the average shell model sum rule strength well when the correlation coefficient of the Hamiltonian with the sum rule operator is small[4]. This is indeed the case for nuclei in the fp shell. The fractional occupancies are evaluated using the spectral distribution methods[5]. The phase space integral 'f' of eq(1) is calculated using finite temperature Fermi Dirac distribution for the electrons. The Fermi strength centroid lies at the Isobaric Analog State whereas the GT centroid lies at a higher energy. We use the expression of Nakayama et al[6] for this. The GT resonance is quite broad and only the tail part of both GT and Fermi resonances is reached through β^- decay. The width of the GT resonance due to nuclear forces (σ_N) is a parameter in our calculation and we use $\sigma_N = 7.5$ MeV. The same model can be used to predict the half-lives $\tau_{1/2}$ for free β^- decay and can be used as a test of the methods. In table I we compare the calculated half lives with the experimental values. This table also gives our predictions for the beta decay rates at a density of 10^8 g/cc and temperature of 4×10^9 K. The variation of the rate with density is seen to be mild. The inclusion of the excited states of the mother nucleus increases the rates by a factor of 3-5. The calculation of such rates for other A> 60 nuclei and other temperatures are in progress. The method can be applied to evaluate electron capture rates as well.

K Kar acknowledges travel support from NSF grant no. : INT 87 - 15411 - A01 -TIFR.

Table I
Rates and half-lives for the β -decay of some fp-shell nuclei

Decaying Nucleus	$\tau_{1/2}$(sec) Calculated	Experimental	$\lambda(sec^{-1})$ at $\rho = 10^8$ g/cc, T = 4×10^{9} °K
Co^{62}	16.4	90	1.79×10^{-1}
Co^{64}	5.4	0.3	4.38×10^{-1}
Ni^{67}	47.9	21	0.70×10^{-2} a)
Cu^{68}	48.9	31	3.08×10^{-2}

a) This includes decay from ground state only, as the excited states are not known experimentally.

References :

1. Bethe H.A. , Brown G.E. , Applegate , J. and Lattimer , J.M. , Nucl Phys A324 (1979) 487 ; Fuller , G.M. , Ap J 252 (1982) 741

2. Fuller , G.M. , Fowler ,W.A. and Newman , M.J. , Ap J 293 (1985) 1 and Ap J 252 (1982) 715

3. Kota , V.K.B. and Kar , K. , Pramana-J Phys (1989) 647

4. Sarkar , S. and Kar , K. , J Phys G14 (1988) L123

5. Kota , V.K.B. and Potbhare , V., Phys Rev C21 (1980) 2637

6. Nakayama , K. , Pio Galeao , A. and Krmpotic , F., Phys Letts B114 (1982) 217

PRECISION MEASUREMENT OF THE HALF-LIFE OF ^{56}CO

D. E. Alburger[1], and C. Wesselborg[1,2]
[1] Brookhaven National Laboratory, Upton, New York, 11973, USA
[2] Inst. für Kernphysik, Justus-Liebig-Universität Giessen, D-6300 Giessen, F.R.G.

ABSTRACT

Gamma rays from a mixed source of ^{56}Co + ^{46}Sc were measured in a Ge(Li) detector. The ^{56}Co/^{46}Sc intensity ratios I_{847}/I_{889} were determined by careful analysis of line shapes and underlying backgrounds. Based on the known ^{46}Sc adopted half-life value of 83.810(10) d, the fit to the intensity ratios corrected for ^{46}Sc decay from 107 runs over an elapsed time of 172 days gave a half-life of 77.29(3) d. By combining this result with other published data we obtain a half-life of $t_{1/2}$=77.27(3) d for ^{56}Co.

INTRODUCTION

Because of its relevance to the light decay rate of supernova 1987A there have been several recent determinations[1-3] of the ^{56}Co half-life, to improve on the accuracy and to resolve some earlier discrepancies in values. The history of measurements to date is given in Ref. 3. Figure 1 shows the reported measurements[1-9] of the half-life of ^{56}Co, and includes the present result. The most recent *Table of Radioactive Isotopes*[10] gives the weighted average of the four measurements made between 1970 and 1980, 77.7(5) d, and the 1987 Nuclear Data Sheets for A=56 adopted[11] 77.12(7) d, taken directly from Lagoutine *et al*.[9]

Of the three recent values for the ^{56}Co half-life, 77.28(4)d (Ref. 1), 77.08(8) d (Ref. 2), and 77.30(9) d (Ref. 3) those of Refs. 1 and 3 agree well, while those of Refs. 1 and 2 differ by nearly twice the sum of their respective uncertainties. In most of the recent work the decay of ^{56}Co was determined relative to radiations from long-lived reference standards, using a high-pressure ionization chamber (Ref. 1) and Ge(Li) detectors (Refs. 2 and 3). As suggested in Ref. 3 we selected ^{46}Sc as a reference source. It decays with an adopted[12] half-life of $t_{1/2}$=83.810(10) d and emits γ rays of 889 and 1121 keV with equal intensity. We selected the 889 keV γ ray for comparison with the 847 keV γ ray of ^{56}Co.

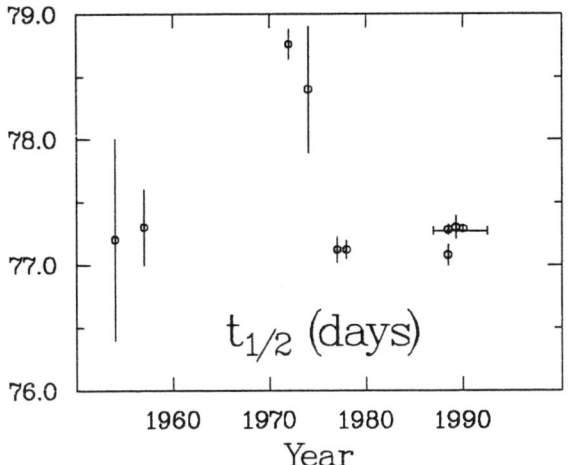

Fig. 1 Measurements of the half-life of ^{56}Co (from Refs. 1-9, and 13). The horizontal bar represents the weighted average 77.27(3) d.

EXPERIMENTAL PROCEDURES AND RESULTS

We outlined our method in Ref. 3. Details of the present experiment can be found in Ref. 13. We clamped the combined sources ^{56}Co + ^{46}Sc in a holder such that the initial intensity of the ^{56}Co 847 keV peak was slightly greater than that of the ^{46}Sc 889 keV peak, and placed it on the axis of a Ge(Li) detector at 13 cm distance. The initial total counting rate was 7100/sec. During the entire 5.7 months of running neither the detector nor the source was moved.

Figure 2 (top and middle part) shows a portion of one of the early spectra (top part expanded by a factor of 70). In the analysis we first shifted each spectrum using the program[14] PAINT so that the 847 and 889 keV lines were at standard peak positions. Only a shift in offset but no gain shift was necessary in the region of interest. We took data over a period of 172 days. We rejected two runs because of peak shifts during the runs. In all remaining 107 runs the peak positions after shifting were constant to better than 0.3 channels. The summed spectra after shifting are shown in the bottom part of Fig. 2.

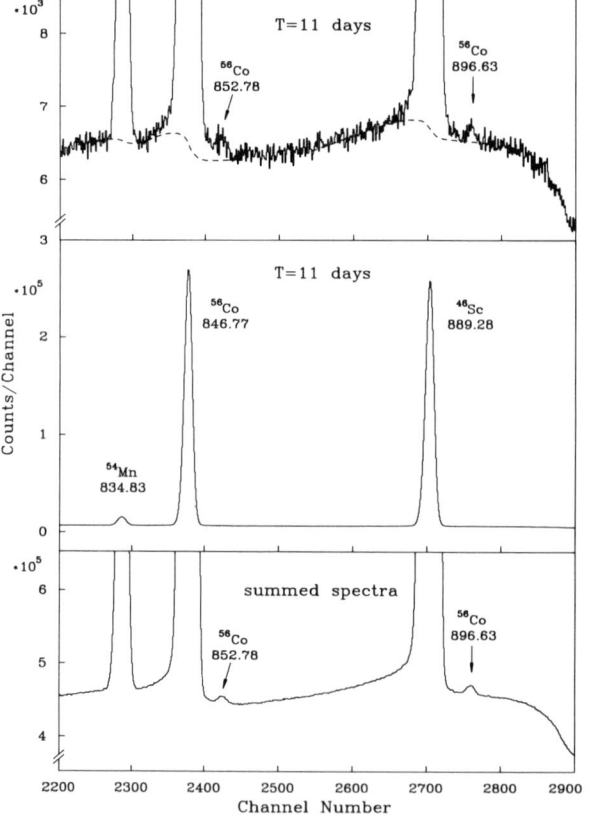

Each peak was fitted using the program LEONE[15] which fits a Gaussian to the peak with exponential high-energy and low-energy tails. The background, as indicated in the top part of Fig. 2, consists of a step function under the peak superimposed on a polynomial function. For the regions of interest, we fitted the peak and background parameters simultaneously. We avoided the region of the new 853 keV γ ray peak[16] of ^{56}Co in the 847 keV peak fit and likewise the region of the ^{56}Co 897 keV peak in the 889 keV peak fit. We determined the exact positions of these weak γ rays from the summed spectra (bottom of Fig. 2).

Fig. 2 Top and middle parts: Spectrum from one of the first Ge(Li) 6 hour runs on a mixed ^{56}Co+^{46}Sc source in the region of the ^{56}Co 847 keV line and the ^{46}Sc 889 keV reference line. The top part illustrates the fitting of the background. The bottom part shows the sum of all individual runs after shifting to standard peak positions.

Initially, the runs were of 6 hours duration. For the very first run the statistical uncertainty in the 847/889 intensity ratio was 0.078%. Towards the end of the experiment the runs were increased to 8 hours.

The I_{847}/I_{889} ratios, after correcting for the decay of ^{46}Sc, are shown in Fig. 3. A fit to all of the points gave a value of 77.29(2) d for the half-life of ^{56}Co. Separate fits to data points for the first and second halves of the series gave 77.34(5) d and 77.22(6) d where the uncertainties are statistical only and include the uncertainty in the ^{46}Sc half-life. Analyses using various background parameters suggested that systematic uncertainties could be present that are comparable to the statistical uncertainties. We therefore combine the statistical and estimated systematic uncertainties and adopt $t_{1/2}$ = 77.29(3) d for the half-life of ^{56}Co. This value agrees very well with the results of Refs. 1 and 3 but is outside the sum of uncertainties when compared with the result of Ref. 2. The weighted average of the four values of Refs. 1 to 3 and the present result, is $t_{1/2}$ = 77.27(3) d.

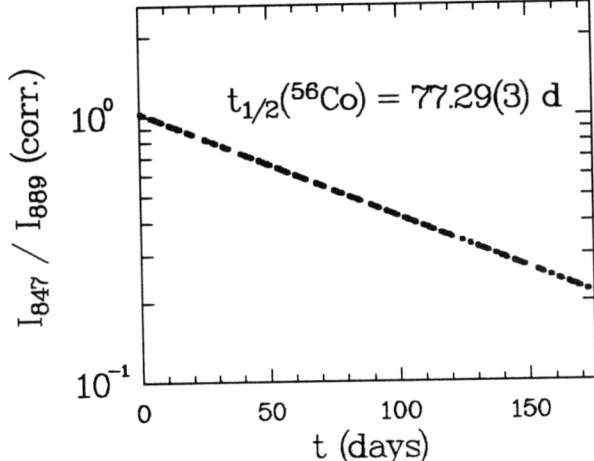

Fig. 3 Decay of the ratio I_{847}/I_{889} corrected for the decay of ^{46}Sc (107 runs over 2.2 ^{56}Co half-lives). Statistical error bars are smaller than the data points.

Research has been performed under contract DE-AC02-76CH00016 with the USDOE. (C.W.) acknowledges support from the Alexander-von-Humboldt Foundation.

1. H. Schrader, Int. J. Appl. Radiat. Isot. 40, 381 (1989).
2. K. T. Lesko, E. B. Norman, et al., Phys. Rev. C 40, 445 (1989).
3. D. E. Alburger, E. K. Warburton, and Z. Tao, Phys. Rev. C 40, 2789 (1989).
4. W. H. Burgus, G. A. Cowan, et al., Phys. Rev. 95, 750 (1954).
5. H. W. Wright, E. I. Wyatt, et al., Nucl. Sci. Eng. 2 427 (1957).
6. J. F. Emery, S. A. Reynolds, et al., Nucl. Sci. Eng. 48, 319 (1972).
7. P. J. Cressy, Nucl. Sci. Eng. 55, 450 (1974).
8. M. E. Anderson, Nucl. Sci. Eng. 62, 511 (1977).
9. F. Lagoutine, J. Legrand, and C. Bac, Int. J. Appl. Radiat. Isot. 29, 269 (1978).
10. E. Browne, and R. B. Firestone, Table of Radiaoactive Isotopes, ed. V. S. Shirley (Wiley, NY, 1986).
11. H. Junde, H. Dailing, et al., Nucl. Data Sheets 51, 1 (1987).
12. D. E. Alburger, Nucl. Data Sheets 49, 237 (1986).
13. D.E.Alburger, and C.Wesselborg, Phys. Rev. C (in press).
14. H. Wolters, Universität Köln, private communication.
15. H. Hanewinkel, Diplomarbeit, Universität Köln (1981); S. Albers, A. Clauberg, et al., Verhandl. DPG(VI) 23, 227 (1988).
16. D. E. Alburger, E. K. Warburton, and Z. Tao, Phys. Rev. C 40, 2891 (1989).

IMPORTANCE OF E2 TRANSITIONS IN ^{175}Lu(n,γ) ISOMER PRODUCTION CALCULATIONS*

M. A. Gardner and D. G. Gardner
University of California, Lawrence Livermore National Laboratory
Livermore, California 94551, USA

ABSTRACT

In the calculation of the isomer production for the ^{175}Lu(n,γ)^{176m,g}Lu reaction we have found that the inclusion of E2 transitions from 62 band-head members of ^{176}Lu (a possible s-process chronometer and/or thermometer) led to an m/total value of 0.82 at E_n = 30 keV, compared to the 0.62 m/total value obtained previously when no E2 transitions from the band heads were considered.

INTRODUCTION

The ^{175}Lu(n,γ)^{176m,g}Lu reaction produces ^{176}Lu both in the ground state(g.s.)(7^-, $t_{1/2}$=3.6 x 10^{10} y) and in an isomeric state at 0.123 MeV(1^-, $t_{1/2}$=3.7 h). In our previous cross-section calculations for this reaction,[1] we had used a set of 291 discrete levels for ^{176}Lu, as modeled by Hoff;[2] this set included 62 rotational band heads. We also used his computed branching fractions, where only dipole transitions were allowed during the depopulation of the discrete levels, even from the band heads. At the incident neutron energy of importance in s-process studies, 30 keV, we calculated an m/total ratio of 0.62; when we considered allowed E2 transitions to take place from some of the band heads, we obtained a value of 0.77. At that time, experimental m/total ratios for this reaction around E_n = 30 keV also differed markedly, ranging from 0.62 to 0.88.

PRESENT CALCULATIONS AND RESULTS

All recent cross-section calculations were made with STAPLUS, our modification of the STAPRE code.[3] We used the same neutron optical model potential and the same discrete levels and level density parameterizations for ^{175}Lu and ^{176}Lu as in our earlier work.[1] At that time, we adjusted the magnitude of the value of the quantity Γγ/D, that is, the total radiation width divided by the average s-wave neutron resonance spacing at the neutron separation energy, in order to obtain agreement with total ^{175}Lu(n,γ)^{176}Lu experimental cross section data available then. This was done in order to deduce an absolute total radiation width and subsequently, an absolute E1 strength function for ^{176}Lu.[4] Revised experimental data have been published and based on it we have reduced the value of Γγ/D in our calculations such that our (n,γ) cross section at E_n = 30 keV is now 8% lower.

We modeled the gamma-ray cascades as follows. The continuum bins were depopulated according to our dipole and E2 strength function systematics, where the E1 strength function has an energy-dependent Breit-Wigner line shape, the M1 strength function is a constant, and the E2 strength is also energy dependent with a Lorentzian line shape. The depopulation of the discrete levels followed the

*Work performed under auspices of US DOE by Lawrence Livermore National Laboratory, contract #W-7405-ENG-48.

prescription of Hoff[2] except in the case of the band heads. We now can approximate the computation of multipole branching fractions, using our PC code, NUSTART. For a given set of discrete levels, the general scheme is to calculate for each radiating level the relative strengths of up to 30 of the strongest transitions to lower-energy levels, conserving angular momentum and parity. Up to 20 dipole and 10 E2 transitions are located; if an M1 and an E2 transition populate the same final level, their strengths are added. The number of transitions is truncated to the 10 most intense transitions and the branching fractions are created by normalizing the total strength to unity. The relative transition strengths are derived from transmission coefficients of the form $CE1 \bullet E\gamma^3$, $CM1 \bullet E\gamma^3$, and $CE2 \bullet E\gamma^5$, where $E\gamma$ is the difference between the initial and final level energies. The coefficients CE1, CM1, and CE2 are constants, except for CE1, which may be either constant or a function of $E\gamma$. These coefficients are equivalent to gamma-ray strength functions; they may be either relative or absolute in magnitude and must be ≥ 0.

While initial studies have shown that NUSTART should not be used to calculate multipole transitions among the complete set of levels for a rigid rotator, we can extract those levels that are the band heads of such a nucleus and use the code to calculate dipole and E2 transitions among these band heads. We can also choose whether the K-quantum number restrictions for change be followed.

In Fig. 1 we show the first 20 of the total 62 band heads modeled for ^{176}Lu.[2] As indicated, our calculated E2 transitions enhance decay to the g.s.(G) or to the isomer(I), compared with results from the dipole-only model. Decay of the 9⁻ level at 0.633 MeV(without symbol) reached the g.s. by dipole transitions. The 4⁻ level at 0.658 MeV was an "isomer" when only dipole transitions were allowed. Two of the five band heads are shown where $J \neq K$. Including allowed E2 transitions from all 62 band heads led to increases of 30% in the calculated production of the isomer, as shown in Fig. 2. Several measurements[5-8] of ^{176m}Lu production are also shown; the most recent are, within the range of the reported experimental uncertainties, in agreement among themselves and with our present calculations. The computed isomer fraction was not sensitive to the magnitude of the E2 strength; a range of 10^3 in the E2 strength gave m/total values that varied by about 5%. Changing the relative magnitudes and shapes of the E1 strength function also led to variations of a few % in the m/total value. A series of calculations in which the 291 discrete levels used in ^{176}Lu were truncated just below a band head showed that, with the inclusion of the E2 transitions, significantly fewer levels were needed to achieve convergence in the computation of the m/total value. Table I. summarizes these findings. The importance of E2 transitions may depend on the band-head structure; we find that preliminary studies of ^{174}Lu and ^{236}Np seem to indicate that the inclusion of E2 transitions does not affect significantly their isomer production cross sections.

In related calculations of isomer production via incident photons, i.e. the ^{176g}Lu $(\gamma,\gamma')^{176m}$Lu reaction, we found that the m/total values above $E\gamma = 0.75$ MeV were increased by factors of 1.2 to 3 with the inclusion of E2 transitions from the ^{176}Lu band heads. Nevertheless, such increases had no discernable effect on the stellar rate of photoproduction of the ^{176m}Lu from the g.s. as a function of the temperature of Planck photon-energy distributions because of the very large magnitudes of such distributions for $E\gamma < 0.75$ MeV relative to those for $E\gamma > 0.75$ MeV.

REFERENCES

1. M. A. Gardner, et al., Intern. Symp. Capture Gamma-Ray Spectroscopy - 1984, Knoxville, TN, AIP Conf. Proceed. #125 (American Institute of Physics, N. Y., 1985) p. 547.

2. R. W. Hoff, et al., Nucl. Phys. A437, 285 (1985).
3. M. Uhl and B. Strohmaier, Rpt. IRK 76/01(1976, updated 1978).
4. D. G. Gardner, et al., Intern. Symp. Capture Gamma-Ray Spectroscopy - 1984, Knoxville, TN, AIP Conf. Proceed. #125 (American Institute of Physics, N. Y., 1985) p. 513.
5. H. Beer and F. Kappeler, Phys. Rev. C21, 534 (1980).
6. F. Stecher-Rasmussen, et al., Intern. Symp. Capture Gamma-Ray Spectroscopy - 1987, Leuven, Belgium, Inst. Phys. Conf. Ser. No. 88 (1988) p. S754.
7. J. R. DeLaeter, et al., J. Astrophys. Astr. 9, 7 (1988).
8. W. R. Zhao and F. Kappeler, personal communication (1989).

Table I. Summary of results of sensitivity studies.

^{176}Lu Levels	Bands	Energy-Dependent E1	Relative Strength Function Values at 1 MeV			$E_n = 30$ keV $\sigma_{n,\gamma}$ to Isomer (b)	m/total
			E1	M1	E2		
291	62	No	1.0	3.2	0.0	0.733	0.612
		Yes	1.0	3.2	0.03	0.930	0.777
		Yes	1.0	3.2	0.32	0.955	0.798
		Yes	5.65	3.2	0.32	0.976	0.815
		Yes	1.0	3.2	32.0	0.982	0.820
		No	1.0	3.2	3.2	0.981	0.819
		Yes	1.0	3.2	3.2	0.981	0.819
194	48	Yes	1.0	3.2	3.2	0.972	0.812
96	29	Yes	1.0	3.2	3.2	0.995	0.830
48	13	Yes	1.0	3.2	3.2	1.058	0.876

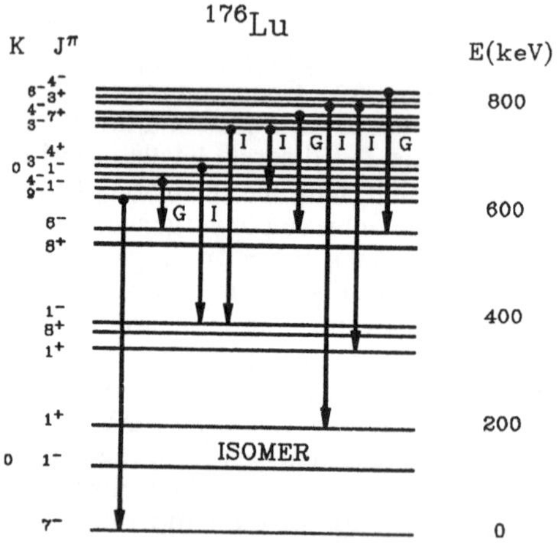

Fig 1. The first 20 of the total 62 modeled[2] band heads for ^{176}Lu and and our calculated E2 transitions. Enhanced decay to the g.s.(G) or to the isomer(I) compared with results from the dipole-only model are indicated.

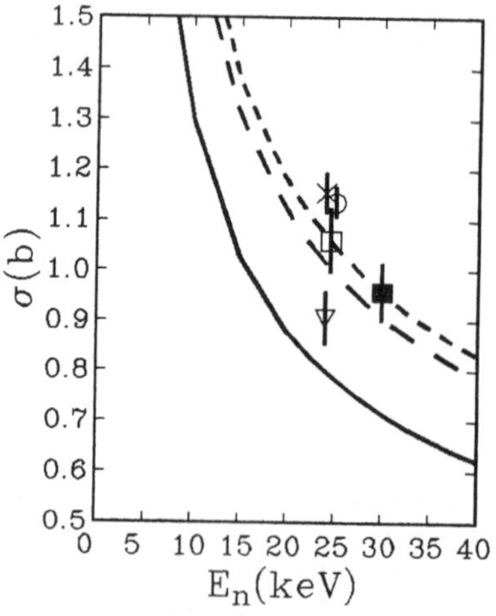

Fig 2. Our calculated ^{175}Lu(n,γ)^{176m}Lu cross sections (short-dashed curve: E2 transitions from band heads allowed, E2/M1 = 1; long-dashed curve: E2/M1 = 0.01; solid curve: no E2 transitions) compared with measurements.[5-8] Open square is a 1/v extrapolation of the datum[7] (filled square).

THE ^{180m}Ta$(\gamma,\gamma')^{180}$Ta CROSS SECTION AT 1.33 AND 4.0 MeV AND ITS ASTROPHYSICAL CONSEQUENCES

Zs. Németh[a,b], F. Käppeler[b], and G. Reffo[c]
[a] Institute of Isotopes, Budapest, Hungary
[b] Kernforschungszentrum Karlsruhe, Karlsruhe, Federal Republic of Germany
[c] ENEA, Bologna, Italy

ABSTRACT

Enriched ^{180m}Ta samples were irradiated by an intense ^{60}Co source of 1.5 PBq and with 4 MeV bremsstrahlung. Using the well known ^{115}In$(\gamma,\gamma')^{115m}$In cross section as a standard, we obtained a cross section of 0.52 mbarn for the ^{180m}Ta$(\gamma,\gamma')^{180}$Ta reaction at 4 MeV, and established an upper limit of 14 nbarn (2σ) at 1.33 MeV. These results mean that ^{180m}Ta can survive at temperatures lower than $\sim 5 \times 10^8$ K, but that it is quickly destroyed via photoexcitation above 7×10^8 K. Accordingly, an s-process production of ^{180m}Ta during stellar helium burning remains a plausible possibility. The respective consequences for the observed abundance of nature's rarest stable isotope are discussed by means of new cross section information.

The origin of ^{180m}Ta, nature's rarest stable isotope, is still an unresolved puzzle of nuclear astrophysics. There were numerous attempts to describe its production by s-, r-, p-, and ν-processes. Since an r-process origin is not supported by recent experiments[1], and since significant p-process contributions are excluded by theoretical studies[2,3], the s-process remains the most promising possibility for producing ^{180m}Ta, which can be investigated experimentally, whereas the ν-process[4] represents a rather speculative alternative.

In the s-process, there are two ways for producing ^{180m}Ta as indicated in Fig. 1 by solid arrows: Beer and Ward[5] suggested a weak beta branch in the decay of the isomer ^{180m}Hf that was later shown to contribute $\sim 20\%$ to the observed ^{180m}Ta abundance[7], whereas Yokoi and Takahashi[8] introduced an s-process branching at ^{179}Hf, which causes a small part of the s-process flow to pass through ^{180m}Ta. The terrestrially stable isotope ^{179}Hf becomes unstable under the physical conditions of the s-process and equilibrizes with the beta decay of ^{179}Ta; finally, neutron captures of the resulting ^{179}Ta nuclei lead to ^{180m}Ta.

The so produced ^{180m}Ta, however, could be destroyed again via (γ,γ') reactions if the metastable state is strongly coupled to the ground state by low-lying activation levels[9]. Our knowledge on the decay scheme of ^{180}Ta^m is poor and the location of the activation levels is unknown. Therefore, the first part of this study was devoted to a search for activation levels by irradiating enriched T$_2$O$_5$ samples (500 mg, 0.26% ^{180m}Ta) with an intense ^{60}Co source of 1.5 PBq and with 4 MeV bremsstrahlung at the Institute of Isotopes, Budapest. The bremsstrahlung was produced by converting the 4 MeV electron beam of 33μA by means of a 1 mm thick platinum converter. In total, six irradiations were carried out; in all cases the cross section of the ^{115}In$(\gamma,\gamma')^{115m}$In reaction was used as a reference standard.

The induced ^{180g}Ta activity was recorded versus time via the hafnium K X-rays, which were easily observable in the spectra. In this way, the background subtraction could be verified by reproducing the 8.1 h half-life of ^{180g}Ta. Since no ^{180g}Ta activity was found after the ^{60}Co irradiations, an upper limit of 14

© 1991 American Institute of Physics

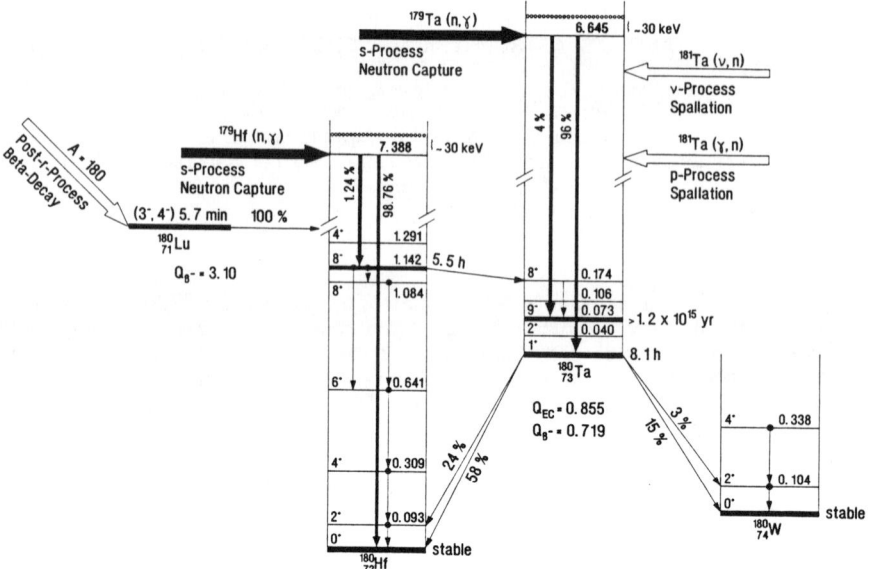

Fig. 1 The level structure at A = 180 and the various production processes of ^{180m}Ta.

nbarns (95 % confidence level) could be deduced for the photodeexcitation cross section of ^{180m}Ta. In the irradiations with 4 MeV bremsstrahlung, a significant activity was observed; the deduced tantalum/indium cross section ratio of 7.4±1. corresponds to an overall ^{180m}Ta photodeexcitation cross section of 0.52±0.20 mbarn. Our results are consistent with those of Richter[9], who found finite ^{180m}Ta $(\gamma,\gamma')^{180g}$Ta cross sections only above 2.7 MeV.

The negative result with the ^{60}Co source indicates the lack of activation levels in ^{180}Ta below 1.33 MeV; this has the consequence that ^{180m}Ta survives at temperatures below $2.5 \cdot 10^8$ K. If the results of Richter[9] are considered as well ^{180m}Ta seems to be safe below $5 \cdot 10^8$ K, which means, that it is practically stable under all reasonable s-process conditions[7]. The large cross section obtained at 4 MeV implies that ^{180m}Ta will be quickly destroyed by photodeexcitation a temperatures above $7 \cdot 10^8$ K.

Since these results strongly support the s-process origin of ^{180m}Ta, the relevant stellar neutron capture cross sections were determined in the second par of this study for the quantitative description of its s-process abundance. Experimen tal studies on the lighter tantalum isotopes are not in sight due to problems with their instability and low natural abundance. Therefore, improved statistical mode calculations were performed with a carefully evaluated *local* parameter systematic based on experimental data for the neighboring isotopes. From this systematics the parameters for the isotopes under investigation can be interpolated, and their reliability can be inferred from the spread of the various data points with respec to the systematic trends versus neutron number[10].

In case of ^{179}Ta and ^{180m}Ta, the stellar cross sections are significantl influenced by capture in thermally populated nuclear states, which occur at rathe low excitation energies. Accordingly, the usual Hauser-Feshbach theory had to b complemented to account for competition with superelastic scattering, where th

Table I. Calculated (n,γ) cross sections for ^{179}Ta and ^{180m}Ta at 25 keV.

Isotope	Nuclear level (keV), I^π	σ(25 keV) (mbarns)	Stellar average for kT = 25 keV (mbarns)
^{179}Ta	g.s. (3.5$^+$) 31 (4.5$^-$) 134 (4.5$^+$)	1248 1184 785	1230 $\pm\,^{370}_{250}$
^{180m}Ta	Isomer (9$^-$) 102 (8$^+$)	2751 2572	2750 ± 540

scattered neutron gains in energy. The calculations were performed in the approximation of the spherical optical model with particular emphasis on a careful parametrization of the level densities. This was achieved by analyzing the neutron resonances of ^{180}Ta, ^{181}Ta, and ^{182}Ta using seven different statistics in an iterative way until convergence to a common average resonance parameter was obtained. Eventually, the final level density parameters for ^{180}Ta, ^{181}Ta, and ^{182}Ta were determined from the simultaneous fit of the discrete levels and of the resonance schemes, and then served for the extrapolation to the corresponding value for ^{179}Ta.

The calculated cross sections for the various relevant excited states in ^{179}Ta and in ^{180m}Ta are given in Table 1 together with the respective Maxwellian averages. Note, that the relatively large cross section for the isomer ^{180m}Ta is due to the suppression of superelastic scattering to the lower states because of the large spin difference. The isomeric ratio was also calculated. We found that only 3.8% of all neutron captures in ^{179}Ta feed the isomer ^{180m}Ta. The present results are estimated to exhibit uncertainties of ~25%, and are significantly smaller compared to previous data used by Yokoi and Takahashi[8].

With these improved cross sections and with typical s-process parameters (n_n = 3.4 · 10^8 cm^{-3} and T_s = 3.3 · 10^8 K or kT_s = 29 keV) one finds, indeed, that ^{180m}Ta can efficiently be produced: Depending on temperature and electron density (that corresponds to the mass density) at the s-process site, up to 100% of the observed ^{180m}Ta abundance can be ascribed to neutron captures on ^{179}Ta. Together with the 20% from the beta decay of ^{180m}Hf (Ref.7) and an additional 10% from the so-called weak component, this result may, therefore, be used to constrain the physical conditions during the s-process.

1. E. Runte, W.-D. Schmidt-Ott, W. Eschner, I. Rosner, R. Kirchner, O. Klepper, and K. Rykaczewski, Z. Phys. A 328, 119 (1987).
2. M. Rayet, N. Prantzos, and M. Arnould, Astron. Astrophys. 227, 271 (1990).
3. W. M. Howard, B. S. Meyer, and S. E. Woosley, Int. Symp. on *Nuclei in the Cosmos*, Baden, Austria, June 18-22 (1990).
4. S. E. Woosley, D. H. Hartmann, R. D. Hoffman, and W. C. Haxton, Ap. J. 356, 272 (1990).
5. H. Beer and R. A. Ward, Nature 291, 308 (1981).
6. S. E. Kellogg and E. B. Norman, Phys. Rev. C 31, 1505 (1985).
7. F. Käppeler, H. Beer, and K. Wisshak, Rep. Prog. Phys. 52, 945 (1989).
8. K. Yokoi and K. Takahashi, Nature 305, 198 (1983).
9. C. B. Collins et al., Phys. Rev. C (in press).
10. G. Reffo, F. Fabbri, K. Wisshak, and F. Käppeler, Nucl. Sci. Eng. 83, 401 (1983), and references therein.

New Facilities Applications, and Related Topics

THE ADVANCED NEUTRON SOURCE

S. Raman and J. B. Hayter
Oak Ridge National Laboratory,* Oak Ridge, Tennessee 37831

ABSTRACT

The Advanced Neutron Source (ANS) is a new user experimental facility planned to be operational at Oak Ridge in the late 1990's. The centerpiece of the ANS will be a steady-state research reactor of unprecedented thermal neutron flux ($\phi_{th} \approx 8 \times 10^{19}$ m^{-2}·s^{-1}) accompanied by extensive and comprehensive equipment and facilities for neutron-based research.

INTRODUCTION

Many research reactors were designed and built in the 1950's and 1960's, culminating with the high-flux reactors at the Brookhaven National Laboratory (HFBR) and at the Oak Ridge National Laboratory (HFIR) in the United States and at the Institute Laue-Langevin (ILL) in France; all of these reactors offer fluxes of $\approx 1.5 \times 10^{19}$ m^{-2}·s^{-1}. However, while new research reactors have come on-line (or are about to do so) since that time in Western Europe, Japan, the Soviet Union, and elsewhere, it is now about 25 years since the last research reactor was constructed in the United States. The Advanced Neutron Source (ANS) is a new multi-purpose research reactor being designed under the leadership of the Oak Ridge National Laboratory (ORNL), with construction scheduled for the mid-1990's.

THE NEUTRON SOURCE

The ANS reactor is being designed to optimize the flux in a variety of neutron beams, both in the central reactor building and in a series of neutron guides extending out into a guide hall. The reactor is of a coaxial, split core design in which the two halves are axially separated and separately cooled. The reactor volume of 67.4 L contains 18 kg of ^{235}U (93% enriched) in the form of involute plates of U_3Si_2 in an Al matrix. The unperturbed peak thermal flux generated in the reflector is expected to be $\approx 8 \times 10^{19}$ m^{-2}·s^{-1} at a power level of 350 MW. The reactor core is cooled by a heavy water primary coolant loop, with coolant upflow. Safety is a primary concern in the reactor design. The reactor is controlled by four independent control rods in the central core channel, with a second independent shutdown system outside the core. Two liquid deuterium cold sources and one graphite hot source will complement the thermal neutron source provided by the heavy water reflector. In addition, a number of irradiation positions in the reflector tank will support programs for transplutonium and other isotopes production, materials irradiation, and analytical chemistry programs.

---

*Operated by Martin Marietta Energy Systems, Inc., under contract DE-AC05-84OR21400 with the U.S. Department of Energy.

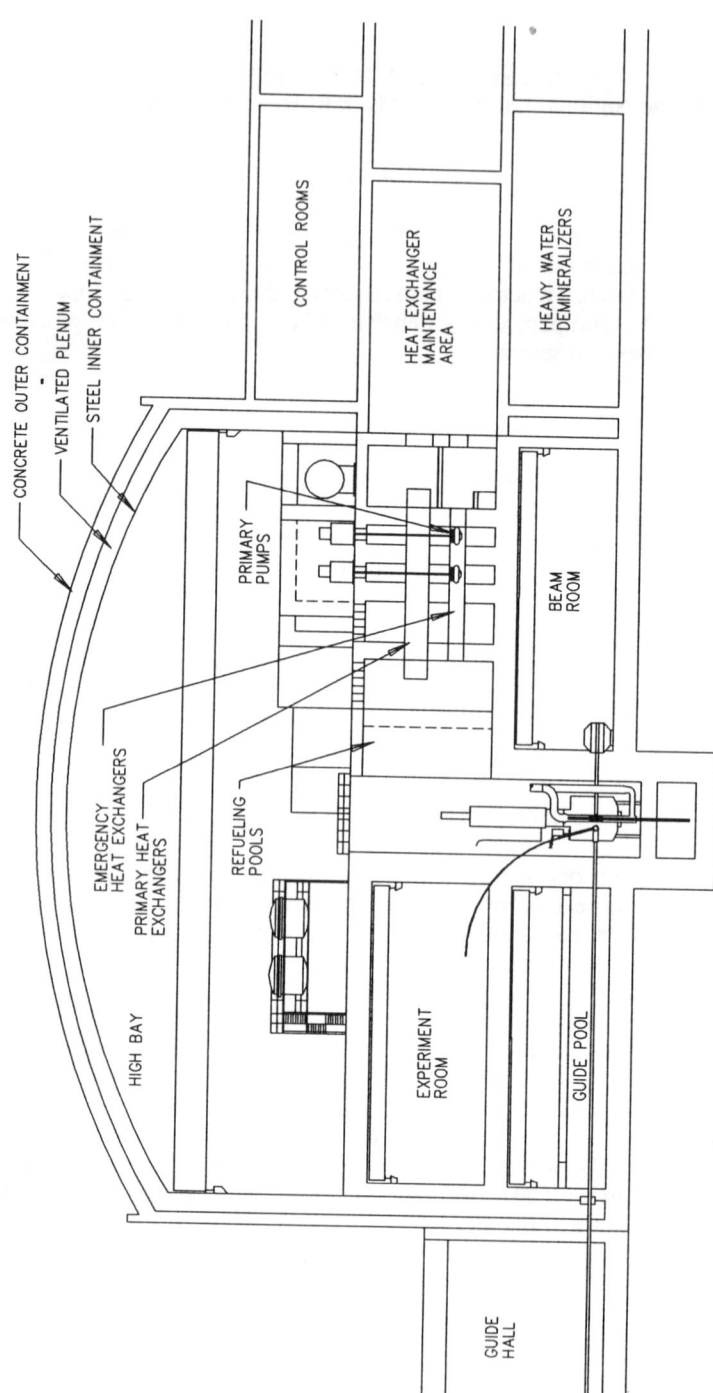

Figure 1. Section through reactor containment building, showing experimental areas.

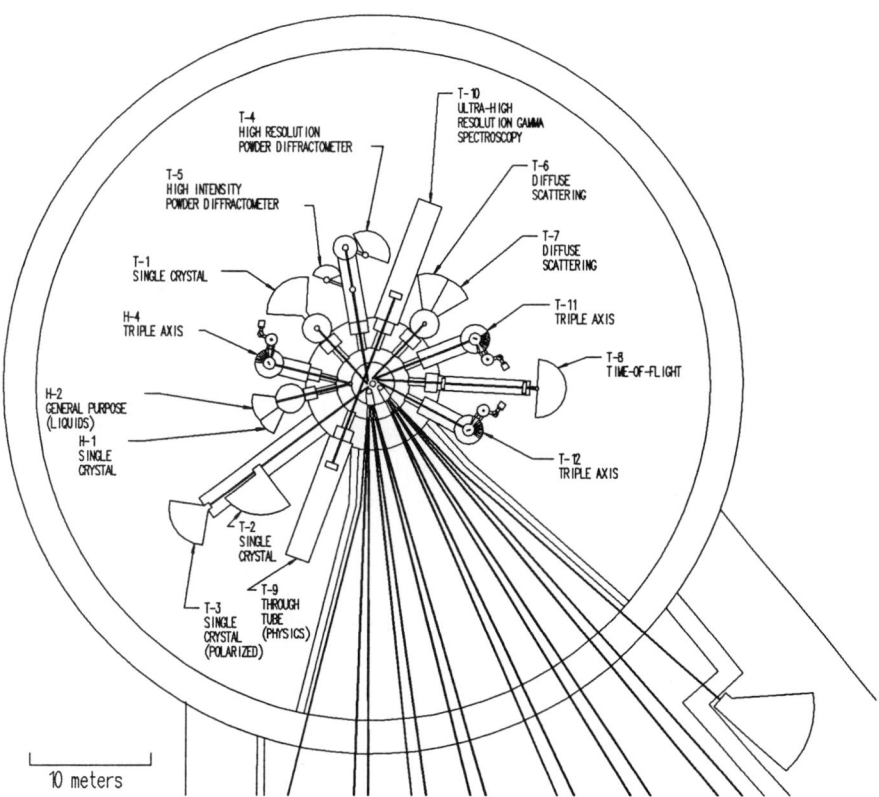

Figure 2. The ground floor beam room

RESEARCH FACILITIES

Three user halls are planned for experimental work at the ANS (Fig. 1). The *Ground Floor Beam Room* (Fig. 2) will provide "conventional" access to thermal and hot neutrons via horizontal tubes terminating at the outside of the biological shielding. Inclined beams and other services, such as rabbit tubes, will terminate in the *Second Floor Experiment Room* (Fig. 3). Very cold and ultracold neutron research will take place on this level, as well as such activities as neutron depth profiling and some of the fundamental and nuclear physics work including an on-line isotope separator. The third main experimental area is the *Neutron Guide Hall* (Fig. 4), which will provide the primary instrumentation for cool and cold neutron research.

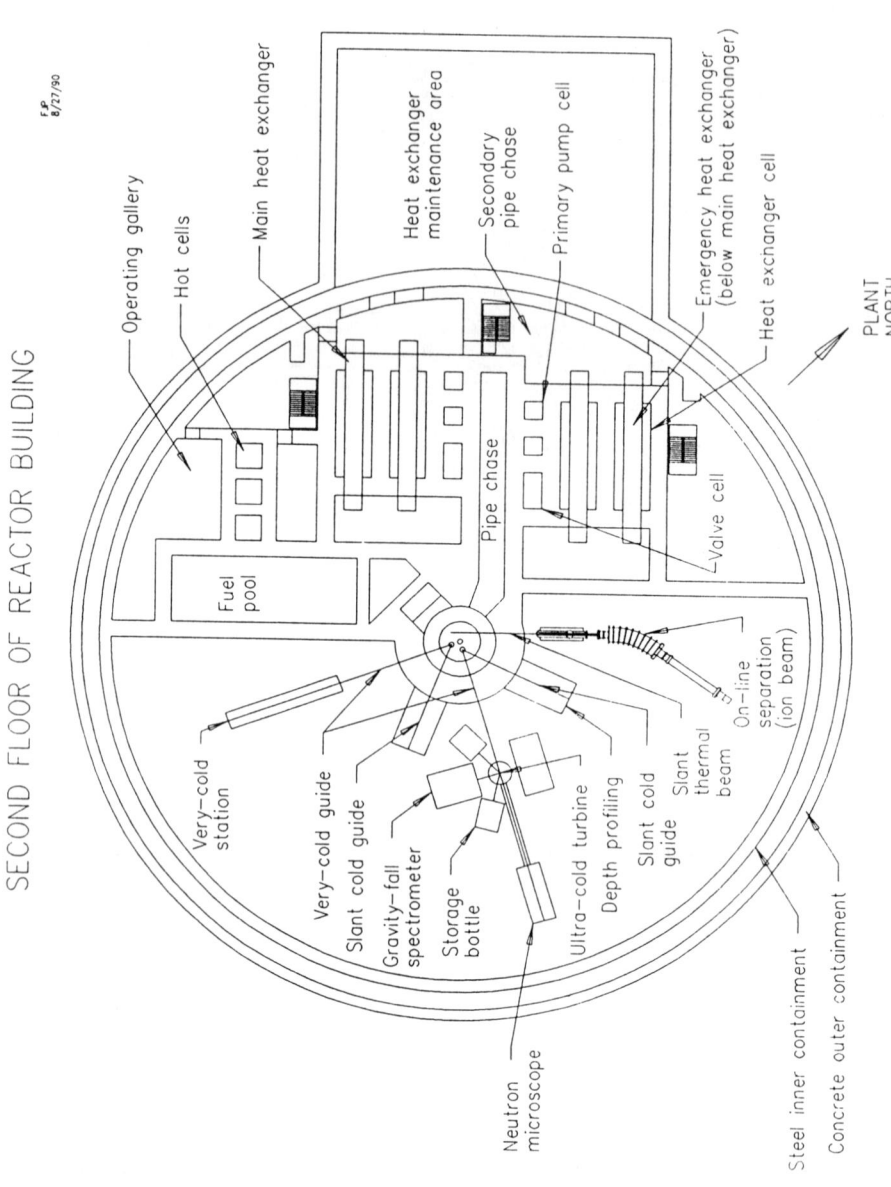

Figure 3. The second floor beam room.

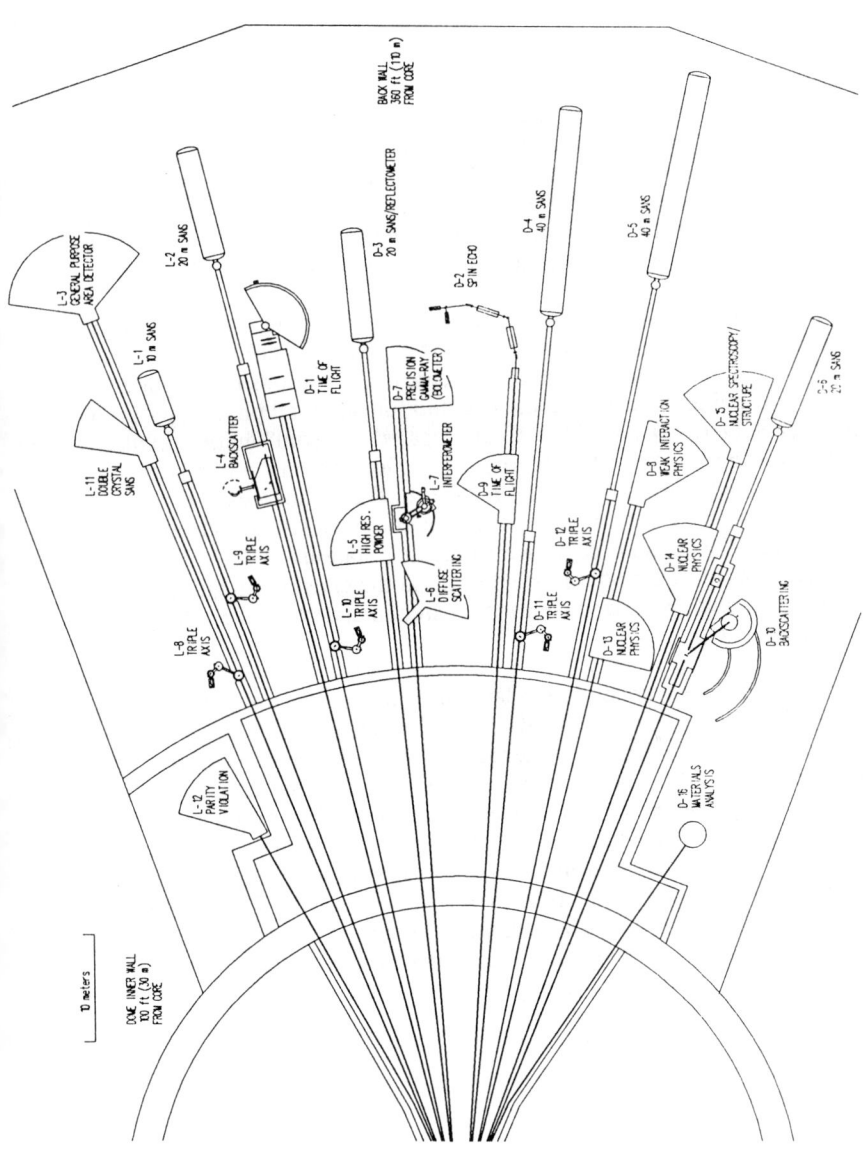

Figure 4. The cold neutron guide hall

NUCLEAR AND FUNDAMENTAL PHYSICS

Under current plans, most of the facilities in the Ground Floor Beam Room (Fig. 2) will be used for a variety of neutron scattering programs. However, the through tube (with end positions designated T-9 and T-10) is intended for nuclear and fundamental physics use. This tube will be located slightly outside the peak thermal flux resulting in a small reduction in the thermal flux but a significant reduction in the fast neutron and gamma flux. Other access to the thermal flux, as well as very cold and ultracold neutrons, will be provided from the second floor (Fig. 3).

The current layout calls for 14 neutron guides extending into the Neutron Guide Hall (Fig. 4) with seven guides pointing at each of two cold sources. A total of 28 experimental stations are located on these guides. Six stations in the guide hall are currently shown as being dedicated to nuclear and fundamental physics programs. Of these, stations D-8, D-13, D-14, and D-15 are located on "cold" guides pointed at cold source CS-1 (not shown). The fifth station, D-7, is located on a guide pointing at cold source CS-2. Although current plans call for identical cold sources, the guides may be customized as necessary, using Ni or supermirror coatings; straight guides will provide access to all "cool" and "cold" wavelengths. Stations D-7 and D-8 are currently indicated as utilizing polarized neutrons at the ends of the guides. Station L-12 provides a clean, cold, polarized beam to an area intended for the safe handling of liquid hydrogen targets.

Nuclear physicists who study fundamental interactions use a wide variety of techniques and facilities, often combined with other fields. In developing a long range plan for nuclear science, the Nuclear Science Advisory Committee (for the U.S. Department of Energy and the National Science Foundation) has identified the ANS as a facility of major importance for carrying out basic experiments on parity violation, time-reversal violation, and the lifetime, electric dipole moment, and beta-decay angular correlation of the free neutron. A facility capable of providing high fluxes of cold and ultracold neutrons has great potential for advancing the field of both basic and applied radiative neutron capture. These research opportunities will be explored in a series of workshops in the coming years. Input from the physics community is being solicited so that the criteria used for the design will enable the ANS to meet the needs for research programs in the 21st century.

Capture Gamma-ray Spectroscopy Using Cold Neutron Beams

C.A. Stone

Department of Chemistry, San Jose State University, San Jose, CA 95192-0101
and
National Institute of Standards and Technology, Gaithersburg, MD 20899

D.F.R. Mildner, R. Zeisler and D.C. Cranmer
National Institute of Standards and Technology, Gaithersburg, MD 20899

Abstract

The availability of cold neutron beams can improve the quality of results obtainable from a capture gamma-ray measurement. Although capture gamma-ray instruments that use cold neutrons are less susceptible to problems that limit the capabilities of thermal instruments, new design parameters become important to consider. We discuss some of the questions that relate to neutron beam handling and present techniques for improving the quality of the capture gamma-ray instrument.

Introduction

Capture gamma-ray instruments are useful for both materials analysis and the characterization of nuclear structure but are limited by a number of problems that restrict their capabilities. The most notable are the large gamma-ray and fast neutron components in neutron beams. Not only do they cause shielding problems, but they induce additional background in the capture gamma-ray spectrum. Gamma rays in the neutron beam can reach the detector by Compton scattering off samples, creating a significant background. Fast neutrons may be moderated and captured in the shielding material, producing gamma rays which increase the background near the detectors. Neutron filters attenuate the gamma rays and fast neutrons in the neutron beam, but with some loss in thermal neutron transmission. Often the small size of the beam tube prevents optimum filters or their cryogenics from being installed.

Spatial restrictions near capture gamma-ray instruments are usually severe. Instruments in a neutron beam facility should be placed close to the neutron source to increase the neutron current density. On the other hand, this increases background problems. Gamma-ray and fast neutron components in the neutron beam are larger and more difficult to filter closer to the neutron source. Neutron scattering instruments, which often compete for the same space as a capture gamma-ray instrument, commonly use cadmium and other strong absorbers for neutron shielding and beam definition. Although the neutron scattering instruments are relatively insensitive to gamma rays produced by these materials, they create a background for the capture gamma-ray instrument. Placing a capture gamma-ray instrument closer to the neutron source thus increases its susceptibility to background problems.

Capture gamma-ray instruments that use cold neutron beams do not suffer so severely from the problems that limit thermal instruments. Spatial restrictions and background problems are not as acute and the cold neutron beams may have higher neutron current densities. As the quality of the neutron beams improve, new parameters become important in optimizing capture gamma-ray instruments. We discuss some of these important design problems.

Quality of Cold Neutron Beams

Cooling a neutron beam to subthermal temperatures has a significant effect on the neutron current density, the background, and the reaction rate at the sample position. As a neutron spectrum is cooled its average wavelength increases. Although neutron guides can be used for thermal neutron beams, they have a much higher efficiency for transmitting cold neutrons. Typical efficiencies are on the order of 99.5% transmission per meter of neutron guide. Neutrons can be transported over long distances with little loss in beam intensity. The advantage of neutron guides is that the divergence angle of neutrons at the sample is increased, effectively transporting the source much closer to the sample position. The entrance to the guide can be placed deep into the beam port where the neutron current density is higher. The combination of the larger initial current density and the high efficiency for transmitting neutrons provides intense neutron beams at the sample position.

While neutron guides cause the effective sample position to move closer to the neutron source, they enable the real sample position to be removed from the source of gamma rays and fast neutrons. Neutron guides have little effect on the transmission efficiency of fast neutrons and gamma rays, which are reduced by the $1/r^2$ factor. Hence, the effect of this background on the capture gamma-ray measurement is much reduced with the sample further from the neutron source.

The longer neutron wavelengths available from a cold source result in a larger probability of interaction between the neutron and a nucleus in the sample. Absorption cross sections increase linearly with wavelength, increasing the reaction rate and improving the sensitivity in a capture gamma-ray measurement. Figure 1 shows the relative gain in the total reaction rate as a function of the Maxwellian temperature of the neutron beam. This has been calculated by folding a $1/v$ cross section into the Maxwellian probability distribution and integrating over all neutron wavelengths. The reaction rate is normalized to 1.0 at a Maxwellian temperature of 300 K. It is easy to see the benefits of cold neutrons.

Many elements have free atom cross sections that are several barns in magnitude, often larger than the corresponding absorption cross sections. Hydrogen (20.5 b), nickel (17.8 b), and lead (11.3 b) are examples of elements with large free atom cross sections. The sample under irradiation can therefore be an efficient neutron scatterer. Since these scattered neutrons can be captured in nearby materials, the sample becomes a source of gamma-ray background, and, more importantly, a radiological concern.

Filters for Cold Neutron Beams

Although neutron guides have little effect on the transmission of gamma rays and fast neutrons, their intensity may still be high enough to require filtering. Some facilities include gamma-ray filters to reduce the gamma-ray heating of the cold source moderator. The ELLA cold source facility at KFA Jülich looks directly at fuel elements in the FRJ-2 reactor. 42 cm of He-cooled single crystal Bi attenuate the gamma-ray and fast neutron beam. This provides neutron beams that require no further filtering in the guide hall. Gamma-ray heating is not as severe at the National Institute of Standards and Technology (NIST) cold neutron research facility. The NBSR reactor at NIST has a split core and the cold source moderator does not look directly at fuel elements. With the current cold source, about 8 cm of Bi (at 350 K) are needed to reduce the gamma-ray heating problems. However, this is not enough Bi to reduce the radiological problems of gamma rays and fast neutrons in the guide hall.

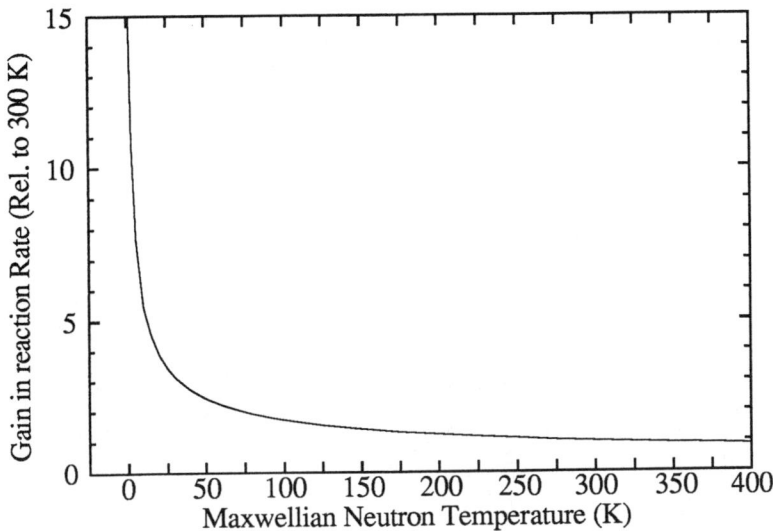

Figure 1. Gain in reaction rate versus Maxwellian neutron temperature. Reaction rates are calculated for a 1/v neutron absorber with an incident Maxwellian neutron spectrum and are shown relative to 1.0 at a Maxwellian temperature of 300 K.

Our knowledge of single crystal neutron filters is relatively poor. Few data are available on total neutron cross sections for most materials. Many of the measurements are old and the data show evidence of crystal imperfection. Since crystal technology has improved during the last decade, results from previous measurements may not reflect the performance achievable today. Results are often reported using figures of merit, primarily to overcome limited data. These figures of merit are valid over a small range of wavelengths but are not applicable to absorption experiments which use all available thermal and subthermal neutrons. Furthermore, these figures of merit may be valid with thermal neutrons but are not necessarily so with cold neutrons. Because of this incomplete knowledge, it is difficult to design filters for cold neutron beams.

Freund[1] has described a method of calculating total neutron cross sections for single crystal materials. Results from his calculations agree with the measured total neutron cross section of silicon, a well-characterized neutron filter. The calculation also reproduces trends in the total neutron cross section for crystals with few available data. The method described by Freund provides a means of comparing the performance of single crystal neutron filters.

We have calculated the properties of single crystal neutron filters using the prescription given by Freund. Figure 2 shows the total neutron cross section for five commonly used crystal filters cooled to 77 K. From this figure, it can be concluded that silicon should be used for thermal energies and beryllium for cold neutrons. The attenuation of the neutron beam at any given energy is dependent on the total neutron cross section at that energy, the number density for the crystal, and the length of the crystal. The desired performance is relative to the epithermal or fast neutron attenuation, and it is important to compare attenuation curves for crystals whose length is normalized to a common attenuation in the epithermal region. Figure 3 shows normalized attenuation curves for these crystals.

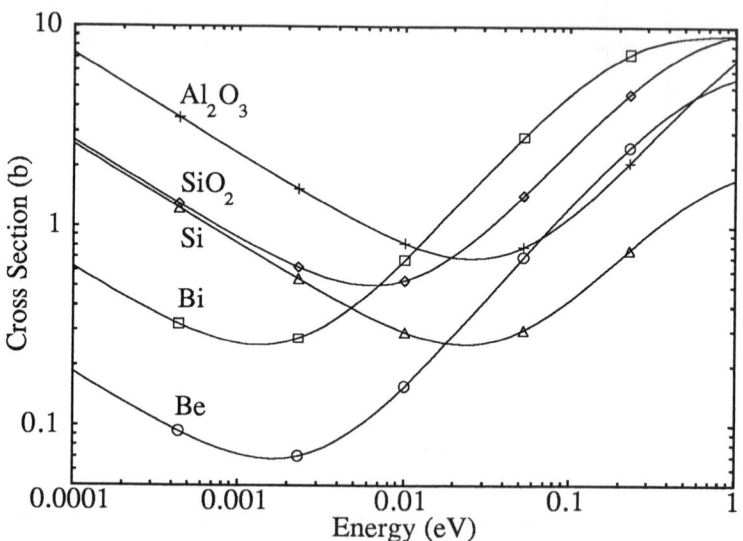

Figure 2. Total neutron cross sections for several neutron filters, calculated using the prescription by Freund.[1] Symbols are placed on the curves for clarity: circles for Be; squares for Bi; diamonds for quartz; triangles for silicon; and crosses for sapphire.

The crystal lengths are: 1.0 cm for Be, 2.9 cm for Bi, 3.6 cm for quartz, 7.5 cm for silicon, and 2.3 cm for sapphire. Sapphire has the lowest neutron attenuation in the thermal region but a larger one at lower neutron energies, due to the Al absorption cross section. Below about 0.025 eV the neutron attenuation in Be is smaller than that for sapphire. Beryllium, though, is a polycrystalline material and has a Bragg cutoff at 5 meV (4 Å). Neutrons with energies larger than 5 meV are scattered from the beam. If the neutron beam has a significant current density above 5 meV, these neutrons will be removed from the beam and the resulting beam intensity will be lower than for sapphire.

Bismuth is not an ideal neutron filter but, because of its high atomic number, it is an essential gamma-ray filter. The problem with Bi is the low Debye temperature of the crystal, allowing creation of phonons in the crystal lattice. If Bi can be diluted in a crystal lattice, creating a material with a higher Debye temperature and a high Bi number density, the neutron transmission will improve. One possible candidate is single crystal bismuth silicate (also known as BSO). The properties of bismuth silicate are similar to bismuth germanate and large crystals should be easily grown.

Thermal (Cold) Neutron Shielding

An ideal neutron shielding is one that rapidly attenuates the neutron beam without creating additional background. The (n,γ) reaction is the dominant absorption reaction for most nuclides. Cadmium and gadolinium are examples of elements with large absorption cross sections, and these elements are commonly used for neutron attenuation. These elements, though, effectively convert the neutron beam into a gamma-ray background. Each produces a large number of gamma rays whose intensity is spread over a wide range of energies. This gamma-ray background is difficult to shield and affects results across the gamma-ray spectrum.

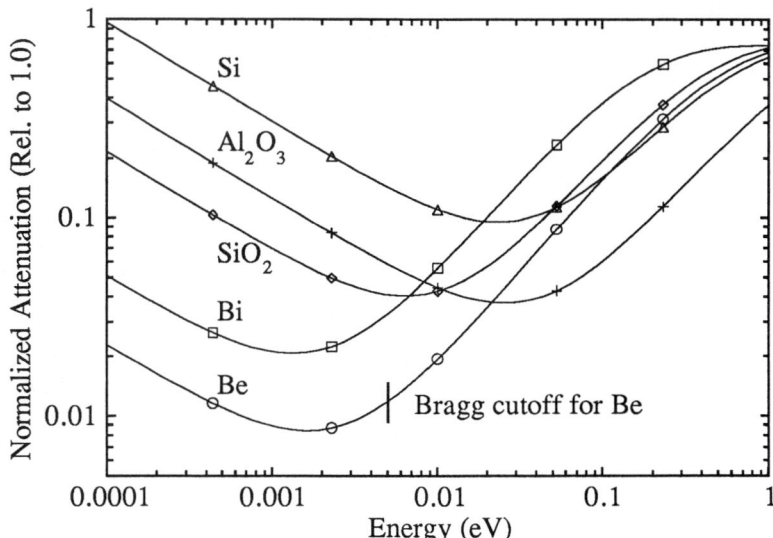

Figure 3. Normalized attenuation coefficients for several neutron filters. These curves were generated by normalizing the crystal lengths to give a common attenuation in the epithermal region. Symbols are placed on the curves for clarity: circles for Be; squares for Bi; diamonds for quartz; triangles for Si; and crosses for sapphire.

Boron has been useful for neutron shielding because of the dominant (n,α) charged particle reaction. The capture gamma-ray reaction is only 0.01% of the total neutron absorption reaction. Much of the $^{10}B(n,\alpha)$ reaction populates a low-lying, 477-keV level in ^{7}Li. A 477-keV gamma ray is emitted, creating a soft gamma-ray background. This background is easier to shield than that from cadmium and gadolinium, and it effects primarily the low-energy portion of the gamma-ray spectrum.

Isotopically separated ^{6}Li is a more attractive neutron shield. The nuclide ^{6}Li has a large absorption cross section, 941 b (at 2200 m/s velocity), and the primary reaction is the (n,α) charged particle reaction. The partial absorption cross section for the (n,γ) reaction is only 39 mb. Isotopically separated ^{6}Li is commercially available, even in large amounts, with isotopic enrichments of over 95%. It is feasible to use ^{6}Li as a shielding material in capture gamma-ray instruments.

A disadvantage of using ^{6}Li as a shielding material has been the lack of a suitable host matrix. It is often incorporated into hydrogenous materials. McGuire[2] has constructed shielding using lithium carbonate mixed with an epoxy, and Failey and Anderson[3] have described one constructed by adding lithium carbonate to a mixture of paraffin and garlic salt (sic). The hydrogen in these materials creates a substantial background at the instrument. Fused ^{6}Li$_2$CO$_3$ does not have the large background problems found with hydrogenous materials, but it is prone to attack by moisture and has poor structural properties. Sintered ^{6}LiF is better. It is resistant to moisture and is less brittle than lithium carbonate. This material, though, is not easily formed into irregular shapes and cannot be machined easily.

We have developed a highly-doped ^{6}Li silicate glass[4,5] at NIST. Glasses have been prepared with ^{6}Li$_2$O concentrations of up to 18 weight percent, a thermal neutron attenuation of 76% per mm of glass. The neutron attenuation is greater than 95% per mm for a neutron beam with a 60-K Maxwellian temperature. Long-term irradiation studies have been performed on the glass and they demonstrate that the glass can withstand irradiations of greater than 10 hours in a neutron field of 8×10^{12} neutrons/cm^{2}s.

Structural properties of the ^{6}Li silicate glass are similar to those of other silicate glasses. It is resistant to moisture and is not susceptible to crystallization at ^{6}Li$_2$O concentrations of less than 18 weight percent. The glass can be poured into a variety of shapes and sizes and can be machined using ordinary glass machining techniques into shapes with close tolerances.

The ^{6}Li silicate glass is an ideal neutron shielding material in cold neutron facilities. Shutters and beam stops can be constructed using one-cm thick blocks of glass. Collimators can be constructed by milling a hole into the glass. Shielding that covers a larger area may require several pieces of glass to be joined together. In such an application, steps are machined into the glass, thereby preventing neutron streaming along a joint. We have formed thin plates of glass by pressing the glass melt between two heated pieces of aluminum. These glass plates are useful as neutron shields for the detectors and other places in the capture gamma-ray instrument where only partial shielding is necessary. An especially important feature of the ^{6}Li silicate glass is its compatibility with the glass used in neutron guides. Pieces of the glass can be machined and incorporated into parts of the guide itself.

Neutron Beam Handling

Although straight guide tubes can transport neutrons over large distances with little loss of intensity, they often view the neutron source directly so that large, unwanted fast neutron and gamma-ray fluxes also propagate down the tube. This can be circumvented using a uniformly curved tube with a radius of curvature which is much greater than the width of the guide. In practice curved guides are constructed in relatively short straight sections, successively misaligned slightly from each other. This small polygon approximation to an arc introduces only small extra losses into guide systems. The spatial distribution of transmitted neutrons at the guide exit is asymmetric, particularly for wavelengths less than that characteristic of the curved guide. This development can be duplicated on a much smaller scale to give a curved, totally-reflecting multichannel system for focusing neutrons by the superposition of intensities transmitted through many thin layer films. This concept offers the best promise for developing an analytical probe for materials research using cold neutrons.

The neutron current density at the sample can also be increased by using a converging guide. This can be particularly useful as the last section of a long straight guide. An obelisk shaped focusing guide coated with nickel-titanium supermirror which reduces the guide cross section by a factor of three has been installed on the capture gamma-ray instrument[6] at ELLA. The greatest gains in neutron current density are found at the longest wavelengths (gain of 3 at 10 Å). The gain is obtained at the expense of increased divergence of the beam. Further increases will be available with improved supermirror coatings. Increases of perhaps an order of magnitude would be expected on well-designed instruments.

An alternative idea is to diffract a beam of neutrons using a large mosaic pyrolytic graphite monochromator. The sample will be placed in the diffracted beam away from the direct gamma-ray and fast neutron radiation. For example, the neutron beam diffracted at 90° has a wavelength of around 4.75 Å, which is useful in a beryllium filtered beam ($\lambda > 4$ Å). The mosaic of the crystal should be matched to the divergence of the incoming beam, both of which determine the wavelength spread of the diffracted beam.

References

1. A.K. Freund, Nucl. Instr. Meth. **213**, 495 (1983).
2. S.C. McGuire, Ph.D. Thesis, Cornell University, 1979.
3. M.P. Failey, D.L. Anderson, Woodberry Laboratories Progress Report, Hyattsville, MD, 1977.
4. C.A. Stone, R. Zeisler, D.H. Blackburn, D.A. Kauffman, D.C. Cranmer, In *NIST (U.S.) Tech Note*, O'Conner, C. (Ed.), pg. 124 (in publication).
5. C.A. Stone, R. Zeisler, D.H. Blackburn, D.A. Kauffman, D.C. Cranmer, I, Olmez, to be submitted to Nucl. Instr. Meth.
6. M. Rossbach, O. Schärpf, W. Kaiser, W. Graf, A. Schirmer, W. Faber, J. Duppich and R. Zeisler, Nucl. Instr. Meth. **B235**, 181 (1988).

CONSIDERATIONS IN UPGRADING INTERMEDIATE FLUX REACTORS BY THE ADDITION OF COLD NEUTRON BEAMS

David D. Clark
Cornell University, Ithaca, NY 14853

ABSTRACT

Arguments are presented that the addition of cold neutron beam facilities at intermediate flux reactors is an effective and feasible way to upgrade their value as research facilities by expanding the range of topics in which they can carry on competitive and useful research. These include prompt gamma-ray analysis, basic level scheme studies with capture gamma-ray and conversion electron spectroscopy in which very pure or very low background beams are required, development and testing of new methods and instruments for research with cold neutrons, and pedagogical experiments that demonstrate quantum mechanical interference and related phenomena at the macroscopic level. Reasons are suggested that may explain why the only US reactors with cold sources are at Brookhaven National Laboratory (BNL) and the National Institute of Standards and Technology (NIST). It is also noted that only the latter has a system of cold neutron beams comparable to those at European laboratories. Design considerations for installing cold neutron beams at intermediate flux reactors, including safety and operational simplicity, are discussed in general terms. Some examples of specific design compromises are drawn from the Cornell project that has been underway for several years.

INTRODUCTION

There are several reasons for limiting the scope of this paper to moderate power reactors from (say) 100 kW to 1 or 2 MW. First, there are only a few higher power research reactors in the U. S., and except for the University of Missouri at Columbia and MIT they are at national laboratories (BNL, Oak Ridge, NIST), operate with large professional staffs, and because of their size are expensive to operate. They are used primarily for research that requires the highest and most continuous beam fluxes that are attainable -- mostly for atomic structure studies using neutrons as probes. A second reason to exclude discussion of cold source design problems at the larger reactors is that although the problems are similar in nature to those for intermediate power, the different magnitudes mean that the design criteria are distinctly different.

Currently only two U.S. reactors have cold sources: BNL, which uses liquid hydrogen, and NIST, which uses heavy water ice now, and plans to use liquid hydrogen. (I would like to be able to say that there is a third -- Cornell -- but we have only operated one model in test runs and are now incorporating the lessons learned then into the final modified design.)

Why should there be only two U. S. cold sources? I can't speak for others of course but I can suggest two reasons. The first is safety considerations, both substantive and regulatory, against having liquid hydrogen in or near a reactor core on a university campus. The second is a common misconception that the only use of cold sources at NIST, BNL, and at the large reactors in Europe such as at Munich, Julich, and Grenoble is to provide copious continuous fluxes of long wavelength neutrons for condensed matter research in physics, chemistry, and biology.

I think neither reason is sufficient to keep at least some moderate power facilities from considering installation of cold neutron beams. At Cornell we have been working on this idea for a few years without external funding[1,2,3] and now with a modest DOE grant are moving to put our concept to the test.

In the following I will not confine myself to the Cornell project parameters but try to speak in more general terms. First let me counter the use argument with reasons why cold neutrons are useful even at modest fluxes and less than 100% duty cycle. Then I will discuss the design considerations, including the prime one of safety.

REASONS FAVORING INSTALLATION OF A COLD NEUTRON BEAM

If one forgoes the idea of competitive condensed matter research there are still other areas in which a pure cold neutron beam permits important research. Heading the list is prompt gamma analysis, sometimes called prompt gamma-ray neutron activation analysis (PGNAA). As is well known, this complements instrumental neutron activation analysis (INAA), which is a principal activity at most intermediate flux reactors, by providing analysis of elements INAA can't touch, such as H, B, C, N, and others, and does so using very much the same detectors, data handling electronics, and interpretation methods as INAA. But it does require an external beam as free as possible from gamma rays and fast neutrons, which are all too abundant in the core.

PGNAA is of course a technique that is useful at high flux reactors also. NIST, which already has a thermal PGNAA beam, will shortly have a setup operating at one of their cold neutron guides, and one exists at Julich.

There are other beam experiments -- well-known to those at this symposium -- of basic studies of level schemes, isomeric states, and so forth with capture gamma-ray and capture conversion-electron spectroscopy, especially for those cases where clean beams are important.

Another use is to develop and test ideas for instrumentation and methods involving cold neutrons before undertaking full-scale experiments at a high flux reactor such as at NIST or ILL. Several European reactors such as the one in Vienna have played this feeder or staging role with respect to Grenoble. One can also provide initial training to students before they join the large scale experiments.

An area of pedagogical value is that of experiments that demonstrate quantum mechanical effects such as interference on a macroscopic scale.

A project we plan to pursue which does not require a high purity beam but can benefit from it is to explore the possibility of extending neutron depth profiling in its current form to many more elements and greater depths by using conversion electrons instead of heavier charged particles as the probes whose energy losses before emerging from the surface of a sample are directly related to the depth from which they originate. (See Ref. 3.) These can be conversion electrons emitted promptly upon neutron capture in a beam, or can be ones emitted from nuclei exhibiting delayed radioactivity.

All of these are important uses that do not require high and/or continuous flux but can upgrade an intermediate flux reactor and increase its utilization and value for research.

MAXIMIZING BEAM PURITY

How do we get as pure as possible a thermal or subthermal beam? Let me start at the using end -- the target -- and work back toward the reactor core, covering some of the design choices at each point. I will assume we deal with a horizontal beam with the necessary components installed within a beam plug inserted into an existing beam port. This avoids altering the reactor.

The first step is to reduce the ambient background due to neighboring experiments by locating the exit of the beam as far as possible from those other experiments. Here moderate power reactors have a distinct advantage. We generally have fewer beams and fewer users, and very likely do not have to share the beam with others.

We wish to filter out the fast neutrons and gamma rays from the core. There are various filters; e.g., single crystals, warm or cold, that let the slow neutrons with their longer wavelengths slither through without encountering Bragg planes but do incoherently scatter the fast neutrons and gamma rays. There are polycrystalline filters such as Be that Bragg-scatter all except the longest wavelength neutrons. But the most effective filter is the totally reflecting neutron guide, first made in a practical form by the group at the Technical University of Munich.[4] For those neutrons that meet the critical angle requirement, practical transmission is about 99% per meter. A straight guide passes the slow neutrons while fast neutrons and gamma rays fall off as the inverse square. To remove even those, or in cases where a long guide of 50 or more meters length isn't feasible, the guide elements can be laid out on a curve. In that case there is no direct line of sight into the core; however, the flux at the exit is not uniformly distributed across the dimension in the plane of curvature.

But the critical angles for total reflection are very small, (15.5 milliradians for 1 meV neutrons on nickel), so the fraction of isotropic neutrons at the guide entrance that is transmitted is very small. The critical angle for total reflection is proportional to $E^{-1/2}$ so the solid angle for acceptance into the guide is proportional to $1/E$. There is therefore a large payoff in chilling the neutrons to a low temperature by inserting a cold moderator block between the core and the guide.

REDUCING THE NEUTRON TEMPERATURE

Cold sources were developed first in England in the 1960's. (Examples of early work are in papers by Webb and by VanDingenen.[5]) Cold sources consist of hydrogen or hydrogenous materials at cryogenic temperatures. The hydrogen nuclei cannot be treated as free particles when the neutrons have energies less than several eV, so the criterion for effective slowing down of neutrons of ordinary thermal energies is that the structures in which the hydrogen is bound must have states of excitation that are as low as possible, such as a few meV. Substances that have been used are liquid hydrogen, frozen heavy water, frozen methane, and frozen mesitylene -- and others.

The best overall choice for a cold moderator is liquid hydrogen. This was recently restated by Egelstaff[6] at the International Workshop on Cold Neutron Sources in Los Alamos this March, where designers and users of cold sources for reactors and for pulsed accelerators discussed this and many other aspects of cold source design.

For laboratories with intermediate flux reactors, and especially those under U.S. Nuclear Regulatory Commission rules, liquid hydrogen presents unacceptable safety hazards of sudden pressure increases if cooling is lost and of fire or explosion if the hydrogen escapes into the air. Methane has similar problems. Heavy water is quite safe but it is expensive; large volumes are needed and it must be protected from contamination by ordinary water. Mesitylene is liquid at room temperature and is not highly flammable. Its properties have been investigated by Utsuro et al in Japan.[7] Their results show that it performs quite well as a cold moderator.

Another consideration in the choice of moderator is radiation damage effects from fast neutrons and gamma rays. Possible consequences include production of free hydrogen and of polymers. Another type of consequence is stored energy in the solids; that is, vacancies and interstitial atoms caused by the intense radiation can be frozen in metastable states that will release energy abruptly when the solid is warmed and/or melts, a situation similar to that of Wigner energy in graphite. Liquid hydrogen shows no effects. According to Utsuro[8] mesitylene experiences slight radiolysis with the production of hydrogen and methane in small amounts under the conditions of radiation dose and temperature that he used. Solid methane when irradiated at 4K shows a strong stored energy effect.[9]

The best choice for an intermediate flux reactor cold source appears to be mesitylene. (It is our choice at Cornell.) This remark deserves amplification, as it involves both a more detailed characterization of cold moderator properties and a deliberate design compromise. With none of the moderators can one expect the cold spectrum to be a Maxwellian at the physical temperature of the moderator, since even the smallest changes in excited levels in the moderator are finite. In the case of 1,3,5-trimethylbenzene (the long name of mesitylene) the lowest-lying excitations have been deduced to be the libration of the whole molecule in the solid and the rotation of the methyl radicals at alternate corners of the benzene ring; both excitations occur at about 7 meV.[7,10] Measurements by Utsuro et al[7] show that the spectrum changes only slightly when the physical temperature is lowered from 40K to 25K. This means, conveniently, that there will be no strong adverse effect on the spectral yield if factors such as nuclear heating and heat removal that affect the source temperature cause it to be as large as 40K.

COOLING THE COLD SOURCE

How can one achieve and maintain the cold temperature? Here again safety and simplicity of operation are important for the moderate power facility. One could circulate liquid hydrogen or helium from outside the shielding to the mesitylene chamber, but those materials present potential hazards. One choice, which we at Cornell prefer, is solid conductors of copper, like the cold finger in a Ge(Li) detector, between the chamber and a cryogenic refrigerator outside the reactor shield. These will be long fingers -- about 3 meters in our case. An alternative choice would be heat pipes of suitable design. This possibility is being pursued by Wehring and Emoto at the University of Texas at Austin.[11]

DESIGN PROBLEM 1: OPTIMIZING SOURCE GEOMETRY

The system I have outlined up to this point does meet the criteria of safety and simplicity of operation. Within this broad outline what are some of the more difficult design problems? I will start this time at the core end of the system, which is where most of the problems arise.

First is the optimization of the size and shape of the cold source itself. There is no agreement in the literature on source geometry. It should be as small and as light as possible to keep the thermal load due to heating by gamma rays and fast neutrons low. It should be large enough that most of the neutrons of the reactor spectrum will be moderated to the cold temperature, but not larger than that or excessive absorption of the cold neutrons will occur. This means that the linear dimensions should be only a few times the scattering mean free paths, which are about 0.3 cm at 300K and about 0.25 cm at 40K. The obvious method for computing the optimum geometry is Monte Carlo simulation. Unfortunately there are no data on the double differential cross section for mesitylene; however, the total neutron scattering cross section vs neutron energy for mesitylene at 10K has been measured and interpreted by Utsuro.[10] He divides the energy scale into four intervals, in each of which the cross section varies linearly with the reciprocal of the neutron energy. Furthermore, in each interval the struck protons have a constant effective mass, different from one unit in all except the highest energy interval, and are in motion with energies described by the excitation levels of three-dimensional harmonic oscillators, a different one for each energy interval. These masses and oscillators model the conditions that the protons are neither free nor at rest. From this we constructed a simple model of mesitylene for Monte Carlo simulation of the histories of warm neutrons isotropically incident on the surfaces of a circular disk and tallied the energy and angle of neutrons leaving a prescribed area on the downstream face of the disk towards the entrance of the neutron guide. We varied the radius and thickness of the disk and also studied the effect of a rectangular re-entrant hole facing the guide. These calculations were run with Turbo Pascal under MS-DOS on a personal computer (DEC Rainbow 100B with 80286 and math coprocessor chips). Broad maxima in neutron yield and in the degree with which their spectrum matched the physical temperature were found, indicating that more realistic models and more complex calculations were not warranted. The program and results have been reported at an American Nuclear Society meeting[12] and are further described in an article in preparation for submission to Nuclear Science and Engineering. We selected a simple disk, 3.75 cm in radius and 3.0 cm thick without a re-entrant hole.

DESIGN PROBLEM 2: NUCLEAR HEATING

The next and toughest problem was (and is) nuclear heating of the aluminum chamber containing the mesitylene, of the mesitylene itself, and even of the cold fingers that remove heat from the source. These are significant even in an intermediate flux reactor. It was necessary to make direct measurements of the heating rates, due primarily to gamma-ray flux, in the intended location of the source, 15 to 18 cm from the edge of the fuel-moderator region of the core. We used thermocouples on thermally isolated samples of construction material. At that location, which is within a beamport that penetrates all the way through the 30-cm thick graphite reflector up to the core edge, we measured rates that when extrapolated to 500 kW, our maximum power, corresponded to 4 mW/g, or 1.4 Mrad/h, even with 7.6 cm of lead shielding between the core edge and the cold chamber position. The neutron fluxes at that location, measured by Au and Ni foil activation, were 10^{12} n/cm^{2}s thermal and 1.7×10^{10} fast, when referred to 500 kW. These radiation levels have a severe impact on the design of the heat removal system.

The critical parameters here are the thermal conductivity of the copper rods in the temperature range of 20K to 40K, the thermal load imposed by heating of the chamber and the mesitylene, the thermal load imposed by the heating of the rods themselves, fast-neutron radiation damage that causes the conductivity of the rods to decrease with exposure, and the degree to which the decrease can be reversed by annealing at room temperature when the source is not in use. Problems of this sort are not unique to cold sources in reactor environments. They are encountered also in the design of cryogenic magnets for fusion reactors. A large number of measurements have been reported and more are in progress. A standard cryogenic conductor is OFHC copper with a conductivity of about 10 W/cm-K at around 20K. But there are specialty coppers, either six-nines pure or oxygen-annealed, with conductivities 10 or more times higher. Defect production by fast neutrons, which is about the same in any type of copper, can cause conductivity reductions of several tens of percent, even of 50%, but after annealing by returning to room temperature, much of the reduction is recovered, leaving a net reduction of a few percent per operating cycle. (I should remark at this point that the duty cycle of the cold beam facility at a moderate power reactor is likely to be no more than 50% because operating crews and even research teams are not continuously available.) The useful life of the rods will end when the conductivity becomes too small to maintain a satisfactory temperature in the source. In this problem the fast neutron and gamma-ray fluxes, the gamma-ray shielding, and the operating cycle all affect the source temperature and must be factored carefully into the design along with the properties of the copper. Although this part of the system design is certainly complex, it appears to be soluble in a number of different ways because many of the factors can be controlled. In the Cornell case we believe that annealed OFHC copper is suitable but we are currently reviewing the literature[13] on the properties, including radiation sensitivity and commercial availability, of the various coppers and copper alloys to find if there is a better one for out purposes.

Heat pipes may be a good alternative in this situation, but they have their own problems. Before cooling down they would be at several atmospheres internal pressure and leaks could cause difficulties. Care needs to be taken during operation to avoid clogging the system by freezing the working gas, which in the case of neon, the most appropriate one for this application, occurs at 24.5K, only 2.6 degrees below the liquefication temperature of 27.1K.

Thermal radiation shields for the chamber and the cold fingers will be desirable. In some regions passive shields (i.e., multiple insulated reflecting layers) may be sufficient; in others active shields (i.e., conducting material connected to the higher temperature stage of the cryogenic refrigerator) may be preferable. Materials such as aluminized Mylar deteriorate rapidly in high gamma-ray fluxes, and another material must be chosen.,

IN CONCLUSION

There are a number of smaller, less serious design problems. The beam plug must include biological shielding around the guide and other penetrations. Neutron and gamma-ray shielding around guide is needed along most of its length, and the inside of the guide has to be evacuated to avoid neutron scattering by air. There should be one or more beam shutters, and at the exit of the guide an evacuated target chamber, detector shields, target holders and a beam stop are needed. Many of those items are easier to construct for a pure subthermal beam than for the usual mixed radiation beam.

We end with the conclusion that a cold neutron beam is a useful and feasible addition to an intermediate flux reactor. Although there are design problems, they are interesting and not insurmountable.

Our description has brought us back to the target end, where we look forward to happy hunting with cold neutrons.

ACKNOWLEDGMENTS

The work reported here is a group effort and it is a pleasure to acknowledge the contributions of the many people involved. Lydia Young was the first graduate student to receive a Ph.D. for work on the project, and Takashi Emoto was the second. Other graduate students associated with the project are Carol Ouellet and Alexander Atwood, currently, and Elissa Pekrul, previously. J. Scott Berg, as an undergraduate, wrote the major part of the Monte Carlo program for moderation in mesitylene. We also wish to thank H. C. Aderhold, Paul Craven, and R. J. Reposh of the Ward Laboratory staff for their very considerable help. We are grateful to W. L. Whittemore of General Atomics for calling our attention to the work of Utsuro and his colleagues. I wish to express appreciation to M. S. Nelkin, K. B. Cady, and M. Utsuro for useful conversations about the project, but absolve them of responsibility for any errors that may be present. Finally, the project is currently supported in part by U.S. DOE grant DE-FG07-89ER128907.

REFERENCES

1. Lydia Young and David D. Clark, Atomkernenergie Kerntechnik 44 (Supplement), 383 (1984).
2. D. D. Clark and T. Emoto, in Capture Gamma-Ray Spectroscopy 1987, K. Abrahams and P. Van Assche, eds., Inst. Phys. Conf. Ser. No. 88, (Institute of Physics, Bristol and Philadelphia, 1988), p. S596 and T. Emoto, Ph.D. thesis, Cornell University, Ithaca, NY 1990 (unpublished).
3. David D. Clark, Takashi Emoto, Carol G. Ouellet, Elissa Pekrul, J. Scott Berg, in 50 Years with Nuclear Fission, J. W. Behrens and A. D. Carlson, eds., (American Nuclear Society, 1989), p. 855.
4. Josef Christ and Tasso Springer, Nukleonik 4, 23 (1961); H. Maier-Leibnitz and T. Springer, Reactor Science and Technology 17, 217 (1963).
5. F. J. Webb, Nuclear Sci. and Eng. 9, 120 (1961); W. Van Dingenen, Nucl. Instrum. and Meth. 16, 116 (1962).
6. P. Egelstaff, private communication. (The proceedings of the Workshop will appear as a Los Alamos report.)
7. Masahiko Utsuro, Masaaki Sugimoto, and Yoshiaki Fujita, Annu. Rep. Res. Reactor Inst. Kyoto Univ. 8, 17 (1975).
8. Masahiko Utsuro and Masaaki Sugimoto, J. of Nucl. Sci. and Tech. 14 (5) 390, 1977).
9. John M. Carpenter, Nature (London) 330, 358 (1987).
10. Masahiko Utsuro, J. Phys. C: 9, L171 (1976).
11. T. Emoto, private communication
12. David D. Clark, Takashi Emoto, Carol G. Ouellet, Elissa Pekrul, and J. Scott Berg, Trans. Am. Nucl. Soc. 61, 93 (1990).
13. For example, see M. W. Guinan, P. A. Hahn, C. E. Klabunde, and R. R. Coltman, J. Nucl. Mater. 155-157, 1315 (1988).

EXPERIMENTS WITH (n,γ)-(γ,e$^+$e$^-$) CONVERTERS

B.Krusche

Institut Laue-Langevin,156X Centre de Tri, 38042 Grenoble Cedex,France *

ABSTRACT

Intense positron sources from pair production of high energy prompt γ-rays from neutron capture reactions are described. A titanium-platinum target placed in a thermal neutron flux of 3.3×10^{14} n/cm^{2}s was used in conjunction with the BILL beta spectrometer as a high intensity, monochromatic and tunable positron source. The system yielded 2.5×10^6 e$^+$/s in 10cm of the focal plane with a momentum dispersion of $\Delta p/p = 1.5 \times 10^{-3}$ per cm. It was applied for the search for neutral resonances superimposed on the excitation function of elastic e$^+$e$^-$ (Bhabha) scattering. These resonances had been proposed as an explanation of the GSI positron lines from heavy ion reactions. No such resonances were found and very stringent limits were set for their intrinsic lifetimes. In addition, the possibility of constructing a very intense source of moderated positrons based on the same principle is discussed.

INTRODUCTION

During the last four years the development of high intensity positron sources has been undertaken at the ILL [1]. The interest in this sources was prompted by the puzzling observation of correlated narrow positron and electron lines in heavy ion collisions at the GSI, Darmstadt [2,3]. These monoenergetic e$^+$e$^-$-pairs could have been evidence for the existence and decay of a series of exotic neutral resonant states with invariant masses between 1.4 and 1.8 MeV/c^2. Lifetime limits of the emitting source between 10^{-20}s and 10^{-10}s were derived from the heavy ion experiments. These lifetimes are still large compared to the typical collision times and suggest the decay of a _free_ neutral object. Thus the invariance under time reversal requires that resonances should show up in the inverse e$^+$e$^-$-scattering process. However it was shown [4] that the constraints derived from the very precisely measured (g-2)-factor of the electron implies that the maximum resonant contribution cannot exceed 1% of the elastic e$^+$e$^-$-scattering cross section. Consequently the excitation function of Bhabha scattering must be studied with uncertainties well below the per cent level. This highlights the importance of a monoenergetic, tunable positron source of high intensity and high stability.

The source described here is based on the conversion chain neutrons - gamma rays - positrons in combination with a magnetic spectrometer as momentum selector. This setup was used for high sensivity measurements of the excitation function for Bhabha scattering [5,6].

Based on the experience with this source it was suggested to construct a very high intensity source of moderated positrons with applications mainly in atomic, solid state and fundamental physics. Measurements of the fast positron flux from such sources were carried out at the BILL spectrometer and the results were compared to numerical simulations, which help to estimate the source parameters.

PRINCIPLE OF THE POSITRON SOURCES

The positrons are produced by e$^+$e$^-$-pair creation from thermal neutron capture γ-rays. The energy release in the capture reaction of typically 6 - 10 MeV is normally shared between several γ-ray transitions cascading from the capture to the ground state.

*present address: II Physikalisches Institut, Justus Liebig Universität, 6300 Gießen, W.-Germany

A significant fraction of the neutrons can be captured with a target nucleus of high neutron capture cross section. Thus the produced number of γ-rays can be similar to, or even higher than, the available number of neutrons. The positrons are produced when the γ-rays undergo pairproduction in the (n,γ)-target itself and/or in a separate converter material. An element of high nuclear charge Z is favourable since the pair production cross section is approximately proportional to Z^3. The positron energy spectrum depends primarly on the energy distribution of the γ-rays from the (n,γ)-reaction. However it is modified by the energy loss and scattering of the positrons within the source.

THE TITANIUM-PLATINUM SOURCE FOR MeV POSITRONS

The most prominent e^+e^--line from heavy ion experiments corresponds to an invariant mass of about 1.83 MeV/c^2. The center of mass energy in the Bhabha scattering reaction must equal the rest mass to produce this object at resonance. This requires positrons with kinetic energies of $\approx$2 MeV. In order to efficiently produce such positrons by pair production, the γ-rays should have roughly twice the positron energy plus one MeV for the e^+e^- rest masses, i.e. about 6 MeV. The reaction ^{48}Ti(n,γ)^{49}Ti with $\sigma = 7.8$ barns provides such γ-rays. The dominant γ-rays have energies of 6760 keV, 6556 keV and 6418 keV and the sum of their intensities amounts to 0.8 photons per captured neutron. The natural abundance of ^{48}Ti in titanium is 73.8% and all (n,γ)-reactions from Ti isotopes lead to stable or very short lived daughter isotopes. Thus no long lived activity is produced by the irradiation with thermal neutrons. In addition titanium has a high melting point (1660 °C) and can be machined easely. The titanium can be covered with a high Z material to improve the pair production rate. Here again the coice is limited by the melting point and the activation by thermal neutron capture. Natural platinum (melting point 1772 °C) is the material with the highest Z that can be used in large quantities at the BILL irradiation position.

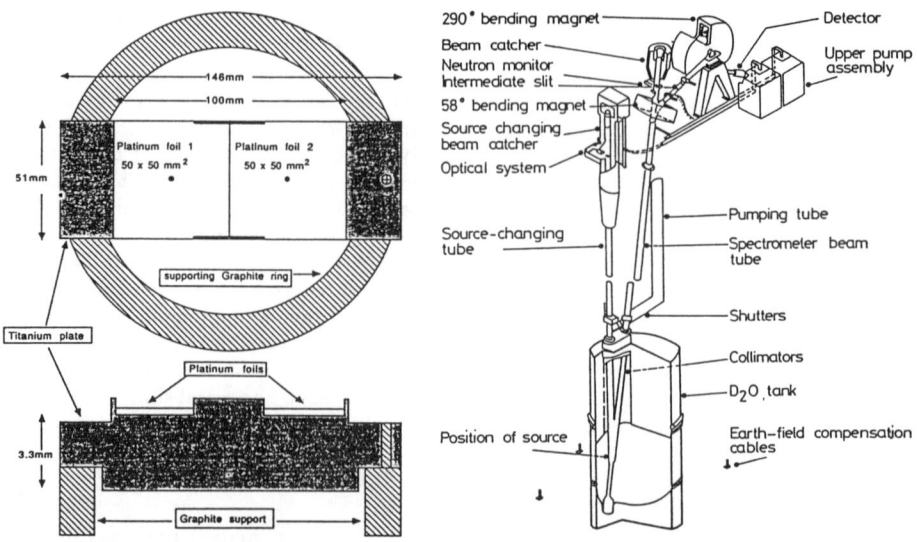

Fig. 1a: The titanium platinum source for MeV positrons

Fig. 1b: The BILL spectrometer

Three different targets have been used so far for the investigation of Bhabha scattering, each improving the intensity of the beam. The first was a titanium plate of 3 mm thickness (target (1)), the second a 2 mm Ti plate covered with a 0.25 mm platinum foil (target (2)).

The last one (target (3)) which is shown in figure 1a was made of a 3.3 mm titanium plate (surface area 50×140 mm^2) covered with a 0.25 mm thick platinum foil (surface area 10×5 cm^2). They were exposed to a thermal neutron flux of 3.3×10^{14} n/cm^{2}s at the inpile site of the beta spectrometer BILL [7].

The BILL spectrometer (see figure 1b) views the target from a distance of 14 m under a solid angle of 3×10^{-6} of 4π. The two double focussing iron magnets momentum analyse the positrons and focus them onto the focal plane of the spectrometer. The dispersion in the focal plane is $\Delta p/p = 1.5 \times 10^{-3}$ per cm, the geometrical reduction of the image of the target is 0.1 and 0.07 in the axial and radial dimensions, respectively.

An important advantage of this setup is the high stability of the source. The intensity follows directly the reactor neutron flux, which is stable to about 0.1%. Even more important than the time stability is the geometrical stability of the beam. The spectrum can be easily scanned by changing the magnetic fields of the spectrometer without moving the position of the beam in the focal plane.

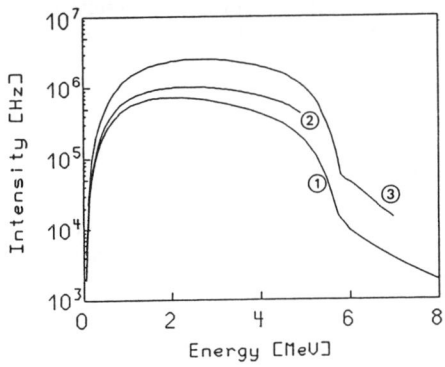

The spectra of the sources were measured in an 88 mm segment of the focal plane by a 32 wire proportional counter. Due to the dispersion of the spectrometer the measured intensities correspond to a relative momentum window of $\Delta p/p = 1.32 \times 10^{-2}$ which translates into an energy window of 32 keV at 2 MeV. The measured spectra (see figure 2) show a pronounced step in intensity around 5.5 MeV. This corresponds to the intense γ-rays from ^{49}Ti with energies between 6.4 and 6.8 MeV. The most intense source gave about 2×10^6 positrons per second at energies between 1.5 MeV and 4 MeV.

Fig 2: Measured positron spectra from target (1)-(3).

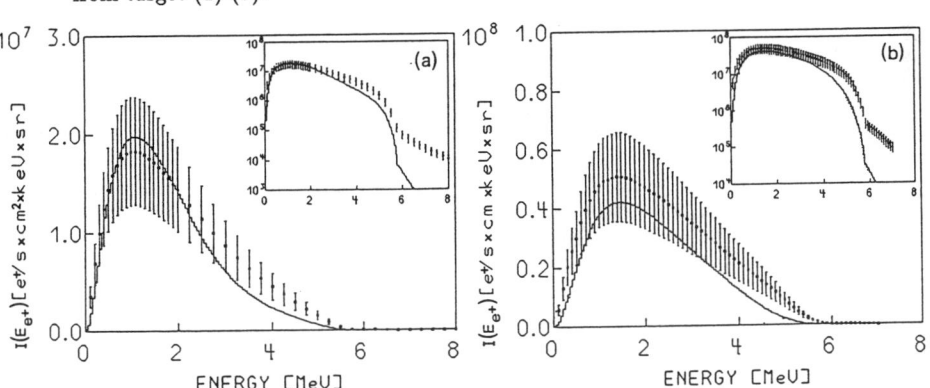

Fig 3: Comparision of measured spectra (error bars) from target (1) (figure 3a) and target (3) (figure 3b) with the results of a numerical simulation (solid lines).

The measured intensities were transformed to a constant energy window and calibrated with the target mass, the target surface area, the neutron flux, the measuring time and the known spectrometer efficiency. The result was compared to the prediction of numerical simulations based on the absorption of an isotropic neutron flux, the γ-ray spectra from thermal neutron capture in the target material, the cross sections for total γ-absorption, pair effect and Compton scattering, the energy distribution of compton scattered γ-rays, the energy distribution of positrons produced by pair formation of a photon with known energy, the energy loss and the small angle multiple scattering of positrons in the target (for details see ref. [1]).

The measured and calculated spectra are shown in figure 3, both for an isotropic neutron flux of 3.3×10^{14} n/cm^2. The systematic uncertainties (≈ 30 %) of the measured spectra stem from the uncertainty in the target surface area seen by the spectrometer and the solid angle covered by the spectrometer. The agreement between measurment and simulation is very satisfactory.

Simulations of even more massive targets show that the positron flux cannot be improved by much more than a factor of two due to γ-ray and positron self absorption. However a factor of two would require substantially thicker targets of more than 10 mm titanium. The thickness is limited by the γ-heating of the target which is suspended in vacuum. The heat input must be carried away by thermal radiation into the beam tube. The results of the simulations together with the measured thermal emissivity of $\epsilon \approx 0.5$ suggest a target temperature of ≈ 800 °C for the present target but about 1500 °C for a twice as intense target. We thus conclude, that the present source is already very close to the optimum that can be achieved in the BILL irradiation position.

THE BHABHA SCATTERING EXPERIMENTS

For the search of the neutral resonances two different experimental setups have been used. The first [5] was optimized for a very sensitive measurement of the elastic Bhabha scattering excitation function to detect small resonances on top of it. The second [6] was a shadow method used to suppress the elastic events and gain in sensitivity for the decay of long lived neutral particles. The two types of experiments will be discussed below.

HIGH SENSIVITY MEASURMENT OF THE EXCITATION FUNCTION FOR BHABHA SCATTERING

The experimental setup is shown in figure 4. A beryllium foil (10 × 100 cm^2 and 4.6 mg/cm^2 thick) was suspended as a scatterer along the focal plane of the spectrometer. Beryllium was chosen because of the relative low average binding energy of its electrons. Bound electrons exhibit a momentum distribution due to their localisation which increases with growing atomic binding energy. The corresponding spread out of the available center of mass energy in the scattering process leads to a Doppler broadening of resonances. This broadening is of the order of 20-30 keV (FWHM) for the low Z beryllium.

The detector device was made of two arrays of four high resolution Si(Li) detectors each with an area of 20×20 mm^2 and an effective thickness of 2 mm. They achieved an energy and time resolution of 7 keV and 4 ns respectively when cooled down to about -50 °C. The detectors were placed at a distance of 85 mm behind the scattering foil and covered mean scattering angles around 30° symmetrically to the beam direction (see figure 4). The angle acceptance of one detector row was reduced by apertures by about 20 % to achieve 100 % coincidence efficiency and to avoid a dependence of the coincidence efficiency on slight misalignments. The opposite detector rows were shielded against each other by a low Z material to avoid coincident background from Mott scattered positrons which are registered in one detector and subsequently reach an opposite detector by backdiffusion.

Bhabha scattering was observed by requiring kinematic coincidences between the scattered positron and the recoil electron recorded by detectors from the opposite arrays. In addition, single events from elastic scattering of the positrons at the beryllium nuclei (Mott scattering) were recorded scaled down for normalisation purposes.

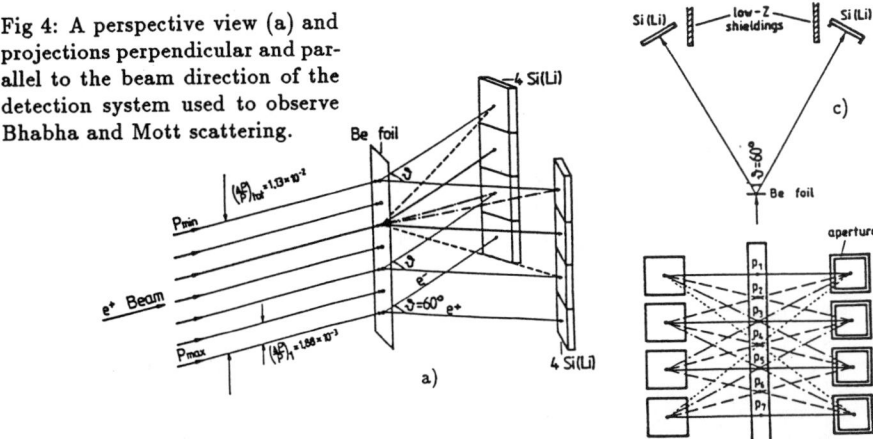

Fig 4: A perspective view (a) and projections perpendicular and parallel to the beam direction of the detection system used to observe Bhabha and Mott scattering.

The measurment of the Bhabha excitation function was carried out in two experiments each taking about two weeks of beam time with in total ten independent runs. For each run the spectrometer field was scanned between 2.13 and 2.38 MeV in steps of 5 keV. The measuring time was about 40 min per step per scan.

The momentum interval of the positron beam was limited by a diaphragm in the intermediate image of the spectrometer to $\Delta p/p = 1.13 \times 10^{-2}$, i.e. ≈ 30 keV at 2.3 MeV. However this relatively large energy window does not define the actual energy resolution for Bhabha events. The two detector arrays allow 16 different kinematical coincidences which are related to scattering processes in seven different areas on the Be foil (see figure 4). Due to the momentum dispersion along the foil, each of this areas belongs to a momentum bin of $\Delta p/p = 1.88 \times 10^{-3}$ or ≈ 5 keV at 2.3 MeV. In this way seven different initial energies are simultanously measured for each setting of the magnets. This information was taken into account for the construction of the Bhabha sum spectra. For Mott scattering however the energies are averaged over the full 30 keV range.

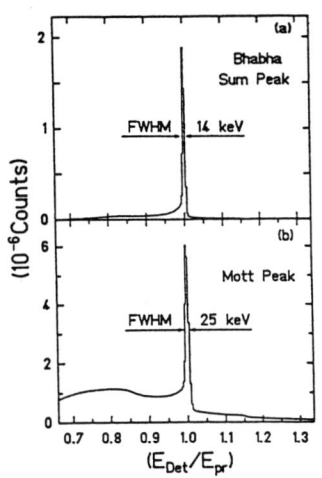

Fig 5: Measured Bhabha sum and Mott peak

The quality of the measured spectra is illustrated in figure 5. Here the measured energies of the sum coincidence and single events (E_{Det}) were divided event by event by the incident positron energy (E_{pr}) which was obtained from the spectrometer calibration. The widths of the peaks reflect the energy spread of the incident positron beam, the energy loss straggling in the target and the detector resolution. As expected the Bhabha sum peak is narrower due to the smaller effective spread of the positron beam energy. The excitation functions for Bhabha and Mott scattering were derived from the energy spectra by an integration of the full-energy peaks in the window $0.991 \leq E_{Det}/E_{pr} \leq 1.009$. They are displayed in figure 6a. Within the statistical accuracy of 0.25 % (Bhabha) and 0.11 % (Mott) both excitation functions are smooth which demonstrates the excellent stability of the beam. In order to account for remaining little fluctuations the Bhabha scattering events were normalised to the Mott events for each measuring point. The result is shown in figure 6b together with the relative and standard deviations from a fitted second order polynomial. No significant deviations from a smooth behaviour occur on a level of statistical accuracy of 0.25 %.

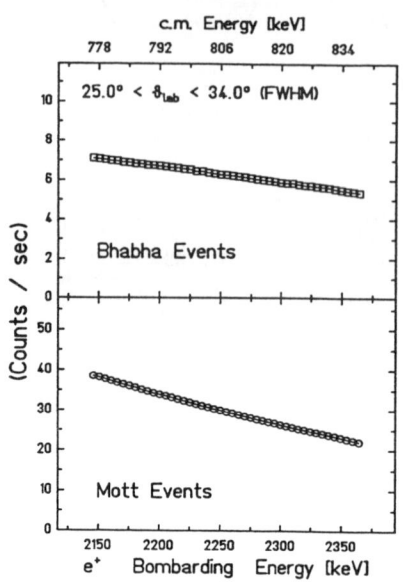

Fig 6a: Total counting rate per 5 keV step for Bhabha and Mott events

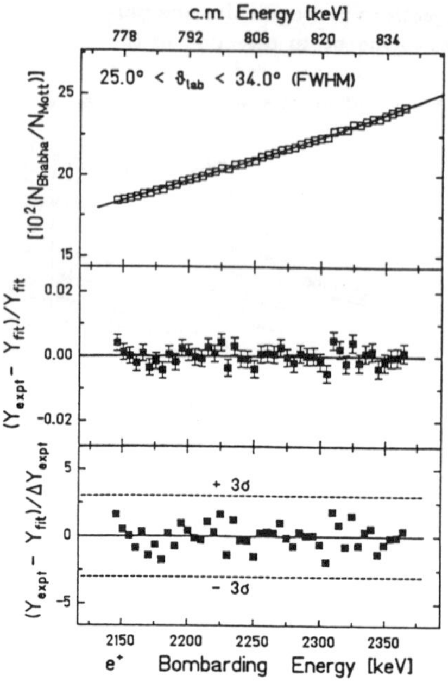

Fig 6b: Ratio of Bhabha to Mott scattering and relative and standard deviations from fitted curve

The maximum height δ of a resonance with Lorentzian shape and a width fixed to the experimental resolution can be determined relative to the Bhabha cross section by a least squares procedure. This maximum height then limits the energy integrated total cross section of a hypothetical resonance via: [5] (in this work * refers to quantities in the c.m.-system):

$$(\sigma_R \Delta E^\star)_{exp} \leq 4\pi\delta \left(\frac{d\sigma_B}{d\Omega^\star}\right)_{eff} \Delta E^\star_{exp} \qquad (1)$$

Here the effective Bhabha cross section $\left(\frac{d\sigma_B}{d\Omega^\star}\right)_{eff}$ must be deduced by proper weighting the differential cross section for Bhabha scattering with the angle dependent detection efficiency [5]. The experimental width $\Delta E \approx$ 28 keV is dominated by the contribution from the Doppler broadening of the bound electrons ($\approx$27 keV) and includes the energy spread of the incident positrons (5 keV) as well as the energy loss straggling. Thus we have $\Delta E^\star_{exp} \approx$ 7.8 keV in the c.m.-system. For a c.m.-energy of 810 keV this yields a limit of

$$(\sigma_R \Delta E^\star)_{exp} < 6.3 \text{ beV}$$

An upper limit of the intrinsic resonance width $\Gamma^\star_R$ can then be obtained using the Breit-Wigner formula and assuming the e^+e^--decay channel to be the dominant one by:

$$(\sigma_R \Delta E^\star)_{exp} = \frac{2\pi^2(2J+1)\hbar^2}{M_R^2 - 4m_0^2}\Gamma^\star_R \qquad (2)$$

where M_R stands for the invariant mass of the resonance. For M_R = 1.832 MeV/c^2 corresponding to the most prominent GSI line, we arrive at $\Gamma^\star_R \leq$ 1.9 meV or $\tau^\star \geq 3.5 \times 10^{-13}$s for a spinless resonance.

The lower lifetime limit of $\tau^* \geq 10^{-20}$s from the heavy ion experiments was thus improved by more than seven orders of magnitude. The new limit was already more stringent than the one derived from precision measurments of the (g-2)-factor of the electron. Furthermore it is independent of assumptions about the internal structure of the hypothetical resonance because it is measured on mass shell. However a lifetime gap between 3.5×10^{-13}s and 10^{-10} was still left.

SEARCH FOR LONG-LIVED NEUTRAL RESONANCES IN BHABHA SCATTERING AROUND 1.8 MeV/c²

The accuracy of the experiments discussed above was only limited by statistical uncertainties, systematical errors were negligible. However a much longer measuring time would be neccessary to improve significantly the statistical accuracy because further improvements of the intensity of the positron source are hampered by heating problems.

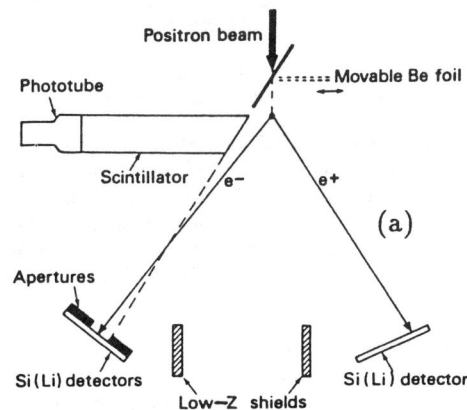

Fig 7: Experimental setup for the search of long lived resonances (see text)

However a different experimental technique can be used for neutral particles with lifetimes above 10^{-13}s. The experiment was modified to the setup shown in figure 7. The positron beam was focussed as before onto the 4.6 mg/cm² thick beryllium foil and coincidences between the two arrays of silicon detectors were recorded. However the foil was now tilted at an angle of 30° relative to the positron beam and a plastic scintillator was interposed between the foil and one of the detector arrays. The scintillator acted as a passive and active shield to suppress the coincident events from elastic Bhabha scattering within the foil. A hypothetical neutral particle would be produced with a velocity of $\approx 0.8c$ and would leave the foil within 5×10^{-14}s. Thus for the lifetime range of interest it would decay outside the foil and could trigger a coincident event.

The excitation function was measured similar to the previous experiment in four indepent runs with a measuring time of 45 minutes per 5 keV step in each run. The calibration of the Si(Li) detectors was checked at the beginning and end of each run with Bhabha and Mott scattering from a second Be-foil which could be moved in perpendiculary to the positron beam. The Bhabha peak observed from the calibration foil was used to extract the position and width of the Bhabha window for the analysis of the measurement with the tilted foil.

The total coincident rate in the proper sum energy window from all four runs is shown in figure 8a. A small background remains which is mainly due to Bhabha events that deposited less than 60 keV (lower threshold) in the scintillator. The data does not exhibit any resonance like structure and the maximum peak height was found to be ≤ 25 % (95 % CL) relative to the fitted curve, i.e. ≈ 5 counts at 810 keV c.m.-energy. An upper bound of the energy integrated total cross section $(\sigma_R \Delta E^*)_{exp}$ can be determined with equ. (1). Here $\delta = (N_P/N_B)/\epsilon(\tau^*)$ is calculated from the limit on the number of particle decays ($N_P \approx 5$), the number of Bhabha events that would be detected without the scintillator (N_B) and the dectection efficiency $\epsilon(\tau^*)$ for e^+e^--pairs from long lived particles relative to the efficiency for Bhabha counts with the shield removed. The number of Bhabha events $N_B \approx 5.4 \times 10^4$ was obtained from the Mott rate observed in the unshielded detector row and the Bhabha to Mott ratio from the previous experiment. The realtive efficiency $\epsilon(\tau^*)$ was calculated numerically (see ref. [6]).

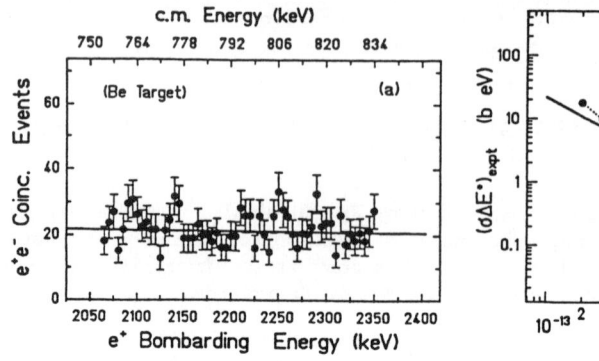
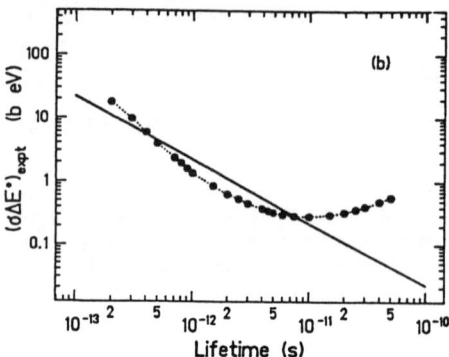

Fig 8a: Number of coincident events in the total count time of 10800 s

Fig 8b: Upper limit on total cross section and unitarian limit (solid line)

The limit obtained for $(\sigma_R \Delta E^*)_{exp}$ (95 % CL) is compared in figure 8b to the total cross section calculated in the unitarian limit equ.(2) for a spinless resonance of an invariant mass $M_R = 1.832$ MeV/c². Obviously lifetimes between 4.5×10^{-13}s and 7.5×10^{-12}s are excluded.

In conclusion the experiments of both types produced no evidence for neutral resonances around an invariant mass of 1.832 MeV/c². If the decay of an unknown neutral particle is responsible for the line found in the heavy ion experiments its lifetime must fit into the narrow range 7.5×10^{-12}s $< \tau^* < 10^{-10}$s.

A SOURCE FOR MODERATED POSITRONS

Research with low energy positrons covers a wide field of investigations in solid state physics, atomic physics and particle physics [8]. A serious drawback is the low positron flux that is currently available, consequently a new positron source is highly desirable. The proposed production method is similar to those at LINACS [9], where high energy electrons produce bremsstrahlung which then creates positrons by the pair effect.

The platinum-titanium source is not optimal for this kind of application because the capture cross section of titanium is rather low. A layer thick enough to absorb the neutron flux would give rise to high self absorption losses. The shape of the γ-ray spectrum is also not favourable, because the very strong high energy γ-rays lead to a hard spectrum with low γ-ray multiplicity.

A much better choice is a cadmium-tungsten source. Thermal neutron capture in natural cadmium is completly dominated by the ^{113}Cd(n,γ)^{114}Cd reaction with a cross section of ≈ 26000 barns in the thermal neutron flux of the ILL reactor. Already a layer of 50 mg/cm² of natural cadmium absorbs thermal neutrons completely. Furthermore ^{113}Cd has a relatively high neutron binding energy (9041 keV) which is available for the cascade γ-rays. Tungsten is very efficient for the conversion of γ-rays to e^+e^--pairs and tungsten is a common moderator for positrons, so production and moderation of the positrons are performed in the same material.

The spectra from five different Cd-W sources in the few MeV range have been measured with the BILL spectrometer. The targets were made of about 2 mg natural cadmium oxide sedimented onto a 1×1 cm² area of a nickel foil. The cadmium was covered with tungsten foils of 0.025 mm or 0.1 mm thickness. The surfaces of the targets were viewed under a 7° angle with respect to the beam axis and at approximately 45°.

The spectra were measured three times from 20 keV to 6 MeV in steps of 10 keV during a period of 50-60 hours. The intensity of the ^{113}Cd(n,γ) reaction decreased during this time by almost an order of magnitude due to radioactive burnup. This allowed the elimination of the background from positrons produced e.g. from capture in other Cd, W and Ni isotopes. A target without tungsten covering was used to estimate the background from internal pair production in ^{114}Cd. The burnup corrected count rates were then converted to a constant energy window etc. in the same way as for the titanium-platinum spectra. The normalisation was taken from the measured count rate (uncovered target) of the K-conversion line of the 1305 keV E0 transitions since this intensity is known to be 2.33 e^- per 1000 captured neutrons.

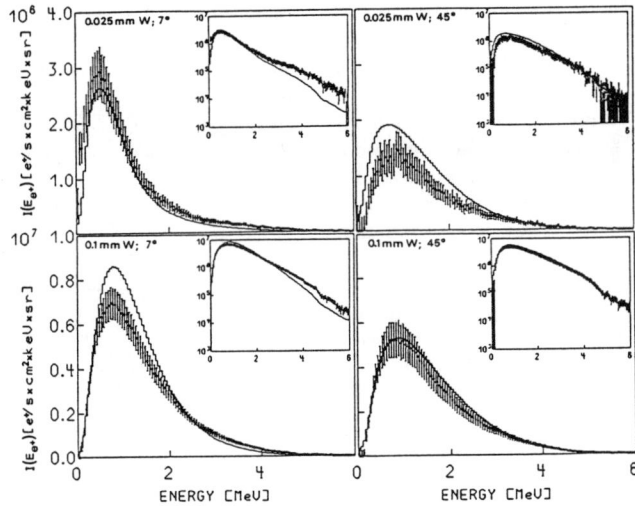

Fig 9: Measured and simulated positron spectra from Cd-W targets for 0.025 and 0.1 mm thick tungsten foils and a captured neutron flux of 10^{13}n/cm^2. The targets were inclined by 7 ° or 45 ° with respect to the position perpendicular to the beam. The points with error bars show the measured spectra, the full lines indicate the result of the simulation. The insets show the same data in a logarithmic scale.

The measured spectra are compared in figure 9 to a numerical simulation carried out with the code POSGEN [1]. The indicated errors are mainly systematic ones (normalisation etc), The absolute intensity scale and the shape of the spectra are quite well reproduced. It should be noted, that the spectra for the thicker foils or the larger angles appear to be harder. This effect is entirely due to small angle multiple scattering.

The simulations can be used to compute the produced, absorbed and emitted positron flux depending on the thickness of the tungsten layer. We can then consider a stack of n tungsten foils enclosed between two cadmium layers to get an approximation of the possible source strength. The simulations predict (for details see ref [1]) an intensity of moderated positrons of $I[e^+/s] \approx 10^{-5} \times R[n/cm^2 s] \times A[cm^2] \times n$ where R is the neutron capture rate in each cadmium layer and A the surface area of each tungsten foil. Here it is assumed that the total thickness of the tungsten layer is between 2 - 5 g/cm^2 and that positrons stopped in a surface layer of about 500 Å can diffuse out ouf the foils. For instance with 20 separate W foils of area 10×10 cm^2 and R = 10^{13} n/cm^2 we arrive at a moderated e^+ intensity of 2×10^{11} e^+/s.

The thermal neutron flux is available over a very large area and therefore the intensity of the source is mainly limited by the requirement that all produced positrons must be collected in a small beam. A possible layout is shown in figure 10a. The source is illustrated for the proposed site in the light water pool of the ILL High Flux Reactor. The capture neutron flux at this side was estimated to about 4×10^{12} n/cm^2 from a gold activation measurement. The vacuum tube containing the tungsten converter is surrounded by cadmium. The positrons are produced and moderated in the tungsten vanes and focussed by an electrical field in form of an "electron gun" onto a remoderator.

With the estimate given above, the measured capture flux and a typical remoderation efficiency of 0.15 one can roughly approximate the moderated positron flux in the transport solenoid to be of the order of 10^{10} e$^+$/s.

A still higher intensity would be possible if positrons from an even larger source area can be focussed. However, the electrostatic focussing is limited by the initial transversal momentum of the positrons. This problem might be overcome by the use of a drift tube as shown in figure 10b. In this case the positrons drift along the electric field lines without accelaration and below the threshold for positron formation, keeping the annihilation probability rather low. Even though certainly a number of problems arise for the use of such a drift chamber in the presence of the high radiation level inside the reactor pool, it seems worthwhile to be further investigated.

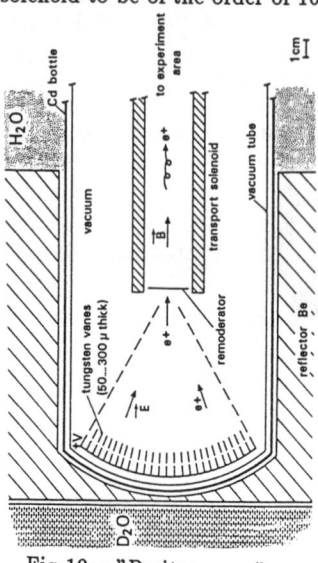

Fig 10a: "Positron gun"

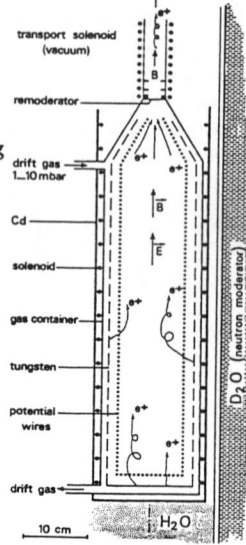

Fig 10b: Drift tube

CONCLUSIONS

It was demonstrated that the chain thermal neutrons - capture γ-rays - e$^+$e$^-$ pairs is a potentially powerful method for intense sources of both MeV and moderated positrons. With some development on the layout of such sources a stable and high quality beam of moderated positrons should be feasible.

The high energy source was applied for the sensitive search of neutral resonances in Bhabha scattering. These experiments have set the most stringent bounds for new particles around invariant masses of $\approx$ 1.8 MeV/c^2 independent of assumptions about their internal structure. The liftime τ^* of this hypothetical particles was limited to the small range of 7.5×10^{-12}s $< \tau^* < 10^{-10}$s.

ACKNOWLEDGEMENTS

The presented work was done in close collaboration with K.Schreckenbach, ILL. The Bhabha scattering experiments were carried out by an ILL - GSI collaboration: K.Schreckenbach, S.Judge, and B.Krusche (ILL), H.Tsertos, C.Kozhuharov, P.Kienle and P.Armbruster (GSI). This contribution was supported by the BMFT.

REFERENCES

[1] B.Krusche, K.Schreckenbach, NIM, in press
[2] T.Cowan et al., Phys.Rev.Lett 56, 444 (1986)
[3] E. Berdermann et al., Nucl.Phys. A448, 683 (1980)
[4] J.Reinhardt et al., Z.Phys. A327, 367 (1987)
[5] H. Tsertos et al., Rev.Lett. B207, 273 (1988) and Z.Phys. A331,103 (1988) and Phys.Rev. D40,1397(1989)
[6] S.M.Judge et al.,Phys.Rev.Lett.65,972(1990)
[7] W. Mampe et al., NIM 154, 127 (1978)
[8] e.g. J.A.W. Humberstone and E.A.G. Armour (eds), "Atomic Physics with Positrons", Nato Advanced Studies Inst. Series B, Phys. Vol 169 (plenum press 1987); L. Dorikens - van Praet, M. Dorikens and D. Segers (eds), "Positron Annihilation" (World Sci., 1988)
[9] F. Ebel et al., NIM 272, 626 (1988)

EXPERIMENTAL ASSESSMENT OF THE PERFORMANCE OF A PROPOSED LEAD SLOWING-DOWN SPECTROMETER AT WNR/PSR*

M. S. Moore, P. E. Koehler, A. Michaudon, and A. Schelberg**
University of California, Los Alamos National Laboratory
Los Alamos, NM 87545 USA

Y. Danon, R. C. Block, and R. E. Slovacek
Rensselaer Polytechnic Institute
Troy, NY 12181 USA

R. W. Hoff and R. W. Lougheed
University of California
Lawrence Livermore National Laboratory
Livermore, CA 94551 USA

ABSTRACT

In November 1989, we carried out a measurement of the fission cross section of ^{247}Cm, ^{250}Cf, and ^{254}Es on the Rensselaer Intense Neutron Source (RINS) at Rensselaer Polytechnic Institute (RPI). In July 1990, we carried out a second measurement, using the same fission chamber and electronics, in beam geometry at the Los Alamos Neutron Scattering Center (LANSCE) facility. Using the relative count rates observed in the two experiments, and the flux-enhancement factors determined by the RPI group for a lead slowing-down spectrometer compared to beam geometry, we can assess the performance of a spectrometer similar to RINS, driven by the Proton Storage Ring (PSR) at the Los Alamos National Laboratory. With such a spectrometer, we find that it is feasible to make measurements with samples of 1 ng for fission 1 μg for capture, and of isotopes with half-lives of tens of minutes. It is important to note that, while a significant amount of information can be obtained from the low resolution RINS measurement, a definitive determination of average properties, including the level density, requires that the resonance structure be resolved.

I. THE SLOWING-DOWN SPECTROMETER AT RPI

The Rensselaer Intense Neutron Source (RINS) is generally recognized as the highest intensity neutron spectrometer in the world.[1] It consists of a $(1.8 \text{ m})^3$ block of highly pure lead, in the center of which is placed a neutron production target so that the spectrometer can be driven by electron pulses from the RPI electron linear accelerator. Four penetrations are provided for the insertion of experiments; these are horizontal channels located 60 cm from each of four faces of the block of lead.

The earliest use of a lead slowing-down spectrometer (SDS) was reported at the first Geneva conference (1955) by Bergmann et al.,[2] who also calculated the performance characteristics. In a lead SDS, the neutron pulse is time and energy

---

*Work performed under the auspices of the U.S. Department of Energy by the Los Alamos National Laboratory under contract number W-7405-ENG-36.

**Present address EG&G, Inc., P. O. Box 809, Los Alamos, NM 87544.

focused in the slowing-down process (lower energy neutrons make fewer collisions per unit time, and thus lose less energy than higher energy neutrons in the pulse). The neutron intensity is much higher than in a standard time-of-flight experiment because the sample can interact with the same neutrons many times in the slowing-down process. A rough rule-of-thumb based on the experience of the RPI group[3] suggests that the SDS gives an intensity about 10,000 times that of a standard time-of-flight measurement at the equivalent flight path, 5.6 m, which is characteristic of the SDS made of solid lead. The energy dependence of the enhancement factor is shown in Table I.

Table I. Energy dependence of the flux enhancement of a lead slowing-down spectrometer over that of a conventional time-of-flight experiment at a 5-meter flight path, as determined by the RPI group.[3]

Neutron Energy (eV)	Φ_{RINS}/Φ_{TOF}
1000	5800
100	7800
10	10500
1	14300

According to Bergmann's calculations, the resolution of the SDS should be about 30% in energy. In practice, the RPI group found that 35-40% in energy is to be expected, because of the presence of light element contaminants. Even a small amount of hydrogen degrades the resolution to the point that the instrument is no longer useful as a spectrometer.[4]

While Bergmann et al., used the SDS for radiative capture cross-section measurements, the RPI group has concentrated on fission. One of the early triumphs was the first demonstration of subthreshold fission in (^{238}U + n).[5] More recently, they have completed a series of measurements of the resonance fission cross sections of the curium isotopes,[6] using a special hemispherical fission chamber suggested by N. W. Hill[7] to minimize alpha pileup.

One problem the RPI group has experienced involves target cooling. A hydrogenous coolant such as H_2O cannot be used, so in RINS the target is cooled by flowing helium gas. However, bremsstrahlung heating of the lead limits the beam power on target to <1 kW. In an experiment in which the background is dominated by radiation from the sample (alpha pileup or spontaneous fission in a fission measurement, or gamma emission in a capture measurement), the signal-to-noise ratio is maximized if the beam power per pulse is as high as can be achieved, with the repetition rate adjusted to give the maximum allowable average power. In a recent typical measurement,[8] the RPI linac operated at 60 MeV with an average current of 15 μA (giving 0.9 kW on target), achieved with 200 ns wide electron bursts at a repetition rate of 90 Hz. Under these conditions, it took about 24 hours of data collection to discover the 0.53 eV resonance in ^{250}Cf, using a 150 ng sample.

II. A PROTON-DRIVEN SLOWING-DOWN SPECTROMETER

When we first proposed adding a Proton Storage Ring (PSR) to the Weapons Neutron Research (WNR) facility at the Los Alamos National Laboratory, we recognized that a lead slowing-down spectrometer driven by bursts of 800 MeV protons would have some unique features. A proton spallation target is much more efficient at producing neutrons than is an electron bremsstrahlung target: we estimate 10-15 neutrons produced per 800 MeV proton on a tungsten target, compared with 0.01-0.02 neutrons for each 60 MeV electron on target. So if the performance of the lead SDS is limited by the problem of average heat removal from the target, we expect a factor of 50-100 higher average neutron population in a proton-driven spectrometer. However, if the problem is primarily due to gamma heating of the lead, the improvement factor could be much larger than this estimate because the gamma pulse is very much smaller with a proton-spallation source. We also expect a dramatic improvement in the signal-to-noise ratio with a proton-driven SDS at WNR/PSR. The proton storage ring, currently used to provide neutrons for the Los Alamos Neutron Scattering Center (LANSCE) at Los Alamos, typically delivers 50-60 μA of 800 MeV protons in 250 ns wide pulses at 20 Hz.

In summary, based on the foregoing performance calculations, we might expect the average performance of a proton-driven lead slowing-down spectrometer at WNR/PSR to show an enhancement of a factor of 50-100 in flux, compared with RINS, at the same average power on target. The signal-to-noise ratio, if the noise is dominated by sample activity such as spontaneous fission, alpha pileup, or gamma radiation associated with sample decay for a capture measurement, should show an enhancement of a factor of about 10^4 over that at RINS. If this conclusion is correct, it suggests that we could consider making fission cross-section measurements on samples that are less than 1/10000 as large as those currently being done (~1 μg) or on samples with more than 10000 times the specific activity.

III. AN EXPERIMENTAL TEST

In November 1989, we fielded a hemispherical fission chamber with microgram sized deposits of ^{235}U, ^{247}Cm, ^{250}Cf, and ^{254}Es for a measurement in the RINS slowing-down spectrometer.[8] Eight months later (almost exactly one half-life for ^{254}Es) we fielded the same chamber in a measurement in conventional geometry at the LANSCE facility. The two experimental configurations are compared in Table II. If the calculations and the conclusions reached in the previous sections are correct (that a PSR-driven SDS should have 10^4 times the flux of RINS), and if the experience at RPI summarized in Table I obtains for PSR/WNR (that the SDS flux is about 10^4 times that of a conventional experiment), then in our conventional experiment at LANSCE we should see the same count rate per burst as we saw with the RINS spectrometer, if we correct to the equivalent flight path.

We choose the 1.25 eV resonance in (^{247}Cm +n) to make the comparison; it is very nearly resolved in the RINS measurement, and thus allows a reasonable comparison to be made. The raw count rate comparisons are given in Figs. 1 and 2. From these data, it is evident that the raw count rates per burst are nearly the same, 0.21 c/burst in this resonance with RINS and 0.29 c/burst at LANSCE. If one corrects to the equivalent flight path, by multiplying the resonance area observed at LANSCE by $(7.25/5.65)^2 = 1.65$, one can conclude that, at a 5.65 m station at LANSCE, the neutron intensity should be roughly a factor of 2 larger than with RINS.

The two sets of data were analyzed in the same way: the ^{235}U fission count rate, measured at the same time in the experiment, was used to derive a flux shape and magnitude. This then permits the fission cross sections for the ^{247}Cm, ^{250}Cf, and ^{254}Es samples to be determined. The cross section comparisons are shown in Figs. 3, 4, and 5 for ^{247}Cm, ^{254}Es, and ^{250}Cf, respectively. Resonance parameters from area analysis are compared for ^{247}Cm and ^{250}Cf in Table III. For ^{254}Es, no distinct resonances were seen in either measurement.

Table II. Comparison of the November 1989 RINS experimental determination of the fission cross section of ^{247}Cm, ^{250}Cf, and ^{254}Es with the July 1990 LANSCE determination.

	RINS	LANSCE
Geometry	Pb SDS	Conventional
Flight path	5.65 m[a]	7.25 m
Beam	60 MeV e$^-$	800 MeV H$^+$
Target	He-cooled Ta	H$_2$O-cooled W
Pulse width	200 ns	125 ns FWHM
Repetition rate	90 Hz	20 Hz
Average power	0.9 kW	55 kW
Duration of expt.	19 hours[b]	96 hours
	(6.2 · 10^6 bursts)	(5.9 · 10^6 bursts)
Sample sizes:		
^{247}Cm	3.16 µg	3.16 µg
	(30.86 f/s)	(30.6 f/s)
^{250}Cf	0.149 µg	0.139 µg
	(441.9 f/s)	(412.6 f/s)
^{254}Es	0.237 µg[c]	0.117 µg
	(13.03 f/s)[d]	(338.3 f/s)[d]

[a]5.65 m is the equivalent flight path for a solid lead SDS.
[b]19 hours is a partial run, used for the present comparison.
[c]A recalculation of the sample size in November 1989, based on the observed spontaneous fission rate in July 1990 for a 276 day ^{254}Es half-life, gives 0.237 µg instead of the 0.21 µg of Ref. 8).
[d]The increase from 13 to 338 fissions per second is due to the ^{250}Cf content.

    The resolution associated with the LANSCE measurement is much better than that with RINS, although not as good as that associated with the 1969 Physics-8 nuclear explosion measurement of Moore and Keyworth.[9] The resolution of our LANSCE measurement should be sufficient to give a reasonably complete set of parameters below 20 eV, however, where no data were obtained on Physics-8 because the moderator was moving up the line-of-sight pipe faster than the escaping neutrons.

    A surprising feature of the present comparison of the two facilities and the data sets we obtained is that the low resolution lead SDS gives a reasonable, although not complete, picture of the average resonance properties of fissionable nuclei. In the LANSCE measurement, we confirmed the parameters of the lowest energy resonance in ^{250}Cf. We confirmed a net average cross section of ~4 barns between 16 and 80 eV in ^{250}Cf, but saw no additional structure. We confirmed the average cross section of ^{254}Es, and seem to see the same little wiggles, but no distinct resonances. In this sense, we confirmed the conjecture we made earlier[8] that the fission cross section of ^{254}Es has no resolvable resonances. However, for ^{247}Cm we observed many more resonances in the LANSCE measurement, enough to be able to extract the level density below 20 eV and to complete the multilevel analysis[9] in this region.

    In conclusion, we feel that the present two experiments provide a sufficient data base to enable us to project what could be done with a lead slowing-down

M. S. Moore et al. 957

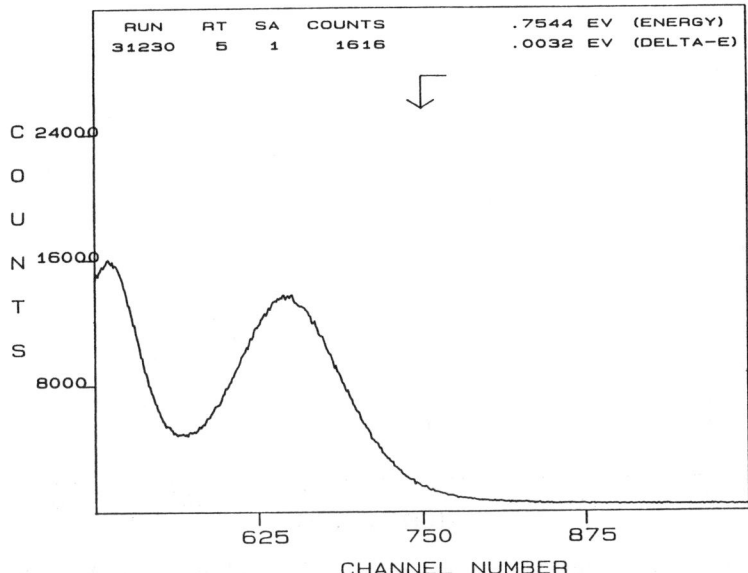

Fig. 1. Raw data plot over the 1.25 eV resonance in (^{247}Cm + n) as obtained with the RINS on 2-3 November 1989. This figure represents a summation of 19 one-hour runs (runs 312-330) at 90 Hz, or 6.2 · 10^6 bursts. The integral over this resonance is about 1.3 · 10^6 counts, giving 0.21 count/burst.

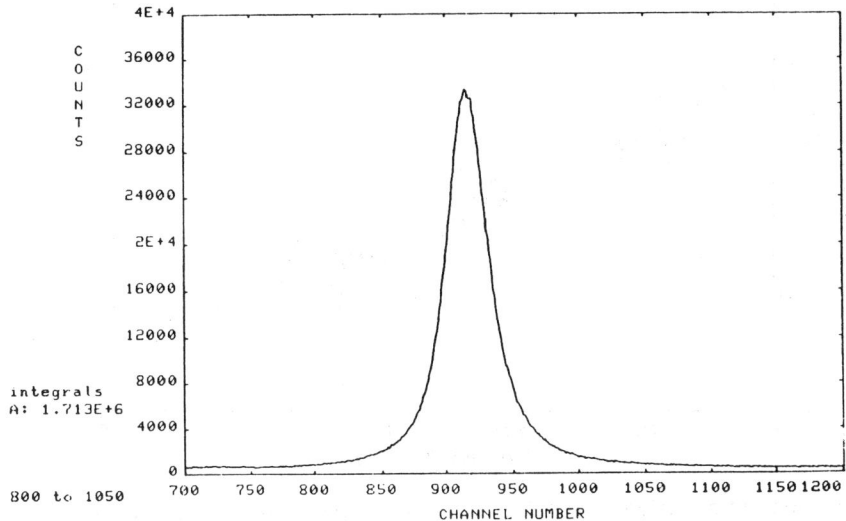

Fig. 2. Raw data plot over the 1.25 eV resonance in (^{247}Cm + n) as obtained in the conventional time-of-flight measurement at a 7.25 m flight path at LANSCE on 26-30 July 1990. This figure represents a summation of four 24-hour runs taken at 20 Hz, or 5.9 · 10^6 bursts. The integral over this resonance is about 1.7 · 10^6 counts, giving 0.29 count/burst.

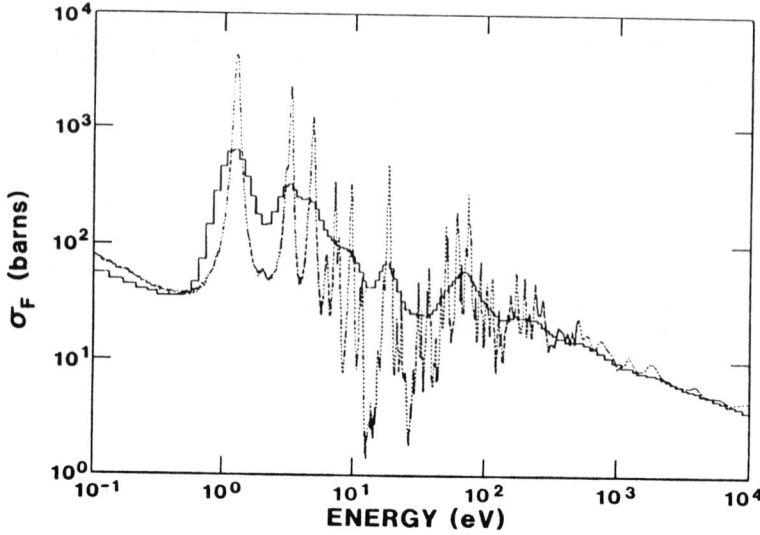

Fig. 3. Fission cross section of (^{247}Cm + n) as determined with the RINS, shown as the solid line, and in conventional geometry at LANSCE, shown as the dotted line, from 0.1 eV to 10 keV.

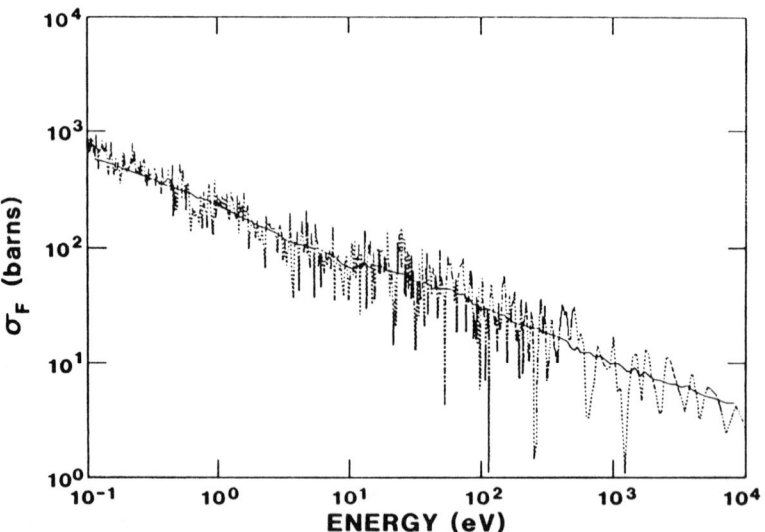

Fig. 4. Fission cross section of (^{254}Es + n) as determined with the RINS, shown as the solid line, and in conventional geometry at LANSCE, shown as the dotted line, from 0.1 eV to 10 keV. The RINS results have been scaled by 0.89, in accordance with a more recent determination of sample size as discussed in the text and in Table I.

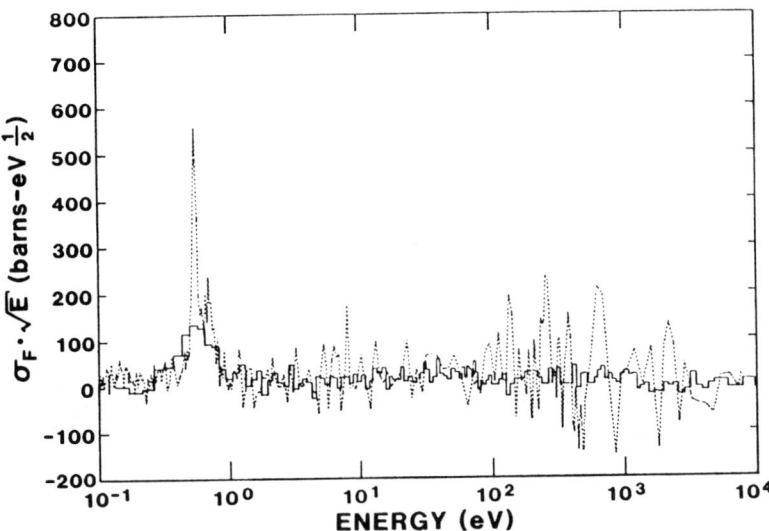

Fig. 5. Fission cross section of (^{250}Cf + n) as determined with the RINS, shown as the solid line, and in conventional geometry at LANSCE, shown as the dotted line, from 0.1 eV to 10 keV. The appearance of the 0.53 eV resonance as a possible doublet in the LANSCE results is considered to be spurious.

Table III Comparison of resonance parameters for ^{250}Cf and ^{247}Cm, as determined in the two experiments.

E_o	A_f (RINS) (b-eV)	A_f (LANSCE) (b-eV)
0.53 (^{250}Cf)	71 ± 9	80 ± 5
1.246	427 ± 34	636
2.97 + 3.18	444 ± 70	579
9.56 + 9.92	141 ± 70	182
18.24 multiplet	332 ± 46	463

spectrometer at WNR driven by 800 MeV proton pulses from the PSR. For each pulse, we expect $2 \cdot 10^4$ times as many neutrons in the SDS as are provided by RINS. How many fissionable isotopes are there whose average properties are of interest but which can be produced only in very limited (i.e., ng) quantities ($2 \cdot 10^{12}$ atoms)? There are perhaps 30 transactinium isotopes that have not been measured, which could be given consideration. Let us choose one of these, ^{235m}U, as an illustrative example.

There have been three recent measurements of the thermal and subthermal fission cross sections of 26 min ^{235m}U.[10-12] This 1/2+ isomer is produced in the alpha decay of ^{239}Pu. In the measurements of d'Eer et al.,[12] samples of $8 \cdot 10^9$

atoms were prepared by collecting recoil atoms from ^{239}Pu decay in a ^{235m}U generator, a 0.5 m D Al sphere coated on the inside with PuF_4. Mostovoi and Ustroyev[10] prepared somewhat larger samples ($2 \cdot 10^{10}$ atoms) in the same way--their cylindrical generator was larger. In the measurement by Talbert et al.,[11] the samples were still larger ($2 \cdot 10^{11}$ atoms) and were prepared by consecutive radiochemical separation of Pu and U.

All three measurements gave roughly the same result: the thermal (2200 m/sec) cross section of ^{235}U and ^{235m}U are about equal. However, this does not answer the question of primary interest: how do the reaction rates compare for fast neutrons? This question can be answered by determining average resonance parameters for ^{235m}U. A sample of $2 \cdot 10^{11}$ atoms is large enough to permit a measurement on a lead SDS at WNR/PSR: if the resonance fission cross section is comparable to that of ^{235}U, we should expect to observe 0.12 ^{235m}U counts/pulse with a fresh sample or 3700 counts between 1 eV and 80 keV for each sample fielded, if the SDS can be driven at 20 Hz. Here the estimate is made by assuming that event addresses are stored with a real-time clock address included and that data are taken for up to ten half lives; this permits one to use the sample until it is completely decayed, weighting each event as being due to ^{235}U or ^{235m}U fission by Bayes' equation.

This experiment can be considered as an effective lower limit on sample size and half-life for the SDS at WNR/PSR. While 3700 events are not an overwhelmingly large statistical sample for the determination of average properties, we feel that it would be adequate, noting that d'Eer at al[12] analyzed 1402 events in their measurement with thermal neutrons.

REFERENCES

1. R. E. Slovacek, D. S. Cramer, E. B. Bean, J. R. Valentine, R. W. Hockenbury, and R. C. Block, Nucl. Sci. Eng. **62**, 455 (1977).
2. A. A. Bergmann, A. I. Isakov, I. D. Murin, F. L. Shapiro, I. V. Shtranikh, and M. V. Kazarnovsky, Proc. 1st Int. Conf. Peaceful Uses At. Energy **4**, 135 (1955).
3. R. C. Block, R. E. Slovacek, Y. Nakagome, and R. W. Hoff, in "50 Years with Nuclear Fission," p. 354 (1989).
4. D. R. Harris, F. Rodriguez-Vera, N. Abdurahman, Y.-D. Lee, R. E. Slovacek, and R. C. Block, Trans. Am. Nuc. Soc. **61**, 73 (1990).
5. R. C. Block, R. W. Hockenbury, R. E. Slovacek, E. B. Bean, and D. S. Cramer, Phys. Lett. **31**, 247 (1973).
6. H. T. Maguire, Jr., C. R. S. Stopa, R. C. Block, D. R. Harris, R. E. Slovacek, J. W. T. Dabbs, R. S. Dougan, R. W. Hoff, and R. W. Lougheed, Nuc. Sci. Eng. **89**, 293 (1985).
7. J. W. T. Dabbs, N. W. Hill, C. E. Bemis, and S. Raman, "Nuclear Cross Sections and Technology," NBS Spec. Pub. 425, I, 81 (1975).
8. Y. Danon, R. E. Slovacek, R. C. Block, R. W. Lougheed, R. W. Hoff, and M. S. Moore, Trans. Am. Nuc. Soc. **61**, 401 (1990).
9. M. S. Moore and G. A. Keyworth, Phys. Rev. **C3**, 1656 (1971).
10. V. I. Mostovoi and G. I. Ustroyev, At. En. **5**, 241 (1984).
11. W. Talbert, J. Starner, S. Balestrini, M. Attrep, D. Efurd, and F. Roensch, Phys. Rev. **C36**, 1896 (1987).
12. A. D'Eer, C. Wagemans, M. Neve de Mevergnies, F. Gonnenwein, P. Geltenbort, M. S. Moore, and J. Pauwels, Phys. Rev. **C38**, 1270 (1988).

USE OF N-CAPTURE γ-RAYS FOR STUDIES IN CONDENSED MATTER PHYSICS

Raymond Moreh

Nuclear Research Center - Negev, Beer-Sheva 84105, Israel, and
Ben-Gurion University of the Negev, Beer-Sheva 84120, Israel

ABSTRACT

The use of n-capture gamma rays for studies of some topics in condensed matter physics is reviewed. These include the study of Debye temperatures of some metallic elements and molecular gases. Properties of 2-dimensional systems such as adsorbed gases on graphite and also of graphite intercalated compounds are discussed in detail. In particular, the zero-point energies of the vibrational and librational potentials of the molecules with respect to the adsorbing surfaces was determined.

INTRODUCTION

Photon beams obtained from the (n,γ) reaction are being used for studies in condensed matter physics. In such studies, the nuclear resonance photon scattering (NRPS) method is employed in which one of the γ-lines of the beam overlaps by chance a nuclear level in any isotope of the scatterer. In this technique, one monitors the Doppler width, $\Delta_r = E_r(2kT_r/M_rc^2)^{1/2}$, of a nuclear level of energy E_r in the scatterer of mass M_r, where T_r is the effective temperature of the scatterer[1-6] and plays a central role in the present technique. T_r is also of importance in scattering of epithermal neutrons from spallation neutron sources for studying high temperature superconductors[7]. The value of T_r enters into the expression of the Doppler width and is influenced by the effect of the binding potentials of the atomic environment through T_r. The binding potentials usually modify the instantaneous velocity of the scattering atom which in turn influence Δ_r. Small variations in Δ_r can be precisely monitored using nuclear resonance scattering as illustrated in Fig. 1. It depicts the Doppler-broadened shapes of a nuclear level and a γ-line of the (n,γ) reaction; the two lines have a large separation distance and overlap through their tails.

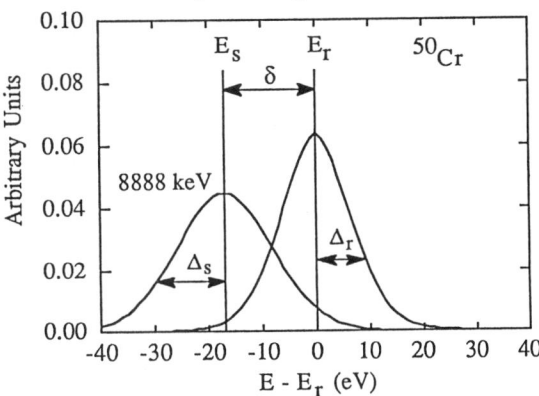

Fig. 1. Doppler broadened shapes of a line from the Fe(n,γ) reaction at E_s= 8888 keV (represented by a Gaussian) and a level E_r in ^{50}Cr (represented by a ψ-function)[8], separated by an energy δ = 17 eV. The cross section σ_s is proportional to the ooverlap integral.

© 1991 American Institute of Physics

This situation ensures that the scattering cross section σ_s becomes sensitive to small changes in the Doppler broadening of both line shapes as explained in detail in Ref. 1. Thus σ_s is strongly dependent on the kinetic energy component of the scatterer (including that of its zero-point motion). At low temperatures, the zero-point contribution constitutes in most cases a substantial part of the effective temperature. In anisotropic systems, this effect could be dependent on direction. This may be illustrated by noting that the scattered intensity of 6324 keV photons from ^{15}N in the form of Na^{15}NO$_3$ single crystal increases by 43% when the orientation of the nitrate planes is changed from a perpendicular geometry (relative to the photon beam) to a parallel geometry[2].

THE PHOTON BEAM AND EXPERIMENTAL METHOD

The photon beam is generated from the (n,γ) reaction on discs of Fe or Cr placed along a tangential beam tube and near the reactor core. Typical intensities obtained per γ-line ranged between 10^4 and 10^6 photons/cm^2.s on the target position with energies in the range 6 to 9 MeV and Doppler widths of around 10 eV. The scattered radiation was detected using either 12.5 cm x 12.5 cm NaI detectors or Ge(Li) or pure Ge detectors. In most cases the scattered radiation was measured as a function of T between 300 K and 10 K.

Pure gas samples were inserted inside a spherical stainless steel cell 2 mm thick and 20 mm internal diameter designed to sustain pressures of around 4000 psi. We thus obtained a large amount of the scatterer (~ 500 mg) in a small region of space over which the photon beam intensity is homogeneous.

The adsorption sample consisted of a thin-walled, 4 cm x 4 cm stainless steel cylinder containing 47 gm Grafoil (a commercial form of graphite with large specific adsorbing area) in the form of 82 parallel sheets. Gas coverage ranging between submonolayer and 1.5 monolayers were used. The coverage was determined by measuring the vapor pressure at 77.3 K. The nitrogen-15 amounts in the

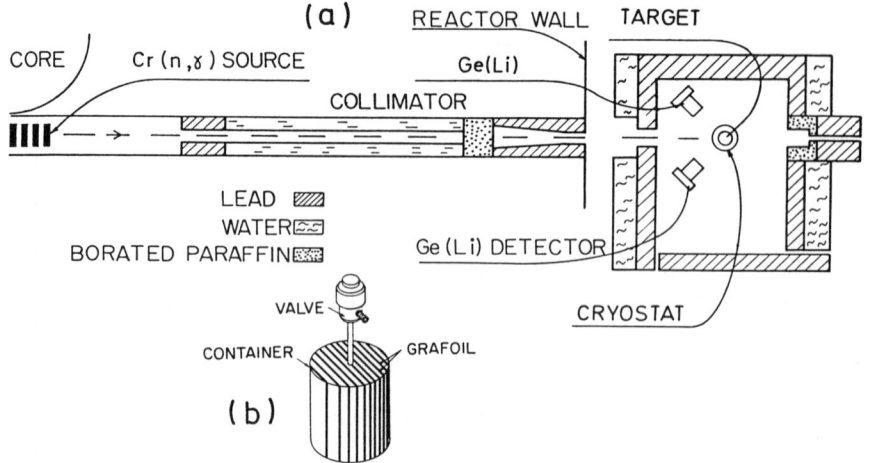

Fig. 2. Schematic diagram of: (a) experimental system. (b) Adsorption target showing Grafoil sheets inside a stainless steel cell.

TABLE I. Resonance scatterers which can be used for solid state applications. Parantheses indicate uncertainties.

Source	Target	Energy (keV)	Γ_o (eV)	Γ (eV)	δ (eV)
Cr	^{15}N	6324[a]	3.0 ± 0.2	3.0 ± 0.2	29.5 ± 2.0
V	^{48}Ti	6600[b]	.64 ± .10	.86 ± .20	15.3 ± .9
Fe	^{50}Cr	8888[c]	.68 ± .18	.75 ± .20	17.0 ± .5
Co	^{55}Mn	7491[d]	.08 ± .04	.45 ± .25	17.0 ± 1.0
Cr	^{56}Fe	8512[e]	(.03 ± .01)	(.09 ± .03)	(23 ± 3)
Fe	^{62}Ni	7646[f]	.31 ± .03	.48 ± .05	14.0 ± 1.0
Cr	^{68}Zn	7362[g]	1.6 ± .2	1.9 ± .1	26.0 ± 1.0

[a]) Ref. 1. [b]) Ref. 4. [c]) Ref. 6. [d]) Ref. 9.
[e]) Ref. 8. [f]) Ref. 3. [g]) Ref. 5.

samples were around 200 mg. The target was placed in a cryostat which could be rotated from a position where the beam and the grafoil planes were parallel, to a perpendicular geometry. The cross section ratios (σ_a/σ_c) with the photon beam parallel (σ_a) and perpendicular (σ_c) to the grafoil planes were measured against T.

RESONANCE SCATTERERS

Table I lists the parameters of some strong accidental resonance scatterers which are good candidates for the present studies. Not any resonance scatterer is suitable for solid state work. Its mass should be small and the value of δ should be large. The following conditions should be fulfilled: (1) $\delta > \Delta_s$, Δ_r. (2) $\Delta_r > \Gamma$, and (3) The scattering signal should be strong to ensure good statistics.

SCATTERING CROSS SECTIONS

Theoretically, the scattering cross section is given by the overlap integral of a gaussian F(E) and a ψ-function. The result

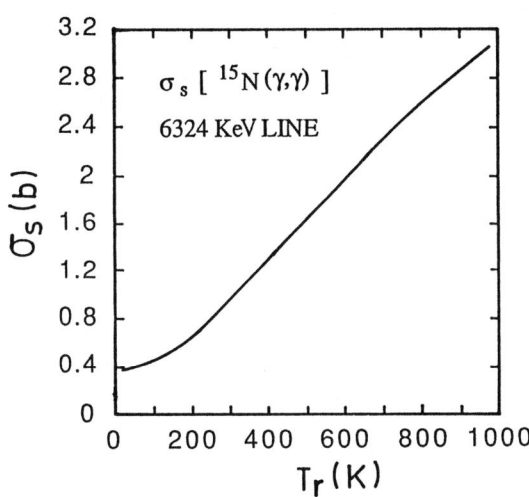

Fig. 3. Calculated scattering cross section from the 6324 keV level in ^{15}N versus T_r obtained at a γ source temperature 240 K, with the following parameters of the resonance process: $\Gamma = \Gamma_o = 2.9$ eV; $\Delta_s = 9.2$ eV; $\delta = 29.5$ eV.

may be expressed by another ψ-function, as follows[10-12]:

$$\sigma_s = C \psi(x_o, t_o) \quad (1)$$

where
$$\psi = (\pi t)^{1/2} \int_0^\infty \frac{e^{-(x_o-z)^2/4t_o}}{1+z^2} dz \quad (2)$$

with $x_o = 2|E_r - E_s|/\Gamma = 2\delta/\Gamma$, $t_o = (\Delta_s^2 + \Delta_r^2)/\Gamma^2$, and C is a constant depending on the parameters of the resonance level. Thus $\sigma_s = f(T_r)$ which means that σ_s depends on T_r through Δ_r. From the above and using Table I, the value of T_r can be determined if σ_s is measured. Fig. 3 shows a plot of the calculated values of σ versus T for the 6324-keV level in ^{15}N.

DEBYE TEMPERATURES - METALS

Using the measured data of σ_s versus T, in the 300 K to 10 K range, the Debye temperatures Θ for metallic targets, were deduced. This was done using the following procedure: (i) Calculate the scattered intensity versus T between 10 K and 300 K. This requires a knowledge of the precise geometry of the beam-target-detector system and is carried out for several assumed values of the Debye temperature. (ii) The T_r values corresponding to the assumed Θ are calculated using the Lamb formula[10,11]:

$$T_r/T = 3 \left(\frac{T}{\Theta}\right)^3 \int_0^{\Theta/T} x^3 \left(\frac{1}{e^x - 1} + \frac{1}{2}\right) dx \quad (3)$$

where $x = h\nu/kT$. (iii) By comparing the resulting curve with the measured data, the value of Θ is deduced by a best fit procedure. This method of deducing Θ was explained in detail in Ref. 3. Table II lists the results together with the corrected values for natural elements which were deduced by assuming a $M^{-1/2}$ correction to allow for the different atomic masses. The present measurement of Θ is sensitive to all the vibrational modes of the lattice and weights more strongly the high-frequency phonon modes; it differs from the x-ray measurements which are more sensitive to the low-frequency modes. Table II compares the present values of Θ with those deduced from the phonon spectrum $g(\nu)$, obtained from inelastic neutron diffraction[13]. To do this, we employ the expression:

$$T_r = \frac{1}{k} \int_0^\infty h\nu \, g(\nu) \left(\frac{1}{\exp(h\nu/kT) - 1} + \frac{1}{2}\right) d\nu \quad (4)$$

where $g(\nu)$ is normalized so that $\int_0^\infty g(\nu) d\nu = 1$.

Table II shows that the values of Θ obtained by the present technique are higher by about 10% than those deduced from the phonon spectra. This is probably due to some vibrational modes of the lattice which are monitored by the NRPS technique but were missed by inelastic n-diffraction.

DEBYE TEMPERATURES - MOLECULAR GASES

Here, we consider only molecular gases containing nitrogen (such as N_2, NO, N_2O and NH_3). This is because ^{15}N is the only

TABLE II. Metallic Debye temperatures for some isotopes determined in the present work together with those obtained by integrating over the phonon spectra as in eq. (7). The values of Θ corrected for the composition of the natural elements are also listed.

Isotope	NRPS Θ(K) Isotope	NRPS Θ(K) Element	PHONONS[a] Θ(K) Element
^{48}Ti	420 ± 20[b]	420 ± 20	360
^{50}Cr	520 ± 20[c]	510 ± 20	464
^{62}Ni	420 ± 15[d]	431 ± 15	384
^{68}Zn	235 ± 10[e]	240 ± 10	230
^{208}Pb	104 ± 6[d]	104 ± 6	93

[a]) Ref. 10. [b]) Ref. 4. [c]) Ref. 8. [d]) Ref. 3. [e]) Ref. 5.

known resonance scatterer of light mass (see Table I). We first calculate T_r of ^{15}N for each of the above gases and consider as an example the case of N_2 assumed to be in a gaseous form. Here, T_r should reflect the actual **kinetic** energy of the two atoms of the molecule which together has six kinetic degrees of freedom: three translational (3kT/2), two rotational (kT), and one vibrational with frequency ν, thus[1]:

$$6kT_r/2 = 3kT/2 + kT + 0.5(\frac{h\nu}{e^{h\nu/kT}-1} + \frac{h\nu}{2}) \quad (5)$$

As a more involved example we discuss the case of the amonia molecule[11] assumed to be in a gaseous form. The effective temperature of the ^{15}N-atom in ^{15}NH$_3$ is given by[14]:

$$3kT_r/2 = S_t(3kT/2) + S_r(kT) + 0.5 \sum_1^6 S_j k\alpha_j \quad (6)$$

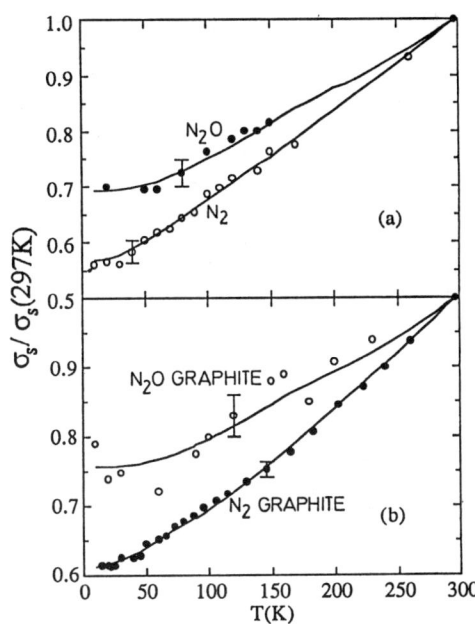

Fig. 4. (a) Scattered intensity ratios from pure gases. The solid lines were calculated using the Einstein temperatures of Table III. (b) Same as in (a) but for gases adsorbed on non-oriented graphite.

TABLE III. Measured Einstein and Debye temperatures of some molecular solids. The Einstein temperature of the molecule-graphite interaction is listed. Uncertainties are ± 15%.

Target	Present work			Literature
	Molec-Graph (K)	Einstein (K)	Debye (K)	Debye (K)
$^{15}N_2$	235[a]	58[a]	92	81[e]
^{15}NO	448[b]	147[b]		
$^{15}N^{14}NO$	432[c]	173[c]	215	
$^{15}NH_3$			241 ± 8[d]	235[d]

[a]) Ref. 16.   [b]) Ref. 17.   [c]) Ref. 18.   [d]) Ref. 14.   [e]) Ref. 19.

S_t, S_r, and S_j (j = 1..6) are the energy fractions shared by the N-atom in the translational, rotational, and the six internal modes of vibration (of frequencies v_j) respectively. S_t = 0.8323, is equal to the ratio of the of the atomic masses of the N-atom to that of amonia. Similarly, S_r = 0.036 is obtained by considering the moments of inertia around the three principal axes of the molecule where the precise geometry, and the normal frequencies v_j were taken from the literature[15]. S was calculated using method of infra-red spectroscopy[14]; α_j (j = 1,..,6) are defined as:

$$\alpha_j = (h v_j/k)([\exp(h v_j/kT) - 1]^{-1} + 1/2) \qquad (7)$$

Assuming a Debye model to describe the vibrations and rotations of the entire amonia molecule in the solid phase, the expression of T_r of the N-atom in amonia may be written as:

$$T_r = S \, 3\frac{T^4}{\Theta^3} \int_0^{\Theta/T} x^3 \left(\frac{1}{e^x - 1} + \frac{1}{2} \right) dx + (\Sigma_1^6 S_j \alpha_j)/3 \qquad (8)$$

S = S_t + 2S_r/3 = 0.8563, obtained from eq. (7). The last equation was used to calculate T_r versus T. Using this calculated curve and the measured data of σ_s versus T (Fig. 4), the value of Θ was determined as explained above and in Ref. 14. The vibrations and rotations of NH_3 in the solid phase may also be described using a single Einstein frequency f and a similar expression for T_r can be deduced. The same procedure was carried out for other gases. Fig. 4 displays some measured data and the fitted curves; Table III lists the Einstein and Debye temperatures derived in this manner. The Θ values for N_2 and NH_3 agree nicely with the literature[19,20].

MOLECULAR ADSORPTION ON GRAPHITE

The adsorption process is caused by Van der Waals forces acting between a surface such as graphite (G) and gas molecules. To study such a system we first define the effective temperatures of a N-atom of a gas such as nitrogen adsorbed on **fully oriented graphite**. To do so we use a simple model which describes the G-nitrogen system using 2 Einstein oscillators only: (i) A single frequency f_1 representing both the out-of-plane vibration and libration of the entire molecule with respect to the G-plane; (ii)

Fig. 5. Relative measured ratios σ_a/σ_c of scattered intensities vs T for gases adsorbed on Grafoil. The solid lines are drawn through the measured points to lead the eye.

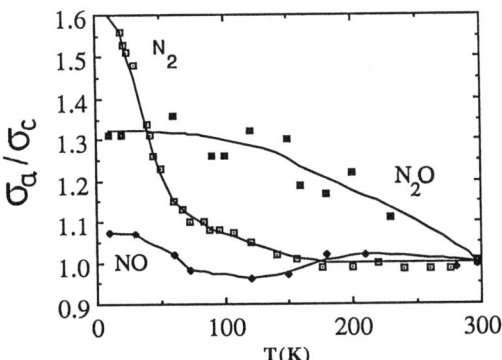

Another frequency f_2 representing both the in-plane libration and the two in-plane degenerate vibrations of the entire molecule. We further assume that f_1 is identical to the Einstein frequency of a pure nitrogen molecular solid (Table III). For an N_2 molecule inclined at an angle ϑ with respect to the G-plane, the effective temperature T_a of the N-atom parallel to the G-plane may be obtained by projecting the kinetic energy of the N-atom along the G-plane. T_c is defined in a similar manner, but perpendicular to the G-plane. The resulting values of T_a and T_c are given by[16]:

$$T_a = (a_0\cos^2\vartheta + a_1\sin^2\vartheta + 3a_2)/4 \qquad (9)$$
$$T_c = (a_0\sin^2\vartheta + a_1[1+\cos^2\vartheta])/2 \qquad (10)$$

where a_j ($j = 1,2$) are defined in a similar manner to that of α_j of eq. (8) with f_j replacing ν_j. Similar calcualtions of T_a and T_c were carried out for other gas-Grafoil systems[17,18].

From the above, one can calculate the T_r of the N-atom for the case of a nitrogen molecule adsorbed on <u>non-oriented</u> graphite:

$$T_g = (2T_a + T_c)/3 = (a_0 + 2a_1 + 3a_2)/6 \qquad (11)$$

In practice T_a and T_c were corrected to account for the actual structure of **Grafoil** (a commercial form of partially oriented graphite) known to contain a randomly oriented fraction ($f = 0.44$) of crystallite surfaces with a mosaic spread of half width at half maximum angle $\phi_0 = 15°$. The corrected values[15], were converted to cross sections to obtain the ratios σ_a/σ_c.

Experimentally, the ratios σ_a/σ_c were measured as a function of T for N_2, N_2O and NO adsorbed on Grafoil for 1 monolayer coverage; the results are displayed in Fig. 5. By comparing these data with the calculated values the Debye and Einstein temperatures of the gas-graphite system were determined (Table III) and also the tilt angles of the molecules relative to the G-surface (see below).

TESTING PHONON SPECTRA

It is important to point out that the above data can be used for testing the reported phonon spectra[19-22] of solid nitrogen and

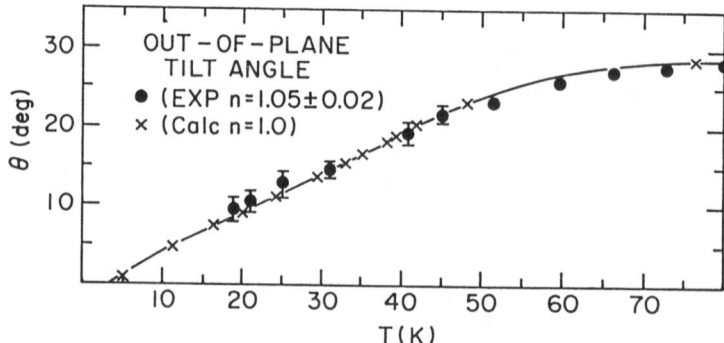

Fig. 6. Measured tilt angles of the $^{15}N_2$ molecules with respect to the G-planes for a coverage n = 1.05 ± 0.02 commensurate monolayers. The solid line is drawn through the calculated points (x) obtained using molecular-dynamic simulations[25-27] for n = 1.00.

amonia. This test may be done by comparing the values of θ deduced by the present method and that from the phonon spectra using eq. (4). The results are shown in Table III which reveals a very nice agreement. This test can also be carried out by comparing the value of T_r at T = 0 K. For nitrogen, we get: T_r = 29 K at T = 0 K, to be compared with T_r = 26 K and 30 K using the phonon spectra of ref. 18 and 19 respectively. This is a remarkable agreement. However, when a similar test was carried out for the case of the Grafoil-nitrogen system, a huge deviation was observed. From Fig. 5 we obtained: T_r = 117 K at T = 0 K. This value is a factor of 2.7 larger than, T_r = 43 K, deduced from the phonon spectra of ref. 19. This large deviation is far higher than the ∼ 10% deviation observed in the other systems studied hitherto (see Tables II and III). The only possible explanation which we could forsee is that in the N_2-Grafoil system there are other high energy vibrational modes which were missed by the n-studies and are monitored by the present technique. In a recent paper[23] which carried out a renewed study of this system using inelastic n-scattering, the acoustic-phonon branches were found to be a factor of 2 larger than the expected values. However, this large deviation at the low-frequency modes did not change appreciably the resulting value of T_r at T = 0 K and the large deviation factor of 2.7 remained unexplained.

TILT ANGLES

Fig. 6 show the results of the tilt angles versus T for the Grafoil-nitrogen system as deduced from the σ_a/σ_c ratios. The figure also shows the calculated values of the tilt angle as obtained using molecular-dynamics simulations[25-27] for a coverage of 1 monolayer. An excellent agreement is observed which illutstrates some of the novel uses of the NRPS technique. It should also be emphasized that the present method can determine the tilt angle even when the molecules are in the vapor phase. All other techniques require the system to be in a solid phase.

A consideration of Fig. 5 shows a remarkable similarity in the behaviour of the N_2-grafoil and the N_2O-grafoil systems: At low T, the molecules lie almost flat on the G-surface with the average tilt angle ϑ increasing slowly with T and becoming almost randomly tilted at T >> Θ. In fact, ϑ is almost the same for both gases if the temperature is measured in units of Θ of the adsorbed gas. The behaviour of the NO-grafoil system is entirely different because the NO molecules form dimers and at low T tilts at a large angle (ϑ ≈ 21°) relative to the G-surface[17]. This result is in contrast with that deduced by n-diffraction in which a study of the structure of the NO-grafoil system was made[27].

GRAPHITE INTERCALATED COMPOUNDS

Here we give an example in which the present technique is used for the determination of the structural properties of some graphite intercalated compounds (GIC): $C_{5n}(HNO_3)$ and the residue compounds $C_{7n}(HNO_3)$, (n = 1,2,...) both prepared from highly oriented pyrolitic graphite. These are highly anisotropic compounds[28] with the electric conductivity changing by a factor ~ 10^5 along and perpendicular to the graphite planes. The $C_{5n}(HNO_3)$ GICs are unstable upon exposure to air turning gradually (within a year or two) to residue compounds of the form $C_{7n}(HNO_3)$.

Our interest in these compounds was to determine the orientation of the NO_3 molecules relative to the graphite planes in two samples: a 'normal' stage-2 sample, $C_{10}(HNO_3)$, and a stage-3 residue sample having the formula $C_{19}(HNO_3)$. The orientation was determined by relying on the fact that σ_s is known to increase by 43% when the orientation of the nitrate planar molecule changes from a parallel to a perpendicular geometry relative to the photon beam[2]. Experimentally, we found that the measured ratios of scattered intensities parallel and perpendicular to the graphite planes were: $R = \sigma_a/\sigma_c = 0.86 \pm 0.02$ and $R = 1.36 \pm 0.03$ for the

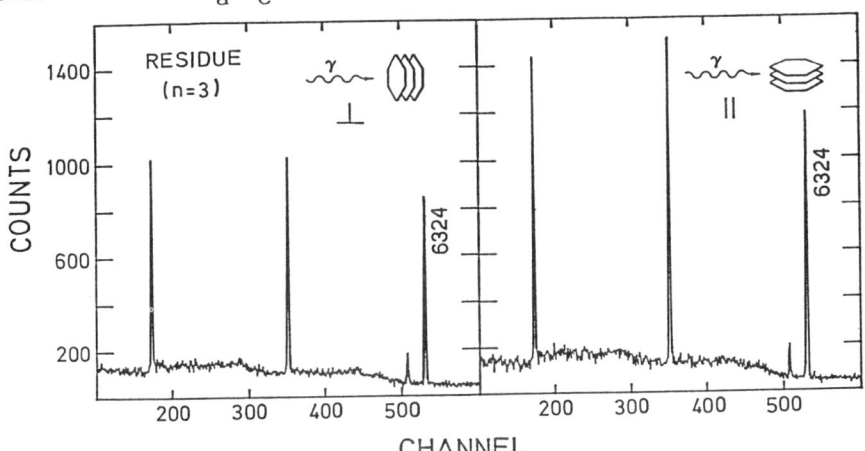

Fig. 7. Scattered spectra from the 'residue' sample $C_{19}(H^{15}NO_3)$ obtained with a 100 cc GeLi detector, at T = 297 K, with the photon beam parallel and perpendicular to the graphite planes. The intensity ratio is $\sigma_\perp/\sigma_\parallel = 1.36$.

normal and the residue samples respectively. Fig. 7 show the scattered spectra for the residue sample from which the tilt angles of the nitrate molecules with respect to the graphite planes were deduced: $\vartheta = 82 \pm 8°$, and $\vartheta = 13 \pm 5°$ respectively. This means that the nitrate molecules are nearly perpendicular to the graphite planes in the stage-2 compound and nearly parallel to the plane in the stable residue compound.

It is important to note that another GIC, namely C_8K was studied using a different version of the present technique which employs a bremsstrahlung beam from an electron linac[30]. With this technique, it was possible to study the directional binding properties and hence the Debye temperatures of the K-atoms in directions parallel and perpendicular to the graphite planes.

ACKNOWLEDGMENTS

We would like to thank Dr. O. Shahal and D. Levant for many helpful discussions. This work was supported by the US - Israel Binational Science Foundation (BSF), Jerusalem, Israel.

REFERENCES

1. R. Moreh, O. Shahal, V. Volterra, Nuc. Phys. A262, 221 (1976).
2. O. Shahal, R. Moreh, Phys. Rev. Lett. 40, 1714 (1978).
3. R. Moreh, O. Shahal, I. Jacob, Nuc. Phys. A228, 77 (1974).
4. I. Jacob, R. Moreh, O. Shahal, A. Wolf, Phys. Rev. B35, 8 (1987).
5. R. Moreh, O. Shahal, Nuc. Phys. A491, 45 (1989).
6. R. Moreh, S. Shlomo, A. Wolf, Phys. Rev. C2, 1144 (1970).
7. H. Rauh, N. Watanabe, Phys. Lett. 100A, 244 (1984); S. Ikeda, N. Watanabe, Phys. Lett. A121, 34 (1987).
8. D. Levant, R. Moreh, O. Shahal, (To be published).
9. J. Tenenbaum, R. Moreh, A. Nof, Nuc. Phys. A218, 95 (1974).
10. W.E. Lamb, Phys. Rev. 55, 190 (1939).
11. R. Moreh, Nuc. Inst. Meth. 166, 45 (1979).
12. B Arad, G. Ben-David, Rev. Mod. Phys. 45, 230 (1973).
13. W. Kress, in Landolt-Bornstein, New Series, Group III, 13B, 174 (1973).
14. R. Moreh, O Shahal, Phys. Rev. B42, 913 (1990).
15. G. Herzberg, Infrared and Raman Spectra (Van Nostrand Reinhold, N.Y., 1968).
16. R. Moreh, O. Shahal, Surf.Sci. 177, L963 (1986).
17. R. Moreh, O. Shahal, Mol. Phys. 65, 279 (1988).
18. R. Moreh, O. Shahal, Phys.Rev. B40, 1926 (1989).
19. T.A. Scott, Phys. Rep. 27C, 89 (1976).
20. P.S. Goyal, B.A. Dasannacharya, C.L. Thaper, P.K. Iyengar, Phys. Status. Solidi (b) 50, 701 (1972).
21. G. Cardini and S.F. O'Shea, Surf Sci. 154, 231 (1985).
22. J.K. Kjems and G.Dolling, Phys. Rev. B11, 1639 (1975).
23. F.Y. Hanson, V.L.P. Frank, H. Taub, L.W. Bruch, H.J. Lauter, J.R. Dennison, Phys. Rev. Lett. 64, 764 (1990).
24. J. Talbot, D.J. Tildesley and W.A. Steele, Far. Disc. Chem. Soc. 80, 1 (1985).
25. A.D. Migone et al., Phys. Rev. Lett. 51, 192 (1983).

26. K.D. Minor, M.H.W. Chan, and A.D. Migone, Phys. Rev. Lett. 51, 1465 (1983).
27. J. Suzanne et al., Phys. Rev. Lett. 41, 760 (1978).
28. R. Moreh, O. Shahal, Sol. St. Commun. 43, 529 (1982).
29. R. Moreh, O. Shahal, G. Kimmel, Phys. Rev. B33, 5717 (1986).
30. R. Moreh, W.C. Sellyey, D.C. Sutton, H. Zabel, Phys. Rev. B35, 821 (1987).

INTERNAL OXIDATION OF Sb AND In IN SILVER STUDIED WITH NUCLEAR REACTION ANALYSIS/CHANNELING COMBINED WITH HFI MEASUREMENTS

D.O. Boerma
Laboratorium voor Algemene Natuurkunde and Materials Science Centre, University of Groningen, Westersingel 34, 9718 CM Groningen, The Netherlands

ABSTRACT

The internal oxidation of dilute solid solutions of Sb and In in single crystals of silver was studied with nuclear reaction analysis, depth profiling and channeling using the $^{18}O(p,\alpha)^{15}N$ reaction. Combining the results obtained for the O concentration depth profiles and the O lattice sites with results of hyperfine interaction measurements, a detailed picture of the internal oxidation process was obtained. It was shown that isolated molecules of SbO_2 and InO_2 embedded in the Ag matrix are created, which diffuse as an entity at elevated temperatures to form an oxide layer at the surface or to form precipitates in the Ag matrix. At still higher temperatures these oxides decompose and the Sb or In redissolves in the silver. The information obtained for this model system may be important, among other things, to understand the formation of buried nitride layers in metals or silicide layers in silicon by high-energy implantation.

INTRODUCTION

Since more than 20 years nuclear methods are widely used for materials analysis. Especially Ion Beam Analysis (IBA)[1] is used in many laboratories in its various versions. Rutherford Backscattering (RBS), Nuclear Reaction Analysis (NRA) and Elastic Recoil Detection (ERD) are used to measure absolute concentrations of impurity atoms as a function of depth. With Particle Induced X-ray Emission (PIXE) trace element concentrations can be determined[2]. When the above methods are combined with channeling[1], impurity lattice sites can be determined in single crystals. In many works IBA has been successfully combined with other methods like electron spectroscopy, EXAFS, X-ray diffraction, TEM and also with Mössbauer Spectroscopy (MS) and Time-dependent Perturbed Angular Correlations (TDPAC).

In this contribution we chose to present results on the internal oxidation of 5-sp impurities in silver. Here IBA is combined with other nuclear technqieus, namely MS and TDPAC. Part of the results have been published elsewhere[3-5]. As will be shown, isolated Sb and In atoms embedded in an Ag matrix can react with oxygen diffusing interstitially into the matrix. In comparison with other noble metals the solubility of O in Ag is high. In the process, called internal oxidation, isolated oxide molecules are formed. In general the properties of isolated molecules are strongly influenced by the boundary conditions imposed by the host material. The binding enthalpy and the geometry can be different as compared to the same molecules in vacuum or in the bulk. It is clear that isolated molecules embedded in a host matrix can give rise to a new kind of chemistry. At elevated temperatures the molecules may precipitate to form small grains in the material or to form a layer at the surface of the material. The precipitates in the material have a profound influence on the mechanical properties since they interact with moving dislocations. The electric and thermal conductivity of high purity copper and silver can be improved by internal oxidation. Isolated transition metal impurities that act as effective electron scatterers are rendered harmless through precipitation in the form of oxides[6]. The formation of precipitates after internal oxidation of impurities in noble metals has been studied extensively with standard metallurgical techniques[7].

The theoretical understanding of internal oxidation and subsequent precipitation is still in its infancy since the necessary (cluster) calculations are complicated. A better understanding could contribute to other important areas such as the formation by high-energy implantation of buried nitride layers[8] in iron or silicide layers in silicon[9].

Many works have been published on internal oxidation studies with MS[10-12] or with TDPAC[13-15] alone, which will not be discussed here.

EXPERIMENTAL METHODS AND DATA ANALYSIS

Single crystals of pure Ag were implanted with doses of a few times 10^{15} atoms/cm^2 of ^{121}Sb at an energy of 110 keV. A small dose ($\simeq 10^{13}$ at./cm^2) of radioactive ^{119}Sb atoms was co-implanted with the same energy. After implantation the samples were annealed for 5 min at 823 K in H$_2$ gas to recover the lattice damage and to diffuse the Sb to a mean depth of $\simeq$ 150 nm. The maximum concentration is then only 0.5 at.%. In the same way samples implanted with ^{115}In and a small dose of radioactive ^{111}In were produced. To form the oxides the samples were heated for several minutes to 550 K in $\simeq$ 100 mbar ^{18}O$_2$.

Mössbauer or TDPAC spectra were taken after the various preparation steps. For MS the gamma rays from the decay of the 23.9 keV level (τ = 17.8 ns) in ^{119}Sn were used. This level is fed (100%) in the EC decay of ^{119}Sb. The resonant, recoilless absorption of the gamma rays in Ca^{119}SnO$_3$ was detected as a function of absorber velocity by counting the conversion electrons produced during the absorption[16]. The isomer shifts with respect to Ca^{119}SnO$_3$, determined from the peak positions in the Mössbauer spectra, are a measure for the electron density at the nucleus. The widths of the peaks are a measure for (the distribution of) the Electric Field Gradients (EFG) at the nucleus. For the TDPAC measurements[17] four NaI(Tl) detectors at relative angles $\theta_1 = 180°$ and $\theta_2 = 90°$ were used. Four coincidence spectra were recorded simultaneously of gamma rays of a cascade via the 245 keV level ($\tau = 85$ ns) of ^{111}Cd. The cascade is fed by EC decay of ^{111}In. The coincidence spectra were combined to form the ratio

$$R(t) = 2 \frac{I(\theta_1,t) - I(\theta_2,t)}{I(\theta_1,t) + 2I(\theta_2,t)}$$

where $I(\theta,t)$ is the coincidence count rate for a pair of detectors and t is the time after excitation of the 245 keV level. The (distribution of) EFG's at the site of the nucleus can be extracted from a Fourrier analysis of the time spectrum R(t).

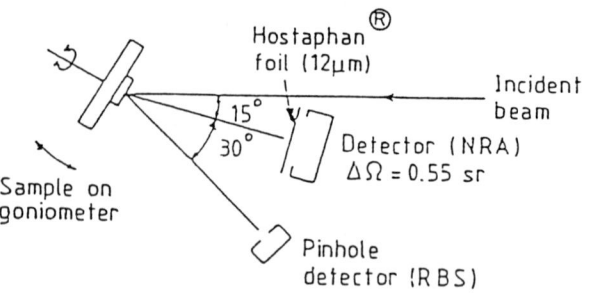

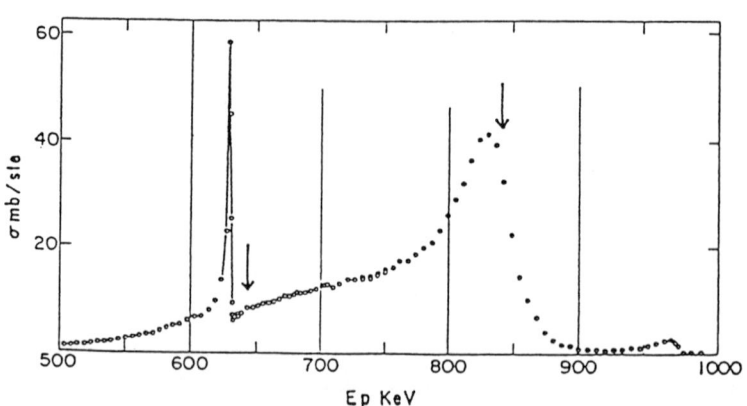

Fig. 1. *Schematic top-view of the set-up for NRA/depth profiling/channeling measurements. The cross section (ref. 18) of the $^{18}O(p,\alpha)^{15}N$ reaction is also given. The energies chosen for the present measurements are indicated by arrows.*

In both MS and TDPAC the spectra are determined by the electronic structure around the daughter nuclei of the probes. The time for electronic rearrangement in metals is very short compared to the life time of the MS state or the TDPAC state. The time needed for atomic rearrangement is long compared to the life time of these states. So the equilibrium charge distribution around the daughter nuclei is probed for atomic positions originally occupied by the probe atoms.

For the NRA measurements the reaction $^{18}O(p,a)^{15}N$ (Q = 3.970 MeV) was used. The cross section of this reaction as determined by Amsel and co-workers[18] is given in fig. 1, together with the schematic experimental set-up. For depth profiling the proton energy was set to 840 keV, the upper limit of a region between 820 keV and 840 keV where the energy dependence of the cross section is relatively weak. The a-particle peak has an energy of more than three times the maximum energy of the protons scattered from Ag. The spectra are taken using three surface barrier detectors with a total solid angle of 0.06 sr and with 12 keV resolution at backward angles. The depth distribution of ^{18}O is derived from the energy distribution of the a particles, using stopping power tables. With this set-up a measuring time of 12 h. is needed to achieve a sensitivity of 0.05 at% in the determination of the ^{18}O concentration profile. For the channeling experiments the beam energy was set to 640 keV. At this energy the scattered protons are stopped in a 12 mm thick Hostaphan foil placed in front of a large (W = 0.55 sr) silicon detector. The a particles penetrate the foil, so that only a particles are detected. Simultaneously the scattered protons are detected with a pin-hole silicon detector (see fig. 1). Angular scans through the three major crystallographic directions were taken by rotating the crystal mounted on a two-axes goniometer. At each selected angle the yield of protons in a window coinciding with a depth interval of 0-250 nm and of all a particles was measured. The angular scans for Ag and ^{18}O thus obtained were compared to computer simulations. In the simulations[19] the trajectories of = 4000 protons are calculated and the probaiblity for scattering or a nuclear reaction is calculated as a function of depth, taking into account the stopping, the thermal vibrations, and the scattering cross section as a function of depth. For ^{18}O the calculations were performed for many assumed lattice positions taking into account the depth profile. The measured scans were fitted with one or more calculated scans to determine the (mixture) of sites occupied by the ^{18}O atoms.

INTERNAL OXIDATION OF Sb IN Ag

In fig. 2 MNssbauer spectra are shown for a sample implanted with $4:10^{15}$ Sb/cm^2 after annealing and after various oxidation and annealing steps. The single line observed for the non-oxidized sample (fig. 2a) indicates a

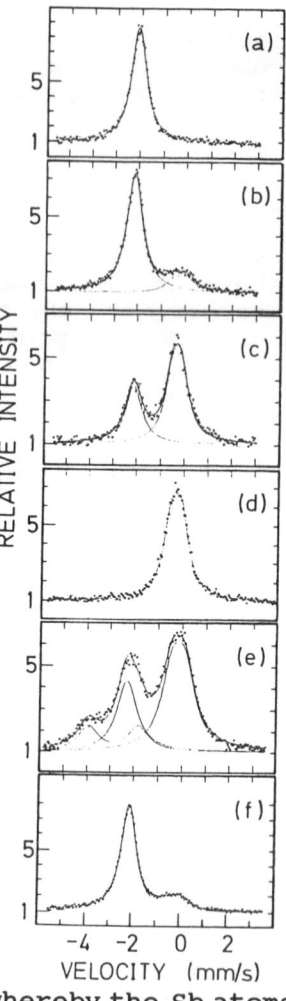

Fig. 2. ^{119}Sb Mössbauer spectra after various oxidation and anneal treatments:
(a) as implanted with $4 \cdot 10^{15}$ Sb/cm^2
(b) annealed for 10 min at 723 K in H_2. Some oxidation is visible.
(c) oxidized for 5 min at 550 K in 100 mbar $^{18}O_2$
(d) oxidized for 10 min at 550 K in 100 mbar $^{18}O_2$
(e) oxidized for 7 min at 550 K in 250 mbar $^{18}O_2$ (68% oxidized) and annealed for 30 min at 773 K
(f) as (e) and annealed for 30 min at 873 K.

perfect solid solution of Sb atoms in Ag, whereby the Sb atoms occupy substitutional sites. Some oxidation occurred in this case. For the partially oxidized samples (figs. 2b and 2c) a second line is observed with an isomer shift S = 0.32(3) mm/s, which is close to the value S = 0.06 mm/s observed for bulk SnO$_2$. For the fully oxidized sample the substitutional line has disappeared. From the relative intensities of the two lines the fraction of oxidized Sb atoms can be derived. The same observations have been made[20] for a sample implanted with only $4 \cdot 10^{13}$ Sb atoms. This indicates that isolated SbO$_n$ molecules are formed, i.e. no clustering occurred. In fig. 3 the concentration depth profiles for ^{18}O are shown for a fully and a partially oxidized sample. From the known implantation dose, the oxidized fraction, and the observed ^{18}O concentration, it can be concluded that SbO$_2$ molecules are formed exclusively in both cases, in agreement with the observation of a single Sn^{4+} line in the Mössbauer spectra. The ^{18}O depth profile for the fully oxidized sample is equal to the Sb depth profile calculated from the known diffusion coefficient. The shape of the depth profile for the partially implanted sample is remarkable. With a simple model due to Wagner[21], in which it is assumed that in-diffusing O atoms or molecules encountering Sb atoms within one interatomic distance are trapped, a sharp oxidation front is predicted. This means that all Sb atoms up to a certain depth must be oxidized, which is clearly not the case. An implication is that O is diffusing in the atomic state, as expected, and not in the molecular state. It also implies that SbO molecules are only marginally stable. The oxidation can be modelled[5] to take place in a two-step process in which SbO_2 is formed if O is trapped at an SbO molecule during its short life time. By solving the appropriate rate equations a value for the binding enthalpy for SbO of H = 0.6 eV can be found[5] with which the observed depth profile can be fitted as indicated in the figure.

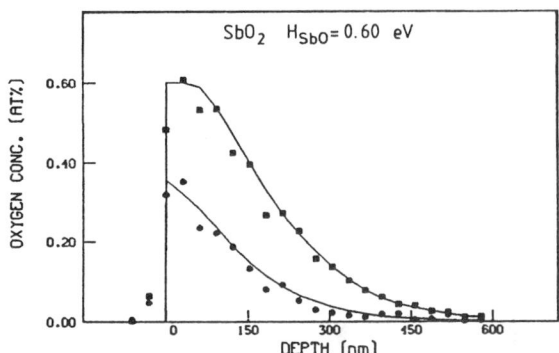

Fig. 3. Oxygen concentration depth profiles for a fully and partially (40%) oxidized sample. The solid lines are calculated results (see text).

The geometry of the SbO_2 molecules was studied by channeling measurements on a fully oxidized sample. The angular scans through the three major axes are shown in fig. 4, together with the computer simulations. It appears that some damage is present in the single crystal, as evidenced by the constant fraction needed to fit the Ag scans. The ^{18}O scans could be fitted with one single ^{18}O position located 0.3(1)Å (measured along a <111> direction) from the octahedral site. Assuming that the Sb atoms occupy substitutional sites after oxidation, a model for the geometry of the SbO_2 molecules as depicted in fig. 4 can be made. By choosing the shifts from the octahedral sites in specific <111> directions, the Sb-O bond length can be fitted to the expected value.

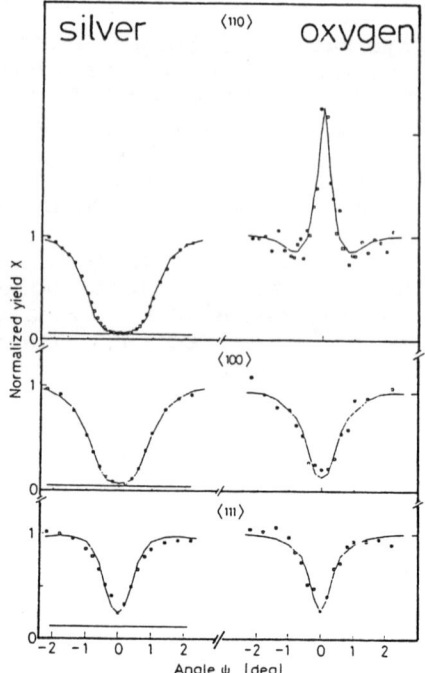

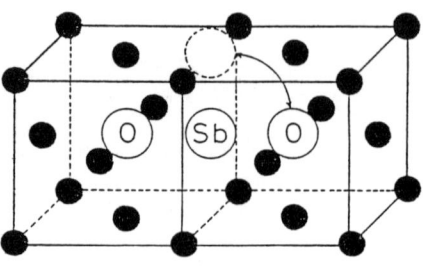

● Ag

Fig. 4. Angular scans through the major axes of a silver single crystal with isolated SbO_2 molecules. The lines represent fits with computer simulated data. In the lower part possible arrangements of the isolated SbO_2 molecules in silver are given, with the Sb atom assumed to occupy a substitutional site.

To study the thermal stability of the molecules, a sample with a 68% oxidized fraction was annealed at temperatures of 773 K and 873 K. The Mössbauer spectra observed after these annealings are given in fig. 3e and 3f. As indicated, the 4^+ component decreases and a doublet closely resembling the lines observed for bulk Sb_2O_3 appears at 773 K. The measured ^{18}O depth profile indicates that most oxygen is concentrated at the surface and that 40% of the oxygen is lost. A straightforward interpretation of these data is that during annealing at 773 K SbO_2 moleucles precipitate at the surface and form a thin Sb_2O_3 layer. Since such a large fraction of ^{18}O is retained, it is probable that the SbO_2 molecules diffuse as an entity. After annealing at 883 K all ^{18}O has disappeared as observed with NRA. The Sb has redissolved in the bulk as revealed by the observed substitutional fraction in the Mössbauer spectrum (fig. 3f).

INTERNAL OXIDATION OF In IN Ag

Similar experiments were performed to study the internal oxidation of In in Ag. The results are also similar, with a few exceptions. In fig. 5 TDPAC spectra are shown after the various treatments of the sample. In fig. 5a the spectrum for the non-oxidized case shows an unperturbed fraction, indicating a perfect substitutional solution of In in Ag. In spectra b and c the partial and full oxidation with ^{18}O is revealed by the growth of a component named 2a in the literature. The ^{18}O depth profiles measured for these cases are very similar to the profile shown in fig. 3, indicating that only InO_2 molecules are formed and that InO is unstable ($H_{SbO} \simeq 0.6$ eV). After annealing the sample at 750 K a new feature was observed. The TDPAC spectrum has a different shape. The stoichiometry of the new complex formed (called 2b) as measured with NRA is still InO_2. Channeling measurements were done to determine the geometry of the complexes 2a and 2b. The measurements and the fits with computer simulations are given in fig. 6. Again some damage is observed in the Ag lattice. The ^{18}O scans for complex 2a can be fitted fairly well assuming again one near-octahedral site for the ^{18}O atoms, shifted 0.5Å into a <111> direction away from the true octahedral site. The ^{18}O scans for the complex 2b are less pronounced. They can be fitted with a near substitutional fraction of 40%, displaced some 0.4 Å from the true substitutional site, and a "random" fraction of 60%. Random means that the ^{18}O fraction occupies many different sites, giving rise to a constant contribution to the angular scan. The results can be interpreted in a model as depicted in fig. 6. It is clear that in the formation of the new complex during annealing at 750 K vacancies must be trapped to allow the ^{18}O atoms to occupy near-substitutional sites. The TDPAC spectrum observed after annealing at 873 K is characteristic for bulk In_2O_3 (fig. 5e). This time the ^{18}O depth profile does

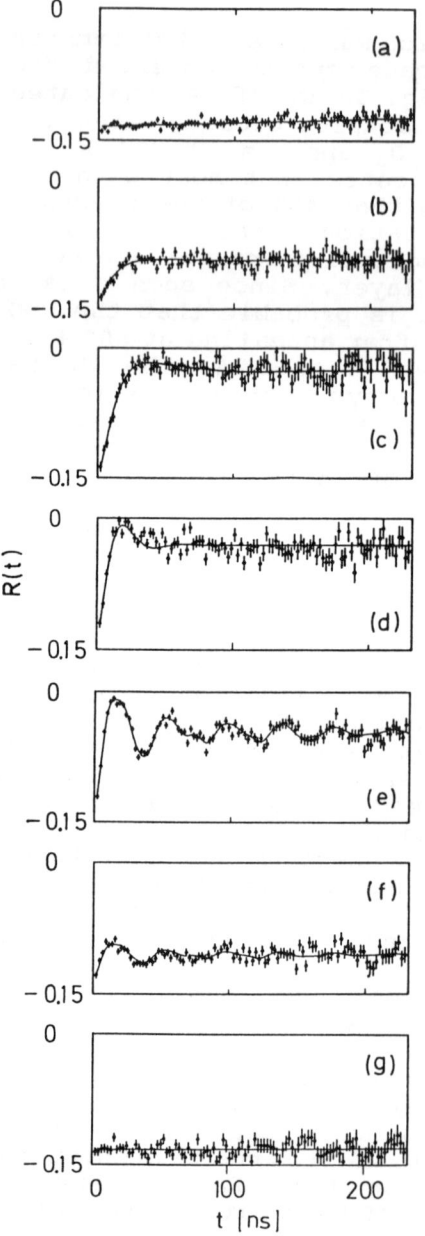

Fig. 5. ^{111}In TDPAC spectra after various oxidation and anneal treatments
(a) impl. with $4 \cdot 10^{15}$ In/cm^2 + annealed, 10 min, 723 K in H_2
(b) oxidized for 1 min at 550 K in 50 mbar $^{18}O_2$
(c) oxidized for 10 min at 550 K in 300 mbar $^{18}O_2$
(d) as (c) and annealed for 10 min at 750 K in 400 mbar $^{18}O_2$
(e) as (d) and annealed for 10 min at 873 K
(f) as (e) and annealed for 10 min at 1023 K
(g) as (f) and annealed for 10 min at 1043 K.

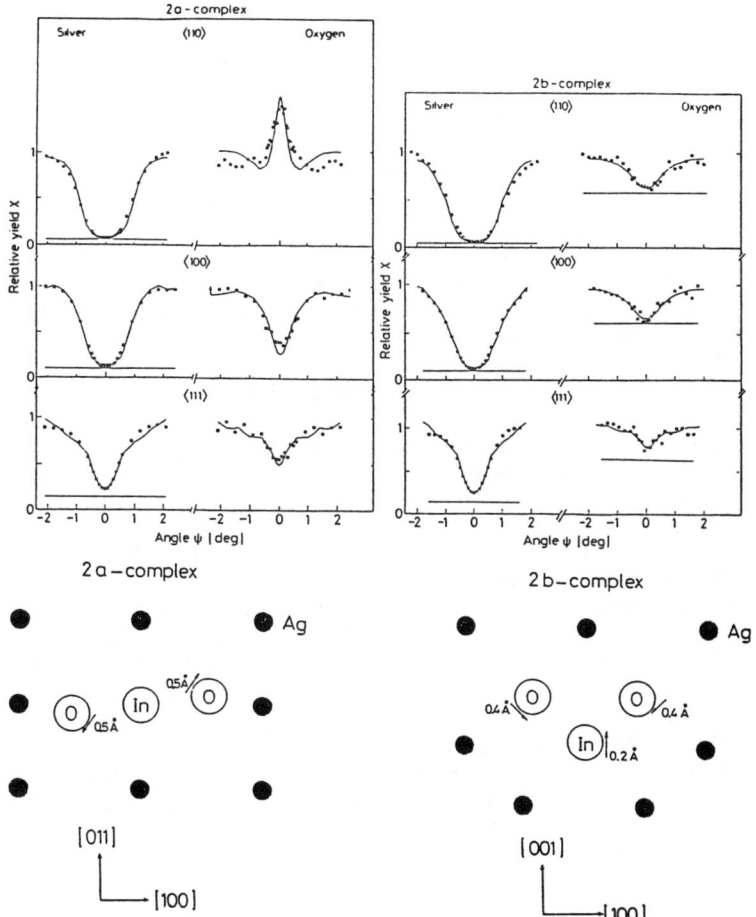

Fig. 6. Angular scans through the major axes of a silver single crystal with InO_2 molecules in complex 2a (left) and complex 2b (right). The lines represent fits with computer simulated data. In the lower part possible arrangements of the 2a and 2b complexes are depicted.

not change its shape, indicating that small precipitates of In_2O_3 are formed in the bulk. During annealing at 1043 K these clusters break up. The ^{18}O is released and the In dissolves again in the Ag matrix (figs. 5f and 5g).

CONCLUSIONS

A very detailed picture of the internal oxidation process for Sb and In in silver could be deduced by combining the MS and TDPAC results with results obtained with NRA concentration depth profiling and channeling of ^{18}O. Only a few details in this phenomenological picture remain unclear. These points are discussed elsewhere[3].

It is clearly illustrated that molecules in metals are very different from molecules in vacuum or in the bulk. For example SbO and InO molecules found to be unstable in Ag have binding enthalpies around 2 eV in vacuum. Bulk SbO_2 is not a stable compound, whereas SbO_2 molecules were found to be stable in Ag.

This work was performed as part of the research program of The "Stichting voor Fundamenteel Onderzoek der Materie" (FOM) with financial support from the "Nederlandse Organisatie voor Wetenschappelijk Onderzoek" (NWO).

REFERENCES

1. See for a recent review on IBA:
D.O. Boerma, Nucl. Instr. and Meth. B50, 77 (1990).
2. D.O. Boerma, E.P. Smit and N. Roosnek, Nucl. Instr. and Meth. B36, 60 (1989).
3. W. Segeth, D.O. Boerma, L. Niesen and P.J.M. Smulders, Phys. Rev. B39, 10725 (1989).
4. W. Segeth, D.O. Boerma, L. Niesen, J.R. Heringa and A. van Veen, Z. Phys. B - Condensed Matter 73, 43 (1988).
5. G.L. Zhang, H. Andreasen, D.O. Boerma, L. Niesen and W. Segeth, Phys. Rev. B39, 1068 (1989).
6. A.C. Ehrlich, J. Mat. Sci. 9, 1064 (1974).
7. R.A. Rapp, Corrosion 21, 382 (1965).
8. A.M. Vredenberg, F.Z. Cui, F.W. Saris, N.M. van der Bers and P.F. Colijn, Mat. Sci. and Eng. A115, 297 (1989).
9. A.E. White, K.T. Short, R.C. Dynes, R. Hull and J.M. Vandenberg, Nucl. Instr. and Meth. B39, 253 (1989)
10. H.H. Podgurski and G.P. Huffmann, Nature Phys. Sc. 237, 77 (1972).
11. G.P. Huffmann and H.H. Podgurski, Acta Metall. 21, 449 (1973).
12. F.H. Sanchez, R.C. Mercader, A.F. Pasquevich, A.G. Biblioni and A. Lopez-Garcia, Hyp. Int. 20, 295 (1984).
13. J. Desimoni, A.G. Biblioni, L. Mendoza-Zelis, A.F. Pasquevich, F.H. Sanchez and A. Lopez-Garcia, Phys. Rev. B28, 5739 (1983).
14. W. Bolse, P. Wodniecki, H. Schröder, M. Uhrmacher and K.P. Lieb, Nucl. Instr.and Meth. B2, 706 (1984).
15. D. Wegner, M. Uhrmacher and K.P. Lieb, Z. Phys. B68, 461 (1987).
16. G. Weyer, Nucl. Instr. and Meth. 186, 201 (1981).
17. F. Pleiter and C. Hohenemser, Phys. Rev. B25, 106 (1982).
18. G. Amsel and D. Samuel, Anal. Chem. 39, 1689 (1967).
19. P.J.M. Smulders and D.O. Boerma, Nucl. Instr. and Meth. B29, 471 (1987).
20. L. Niesen, Hyp. Int. 35, 587 (1987).
21. C. Wagner, Z. Electrochem. 63, 772 (1959).

THIN FILM ANALYSIS BY MEANS OF RESONANT PROTON CAPTURE REACTIONS

K.P. Lieb, W. Bolse, T. Corts, T. Kacsich,
A. Kehrel and M. Uhrmacher,
II. Physikalisches Institut, Universität Göttingen,
D-3400 Göttingen, Fed. Rep. Germany

ABSTRACT

Resonant capture reactions are a versatile tool to measure concentration profiles of light elements in near-surface layers. They allow us to monitor, with nanometer depth resolution, atomic transport processes caused by thermal diffusion, ion irradiation, nucleation and chemical forces. We have investigated properties of ion-implanted super saturated Na solutions in metals by using the 309 keV ^{23}Na(p,γ) resonance. Results on Na-nucleation, outdiffusion and hydrogen decoration are presented; the H-profiles were measured with the resonant ^{1}H(^{15}N,$\alpha\gamma$) reaction. The second example refers to 30-300 nm magnetron sputtered TiN and Cr$_2$N hard coatings irradiated with 100-900 keV Ar, Kr and Xe ions. The resonant ^{15}N(p,$\alpha\gamma$) reaction served to monitor the composition of the coatings and ion-induced surface sputtering and interface mixing effects.

INTRODUCTION

Solid state reactions often involve atomic displacement processes which may be driven by thermochemical forces or initiated by ion-beam irradiations. The relaxation of the highly non-equilibrium state produced in the ion-induced collision cascades can lead to macroscopic modifications and, therefore, is of technological importance in the fabrication of coatings and layered structures. An understanding of the transport mechanisms within the films and through the interfaces relies, to large extent, on the results of high resolution measurements of the concentration profiles of the components.

Ion-beam assisted analyzing methods such as Rutherford backscattering and channeling spectroscopy (RBS), proton and heavy ion induced X-ray emission (PIXE, HIXE), ion sputtering are commonly used to depth profile heavy elements in subsurface layers. Activation analysis via neutron capture reactions is used for trace element detection. However, most of these methods do not apply for depth profiling the concentration of light elements in a high-Z matrix. Under these conditions, resonant nuclear reaction analysis (RNRA) offers a very useful and elegant alternative: proton induced reactions on Z≤20 isotopes are dominated by narrow resonances, well separated in energy, and thus provide precise depth markers.

For a number of resonances in proton induced reactions on ^{15}N, ^{19}F, ^{23}Na, ^{24,25,26}Mg and ^{27}Al below 530 keV, we have recently measured the resonance energies E_R and resonance widths Γ with high

© 1991 American Institute of Physics

precision, by means of the Göttingen ion implanter IONAS[1-3]. Table I summarizes the resonance parameters and the effective energy spread W of the proton beam which is made up by the (very small) high voltage ripple of IONAS and the Doppler broadening width, caused by the thermal vibrations of the capturing nuclei in the target[1].

Table I: Resonance energies (E_R), widths (Γ) and beam spreads (W) of some proton-induced capture reactions in light nuclei [1-5]

Reaction	E_R (keV)[a]	Γ (eV)[a]	W (eV)[b]	Ref.
$^{15}N(p,\alpha\gamma)$	429.57±0.09	124±17		2
$^{19}F(p,\alpha\gamma)$	223.99±0.07	985±20		1
	340.46±0.04	2340±40		
	483.91±0.10	903±30		
$^{23}Na(p,\gamma)$	250.90±0.20	<20 [4]	68±5	1
	308.75±0.06	<20 [4]	64±5	
	512.1 ±0.1	<46		
$^{24}Mg(p,\gamma)$	222.89±0.08	<32	52±5	1
$^{25}Mg(p,\gamma)$	316.16±0.11	<37	68±5	3
	389.24±0.11	< 4	86±6	
	434.85±0.12	<44	82±6	
	496.75±0.12	<51	95±7	
$^{26}Mg(p,\gamma)$	292.06±0.09	<37	69±5	1
	338.4 ±0.1	<40	79±5	
$^{27}Al(p,\gamma)$	222.82±0.10	<34	57±5	1
	293.08±0.08	$59(^{+12}_{-16})$	68±5	
	326.97±0.05	<38	73±5	
	405.44±0.10	<42	87±5	
	445.75±0.15	---	96±6	
	504.88±0.15	<71 [5]	119±7	
	506.9 ±0.2	< 1.5 [5]	115±7	

[a] E_R, Γ given in the lab system, [b] IONAS, at room temperature

SUPER SATURATED SODIUM SOLUTIONS IN METALS

Na atoms are oversized and therefore essentially insoluble in metals; however, Na can be introduced by ion implantation in rather large concentrations[6]. The storage and diffusivity of Na in (radiation damaged) metals is important for the fabrication of ^{22}Na sources in positron annihilation spectroscopy and thermonuclear reaction studies as well as for the design of Na-cooled breeder reactors. We have investigated the transport processes of implanted Na in Al, Cr, Fe, Mo, Ni, Ta and AISI 316 Ti austenitic steel[7-10].

The 10 μA 250 keV Na⁺ ions were implanted into the water-cooled, 10x10x2 mm³ polycristalline samples up to a fluence of 1.5 $10^{17}/cm^2$; this corresponds to a maximum Na concentration of about 8 at%. Each sample was vacuum annealed at $4*10^{-6}$ mbar for 30 min, removed within 30 s from the hot region of the oven (to freeze-in the actual Na profile) and then cooled down to room temperature in about 20 min. The Na concentration profiles were determined by scanning the energy of the 20-25 µA proton beam over the 309 keV $^{23}Na(p,\gamma)^{24}Mg$ resonance and measuring the yield of the decay radiation of the 11.69 MeV capture state in ^{24}Mg by means of a lead shielded NaI(Tl) detector, 16 cm in diameter and 14 cm long. Further details on sample preparation, annealing and depth profiling are given in Refs. 7,8.

Fig. 1 illustrates Na concentration profiles in Al[7] and Fe[9], while fig. 2 shows the Na fraction retained in the sample after a single 30 min annealing at temperature T_a. The Na concentration at depth x is proportional to the γ-count rate $N(E_P)$ and the stopping power (dE/dx) at the incident proton energy E_P (eventually modified by the implanted Na). The minimum detectable Na concentration lies around 0.1 at%. The depth scale itself is approximately given by $x=(E_P-E_R)(dE/dx)^{-1}$. The depth resolution $\Delta(x)$ depends on the resonance width Γ, the energy spread W and the energy straggling $\delta E(x)$ of the beam along its path in the sample: $\Delta(x) = \{\Gamma^2+W^2+\delta E(x)^2\}^{1/2}(dE/dx)^{-1}$. Beyond some 30 nm below the surface, the energy straggling $\delta E(x) \propto E_P^{1/2}$ is the dominant contribution. At 250 keV ion energy, the Na implantation profile in Al has a projected range of $R_P=423\pm5$ nm and a FWHM of 270±9 nm, and the RNRA depth resolution is $\Delta(R_P)\approx 77$ nm[7].

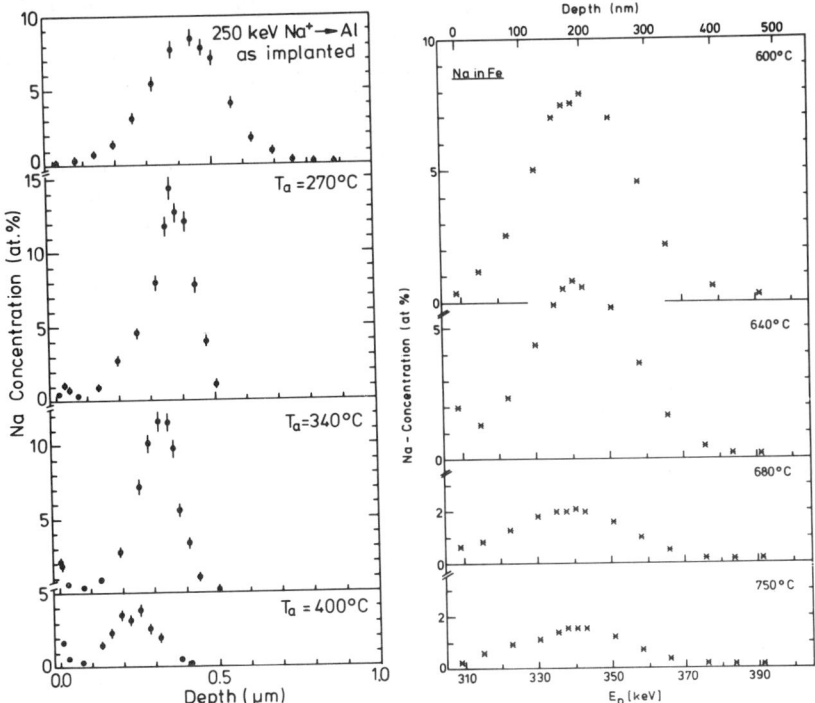

Fig. 1: Na-concentration profiles in Al and Fe as-implanted and after 30 min annealings at the temperature T_a [7,9].

In all matrices studied so far Na diffuses, within a rather narrow temperature range, to the surface from where it evaporates. As illustrated for the Na<u>Al</u> case, this diffusion process occurs in several steps: first Na moves into the region of <u>correlated radiation damage</u> where it forms <u>precipitates</u> (note the reduced FWHM of the 270 and 340°C profiles in fig. 1). These precipitates then dissolve at higher temperature, and the retained Na content f_r drops to nearly zero (fig. 2). It turns out that the essential parameters in the <u>outdiffusion</u> process are the melting temperature T_m and lattice structure of the matrix: If we plot the retained fraction f_r versus the reduced temperature T_a/T_m, we find consistently that half the Na content has left the matrix at $0.53(2) T_m$ in the bcc-type metals and at $0.70(2) T_m$ in the fcc-type matrices[11] (fig. 3). This two-step process is not unexpected as outdiffusion involves thermally activated vacancies and precipitation evidently involves radiation induced vacancies. The influence of radiation induced defects has been proven by pre-irradiating Na-doped Al samples with He⁺ or Al⁺ ions[7].

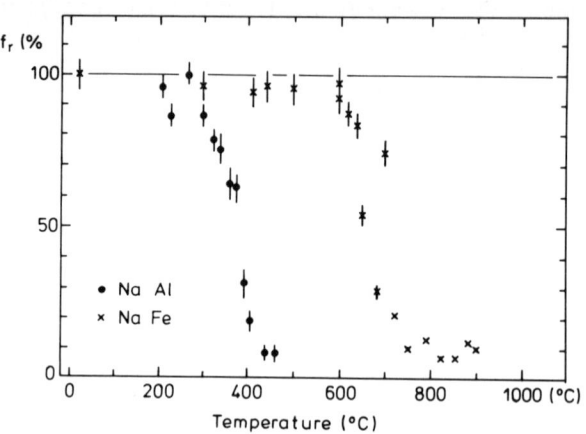

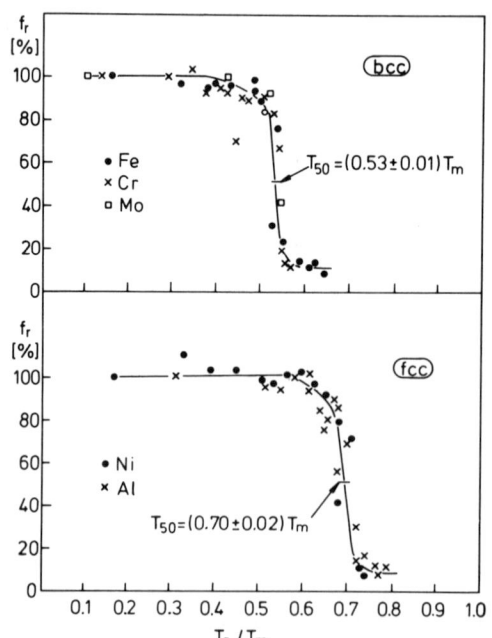

Fig. 2: Fraction f_r of Na retained in Al and Fe versus annealing temperature [7,9].

Fig. 3: Retained fraction f_r versus T_a/T_m.

Further insight into the stabilisation of the Na precipitates has been recently gained by Kehrel et al.[10] in their detailed study of the role of implanted and solved hydrogen. Note that the RNRA experiments are performed at room temperature where hydrogen is mobile in some of the matrices (e.g. in Fe and Ta), but not in Ni and Mo. One might therefore expect that H introduced during the (p,γ) RNRA, can diffuse towards the Na profile and either "decorate" the radiation-induced defects and/or the interface between of the Na precipitates and the matrix, or even may form Na-hydride. H-profiling was done via the resonant ^{1}H(^{15}N,αγ) reaction at the 6.34 MeV resonance, at the University of Helsinki tandem accelerator. It was demonstrated that H-trapping, indeed, takes place at Na damage, but that no Na-hydride is formed. Fig. 4 illustrates the measured H-profiles in a Na-doped Fe sample annealed at 250°C (upper part), and in a Na-doped Mo sample annealed at 300°C (lower part). The samples were post-irradiated with either $2 \cdot 10^{18}$ 300-400 keV protons/cm² (cercles) or with $1 \cdot 10^{17}$ 100 keV protons/cm² (dots). The full lines indicate the measured Na-profiles. It is seen that defect-hydrogen formation is more effective in regions where the damage due to the deposited recoil energy of the Na-implantation is low (at the far side) than in the peak region of the damage energy near the maximum of the Na-profile). Most of the hydrogen bound to these defects was solved in the samples and did not originate from the RNRA (p,γ) reaction.

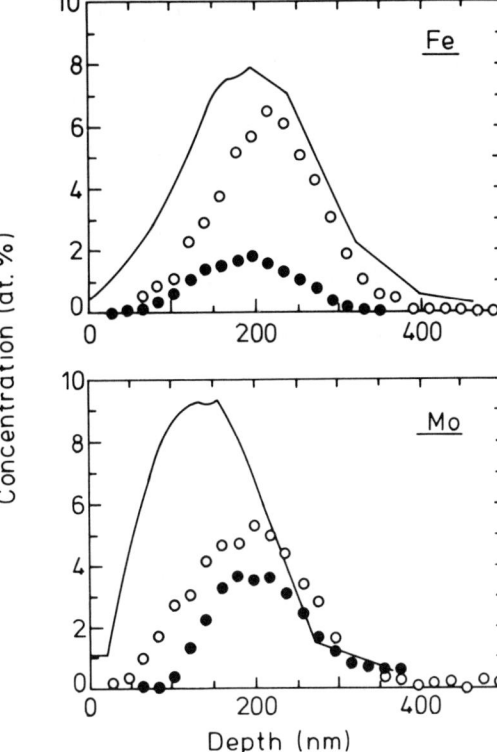

Fig. 4: Na-profiles (full lines) and H-profiles (dots, circles) in Fe (a) and Mo (b) post-irradiated with up to $2 \cdot 10^{18}$ protons/cm² and annealed for 30 min at 520 K (Fe) and 670 K (Mo) [10].

NITROGEN PROFILING IN ION-IRRADIATED TiN AND Cr_2N COATINGS

Thin nitride films are of technological importance as hard coatings which protect metallic surfaces against wear and corrosion. Ion beam irradiation has been used to improve the adhesion of these coatings to the substrates, either <u>during</u> the deposition phase of the films in a process called Ion Beam Assisted Deposition (IBAD) or <u>after</u> deposition. We have studied in great detail and by a number of techniques, the atomic transport processes in 30-300 nm thick TiN and Cr_2N films which were produced via magnetron sputtering on various substrates and irradiated with 100-900 keV noble gas ions (Ar, Kr, Xe). We were mostly interested in
- the stoichiometry of the as-deposited films[12],
- the surface sputtering and interface mixing processes as function of the ion energy and mass[13-18], and
- the balance between improved adhesion due to ion mixing, and film destruction due to blisters of the implanted gas[19,20].

A summary of our results for TiN coatings is given in refs. 14,16 and for Cr_2N coatings in ref. 18.

The N-content of the films was depth profiled with the resonant $^{15}N(p,\alpha\gamma)$ reaction[2] at 340 keV using a 20 µA proton beam, in energy steps of 0.1-1.0 keV and for a total beam charge of 3 mC at each energy. The other components of the films and/or substrates were analyzed via RBS and/or PIXE[17]. The three analyzing methods complement each other in a favorable way[14]: PIXE at few hundred keV proton energies gives very precise data on the overall film thickness and thus on the ion sputtering process. The RBS spectra, taken at 900 keV α-energy in two high-resolution surface barrier detectors at ±165°, exhibit the Ti and Cr concentration profiles and in the case of light-element substrates such as AlMg3, Al_2O_3 and Si.

Stoichiometry and mechanical properties of the hard coatings:
Fig. 5 illustrates γ-ray yield curves for an initially 42 nm and 140 nm Cr_2N film bombarded with $5*10^{16}$ Xe^+ ions/cm^2 at 350 keV[18]. Note the dip in the 140 nm yield curve which is a measure of the implanted Xe distribution; the horizontal dashed line represents the γ-yield for stoichiometric Cr_2N. Xe irradiation of the 42 nm layer evidently has changed the composition of the film and/or sputtered it off partially over the 8x8 mm^2 beam spot.

Studies on the stoichiometry and some mechanical properties of TiN_x coatings produced by DC magnetron sputtering of Ti in an Ar/N_2 atmosphere have been reported by Bolse et al.[12]. Fig. 6 illustrates the γ-scan of thick TiN_x coatings as function of the N_2-component in the mixing gas (20-30%) and the temperature of the steel substrate (320K, 470K). Stoichiometric TiN films can be produced either at low N_2-flux and higher substrate temperature or at higher N_2-flux and room temperature. Fig. 7 illustrates how the Ti/N concentration in the films relates to their color, critical load L_c and microhardness HK: films of optimal properties are indeed produced for the correct 1:1 atomic concentration ratio.

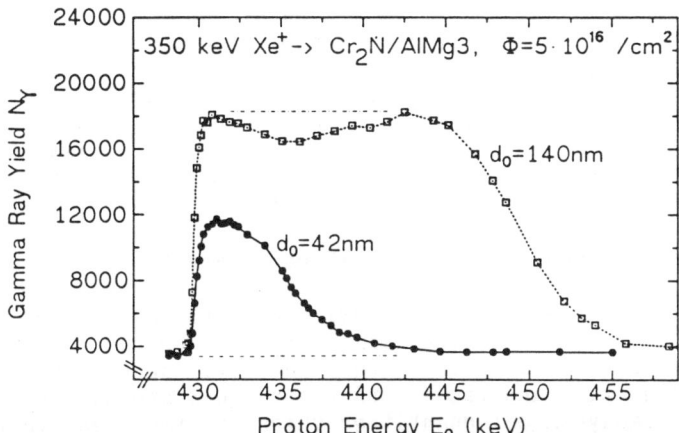

Fig. 5: N-profiles of initially 42 nm and 140 nm thick Cr_2N films irradiated with 350 keV Xe^+-ions to a fluence of $5*10^{16}/cm^2$.

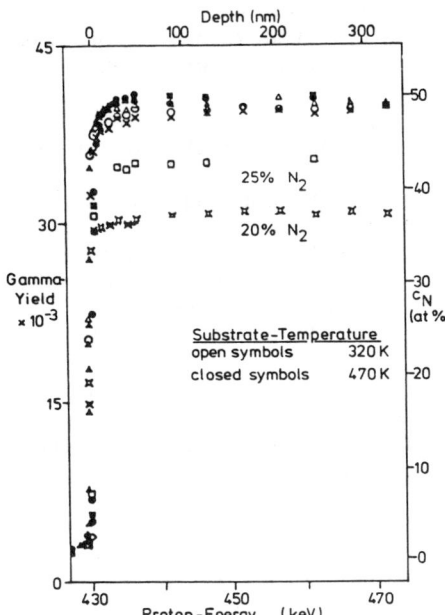

Fig. 6: $^{15}N(p,\alpha\gamma)$ yields of TiN_x films as function of the N_2 flux and substrate temperature during magnetron sputtering.

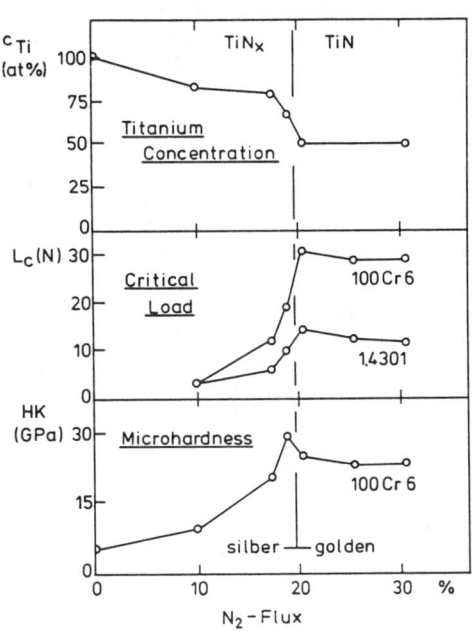

Fig. 7: Ti-content c_{Ti}, micro-critical load L_c and hardness HK of TiN_x films as function of the N_2-flux during magnetron deposition [12].

Interface mixing: The ion mixing process in a bi-layer involves the transport of atoms from the coating through the interface to the substrate and vice versa. It is a consequence of the kinetic energy deposition of the primary ion "near" the interface, via atomic collisions. Many authors[21-23] have discussed various substages of this complicated process and their time scales: primary collision, collision cascade (and possible formation of a thermal spike), relaxation (eventually followed by radiation enhanced diffusion,..).

The experimental quantity which characterizes the interface broadening effect is the mixing rate k defined as $k=\Delta\sigma^2/\Phi$. Here $\Delta\sigma^2$ denotes the increase of the variance of the N profile and Φ the incoming ion fluence. The mixing rate evidently depends on the parameters of the ballistic energy transfer, such as the mass ratio between projectile and target and the nuclear stopping power S_n, but also on thermodynamical quantities which are important in the relaxation process, such as the cohesive and mixing energy[21]. Fig. 8 illustrates the variation of the $^{15}N(p,\alpha\gamma)$ yield measured for an initially 30 nm thick TiN layer on stainless steel and irradiated with up to $4*10^{16}/cm^2$ $^{84}Kr^+$ ions at 150 keV[16]. One finds an error-function type profile at the back edge the variance of which increases linearly with Φ and gives $k = 0.12\pm0.02$ nm^4. The energy loss of the Kr ions at the interface as calculated with the program TRIM[24] is $F_D=2.3$ keV/nm, resulting in a mixing efficiency of $k/F_D=5\pm1$ A^5/eV.

We now like to mention some important new results of these experiments:
1. Generally, the mixing efficiency of TiN compound layers[14-16] is by an order of magnitude smaller than that of metallic Ti layers on the same substrates (for Ti: $k/F_D\approx67$ A^4/eV^{25}). We attribute this difference to the very high binding energy of TiN which prevents dissociation of TiN in the ballistic phase and favors the formation of TiN in the relaxation phase of the collision over that of any other compound or intermetallic phase.
2. One of the essential questions in ion beam mixing is the the problem of the shape and size of the zone affected by the primary collision cascade. We have tackled this problem[13,18] by measuring, for many TiN and Cr_2N films, the mixing rate k as function of the projectile mass and the ratio between the film thickness d and projected ion range R_P. Fig. 9 illustrates k/k_{max} values versus d/R_P for TiN/AlMg3 bilayers irradiated with Ar and Xe ions[13], and for Cr_2N/AlMg3 bilayers irradiated with Xe ions[18]. One first notes that the mixing rate for Ar bombardment produces a rather narrow distribution of k having a maximum at $k/R_P=0.7$. The Xe irradiations produce much broader distributions. Even for $d/R_P\geq2$, considerable ion mixing is observed. This indicates that the collision cascade zone extends far beyond the projected range of the primary ion.
3. Usually, the mixing rate k increases with the energy F_D deposited at the interface. However, no simple scaling of k with either F_D (as predicted by the collisional model[22]) or F_D^2 (as suggested by the thermal spike model of Johnston et al.[21]) is found.

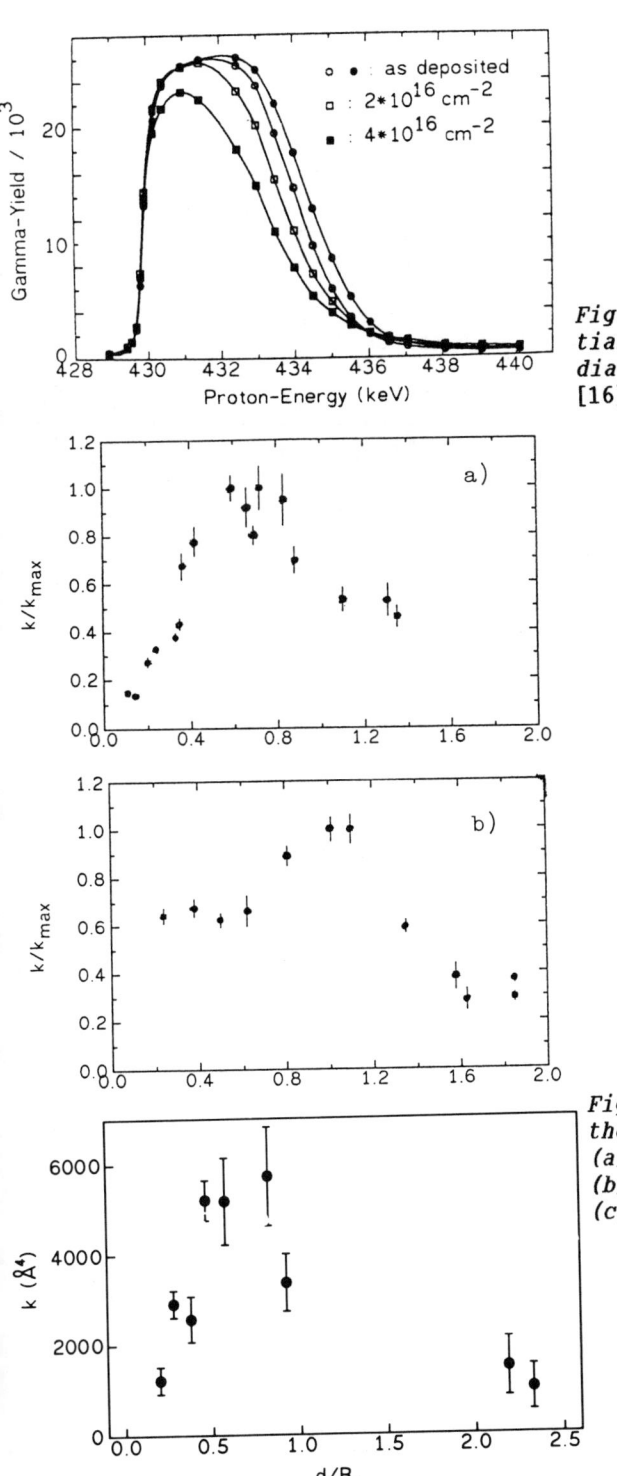

Fig. 8: N-profiles of initially 30 nm TiN films irradiated with 150 keV Kr ions [16].

Fig. 9: Mixing rate k versus the parameter d/R_p [13,19]:
(a) TiN layers + Ar ions;
(b) TiN layers + Xe ions;
(c) Cr_2N layers + Xe ions.

Sputtering: The sputter yield Y, defined as the number of target atoms removed per incoming ion, depends on the nuclear stopping power of the ions at the surface, S_n, and the surface binding energy, U_b. For mono-elemental targets, the Sigmund-Winterbon theory[26,27] gives the estimate $Y = 0.042\ \alpha(A_2/A_1)\ S_n/U_b$, where the factor $\alpha(A_2/A_1)$ depends on the ion and target mass numbers and is given in ref. [28]. Fig. 10 compares sputter yields of 150-700 keV ^{84}Kr ions impinging onto 30 nm and 130 nm TiN films deposited onto stainless steel substrates[16]. The data were obtained by PIXE (upper part) and RNRA and RBS (lower part). Note that RBS and PIXE determine the Ti sputter yield, while RNRA measures the N sputter yield. Fig. 10 is representative in the sense that PIXE can provide much more precise sputter yields than RBS in cases where the components of the film (Ti) and substrate (Fe,Cr,Ni) and the ions (Kr) have similar masses. This prevents the unambiguous deconvolution of the RBS spectra. The solid line gives the theoretical prediction for $\alpha=0.13$ and $U_b=13.27$ eV/TiN, the formation enthalpy to build solid TiN out of its atomic components.

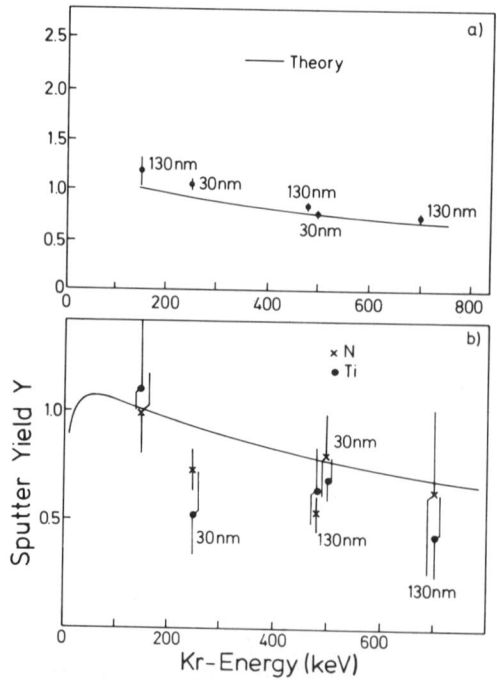

Fig. 10: Sputter yields Y of Kr-irradiated TiN films obtained by PIXE (a) and RBS and RNRA (b). The full line is the Sigmund-Winterbon [28] prediction discussed in the text [16].

The authors are most grateful to Dr. F.J. Bergmeister, Battelle Institut, Frankfurt, for providing the hard coatings, to D. Purschke for their help with the IONAS experiments and to Prof. J. Keinonen and W. Müller for the permission to quote unpublished results. This work was partially supported by Deutsches BMFT.

REFERENCES

1. M. Uhrmacher, et al., Nucl. Instr. Meth. $\underline{B9}$, 234 (1985).
2. T. Osipowicz, K.P. Lieb, S. Brüssermann,
   Nucl. Instr. Meth. $\underline{B18}$, 232 (1987).
3. F.J. Bergmeister, K.P. Lieb, K. Pampus, M. Uhrmacher,
   Z. Phys. $\underline{A320}$, 693 (1985).
4. S. Wagner, M. Heitzmann, Zeitschr. f. Naturf. $\underline{15a}$, 74 (1960).
5. A. Ritz, et al., Nucl. Phys. $\underline{43}$, 229 (1963).
6. S. Seuthe, et al., Nucl. Instr. Meth. $\underline{A260}$, 33 (1987);
   Nucl. Phys. $\underline{A514}$, 471 (1990).
7. N. Scapellato, M. Uhrmacher, K.P. Lieb, J. Phys. $\underline{F18}$, 677 (1988).
8. A. Kehrel, K.P. Lieb, M. Uhrmacher,
   Mat. Sci. Eng. $\underline{A115}$, 43 (1989).
9. A. Kehrel, K.P. Lieb, N. Scapellato, M. Uhrmacher,
   J. Nucl Mat. $\underline{173}$ (1990), in press.
10. A. Kehrel, J. Keinonen, K.P. Lieb, M. Uhrmacher,
    Appl. Phys. A, submitted.
11. M. Uhrmacher, K.P. Lieb, to be publ.
12. W. Bolse, et al., in "Plasma Surface Engineering" (DGM Informationsgesellschaft, Oberursel, 1989) p. 767.
13. W. Müller, W. Bolse, K.P. Lieb, submitted for publication;
    W. Müller, diploma thesis, Göttingen, 1990, unpubl.
14. K.P. Lieb, et al., Nucl. Instr. Meth. $\underline{B50}$, 10 (1990).
15. W. Bolse, et al. Thin Solid Films $\underline{174}$, 139 (1989).
16. T. Corts, W. Bolse, T. Osipowicz, K.P. Lieb,
    Appl. Phys. $\underline{A51}$ (1990) in press.
17. T. Osipowicz, T. Corts, K.P. Lieb, F.J. Bergmeister,
    Nucl. Instr. Meth. $\underline{B50}$, 238 (1990).
18. W. Bolse, T. Kacsich, F.J. Bergmeister, Mat. Sci. Eng., in press.
19. T. Weber, W. Bolse, K.P. Lieb, Nucl. Instr. Meth. $\underline{B50}$, 95 (1990)
20. W. Bolse, T. Weber, T. Flottmann, Mat. Sci. Eng., to be publ.
21. W.L. Johnson, Y.T. Cheng, M. van Rossum, M.-A. Nicolet,
    Nucl. Instr. Meth. $\underline{B7/8}$, 657 (1985).
22. P. Sigmund, A. Gras-Marti, Nucl. Instr. Meth. $\underline{182/183}$, 25 (1981).
23. R.S. Averback, S. Seidman, Mat. Sci. Forum $\underline{15-18}$, 963 (1981).
24. Program TRIM89, in J.F. Ziegler, G. Cuomo, J.P. Biersack, "The Stopping and Ranges of Ions in Matter" (Plenum, New York, 1989).
25. W. Bolse, T. Weber, W. Lohmann, Nucl. Instr. Meth. $\underline{50}$, 416 (1990)
26. P. Sigmund, Phys. Rev. $\underline{184}$, 393 (1969).
27. K.B. Winterbon, in "Ion Implantation Ranges and Energy Deposition Distributions" (Plenum, New York, 1975).
28. H.H Andersen, Nucl. Instr. Meth. $\underline{B33}$, 466 (1988); and in "Sputtering by Particle Bombardment", vol. 1, R. Behrisch, Ed. (Springer, Berlin, 1981).

ELEMENTAL MAPPING OF PLANETARY SURFACES USING GAMMA-RAY SPECTROSCOPY*

Robert C. Reedy
Space Science and Technology Division, Mail Stop D438
Los Alamos National Laboratory, Los Alamos, NM 87545, USA

ABSTRACT

The gamma rays escaping from a planet can be used to map the concentrations of various elements in its surface. In a planet, the high-energy particles in the galactic cosmic rays induce a cascade of particles that includes many neutrons. The γ rays are made by nuclear excitations induced by these cosmic-ray particles and their secondaries (especially neutron capture or inelastic-scattering reactions) and by the decay of the naturally-occurring radioelements. After a short history of planetary γ-ray spectroscopy and its applications, the γ-ray spectrometer planned for the Mars Observer mission is presented. Laboratory experiments that simulate the cosmic-ray bombardments of planetary surfaces or measure cross sections for the production of γ rays are reviewed. Theoretical calculations for the processes that make and transport neutrons and γ rays are discussed. The emphasis here is on studies of Mars and on new ideas, concepts, and problems that have arisen over the last decade, such as Doppler broadening and peaks from neutron scattering with germanium nuclei in a high-resolution γ-ray spectrometer.

INTRODUCTION

The energies and intensities of γ rays escaping from a planet with very little or no atmosphere can be used to determine and map the concentrations of various elements in the top few tens of centimeters of the surface. In the planet, the high-energy ($\sim$0.1–10 GeV/nucleon) particles in the galactic cosmic rays induce a cascade of particles that includes many neutrons.[1] The γ rays used for elemental mapping are made by the decay of the naturally-occurring radioactive elements (K, U, and Th) and by nuclear excitations induced by cosmic-ray particles (mainly by neutron capture or inelastic-scattering reactions).[2,3] Certain elements, such as hydrogen, carbon, samarium, and gadolinium, can strongly affect the spectrum of neutrons in a planet and thus can be sensed indirectly. Hydrogen concentration-versus-depth profiles can be determined from neutron spectra[4,5] or γ rays from other elements.[6,7] The Earth's atmosphere is so thick ($\approx$1000 g/cm^2) that few cosmic-ray particles reach the surface, and it also prevents γ rays made in the surface from traveling very far. Thus γ-ray spectroscopy on the Earth has been limited to searches for natural radioactivity (e.g., U) and to studies of the cosmic-ray-produced γ rays near the top of the atmosphere. Planetary γ-ray spectroscopy as considered here refer to solid objects with no or thin atmospheres, such as the Moon, Mars, asteroids, and comets.

---

* This work was supported by NASA and done under the auspices of the U.S. Department of Energy.

The elements that are of interest to future planetary missions are listed in Table I with their abundances in the Moon, Mars, and a CI-chondritic meteorite. The last composition could be present in asteroids and, with additional amounts of volatile hydrogen, carbon, nitrogen, and oxygen, in a comet. Not listed in Table I are a few minor elements that are enriched in evolved objects like the Moon relative to chondrites, such as Sr, Zr, Ba, Gd, Sm, Th, and U. The only γ-ray-producing reactions of interest with these minor elements are the neutron-capture ones with Gd and Sm. Also not included is the martian atmosphere, which would add ^{40}Ar to the list, although it is mainly CO_2 and some N_2. In the Moon, the darker, mare regions are enriched in Fe and Ti, while the lighter colored, highland areas have enhanced amounts of Al and Ca. Considerable variability in elemental concentrations is also expected at Mars.

Planetary γ-ray spectroscopy was proposed around 1960 by James Arnold and colleagues.[8] However, planetary missions with such instruments have been very rare. On the Apollo 15 and 16 missions in 1971 and 1972, respectively, NaI(Tl) γ-ray spectrometers were flown, and spectra were accumulated over about 22% of the Moon's surface. Maps of iron, titanium, magnesium, and natural radioactivity were produced from the Apollo γ-ray data.[9,10] Most papers on planetary γ-ray spectroscopy date back to the Apollo era. Future missions will use advanced technologies for γ-ray spectroscopy.

A high-purity-germanium γ-ray spectrometer is scheduled to be launched in 1992 on Mars Observer and others will probably be on lunar orbiters and Soviet martian orbiters. A γ-ray spectrometer in a penetrator[11] has been proposed for the comet rendezvous part of the Comet Rendezvous Asteroid Flyby (CRAF) mission. The greatly improved detection capabilities (such as the high resolution for γ rays) and new targets (e.g., Mars with its thin atmosphere) have been changing our ideas for planetary γ-ray spectroscopy considerably since the

Table I. Elemental compositions of a few objects of interest to planetary γ-ray spectroscopy (mg-element/g-total).

Element	Average Moon[3]	Mars (Dust)[7]	CI chondrite[12]
H	0.04	~1.	20.2
C	–	10.	34.5
N	–	1.	3.2
O	435.	442.	464.
Na	3.5	9.	5.0
Mg	40.	40.	98.9
Al	110.	42.	8.7
Si	200.	226.	106.4
P	0.6	–	1.2
S	0.7	31.	62.5
Cl	0.02	9.1	0.7
K	1.2	1.2	0.56
Ca	100.	44.	9.3
Ti	14.	4.1	0.4
Cr	1.0	1.4	2.7
Mn	0.8	3.7	2.0
Fe	90.	135.	190.4
Co	–	–	0.5
Ni	0.4	–	11.0

Apollo missions. The Mars Observer Gamma-Ray Spectrometer (GRS) is described below. Also discussed are the results of some simulation experiments and preliminary results for γ-ray and neutron calculations for Mars. The needs for data on reactions producing γ rays are reviewed. The new instruments will produce significantly improved measurements, but they also require additional studies and calculations to anticipate possible complications arising from their use and to fully utilize the measured γ-ray spectra to get elemental abundances and distributions.

THE MARS OBSERVER GAMMA-RAY SPECTROMETER

The Mars Observer mission is scheduled to be launched in September 1992 and placed in a polar orbit to globally map Mars.[13] The mission's objectives include determination of the elemental composition of surface features and studies of volatile materials (especially water and carbon dioxide), problems that can be addressed by a GRS on the Mars Observer spacecraft. The martian atmosphere can also be studied by the γ rays measured from orbit.[14] From a knowledge of the present state of Mars obtained by the experiments on Mars Observer, its origin and evolution can be inferred and Mars can be compared with its sister planets, the Earth and Venus.[13] A similar spacecraft, the Lunar Observer, has been proposed to do similar studies at the Moon, but such a lunar mapping spacecraft is not yet an approved mission.

The Mars Observer γ-ray detector will be a high-purity n-type germanium (hpGe) coaxial diode with a 56-mm diameter and a 56-mm length. It will be cooled to $\lesssim$100 K by a passive radiator. The hpGe will be surrounded by a plastic scintillator, and the GRS's electronics will reject signals in the hpGe detector that are in coincidence with a signal in the plastic scintillator, eliminating background signals from the passage of charged cosmic rays through the hpGe detector. Signals from the hpGe for energies from ~0.2 to ~10 MeV will be processed in a pulse height analyzer. Below and above ≈2.4 MeV, the spectra will have ≈0.6 and ≈1.2 keV per channel, respectively. An entire γ-ray spectrum (≈10,000 channels) will be transmitted every ~20 seconds.

Thermal ($\lesssim$0.1 eV) and epithermal ($\simeq$1-1000 eV) neutrons will be detected using a boron-loaded plastic scintillator for the anti-coincidence shield. The $^{10}B(n,\alpha)^7Li$ reactions induced by these neutrons in the borated plastic will produce a unique signal in the scintillator's output.[15] Because the spacecraft moves at a velocity slightly faster (3.4 km/s) than that of a thermal neutron, the neutron count rates in each of the four faces of the anti-coincidence shield (which is pyramid shaped and fixed relative to the spacecraft's velocity) will be used as a Doppler filter to determine the fluxes and spectral shapes of thermal and epithermal neutrons.[16]

SIMULATION EXPERIMENTS

Several experiments have been done recently at accelerators to simulate the processes that produce γ rays in a planet's surface, and more are planned by the Mars Observer GRS team. In one series of irradiations, thick targets were bombarded with 6-GeV protons,[17] simulating the cascade of galactic-cosmic-ray particles in a large solid target. The spectra of γ rays measured in front of thick iron targets showed many narrow lines for γ rays whose relative fluxes were in good agreement with theoretical calculations.[17]

As neutrons dominate the production of most γ rays,[2,3] another series of irradiations was done using neutrons from a 14-MeV neutron generator.[18,19] The concrete in the room around the neutron generator moderated many neutrons and produced neutrons with a continuum of energies from 14 MeV to thermal. The γ-ray spectrum from the irradiation of an aluminum target[19] is shown in Fig. 1. The relative fluxes of γ rays made by Fe(n,γ) reactions were in good agreement with calculated planetary γ-ray fluxes that only considered production by thermal neutrons.[3] Because the spectrum of neutrons in the simulation had an epithermal/thermal neutron ratio similar to that in the Moon, this agreement shows that thermal yields are good for calculating fluxes of neutron-capture-produced γ rays.[18]

Several aspects of the results for the simulations with $\leq$14-MeV neutrons were different from our experience with the low-resolution Apollo γ-ray data. As marked with diagonal lines in the top part of Fig. 1, five peaks with unusual shapes were observed. These peaks are shaped normally at their low-energy sides, and their energies correspond to those for the de-excitation of low-lying levels in germanium isotopes. The high-energy sides of these peaks extend for ~50 keV, and are caused by the summing of the recoil energy from a Ge(n,n') reaction with the de-excitation γ ray.[18] Except for the peak at and above 834 keV, these sawtooth-shaped peaks made in a Ge detector should not interfere with the major γ rays expected from a planetary surface. The high-energy tail above 834 keV will be under the inelastic-scattering peaks from Al and Fe at 844 and 847 keV, respectively. Also marked in Fig. 1 is the peak at 4.438 MeV from the ^{12}C(n,nγ)^{12}C reaction (and also probably including some γ rays from the ^{16}O(n,n$\alpha\gamma$)^{12}C reaction), which has a width of 53 keV compared to the 5-keV width of an adjacent γ ray from Si. This Doppler broadening of the major carbon inelastic-scattering γ ray and the low cross section for the ^{12}C(n,γ) reaction will make the detection of carbon in a planetary surface by high-resolution γ-ray spectrometers difficult.

NUCLEAR DATA FOR GAMMA-RAY-PRODUCING REACTIONS

In my 1978 paper on the fluxes of γ rays expected from planetary surfaces,[3] I noted that cross sections for the production of γ rays by nonelastic-scattering reactions were often scarce. They still are scarce. Usually the highest neutron energy used in measuring cross sections for the production of nonelastic-scattering γ rays is below 20 MeV, and often cross sections have not been measured at some energies, such as from $\approx$6-13 or >15 MeV.[20] Few γ-ray-production cross sections have been measured for the proton energies of interest (hundreds of MeV to several GeV). Recently several irradiations have been done with neutrons having energies $\leq$78 MeV[21] and up to several hundred MeV,[22] and more irradiations are planned to measure such cross sections.

The lack of good cross sections for nonelastic-scattering reactions limits our ability to calculate the leakage fluxes of such γ rays, especially for those γ rays that are made by high-energy particles and interfere with inelastic-scattering γ rays. For example, the production of 1.369-MeV γ rays by the ^{28}Si(n,n$\alpha\gamma$)^{24}Mg reaction strongly interferes with the signal from the ^{24}Mg(n,nγ)^{24}Mg reaction that is used to determine magnesium concentrations.[3] Such interferences by nonelastic-scattering reactions occur for many elements, such as the 4.438-MeV γ ray from ^{12}C also being produced in high fluxes from

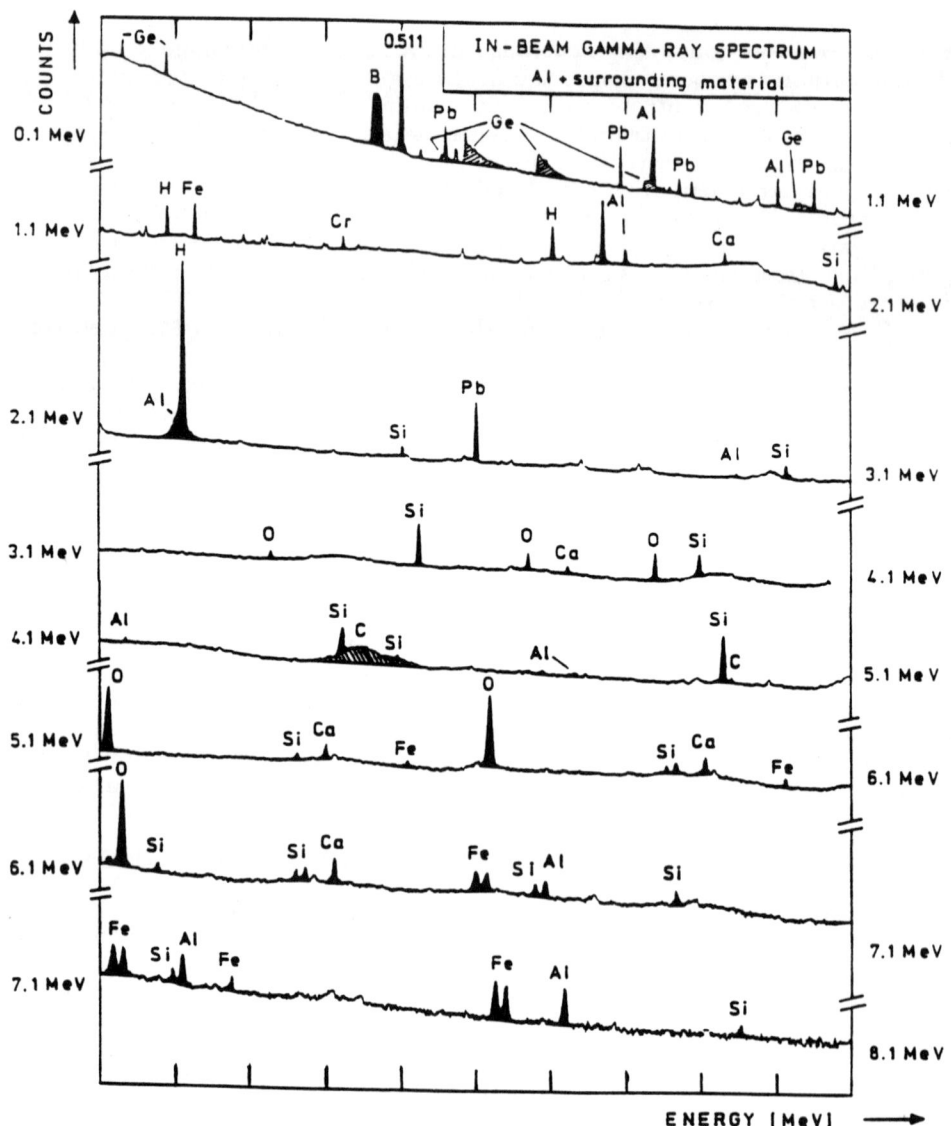

Fig. 1. The γ-ray spectrum observed from aluminum irradiated with 0 to 14-MeV neutrons.[19] Most γ rays were produced in the concrete around the 14-MeV neutron generator and in the material (such as lead and borated paraffin) surrounding the Ge detector. Shaded are the five asymmetric Ge peaks from 596 to 1040 keV and the Doppler-broadened peak at 4.438 MeV from the $^{12}C(n,n\gamma)^{12}C$ reaction (and also probably the $^{16}O(n,n\alpha\gamma)^{12}C$ reaction).[18]

^{16}O or the 1.434-MeV γ ray from ^{52}Cr being overwhelmed by reactions with ^{56}Fe.[3] Where such nonelastic-scattering reactions with heavier nuclei interfere with inelastic-scattering γ rays, γ rays made by neutron-capture reactions often can be used instead,[3] although this is not always the case (e.g., carbon). The γ rays made with elements both in the atmosphere and surface, such as oxygen, can often be distinguished by the fact that the width of a γ-ray line from a solid is almost always narrow while that from a gas usually is Doppler broadened.[14,23] The elements that are generally well mapped by inelastic-scattering γ rays include carbon, oxygen, magnesium, aluminum, silicon, calcium, and iron.

Many elements have sufficiently large cross sections for the capture of thermal neutrons and γ rays emitted in large enough yields that they can be studied with such γ rays. The fluxes of thermal neutrons in a planet can be considerably modified by light elements, such as hydrogen or carbon, which rapid moderate the neutron spectrum.[5,6] The total cross section for the capture of thermal neutrons also can affect the flux and energy spectrum of low-energy neutrons in a planet's surface.[6] While the fluxes of low-energy neutrons can be calculated well by various codes[4,5,6,7,11,24] for neutron transport, the composition of many surface features may not be well determined. Undetected elements, such as gadolinium and samarium, can seriously affect the equilibrium neutron distributions.[4,6] Thus elemental ratios will usually be obtained initially from planetary γ-ray spectra. As iron and silicon have γ rays made at good rates by both inelastic-scattering and neutron-capture reactions, their γ rays can be used to normalize elemental ratios obtained from both types of reactions.

There is a need for better yields for the production of γ rays by the capture of thermal neutrons. The status of the data for the yields of neutron-capture γ rays for planetary GRS applications was reviewed over a decade ago[3] and indicated the need for additional data. Existing compilations[25,26] of the yields of neutron-capture γ rays have missed many measurements in the literature, and there are not good published γ-ray yields for some of the elements listed in Table I. The elements well mapped by the neutron-capture γ rays include hydrogen, aluminum, silicon, calcium, titanium, chromium, iron, and nickel.

RECENT CALCULATIONS OF MARTIAN NEUTRON AND GAMMA-RAY LEAKAGE FLUXES

In planning for the Mars Observer Gamma-Ray Spectrometer experiment, calculations recently have been done for the production and transport of neutrons[5] and γ rays[7] in the martian surface and atmosphere. The attenuation of γ rays from the surface by the martian atmosphere[27] needs to be considered in such calculations. All of these calculations included a $\approx$16-g/cm^2-thick atmosphere (95.7% CO_2, 2.7% N_2, and 1.6% ^{40}Ar) and used a martian-surface composition estimated from chemical analyses by the Viking landers and of the "martian" (SNC) meteorites (the shergottites, nakhlites, and Chassigny).[5,7] Much of the emphasis in these calculations has been on the highly-variable amounts of volatiles (H_2O and CO_2) that can be present in or on the martian surface. The equilibrium distributions of neutrons in Mars were calculated using the ONEDANT[5,24] and the ANISN[7] neutron-transport codes. The ONEDANT code was modified to include the effects of gravity and the neutron's beta decay.[24] Neutrons that escape Mars with E $\leq$ 0.132 eV are gravitationally bound, although some neutrons beta decay before returning to the planet. Neutron-transport calculations done with and without the effects of gravity showed that gravity increased the flux of neutrons at the top of the martian atmosphere by $\approx 29\%$, but that the neutron-flux increase at the top of the martian surface due to gravity was only a few percent.[24]

These calculations[5,7,24] indicate that the martian atmosphere and the presence of H_2O in or CO_2 on the martian surface significantly affect the distributions of neutrons. Hydrogen rapidly thermalizes neutrons and shifts the peak of their depth distribution towards the surface. Because of its low absorption cross section, CO_2 builds a large reservoir of low-energy neutrons that can leak back into the surface.[5,24] The neutron count rates expected in the GRS's anti-coincidence shield are high enough to allow a rapid determination of the concentrations of H_2O and CO_2 in and on the surface from the observed fluxes of the thermal and epithermal neutrons.[5] The depth that H_2O is below the surface can often be determined from the neutron[5] and γ-ray[7] data. The measured γ-ray and neutron leakage fluxes can be used together to get additional information on the concentration and stratigraphy of H_2O and CO_2 in the top meter of the martian surface.

The fluxes from the ANISN neutron-transport calculations were used to determine the production rates of γ rays by nonelastic-scattering and neutron-capture reactions.[7] The γ rays made by these reactions and by the natural decay of K, Th, U, and their daughters were transported through the martian surface and atmosphere to get fluxes at the spacecraft. The γ rays most suitable for detecting the expected major elements and the radioelements have strong enough leakage fluxes that they should be detectable with integration times of hours to several hundred hours. The measurement of these elements will aid in the determination of the major rock types present and of the degree of local and global refractory enrichment. Major types of volcanic materials, especially in the southern highlands of Mars unsampled by Viking, should be identified. Readily detectable in γ-ray spectra will be S and Cl, which might be present in surface precipitates or subsurface brines. Besides elemental abundances, the γ-ray data can also be used to study the distribution of hydrogen and CO_2 in Mars by comparing γ rays made by both inelastic-scattering and neutron-capture reactions with one element.[7]

The data obtained from both the γ-ray and neutron modes of the Mars Observer GRS will complement each other, and their use together will considerably improve the scientific return. For example, the elemental results from the γ-ray spectra are needed to help interpret the transport of the leakage neutrons. As neutrons are the major source of most γ rays, direct measurement of the neutron leakage flux can aid in interpreting the γ-ray data. The measured neutron and γ-ray fluxes also can help to infer the presence of certain elements not directly observed in the γ-ray spectra, such as relatively high amounts of strongly-neutron-absorbing Gd and Sm. The intensity of the leakage thermal and epithermal neutrons can be used with the flux ratio for the neutron-capture γ rays from hydrogen and a major element like Si or Fe to determine the concentration and stratigraphy of H_2O in the top ~ 100 g/cm^2 of the martian surface or the thickness of CO_2 frost on the martian surface. Such studies of volatiles will be very important not only for studies of Mars but also for comets and possibly for asteroids and the polar regions of the Moon.

SUMMARY

Future planetary missions to the terrestrial planets and to small bodies (comets and asteroids) will have as one of their major objectives the determination of their chemical compositions. Gamma-ray spectroscopy is an excellent method for orbital or in-situ elemental studies of these objects. A gamma-ray spectrometer is scheduled to fly on the Mars Observer mission and is a prime candidate for the Lunar Observer mission. Our ideas for these missions have changed considerably since the time of the Apollo missions with their NaI(Tl) γ-ray spectrometers. New detectors (e.g., high-purity germanium) and techniques (Doppler-filter neutron spectroscopy) are available, and the new targets are different in many ways from the Moon (atmospheres[28] and volatiles). Nuclear data is critical to the planning for and analysis of measurements from such missions. As discussed above, much work, including simulation experiments and theoretical calculations, is being done in planning for upcoming missions. There is also a strong need for more nuclear data for the production of γ rays by inelastic- and nonelastic-scattering reactions and by neutron-capture reactions.

Several problems have been identified with these future γ-ray spectroscopy experiments. The hpGe detectors can be fairly easily damaged by cosmic radiation. Experiments to understand how and when such radiation damage occurs are being done with the goal of minimizing such radiation effects. The high resolution of hpGe γ-ray spectrometers increases our ability to measure concentrations of most elements[29,30] but means that we must be careful of effects such as Doppler broadening and interferences to major γ-ray lines from other sources. The laboratory irradiations and theoretical calculations are important, especially now that we will be going to objects for which we have no "ground truth," which we had at the Moon, to normalize our measurements.[9,31] At present, there appear to be no serious problems to the use of γ-ray spectroscopy for elemental determinations of a variety of solar-system objects. Additional laboratory measurements, such as cross sections or yields for γ-ray production, will be very valuable for such exploration missions and will help to produce good elemental maps for planetary objects like Mars and the Moon.

ACKNOWLEDGMENTS

Many results presented here represent the efforts of other Mars Observer Gamma-Ray Spectrometer team members, including J. R. Arnold, W. V. Boynton, J. Brückner, D. M. Drake, P. A. J. Englert, L. G. Evans, W. C. Feldman, E. L. Haines, A. E. Metzger, S. W. Squyres, J. I. Trombka, and H. Wänke.

REFERENCES

1. R. C. Reedy and J. R. Arnold, J. Geophys. Res. 77, 537 (1972).
2. R. C. Reedy, J. R. Arnold, and J. I. Trombka, J. Geophys. Res. 78, 5847 (1973).
3. R. C. Reedy, Proceedings of the 9th Lunar and Planetary Science Conference (Pergamon, N. Y., 1978), p. 2961; also Gamma Ray Spectroscopy in Astrophysics, NASA-TM-79619, p. 98 (1978).
4. R. E. Lingenfelter, E. H. Canfield, and W. N. Hess, J. Geophys. Res. 66, 2665 (1961).

5. D. M. Drake, W. C. Feldman, and B. M. Jakosky, J. Geophys. Res. 93, 6353 (1988).
6. J. R. Lapides, Planetary Gamma-Ray Spectroscopy: The Effects of Hydrogen and the Macroscopic Thermal-Neutron Absorption Cross Section on the Gamma-Ray Spectrum, (Ph.D. Thesis, University of Maryland, 1981).
7. L. G. Evans and S. W. Squyres, J. Geophys. Res. 92, 9153 (1987).
8. M. A. Van Dilla, E. C. Anderson, A. E. Metzger, and R. L. Schuch, IRE Trans. Nucl. Sci. NS-9, 405 (1962).
9. M. J. Bielefeld, R. C. Reedy, A. E. Metzger, J. I. Trombka, and J. R. Arnold, Proceedings of the 7th Lunar Science Conference (Pergamon, N. Y., 1976), p. 2661.
10. M. I. Etchegaray-Ramirez, A. E. Metzger, E. L. Haines, and B. R. Hawke, J. Geophys. Res. 88, A529 (1983).
11. L. G. Evans, J. I. Trombka, and W. V. Boynton, J. Geophys. Res. 91, D525 (1986).
12. E. Anders and N. Grevesse, Geochim. Cosmochim. Acta 53, 197 (1989).
13. A. L. Albee and D. F. Palluconi, Trans. Am. Geophys. Union (EOS) 71, 1099 (25 September 1990).
14. R. C. Reedy, Nuclear Spectroscopy of Astrophysical Sources (AIP Conf. Proc. 170, N.Y., 1988), p. 203.
15. D. M. Drake, W. C. Feldman, and C. Hurlbut, Nucl. Instrum. & Methods A247, 576 (1986).
16. W. C. Feldman and D. M. Drake, Nucl. Instrum. & Methods A245, 182 (1986).
17. A. E. Metzger, R. H. Parker, and J. Yellin, J. Geophys. Res. 91, D495 (1986).
18. J. Brückner, H. Wänke, and R. C. Reedy, J. Geophys. Res. 92, E603 (1987).
19. J. Brückner, Neutronen-Induzierte Gamma-Spektroskopie: Beitrag zur Chemischen Fernerkundung von Planetaren Oberflächen (Doctoral Dissertation, Johannes Gutenberg-Universität, Mainz, FRG, 1984).
20. R. C. Reedy, Radiat. Eff. 94, 259 (1986).
21. P. Englert, J. Brückner, and H. Wänke, J. Radioanal. Nucl. Chem., Articles 112, 11 (1987).
22. R. O. Nelson et al., this Symposium (October 1990).
23. R. Ramaty, B. Kozlovsky, and R. E. Lingenfelter, Astrophys. J. Suppl. Ser. 40, 487 (1979).
24. W. C. Feldman, D. M. Drake, R. D. O'Dell, F. W. Brinkley, Jr., and R. C. Anderson, J. Geophys. Res. 94, 513 (1989).
25. M.A. Lone, R. A. Leavitt, and D. A. Harrison, Atomic Data Nucl. Data Tables 26, 511 (1981).
26. J. K. Tuli, Thermal Neutron Capture Gamma-Rays, BNL-NCS-51647 (1983).
27. A. E. Metzger and J. R. Arnold, Appl. Opt. 9, 1289 (1970).
28. A. E. Metzger and E. L. Haines, J. Geophys. Res., in press (1990).
29. E. L. Haines, J. R. Arnold, and A. E. Metzger, IEEE Trans. Geosci. Electron. GE-14, 141 (1976).
30. A. E. Metzger and D. M. Drake, J. Geophys. Res. 95, 449 (1990).
31. A. E. Metzger, E. L. Haines, R. E. Parker, and R. G. Radocinski, Proceedings of the 8th Lunar Science Conference (Pergamon, N. Y., 1977), p. 949.

NEUTRON FOCUSING USING CONVERGING GUIDES

D.F.R. Mildner
National Institute of Standards and Technology,
Gaithersburg, Maryland 20899

ABSTRACT

The neutron current density can be increased by using a converging guide; this can be particularly useful as the last section of a long straight guide. We give the expected flux gains for such a system. This increased density is obtained at the expense of increased divergence of the beam, which is unimportant in neutron absorption techniques, such as prompt gamma activation analysis and neutron depth profiling. These experiments can benefit from improved detection limits available when the neutron beam is focused by a neutron optical element such as a converging guide.

INTRODUCTION

Neutron guides using the principle of total reflection are used to transport neutrons beams over considerably large distances, by translating the neutron source density at the guide entrance to the guide exit with relatively little loss in transmission. This enables experiments to be placed in regions where the radiation backgrounds are much lower. Frequently a nickel surface coated on float glass is used as the guide material, though a 40% increase in transmission is obtained with isotopically pure ^{58}Ni. Further increases in transmission may be obtained using multilayer coatings, composed of two materials of high and low neutron scattering density, such as nickel-titanium. These give a critical angle at least twice that of natural nickel, and efforts are in progress to extend the reflectivity curve even further. A diagrammatic representation of neutron trajectories within the guide in each of the two transverse directions of the guide is useful for deriving analytical expressions in the small angle approximation which describe the transmission properties of the guide[1]. When the guide entrance is fully illuminated, it is easier to consider the acceptance at the exit of the converging guide. When the illumination is restricted, as from the exit of a long straight guide, it is more convenient to consider the acceptance at the entrance.

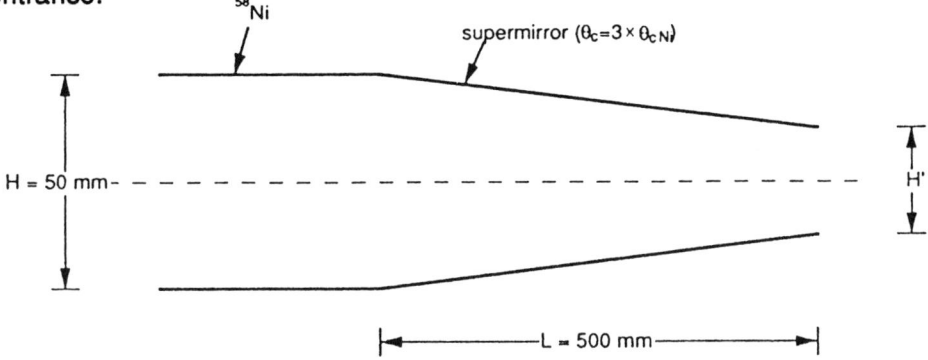

Fig 1. Schematic diagram of the guide configuration

We define a one-dimensional gain G for the converging focusing guide as the ratio of the current density of neutrons (transmitted acceptance per unit dimension) leaving the converging guide system to the current density of neutrons (acceptance area per unit dimension) incident on the system. Using the acceptance technique, Anderson[2] has indicated that the converging guide gain as a function of wavelength shows discontinuities in the first derivative. Expressions for the functional form of the gain for the focusing guide with limited acceptance have been given[3]. The greatest gains are achieved when the critical angle of the converging guide is much greater than than the divergence of the incoming beam.

Intensity gains for the converging guide[4] are limited to perhaps an order of magnitude on account of the increase in the angle that the neutron makes with the device axis upon successive reflections. This increase occurs until either the neutron is transmitted by the system, or the glancing angle is greater than the critical angle for reflection, in which case the neutron is lost. The output of the converging guide may be increased by using a material of greater critical angle. Relationships have been derived[3] for the gain of the converging guide as a function of the converging guide dimensions, the incoming beam divergence, and the critical angle of the converging guide surface material. These analytical expressions are in agreement with measurements[5] for the converging guide of nickel-titanium supermirrors.

RESULTS

We estimate the expected gain in the cold neutron flux by the use of a converging guide with a supermirror coating for cold neutron prompt gamma activation analysis at the exit of NG7 guide on the cold source at the NIST reactor. Consider a ^{58}Ni guide with a cross section 50 x 50 mm^2, followed by a converging guide of length L = 500 mm (see Figure 1). Assume that this converging guide is a truncated square cone, so that in the small angle approximation all calculations can be performed for one transverse direction, and the overall gain is simply the square of the one dimensional gain G of the neutron current density. Assume also that the converging guide is coated with nickel-titanium supermirror for which the critical angle is three times that of ordinary nickel. This should be available on a large scale within a short time.

The critical angle per unit wavelength for various materials are as follows:
Ni $\quad \gamma_c$ = 1.73 mrad Å^{-1}
^{58}Ni γ_c = 2.04 mrad Å^{-1}
Ni-Ti supermirror
$\quad \gamma_c$ = 5.19 mrad Å^{-1}

We assume that the reflectivity is unity for grazing angles below the critical angle, and

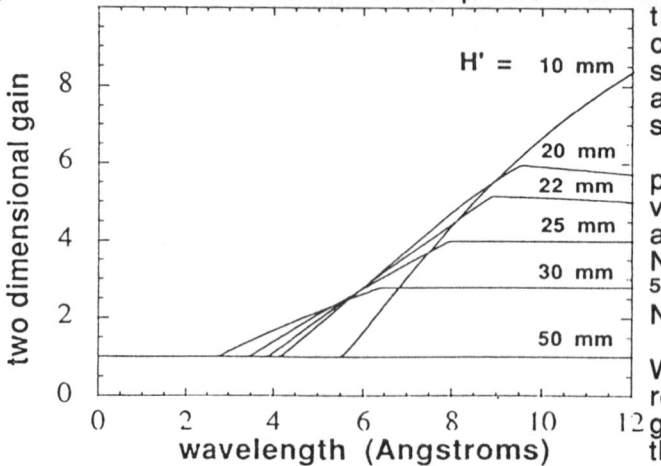

Fig 2. The two dimensional gain G^2 as a function of λ

zero above. We take the geometry of the converging guide as:
entrance height   H = 50 mm,
exit height   H' = variable,
and length   L = 500 mm.

We define angles $\psi_0 = (H + H')/2L$, and $\psi_1 = (H - H')/2L$. For each value of H', we find f such that $(f-1)/f\, \psi_0 < \psi_1 \leq f/(f+1)\, \psi_0$. The value of f enables the sketching of a (θ, θ_c) diagram[3] depicting the regions for the various functional forms for the gain G. These diagrams are solely dependent on the geometry, and **not** on the materials of the coatings. The line $\theta_c = m\theta$ gives the locus of different combinations of guide surfaces, where m is given by the ratio of the critical angles of the two coatings; here, m= 5.19/2.04 = 2.54. Note also that the various angles ψ_0, ψ_1, θ_c and θ can be placed on a wavelength scale using $\theta = \gamma_c \lambda$, where $\gamma_c = 2.04$ mrad Å^{-1}.

The two dimensional gain G^2 is shown in Figure 2. At short wavelengths the gain is unity. It then rises at longer wavelengths, until the divergence angle becomes too great to be reflected within the converging guide, at which point the gain levels off or even drops. Except for certain cases, the distribution of neutron trajectories at the exit is neither uniform in position nor in angle.

Since the output of the straight guide varies as the square of the critical angle, or as λ^2, we define a figure of merit as $(G/\lambda)^2$ for the converging guide. The value of this figure of merit as function of λ is shown in Figure 3. Since the NG7 will have a beryllium filter, values below 4 Å are useless. Though absorption cross sections vary as λ, the near Maxwellian spectrum of the cold source will be falling rapidly at long wavelengths, so that the greatest gains will be achieved in the region 4-6 Å.

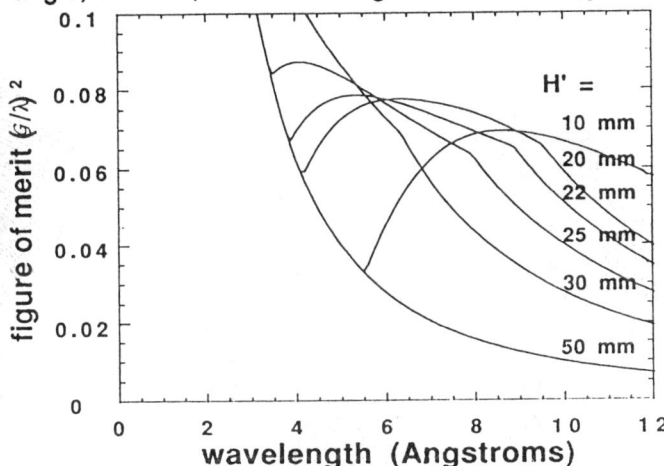

Fig 3. The figure of merit $(G/\lambda)^2$ as a function of λ

REFERENCES

1. D.F.R. Mildner, Nucl. Instrum. & Meth. **200**, 167-173 (1982).
2. I.S. Anderson, S.P.I.E. Proc. **983**, 84-92 (1989).
3. D.F.R. Mildner, Nucl. Instrum. & Meth., submitted (1990).
4. M. Rossbach, O. Schärpf, W. Kaiser, W. Graf, A. Schirmer, W. Faber, J. Duppich and R. Zeisler, Nucl. Instrum. & Meth. **B35**, 181-190 (1988).
5. J.R.D. Copley and C.F. Majkrzak, S.P.I.E. Proc. **983**, 93-104 (1989).

BORON NEUTRON CAPTURE THERAPY

F. Stecher-Rasmussen[1], R.L. Moss[2] and M.W. Konijnenberg[3]

[1]Netherlands Energy Research Foundation ECN, Petten, The Netherlands
[2]Commission of the European Communities, Joint Research Centre, Petten, The Netherlands
[3]Netherlands Cancer Institute, Amsterdam, The Netherlands

Boron Neutron Capture Therapy is a cancer therapy based on selective uptake of boronated compounds in tumorous tissue and subsequent irradiation with neutrons. Through the $^{10}B(n,\alpha)^7Li$ reaction the capture of thermal neutrons results in the emission of densely ionizing α- and Li-particles, thus releasing a cell-destructing field over the volume of the malignant tissue. Within the frame of a European Concerted Action Project an intermediate energy neutron filter has been installed in the Large Neutron Facility[1] of the High Flux Reactor in Petten. It is the aim of the European Collaboration to arrive at the first clinical trials on brain tumours using neutrons from this facility in a couple of years.

Apart from the $^{10}B(n,\alpha)$ reaction the intermediate energy neutrons interact with the brain matter through neutron scattering, the (n,γ) process and the (n,p) process. All reaction products contribute to the total dose in the target. The neutron filter, which consists of Cd, Al, Ti, S and liquid Ar, has been optimized to yield a neutron beam which produces a tolerable dose in healthy tissue and a lethal dose in cancer tissue. Monte Carlo transport calculations, tracking neutrons and photons from the reactor core through the filter configuration into a tumourized brain model, demonstrate the dose gain in the tumour area.

To simulate a therapy situation a headlike phantom has been computer modelled. An intermediate-energy neutron beam from the therapy facility, as resulting from Monte Carlo calculations (MCNP) of the filter configuration, is impinging on the phantom. By neutron transport calculations (DOT 3.5), taking into account all relevant nuclear reactions in the phantom, the dose distributions resulting from the interaction between phantom and beam are calculated. These calculations do not only give insight into the contribution to the total dose from the various reactions - and the therapy beam itself - but also allow for feedback to the beam characteristics, thus yielding information about the optimal parameters of the beam filter.

A tumour model has been included in the calculations and the results, an example hereof is given in fig. 1, clearly show the boron concentration dependent dose enhancement within the tumour volume. The model in fig. 1 includes 3 ppm of ^{10}B in normal tissue and 30 ppm in the tumour.

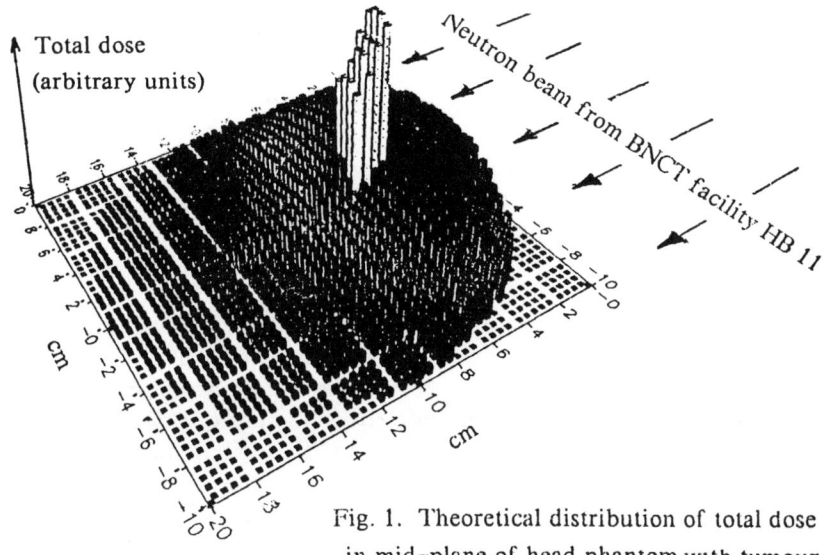

Fig. 1. Theoretical distribution of total dose in mid-plane of head phantom with tumour.

A physical phantom is under construction. This phantom will serve two purposes: i/ verification of the calculations and ii/ the study of the biological effects of the Petten BNCT beam. Extensive dose mapping using as well physical metrology techniques as biological dosimeters (cell cultures).

To allow for comparison with phantom experiments at other centres recommendations for phantom standards are discussed.
The calculational model and the experimental phantom should finally merge into a basis for development of an integrated treatment planning system capable of handling all relevant parameters.

The determination of small concentrations of boron (1 to 50 ppm) is important for the pre-clinical and for the clinical studies. For this purpose a capture gamma-ray spectroscopy facility with low background has been developed at a thermal neutron beam of the HFR.

Finally, the use of capture gamma-ray spectroscopy in the dose monitoring system at the clinical facility will be discussed.

---

[1]F. Stecher-Rasmussen, in "Proc. 5th Int. Symp. on Capture Gamma-Ray Spectroscopy and Related Topics", Knoxville, USA, Sept. 10-14, 1984 (AIP, Conf. Proc. No. 125, 1985), p. 933.

DESIGN AND APPLICATION OF IN SITU PROMPT-GAMMA
PROBE AS A SALINOMETER

Jiunn-Hsing CHAO and Chien CHUNG

Institute of Nuclear Science, National Tsing Hua University,
Hsinchu 30043, Taiwan, Republic of China

ABSTRACT

In situ survey of river salinity using a portable neutron-gamma probe was conducted at northern Taiwan. The probe, consisting of a 40 μg ^{252}Cf neutron source and a HPGe detector, was mounted on board a fishing boat to survey the salinity at river delta. The variations of water salinity along the river were mapped according to the tidal cycle. Detecting range, radiation doses, characteristics of the in situ prompt gamma spectrum, and its potential application as a salinometer are discussed.

INTRODUCTION

Application of in situ survey using neutron-capture prompt gamma-ray activation analysis(PGAA) technique to water pollutant monitoring has been investigated in our previous studies[1,2]. In this work, we describe the feasibility of this techniqe employed in rapid measurement of water salinity. The salinity, defined as total dissolved salt in water, can be measured in terms of chlorine concentration by detecting the 6111 keV photopeak, one of major chlorine prompt gamma rays. A HPGe-^{252}Cf probe, assembled and mounted on a fishing boat to survey the salinity of Tamsui River at northern Taiwan, provides an alternative method for the in situ measurement of water salinity in field operation.

APPLICATION

The HPGe-^{252}Cf probe, in connection with a gamma-ray spectroscopic system, is assembled and mounted on a fishing boat, which is propelled by a 18 hp outboard engine for in situ operation from Tamsui River mouth all the way upstream until water salinity below the detection limit of the probe. Neutron source locked in a bamboo bar was submerged in water from the port side while the HPGe detector was positioned on board right above the source, as shown in Fig. 1. The

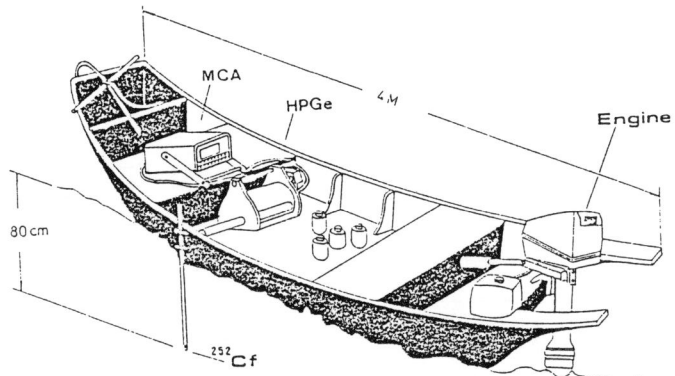

Fig. 1. In situ PGAA setup on board a fishing boat.

portable HPGe detector with 2-liter liquid nitrogen in dewar enables it to operate for 24 hours continuously. A portable CANBERRA-10 PLUS multichannel analyzer(MCA) with built-in amplifier, high-voltage power supply, and rechargable Ni-Cd batteries, is used in connection with the HPGe detector. The in situ survey began at river mouth all the way upstream to map the variations of salinity in river water at different sampling stations. The river water in each measurement was also sampled in 1-liter plastic bottle and subject to neutron activation analysis(NAA) using 1 MW Tsing Hua Open-pool Reactor(THOR) facility.

RESULTS AND DISCUSSION

The salinity of river water varies from that of fresh water to normal marine content, depending on the rate of freshwater from inland and the rate of tidal movement. In situ PGAA survey began at the center along the river, carried out 500 m from the river mouth to 24 km inland. The maximum and minimum salinities at these locations were measured and recorded. The highest salinity found to be 27.4 ‰ at river mouth and lowest at 0.08 ‰ at 24 km inland. Difference between maximum and minimum salinity becomes significantly large at 6 km inland, this is due primarily to the local discharge of fresh water and sewage, further reducing the salt content at low tide measurement. In Fig. 2, variation of salinity according to the tidal movement is plotted at 6 km inland at fishing port based on rapid 10-min in situ measurements. The effective detecting range, as described in our previous work[2], is calculated to be 105 cm depth with 75 cm radius around the submerged neutron source, totaling 1.86 tons of water. Therefore, more represntative concentration of salinity can be obtained by the in situ PGAA than other field techniques.

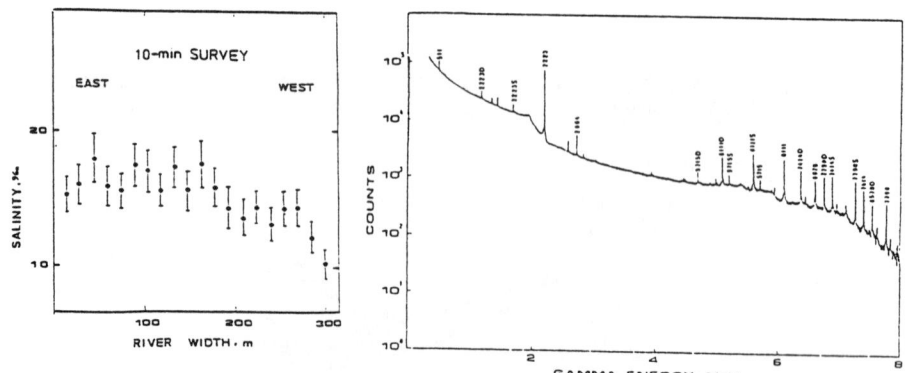

Fig. 2. Variation of salinity at 6 km inland as a function of tidal cycle.

Fig. 3. Prompt gamma-ray spectrum taking in situ at Tamsui River. Counting time is one hour; 1.5 km inland.

Almost all neutrons from ^{252}Cf source are thermalized and confined in water near the source, little neutron dose contributes to the workers on board. The neutron-induced prompt gamma rays, mainly the medium energy 2223 keV from H(n,γ) and high energy chlorine prompt gamma rays, as shown in Fig. 3, inducing only 0.4 mR/hr around the detector, and further reducing to 0.15 mR/hr one meter from the detector on board.

CONCLUSION

Many of the densely populated areas in Taiwan are located near estuary at coastal plain, where municipal and industrial wastewater discharge into the river and alter composition of salt water. This in situ PGAA technique is served as an attractive alternative to traditional salinometer, which in turn cannot resolve sodium chloride from a wide variety soluble metal salts in water discharged from factories. Some pollutants with detectable prompt gamma rays of high sensitivity can be identified simultaneously by measuring their characteristic high-energy prompt gamma rays.

REFERENCES

1. C. Chung and T. C. Tseng, Nucl. Instr. and Meth. A267 (1988)223.
2. J. H. Chao and C. Chung, Nucl. Instr. and Meth. in press(1990).

PHOTOEXCITATION OF ^{50}Cr, ^{53}Cr AND THE DEBYE TEMPERATURE OF CHROMIUM

D. LEVANT*, R. MOREH*,⁺ AND O. SHAHAL*
*Ben-Gurion University of the Negev, Beer-Sheva 84105, Israel and
⁺Nuclear Research Center - Negev, Beer-Sheva 84120, Israel.

ABSTRACT

The random photoexcitation process of n-capture γ-rays was used for studying the decay properties of the 8888 keV level in ^{50}Cr and a level at 7646 keV in ^{53}Cr. The γ source was obtained from the Fe(n,γ) reaction. The scattering cross section σ_s from the 8888-keV level was measured in the range 297 K - 10 K and the Debye temperature of metallic chromium was found to be $\Theta = 510 \pm 20$ K. The result is compared with that obtained by other methods.

Photon beams based on thermal n-capture were used for several diverse studies in nuclear and condensed matter physics[1]. Here we used the random photoexcitation method to study the decay of the 8888 keV level[2] in ^{50}Cr and another resonance level in ^{53}Cr at 7646 keV. In addition, we measured the cross section σ_s of the scattered 8888 keV line as a function of temperature in order to determine the Debye temperature Θ of metallic chromium.

The present work was motivated by the fact that the values of Θ for metallic Cr, reported in the literature[3,4] differ markedly and are in the range 400 K to 670 K. A precise knowledge of Θ is important because any uncertainty in Θ may introduce larger uncertainties in resonance scattering measurements in which metallic Cr is used as a γ source[5].

The possibility of measuring Θ by the present method depends on the fact that the energy separation δ between the incident γ line and the nuclear level happens to be large, namely, $\delta = 18$ eV and larger than the Doppler widths Δ_s and Δ_r of the γ line and the resonance level[6]. In addition, Δ_r is much larger than Γ. These are the two requirements for a precise measurement of Θ.

Experimentally, the γ source was obtained from the (n,γ) reaction on four metallic iron discs placed along a tangential beam tube and near the core of the IRR-2 reactor[2]. The intensities of the 8888 keV and the 7646 keV lines were around 10^5 and 10^6 photons/cm^2.s on the target position. The resonant scattered spectrum (Fig. 1) shows elastic and inelastic lines scattered by ^{50}Cr and ^{53}Cr (having natural abundances of 4.3% and 9.5%). The identification of ^{50}Cr was made by using a separated isotope[2] while ^{53}Cr was identified from the inelastic lines. Fig. 2 shows a decay scheme of the two levels constructed from the known excited levels in ^{50}Cr and ^{53}Cr.

To get good statistics in a relatively short running time a 12.5 cm x 12.5 cm NaI detector was used for measuring the scattered intensity versus T. A 160 gm rectangular metallic sample (7.5 cm x 4 cm) was placed perpendicular to the beam and inside a Displex cryostat which varied the temperature of the target between 12 K to 297 K. The geometry of target-detector-beam system is important

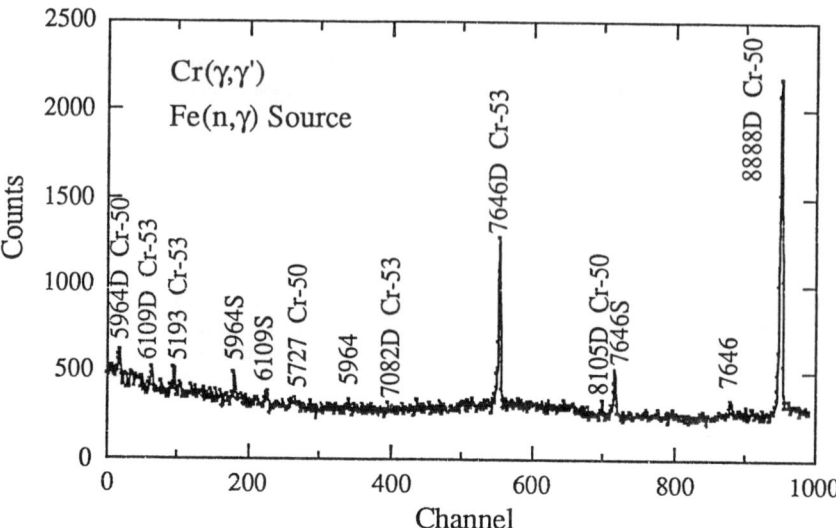

Fig. 1. Scattered γ-spectrum from a 10 g/cm² thick Cr target as measured by a 30 cm³ GeLi detector placed 25 cm away from the target and making an angle of 130° with respect to the beam. D and S denote double and single escape peaks respectively. A Fe(n,γ) source was used. The scattering isotopes are indicated.

because of nuclear self-absorption in the target. The running time per data point was about 12h.

Fig. 3 shows the measured scattered intensities of 8888 keV line versus T relative to 297 K. Each data point was obtained by integrating the counts in the energy region 7.8 MeV to 9.0 MeV which is due to ^{50}Cr only. The background was subtracted using an Fe sample of identical geometry to the Cr sample.

The value of Θ was determined by searching for the values of both δ and Θ which best fitted the data points of Fig. 3. The calculations involved were explained in detail in Ref. 6; the relevant parameters are given in the caption to figure 3 (see also Ref. 2). The results for ^{50}Cr, are: Θ = 520 ± 20 K; δ = 17.0 ± 0.5 eV. Applying a $M^{-1/2}$ correction for atomic mass, we obtain for metallic Cr of natural abundance: Θ = 510 ± 20 K. It should be noted that this δ is more precise than δ = 18 ± 1 eV, reported previously[2]. This Θ is higher by ≈ 10% than Θ = 464 K obtained by integrating over the phonon spectrum[7] of metallic Cr and conforms to the behaviour of Θ of other metallic scatterers as deduced by the present technique[8]. This could probably be due to the fact that

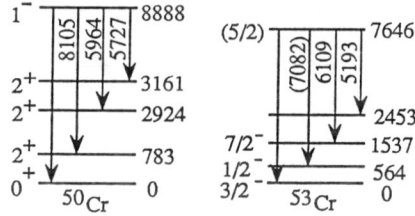

Fig. 2. Decay scheme of the 8888 keV level of ^{50}Cr and the 7646 keV level of ^{53}Cr.

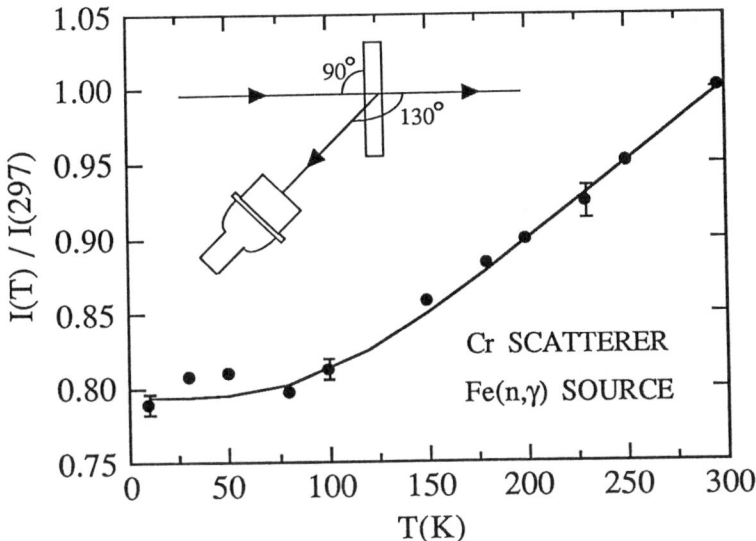

Fig. 3. Scattered intensities versus T (relative to 297 K) from metallic Cr (5.3 g/cm² thick). The solid line represents the best fit calculated curve obtained using δ = 17 eV, Γ = 0.75 eV, Γ_0 = 0.68 eV, Δ_s = 11.75 eV and Θ = 520 K for metallic ^{50}Cr; the target-detector geometry is shown. Δ_s corresponds to an effective temperature, T_s = 519 K, obtained using an iron source temperature of T = 498 K with Θ = 460 K. Typical errors are indicated.

with the present technique one can in principle view all vibrations of the lattice while with the inelastic neutron scattering from which the phonon spectra are deduced, some vibrational modes occuring at the high energy end could be missed.

This work was supported by the US - Israel Binational Science Foundation (BSF), Jerusalem, Israel.

REFERENCES

1. R. Moreh, in Neutron-Capture Gamma-Ray Spectroscopy, (Inst. of Physics Conf. Ser. No. 62, Bristol and London, 1982) p. 575.
2. R. Moreh, S. Shlomo and A. Wolf, Phys. Rev. C2, 1144 (1970).
3. E.S.R. Gopal, in Specific Heats at low Temperatures: Plenum, N.Y. 1966, p. 33.
4. K. A. Gschneider, Sol. St. Phys. 16, 275 (1964).
5. R. Moreh, O. Shahal, Surf. Science 177, L963 (1986); Phys. Rev. B40, 1926 (1989).
6. R. Moreh, O. Shahal and I. Jacob, Nucl. Phys. A228, 77 (1974).
7. W. Kress, in Landolt-Bornstein, New Series, Group III, 13B, 174 (1973).
8. R. Moreh (see contribution to this Proceedings).

ELEMENTAL ANALYSIS OF COAL BASED ON SPECTROMETRY OF PROMPT GAMMA RAYS FROM NEUTRON INDUCED REACTIONS

S.Pospíšil, Z.Janout, J.Koníček
Czech Technical University
Faculty of Nuclear Science and Physical Engineering
115 19 Praha 1, Břehová 7, Czechoslovakia
M.Vobecký
Czechoslovak Academy of Sciences
Institute of Nuclear Biology and Radiochemistry
142 20 Praha 4, Vídeňská 1083, Czechoslovakia

ABSTRACT

The non-destructive determination of technologically important elements (Fe,S,Si,C) in large-scale sample of coal, based on radiative capture and inelastic scattering of neutrons, is described. The standard sample was represented by amount of bulk coal in four 280x280x140 mm^3 polyethylene vessels. The total sample mass was about 35 kg. For the measurement based on radiative capture, the ^{252}Cf source with total emission of 2.4x10^7 s^{-1} located in D$_2$O moderator was used. The ^{241}AmBe source with total emission of 2.4x10^7 s^{-1} was used in case of inelastic scattering. The sample surrounded the neutron source. From outside, the sample was surrounded by a layer of D$_2$O. Heavy water was in an aluminium vessel. A Ge(Li) or BGO detector was used. The detector was located under the neutron source and from the direct source radiation was shielded by a bismuth bar. The source-sample-detector setup thus had a geometry near cylindrical. The supporting structure was made of wood. In order to provide calibration characteristics of the analyzer, reference large-scale brown and bituminous coal samples were prepared by means of chemical analysis. Glass sand was chosen as a material free of iron, sulphur and carbon. On the basis of the known parameters of the Ge (Li) detector and cross sections and gamma-ray intensities for individual nuclides, the tables of the expected experimental analytical sensitivities for particular elements, contained in a different types of coal, were prepared. The main analytical lines were selected as follows: Fe (7645 and 7631 keV), S (5421 keV), Si (4935 and 3539 keV) from (n,gamma); Si (1779 keV), C (4438 keV) from (n,n'gamma). In exposure times of up to 1h, the detection limits of 0.34 % and 0.64 % and accuracies of 0.25 % and 0.40 % were achieved in case iron and sulphur, respectively. For silicon determination, in exposure times of up to 2h, the detection limits of 1.3 % and 0.9 % and accuracies of 0.66 % and 0.40 % were achieved in case of (n,gamma) and (n,n'gamma) processes, respectively. In case of carbon determination, an exponential curve was analytical response to the carbon content. In exposure times of up to 1h, the detection limit was found 3.3 %, accuracy of carbon determination at 40 % carbon content was estimated 4.5 %.

Participants

Dr. Kees Abrahams
FOM-ECN
Nuclear Structure Group
P.O. Box 1
NL-1755 ZG Petten
Netherlands

Prof. Y.K. Agarwal
Tata Inst. of Fundamental Research
Homi Bhabha Road
Colaba
Bombay, 400 005
India

Dr. Irshad Ahmad
Argonne National Laboratory
9700 S. Cass Avenue, D-203
Argonne, IL 60439

Dr. John D. Anderson
Lawrence Livermore Nat'l Laboratory
P.O. Box 808, L-295
Livermore, CA 94551

Dr. A. Aprahamian
University of Notre Dame
Department of Physics
Notre Dame, IN 46556

Prof. G. Ardisson
Universite de Nice
Institut de Physique Nucleaire
BP 1
F-06034 Nice Cedex
France

Dr. Martin K. Balodis
Latvian SSR Academy of Sciences
Physics Institute, Riga
229021 Salaspils 1
USSR

Dr. John A. Becker
Lawrence Livermore Nat'l Laboratory
P.O. Box 808, L-280
Livermore, CA 94551

Prof. Frantisek Becvár
Charles University
Mathematics and Physics
V. Holesovickách 2
CS18000 Prague 8
Czechoslovakia

Dr. Tamás Belgya
Hungarian Academy of Sciences
Institute of Isotopes
P.O. Box 77
H-1525 Budapest
Hungary

Dr. Yair Birenbaum
Nuclear Research Center-Negev
Physics Department, N.R.C.N.
P.O. Box 9001
Beer-Sheva, 84190
Israel

Dr. Marshall Blann
Lawrence Livermore Nat'l Laboratory
P.O. Box 808, L-289
Livermore, CA 94551

Dr. Stewart Bloom
Lawrence Livermore Nat'l Laboratory
P.O. Box 808, L-297
Livermore, CA 94551

Prof. Dik O. Boerma
University of Groningen
LAN & Mat'ls Science Center
Westersingel 34
NL-9718 DB Groningen
Netherlands

Dr. Paul Bonche
SPhT-CEN Saclay
F-91191 Gif sur Yvette
France

Dr. Charles D. Bowman
Los Alamos National Laboratory
Physics Division
MS H803
Los Alamos, NM 87545

Participants

Dr. Hans Börner
Institut Laue-Langevin
156X Centre de Tri
F-38042 Grenoble
France

Mr. M.J. Brinkman
Lawrence Livermore Nat'l Laboratory
P.O. Box 808, L-280
Livermore, CA 94551

Prof. Dennis G. Burke
McMaster University
Physics Department, GSB 105
1280 Main Street West
Hamilton, Ont. L85 4Kl
Canada

Dr. Thomas W. Burrows
Brookhaven National Laboratory
Nat'l Nuclear Data Center
Building 197D
Upton, NY 11973

Dr. Richard F. Casten
Brookhaven National Laboratory
Department of Physics
Upton, NY 11973

Dr. Richard R. Chasman
Argonne National Laboratory
9700 5. Cass Avenue
Argonne, IL 60439

Dr. Robert E. Chrien
Brookhaven National Laboratory
Department of Physics
510A
Upton, NY 11973

Prof. Jolie A. Cizewski
Rutgers University
Department of Physics
P.O. Box 849
Piscataway, NJ 08855

Prof. David D. Clark
Cornell University
Ward Laboratory
Ithaca, NY 14853

Dr. Francesco Corvi
CEC-JRC
Central Bureau for Nuclear Meas.
Steenweg naar Retie
B-2440 Geel
Belgium

Prof. Paul D. Cottle
Florida State University
Department of Physics
Tallahassee, FL 32306

Mr. Roderick Dayton
Bicron Corporation
12345 Kinsman Road
Newbury, OH 44065

Dr. Caroline De Coster
University of Gent
Inst. for Nuclear Physics
Proeftuinstraat 86
B-9000 Gent
Belgium

Dr. Daniel J. Decman
Lawrence Livermore Nat'l Laboratory
P.O. Box 808, L-396
Livermore, CA 94551

Dr. Maynard S. Dewey
National Institute of
Standards and Technology
Quantum Metrology Group
Building 221, Room A-141
Gaithersburg, MD 20899

Dr. Frank S. Dietrich
Lawrence Livermore Nat'l Laboratory
P.O. Box 808, L-289
Livermore, CA 94551

Participants

Prof. J.P. Draayer
Louisiana State University
Department of Physics
Baton Rouge, LA 70803

Dr. Takashi Emoto
University of Texas
Nuclear Engineering Teaching Lab.
Balcones Research Center
Austin, TX 78758

Dr. Gerald Feldman
Duke University and TUNL
Department of Physics
Durham, NC 27706

Prof. Gerald Fink
Ohio University
Department of Physics
Accelerator Laboratory
Athens, OH 45701

Dr. Hans Frisk
Lawrence Berkeley Laboratory
Nuclear Science Division
Berkeley, CA 94720

Dr. Donald G. Gardner
Lawrence Livermore Nat'l Laboratory
P.O. Box 808, L-234
Livermore, CA 94551

Dr. Maureen A. Gardner
Lawrence Livermore Nat'l Laboratory
P.O. Box 808, L-234
Livermore, CA 94551

Dr. Paul E. Garrett
McMaster University
Physics Department, GSB105
1280 Main Street West
Hamilton, Ont. L8S 4Kl, Canada

Dr. William Gelletly
Daresbury Laboratory
Science & Engineering Research Coun.
Daresbury, Warrington
Cheshire, WA4 4AD, UK

Dr. Ronald L. Gill
Brookhaven National Laboratory
Department of Chemistry
Building 510A
Upton, NY 11973

Dr. Joseph N. Ginocchio
Los Alamos National Laboratory
T-S, MS B-283
P.O. Box 1663
Los Alamos, NM 87545

Prof. Chris R. Gould
North Carolina State Univ.
Los Alamos National Laboratory
MS-H846
Los Alamos, NM 87545

Dr. Zhendi Guo
Rijksuniversiteit Utrecht
R.J. Van de Graaff Laboratory
Postbus 80.000
NL-3508 TA Utrecht
Netherlands

Dr. Luisa Hansen
Lawrence Livermore Nat'l Laboratory
P.O. Box 808, L-289
Livermore, California 94551

Prof. Hanns-Ludwig Harney
Max Planck Inst. Kernphysik
Postfach 10 39 80
D-6900 Heidelberg
Germany

Dr. Evans Hayward
National Institute of
Standards and Technology
Gaithersburg, MD 20899

Dr. Ralf-Dieter Heil
Justus-Liebig-Universitat Giessen
Institut fur Kernphysik
Leihgesterner Weg 217
D630 Giessen
Germany

Dr. Eugene A. Henry
Lawrence Livermore National Laboratory
P.O. Box 808, L-234
Livermore, CA 94551

Prof. Kris Heyde
University of Gent
Institute for Theoretical Physics
Proeftuinstraat 86
B-9000 Gent
Belgium

Prof. Yu-Kun Ho
Fudan University
Nuclear Science Department
Shanghai, China

Dr. Antje Hoering
Max Planck Institut Kernphysik
Postfach 10 39 80
D-6900 Heidelberg
Germany

Dr. Richard W. Hoff
Lawrence Livermore National Laboratory
P.O. Box 808, L-234
Livermore, CA 94551

Dr. W. Michael Howard
Lawrence Livermore Nat'l Laboratory
P.O. Box 808, L-297
Livermore, CA 94551

Dr. Friedrich Hoyler
University of Tuebingen
Department of Physics
Morgenstelle 14
D-7400 Tuebingen
Germany

Prof. Bo Höistad
Uppsala University
Department of Radiation Sciences
P.O. Box 535
S-75121 Uppsala
Sweden

Dr. Masayuki Igashira
Tokyo Institute of Technology
Research Lab. for Nuclear Reactors
2-12-10-0kayama, Meguro-ku
Tokyo, 152
Japan

Prof. Ashok Kumar Jain
University of Roorkee
Physics Department
Roorkee, 247667
India

Dr. Jan Jolie
Institut Laue-Langevin
Avenue des Martyrs
B.P. 156X
F-38042 Grenoble
France

Dr. Sylvian Kahane
Nuclear Research Center-Negev
Physics Department
P.O. Box 9001
Beer-Sheva, 84-190
Israel

Dr. Kamales Kar
Saha Institute of Nuclear Physics
TNP Division
92, A.P.C. Road
Calcutta, 700-009
India

Dr. Franz Käppeler
Kernforschung. Karlsruhe
Institut fur Kernphysik 1
Postfach 3640
D-7500 Karlsruhe 1
Germany

Prof. Juhani Keinonen
University of Helsinki
Accelerator Laboratory
Hameentie 100
SF-00550 Helsinki
Finland

Participants

Prof. Jean Kern
University of Fribourg
Physics Department
Perolles
CH-1700 Fribourg
Switzerland

Dr. Ernest G. Kessler
National Institute of
Standards and Technology
A141, Physics Building
Gaithersburg, MD 20899

Dr. Michael Z. Kirson
Weizmann Institute of Science
Nuclear Physics Department
Rehovot, 76100, Israel

Prof. Hideo Kitazawa
Tokyo Institute of Technology
Research Lab. for Nuclear Reactors
2-12-10-0kayama, Meguro-ku
Tokyo, 152
Japan

Prof. Dr. Hans V. Klapdor-Kleingrothaus
Max Planck Inst. Kernphysik
Postfach 10 39 80
D-6900 Heidelberg 1
Germany

Dr. Jorg Klora
TU Munchen
Physik-Dept. E-18
James-Franck-Str.
D-8046 Garching
Germany

Dr. Paul E. Koehler
Los Alamos National Laboratory
P17, MS-H803
Los Alamos, NM 87545

Prof. I.A. Kondurov
Leningrad Nuclear Physics Inst.
Gatchina
Leningrad, 188350
USSR

Dr. Jiri A. Kopecky
Netherlands Energy Research Found.
Physics Department
P.O. Box 1
NL-1755-ZG Petten
Netherlands

Prof. V. Krishna Brahmam Kota
Physical Research Laboratory
Ahmedabad, 380 009
India

Dr. David Krofcheck
Lawrence Livermore Nat'l Lab.
P.O. Box 808, L-280
Livermore, CA 94551

Dr. Bernd Krusche
Justus-Liebig-Universitaet Giessen
II Physikalisches Institut
Henrich-Buff-Ring 16
D-6300 Giessen
Germany

Dr. Andreas Kuhnert
Lawrence Livermore Nat'l Laboratory
Nuclear Chemistry Division
P.O. Box 808, L-396
Livermore, CA 94551

Dr. Jaana Kumpulainen
University of Jyväskylä
Department of Physics
Seminaarinkatu 15
SF-40100 Jyväskylä
Finland

Dr. Dimitri Kusnezov
Michigan State University
NSCL
East Lansing, MI 48824-1321

Ms. Ruth-Mary Larimer
Lawrence Berkeley Laboratory
Building 88
1 Cyclotron Road
Berkeley, CA 94720

Dr. Robert D. Lawson
Argonne National Laboratory
Building 314
9700 S. Cass Avenue
Argonne, IL 60439

Dr. K.T. Lesko
Lawrence Berkeley Laboratory
Building 88
Berkeley, CA 94720

Dr. Amiram Leviatan
Los Alamos National Laboratory
T-5, MS B283
Los Alamos, NM 87545

Prof. Dr. K. P. Lieb
University of Göttingen
II. Physik Institut
Bunsenstr. 7-9
D-3400 Gottingen
Germany

Dr. Helmut Lindner
TU Munchen
Physics Department, E18
James-Franck-Str.
D-8046 Garching
Germany

Dr. Ronald W. Lougheed
Lawrence Livermore Nat'l Laboratory
P.O. Box 808, L-232
Livermore, CA 94551

Mr. Daniel Love
Bicron Corporation
12345 Kinsman Road
Newbury, OH 44065

Dr. J.E. Lynn
Los Alamos National Laboratory
Neutron Scattering Center
LANSCE, MS H-805
Los Alamos, NM 87545

Dr. Douglas R. Manatt
Lawrence Livermore Nat'l Laboratory
Nuclear Chemistry Division
P.O. Box 808, L-233
Livermore, CA 94551

Dr. Lloyd G. Mann
Lawrence Livermore Nat'l Laboratory
P.O. Box 808, L-233
Livermore, CA 94551

Dr. Grant J. Mathews
Lawrence Livermore Nat'l Laboratory
P.O. Box 808, L-405
Livermore, CA 94551

Mr. Gary Mattesich
Canberra Nuclear Products Group
Golf and Meacham Roads
Schaumberg, IL 60196

Dr. Ulrich Mayerhofer
TU Munchen
Physics Department, E18
James-Franck-Str. 1
D-8046 Garching
Germany

Dr. David F.R. Mildner
National Institute of
Standards and Technology
235/B108
Gaithersburg, MD 20899

Dr. Gabor Molnár
Hungarian Academy of Sciences
Institute of Isotopes
P.O. Box 77
H-1525 Budapest
Hungary

Prof. Alfonso Mondragón
UNAM
Instituto de Fisica
Apartado Postal 20-364
Mexico, D.F., 01000
Mexico

Participants

Dr. Michael S. Moore
Los Alamos National Laboratory
P-15, MS D-406
Los Alamos, NM 87545

Prof. R. Moreh
Ben Gurion University
Department of Physics
P.O. Box 653
Beer-Sheva, 84-105
Israel

Prof. Steven A. Moszkowski
University of California
Department of Physics
405 Hilgard Avenue
Los Angeles, CA 90024

Dr. Naoki Mukai
Tokyo Institute of Technology
Research Lab. for Nuclear Reactors
2-12-10-0kayama, Meguro-ku
Tokyo, 152
Japan

Dr. M.G. Mustafa
Lawrence Livermore Nat'l Laboratory
P.O. Box 808, L-234
Livermore, CA 94551

Dr. Andreas Müller
Max Planck Inst. Kernphysik
Postfach 10 39 80
D-6900 Heidelberg
Germany

Prof. Yasuki Nagai
Tokyo Institute of Technology
Department of Applied Physics
Oh-okayama 2-12-1, Meguro
Tokyo, 152
Japan

Dr. M.N. Namboodiri
Lawrence Livermore National Laboratory
P.O. Box 808, L-234
Livermore, CA 94551

Prof. R.A. Naumann
Princeton University
Department of Physics
Jadwin Hall
Princeton, NJ 08544

Dr. Ronald O. Nelson
Los Alamos National Laboratory
MS H803
Los Alamos, NM 87545

Dr. Eric B. Norman
Lawrence Berkeley Laboratory
1 Cyclotron Road
Building 88
Berkeley, CA 94720

Dr. H.A. O'Brien
Los Alamos National Laboratory
P.O. Box 1663, MS-D449
Los Alamos, NM 87545

Prof. Pavel Oblozinsky
Slovak Academy of Sciences
Electro-Physical Research Centre
Dubravska Cesta 9
CS-84228 Bratislava
Czechoslovakia

Prof. Takaharu Otsuka
University of Tokyo
Department of Physics
7-3-1 Hongo Bunkyo-ku
Tokyo, 113
Japan

Prof. Vladimir Paar
University of Zagreb
Theoretical Physics Department
Marvlicev TRG 19
YU-41000 Zagreb
Yugoslavia

Prof. W.R. Phillips
University of Manchester
Department of Physics
Oxford Road
Manchester, M13 9PL, UK

Prof. Yu. P. Popov
Joint Institute for Nuclear Research
Laboratory of Neutron Physics
Head Post Office, Box 79
Moscow, 101000, USSR

Dr. Carl H. Poppe
Lawrence Livermore Nat'l Laboratory
Nuclear Chemistry Division
P.O. Box 808, L-231
Livermore, CA 94551

Prof. Peter T. Prokofjev
Latvian SSR Academy of Sciences
Institute of Physics, Riga
Salaspils 1, 229021
USSR

Prof. Philippe Quentin
Universite de Bordeaux I
Laboratory de Physique Theorique
Rue du Solarium
F-33170 Gradignan
France

Dr. S. (Ram) Raman
Oak Ridge Nat'l Laboratory
Building 6010
P.O. Box 2008
Oak Ridge, TN 37831

Dr. Sathyavathi Ramavataram
Brookhaven National Laboratory
Dept. of Nuclear Energy
Building 197D
Upton, NY 11973

Prof. Gianluca Raselli
University of Pavia
Nuclear Physics Dept.
via A. Bassi 6
27100 Pavia, Italy

Prof. John O. Rasmussen
Lawrence Berkeley Laboratory
1 Cyclotron Road
MS 70A-3307
Berkeley, CA 94720

Dr. Robert C. Reedy
Los Alamos National Laboratory
Group SST-8
MS D-438
Los Alamos, NM 87545

Dr. Gianni Reffo
ENEA
C.R.E. "E. Clementel"
V. Le Ercolani 8
I-40138 Bologna
Italy

Dr. C.W. Reich
EG&G Idaho, Inc.
P.O. Box 1625
Idaho Falls, ID 83415

Dr. L. Robinson
Oak Ridge Nat'l Laboratory
Martin Marietta Energy Systems
P.O. Box 2008
Oak Ridge, TN 37831

Dr. Steve J. Robinson
Institut Laue-Langevin
Avenue des Martyrs
B.P. 156X
F-38042 Grenoble
France

Dr. Gert Rohr
CBNM-Euratom
CEC/JRC
Steenweg naar Retie
B-2440 Geel
Belgium

Dr. A.P. Serebrov
Leningrad Nuclear Physics Inst.
Gatchina
Leningrad, 188350
USSR

Dr. E.I. Sharapov
Joint Institute for Nuclear Research
Laboratory of Neutron Physics
Dubna, 141980, USSR

Prof. R.K. Sheline
Florida State University
Department of Chemistry
Nuclear Research Building
Tallahassee, FL 32306

Dr. Adnan Shihab-Eldin
Lawrence Berkeley Laboratory
1 Cyclotron Road
MS 70A-3307
Berkeley, CA 94720

Prof. John F. Shriner
Tennessee Technological Univ.
Department of Physics
P.O. Box 5051
Cookeville, TN 38505

Dr. Kornelius Sistemich
Forschungszentrum Julich GmbH
Institut fur Kernphysik
Postfach 19 13
D-5170 Julich 1, Germany

Prof. Vadim G. Soloviev
Joint Institute for Nuclear Research
Laboratory of Theoretical Physics
Head Post Office, Box 79
Moscow, 101000, USSR

Prof. Brian M. Spicer
University of Melbourne
School of Physics
Parkville
Victoria, 3052, Australia

Dr. Finn Stecher-Rasmussen
Netherlands Energy Research Found.
ECN
P.O. Box 1
NL-1755 ZG Petten
Netherlands

Prof. Craig A. Stone
San Jose State University
Dept. of Chemistry
1 Washington Square
San Jose, CA 95192-0163

Dr. Chavdar H. Stoyanov
Inst. Nuclear Res. & Nuclear Energy
Boul. Lenin 72
1784 Sofia
Bulgaria

Mr. Mark A. Stoyer
Lawrence Berkeley Laboratory
Nuclear Science Division
Building 70A-3307
Berkeley, CA 94720

Dr. Gordon L. Struble
Lawrence Livermore Nat'l Laboratory
Nuclear Chemistry Division
P.O. Box 808, L-233
Livermore, CA 94551

Mr. Bhaskar Sur
Lawrence Berkeley Laboratory
Building 88
1 Cyclotron Road
Berkeley, CA 94720

Prof. D. Ronald Tilley
North Carolina State University
Department of Physics
P.O. Box 8202
Raleigh, NC 27695

Dr. Kenneth S. Toth
Oak Ridge Nat'l Laboratory
P.O. Box 2008, MS 6371
Building 6000
Oak Ridge, TN 37831

Prof. Fumikatsu Uesawa
Tokyo Institute of Technology
Research Lab. for Nuclear Reactors
2-12-10-0kayama, Meguro-ku
Tokyo, 152, Japan

Prof. Dr. Mario Uhl
University of Vienna
Inst. f. Radiumforschung/Kernphysik
Boltzmannstrasse 3
A-1090 Vienna, Austria

Dr. Pieter H.M. Van Assche
SCK/CEN
Nuclear Energy Centre
Mol and Leuven University
B-2400 Mol
Belgium

Prof. Cornelius Van der Leun
Rijksuniversiteit Utrecht
R.J. Van de Graaff Laboratory
Postbus 80.000
NL-3508 TA Utrecht
Netherlands

Dr. Jeffrey R. Vanhoy
U. 5. Naval Academy
Department of Physics
Annapolis, MD 21402

Dr. Michel Vergnes
Institut de Physique Nudeaire
F-91406 Orsay
France

Prof. P. von Brentano
Universitat Koln
Institut fur Kernphysik
Zulpicher Strasse 77
D-5000 Koln 41
Germany

Prof. Dr. Till von Egidy
TU Munchen
Physics Department, E18
D-8046 Garching
Germany

Dr. Tzu-Fang Wang
Lawrence Livermore Nat'l Laboratory
P.O. Box 808, L-397
Livermore, CA 94551

Dr. Ernest Warburton
Brookhaven National Laboratory
Department of Physics
Upton, NY 11973

Dr. Richard A. Ward
Lawrence Livermore Nat'l Laboratory
P.O. Box 808, L-023
Livermore, CA 94551

Dr. David D. Warner
SERC Daresbury Laboratory
Daresbury
Warrington, WA4 4AD
UK

Ms. Kristin Wedding
Lawrence Berkeley Laboratory
Building 88
1 Cyclotron Road
Berkeley, CA 94720

Dr. Morton 5. Weiss
Lawrence Livermore Nat'l Laboratory
P.O. Box 808, L-297
Livermore, CA 94551

Prof. Henry R. Weller
Duke University and TUNL
Physics Department
Durham, NC 27706

Dr. Steve A. Wender
Los Alamos National Laboratory
Physics Department
MS H803
Los Alamos, NM 87545

Dr. Christopher Wesselborg
Brookhaven National Laboratory
Physics Department 510A
Upton, NY 11973

Prof. Donald H. White
Western Oregon State College
Department Natural Sciences
345 N. Monmouth
Monmouth, OR 97361

Dr. Roger M. White
Lawrence Livermore Nat'l Laboratory
P.O. Box 808, L-298
Livermore, CA 94551

Prof. Michael C.F. Wiescher
University of Notre Dame
Department of Physics
Notre Dame, IN 46556

Dr. John F. Wild
Lawrence Livermore Nat'l Laboratory
P.O. Box 808, L-232
Livermore, CA 94551

Prof. Fred K. Wohn
Iowa State University
Department of Physics
Ames, IA 50011

Dr. Alex Wolf
Nuclear Research Center-Negev
Physics Department
P.O. Box 9001
Beer-Sheva, 84-190
Israel

Prof. John L. Wood
Georgia Institute of Technology
School of Physics
Atlanta, GA 30332

Prof. Steven W. Yates
University of Kentucky
Department of Chemistry
Lexington, KY 40506

Author Index

(Speakers are identified by bold type)

A

Abrahams, K., 764
Achiha, Y., 627
Adams, G. S., 733
Adarkar, A., 165
Afanasjev, A. V., 253, 446
Ahmad, I., 339
Alburger, D. E., 909
Alderliesten, C., 404
Alfimenkov, V. P., 787, 811
Ando, T., 900
Andrzejewski, J., 636, 808
Antonov, A. D., 808
Anze, H., 624
Ardisson, G., 470
Azaiez, F., 523
Aziz, S. M., 437

B

Baader, H. A., 421
Babeliowsky, T., 906
Balbes, M. J., 587
Baldo-Ceolin, M., 805
Balodis, M. K., 253, 263, 446
Barfield, A. F., 134
Barnes, F. L., 549
Barreau, G., 421
Barrett, B. R., 134
Baum, E. M., 434
Bazan, G., 827
Beausang, C. W., 546
Becker, J. A., 523, 546, 549
Becvar, F., 287
Beer, H., 850
Beer, W., 263
Beitins, M. R., 494
Belgya, T., 434
Benetti, P., 805
Berrier, G., 461
Bingham, C., 473
Birenbaum, Y., 620

Bitter, T., 805
Blann, M., 723
Block, R. C., 953
Bloom, S. D., 122
Bobisut, F., 805
Boerma, D. O., 972
Bogdanović, M., 421, 458
Bolse, W., 983
Bonche, P., 511
Bondarenko, V. A., 249, 491
Boneva, S. T., 485, 488, 494
Booten, J. G. L., 407
Börner, H. G., 173, 189, 199, 257, 396, 407, 421, 440, 455, 476
Borzakov, S. B., 811
Bowman, C. D., 747, 756
Bowman, J. D., 747
Brant, S., 150, 263, 425, 672
Breitig, D., 421
Brenner, D. S., 418
Brillard, L., 470
Brinkman, M. J., 523, 546, 549
Brissot, R., 421
Brown, J. D., 437
Buccino, S. G., 437
Burke, D. G., 279
Büscher, M., 367
Bush, J. E., 747

C

Calligarich, E., 805
Casten, R. F., 67, 140, 173, 263, 418
Cejnar, P., 287
Chadraabal, I., 636
Chalupka, A., 263
Chao, J.-H., 1008
Chasman, R. R., 339, 499
Chiba, S., 579
Chrien, R. E., 287
Chung, C., 1008
Cizewski, J. A., 173, 323, 523, 546

C

Clark, D. D., 936
Cline, D., 473
Colvin, G., 476
Copnell, J., 210
Corts, T., 983
Corvi, F., 906
Cottle, P. D., 98, 437
Coveca, C., 263
Cowan, J. J., 827
Cranmer, D. C., 929

D

Dahlgren, S., 733
Dalmasso, J., 470
Danon, Y., 953
Davis, E. D., 797
De Coster, C., 3, 105
De Sutter, B., 367
Decman, D. J., 415
Degener, A., 479
Deleplanque, M. A., 523, 546
Delheij, P. P. J., 747
Deloncle, I., 55
Deslattes, R. D., 173, 199
Dewey, M. S., 173, 199, 396, 774
Diamond, R. M., 523, 546
Dietrich, F. S., 742
Ding, D., 630
Dobaczewski, J., 511
Dodge, W. R., 595
Dolfini, R., 805
Doll, P., 617
Dombrádi, Zs., 150, 425
Dousse, J.-Cl., 263
Draayer, J. P., 30
Drake, D. M., 697
Draper, J. E., 523, 546
Drosg, M., 697
Dubbers, D., 805
Durham, F. E., 437

E

Eder, R., 263
Eisert, F., 805
El-Muzeini, P., 805
Endt, P. M., 404
Erozolimskii, B. G., 814

F

Faust, H., 421
Fazekas, B., 434
Feldman, G., 587, 595
Fényes, T., 425
Fink, G., 617
Fioni, G., 906
Flocard, H., 511
Föhl, K., 257
Ford, C. E., 431
Fortier, S., 461
Frankle, C. M., 747
Friedman, A. M., 339
Friedrichs, H., 479
Frisk, H., 665
Fukuda, T., 900, 903

G

Gardner, D. G., 263, 882, 914
Gardner, M. A., 263, 882, 914
Garrett, P. E., 279
Gatenby, R. A., 434
Gelletly, W., 308
Genoni, M., 805
Giacobbe, P., 263
Gibin, D., 805
Gigli Berzolari, A., 805
Gill, R. L., 263, 375, 418
Ginocchio, J. N., 82, 137
Gledenov, Yu. M., 636, 808
Gobrecht, K., 805
Goel, A., 153, 162
Görres, J., 840
Gould, C. R., 747, 756
Greene, G. L., 173, 199
Gudkov, V. P., 787
Guenther, P. T., 579
Guglielmi, A., 805
Guidotti, R., 639
Guidry, M. W., 473
Guo, Z., 404
Guseva, T. V., 253

H

Haase, D. G., 747
Hagn, E., 263
Haight, R. C., 697

Halbert, M. L., 473
Han, X. L., 473
Harney, H. L., 797
Hauber, S., 617
Haupenthal, M., 617
Hayter, J. B., 923
Hayward, E., 595
Heenen, P. H., 511
Heil, R. D., 234, 479
Helmer, K., 473
Henry, E. A., 523, 546, 549
Henry, R. G., 323
Hensley, D., 473
Herman, M., 742
Hernández, E., 604
Heyde, K., 3, 105, 143
Hill, J. C., 418
Hiller, H., 455, 464, 467
Ho, L., 630
Ho, Y. -K., 614, 705
Ho Bom, L., 636
Hoff, R. W., 159, 263, 476, 953
Höistad, B., 733
Holcomb, J. W., 437
Honma, M., 543
Horiguchi, M., 624
Höring, A., 742
Hoyler, F., 257, 412, 476
Huang, Z., 630
Hungerford, M. A., 263
Hussonnois, M., 470

I

Igashira, M., 624, 627, 900, 903
Isaksson, S., 733

J

Jacobsen, E. R., 437
Jain, A. K., 153, 156, 159, 162
Janout, Z., 1014
Johnson, E. L., 434
Johnson, T. D., 437
Jolie, J., 143, 146, 173, 189, 210, 396, 412, 440
Jolos, R., 234
Judge, S., 440, 455
Julin, R., 428
Jung, A., 479

K

Kacsich, T., 983
Kahane, S., 620
Kalifa, J., 461
Kantele, J., 428
Käppeler, F., 850, 917
Kar, K., 909
Katriel, J., 168
Kavka, A. E., 473
Kehrel, A., 983
Keinonen, J., 383
Kemper, K. W., 437
Kern, J., 263
Kernan, W. J., 473
Kerr, S., 421
Kessler, E. G., 173, 199
Kessler, M., 805
Khang, P. D., 485, 491
Kharitonov, A. G., 787
Khiem, L. H., 249, 485, 491
Khitrov, V. A., 249, 482, 485, 488, 491, 494
Kholnov, Yu. V., 249, 485, 491
Khrykin, A. S., 811
Kincaid, R. W., 473
Kinkel, U., 805
Kirson, M. W., 82
Kitamura, S., 624
Kitazawa, H., 624, 627, 900, 903
Klages, H. O., 617
Klapdor-Kleingrothaus, H. V., 870
Klay, N., 850
Klemt, E., 805
Klora, J., 455, 464, 467
Kneissl, U., 234, 479
Knudson, J., 747
Koch, R., 421
Koehler, P. E., 892, 953
Kondurov, I. A., 263, 271, 421
Koníček, J., 1014
Konijnenberg, M. W., 764, 1006
Kononenko, I. V., 263
Kopecky, J., 287, 607
Korten, W., 523
Kota, V. K. B., 685
Kotlinski, B., 473
Kramer, N. D., 263
Kramer, L. H., 587, 595
Krieger, S. J., 511
Krips, W., 418

Krusche, B., 173, 257, 412, 455, 610, 943
Kubono, S., 900
Kuhnert, A., 523
Kuida, I. A., 814
Kulik, V. D., 485
Kumpulainen, J., 428
Kusnezov, D., 90
Kuvaga, I. L., 249, 491
Kuznetsov, I. A., 814

L

Landberg, C., 733
Lanier, R. G., 672
Larson, D. C., 697
Laveder, M., 805
Lawson, R. D., 579
Laymon, C. M., 697
Le Du, J. F., 470
Lee, C. S., 323
Lee, Y., 627
Lesko, K. T., 860
Levant, D., 1011
Leviatan, A., 82, 137
Liang, M., 367
Libert, J., 55
Libman, V. A., 263, 455
Lieb, K. P., 173, 393, 396, 399, 407, 983
Lindenstruth, S., 479
Lindner, H., 455, 464, 467
Lippert, W., 805
Litvinsky, L. L., 455
Liu, X. T., 473
Loginov, Yu. E., 263, 271
Lougheed, R. W., 953
Lu, Z. H., 614
Lushikov, V. I., 787
Lynn, J. E., 555

M

Mac Mahon, T. D., 421
Mach, H., 418
Maino, G, 639
Maison, J. M., 461
Malov, L. A., 494
Manatt, D. R., 546, 549

Mantica, Jr., P. F., 431
Mareev, Yu. D., 811
Margraf, J., 479
Mathews, G. J., 827
Mattioli, F., 805
Mauri, A., 906
Mauri, F., 805
Mayerhofer, U., 455, 464, 467
Meyer, J., 511
Mezzetto, M., 805
Michaelsen, S., 393, 399
Michaudon, A., 953
Mildner, D. F. R., 929, 1003
Mitchell, G. E., 655, 747
Mitrikov, M. P., 636, 808
Mitsunari, T., 421
Mo, C. Z., 614
Molnár, G., 227, 367, 418, 434
Moltz, D. M., 347
Mondragón, A., 604
Moore, M. S., 953
Moreh, R., 961, 1011
Morrison, I., 114
Moss, R. L., 1006
Mostovoi, Ju. A., 814
Moszkowski, S. A., 122
Moszyǹski, M., 418
Motoyama, S., 903
Mukai, N., 627, 900, 903
Müller, A., 797
Müller, G., 479
Murzin, A. V., 263, 455
Mustafa, M. G., 672
Muto, K., 627

N

Nagai, Y., 900, 903
Nakada, H., 131
Naumann, R. A., 165
Nelson, R. O., 697, 723
Nesvizhevsky, V. V., 787
Németh, Zs., 917
Nilsson, E., 733
Nitschke, J. M., 347
Novoselsky, A., 168

O

Obložinský, P., 714
O'Brien, H. A., 892
Oda, T., 627
Ohm, H., 367
Ohsaki, T., 900
Okunev, I. S., 808
Otsuka, T., 114, 131, 543
Owens, M. L., 437

P

Paar, V., 150, 263, 425, 672
Paffrath, U., 367
Passoja, A., 428
Pavlik, A., 697
Penttilä, S., 747
Peskov, B. G., 808
Phillips, W. R., 355
Piazzoli, A., 805
Pikelner, L. B., 811
Pitz, H. H., 234, 479
Popov, Yu. P., 249, 298, 482, 485, 491, 494, 636, 756, 808
Pospíšil, S., 1014
Postma, H., 747, 756
Prokofjev, P. T., 249, 263, 446, 491, 494
Puglierin, G., 805

Q

Quentin, P., 55

R

Raemy, A., 263
Raman, S., 555, 923
Rappoldi, A., 805
Raselli, G. L., 805
Rasmussen, J. O., 473
Ray, A., 909
Reber, E. L., 437
Reedy, R. C., 994
Reffo, G., 742, 917
Resler, D. A., 122
Rezvaya, G. L., 491, 494
Richter, A., 234

Roberson, N. R., 747
Robertson, J. D., 347
Robinson, S. J., 173, 189, 210, 257, 396, 399, 407, 440, 443, 455
Rohr, G., 572
Rosier, L., 461
Rotbard, G., 461
Roy, N., 546

S

Saporetti, F., 639
Sarkar, S., 909
Sauvage, J., 55
Scannicchio, D., 805
Schacht, H., 479
Schatz, G., 850
Schechter, H., 473
Scheerer, H. J., 263
Schelberg, A., 953
Schieler, H., 617
Schillaci, M. E., 723
Schillebeeckx, P., 257, 440
Schlitt, B., 479
Schmidt, H. H., 263
Schreckenbach, K., 421, 476
Schumacher, M., 479
Schwitz, W., 263
Sconza, A., 805
Sebe, T., 131
Sedyshev, P. V., 636
Seemann, U., 479
Seestrom, S. J., 747, 756
Serebrov, A. P., 787
Seyfarth, H., 421
Shahal, O., 1011
Sharapov, E. I., 633, 756, 811
Sharon, Y. Y., 165, 437
Sheline, R. K., 159, 533
Shi, X., 630
Shihab-Eldin, A. A., 473
Shriner, Jr., J. F., 655
Shul'gina, E. V., 808
Shvetsov, V. N., 787
Simić, J., 421, 458
Simonova, L. I., 263, 491, 494
Sistemich, K., 367, 418
Skoy, V. R., 633, 811
Slovacek, R. E., 953
Smend, F., 617

Smith, A. B., 579
Soloviev, V. G., 16
Sood, P. C., 159, 533
Stecher-Rasmussen, F., 1006
Stepanenko, I. V., 814
Stephens, F. S., 523, 546
Stock, R., 479
Stoeffl, W., 415
Stone, C. A., 431, 929
Stoyer, M. A., 473
Strelkov, A. V., 787
Stuckey, K. A., 437
Sukhovoj, A. M., 249, 482, 485, 488, 491, 494
Sumbaev, S. O., 787
Sunko, D. K., 672
Sushkov, P. A., 263, 271, 421
Szymanski, J. J., 747

T

Tabor, S. L., 437
Takeda, K., 900, 903
Taldaev, R. R., 787
Tambergs, J. J., 253
Thielemann, F. K., 840
Thun, J., 733
Tilley, D. R., 587
Toth, K. S., 347
Trieb, H., 464, 467
Trzaska, W. H., 428

U

Uesawa, F., 900, 903
Uhl, M., 607
Uhrmacher, M., 983
Ulbig, S., 173, 396, 407

V

Väärämäki, J., 428
van der Leun, C., 404, 645
Van Isacker, P., 461
Van Neck, D., 3
Varlamov, V. E., 787
Vascon, M., 805
Vasilieva, E. V., 485, 494
Vasilyev, A. V., 787

Ventura, A., 639
Veres, Á., 434
Vergnes, M., 461
Verho, E., 428
Vernotte, J., 461
Vesna, V. A., 808
Visentin, L., 805
Vobecký, M., 1014
Vogt, E., 473
Vojnov, A. V., 488
von Brentano, P., 234, 244, 479
von Egidy, T., 263, 393, 455, 464, 467, 655
Vonach, H., 697

W

Walter, A., 464, 467
Walters, W. B., 431
Wang, T. F., 523
Warburton, E. K., 331
Warner, D. D., 46
Waroquier, M., 3
Wasson, A., 756
Weidenmüller, H. A., 742
Weiss, M. S., 511
Weller, H. R., 587, 595
Wender, S. A., 697, 723, 756
Werner, R., 805
Wervelman, R., 764
Wesselborg, C., 234, 479, 911
White, D. H., 476
Wicke, G. D., 617
Wiescher, M., 840
Williams, J. Z., 587
Wilmarth, P. A., 347
Wohn, F. K., 418
Wolf, A., 140, 620
Womble, P. C., 437
Wu, C. Y., 473

Y

Yates, S. W., 218, 339, 434, 523
Yoo, S. H., 747
Young, P. G., 697
Yuan, V. W., 747

Z

Zech, E., 263
Zeisler, R., 929
Zhu, L., 630
Zhu, X., 747
Ziegeler, L., 393
Zilges, A., 234, 479
Zimmerman, B. E., 431

AIP Conference Proceedings

		L.C. Number	ISBN
No. 152	Heavy Ion Inertial Fusion (Washington, DC, 1986)	86-73185	0-88318-352-8
No. 153	Physics of Particle Accelerators (SLAC Summer School, 1985) (Fermilab Summer School, 1984)	87-70103	0-88318-353-6
No. 154	Physics and Chemistry of Porous Media—II (Ridge Field, CT, 1986)	83-73640	0-88318-354-4
No. 155	The Galactic Center: Proceedings of the Symposium Honoring C. H. Townes (Berkeley, CA, 1986)	86-73186	0-88318-355-2
No. 156	Advanced Accelerator Concepts (Madison, WI, 1986)	87-70635	0-88318-358-0
No. 157	Stability of Amorphous Silicon Alloy Materials and Devices (Palo Alto, CA, 1987)	87-70990	0-88318-359-9
No. 158	Production and Neutralization of Negative Ions and Beams (Brookhaven, NY, 1986)	87-71695	0-88318-358-7
No. 159	Applications of Radio-Frequency Power to Plasma: Seventh Topical Conference (Kissimmee, FL, 1987)	87-71812	0-88318-359-5
No. 160	Advances in Laser Science–II (Seattle, WA, 1986)	87-71962	0-88318-360-9
No. 161	Electron Scattering in Nuclear and Particle Science: In Commemoration of the 35th Anniversary of the Lyman-Hanson-Scott Experiment (Urbana, IL, 1986)	87-72403	0-88318-361-7
No. 162	Few-Body Systems and Multiparticle Dynamics (Crystal City, VA, 1987)	87-72594	0-88318-362-5
No. 163	Pion–Nucleus Physics: Future Directions and New Facilities at LAMPF (Los Alamos, NM, 1987)	87-72961	0-88318-363-3
No. 164	Nuclei Far from Stability: Fifth International Conference (Rosseau Lake, ON, 1987)	87-73214	0-88318-364-1
No. 165	Thin Film Processing and Characterization of High-Temperature Superconductors (Anaheim, CA, 1987)	87-73420	0-88318-365-X
No. 166	Photovoltaic Safety (Denver, CO, 1988)	88-42854	0-88318-366-8
No. 167	Deposition and Growth: Limits for Microelectronics (Anaheim, CA, 1987)	88-71432	0-88318-367-6
No. 168	Atomic Processes in Plasmas (Santa Fe, NM, 1987)	88-71273	0-88318-368-4
No. 169	Modern Physics in America: A Michelson-Morley Centennial Symposium (Cleveland, OH, 1987)	88-71348	0-88318-369-2
No. 170	Nuclear Spectroscopy of Astrophysical Sources (Washington, DC, 1987)	88-71625	0-88318-370-6

No.	Title		
No. 171	Vacuum Design of Advanced and Compact Synchrotron Light Sources (Upton, NY, 1988)	88-71824	0-88318-371-4
No. 172	Advances in Laser Science–III: Proceedings of the International Laser Science Conference (Atlantic City, NJ, 1987)	88-71879	0-88318-372-2
No. 173	Cooperative Networks in Physics Education (Oaxtepec, Mexico, 1987)	88-72091	0-88318-373-0
No. 174	Radio Wave Scattering in the Interstellar Medium (San Diego, CA, 1988)	88-72092	0-88318-374-9
No. 175	Non-neutral Plasma Physics (Washington, DC, 1988)	88-72275	0-88318-375-7
No. 176	Intersections Between Particle and Nuclear Physics (Third International Conference) (Rockport, ME, 1988)	88-62535	0-88318-376-5
No. 177	Linear Accelerator and Beam Optics Codes (La Jolla, CA, 1988)	88-46074	0-88318-377-3
No. 178	Nuclear Arms Technologies in the 1990s (Washington, DC, 1988)	88-83262	0-88318-378-1
No. 179	The Michelson Era in American Science: 1870–1930 (Cleveland, OH, 1987)	88-83369	0-88318-379-X
No. 180	Frontiers in Science: International Symposium (Urbana, IL, 1987)	88-83526	0-88318-380-3
No. 181	Muon-Catalyzed Fusion (Sanibel Island, FL, 1988)	88-83636	0-88318-381-1
No. 182	High T_c Superconducting Thin Films, Devices, and Application (Atlanta, GA, 1988)	88-03947	0-88318-382-X
No. 183	Cosmic Abundances of Matter (Minneapolis, MN, 1988)	89-80147	0-88318-383-8
No. 184	Physics of Particle Accelerators (Ithaca, NY, 1988)	89-83575	0-88318-384-6
No. 185	Glueballs, Hybrids, and Exotic Hadrons (Upton, NY, 1988)	89-83513	0-88318-385-4
No. 186	High-Energy Radiation Background in Space (Sanibel Island, FL, 1987)	89-83833	0-88318-386-2
No. 187	High-Energy Spin Physics (Minneapolis, MN, 1988)	89-83948	0-88318-387-0
No. 188	International Symposium on Electron Beam Ion Sources and their Applications (Upton, NY, 1988)	89-84343	0-88318-388-9
No. 189	Relativistic, Quantum Electrodynamic, and Weak Interaction Effects in Atoms (Santa Barbara, CA, 1988)	89-84431	0-88318-389-7
No. 190	Radio-frequency Power in Plasmas (Irvine, CA, 1989)	89-45805	0-88318-397-8
No. 191	Advances in Laser Science–IV (Atlanta, GA, 1988)	89-85595	0-88318-391-9
No. 192	Vacuum Mechatronics (First International Workshop) (Santa Barbara, CA, 1989)	89-45905	0-88318-394-3

No. 193	Advanced Accelerator Concepts (Lake Arrowhead, CA, 1989)	89-45914	0-88318-393-5
No. 194	Quantum Fluids and Solids—1989 (Gainesville, FL, 1989)	89-81079	0-88318-395-1
No. 195	Dense Z-Pinches (Laguna Beach, CA, 1989)	89-46212	0-88318-396-X
No. 196	Heavy Quark Physics (Ithaca, NY, 1989)	89-81583	0-88318-644-6
No. 197	Drops and Bubbles (Monterey, CA, 1988)	89-46360	0-88318-392-7
No. 198	Astrophysics in Antarctica (Newark, DE, 1989)	89-46421	0-88318-398-6
No. 199	Surface Conditioning of Vacuum Systems (Los Angeles, CA, 1989)	89-82542	0-88318-756-6
No. 200	High T_c Superconducting Thin Films: Processing, Characterization, and Applications (Boston, MA, 1989)	90-80006	0-88318-759-0
No. 201	QED Stucture Functions (Ann Arbor, MI, 1989)	90-80229	0-88318-671-3
No. 202	NASA Workshop on Physics From a Lunar Base (Stanford, CA, 1989)	90-55073	0-88318-646-2
No. 203	Particle Astrophysics: The NASA Cosmic Ray Program for the 1990s and Beyond (Greenbelt, MD, 1989)	90-55077	0-88318-763-9
No. 204	Aspects of Electron–Molecule Scattering and Photoionization (New Haven, CT, 1989)	90-55175	0-88318-764-7
No. 205	The Physics of Electronic and Atomic Collisions (XVI International Conference) (New York, NY, 1989)	90-53183	0-88318-390-0
No. 206	Atomic Processes in Plasmas (Gaithersburg, MD, 1989)	90-55265	0-88318-769-8
No. 207	Astrophysics from the Moon (Annapolis, MD, 1990)	90-55582	0-88318-770-1
No. 208	Current Topics in Shock Waves (Bethlehem, PA, 1989)	90-55617	0-88318-776-0
No. 209	Computing for High Luminosity and High Intensity Facilities (Santa Fe, NM, 1990)	90-55634	0-88318-786-8
No. 210	Production and Neutralization of Negative Ions and Beams (Brookhaven, NY, 1990)	90-55316	0-88318-786-8
No. 211	High-Energy Astrophysics in the 21st Century (Taos, NM, 1989)	90-55644	0-88318-803-1
No. 212	Accelerator Instrumentation (Brookhaven, NY, 1989)	90-55838	0-88318-645-4
No. 213	Frontiers in Condensed Matter Theory (New York, NY, 1989)	90-6421	0-88318-771-X 0-88318-772-8 (pbk.)
No. 214	Beam Dynamics Issues of High-Luminosity Asymmetric Collider Rings (Berkeley, CA, 1990)	90-55857	0-88318-767-1
No. 215	X-Ray and Inner-Shell Processes (Knoxville, TN, 1990)	90-84700	0-88318-790-6

No. 216	Spectral Line Shapes, Vol. 6 (Austin, TX, 1990)	90-06278	0-88318-791-4
No. 217	Space Nuclear Power Systems (Albuquerque, NM, 1991)	90-56220	0-88318-838-4
No. 218	Positron Beams for Solids and Surfaces (London, Canada, 1990)	90-56407	0-88318-842-2
No. 219	Superconductivity and Its Applications (Buffalo, NY, 1990)	91-55020	0-88318-835-X
No. 220	High Energy Gamma-Ray Astronomy (Ann Arbor, MI, 1990)	91-70876	0-88318-812-0
No. 221	Particle Production Near Threshold (Nashville, IN, 1990)	91-55134	0-88318-829-5
No. 222	After the First Three Minutes (College Park, MD, 1990)	91-55214	0-88318-828-7
No. 223	Polarized Collider Workshop (University Park, PA, 1990)	91-71303	0-88318-826-0
No. 224	LAMPF Workshop on (π, K) Physics (Los Alamos, NM, 1990)	91-71304	0-88318-825-2
No. 225	Half Collision Resonance Phenomena in Molecules (Caracus, Venezuela, 1990)	91-55210	0-88318-840-6
No. 226	The Living Cell in Four Dimensions (Gif sur Yvette, France, 1990)	91-55209	0-88318-794-9
No. 227	Advanced Processing and Characterization Technologies (Clearwater, FL, 1991)	91-55194	0-88318-910-0
No. 228	Anomalous Nuclear Effects in Deuterium/Solid Systems (Provo, UT, 1990)	91-55245	0-88318-833-3
No. 229	Accelerator Instrumentation (Batavia, IL, 1990)	91-55347	0-88318-832-1
No. 230	Nonlinear Dynamics and Particle Acceleration (Tsukuba, Japan, 1990)	91-55348	0-88318-824-4
No. 231	Boron-Rich Solids (Albuquerque, NM, 1990)	91-53024	0-88318-793-4
No. 232	Gamma-Ray Line Astrophysics (Paris–Saclay, France, 1990)	91-55492	0-88318-875-9
No. 233	Atomic Physics 12 (Ann Arbor, MI, 1990)	91-55595	088318-811-2
No. 234	Amorphous Silicon Materials and Solar Cells (Denver, CO, 1991)	91-55575	088318-831-7
No. 235	Physics and Chemistry of MCT and Novel IR Detector Materials (San Francisco, CA, 1990)	91-55493	0-88318-931-3
No. 236	Vacuum Design of Synchrotron Light Sources (Argonne, IL, 1990)	91-55527	0-88318-873-2
No. 237	Kent M. Terwilliger Memorial Symposium (Ann Arbor, MI, 1989)	91-55576	0-88318-788-4